Lecture Notes in Computer Science 3794

Commenced Publication in 1973
Founding and Former Series Editors:
Gerhard Goos, Juris Hartmanis, and Jan van Leeuwen

Xiaohua Jia Jie Wu Yanxiang He (Eds.)

Mobile Ad-hoc and Sensor Networks

First International Conference, MSN 2005
Wuhan, China, December 13-15, 2005
Proceedings

Volume Editors

Xiaohua Jia
City University of Hong Kong, Department of Computer Science
Tat Chee Avenue, Kowloon Tong, Hong Kong SAR
E-mail: csjia@cityu.edu.hk

Jie Wu
Florida Atlantic University
Department of Computer Science and Engineering
Boca Raton, FL 33431, USA
E-mail: jie@cse.fau.edu

Yanxiang He
Wuhan University, Computer School
Wuhan, Hubei 430072, P.R. China
E-mail: yxhe@whu.edu.cn

Library of Congress Control Number: 2005937083

CR Subject Classification (1998): E.3, C.2, F.2, H.4, D.4.6, K.6.5

ISSN 0302-9743
ISBN-10 3-540-30856-3 Springer Berlin Heidelberg New York
ISBN-13 978-3-540-30856-0 Springer Berlin Heidelberg New York

Springer is a part of Springer Science+Business Media

springeronline.com

Printed in Germany

Typesetting: Camera-ready by author, data conversion by Scientific Publishing Services, Chennai, India
Printed on acid-free paper SPIN: 11599463 06/3142 5 4 3 2 1 0

Preface

MSN 2005, the First International Conference on Mobile Ad-hoc and Sensor Networks, was held during 13-15 December 2005, in Wuhan, China. The conference provided a forum for researchers and practitioners to exchange research results and share development experiences. MSN 2005 attracted 512 submissions (including the submissions to the Modeling and Security in Next Generation Mobile Information Systems (MSNG) workshop), among which 100 papers were accepted for the conference and 12 papers were accepted for the workshop.

We would like to thank the International Program Committee for their valuable time and effort in reviewing the papers. Special thanks go to the conference PC Vice-chairs, Ivan Stojmenovic, Jang-Ping Sheu and Jianzhong Li for their help in assembling the International PC and coordinating the review process. We would also like to thank the Workshop Chair, Dongchun Lee, for organizing the workshop.

We would like to express our gratitude to the invited speakers, Laxmi Bhuya, Lionel Ni and Taieb Znati, for their insightful speeches. Finally, we would like to thank the Local Organization Chair, Chuanhe Huang, for making all the local arrangements for the conference.

December 2005

Xiaohua Jia
Jie Wu
Yanxiang He

Organization

Steering Co-chairs

Lionel Ni, Hong Kong University of Science and Technology, HKSAR
Jinnan Liu, Wuhan University, PRC

General Co-chairs

Taieb Znati, University of Pittsburgh, USA
Yanxiang He, Wuhan University, PRC

Program Co-chairs

Jie Wu, Florida Atlantic University, USA
Xiaohua Jia, City University of Hong Kong, HKSAR

Program Vice Chairs

Ivan Stojmenovic, University of Ottawa, Canada
Jang-Ping Sheu, National Central University, Taiwan
Jianzhong Li, Harbin Institute of Technology, PRC

Publicity Chair

Makoto Takizawa, Tokyo Denki University, Japan
Weifa Liang, The Australian National University, Australia
Jiannong Cao, Hong Kong Polytechnic University, HKSAR

Publications Chair

Hai Liu, City University of Hong Kong, HKSAR

Local Organization Chair

Chuanhe Huang, Wuhan University, PRC

Program Committee

Amiya Nayak (University of Ottawa, Canada)
Cai Wentong (Nanyang Technological University, Singapore)
Chih-Yung Chang (Tamkang University, Taiwan)
Christophe Jelger (University of Basel, Switzerland)
Cho-Li Wang (The University of Hong Kong, HKSAR)
Chonggang Wang (University of Arkansas, USA)
Chonggun Kim (Yeungnam University, Korea)
Chun-Hung Richard Lin (National Sun Yat-Sen University, Taiwan)
Chung-Ta King (National Tsing Hua University, Taiwan)
David Simplot-Ryl (University of Lille, France)
Duan-Shin Lee (National Tsing Hua University, Taiwan)
Dong Xuan (Ohio State University, USA)
Eric Fleury (INRIA, France)
Fei Dai (North Dakota State University, USA)
Geyong Min (University of Bradford, UK)
Guangbin Fan (The University of Mississippi, USA)
Guihai Chen (Nanjing University, PRC)
Guojun Wang (Central South University, PRC)
Han-Chieh Chao (National Dong Hwa University, Taiwan)
Hong Gao (Harbin Institute of Technology, PRC)
Hongyi Wu (University of Louisiana at Lafayette, USA)
Hsiao-Kuang Wu (National Central University, Taiwan)
Hu Xiaodong (Chinese Academy of Science, PRC)
Ingrid Moerman (Ghent University, Belgium)
Isabelle Guerin Lassous (INRIA, France)
Isabelle Simplot-Ryl (University of Lille, France)
Jaideep Srivastava (University of Minnesota, USA)
Jean Carle (University of Lille, France)
Jehn-Ruey Jiang (National Central University, Taiwan)
Jiang (Linda) Xie (The University of North Carolina at Charlotte, USA)
Jie Li (University of Tsukuba, Japan)
Jiangliang Xu (Baptist University of Hong Kong, HKSAR)
Jiangnong Cao (Hong Kong Polytechnic University, HKSAR)
Jianping Wang (Georgia Southern University, USA)
Justin Lipman (Alcatel Shanghai Bell, PRC)
Jyh-Cheng Chen (National Tsing Hua University, Taiwan)
Kuei-Ping Shih (Tamkang University, Taiwan)
Kui Wu (University of Victoria, Canada)
Lars Staalhagen (Technical University of Denmark, Denmark)
Li-Der Chou (National Central University, Taiwan)
Li Xiao (Michigan State University, USA)
Liansheng Tan (Central China Normal University, PRC)
Mei Yang (UNLV, USA)
Min Song (Old Dominion University, USA)

Minglu Li (Shanghai Jiao Tong University)
Natalija Vlajic (York University, Toronto, Canada)
Ning Li (University of Illinois, USA)
Pedro M. Ruiz (University of Murcia, Spain)
Qingfeng Huang (Palo Alto Research Center, Palo Alto, USA)
Rong-Hong Jan (National Chiao Tung University, Taiwan)
Ruay-Shiung Chang (National Dong Hwa University, Taiwan)
Scott Huang (City University of Hong Kong, HKSAR)
Sunghyun Choi Seoul (National University, South Korea)
Suprakash Datta (York University, Canada)
Timothy Shih (Tamkang University, Taiwan)
Tracy Camp (Colorado School of Mines, Golden, USA)
Vojislav Misic (University of Manitoba, Canada)
Wan Pengjun (Illinois Ins. of Technology, USA)
Weifa Liang (The Australian National University, Australia)
Weijia Jia (City University of Hong Kong, HKSAR)
Weili Wu (University of Texas at Dallas, USA)
Wenzhan Song (Washington State University, Vancouver, USA)
Winston Seah (Institute for Infocomm Research, Singapore)
Wu Zhang (Shanghai University, PRC)
Xiao Bin (Hong Kong Polytechnic University, HKSAR)
Xiao Chen (Texas State University, USA)
Xiaobo Zhou (University of Colorado at Colorado Springs, USA)
Xiaoyan Maggie Cheng (University of Missouri, Rolla, USA)
Xiuzhen Susan Cheng (George Washington University, USA)
Xuemin Shen (University of Waterloo, Canada)
Xueyan Tang (Nanyang Technological University, Singapore)
Yang Xiao (The University of Memphis, USA)
Yao-Nan Lien (National Chengchi University, Taiwan)
Yen-Wen Chen (National Central University, Taiwan)
Yongbing Zhang (University of Tsukuba, Japan)
Yu-Chee Tseng (National Chiao Tung University, Taiwan)
Yuh-Shyan Chen (National Chung Cheng University, Taiwan)
Yu Wang (University of North Carolina, Charlotte, USA)
Zhen Jiang (West Chester University, USA)

Table of Contents

An Overlapping Communication Protocol Using Improved Time-Slot Leasing for Bluetooth WPANs

Yuh-Shyan Chen[1], Yun-Wei Lin[1], and Chih-Yung Chang[2]

[1] Department of Computer Science and Information Engineering, National Chung Cheng University, Taiwan
[2] Department of Computer Science and Information Engineering, Tamkang University, Taiwan

Abstract. In this paper, we propose an overlapping communication protocol using improved time-slot leasing in the Bluetooth WPANS. One or many slave-master-slave communications usually exist in a piconet of the Bluetooth network. A fatal communication bottleneck is incurred in the master node if many slave-master-slave communications are required at the same time. To alleviate the problem, an overlapping communication scheme is presented to allow slave node directly and simultaneously communicates with another slave node to replace with the original slave-master-slave communication works in a piconet. This overlapping communication scheme is based on the improved time-slot leasing scheme. The key contribution of our improved time-slot leasing scheme additionally offers the overlapping communication capability and we developed an overlapping communication protocol in a Bluetooth WPANs. Finally, simulation results demonstrate that our developed communication protocol achieves the performance improvements on bandwidth utilization, transmission delay time, network congestion, and energy consumption.

Keywords: Bluetooth, time-slot leasing, WPAN, wireless communication.

1 Introduction

The advances of computer technology and the population of wireless equipment have promoted the quality of our daily life. The trend of recent communication technology is to make good use of wireless equipments for constructing an ubiquitous communication environment. Bluetooth[2] is a low cost, low power, and short range communication technology that operates at 2.4GHz ISM bands.

A master polls slaves by sent polling packets to slaves using round robin (RR) scheme within the piconet. The master communicates with one slave and all other slaves must hold and wait the polling packet, so the transmission of other slaces is arrested. This condition is called the "transmission holding problem." To reduce the "transmission holding problem", one interest issue is how to develop a novel scheme which can effectively solve the "transmission holding" problem under the

X. Jia, J. Wu, and Y. He (Eds.): MSN 2005, LNCS 3794, pp. 1–10, 2005.

fixed-topology situation. To satisify that purpose, Zhang *et al.* develops a time-slot leasing scheme (TSL) [5]. Zhang *et al.*'s time-slot leasing scheme provides a general mechanism to support the direct slave-slave communication, but the master node uses round-robin (RR) mechanism to check slave node intended to send or receive data. Other slave node waits the polling time and gains the transmission holding time. More recently, Cordeiro *et al.* proposed a QoS-driven dyamic slot assignment (DSA) schedule scheme [1] to more efficiently utilize time-slot leasing scheme. With TSL and DSA scheme, the tramnsission holding problem still exists. Effort will be made to effectively reduce the the tramnsission holding problem under the fixed topology structure.

In this paper, we propose an overlapping communication protocol using improved time-slot leasing in the Bluetooth WPANS. The overlapping communication scheme is based on the improved time-slot leasing scheme which modified from the original time-slot leasing scheme, while the original time-slot leasing scheme only provides the slave-to-slave communication capability. The key contribution of our improved time-slot leasing scheme additionally offers the overlapping communication capability and we developed an overlapping communication protocol in a Bluetooth WPANs. Finally, simulation results demonstrate that our developed communication protocol achieves the performance improvements on bandwidth utilization, transmission delay time, network congestion, and energy consumption.

This paper is organized as follows. Section 2 describes the basic idea of our new scheme. The new communication protocol is presented in Section 3. The performance analysis is discussed in Section 4. Section 5 concludes this work.

2 Basic Idea

The transmission holding problem is originated from the drawback of the master/slave model. In a piconet, since the slave may transmit packets only if it receives the polling packet from master. As a result, when there are many salves have to transmit data, slaves must hold its transmission until receving the polling packet, as shown in Fig. 1(a). To solve the transmission holding problem, a time-slot leasing (TSL) approach [5] has been proposed. Slaves can directly transmit packets to each others without the master relaying. Using TSL approach, the waiting time of the other holding slaves are reduced. Unfortunately, the effect of transmission holding problem is reduced, but it still exists. To slove the transmission holding problem completely and overcome the drawback of RR scheme, an overlapping communication scheme is investigated in this work to offer the overlapping communication capability for multi-pair of devices within a piconet. With the overlapping communication scheme, Bluetooth device can simultaneously and directly communicate with each other. The performance of communication will be improved.

In the following, we describe the main contribution of our scheme, compared to time-slot leasing (TSL) scheme [5] and QoS-driven dyamic slot assignment (DSA) scheme [1]. The frequency-hopping spread spectrum is used in the Bluetooth network. Let $C(x)$ denote the used channel for time slot x, and $\overrightarrow{\alpha\beta}$ denote slave node α sends data to slave node β. An example is shown in Fig. 1(b), $\overrightarrow{S_1S_2}$,

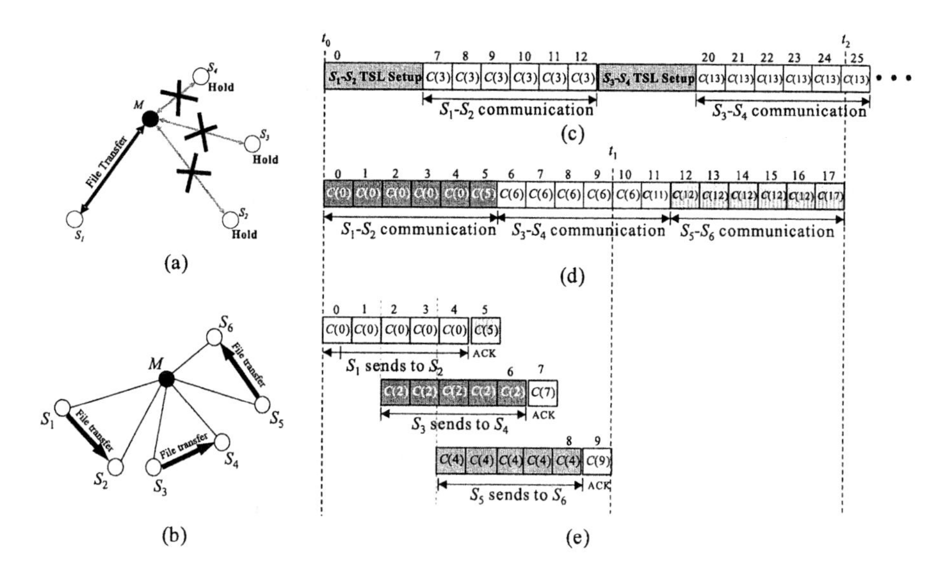

Fig. 1. (a) Transmission holding problem (b) Three communication requests (c) TSL (d) DSA (e) Overlapping communication protocol

$\overrightarrow{S_3S_4}$, and $\overrightarrow{S_5S_6}$ simultaneously occur in a piconet. The transmission holding problem is heavily occurred in master node M. Fig. 1(c) shows that the time cost is more than $t_2 - t_0$ by using TSL scheme for the $\overrightarrow{S_1S_2}$, $\overrightarrow{S_3S_4}$ and $\overrightarrow{S_5S_6}$. With the same works of $\overrightarrow{S_1S_2}$, $\overrightarrow{S_3S_4}$ and $\overrightarrow{S_5S_6}$, time cost is obviously reduced to $t_2 - t_0$ by using the DSA scheme as illustrated in Fig. 1(d). However, the time cost will be improved by using our overlapping communication scheme. Fig. 1(e) illustrates that the works of $\overrightarrow{S_1S_2}$, $\overrightarrow{S_3S_4}$, and $\overrightarrow{S_5S_6}$ can be accomplished in time $t_1 - t_0$.

To explain the frequency hopping technology, every time slot during the transmission adopts the different channel, we let channel $FH(x)$ denote the frequency used at time slot x. From the sepcification of the Bluetooth system 1.2 [2], the consecutive five time slots keep the same channel if using a DH5 packet. The rule is same for packets DH3, DM3, and DM5. Let channel $C(x)$ denote a Bluetooth device sends a packet at time slot x using the channel $C(x)$. If a device sends a DH1 packet at time slot x, then we have the result of $C(x) = FH(x)$, $C(x+1) = FH(x+1)$, $C(x+2) = FH(x+2)$, $C(x+3) = FH(x+3)$, $C(x+4) = FH(x+4)$, where $C(x) \neq C(x+1) \neq C(x+2) \neq C(x+3) \neq C(x+4)$. But if a device sends a DH5 packet at time slot x, then five connective time slots use the same channel, $C(x) = C(x+1) = C(x+2) = C(x+3) = C(x+4) = FH(x)$. Observe that channels $F(x+1)$, $F(x+2)$, $F(x+3)$, and $F(x+4)$ in the original frequency hopping sequence are free. Efforts will be made to significantly improve the throughput in a scatternet by using our new overlapping communication scheme. This work is achieved by developing intra-piconet and inter-piconet overlapping protocols, which are presented in the following sections. The intra-piconet overlapping communication protocol is shown in Figs. 2(a)(b). The data transmission of $\overleftrightarrow{M_1S_3}$

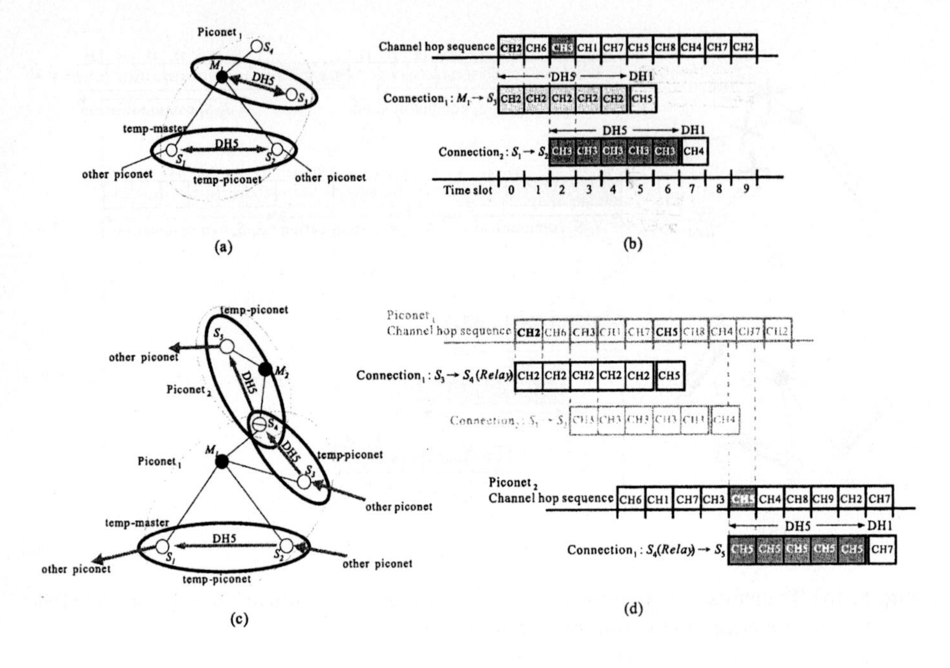

Fig. 2. The concept of overlapping communication scheme

and $\overleftrightarrow{S_1S_2}$ can be overlapped. In addition, Fig. 2(c)(d) illustrates the overlapping condition for performing the intra-piconet overlapping communication protocol.

3 Overlapping Communication Protocol

To significantly overcome the "transmission holding problem", an overlapping communication protocol is presented. The overlapping communication protocol is divided into two phases; (1) *queuing scheduling*, (2) *overlapping time-slot assignment.* The details descrilbe in the following.

3.1 Queuing Scheduling Phase

The queuing scheduling phase is to achieve the overlapping communication schedule. To complete the overlapping communication, master initially forms *data-flow matrix* and *queuing table.* This work can be done in the *BTIM* window as shown in Fig. 3(c). During a schedule interval, each source node (Bluetooth device) in a piconet just can transmit data to one destination node. The amount of data transmission of all pair of source-destination nodes is kept in the master node, and can be stored in a *data flow matrix* $DM_{m\times m+1} = \begin{bmatrix} D_{10} & \cdots & D_{1m} \\ \vdots & D_{ij} & \vdots \\ D_{m0} & \cdots & D_{mm} \end{bmatrix}$, where D_{ij} denote the amount of data transmission from node i to node j, where

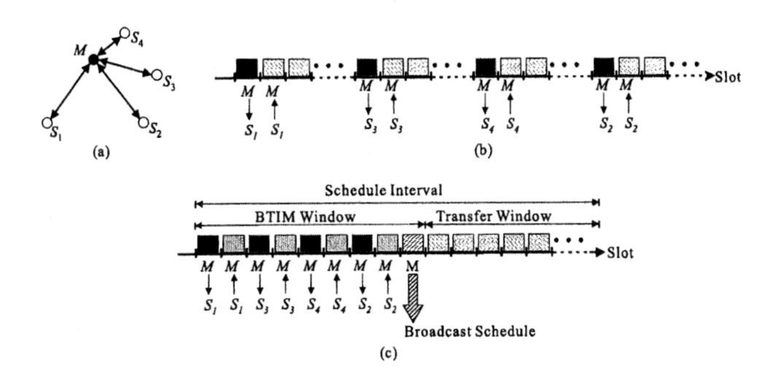

Fig. 3. (a) Master polling (b) Polling and transmission in origin piconet (c) The structure of BTIM

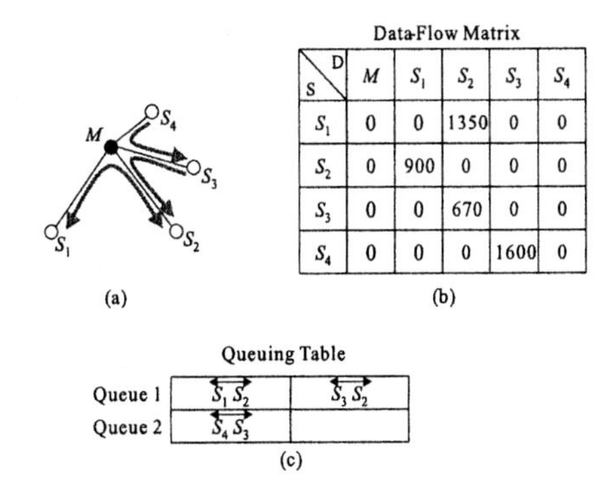

S \ D	M	S1	S2	S3	S4
S1	0	0	1350	0	0
S2	0	900	0	0	0
S3	0	0	670	0	0
S4	0	0	0	1600	0

Queuing Table		
Queue 1	$\overleftrightarrow{S_1S_2}$	$\overleftrightarrow{S_3S_2}$
Queue 2	$\overleftrightarrow{S_4S_3}$	

Fig. 4. (a) An example of piconet communication, (b) flow matrix, and (c) n-queue

$1 \leq i \leq m$, $0 \leq j \leq m$, and m is the number of slave nodes in a piconet. Observe that node j is the master node if $j = 0$. Example is shown in Fig. 4(b).

Master node further calculates a *queuing sequence* based on data-flow matrix $DM_{m \times m+1}$. Our overlapping communication scheme is to utilize the long packet, since the high utilization of the long packet can significantly increase the chance of overlapping communication.

Before describing the overlapping time-slot assignment operation, we define the following notations. First, a link with a greater number of data has the higher priority for the data transmission. Therefore, we define priority function $pri(\overleftrightarrow{ij})$ of link $\overleftrightarrow{ij}$ as

$$pri(\overleftrightarrow{ij}) = \frac{D_{ij} + D_{ji}}{|D_{ij} - D_{ji}|}.$$

A link queue Q is defined to record $m+1$ pairs of source-to-destination links $\overleftrightarrow{i_0j}, ..., \overleftrightarrow{i_kj}, \overleftrightarrow{i_{k+1}j}, ...,$ and $\overleftrightarrow{i_mj}$ in a piconet, where all of these links have the same destination node j. Further, let $Q = \{\overleftrightarrow{i_0j}, \cdots, \overleftrightarrow{i_kj}, \overleftrightarrow{i_{k+1}j}, \cdots, \overleftrightarrow{i_mj}\}$, where $pri(\overleftrightarrow{i_kj}) > pri(\overleftrightarrow{i_{k+1}j})$ and $0 \leq k \leq m-1$. For instance, $\{\overleftrightarrow{S_1S_2}, \overleftrightarrow{S_3S_2}\}$.

Assumed that there are n link queues $Q_1, Q_2, \cdots, Q_q, \cdots, Q_n$, where each Q_q has differnt destination node, and $1 \leq q \leq n$, where n is real number. For instance, $Q_1 = \{\overleftrightarrow{S_1S_2}, \overleftrightarrow{S_3S_2}\}$ and $Q_2 = \{\overleftrightarrow{S_4S_3}\}$. Given $Q_q = \{\overleftrightarrow{i_0j_q}, \cdots, \overleftrightarrow{i_kj_q}, \overleftrightarrow{i_{k+1}j_q}, \cdots, \overleftrightarrow{i_mj_q}\}$, we denote $MAX(Q_q) = \underset{k=0}{\overset{m}{MAX}}\ pri(\overleftrightarrow{i_kj_q}) = pri(\overleftrightarrow{i_0j_q})$. All of

$Q_1, Q_2, \cdots, Q_q, \cdots, Q_n$ can be combined into a queuing sequence $= \begin{pmatrix} Q_1 \\ \vdots \\ Q_q \\ Q_{q+1} \\ \vdots \\ Q_n \end{pmatrix}$,

where $MAX(Q_q) > MAX(Q_{q+1})$, and $1 \leq q \leq n$. Example is shown in Fig. 4(a).

3.2 Overlapping Time-Slot Assignment Phase

The queuing scheduling phase determines the appropriate transmission order which records in a queuing sequence. This queuing sequence is used in the overlapping time-slot assignment phase to assign suitable time-slots for each transmission. In the overlapping time-slot assignment phase, there are two conditions occurred, collision-free and collision detection. In the following, we introduce collision-free and collision detection.

In the collision-free condition, the overlapping time-slot assignment phase assign suitable time-slots to each transmission $\overleftrightarrow{ij}$ according to the order in the queue sequence. Q_q processes before Q_{q+1} in queue sequence. Each Q_q starts to process at time slot $2k$, and $q \leq k \leq n$. For higher bandwidth utilization, DH5 packet type is the prior choice to transmit data. Overlapping communication scheme uses the appropriate packet type improves bandwidth and decrease the energy consumption. We let n_DH5 and n_DH3 denote the amount of DH5 and DH3 packet type respectively, D_{ij} denote the amount of data transmission from node i to node j, and $Excess$ denote the excess part of transmission. For each transmission, master arranges approoriate packet type by a packet type distribution rule which describes as following:

$$n\_DH5 * 339 + n\_DH3 * 183 = D_{ij} + min(Excess)$$

The packet amount of DH5 and DH3 is 339 bytes and 183 bytes, respectively. A set of packet type distribution $PD_{ij} = \{pd_{ij_0}, pd_{ij_1}, ..., pd_{ij_s}, ..., pd_{ij_p}\}$ is used to record $p+1$ packet type distribution from node i transmitting to node j, where pd_{ij_s} denotes the tth transmission packet type form i to j and $0 \leq s \leq p$. The value of p is $n\_DH5 + n\_DH3$, where n_DH5 and n_DH3 are related to pd_{ij_s}. For example in Fig. 4(b), $D_{S_1S_2}$ is 1350 bytes composed of four DH5 packets ($n\_DH5 = 4$), $PD_{S_1S_2} = \{5, 5, 5, 5\}$ and $D_{S_2S_1}$ is 900 composed of three DH5 packets bytes ($n\_DH5 = 3$), $PD_{S_2S_1} = \{5, 5, 5\}$. Master gets a PD_{ij} for link $\overleftrightarrow{ij}$ by packet type distribution rule. Set PD_{ij} is used to predict the occupied time-slots by link $\overleftrightarrow{ij}$.

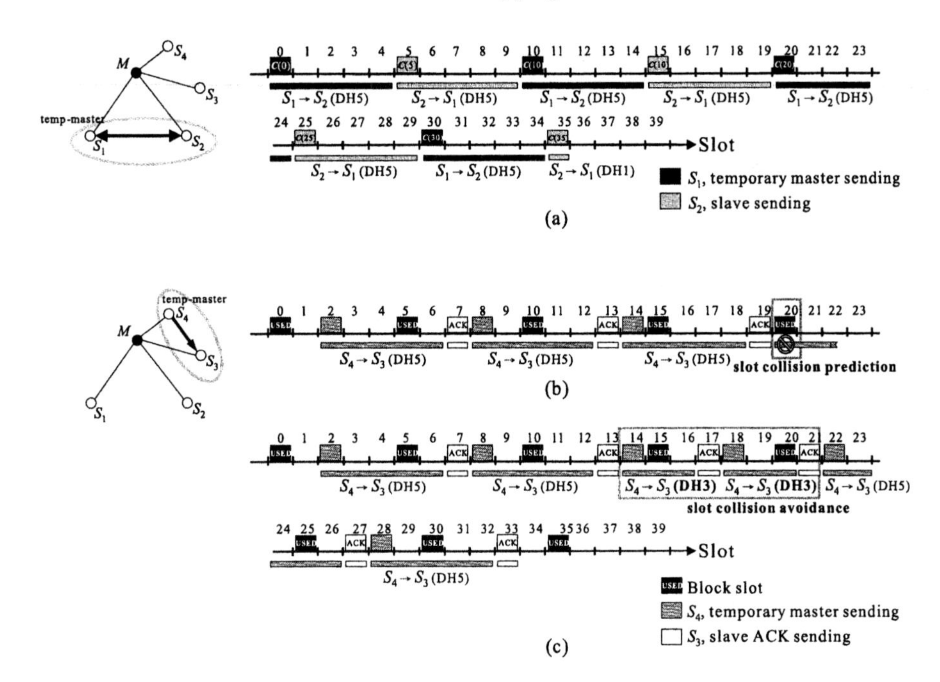

Fig. 5. (a) the block slot calculation, (b) collision occurred, (c) packet type changing

According the packet type distribution, the overlapping time-slot assignment phase assigns suitable time-slot to use improved time-slot leasing for data transmission. A time-slot WS_{ij} which is defined as the i's wake up time-slot between link $\overleftrightarrow{ij}$. A set of time-slot offset $SO_{ij} = \{so_{ij_0}, so_{ij_1}, ..., so_{ij_t}, ..., so_{ij_q}\}$ is defined to record $q+1$ time-slot offset from node i transmitting to node j, and $0 \leq t \leq q$. The so_{ij_t} denotes the tth time-slot offset transmission from i to j. SO_{ij} is accumulated form PD_{ij} and PD_{ji}, which is used to predict the occupied time-slot offset by link $\overleftrightarrow{ij}$. For using improve time-slot leasing, master announces each slave WS_{ij} and SO_{ij}. Slaves wake up at assigned slot WS_{ij} and obey the set of time-slot offset SO_{ij} to transmit data.

About overlapping communication scheme, master assigns the roles, WS_{ij}, and SO_{ij} to each transmission $\overleftrightarrow{ij}$ at BTIM. We consider two roles, temp-master and slave and if $D_{ij} > D_{ji}$ then i is temp-master and j is slave. For a transmission from temp-master i to slave j, master assigns time-slot to link $\overleftrightarrow{ij}$ as follows.

$$SO_{ji} : so_{ji_t} = so_{ij_t} + pd_{ij_s}$$
$$SO_{ij} : so_{ij_{t+1}} = so_{ji_t} + pd_{ji_s},$$

where the so_{ij_0} initially sets to 0. During computing SO_{ij}, if master runs out of the PD_{ij} set, $pd_{ji_{p+t}}$ sets to 1 for ACK responding. Master assigns the nearest free time slot as WS_{ij} and $WS_{ji} = WS_{ij} + pd_{ij_1}$. $WS_{ij} + SO_{ij}$ is the real time-slot. For example about $\overleftrightarrow{S_1S_2}$ in Fig 4(c), $D_{S_1S_2}$ is 1350 bytes, $SO_{S_1S_2} = \{0, 10, 20, 30\}$,

$D_{S_2S_1}$ is 900 bytes, $SO_{S_2S_1} = \{5, 15, 25, 35\}$, $WS_{S_1S_2} = 0$, $WS_{S_2S_1} = 5$, and the all transmission process of $\overleftrightarrow{S_1S_2}$ as shown in Fig. 5(a).

In the collision detection condition, some occupied time-slots have been assigned again to transmission possibly. Before transmission, master should detect this condition first. A set of used time-slot $US = \{us_0, us_1, ..., us_k\}$ is defined to record $k+1$ used time-slots in a piconet. Master add $WS_{ij} + SO_{ij}$ into US after assigned time-slots for i. If master assigns time-slots including in US to slaves, we called *time-slot collision.* Master uses the following equation to check the collision status.

$$(WS_{ij} + SO_{ij}) \bigcap US \neq \{\ \phi\ \}$$

If the equation is ture, the problem of time-slot collision is detected. For example as shown in Fig. 5(b), master checks $(WS_{S_4S_3} + SO_{S_4S_3}) \bigcap BS = \{20\}$. The result set is non-empty. Collision is detected in assigned time-slots for $\overrightarrow{S_4S_3}$.

In the following, we describe how to solve the time-slot collision problem. When the time-slots is collision, master adapts the following rules to avoid the collision.

S1: /* Avoidance for DH5 packet */
If the tth collision packet type is DH5, master change the packet to two DH3 packets. Master changes the $pd_{ij_t} = \{5\}$ to $pd_{ij_t} = \{3, 3\}$.
S2: /* Avoidance for DH3 packet */
If the tth collision packet type is DH3, master change the packet to one DH5 packet. Master changes the $pd_{ij_t} = \{3\}$ to $pd_{ij_t} = \{5\}$.
S3: /* Avoidance for DH1 packet */
If the tth collision packet type is DH1, master change the packet to one DH3 packet. Master changes the $pd_{ij_t} = \{1\}$ to $pd_{ij_t} = \{3\}$.

After master applying above rules to the collision condition of $\overrightarrow{S_4S_3}$, the transmission process has shown in Fig. 5(c). After the schedule arrangement, master informs all slaves by a schedule-assignment packet. If a device does not do anything, it changes to sleep mode to save energy until next BTIM.

4 Experimental Results

In simulation, we investigate the performance of OCP protocol. To compare with three algorithms, RR, TSL [5] and DSA [1], we have implemnted OCP protocol and others using the Network Simulator (ns-2) [3] and BlueHoc [4]. The performance metrics of the simulation are given below. The simulation parameters are shown in table 1.

- *Average Holding Time*: the time period of packet is arrested by other transmission.
- *Transmission Delay*: the latency from a souurce device to a destination device.
- *Throughput*: the number of data bytes received by all Bluetooth devices per unit time.

Table 1. The detail simulation parameters

Parameters	Value
Number of device	$2 \geq N \geq 70$
Network region	100m $\times$100m
Radio propagation range	10m
Mobility	No
Schedule Interval	64 time slots
Packet type	DH1 or DH3 or DH5

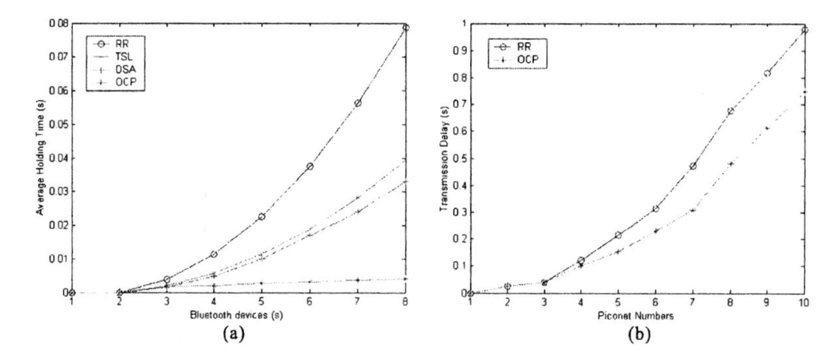

Fig. 6. Performance of comparison about (a) average of holding time vs. Bluetooth devices (b) transmission delay vs. piconet numbers

4.1 Performance of Average Holding Time and Transmission Delay

In a piconet, Fig. 6(a) indicates that the holding time of TSL is half of RR approximately. DSA uses TSL more efficiently and decreases the holding time more than TSL. OCP uses the overlapping communication scheme. Therefore, there are several communication pairs transmission simultaneously and the average holding time is almost reduced to two slots.

In scatternet communication using RR, relay can transmit or receive data only by master polling. If master routes packets to relay and relay can not transmit packets to another master immediately, the transmission delay time grows. In OCP, a relay can have more high priority to decide the transmission and receive time. As shown in Fig. 6(b), OCP can decrease the transmission delay in scatternet communication.

4.2 Performance of Throughput

In a piconet, RR keeps one slave transmits data to another slave by master relay. TSL adopts the slave-to-slave direct communication without master relay. As shown in Fig. 7(a), by the time passing, the total throughput is more than RR. DSA uses the TSL more efficiently. Therefore, the total throughput is more than TSL. OCP uses the overlapping communication scheme, there are several communication pairs transmitting at the same time, hence the throughput is more than DSA.

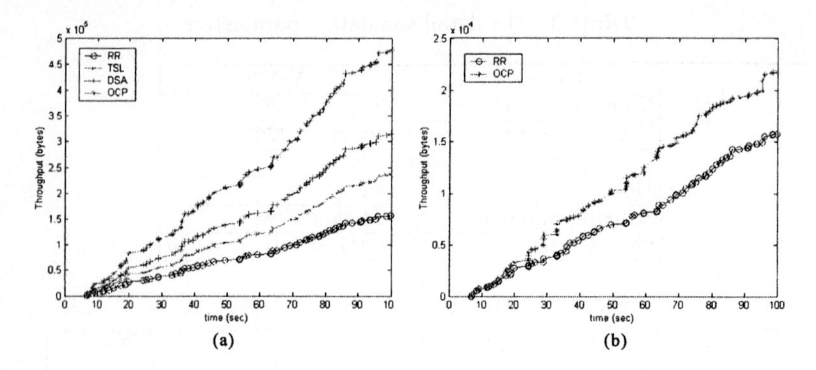

Fig. 7. Performance of throughput vs. seconds in (a) a piconet (b) a scatternet

In a scatternet, OCP decreases the holding time because the relay has more high priority to decide the transmission time and transmits to another relay directly by overlapping communication scheme. As shown in Fig. 7(b), the total throughput of OCP is more than RR in scatternet communication.

5 Conclusion

In this paper, we have proposed an efficient overlaping communication protocol to address the transmission holding problem. We realize OCP by using improved time-slot leasing. By OCP, devices can communicate simultaneously under the same frequence hopping sequence, and there are less packet delay time and low probability of network congestion. In scatternet, OCP is able to advance the data packet transmission between piconets. OCP has improvement in bandwidth utilization and transmission delay.

References

1. C. Cordeiro, S. Abhynkar, and D. Agrawal. "Design and Implementation of QoS-driven Dyamic Slot Assignment and Piconet Partitioning Algorithms over Bluetooth WPANS". *INFOCOM 2004 Global Telecommunications Conference*, pages 27–38, April 2004.
2. Bluetooth Special Interest Group. *"Specification of the Bluetooth System 1.2"*, volume 1: Core. http://www.bluetooth.com, March 2004.
3. VINT Project. "Network Simulator version 2 (ns2)". Technical report, `http://www.isi.edu/nsnam/ns` , June 2001.
4. IBM research. "BlueHoc, IBM Bluetooth Simulator". Technical report, http://www-124.ibm.com/developerworks/opensource/bluehoc/, February 2001.
5. W. Zhang, H. Zhu, and G. Cao. "Improving Bluetooth Network Performance Through A Time-Slot Leasing Approach". *IEEE Wireless Communications and Networking Conference (WCNC'02)*, pages 592–596, March 2002.

Full-Duplex Transmission on the Unidirectional Links of High-Rate Wireless PANs*

Seung Hyong Rhee[1], Wangjong Lee[1], WoongChul Choi[1], Kwangsue Chung[1], Jang-Yeon Lee[2], and Jin-Woong Cho[2]

[1] Kwangwoon University, Seoul, Korea
{shrhee, kimbely96, wchoi, kchung}@kw.ac.kr
[2] Korea Electronics Technology Institute, Sungnam, Korea
{jylee136, chojw}@keti.re.kr

Abstract. The IEEE 802.15.3 WPAN(Wireless Personal Area Network) has been designed to provide a very high-speed short-range transmission capability with QoS provisions. Its unidirectional channel allocations for the guaranteed time slots, however, often result in a poor throughput when a higher layer protocol such as TCP requires a full-duplex transmission channel. In this paper we propose a mechanism, called *TCP transfer mode*, that provides the bidirectional transmission capability between TCP sender and receiver for the channel time allocations(CTAs) of the high-rate WPAN. As our scheme does not require additional control messages nor additional CTAs, the throughput of a TCP connection on the high-rate WPAN can be greatly improved. Our simulation results show that the proposed scheme outperforms any possible ways of TCP transmission according to the current standard of the WPAN.

1 Introduction

The emerging high-rate wireless personal area network (WPAN) technology, which has been standardized[1] and being further enhanced by the 15.3 task group in IEEE 802 committee, will provide a very high-speed short-range transmission capability with quality of service (QoS) provisions. Its QoS capability is provided by the channel time allocations using TDMA; if a DEV (device) needs channel time on a regular basis, it makes a request for isochronous channel time. Asynchronous or non-realtime data is supposed to use CAP (Contention Access Period) which adopts CSMA/CA for the medium access.

Among the high-rate applications expected to be prevalent in a near future, the high-quantity file transfer using TCP will also occupy a large portion of the traffic transmitted in the WPAN environment. The unidirectional channel allocations for the guaranteed time slots, however, often result in poor throughput because a TCP connection requires a full-duplex transmission channel. In order

* This work has been supported in part by the Research Grant of Kwangwoon University in 2004, and in part by the Ubiquitous Autonomic Computing and Network Project, the MIC 21st Century Frontier R&D Program in Korea.

X. Jia, J. Wu, and Y. He (Eds.): MSN 2005, LNCS 3794, pp. 11–20, 2005.

to transmit the TCP traffic according to the current standard of the high-rate WPAN, one of the following three methods can be adopted. First, it can be transmitted during CAP. However, as the duration of the CAP is determined by the piconet coordinator (PNC) and communicated to the DEVs via *beacons*, it is very hard for the DEVs to estimate the available bandwidth for a TCP connection. Second, the TCP connection may request a guaranteed time slot and use it for the bidirectional TCP data and acknowledgment packets. Clearly it will cause frequent collisions at the MAC layer between TCP sender and receiver, and thus significantly degrade the transmission performance. Finally, they may request two CTAs, one for the TCP data and another for the TCP acknowledgment. Due to the dynamic nature of the TCP flow control, it is very hard to anticipate or dynamically allocate the size of the CTAs.

Recently, a lot of work has been done by many researchers in the area of the high-rate WPAN. However, few attempts have been made at the problem of non-real time TCP transmission on the unidirectional link so far. In this paper, we propose *TCP transfer mode* which is a mechanism that provides the bidirectional transmission capability between TCP sender and receiver on the guaranteed time slots of the high-rate WPAN. If a CTA is declared to be in the TCP transfer mode, the source DEV alternates between transmit mode and receive mode so that the destination DEV is able to send data (TCP ACK) in the reverse direction. Our mechanism is transparent to higher-layer entities: the source DEV regularly makes transitions between transmit and receive mode, and the destination DEV sends data only when the CTA is in the TCP mode. In addition, as our scheme does not require additional control messages nor additional CTAs, the throughput of a TCP connection on the high-rate WPAN can be greatly improved.

The remaining part of this paper is organized as follows. After introducing related works and the high-rate WPAN protocol in chapter 2, we describe the three methods of TCP transmission under the current standard in chapter 3. In chapter 4, we propose a new transmission mode that allows bidirectional TCP transfer on the guaranteed time slots. Simulation results are provided and discussed in chapter 5, and finally chapter 6 concludes the paper.

2 Preliminaries

2.1 IEEE 802.15.3 High-Rate WPAN

The IEEE 802.15.3 WPAN has been designed to provide a very high-speed short-range transmission capability with QoS provisions[7,8]. Besides a high data rate, the standard will provide low power and low cost solutions addressing the needs of portable consumer digital images and multimedia applications. Figure 1 shows several components of an IEEE 802.15.3 piconet. The piconet is a wireless ad hoc network that is distinguished from other types of networks by its short range and centralized operation. The WPAN is based on a centralized and connection-oriented networking topology. At initialization, one device (DEV) will be required to assume the role of the coordinator or scheduler of the piconet. It is

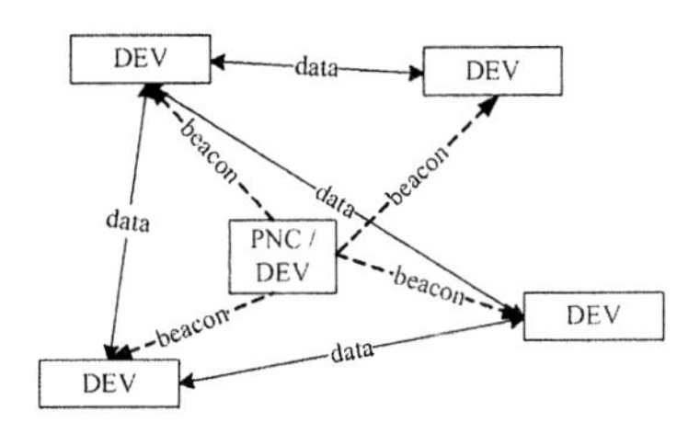

Fig. 1. IEEE 802.15.3 piconet components

called PNC (piconet coordinator). Its duty includes allocating network resources, admission control, synchronization in the piconet, providing quality of services, and managing the power save mode.

The superframe of the piconet consists of several periods as follows. In the first period, the PNC transmits a beacon frame which contains all the necessary information to maintain the piconet. All the DEVs in the piconet should receive the beacon and synchronize their timer with the PNC. The beacon frame is used to carry control information and channel time allocations to the entire piconet. In the second period, CAP (Contention Access Period) can be allocated optionally for the purposes of association request/response, channel time request/response and possible exchange of asynchronous traffic using CSMA method. CTA period is in the third period and is the most part of the superframe. This period is used for isochronous streams and asynchronous data transfer. Using a TDMA mechanism, the period allocates guaranteed time slots for each DEV. All transmission opportunities during the CTA period begin at predefined times, which is relative to the actual beacon transmission time and extends for predefined maximum durations. Those allocation information is communicated in advance from the PNC to the respective devices using the traffic mapping information element conveyed in the beacon. During its scheduled CTA, a DEV may send arbitrary number of data frames with the restriction that aggregate duration of these transmissions does not exceed the scheduled duration limit.

2.2 Related Works

Recently, a lot of work has been done by many researchers in the area of the high-rate WPAN. However, few attempts have been made at the problem of non-real time TCP transmission on the unidirectional link so far. In [2], the authors proposed a MAC protocol that enhances the TCP transmission mechanism in TDMA-based satellite networks. This is an approach of TCP throughput

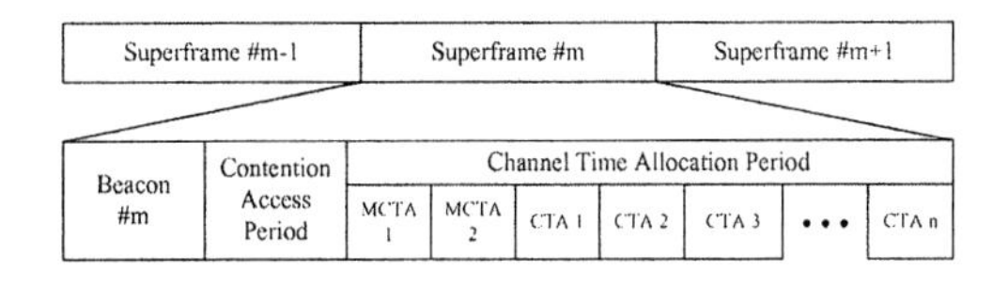

Fig. 2. IEEE 802.15.3 superframe

enhancement using the modified MAC protocol. Similarly, [3] proposes a mechanism for enhancement of TCP transfer via satellite environment. [4] proposes a MAC layer buffering method to improve handoff performance in the Bluetooth WPAN system in order to improve the TCP performance. They can minimize the negative effects of the exponential backoff algorithm and prevent duplicate packets during handoffs. In [5], although authors are not concerned with TCP transmission, they proposed an *application-aware* MAC mechanism which considers the status of higher layer. To the best of our knowledge, however, there have been no research in which the MAC layer supports efficient TCP transfer in the high-rate WPAN.

3 TCP Transmissions for the WPAN

In this chapter, we describe three possible methods of TCP transmission with the MAC protocol of the current WPAN standard which contains no mention on higher layer protocols. The performance of TCP transmission using each possible method will be discussed and compared. Except the three methods discussed in this chapter, TCP traffic can be transmitted during CAP. However, as the duration of the CAP is determined by the PNC and communicated to the DEVs via the beacon, it is very hard for the devices to estimate the available bandwidth for the TCP connection. We will consider only the methods of using CTAP in this paper. In order to transmit the TCP traffic using CTAP according to the current standard of the high-rate WPAN, one of the three methods in this chapter can be adopted.

Figure 3 shows a TCP transmission process using immediate ACK policy in high-rate WPAN. First, TCP data packet comes from the higher layer and is processed at the MAC layer. The sender DEV sends the MAC frame to the receiver DEV via wireless interface. At the MAC layer of TCP receiver, it receives the MAC frame and sends a MAC ACK to the sender. The TCP sink that accepted TCP data frame sends a TCP ACK packet to the TCP sender. The TCP sender that received this TCP ACK packet sends MAC ACK frame for the TCP ACK. TCP sender transmits next TCP data packet to the receiver when it received TCP ACK. All TCP transmissions are achieved by this way in the high-rate WPAN.

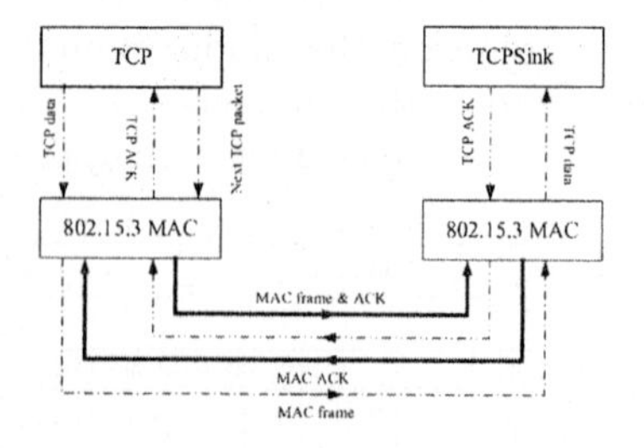

Fig. 3. TCP transmission in IEEE 802.15.3 high-rate WPAN

3.1 TCP Transmission on a Single CTA

According to the current MAC protocol, one may use a single (unidirectional) CTA for a TCP connection. Clearly, this will result in a poor throughput because the TCP connection requires a full-duplex transmission channel. TCP traffic can not be transmitted between the sender and receiver using this method, because the CTA is defined as unidirectional and the TCP receiver has no way of sending TCP ACKs to the sender. Thus transmission of transport layer ACK is impossible and the connection can not be maintained. Figure 4 depicts the case where a single CTA is allocated to the TCP sender and the TCP receiver has no way of sending back the ACKs.

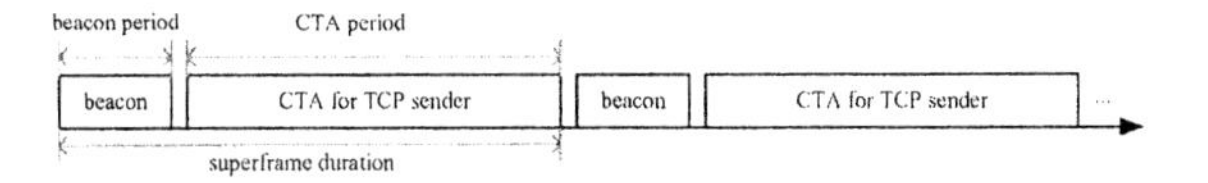

Fig. 4. Single CTA in the high-rate WPAN

3.2 Allocating Two CTAs for TCP Data/ACK

The PNC may allocate extra CTA for the TCP receiver, so that the receiver is able to send back the necessary ACKs. In this method, the TCP sender transmits data during its own CTA, and the receiver transmits ACKs also during its allocated CTA. The problem here is that, due to the dynamic nature of the TCP flow control, it is very hard to anticipate or dynamically allocate the size of those CTAs. In addition, TCP sender waits an ACK packet after sending data up to the window size, and the receiver can not send an ACK before its CTA comes. Therefore, this method may waste the two CTAs because exact channel time allocation is almost impossible due to the dynamic property of the TCP connection.

Figure 5 explains this method: TCP sender is assumed to transmit data packets during CTA1, and the receiver sends ACK packets during CTA2. In this method, the throughput can be very different according to the ratio of the durations of the two CTAs. If the ratio of the allocated CTAs does not consider the current status of the TCP connection, those CTAs can be seriously wasted. The problem of using two separate CTAs for a TCP connection is that it is extremely hard to adjust the ration of the two CTAs according to the dynamics of a TCP connection.

3.3 Sharing a Single CTA

TCP connection may request a single guaranteed time slot and use it for the bidirectional TCP data and acknowledgment packets. A single CTA is shared between TCP sender and receiver. That is, the sender send TCP data to the

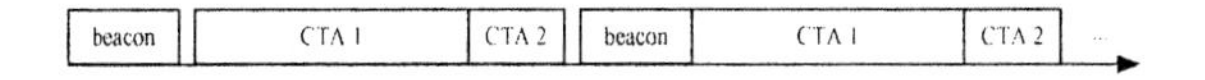

Fig. 5. Two CTAs for a TCP connection

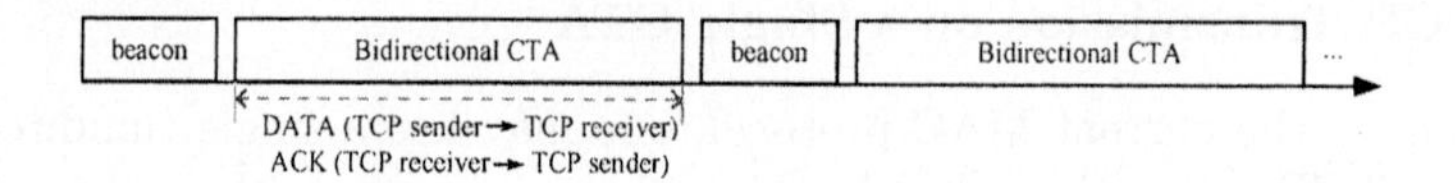

Fig. 6. Sharing a single CTA between two device

receiver and the receiver send TCP ACK to the sender during a single CTA. Clearly it will cause frequent collisions at the MAC layer between TCP sender and receiver, and significantly degrade the transmission performance.

4 TCP Transfer Mode

In this chapter, we propose *TCP transfer mode* which can maintain the throughput of a TCP connection without collision in a single CTA that is shared between TCP sender and receiver. The sender device informs PNC that it will send TCP data when it makes a channel time request. For this purpose, we have defined TCP Enable bit using the reserved bits in 15.3 MAC header. The PNC responds to the request, and then broadcast the beacon frame with the information on the newly allocated CTA for the TCP connection. The CTA information contains the stream index field which tells that the CTA is allocated to a TCP connection and the CTA will be used in the *TCP Transfer mode.* The TCP stream index and CTA block are explained in Figure 7.

There are three kinds of ACK policy in the IEEE 802.15.3 high-rate WPAN: Immediate-ACK, No-ACK and Delayed-ACK policy. We consider only Immediate-ACK and No-ACK policy in this paper. We describe the TCP transfer mode with the No-ACK case first. The (TCP) sender changes its radio interface from TX mode to RX mode immediately after it sends a (TCP) data. Then it senses a frame in the reverse direction, which is possibly TCP ACK from the TCP receiver, during the SIFS (short inter frame space). TCP sender will be transmitting a TCP data continually if channel is idle. If TCP sender receives a TCP ACK from TCP receiver, TCP sender maintains the radio interface status as RX. TCP sender received whole ACK then sends next TCP data after waiting for a SIFS time. There is a figure in below that explains this operation of proposed TCP transfer mode. This TCP transfer mode support transmission without collision.

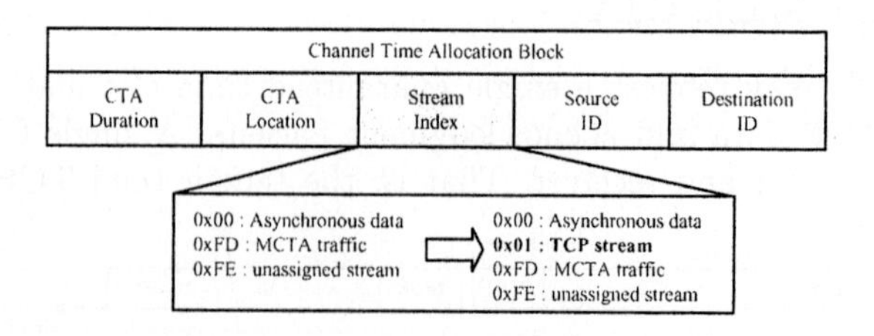

Fig. 7. Stream Index field and value in CTA block

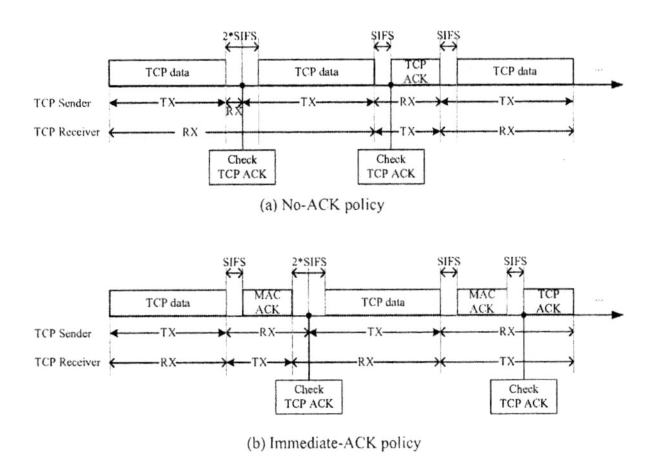

Fig. 8. Difference of ACK policies: (a) No-ACK policy (b) Immediate-ACK policy

Table 1. Simulation Environment

Attribute	Value
Bandwidth	100Mbps
Number of flows	1, 2, 3
CAP duration	4000us
CTAP duration	4000us
CTA duration	3500:500, 3000:1000, 2500:1500
MAC ACK policy	Immediate-ACK policy
TCP packet size	1024, 2048, 4096 byte
TCP window size	20
Error rate	0, 25, 50 %

5 Performance Evaluations

5.1 Simulation Environment

We have implemented our *TCP transfer mode* in the ns-2 network simulator with the CMU wireless extension[9,10]. Our implementation includes beacon transmission, channel time management and ACK policies for the 15.3 WPAN. The parameters used for the simulation are summarized in Table 1. We assume that all the DEVs are fixed during the simulation, and are associated to the piconet before the simulation starts. Moreover, there are no control frames and management overhead except the beacon transmission. We have chosen the channel bit rate as 100Mbps in order to allow enough increasing of TCP window size. For the purpose of comparison, we use a same amount of time for the CAP and CTA duration. Finally, we have use various number of flows and adopts different error rates in order to verify the performance of our mechanism under diverse environments.

5.2 Simulation Results Without Channel Errors

In this section we evaluate the performance of the proposed TCP transfer mode in an *error-free* channel, in which no errors are assumed during the transmission of MAC frames. The effect of proposed mechanism can be verified in figure 9(a). In the case of single CTA method, although the entire bandwidth is 100Mbps, the aggregate throughput is saturated with about 28Mbps. This throughput degradation is due to the collisions between TCP sender and receiver. Since MAC layer can not distinguish TCP data and ACK packets, it transmits MAC frames whenever its queue is backlogged and the wireless medium is idle. Thus, collisions between peer DEVs are inevitable and the throughput is degraded.

By using *TCP transfer mode*, we can achieve the throughput of 38Mbps in the simulation. This performance is due to the MAC entities' ability that avoids collisions by checking the frame transmissions during the inter-frame space. Thus, the peer entities use the channel bidirectionally in the mode without packet loss and retransmission. TCP sender transmits TCP data packets as much as TCP window at once. This effect should be distinguished with that of TCP transmission in CAP periods: PNC can not determine the required length of the CAP for a TCP connection, and RTS/CTS frames are required in the CAP.

In figure 9(b), we have used two flows in the piconet. As the number of flows is increased, the channel times allocated to a connection is decreased. Thus the throughput of each TCP connection is also decreased regardless of the mechanisms. The graph shows that the throughput is reduced by half compared that of figure 9(a). In this case, however, TCP transfer mode still outperforms the other methods. Figure 11(a) shows the throughput of a TCP connection when the number of flows in the piconet varies. As the duration of a superframe is fixed (4000μsec) during the simulations, a large number of flows means a small CTA for a connection, and thus a low throughput. For example, the duration of a CTA is 2000μsec for 2 flows and 1333μsec for 3 flows. Table 2 compares the TCP throughput according to the number of flows under different mechanisms. For any number of flows, TCP transfer mode shows a higher throughput than other methods.

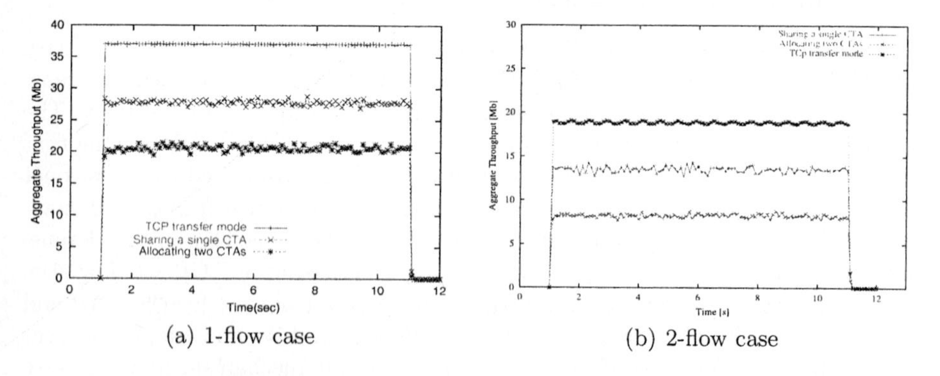

(a) 1-flow case (b) 2-flow case

Fig. 9. TCP throughput in a 15.3 WPAN

Table 2. Comparison of throughput for different mechanisms

Flows	CAP	Two CTAs	*TCP Transfer Mode*
1	28 Mbps	21 Mbps	38 Mbps
2	13.5 Mbps	8.3 Mbps	18.8 Mbps
3	8.7 Mbps	-	12.7 Mbps

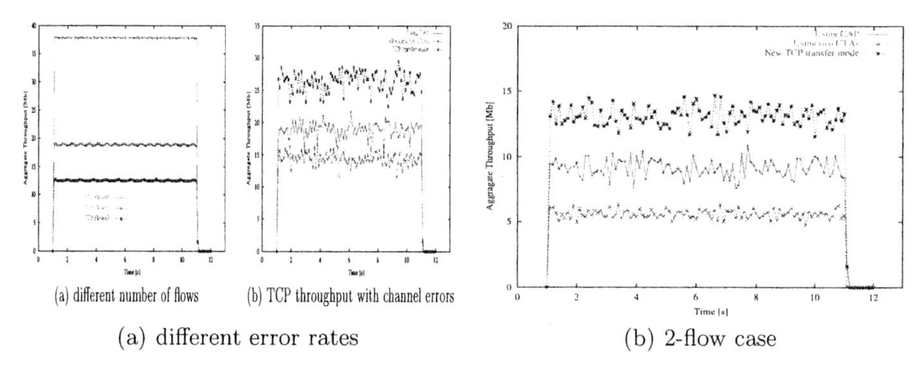

(a) different error rates (b) 2-flow case

Fig. 10. TCP transfer mode with frame error

5.3 Simulation Results with Channel Errors

In this section we evaluated the performance of the proposed TCP transfer mode in the wireless channel with frame errors. We assume that the frames are corrupted randomly according to a uniform distribution and, once a frame error occurs, the TCP receiver can not interpret the frame. Figure 11(b) shows the

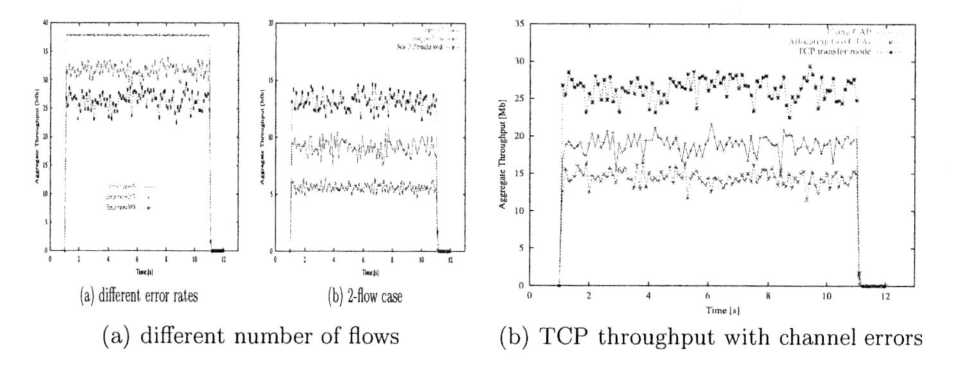

(a) different number of flows (b) TCP throughput with channel errors

Fig. 11. Simulation Results

Table 3. Performance of the mechanisms with frame errors

Error rate	CAP	Two CTAs	*TCP Transfer Mode*
0%	28 Mbps	21 Mbps	38 Mbps
25%	23 Mbps	18 Mbps	31 Mbps
50%	18 Mbps	14.5 Mbps	24.5 Mbps

simulation result when 50% of frame error is assumed in the wireless medium. The performances of three methods are depicted in the graph, and again, TCP transfer mode shows the best performance. Note that the graphs shows fluctuations as the frame errors cause the frame retransmissions and variation of TCP window sizes. In figure 10(a), we simulate the performance of TCP transfer mode for different rates of frame errors to examine the influence of channel errors. The throughput of a TCP connection decreases as the error rate increases. Table 3 compares the performance of different mechanisms discussed in this paper when there are non-zero values of error rates. In any cases our TCP transfer mode shows better throughput than those of other mechanisms. Finally, figure 10(b) depicts the performance for two TCP flows and 50% of frame errors in the piconet. Again in this simulation, TCP transfer mode outperforms the other methods.

6 Conclusion

In this paper we have proposed an efficient TCP transmission method that provides the bidirectional transmission capability between TCP sender and receiver for the channel time allocations of the IEEE 802.15.3 high-rate WPAN. Our scheme does not require additional control messages nor additional CTAs and it can be implemented with a minor change of current standard. We have described three possible methods of TCP transmission with the MAC protocol of the current WPAN standard, and compared the performance of our mechanism with those of the three methods under various simulation environments. Our extensive simulation shows that our proposed mechanism greatly improves the throughput of a TCP connection in the piconet regardless of the number of flows or the error rates.

References

1. Wireless Medium Access Control and Physical Layer Specifications for High Rate Wireless Personal Area Networks, IEEE standard, Sep. 2003
2. J. Zhu, S. Roy: Improving TCP Performance in TDMA-based Satellite Access Networks: ICC2003 IEEE(2003)
3. J. Neale, A. Mohsen: Impact of CF-DAMA on TCP via Satellite Performance: GLOBECOM2001 IEEE(2001)
4. S. Rhee *et al.*: An Application-Aware MAC Scheme for A High-Rate Wireless Personal Area Network: IEEE WCNC2004 (2004)
5. H. Balakrishnan, S. Seshan, E. Amir, R.Kantz: Improving TCP/IP Performance Over Wireless Networks: ACM MOBICOM 95(1995)
6. J. Karaoguz: High-rate Wireless Personal Area Networks: IEEE Communication magazine(2001)
7. P. Gandolfo, J. Allen: 802.15.3 Overview/Update: The WiMedia Alliance(2002)
8. The CMU Monarch Project: Wireless and mobile extension to ns Snapshot Release 1.1.1: Carnegie Mellon University(1999)
9. K. Fall, : The ns Manual, UC Berkeley, LBL, USC/ISI, and Xerox PARC(2001)

Server Supported Routing: A Novel Architecture and Protocol to Support Inter-vehicular Communication

Ritun Patney, S.K. Baton, and Nick Filer

School of Computer Science, University of Manchester,
Manchester, United Kingdom M13 9PL
{patneyr, S.K.Barton, nfiler}@cs.man.ac.uk

Abstract. A novel architecture and multi-hop protocol to support inter-vehicular communication is proposed. The protocol uses 'latency' as a metric to find routes. We introduce the concept of a Routing Server (RS), which tries and keeps up-to-date information about the state of the network, i.e. the network topology, and the latency associated at each node averaged over a short time. Route discovery is now carried out by sending a Route Request packet directly to the RS, rather than flooding the network. We call this model 'Server Supported Routing (SSR)'. A simulation study is carried and the performance of SSR compared with Dynamic Source Routing (DSR) protocol. It is found that SSR performs better than DSR over longer hops, faster rate of topology change, and high offered load in the network. The study also reveals that DSR is much more sensitive to changes in the network than SSR.

1 Introduction

The demand for mobile broadband services is increasing and soon vehicles will be enabled with broadband aware devices. Users will expect similar quality of service as in wired networks. However, current 3G networks, at vehicular traffic speeds, can only provide a transmission rate of 144 Kbps [1]. Wireless Local Area Networks (LAN) support data rates of up to 54Mbps [2], but such high data rates are restricted by very short transmission ranges (50-80m). One solution is to provide an Access Point (AP) at every 50m along all major road networks. However, such a solution will be too expensive to implement. Using a pure ad-hoc based solution tends to give a very low throughput for longer paths [3].

In this work we present a new idea, which, we believe, has not been proposed before in the literature. The solution incorporates a novel architecture along with changes to current multi-hop ad-hoc routing protocols. The model is motivated from the work carried out by Lowes [4]. Lowes measures the maximum achievable performance a routing protocol can reach using simulation studies. For this, his work implements a piece of software called the 'magic genie routing protocol'. This software sits above all routers in the simulation environment. All nodes, at periodic intervals of time, keep passing their dynamic state, e.g. latency experienced by the packets being forwarded (averaged over a short time), current buffer size, etc. to the magic genie software. The software is also kept aware of the current network

X. Jia, J. Wu, and Y. He (Eds.): MSN 2005, LNCS 3794, pp. 21 – 30, 2005.

topology. The best path (dependent on the desired metric - hop count, latency, buffer size, etc.) from a source to a destination can now be found using the Dijkstra or Bellman-Ford [5] algorithm. When a node now has data for a destination, it simply asks the magic genie for the best available route.

In our model, we introduce the concept of a Routing Server (RS), which is analogous to the magic genie routing server. We call our model 'Server Supported Routing' (SSR), and the network uses latency as the routing metric. We compare its performance with Dynamic Source Routing (DSR) protocol [6]. The remaining paper is organized as follows. In section 2, we describe the proposed model. Section 3 enumerates some details of implementation. In section 4, we present our results and analysis them. We finally conclude with a short discussion.

2 Approach and Proposed Model

Fig. 1 provides an insight into the proposed architecture. The Routing Server (RS) is connected with immobile nodes, which we call Routing Server Access Points (RSAP). These are provided at regular distances (which can be much greater than the 50-80m limit for infrastructure based WLANs), and enable communication between the vehicles and the RS. Apart from RSAPs being immobile, they have the same functionality as any other vehicle. In the following discussion, a 'node' refers to the vehicles as well as the RSAPs. We assume the physical wireless links are bi-directional. The communication between the RSAP and RS is assumed to be done via underground cables. Furthermore, each vehicle is considered as a single network node, and the road network has no junctions and stoppages.

2.1 Overview

Every node, at all times, tries to maintain a valid route to the RS. The route is based on the least latency a packet would experience to reach the RS, and contains the list of all nodes on this route, but itself. We refer to this route as R_n, which is defined below.

Let S = {n, n-1,...., 1} be a set of ordered nodes in a path. Then R_n is the subset {n-1,...., 1} of S, and where node numbered '1' is the RS.

Furthermore, every node also maintains a list of its current neighbours, and the average latency experienced by packets forwarded by it over time t (L_t). The neighbour list plus the latency at its end is referred to as 'Routing Information Set (RIS)'. Periodically, every node, sends its RIS to the RS on R_n. In turn, the RS uses the RIS packets received from different nodes to maintain a link state graph of the network, along with the average latency experienced by packets at each node.

2.2 Building the RIS and Maintaining R_n

The RS and every node periodically broadcast a RSBROADCAST (Routing Server Broadcast) packet on all interfaces. The packet is used by the receiving nodes to

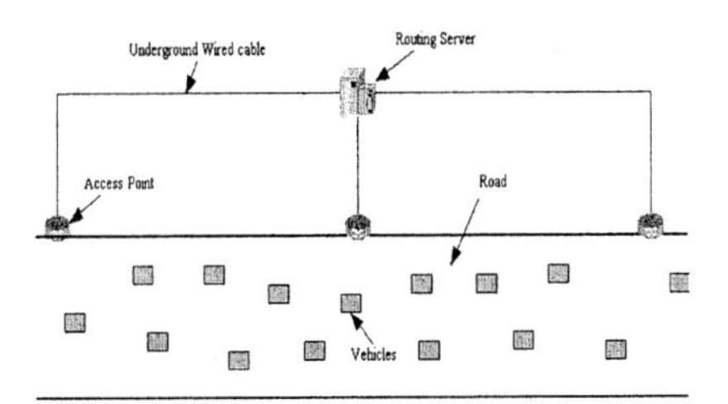

Fig. 1. Architecture for SSR

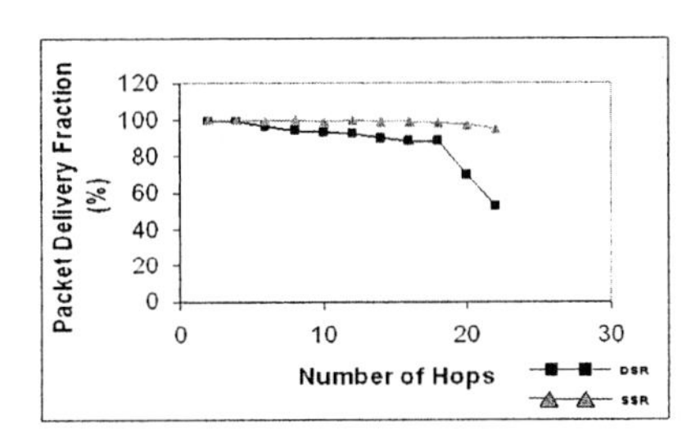

Fig. 2. Impact of hop count on p.d.f

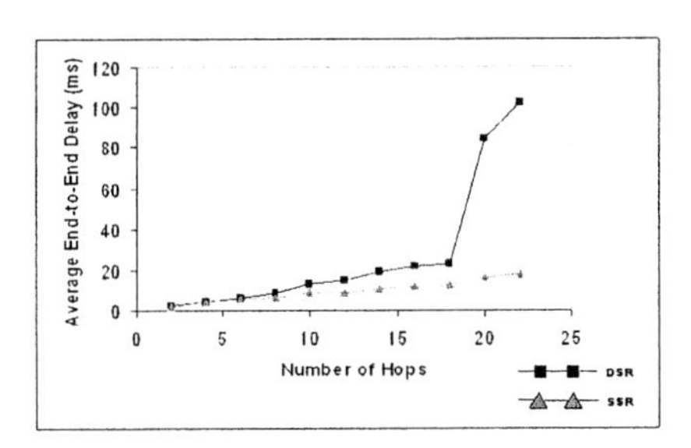

Fig. 3. Impact of hop count on delay

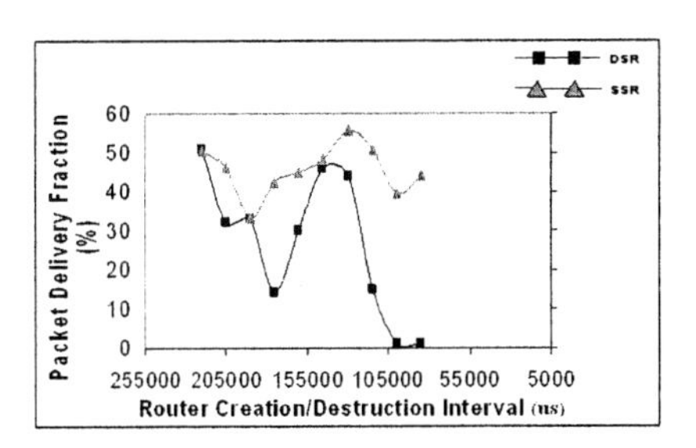

Fig. 4. p.d.f vs rate of topology change

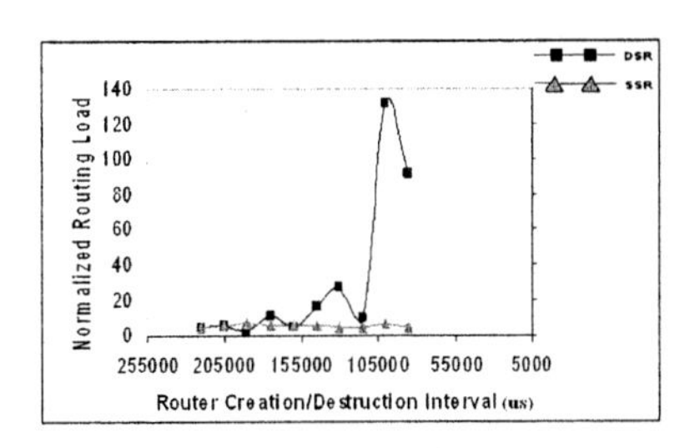

Fig. 5. Routing Load vs rate top. change

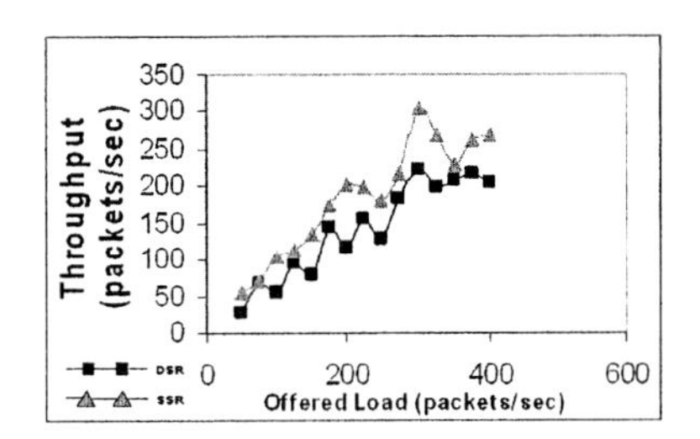

Fig. 6. Throughput vs offered load

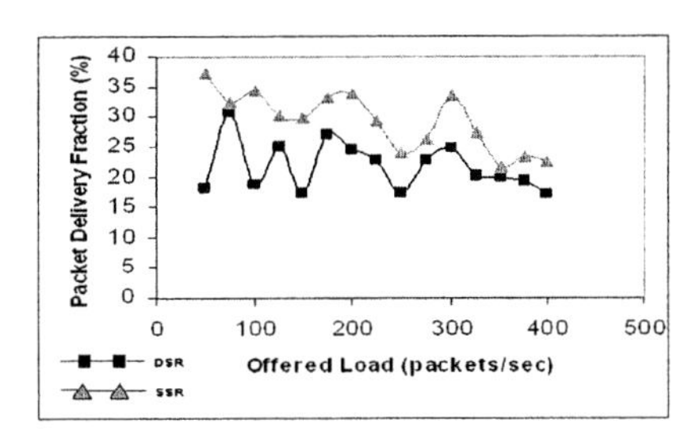

Fig. 7. p.d.f vs offered load

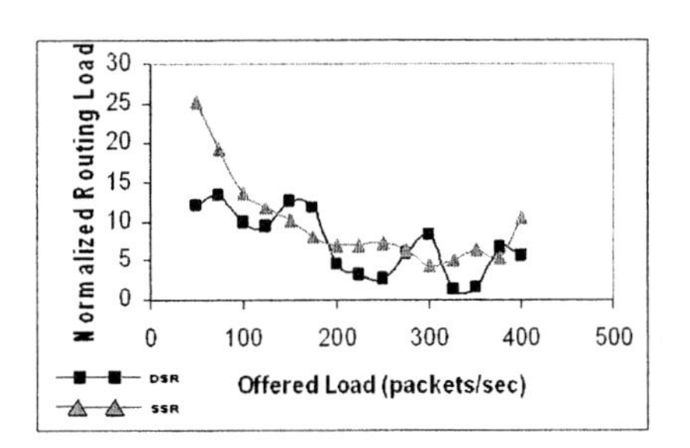

Fig. 8. Routing load vs offered load

maintain an updated R_n as well as to maintain a list of their neighbours. A node broadcasts a RSBROADCAST packet only if it itself has a valid route to the RS. The originator of this packet fills up the two fields of RSBROADCAST in the following manner:

1. Route to the RS (R_n) being currently maintained by the originator – The set S, obtained by adding itself to R_n.
2. Latency – This field indicates the latency a packet forwarded on S would experience. It is the sum of the average latencies at each node in the set S. The RS fills in a value of 0 before originating such a packet. Each node adds its own latency to the latency of the path R_n before filling this field.

On reception of a RSBROADCAST packet, a node does the following:

1) Checks the route to the RS contained in the received packet. If this node's address already appears in the route, then it silently discards the packet.
2) Else, if the current Rn passes through the originator of RSBROADCAST, then it updates Rn to use the new route.
3) Else, if the node does not have a valid or fresh enough Rn, it updates Rn with the route in the received packet.
4) Else, it compares the latency associated with its current Rn, with the latency in the received packet. It replaces Rn with the new route if the latter claims a lesser latency.

Apart from the above, it always adds the originating node as its neighbour, and associates a time out for it. This timeout indicates the time by which it should receive the next RSBROADCAST packet for the node to qualify as its neighbour. A similar timeout value is also associated with R_n, after which the route is no longer valid.

2.3 Creating a Link Cache at RS

RIS packets contain the originating node's neighbours, and its L_t. This is used by the RS to update its maintained network state. The originating node of the RIS packet, along with each of its neighbour forms a link, and each node's latency is stored as extracted from the RIS packet. The RS also associates each node and link with a validity time. Both are removed if the time expires, and no information has been received about them recently enough. In case, a node is removed, all links associated with the node are also removed.

While adding a link, it maybe possible that the node at the other end of the link does not exist in the cache (no RIS packet received from it recently enough), the RS still adds this link to its cache since we have assumed bi-directional links. However, since no information has been explicitly received from the node in question, a high enough latency is associated with this node such that routes through it will only be chosen in case of no other route being possible between a source and a destination without its inclusion.

2.4 Route Discovery

A node requiring a route to a destination, unicasts (thus preventing the overhead associated with flooding) a Route Request (RREQ) packet to the RS; the packet is source routed with route Rn. The model was first created such that hop-by-hop routing is used by RREQ packets, but however it was observed that due to the asynchronous change of Rn at different nodes, some RREQ packets were looping between nodes, and never reached the RS.

On receiving a RREQ, the RS calculates a route based on the least latency and sends a Route Reply (RREP) back to the source node on the same route on which the RREQ packet had arrived. In our model, the source uses this route either for a fixed time, or to forward a fixed number of packets, whichever limit is reached first. Route discovery is done again once the limit is reached.

2.5 Route Maintenance

We have eliminated RERR packets completely as opposed to DSR. The following discussion summarizes the situations where a RERR would have been sent by DSR, and how our model avoids them.

While forwarding a packet (with a source route) to its next hop, if the link layer is not able to deliver the packet, it informs the network layer. In DSR, the network layer initiates a RERR for the originator of the failed packet (one RERR per one failed packet). In our network layer, route maintenance is inherent in the design, which is explained as follows:

1) A node re-starts a route discovery when it has used a route a maximum number of times (hence new routes are fetched periodically).
2) Nodes do not transmit RIS and RSBROADCAST packets if they do not have a fresh enough route to the RS (thus avoiding route failures).
3) The RREQs are only generated by a node only if it has a valid route to the RS. Furthermore, the RS does not reply back with a RREP if it does not find a route in its route cache between the source and the destination.
4) Links and routes time out and become invalid after a certain period of time unless new information is obtained about them within the time out period.

All this makes it unnecessary for RERRs to be propagated. The non-generation of RERRs also saves network bandwidth.

In DSR, while forwarding a RERR, a failure of another link may cause more RERRs to flow. Also, if a node generating a RERR does not have a route to the source node; it starts its own route discovery for the source, causing greater routing overhead.

3 Implementation

We chose Dynamic Source Routing (DSR) as the basis for comparison because SSR has been formed after studying and modifying DSR in detail; some qualities like source routing for data packets, and using the same path for forwarding RREP packets on which RREQ arrived, are the same for both. We simulate SSR and DSR on a discrete event simulator [7]. The simulator has been developed by the Mobile Systems Architecture Research Group, School of Computer Science, University of Manchester, UK [8]. The simulations have been carried out in Windows environment, with Java as the programming language. The machine used was Intel based with 256MB of memory.

3.1 Modeling the Protocol Stack

The OSI reference model [9] divides the protocol stack into layers, each having a different functionality. For our simulations, we model the physical layer, MAC layer, and the network layer. The MAC layer used for simulations is Aloha [10]. In Aloha, each node transmits whenever it has data to transmit. In case of a collision, the frame is re-transmitted after a random amount of time. The randomness is necessary; as otherwise, the frames will keep colliding over and over again.

3.1.1 Traffic Simulation

Traffic between a source and a destination is simulated by attaching a 'traffic stream' object between the pair (rather than nodes having an application layer which generates packets). The stream decides when the source generates a packet. The stream makes a function call on the node's network layer to make it generate a packet. The stream consists of the start time of the first packet, the rate at which packets are to be generated, and number of packets to generate per call.

3.1.2 Network Layer Design

We implement part of the DSR protocol. The basic route discovery mechanisms (RREQ and RREP) are implemented. The RREP is sent after reversing the route on which the RREQ arrives. The route maintenance (RERR) mechanism is also implemented. However, the extensions to the DSR protocol are not implemented. These include caching overhead routing information, support for unidirectional links, expanding route discovery, packet salvaging, automatic route shortening, etc.

3.1.2 Network Layer Design

We implement part of the DSR protocol. The basic route discovery mechanisms (RREQ and RREP) are implemented. The RREP is sent after reversing the route on which the RREQ arrives. The route maintenance (RERR) mechanism is also implemented. However, the extensions to the DSR protocol are not implemented. These include caching overhead routing information, support for unidirectional links, expanding route discovery, packet salvaging, automatic route shortening, etc. Additional functionality required to simulate SSR (like unicasting of RREQ, etc.) has been added to the DSR design. This allows the program to be run in both modes (DSR as well as SSR) and can be set from an external parameter.

3.2 Topology Simulation

We model node mobility by creating and destroying vehicles on a random basis. The reasoning behind this is that a link is only important at the time when a packet is being forwarded on it. For all simulations, the number of vehicles at any time is kept constant, i.e. for one vehicle destroyed; another is created at a random location (by choosing a random set of co-ordinates).

4 Simulations

The following metrics are measured and evaluated:

1) Packet Delivery Fraction (P.D.F) – The ratio of the data packets received by the destinations to those generated by the sources.

2) Throughput – Number of packets received per second at all destinations. We plot the average throughput of a stream. It is plotted in terms of packets received per second.
3) Normalized Routing Load – Number of routing packets transmitted per application packet received. A single transmission (hop-wise) of a routing packet by each node is counted as one transmission. This is summed for the whole network, and divided by the number of packets received at all destinations.
4) Average end-to-end delay – This is the delay experienced by a packet from creation until being successfully received at the destination. It includes re-transmission delays at the MAC layer, and the route discovery delays. It is averaged over all streams.

The metrics are plotted against the offered load per traffic stream, number of hops, and the rate of topology change, depending on the experiments.

4.1 Configuration

For all the simulations, we use fixed packet sizes of 512 bytes. All simulations are started with 225 vehicles. The number of access points for all simulations has been kept at 8 (making the road length of about 6000m). The channel bandwidth (bit transmission rate) is configured at 54 Mbits/s.

All simulations are 40s long. Traffic streams are created between 10 and 15s. The vehicles are created and destroyed starting from 1s. The network is found to stabilize from 25s, and all measurements are done from 25 to 40s.

4.2 Experiments

Three different sets of experiments are performed. In the first, two traffic streams are run in the opposite direction between the same set of nodes. The number of hops is varied in increments of 2 and the packet generation rate is kept at 300 packets/sec. This is done on a static topology. We measure the packet delivery fraction and the end-to-end delay (fig. 2, and fig. 3). The packet delivery fraction in DSR falls rapidly with increase in number of hops, whereas in SSR, it falls slowly. This in a way is consistent with the theory that probability of a packet drop increases with increasing path lengths [11]. The end-to-end delay in DSR, for longer path lengths, is seen to be much higher than for SSR. Collision of frames leads to re-transmissions, which adds to their delay. This delay also adds to the delay of the buffered frames (which will have to wait till this frame finishes delivery or is dropped). Furthermore, a frame colliding more than a specific number of times is ultimately dropped by the MAC layer, leading to greater packet loss. These problems are exacerbated with increasing of intermediary nodes (longer routes).

In SSR, because routes are acquired on a regular basis, data is sent on different routes (may not be completely disjoint routes). Some of these routes may also pass through the RS (a wired link not suffering from the problems of wireless links). For packets routed via the RS, these problems come into existence only for the hops from an AP to the destination and from the source to the AP. SSR is therefore able to give a reasonably steady performance over varying hop counts.

In the second set of experiments, we run 3 traffic streams, between arbitrarily chosen sources and destinations. The pairs are chosen such that they are at least 6 hops apart. The packet generation rate is kept constant at 300 packets per second, and the rate of vehicles being created and destroyed is varied. We measure packet delivery fraction, and the normalized routing load (fig. 4, and fig. 5).

It is found that the packet delivery fraction for DSR is very poor as compared to SSR for this high rate of topology change. This can be attributed to the following:

1. In DSR, to obtain a route, RREQ and RREP packets need to traverse between the source and the destination. The RREP is returned on the route on which the RREQ arrives. Some intermediary nodes may be destroyed while the route is being discovered, thus breaking the individual links, and thus the path on which the RREP packet was to be forwarded. This may lead to failure of the source receiving any RREP. This makes the source re-start the route discovery, and after a certain number of retries, all packets for this destination are dropped.
2. In DSR, data packets are also source routed. Destroying a node which falls on the route on which data packets are being forwarded, will result in failure of some data packets to reach the destination. This will also result in the generation of RERR packets, not only causing a higher routing load, but also greater interference at the physical layer. Furthermore, it will take time before the source node gets a RERR packet. In the meantime, the source keeps forwarding data packets on the same failed route.

In SSR, the RREQ is not sent all the way to the destination but only to the RS through the nearest AP (which also means fewer hops generally). For data packets forwarded to their destinations through the RS, the probability of them getting lost due to intermediate nodes dying is less (could happen over routes between the source to AP and AP to the destination).

The normalized routing load in SSR is more or less the same irrespective of the topological changes. All nodes keep sending RIS and RSBROADCAST packets even if they do not have any data to transmit or receive. The only other routing load is due to the RREQ and the RREP packets. RREQ is not broadcasted but unicasted to the RS and RERR packets are not generated in the event of a link fail.

To stress the above discussed points further, the variation in the graphs for DSR is much more than for SSR. This implies that DSR is more likely to be affected by changes in the links than SSR.

It is interesting to note that at slower rates of topology change, performance of DSR and SSR is quite similar. A slower rate means the network is closer to a static network. DSR may find multiple routes to a destination for every RREQ it sends (since a RREQ is broadcasted and RREP is generated for each RREQ received). Therefore it may use a second route available on receiving a RERR, thus reducing the routing load and improving the packet delivery fraction.

In the last set of experiments, we run 3 traffic streams, chosen arbitrarily (with a minimum hop separation of 6), and vary the rate of packet generation of each stream. To make it fairly challenging for the protocols, we keep the rate of router creation/destruction constant at 8.7 routers per second. With an even faster rate of router creation and destruction, the throughput of DSR starts to approach zero. This happens because no route discovery process (sending of RREQ and receiving a RREP) is able to complete with such a fast rate of topology change, and thus not providing a basis for meaningful analysis and comparison with SSR. The number of packets generated per second is varied from 50 to 400 packets per second (in increments of 25 packets per second). We measure the throughput (packets/sec), normalized routing load, and packet delivery fraction (fig. 6, fig. 7, and fig. 8). The whole process is repeated with traffic streams run between a different set of sources and destinations. The performance metrics are then averaged for the two.

SSR performs better than DSR when it comes to packet delivery fraction and the average throughput of each stream. It is however interesting to note that the throughput increases with increase in data rate, whereas the percentage of packets delivered decreases. This means that although both protocols, with increasing data rates, are able to get more number of packets to the destination with in the same time, they are not able to achieve this proportionally to the increase in data rate. This needs to be studied in more detail in future work.

The normalized routing load decreases for SSR with increasing data rates. This is because, as discussed above, the routing load in the SSR architecture remains almost the same. The graph falls due to the increase in throughput (routing load per packet received falls). It generally falls for DSR (increased throughput) too, though it is more, which again indicates the sensitivity of DSR to changes in network conditions (offered load in this case).

5 Conclusion and Future Work

It is observed that though DSR claims to use least hop count as a metric, the routes found are actually based on the round trip times for the RREQ and RREP packets. These packets are delayed by different amounts at different nodes; due to the instantaneous state of various parameters at these nodes e.g. transmit buffer size, etc. The back-off algorithm of the MAC layer also adds to this. Thus, the routes which a source learns via its route discovery are not based on any formal criterion. Furthermore, the source chooses a route amongst these on a purely random basis, which may not be the best route from itself to the destination.

The chosen route also remains fixed either for the entire communication session or until the route breaks. A node gets no indication of a better route existing during a session. Whereas, in SSR, it is the RS that provides the best possible route based on the state of the whole network. A node also refreshes this route periodically. This enables it to be aware of the best available route between itself and the destination at most times. These reasons also eliminate the need to generate RERR packets.

The observation is that the SSR model performs better than DSR (in terms of higher throughput and average latency) with increasing hop count. SSR seems to be more stable with changing network parameters, including link failures and traffic conditions. The routing load for SSR does not change very much with varying network conditions (NRL decreases as throughput increases); whereas in DSR, it fluctuates, depending on the dynamic state of the network. SSR also enables remote machines to initiate communication sessions with vehicles via the RS, which is not possible with DSR.

The model needs to be simulated with other MAC protocols like MACA [12]. The possibility of making a fresh routing decision at each or some other intermediary nodes should be explored. Furthermore, a node should maintain the latency metric for each link as opposed to maintaining an average latency over all neighbour links. To improve the models efficiency, a prediction algorithm is needed, which is able to calculate the status of a wireless link in the near future. As suggested from the simulation studies, having the RSAPs form a backbone routing infrastructure may lead to a much better performance. In such a scenario, data packets only traverse over wireless links between the source/destination and the nearest RSAP.

References

1. Y. Lin, and I.Chlamtac. Wireless and Mobile Network Architectures. John Wiley and Sons, 2001, Chapter 21.
2. LAN MAN Standards Committee of IEEE Computer Society. Wireless LAN Medium Access Control (MAC) and Physical (PHY) Layer Spec. IEEE Std. 802.11, IEEE, June 99.
3. J. P. Singh, N. Bambos, B. Srinivisan, and D. Clawain. Wireless LAN Performance under Varied Stress Conditions in Vehicular Traffic Scenarios. IEEE Vehicular Technology Conference, vol. 2, pp. 743 - 747, Vancouver, Canada, Fall 2002.
4. James A. Lowes. Ad-hoc Routing with the MSA Simulation Engine. Research Library, Department of Computer Science, University of Manchester, UK, 2003.
5. Thomas H. Cormen. Introduction to Algorithms. MIT Press, 2001.
6. David B. Johnson, David A. Maltz, and Yin-Chun Hu. Dynamic Source Routing Protocol for Mobile Ad-hoc Networks. IETF Internet Draft, draft-ietf-manet-dsr-10.txt, July 2004.
7. Peter Ball. Introduction to Discrete Event Simulation. DYCOMANS workshop on 'Management and Control: Tools in Action', Algarve, Portugal, May 1996.
8. Stephen Q. Ye. A New Packet Oriented Approach to Simulating Wireless Ad-hoc Network Protocols in Java. Research Library, Department of Computer Science, University of Manchester, UK, 2002.
9. Andrew S. Tanenbaum. Computer Networks. Prentice-Hall, third edition, 2001.
10. N. Abramson. Development of the ALOHANET. IEEE Transactions on Information Theory, vol. IT-31, March 1985, pp. 119-123.
11. D. Bertsekas, R. Gallager. Data Networks. Prentice-Hall, second edition, 2000.
12. Phil Karn. MACA – A New Channel Access Protocol for Packet Radio. In Proceedings of the ARRL/CRRL Amateur Radio Ninth Computer Networking Conference, 1990, pp. 134-140.

Energy-Efficient Aggregate Query Evaluation in Sensor Networks

Zhuoyuan Tu and Weifa Liang

Department of Computer Science,
The Australian National University,
Canberra, ACT 0200, Australia
{zytu, wliang}@cs.anu.edu.au

Abstract. Sensor networks, consisting of sensor devices equipped with energy-limited batteries, have been widely used for surveillance and monitoring environments. Data collected by the sensor devices needs to be extracted and aggregated for a wide variety of purposes. Due to the serious energy constraint imposed on such a network, it is a great challenge to perform aggregate queries efficiently. This paper considers the aggregate query evaluation in a sensor network database with the objective to prolong the network lifetime. We first propose an algorithm by introducing a node capability concept that balances the residual energy and the energy consumption at each node so that the network lifetime is prolonged. We then present an improved algorithm to reduce the total network energy consumption for a query by allowing group aggregation. We finally evaluate the performance of the two proposed algorithms against the existing algorithms through simulations. The experimental results show that the proposed algorithms outperform the existing algorithms significantly in terms of the network lifetime.

1 Introduction

Wireless sensor networks have attracted wide attention due to their ubiquitous surveillant applications. Recent advances in microelectronical technologies empower this new class of sensor devices to monitor information in a previously unobtainable fashion. Using these sensor devices, biologists are able to obtain the ambient conditions for endangered plants and animals every few seconds. Security guards can detect the subtle temperature variation in storage warehouses in no time. To meet various monitoring requirements, data generated by the sensors in a sensor network needs to be extracted or aggregated. Therefore, a sensor network can be treated as a database, where the sensed data periodically generated by each sensor node can be treated as a segment of a relational table. During each time interval, a sensor node only produces a message called a tuple (a row of the table). An attribute (a column of the table is either the information about the sensor node itself (e.g., its id or location), or the data detected by this node (e.g., the temperature at a specific location). There is a special node in the network called the *base station* which is usually assumed to have constant energy supply. The base station is used to issue the queries by users and collect the aggregate results for the whole network. In such a sensor

X. Jia, J. Wu, and Y. He (Eds.): MSN 2005, LNCS 3794, pp. 31–41, 2005.

network, users simply specify the data that they are interested in through the various SQL-like queries as follows, and the base station will broadcast these queries over the entire network.

```
SELECT      {attributes, aggregates}
FROM        sensors
WHERE       condition-of-attributes
GROUP BY    {attributes}
HAVING      condition-of-aggregates
DURATION    time interval
```

To respond to a user aggregate query, the network can process in either centralized or in-network processing manner. In the centralized processing, all the messages generated by the sensor nodes are transmitted to the base station directly and extracted centrally. However, this processing is very expensive due to the tremendous energy consumption on the message transmission. By virtue of the autonomous, full-fledged computing ability of sensor nodes, the collected messages by each sensor node can also be filtered or combined locally before transmitted to the base station, which is called *in-network aggregation*. In other words, a tree rooted at the base station and spanning all the sensor nodes in the network will be constructed for the data aggregation. Data collected by each node is aggregated before being transmitted to its parent. Ultimately the aggregate result will be relayed to the base station. To implement data aggregation, each node in the routing tree will be assigned into a group according to the distinct value of a list of group by attributes in a SQL-like query. Messages from different nodes are merged into one message at an internal node if they belong to the same group [10]. For example, if we pose a query of "the average temperature in each building", each sensor node will first generate its own message and collect the messages from its descendants in the tree, and then use `SUM` and `COUNT` functions in SQL to compute the average temperature for each group (each building) before forwarding the result to its parent. In the end, all the messages in the same building will be merged into one message so that the transmission energy consumption will be dramatically reduced. Therefore, the number of messages finally received by the base station is equal to the number of buildings.

Related work. *Network lifetime* is of paramount importance in sensor networks, because one node failure in the network can paralyse the entire network. Network lifetime of a wireless sensor network can thus be defined as the time of the first node failure in the network [1].

To improve the energy efficiency and prolong the network lifetime, several existing protocols for various problems have been proposed in both ad hoc networks and sensor networks [1, 2, 3, 4, 5, 6, 7, 11]. For example, in ad hoc networks Chang and Tassiulas [1, 2] realized a group of unicast requests by discouraging the participation of low energy nodes. Kang and Poovendran [6] provided a globally optimal solution for broadcasting through a graph theoretic approach. While in sensor networks, Heinzelman *et al* [3] initialized the study of data gathering by proposing a clustering protocol `LEACH`, in which nodes are grouped into a number of clusters. Within a cluster, a node is chosen as the cluster head which will be used to gather and aggregate the data for other members and forward the aggregated result to the base station directly. Lindsey and Raghavendra [7] provided an improved protocol PEGASIS using a chain concept, where all the nodes in the network form a chain and one of the nodes is chosen as the chain head in turn to report the aggregated result to the base station. Tan and Kórpeoğlu [11] provided a protocol PEDAP for the data gathering problem, which constructs a minimum spanning tree (MST) rooted at the base station to limit the total energy consumption. Kalpakis *et al* [5] considered a generic data gathering problem with the objective to maximize the network lifetime, for which they proposed an integer program solution and a heuristic solution.

This paper provides the evaluation of an aggregate query in a sensor network with an objective to prolong the network lifetime. The pervasive way to do this is to apply the in-network aggregation to proceed the query evaluation, which has been presented in [8], where the information-directed routing is proposed to minimize the transmission energy consumption while maximizing data aggregation. Yao and Gehrke [12] have generated efficient query execution plans with in-network aggregation, which can significantly reduce resource requirements. Apart from these, the query semantics for efficient data routing has been considered in [9] to save transmission energy, in which a semantic routing tree (SRT) is used to exclude the nodes that the query does not apply to. Furthermore, group aggregation has been incorporated into the routing algorithm GaNC in [10], where the sensor nodes in the same group are clustered along the same routing path with the goal of reducing the size of transmitted data. However, an obvious indiscretion in some of the routing protocols, such as MST and GaNC, is that a node is chosen to be added into the tree without taking into account its residual energy during the construction of the routing tree. As a result, the nodes closer to the root of the routing tree will exhaust their energy rapidly due to the fact that they serve as relay nodes and forward the messages for their descendants in the tree. Thus, the network lifetime is shortened.

Our contributions. To evaluate an aggregate query in a sensor network, we first propose an algorithm by introducing the node capability concept to balance the residual energy and the energy consumption of each node in order to prolong the network lifetime. We then present an improved algorithm which allows the group aggregation to reduce the total energy consumption. We finally conduct the experiments by simulation. The experimental results show that the proposed algorithms outperform the existing ones.

The rest paper is organized as follows. Section 2 defines the problem. Section 3 introduces the node capability concept and a heuristic algorithm. Section 4 presents an improved algorithm. Section 5 conducts the experiments. Section 6 concludes.

2 Preliminaries

Assume that a sensor network consists of n homogeneous energy-constrained sensor nodes and an infinite-energy-supplied base station s deployed over an interested region. Each sensor periodically produces sensed data as it monitors its vicinity. The communication between two sensor nodes is done either directly (if they are within the transmission range of each other) or through the relay nodes. The network can be modeled as a directed graph $M = (N, A)$, where N is the set of nodes with $|N| = n+1$ and there is a directed edge $\langle u, v\rangle$ in A if node v is within the transmission range of u. The energy consumption for transmitting a m-bit message from u to v is modeled to be $md^{\alpha}_{v,u}$, where $d_{v,u}$ is the distance from u to v and α is a parameter that typically takes on a value between 2 and 4, depending on the characteristics of the communication medium. Given an aggregate query issued at the base station, the problem is to evaluate the query against the sensor network database by constructing a spanning tree rooted at the base station such that the network lifetime is maximized. We refer to this problem as *the lifetime-maximized routing tree problem* (`LmRTP` for short).

3 Algorithm `LmNC`

In this section we introduce the node capability concept and propose a heuristic algorithm called the `Lifetime-maximized Network Configuration (LmNC)` for `LmRTP` based on the capability concept.

Capability concept. Given a node v, let $p(v)$ be the parent of v in a routing tree. The energy consumption for transmitting a m-bit message from v to $p(v)$ is $E_c(v, p(v)) = md^{\alpha}_{v,p(v)}$, where $d_{v,p(v)}$ is the distance between v and $p(v)$. Let $E_r(v)$ be the residual energy of v before evaluating the current query. Assume that the length of the message sensed by every node is the same (m-bit), then the capability of node v to $p(v)$ is defined as

$$C(v, p(v)) = E_r(v)/E_c(v, p(v)) - 1 = E_r(v)/md^{\alpha}_{v,p(v)} - 1. \tag{1}$$

If v has k descendants in the routing tree, then the energy consumption at v to forward all the messages (its own generated message and the messages collected from its descendants) to its parent $p(v)$ will be $(k+1)md^{\alpha}_{v,p(v)}$, given that there is no data aggregation at v. If after this transmission, v will exhaust its residual energy, then $E_r(v) = (k+1)md^{\alpha}_{v,p(v)}$. From Equation (1), it is easy to derive that $k = E_r(v)/md^{\alpha}_{v,p(v)} - 1 = C(v, p(v))$. So, if there is no aggregation at v, the capability of node v to $p(v)$, $E_c(v, p(v))$, actually indicates the maximum number of descendants that it can support by its current residual energy.

Algorithm description . Since a node with larger capability can have more descendants in the routing tree (if data aggregation is not allowed), it should be placed closer to the tree root to prolong the network lifetime. Based on this idea, we propose an algorithm `LmNC`, where each time a node with the maximum capability is included into the current tree. Thus, the nodes are added one by one until all the nodes are included in the tree. The motivation behind this algorithm is that adding the node with the maximum capability innately balances the node residual energy $E_r(v)$ and the actual energy consumption for transmitting a message to its parent $E_c(v, p(v))$ (as the definition of the node capability), so that the network lifetime is prolonged. Specifically, we denote by T the current tree and V_T the set of nodes included in T so far. Initially, T only includes the base station, i.e. $V_T = \{s\}$. Algorithm `LmNC` repeatedly picks a node v ($v \in V - V_T$) with maximum capability to u ($u \in V_T$) and adds it into T with u as its parent. The algorithm continues until $V - V_T = \emptyset$. The detailed algorithm is given below.

Algorithm. Lifetime_Efficient_Network_Configuration (G, E_r)
/* G is the current sensor network and E_r is an array of the residual energy of the nodes */
begin
1. $V_T \leftarrow \{s\}$; /* add the base station into the tree */
2. $Q \leftarrow V - V_T$; /* the set of nodes which is not in the tree*/
3. while $Q \neq \emptyset$ do
4. $\quad C_{\max} \leftarrow 0$; /* the maximal capability of nodes in the tree */
5. $\quad$ for each $v \in Q$ *and* $u \in V_T$ do
6. $\quad\quad$ compute $C(v, u)$;
7. $\quad\quad$ if $C_{\max} < C(v, u)$
8. $\quad\quad\quad$ then $C_{\max} \leftarrow C(v, u)$;
9. $\quad\quad\quad\quad$ $added\_node \leftarrow v$;
10. $\quad\quad\quad\quad$ $temp\_parent \leftarrow u$;
11. $\quad$ $p(added\_node) \leftarrow temp\_parent$;
$\quad$ /* set the parent for the node with maximum capability */
12. $\quad$ $V_T \leftarrow V_T \cup \{added\_node\}$; /* add the node into tree */
13. $\quad$ $Q \leftarrow Q - \{added\_node\}$;

end.

Note that, although there have been several algorithms for `LmRTP` considering the residual energy of nodes during the construction of the routing tree (including [1]), they failed to consider the actual transmission energy consumption from a node to its parent. This can be illustrated by the following example. Assume that there is a partially built routing tree and a number of nodes to be added into the current tree. Node v_i has the maximum residual energy among the nodes out of the tree, while the distance between v_i and its parent is much longer than that between another node v_j and its parent. Now, if node v_i is added into the tree, it will die easily in the further tree construction because of the enormous transmission energy consumption from v_i to its parent. Therefore, although v_i has more residual energy than v_j at the moment, the maximum number of the messages transmitted by v_i to its parent is less than that by v_j. We thus conclude that the lifetime of node v_i is shorter than that of node v_j.

4 Improved Algorithm `LmGaNC`

Although algorithm `LmNC` manifests the significant improvement on the network lifetime for `LmRTP`, the total energy consumption for each query is hardly considered during the construction of the routing tree. Because the node with maximum capability may be far away from its parent and the excess transmission energy consumption by the node will be triggered. In this section we present an improved algorithm called `Lifetime-maximized Group-aware Network Configuration` (`LmGaNC`) by allowing group aggregation to reduce the total energy consumption.

Algorithm description. Since group aggregation is able to combine the messages from the same group into one message, incorporating the nodes of the same group into a routing path will reduce the energy consumption and maximize the network lifetime, because the messages drawn from these nodes will contain fewer groups. With this idea Sharaf *et al* provided a heuristic algorithm (in [10]) to construct an energy-efficient routing tree. Further incorporating this idea into algorithm `LmNC`, we propose an improved algorithm `LmNC` as follows.

Algorithm. Lifetime_Efficient_Network_Configuration (G, E_r)
/* G is the current sensor network and E_r is an array of the residual energy of the nodes */
begin
1. $V_T \leftarrow \{s\}$; /* add the base station into the tree */
2. $Q \leftarrow V - V_T$; /* the set of nodes which is not in the tree*/
3. while $Q \neq \emptyset$ do
4. $C_{\max} \leftarrow 0$; /* the maximal capability of nodes in the tree */
5. for each $v \in Q$ and $u \in V_T$ do
6. compute $C(v, u)$;
7. if $C_{\max} < C(v, u)$
8. then $C_{\max} \leftarrow C(v, u)$;
9. $added\_node \leftarrow v$;
10. $temp\_parent \leftarrow u$;
11. $p(added\_node) \leftarrow temp\_parent$;
/* set the parent for the node with maximum capability */
12. $d_{min} \leftarrow \infty$; /* minimum distance to choose */
13. for each $u' \in V_T$ and $u' \neq temp\_parent$ do
14. if $group\_id(u') = group\_id(added\_node)$ and $d_{added\_node,u'} < d_{min}$
15 and $d_{added\_node,u'} \leq df * d_{added\_node,temp\_parent}$
16. then $p(added\_node) \leftarrow u'$;
17. $V_T \leftarrow V_T \cup \{added\_node\}$; /* add the node into tree */
18. $Q \leftarrow Q - \{added\_node\}$;

end.

Algorithm `LmGaNC` is similar to algorithm `LmNC`. The difference is that, during the construction of the routing tree, a child with the maximum capability chosen by `LmNC` will keep checking if there is a node in the same group as itself in the current tree in terms of `LmGaNC`. We call this node a better parent. If yes, the child will switch to this better parent. If there are more than one better parent to choose from, the closest one will be chosen. Notice that choosing a better parent far away will cause the extra transmission energy consumption. So, a concept of *distancefactor* (*df*) is employed, which is the upper bound of the distance between a child and its selected parent. For example, if df=1.5, then we only consider the parent whose distance to the child is at most $df * d_{v,u} = 1.5 d_{v,u}$, where $d_{v,u}$ is the distance between v and its current parent u. Energy reduction brought by algorithm `LmGaNC` is demonstrated by the following example. In Figure 1, we have a partially built routing tree (see Fig. 1(a)).

Assume that black nodes 2 and 7 belong to Group 1, shaded node 6 belongs to Group 2, and the rest belong to Group 3. The numbers of messages are as shown in the figure (depending on the number of various groups in the subtree). Under algorithm LmNC, shaded node 8 (in Group 2) has maximum capability to its parent node 7 (see Fig. 1 (b)). In order to forward one message originally from node 8, all the nodes in the path from node 8 to the root, except the root 1, have to consume extra energy for this transmission. While the improved algorithm LmGaNC allows node 8 to switch to a better parent (node 6) which is in the same group, assuming without violation of the distance factor. As a result, none of the nodes, except node 8 itself, needs to forward one extra message for node 8, so that energy can be saved. Here, after applying group aggregation,

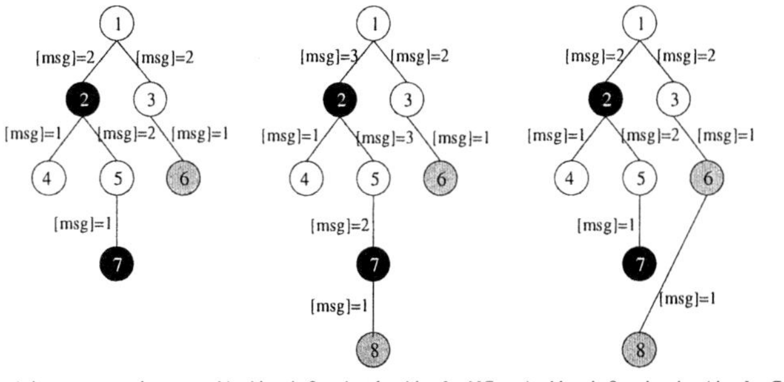

a) the current routing tree
b) add node 8 under algorithm LmNC
c) add node 8 under algorithm LmGaNC

Fig. 1. Benefit of algorithm LmGaNC

a node capability indicates the number of its descendants in different groups (excluding the group of the node itself) rather than the total number of its descendants, because the messages from the descendants in the same group can be merged into one message only.

Effect of conditional data aggregation. The Group-By clause can divide an aggregate query's result into a set of groups. Each sensor node is assigned to one group according to its value of Group-By attributes. In practice, however, this clause may not be enough to answer the query like "what is the average temperature in each room at level 5". To match the query condition, we normally employ the *where* clause originated from the SQL language to further reduce the energy consumption delivered by algorithm LmGaNC. So, in the above case, the condition clause "WHERE Level_no.=5" will be imposed on the query. Before each sensor node transmits its sensed data to its parent, it will check whether the data matches the query condition. If not, then the node just transmits a bit of notification information with value 0 rather than its original data to its parent, so that its parent will not keep waiting for the data from that child. This is especially true under the aggregation schema of Cougar [12], where each node holds a waiting list of its children and will not transmit its data to its parent until it hears from all the nodes on the waiting list. Since the size of the data is shrunk into only 1 bit given the mismatch, the total transmission energy consumption will be further reduced. However, even if a node matches the query condition, whether its residual energy can afford the message transmission is still questionable. One possible solution for this is that the node checks if it has sufficient residual energy to complete the transmission. If not, it will send a bit of notification information with value 1 instead of its original data to its parent to indicate the insufficiency of its residual energy.

5 Simulation Results

This section evaluates algorithms `LmNC` and `LmGaNC` against the existing algorithms including `MST (Minimum Spanning Tree)`, `SPT (Shortest Path Tree)` and `GaNC`. The experimental metrics adopted are the network lifetime and the total energy consumption, based on different numbers of groups, various distance factors, and with and without the *where* condition clause. We assumed that the network topologies are randomly generated from the *NS-2* network simulator with the nodes distributed in a 100×100 m^2 region and each sensor is initially equipped with 10^5 μ-*Joules* energy. For each aggregate query, we assign a "Group_id" for each node randomly and take the average of the experimental results from 30 distinct network topologies for each network size.

Performance analysis of the proposed algorithms. Before we proceed, we reproduce an existing heuristic algorithm called `GaNC` [10] for the concerned problem, which will be used as the benchmark. Algorithm `GaNC` with the group aggregation concept is derived from a simple `First-Heard-From (FHF)` protocol where the nodes always select the first node from which they hear as their parents after the query specification is broadcast over the network. The main difference between `GaNC` and `FHF` is that the child under `GaNC` can change to a better parent in the same group within the given distance factor. The simulation results in Figure 2(a) show that the network lifetimes delivered by algorithms `LmNC` and `LmGaNC` significantly outperform the ones delivered by `MST`, `SPT` and

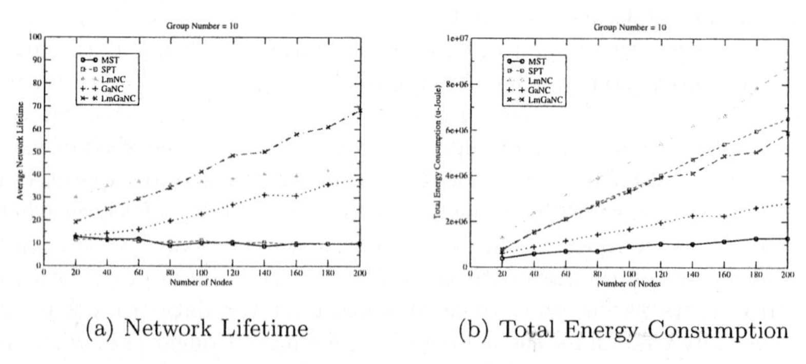

(a) Network Lifetime (b) Total Energy Consumption

Fig. 2. Performance comparison among various algorithms

GaNC. Figure 2(b) shows that algorithm LmGaNC gracefully balances the total energy consumption of LmNC with a slight shortening on the network lifetime when the number of nodes in the network is less than 80.

Sensitivity to the number of groups. In comparison to algorithm GaNC, the average network lifetime under LmGaNC is more sensitive to the number of groups. Figure 3(a) indicates both algorithms manifest their lifetime improvements by 50% approximately, when the number of groups is decreased from 10 to 5. The reasons behind is as follows. On one hand, fewer groups mean that more sensor nodes will be in the same group, and thus the possibility of message suppression under group aggregation will be enhanced. On the other hand, fewer groups make a child node have more chances to switch to a better parent in the same group, so less transmission energy will be consumed.

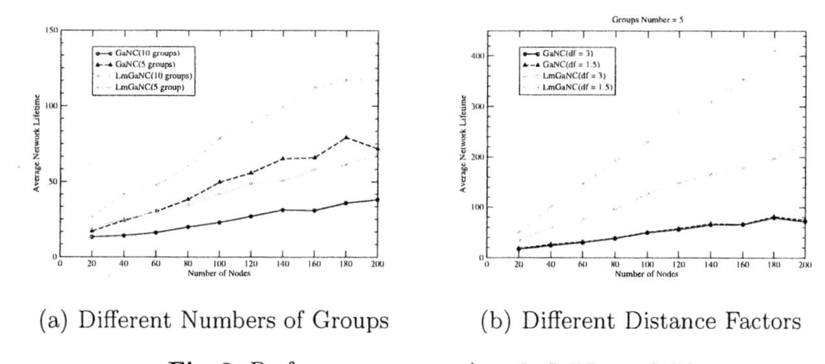

(a) Different Numbers of Groups (b) Different Distance Factors

Fig. 3. Performance comparison LmGaNC vs GaNC

Sensitivity to the distance factor. As discussed earlier, the distance factor is introduced to limit the maximum distance that is acceptable when a child node switches to a better parent in the same group. It avoids unnecessary energy dissipation resulting from this switching. As such, the smaller the distance factor is, the less the energy dissipation will be, therefore, the longer the network can endure. Figure 3(b) shows that when distance factor is decreased from 3 to 1.5, algorithm LmGaNC exhibits its sensitivity immediately and the network lifetime is significantly prolonged, while GaNC reacts much more rigidly.

Sensitivity to the where condition clause. The experiments here aim to further reduce the energy consumption of evaluating an aggregate query through allowing the nodes that mismatch the query condition to send a 1-bit notification to their parents instead of the sensed data. Figure 4 (a) and (b) illustrate the effects of the *where* clause in an aggregate query on both the network lifetime and the total energy consumption. The experimental results show that the network lifetime increases by more than 50% while the total energy consumption only goes up by around 25% under LmGaNC.

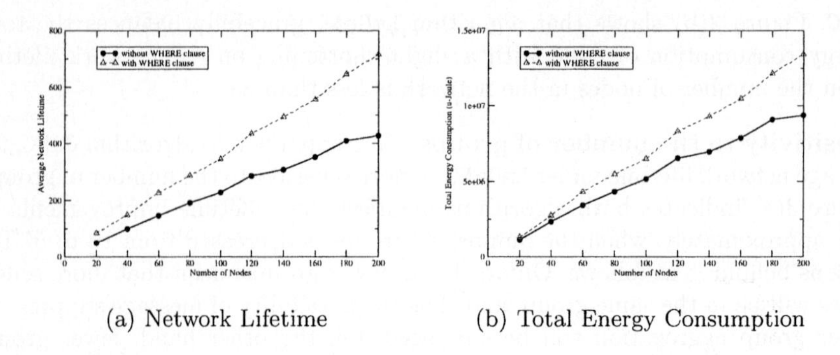

(a) Network Lifetime (b) Total Energy Consumption

Fig. 4. Performance of `LmGaNC` with the `where` condition clause

6 Conclusions

This paper considered the aggregate query evaluation in a senor network database by exploring the node capability concept. Based on this concept we first proposed a heuristic algorithm to prolong the network lifetime, then presented an improved algorithm by incorporating group aggregation to reduce the total energy consumption. We finally conducted experiments to evaluate the performance of the proposed algorithms against those of the existing ones. The experimental results showed that the proposed algorithms outperform the existing algorithms.

Acknowledgment. It is acknowledged that the work by the authors was supported by a research grant from the Faculty of Engineering and Information Technology at the Australian National University.

References

1. J-H Chang and L. Tassiulas. Energy conserving routing in wireless ad hoc networks. *Proc. INFOCOM'00*, IEEE, 2000.
2. J-H Chang and L. Tassiulas. Fast approximate algorithms for maximum lifetime routing in wireless ad hoc networks. *IFIP-TC6/European Commission Int'l Conf.*, Lecture Notes in Computer Science, Vol. 1815, pp. 702–713, Springer, 2000.
3. W. R. Heinzelman, A. Chandrakasan and H. Balakrishnan. Energy-efficient communication protocol for wireless microsensor networks. *Proc. 33th Hawaii International Conference on System Sciences*, IEEE, 2000.
4. C. Intanagonwiwat, D. Estrin, R. Govindan, and J. Heidemann. Impact of network density on data aggregation in wireless sensor networks. *Proc. 22nd International Conference on Distributed Computing Systems (ICDCS'02)*, IEEE, 2002.
5. K. Kalpakis, K. Dasgupta and P. Namjoshi. Efficient algorithms for maximum lifetime data gathering and aggregation in wireless sensor networks. *Computer Networks*, Vol. 42, pp. 697–716, 2003.
6. I. Kang and R. Poovendran. Maximizing static network lifetime of wireless broadcast ad hoc networks. *Proc. ICC'03*, IEEE, 2003.

7. S. Lindsey and C. S. Raghavendra. PEGASIS: Power-efficient gathering in sensor information systems. *Proc. Aerospace Conference*, IEEE, pp. 1125–1130, 2002.
8. J. Liu, F. Zhao, and D. Petrovic. Information-directed routing in ad hoc sensor networks. *Proc. 2nd international conference on wireless sensor networks and applications*, ACM, 2003.
9. S. Madden, M. Franklin, J. Hellerstein and W. Hong. The design of an acquisitional query processor for sensor networks. *Proc. of SIGMOD'03*, ACM, 2003.
10. M. A. Sharaf, J. Beaver, A. Labrinidis and P. K. Chrysanthis. Balancing energy efficiency and quality of aggregate data in sensor networks. *J. VLDB*, Springer, 2004.
11. H. Ó. Tan and İ. Kórpeoğlu. Power efficient data gathering and aggregation in wireless sensor networks. *ACM SIGMOD Record*, Vol. 32, No. 4, pp. 66–71, 2003.
12. Y. Yao and J. Gehrke. Query processing in sensor networks. *Proc. 1st Biennial Conf. Innovative Data Systems Research (CIDR'03)*, ACM, 2003.

Data Sampling Control and Compression in Sensor Networks*

Jinbao Li[1,2] and Jianzhong Li[1,2]

[1] Harbin Institute of Technology, 150001, Harbin, China
[2] Heilongjiang University, 150080, Harbin, China
jbLi@hlju.edu.cn, lijzh@hit.edu.cn

Abstract. Nodes in wireless sensor networks have very limited storage capacity, computing ability and battery power. Node failure and communication link disconnection occur frequently, which means weak services of the network layer. Sensing data is inaccurate which often has errors. Focusing on inaccuracy of the observation data and power limitation of sensors, this paper proposes a sampling frequency control algorithm and a data compression algorithm. Based on features of the sensing data, these two algorithms are combines together. First, it adjusts the sampling frequency on sensing data dynamically. When the sampling frequency cannot be controlled, data compression algorithm is adopted to reduce the amount of transmitted data to save energy of sensors. Experiments and analysis show that the proposed sampling control algorithm and the data compression algorithm can decrease sampling times, reduce the amount of transmitted data and save energy of sensors.

1 Introduction

Recent advancement in digital-electronics, micro-processors and wireless technologies enable the creation of small and cheap sensors which has processor, memory and wireless communication ability. This accelerates the development of large scale sensor networks. In sensor networks, various sensors which have different functions are distributed to some given area to collect, monitor and process information. Sensor networks integrate sensor technique, computer technique, and distributed information processing technique and communication technique. Sensors are used to collect, obtain or monitor their surroundings, process the information to get detail and accurate information about the area of sensor networks. For example, by obtaining geographical features such as hardness and humidity of the jungle of enemies in battlefield, the blue print of b'attle can be made. There are a lot of attractive features of sensors such as small, cheap, flexible, movable and wireless communication ability, so sensor networks can be used to obtain detailed and reliable information in any time, location, hypsography or environment. In military affairs, sensor networks can be used to

* Supported by the National Natural Science Foundation of China under Grant No.60473075 and No.60273082; the Natural Science Foundation of Heilongjiang Province of China under Grant No.ZJG03-05 and No.QC04C40.

X. Jia, J. Wu, and Y. He (Eds.): MSN 2005, LNCS 3794, pp. 42 – 51, 2005.

monitor the action of enemies and the existence of dangerous features such as poison gas, radiation, exploder, etc. In environment monitoring, sensors can be set at plain, desert, mountain region or seas to monitor and control changes in the environment. In traffic applications, sensors can be used to monitor and control the traffic in freeway or crowed area in cities. Sensors can also be used in security supervising of large shopping center, carport and other devices and in supervising the occupying of parking spaces in parks.

Sensors generally have limited processing, storing and communication ability. They connected with each other by wireless network. Sensors can move, which makes the topology structure of the network change. They communicate with each other by Ad-Hoc manner. Each sensor can act as a router and has the ability of search, localization and reconnection dynamically. Sensor networks is a special kind of wireless Ad-Hoc network, which has features such as frequent moving, connection and disconnection, limitation of power, large distributed area, large amount of nodes, limitation of own resources [1]. The reliability of communication is weak and the power is limited in sensor networks. Each node may be invalid at any moment. The network layer can provide weak service. Each sensor has limited storage capacity, computing ability and battery power. There are errors in the measured value of sensors, so the observation data are not accurate.

This paper focuses on data management and query processing [2,6,8,9,10,17] in sensor networks. Aimed at the inaccuracy of sensing data, a sampling frequency control algorithm and a data compression algorithm are proposed, which are suitable for approximate query in sensor networks. By controlling the sampling of nodes, the sampling frequency control algorithm can reduce sampling frequency and decrease power consumption. The data compression algorithm well uses the limited storage capacity and computing ability, compresses the sampling data through a compression algorithm, which needs few computing. Through these two algorithms, this paper reduces sampling frequency and amount of transmitted data. The power is thus saved. The paper is organized as follows. Section 1 introduces sensor networks. Section 2 is the related work. Section 3 proposes the sampling frequency control algorithm and the data compression algorithm. Then the experiments and analysis is given in section 4. The last section is the conclusion.

2 Related Works

Energy saving is an important optimization target in sensor networks. Each sensor is power by battery. Battery cannot be replaced when it is exhausted because of the deserted or dangerous environment it lies. So the limited power of sensors should be efficiently used to prolong the lifetime of sensor networks. S. Madden and Yao.Y proposed a clustering method to reduce communication cost, which saved a lot of energy [3,4,6,7,8,17]. This method first constructs a aggregating tree. Before non-leaf node transmits data to its parent, it aggregates the data in its subtree and tramsmits the results to its parent. It not only reduces the communication cost by aggregation, but also gets more accurate query results.

In TAG[3], base station is used as root node in sensor networks. When receiving an aggregation operation, a query tree is constructed. Each node in the tree combines data received from its subtree with its own data and transmits the result to its parent. In sensor networks, node invalidation frequently occurs. The communication line also fails frequently because of environment, packet conflict, lower Signal-to-Noise, etc. If a node is invalid and the message cannot be transmitted to its parent, the aggregation from its subtree will be lost. If failure node is near the root node, the failure will effect the aggregation dramatically.

Fault tolerance of query processing on stream data from sensors is investigated by Olson[13]. Restricted by a given accuracy, it processes continues query on a central data stream processor. The data stream processor sets a filter on each remote data source. Thus, the limitation of fault tolerance is dynamically distributed to data sources. For each data source, data should be transmitted only when the current value(compared with the last value) is beyond the threshold of its filter. Centrolized processing is adopted in this method, so it cannot be used in sensor networks directly. Mohamed.A etc. investigate how to implement approximate query processing by setting filters on sensors when query accuracy is satisfied [11][12]. Based on Olson's work, A. Deligiannakis etc. research on distribution problem below a given threshold of fault tolerance in sensor networks [12]. They extend the idea of using filter to reduce the transmission cost to sensor networks. By increasing fault tolerance threshold and using residual scheme, energy of sensors and bandwidth of networks are obviously saved at the cost of accuracy. The lifetime of network is extended. R. Cheng researched on probabilistic query on inaccurate data [14]. He gave the definition, classification, query processing algorithm and query evaluation method of inaccurate data. Lazaridis introduced a data transmission method in sensor networks which can reduce transmission cost [15]. This method first piecewise constantly compresses raw data on each sensors. Only if the compressed data beyonds the threshold of fault tolerance, it is transmitted to base station. Considine etc. implement approximate in-network aggregation in sensor network database through small sketches[16]. They extend the copy-sensitive sketches, which is suitable for aggregation query. They use the sketches of that method to produce accurate results which have lower communication cost and lower computing cost.

3 Sampling Frequency Control and Data Compression

Sending and receiving data are the most energy consuming operations. In Mica motes system in Berkeley, the energy cost by transmitting a bit through wireless network is equal to the energy cost by executing 1000 instructions by CPU [5]. This observation shows that to reduce the energy cost when transmits data, we should try to avoid communications between nodes. Besides sending and receiving data, sampling is another energy consuming operation [5]. In the situation that sampling accuracy is satisfied, if we can control sampling and reduce the sampling frequency, not only the energy used to sample data can be saved, but also the amount of transmitted data will be reduced. Affected by node invalidation, wireless communication uncertainty and power restriction, the obtaining, processing and transmission of sensing data often had errors. Sensing data are uncertainty to some extent [14][15]. At the same time, user queries often need not accurate results. Query on sensing data is a kind of query on uncertain data. Some errors are allowable in query results. For example, counting the number of rooms, the temperature of which is beyond 30°C, in every 3 seconds. The error is defined below 10%.

In applications, observation data from sensors are often continues data, such as temperature, humidity, etc. Data from one sensor often lies in a stable range during a given period. For sensors that the measured data changes continuously, the measured data may fluctuate slightly in some successive periods and may fluctuate obviously but in linearity in some other periods. Figure 1 gives an example. Though data in areas $[t_1,t_2]$, $[t_3,t_4]$, $[t_4,t_5]$ and $[t_6,t_7]$ are all different, they fluctuate slightly in each area. Data in area $[t_2,t_3]$ and $[t_5,t_6]$ fluctuate obviously, but they change in linearity.

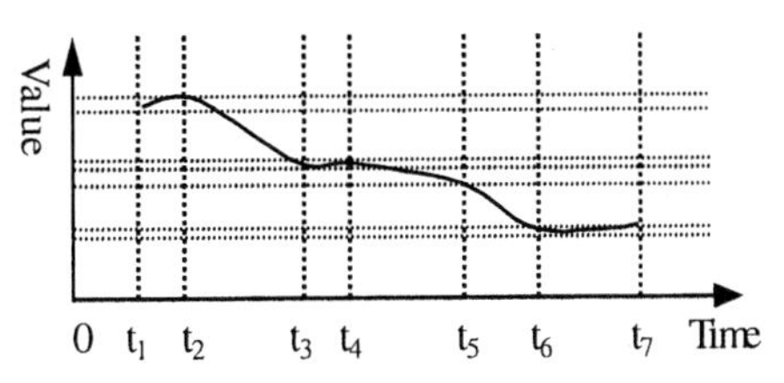

Fig. 1. Time series of sensing data

3.1 Sampling Frequency Control

For sensors that observation data changes continuously, we can use the *n* latest measured value of sensing data to predict the time interval △t, in which the measured data within a given error bound. If △t is bigger than the sampling cycle, we can get next sample data at time t_{N+1}=△t+ t_N. This prediction method can control the sampling frequency and reduce energy cost when sampling.

Linear Regression model is used in this paper to predict the steady time interval △t of measured data within given error bound. Suppose the prediction of sensing data conforms to unitary Linear Regression model, the prediction function can be represented as formula 3.1.

$$v = a + bt \tag{3.1}$$

where, *v* is the value of sensing data, *t* is the sampling moment, *a* and *b* are regression coefficients.

Till current time *N*, the time series of sensing data is supposed to be $<(v_i, t_i)>$, $0<i<=N$. Regression coefficients *a* and *b* can be determined by Least Square method, which is described as follows.

$$b = \frac{\sum (v_i - \bar{v})(t_i - \bar{t})}{\sum (t_i - \bar{t})^2} \tag{3.2}$$

$$a = \frac{1}{N}(\sum v_i - b\sum t_i) \tag{3.3}$$

$$\bar{t} = \frac{1}{N}\sum t_i, \bar{v} = \frac{1}{N}\sum v_i$$

Only when *v* and *t* are linearly correlated with each other, formula 3.1 can be used to predict *v* and △t. Thus, the relationship between *v* and *t* should be verified first. If they are not linearly correlated, *b*=0 in formula 3.1. Otherwise, *v* and *t* are linearly correlated. Verifying the relationship between *v* and *t* is a hypothesis test: H_0: *b*=0.

Consider 2 statistical variables: $Q_e = \sum_{i=1}^{N}(v_i - \hat{v}_i)^2$ and $Q_R = \sum_{i=1}^{N}(\hat{v}_i - \bar{v})^2$, where $\hat{v}_i$ is the prediction value of v_i. When H_0 is true, statistical variable F satisfies the following condition.

$$F = \frac{Q_R}{Q_e(N-2)} \sim F(1, N-2) \tag{3.4}$$

F-examination can be used to do Significance Test of formula 3.1. For a given significant levelα, if the observation data of F is bigger than $F_\alpha(1,N\text{-}2)$, H_0 is rejected and say the linear correlation between v and x is significant. If F is smaller than $F_\alpha(1,N\text{-}2)$, H_0 is accepted and say there is no linear correlation between v and x. That is to say, if the statistical variable F determined by sampling is bigger than $F_\alpha(1,N\text{-}2)$, the Linear Regression model exists.

Suppose T is the sampling frequency of sensing data, sampling accuracy is $\delta > 0$. That is, the tolerant error of sensing data is $\pm\delta$. Formula 3.1 and 3.5 are used to predict the moment when next value v_{N+1} will be produced. That is, they are used to predict t_{N+1} when v_{N+1} is produced under the condition $|v_{N+1}\text{-}v_N| < \delta$. Time t_{N+1} is calculated as formula 3.5.

$$t_{N+1} = \frac{1}{b}\left[v_N + \delta - t_{\frac{\alpha}{2}}(N-2)\sigma^* \sqrt{1 + \frac{1}{N} + \frac{(t_N - \bar{t})^2}{\sum_{i=1}^{N}(t_i - \bar{t})^2}} - a \right], \quad b > 0$$

$$\text{or } t_{N+1} = \frac{1}{b}\left[v_N - \delta + t_{\frac{\alpha}{2}}(N-2)\sigma^* \sqrt{1 + \frac{1}{N} + \frac{(t_N - \bar{t})^2}{\sum_{i=1}^{N}(t_i - \bar{t})^2}} - a \right], \quad b < 0 \tag{3.5}$$

$$\text{Where } \sigma^* = \sqrt{\frac{Q_e}{N-2}}$$

To get more accurate prediction time interval, more observed nodes should be used to construct Regression function. Sampling window is enlarged in this paper. This method is divided into 3 steps. The first step, determines the minimal window size in which every two samples satisfy linear correlation, say W_{min}, to verify the significance; The second step, if the first step satisfy linear correlation, W_{min} is enlarged β times to verify the significance; The third step, if the second step satisfy linear correlation $W = \beta * W_{min}$, otherwise $W = W_{min}$.

If the Linear Regression model exists and the prediction value $\Delta t = t_{N+1} - t_N > T$, t_{N+1} is the next sampling moment. If the prediction value $\Delta t \le T$ or the linear regression model in the first step not exists, the general sampling frequency is adopted.

Algorithm 1 controls sampling frequency and data transmission by prediction. It is executed in each sensor.

In some time interval, if sample data are linearly correlated, prediction and sampling frequency control method are used to control sampling rate. Otherwise, the method proposed in section 3.2 will be used to compress sample data. This compression method needs few computations. This paper mainly proposes these two methods to reduce the sampling frequency and the amount of transmitted data. Battery power is dramatically saved.

3.2 Data Compression Algorithm

Sensors produce a lot of continuous sensing data, but have very limited storage capacity. As a result, not all the sensing data can be stored in sensors. The storage of a sensor can cache or store only part of the data. To make full use of storage capacity of sensors, compression method can be used to compress sensing data. Thus, more data can be cached in sensor storage and used for query processing and data prediction.

The region of sensor networks, say D, are divided into some areas based on a given error ε. That is, 2ε is the length of each data area and D is divided into $D/(2\varepsilon)$ areas. Suppose the given error is 5, region D is [20,99], ranges of all data areas should be [20,29], [30,39], … , [90,99]. After the data areas are determined, we assign the sample data of each sampling cycle to corresponding data area. For example, data from cycle 1 to cycle 8 are all belong to data area [20,30], so cycle range [1,8] is stored instead of storing 8 sample data. Data from cycle 9 and cycle 11 belong to data area [30, 40], while cycle 10 belongs to data area [50,60]. Cycles [9,9], [10,10] and [11,11] have to be assigned to corresponding data areas separately as

Algorithm 1. Precision Based Sampling and Transmit Algorithm (PBSA)

Input:
Sampling period T,
Tolerance error δ,
α, β, W_{min},
MPC time series S=
$<s[t_1], s[t_2], \ldots>$ with tolerance $\varepsilon_i < \delta$

Output:
(1) $b = \frac{\sum (v_i - \bar{v})(t_i - \bar{t})}{\sum (t_i - \bar{t})^2}$, constructs Linear Regression equations, gets values of regression coefficients a and b;
(2) $a = \frac{1}{N}(\sum v_i - b\sum t_i)$
(3) $v = a + bt$,
(4) n= W_{min}
(5) if $F = \frac{Q_R}{Q_e(n-2)} > F_\alpha(1, n-2)$
(6) n=β*W_{min}
(7) Reconstructs Linear Regression equations and gets the values of a and b;
(8) if $F = \frac{Q_R}{Q_e(n-2)} > F_\alpha(1, n-2)$
(9) n=β*W_{min}
(10) else
(11) n=W_{min}
(12) end
(13) Calculates t_{N+1} from formula 3.5
(14) t_{N+1}=max(t_N+T, t_{N+1})
(15) else
(16) t_{N+1}= t_N+T
(17) end
(18) if abs(v_{N+1}-v_N)>δ
(19) SnedToParent(v_{N+1}, t_{N+1})
(20) end

Table 1. Example of DCA algorithm

Data area	cycle	…	cycle
[20,29]	[1,8]	[26,32]	…
[30,39]	[9,9]	[11,11]	…
[50,59]	[10,10]	[40,47]	…
…	…	…	…

(a)

Data area	cycle	cycle
[20,30]	[1,3]	[6,6]
[40,50]	[4,4]	
[60,70]	[5,5]	[7,10]

(b)

Data area	cycle	cycle
[20,30]	[1,3]	[6,6]
[40,50]	[4,4]	*[11,11]*
[60,70]	[5,5]	[7,10]

(c)

Data area	cycle	cycle
[20,30]	[1,3]	[6,6]
[40,50]	[4,4]	*[11,12]*
[60,70]	[5,5]	[7,10]

(d)

shown in figure 1. Following this rule, when sending data to sink, only cycles and the corresponding data areas are sent. Partial of the raw data instead of the whole data are sent to sink. Data are compressed in this manner.

To implement the above compression method, buffers should be set on each node, which stores data shown in table 1(a). The cycles corresponding to each data area are maintained as a circular queue. The cycle of the current sample data is inserted into the queue, which corresponds to the data area it belongs to. If the sequential number of data area corresponding to the current cycle is the same as the last one, data need not be sent out. The only thing should be done is increasing the upper bound of the last cycle range corresponding to that data area.

```
Algorithm 2. Data Compress Algorithm (DCA) ,it is executed in each
sensor
Input:
  Sampling period T,
  Tolerance error ε,
  Sample series v_i, t_i
Output: Compressed Sample series
   (1) Divide the observation region D of sensor networks
       into D/(2ε) data area; construct a table T to store
       these data areas;
   (2) Obtain the current observation data v_i, t_i
   (3) if |v_i-v_i-1|<=ε then
   (4)     update t_i-1 to t_i
   (5)     do not send data to sink;
   (6) else
   (7)     insert t_i into table T ;
   (8)     Send t_i and the number of
       data area corresponding to v_i
       to sink;
   (9) End
```

Sink can decompress the coming data based on D and ε. If it received nothing at a cycle, sink will use the sample data of last cycle.

Table 1(b) shows the sending buffer of a node. If sampling data of the node in cycle 11 is 44, which is in data area [40,50], cycle [11,11] should be written into the buffer of data area [40,50] as shown in table 1(c). Suppose the sample data in cycle 12 is 48, which is also in data area [40,50], cycle [11,11] should be updated to [11,12]. Because the data areas of the two cycles are the same, data need not be sent to sink in this cycle, which is shown in table 1(d).

4 Experiments and Analysis

In the experiments, our SCA&DCA algorithm is compared with TAG and TiNA. In the simulated environment, 25 sensors are randomly put to a 100*100 region. Communication radius of each sensor is 10.

All data in the experiments are either real data or compositive data from Tropical Atmosphere Ocean Project[18]. We use the temperature from Tropocal Atmosphere Ocean Project as the observation data of sensors in the experiments. 25 sensors and 100 cycles are simulated.

The experiments are divided into two parts: (1) comparing data transmission rates of algorithms SCA&DCA, TiNA and TAG; (2) comparing power consumed in algorithms SCA&DCA, TiNA and TAG.

Figure 2 shows the data transmission rate of the three algorithms in different tolerance error. It can be seen from the figure that data transmission rate of algorithm TAG is 100%. It is because it sends raw data directly in each cycle. Algorithms SCA&DCA and TiNA allow some errors so that the data transmission rate is lower. With the increasing of tolerance error, data transmission rate of the two algorithms both decrease. The data transmission rate of SCA&DCA algorithm is a bit lower than TiNA because SCA&DCA not only reduces the amount of data transmitted through data compression, but also reduces the times of sampling through sampling frequency control algorithm dynamically. Data transmission cost is reduced further more.

Sending and receiving power of Chip CC1000 used in Mica2 is 15mW. Sending or

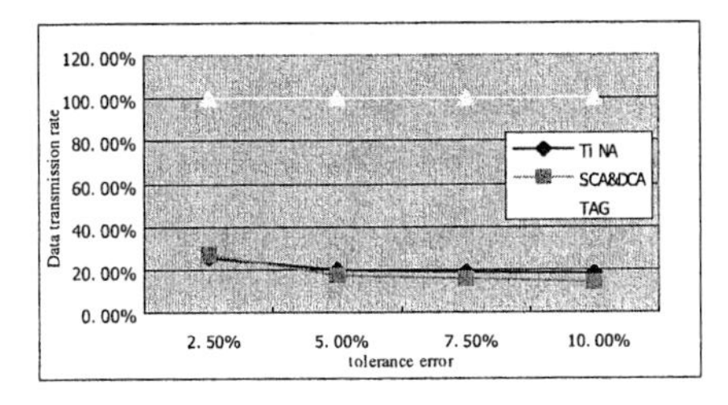

Fig. 2. Data transmission rate of different algorithms

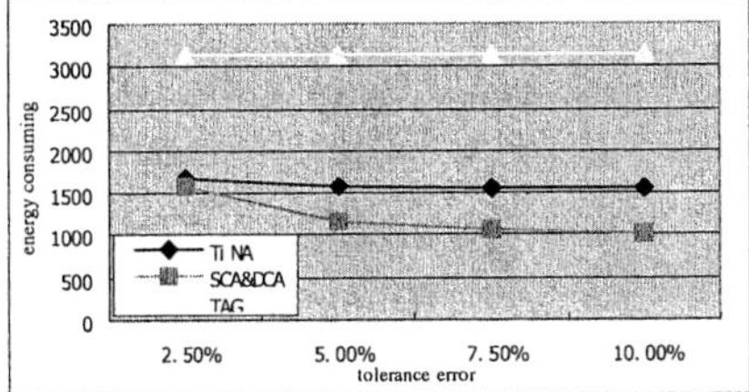

Fig. 3. Energy consuming when sampling frequency is 2048ms

receiving a packet will cost 27ms. Thus, 0.4mJ energy will be consumed when sending or receiving a packet in Mica2. Because the energy consumed by sending a bit wirelessly is equal to the energy of executing 1000 instructions of CPU in Mica2, the computing cost can be omitted.

Different sensors consume different amount of energy when sampling data. Table 2 presents the energy consumed by different sensors when sampling data[19].

Temperature sensors are used as examples in this paper to test the energy consumed by different algorithms. Sampling one bit will consume 0.5mJ energy for a temperature sensor.

Table 2. Energy Consumed by Different Sensors

Sensor	Energy Per Sample (@3V), mJ
sun radiancy	0.525
air pressure	0.003
temperature and humidity	0.5
voltage	0.00009

In different error bound, energy consumed by TiNA and TAG can be expressed by the following equation:

$$E_{TiNA\text{-}TAG} = R_{send} \times E_{sr} \times 2 + E_{sample}$$

where R_{send} is data sending rate in TiNA and TAG, E_{sr} is the energy consumed when sending and receiving a packet, E_{sample} is the energy consumed by sampling.

Energy consumed by DCA&LCA is described as follows:

$$E_{DCA\&LCA} = (R_{DCAsend} + R_{SCAsend}) \times E_{sr} \times 2 + R_{SCAsend} \times E_{sample}$$

where $R_{DCAsend}$ is data sending rate of DCA, $R_{SCAsend}$ is data sending rate of SCA , E_{sr} is the energy consumed when sending and receiving a packet, E_{sample} is the energy consumed by sampling.

Figure 4 to figure 7 show the energy consumed in SCA&DCA, TiNA and TAG with different sampling rates. It can be seen from these figures that the energy consumed by SCA&DCA is always lower than the other two algorithms. With the increasing of error bound, the decreasing of energy consumed in SCA&DCA is faster than the other two algorithms. SCA&DCA cuts down the amount of data transmitted by data compression and further saved energy when sending and receiving data. At the same time, it cuts down the times of sampling by controlling sample frequency dynamically and saved energy when sampling. TiNA and TAG cannot reduce energy consumed when sampling, so it is not as good as SCA&DCA. These figures also show that with the reducing of sampling cycle, energy consumed in SCA&DCA, TiNA and TAG all decrease obviously. But, the decreasing speed of SCA&DCA is faster than the others and energy consumed by it is the lowest. All these indicate that SCA&DCA has the best performance among the three algorithms.

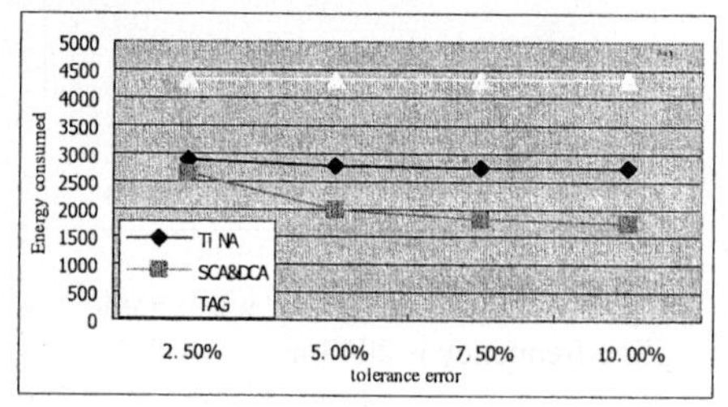

Fig. 4. Energy consumed when sampling frequency is 1024ms

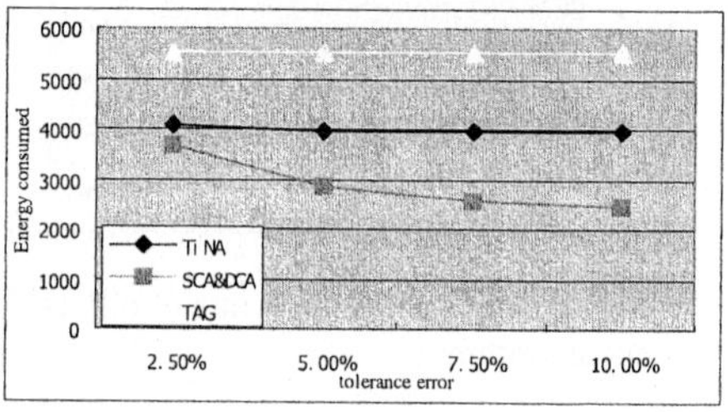

Fig. 5. Energy consumed when sampling frequency is 512ms

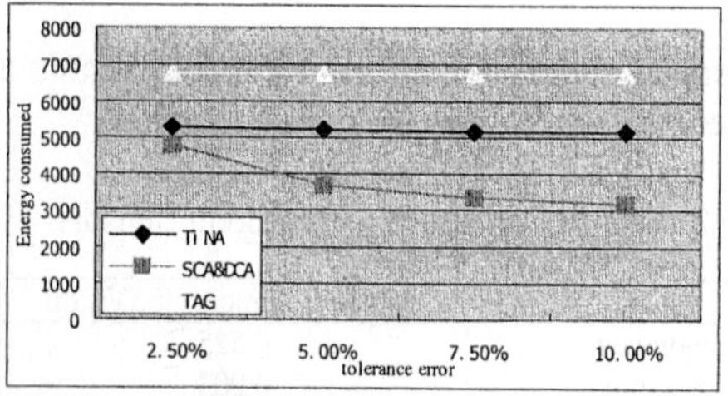

Fig. 6. Energy consumed when sampling frequency is 256ms

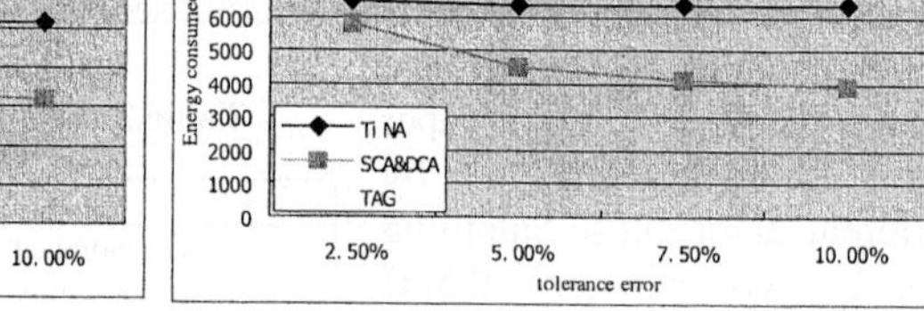

Fig. 7. Energy consumed when sampling frequency is 128ms

5 Conclusion

Power saving is an important optimization target in sensor networks. Limited storage capacity, communication ability, computing ability and the limited battery power are main restrictions in sensor networks. Sampling frequency control and data compression algorithms proposed in this paper take these restrictions into account. In a given tolerance error, based on the features of sensing data, sampling frequency is dynamically controlled and sensing data are compressed. Times of sampling and the amount of transmitted data are dramatically reduced in this method. The limited energy of sensors is thus largely saved.

References

1. Ian F. Akyildiz, Weilian Su, Yogesh Sankarasubramaniam and Erdal Cayirci, A Survey on Sensor Networks, *IEEE Communications Magazine*, vol. 40, no. 8, Page(s): 102 - 114. August 2002.Rentala P, Musunuri R, Gandham S, Saxena U. Survey on sensor networks. Technical Report, UTDCS-33-02, University of Texas at Dallas, 2002.
2. Bonnet P, Gehrke JE, Seshadri P. Towards sensor database systems. In: Tan K-L, Franklin MJ, Lui JCS, eds. Proceedings of the 2nd International Conference on Mobile Data Management. Hong Kong: Springer-Verlag, 2001. 3~14.
3. Yao Y, Gehrke J. The cougar approach to in-network query processing in sensor networks. SIGMOD Record, 2002,31(3):9~18.
4. Gerhke J. COUGAR design and implementation. http://www.cs.cornell.edu/database/cougar/.
5. S. Madden, M. J. Franklin, J. M. Hellerstein, and W. Hong.. The design of an acquisitional query processor for sensor networks. In: Halevy AY, Ives ZG, Doan AH, eds. Proceedings of the SIGMOD Conference. New York: ACM Press, 2003. 491~502.
6. S. Madden, M. J. Franklin, J. M. Hellerstein, and W. Hong. Tag:A Tiny Aggregation Service for ad hoc Sensor Networks. In OSDI Conf., 2002.
7. University of California at Berkeley. TinyDB. http://telegraph.cs.berkeley.edu/tinydb/.
8. Yong Yao Johannes Gehrke, Query Processing for Sensor Networks, Proceedings of 1st Biennial Conference on Innovative Data Systems Research (CIDR 2003), Jan 2003, Asilomar, CA.
9. S. Madden, M. J. Franklin, J. M. Fjording the stream: An architechture for queries over streaming sensor data. In: Ceri S, di Milano P, eds. *Proceedings of the ICDE Conference.* Los Alamitos: IEEE Computer Press, 2002. 555~666.
10. Jonathan Beaver, Mohamed A. Sharaf, Alexandros Labrinidis, and Panos K. Chrysanthis, Power-Aware In-Network Query Processing for Sensor Data. *In the Proceedings of the 2nd Hellenic Data Management Symposium*, September 2003.
11. Mohamed A. Sharaf, Jonathan Beaver, Alexandros Labrinidis, and Panos K. Chrysanthis, TiNA: A Scheme for Temporal Coherency-Aware in-Network Aggregation. In the Proceedings of the 3rd ACM MobiDE Workshop, September 2003.
12. A. Deligiannakis, Y. Kotidis, N. Roussopoulos.Hierarchical in-Network Data Aggregation with Quality Guarantees. E. Bertino et al. (Eds.): *In Proceedings of the 9th International Conference on Extending DataBase Technology (EDBT)*, March, 2004.p658-675.
13. C. Olston, J. Jiang, and J.Widom. Adaptive Filters for Continuous Queries over Distributed Data Streams. In *ACM SIGMOD Conference*, pages 563–574, 2003.
14. R. Cheng, D. V. Kalashnikov, and S. Prabhakar. Evaluating Probabilistic Queries over Imprecise Data. In *ACM SIGMOD Conference*, pages 551–562, 2003.
15. Iosif Lazaridis, Sharad Mehrotra: Capturing Sensor-Generated Time Series with Quality Guarantees. *Proc. of the Intl. Conf. on Data Engineering (ICDE 2003)*, page 429-441
16. J. Considine, F. Li, G. Kollios, J. Byers. Approximate Aggregation Techniques for Sensor Databases. Proceedings of 20th *IEEE International Conference on Data Engineering (ICDE2004)*, Boston, MA, March 30 - April 2, 2004.
17. Samuel Madden. The Design and Evaluation of a Query Processing Architecture for Sensor Networks Ph.D. Thesis. UC Berkeley. Fall, 2003A. Deligiannakis, Y. Kotidis, N.
18. M. J. McPhaden. Tropical atmosphere ocean project. *Pacific marine environmental laboratory*. http://www.pmel.noaa.gov/tao/.

Worst and Best Information Exposure Paths in Wireless Sensor Networks

Bang Wang, Kee Chaing Chua, Wei Wang, and Vikram Srinivasan

Department of Electrical and Computer Engineering (ECE),
National University of Singapore (NUS), Singapore
{elewb, eleckc, g0402587, elevs}@nus.edu.sg

Abstract. This paper proposes the concept of information exposure for a point and for a path based on estimation theory. The information exposure of a point is defined as the probability that the absolute value of estimation error is less than some threshold; and the information exposure for a path is the average information exposure of all points along the path. The higher the information exposure, the higher the confidence level that some information of a target is exposed and the better the target is monitored. An approximation algorithm is proposed to solve the problem of finding the worst (best) information exposure path in wireless sensor networks, and its performance is evaluated via simulations. Furthermore, a heuristic for adaptive sensor deployment is proposed to increase the information exposure of the worst information exposure path.

1 Introduction

Recently, *wireless sensor networks* (WSNs) that consist of a large number of sensors each capable of sensing, processing and transmitting environmental information have attracted a lot of research attention [1]. WSNs can be used for example in environmental monitoring, military surveillance, space exploration, etc. A fundamental issue in a WSN is the *coverage problem* [2][3]. In general, coverage is used to determine how well an area (points, lines or regions) is monitored or tracked by sensors. Many algorithms and solutions have been proposed to determine the points, lines and region coverage based on a unit disk sensing model such as in [4][5].

In target tracking, when a target passes through a sensor field, the target may choose a path on which it is least likely to be detected [6][7][8]. In [6], a *maximal breach path* (*maximal support path*) is defined as the path on which the distance from any point to the closest sensor is maximized (minimized). If the target moves along the maximal breach path, it is least likely to be detected by the closest sensor. In [7], a minimum exposed path is defined as the path on which each point is monitored by a number of sensors and the total sensing intensity is minimized. This model considers that each point of the path may be monitored by more than one sensor. The target experiences the least sensing intensity along the minimal exposed path, and hence it is least likely to be detected.

X. Jia, J. Wu, and Y. He (Eds.): MSN 2005, LNCS 3794, pp. 52–62, 2005.

In [8], the *exposure* of a path is defined to be the net probability of detecting the target when the target moves along the path. It not only considers using one or more sensors to monitor a point, but also applies *value fusion* to determine the detection probability of the point. However, the model considers only the detectability. In some cases, one may need not only to detect the existence of the target, but also to estimate some parameter values of the target. For example, when a tank traverses a field, obtaining the emitted energy amplitude of the tank (or the seismic energy amplitude) may not only help to determine the existence of the tank but also the model of the tank.

The above motivates us to propose an *information exposure* model based on parameter estimation. In particular, when a number of sensors are used to monitor a point and to estimate the value of a parameter at this point, the information exposure of this point can be defined as the probability that the absolute estimation error is less than some threshold. If the information exposure is high, information about the target at this point can be estimated (exposed) with a high confidence level. Hence, the higher the information exposure, the better the point is monitored. The information exposure for a path is then defined to be the average of the information exposure of all points on the path, and the worst (best) information exposure path problem is defined as finding the path with the minimal (maximal) information exposure. We propose an algorithm to approximately solve this problem by converting it to a problem of finding a minimal mean weighted path in a weighted graph. Numerical examples are provided to illustrate the proposed concepts and the algorithm, and a heuristic is proposed for adaptive sensor deployment.

The rest of the paper is organized as follows. Section 2 introduces the concepts of point and path information exposure based on estimation theory. The approximation algorithm to solve the worst (best) information exposure path problem is proposed in Section 3 and evaluated in Section 4. Section 5 concludes the paper with some remarks.

2 Information Exposure Path

2.1 Target Parameter Estimation

Consider a field with a number of sensors deployed to detect a target passing through the sensor field. The target emits a signal with amplitude θ which is measured by the sensors. Consider a snapshot of the sensor field and the target, and let $d_k, k = 1, 2, ...,$ denote the distance between a sensor k and the target at the snapshot. The parameter θ is assumed to decay with distance, and at distance d it is θ/d^α, where $\alpha > 0$ is the decay exponent. The above sensing model has been used in [7] for determining path exposure. The measurement of the target signal amplitude, x_k, at a sensor k may also be corrupted by an additive noise, n_k. Thus,

$$x_k = \frac{\theta}{d_k^\alpha} + n_k, k = 1, 2, ..., \quad (1)$$

The above measurement model has also been used in [8] to detect the presence of a target. Unlike the use of value fusion for target detection in [8], we use the estimate of the target parameter θ based on the corrupted measurements. Let $\hat{\theta}$ and $\tilde{\theta} = \hat{\theta} - \theta$ denote the estimate and the estimation error, respectively. A commonly used performance criterion is to minimize the *mean squared error* (MSE) of an estimator, i.e., to minimize $\mathbb{E}[\tilde{\theta}^2]$.

The measurement given by (1) can be written in matrix format for K sensors as

$$\mathbf{X} = \mathbf{D}\theta + \mathbf{N}, \tag{2}$$

where $\mathbf{X} = (x_1, x_2, ..., x_K)^T$, $\mathbf{D} = (d_1^{-\alpha}, d_2^{-\alpha}, ..., d_K^{-\alpha})^T$, and $\mathbf{N} = (n_1, n_2, ..., n_K)^T$. The additive noises are assumed to be spatially uncorrelated white noise with zero mean and σ_k^2 variance, but otherwise unknown. The covariance matrix of the noises $\{n_k : 1, 2, ..., K\}$ is given by

$$\mathbf{R} = \mathbb{E}[\mathbf{N}\mathbf{N}^T] = \text{diag}[\sigma_1^2, \sigma_2^2, ..., \sigma_K^2]. \tag{3}$$

A well-known *best linear unbiased estimator* (BLUE) [9] can be applied to estimate $\hat{\theta}_K$ and to achieve a minimum MSE. According to BLUE, when K measurements are available, the estimate $\hat{\theta}_K$ of the original signal θ is given as

$$\hat{\theta}_K = \left[\mathbf{D}^T\mathbf{R}^{-1}\mathbf{D}\right]^{-1}\mathbf{D}^T\mathbf{R}^{-1}\mathbf{X}. \tag{4}$$

The MSE of BLUE is given as

$$\mathbb{E}[(\theta - \hat{\theta}_K)^2] = (\mathbf{D}^T\mathbf{R}^{-1}\mathbf{D})^{-1}, \tag{5}$$

and the estimation error $\tilde{\theta}$ is given as

$$\tilde{\theta}_K = \hat{\theta}_K - \theta = [\mathbf{D}^T\mathbf{R}^{-1}\mathbf{D}]^{-1}\mathbf{D}^T\mathbf{R}^{-1}\mathbf{N}. \tag{6}$$

2.2 Information Exposure

In general, the estimate of a parameter at a point where a target is present is different from the estimate of the same point without the target. This is because the signal energy of the target is normally larger than the background noise energy. For example, the seismic vibrations caused by a tank are greater than the background noise. If the estimation error is small, not only the target can be claimed to be detected but also the target parameter can be obtained within some confidence level. Note that the estimation error $\tilde{\theta}_K$ given by (6) is a random variable with zero mean (due to zero mean uncorrelated noises) and variance denoted as $\tilde{\sigma}_K^2$. This motivates use to define the *information exposure of a point* as follows.

Definition 1. *K sensors cooperate to estimate a parameter at a point p that has distances $d_k, k = 1, 2, ...K$ to these sensors. The information exposure of this point is defined as the probability that the absolute value of estimation error is less than some threshold, A, i.e.,*

$$I(p, K) = \Pr[|\tilde{\theta}_K| \leq A]. \tag{7}$$

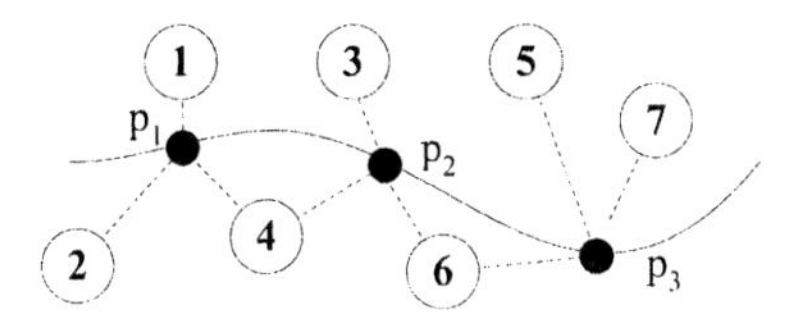

Fig. 1. Illustration of path exposure. Each point is monitored by three sensors.

The information exposure of a point depends on the number of sensors doing the estimation as well as their distances to the point. According to the definition, the larger $I(p, K)$ is, the lower the probability that the estimation error $\tilde{\theta}_K$ takes on a value by more than A, and hence the better point p is monitored. When the target is at location $p(t)$ at time t, we use $I(p(t), K)$ to denote the information exposure of this point. Suppose a target is moving in the sensor field from point $p(t_1)$ to point $p(t_2)$ along the path (or curve) $P(t_1, t_2)$. We now define the information exposure of the path along which the target moves as follows.

Definition 2. *Suppose that every point of a path $P(t_1, t_2)$ is monitored by K sensors. The information exposure for the path $P(t_1, t_2)$ along which a target moves during the time interval $[t_1, t_2]$ is defined as:*

$$\Phi(P(t_1, t_2), K) = \frac{1}{||P(t_1, t_2)||} \int_{t_1}^{t_2} I(p(t), K) \left| \frac{dP(t_1, t_2)}{dt} \right| dt, \tag{8}$$

where $||P(t_1, t_2)||$ is the Euclidean length of the path (i.e., the number of points on the path), $I(p(t), K)$ is the information exposure of point $p(t)$ and $|\frac{dP(t_1,t_2)}{dt}|$ is the element of curve length.

From the above definition, the path information exposure is an average for all path points' information exposures. Hence it can be considered as a measure of the average target estimation ability of the path. Fig. 1 illustrates the path information exposure. The path consists of three points each monitored by three sensors and hence the information exposure for the path is $\frac{1}{3} \sum I(p_i, 3)$. We compare the path information exposure to the path exposure defined in [7] and in [8]. The definition in [7] only considers simple processing for the measurements, viz., summing up the decayed parameters. It does not consider the effect of noise in the measurements. The definition in [8] applies value fusion to determine the detection probability of a point. In contrast, our definition includes advanced processing for noise corrupted measurements, i.e., the BLUE estimator, and provides the estimation of some target parameter as well as the detectability of a target.

3 Best and Worst Information Exposure Paths

Assume a rectangular sensor field in which each point is monitored by exactly K sensors. Let P denote a path from a point located at the west side of the sensor field to a point located at the east side of the sensor field, and let $\mathcal{P}$ denote

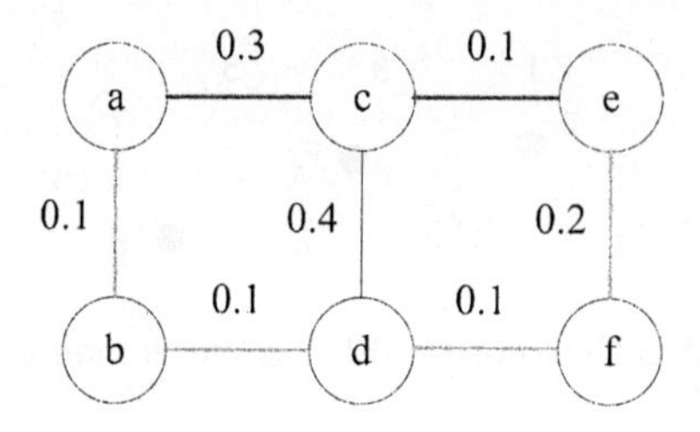

Fig. 2. Illustration of the difference between the mean weighted path problem and the shortest path problem. The path $< a, c, e >$ is the shortest path from a to e, however, the mean weighted path is $< a, b, d, f, e >$.

the set of such paths. The best (worst) information exposure path P_B (P_W) is defined to be the path with the largest (smallest) information exposure among all paths in $\mathcal{P}$ traversing from west to east in the sensing field.

In general, the best (worst) information exposure path can be fairly arbitrary in shape. An optimal solution may be very hard, , if not impossible, to obtain. Instead, we propose an algorithm to approximately solve the problem. The basic idea is to use a grid approach to transform the problem from a continuous domain to a discrete domain, and to restrict the best (worst) information exposure path search only along the grid. The formation of a grid again can be rather arbitrary and higher-order grids [7] can also be used. However, regular polygons are suggested to form a regular grid to simplify the computation of the average path information exposures. In general, the smaller the side of a regular polygon, the closer the approximation to an optimal solution; however, the computation becomes more complicated and takes longer time. The construction of the grid needs not to include the specified start and end points as the vertices of some polygons. If the start and end points are not on the grid, we can use their nearest polygon vertices to approximate them. Each side of a regular polygon is assigned a weight that represents the information exposure of this side. The target is restricted to move only along the sides of the polygons. Let graph $G(V, E)$ denote the grid, where V is the set of vertices and E the set of edges. Let $w(v_i, v_j)$ denote the weight of the edge connecting vertices v_i and v_j. The problem to find the best (worst) information path is then converted to find the maximum (minimum) mean weighted path connecting the start point p_s and the end point p_e in the graph $G(V, E)$, i.e., to find a path $P =< p_s = v_1, v_2, ..., v_j = p_e >$ with the largest (smallest) mean weight $w(P) = \frac{1}{j} \sum_{i=1}^{j} w(v_j, v_{j+1})$. We note that the best (worst) information exposure path may not be unique in the graph.

The above maximum (minimum) mean weighted path (abbreviated as max-/min-MWP hereafter) problem is different from the classical *single source shortest path* problem in that the latter is to find a path with the smallest path weight $\sum_{i=1}^{j} w(v_i, v_{i+1})$. Fig. 2 presents an example to illustrate that the min-MWP and the shortest path may not be the same in a graph. However, we note that the classical Dijkstra's algorithm (see e.g., [10], page 595) to solve the shortest path problem can also be applied to solve the mean weighted path problem with some modifications. The modifications include introducing variables $\mathbf{l}[v_i]$ to

Table 1. Procedure for Finding Worst Information Exposure Path

Notations:

S	the set of vertices whose min-MWP from the source have already been determined.
Q	the remaining vertices.
$\mathbf{d}$	the array of best estimates of the min-MWP to each vertex.
$\mathbf{l}$	the array of the path lengths from the source to each vertex.
$\mathbf{p}$	the array of predecessors for each vertex.

Find_Min_MWP

(1) generate a suitable regular grid;
(2) initiate the graph $G(V, E, v_s, v_e)$;
(3) compute information exposure for all edges;
(4) initiate $\mathbf{l}[v] = 0$ for all vertices;
(5) initiate $\mathbf{d}[v] \leftarrow \infty$ and $\mathbf{p}[v] \leftarrow nil$ for all vertices;
(6) initiate $\mathbf{d}[v_s] = 0$ for the start vertex;
(7) initiate $S \leftarrow \emptyset$ and $Q = V$;
(8) **while** $v_e \notin S$
(9) $\quad u \leftarrow$ ExtractMin(Q)
(10) $\quad S = S \cup \{u\}$
(11) $\quad \mathbf{l}[u] = \mathbf{l}[u] + 1$
(12) $\quad$ **for** each vertex v in $Adjacent(u)$
(13) $\quad\quad z = \frac{\mathbf{l}[u]-1}{\mathbf{l}[u]}\mathbf{d}[u] + \frac{1}{\mathbf{l}[u]}w(u, v)$
(14) $\quad\quad$ **if** $\mathbf{d}[v] > z$
(15) $\quad\quad\quad \mathbf{d}[v] = z$
(16) $\quad\quad\quad \mathbf{p}[v] = u$
(17) $\quad\quad\quad \mathbf{l}[v] = \mathbf{l}[u]$
(18) $\quad\quad$ **endif**
(19) $\quad$ **endfor**
(20) **endwhile**
(21) $min\_exposure = \mathbf{d}[v_e]$

denote the path length (i.e., the number of edges) from the start vertex to the vertex v_i and modifying the computation of the RELAX algorithm ([10], page 586) to set predecessor vertex weight as the information exposure from the start vertex to the vertex in consideration. That is, the modified Dijkstra algorithm uses the average edge weight instead of the total edge weight to compute the vertex weight for predecessors.

The procedure for finding the worst information exposure path is given in Table 1. Lines (1) to (7) are the initialization part of the algorithm. The edge weight can be computed by integrating the point information exposure of all the points of the edge or by simply using the point information exposure of the median point of the edge. Lines (5) to (7) are the same initialization part as in Dijkstra's algorithm. Since we only need to find a path connecting the start vertex to the end vertex, the condition for the **while** loop is changed accordingly. Line (11) is added to update the counter of the number of edges from the start vertex to the current vertex. Lines (12) to (19) are the modified RELAX algorithm. The RELAX algorithm ([10], page 586) computes $z = \mathbf{d}[u] + w(u, v)$ as the total weight from the start vertex to the vertex in consideration; while line (13) computes the mean weight. Since the grid is regular and all edges have equal lengths, $\mathbf{l}$ is as simple as a counter. Furthermore, line (17) is added to update the counter for the vertex in consideration as its predecessor's counter. Finally, the vertices in the predecessors $\mathbf{p}[v_e]$ and the edges connecting these vertices provide the min-MWP path. The procedure in Table. 1 can also be applied to compute the max-MWP with the following modifications. Line (5) should be modified to initialize $\mathbf{d}[v] = -\infty$; line (9) changed to '$u \leftarrow ExtractMax(Q)$'; line (13) changed to '**if** $\mathbf{d}[v] < z$', and finally line (21) changed to '$max\_exposure = \mathbf{d}[v_e]$'.

4 Numerical Examples

This section presents some numerical examples. For simplicity, we consider a special case where all noises are Gaussian. Since all noises are Gaussian and independent, the sum of these Gaussian noises is still Gaussian with zero mean and variance $\tilde{\sigma}_K^2 = \sum_{k=1}^{K} a_k^2 \sigma_k^2$, where $a_k = B_K / d_k^{\alpha} \sigma_k^2$ and $B_K = \left(\sum\limits_{k=1}^{K} 1/d_k^{2\alpha} \sigma_k^2 \right)^{-1}$. We further assume that all noises have the same variance, i.e., $\sigma_k^2 = \sigma^2$ for all $k = 1, 2 \ldots$. Via some algebra, the information exposure for a point is given by

$$I(p, K) = \int_{-A}^{A} \frac{1}{\sqrt{2\pi \tilde{\theta}_K}} \exp\left(-\frac{\tilde{\theta}_K^2}{2\tilde{\sigma}_K^2} \right) d\tilde{\theta}_K = 1 - 2Q(\frac{A}{\tilde{\sigma}_K}),$$

where $Q(x)$ is defined as $Q(x) = \frac{1}{\sqrt{2\pi}} \int_x^{\infty} \exp(-\frac{t^2}{2}) dt$, $\tilde{\sigma}_K = \sigma \sqrt{C_K}$, and $C_K = (\sum_{i=1}^{K} 1/d_i^{2\alpha})^{-1}$. For simplicity, A can be set as $\beta\sigma$, $\beta > 0$. In this paper, we set $\beta = 0.5$.

Consider a 10×10 grid sensor field with 10 sensors at locations marked by red disks, as shown in Figs. 3 and 4. The side length of a square is set as a unit. The start and end points are marked by the blue circle, and located at the left and right sides of the grid, respectively. For simplicity, the information exposure of the middle point of each segment is used as the path information

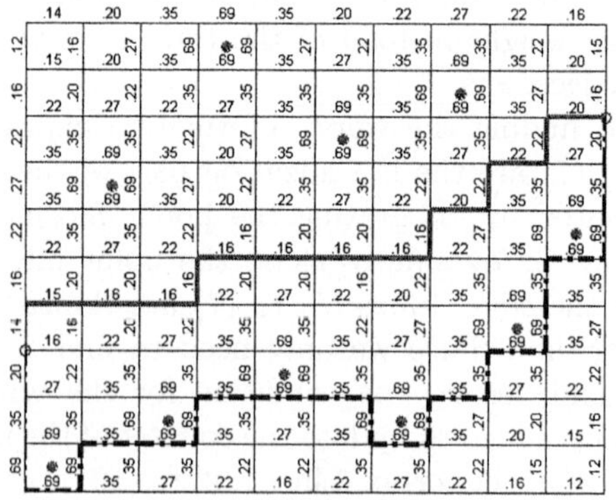

Fig. 3. Worst and best information exposure paths when $K = 1$. The red line is the Min-MWP and blue dashed line is the Max-MWP.

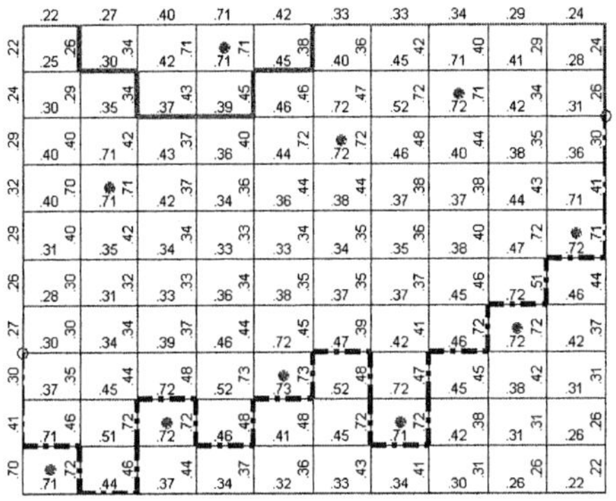

Fig. 4. Worst and best information exposure paths when $K = 10$. The red line is the Min-MWP and blue dashed line is the Max-MWP.

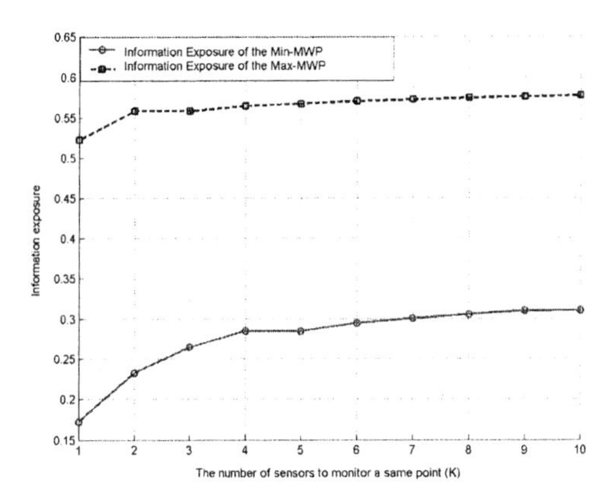

Fig. 5. Path information exposure as a function of the number of sensors for estimation of a same point

of the whole segment. The sensors selected to monitor a point are the ones closest to the monitored point, which is the most efficient selection as shown in [11]. Furthermore, we set $\alpha = 1.0$ in the computations. Figs. 3 and 4 show the computed worst and best information exposure paths when using 1 and 10 sensors to monitor each segment, respectively. The worst and best information exposures for $K = 1$ are 0.17 and 0.52 respectively, and for $K = 10$ are 0.31 and 0.58 respectively.

In general, to increase the path information exposure, one can use more sensors to monitor a point; or deploy more sensors in the field; or do both. Fig. 5 plots the path information exposures for min-MWP and max-MWP when using different numbers of sensors to monitor a point. The sensor field and the start and end points are the same as in Fig. 3. As expected, the path information exposure increases when using more sensors to monitor a point. This is because the estimation accuracy of a point can be improved by using more sensors for esti-

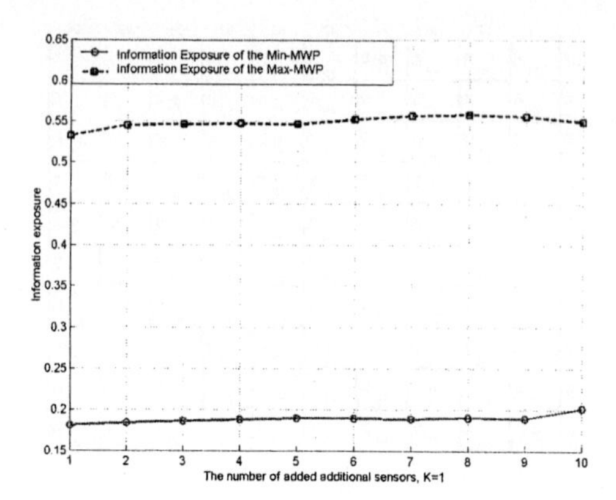

Fig. 6. Path information exposure as a function of the number of added additional sensors. $K = 1$.

mation. However, we note that there might not exist a monotonically increasing relationship between the information exposure of min-MWP and K. The information exposure for the min-MWP has improved $(0.31 - 0.17)/0.17 = 82\%$ when using $K = 10$ sensors compared with that using $K = 1$ sensor.

Next we evaluate the impacts to information exposure when randomly adding more sensors to the field. The same sensor deployment as in Fig. 3 is used, and 1 to 10 additional sensors are randomly added to the field. However, each point is monitored by only one sensor, i.e., $K = 1$. Fig. 6 plots the information exposure against the number of additional sensors. Note that a value of 1 in the x-axis indicates a total of 11 sensors in the field, 10 of which are deterministically distributed as in Fig. 3 and 1 randomly distributed. The information exposure is obtained over 20 runs of simulations. It is observed that the improvement of information exposure for min-MWP is reduced. The information exposure for the min-MWP has improved $(0.21 - 0.18)/0.18 = 17\%$ when randomly adding $K = 10$ sensors compared with that adding $K = 1$ sensor. This observation suggests that using more sensors to collaboratively monitor a point is better in terms of increased information exposure than having only one sensor do the monitoring despite having a larger total number of sensors in the field.

Motivated by this observation, we now propose a heuristic to adaptively deploy sensors so as to increase the information exposure of min-MWP. Let $add(K)$ denote the method of increasing the number of sensors by one to monitor a point, and $add(Nr)$ denote the method of deploying a new sensor to the field. To increase information exposure, a new sensor should be deployed as close as possible to min-MWP. In this paper, a new sensor is added to the center of a square that has at least one side as part of min-MWP and whose total information exposure of the four sides is the smallest. For example, if we add a new sensor to Fig. 3, it is added to the center of the square located at the first column and the fourth row (from bottom to above). Assume that at first the field is randomly deployed

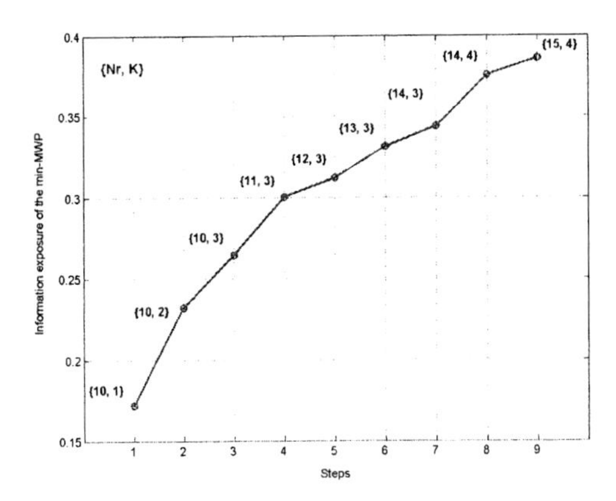

Fig. 7. Information exposure of the min-MWP by using the heuristic method

Table 2. Pseudo-codes for the heuristic sensor deployment

A Heuristic Deployment Method
(1) set $K = 1$ and $\Phi(P_0, 0) = 0$
(2) compute $\Phi(P_1, 1)$
(3) **while** $\Phi < \Phi_{target}$
(4) **if** $\Phi(P_K, K) - \Phi(P_{K-1}, K-1) < \Delta\Phi$, $add(Nr)$
(5) **else** $add(K)$; **endif**
(6) compute new Φ
(7) **endwhile**

with Nr sensors. The pseudo-codes for the heuristic are given in Table. 2. In the table, Φ_{target} and $\Delta\Phi$ are two predefined thresholds. $\Delta\Phi$ relate to the relative cost between using $add(K)$ and using $add(Nr)$. We still use the sensor deployment in Fig. 3 to evaluate the proposed heuristic, and set $\Delta\Phi = 0.03$. Fig.7 shows the information exposure of min-MWP computed by the heuristic. It is observed that the information exposure increases faster than that shown in Fig. 5 (Fig. 6).

5 Concluding Remarks

We have proposed the concept of the best and worst information exposure paths for WSNs based on parameter estimation theory. The proposed information exposure can be used as a measure of the goodness of sensor deployment or coverage. An algorithm has been proposed to find the worst/best information exposure path and has been evaluated by simulations. Furthermore, a heuristic has been proposed for sensor deployment to increase the minimum information exposure.

References

1. Akyildiz, I.F., Su, W., Sankarasubramaniam, Y. and Cayirci, E.: Wireless Sensor Networks: A Survey. *Computer Networks, Elsevier Publishers* (2002) vol. 39, no. 4, 393–422
2. Cardei, M., and Wu, J.: Energy-Efficient Coverage Problems in Wireless Ad Hoc Sensor Networks, *Handbook of Sensor Networks, (Ilyas, M. and Mahgoub, I. Eds) chapter 19, CRC Press* (2004)
3. Huang, C.-F., and Tseng, Y.-C.: A Survey of Solutions to The Coverage Problems in Wireless Sensor networks, *Journal of Internet Technology*, (2005) vol. 6, no. 1
4. Huang, C.-F. and Tseng, Y.-C.: The Coverage Problem in A Wireless Sensor Network, *ACM international workshop on wireless sensor networks and applications (WSNA)*, (2003) pp. 115–121
5. Wang, X., Xing, G., Zhang, Y., Lu, C., Pless, R., and Gill, C.: Integrated Coverage and Connectivity Configuration in Wireless Sensor Networks, *In ACM International Conference on Embedded Networked Sensor Systems (SenSys)*, (2003) 28–39
6. Meguerdichian, S., Koushanfar, F., Potkonjak M., and Srivastava, M. B.: Coverage Problems in Wireless Ad-hoc Sensor Networks, *IEEE Infocom*, (2001) vol. 3, 1380–1387
7. Meguerdichian, S., Koushanfar, F., Qu, G., and Potkonjak, M.: Exposure in Wireless Ad Hoc Sensor Networks, *ACM International Conference on Mobile Computing and Networking (MobiCom)*, (2001), 139–150
8. Clouqueur, T., Phipatanasuphorn, V., Ramanathan, P., and Saluja, K. K: Sensor Deployment Strategy for Target Detection, *First ACM International Workshop on Wireless Sensor Networks and Applications(WSNA)*, (2002) 42–48
9. Mendel, J. M.: *Lessons in Estimation Theory for Signal Processing, Communications and Control*, Prentice Hall, Inc, (1995)
10. Cormen, T. H., Leiserson, C. E., Rivest, R. L., and Stein, C., *Introduction to Algorithms*, The MIT Press; 2nd Edition, 2001.
11. Wang, B., Wang, W., Srinivasan, V., and Chua, K. C.: Information Coverage for Wireless Sensor Networks, *accepted by IEEE Communications Letters*, 2005.

Cost Management Based Secure Framework in Mobile Ad Hoc Networks

RuiJun Yang[1,2], Qi Xia[1,2], QunHua Pan[1], WeiNong Wang[2], and MingLu Li[1]

[1] Department of Computer Science and Engineering,
Shanghai Jiao Tong University, Shanghai, China
rjyang@sjtu.edu.cn
[2] Network Information Center,
Shanghai Jiao Tong University, Shanghai, China

Abstract. Security issues are always difficult to deal with in mobile ad hoc networks. People have seldom studied the costs of those secure schemes respectively and for some secure methods designed and adopted before-time, their effects are often investigated one by one. In fact, when facing certain attacks different methods would response individually and it would result in some waste of resource, which may be worthless from the viewpoint of whole network. Making use of the cost management idea, we analyze the costs of secure methods in mobile ad hoc networks and introduce a secure framework. Under the framework not only the network systems own tasks can be finished in time but also the whole network's security costs can be decreased. We discuss the process of security costs computation at each mobile node and in certain nodes groups. We then exemplify the DoS attacks and costs computation among defense methods. The results show that more secure environment can be achieved based on the framework in mobile ad hoc networks.

1 Introduction

Mobile ad hoc networks are an active area of current researches and until recently the focus has been on issues such as routing [4], security [6] and data management. As for network security people have proposes many kinds of secure mechanisms such as SAR [3], SEAD [4] and so on. The resource consumptions of secure mechanisms are always large in mobile ad hoc networks so the cost is very high when multi-secure methods collaborate to resist the attacks and system weaknesses. Sometimes the resources such as bandwidths, power and media storages are wasted or inefficiently applied. From the view of whole ad hoc networks, the costs of security techniques should be taken into account besides finishing the main networks tasks.

There are some reasons for the high costs about security mechanisms. Firstly, the own characteristics of mobile ad hoc network account for the main factor. Mobile ad hoc networks are the resource limited systems themselves and each component mobile nodes have finite usable resources either power, bandwidths or processing ability and memories though their capacities are determined by

X. Jia, J. Wu, and Y. He (Eds.): MSN 2005, LNCS 3794, pp. 63–72, 2005.

the certain applications. Nevertheless the restricted resource in each node is a big weakness for mobile ad hoc network. So the cost issues should be considered both in network own basic functions design and other appended mechanisms.

The second reason is related with the working mechanisms of mobile ad hoc network. At the beginning of ad hoc network protocols designing, people have not brought into the security problems. And some secure mechanisms are developed and attached to the older protocols when facing more and more security leaks and network attacks. Even added into the network protocols or integrated as a secure module, these operations are dealt for themselves and the costs of multi nodes or multi mechanisms united defense again attacks are rarely considered. So the working mechanisms related security costs should also be noticed and paid more attention.

The assumptions about radio propagations may be one of the reasons about little attention for security costs. Those secure schemes are declared that they may help the system to satisfy some secure requirements though the results are just get through some simulations. It is simulation's radio propagation assumptions that make things to be different. Those simulations are executed under nearly perfect assumptions and outcomes may be subverted causing larger costs when they were done in more realistic scenarios. It is certain that simulations and designing results may run a risk in testifying secure protocols. [1]points out the simplistic radio models may lead to manifestly wrong results. [2]indicates that theoretical results on the capacity of ad hoc networks are still based on some simplified assumptions and the derivation of new results by considering critical factors.

The rest of this paper is organized as follows: In section 2, we discuss the existing secure schemes and attacks from layer to layer and attempts a detailed presentation about secure costs in each nodes or each mechanism. In section 3, we provide a secure framework based on cost management and describe the costs computing process at each node or in some nodes groups. The analysis about DoS attacks and defenses under the proposed secure framework can be found in section 4 and some secure mechanisms reconfiguration's effects are also presented. At last we conclude this paper in section 5 with a stating about existing issues and some future works.

2 Analysis of Network Attacks and Security Mechanisms

The fundamental requirements of computer security like confidentialityintegrity, authentication and non-repudiation are also valid when protection of correct network behaviors is to be considered in Mobile ad hoc networks. The characteristics of Mobile ad hoc networks make them vulnerable to various forms of attacks, which can be classified into different categories according to different standard. One of the classifications is passive attacks and active attacks. Passive attacks are the attacks in which an attacker just eavesdrops on the network traffic to get more useful information for future attacks. Active attacks are the attacks in which an attacker actively participates in disrupting the normal operation of network protocols [6].

For different layers and different types of attacks, people have made great corresponding efforts and produced lots of secure mechanisms and in [5] an adaptive secure framework has been proposed which take into account the structures of security but have little knowledge on how to cooperation among multi-security methods.

The attacks are always restricted within certain network applications and when facing aggression the secure mechanisms have to do their own bests they can to cope with hardness with little thinking about the cooperation among them. It is believable that attacks would happen at almost all layers from one time to another time with different styles. Facing combined attacks, people have developed combined secure mechanisms to resist them and prevent system from falling. Actually there may be many methods in each layer.

When facing secure problems bursting at more than two layers simultaneously, single layer protocol can not dealt with them itself and there is need for cooperation among several layers' secure mechanisms. But protection against one type of attacks may weaken the network against a second type of attacks and finding the right balance is extremely difficult. Another problem exists that having two separate protocols at each layer performing similar functions in an uncoordinated manner may lead to large overhead and make the cost high.

Some secure mechanisms begin to run before encountering attacks and others startup while suffering aggressions. So it is hard to compute the cost of secure schemes in mobile ad hoc networks. The composing parts of whole running cost at one node are CPU occupancy factor, power consumption, bandwidth occupancy factor and memory utilizations all of which are the expenditure of network running and are necessary for the attainment of finishing tasks. It is necessary for mobile nodes to quantify the running cost and for each node$i$, its running cost C_i is the weighted sum of four factors corresponding costs:

$$C_i = \alpha C_{i-CPU} + \beta C_{i-Memmory} + \delta C_{i-Battery} + \gamma C_{i-Band} \tag{1}$$

here C_{i-CPU}, $C_{i-Memory}$, $C_{i-Battery}$,C_{i-Band} are the corresponding CPU occupancy factor, power consumption, bandwidth occupancy factor and memory utilizations$\alpha, \beta, \delta, \gamma$are the weighted coefficients representing their own importance. Every secure mechanism has its own secure cost $C_{security}$. Before the working of secure mechanism, the running cost of normal network protocol is $C_{pre-secure-mechanism}$ and after secure mechanism adopted, the running cost is $C_{post-secure-mechanism}$ so:

$$C_{security} = C_{post-secure-mechanism} - C_{pre-secure-mechanism} \tag{2}$$

Once the cross layer secure mechanisms can cooperate to resist attacks, their combined cost should be small than the sum of every sole mechanism cost. For the combined secure mechanisms there are many algorithms and their costs is not the simple sum of nodes $C_{security}$. It needs to propose an adaptive secure framework to deal with that complex cost computing.

3 Secure Framework Based on Cost Management

There are some cross layer secure framework aiming at deal with complex weaknesses and attacks in mobile ad hoc networks. Commonly secure mechanisms have not taken into account the cost of security. Based on cost management, there are several principles for designing: 1) the cost management should be executed from the view of the whole network. 2) It should calculate secure scheme cost of every layer. 3) It should increase the validity of the cost management to get the greatest secure benefits at the expense of the smallest secure costs. We should forecast, analyze, calculate and harmonize secure costs among nodes effectively.

When facing threats, system can configure own limited resource according to the results feedback by the different stage of cost management. Every node firstly calculates its own cost, judge whether it needs support from other nodes. If does, it should collect the involved nodes' costs according to corporations in nodes group resisting attacks. After the simple calculation, system judges and decides how to reconfigure secure mechanisms in distributed style. Then system distributes the safeguard tasks among nodes according to predefined principles with the nodes' resource effectively utilizing. As shown in Fig. 1, we propose a secure framework based on cost management. In the secure framework, there are four parts. Besides the normal ad hoc network protocols' running module, the secure mechanisms configuration module and reconfiguration module are also important for security costs effects of the whole network. Both them are centered with cost management module.

Firstly, some secure mechanisms such as secure routing protocol, authorization and authentication should be set in the network nodes in order to prevent some known attacks. While the network running, some attacks may be found and related secure schemes would be adopted to resist them. Before their working, we should compute the costs budgets of security and compute the costs of

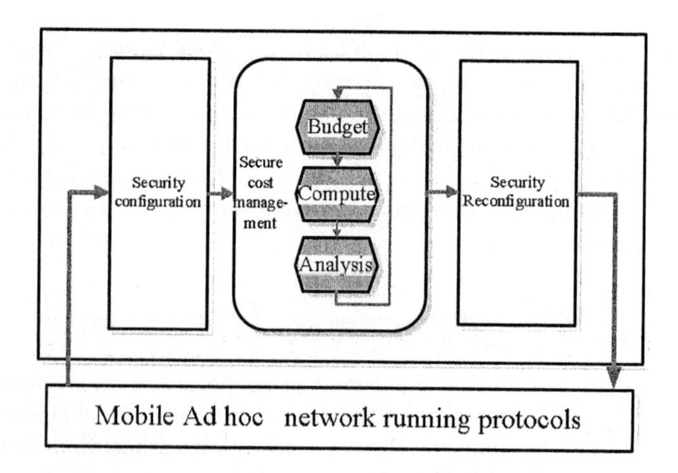

Fig. 1. Cost Management Based Secure Framework

them again after some while or the stop of secure schemes. These processes are executed more times when facing large mounts of attacks. The changes between the budgets and real costs can be fed back and used for analysis of secure mechanisms reconfigurations. At the same time, resistant nodes or nodes groups can make use of the feedback information to evaluate the whole effectiveness of costs and reconfigure the secure mechanisms according to that information. The flow of costs computation under secure framework is shown below:

3.1 Each Mobile Node Solely Computes Its Own Security Costs

For the node in whole network, there are corresponding network operating costs from application, transport, network, to MAC and physical layer, which are shown in Table. 1. As mentioned in former section, the normal running cost of each mobile node is shown in formula 1.: every layer's running costs are C_{iA},C_{iT},C_{iN},C_{iM},C_{iP},which all can be computed according to

$$C_{i\psi} = \alpha_{\psi} C_{i\psi-CPU} + \beta_{\psi} C_{i\psi-Memmory} + \delta_{\psi} C_{i\psi-Battery} + \gamma_{\psi} C_{i\psi-Band} \quad (3)$$

ψ means certain layer. After computed C_{iA},C_{iT},C_{iN},C_{iM},C_{iP}, the results are stored in a vector called cost vector:

$$\hat{C}_i = (\, C_{iA},\, C_{iT},\, C_{iN},\, C_{iM},\, C_{iP}\,) \quad (4)$$

As mentioned above, every secure mechanism' cost is shown in formula 2. For certain layer secure mechanism, the costs C_{iA},C_{iT},C_{iN},C_{iM},C_{iP}, always change before and after secure mechanisms adoption. And they can be computed:

$$C_{i\psi-security} = C_{i\psi-post-secure-mechanism} - C_{i\psi-pre-secure-mechanism} \quad (5)$$

ψ in $C_{i\psi-security}$ indicate some layer of network protocol. The security weaknesses occur indeterminably in some layers and we use one vector to distinguish the security related costs change from the changes due to normal system running when computing the secure mechanisms costs.

$$\hat{O}_i = (o_{iA}, o_{iT}, o_{iN}, o_{iM}, o_{iP}), o_{ik}, k \in A, T, N, M, P \quad (6)$$

Table 1. Normal Running Cost of Each Layer Protocol

iNode	*A(application)*	*T(transfer)*	*N(network)*	*M(MAC)*	*P(physical)*
1	C_{1A}	C_{1T}	C_{1N}	C_{1M}	C_{1P}
2	C_{2A}	C_{2T}	C_{2N}	C_{2M}	C_{2P}
3	C_{3A}	C_{3T}	C_{3N}	C_{3M}	C_{3P}
⋮	⋮	⋮	⋮	⋮	⋮
i	C_{iA}	C_{iT}	C_{iN}	C_{iM}	C_{iP}
⋮	⋮	⋮	⋮	⋮	⋮
n	C_{nA}	C_{nT}	C_{nN}	C_{nM}	C_{nP}

The values of o_{iA},o_{iT},o_{iN},o_{iM},o_{iP} can only are 0 or 1 respectively representing whether adopting secure schemes. If not, the results in each factors should not be included in the $C_{i\psi-security}$. That is to say, for node i, its security cost can be computed though the non-zero values in the vector $\widehat{CO}_i$:

$$\widehat{CO}_i = \hat{C}_i \times \hat{O}_i = (C_{iA}, C_{iT}, C_{iN}, C_{iM}, C_{iP}) \times (o_{iA}, o_{iT}, o_{iN}, o_{iM}, o_{iP}) \quad (7)$$

The values of o_{iA},o_{iT},o_{iN},o_{iM},o_{iP} can be re-written by its seated node. The whole costs of one node is:

$$C_{i-security} = \sum C_{i\phi} o_{i\phi} \quad (8)$$

Here ϕ represents the layer which value in $\widehat{CO}_i$ is nonzero. When the node computing its own costs, it can also calculate the costs ratio which indicates the secure mechanisms costs proportion to the whole running costs.

$$\eta_i = C_{i-security}/C_i = \sum C_{i\phi} o_{i\phi} \Big/ (C_{iA} + C_{iT} + C_{iN} + C_{iM} + C_{iP}) \quad (9)$$

3.2 Local Group Costs Computation

Every small period, some nodes can make up of one group in some small area and the principal node, which has largest quantity of usable resource, is charge with the local group costs computation.

Selection of The Principal Node: When one node has the largest resource to utilization comparing with neighboring nodes, it can be made use of computing the group cost around it. In order to make the correct choice of principal node, the neighboring nodes have to exchange their information about usable resource. These exchanging also make some costs for themselves so the size of local group should not be too large.

Setup for Size of Group and Computing Period: The local group cost computing should be executed periodically. Once the node finishing its own first time secure cost computing, it can send one indication about its attempt to call local group cost computing. When more than half of several adjacent nodes sense the attempts from each other they can select the principal node and begin the first-time group cost computation. The principal node only include those nodes inside one-hop in its current group. After group cost calculating, the principal node feeds back some cost information to its group's nodes. One node cannot take part in another group cost computation when it has already been participating one computation.

Local Groups Cost Computation: The principal node requires nodes in its group to send their own secure costs and compute the group costs:

$$C_{group} = \sum_{i=1}^{m} C_{i-security} \quad (10)$$

At the same time, it can also calculate the group secure costs ratio used in future security analysis. Here m is the number of nodes in the group.

$$\eta_{group} = \frac{\sum_{i=1}^{m} C_{i-security}}{\sum_{i=1}^{m} C_i} = \frac{\sum_{i=1}^{m} \sum C_{i\phi} o_{i\phi}}{\sum_{i=1}^{m} C_{iA} + C_{iT} + C_{iN} + C_{iM} + C_{iP}} \tag{11}$$

For the whole mobile ad hoc network it is very hard to share the security information so it is too difficult to implement the global security costs and co-operations among nodes. Some models should be proposed to predict the execution time of multiple jobs on each mobile node with varied secure mechanisms costs.

3.3 Global Secure Costs Compute

Every larger period, the former principals can set up a large-scale secure cost computation in larger area. For the mobility of nodes it is too hard to compute global secure costs.

We can calculate secure in larger area than steps in 3.2. The tasks are still charged by principals and it needs them to store some local secure costs and cost ratios and to exchange them among nodes. These principal nodes should send indications to the same type of nodes outside two-hop about the intension of calculating secure costs in larger area. Those indications only can be accepted and calculated instantly providing the involved groups' security association to system global cost analysis. The method of computing is similar with step2's but only two types of parameters are used here. For example, node B receives two local secure cost related values $\eta_{Group-A}$ and $C_{Group-A}$ from node A (here $C_{Group-A} = \sum C_{i-security}, i \in Group - A$) it can combine them with its own parametersgroup-B and $C_{Group-B}$. Then it can calculate the security association $\Phi_{Group-A,Group-B}$ in larger area around link A-B

$$\Phi_{Group-A,Group-B} = \frac{(C_{Group-A} + C_{Group-B})}{[(C_{Group-A}/\eta_{Group-A}) + (C_{Group-B}/\eta_{Group-B})]} \tag{12}$$

Here we only make the simple cost management and through several larger areas secure cost computations the approximate global secure costs can be obtained. For more complex global secure cost management algorithms it needs more efforts to study them in the future. In mobile ad hoc networks there are some problems in secure costs share and cross layers secure mechanisms cooperation among nodes because the mobile nodes keep moving and the nodes group changing. So we can only remind system about secure conditions through checking the secure cost ratios and secure costs in short times.

4 Example of Security Configuration Under Secure Framework

In the implementation of the proposed secure framework, each node has to be charge with its own tasks when facing attacks in order to protect system using

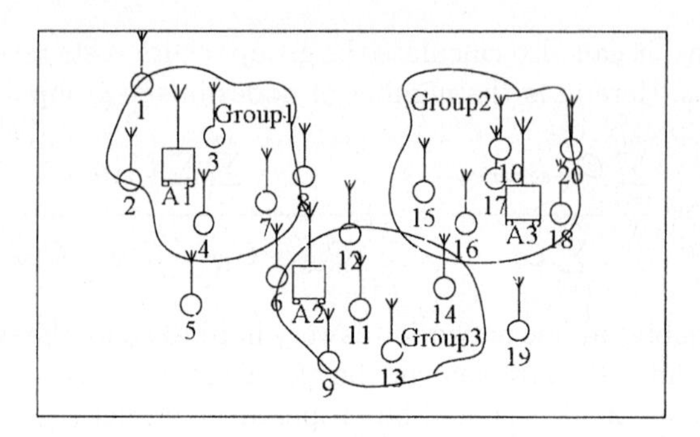

Fig. 2. Scenario of Example of DoS Attacks and Defenses

rational schemes. Here we exemplify the DoS attacks and protections as the illustration of our cost management based secure framework.

DoS attacks aim to prevent access to network resources and it is difficult to protect against. It can target different layers and there is much difference between various types of DoS attacks. Traffic patterns generated by an attacking node, its location in the network, availability of other compromised nodes, and availability of routing information are key factors in determining the efficacy of the DoS [7]. In fact, the inherent computational costs enable other DoS attacks at other layers on the server performing some protections. An appropriate secure mechanism costs' balance should be introduced for the defense against DoS attacks. As shown in Fig. 2,let's suppose an ad hoc network contain twenty nodes numbered from 1 to 20 and three attack nodes named A_1,A_2,and A_3. Nodes around the attack nodes organize randomly three nodes groups. It will set up two big local secure conjunctions between the node 3 and 11 as well as the node 11, 17.First,we calculate every security cost of the twenty nodes marked as C_1,C_2,... C_20. Then, we compute the group costs of $Group_1 Group_2 Group_3$ and proportion of security cost to whole running cost $\eta_{Group-1}$,$\eta_{Group-2}$,$\eta_{Group-3}$, centered the node 2,11 and 17 which are respond for the computing tasks. At last, the security association$\Phi_{Group1,Group3}$ between node 1 and node 11 as well as$\Phi_{Group2,Group3}$ between node 11, 17 are figured out.

If system has detected attacks and defense actions are deployed, the costs of security will turn larger. At this time, system should estimate the invalid costs. For the residual resources there are three kinds of scenarios including satisfying the requirements of system running, completing parts of mission ineffectively, and losing basic functions. For each instance we should adjust the costs in time.

These data can be used to collocate their secure scheme of the nodes in the group. For example, at some time in the $Group_1$, the attacker A_1 sends Exhaustion attack to node 1, the node 1 prefabricates small frames secure scheme, while the nodes 2 and 4 startup client puzzles secure scheme to protect the flooding attack from A_1. If A_1 make misdirection to node 7 and 8, which we assume the two nodes have prefabricated authorization. So, because the node 3 has not been attacked from A_1, it has so many sources to use to startup the secure cost

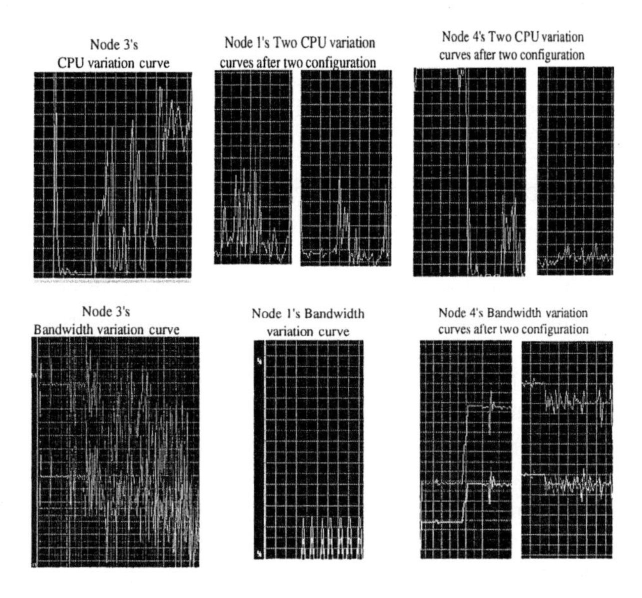

Fig. 3. CPU and Band Utilization

calculation of $Group_1$. It can be harmonize every node to relocate the secure mechanisms by calculating every node's secure cost and ratio.

Through simulating the secure mechanisms reconfiguring of the nodes in $Group_1$, we get some graphs of CPU and band utilization ratio shown in Fig. 3. The occupancy factor of CPU of **Node3** is small at the beginning. It increases after starting up the calculation of the costs. At first, the band is transferring the data normally. When the other nodes are attacked, it has less data to transfer with them so that its band is decreased. Because the **Node7**,**Node8** have prefabricated secure schemes, their bands and CPUs are changing when attacked. And then, they can resist the attacks through their own secure schemes, which result in that changing is smoothly. But for the band of **Node1** attacked, it is stable straightly and keeps in a low bandwidth-utilizing ratio, so it could not be adjusted. But because it uses small frames, its occupancy factor of CPU will be high. We can modify it through its secure scheme to decrease its CPU occupancy factor. **Node2** and **Node4** have the same conditions with facing flooding attacks, because they adapt the client puzzle methods the CPU keep working in a high occupancy factor and the bandwidths utilizing are also too large. At this time, the secure costs become too abnormally high and the system running is affected badly. For some attacks we should not care them and make the CPU's overload mitigating based on the sacrifice of bandwidths which can satisfy the basic data transferring requirements though sill keeping higher.

Above figures tell us that local group's secure costs can be used to enhance the effects of security costs in the mobile ad hoc networks and to improve the availabilities of nodes in larger areas. For the more complex applications and implementations of secure framework such as the costs computation of whole networks nodes, we still keep on studying and in the future may introduce some QoS's evaluating methods to detect the effectiveness of secure mechanisms reconfigurations schemes under this secure framework.

5 Conclusion and Future Works

In this paper we analyze the radio propagation assumptions from the viewpoint of security costs, then propose a secure framework based on cost management. After presenting the implementation of costs computing, we provide an example about DoS attacks and defenses. The results show that the mobile ad hoc network shortcoming of node's limited resources can be overcome to some extent and the effects of security costs can be improved with increasing the availability of ad hoc networks in more realistic scenarios.

There are still some existing problems left for us to work over in the future. The weighted coefficients in the weighted sum of every layer secure mechanisms costs are hard to choose suitably. It should be seasoned with the certain application security requirements and now we give them values by our experiences. It should be determined self-adaptively in the future. It is difficult to calculate the costs of cross layers secure mechanisms for there is some counteracts in the co-operations. The modularization of each secure mechanism costs computing should be considered in the future work under our proposed secure framework.

References

1. Calvin Newport, Simulating mobile ad hoc networks: a quantitative evaluation of common MANET simulation models, Dartmouth College Computer Science Technical Report TR2004-504, June 16, 2004.
2. Ian F. Akyildiz, Xudong Wang, and Weilin Wang, Wireless mesh networks: a survey, Computer Networks (to be published in 2005), from www.elsevier.com/locate/comnet
3. S. Yi, P. Naldurg, R. Kravets, A security-aware ad hoc routing protocol for wireless networks, in: The 6th World Multi-Conference on Systemic, Cybernetics and Informatics (SCI 2002), 2002.
4. Y. -C. Hu, D.B. Johnson, A. Perrig, SEAD: secure efficient distance vector routing for mobile wireless ad hoc networks, in: Proceedings of the 4th IEEE (WMCSA 2002), IEEE, Calicoon, NY, June 2002, pp. 3-13.
5. Shuyao Yu, Youkun Zhang, Chuck Song, Kai Chen, security architecture for Mobile Ad Hoc Networks, Proc. of APAN'2004.
6. Lakshmi V. and D. P. Agrawal, Strategies for enhancing routing security in protocols for mobile ad hoc networks, J. Parallel Distributed Computing. 63 (2003) 214-227
7. Imad Andy JeanPierre, Hubaux, y and Edward W. Knightlyz, Denial of Service Resilience in Ad Hoc Networks, MobiCom'04, Sept. 26Oct.1, 2004, Philadelphia, Pennsylvania, USA.

Efficient and Secure Password Authentication Schemes for Low-Power Devices*

Kee-Won Kim, Jun-Cheol Jeon, and Kee-Young Yoo**

Department of Computer Engineering,
Kyungpook National University,
Daegu, Korea, 702-701
{nirvana, jcjeon33}@infosec.knu.ac.kr, yook@knu.ac.kr

Abstract. In 2003, Lin et al. proposed an improvement on the OSPA (optimal strong-password authentication) scheme to make the scheme withstand the stolen-verifier attack, using smart card. However, Ku et al. showed that Lin et al.'s scheme is vulnerable to the replay and the denial of service attack. In 2004, Chen et al. proposed a secure SAS-like password authentication schemes. Their schemes can protect a system against replay and denial-of-service attacks. In this paper, we propose two efficient and secure password authentication schemes which are able to withstand replay and denial-of-service attacks. The proposed schemes are more efficient than Chen et al.'s schemes in computation costs. Moreover, the proposed schemes can be implemented on most of target low-power devices such as smart cards and low-power Personal Digital Assistants in wireless networks.

Keywords: password authentication, low-power device, mutual authentication, wireless network.

1 Introduction

The password authentication scheme is a method to authenticate remote users over an insecure channel. A variety of password authentication schemes have been proposed [1, 2, 3, 4, 5, 7, 9].

Lamport [1] proposed a one-time password authentication scheme using a one-way function, but this scheme has two practical difficulties: the high hash overhead and the requirement of resetting the verifier. Thereafter, many strong-password authentication schemes have been proposed, e.g., CINON [2] and PERM [3]. Unfortunately, none of these earlier schemes is both secure and practical.

In 2000, Sandirigama et al. [4] proposed a simple and secure password authentication scheme, called SAS. However, Lin et al. [5] showed that the SAS suffers from vulnerability to both replay and denial-of-service attacks and proposed an

* This work was supported by the Brain Korea 21 Project in 2005.

** Corresponding author.

X. Jia, J. Wu, and Y. He (Eds.): MSN 2005, LNCS 3794, pp. 73–82, 2005.

optimal strong-password authentication scheme, called OSPA, to enhance the security of SAS. Chen and Ku [6] pointed out that SAS and OSPA are vulnerable to stolen-verifier attacks. In 2003, Lin et al. [7] proposed an improved scheme to enhance the security of OSPA. Ku et al. [8] showed that Lin et al.'s scheme [7] is vulnerable to replay and denial-of-service attacks. In 2004, Chen et al. [9] proposed secure SAS-like password authentication schemes. They proposed not only the unilateral authentication but also the mutual authentication. Their schemes can protect a system against replay and denial-of service-attacks.

In this paper, we propose two efficient and secure password authentication schemes, that can withstand replay and denial-of-service attacks. Moreover, the proposed schemes are more efficient than Chen et al.'s schemes, and can be implemented on most of target low-power devices such as smart cards and low-power Personal Digital Assistants in wireless networks.

The remainder of this paper is organized as follows. The proposed schemes are presented in Section 2. In Section 3, we discuss the security and computation costs of the proposed schemes. Finally, we state the conclusions of this paper in Section 4.

2 The Proposed Schemes

In this section, we propose two schemes to withstand denial-of-service attacks and replay attacks. The proposed schemes are the unilateral authentication scheme(Method 1) and the mutual authentication scheme(Method 2).

2.1 Notations

The following notations are used throughout this paper.

- U denotes the low-power client.
- S denotes the server.
- E denotes the adversary.
- ID denotes the identity of U.
- PW represents the password of U
- N, N', r and r' denote random nonces.
- h denotes a one-way hash function. $h(x)$ means x is hashed once. $h^2(x)$ means x is hashed twice.
- $\oplus$ denotes a bitwise XOR operation.
- $||$ denotes a string concatenation.
- x denotes the secret key of S.

2.2 The Proposed Scheme with Unilateral Authentication (Method 1)

Registration Phase

Suppose a new user U wants to register with a server S for accessing services. The registration phase is shown in Fig. 1. The details are presented as follows:

Step (R1) U sends his identity ID and password P through a secure channel.

Step (R2) S selects a random nonce N and computes $X = h(x||ID)$, $vpw\_src = h(P \oplus N)$ and $vpw = h^2(P \oplus N)$, where x is the secret key of the server.

Step (R3) S stores vpw into the database and issues a low-power device storing $\{X, vpw\_src, N, h(\cdot)\}$ to U through a secure channel.

Authentication Phase

The authentication phase is shown in Fig. 2. The details are presented as follows. If the user U wants to login, U enters the identity ID and password P in his low-power device, then the low-power device will perform the following operations.

Step (A 1) Verify $h(P \oplus N)$ and vpw_src. If they are not equal, the low-power device terminates this session.

Step (A 2) Select a new random nonce N' and compute

$$new\_vpw\_src = h(P \oplus N'), \tag{1}$$

$$C_1 = X \oplus new\_vpw\_src, \tag{2}$$

$$C_2 = vpw\_src \oplus h(new\_vpw\_src). \tag{3}$$

Step (A 3) Send $\{ID, C_1, C_2\}$ to the server as a login request.

Upon receiving the login request $\{ID, C_1, C_2\}$ from U, the server will perform the following operations:

Step (A 4) Check the format of ID.

Step (A 5) Compute

$$X' = h(x||ID), \tag{4}$$

$$new\_vpw\_src' = C_1 \oplus X' = h(P \oplus N'), \tag{5}$$

$$vpw' = h(C_2 \oplus h(new\_vpw\_src')). \tag{6}$$

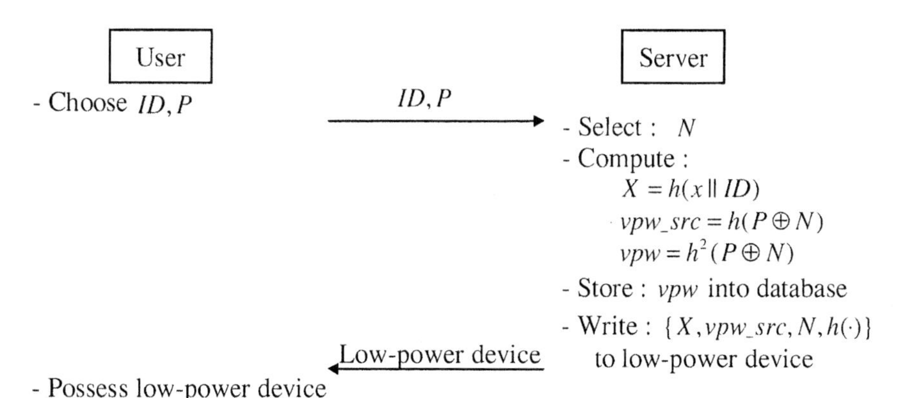

Fig. 1. The registration phase of Method 1 and Method 2 of the proposed schemes

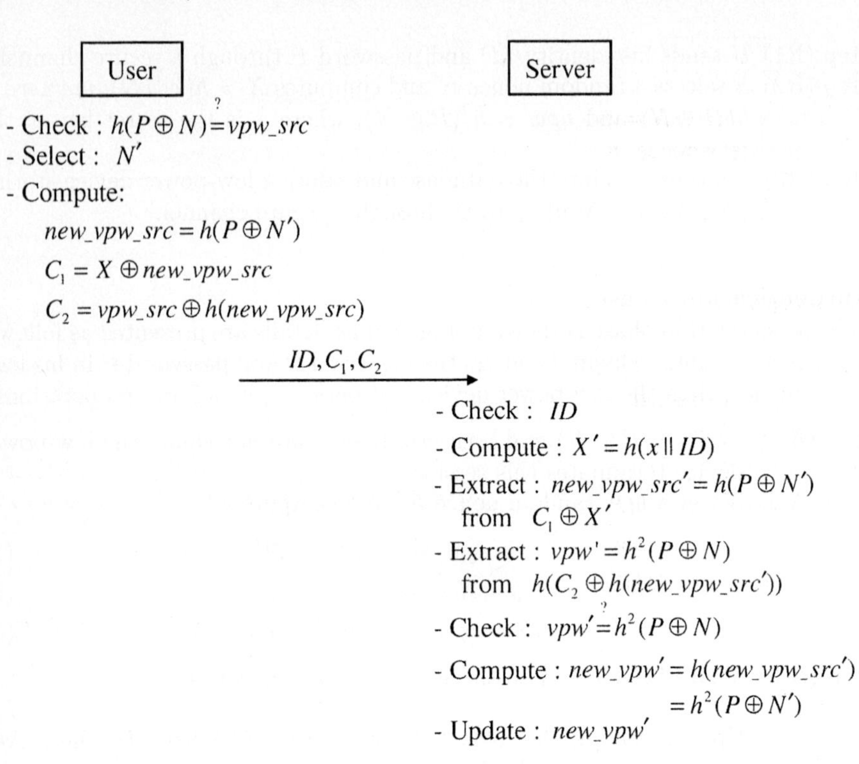

Fig. 2. The authentication phase of Method 1

and check whether vpw' is equal to the stored verifier $vpw = h^2(P \oplus N)$. If it holds, then the server accepts the login request.

Step (A 6) Compute $new\_vpw' = h(new\_vpw\_src')$ and updates $vpw = h^2(P \oplus N)$ with $new\_vpw' = h^2(P \oplus N')$ for the next authentication session.

2.3 The Proposed Scheme with Mutual Authentication(Method 2)

To provide the mutual authentication, we propose Method 2 based on Method 1. The registration phase is the same as that of Method 1 and omitted here.

Authentication Phase

The authentication phase is shown in Fig. 3. The details are presented as follows. If the user U wants to login, U enters the identity ID and password P in his low-power device, then the low-power device will perform the following operations.

Step (A 1) Verify $h(P \oplus N)$ and vpw_src. If they are not equal, the low-power device terminates this session.

Step (A 2) Generate a random nonce r, where r is used to identify this transaction uniquely.

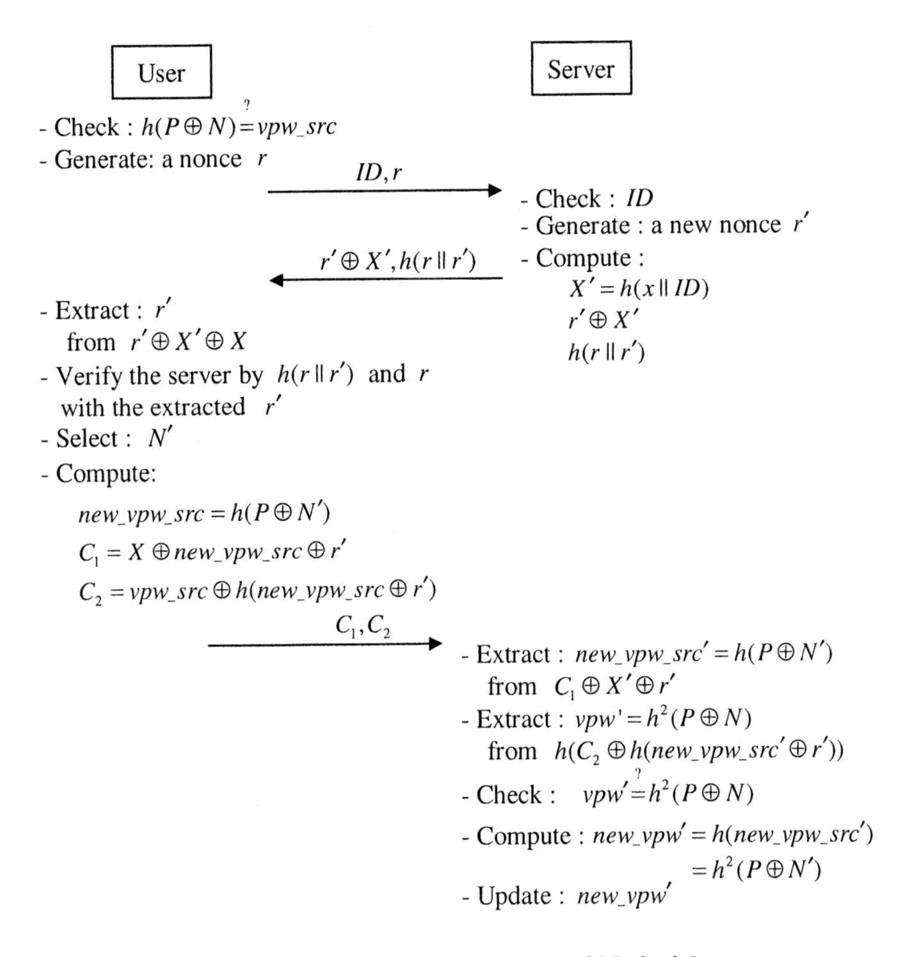

Fig. 3. The authentication phase of Method 2

Step (A 3) Send $\{ID, r\}$ to S.

Upon receiving the message $\{ID, r\}$ from U, the server will perform the following operations:

Step (A 4) Check the format of ID.

Step (A 5) Generate a random nonce r' and compute $X' = h(x||ID)$, $r' \oplus X'$ and $h(r||r')$.

Step (A 6) Send $\{r' \oplus X', h(r||r')\}$ to U.

Upon receiving the message $\{r' \oplus X', h(r||r')\}$ from S, the low-power device will perform the following operations:

Step (A 7) Extract r' from $r' \oplus X' \oplus X$ and verify $h(r||r')$ using r and r' to authenticate the remote server.

Step (A 8) Select a new random nonce N' and compute

$$new\_vpw\_src = h(P \oplus N'), \tag{7}$$
$$C_1 = X \oplus new\_vpw\_src \oplus r', \tag{8}$$
$$C_2 = vpw\_src \oplus h(new\_vpw\_src \oplus r'). \tag{9}$$

Step (A 9) Send $\{C_1, C_2\}$ to the server as a login request.

Upon receiving the login request $\{C_1, C_2\}$ from U, the server will perform the following operations:

Step (A10) Compute

$$new\_vpw\_src' = C_1 \oplus X' \oplus r' = h(P \oplus N'), \tag{10}$$
$$vpw' = h(C_2 \oplus h(new\_vpw\_src' \oplus r')). \tag{11}$$

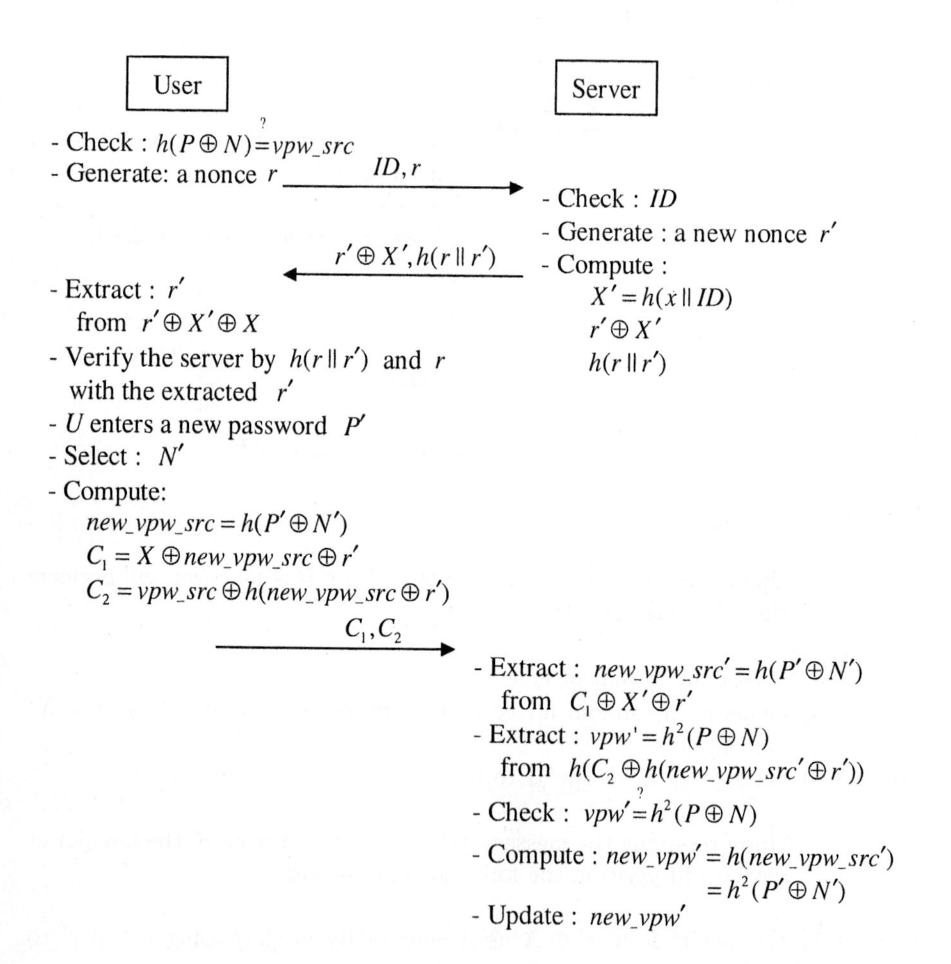

Fig. 4. The password change phase of the proposed scheme

and check whether vpw' is equal to the stored verifier $vpw = h^2(P \oplus N)$. If it holds, then the server accepts the login request.

Step (A11) Compute $new\_vpw' = h(new\_vpw\_src')$ and updates $vpw = h^2(P \oplus N)$ with $new\_vpw' = h^2(P \oplus N')$ for the next authentication session.

Password Change Phase

The password change phase is shown in Fig. 4. The details are presented as follows. If the user U wants to change his old password P to a new password P', he only needs to perform the procedures below. The password change phase is much the same as the authentication phase, except for Steps (A8), (A10) and (A11).

After executing from Step (A1) to Step (A7), U's low-power device card executes Step (A'8), as below.

Step (A′ 8) U enters a new password P'. The low-power device selects a new random nonce N' and computes

$$new\_vpw\_src = h(P' \oplus N'), \tag{12}$$

$$C_1 = X \oplus new\_vpw\_src \oplus r', \tag{13}$$

$$C_2 = vpw\_scr \oplus h(new\_vpw\_src \oplus r'). \tag{14}$$

After executing Step (A'8), U's low-power device executes Step (A9).

Upon receiving the login request $\{C_1, C_2\}$ from U, the server will perform the following operations:

Step (A′10) Compute

$$new\_vpw\_src' = C_1 \oplus X' \oplus r' = h(P' \oplus N'), \tag{15}$$

$$vpw' = h(C_2 \oplus h(new\_vpw\_src' \oplus r')). \tag{16}$$

and check whether vpw' is equal to the stored verifier $vpw = h^2(P \oplus N)$. If it holds, then the server accepts the login request.

Step (A′11) Compute $new\_vpw' = h(new\_vpw\_src')$ and update $vpw = h^2(P \oplus N)$ with $new\_vpw' = h^2(P' \oplus N')$ for the next authentication session.

3 Security and Efficiency of the Proposed Schemes

In this section, we examine the security and the efficiency of the proposed schemes.

3.1 Security

In the following, we analyze the security of the proposed schemes.

Password Guessing Attack
There are only two instances including the password P: the login message $\{C_1, C_2\}$ and the verifier $h^2(P \oplus N)$ stored by the server. If an adversary E intercepts $C_1 = h(x||ID) \oplus h(P \oplus N')$(or $C_1 = h(x||ID) \oplus h(P \oplus N') \oplus r'$, in Method 2) and $C_2 = h(P \oplus N) \oplus h(h(P \oplus N'))$(or $C_2 = h(P \oplus N) \oplus h(h(P \oplus N') \oplus r')$, in Method 2), it is infeasible to guess the user's password without knowing x, N, N', and r' because E has no feasible way to ascertain the password. Suppose adversary E has stolen the verifier $h^2(P \oplus N)$. E cannot guess the password without knowing N.

Replay Attack
In the proposed schemes, the server extracts $vpw' = h^2(P \oplus N)$ from C_2 using $new\_vpw\_src' = h(P \oplus N')$ obtained by C_1. Then the server checks whether vpw' is equal to the stored verifier $h^2(P \oplus N)$ or not. If C_1 or C_2 are replaced with a value in any previous session, the verification equation $vpw' \stackrel{?}{=} h^2(P \oplus N)$ does not hold. Then, the server rejects the login request. Therefore, adversary E cannot login to the remote sever by replaying the previous login request.

Impersonation Attack
In Method 1 and Method 2, if adversary E wants to impersonate the user, E must compute a valid $\{C_1, C_2\}$. Because E has no idea about P, N, r', and the server's secret key x, E cannot forge a valid $\{C_1, C_2\}$. Therefore, E has no chance to login by launching the impersonation attack.

In Method 2, if adversary E wants to impersonate the server, he must send a valid $\{r' \oplus h(x||ID), h(r||r')\}$ to the user. Because E has no idea about the server's secret key x, he cannot compute $h(x||ID)$ to forge a valid $\{r' \oplus h(x||ID), h(r||r')\}$. Therefore, the proposed schemes can resist impersonation attacks.

Stolen-Verifier Attack
Assume that adversary E has stolen verifier $h^2(P \oplus N)$. To pass the authentication in the proposed schemes, E must have $h(x||ID)$ and $h(P \oplus N)$ to compute $\{C_1, C_2\}$. The adversary cannot compute $h(x||ID)$ because he does not know the system secret key x. E cannot derive $h(P \oplus N)$ from $h^2(P \oplus N)$ because $h(\cdot)$ is a one-way hash function. Therefore, the proposed schemes can resist stolen-verifier attacks.

Denial-of-Service Attack
In the proposed schemes, the server extracts $vpw' = h^2(P \oplus N)$ from C_2 using $new\_vpw\_src' = h(P \oplus N')$ obtained by C_1. Then, the server checks whether vpw' is equal to the stored verifier $h^2(P \oplus N)$. If C_1 or C_2 is replaced with another value, the verification equation $vpw' \stackrel{?}{=} h^2(P \oplus N)$ does not hold. Then, the server rejects the login request and does not update the verifier. Therefore, the proposed schemes can resist denial-of-service attacks.

Table 1. Efficiency comparison between Chen et al.'s schemes and the proposed schemes

	Chen et al.'s schems [9]		The proposed schemes	
	Method 1	Method 2	Method 1	Method 2
Computation cost of registration	$2T_{H_u} + 1T_{H_s}$	$2T_{H_u} + 1T_{H_s}$	$3T_{H_s}$	$3T_{H_s}$
Computation cost of authentication	$7T_{H_u} + 5T_{H_s}$	$7T_{H_u} + 5T_{H_s}$	$3T_{H_u} + 3T_{H_s}$	$4T_{H_u} + 5T_{H_s}$

Method 1 : the unilateral authentication
Method 2 : the mutual authentication
T_{H_u} : the time for performing a one-way hash function by the user
T_{H_s} : the time for performing a one-way hash function by the server

3.2 Efficiency

We compare the proposed and related schemes in terms of computation costs. Table 1 shows the efficiency comparison of the proposed and related schemes in the registration and authentication.

As shown in Table 1, the proposed scheme is more efficient than Chen et al.'s schemes in computation costs. Therefore, the proposed schemes are efficient enough to be implemented on most of target low-power devices in wireless networks.

4 Conclusions

In this paper, we have proposed an improvement on Lin et al.'s scheme that can withstand replay attacks and denial-of-service attacks. Moreover, we have proposed mutual authentication based on unilateral authentication of the proposed scheme. The proposed scheme is more efficient than Chen et al.'s schemes in computation costs. Therefore, the proposed schemes are efficient enough to be implemented on most of the target low-power devices such as smart cards and low-power Personal Digital Assistants in wireless networks.

References

1. L. Lamport, Password Authentication with Insecure Communication, Communications of ACM, Vol.24, No.11, pp.770–772, 1981.
2. A. Shimizu, A Dynamic Password Authentication Method by One-way Function, IEICE Transactions, Vol.J73-D-I, No.7, pp.630–636, 1990.
3. A. Shimizu, T. Horioka, H. Inagaki, A Password Authentiation Methods for Contents Communication on the Internet, IEICE Transactions on Communications, Vol.E81-B, No.8, pp.1666–1673, 1998.
4. M. Sandirigama, A. Shimizu, M.T. Noda, Simple and Secure Password Authentication Protocol (SAS), IEICE Transactions on Communications, Vol.E83-B, No.6, pp.1363–1365, 2000.

5. C.L. Lin, H.M. Sun, T. Hwang, Attacks and Solutions on Strong-password Authentication, IEICE Transactions on Communications Vol.E84-B, No.9, pp.2622-2627, 2001.
6. C.M. Chen, W.C. Ku, Stolen-verifier Attack on Two New Strong-password Authentication Protocols. IEICE Transactions on Communications, Vol.E85-B, No.11, pp.2519–2521, 2002.
7. C.W. Lin, J.J. Shen, M.S. Hwang, Security Enhancement for Optimal Strong-password Authentication Protocol, ACM Operating Systems Review, Vol.37, No.2, pp.7–12, 2003.
8. W.C. Ku, H.C. Tsai, S.M. Chen, Two Simple Attacks on Line-Shen-Hwnag's Strong-password Authentication Protocol, ACM Operating Systems Review, Vol.37, No.4, pp.26–31, 2003.
9. T.H. Chen, W.B. Lee, G. Horng, Secure SAS-like Password Authentication Schemes, Computer Standards and Interfaces, Vol.27, No.1, pp.25-31, 2004.

Improving IP Address Autoconfiguration Security in MANETs Using Trust Modelling

Shenglan Hu* and Chris J. Mitchell

Information Security Group, Royal Holloway, University of London
{s.hu, c.mitchell}@rhul.ac.uk

Abstract. Existing techniques for IP address autoconfiguration in mobile ad hoc networks (MANETs) do not address security issues. In this paper, we first describe some of the existing IP address autoconfiguration schemes, and discuss their security shortcomings. We then provide solutions to these security issues based on the use of trust models. A specific trust model is also proposed for use in improving the security of existing IP address autoconfiguration schemes.

1 Introduction

IP address autoconfiguration is an important task for zero configuration in ad hoc networks, and many schemes have been proposed [5, 6, 11]. However, performing IP address autoconfiguration in ad hoc networks securely remains a problem. Most existing schemes are based on the assumption that the ad hoc network nodes will not behave maliciously. However, this is not always a realistic assumption, since not only may some malicious nodes be present, but ad hoc network nodes are often easily compromised.

Of the existing IP address autoconfiguration schemes, we focus here on *requester-initiator schemes*. In such a scheme, a node entering the network (the *requester*) will not obtain an IP address solely by itself. Instead, it chooses an existing network node as the *initiator*, which performs address allocation for it.

The rest of this paper is organised as follows. In Section 2, we review existing requester-initiator schemes, and analyse their security properties. In Section 3, we describe how to improve the security of these schemes using trust models. In Section 4, a new trust model is proposed which can be used to secure the operation of requester-initiator schemes, and an analysis of this trust model is given. Finally, a brief conclusion is provided in section 5.

2 Requester-Initiator Address Allocation Schemes

We first briefly review two existing requester-initiator schemes. We then use them to illustrate a variety of security issues that can arise in such schemes.

* The work of this author was sponsored by Vodafone.

X. Jia, J. Wu, and Y. He (Eds.): MSN 2005, LNCS 3794, pp. 83–92, 2005.

Nesargi and Prakash proposed the MANETconf scheme [5], a distributed dynamic host configuration protocol for MANETs. In this scheme, when a new node (the requester) joins an ad hoc network, it broadcasts a request message to its neighbour nodes. If the requester is the only node in the network, then it becomes an initiator; otherwise, it will receive a reply message from one or more of its neighbour nodes. It then selects one of its reachable neighbour nodes as an initiator. Each node in the network stores the set of addresses currently being used, as well as those that can be assigned to a new node. The initiator selects an IP address from the available addresses and checks the uniqueness of the address by broadcasting a message to all network nodes. If the IP address is already being used, then the initiator selects another IP address and repeats the process until it finds a unique IP address which it allocates to the requester.

Fazio, Villri and Puliafito [2] proposed another requester-initiator scheme for IP address autoconfiguration based on MANETconf [5] and the Perkins-Royer-Das [6] scheme. In this protocol, a NetID is associated with each ad hoc network, and each node is therefore identified by a (NetID, HostID) pair. When a new node (a requester) joins an ad hoc network, it randomly selects a 4-byte HostID and requests the initiator to allocate it a unique address. The initiator randomly chooses a candidate IP address and broadcasts it to all other nodes to check for uniqueness. When the initiator finds a unique address, it sends the requester this address and the NetID of the network. The NetID is used to detect merging of networks. When partitioning happens, the NetID of each part will be changed.

In all requester-initiator schemes, including those briefly described above, IP address allocation for new nodes depends on the correct behaviour of the initiator and other existing nodes. However, in reality, malicious nodes may be present in an ad hoc network, potentially causing a variety of possible problems. We now note a number of potential security problems.

Firstly, if a malicious node acts as an initiator, it can deliberately assign a duplicate address to a requester, causing IP address collisions. In schemes where a node stores the set of addresses currently being used, it can also trigger IP address allocations for nodes that do not exist, thereby making IP addresses unavailable for other nodes that may wish to join the MANET. This gives rise to a serious denial-of-service attack.

Secondly, a malicious node can act as a requester and send address request messages to many initiators simultaneously, who will communicate with all other nodes in order to find a unique address. This will potentially use a lot of the available bandwidth, again causing a denial-of-service attack.

Thirdly, a malicious node in the network could claim that the candidate IP address is already in use whenever it receives a message from an initiator to check for duplication. As a result no new nodes will be able to get an IP address and join the network. It can also change its IP address to deliberately cause an IP address collision, forcing another node to choose a new IP address. This could lead to the interruption of that node's TCP sessions.

The main purpose of this paper is to consider means to address these threats. We start by showing how a trust model satisfying certain simple properties can be used to reduce these threats. Note that all three threats identified above result in denial-of-service attacks — this provides us with an implicit definition of what we mean by a malicious node, i.e. a node seeking to deny service to other network nodes; the nature of a malicious node is discussed further below.

3 Solutions Based on Trust Models

According to clause 3.3.54 of ITU-T X.509 [9], trust is defined as follows: "Generally an entity can be said to 'trust' a second entity when the first entity makes the assumption that the second entity will behave exactly as the first entity expects". In this paper, trust (in the form of a *trust value*) reflects the degree of belief that one entity has in the correctness of the behaviour of another entity. It is dynamic, and reduces if an entity misbehaves, and vice versa.

For the moment we do not assume any particular method of computing trust values; we simply suppose that such a method has been selected. We use the following terminology. For any nodes A and B in an ad hoc network, the trust value held by A for B, i.e. the level of trust A has in B, is a rational (floating point) value denoted by $T_A(B)$. Every node A has a *threshold trust value*, denoted by T_A^*. That is, A deems B as trustable if and only if the trust value that A currently assigns to B is at least its threshold trust value, i.e. $T_A(B) \geq T_A^*$; otherwise node A will regard node B as a potentially malicious node.

Each node maintains its own threshold value T_A^*, i.e. different nodes may choose different trust thresholds. Hence the definition of malicious node may vary from node to node, depending on local policy. Every node also keeps a *blacklist*. Whenever it finds another node for which its trust value is lower than its threshold trust value, it deems this node a malicious node and adds this node to its *blacklist*. It will regularly calculate its trust values for the nodes in its *blacklist* and update its *blacklist* based on these new trust values. Except for the messages used for calculating trust values, it will ignore all other messages from nodes in its *blacklist* and will not route any other messages to these nodes.

One underlying assumption in this paper is that the number of malicious nodes in an ad hoc network is small. We also assume there is a trust model available with the following two properties, where the neighbour nodes of a node are those nodes that are within direct transmission range. (1) Any node can make a direct trust judgement on its neighbour nodes based on the information it gathers in a passive mode. If one of its neighbour nodes is malicious, it can detect the misbehaviour of this malicious node. It maintains trust values for all its neighbour nodes and regularly updates them. (2) Any node is able to calculate the trust values of non-neighbour nodes based on the trust values kept by itself and/or other nodes. We now show how such a trust model can be used to improve the security of any requester-initiator scheme.

3.1 Choosing a Trustable Node as the Initiator

When a node N joins a network, it broadcasts a request message $Neighbour\_Query$ containing T_N^* to its neighbour nodes. If the requester is the only node in the network, then it becomes an initiator; otherwise, each of the other nodes receiving a *Neighbour_Query* message will check the trust values it hold for its neighbour nodes, and send N a reply message *InitREP*, containing identifiers of the nodes for which the trust values it holds are greater than or equal to T^*_N. Once N has received *InitREP* messages from its neighbour nodes, it combines these messages and chooses as its *initiator* the responding neighbour node which appears in the most received *InitREP* messages. A malicious node is unlikely to appear in *InitREP* messages generated by honest neighbour nodes. Hence, given our assumption that a majority of the nodes in the network are honest nodes, the probability that a malicious node will be chosen as an initiator is low.

3.2 Checking for Duplication of the Candidate Address

When an initiator A chooses a candidate IP address for a new node, it broadcasts an *Initiator_Request* message to check for duplication. In order to protect a requester-initiator scheme against a DoS attack caused by malicious nodes claiming the possession of arbitrary candidate IP addresses chosen by initiators, the trust model can be used to discover possible malicious nodes. If initiator A receives a reply message *Add_Collision* from an existing node (node B, say) indicating that B is already using the candidate IP address and B is not in A's *blacklist*, then A will react to this reply as follows.

A either maintains a trust value for B or can calculate one (if B is not a neighbour node). If A's trust value for B is greater than or equal to T^*_A, then A believes that the candidate IP address is already being used. Node A will then choose another candidate IP address and repeat the procedure to check for duplication. Otherwise, A deems B a malicious node. In this case, A adds B to its *blacklist* and ignores this *Add_Collision* message. A also broadcasts a *Malicious_Suspect* message about B to all other nodes. When it receives A's message, each node uses its trust value for node A (if necessary calculating it) to decide if it should ignore A's message. If A is deemed trustworthy, then it will calculate its trust value for B. If its newly calculated trust value for B is lower than its threshold acceptable trust value, then it adds B to its *blacklist*. As a result, misbehaving nodes will be permanently excluded from the network.

3.3 Dealing with Possible Address Collisions

Suppose node E detects a collision of IP addresses between two existing nodes. It will then send a message about the collision to both of them. Any node (node F, say) receiving such a message considers its trust value for node E (calculated if necessary). If $T_F(E) \geq T^*_F$, then F will assign itself a new IP address. Otherwise, F will keep its current IP address and add E to its *blacklist*.

3.4 Brief Analysis

The above protocol enhancements significantly improve the security of requester-initiator schemes. Only trusted nodes will be chosen as initiators for address allocation. Malicious nodes will be detected and isolated with the cooperation of other network nodes. The DoS attack caused by a malicious node claiming the possession of candidate IP addresses is prevented. Only minor computational costs will be incurred if the number of malicious nodes is small.

However, when a malicious node acts as a requester and simultaneously asks many initiators for IP address allocation, each initiator will treat this malicious node as a new node and will not be able to calculate a trust value for it (since there will no history of past behaviour on which to base the calculations). This kind of attack cannot be prevented by using the trust model approach described in this paper. Some other method outside the scope of trust modelling is required.

4 A Novel Trust Model

In the solutions described in section 3, we assume that there is a trust model by which the trust values between any two nodes can be calculated. Many methods have been proposed for trust modelling and management, see, for example [3, 10]. Unfortunately, none of them has all the properties discussed in Section 3. Thus, they cannot be straightforwardly adopted for use in our scheme. In this section we propose a trust model specifically designed to be used in this environment.

In our trust model, each trust value is in the range 0 to +1, signifying a continuous range from complete distrust to complete trust, i.e. $T_A(B) \in [0, +1]$. Each node maintains a *trust table* in which it stores the current trust value of all its neighbour nodes. When a new node joins a network, it sets the trust values for its neighbour nodes in its trust table to an initial value T_{init}, and dynamically updates these values using information gathered. A node computes its trust value for another node using one of the following two methods, depending on whether or not the other node is a neighbour node.

4.1 Calculating Trust Values Between Neighbour Nodes

If B is a neighbour node of A, then A calculates $T_A(B)$ based on the information A has gathered about B in previous transactions, using so-called passive mode, i.e. without requiring any special interrogation packets. Here we adopt the approach of Pirzada and McDonald [10] to gather information about neighbour nodes and to quantify trust. Potential problems could arise when using passive observation within a wireless ad hoc environment [8, 12]. The severity of these problems depends on the density of the network and the type of Medium Access Control protocol being used. This is an open issue which needs further research. In our paper, we assume that these problems will not occur.

Information about the behaviour of other nodes can be gathered by analysing received, forwarded and overheard packets monitored at the various protocol layers. Possible events that can be recorded in passive mode are the number and accuracy of: 1) Frames received, 2) Streams established, 3) Control packets forwarded, 4) Control packets received, 5) Routing packets received, 6) Routing packets forwarded, 7) Data forwarded, 8) Data received. We also use the following events: 9) The length of time that B has used its current IP address in the network compared with the total length of time that B has been part of the network, and 10) How often a collision of B's IP address has been detected.

The information obtained by monitoring all these types of event is classified into n ($n \geq 1$) trust categories. Trust categories signify the specific aspect of trust that is relevant to a particular relationship and is used to compute trust for other nodes in specific situations. A uses the following equation, as proposed in [10], to calculate its trust for B:

$$T_A(B) = \sum_{i=1}^{n} W_A(i) T_{A,i}(B)$$

where $W_A(i)$ is the weight of the ith trust category to A and $\sum_{i=1}^{n} W_A(i) = 1$; $T_{A,i}(B)$ is the situational trust of A for B in the ith trust category and is in the range [0,+1] for every trust category. More details can be found in [10].

Each node maintains trust values for its neighbour nodes in a trust value table, and regularly updates the table using information gathered. If a neighbour node moves out of radio range, the node entry in the trust table is kept for a certain period of time, since MANETs are highly dynamic and the neighbour node may soon be back in the range. However, if the neighbour node remains unreachable, then the entry is deleted from the trust table.

4.2 Calculating Trust Values for Other Nodes

If B is not A's neighbour node (as may be the case in multi-hop wireless ad hoc networks), A needs to send a "Trust Calculation Request" to B, and B returns to A a "Trust Calculation Reply", which contains trust values for all nodes in the route along which it is sent back to A, as follows.

We adopt the *Route Discovery* method proposed in the Dynamic Source Routing Protocol (DSR) [7] to send A's Trust Calculation Request (TCReq) message to B. Node A transmits a TCReq as a single local broadcast packet, received by all nodes currently within wireless transmission range of A. The TCReq identifies the initiator (A) and target (B), and also contains a unique request identifier, determined by A. Each TCReq also contains a *route record* of the address of each intermediate node through which this particular copy of the TCReq has been forwarded. When a node receives this TCReq, if it is not the target of the TCReq and has recently seen another TCReq from A bearing the same request identification and target address, or if this node's own address is already listed in the route record, it is discarded. Otherwise, it appends its own address to the route record in the TCReq and propagates it by transmitting it as a local broadcast packet. The process continues until the TCReq reaches B.

When B receives the TCReq message, a route from A to B has been found. Node B now returns a Trust Calculation Reply (TCReply) to node A along the reverse sequence of nodes listed in the *route record*, together with a copy of the accumulated route record from the TCReq. In our trust model, the TCReply also contains a *trust value list* of the trust value for each node in the route record,

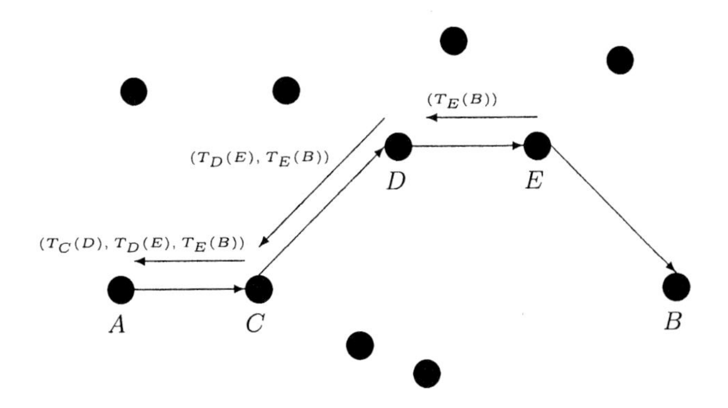

Fig. 1. A route from A to B

as calculated by its predecessor in the route. That is, whenever a node in this route receives a TCReply, it forwards the message to the preceding node on the route and appends to the trust value list its trust value for the succeeding node on the route, i.e. the node which forwarded the TCReply to it.

For example, as shown in Figure 1, when a TCReq is sent along the route $A \rightarrow C \rightarrow D \rightarrow E \rightarrow B$ and arrives at B, a route is found, and the route record is $ACDEB$. B will send a TCReply back to A along the route $B \rightarrow E \rightarrow D \rightarrow C \rightarrow A$. The immediate nodes E, D, and C append $T_E(B)$, $T_D(E)$, $T_C(D)$ to the trust value list respectively when they send back the TCReply. Therefore, when node A receives the TCReply, it will obtain a trust value list ($T_E(B)$, $T_D(E)$, $T_C(D)$); it will also have $T_A(C)$ from its own trust value table. Thus, A finds a route to B and also trust values for all the nodes in this route.

Following the above procedure, A may find one or more routes to B. Since there might be malicious nodes present, A will check if each route is a *valid route*, i.e., if all the trust values in the TCReply message are above A's threshold acceptable trust value T_A^*. Observe that when a malicious node receives a TCReply, it can change any of the trust values in the trust value list it receives, which only contains trust values for succeeding nodes on the route. However, it cannot change the trust value of any other node for itself. Suppose the route received by A consists of nodes $A = N_0, N_1, \ldots, N_i = B$. If a certain trust value $T_{N_m}(N_{m+1})$ on this route is below T_A^*, then either node N_{m+1} is a malicious node or another node $N_k (k \leq m)$ is a malicious node and has intentionally changed the trust value list when it received the TCReply. Therefore, when A obtains the trust list, it will learn that there is at least one malicious node in the route, and this route is therefore regarded as an *invalid route.*

If there is no valid route from A to B, A cannot calculate its trust value for B. In order to prevent certain attacks (see below), A is obliged to regard B as a potentially malicious node. Given that malicious nodes are rare, this situation is unlikely to occur frequently. If there does exist at least one valid route from A to B, then we can calculate the trust value of node A for node B based on the weighted average of the trust values of the nodes preceding node B on all valid routes for node B, and the weight of each route is based on the trust rating of all the intermediate nodes on each valid route. Suppose the route R is a valid route from A to B, denoted by:

$$N_{0,R} \rightarrow N_{1,R} \rightarrow N_{2,R} \rightarrow \ldots \rightarrow N_{i-2,R} \rightarrow N_{i-1,R} \rightarrow N_{i,R}$$

where $N_{0,R} = A$ and $N_{i,R} = B$. A can then calculate the *trust weight* for route R by computing the geometric average of the trust values listed in the TCReply in route R (all these trust values are above T_A^* given that R is a valid route):

$$W_{A,B}(R) = \prod_{j=0}^{i-2} T_{N_{j,R}}(N_{j+1,R})^{\frac{1}{i-1}}$$

A's trust value for node B is computed as the weighted arithmetic average of the trust values of A for B on all valid routes, $R_1, R_2, \ldots, R_g$, say.

$$T_A(B) = \frac{1}{\sum_{h=1}^{g} W_{A,B}(R_h)} \sum_{h=1}^{g} (W_{A,B}(R_h) T_{N_{i-1,R_h}}(B))$$

Node A only calculates its trust value for B when needed, and A will not store this trust value in its trust table. This is particularly appropriate for highly dynamic ad hoc networks.

4.3 Analysis

An underlying assumption for the scheme described above is that a node is considered malicious if it does not adhere to the protocols used in the network. Under this definition, two types of malicious node can be identified. Firstly, a node may be malicious all the time, i.e. it will behave maliciously when interacting with other nodes for all types of network traffic; it may also be malicious in its behaviour with respect to the trust model by sending incorrect trust values to requesting nodes. This type of malicious node behaviour can be detected by its neighbours, and thus we can expect that an honest neighbour will maintain a low trust value for such a node.

Alternatively, a node can behave honestly for all network interactions, and only behave maliciously with respect to trust model functionality. In our scheme, malicious nodes of this type will not be detected by other nodes, and the calculation of trust values will potentially be affected by these nodes. This is therefore an example of a vulnerability in our trust model approach. Hence, in the analysis below, we assume that all malicious nodes are of the first type, and we can assume that a misbehaving node will be detected by its neighbour nodes.

The trust value of any node A for any other node B in the network can be calculated. We claim that the existence of a small number of malicious nodes will not affect the calculation of the trust values. First note that, if B is A's neighbour node, A calculates its trust value based on information it gathers itself, which will not be affected by the existence of malicious nodes. Otherwise, A calculates its trust value for B based on the trust values listed in the TCReply in one or more routes from A to B.

If there are malicious nodes in a route R from A to B, then there will be a unique malicious node $N_{k,R}$ 'closest' to A on this route – all nodes on this route that are closer to A can therefore be trusted not to modify the trust value list. Consider the route R:

$$A = N_{1,R} \to \cdots \to N_{k-1,R} \to N_{k,R} \to N_{k+1,R} \to \cdots \to N_{i,R} = B$$

where $N_{k,R}$ is malicious and $N_{k-1,R}$ is honest. When the TCReply message from B to A is forwarded to node $N_{k,R}$ by node $N_{k+1,R}$, node $N_{k,R}$ appends $T_{N_{k,R}}(N_{k+1,R})$ to the trust value list and send this TCReply message to node $N_{k-1,R}$. Since node $N_{k,R}$ is a malicious node, it may deliberately modify (lower or raise) some of the trust values in the trust value list, i.e. any of $(T_{k,R}(N_{k+1,R})$, $T_{N_{k+1,R}}(N_{k+2,R})$, $\cdots$, $T_{N_{i-1,R}}(N_{i,R}))$. Moreover, other malicious nodes on this route may collude and raise each other's trust values. However, $T_{N_{k-1,R}}(N_{k,R})$ should be very low, since the maliciousness of $N_{k,R}$ will be detected by its honest neighbour node $N_{k-1,R}$. When A receives the TCReply message, A will regard this route as invalid, and will not use this route to calculate its trust value for B. Hence all intermediate nodes in a valid route must be honest.

The number of malicious nodes that our trust model can tolerate varies depending on the network topology. Our scheme requires at least one valid route from A to B in order to calculate the trust value of A for B. If at any time, no valid route from A to B is found, A cannot calculate a trust value for B. In this case, if A regards B as an honest node, a malicious node B can attack our trust model by modifying the *route record* when B receives a request message, or by just ignoring this request message to prevent A from finding a valid route to B. Thus, as mentioned in Section 4.2, if A cannot calculate a trust value for B, then A must treat B as as a potentially malicious node. If this trust model is used in the scheme described in Section 3, A adds B to its *blacklist*. However, a MANET is highly dynamic. A can calculate its trust value for B and update its *blacklist* frequently as long as a new valid route from A to B can be found when the topology of the network changes. All honest nodes will adhere to our trust model and make sure that the TCReply messages are sent back correctly.

The model suffers from Sybil attacks, where a node fraudulently uses multiple identities. We can use other methods to prevent Sybil attacks; for example, it may be possible to use trusted functionality in a node to provide a unique node identity. A detailed discussion of such techniques is outside the scope of this paper. We also ignored the problem posed by malicious nodes impersonating honest nodes. This problem cannot be completely overcome without using origin authentication mechanisms. We assume that, in environments where impersonation is likely to be a problem, an authentication mechanism is in place, i.e. fake TCReply messages sent by a malicious node will be detected. Many protocols and mechanisms for authentication and key management are available [1]. However, providing the necessary key management to support a secure mechanism of this type is likely to be difficult in an ad hoc environment. As mentioned above, trusted functionality, if present in a device, may help with this problem. Possible solutions to these issues will be considered in future work.

5 Conclusion

This paper focuses on IP address autoconfiguration in adversarial circumstances. The main contribution of this paper is to use a trust model to provide a number of enhancements to improve the security of requester-initiator schemes for IP address autoconfiguration in an MANET. It also gives a new trust model which can be used in these enhancements. Nevertheless other possible trust models with different trust quantification methods can also be applied in our solutions, as long as they satisfy the properties described in Section 3.

References

1. Boyd, C., Mathuria, A.: Protocols for Authentication and Key Establishment. Springer-Verlag, June 2003
2. Fazio, M., Villri, M., Puliafito., A.: Autoconfiguration and maintenance of the IP address in ad-hoc mobile networks. In: Proc. of Australian Telecommunications, Networks and Applications Conference, 2003
3. Huang, C., Hu, H.P., Wang, Z.: Modeling Time-Related Trust. In Jin, H., Pan, Y., Xiao, N., Sun, J. eds.: Proceedings of GCC 2004 International Workshops. Volume 3252 of Lecture Notes in Computer Science., Springer-Verlag (2004) 382–389
4. Mezzetti, N.: A Socially Inspired Reputation Model. In Katsikas, S.K. et al. eds.: Proceedings of 1st European PKI Workshop. Volume 3093 of Lecture Notes in Computer Science., Springer-Verlag (2004) 191–204
5. Nesargi, S., Prakash, R.: MANETconf: Configuration of Hosts in a Mobile Ad Hoc. Network. In: Proceedings of INFOCOM 2002, Volume 2, IEEE 2002 1059–1068
6. Perkins, C.E., Royer, E.M., Das, S.R.: IP Address Autoconfiguration in Ad Hoc Networks. 2002, Internet Draft: draft-ieft-manet-autoconf-00.txt
7. Johnson, D.B., Maltz, D.A., Hu, Y.C.: The Dynamic Source Routing Protocol for Mobile Ad Hoc Networks. (DSR). 2004, IETF Draft: draft-ietf-manet-dsr-10.txt
8. Marti, S., Giuli, T.J., Lai, K., Baker, M.: Mitigating routing misbehavior in mobile ad hoc networks. In Pickholtz, R., Das, S., Caceres, R., Garcia-Luna-Aceveseds, J. eds.: Proceedings of the Sixth Annual International Conference on Mobile Computing and Networking, ACM Press (2000) 255–265
9. International Telecommunication Union: ITU-T Recommendation X.509 (03/2000), The directory – Public-key and attribute certificate frameworks, 2000
10. Pirzada, A.A., McDonald, C.: Establishing Trust in Pure Ad hoc Networks. In: Proceedings of the 27th Conference on Australasian Computer Science, volume 26, Australian Computer Society, Inc. (2004) 47–54
11. Thomson, S., Narten, T.: IPv6 stateless address autoconfiguration. Dec. 1998, IETF RFC 2642
12. Yau, P, Mitchell, C.J.: Reputation methods for routing security for mobile ad hoc networks In: Proceedings of SympoTIC '03, IEEE Press (2003) 130–137

On-Demand Anycast Routing in Mobile Ad Hoc Networks

Jidong Wu

Zhejiang University, Hangzhou, Zhejiang, China 310027

Abstract. Anycast allows a group of nodes to be identified with an anycast address so that data packets destined for that anycast address can be delivered to one member of the group. An approach to anycast routing in mobile ad hoc networks is presented in the paper. The approach is based on the Ad Hoc On-demand Distance Verctor Routing Protocol (AODV) and is named AODVA Anycast Routing Protocol (AODVA). AODVA extends AODV's basic routing mechanisms such as on-demand discovery and destination sequence numbers for anycast routing. Additional mechanisms are introduced to maintain routes for anycast addresses so that no routing loops will occur. Simulations show that AODVA achieves high packet delivery ratios and low delivery delay for data packets destined for anycast addresses.

Keywords: Routing protocols, anycast, ad hoc networks, mobile networking.

1 Introduction

Anycast has been developed in the context of IPv6 [1]. With anycast, a group of nodes providing the same service in the network is identified via a so-called anycast address. Data packets destined for that anycast address are then delivered to any one of this group. Many network applications could benefit from the use of anycast. For example, by using anycast, a node is able to communicate with one of rendezvous points so that this node can integrate itself into a multicast tree used for support multicast routing [2].

Anycast may have many potential applications in mobile ad hoc networks, which are infrastructure-less and self-organized. For example, anycast could be used to improve service resiliency or to dynamically discover services available. Mobile ad hoc networks are able to operate completely autonomously, not needing support from a fixed infrastructure. However, their topology and structure could change frequently. Consequently, services in mobile ad hoc networks should be constructed in distributed fashion to avoid "a single point failure". Furthermore, duplicated servers might also be of more interest. For instance, some critical databases might be duplicated in the network so that a high level of service availability can be provided in spite of intermittent connectivity and node failure in ad hoc networks. Anycast provides an elegant solution to the deployment of such distributed services: the nodes providing the same service can be treated just as a group of the nodes, and that group can be identified with an anycast address.

X. Jia, J. Wu, and Y. He (Eds.): MSN 2005, LNCS 3794, pp. 93–102, 2005.

This paper presents an on-demand anycast routing protocol, which is based on the Ad hoc On-demand Distance Vector Routing Protocol (AODV) [3, 4] and is named AODV Anycast Routing Protocol (AODVA). Additional mechanisms are introduced to deal with anycast routing, although the most basic mechanisms of AODV are still applied to AODVA as well.

The remainder of the paper is organized as follows. AODVA is described in detail in Section 2. Simulation results are presented in Section 3. Section 4 discusses related work, and Section 5 concludes the paper.

2 AODV Anycast Routng Protocol (AODVA)

Similarly to AODV, AODVA assumes that communication among nodes in the network is bi-directional and does not need to be reliable. Each node has a unique address. Additionally, it is assumed that anycast addresses have no specific prefix, since anycast addresses in IPv6 are not distinguished from unicast addresses in the address format, as opposed to multicast addresses [1].

Nodes possessing the same anycast address, called *anycast destinations* for that anycast address, form a so-called *anycast destination group* of that address. A data packet destined to an anycast address, called an *anycast data packet*, can be delivered to any member of the anycast destination group identified by that anycast address. However, it is preferable that it is delivered to the nearest member according to the distance metric used by the routing protocol. The route that an anycast data packet is forwarded is called an *anycast route*. An anycast destination used by the routing protocol for calculating the anycast route at a node is called an *anycast peer* of that node, and route calculation takes place independently at each node.

Each node maintains an anycast routing table, which keeps the routing information of anycast addresses of interest. An entry in the anycast routing table contains an anycast address, the successor node of the anycast route for that address, the unicast address of the anycast peer, the anycast peer's destination sequence number (DSN), the distance to the anycast peer (in hops), and the expiration time of the entry. The fields of anycast address and the successor are looked up for forwarding anycast data packets. AODVA uses a pair consisting of the unicast address of an anycast peer and this anycast peer's destination sequence number to identify the freshness of the associated anycast routing information.

Fig. 1 illustrates an example. Nodes A and B represent two anycast destinations with anycast address X. Nodes J,K and L select node A as their anycast peer, while M and N select B as their anycast peer. Furthermore, in Fig 1 each node is associated with a list of labels $[w, x, y, z]$. The first label w represents the anycast address in question; x is the anycast peer of the node; the anycast peer has a DSN equal to y and is located z hops away from the node. For example, node K has node A as its anycast peer for the anycast address X. Besides, Node K knows at the moment that the anycast peer A has the DSN equal to 1 and is located 6 hopes away from node K. Furthermore, *upstream nodes* and *down-*

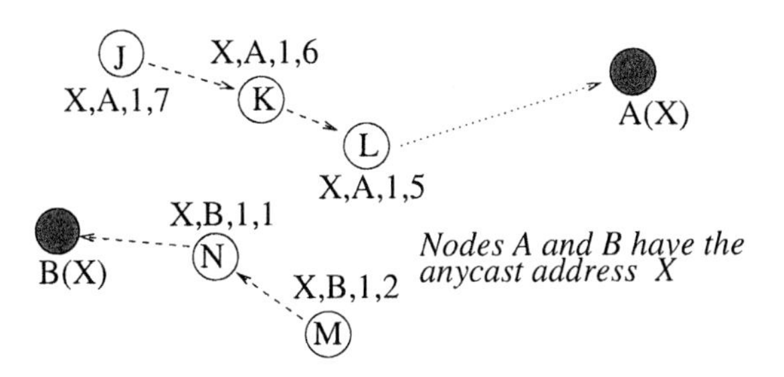

Fig. 1. Anycast routes and anycast peers

stream nodes on an anycast route can be distinguished. For example, nodes K and L be located on the same anycast route with node L being located closer to the anycast destination A. Hence, node L is located downstream from node K and node K is located upstream from node L.

2.1 Anycast Route Discovery and Maintenance

When a node has a data packet destined for an anycast address, and there is no known route or the route is broken, the node initiates a route discovery. If the node recognizes the address as an anycast address, it broadcasts an anycast route request (AREQ) message to all its neighbors. An anycast route request includes, besides the anycast address in question, the unicast address and the DSN of the known anycast destination. Otherwise, it broadcasts a route request (RREQ) message as if the anycast address ware an unicast address.

When an intermediate node receives a RREQ or AREQ for the first time and it has a valid route to the anycast address in question, it sends an anycast Route Reply (AREP) message back to the RREQ or AREQ originator in the following cases:

- if the DSN it stored is not smaller than the one contained in the AREQ, and both DSNs are associated with the same anycast destination.
- if it has a path to another anycast destination.

If the conditions above is not satisfied, the intermediate node records the node from which it receives the RREQ or AREQ in order to set up a reverse route so that it can forward AREPs later. Furthermore, it stores the necessary information from the received RREQ or AREQ in order to detect duplicated RREQs or AREQs later. Then, the RREQ or AREQ is re-broadcast to its neighbors.

If an anycast destination with the anycast address in question receives the RREQ or AREQ, it always replies with an AREP, which includes its unicast address, DSN and expiration time associated with this entry in the routing table. When a node receives an AREP, it updates its routing table as described below:

- The AREP reports an anycast route to the same anycast peer, but with a greater DSN, or with an equal DSN but a lower distance. In this case, as in

unicast routing in AODV, the node updates the routing information such as the DSN and the distance. The node transmitting the AREP is selects as the new successor, but the anycast peer is not changed. We call this case a *peer refresh.*
- The AREP has a DSN associated with another anycast peer, but the AREP was received from its current successor. In this case, the node updates only the information about the anycast peer, and does not change its successor node. We call such a change a *peer revision.*
- The AREP reports an anycast route with a shorter distance, and comes from a node other than its current successor. This route, however, leads to an anycast destination other than its current anycast peer. To determine the freshness of the routing information reported, the node initiates an *anycast echo procedure.* In case of a positive result a so-called *peer switch* is performed. As a result, the node has a new successor node as well as information about a new anycast peer.

If the node is an intermediate node, it modifies the AREP or creates a new AREP and sends the AREP to the next node along the reverse route. In the case of peer refreshes or peer revisions, it increases the distance contained in the received AREP. In the case of peer switches, it creates a new AREP using the new destination's information. Finally, the AREP arrives at the RREQ or AREQ originator, and the route to the anycast destination is then established.

2.2 Anycast Echo Procedure

In case of a peer switch, a node should ensure that no routing loop is introduced. A route loop will occur if a node uses out-of-date routing information and selects a upstream node as its new successor of an anycast route. Therefore, an anycast echo procedure is used so that nodes on an anycast route can update the corresponding routing information. Four routing messages are defined in AODVA for the anycast echo procedure: Update Request (UREQ), Update Reply (UREP), Update Clear (UCLR), and Update Release (UREL).

The node, which initiates an anycast echo procedure, sends a UREQ to the node that has reported a shorter route leading to another anycast destination. As a matter of convenience we call the two nodes the *echo-issuer* and the *echo-relay*, respectively. The UREQ is destined for the anycast address in question.

Two tables are used for anycast echo procedures: a pending reply table and an echo request table. The pending reply table is only used by the echo-issuer. An entry in the pending reply table contains the anycast address for which the anycast echo procedure is in progress, the address of the corresponding echo-relay, and an expiration timer for receiving a UREP.

An echo request table is used by all nodes which receive and propagate UREQs. An entry contains the anycast address for which an anycast echo procedure is in progress, a *list of backward nodes* from which the node receives UREQs, and an expiration timer for receiving an UREP. The backward nodes determine the reverse route to the echo issuer. This route is used for forwarding

UREPs. A node first checkes its pending reply table and its echo request table before initiating an anycast echo procedure, so that it does not initiate multiple anycast echo procedures for the same anycast address.

Interaction of UREQs and UREPs. An echo-issuer sends a UREQ to an echo-relay at the beginning of the anycast echo procedure. It inserts an entry in its pending reply table. It does not change the routing entry for an anycast address until it receives the corresponding UREP if this anycast address is listed in the pending reply table.

When an intermediate node receives a UREQ, it adds a new entry to its echo request table if no anycast echo procedure for this anycast address is in progress. If a procedure is already in progress, it only needs to add the new node to the backward node list in the already existing entry. Then the UREQ is forwarded. It does not change the routing entry for an anycast address until it receives the corresponding UREP, if this anycast address is listed in its echo request table.

The UREQ finally arrives at an anycast destination. The destination increases its DSN, and returns a UREP along the reverse route. The UREP includes the anycast address in question, the unicast address of the anycast destination, its DSN, and the hop count to the destination. An intermediate node receiving the UREP updates its own routing table, increases the hop count contained in the UREP, forwards it to the next node in the backward node list, and deletes the corresponding entry in the echo request table.

After the UREP arrives at the echo-issuer, the echo-issuer selects the echo-relay as its new successor for the anycast address in question, updates its routing table accordingly, and deletes the corresponding entry in the pending reply table. Furthmore, If the corresponding timer in the pending reply table expires, the anycast echo procedure is ended and the corresponding table entry is deleted.

Dealing with Stall Routing Information. When a UREQ arrives at an echo-issuer, it determines if it is the originator of this UREQ. If this is the case, it terminates the anycast echo procedure. Furthermore, it returns a UCLR to the node from which it received the UREQ. The UCLR contains similar information to a UREP. Nodes that receive a UCLR update their routing table and delete the corresponding entries in the echo request table. Fig. 2 shows an example of

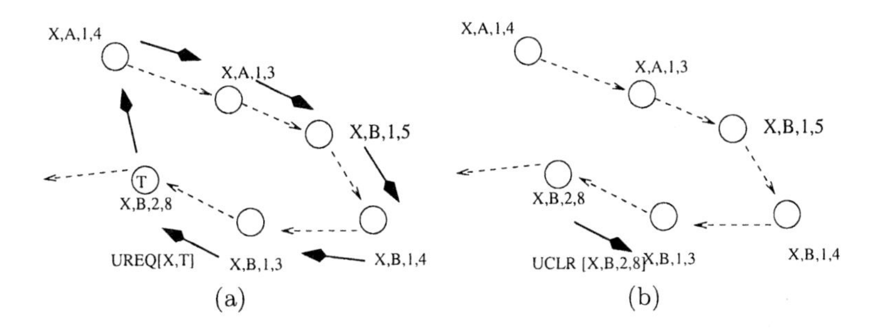

Fig. 2. Detecting a path between the echo-issuer and echo-relay

such a case. Node T is the echo-issuer. Since T is located on the anycast route, it receives its own UREQ. T, thus, returns a UCLR to terminate the anycast echo procedure.

Dealing with Simultaneous Anycast Echo Procedures. More than one anycast echo procedure may be active for the same anycast address. If a node other than an echo-issuer receives another UREQ from a different echo-issuer, it can add the node from which it received the UREQ into the backward node list in its echo request table. The UREQ is not forwarded. The UREP received later must be forwarded to all nodes in the backward node list.

However, if an echo-issuer receives a UREQ for the same anycast address from another echo-issuer, it returns a UREL to that echo-issuer, which, as a consequence, terminates its anycast echo procedure. A UREL contains the anycast address in question, and the unicast address, the DSN and the distance to the anycast peer. The UREL passes along the reverse route from which the UREQ is forwarded. It updates the routing tables and the echo request table.

3 Simulative Analysis of On-Demand Anycast Routing

In this section, simulation results of AODVA are presented and discussed. AODVA is implemented in the network simulator *ns-2* as an extension of the AODV agent. The goal of the simulations is to evaluate the performance of AODVA for support anycast routing. and the following metrics are considered [5, 6]:packet delivery ratio, end-to-end delay of packets, and control overhead.

Packet delivery ratio is the quotient of the number of packets sent by sources and the number of packets received by the anycast destination. This ratio reflects the effectiveness of the algorithm. End-to-end delay of packets is the average value of time which packets take to reach anycast destination nodes. Delay of a packet is calculated as time interval between the time at which packet received by a data sink and sent by a data source, respectively. Control overhead is the amount of all routing packets sent in the simulation. Packets that are forwarded across multiple hops are considered on a per-hop basis and not on a per-path basis. We also consider the normalized control overhead, which is the average number of routing messages sent per data packets.

3.1 Simulation Configurations

The setup used in the simulations is comparable to the one used for performance evaluation of unicast routing protocols [5]. In each simulation experiment, 50 wireless nodes move over a $1500m$ x $300m$ space for 900 seconds of simulated time. Nodes move according to the model "random waypoint"[5], in which the characteristics of node movement are defined by two parameters: *speed* and *pause time*. In the simulations, the speed of node movement is randomly selected and is between 0.1 and 10m/second. Seven different pause times are selected: 0, 30, 60, 120, 300, 600, and 900 seconds. For each value of pause time, ten runs of simulations have been conducted. Constant bit rate (CBR) traffic with 20 sources is used in the simulation, for CBR traffic is popularly used for evaluating routing protocols in the literature [5, 6]. Without losing generality, in the simulation only one anycast address is used. The anycast destination group with 3, 5, and 7 destinations (sinks) has been used in simulations, respectively.

For the purpose of comparison, an "ideal" algorithm, called "omniscient anycast routing" (OMNIACAST) here, has been implemented, and the corresponding simulations have been performed . In the OMNIACAST scheme, the node's routing agent is able to access the global data structures used by the simulator. From these global data structure, the routing agent is able to read the information about anycast group membership and the network topology, and to calculate paths to anycast destination directly. No routing messages are needed to be exchanged.

3.2 Discussion of Results

The results of both AODVA and OMNIACAST are plotted in Fig. 3. A number enclosed in parentheses indicates the numbers of anycast sinks. As seen from Fig. 3.a, AODVA achieves high packet delivery ratios. The ratios are above 95.5% in the simulations. The delivery ratio increases as the pause time increases, that is, the delivery ratio increases as the degree of mobility in the network reduces. Differences of delivery ratio between AODVA and OMNIACAST become smaller when the pause time increases. It is observed from the figure, even in the case of OMNIACAST, the delivery ratios are only between 99.43 and 99.99%, for mobility and wireless transmission collisions result in a non-hundred percent delivery ratio.

As seen from Fig. 3.b, as mobility in the network reduces, the average end-to-end delay reduces for AODVA. On the other hand, it is relatively stable for OMNIACAST. Similar to the behavior of delivery ratios, differences of end-to-end delay between AODVA and OMNIACAST become smaller when the pause time increases. In the following, the impact of anycast group size is discussed.

Fig. 4 shows performace mertics as a function of both mobility and the size of anycast group. As seen from Fig. 4.a, the packet delivery ratios of AODVA

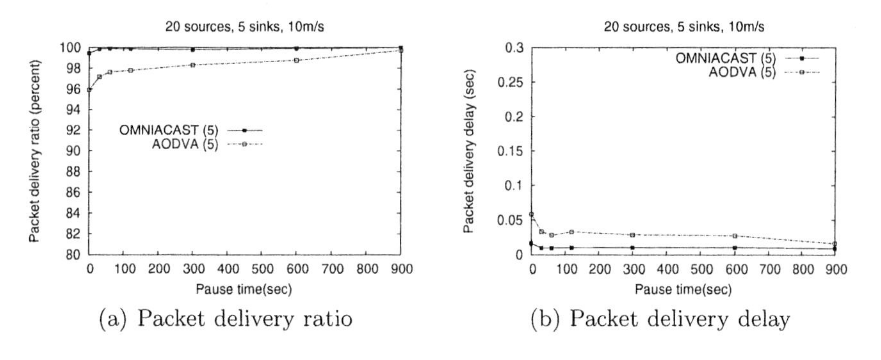

(a) Packet delivery ratio

(b) Packet delivery delay

Fig. 3. Simulation results with 20 flows, 5 anycast sinks

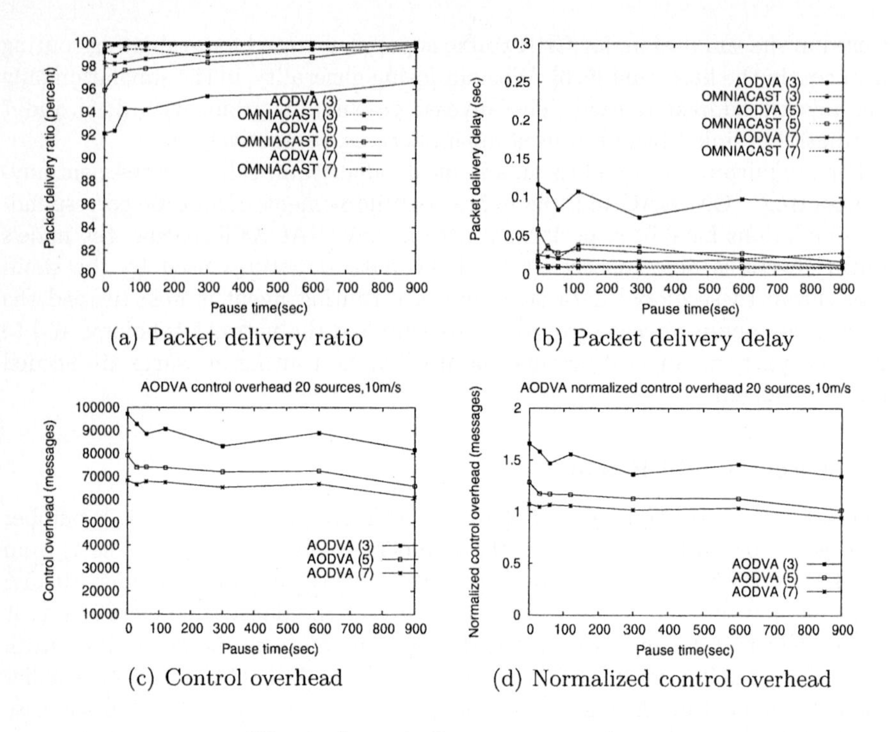

(a) Packet delivery ratio

(b) Packet delivery delay

(c) Control overhead

(d) Normalized control overhead

Fig. 4. Impact of anycast group size

are high, namely, they are all above 92.1%. The packet delivery ratio increases as the size of anycast group increases. The more anycast destinations, the closer is the delivery ratio of AODVA to that of OMNIACAST.

As seen from Fig. 4.b, the average end-to-end delay reduces quickly as the size of anycast group increases. For example, the delay in the cases of 5 anycast destinations reduces dramatically compared with that in the case of 3 anycast destinations. This is caused by the reduction of network load. First, as shown in Fig. 4.c, the amount of routing messages decreases as the number of the anycast sinks increases. Second, it was observed in the simulations that more data packets reach the anycast destinations through short paths with the increase of the size of anycast group.

Fig. 4.c and Fig. 4.d show that the increase of the size of anycast group reduces control overhead. The main reason is that less route discovery was issued when the size of anycast group increases. It was observed in the simulations that route breaks occurred more frequently in the cases of a smaller size of anycast group; the number of the route requests sent in the case of 3 sinks is about 150% of that in the case of 5, which is nearly 150% of that in the case of 7 anycast sinks. The two figures also show that control overhead tends to reduce as the mobility of nodes reduces, just as expected. It is interesting to see that in the case of stationary scenarios with 5 or 7 anycast sinks, the normalized control overheads are near or even under 1. The corresponding lines in the two figures have similar shapes. It is not an incident. In the calculation of the normalized control overhead, the numbers of received data packets are nearly equal, as in all three cases there exist 20 CBR sources.

As shown in Fig. 4, the simulation results with AODVA and three anycast sinks exhibit some irregularities at the point of pause time 60 seconds. This is due to the movement scenarios used in the simulation. It was observed that there are less temporary unreachable nodes in the movement scenarios with 60 seconds than in those with 120 seconds, although the former exhibits more link breaks than the latter. Consequently, in the case of three anycast sinks, the AODVA routing performance at the point of pause time 60 seconds is a bit better than that at the point of pause time 120 seconds. But with the increase of the size of anycast group (e.g., with 5 or 7 anycast sinks), the simulation results are less influenced by the irregularity contained in the movement scenarios.

4 Related Work

In [7], the members of an anycast group are treated as one "virtual node", so that anycast routing could be integrated and unified into unicast routing mechanisms. With such an approach, in the case of AODV, a "virtual destination sequence" for a virtual node has to be agreed on or coordinated by all anycast group members. One way to do this was proposed in [8]. Each anycast destination node maintains, in addition to the destination sequence number, an *anycast sequence numbers* for each anycast address it possesses. When answering route requests, it updates its local anycast sequence number so that the anycast sequence number is greater by one than the maximum of it local anycast sequence number and the one contained in route requests. The approach proposed in this paper, on contrast, does not need a common destination sequence member for a anycast address.

An extension of AODV for anycast was proposed in [9]. In that extension, route request messages are extended to include anycast group-IDs, and route discovery for anycast groups is conducted similarly for unicast addresses. Compared with that extension, the approach developed in this paper provides some special procedures to ensure loop freedom of anycast routes.

5 Conclusion

Anycast has many potential applications in mobile ad hoc networks. In this paper an ad hoc on-demand anycast routing protocol called AODVA is proposed. AODVA extends the well-established ad hoc routing protocol AODV for supporting anycast routing. The simulations have shown that AODVA achieves good performance.

Acknowledgments

Main work of this paper have been done at the Institute of Telematics, University of Karlsruhe, Germany. The work was supported by the IPonAir project funded by the German Federal Ministry of Education and Research.

References

1. Hinden, R., Deering, S.: IP version 6 addressing architecture. RFC 2373 (1998)
2. Kim, D., Meyer, D., et al.: Anycast rendevous point (RP) mechanism using protocol independent multicast (PIM) and multicast source discovery protocol (MSDP). RFC 3446 (2003)
3. Perkins, C.E., Royer, E.M.: Ad hoc on-demand distance vector routing. In: Proceedings of the 2nd IEEE Workshop on Mobile Computing Systems and Applications, New Orleans, LA (1999) 90–100
4. Perkins, C., Belding-Royer, E., Das, S.: Ad hoc on demand distance vector (AODV) routing. RFC 3561 (2003)
5. Broch, J., Maltz, D.A., Johnson, D.B., Hu, Y.C., Jetcheva, J.: A performance comparison of multi-hop wireless ad hoc network routing protocols. In: Proceedings of ACM/IEEE MOBICOM. (1998) 85–97
6. Das, S.R., Perkins, C.E., Royer, E.M.: Performance comparison of two on-demand routing protocols for ad hoc networks. In: Proceedings of the IEEE Conference on Computer Communications (INFOCOM), Tel Aviv, Israel (2000) 3–12
7. Park, V., Macker, J.: Anycast routing for mobile services. In: Proc. Conference on Information Sciences and Systems (CISS) '99. (1999)
8. Gulati, V., Garg, A., Vaidya, N.: Anycast in mobile ad-hoc networks. Course project `http://ee.tamu.edu/~vivekgu/courses/cs689mobile_report.ps` (2001)
9. Wang, J., Zheng, Y., Jia, W.: An aodv-based anycast protocol in mobile ad hoc network. In: Personal, Indoor and Mobile Radio Communications, 2003. PIMRC 2003. 14th IEEE Proceedings on. Volume 1. (2003) 221 – 225

MLMH: A Novel Energy Efficient Multicast Routing Algorithm for WANETs*

Sufen Zhao[1], Liansheng Tan[1], and Jie Li[2]

[1] Department of Computer Science,
Central China Normal University, Wuhan 430079, PR China
{S.Zhao, L.Tan}@mail.ccnu.edu.cn
[2] Graduate School of Systems and Information Engineering,
University of Tsukuba, Tsukuba Science City, Japan
lijie@cs.tsukuba.ac.jp

Abstract. Energy efficiency is of vital importance for wireless ad hoc networks (WANETs). In order to keep the nodes active as long as possible, it is essential to maximize the lifetime of a given multicast tree. However, hop count is generally an important metric for WANETs, and any efficient routing protocol should have low hop count. The problem of generating the optimized energy efficient routing protocol for WANETs is NP-hard. Any workable heuristic solution is, therefore, highly desirable in this case. To take into account the tradeoff between the lifetime and the hop count in routing of multicast tree in WANETs, this paper defines a new metric termed energy efficiency metric (EEM) function. Theoretical analyses show that it is efficient to fully characterize the energy efficiency of WANETs. A distributed routing algorithm called Maximum Lifetime and Minimum Hop-count (MLMH) is then proposed with the aim that extends the lifetime while minimizes the maximal hop count of a source-based multicast tree in WANETs. Simulation results give sound evidence that our algorithm achieves a balance between the hop count and the lifetime of the multicast tree successfully.

1 Introduction

For a wireless ad hoc network (WANET), each node in the network cooperates to provide networking facilities to various distributed tasks. So nodes are usually powered by limited source of energy. However, the set of network links and their capacities is not pre-determined because it depends on factors such as distance between nodes, transmission power, hardware implementation and environmental noise. Thus, it is different from a wired network. Due to the fact that the devices are dependent on battery power, it is important to find energy efficient routing paths for transmitting packets. In order to find optimal routing paths for WANETs, it is necessary to consider not only the cost of transmitting a packet, but also that of receiving, and even discarding

* The research of Sufen Zhao and Liansheng Tan has been supported by National Natural Science Foundation of China under Grant No.60473085. The work of Jie Li has been supported in part by JSPS under Grant-in-Aid for Scientific Research.

X. Jia, J. Wu, and Y. He (Eds.): MSN 2005, LNCS 3794, pp. 103 – 112, 2005.

a packet. From this viewpoint, the proportions of broadcast and point-to-point traffic used by the protocol should be considered carefully.

However, in many ad hoc networks, the metric of actual interest is not the transmission energy of individual packets, but the total operational lifetime of the network. From a conceptual viewpoint, power-aware routing algorithms attempt to distribute the transmission load over the nodes in a more egalitarian fashion, even though such distribution drives up the total energy expenditure. However, if the total energy of the nodes is used up, the links will break down and the data transmission will fail. So maximizing the total operational lifetime of the network is more important.

On the other hand, hop count of the multicast tree also plays an important role in WANETs. It is well known that an ad hoc wireless network is unreliable and its links can break down at any time. A longer routing path may take more risks. Wireless links, typically perform link-layer re-transmissions, and therefore choosing a path with a very large number of short hops can be counter-productive. In fact, as the hop count increase, the resulted increase in the total number of re-transmissions could not be neglected. So an appreciated routing protocol should have short routing path.

The energy-efficiency issue in wireless network design has been received significant attention in the past few years. In the past, several heuristic energy-efficient multicasting algorithms were proposed. These algorithms include Shortest Path Tree (SPT) algorithm, Minimum Spanning Tree (MST) algorithm and Broadcasting Incremental Power (BIP) algorithm [5]. The MST heuristic applies Prim's algorithm to obtain a MST, and then broadcasting messages rooted at the source node. The SPT heuristic applies Dijkstra's algorithm to obtain a shortest path tree rooted at the source node. The BIP heuristic is a different version of Dijkstra's algorithm for SPT. All these minimum total energy protocols can result in rapid depletion of energy at intermediate nodes; possibly leading to network getting partitioned and interruption to the multicast service. In [2], Wang and Gupta proposed an algorithm called L-REMiT to maximize lifetime for WANETs. L-REMiT extends lifetime of a multicast tree through extending the lifetime of the bottleneck node in the multicast tree. Simulation results presented in [2] show the promising performance of L-REMiT. However, it does not consider hop count while switching the parent for the bottleneck node's child.

To design energy efficient routing algorithm is NP-hard [3], an efficient solution is highly desirable. In this paper, we focus on multicast routing protocols for WANETs, and propose a novel method called Maximum Lifetime and Minimum Hop-count (MLMH) for maximizing the lifetime of source-based multicast trees in WANET. MLMH algorithm defines a new metric termed energy efficiency metric (EEM) function. The EEM function is the weighted summation of relative increment for lifetime and hop count, and it is shown to be efficient to fully characterize the energy efficiency of WANETs. Theoretical results show that our algorithm improves SPT protocol significantly.

The remainder of this paper is organized as follows. Section II describes the system model. Section III presents the energy efficiency problem definition. Section IV proposes the novel algorithm MLMH to maximize the lifetime of source-based multicast tree. Section V presents the simulation results that demonstrate the superiority of our algorithms. Finally, the conclusion is given in Section VI.

2 System Model

The wireless ad hoc network consists of several mobile hosts (nodes) and routers. The topology of the wireless ad hoc network is modeled by an undirected graph $G = (V, E, W_x, W_e)$, where V is the set of nodes, E is the set of full-duplex communication links, where $|V| = n$ and $|E| = m$. W_x: $V \rightarrow R^+$ is the node weight function, and W_e: $E \rightarrow R^+$ is the edge weight function. Here, we assume W_x defines the residual energy of node x, and the Euclidean distance between two neighbor nodes i and j is defined by W_e, naming $d_{i,j}$, and node i and node j are within the transmission range of each other when they keep their power at this level. Let $P_{i,j}$ be the minimum energy needed for link between nodes i and j for a data packet transmission. Here, we assume all the packets are of the same size. Therefore, from [1] we can know that:

$$P_{i,j} = K\,(d_{i,j})^{\theta} + C, \tag{1}$$

where K is a constant upon the properties of the antenna. θ is the propagation loss exponent, and it's value is typically between 2 and 4. C is a fixed component that accounts for overheads of electronics and digital processing. For long range radios, $C << K(d_{i,j})^{\theta}$, so we may ignore it. However, for short-range radios, C is not negligible.

In order to simplify the problem, we have following assumptions:

- Every node has local view of the network and knows the physical distance between itself and its neighbor nodes. This can be easily implemented by some distance estimation method.
- Every node can dynamically change its transmission power level.
- Each data packet is of the same size.
- Every node does not move during the transmitting process.

We now define **neighbors** of a node i are the nodes within the transmission range of node i, and **tree neighbors** of node i are those neighbors of node i which also belong to multicast tree T. In multicast tree T, there is a source node, namely s, and it is the root of the multicast tree. Every leaf node is a destination node. We take Fig.1 for an example. We define node 4 as the source node, nodes 1, 2, 5, 6, 7, 9, 10, 11, and node 12 are multicast group members; node 3 and node 8 are not the group members. The value above the link is the distance between two tree neighbor nodes.

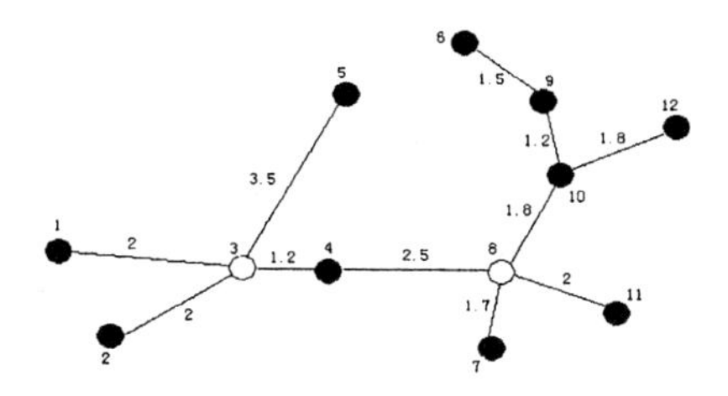

Fig. 1. A Multicast Tree Example

3 Problem Definition

As we have mentioned above, we assume the residual battery energy at a certain instance of time at node i is E_i. In a source based multicast tree, the energy consumption at every node in a tree is determined by the distance to its children nodes. For example, the energy consumed at node 3 is $max\ (P_{3,1}, P_{3,2}, P_{3,5})$. Let r_i be the distance from node i to it's farthest child. The energy consumption per packet of node i in a multicast tree T is defined as $P(T, i)$:

$$P(T,\ i) = \begin{cases} E_{elec} + Kr_i^{\theta} & \text{if } i \text{ is the source node;} \\ E_{elec} + Kr_i^{\theta} + E_{recv} & \text{if } i \text{ is neither the source nor a destination node;} \\ E_{recv} & \text{if node } i \text{ is a destination node,} \end{cases} \quad (2)$$

where E_{elec} is a distant independent constant that accounts for the overheads when transmitting a packet. E_{recv} denotes the energy cost of receiving a data packet.

Next, we define the lifetime metric. The lifetime of a node is the maximum number of packets which can be transmitted by the node. From above, we know the node i's lifetime is as:

$$L(T,i) = \frac{E_i}{P(T,i)}, \quad (3)$$

where E_i is the residual energy for node i. Let the node with minimum lifetime in a multicast tree T be the bottleneck node for T. Thus, the lifetime of a multicast tree T is the lifetime of the bottleneck node, i.e.,

$$LT(T) = \min_{\forall i \in T} LT(T,i) = \min_{\forall i \in T} \frac{Ei}{P(T,i)}. \quad (4)$$

The problem of maximizing the lifetime of a multicast tree is equivalent to that of maximizing the lifetime of the tree's bottleneck node.

4 Proposed MLMH Protocol

In this section, we propose the novel algorithm MLMH and analyze its performance.

4.1 Algorithm Description

Our algorithm tries to improve the lifetime of bottleneck nodes in the initial multicast tree by changing bottleneck nodes' children nodes, so that the multicast tree's lifetime is lengthened. The algorithm uses SPT or MST as the initial tree. From above, we know that if we change the bottleneck node, then we get another multicast tree T', and the lifetime of T' is different from the tree T. Let the bottleneck node of T be i, the farthest child of i is x, and the node j is the node x's new parent in T'. The process of above is noted as $Change_x^{i,j}$. The difference between the two trees' lifetime is

$$LT\_gain(T,x,i,j) = LT(T') - LT(T).$$

Then, the relative lifetime gain is defined as

$$rLT\_gain(T,x,i,j) = \frac{LT\_gain}{LT(T)} = \frac{LT(T') - LT(T)}{LT(T)} = \frac{LT(T')}{LT(T)} - 1 \, . \quad (5)$$

Furthermore, due to the fact that every leaf node is a destination node, we define the **maximum hop count** of a tree to be the depth of the tree, *i.e.*, the length from the source node to the farthest leaf node. We sign $HT(T)$ as the maximum hop count of tree T, $HT(T')$ is the maximum hop count of tree T'. When tree T switches to T', the relative maximum hop count difference is defined as

$$rHT\_price(T,x,i,j) = \frac{HT(T') - HT(T)}{HT(T)} = \frac{HT(T')}{HT(T)} - 1 \, . \quad (6)$$

While $LT\_gain(T, x, i, j) > 0$, we define the EEM function as

$$F(T,x,i,j) = \alpha \times rLT\_gain(T,x,i,j) - \beta \times rHT\_price(T,x,i,j) \, , \quad (7)$$

where $\alpha + \beta = 1$. The above EEM function accounts for both impacts of the relative lifetime gain and the relative maximum hop count difference on performance. The two parameters α and β can be chosen on the basis of the following thumb-rule: larger value of α indicates the importance of the relative lifetime gain; while larger value of β indicates the importance of relative maximum hop count difference. Let M be the set of the neighbors of the node x. A node j^* is selected as the new parent of node x if and only if

$$F(T,x,i,j^*) = \max_{j \in M} F(T,x,i,j) \, , \quad (8)$$

On the basis of the above analyses, we propose a heuristic algorithm described as followings:

Step 1. Construct the initial multicast tree: The multicast tree is build as SPT, because it is proved energy efficient. We assume that after building tree T, every node can know it's parent and children nodes with respect to the source node s.
Step 2. Select the bottleneck node: After building the multicast tree, the source node s requires all the nodes except leaf nodes in T to select the minimum lifetime node in a bottom up manner from the reciprocal second layer nodes. This process is described using mathematical induction method as Fig. 2.
Step 3. Change the bottleneck node: This process is referenced as the selection function (8). After the bottleneck node x is selected, x needs to look for its neighbor list M. If neighbor j isn't in the subtree of node x, we consider the case if the node x change it's parent from node i to node j. Firstly the value of LT_gain function should be computed. If the value of LT_gain is larger than zero, then we compute the value of the function *EEM*. For every neighbor in the neighbor list of x, this procession is repeated. Finally, the node that has the largest F value will be found out to be the new parent for x. If the lifetime gain of every neighbor is less than zero, then stop; else go to step 2 and start next repetition.

```
Bottleneck_Node_Selection(T)
  /* find a bottleneck node in tree T. */
  MaxEnergy = - ∞ ;
  Bottlenecknode = NULL;
  for i=1:n
     if child(i)=NULL then continue;
      else
         computemaxdist(maxdist(i), maxnode(i));
         computeenergy(energy(i));
         if energy(i) > MaxEnergy
             MaxEnergy = energy(i);
             Bottlenecknode=i;
         endif
       endif
  endfor
  return bottlenecknode;
```

Fig. 2. Algorithm for selecting the bottleneck node

4.2 Performance Analysis

The worst case complexity of MLMH algorithm is not very high. Next, we analyze the complexity of it.

Theorem 1. *MLMH has O(n*R) time complexity, where n is the number of nodes in the network and R is the number of rounds performed.*

Proof: The complexity of bottleneck node selection is $O(n)$. The complexity of changing a node's parent is $O(1)$. The complexity of a round in which a tree switching is $O(\delta_{max})$, whereδ_{max} is the maximum number of neighbors for every node's coverage area. Thus, the complexity of our algorithm is

$$O(R * (O(n) + O(\delta_{max}))) = O(n * R).$$ ∎

How to select the value of the parameter α and β in the optimized function (7) have great impact on the performance of the algorithm. Next, we study the special case with α is equal to 1. From the following observation, one finds our algorithm is actually a generalization to the L-REMIT algorithm.

Theorem 2. *When α is equal to 1, MLMH degenerates to the L-REMIT algorithm.*

Proof: If α is equal to 1, β will be zero. Then *EEM* function is equal to *rLT_gain(T, x, i, j)*. Because

$$rLT\_gain(T,x,i,j) = \frac{LT(T')}{LT(T)} - 1$$

and

$$F(T,x,i,j^*) = \max_{j \in M} F(T,x,i,j),$$

then

$$j^* = \arg\max_{j \in M} F(T, x, i, j) = \arg\max_{j \in M} LT\_gain(T, x, i, j) \cdot$$

Thus, in this case the MLMH algorithm is equivalent to L-REMiT algorithm. ■

Theorem 2 signifies that selection of the parameters in the optimized function is very important. Suitable selection of the parameters results in good performance. Furthermore, one notes that if α is equal to 0, MLMH algorithm does definitely not degenerate to SPT. They can be different from each other even in this case.

4.3 An Example

In Fig. 1, we assume that $K = 1.5$, $\theta = 2$, $E_{elec} = 1$, $E_{recv} = 2$, $R = 3.5$, $\alpha = 0.7$ and $E_i =$ 150 unit. The transmission range of every node is 4 meters. There are twelve nodes in the multicast tree. let node 4 be the source node, the residual nodes are destination nodes except node 3 and node 8. Bottleneck node is selected as node 3. And $LT(T, 3) = 150/(1+1.5\times 3.5^2+2) \approx 7.02$. The node 5 is the node 3's farthest child. So we look for the neighbor list of node 5, and we find node 6, 8, 9, 10 are in the list. If we change node 5's parent to node 6, T and T' denote the multicast tree before and after $change_5^{3,6}$. Assume that the distance between Node 5 and Node 6 is 1.6, namely $d_{5,6}$, $d_{5,9} = 2.6$, $d_{5,10} = 3.5$, $d_{5,8} = 3.6$, then after $change_5^{3,6}$, we compute the lifetime of the new tree T', it is: $LT(T, 4) = 150 / (1 + 1.5\times 2.5^2 + 2) = 12.12$. The maximum hop count of tree T is 4, and the maximum hop count of T' is 5. Then, using formula (7), we compute the value of *EEM* function is: $F(T, 5, 3, 6) = 0.4335$. And we also compute $LT\_gain(T, 5, 3, 9) = 11.41$, $F(T, 5, 3, 9) = 0.4387$, $LT\_gain(T, 5, 3, 10) = 0$. Thus, according to formula (8), node 9 should be as the node 5's new parent. In Fig.1, the branch between nod 5 and node 3 should be deleted, and branch between node 5 and node 9 will be added.

5 Performance Evaluation

The computer simulation is conducted for comparing our protocol with SPT. And we use two different energy models: ① long-range radios: in this model, the nodes are distributed in 500×500 unit area. The parameters are as following: $E_i = 10000$, $R = 150$, $E_{elec} = 10$, $E_{recv} = 20$, $K = 1$, $\theta = 2$. ② short-range radios: in this model, the nodes are distributed in 100×100 unit area. The parameters are: $E_i = 10000$, $R = 30$, $E_{elec} = 0.1\times Kr_i^{\alpha}$, $E_{recv} = 0.2 \times Kr_i^{\alpha}$, $K = 1$, $\theta = 3$. In these two models, we assume the network

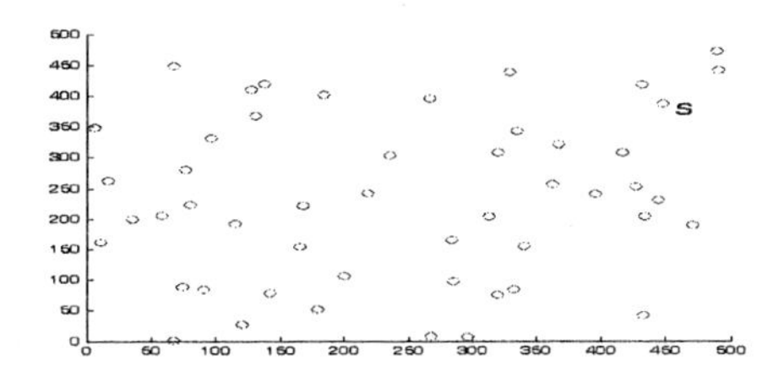

Fig. 3. A wireless ad hoc network model

graph G is connected, and all the nodes are distributed randomly in the area. In our experiments, we assume that once a path is built, the routing process will succeed, and we do not consider that the re-transmissions due to the large hop count of the tree. Simulation shows that MLMH performs well even in this condition.

Consider a wireless ad hoc network model shown in Fig. 3 as a specific example. In Fig. 3, all nodes are distributed randomly in 500 × 500 unit area. We assume the transmission range of every node is 150 units. Next, we produce an undirected graph G that is described in Fig. 4. After specifying a source node S, according to Dijkstra's algorithm, the algorithm produces a SPT, which is shown in Fig. 5. Our algorithm improved the SPT algorithm; the results are presented in Fig. 6, Fig. 7, Fig. 8 and Fig. 9. In our experiments, we have chosen the parameter α to be 0.3, 0.5, 0.7, and 0.9.

Fig. 6 shows the contrast of lifetime of the multicast tree for these two algorithms. We carry out the simulation for the network model with the number of its nodes is 50, 60, 80, and 100. One observes that the lifetime of the multicast tree is decreasing when nodes are densely distributed, and the proposed algorithm MLMH has longer lifetime than that of SPT. Fig. 7 presents the comparison of maximum hop count for these algorithms. Upon observations of the simulation results displayed in Figures 6 and 7, one concludes that, either in long or short radio range, the MLMH algorithm gains about 20% lifetime comparable with SPT in average. However, the average maximum hop count in MLMH is only about 4% longer than SPT. In the case that α is equal to 0.9, both the lifetime and the maximum hop count arrive the maximum value. Thus, let the value of α be in the range of 0.1 and 0.9 is always suitable.

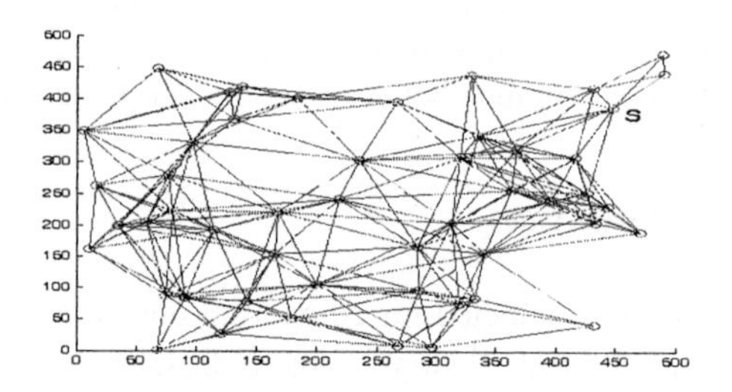

Fig. 4. The topology derived using maximum transmission power

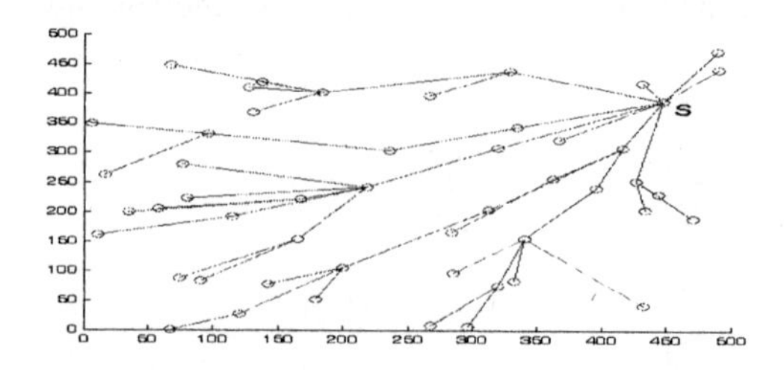

Fig. 5. A shortest path tree for the network model

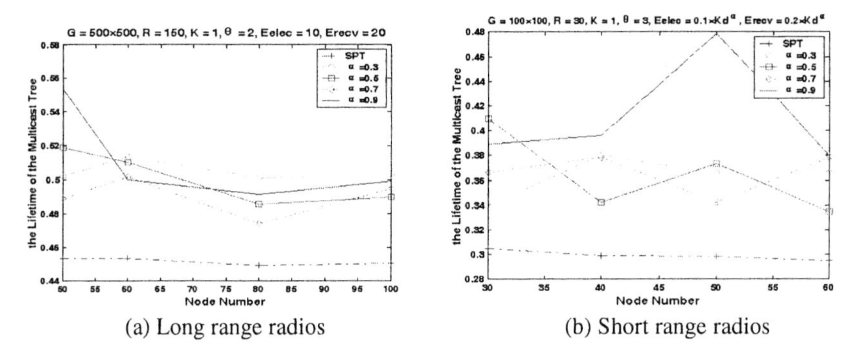

(a) Long range radios (b) Short range radios

Fig. 6. Lifetime of the multicast tree

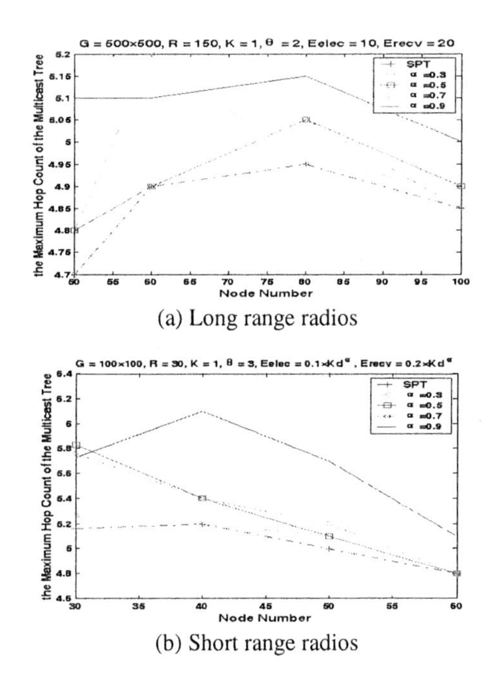

(a) Long range radios

(b) Short range radios

Fig. 7. Maximum hop count of the multicast tree

The whole simulations run 15 times for each configuration. All data points are averaged with 90% confidential intervals, which are approximately within 5 to 10 percent of the average.

6 Conclusions

We have studied the tradeoff between lifetime and maximum hop count of the multicast tree on wireless ad hoc network, and proposed a new algorithm termed MLMH. Both theoretical analyses and simulation results have demonstrated that our algorithm outperforms the known SPT algorithm. The main merit of the proposed algorithm lies in the fact that it is able to obtain a certain balance between lifetime and maximum hop count of a multicast tree.

For the sake of simplicity, we have ignored re-transmissions, which are due to the large hop count of a multicast tree. However, if this is taken into account, the performance of the simulation results will certainly be improved significantly. Our future work would include in generalizing the present approach to maximize the lifetime of mobile ad hoc networks (MANETs).

References

1. Archan Misra, Suman Banerjee, "MRPC: Maximizing network lifetime for reliable routing in wireless environments", in *Proceedings of WCNC 2002 - IEEE Wireless Communications and Networking Conference*, vol. 3, no. 1, pp. 688-694, March 2002.
2. Bin Wang and Sandeep K.S.Gupta, "On Maximizing Lifetime of Multicast Trees in Wireless Ad Hoc Networks", in *Proceedings of the 2003 international Conference on Parallel Processing* (*ICCP'03*), 2003, pp.333-340, Kaohsiung, Taiwan, October 06-09.
3. A. E. F. Clementi, P. Crescenzi, P. Penna, G.Rossi, and P. Vocca. On the complexity of computing minimum energy consumption broadcast subgraphs. In *Proceedings of 18th Annual Theoretical Aspects of Comp. Sc.* (*STACS*), vol. 2010, pp. 121-131, Springer-Verlag, 2001.
4. Deying Li, Xiaohua Jia, "Energy efficient broadcast routing in static ad hoc wireless networks", *IEEE Transactions on Mobile Computing*, vol. 3, no.2, pp. 144-151, April- June 2004.
5. Jeffrey E. Wieselthier, Gam D. Nguyen, Anthony Ephremides, "On the construction of energy-efficient broadcast and multicast trees in wireless networks", in *Proceedings of IEEE INFOCOM 2000 - The Conference on Computer Communications*, no. 1, pp. 585-594, March 2000.
6. Julien Cartigny, David Simplot and Ivan Stojmenovie, "Localized minimum-energy broadcasting in ad-hoc networks", in *Proceedings of IEEE INFOCOM 2003 - The Conference on Computer Communications*, vol. 22, no. 1, Mar 2003, pp. 2210-2217.
7. Laura Marie Feeney, Martin Nilsson, "Investigating the energy consumption of a wireless network interface in an ad hoc networking environment", in *Proceedings of IEEE INFOCOM 2001 - The Conference on Computer Communications*, no. 1, pp. 1548-1557, April 2001.
8. M. Cagalj, J. P. Hubaux and C. Enz, "Minimum-energy broadcast in all-wireless networks: NP-Completeness and distribution issues", in *Proceedings of ACM MobiCom 2002*, pp. 172-182, Atlanta, Sept. 2002.
9. Morteza Maleki, Massoud Pedram, "Lifetime-aware multicast routing in wireless ad hoc networks", in P*roceedings of WCNC 2004 - IEEE Wireless Communications and Networking Conference*, vol. 5, no. 1, pp. 1305-1311 March 2004.
10. Ning Li, Jennifer C. Hou, Lui Sha, "Design and analysis of an MST-based topology control algorithm", in *Proceedings of IEEE INFOCOM 2003 - The Conference on Computer Communications*, vol. 22, no. 1, pp. 1702-1712, Mar 2003.
11. Sorav Bansal, Rajeev Shorey, Archan Misra, "Comparing the routing energy overheads of ad-hoc routing protocols", in *Proceedings of WCNC 2003* - IEEE Wireless Communications and Networking Conference, vol. 4, no. 1, pp. 1155-1161, Mar 2003.
12. Weifa Liang and Yang Yuansheng, "Maximizing Battery Life Routing in Wireless Ad Hoc Networks", in *Proceedings of the 37th Hawaii International Conference on System Sciences 2004*.

ZBMRP: A Zone Based Multicast Routing Protocol for Mobile Ad Hoc Networks

Jieying Zhou[1], Simeng Wang[1], Jing Deng[2], and Hongda Feng[1]

[1] Dept. of Electronics and Communication Engineering,
School of Information Science & technology,
Zhongshan University, Guangzhou, 510275, P. R. China
isszjy@zsu.edu.cn

[2] Dept. of Computer Science, University of New Orleans,
2000 Lakeshore Dr., New Orleans, LA 70148, U.S.A
jing.deng@ieee.org

Abstract. In this paper, we propose a Multicast Routing Protocol termed ZBMRP (Zone Based Multicast Routing Protocol) for Mobile Ad Hoc Networks (MANETs). ZBMRP applies on-demand procedures to dynamically establish mesh-based multicast routing zones along the path from the multicast source node to the multicast receivers. Control packet flooding is employed inside multicast zones, thus multicast overhead is vastly reduced, and good scalability can be achieved. It will also be easier to secure multicast routing. ZBMRP fits well for MANETs where bandwidth is limited, topology changes frequently, power is constrained and security problem is serious. Simulation results are presented to support our claim.

1 Introduction

Mobile Ad hoc Networks (MANETs) is the cooperative engagement of a collection of wireless mobile nodes that also performs as routers. Nodes in MANETs communicate with each other through multi-hop transmission that does not need any existing infrastructure or communication supporting center. Topologies of MANETs may change quickly. Nodes in MANETs often perform a given task with other nodes together. This often leads to sending the same information to a group of members. Instead of sending data packets to each of the group members individually as unicast technique does, multicast technique allows the source node to send packets to a group of nodes as a single entity that will be sent only once on the shared path, greatly conserving network bandwidth on the shared path of the group member.

Multicast is one of the key techniques for group communication in MANETs. Among the existed multicast techniques proposed for MANETs, On Demand Multicast Routing protocol, ODMRP [1], stands as a good example. ODMRP is a mesh-based multicast protocol in which a collection (mesh) of nodes forwarding multicast packets is created between the senders and receivers. The main disadvantage of ODMRP is its excessive overhead incurred in keeping the forwarding group current and in the global flooding of the JOIN-REQUEST packets frequently [2]. Therefore, ODMRP may suffer scalability issue.

X. Jia, J. Wu, and Y. He (Eds.): MSN 2005, LNCS 3794, pp. 113 – 122, 2005.

To overcome the disadvantage of ODMRP, we propose a new hybrid multicast routing protocol termed ZBMRP (Zone-Based Multicast Routing Protocol) for MANETs. In ZBMRP, when a node has multicast packets to send but no route information is available, it starts to create a forwarding mesh in the entire network as ODMRP does. Then, it creates multiple mesh-based routing zones, including source and branch zones, along the route from source node to multicast receiver nodes according to the distribution of source node, receiver nodes and forwarding group nodes in the forwarding mesh. Zone leaders are selected according to FDW (First Declaration Wins) principle which is responsible for creating and maintaining zones periodically. Inside each zone, a mesh-based multicast routing strategy similar to ODMRP is used. Zone size and the number of zones can be decided according to the network size and multicast nodes distribution. Tunneling technology is employed to deliver multicast packets among zones and other sporadic multicast receivers that are not included in any zone in which multicast packets are encapsulated in the unicast packet for transmission. Since control packets flooding is restricted inside multicast zones, multicast overhead will be vastly reduced, and good scalability can be obtained.

ZBMRP scheme integrates three advanced techniques: mesh-based, on demand, and zone based such as ODMRP[1], AODV[3], ZRP [4] respectively and has the characteristics of adaptive to network topology change, robust to nodal mobility, and good scalability. ZBMRP will provide adequate multicast service to MANETs where bandwidth is limited, topology changes frequently, and power is constrained.

The rest of this paper is organized as follows. Section 2 describes the operational details of ZBMRP. Section 3 gives a description of simulation environment for ZBMRP. Simulation results of ZBMRP are reported in Section 4. Section 5 is conclusions for ZBMRP.

2 Zone Based Multicast Routing Protocol Overview

2.1 Mesh Created in the Entire Network in the Establishment Phase

When a multicast source has packets to send but cannot find any route and group membership information, it broadcasts a member advertising and route request packet, termed RREQ, to the entire network. Only multicast receivers send back a route reply message, called RREP, in order to allow the source node to get the current routing information. Then ZBMRP uses the same strategy as ODMRP does [1] to establish a mesh of nodes for forwarding packets between a multicast source and receivers. The mesh is created using the forwarding group concept. The forwarding group is a set of nodes that are in charge of forwarding multicast packets. It supports shortest paths between any member pairs. A multicast receiver may serve as a forwarding group node if it is on the path between a multicast source and another receiver. In ZBMRP, we further classify forwarding group nodes into two categories: FG-F and FG-B. FG-F means a forwarding group node which only forwards packets to one other node. FG-B is a forwarding group node that should forward packets to more than one node (i.e., forwarding branches), as node C shown in Fig. 1.We call source node, FG-B, and multicast receivers as ZANs (Zone-Associated Nodes). For example, in Fig. 1, node B, C, R_1, R_6, are the nearest downstream ZANs of source node S. Node A is a FG-F node but not a ZAN.

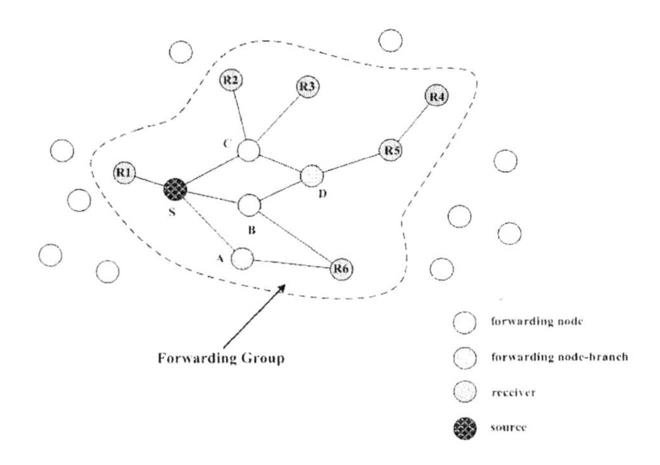

Fig. 1. Illustration of forwarding group nodes

2.2 Source Zone Creation

After the forwarding mesh between a multicast source and receivers is established in the entire network, mesh-based multicast routing zones are created according to the distribution of source node, FG-Bs, and multicast receiving member nodes, i.e. ZANs.

Multicast source node will firstly establish mesh-based multicast zone, named source zone. It collects the information of its nearest downstream ZANs from RREP messages that it receives. Such information includes IP address and distance (in terms of hop counts). If the source's nearest downstream ZAN is far away from itself, e.g., more than 3 hops away, and relatively sparse, then the source node will not establish source zone (or we can say the zone size is 0). It just tunnels multicast packets in the unicast packets to its nearest downstream ZANs. If the source node finds many ZANs within N hops, then it establishes source zone with a zone radius of N. Source node becomes the leader of this source zone. A zone leader is in charge of constructing and maintaining a zone. When N is no less than the size of the entire network, ZBMRP becomes ODMRP.

After source node selects the size of the source zone as N, it sets the TTL of the periodically flooded RREQ packet to be N, puts the zone ID, i.e. the IP address of the zone leader, in the reserved field of IP header, and then sends the IP multicast packet out. Inside the source zone, forwarding mesh-based routing strategy is used. Source node and ZANs inside zone communicate through a mesh of forwarding group nodes. For the sporadic ZANs of source node that is outside source zone, the source node will tunnel packet to them.

FG-B receiving packets from the source node with zone information just joins the source zone as a normal zone member, it will not build zone by itself. They find out whether their ZANs are in the source zone based on the source zone size (optimal zone size depends largely on node density and traffic load and its study is beyond the scope of this work). FG-B tunnels packets to its ZANs are that are outside of source zones if there are any.

If a downstream ZAN receives the same packets from both multicast source node and an upstream FG-B, it will send a message packet N-Tunnel (Not to Tunnel packet) to its upstream FG-B to notify the FG-B to stop sending packets to it.

2.3 Branch Zone Creation

If a FG-B gets data packets without zone information, which means it gets the packets through tunneling, it is outside upper level zone, then it has to create and maintain its mesh based routing zone according to the distribution of its nearest ZANs. We call this kind of zone as branch zone. Other FG-Bs inside this zone just join it and will not create the zone of itself as in FG-B IN source zone does. It is a kind of FDW (First Declare Win) strategy. This kind of work will continue until the far end of the network. There are no two or more FG-Bs to contend for building a same branch zone.

If a ZAN resides in multiple zones, it will receive multiple copies of the same multicast packet from several zone leaders with different zone ID. It then just discards the replicate packet. Note that, if this ZAN received a replicate packet tunneled from its upstream ZAN, it needs to send an N-Tunnel message packet to its upstream ZAN to save bandwidth. Upon receiving data packets without zone information, FG-B will start establish its zone. Thus along with the data packets delivering, multiple mesh based routing zones will be established along the path from the source node to multicast receivers.

In Fig. 2, we show an example of several zones created. Node S is the source node. The source zone has a zone radius of 1. The branch zone created by node C has a zone radius of 2. Note that the dashed lines represent tunneling transmissions. Therefore, nodes R2 and R3 receive their multicast data through tunneling.

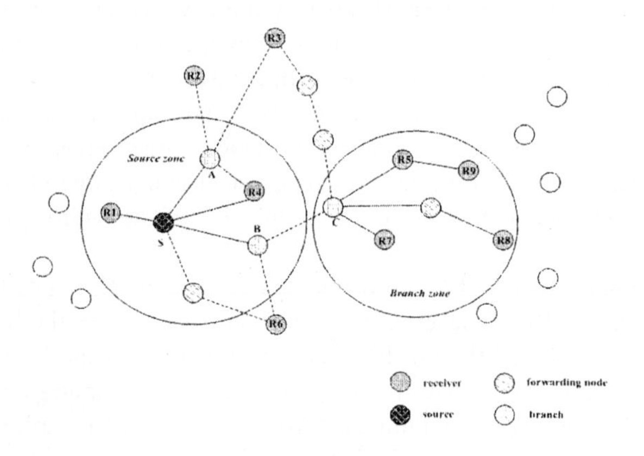

Fig. 2. Zone creation

2.4 Zone and Route Maintenance

Zone leader periodically broadcasts RREQ message inside zone with TTL equal to N (zone radius). Every node inside zone that receives RREQ forwards RREQ until TTL becomes zero. It also sends RREP back to zone leader. With these, zone leader updates mesh routing inside zone.

For the sporadic ZANs that are not included in any zone, the upstream ZAN of it tunnels packets to it unless the upstream ZAN receives explicit N-Tunnel messages from these sporadic ZANs. A periodic positive update of tunneling membership is also possible. In this technique, the downstream nodes periodically send the upstream nodes a Tunnel message. Reception of such message indicates validity of the tunnel.

2.5 New Node Joining the Multicast Group

During the process of multicasting, if a new node wants to join the multicast group, it explicitly generates a RREP-R packet that will be broadcasted to its neighbors. This RREP-R message will be forwarded until it is received by a forwarding group node, or source node. Then this node will be added to the multicast group, through a zone or a forwarding route.

If the RREP-R is terminated at the normal forwarding group node with no branch, i.e. FG-F, then this FG-F node will become as a FG-B node which will update this information to its upstream ZAN and responsible for forwarding packets to this new group member.

2.6 Multicast Group Member Leaving the Group

Only the multicast group member node outside any zone needs to send an N-Tunnel message to its upstream ZAN to tell its leave. Other group member nodes need only to stop sending back any RREP or RREQ (for source node). It is a kind of soft leaving mechanism.

2.7 Process for Link Breakage

If the link for forwarding breaks, the downstream receiver needs to send RREP-R packet as a new node does to join the multicast group again, also to inform the downstream ZANs about this state. If one link inside mesh-based zone breaks, the packet may be sent to the receiver members through possible redundancy route. Besides, zone leader will broadcast RREQ packets periodically inside zone. The receiver with link breakage can send RREP packets after it receive the new RREQ to update route information with zone leader.

2.8 Unicast Capability

Unicast can be seen as a special case of ZBMRP, i.e. only one group receiver with unicast IP address. No zone will be created. Source node broadcasts RREQ to the entire network, unicast receiver sends back RREP and activates the nodes to join the forwarding group along the reverse route. After the unicast route is established, the source can send packets to the receiver until link breakage information is received which will trigger route finding again, i.e. sending RREQ and RREP again.

2.9 Tables Used in ZBMRP

Nodes build and maintain following tables on demand. First one is membership table for recording which groups they are joining. Second one is message buffer for saving recently received packet which will be used for checking out whether a replicate packet is received. Third one is routing table in which every entry records multicast address, forwarding flag, pointer pointing to its Nearest ZANs list, flag indicating multicast route valid or not, multicast route expire time, flag indicating zone created or not, zone ID, zone radius. Fourth one is the Nearest ZANs list for recording the information about the downstream nearest ZANs which will be used to establish mesh-based multicast routing zones.

3 Description of Simulation Environment

Simulations have been implemented in OPNET (version 10.5.a). In the simulations, a free space propagation model with a threshold cutoff is used in our experiments. In the free space model, the power of a signal attenuates as $1/r^2$, where r is the distance between radios. In the radio model, we assume the ability of a radio to lock on to a sufficiently strong signal in the presence of interfering signals, i.e., radio capture. If the capture ratio (the minimum ratio of an arriving packet's signal strength relative to those of other colliding packets) is greater than the predefined threshold value, the arriving packet is received while other interfering packets are dropped. The IEEE 802.11 Distributed Coordination Function (DCF) is used as the medium access control protocol. The scheme used is Carrier Sense Multiple Access/Collision Avoidance (CSMA/CA) with acknowledgments.

In our simulations, we investigated the performance of the ZBMRP scheme in networks with two different nodal densities: one network with 50 mobile hosts placed randomly within a 1000x1000 area; the other with 100 mobile hosts in the same network area. Radio propagation range for each node is 300 meters and channel capacity is 2 Mbps. Each simulation lasts 600 seconds of simulation time. Multiple runs with different seed numbers are conducted for each scenario and collected data is averaged over those runs. We are still working on other simulation scenarios and will report new results as they become available.

The source node sends data at the rate of 2 packets per second. The size of data payload is 512 bytes. The member nodes of the multicast group are chosen randomly. Members join the multicast group in the beginning of the simulation and remain as members throughout the simulation. Random waypoint mobility model [1] is used. A node randomly selects a destination and moves towards that destination at a predefined speed. Once the node arrives at the destination, it stays in its current position for a pause time between 0 and 10 seconds. After being stationary for the pause time, it selects another destination and repeats the same process. Mobility speed is varied from 0 km/hr to 72 km/hr.

We choose the following ZBMRP parameters in our simulations: The multicast route entry lifetime is 10 s. The number of multicast group is 1. The period of sending RREQ packet is 3 s. The ratio of ZANs to be included into a zone is 0.8 which means at least 80 percent of the nearest ZANs are included in the zone.

4 Simulation Results and Discussions

4.1 Packet Delivery Ratio for Different Node Mobility

Figure 3 shows the packet delivery ratio of ZBMRP as a function of node mobility speed. Packet Delivery Ratio is the number of data packets delivered to multicast receivers over the number of data packets supposed to be delivered to multicast receivers. We assumed 20 multicast receivers exist among the 50 (or 100) network nodes. As confirmed by Fig. 3, packet delivery ratio decreases as nodal speed increases. This is due to the higher probability of link breakage and topology change, which cause more multicast control packets to be transmitted, lowering the overall data delivery ratio. As the nodal density doubles, the packet delivery ratio only lowers slightly, indicating the good scalability of the ZBMRP scheme. Overall, a relatively high packet delivery ratio can be obtained.

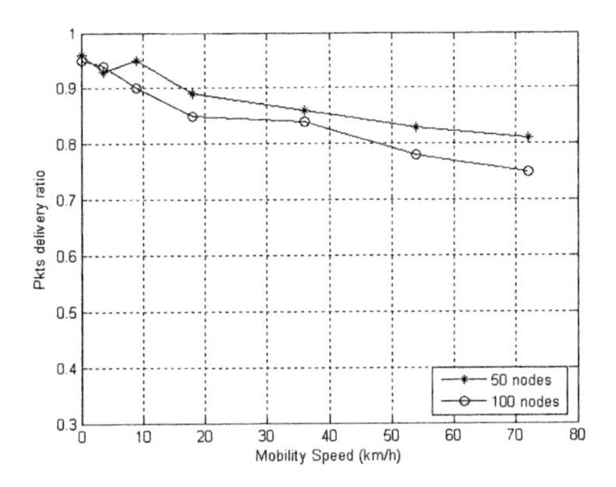

Fig. 3. Packet delivery ratio for different node mobility

4.2 Number of Data and Control Packets Transmitted Per Data Packet Delivered

Figure 4 shows the average number of total packets transmitted per data packet delivered. Total packets include data and control packets. Since most Medium Access Control (MAC) schemes used in MANETS are contention-based, it is crucial to be able to send one data packet with as less control packets as possible. When nodes contend less for the channel access, the probability of successful delivery of packets in a short time becomes higher. As suggested by Fig. 4, the average number of packet transmitted per data packet delivered maintain relatively in the range of 1.2-1.5, although it climbs up as the node mobility increases. Total packets sent in the network with 100 nodes are a little more than in the network of 50 nodes. Therefore, the control packet overhead introduced by the ZBMRP scheme is relatively low in both 50 nodes and 100 nodes networks which show good scalability of ZBMRP also.

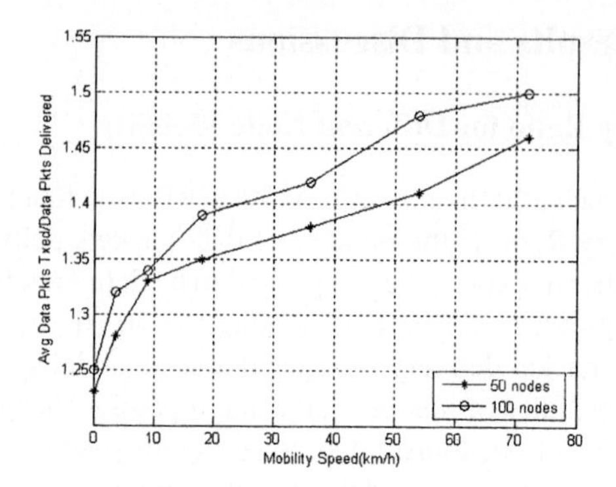

Fig. 4. Number of Total Packets Transmitted

4.3 Number of Control Bytes Transmitted Per Data Byte

The average number of control bytes transmitted per data byte delivered is shown in Fig. 5. Here, we choose to use a ratio of control bytes transmitted to data byte delivered to investigate how efficiently control packets are utilized in delivering data. In addition to bytes of control packets (e.g., RREQ, RREP, RREP-R, N-Tunnel), bytes of data packet headers are included in calculating control bytes transmitted. Thus, only bytes of the data payload contribute to the data bytes delivered.

We can find that the control overhead increases with the node mobility speed add. Two aspects results in the control overhead increment. One is the mesh-based routing which leads to redundancy route with more control overhead. Another one is link breaks more frequently as the nodes move faster which leads to more control packages, i.e., RREP-R and RRER will be sent. To deliver packets reliably to the destination, some control packets have to be sent. Protocol design has to make some compromise between efficiency and reliability. Fig.5 shows that ZBMRP gets high reliability with relative low control overhead in both 50 nodes and 100 nodes networks. It means good scalability also.

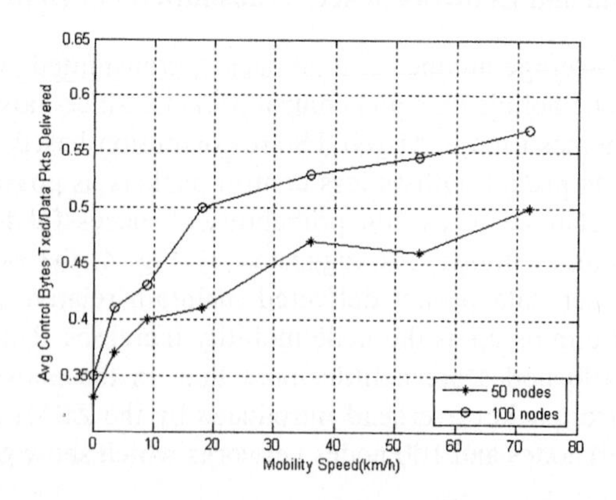

Fig. 5. Number of control bytes transmitted per data byte

4.4 Multicast Group Size

One other indication of the performance of a multicast scheme is its deliverability as the number of multicast receivers of the same multicast group increases. A good multicast scheme should scale well even if a wide range of number of receivers "tap" to the multicast group. We present our simulation results of the ZBMRP scheme in this respect in Fig. 6. We can see that the packet delivery lowers slightly as receiver number increases from 1 to 15 (with 50 or 100 network nodes)□then climbs up slowly as receiver number increases from 15 to 30. When the receiver number equals to 1, it is similar to unicast which has high packet delivery ratio. When the number of multicast group member increases, but is not too much, mesh is established with few redundancies, packets are lost more frequently. When the number of receiver member is big, more redundancy routes exist. Packet delivery ratio becomes high also. Fig.6 shows packet delivery ratio does not change much with group receiver numbers changes.

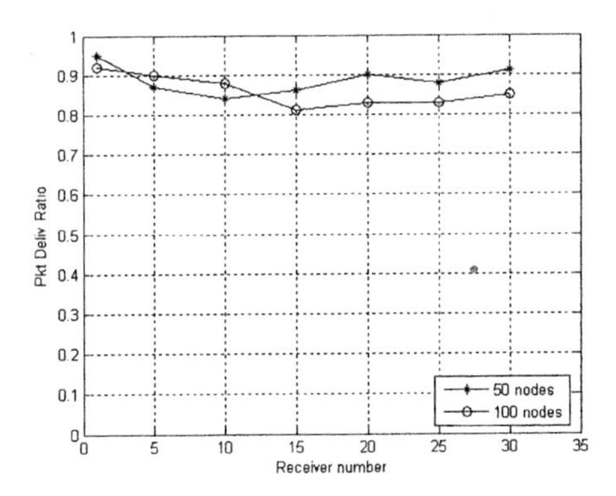

Fig. 6. Packet delivery ratio changed with group receiver numbers

5 Conclusions

In this paper, we have proposed a new multicast scheme, termed Zone-Based Multicast Routing Protocol (ZBMRP). It builds mesh-based multicast routing zones along the source node to multicast receivers on demand. Tunneling technology with link breakage repairing is used to deliver multicast packets between zones or levels which is not mesh based but a kind of tree. Only with this the overhead may be reduced a lot. And it is still robust for we have adopted link breakage repairing technique for it. A compromise of tree and mesh according the real network situation will get the best performance for ZBMRP. We will continue to research on how to do the compromise in the future.

With cohesive integration of the mesh-based, on demand, zone-based and tunneling techniques, ZBMRP scheme strives to provide adequate multicast service to MANETs where bandwidth is limited, topology changes frequently, power is constrained and security problem is serious. ZBMRP restricts RREQ control message flooding inside zones, thus get good scalability. ZBMRP uses RRER and RREP-R control message to repair link breakages and get good packets delivery ratio. ZBMRP lets only receiver member initiating to send back RREP message and get latest routing information. With zone structure, it will also be easier for ZBMRP to secure multicast routing compared with other scheme. Our simulation results show that ZBMRP scheme performs quite well in terms of data delivery ratio, overhead, and sensitivity to node speeds in different network scales and different multicast group sizes as well. ZBMRP can work well with unicast protocol such as AODV, etc. ZBMRP can also work without unicast protocol support.

In future work, we will further improve ZBMRP protocol. We will investigate how to secure our ZBMRP multicast routing protocol. We will also compare ZBMRP scheme with other related schemes, e.g., ODMRP. We will analyze ZBMRP scheme theoretically as well.

Acknowledgment

This paper is financially supported in part by National Natural Science Foundation of China grant No. 90304011, titled: "Study of Key Technologies for Network Space-time Behavior and 2008 Olympic Game Network Security Problem".

References

1. M. Gerla, S.-J. Lee, and W. Su, "On-Demand Multicast Routing Protocol (ODMRP) for Ad Hoc Networks", Internet draft, draft-ietf-manet-odmrp-02.txt, 2000.
2. Vijay Devarapalli, Deepinder Sidhu, "MZR: A Multicast Protocol for Mobile Ad Hoc Networks", IEEE 2001, 886-891
3. Ad hoc On-Demand Distance Vector (AODV) Routing, RFC3561, July 2003.
4. Z. J. Haas, M. Pearlman, P. Samar. "The zone routing protocol (ZRP) for Ad hoc networks," IETF Internet Draft , Version 4 , July 2002.

A Survey of Intelligent Information Processing in Wireless Sensor Network

Xiaohua Dai, Feng Xia, Zhi Wang, and Youxian Sun

National Laboratory of Industrial Control Technology,
Zhejiang University, Hangzhou 310027, China
daixh@iipc.zju.edu.cn

Abstract. The main goal of wireless sensor networks is to collect information from the target environment through a large number of micro sensor nodes. However, the constrained resource on each senor node brings great challenges to the information processing procedure in sensor networks. An overview and some open issues of the intelligent information processing approaches in wireless sensor networks are given in this paper to meet these challenges, which include in-network data aggregation and data compression algorithms, distributed database storage and query strategies, and multi-sensor data fusion rules. The advantages and disadvantages of every approach are also discussed.

1 Introduction

Recent advances in sensor, micro-electro-mechanical system (MEMS), wireless communications and distributed signal processing technology have enabled the development of low-cost, low-power, multifunctional sensor nodes, which are small in size and communicate with each other in short distances. These tiny sensor nodes, which consist of sensing, data processing, and communicating components, result in the idea of sensor networks based on collaborative effort of a large number of nodes. A typical wireless sensor network (WSN) is composed of a large number of sensor nodes, which are randomly dispersed over the interested area, picking up the signals by all kinds of sensors and the data acquiring unit, processing and transmitting them to the sink node [1]-[3]. Since the WINS [4] sponsored by DARPA, many similar research plans were proposed, such as Smart Dust [5], uAPMS [6], SensIT [7] and SeaWeb [33]. Now, WSNs are widely used for not only military but also many other fields, e.g., precise agriculture, environments monitoring, health, intelligent home/transport and so on.

The basic role of a WSN is to collect information from the environment by many micro sensor nodes. The sensor nodes typically have finite battery life and nodes failures can lead to a loss of data or network partition. Therefore, it is important to minimize the energy usage of each sensor node and then to prolong the system lifetime. However, the characteristics of WSN bring many challenges, such as the ultra large number of sensor nodes, dense deployment, frequently changing topology structure, and the most important but not the last one, the limited resources, including computation, storage and communication capability. All these require the algorithms and protocols to be energy-efficient, scalable well and robust.

X. Jia, J. Wu, and Y. He (Eds.): MSN 2005, LNCS 3794, pp. 123 – 132, 2005.

Several researchers have done much work to meet the above requirements, taking the intelligent processing schemes, such as in-network data aggregation/compression, distributed database storage/query and multi-sensor data fusion.

The main goal of this paper is to give an overview of the intelligent information processing of WSN. Furthermore, we try to give some open issues for future work, and point out the advantages/disadvantages of the algorithms. The remainder of this paper is organized as follows. Section 2 describes the in-network data aggregation scheme. Section 3 points out the development of data compression/decompression approach. Distributed database storage/query and multi-sensor data fusion issues are addressed in Section 4 and 5, respectively. Section 6 concludes this paper.

2 In-Network Data Aggregation

In-network data aggregation is one of the effective approaches that can reduce the communication traffic in WSN. This paradigm shifts the focus from the traditional address-centric paradigm to a more data-centric one, in which the node can look at the content of the data and perform aggregation on multiple input packets to get a single output packet. Generally speaking, it builds and maintains a hierarchical logic structure according to some principle. Sensing data is sent to its parent from the child node, and aggregated at the parent node. So the parent just only transmits useful data to its parent. The process continues repeatedly until the result data is sent to the sink node. The reducing of communication results in energy savings and prolongs the lifetime of each sensor and finally the whole application system lifetime. The data aggregation algorithms can be divided into four classes, i.e., data-centric, energy based, optimal and performance-based one.

2.1 Data-Centric Aggregation

The Directed Diffusion (DD), proposed by Chelermek [8] and Heidenann [9], is an earlier research development for data aggregation, whose major idea is as follows. Data generated by sensor node is named by attribute-value pairs. The sink requests data by sending interests for named data. Data matching the interest is then drawn down toward that node. Intermediate nodes can cache, or transform data and may direct interests based on previously cached data. Interest and data propagation or aggregations are determined by localized interactions, i.e., message exchanges between neighbors or nodes within some vicinity.

As an on-demand query schedule, DD doesn't need to keep global network topology information, and it's very energy-efficient. However, it has some drawbacks. Firstly, it can't be used in those applications that data must be sent continuously to the sink, for example the environment monitoring case. Secondly, the naming mechanism is application-aware and must be defined in advance, which results in very poor reusability. Thirdly, the matching between data and interest may bring extra energy cost. Finally, useful interest message must be flooded throughout the network, which makes it doesn't fit to time-critical applications because of heavy time delay.

2.2 Resident Energy Based Aggregation

Min Ding [12] proposed the EADAT paradigm, namely Energy Aware Distributed Aggregation Tree. It builds and maintains an energy-efficient tree. In order to reduce the amount of broadcast message, turn off the RF units of all leaf nodes. Only the non-leaf nodes aggregate data and reply to external events. Before sending data, every node waits for some time determined by its resident energy. The more the resident energy is, the shorter the time needs to wait, so that the energy consumption of all nodes is almost even. The nodes having more resident energy have priority to become the non-leaf nodes to engage in data aggregation and reply to external events.

By this approach, energy consumption of the non-leaf nodes is aggressively decreased. And the more densely sensor nodes are deployed, the longer the lifetime is, which makes it scale very well. However, one must take into account the proper distribution of leaf and non-leaf nodes, and make them even as possible. Because a non-active node can't work and respond to any events until it is waken up. If there are no sufficient active nodes, the real-time responsibility will be affected seriously.

2.3 Optimal Data Aggregation

Chalermek proposed the Greedy Aggregation algorithm (GA) in [10]. The major idea is to use a GIT, Greedy Incremental Tree, to improve path sharing for more energy savings by adjusting aggregation points. To construct a GIT, a shortest path is established for only the first source to the sink whereas each of other sources is incrementally connected at the closest point on the existing tree. Compared with the Directed Diffusion [8,9], GA algorithm can give more energy savings especially when the network is very dense, and there is not much affection to time delay and robustness.

Jamal N.Al-Karaki [14] argues one exact and three near optimal approximation algorithms for the optimal data aggregation problem. The author uses a hierarchical model that utilizes data aggregation and in-network processing at two levels of the network hierarchy. First, a set of sensor nodes called LAs, i.e., local aggregators is elected to form a fixed virtual routing architecture on which the first level of aggregation and routing was performed. An optimal subset of LAs, called MAs (Master Aggregators), is then selected to perform the second level of aggregation to maximize the network lifetime. One can get the optimal selection of MAs using the Integer Linear Program (ILP). But solving this ILP requires a time that is exponential in the number of LAs, so the author presents three near optimal approximation algorithms for the joint problem of MAs selection and data routing.

Generally speaking, optimal data aggregation approach can ensure the performance because they take some kind of optimal system parameter. However, the thing must be taken into account is that the temporal and spatial complexity must not be too high because of the resource-constraints on sensor node. So we have to tradeoff between the performance and the energy consumption of the optimal algorithm.

2.4 Performance Based Aggregation

Jerry Zhao [11] addresses the effect of continuous aggregation computation on network performance, including packet loss rate, energy consumption level and so on. The proposed tree structure construction algorithm can be used to solve some kind of aggregation function energy effectively. Using this method, one can get good aggregate accuracy resulting from dropping the link of high packet loss rate and asymmetry.

Athanassions Boulis [13] considers the tradeoff between energy savings and the accuracy of results. For periodic aggregation problem, it gives a guide according to the estimate of temporal-spatial relation. It's shown that the more the end user can tolerate for the estimate error, the less the energy consumption, and vice versus.

Ignacio Solis [15] proposes the timing model of aggregation, namely, before sending data to its parent, a sensor node should wait how long so that it can collect sufficient data information from its child nodes. For periodic aggregation problem, the author compares the performance (energy consumption, data accuracy, data freshness and communication traffic) of three typical aggregation methods, namely Periodic Simple Aggregation, Periodic Per-hop Aggregation and Periodic Per-hop Adjusted Aggregation. It's proved that using the PPAA can get less energy consumption and keep the data accuracy and freshness.

Tri Pham [32] argues the principle of DAQ (i.e. Data Aggregation Quality) for evaluating the performance of various DA algorithms. Accordingly, the author proposes two new extended aggregation algorithms, namely E-LEACH and C-PAGASIS. The energy consumption, average DAQ value and network time delay of all the four protocols are compared by simulation. It's shown that although PEGASIS and C-PEGSIS can give more energy savings, the DAQ values are poor.

2.5 Open Issues

The input and output data for the aggregation algorithm must be transmitted by some routing protocol, and as a consequence, the routing protocol must be taken account during the design of the aggregation algorithm. Another issue is the tradeoff between energy consumption and the aggregate performance in the aggregation scheme.

3 Data Compression/Decompression

In-network data aggregation is simple to implement and can decrease data traffic among nodes by reducing redundancy in sensor measurements. However, the extracted data is only a summary of sensor data readings and this limits it to those that can only be formulated with an aggregation functions. However, using data compression approach, the communication traffic is reduced without any loss of intermediate data information. The current in-network data compression algorithms used for WSN fall into three classes, i.e., data correlation based, optimal distributed and pipeline based one.

3.1 Data Correlation Based Data Compression

Jim Chou [16] proposes distributed compressing model for WSN. Among all the sensor nodes, only one node is elected to send raw data to sink and the others only send coded data. After receiving the sensing data, sink node decodes it through the correlations between the compressed and uncompressed data. The key step is to find a good coding algorithm supporting multi-rate compressing and an energy-aware and low complexity correlation-tracking algorithm. Further, Jim Chou [17] gives a simple prediction model to track and character the correlations of sensing data.

Sundeep Pattem [20] addresses the performance of various data aggregation schemes across the range of spatial correlations. The author uses an application-independent measure to quantify the size of compressed information and bit-hop metric to quantify the total cost of join routing with compression. It's revealed that there exists a practical static clustering scheme, which can provide near-optimal performance for a wide range of spatial correlations.

3.2 Optimal Distributed Data Compression Algorithm

Seung Jun Baek [19] proposes a compression algorithm based on greedy, and explores the optimal distributed data compression issue in the case of single sink. It shows that the result is independent of data sources' distribution and only depends on the relative order of aggregation cost. The author proposes a three levels of hierarchical architecture, i.e., the top layer consisting of sink nodes, the middle layer making up of aggregators/compressors, and the bottom layer consisting of sensor nodes. Using reasonable energy metric function, this optimal hierarchy problem can be transformed into a John-Mehl tessellation Problem, which can by solved by using stochastic geometry.

3.3 Pipeline Based Data Compression Algorithm

Tarik Arici [18] presents the PINCO paradigm, namely Pipelined In-Network COmpression Algorithm. The idea is that: Before transmitted, data is cached in the node's buffer and waits for some time according to the user-specified delay bound. After pipelined compression, the sensor data is combined into Group Data (GD), in which many data from different sensor nodes shares the same packet header. Moreover, there may exist redundancy or correlation between different GDs, so that they can be recompressed to save more energy.

3.4 Open Issues

One can find that all the proposed compression algorithms are highly dependent on the specific WSN application, which makes one compression scheme for this application cannot be used to another without too much efforts. Thereby, one future work is to introduce the general compression algorithm to WSN. And another noticeable study is to combine the compression and aggregation into an integrated framework, which has both of the advantages while throws off the disadvantages of them.

4 Distributed Database Storage and Query

WSN can be viewed as a distributed database system. However, unlike the traditional server-based database system, it is impossible to send all the data of all sensor nodes to the centric server for query because of the limited resource of each sensor node and the expensive wireless communication cost. As a result, for WSN, an effective solution is to make the query distributed into all the sensor nodes, and locally query the sensor data. So far, many work have been done for the distributed database storage/query in WSN, including the general query interface design, query responding algorithms and multi-resolution storage.

4.1 General Query Interface

Samuel Madden [21] applies a generic SQL-style query interface for data aggregation, which can execute queries over any type of sensor data while providing opportunities for optimization. The reliability and performance of the query can be improved. It is shown that through this interface grouped aggregates can be efficiently computed.

Xin Li [22] proposes DIM, namely Distributed Index for Multi-dimensional data. DIMs leverage two key ideas, i.e., in-network data centric storage and locality-preserving geographic hash, which allows nodes to consistently hash an event to some location within the network and efficient retrieval of events. DIMs use a unique technique so that the events whose attribute values are close are likely to be stored at the same or nearby nodes. It's shown that DIMs scale quite well with network size.

James Newsome [25] introduces Graph EMbedding for data-centric storage and information processing in WSN. Unlike previous approaches, it does not depend on geographic information. In GEM, a labeled graph is constructed, which can be embedded in the original network topology in an efficient and distributed fashion. In the label graph, each node is given a label that encodes its position. This allows messages to be efficient routed through the network, while each node only needs to know the labels of its neighbors.

4.2 Query Responding Algorithm

Boris Jan Bonfils [23] discusses the optimal placement of the aggregation operators in WSN. The authors view the aggregation function as a tree of query operators such as filters, aggregations and correlations. The existing polynomial algorithms are centralized and cannot be used in WSN. So an adaptive and decentralized algorithm based on neighbor exploration strategy is proposed, which can progressively reduce the placement of operators by walking through neighbor nodes. In the algorithm, the placement of active operators is continuously refined depending on the estimated cost on their associated tentative nodes, and the decisions are taken at the level of each operator. Active nodes only maintain information about their children, so the message exchange cost is low.

Narayanan Sadagopan [24] proposes the ACQUIRE (i.e., ACtive QUery forwarding In sensoR nEtworks) mechanism for efficient querying, especially in the case of one-shot, complex queries for replicated data. In ACQUIRE, an active query is forwarded through the network, and intermediate nodes uses cached local information in order to partially resolve the query. When the query is fully resolved, a completed response is sent directly back to the querying node. By a mathematical modeling approach, the optimal parameter setting can be found.

4.3 Multi-resolution Storage

Deepak Ganesan [26] demonstrates the use of in-network wavelet-based summarization and progressive aging of summaries in support of long-term querying in storage and communication-constrained sensor network for the special case in which the features of interest are not known a priori, e.g., many scientific applications. Through evaluating the performance of its Linux implementation, the authors show that it can achieve low communication overhead for multi-resolution summarization, highly efficient drill-down search over such summaries, and efficient use of network storage capacity through load-balancing and progressive aging of summaries.

4.4 Open Issues

Distributed database storage and query is not a new research issue, but in the case of WSN, there exists much extra difficulty. We argue that distributed data storage and query capability should become a basic service. So how to integrate this function into the operating systems of the sensor nodes is a valuable work.

5 Multi-sensor Data Fusion

One single sensor can only get local, one-sided information from the target environment, which is very limited and incomplete because of the sensor's quality, performance and noise. So it is important to fuse the data from different sensor nodes. However, the data fusion in WSN is not like the traditional one, of which the accuracy and high precision are the focus, and the computation and storage capability of the devices are almost unlimited. The key challenge of data fusion in WSN lies in the very large number of sensor nodes and the constrained resource of each node.

Swapnil [27] proposes the idea of Serial Data Fusion, in which observations and decisions are combined incrementally and a space-filling curve is used to traverse the neighbor nodes. This technique uses only the really necessary sensor observations and can be very communication efficient. The author demonstrates that it performs better, both in terms of detection errors and message cost, compared with those commonly used mechanisms such as parallel fusion with a tree-based aggregation scheme.

In [28], Tsang-Yi Wang presents the scheme that integrates channel coding with the distributed fault-tolerant classification fusion, i.e., the DCFECC approach to solve the distributed classification problem in WSN. It combines both soft-decision decoding and local decision rules without introducing any additional redundancy, in which the soft decoding scheme is utilized to combat channel fading, while the DCFECC fusion structure provides excellent fault-tolerance capability.

Ruixin Niu [34] suggests that using the total number of detections reported by local sensors for hypothesis testing for a WSN. It is shown that given the sufficient large number of sensor nodes, it can achieve very fine system level detection performance even without the knowledge of local sensors' performance and at low SNR (signal to noise ratio).

5.1 Open Issues

Although data fusion has been researched for many years, it has not been taken much attention in the case of WSN. The difficulty results from the fact that it is not practical to define a communication topology that is fixed a priori for performing data fusion in

WSN and the ultra larger number of sensor nodes in WSN. Moreover, the topology is likely to change with time due to node and link failures.

6 Conclusions

Data aggregation and compression are intended for two different WSN applications. In some cases, an application may be characterized by both of them. Therefore, it is necessary to combine the two approaches into one integrated framework, which can be adjusted by some parameters according to the application's specifications.

An important research direction is to integrate the information process and the routing, the topology control, such as in [29], [30]. In [29], the author issues the optimal gateway placement problem subject to minimizing the delay time and communication cost. In [30], it introduces an integrated model for exploring the comprehensive performance of data aggregation and topology control in order to tradeoff between delay time, energy consumption and data accuracy.

How to build a unified sensor network model used to evaluate the protocols for various network layers is also a valued issue.

Acknowledgement. This work is supported in part by NSFC (60434030), Sino-France Advanced Research Program (SI03-02), Sino-Portugal Science And Technology Cooperation Project (2005-2007) and Zhejiang Academician Fund project (Academician Youxian Sun).

References

1. I. F. Akyildiz, W. Su, Y. Sankarasubramaniam, E. Cayirci: Wireless sensor networks: a survey. Computer Networks. 38: 393-422, 2002.
2. L. Cui, H. L. Ju, Y. Miao, T. P. Li, W. Liu, Z. Zhao: Study on Wireless Sensor Networks. Computer Research and Development. Vol. 42, No. 1: 163-174, 2005.
3. H. B. Yu, P. Zeng, Zh. F. Wang, Y. Liang, Zh. J. Shang: Study On Distributed Wireless Sensor Networks Communication Protocols. Journal of Communication. Vol. 25,No. 10: 102-110, 2004.
4. G. J. Pottie, W.J. Kaiser: WIRELESS INTEGRATED NETWORK SENSORS. COMMUNICATIONS OF THE ACM. Vol.43, No. 5: 51-58, May 2000.
5. B. Warneke, M. Last, B. Liebowitz, K. S. J. Pister: Smart Dust: communicating with a cubic-millimeter computer. Computer. Vol. 34, No. 1: 44-51, Jan 2001.
6. uAMPS: http://www-mtl.mit.edu/research/icsystems/uamps
7. SensIT: http://www.sainc.com/sensit/
8. Chalermek, R. Govindan, D. Estrin: Directed Diffusion: A Scalable and Robust Communication Paradigm for Sensor Networks. MOBICOM 2000, August 2000.
9. J. Heidemann, F. Silva, C. Intanagonwiwat, R. Govindan, D. Estrin, D. Ganesan: Building Efficient Wireless Sensor Networks with Low-Level Naming. ACM SIGOPS Operating Systems Review. December 2001. Volume: 35 Issue: 5. 146-159
10. C. Intanagonwiwat, D. Estrin, R. Govindan, and J. Heidemann: Impact of Network Density on Data Aggregation in Wireless Sensor Networks. The 22nd International Conference on Distributed Computing Systems. July, 2002.

11. J. Zhao, R. Govindan, D. Estrin: Computing aggregates for monitoring wireless sensor networks. 2003 IEEE International Workshop on Sensor Network Protocols and Applications, 11 May 2003 139-148
12. M. Ding, X. Zh. Cheng, G. L. Xue: Aggregation tree construction in sensor networks. VTC 2003-Fall. 2003 IEEE 58th , Volume: 4 , 6-9 Oct. 2003 2168-2172
13. A. Boulis, S. Ganeriwal, M. B. Srivastava: Aggregation in sensor networks: an energy-accuracy trade-off. 2003 IEEE International Workshop on Sensor Network Protocols and Applications, 11 May 2003 128-138
14. J. N. Al-Karaki, R. Ul-Mustafa, A. E. Kamal: Data aggregation in wireless sensor networks- exact and approximate algorithms. HPSR. 2004 Workshop on High Performance Switching and Routing, 2004. 241-245
15. I. Solis, K. Obraczka: The impact of timing in data aggregation for sensor networks. 2004 IEEE International Conference on Communications, Vol: 6, 20-24 June 2004. 3640-3645
16. J. Chou, D. Petrovic.K. Ramchandran: Signals,Tracking and exploiting correlations in dense sensor networks. The Thirty-Sixth Asilomar Conference on Systems and Computers, 2002, Volume: 1. 39 - 43
17. J. Chou, D. Petrovic, R. Kannan: A distributed and adaptive signal processing approach to reducing energy consumption in sensor networks. INFOCOM 2003. Twenty-Second Annual Joint Conference of the IEEE Computer and Communications Societies. IEEE , Volume: 2 , 30 March-3 April 2003 1054-1062
18. T. Arici, B. Gedik, Y. Altunbasak, L. Liu: PINCO: a pipelined in-network compression scheme for data collection in wireless sensor networks. ICCCN 2003. The 12th International Conference on Computer Communications and Networks. 20-22 Oct. 2003 539-544
19. J. B. Seung, G. D. Veciana, Xun Su: Minimizing energy consumption in large-scale sensor networks through distributed data compression and hierarchical aggregation. IEEE Journal on Selected Areas in Communications, Volume: 22 , Issue: 6 , Aug. 2004 1130-1140
20. S. Pattern, B. Krishnamachari, R. Govindan: The impact of spatial correlation on routing with compression in wireless sensor networks. IPSN 2004. Third International Symposium on Information Processing in Sensor Networks, 2004. 28-35
21. S. Madden, R. Szewczyk, M. J. Franklin, D. Culler: Supporting aggregate queries over ad-hoc wireless sensor networks. Fourth IEEE Workshop on Mobile Computing Systems and Applications, 2002. 49-58
22. X. Li, Y. J. Kim, R. Govindan, W. Hong: Multi-dimensional Range Queries in Sensor Networks. Proceedings of the 1st international conference on Embedded networked sensor systems. November, 2003.
23. B. J. Bonfils, P. Bonnet: Adaptive and Decentralized Operator Placement for In-Network Query Processiong. LNCS 2634, 47–62, 2003. IPSN 2003, Springer-Verlag Berlin Heidelberg 2003.
24. S. Das, K. Shuster, C. Wu: The ACQUIRE Mechanism for Efficient Querying in Sensor Networks. Proceedings of the first international joint conference on Autonomous agents and multiagent systems. July 2002
25. J. Newsome, D. Song: GEM: Graph EMbedding for Routing and Data-Centric Storage in Sensor Network Without Geographic Information, Proceedings of the 1st international conference on Embedded networked sensor systems. November 2003.
26. D. Ganesan, B. Greenstein, D. Perelyubskiy, D. Estrin, J. Heidemann: An Evaluation of Multi-resolution Storage for Sensor Network. Proceedings of the first international conference on Embedded networked sensor systems. November 2003

27. S. Patil, S. R. Das, A. Nasipuri: Serial Data Fusion Using Space-filling Curves in Wireless Sensor Networks. IEEE SECON 2004. First Annual IEEE Communications Society Conference on Sensor and Ad Hoc Communications and Networks. 182-190
28. T. Y. Wang, Y. S. Han, P. K. Varshney: A combined decision fusion and channel coding scheme for fault-tolerant classification in wireless sensor networks. ICASSP '04. IEEE International Conference on Acoustics, Speech, and Signal Processing. 1073-10766.
29. J. L. Wong, R. Jafari, M. Potkonjak: Gateway placement for latency and energy efficient data aggregation. 29th Annual IEEE International Conference on Local Computer Networks, 2004. (2004) 490 – 497
30. V. Erramilli, I. Matta, A. Bestavros: On the interaction between data aggregation and topology control in wireless sensor networks. IEEE SECON 2004. (2004) 557–565
31. T. Pham, E. J. Kim, M. Moh: On Data Aggregation Quality and Energy Efficiency of Wireless Sensor Network Protocols-Extended Summary. First International Conference on Broadband Networks, 2004. 730-732
32. R. Min, M. Bhardwaj, S. H. Choi, N. Ickes, E. Shih, A. Sinha, A. Wang and A. Chandrakasan: Energy-centric enabling technologies for wireless sensor networks, IEEE Wireless Communications, 28-39, August 2002
33. I. F. Akyildiz, D. Pompili, T. Melodia: Challenges for Efficient Communication in Underwater Acoustic Sensor Networks [J]. ACM Special Interest Group on Embedded Systems Review, http://www.cs.virginia.edu/sigbed/ and http://www.seaweb.org/. 2004, 1(2)
34. R. X. Niu, P. K. Varshney, M. Moore, D. Kalmer: Decision Fusion in a Wireless Sensor Network with a Large Number of Sensors. The 7th International Conference on Information Fusion. June 28 to July 1, 2004 in Stockholm, Sweden. 21-27

Minimum Data Aggregation Time Problem in Wireless Sensor Networks*

Xujin Chen, Xiaodong Hu, and Jianming Zhu

Institute of Applied Mathematics, Chinese Academy of Sciences,
P. O. Box 2734, Beijing 100080, China
{xchen, xdhu}@amss.ac.cn, sduzhujm21@163.com

Abstract. Wireless sensor networks promise a new paradigm for gathering data via collaboration among sensors spreading over a large geometrical region. Many real-time applications impose stringent delay requirements and ask for time-efficient schedules of data aggregations in which sensed data at sensors are combined at intermediate sensors along the way towards the data sink. The *Minimum Data Aggregation Time* (MDAT) problem is to find the schedule that routes data appropriately and has the shortest time for all requested data to be aggregated to the data sink.

In this paper we study the MDAT problem with uniform transmission range of all sensors. We assume that, in each time round, data sent by a sensor reaches exactly all sensors within its transmission range, and a sensor receives data if it is the only data that reaches the sensor in this time round. We first prove that this problem is NP-hard even when all sensors are deployed a grid and data on all sensors are required to be aggregated to the data sink. We then design a $(\Delta - 1)$-approximation algorithm for MDAT problem, where $\Delta + 1$ equals the maximum number of sensors within the transmission range of any sensor. We also simulate the proposed algorithm and compare it with the existing algorithm. The obtained results show that our algorithm has much better performance in practice than the theoretically proved guarantee and outperforms other algorithm.

1 Introduction

Due to existing and emerging applications in various situations, wireless sensor networks (WSNs) have recently emerged as a premier research topic. A wireless sensor network (WSN) consists of a number of small-sized sensor nodes spreading over a geographical area and a sink node where the end user can access data. All nodes are equipped with capabilities of sensing, data processing, and communicating with each other by means of a wireless ad hoc network. A wide range of tasks can be performed by these tiny devices, such as condition-based maintenance and the monitoring of a large area with respect to some given physical quantity.

* This work was supported in part by the NSF of China under Grant No. 70221001 and 60373012.

X. Jia, J. Wu, and Y. He (Eds.): MSN 2005, LNCS 3794, pp. 133–142, 2005.

In contrast to traditional address-centric networks, WSNs are intrinsically data-centric. In some applications of WSN, the end user (data sink) wants to extract information from the sensor field with low latency, and the sheer number of sensor nodes poses unique challenges on time-efficient data aggregation. First, the sensor nodes employ low-power radio transceivers to enable communications. Data (signal) sent by a senor (*sender*) reaches exactly all its neighbors that are sensors or data sink within the transmission range of the sender; sensors far from the data sink have to use intermediate nodes to relay their data. Second, when two or more sensors send data to a common neighbor at the same time, the data *collide* at the common neighbor which will not receive any of these data. Third, the data sent by a sender is received by any its neighbor (*receiver*) at which no collision occurs; the receiver fuses the data received with its own data (possibly null), and stores the fused data as its new data; the time consumed by a single sending-receiving-fusing-storing is typically normalized to one, and parallel sending-receivings are desirable for reducing the network delay. Fourth, with the large population of sensor nodes, it may be impractical or energy consuming to pay attention to each individual nodes in all situations; for instance, the user wound be more interested in querying "what is the highest temperature in the northeast quadrant?"

Motivated by various applications of time-efficient data aggregation (such as battlefield communications), we study in this paper the *Minimum Data Aggregation Time* (MDAT) Problem: Given a WSN in which a distinguished data sink d is interested in data on a subset S of sensors, the goal of MDAT problem is to find a sending-receiving schedule such that all data on S are aggregated to d in a minimum time. In our work, the geometric nature of WSNs is emphasized, and time efficiency is used as performance metric to evaluate data aggregation algorithms.

The remainder of this paper is organized as follows. In Section 2, we first specify the network model and formalize the MDAT problem, and then present some related work and our contributions in this paper. In Section 3, we prove that MDAT problem is NP-hard even for some special case. In Section 4, we propose an approximation algorithm for MDAT problem, and give the theoretical proof of its performance guarantee. We then evaluate the average performance of the proposed algorithm through simulation and compare it with the existing algorithm. In Section 5, we conclude this paper with remarks on future research.

2 Preliminaries

2.1 Model Description

In view of miniature design of sensor devices, we assume that all sensors in WSNs are fixed and homogeneous. More specifically, the WSN under investigation consists of stationary nodes (sensor nodes and a sink node) distributed in a Euclidean plane. Assuming the transmission range of any sensor node is a unit disk (circular region with unit radius) centered at the sensor, we model a

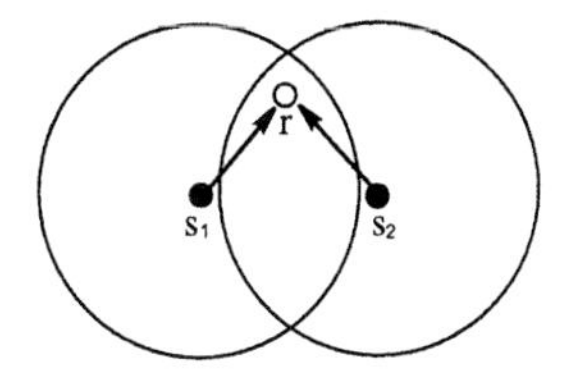

Fig. 1. Description of network model used

WSN as a *unit disk graph* $G = (V, E)$ in which two nodes $u, v \in V$ are considered neighbors, i.e., there is an edge $uv \in E$ joining u and v, if and only if the Euclidean distance between u and v is at most one. Hereafter we reserve symbol G for unit disk graphs modelling WSNs, and Δ for the maximum degree of G. It is always assumed that G is connected.

We assume in this paper that communication is deterministic and proceeds in synchronous *rounds* controlled by a global clock. In each time round

- any node can send data (be a *sender*) or receive data (be a *receiver*) but cannot do both;
- data sent by any sender *reaches* simultaneously all its neighbors;
- a node receives a data only if the data is the only one that reaches it (in this round); and
- each receiver updates its data as the combination of its old data and the data received (according to the aggregation function).

For example, in Fig. 1, two time rounds are required if data on s_1 and that on s_2 need to be aggregated to r which is located within the transmission range of s_1 and that of s_2.

An instance of MDAT problem is denoted by $(G, \mathsf{S}, \mathsf{d})$ in which $G = (V, E)$ models the WSN, the set $\mathsf{S}(\subseteq V)$ consists of nodes possessing data requested by the sink node $\mathsf{d}(\in V)$. The solution of $(G, \mathsf{S}, \mathsf{d})$ is a schedule $\{(S_1, R_1), \ldots, (S_s, R_s)\}$ such that S_r (resp. R_r) is the set of senders (resp. receivers) in round r, $r = 1, \ldots, s$, and all data on S must be aggregated to d within s rounds. For the purpose of energy conserving, we assume that

- each node sends data at most once and the data is received by exactly one receiver (i.e., one of neighbors of the node).

Hence every (S_r, R_r) gives implicitly the 1-1 correspondence between S_r and R_r in a way that $v \in S_r$ corresponds to its receiver in R_r which is the only neighbor of v in R_r. The value s is called the *data aggregation time* of solution $\{(S_1, R_1), \ldots, (S_s, R_s)\}$. The MDAT problem is to find the schedule with minimum data aggregation time $t_{OPT}(G, \mathsf{S}, \mathsf{d})$.

As usual (e.g. in [3]), we assume that each sensor knows its geometric position in the network, which is considered the unique ID of the sensor (the data aggregated may include some of these IDs). We further assume that the sink has global knowledge of all IDs in the WSN. When it needs some data of particular interests at some sensor nodes, it announces, by multicasting, the schedule

$\{(S_1, R_1), \ldots, (S_s, R_s)\}$ which is represented by IDs of senders and receivers. Upon receiving the request, sensor nodes will send their data or receive data from others as specified in the schedule.

2.2 Related Work

Most of works on data aggregation focus on energy efficiency. There are two recent works on time efficiency [1, 7]. They studied a special case of MDAT problem, called *convergecasting problem*, where data on all sensors in the network are required to be sent and aggregated to the data sink. Annamalai et. al [1] proposed a centralized heuristic that constructs a tree rooted at the sink node according to proximity criterion (a node is assigned as a child to the closest possible parent node) and to assign each node a code and a time slot to communicate with its parent node. However the miniature hardware design of nodes in WSNs may not permit employing complex radio transceivers required for spread spectrum codes or frequency bands systems. Additionally, the heuristic is evaluated only through simulations. More recently, Kesselman and Kowalski [7] devised a randomized distributed algorithm for convergecasting in n-node WSNs that has the expected running time $O(\log n)$. An assumption central to their model is that sensor nodes have capability of detecting collisions and adjusting transmission ranges, and that the maximum transmission range might be as large as the diameter of the network. This completes the sensor hardware design and poses challenge to low-transmission-range constraint on sensors.

Given an MDAT instance $(G, \mathsf{S}, \mathsf{d})$, a *shortest path tree* T of $(G, \mathsf{S}, \mathsf{d})$ is a tree in G consisting of shortest paths from d to nodes in S. The *hight* of T, denoted by $h(G, \mathsf{S}, \mathsf{d})$, equals to the length of the longest path in T from d to leaf nodes of T. The following lower bound can be easily obtained by reversing the argument in the estimation of multicasting time in a telephone network [2].

Lemma 1. $t_{OPT}(G, \mathsf{S}, \mathsf{d}) \geq \max\{h(G, \mathsf{S}, \mathsf{d}), \log_2 |\mathsf{S}|\}$ *for any MDAT instance* $(G, \mathsf{S}, \mathsf{d})$.

Unlike traditional telephone network, in WSNs the data aggregation/ convergecast is not simply the reverse of multicast/ broadcast. For example, when the underlying topology G of WSN is a n-node complete graphs, the convergecasting time equals to n, and the broadcasting time and the height of SPT both equal to 1.

2.3 Our Contributions

The first contribution is a new and simple data aggregation model. It is collision-free and does not require any specialized codes. The second contribution is the NP-hardness proof of the corresponding problem. The third contribution is an approximation algorithm proposed for MDAT problem with provable performance guarantee $\Delta - 1$. The algorithm uses a new technique that allows certain flexibility of tree-structures while scheduling parallel transmissions; in other words instead of making a schedule after a SPT is constructed as the existing methods do, we find the data aggregation tree after the schedule is made.

3 NP-Hardness Proof

In order to prove the NP-hardness of the MDAT problem, we apply the result on orthogonal planar drawing and reduction from restricted planar 3-SAT. An *orthogonal planar drawing* of a planar graph H is a planar embedding of H in the plane such that all edges are drawn as sequences of horizontal and vertical segments. A point where the drawing of an edge changes its direction is called a *bend* of this edge. All vertices and bends are drawn on integer points. If the drawing can be enclosed by a box of width g and height g, we call it an embedding with grid size $g \times g$. By a *plane graph* we mean a planar graph together with a planar embedding of it. Given a simple plane graph H on g vertices that is not an octahedron and has maximum degree at most 4, Biedl and Kant [5] presented a linear algorithm which produces an orthogonal planar drawing of H with grid size $g \times g$ such that the number of bends along each edge is at most 2. Using this result, we can prove the following lemma.

Lemma 2. *Let H be a plane graph on g vertices with maximum degree at most 4. Suppose that H is not an octahedron, and let H' be the graph obtained from H by replacing each edge in H with a path of length $120g^2$. Then H' is a unit disk graph and an orthogonal planar embedding of H' of grid size $(40g^2+40g) \times (40g^2+40g)$ can be computed in time polynomial in g.*

As customary $\{x_1, \dots, x_n\}$ and $\{c_1, \dots, c_m\}$ denote, respectively, the set of variables and clauses in a Boolean formula φ in conjunctive normal form, where each clause has at most 3 literals. Associate with φ the formula graph $G_\varphi = (\{x_1, \dots, x_n\} \cup \{c_1, \dots, c_m\}, E_1 \cup E_2)$, where $E_1 := \{x_i c_j : x_i \in c_j \text{ or } \bar{x}_i \in c_j\}$ and $E_2 := \{x_i x_{i+1} : 1 \leq i \leq n-1\} \cup \{x_n x_1\}$. Boolean formula is called *planar* if G_φ is a planar graph. Planar 3-SAT is the problem of finding, if any, a truth assignment that satisfies all clauses in a planar Boolean formula, where each clause has at most three literals. A planar 3-SAT problem is said to be *restricted* if (a) each variable (unnegated or negated) appears in at most three clauses, (b) both unnegated and negated forms of each variable appear; and (c) at every variable node in the planar embedding of G_φ, the edges in E_2 that are incident with node x separate the edges in E_1 incident to x such that all edges representing unnegative appearance are incident to one side of x and all edges representing negative appearance are incident to the other side. It is known that restricted planar 3-SAT is NP-complete [4].

Theorem 1. *The MDAT problem is NP-complete even when the underlying topology is a subgraph of a grid.*

Proof. The proof is by reduction from restricted planar 3-SAT. Given any restricted planar 3-SAT instance φ on n variable and m clauses, from its planar formula graph G_φ, we construct planar graphs G_k for positive integer k as follows.
To every variable x_i, $1 \leq i \leq n$, we associate a rectangle X_i and two node-disjoint paths $\Pi_i, \bar{\Pi}_i$ such that (i) X_i has exactly $10k$ nodes among which equally spaced nodes $p_i, q_i, r_i, s_i, t_i, \bar{s}_i, \bar{r}_i, \bar{q}_i, \bar{p}_i, \bar{t}_i$ are located in cyclic order of X_i, and Π_i (resp.

$\bar{\Pi}_i$) has ends o_i and p_i (resp. $\bar{o}_i$ and $\bar{p}_i$) with $\Pi_i \cap X_i = \{p_i\}$ (resp. $\bar{\Pi}_i \cap X_i = \{\bar{p}_i\}$), and (ii) both Π_i and $\bar{\Pi}_i$ are of length $(6i-5)k-1$. To every clause c_j, we associate a path C_j with ends b_j, c_j and of length $k-1$. All $X_i \cup \Pi_i \cup \bar{\Pi}_i$'s and C_j's are pairwise node-disjoint. For every edge $x_i c_j$ (resp. $\bar{x}_i c_j$) in G_φ, there is a path Υ_{ij} in G_k such that Υ_{ij} has one end c_j and the other end in $\{r_i, s_i\}$ (resp. $\{\bar{r}_i, \bar{s}_i\}$), and for all $1 \le i, i' \le n, 1 \le j, j' \le m$, we have (iii) $\Upsilon_{i,j} \cap (\bigcup_{h=1}^n C_h) = \{c_j\}$, $\Upsilon_{i,j} \cap (\bigcup_{h=1}^n (X_i \cup \Pi_i \cup \bar{\Pi}_i))$ consists of a node in $\{r_i, s_i, \bar{r}_i, \bar{s}_i\}$; (iv) $\Upsilon_{ij} \cap \Upsilon_{i'j'} \neq \emptyset$ iff $\Upsilon_{ij} = \Upsilon_{i'j'}$ or $i \neq i'$ and $\Upsilon_{ij} \cap \Upsilon_{i'j'} = \{c_j\} = \{c_{j'}\}$; and (v) Υ_{ij} is of length $(6i-4)k$ when it has end in $\{r_i, \bar{r}_i\}$ and of length $(6i-3)k$ when it has end in $\{s_i, \bar{s}_i\}$. Finally we add $n-1$ pairwise node-disjoint paths $\Lambda_1, \ldots, \Lambda_{n-1}$ such that (vi) Λ_h has length k and $\Lambda_h \cap (\bigcup_{i,j} (X_i \cup \Pi_i \cup \bar{\Pi}_i \cup C_j \cup \Upsilon_{ij})) = \{t_h, t_{h+1}\}$ for $h = 1, \ldots, n-1$. This completes the construction of $G_k := \bigcup_{i,j} (X_i \cup \Pi_i \cup \bar{\Pi}_i \cup \Lambda_i \cup C_j \cup \Upsilon_{ij})$.

Let G_k^+ be obtained from G_k by adding $2n+m$ pendant edges such that the $2n+m$ degree-one nodes $o_1, \bar{o}_1, o_2, \bar{o}_2, \ldots, o_n, \bar{o}_n, b_1, b_2, \ldots, b_m$ in G_k have degree two in G_k^+. Denote by g the number of nodes in G_1^+ and set $\ell = 120g^2$. It is easy to see that $g < 36n^2 + m$ and that G_ℓ^+ is obtained from G_1^+ by replacing each edge of G_1^+ with a path of length $120g^2$. Since G_φ is a planar graph with maximum degree at least 3, so is G_1. Thus a planar embedding of G_1 might be computed in time polynomial in $n+m$ [6]. By the construction, this planar embedding might be extended to be a planar embedding for G_1^+ in time polynomial in g and hence polynomial in $n+m$. Notice that G_1^+ is a plane graph other than octahedron and that the maximum degree of G_1^+ is at most 4. It follows from Lemma 2 that G_ℓ^+ is a unit disk graph, so is G_ℓ. Moreover from Lemma 2, we deduce that G_ℓ is a subgraph of a grid, and that both the size of G_ℓ and the construction time of G_ℓ are polynomial in $n+m$.

It can be shown that the restricted planar 3-SAT instance φ is satisfiable if and only if the MDAT problem on $(G_\ell, V(G_\ell), t_n)$ has a solution (schedule) which aggregates all data on $V(G_\ell)$ into the sink t_n within $(6n-1)\ell$ rounds. The theorem follows from the NP-completeness of restricted planar 3-SAT. □

4 Approximation Algorithm

In this section we present an approximation algorithm SDA for MDAT that follows the shortest data aggregation philosophy – aggregating data along shortest paths towards the sink. Regarding the performance of SDA, our theoretical analysis proves the worst-case ratio $\Delta - 1$, while our simulation indicates a better average performance.

4.1 Shortest Data Aggregation

Algorithm SDA proceeds by incrementally constructing smaller and smaller shortest path trees rooted at d that span nodes possessing all data of interests. SDA initially sets T_1 to be a SPT of $(G, \mathsf{S}, \mathsf{d})$, and then implements a number of iterations (Step 2-14). Each iteration produces a schedule of a round. In the

rth iteration, T_r is a SPT rooted at d spanning a set of nodes that possess all data aggregated from S till round $r-1$. SDA picks from the leaves of T_r the senders for round r. In Step 4-11, Z^r is initially set to be the set of leaves of T_r excluding d, and the it remains the property that every non-leaf neighbor of a leaf in T_r other than d has a neighbor in Z^r. The leaves of T_r other than d are examined in the decreasing order of the number of their neighbors in G that are non-leaf nodes in T_r. A leaf is eliminated from Z^r if and only if the elimination does not destroy the property of Z^r. When all leaves of T_r other than d have been examined, the remaining nodes in Z^r form the set S_r of the senders in round r. Subsequently, SDA eliminates S_r from its consideration by setting $T_{r+1} := T_r \backslash S_r$ and enters the $(r+1)$th iteration. The following pseudo-code makes our idea more precise. For the simplicity of presentation, we denote by $N_G(U)$ the set $\{v : uv \in E, u \in U, v \in V \backslash U\}$ for $U \subseteq V$, where V is the vertex set of G.

Algorithm SDA SHORTEST_DATA_AGGREGATION

Input MDAT instance $(G, \mathsf{S}, \mathsf{d})$
Output A data aggregation schedule $\{(S_1, R_1), \ldots, (S_s, R_s)\} = SDA(G, \mathsf{S}, \mathsf{d})$

1. $r \leftarrow 1$, $I_1 \leftarrow \mathsf{S}$, $T_1 \leftarrow$ a SPT T of $(G, \mathsf{S}, \mathsf{d})$
2. **while** $T_r \neq \{\mathsf{d}\}$ **do begin**
3. $\quad T_r \leftarrow T_r \backslash (\{\text{leaves of } T_r\} \backslash (\{\mathsf{d}\} \cup I_r))$
4. $\quad Z^r = Z_r \leftarrow \{\text{leaves of } T_r\} \backslash \{\mathsf{d}\}$, $Y_r \leftarrow N_G(Z_r) \cap V(T_r)$
5. $\quad S_r \leftarrow \emptyset$, $R_r \leftarrow \emptyset$
6. $\quad$ **while** $Z^r \backslash S_r \neq \emptyset$ **do begin**
7. $\quad\quad z \leftarrow$ a node in $Z^r \backslash S_r$ of the maximum number of neighbors in Y_r
8. $\quad\quad$ **if** $Y_r \subseteq N_G((Z^r \cup S_r) \backslash \{z\})$ **then** $Z^r \leftarrow Z^r \backslash \{z\}$
9. $\quad\quad\quad$ **else** $y_z \leftarrow$ a node from $Y_r \backslash N_G((Z^r \cup S_r) \backslash \{z\})$
10. $\quad\quad\quad\quad S_r \leftarrow S_r \cup \{z\}$, $R_r \leftarrow R_r \cup \{y_z\}$
11. $\quad$ **end-while**
12. $\quad I_{r+1} \leftarrow I_r \cup R_r$, $T_{r+1} \leftarrow T_r \backslash S_r$
13. $\quad r \leftarrow r+1$
14. **end-while**
15. Output $t_{SDA}(G, \mathsf{S}, \mathsf{d}) = s \leftarrow r-1$ and $SDA(G, \mathsf{S}, \mathsf{d}) \leftarrow \{(S_1, R_1), \ldots, (S_s, R_s)\}$

Next we analyze theoretically the correctness and the performance of SDA. The following lemma can be proved by induction on r.

Lemma 3. *Let $R_0 = S_0 = Z_{s+1} = \emptyset$. All of the following hold for each r with $1 \leq r \leq s$.*

(i) I_r consists of all informed nodes at the beginning of round r.
(ii) T_r is a subtree of $T \backslash (\cup_{i=0}^{r-1} S_i)$ with root d such that $|V(T_r)| \geq 2$ and all data on $\mathsf{S} \backslash V(T_r)$ have been aggregated to nodes in $V(T_r)$ at the beginning of round r (i.e. the end of round $r-1$).
(iii) Y_r and Z_r are nonempty subsets of $V(T_r)$ such that $Y_r \subseteq N_G(Z_r)$.
(iv) $Y_r \subseteq N_G(S_r)$ in Step 12.

(v) There is a 1-1 correspondence between S_r and R_r in a way that every sender $z \in S_r$ corresponds to its receiver $y_z \in R_r$.
(vi) $S_r \subseteq Z_r \subseteq I_r$, $Z_r \setminus S_r \subseteq Z_{r+1}$, $\emptyset \neq R_r \subseteq I_{r+1} \cap Y_r \subseteq V(T_{r+1}) \subseteq V(T_r) \setminus S_r \subsetneq T_r$.

Corollary 1. *(i) $S_1, S_2, \ldots, S_s$ are pairwise disjoint. (ii) $S_r \cap (R_r \cup R_{r+1} \cup \cdots \cup R_s) = \emptyset$ for all $1 \leq r \leq s$. (iii) $T = T_1 \supsetneq T_2 \supsetneq \cdots \supsetneq T_s \supsetneq T_{s+1} = \{\mathsf{d}\}$.*

Lemma 4. *Algorithm SDA solves MDAT problem correctly.*

Proof. The termination of SDA is guaranteed by Corollary 1(iii). From $T_{s+1} = \{\mathsf{d}\}$ and Lemma 3(v) and (vi), it can be seen that T_s is a 2-node tree on $R_s = \{\mathsf{d}\}$ and the only sender in S_s. Since, by Lemma 3(ii), all data on S have been aggregated to $V(T_s)$ at the end of round $s-1$, it is instant that the schedule $\{(S_1, R_1), \ldots, (S_{s-1}, R_{s-1}), (S_s, R_s)\}$ output by SDA aggregates all data on S to d within s rounds. □

We now study the approximation ratio of Algorithm SDA. Denote $h = h(G, \mathsf{S}, \mathsf{d})$. Set $L_i = \{\text{nodes in } T \text{ at } i \text{ hops away from } \mathsf{d}\}$ for every $0 \leq i \leq h+1$; in particular, $L_0 = \{\mathsf{d}\}$ and $L_{h+1} = \emptyset$. Let $T_i = \emptyset$ for all $i \geq s+2$. By induction on i, we have the following lemma.

Lemma 5. *$L_{h+1-i} \cap V(T_{(\Delta-1)i+1}) = \emptyset$ for every $0 \leq i \leq h-1$.*

Theorem 2. *Given any MDAT instance $(G, \mathsf{S}, \mathsf{d})$, Algorithm SDA produces a solution $SDA(G, \mathsf{S}, \mathsf{d})$ that satisfies (i) $t_{SDA}(G, \mathsf{S}, \mathsf{d}) \leq (\Delta-1)h+1$ and (ii) $t_{SDA}(G, \mathsf{S}, \mathsf{d}) \leq (\Delta-1)t_{OPT}(G, \mathsf{S}, \mathsf{d})$.*

Proof. Recall from Lemma 1 that $t_{OPT}(G, \mathsf{S}, \mathsf{d}) \geq h$. If $s = t_{SDA}(G, \mathsf{S}, \mathsf{d}) \leq (\Delta-1)(h-1)+1$ then we are done. So we assume $s > (\Delta-1)(h-1)+1$.

To justify (i), recall from Lemma 5 that $L_2 \cap V(T_{(\Delta-1)(h-1)+1}) = \emptyset$. In turn Lemma 3(ii) gives $V(T_{(\Delta-1)(h-1)+1}) \subseteq L_1 \cup \{\mathsf{d}\}$. Combining $|L_1| \leq \Delta$ and $T_{s+1} = \{r\}$, we deduce from Corollary 1(iii) that $s+1 \leq (\Delta-1)(h-1)+1+|L_1| \leq \Delta(h-1)+2$, which yields (i).

Next we prove (ii). In case of $|L_h| \geq 2$, we have $t_{OPT}(G, \mathsf{S}, \mathsf{d}) \geq h+1$ and (i) implies $s \leq (\Delta-1)t_{OPT}(G, \mathsf{S}, \mathsf{d})$. It remains to consider the case where $|L_h| = 1$. We may assume $\Delta \geq 3$ as otherwise G is a path or a cycle and $s = h = t_{OPT}(G, \mathsf{S}, \mathsf{d})$. It is obvious that $S_1 = L_h$. Let $G' = G \setminus L_h$ and S' consist of the nodes in $\mathsf{S} \setminus L_h$ and the neighbor (parent) of L_h in T. Then $T \setminus L_h$ is a SPT of (G', S', r) that has hight $h-1$; moreover, there is an implementation of SDA on $(G', \mathsf{S}', \mathsf{d})$ which outputs $\{S'_1, S'_2, \ldots, S'_{s-1}\}$ with $S'_i = S_{i+1}$ for all $1 \leq i \leq s-1$. Using (i), we have $s-1 = t_{SDA}(G', \mathsf{S}', \mathsf{d}) \leq (\Delta-1)(h-1)+1$ since the maximum degree of G' is upper bounded by Δ. It follows from $\Delta \geq 3$ that $s \leq (\Delta-1)h \leq (\Delta-1)t_{OPT}(G, \mathsf{S}, \mathsf{d})$, and (ii) is proved. □

4.2 Simulation Study

Deployment of a sensor network can be in random fashion (e.g., dropped from an airplane) or planted manually (e.g., fire alarm sensors in facilities). In the

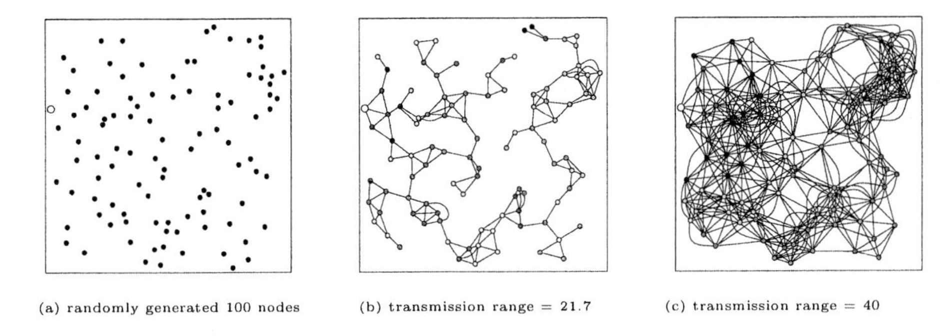

(a) randomly generated 100 nodes (b) transmission range = 21.7 (c) transmission range = 40

Fig. 2. A randomly generated WSN

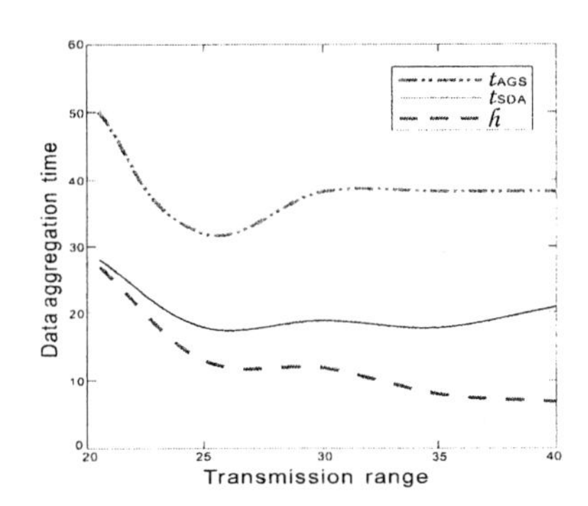

Fig. 3. Simulation results: data aggregation times versus transmission ranges

latter case, WNSs have nice properties including bounded-degree; consequently our theoretical results assure satisfactory approximation for MDAT problem. Thus we focus on the former case. Our test network had 100 nodes which were randomly generated within a 200×200 square region (see Fig. 2(a)). The leftmost node (the larger white node in Fig. 2) was selected as data sink d. We used a simulation technique similar to that in [8]. Our simulation was carried out with deferent transmission ranges varying from 21.7 to 40 such that the number of edges of the unit disk graphs modelling the WSN varies from 167 to 546 (see Fig. 2(b,c)), where 21.7 is the minimum transmission range that guarantees the network connectivity.

For comparison purpose, our simulation was done for convergecast (the case where $V = \mathsf{S} \cup \{\mathsf{d}\}$). Let AGS stand for the convergecast algorithm in [1] using one code. We compare the convergecasting times t_{SDA} and t_{AGS} computed by SDA and AGS with the lower bound $h = h(G, V, \mathsf{d})$ as benchmark, respectively.

It can be seen from Fig. 3 that Algorithm SDA outperforms Algorithm AGS, and the ratio of t_{SDA} to h is much less than the theoretical performance ratio $\Delta - 1$. This highlights the advantages of assigning parallel transmissions according to degree order in our shortest-data-aggregation-based algorithm SDA.

5 Conclusion

In this paper we have studied MDAT problem aiming for time-efficient data aggregation in WSNs. We establish NP-hardness of MDAT problem, and devise an approximation algorithm for MDAT. Our theoretical analysis and simulation results show that the proposed data aggregation achieves time-efficiency both in theory and in practice.

Due to the compactness of sensors' hardware, in the present paper, we make the assumption that all nodes have the same transmission radius. With the advances of techniques, the nodes may have the capability of adjusting their transmission ranges provided the maximum range is bounded. It would be interesting to investigate the topology of unidirectional links and design approximation algorithms for fast data aggregation in WSNs with adjustable transmission ranges and a maximum range bound.

References

1. V. Annamalai, S. K. S. Gupta, and L. Schwiebert, On tree-based convergecasting in wireless sensor networks, *WCNC 2003-IEEE Wireless Communication and Networking Conference*, **4**(1) (2003), 1942-1947.
2. A. Bar-Noy, S. Guha, J. Naor, and B. Schieber, Message multicasting in heterogeneous networks, *SIAM J. Comput.* **30**(2), 347-358.
3. N. Bulusu, J. Heidemann, and D. Estrin, GPS-less low cost outdoor localization for very small devices, *Technical Report 00-729*, Computer Science Department, University of Sourthern California, Apr. 2000.
4. D. Lichtenstein, Planar formulae and their uses, *SIAM J. Comput.* **11** (1982), 329-343.
5. T. Biedl and G. Kant, A better heuristic for orthogonal graph drawings, *Computational Geometry* **9** (1998), 159-180.
6. G. Di Battista, G. Liotta, and F. Vargiu, Spirality of orthogonal representations and optimal drawings of series-parallel graphs and 3-planar graphs, in: *Proc. Workshop on Algorithms and Data Structures, Lecture Notes in Computer Science 709*, Springer, New York, 1993, pp. 151-162.
7. A. Kesselman and D. Kowalski, Fast distributed algorithm for convergecast in ad hoc geometric radio networks, *Proc. 2nd Annual Conf. on Wireless on Demand Network Systems and Serveices* (WoNS'05), January, 2005, St. Moritz, Switzerland.
8. M. Varshnery and R. Bagrodia, Detailed models for sensor network simulations and their impact on network performance, *MSWim'04*, October, 2004, Venezia, Italy.

Mobility-Pattern Based Localization Update Algorithms for Mobile Wireless Sensor Networks

Mohammad Y. Al-laho, Min Song, and Jun Wang

Department of Electrical and Computer Engineering,
Old Dominion University, 231 Kaufman Hall,
Norfolk, VA 23529, USA
{malla003, msong, jwang012}@odu.edu

Abstract. In mobile wireless sensor networks, sensors move in the monitored area at any direction and at any speed. Unlike many other networking hosts, sensor nodes do not have global addresses. Very often they are identified by using a location-based addressing scheme. Therefore, it is important to have the knowledge of the sensor location indicating where the data came from. In this paper, we design three mobility-pattern based localization update algorithms. Specifically, we divide sensor movements into three states, *Pause*, *Linear*, and *Random*. Each state adopts different localization update algorithms. Analytical and simulation results are provided to study the localization cost and location accuracy of the proposed localization-update algorithm in different mobility patterns. The analysis to these results indicates that the localization cost is minimized and the location accuracy is improved.

1 Introduction

In mobile wireless sensor networks, sensors move in the monitored area at any direction and at any speed. Unlike many other networking hosts, sensor nodes do not have global addresses. Very often they are identified by using a location-based addressing scheme [5]. The routing process also requires location-based naming where the users are more interested in querying a location of a phenomenon, rather than querying an individual node [5]. Therefore, it is important to have the knowledge of the sensor location indicating where the date came from.

Many researches have been conducted to solve the localization problem of moving objects. The most famous one is the GPS system. Unfortunately, GPS is not a good choice for mobile wireless sensor networks because of its significant power consumption of the GPS receivers [7]. Intuitively, the localization of mobile wireless sensors can be performed by using prediction techniques to anticipate the possible next step of a sensor node based on the movement pattern model and the history of the movement pattern [4]. Unfortunately, it has been found that prediction techniques are not always accurate [1]. Especially, when the movement pattern is not predictive (not linear) the accuracy decreases significantly. A more reliable method is to integrate the prediction techniques and localization update algorithms. In this case, the localization update algorithms can correct any inaccurate anticipation. This method, however, is rather too complex and needs many calculations and filtering techniques to refine the possible

X. Jia, J. Wu, and Y. He (Eds.): MSN 2005, LNCS 3794, pp. 143 – 152, 2005.

sampling data collected from previous movements. The energy consumption and the memory size restrictions of a sensor node limit the use of these prediction methods.

Another problem for mobile localization schemes in literature is that they impractically simplify the mobility pattern and thus they do not work well when nodes move randomly. The more the mobility pattern deviates from the linear pattern, the more the location accuracy gets inaccurate and the update cost gets higher. It is necessary to differentiate different mobility patterns in the system. For each pattern, a different localization-update scheme is used to get the most accurate results. It is worth to note that one of the critical design issues in mobile sensor localization is simplicity, since wireless sensor nodes have limited power consumption and computational capability due to the small size of a node.

In this paper, we design three mobility-pattern based localization update algorithms. More specifically, we divide sensor movements into three states, *Pause*, *Linear*, and *Random*. Each state adopts different localization update algorithms. The objectives of this design are to 1) manage the localization update period in optimal rates so that the update cost is minimized, and 2) keep the location error low.

The rest of the paper is organized as follows. Section 2 presents the related work in localization update algorithm for mobile wireless sensor networks. In Section 3, we introduce the system models and our mobility-pattern based localization update algorithms. The numerical results are provided in Section 4 to study the update cost and location accuracy of the proposed localization update algorithms in different mobility patterns. Section 5 concludes the paper.

2 Related Work

Location update schemes generally can be classified as 1) time-based—updates are made at fixed time intervals, 2) movement-based—updates are made whenever the number of cell-boundary crossings since the last update exceeds a specified threshold, 3) distance-based—a mobile updates its location whenever its distance from an expected location exceeds a specified value since its last update [9], 4) dead reckoning—a scheme to predict a location based on velocity components calculations from previous locations, and 5) mobility-aware-based—updates are made based on the mobility pattern. As an example of time-based scheme, static fixed rate (SFR) performs the localization periodically with a fixed time period T [1]. The energy consumption of the SFR method is independent of mobility. However, the accuracy of the location varies with the mobility of the sensor.

Dynamic velocity monotonic (DVM) [1] is an example of dead reckoning. In DVM, a sensor node adapts its localization period as a function of its mobility speed. Based on the speed value a node schedule the next time to localize. A parameter α is set to represent the target maximum error. For each localization measurement, the speed value is compared with α to estimate how much time needed to reach the target maximum error if the node continues on the same speed. The next localization is scheduled after that time period. Note that the assumption is for constant velocity between two measurement points and this could affect the scheduling time for localization such that a node reaches values of error higher than the target maximum error (threshold error). To accommodate for extremely low/high speeds, DVM assigns an upper and lower limit for localization periods.

A more complex method is called mobility aware dead reckoning driven (MADRD) [1]. MADRD uses the dead reckoning model (DRM) [3, 8] to calculate the Euclidean distance between the predicted location and the actual current location. If the difference exceeds a predetermined threshold error the process moves into a low confidence state where localization is performed with a higher frequency. Otherwise, the process moves into a high confidence state where localization is performed with a lower frequency. Since MADRD essentially is based the prediction technique, its complexity inherently limits its wide use.

Recently, Song *et al.* proposed an algorithm to track moving objects in location-based services [2]. In their design, the mobility pattern is divided into three types: *Pause*, *Linear*, and *Random*. Different localization update policies are used for different mobility patterns. However, this design in [2] is more suitable for cellular networks. For example, the distance is defined in terms of the number of cells.

3 Mobility-Pattern Based Localization Update Algorithms

3.1 Main Ideas

Motivated by Song's work in [2], we divide the mobility pattern of mobile sensors into three states: *Pause*, *Linear*, and *Random*. A modified time-based scheme is used for *Pause* state since there is no distance or movement involved in a pause. The localization period is increased as the pause time increase. To prevent a long wait time there is a maximum period of wait time that cannot be exceeded. Dead reckoning scheme is most suitable for linear mobility pattern since it works for predictive movement pattern where there is no change of velocity or unpredicted change of direction. The localization period is incremented as long as the distance between the predicted location and real location is within a predefined threshold. The localization is set to the initial localization period if the error exceeds the threshold value.

The most suitable policy for the random movement pattern is distance-based scheme. Our distance-based scheme is different than the one used in [2]. In [2] the distance is between the last update in the last cell and the current cell. The idea in our design is not to actually measure the distance between last update and current position but rather to predict when the node will cross the limit of this distance based on the acceleration value measurement of the node. The acceleration a is measured periodically. The localization period is then measured as

$$\hat{t} = \sqrt{d_{thresh} / |a|} \tag{1}$$

where d_{thresh} denotes the distance limited to localize. After the node localizes, d_{thresh} will be as the radius length of a circle where the node is on the origin. The value of d_{thresh} will depend on the applications. Applications that need more accuracy will have smaller d_{thresh}. For simulation of discrete time intervals, $\hat{t}$ should be converted to discrete time intervals by using (2).

$$n = \left\lceil \hat{t} / timeslot \right\rceil \tag{2}$$

During the n time intervals the locations are estimated as follows,

$$x_t = x_{t-1} + \left(v_x + \hat{v}\right) timeslot \tag{3}$$

$$y_t = y_{t-1} + \left(v_y + \hat{v}\right) timeslot \tag{4}$$

where x_t and y_t are the new estimated position coordinates, and x_{t-1} and y_{t-1} are the previous position coordinates, v_x and v_y are the velocity components. Since a equals to the increment or decrement of velocity per second, a is multiplied by the *timeslot* period to get the increment or decrement in velocity vector for each time interval, $\hat{v} = a \times timeslot$.

3.2 Mobility Models

The mobility model used in this study is Gaussian Markovian Random Waypoint [6]. It is a hybrid of the Random Waypoint and Gaussian Markovian models. In Random Waypoint model the Mobile node chooses a random destination in the simulation area and a speed that is uniformly distributed. When it reaches the chosen destination, it pauses for a specific period of time. Then the node chooses a random destination again. All destinations are randomly chosen from within a predefined area.

In Gaussian Markovian model each node initially is assigned a current speed and direction. The speed (s) and direction (d) are updated at fixed intervals as follows:

$$s_n = \alpha s_{n-1} + (1-\alpha)\overline{s} + \sqrt{1-\alpha^2}\, s_{x_{n-1}} \tag{5}$$

$$d_n = \alpha d_{n-1} + (1-\alpha)\overline{d} + \sqrt{1-\alpha^2}\, d_{x_{n-1}} \tag{6}$$

where α is a value from 0 to 1 and it is the tuning parameter used to vary the randomness: $\alpha = 0$ leads to very random motion, while $\alpha = 1$ leads to completely linear motion; $\overline{s}$ and $\overline{d}$ are the mean value of speed and direction as $n \rightarrow \infty$; $s_{x_{n-1}}$ and $d_{x_{n-1}}$ are two random variables following a Gaussian distribution.

At each time interval the next location is calculated as follows:

$$x_n = x_{n-1} + s_{n-1} \cos d_{n-1} \tag{7}$$

$$y_n = y_{n-1} + s_{n-1} \sin d_{n-1} \tag{8}$$

The idea of Gaussian Markovian Random Waypoint is to use random waypoint at the macro level and use Gaussian Markovian at the micro level to provide a noisy behavior. That is, the mobile node still chooses a random destination location and speed, and then computes the direction of travel. However, it follows the model presented in equations (5) and (6) to travel in this direction in small time steps. The model is initialized with s0 and d0 being equal to the chosen speed and direction. The node travels following this model for an amount of time (T) equal to what it would take for it to reach the current destination using the chosen speed. The mobile node may not reach the destination, but possibly a location close to it depending on the amount of noise. After the time T the node pauses according to a given pause time, chooses a random destination location, and repeats the above process.

In our study, we use the Gaussian Markovian Random Waypoint with some alteration. For *Linear* state we set α value equal to 1 in equations (5) and (6) so it behaves as pure linear movement. However, the dn-1 in (6) should be recalculated to the new location. In *Random* state the α value is set for values between 0.1 and 0.9, because a value of zero will make it a completely random and memoryless model that does not depend on the previous speed and direction. Whenever the mobile node goes in pause time, it automatically switches to the pause state for that amount of time.

3.3 Analytical and System Models

Conditions to move from one state to another will be tested in the current state so the update policy is changed with a state change. Fig. 1 (a) shows the initial and three mobility pattern states. In the actual algorithm the main test to transit from state to another is based on velocity information. A test for the pause state is to have same location detected two times in a row. A same location means the velocity value equals to zero. In *Linear* state velocity should be the same for certain time, which is constant velocity or equivalently acceleration value equals to zero. For *Random* state, the test is to have velocity change which indicates random mobility pattern.

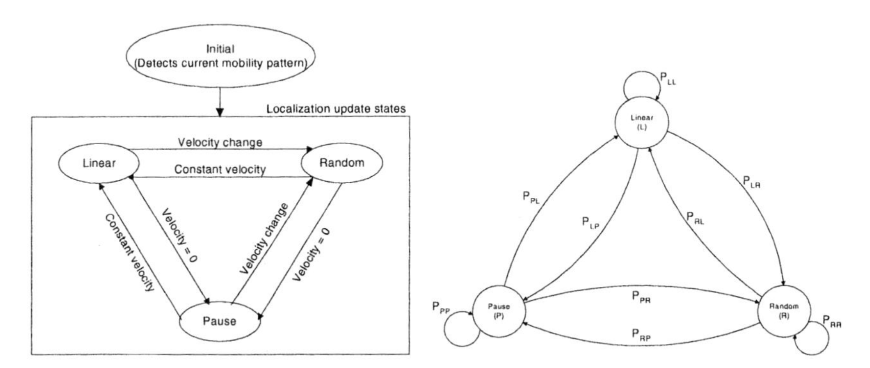

Fig. 1. (a) Mobility pattern transition diagram, (b) state-based mobility model

Fig. 1 (b) shows the state-based mobility model. We have the steady-state transition probability vector π for each state:

$$\pi_P = \frac{P_{LP}P_{RP} + P_{LP}P_{RL} + P_{LR}P_{RP}}{(P_{PR} - P_{LR})(P_{LP} + P_{LR} + P_{RL}) + (P_{LP} + P_{LR} + P_{PL})(P_{RP} + P_{RL} + P_{LR})} \tag{9}$$

$$\pi_R = \frac{P_{PR}P_{LP} + P_{PR}P_{LR} + P_{LR}P_{PL}}{(P_{PR} - P_{LR})(P_{LP} + P_{LR} + P_{RL}) + (P_{LP} + P_{LR} + P_{PL})(P_{RP} + P_{RL} + P_{LR})} \tag{10}$$

$$\pi_L = \frac{P_{PR}P_{RL} + P_{RP}P_{PL} + P_{RL}P_{PL}}{(P_{PR} - P_{LR})(P_{LP} + P_{LR} + P_{RL}) + (P_{LP} + P_{LR} + P_{PL})(P_{RP} + P_{RL} + P_{LR})} \quad (11)$$

Equations (9), (10), and (11) are used to calculate the performance measurement specified in Section 3.5.

3.4 Performance Measurements

The two performance measurements are the localization updates cost C_u and the location imprecision cost C_e. The imprecision cost is the average of the Euclidean distance between the actual location and the estimated location. We have

$$C_u = \sum_{i \in S} \pi_i c_u^i \quad (12)$$

$$C_e = \sum_{i \in S} \pi_i c_e^i \quad (13)$$

where c_u^i and c_e^i are the normalized update cost and imprecision cost for each state, which are defined as follows:

$$c_u^i = \sum_{t=t_0}^{t_{n-1}} \overline{c}_t \Big/ n \quad (14)$$

where n is the number of time slots, $\overline{c}_t$ is 1 if the update occurs at time slot t and 0 otherwise.

$$c_e^i = \sum_{t=t_0}^{t_{n-1}} d_t \Big/ k, \quad i \in \{L, R\} \quad (15)$$

$$c_e^i = \sum_{t=t_0}^{t_{n-1}} d_t \Big/ l, \quad i \in \{P\} \quad (16)$$

where d_t is the Euclidean distance between the estimated location and the real location at time slot t. k is the number of predictions occurred for *Linear* and *Random* states policies. For *Pause* state policy the imprecision cost equation is calculated differently since *Pause* state policy does not have predictions, so in (16) l is the number of updates occurred during *Pause* state duration time.

4 Simulation Results

With the mobility-pattern based localization update algorithms (MBLUA), we make a trade-off between update cost and imprecision cost. We expect that when update cost is low (not updating too much), imprecision cost will increase indicating too much error as a node moves. On the other hand, if update cost is high (update too often), the imprecision cost will decrease indicating a low error.

Our experiment is divided into two parts. The first part concerns the simulation of the node's actual movement. Our model is based on GMRW mobility model [6]. In order to analyze the algorithms, we modify BonnMotion [10], and use our mobility model to generate the workload for different mobility pattern scenarios. The next mobility pattern (the next state) is randomly chosen. However, the probability that the next mobility pattern is *Linear* or *Random* is higher than the probability of *Pause* through the entire simulation. The probability of *Linear* or *Random* is 0.4 for each, and the probability of *Pause* is 0.2. The second part is localization update algorithms, and they are run over the workload generated from the mobility model. Localization update algorithms include algorithm for three states combined using the State-based Mobility Model (SMM), and algorithm for each movement state. The purpose of each state algorithm alone is to compare it with MBLUA.

We use a simulation area of 200 by 200 meters and simulate the mobility model over 10 nodes for 900 seconds. The localization scheme should not affect our proposed model, since the model takes care of the localization period only and it is mobility based dependent.

Fig. 2 (a) and (b) show the trace results for one slow node and one fast node, respectively, including the actual movement trace from modified GMRW model and the MBLUA trace from localization update algorithms for two different speed ranges. For slow node, the speed range is form 0.5 m/s to 1 m/s; for fast node, the speed range is from 2 m/s to 4 m/s. The maximum pause time is 60 seconds. The results indicate a perfect match between actual movement and MBLUA trace when the movement is linear and it tends to have some mismatch when the node turns abruptly. In Fig. 2 (b), the error increases for fast node.

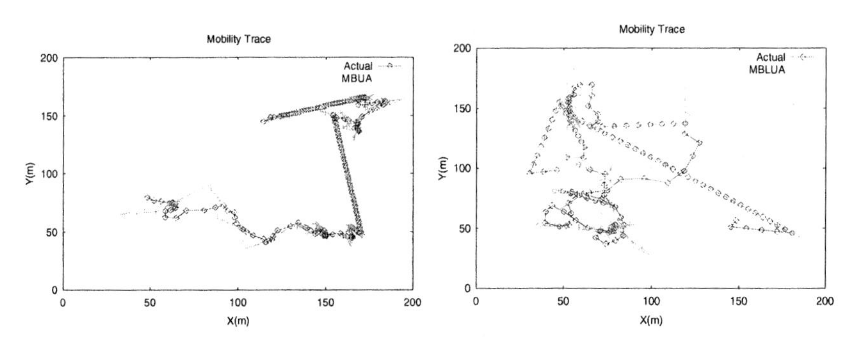

(a) Slow speed (0.5-1 m/s) (Max pause time 60 sec.) (*max_time* = 5 time slots, *Lin_thresh* = 4m, and *d_thresh* = 5m)

(b) Fast speed (2-4 m/s) (Max pause time 60 sec.) (*max_time* = 5 time slots, *Lin_thresh* = 4m, and *d_thresh* = 5m)

Fig. 2. Mobility trace for slow and fast speeds

Table 1 shows the update and imprecision costs for three update policies and our model for the fast and the slow movement with the same parameters. *Pause* state policy has the least update cost but the most imprecision cost. Depending on the mobility motion percentage, if the motion is more linear, then the lowest imprecision cost is *Linear* state policy; and if the motion is more random, then the lowest imprecision cost is *Random* state policy. However, our MBLUA has considerably low update cost with an acceptable imprecision cost between *Linear* and *Random* policies.

Table 1. Policies update and imprecision costs comparison

Policy	Fast node (2-4 m/s)		Slow node (0.5-1 m/s)	
	Update cost	Imprecision cost	Update cost	Imprecision cost
MBLUA	0.373776	10.19647	0.241907	5.388964
Pause-state	0.176768	18.63553	0.176768	10.07666
Linear-state	0.464646	6.349763	0.40404	3.734811
Random-state	0.208122	16.60858	0.177665	8.137531

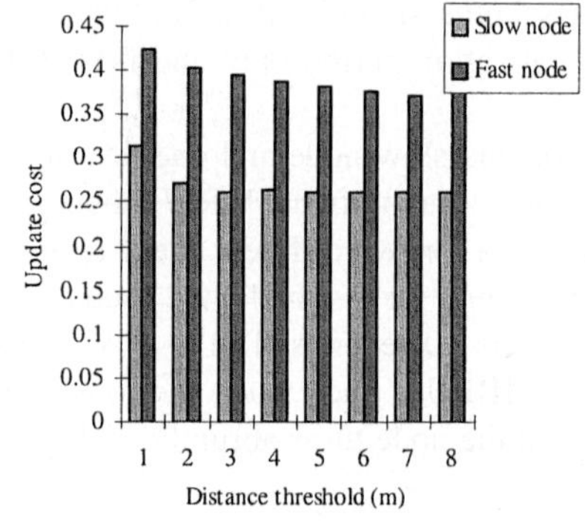

(a) Update cost vs. *d_thresh* (*max_time*=5 time slots and *Lin_thresh*=4m)

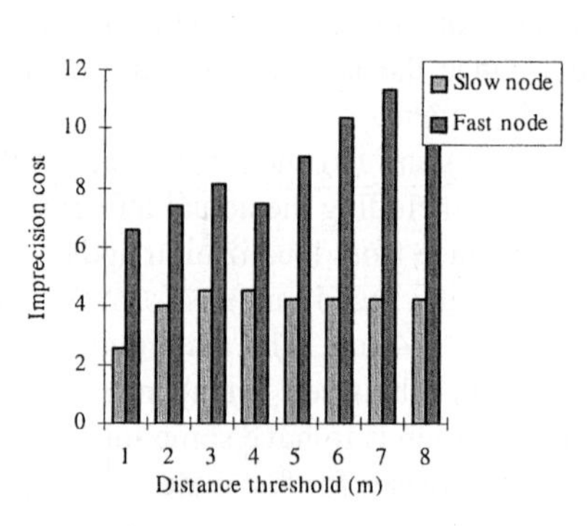

(b) Imprecision cost vs. *d_thresh* (*max_time*=5 time slots and *Lin_thresh*=4m)

Fig. 3. Update and imprecision cost as a function of *d_thresh*

Fig. 3 (a) and (b) show the update and imprecision costs for MBLUA model when varying the *d_thresh*. It was observed that, when *d_thresh* is low, the update cost is high indicating more frequent updates, and the imprecision cost is low since location knowledge is updated more frequent. As *d_thresh* increases, the update cost decreases and the imprecision cost increases, since a larger *d_thresh* value results in larger wait time to localize. We have found that changing *Lin_thresh* only have little impact on the results because *Linear* state policy increase the waiting time to the maximum time as long as the prediction error is less than *Lin_thresh*, which is the case that the error value is close to zero for linear movement.

From the results of Fig. 3, it is observed that any *d_thresh* value greater than 5 does not have effect on the update cost meanwhile imprecision cost does increase. The reason is that the maximum time (*max_time*) to localize is fixed in the simulations. As a result, when *d_thresh* is greater than5, the sensor node will localize after *max_time* period no matter how large *d_thresh* value is, which stabilizes the update cost. Consequently, when the update cost stabilizes, the main factor of changing the imprecision cost is the acceleration value.

Fig. 4 shows the update and imprecision costs for MBLUA model when varying the maximum wait time to localize. With low value of maximum wait time, the update cost is high and the imprecision cost is low since we force the algorithm to localize very often. As the maximum time increases the update cost decreases and the imprecision cost increases because the algorithm will have more possible time to increase the wait time to localize and hence will decrease the localization update frequency. But this causes more error as the wait time increases.

Our experiments show that fast movement has more update and imprecision costs. When a mobile node moves faster, the estimation deviates from the real location in higher rates, which causes larger estimation errors. Moreover, as long as the error value exceeds the predefined error thresholds (*d_thresh* or *Lin_thresh*), a node will localize more frequently.

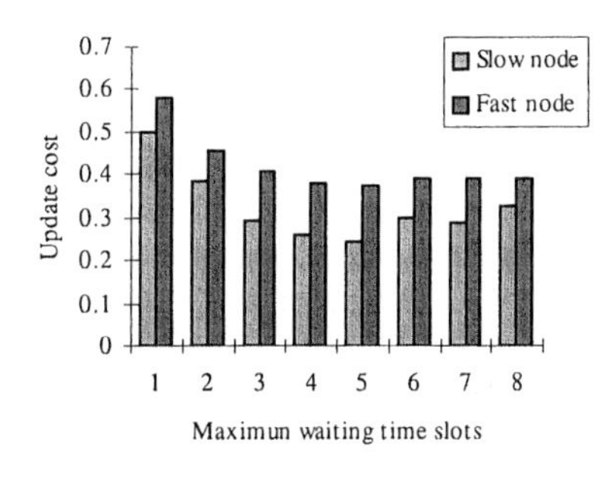

(a) Update cost vs. *max_time* (*Lin_thresh*=4m, *d_thresh*=5m)

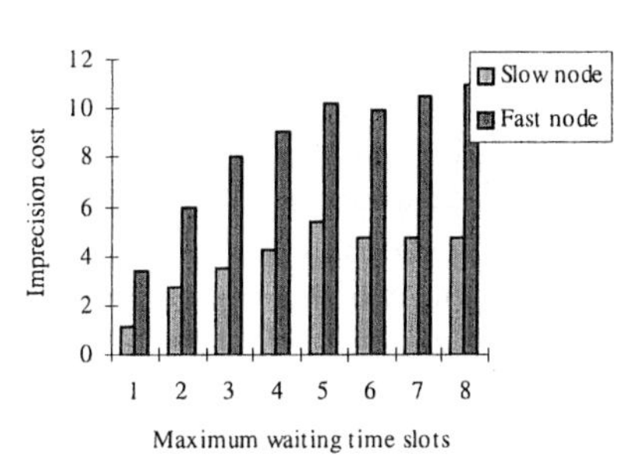

(b) Imprecision cost vs. *max_time* (*Lin_thresh*=4m, *d_thresh*=5m)

Fig. 4. Update and imprecision costs as a function of maximum waiting time slots (*max_time*)

5 Conclusions

We designed three mobility-pattern based localization update algorithms for mobile wireless sensor networks. The main idea is to divide sensor movements into three states, *Pause*, *Linear*, and *Random*. Based on the nature of each movement pattern, different localization update algorithm is adopted at each state. To verify the design, we have performed both mathematical analysis and simulations. The results demonstrate how the localization cost is minimized and how the location accuracy is improved. This research is significant to conserve the power consumption in sensor nodes and to track the location of mobile sensors in a real-time manner.

References

1. S. Tilak, V. Kolar, N. B. Abu-Ghazaleh, and K. D. Kang, "Dynamic Localization Control for Mobile Sensor Networks," *Proc. of IEEE International Workshop on Strategies for Energy Efficiency in Ad Hoc and Sensor Networks*, Apr. 2005.
2. M. Song, K. Park, J. Ryu, and C. Hwang, "Modeling and Tracking Complexly Moving Objects in Location-Based Services," *Journal of Information Science and Engineering*, vol. 20, no. 3, pp. 517–534, May 2004.
3. V. Kumar, and S. R. Das, "Performance of dead reckoning-based location service for mobile ad hoc networks," *Wireless Communications and Mobile Computing Journal*, vol. 4, no. 2, pp. 189–202, Mar. 2004.
4. L. Hu, and D. Evans, "Localization for Mobile Sensor Networks," *Proc. of Tenth Annual International Conference on Mobile Computing and Networking*, pp. 45–57, Sep. 2004.
5. I. F. Akyildiz, W. Su, Y. Sankarasubramaniam, and E. Cayirci, "Wireless sensor networks: a survey," *Computer Networks*, vol. 38, no. 4, pp. 393–422, Mar. 2002.
6. T. Camp, J. Boleng, and V. Davies, "A Survey of Mobility Models for Ad Hoc Network Research," *Wireless Communication & Mobile Computing, Special issue on Mobile Ad Hoc Networking: Research, Trends and Applications*, vol. 2, no. 5, pp. 483–502, Apr. 2002.
7. A. Savvides, C. C. Han, and M. B. Srivastava, "Dynamic fine-grained localization in ad-hoc networks of sensors," Proc. of the 7th ACM/IEEE International Conference on Mobile Computing and Networking (MobiCom), pp. 166–179, Jul. 2001.
8. A. Agarwal, and S. R. Das, "Dead reckoning in mobile ad hoc networks," Proc. of the IEEE Wireless Communications and Networking Conference, Mar. 2003.
9. V. W. S. Wong, and V. C. M. Leung, "An adaptive distance-based location update algorithm for next-generation PCS networks," IEEE J. Select Areas on Communications, vol. 19, no. 10, pp. 1942–1952, Oct. 2001.
10. BonnMotion, http://web.informatik.uni-bonn.de/IV/Mitarbeiter/dewaal/BonnMotion/

Accurate Time Synchronization for Wireless Sensor Networks

Hongli Xu, Liusheng Huang, Yingyu Wan, and Ben Xu

Depart. of Computer Science and Technology, Univ. of Science & Technology of China,
Anhui Province-Most Co-Key Lab. of High Performance Computing and Its Application,
Hefei 230027, P.R. China
hlxu3@mail.ustc.edu.cn, {lshuang, yywan}@ustc.edu.cn

Abstract. Time synchronization is a critical topic in wireless sensor networks for its wide applications, such as data fusion, TDMA scheduling and cooperated sleeping, etc. In this paper, we present an accurate time synchronization (ATS) algorithm using linear least square for sensor networks. Unlike the previous protocols, all nodes aren't synchronized to some reference nodes or sink node, but to a virtual clock. Moreover, each pair of the nodes are synchronized each other. The main advantage of ATS is simple and accurate. The variance of the synchronized drift error is no more than $D_s * \delta / 2\beta^2$, where D_s is the depth of the network, δ is the maximal variance of the link delay and β is the sampling interval. The experiments show the high precision compared with the previous algorithms.

1 Introduction

Recent advances in wireless communications and electronics have enabled to deve- lop the low-cost, low-power, multifunctional sensor nodes that are small in size and communicate in short distance [1]. These tiny nodes, which are capable of sensing, processing and communication, constitute the sensor networks. The main motivation to deploy the sensor network is the requirement for remote surveillance in the desolate environments, such as battling fields and forests.

Time synchronization is an important requirement for all kinds of distributed systems. It is also required in sensor networks for several reasons. First, sensor nodes need to coordinate their operations to achieve a complex task. Data gathering [2] is such an example in which sensed data from sensor nodes are fused into a final result. Secondly, time synchronization helps to preserve the energy, thus prolonging the network's lifetime. For example, the nodes are scheduled to sleep and wake up at the synchronized times [3]. Thirdly, scheduling algorithm such as TDMA is used to share the transmission medium in the different time slots to eliminate transmission colli- sion. Thus, synchronization is a critical part of resource scheduling and management.

Synchronization problem has been deeply studied in traditional distributed system and LAN. Several technologies have been used to provide global synchronization in networks, such as GPS, etc. Complex protocol such as NTP [5] has been to keep the internal clocks ticking in phase. However, the requirements of time synchronization in

X. Jia, J. Wu, and Y. He (Eds.): MSN 2005, LNCS 3794, pp. 153 – 163, 2005.

sensor network differ from the traditional networks. In general, each node is resource-constrained, such as energy and bandwidth etc. Additionally, the sensor networks are often deployed with high density. To operate in such environment, time synchronization should be scalable with the number of nodes. Furthermore, it is required to reach synchronization with high precision among the nodes involved in a task, such as target tracking.

In this paper, we present an accurate time synchronization (ATS) protocol for wireless sensor networks. Each node synchronizes to a virtual clock, which is maintained by two root nodes. In the protocol, the pair of nodes is synchronized to each other using linear least square. The analysis shows that the variance of the node's drift error is no more than $D_s * \delta / 2\beta^2$, where D_s, δ and β are the depth of the net- work, the maximal variance of the link delay and the sampling interval respectively. The experiments show the efficiency and high precision of the ATS protocol.

The rest of the article is organized as follow. The next section introduces some preliminary background for ATS protocol and reviews the related works. Section 3 shows the feasibility of computation of the virtual clock between two nodes. By these analyses, a novel time synchronization protocol is presented for sensor networks in section 4. We analyze the error generated by the protocol in section 5, and give the experimental simulations in section 6. Section 7 concludes the paper.

2 Preliminary Background

2.1 Synchronization Problem

The wireless sensor network consists of *n* wireless sensor nodes, where *n* is the number of nodes. Each node *i* is equipped with a hardware clock $C_i(t)$, which is a function from the real time to the local time. The drift of a perfect clock will equal 1, but all clocks are subject to clock drift. Generally, the clock can be approximated with good accuracy using an oscillator. The hardware clock for node *i* can be modeled as: $C_i(t) = a_i t + b_i$, where a_i and b_i are the drift and offset of this clock from real time *t* respectively. To synchronize, each node *i* keeps two variables, rd_i and ro_i. A new local clock can be computed as $C_i^{'}(t) = a_i^{'} t + b_i^{'} = rd_i \times C_i(t) + ro_i$. Thus the task of synchronization is to compute the rd_i and ro_i for node *i*, so that the new local clocks of all nodes can be synchronized.

Definition 1. two nodes *i* and *j* are synchronized, only if $C_i^{'}(t) = C_j^{'}(t)$.

To measure the synchronization, consider the concept of drift accuracy.

Definition 2. the drift accuracy of the protocol refers to the difference between the maximal and the minimal new drift among the entire network. That is:

$$\text{Drift Accuracy} = Max\{a_i^{'}\} - Min\{a_i^{'}\},\ 1 \le i \le n.$$

2.2 Linear Least Square

Here we introduce some basic concepts of the linear least square method. Given a set of m value-pairs, $\{(x_i, y_i), 1 \le i \le m\}$, there are several methods to fit these data to the function $y = F(x)$. Usually, it is required that the sum of the squares of the deviation of all data is minimized. That is, $R^2 = \sum_{i=1}^{m}(y_i - F(x_i))^2$ is minimal among all functions. It is called linear least square, if F(x) is a linear function. Let $F(x) = ax + b$, R^2 can be formulated as following:

$$R^2 = \sum_{i=1}^{m}(y_i - ax_i - b)^2 \tag{1}$$

In order to minimize R^2, the derivative of equation (1), which is calculated over the variables a and b, should be zero. That is:

$$\frac{\partial}{\partial a}\sum_{i=1}^{m}(ax_i + b - y_i)^2 = 2(a\sum_{i=1}^{m}x_i^2 + b\sum_{i=1}^{m}x_i - \sum_{i=1}^{m}x_i y_i) = 0$$

$$\frac{\partial}{\partial b}\sum_{i=1}^{m}(ax_i + b - y_i)^2 = 2(a\sum_{i=1}^{m}x_i + mb - \sum_{i=1}^{m}y_i) = 0$$

The above two equations can be represented as following:

$$\begin{pmatrix} m & \sum_{i=1}^{m}x_i \\ \sum_{i=1}^{m}x_i & \sum_{i=1}^{m}x_i^2 \end{pmatrix} \times \begin{pmatrix} b \\ a \end{pmatrix} = \begin{pmatrix} \sum_{i=1}^{m}y_i \\ \sum_{i=1}^{m}x_i y_i \end{pmatrix} \tag{2}$$

The variables a and b can be figured out by equation 2 easily. Therefore, we obtain the function F(x).

2.3 Related Works

This section discusses the previous protocols of time synchronization for wireless sensor networks. Among all researchers, Elson and Estrin [6] first address time synchronization problem in sensor networks. For the special features, Elson et al. [7] show that it is a completely different regime for time synchronization in sensor networks. They also point out remarkable differences between sensor networks and traditional networks in synchronization requirements. A comprehensive survey is introduced in [4]. Generally, the problem of time synchronization can be divided into three models.

The simplest model of synchronization only concentrates on maintaining the relative notion of times among nodes. Thus the aim of this model is not to synchronize all local clocks, but to generate a right chronology for event. A scheme for ad hoc networks based on this model is presented in [8]. This algorithm is only initiated when the event takes place in the monitoring area. However, this model cannot be extended to the case where the local time of each node is required, such as target tracking, etc.

A complex model is to maintain relative clocks among all nodes. In this model, though every node keeps an individual clock, these clocks are not synchronized with each other. Instead, every node stores information about the relative offsets between its clock and that of any other nodes in the networks. A scheme based on this model is RBS (Reference Broadcast Synchronization) [9], where sensor nodes periodically send beacons to their neighbors using the network broadcast. Recipients use the message's arrival time as point of reference for comparing their clocks. The offset between any pair of nodes is calculated by exchanging the local timestamps.

The most complex model is the one where each node maintains a clock that is synchronized with a reference node in the network. The aim is to maintain a global time-scale throughout the networks. Some time synchronization protocols [10-14] belong to this scheme. Sichitiu et al. [11] present TINY/MINI-SYNC protocol based on pair-wise synchronization. The exchanged packets are used to estimate the best-fit offset between two nodes. But the authors don't analyze the performance of the protocol theoretically. Lightweight tree-based synchronization [12] is also based on pair-wise synchronization scheme. The authors show that the expected error is zero, and its variance is $4D_s * \delta$. Santashil et al. [14] present a probabilistic method for clock synchronization that uses the receiver-to-receiver synchronization. The protocol can be extended to multi-hop sensors networks by level architecture.

3 Computing the Virtual Clock

Assume that there is a group of p nodes, labeled from 1 to p in turn. Different nodes maintain the different local clocks, which are determined by the clock's drifts and offsets. For real time t_j, the corresponding local time of node i is $T_{i,j}$. If there are m sampling real times, t_1, ..., t_m, it produces $p \times m$ time-pairs, which can be represented by matrix $A_{p \times m}$.

$$A_{p \times m} = \begin{pmatrix} (t_1, T_{1,1}) & (t_2, T_{1,2}) & \cdots & (t_m, T_{1,m}) \\ (t_1, T_{2,1}) & (t_2, T_{2,2}) & \cdots & \cdots \\ \cdots & \cdots & \cdots & \cdots \\ (t_1, T_{p,1}) & (t_2, T_{p,2}) & \cdots & (t_m, T_{p,m}) \end{pmatrix}$$

Each local clock is a linear function of the real time, so the virtual clock is regarded as a linear function of the real time too. Assume that the virtual clock is $V(t) = a_v t + b_v$, where t is the real time. By equation 2, the variables a_v and b_v satisfy that:

$$\begin{pmatrix} C_1 & C_2 \\ C_2 & C_3 \end{pmatrix} \times \begin{pmatrix} b_v \\ a_v \end{pmatrix} = \begin{pmatrix} D_1 \\ D_2 \end{pmatrix} \tag{3}$$

where $C_1 = m * p$, $C_2 = p\sum_{j=1}^{m} t_j$, $C_3 = p\sum_{j=1}^{m} t_j^2$

$$D_1 = \sum_{i=1}^{p}\sum_{j=1}^{m} T_{i,j}, D_2 = \sum_{i=1}^{p}\sum_{j=1}^{m} t_j T_{i,j} \tag{4}$$

Observe the equation (4), D_1 and D_2 can be simplified as following:

$$D_1 = \sum_{i=1}^{p}(a_i \sum_{j=1}^{m} t_j + mb_i) = \sum_{i=1}^{p} a_i \sum_{j=1}^{m} t_j + m\sum_{i=1}^{p} b_i \quad (5)$$

$$D_2 = \sum_{i=1}^{p} \sum_{j=1}^{m} t_j (a_i t_j + b_i) = \sum_{i=1}^{p} a_i \sum_{i=1}^{m} t_j^2 + \sum_{i=1}^{p} b_i \sum_{j=1}^{m} t_j \quad (6)$$

Combined with the above equations, the equation (3) can be solved:

$$b_v = (C_3 D_1 - C_2 D_2)/(C_1 C_3 - C_2^2) = \sum_{i=1}^{p} b_i / p \quad (7)$$

$$a_v = (C_1 D_2 - C_2 D_1)/(C_1 C_3 - C_2^2) = \sum_{i=1}^{p} a_i / p \quad (8)$$

Thus the virtual clock is irrelative to the sampling times. Next, we explain how each node synchronizes to the virtual clock efficiently. Similarly, for node *k*, it can obtain $m \times p$ time-pairs represented by matrix $A'_{p\times m}$.

$$A'_{p\times m} = \begin{pmatrix} (T_{k,1}, T_{1,1}) & (T_{k,2}, T_{1,2}) & \cdots & (T_{k,m}, T_{1,m}) \\ (T_{k,1}, T_{2,1}) & (T_{k,2}, T_{2,2}) & \cdots & \cdots \\ \cdots & \cdots & \cdots & \cdots \\ (T_{k,1}, T_{p,1}) & (T_{k,2}, T_{p,2}) & \cdots & (T_{k,m}, T_{p,m}) \end{pmatrix}$$

Now the task is to compute the relative drift a_g and offset b_g by matrix $A'_{p\times m}$. Thus, the coefficients a_g and b_g satisfy that:

$$\begin{pmatrix} C_1' & C_2' \\ C_2' & C_3' \end{pmatrix} \times \begin{pmatrix} b_g \\ a_g \end{pmatrix} = \begin{pmatrix} D_1' \\ D_2' \end{pmatrix}$$

where $C_1' = m * p$, $C_2' = p\sum_{j=1}^{m} T_{k,j}$, $C_3' = p\sum_{j=1}^{m} T_{k,j}^2$

$$D_1' = \sum_{i=1}^{p} \sum_{j=1}^{m} T_{i,j}, D_2' = \sum_{i=1}^{p} \sum_{j=1}^{m} T_{k,j} T_{i,j} \quad (9)$$

For each node *k*, $T_{k,j} = a_k \times t_j + b_k$. The variables a_g and b_g can be expressed as following: $b_g = b_v - b_k * a_v / a_k$ and $a_g = a_v / a_k$. The new clock of node *k* actually is:

$$C_k(t) = a_g \times T_k + b_g = a_v / a_k \times (a_k \times t + b_k) + (b_v - b_k \times a_v / a_k)$$
$$= a_v t + b_v$$

By the above method, the node *k* can synchronize the local clock to the virtual clock exactly. In this paper, the algorithm only considers the simplest case in which *m* and *p* are both 2. That is, two nodes are synchronized by exchanging two pairs of message each other.

4 Accurate Time Synchronization Protocol

The main idea of the protocol is that the drift variance can be improved by synchronizing between two nodes each other. ATS protocol works in three phases: synchronization tree discovery, local synchronization and global synchronization.

4.1 Synchronization Tree Discovery Phase

The first phase occurs as the synchronization is necessary in sensor networks. In this protocol, the synchronization tree (Sync-tree) is required to contain two root nodes, which is the main difference from the previous tree construction. These two nodes should be connected to generate the global virtual clock. At first, the system appoints one node as the root. Then it selects one neighbor with maximal power as the other root. When one root is failed, the other should select another node as the new root, and broadcast this information in sensor network. The topology can be derived from the traditional spanning tree easily. In order to improve the drift precision, an optimal Sync-tree is one with minimum depth. The synchronization should occur in parallel along all branches so that the procedure finishes in similar time. One type of tree construction that satisfies both properties is width-first tree or level discovery [12]. Time synchronization happens on each edge of Sync-tree. An example of sync- tree topology is shown in Fig. 1.

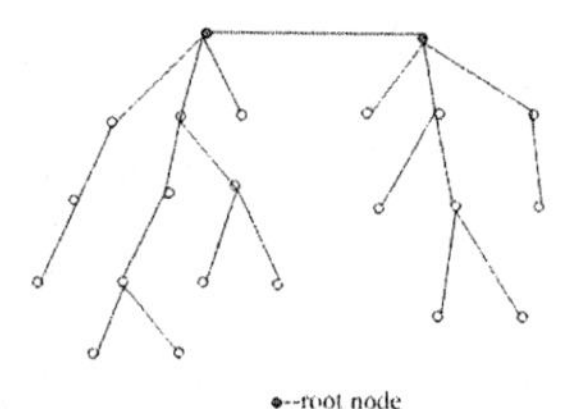

Fig. 1. An example of Sync-Tree Topology

4.2 Local Synchronization

Local synchronization refers to synchronize two connected nodes with each other in the Sync-tree. Each node will do the following steps to synchronize itself to a virtual clock. In the following, assume that node *i* and *j* are connected in the Sync- tree. The formal description of the protocol is given as follows:

1. Node i sends two reference packets to node j at local times $T_{i,1} = t_0$, $T_{i,2} = t_0 + \beta$ respectively. Here, β is called sampling interval.

2. When receiving the *k*-th reference packet from node *i*, the node *j* sends a synch-packet to *i* attached with local time $RT_{j,k}$ immediately.

3. On receiving the *k*-th synch-packet from node *j*, node *i* records the local time $T'_{i,k}$.

4. After receiving all synch-packets, the node *i* handles as following:

4.1 The delay of the link *ij* in the *k*-th sampling is $\Delta_k = (ST_{i,k} - T_{i,k})/2$, where $k = 1,2$.

4.2 Node *i* adjusts the local times $T_{j,k}$ of node *j* corresponding to the sampling time $T_{i,k}$, $T_{j,k} = RT_{j,k} - \Delta_k$, where $k = 1,2$.

4.3 To minimize the drift error, node *i* generates four time-pairs, denoted by A_a, to compute the relative drift a_g using linear least square.

$$A_a = \begin{pmatrix} (-\beta/2, T_{i,1}) & (\beta/2, T_{i,2}) \\ (-\beta/2, T_{j,1}) & (\beta/2, T_{j,2}) \end{pmatrix}, \quad A_b = \begin{pmatrix} (T_{i,1}, T_{i,1}) & (T_{i,2}, T_{i,2}) \\ (T_{i,1}, T_{j,1}) & (T_{i,2}, T_{j,2}) \end{pmatrix}$$

4.4 For the offset, node *i* generates another four time-pairs, denoted by A_b, to compute the relative offset b_g by equation 11, where a_g is known.

By now, the node *i* can compute its new local clock as following: $T_i' = a_g \times T_i + b_g$, where T_i is the local clock of node *i*. The message exchanging for per synchronization is shown in Fig. 2. Similarly, node *j* can be synchronized to this virtual clock with the above procedure.

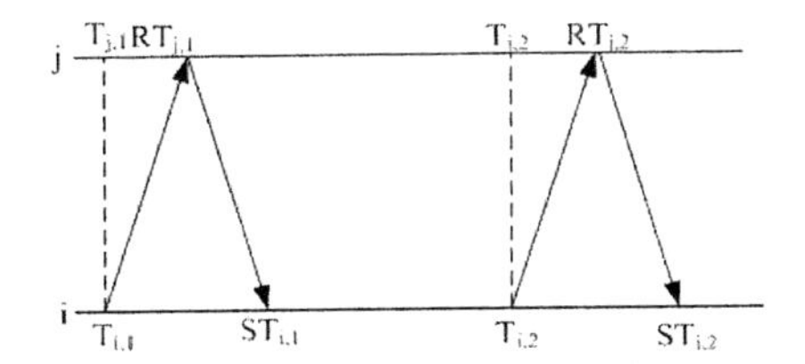

Fig. 2. The message exchanging for local synchronization of node *i*

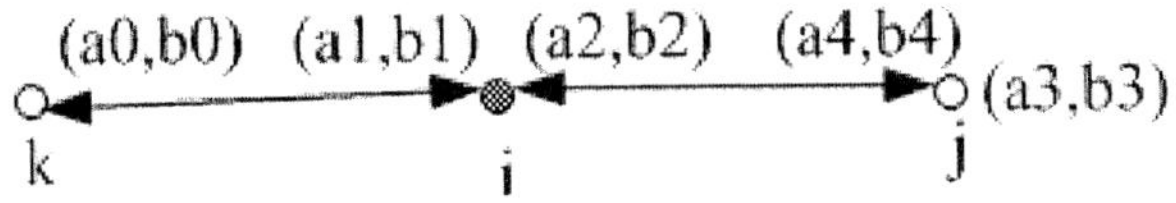

(a0,b0) (a1,b1) (a2,b2) (a4,b4) (a3,b3)
k i j

Fig. 3. Illustration for Global Synchronization

4.3 Global Synchronization

Each two connected nodes in the sync-tree have synchronized to a virtual clock by local synchronization, while different pairs of nodes maintain different virtual clocks. The global synchronization is to let all nodes in the networks keep the unique virtual clock maintained by two root nodes. As shown in Fig. 3, node *i* is connected with nodes *k* and *j* in the sync-tree. This node has synchronized with node *k* and *j* using $(a1, b1)$ and $(a2, b2)$ respectively. Moreover, node *j* has synchronized with node *i* using $(a4, b4)$. Global synchronization is to make node *j* synchronize with *k* by computing the new relative drift and offset $(a3, b3)$.

Node j is connected with node i directly in the sync-tree. Hence, these two nodes keep synchronization both in local and global synchronization. That is:

$$a2*(a_i*t+b_i)+b2=a4*(a_j*t+b_j)+b4 \tag{10}$$

$$a1*(a_i*t+b_i)+b1=a3*(a_j*t+b_j)+b3 \tag{11}$$

From equation (10) and (11), $a3$ and $b3$ can be figured out:

$$\begin{cases} a3=(a1*a4)/a2 \\ b3=a1*(b4-b2)/a2+b1 \end{cases} \tag{12}$$

In this way, node j can resynchronize to another virtual clock using the new drift and offset, $a3$ and $b3$. Hence, all nodes of entire network can be synchronized to the single virtual clock.

5 Error Analysis

This section mainly analyzes the drift error of ATS protocol. The source of clock synchronization error consists of four components: send time, access time, propagation time and receive time [10]. For simplicity, all these errors are denoted as delay error. Without loss of generality, assume that the expectation of the delay error is zero, and its variance is δ. The delay error of the directed link ij in the k-th sampling is denoted as $d_{i,j,k}$. We assume that all error variables are independent.

In ATS protocol, the error of $T_{j,k}$ can be expressed as $\Delta T_{j,k}=(d_{i,j,k}+d_{j,i,k})/2$. Consider the error of drift a_g. By equation 9, $C_1'=4$, $C_2'=0$ and $C_3'=\beta^2$. What's more, the error of D_1' and D_2' are $\Delta T_{j,1}+\Delta T_{j,2}$ and $\beta*(\Delta T_{j,2}-\Delta T_{j,1})/2$ respectively. Thus the error of a_g is:

$$\begin{aligned}\Delta a_g &= (C_1'\Delta D_2'-C_2'\Delta D_1')/(C_1'C_3'-C_2'^2) \\ &= \Delta D_2'/C_3' = (\Delta T_{j,2}-\Delta T_{j,1})/2\beta \\ &= (d_{i,j,2}+d_{j,i,,2}-d_{i,j,1}-d_{j,i,1})/4\beta\end{aligned}$$

The expectation of the drift error is zero, and its variance is $\delta/4\beta^2$. For the global synchronization, assume that the height of the sync-tree is D_s. The variance of the drift error increases along each branch of the tree. By equation 12, assume that the errors of $a1$, $a2$ and $a4$ are $\Delta a1$, $\Delta a2$ and $\Delta a4$ respectively. Thus the error of $a3$ is:

$$\begin{aligned}\Delta a3 &= (a1+\Delta a1)(a4+\Delta a4)/(a2+\Delta a2)-a1*a4/a2 \\ &\approx \Delta a1+\Delta a2+\Delta a4\end{aligned}$$

As we know, $Var(\Delta a2)=Var(\Delta a4)=\delta/4\beta^2$. By recursion, the variance of the drift of entire network is $(2D_s-1)*\delta/4\beta^2 < D_s*\delta/2\beta^2$.

6 Experimental Simulations

This section will evaluate the performance of ATS protocol by comparing with the typical pair-wise synchronization protocols in which each node synchronizes to the father node in the tree [10,12]. The local clock of each node has a fixed drift rate between -50ppm to +50ppm, which is typical for quartz crystals. The error of the link delay is the random Gaussian distribution with a means of 0.1ms and a standard variance of 11 μs . The measurements for the experiments are drift accuracy and absolute time difference among all nodes in the networks. We execute each simulations 50 times, and average the results.

The first two experiments show how the hop number of the network impacts the drift accuracy and time difference, as shown in Fig. 4 and 5. In these two experiments, the sampling interval and elapsed time are fixed, 20s and 86400s respectively. The plotted results show that the performance of both protocols changes similarly. In both figures, the drift accuracy and time difference become larger with the hop number increasing from 1 to 5, which is compatible with the theoretical analysis. For exam- ple, the drift accuracy of the one-hop nodes is about 40% that of five-hop. The results also show that ATS protocol improves about 37% that of typical pair-wise method.

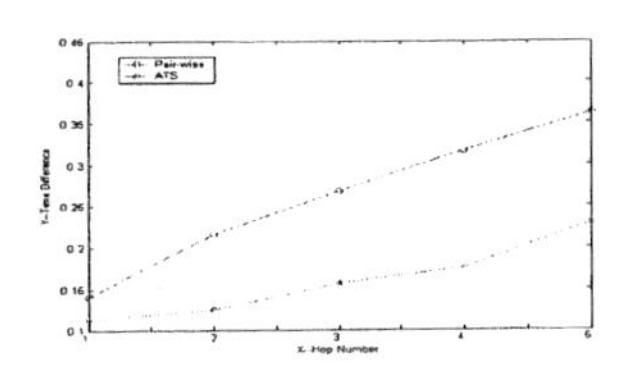

Fig. 4. Hop number vs. Drift accuracy

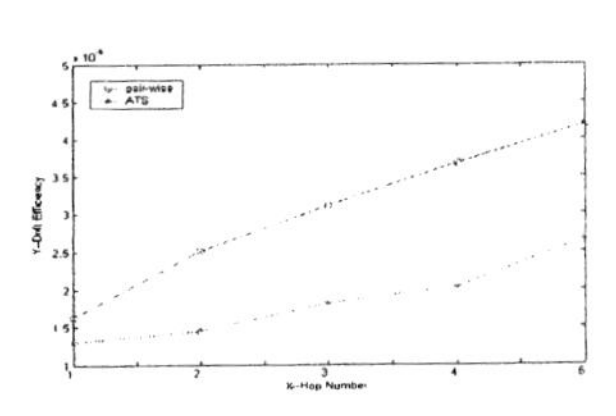

Fig. 5. Hop number vs. Time difference

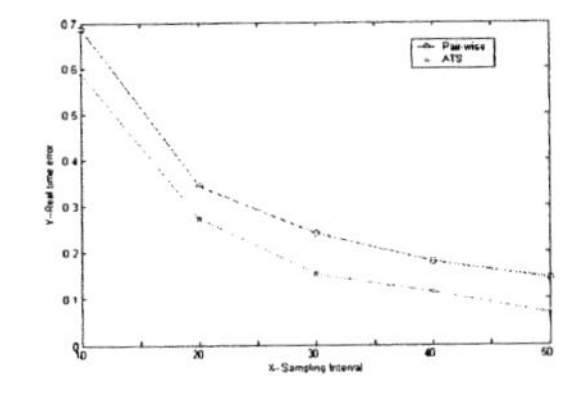

Fig. 6. Sampling interval vs. Drift accuracy

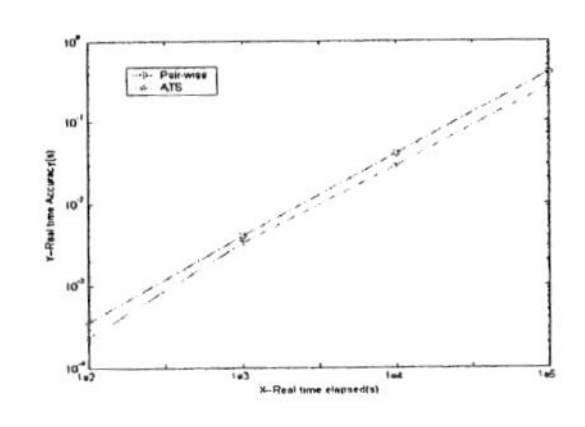

Fig. 7. Sampling Interval vs. Time difference

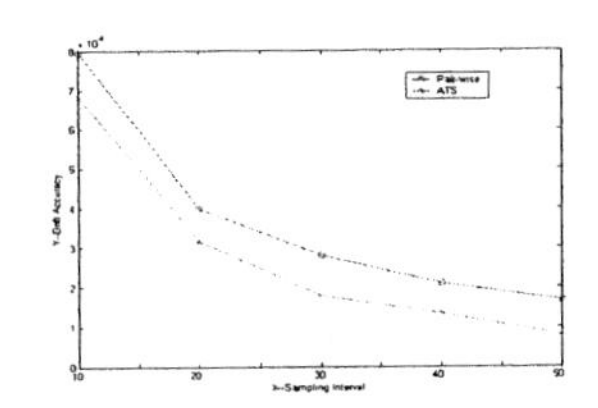

Fig. 8. Elapsed time vs. Time difference

The following two experiments explore how the sampling interval and elapsed time impact the time difference among the network. Also, the simulations are setup on 5-hop networks. First, we change the sampling interval from 10s to 50s, plotted in Fig. 6. The result shows that the drift of the network becomes more precious as the interval is much larger. Time difference is relative to the elapsed time period. Fig. 7 shows the time difference as the sampling interval when the elapsed time is exactly one day, i.e., 86400s. This figure is similar to Fig. 8, that's because the time difference is mainly determined by the drift accuracy. Both figures show that ATS protocol improves about 42% time difference of previous algorithm.

The last experiment studies how the elapsed time impacts the time difference. Fig. 8 displays the simulation results when the elapsed time periods are 10^2 s, 10^3 s, 10^4 s and 10^5 s respectively. The time difference is almost linear as the elapsed time. The reason is that the offset of each clock is too small to impact the time difference. By this figure, the time difference by ATS protocol can be estimated for the certain elapsed time.

7 Conclusions

This paper presents a novel time synchronization protocol using least square method for wireless sensor networks. Unlike the previous protocols, ATS protocol intends to synchronize each node not to a reference node, but to a virtual global clock. The protocol maintains synchronization with high precision, because the variance of the synchronized drift error is no more than $D_s * \delta / 2\beta^2$. The experimental results show the efficiency and precision of ATS protocol. In the future, we will consider the case where the drift of each local clock is not fixed, but varies among an interval, which is more similar to the reality.

References

[1] D. Estrin, R. Govindan, J. Heidemann, and S. Kumar, "Next century challenges: scalable coordination in sensor networks", ACM Mobicom Conference, Seattle, WA, august 1999

[2] W. Heinzelman, A. Chandrakasan, and H. Balakrishnan, "Energy-efficient communication protocol for wireless sensor networks", In Proceedings of the Hawaii Conference on System Sciences, Jan. 2000

[3] Y. Ting, H. Tian, and A. S. John, "Differentiated surveillance for sensor networks", SenSys'03, Nov. 5-7, 2003, Los Angeles, California, USA

[4] S. Fikret, and Y. Bulent, "Time Synchronization in sensor Networks: A survey", IEEE Network, Vol: 18, Issue: 4, p.45-50, July/August 2004,

[5] D. L. Mills, "Internet time synchronization: the network time protocol", In Z. Yang and T. A. Marsland, editors, global states and time in distributed systems. IEEE Computer Society Press, 1994

[6] J. Elson, and D. Estrin, "Time synchronization for wireless sensor networks", in proceeding of the 2001 International Parallel and Distributed Processing Symposium (IPDPS), workshop on Parallel and Distributed Computing Issues in Wireless Networks and Mobile Computing, San Francisco, CA, Apr. 2001

[7] J. Elson and K. Romer, "Wireless sensor networks: A new regime for time synchronization", ACM SIGCOMM Computer Communication Review, Vol 33, 2003

[8] K. Romer, "Time synchronization in ad hoc networks", in proceeding of ACM MobiHoc, Long Beach, CA, 2001

[9] J. Elson, L. Girod, and D. Estrin, "Find-grained network time synchronization using reference broadcasts", in UCLA Technical Report 020008, Feb, 2002

[10] S. Ganeriwal, R. Kumar and M. Srivastava, "Timing sync protocol for sensor networks", ACM SenSys, Loc Angeles, CA, p.1266-73 Nov, 2003

[11] M. L. Sichitiu, and C. Veerarittiphan, "Simple, accurate time synchronization for wireless sensor networks", IEEE WCNC, p.1266-73, Mar. 2003
[12] J. V. Greunen, and J. Rabaey, "Lightweight time synchronization for sensor networks", Proc. 2nd ACM Int'l Conf. Wireless sensor networks and Apps, San Diego, CA, Sep, 2003
[13] Q. Li, and D. Rus, "Global clock synchronization in sensor networks", IEEE 23rd INFOCOM, p. 564-74, Mar. 2004
[14] S. Palchauduri, A. K. Saha, and D. B. Johnson, "Adaptive clock synchronization in sensor net- works", 3rd International Symposium on Information Processing in Sensor Networks, pp.340-8, Apr. 2004

A Service Discovery Protocol for Mobile Ad Hoc Networks Based on Service Provision Groups and Their Dynamic Reconfiguration

Xuefeng Bai[1], Tomoyuki Ohta[1], Yoshiaki Kakuda[1], and Atsushi Ito[2]

[1] Faculty of Information Sciences, Hiroshima City University,
Ozuka-Higashi 3-4-1, Asa-Minami-ku, Hiroshima, 731-3194 Japan
bai@pe.ce.hiroshima-cu.ac.jp, {ohta, kakuda}@ce.hiroshima-cu.ac.jp
[2] KDDI Corporation,
Iidabashi 3-10-10, Chiyoda-ku,
Tokyo, 102-8460 Japan
at-itou@kddi.com

Abstract. Mobile ad hoc networks can be formed by a group of wireless nodes without requiring the use of any preexisting infrastructure. Therefore nodes of such networks can not assume which services exist and where they are hosted. A service discovery protocol that provides automatic discovery of desired services is particularly important to save user from the trouble of configuration and quick access to services. Recently, a service discovery protocol that uses broadcast and introduces the cache function has been proposed. There was a problem of the number of messages increasing with the increase in the number of clients which performs a service discovery request, and becoming the cause of congestion. In this paper, we propose a new service discovery protocol for mobile ad hoc networks, which is characterized by service provision groups and their dynamic reconfiguration for the purpose of reducing the number of control messages, and a simulation study is presented to show the effect of the proposed protocol.

1 Introduction

A mobile ad hoc network[1] is an autonomous system of mobile routers (and associated hosts) connected by wireless links–the union of which form an arbitrary graph. There are no mobility restrictions on these routers and they can organize themselves arbitrarily resulting in rapid and unpredictable change in the network's topology. A mobile ad hoc network may operate in a standalone fashion or may be connected to the internet. The property of these networks that makes it particularly attractive is that they do not require any prior investment in fixed infrastructure. Instead, the participating nodes form their own co-operative infrastructure by agreeing to relay each other's packets. In general, the application that uses the network can be modeled as a client that needs the use of the service provided on other nodes[2]. In an ad hoc network it becomes

X. Jia, J. Wu, and Y. He (Eds.): MSN 2005, LNCS 3794, pp. 164–174, 2005.

difficult to manage, like the fixed network which constituted in a fixed institution, which kind of service is provided on a network, and client cannot grasp which node provided the service beforehand. A service discovery protocol that provides automatic discovery of desired services is particularly important to save user from the trouble of configuration and quick access to services.

In fixed infrastructure networks, some service discovery protocols are available such as SLP (Service Location Protocol)[3], Salutation[4], Jini[5], and UPnP (Universal Plug and Play)[6]. In a service discovery environment, services advertise themselves, supplying details about their capabilities and information which one must know to access the service (e.g., the IP address). Clients (e.g., word processing software) may locate a service by its service type (e.g., printer) and may make an intelligent service selection in case that multiple services of the desired type are available. Generally, it becomes difficult to apply the fixed center server to an ad hoc network [7][8]. As an alternative, a decentralized approach without relying on particular nodes is proposed. A client broadcasts a service request and each node sends a service reply back. This approach can be an expensive process in an ad hoc network that uses limited wireless resources, since it may cause a large number of broadcast messages to be transmitted all over the network whenever clients perform service discovery request. To minimize spreading of the service discovery request messages, a service discovery protocol which introduced the cache function has been proposed[9]. In this protocol, intermediate nodes which transfer a service reply cache the configuration information included in the service reply. However, this protocol has the potential drawback of lowering the number of discoverable services, since the clients can not discover other services whose configuration information is not cached at the respond nodes.

In this paper, we propose a new service discovery protocol for mobile ad hoc networks. The proposed protocol is characterized by dynamic reconfiguration of service provision groups for the purpose of reducing the number of control messages. We evaluate the proposed protocol through a simulation study and compare it with the existing protocol.

The rest of this paper is organized as follows. Section 2 addresses the existing protocol and the problem. Section 3 describes the proposed service discovery protocol in detail. The evaluation results of proposed protocol are shown by simulation experiments in section 4.

2 Existing Protocol and the Problem

2.1 An Example Scenario of Using Service Discovery

Many kinds of applications will be used in ad hoc networks, such as voice communication, file sharing and so on. These applications need to discover with each other before starting communications. Thus there will be many services that have the same service type in the ad hoc networks. Therefore, the service discovery protocol is required to provide discovery of all services of the same service type. Here as an example scenario, consider a traveler at the service station for airplane, hotel, rental car reservation with the protocol of service discovery at an airport(Fig. 1).

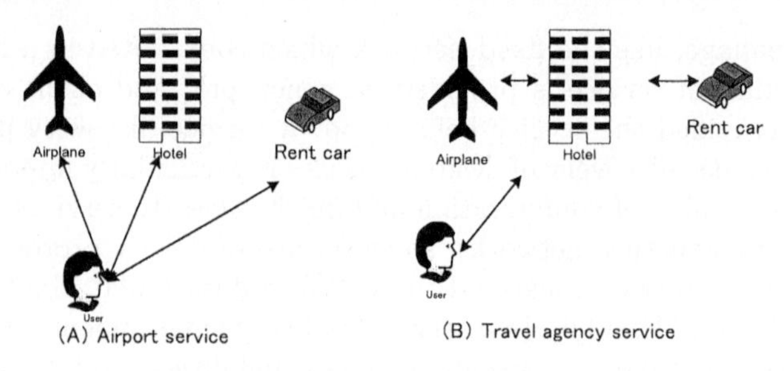

Fig. 1. Scenario of service discovery protocol

The node of the traveler who arrived at the airport has been detected in the ad hoc network by receiving the ad hoc network control message. At this time, the traveler's node does not hold that reservation information, though service exists, offering reservation of an airplane ticket and summary service of lodging. In order to start communication with the node which acquires reservation information and offers service, a service discovery protocol is utilized. The traveler's node specifies the service classification used as the information for discovering the service, and performs the service discovery request. The traveler's node can acquire airplane ticket reservation, lodging, rental car reservation service on an ad hoc network as service reply to this request. From this point, traveler can use these services. In such an ad hoc network, if one service (Fig. 1 (B)), for example, travel agency service exist, traveler which discovers each service will more quickly and the limited wireless resources of ad hoc network can be saved. Therefore, it is required to provide the contents of service dynamically by a user's request.

2.2 Existing Service Discovery

Using centralized nodes is a reasonable approach in fixed infrastructure networks, so existing service discovery protocol such as SLP, Salutation, and Jini mainly use particular nodes which store configuration information of services. It is obvious that an approach relying on such centralized nodes does not work with ad hoc networks, since the ad hoc network must operate without requiring the use of such particular preexisting nodes and/or continuously existing nodes.

(1) Broadcast Method. In SLP, when a client requests service discovery, it broadcasts a service request message specifying a type of the desired service (ex. "lpr", "ftp"). Nodes that provide a service satisfying the specified service type respond with a service reply message including configuration information of the service. Performing such a service discovery can be an expensive process in an ad hoc network which uses limited wireless resources, since it may cause a large number of service request to be transmitted all over the network whenever clients request service discovery[9].

(2) Cache Function. To minimize spreading of the service discovery request messages, a service discovery protocol which introduces the cache function has

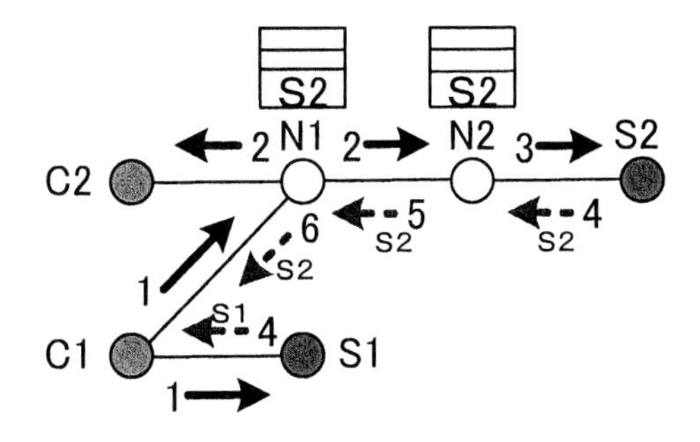

Fig. 2. Process of introduced the cache function

been proposed[8]. As shown in Fig.2, client C1 broadcasts a service request message which specifies the service classification (Fig. 2: 1, 2, 3). S1 and S2 which provide a service of the service classification specified in the service request message send service reply messages by the unicast (Fig. 2: 4, 5, 6). First, S2 unicasts the service reply message to N2 using the routing information implying that the route for delivering the service reply message to C1 passes through N2. Routing information is offered by routing protocols, such as AODV [10]. N2 which received the service reply message carries out the cache of the service information on S2 included in the service reply message, and then unicasts the service reply message to N1 using routing information on C1. N1 as well as N2 uses cache and finally unicasts a service reply message to C1.

When client C2 requests the same service as C1, C2 broadcasts service request messages, and N1 receives this message, because N1 includes the same service information in the cache which C2 requested, so N1 sends a reply message to C2, and does not broadcast service request messages any more. Since the cache function aims at reduction of the number of messages used for a service discovery, the node that has cached the service information also offers service reply, and it can be thus said that the service discovery protocol that introduces the cache function not to broadcast serviced request messages to other nodes is effective. However, in mobile ad hoc networks, there remains a problem that it is difficult to introduce the cache function because of node movement.

3 New Service Discovery Protocol

3.1 Proposed Protocol

The greatest feature of the proposed protocol is service provision group and their dynamic reconfiguration by the service discovery request from users. In the existing protocol (Fig.3 (a)) the user broadcasts a service request message to discover the service for which the user specifies the service classification. When the node received the service request message and cannot offer the service which is consistent with the service classification specified in the received service request message, the node further rebroadcasts the service request message. Reception

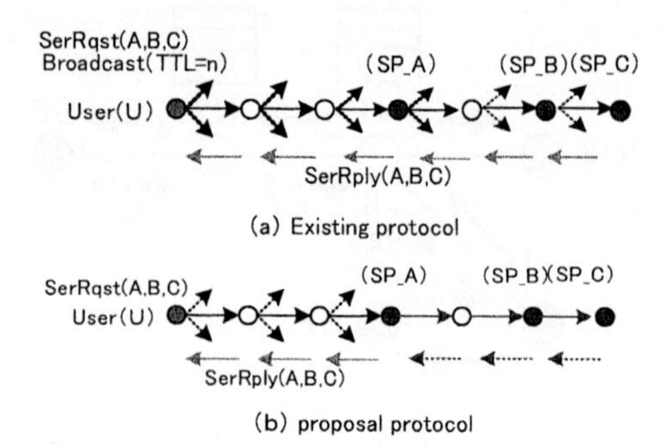

Fig. 3. Service discovery by existing protocol and proposed protocol

and rebroadcasting of this service request message are performed until the node which offers the service specified in the received service request message is found. When the node (that is, service provider) is found, a service reply message including the service information on the service currently offered will reply to the user. In the existing system (Fig. 3 (a)), for the service requests A, B, and C, service providers SP (A), SP (B) and SP (C) will receive the service discovery request message from the user, and have to reply to the user with service reply messages.

The proposed protocol (Fig.3 (b)) as well as the existing protocol broadcasts the service request message for which the user specifies the service classification. In the proposed protocol, when the service provider accepts service request message from a user, a service provision group is dynamically reconfigurated. However, it is necessary to specify beforehand the size of a service provision group and the number of members in the group. When the same user broadcasts service request messages, one of the members in the already constituted service provision group has only to receive this message. As a result, it will reduce the amount of service discovery messages for broadcasting. In the proposed protocol (Fig. 3 (b)), if service providers SP (A), SP (B) and SP (C) constitutes a service provision group, the user will be answered with a service reply message only by one of the service providers SP (A), SP (B) or SP(C) which received the service request message.

Since it is not necessary to look for all services when the user triggers a service discovery request, reduction of the broadcasting messages for service discovery is realized. The reduction amount of broadcasting messages becomes large intensively with increase of the size of an ad hoc network. Moreover, service can be quickly discovered in a shorter time than the existing protocol.

The effect of a service provision group is explained with an example (Fig.4). Suppose that a service provision group G constituted by five kinds of services a, b, c, d, and e is constituted by the service discovery request. The user broadcasts a service request message to G by TTL=n (Fig. 4: 1).The member node d of G which received this service request message send this service request message to the other service provision group members (Fig.4: 2). After checking each member's condition, the member node d answers to the user with a service reply message, so the user can discover the set of services (Fig.4: 3).

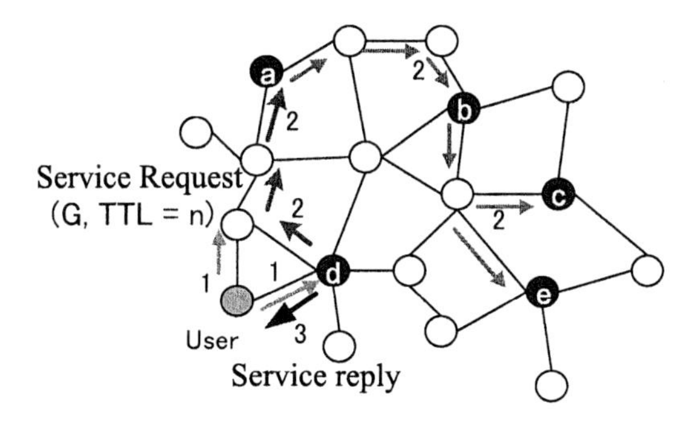

Fig. 4. Process of proposed protocol

3.2 Dynamic Reconfiguration of Service Provision Group

This section shows methods for the initial configuration and reconfiguration of the service provision group in an ad hoc network.

Initial Configuration of a Service Provision Group. Suppose that the service provider SP{SP(1),SP(2),...,SP(n)} respectively offers services S{s(1), s(2),...,s(n)}. In order to discover the services, the user broadcasts request messages of service S (r). Here, let S(r){ s(i),...,s(j),...,s(k)}. When SP (i) who offers s (i) service which is an element of S (r) receives the same request messages of service S (r) q times from two or more users, a service provision group for S (r) is dynamically configured. Here, q is a positive integer and a threshold for triggering configuration of the group. The process is shown below.

1. SP (i) broadcasts Invite message.
2. When service providers which offer service S (r) received Invite message, then they answer to SP (i) with Join message.
3. SP (i) receives Join messages from providers of all services in S (r), and constitutes a service provision group for S (r). During the above process, a tree is constructed for the routing information between group members according to the procedure of MAODV [11].

Reconfiguration of Service Provision Group. In order to accept various service discovery requests flexibly and efficiently, it is necessary to add a new service to the service group and delete an existing service from a service provision group. According to a discovery request of services which are similar to those in the existing service group, service provision group is reconfigurated by performing the addition and deletion of a service in the service provision group.

(1) Addition of Members to Service Provision Group. Suppose that the service provision group G (a, b, c, d, e) exists in an ad hoc network. If a user requests a new service (G+f), the service provision group G needs to accept this request. The additional process is as follows.

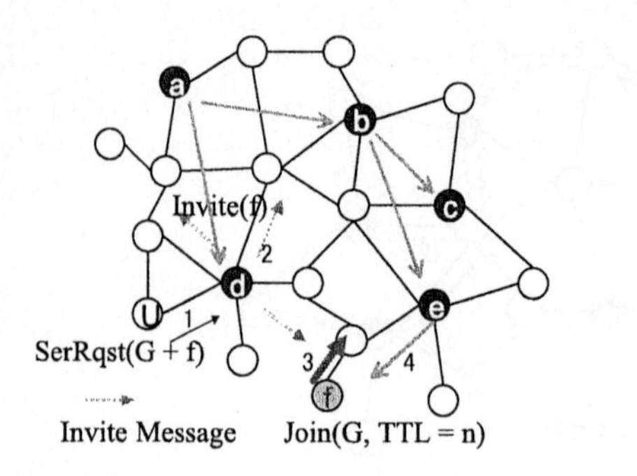

Fig. 5. Addition of group member

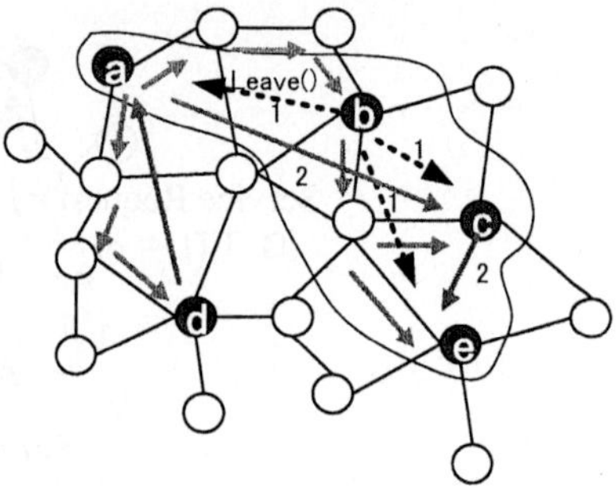

Fig. 6. Deletion of a group member

The user broadcasts a new service request (G+f) (Fig. 5: 1). When the member node d in the service provision group G receives this service request message, it recognizes that service f is not included in the service provision group G. At this time, d broadcasts an Invite (f) message (Fig. 5: 2). When the node f which offers service f receives Invite (f) message, in order to participate in the service provision group G, node f broadcasts Join (G) message by TTL=n > H(d, f)(Fig. 5: 3). Here, H (d, f) is taken as the number of hops between node d and node f. By this process, Join message is delivered to a member node in the service provision group G nearest to f (Fig.5: 4).

(2) Deletion of Service from Service Provision Group. When a member (say b) in a service provision group cannot offer the service, or when it is no longer contained in any service discovery request from users during a predefined time period, it is necessary to reconfigure the service provision group G by deleting b. This process is as follows.

First, the member node b sends Leave message to only its adjacent group members in the tree (Fig.6: 1). Node a which received the Leave message locally reconfigures the tree by sending Invite message only to nodes c and e (Fig.6: 2).

3.3 Disappearance of Service Provision Group

When the service provision group itself has not been used during a predefined long time period, the service provision group disappears automatically.

4 Performance Evaluation

4.1 Simulation Parameters

In this section we evaluate the proposed protocol by means of a simulation study. We implemented a simulator within NS2 [12]. In our simulation, a network is modeled by 50 mobile nodes placed randomly within a 1500 meter x 300 meter area. We used a random waypoint model [13] as the mobility model (that is, the velocity is randomly selected from 0-1 m/s, and the pause time is zero seconds).

Each node has a radio propagation range of 250 meters and the channel capacity is 2 Mb/s. As a medium access control protocol, we used the IEEE802.11 Distributed Coordination Function. Note that each data is an average value which is calculated from the result of ten simulation runs on each combination of parameters.

4.2 Simulation Results

(1) Evaluation of successful service discovery rate (Fig.7)

Fig.7 shows the rate of service discovery when either the proposed protocol with 1-20 users and ten services constituting a service discovery group or the existing broadcast protocol is applied. Here, the (successful) rate of service discovery is defined as the total service reply messages which the users received divided by the total service request messages which the users sent. As shown in Fig.7, comparing the proposed protocol with the broadcast protocol, the rate of service discovery by the broadcast protocol is remarkable low, and the rate of service discovery has become decreased along with increase of the number of users. As opposed to this result, in the proposed protocol, even if the number of users increases, there is no sharp reduction of the rate of service discovery.

(2) Evaluation of response time

When the number of users is fixed to ten, as shown in Fig.8 (a), the response time of service discovery do not change so much with increase of the service number of the service provision group. This is because of the following two reasons: (1) a representative member of the service provision group replies to request of the set of services in the proposed protocol and (2) checking each member's condition is efficiently performed using the tree of the service provision group.

When the number of services is fixed to ten, as shown in Fig.8 (b), it turned out that with increase of the number of users the response time increases gradually. This is because, although the users simultaneously send service request messages to the service provision group, along with increase of the number of

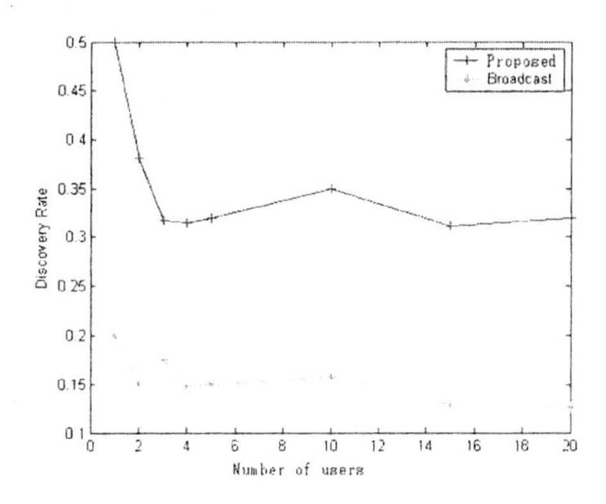

Fig. 7. Simulation results on the rate of service discovery

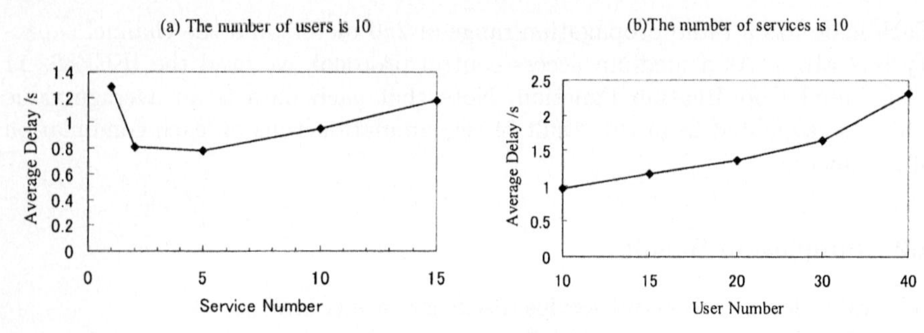

Fig. 8. Simulation results on response time

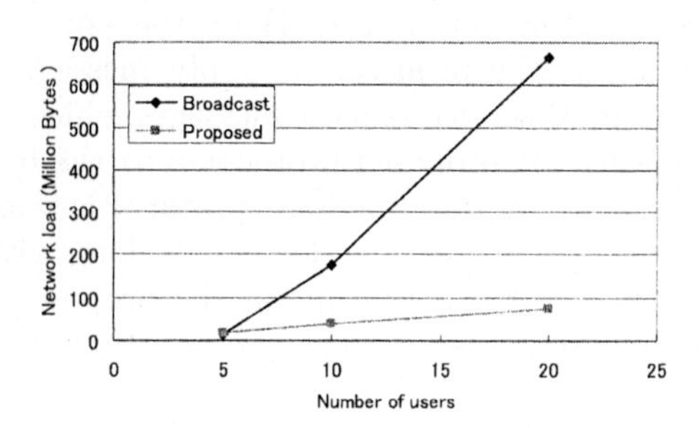

Fig. 9. Simulation results on network load

the users the shortest hop number between the users and the service provision group becomes long.

Similar observations are obtained in cases of other fixed user numbers and fixed service numbers.

(3) Evaluation of network load

To compare the overhead of the broadcast protocol and the proposed protocol, we adopted an application layer packet generation function to generate service requests at regular time intervals. In the simulation environment the following parameters are used: the number of nodes is 50, simulation length is 910 seconds, the number of simulation runs is 10, the service provision group includes 10 kinds of services, and the number of users is 5, 10, 20.

Fig.9 plots the network load (that is, the total number of bytes transmitted at the MAC layer) for the broadcast protocol and the proposed protocol. The network load of broadcast protocol is much higher than that of the proposed protocol with increase of the users. In the broadcast protocol, when each of ten service providers receives service request message, it will broadcast the service request message to other nodes and send a service reply message to the users. On the other hand, in the proposed protocol all the ten service providers will not broadcast service request messages further when one of the service providers received the service request message. Therefore, the network load in the proposed protocol is much lower than that of the broadcast protocol when the number of users becomes larger.

5 Conclusions

This paper has proposed a new service discovery protocol for mobile ad hoc networks, which is based on the service provision groups and their dynamic reconfiguration. Messages among members in the service provision group are delivered through a tree which the service provision group forms and a representative member replies to request of service discovery from users. The proposed protocol is thus effective when users often request a specific collection of services. Even when members of the service provision group increase or decrease, the proposed protocol works effectively owing to the dynamic reconfiguration of the service provision group.

This paper has given simulation results on the rate of service discovery, response time and network load for the proposed protocol. Compared with the typical existing broadcast protocol, the results show that the service discovery rate is high, the response time is short and the network load is low when the number of members in the service provision group is large. Extensive simulation experiments will be conducted in the future.

Acknowledgments

This work was supported in part by the Ministry of Internal Affairs and Communications of Japan under Grant-in-Aid for SCOPE-C (No.052308002), the Ministry of Education, Science, Sports and Culture of Japan under Grant-in-Aid for Scientific Research (C) (No.15500049) and Young Scientists (B) (No.16700075), and Hiroshima City University under their research grants.

References

1. T. Ohta, S. Inoue, Y. Kakuda, "An adaptive multihop clustering scheme for highly mobile ad hoc networks," Proc. 6th IEEE International Symposium on Autonomous Decentralized Systems (ISADS2003) pp.293-300 (April 2003).
2. E. Guttman, et al., "Service Location protocol," Version 2, IETF, RFC 2165, 1999.
3. E. futtman, C. Perkins, J. Veizades, and M. Day, "Service Location Protocol, Version 2," IETF RFC 2608, June 1999. http://www.ietf.org/rfc/rfc2608.txt.
4. The Salutation Consortium, Salutation Architecture Specification, www.salutation.org
5. Sun Microsystems, JINI technology, www.jini.org
6. Universal Plug and Play Forum. Universal Plug and Play Device Architecture. Version 0.91, March 2000.
7. M. Nidd, "Service Discovery in DEAPspace," IEEE Personal Communications, August 2001.
8. L. M. Feeney, B. Ahlgren, and A. Westerlund, "Spontaneous networking: An application-oriented approach to ad hoc networking," IEEE Communication Magazine, 39(6), 2001.
9. C. Bettstetter and C. Renner, "A Comparison of Service Discovery Protocols and Implementation of the Service Location Protocol," Proc. of the Sixth EUNICE Open European Summer School: Innovative Internet Applications, EUNICE 2000, Twente, Netherlands, September, pp.13-15, 2000.

10. B. Raman, P. Bhagwat, S. Seshan, "Arguments for cross-layer optimizations in Bluetooth scatternets," Symposium on Applications and the Internet, pp.176-184, 2001.
11. E. Royer, C. Perkins, "Multicast ad hoc on-demand distance vector (MAODV) routing," Internet-draft, July 2000, draft-ietf-manet-maodv-00.txt.
12. USB/LBNL/VINT Network Simulator ns (Ver.2), http://www.isi.edu/nsnam/ns.
13. J. Broch et al., "A performance comparison of multihop wireless ad hoc network routing protocols", Proc. IEEE/ACM Mobicom'98, 1998.

Population Estimation for Resource Inventory Applications over Sensor Networks

Jiun-Long Huang

Department of Computer Science,
National Chiao Tung University, Hsinchu, Taiwan
jlhuang@cs.nctu.edu.tw

Abstract. The growing advance in wireless communications and electronics makes the development of low-cost and low-power sensors possible. These sensors are usually small in size and are able to communicate with other sensors in short distances wirelessly. A sensor network consists of a number of sensors which cooperates with one another to accomplish some tasks. In this paper, we address the problem of resource inventory applications, which means a class of applications involving population calculation of a specific species or object type. To reduce energy consumption, each sensor only reports the number of sensed objects to the server, and the server will estimate the object number according to the received reports of all sensors. To address this problem, we design in this paper a population estimation algorithm, called algorithm Estimation, to estimate the object numbers. Several experiments are conducted to measure the performance of algorithm Estimation. The experimental results show that algorithm Estimation is able to obtain closer approximations of object numbers than prior algorithms.

Keywords: Sensor network, resource inventory, population estimation.

1 Introduction

The growing advance in wireless communications and electronics makes the development of low-cost and low-power sensors possible. These sensors are usually small in size and are able to communicate with other sensors in short distances wirelessly. A sensor network [1] consists of a number of sensors which cooperates with one another to accomplish some tasks. Sensors can be deployed either in a random or in a predetermined manner. Since being self-organized, sensors are able to form a sensor network automatically. Due to the characteristics of wireless communication and configuration-free deployment, sensor networks are suitable for various application areas including inventory management, product quality monitoring and disaster area monitoring [1][5]. Hence, sensor networks have attracted a significant research attention, including hardware and operating system design [8][13], localization [2][10], data aggregation methods [3][6][9] and applications of sensor networks [11][14].

X. Jia, J. Wu, and Y. He (Eds.): MSN 2005, LNCS 3794, pp. 175–184, 2005.

The authors in [7] described one kind of applications[1], resource inventory applications, which uses sensor networks to calculate object numbers. By resource inventory, it means a class of applications involving population calculation of a specific species or object type. To calculate population of objects, a number of sensors are deployed in a plane and each sensor is able to sense the number of objects within its sensing region. These sensors then form a sensor network and users are able to query the number of objects sensed by the sensor network via a server. For the sake of simplicity, in this paper we use "object number" to indicate the number of objects sensed by a sensor network. Since sensors are usually powered by batteries, energy conservation becomes an important issue in the design of sensor networks. To reduce energy consumption, the authors in [7] suggested that each sensor only reports the number of sensed objects to the server[2], and the server will estimate the object number according to the received reports of all sensors.

Since the sensors may be deployed randomly, the sensing regions of these sensors may be overlapped with one another. This phenomenon results in the difficulty of obtaining the exact object number due to the reason that one object may be sensed by more than one sensor. Fortunately, knowing the ranges of object numbers is still useful enough in many applications [7]. As a consequence, the authors in [7] proposed an energy-conserved scheme which is able to obtain the lower bounds and the upper bounds of the exact object numbers. For convenience, we name the scheme proposed in [7] as scheme SDARIA (i.e., the acronym of the title of [7]). Although being shown to be able to conserve much energy than other schemes [7], the lower bounds and the upper bounds obtained by scheme SDARIA are not informative. In our experiments, the upper bounds obtained by scheme SDARIA are around 170% $\sim$ 250% of the exact numbers, while the lower bounds are around 60% $\sim$ 80% of the exact numbers. Such high error rates make users not able to get enough information about the exact object numbers.

In view of this, we design in this paper a population estimation scheme, called algorithm Estimation, to estimate the object numbers. Specifically, algorithm Estimation first partitions the plane into several disjoint grids, and identifies the full and partial grids of each sensor. Algorithm Estimation then estimates the object number of each grid, and finally estimates the overall object numbers according to the estimated object number of each grid. Several experiments are conducted to measure the performance of algorithm Estimation. The experimental results show that algorithm Estimation is able to obtain closer approximations of object numbers than scheme SDARIA.

The rest of this paper is organized as follows. Section 2 gives a description of resource inventory applications and an overview of scheme SDARIA. The design of algorithm Estimation is given in Section 3. Section 4 shows the performance study of algorithm Estimation. Finally, Section 5 concludes this paper.

---

[1] Interested readers can refer to [7] for more examples of resource inventory applications.

[2] The detailed architecture of resource inventory applications is given in Section 2.

2 Preliminaries

2.1 Related Work

The system architecture of resource inventory applications proposed in [7] is shown in Figure 1. To calculate the number of objects, a number of sensors are deployed in a plane and each sensor is able to sense the number of objects within its sensing region. These sensors then form a sensor network and users are able to query the number of objects via a server. For energy conservation, each sensor only reports the number of sensed objects to the server when the server requests all sensors to report their sensing status. Similar as [7], we assume that the server is able to get the sensing regions of all sensors. The sensing region of each sensor can be obtained from manual measurement by human or automatic measurement by sensors when they are equipped with GPS [4].

Basically, scheme SDARIA comprises the following two phases.

- *Data aggregation phase:* In data aggregation phase, each sensor reports the number of sensed objects to the server via the sink. After receiving the reports of all sensors, the server transforms the received reports into the corresponding snapshot graph $G = \{V, E\}$. The transformation procedure is as follows. Each sensor is modelled as a vertex and there is an edge between two sensors if these two sensors' sensing regions are overlapped with each other. Figure 2 shows an example of the transformation between the reports and the corresponding snapshot graph.
- *Population estimation phase:* In population estimation phase, the server estimates the object number according to the snapshot graph G. In scheme SDARIA, two algorithms are proposed to obtain the upper bound and the lower bound of the exact object number.

Experimental results in [7] showed that the data aggregation method used in scheme SDARIA is able to greatly reduce the power consumption of sensors. Therefore, in this paper we adopt the data aggregation method used in scheme SDARIA and focus on population estimation phase. In scheme SDARIA, upper

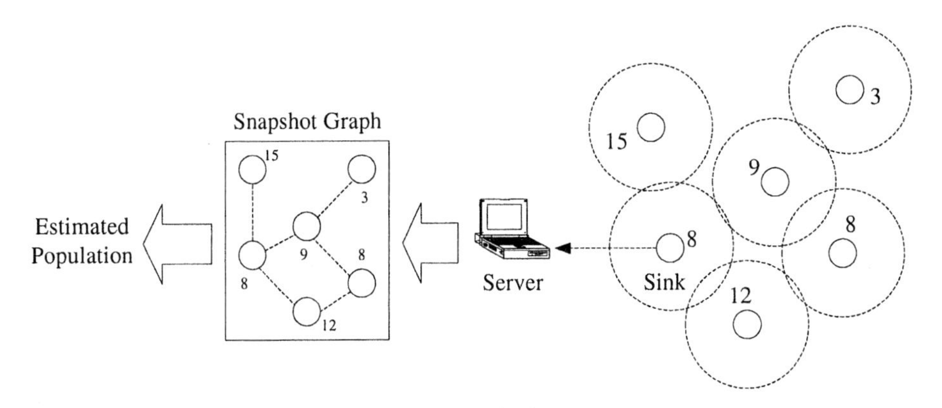

Fig. 1. System Architecture

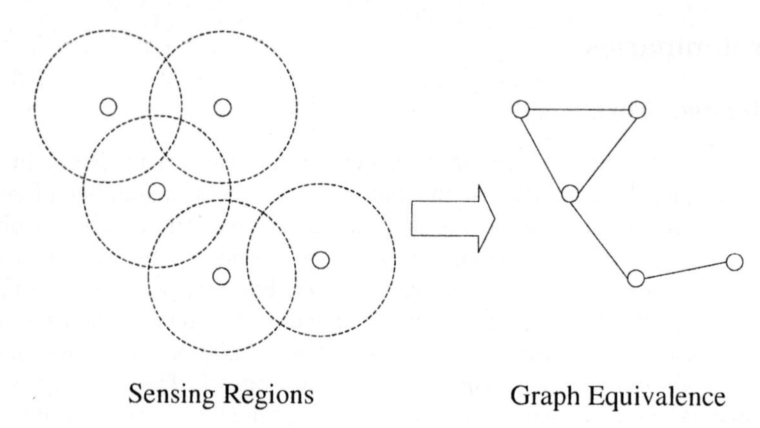

Fig. 2. Transformation between sensing reports and the corresponding snapshot graph

bound calculation is modelled as finding a subset of V which includes solely the vertices whose sensing regions cannot be replaced by the combinations of other vertices' sensing regions. For the interest of space, we do not describe the details of the upper bound calculation algorithm in this section. Interested readers can refer to [7] for details.

3 Design of Population Estimation Algorithm

Although we can obtain the lower bounds and the upper bounds of object numbers, respectively, by the lower bound and the upper bound calculation algorithms used in scheme SDARIA, these bounds do not give users enough information about the exact object numbers. For example, in our experiments, the obtained upper bounds are around 190% $\sim$ 210% of the exact object numbers, and the lower bounds are around 60% $\sim$ 80% of the exact object numbers. To address this problem, we propose a grid-based population estimation algorithm, called algorithm Estimation, to obtain close approximations of exact object numbers.

Before designing algorithm Estimation, we first partition the plane into several non-overlapped grids. Consider the sensing region of a sensor shown in Figure 3. A grid g is called the *full* grid of a sensor s if all the area of grid g is covered by the sensing region of sensor s. Similarly, a grid g is called the *partial* grid of a sensor s if only part of the area of grid g is covered by the sensing region of sensor s. A grid g is called the *overlapped* grid of sensor s if grid g is a full or a partial grid of sensor s. In Figure 3, the full and partial grids of the sensor are marked as 'F' and 'P', respectively.

To facilitate the estimation of the exact object numbers, we have the following assumptions.

1. The objects sensed by sensor s is uniformly distributed in the sensing region of sensor s.
2. The objects in a grid g is uniformly distributed in the area of grid g.

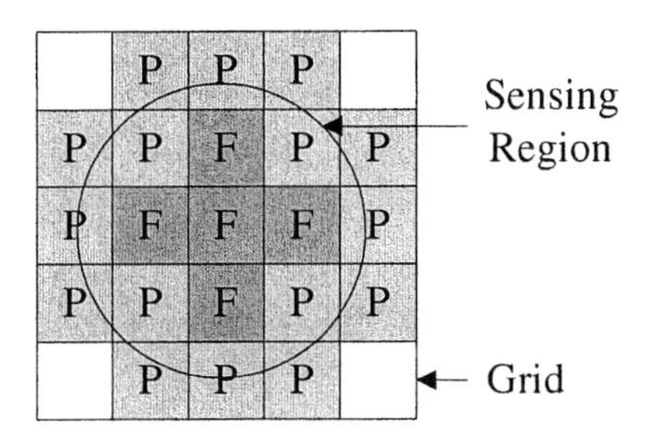

Fig. 3. Full and partial grids

3. On average, for a partial grid g of a sensor s, only half of the area of grid g is covered by the sensing region of sensor s.

Note that these assumptions are made only to guide the design of algorithm Estimation, and are not the limitations of algorithm Estimation. They will be relaxed in the experiments in Section 4.

The procedure of algorithm Estimation is as follows. Initially, each sensor is marked as UNSELECTED. Algorithm Grid-Estimation is an iterative algorithm and selects one sensor marked as UNSELECTED in each iteration. Consider the case that sensor s is selected. Denote the number of objects sensed by sensor s as $s.ObjNo$, and let $full$ and $partial$ be the sets of the full grids and the partial grids, respectively, of sensor s. By Assumption 3, we first calculate the *equivalent* number of full grids by considering one partial grid as $\frac{1}{2}$ full grid. Hence, the equivalent number of the full grids of sensor s is $|full| + \frac{|partial|}{2}$. By Assumption 1, $\frac{s.ObjNo}{|full|+\frac{|partial|}{2}}$ objects are expected in each full grid of sensor s. In addition, by Assumption 2 and Assumption 3, $\frac{s.ObjNo}{|full|+\frac{|partial|}{2}}$ objects are expected in each partial grid of sensor s. Hence, sensor s suggests that $\frac{s.ObjNo}{|full|+\frac{|partial|}{2}}$ objects are expected in each of its overlapped grids. Finally, sensor s is marked as SELECTED. The above procedure repeats until all sensors are marked as SELECTED. Since the expected object number in grid g may be suggested by several sensors, the expected object number in grid g is determined as the average of possible object numbers suggested by sensors. Finally, the estimated object number is determined as the summation of the expected object numbers of all grids. The algorithmic form of the proposed population algorithm is as follows.

Algorithm Estimation

1: **for** each grid g **do** /* Initialization */
2: $\quad g.AvgObjNo \leftarrow 0$
3: $\quad g.SensorNo \leftarrow 0$
4: **end for**
5: Mark all sensors as UNSELECTED
6: **while** (at least one sensor is marked as UNSELECTED) **do**

7: Pick one sensor which is marked as UNSELECTED
8: $full \leftarrow$ full grids of sensor s
9: $partial \leftarrow$ partial grids of sensor s
10: $expectation = \frac{ObjNo}{|full| + \frac{|partial|}{2}}$
11: **for** each grid g in $full \bigcup partial$ **do**
12: $g.AvgObjNo \leftarrow$
$\frac{1}{g.SensorNo+1} \times (g.AvgObjNo * g.SensorNo + expectation)$
13: $g.SensorNo \leftarrow g.SensorNo + 1$
14: **end for**
15: Mark sensor s as SELECTED
16: **end while**
17: $total \leftarrow 0$
18: **for** each grid g **do**
19: $total \leftarrow total + g.AvgObjNo$
20: **end for**
21: return $total$

4 Performance Evaluation

4.1 Simulation Model

Similar as [7], the sensors are uniformly placed in to a 500×500m plane and the sensing radius of each sensor is set to 100m. We use GSTD tool [12] to generate the synthetic datasets used in this simulation. We synthesize the locations of objects by two distributions: uniform distribution and Gaussian distribution with standard deviation 50, and they are shown in Figure 4a and Figure 4b, respectively. Since focusing on population estimation phase, in data aggregation phase, we adopt the data aggregation method used in scheme SDARIA. In addition to algorithm Estimation, we also implement scheme SDARIA for comparison purposes. We implement both schemes in C++ and the simulation is executed in a PC with one Pentium III 500MHz CPU and 512MB memory. The default system parameters are listed in Table 1.

Since both schemes use the same data aggregation method, the energy consumption of the sensors in both schemes is the same. Hence, we take accuracy and execution time as the performance metrics of both schemes. Error rate, which is define as below, is taken to measure the accuracy of population estimation.

$$\text{Error rate} = \frac{\text{Estimated object number-Exact object number}}{\text{Exact object number}}$$

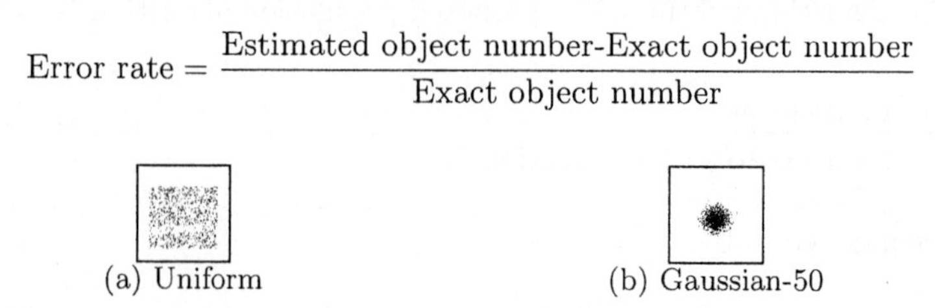

(a) Uniform (b) Gaussian-50

Fig. 4. Datasets

Table 1. Default system parameters

Parameter	Value
Number of sensors (V)	30
Number of objects	2000
Sensing radius	100m
Grid size (Grid size ratio=2%)	2m

Table 2. Description of curves

Curve	Result
Lower bound	Lower bound obtained by scheme SDARIA
Estimation	Estimated object number
Upper bound	Upper bound obtained by scheme SDARIA

Note that error rate smaller than zero indicates that the estimated object number is smaller than the actual object number. The accuracy is higher when error rates are closer to zero. That is, the accuracy is high when the absolute values of error rates are small.

We now describe the meaning of the curves which will appear in the following experimental results. Lower bound represents the lower bounds obtained by scheme SDARIA. The lower bound calculation algorithms used in scheme SDARIA is a brute force-based algorithm to obtain the optimal solutions of maximal independent set problem [7]. Cureve Estimation represents the results of algorithm Estimation, while curve Upper bound represents the upper bounds obtained by scheme SDARIA. A summary of these curves as given in Table 2.

4.2 Effect of Sensor Number

This experiment is conducted to measure the effect of the number of sensors. Figure 5 shows the error rates and execution time of all algorithms with the number of sensors varied. The number of sensors is set from 30 to 180.

As observed in Figure 5a and Figure 5b, the error rates of upper bound calculation algorithm range from 80% to 205%. Note that the lower bound calculation algorithm used in scheme SDARIA is of high complexity, and hence, only the results with 30 sensors are shown in Figure 5a and Figure 5b. In our experiment, the execution time of the lower bound calculation algorithm is longer than six hours when the number of sensors is larger than 30. In addition, the error rates of algorithm Estimation are between -2.9% and 11.25% in this experiment. Since the absolute values of the error rates of algorithm Estimation are smaller than those of the lower bound and the upper bound calculation algorithms used in scheme SDARIA, algorithm Estimation is able to give users closer estimations than the lower bound and the upper bound calculation algorithms. Figure 5c

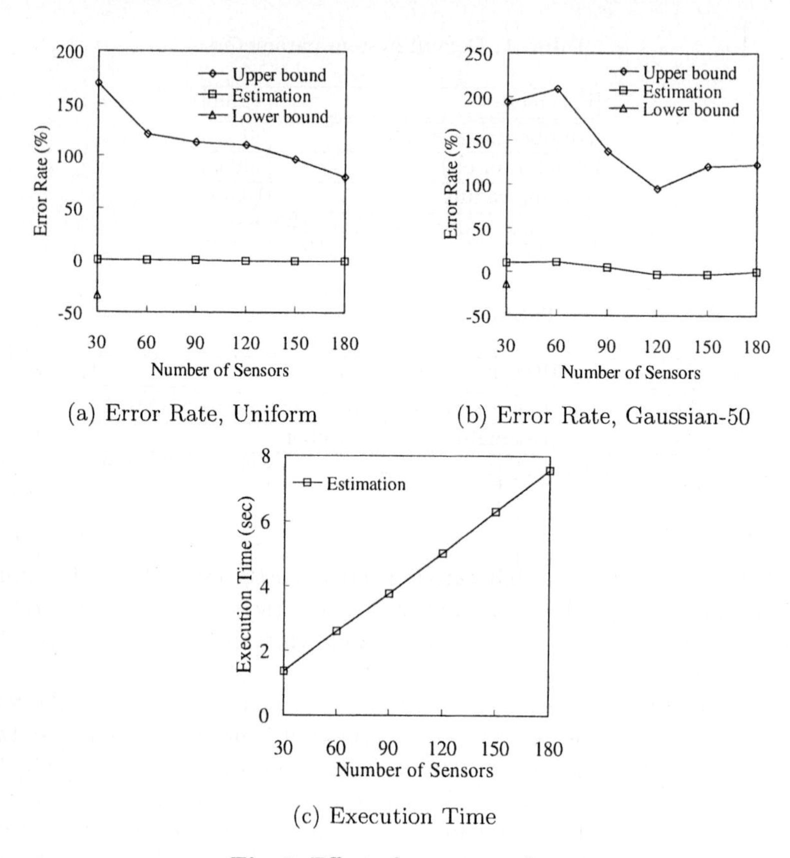

(a) Error Rate, Uniform

(b) Error Rate, Gaussian-50

(c) Execution Time

Fig. 5. Effect of sensor number

shows the execution time of algorithm Estimation. The execution time of algorithm Estimation increases linearly as the number of sensors increases. This result agrees to the complexity analysis of algorithm Estimation in Section 3.

4.3 Effect of Object Number

This experiment is to investigate the effect of the number of objects. Figure 6 shows the error rates of all algorithms by setting object number from 500 to 5000. Since the execution time of all algorithms are not affected by object number and the distribution of objects, we only show the error rates of all algorithms in this subsection. As shown in Figure 6, the error rates of all algorithms are affected by the distribution of objects. Since being designed under the premise that objects are distributed uniformly, algorithm Estimation performs very well in dataset Uniform. It is also observed that the error rates of algorithm Estimation slightly increase as the number of objects increases. In this experiment, the error rates of algorithm Estimation increase from -0.81% to 0.1% and from 8% to 10.6%, respectively, in datasets Uniform and Gaussian-50. As shown in Figure 6, the error rates of the lower bound calculation algorithm in datasets Uniform and Gaussian-50 range from -33.94%

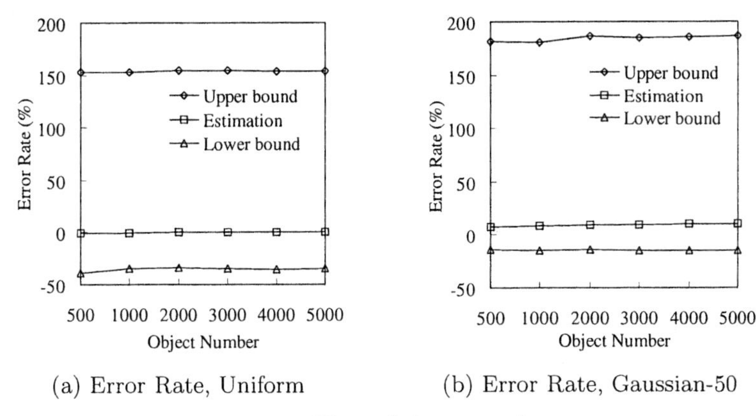

(a) Error Rate, Uniform (b) Error Rate, Gaussian-50

Fig. 6. Effect of object number

to -38.91% and from -14.2% to -15.1%, respectively. We observe that the absolute values error rates of the lower bound calculation algorithm become smaller when the distribution of objects becomes centralized.

5 Conclusion

In this paper, we addressed the problem of resource inventory applications over wireless sensor networks. To reduce energy consumption, each sensor reports only the number of sensed objects to the server, and the server will estimate the object number according to the received reports of all sensors. In view of this, we designed algorithm Estimation to estimate the object numbers. Several experiments were conducted to measure the performance of algorithm Estimation. The experimental results showed that algorithm Estimation was able to obtain close approximations of object numbers in reasonable execution time. In our experiments, the approximations of algorithm Estimation were around 95% $\sim$ 115% of the exact object numbers. This result showed that algorithm Estimation is more suitable for practical use than prior schemes.

References

1. I. F. Akyildiz, W. Su, Y. Sankarasubramaniam, and E. Cayirci. A Survey on Sensor Networks. *IEEE Communications Magazine*, August 2002.
2. K. Chintalapudi, R. Govindan, G. Sukhatme, and Amit Dhariwal. Ad-Hoc Localization Using Ranging and Sectoring. In *Proceedings of the IEEE INFOCOM Conference*, March 2004.
3. J. Considine, F. Li, Kollios G, and J. Byers. Approximate Aggregation Techniques for Sensor Databases. In *Proceedings of the 20th IEEE International Conference on Data Engineering*, March-April 2004.
4. D. M. Doolin, S. D. Glaser, and N Sitar. Software Architecture for GPS-enabled Wildfire Sensorboard. In *TinyOS Technology Exchange*, February 2004.

5. S. D. Glaser. Some Real-World Applications of Wireless Sensor Nodes. In *Proceedings of SPIE Symposium on Smart Structures and Materials*, March 2004.
6. C. Intanagonwiwat, D. Estrin, R. Govindan, and J. Heidemann. Impact of Network Density on Data Aggregation in Wireless Sensor Networks. In *Proceedings of the 22nd IEEE International Conference on Distributed Computing Systems*, July 2002.
7. T.-H. Lin and P. Huang. Sensor Data Aggregation for Resource Inventory Applications. In *Proceedings of the IEEE Wireless Communications and Networking Conference*, March 2005.
8. I. Locher, S. Park, M. Srivastava, A. Chen, R. Muntz, and S. Yuen. A Support Infrastructure for the Smart Kindergarten. *IEEE Pervasive Computing Magazine*, April-June 2002.
9. S. Madden, M. J. Franklin, J. M. Hellerstein, and W. Hong. TAG: A Tiny Aggregation Service for ad hoc Sensor Networks. In *Proceedings of the 5th Symposium on Operating Systems Design and Implementation*, December 2002.
10. Y. Shang and W. Ruml. Improved MDS-Based Localization. In *Proceedings of the IEEE INFOCOM Conference*, March 2004.
11. G. Simon, M. Maroti, A. Ledeczi, G. Balogh, B. Kusy, A. Nadas, G. Pap, J. Sallai, and K. Frampton. Sensor Network-based Countersniper System. In *Proceedings of the 2nd ACM International Conference on Embedded Networked Sensor Systems*, November 2004.
12. Y. Theodoridis, J. R. O. Silva, and M. A. Nascimento. On the Generation of Spatialtemporal Datasets. In *Proceedings of the 6th International Symposium on Large Spatial Databases*, July 1999.
13. TinyOS Community Forum. http://www.tinyos.net/.
14. N. Xu, S. Rangwala, K. K. Chintalapudi, D. Ganesan, A. Broad, R. Govindan, and D. Estrin. A Wireless Sensor Network for Structural Monitoring. In *Proceedings of the 2nd ACM International Conference on Embedded Networked Sensor Systems*, November 2004.

Segmented Broadcasting and Distributed Caching for Mobile Wireless Environments*

Anup Mayank and Chinya V. Ravishankar

Department of Computer Science & Engineering,
University of California, Riverside,
Riverside, California, 92521
{mayank, ravi}@cs.ucr.edu

Abstract. Broadcast data dissemination is well-suited for mobile wireless environments, where bandwidth is scarce, and mutual interference must be minimized. However, broadcasting monopolizes the medium, precluding clients from performing any other communication. We address this problem in two ways. First, we segment the server broadcast, with intervening periods of silence, during which the wireless devices may communicate. Second, we reduce the average access delay for clients using a novel cooperative caching scheme. Our scheme is fully decentralized, and uses information available locally at the client. Our results show that our model prevents the server from monopolizing the medium, and that our caching strategy reduces client access delays significantly.

Keywords: Mobile Ad-hoc networks, Broadcasting, Cooperative Caching.

1 Introduction

Mobile users are increasingly interested in a range of information, such as stock quotes, pricing information at shopping malls, weather, news and traffic information, schedules at bus stands, railway stations, and airports. Push-based information dissemination can be much more effective and scalable than the pull model in such applications. Its advantages are well-known [20, 19, 22, 2, 8, 17, 14, 9]. In contrast, the pull model requires clients to send each request to the server, and increases traffic and resource consumption at both clients and the server [17].

Broadcasting is a good approach for disseminating information in mobile wireless systems. However, when the medium is shared, as in the 802.11 protocol suite, broadcasting monopolizes the medium, preempting clients from performing any other communication. Blindly interrupting the broadcast schedule to create slots for clients to communicate would increase client wait times proportionally, degrading an important performance metric. Although wireless protocols make several channels available for communication, it is better to treat the set of channels as a common bandwidth resource. We show how to share each channel between the broadcast program and other communications.

* Supported by grants from Tata Consultancy Services, the DiMI program of the University of California, and contract F30602-01-2-0535 of DARPA's FTN program.

X. Jia, J. Wu, and Y. He (Eds.): MSN 2005, LNCS 3794, pp. 185–196, 2005.

Work to date has arbitrarily assumed that clients perform no other communication besides listening to the broadcast. It has therefore focused merely on mechanisms to reduce client access times. Some approaches, for example, reduce latency by repeating popular items several times in a broadcast cycle [14, 9]. Others use client feedback [17] to reduce broadcast cycle length. Such approaches are inadequate since clients are still held captive by the broadcast server.

1.1 Broadcast Scenarios and Caching

Consider a hypothetical scenario with a number of mobile users with PDAs in a shopping mall. To help customers and improve sales, the mall broadcasts a variety of information such as mall maps, store names, prices, sales under way, advertisements, and so on. Customers use their PDAs to listen to the broadcast, and wait for information relevant to their own shopping objectives (see Figure 1).

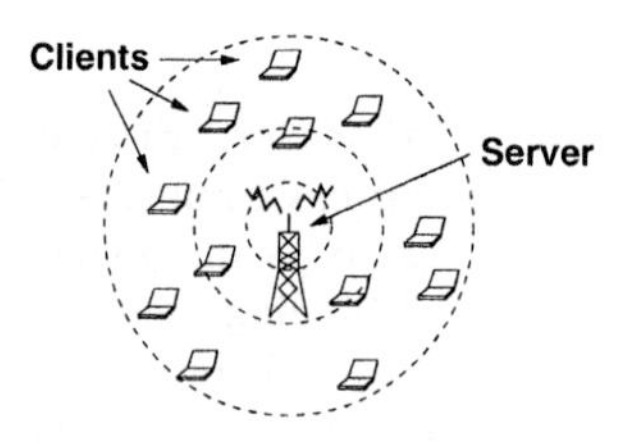

Fig. 1. Broadcasting

However, if the mall broadcasts constantly, it will effectively be ''jamming'' the medium, making any other communication impossible for user PDAs. The mall will surely lose customers and revenues. Interrupting the broadcast to allow PDAs to communicate increases wait latencies, prolong wait times, aggravate clients, and also lose revenues for the mall.

Clients can also reduce latency by caching data. Local caching methods such as prefetching [14] do reduce average access latency, but are limited by the amount of storage at each node. It would be better for caches to cooperate.

With cooperative caching, it suffices for each node to cache only a subset of items. When a desired item is absent from its local cache, a node tries to locate it in one of the other caches, using a lookup mechanism. Work already exists on cooperative caching in mobile wireless systems [9, 5, 7, 4]. Unfortunately, such approaches require caches to exchange messages to maintain global state, which is impossible when the broadcast server monopolizes the medium. Besides, message exchanges introduce considerable overhead. We do not see such models as appropriate for the environments we consider.

1.2 Our Contributions

We address these issues for mobile ad-hoc networks using 802.11-style protocols, where a shared medium is used for all communication. There are two interesting aspects to our approach. First, it segments the server broadcasts, dividing the broadcast cycle into a fixed number of segments interleaved with periods of silence, during which clients can communicate among themselves. Second, we propose a fully decentralized cooperative caching mechanism for use by clients, which requires no message exchanges for maintaining information on cache contents. Earlier work on caching for push-based dissemination are either non-cooperative [14] or are based on information exchanges between mobile nodes [9] for maintaining cache states. We show that our mechanism greatly reduces average object access delays, and is both fault tolerant and scalable.

2 Related Work

In the COCA [7] architecture, each mobile node is in a High Activity (HA) or Low Activity (LA) state. When its state changes from HA to LA, it retrieves and caches a set of less frequently accessed objects. More frequently accessed items would be already cached in the local cache of HA nodes, which can access replicas of less frequently accessed data items cached nearby. They do not explicitly address the issue of the server blocking out client communications.

A scheduling algorithm for correlated data is proposed in [21], so that data is accessed in a group. A caching strategy is used at the client to improve the performance of the scheduling mechanism. However, their caching approach involves no cooperation among clients. Their broadcast is continuous, and preempts all client communications.

In the CachePath and CacheData [5] mechanisms, caching is done at the routing layer. When a node routes a data item to a particular node, it either caches the path to reach the cached data item or caches the data item depending on the node's distance to server, caching node and route stability.

In GCLP [15], the server periodically sends information about its contents and physical location to a set of selected nodes called content location servers. A client locates an object by sending a query along suitable routes. When the query reaches a content location server, it returns the name of the nearest content server to the client. The client can obtain the requested object from the server.

The scheme in [13] allocates data to the servers, based on the movement pattern of users, so that a mobile user can obtain most recent replica from a nearby server, instead of sending requests to multiple hop away main server.

In [14], the local cache at the client end is used to store data items prefetched from the broadcast cycle. This scheme considers only the single cache at each client. It is not suitable if the cache size is small and total number of broadcasted data items is large.

The approach in [9] uses a cooperative caching approach among mobile clients in broadcast based information system. Mobile clients rely heavily on message exchanges among themselves for caching of data items. Unfortunately [9] uses a continuous broadcast based model, and it is unclear how clients are to communicate among themselves in a wireless environment.

3 Hash-Based Cooperative Caching

Let $C_1, \ldots, C_n$ be the names (or IDs) of a set of cooperating caches. Cooperative caching will be most effective if any given object O_k were never cached at more than one of these caches, and if each client could independently determine the identity of this cache. The Higest Random Weight (HRW) approach [16] was the first to show how to use hashing to allow clients to agree, with no communication, about which cache should hold an object O_k. A related idea appeared subsequently under the name of consistent hashing [10].

The HRW approach [16] uses a hash function H as follows. Each client computes the n hash values $H(C_1, O_k), \cdots, H(C_n, O_k)$ independently, and selects π_k, the cache that that yields the highest hash value. The crucial idea in HRW is to always cache the object O_k only at π_k, which is called the *prime cache* for O_k in the group $C_1, \cdots, C_n$. Conversely, all objects O_k that have π_k as their prime are called *prime objects* for π_k.

Since all clients apply HRW using the same H and to the same cache cluster $C_1 \ldots, C_n$, each client will independently select the same π_k as the prime for any given O_k. Since O_k is cached only at π_k, there is no object duplication across the C_i, thereby maximizing hit rates. The work in [16] discusses suitable H, and shows that HRW is very efficient and effective, and that it randomizes very well, so that every cache is likely to be the prime for the same number of objects.

HRW is also very robust in the face of cache failures. If the clients find that the prime π_k for some object O_k has become inaccessible, they simply pick the cache π'_n that yields the next highest hash value. Now, any request assigned to π_k becomes automatically reassigned to π'_k at each client. Since HRW is randomizing, each of the remaining proxies receives an equal share of these reassignments, ensuring that cache loads continue to be balanced.

When a cache C_i comes back up or is added to the cluster, the objects reassigned to it are exactly those which yield a higher hash value for C_i than any other cache in the cluster. Thus, HRW ensures that the fewest possible number of objects are reassigned in the case of cache failures or cache addition.

4 Segmented Broadcasting and Hash-Based Caching

Our approach divides the broadcast cycle into segments, with intervening periods of silence. The server broadcasts a subset of objects in each segment, but yields the medium to clients during the inter-segmental silences. With longer silences, the clients will have more use of the medium, but must tolerate longer wait times for receiving objects from the broadcast. The segment size and silence interval are parameters to be adjusted according to the number of objects to be broadcasted, average object size, the client interest patterns, and waiting times considered reasonable. There will be frequent intervals

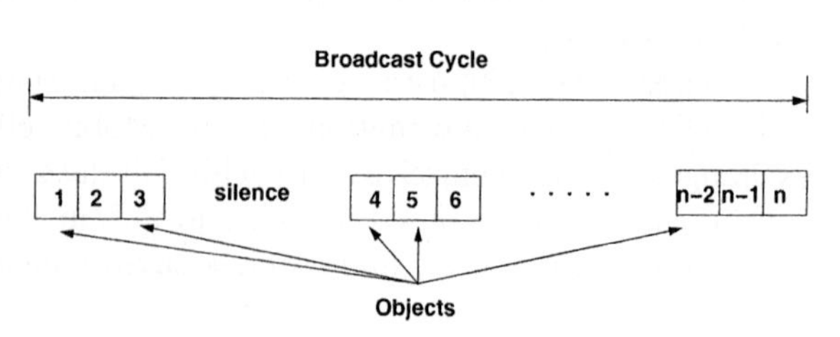

Fig. 2. Segmented broadcast Model

when the medium is free, even with large broadcast data sets. Clients can use these intervals for communication, and for obtaining desired objects from peers.

4.1 Cooperative Caching Using Local Information

We use cooperative caching to reduce access delays, and to allow clients to obtain data items from other clients in their neighborhood, rather than wait for the server to broadcast it. To be effective, cooperative caching must minimize object replication across caches, and incur low communication overhead. This can be tricky to accomplish in mobile ad-hoc systems, since mobile ad-hoc environments are dynamic, with nodes joining and leaving the system as they please.

A node may not know all other nodes in the system, but it usually knows its 1-hop neighborhood. It is typical, for instance [15], for 1-hop neighbors to exchange periodic hello messages. Moreover, 1-hop neighborhoods are likely to remain stable over short or medium durations, as in the case of passengers waiting for a particular flight or train, or customers near a particular exhibition stall.

Each node has a cache of limited size for holding a subset of the objects broadcasted by the server, so that replication within a neighboring set of mobile nodes must be minimized. It has been typical, as in [9], for caches to control replication by communicating their contents to each other and agreeing on a global caching policy. However, this method requires a great deal of communication, and is unrealistic for power- and bandwidth-limited wireless systems.

HRW Search of 1-Hop Neighborhoods. When node n_i needs an object O, it first checks its own cache. If O is not found, n_i determines the next broadcast time for O from the broadcast schedule. (It is typical [14, 9] for the server to periodically disseminate its schedule for broadcasting objects. This schedule can be cached and shared with other clients.) If this delay is too high, n_i tries to locate O in its peer caches as follows.

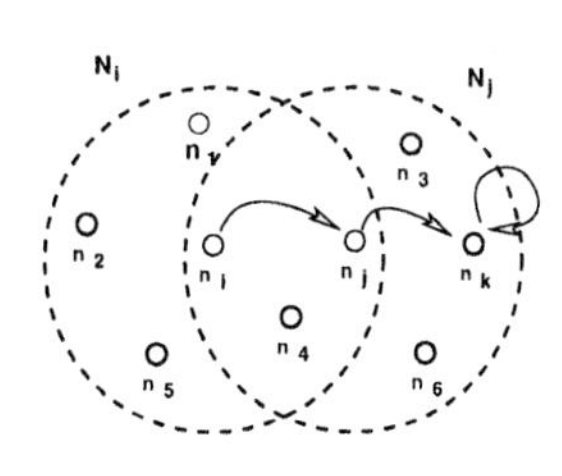

Fig. 3. Object search

Let the set $N_i = \{n_{i1}, \ldots, n_{ik}\}$ be the set of 1-hop neighbors of n_i. Node n_i applies HRW, and selects the node n_j yielding the highest hash value. From n_i's perspective, n_j is the prime for O in the set N_i, so that O would be at n_j if it were in the set N_i. Therefore, n_i sends n_j a request for the object O. If n_j has O, it returns O to n_i. Otherwise, n_j continues the search, using HRW on its own 1-hop neighborhood N_j.

As the search continues, n_j may itself turn out to be the prime cache for O within its own neighborhood N_j, in which case the search bottoms out. Such a search can be continued until a node is prime for its own neighborhood, (as is n_k in Figure 3), or for a fixed number of hops from n_i.

We choose to continue the search through n_j rather than some randomly chosen $n_k \in N_i$ for a specific reason. Since n_j is the prime for O in the neighborhood N_i, subsequent requests for O in N_i will also be sent to n_j. Continuing the search through n_j allows n_j to cache O, increasing future hit rates.

4.2 Caching of Objects and Cache Replacement

Each object O_i obtained by n_j is cached at n_j. If n_j's cache is full, a victim object is selected, as explained below, and evicted to make space for O_i. If a node receives a request for object O, it is likely to soon receive another request for O, due to locality effects. Hence, nodes cache all arriving objects, even if they are not prime objects.

Cache Replacement Mechanism. Since wireless devices have limited storage capacity, the cache replacement policy can have a major influence on performance. Our cache replacement policy favors popular and prime objects, increasing hit rates and reducing average access latency.

Our replacement policy tries to preserve prime objects, since they are likely to be requested by peers. When a cache must make space, it will first evict non-prime objects, starting with those having the lowest access probability. If no non-prime objects are present, some prime object must be evicted. Prime objects are also evicted starting with those having the lowest access probability.

The access probability function of an object O_i is calculated as

$$\psi(O_i) = \min\{1, \frac{T_{avg}}{T_r - T_c}\},$$

where T_{avg} is the cumulative average request arrival interval of the object, T_r is its last reference time, and T_c is the current time. T_{avg} is recomputed as

$$T_{avg}^{new} = \beta * T_{avg}^{old} + (1 - \beta)(T_r - T_c),$$

where β is a positive constant less than 1. In our experiments we have used $\beta = 0.5$. Our access probability function is similar to those used in [18, 6].

4.3 Analysis of Search Method

As in Figure 3, let n_i and n_j have 1-hop neighborhoods N_i and N_j respectively. Let a request for object O originate at n_i, and let the application of HRW yield $n_j \neq n_i$ as prime in N_i, and n_k as prime in N_j. Since HRW orders nodes linearly, n_j and n_k can not both be in N_i, unless $n_j = n_k$. The probability that $n_j = n_k$ is high when the neighborhoods N_i and N_j share most of their nodes, that is, when $N_i \cap N_j$ has higher cardinality than $N_i \setminus N_j$ and $N_j \setminus N_i$.

Let x be the distance between n_i and n_j, and let R be the transmission range of a node. Then the area of intersection $A(x)$ of 1-hop neighborhoods is

$$A(x) = 2R^2 \cos^{-1}\left(\frac{x}{2R}\right) - \frac{x\sqrt{4R^2 - x^2}}{2}$$

If $A = \pi R^2$ is the area of a single 1-hop neighborhood, the fraction of overlap area at any separation x is $A(x)/A$. Since x can vary in the range $(0, R)$, we can obtain the expected fraction of overlap area over the range $(0, R)$ as

$$\frac{1}{R}\int_0^R \frac{A(x)}{\pi R^2}\,dx = \left| \frac{(4R^2 - x^2)^{\frac{3}{2}}}{6\pi R^3} - \frac{4}{\pi}\sqrt{1 - \frac{x^2}{4R^2}} + \frac{2x}{\pi R}\cos^{-1}\left(\frac{x}{2R}\right) \right|_0^R \approx 0.6884$$

If nodes are uniformly distributed, the number of nodes in a region is roughly proportional to its area. Consequently, as the search progresses, the expected overlap between two successive 1-hop neighborhoods is more than 68%. For a given object O, n_j will be picked randomly in N_i, since HRW is randomizing. Thus, the probability $\Pr[n_k = n_j]$ of the HRW search process stopping at any given step exceeds 0.68. If we model the search as a series of Bernoulli trials, each with a probability $p = 0.6884$ of success, the expected number of trials to success is given by the mean of the Geometric distribution $G(k) = (1-p)^k p$, which is simply $\frac{1}{p} = \frac{1}{0.6884} \approx 1.45$. Thus, the search is likely to bottom out quickly when the node distribution is uniform and sufficiently dense.

5 Experiments and Results

Our goal is to minimize the average *access delay* for clients, that is, the time to retrieve a document, either from the peer nodes or from broadcast by the server. Our experiments show that our mechanism greatly reduces access delays.

We used ns-2 (version 2.26) [1] in our experiments. N mobile nodes were randomly dispersed in a $2000m \times 2000m$ square, each moving according to the random waypoint model [3]. A stationary server with a range covering this entire region broadcasted a set of D objects at a bandwidth of 11 Mbps. At each step, a node moved to a random destination at a randomly chosen velocity between $(0, 2.0)$ m/s, and remained there for a pause time of 1 minute.

The bandwidth between mobile nodes was 2Mbps, and the communication range was 250 m. Requests were generated by randomly selecting a node n_i, and having it generate a request for an object O_j according to the Zipf popularity model [23]. O_j will ultimately be retrieved from peer nodes or from the broadcast. Requests were distributed exponentially with an average rate of 10 requests per second. For each set of experiments, 10,000 requests were generated and average delay between request generation and object retrieval was computed.

5.1 Hop Count and Hit Rates

We first evaluated how the hit rate increased with the number of hops. This is an important metric, since going to nodes farther away also increases access delays and traffic. We varied the Zipf parameter α, but kept all other parameters at their default values (Table 1).Figure 4 shows the cumulative object hit rate as we go h hops away from the source. We see that hit

α	h=0	h=1	h=2	h=3	h≥4
1.0	19.6	60.7	68.8	69.8	70.0
0.8	13.5	46.0	52.4	53.5	53.7
0.6	9.1	32.1	36.8	37.6	37.7
0.4	5.6	21.1	24.5	25.1	25.2

Fig. 4. Hit rates

Table 1. Simulation parameters

Parameter	Range	Default
# Objects (D)	500–5000	1000
# Mobile nodes (N)	25–200	100
Node velocity (V)	0.0–2.0 m/s	1.0 m/s
Cache size (% of total object size)	5%–40%	20%
Zipf parameter	0.0–1.0	0.7
Segment size (# objects)	100–500	100
Inter segment delay	10–120 seconds	60 seconds

rates are quite high for small h, especially when α is high. It also suffices to go 3 hops, since the increase in hit rates for more hops is marginal. Our search mechanism is extremely efficient. We can achieve even better performance by increasing the cache size and number of nodes.

5.2 Effect of Cache Size

Larger caches allow mobile nodes to store more objects, thus more objects are served from the local and peer cache. This leads to a reduction in average access delay observed by nodes, and this effect is quite obvious in Figure 5(a).

5.3 Effect of Zipf Parameter

The Zipf parameter α is a measure of skew in object popularity. Higher α indicates that some objects are requested more frequently than others. Caching popular objects minimizes average access delays.

Our caching mechanism favors caching prime and popular objects, reducing average access delays. This effect is clearly shown in 5(b). Average access delay decreases from 108.15 seconds to 72.58 seconds, as the value of α is changed from 0.0 to 1.0, whereas without caching average access delay is around 303 seconds.

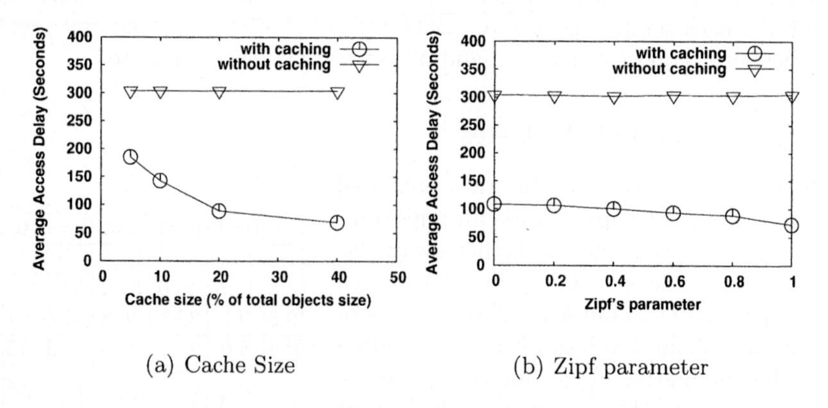

(a) Cache Size

(b) Zipf parameter

Fig. 5. Effects of cache size and Zipf parameter

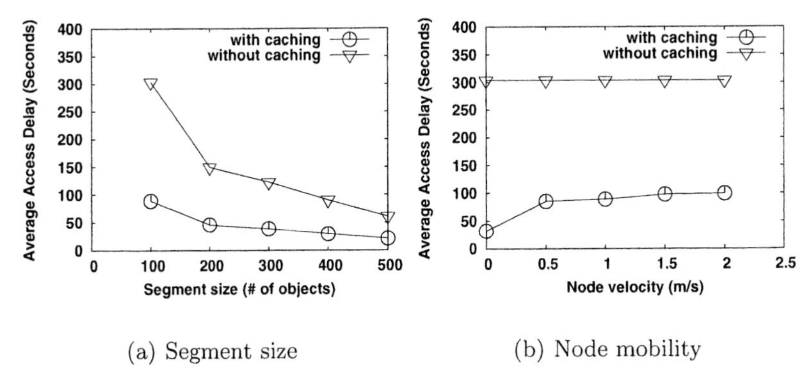

(a) Segment size (b) Node mobility

Fig. 6. Effects of segment size and node mobility

5.4 Effect of Segment Size

Segment size measures the number of objects in each segment. The total number of segments in a broadcast cycle decreases as segment size increases, so that the number of inter-segment silence periods decreases. Our inter-segment silences are greater than the broadcast time of a segment, so that the broadcast cycle length decreases with an increase in segment size.

Increasing segment size has both positive and negative effects. Larger segments result in lower average access delay (see Figure 6(a)), but the wireless channel is occupied for a longer period of time by the broadcast server, which may result in longer wait times for inter-client communications. Segment size is a tunable parameter, and can be set for optimum performance.

5.5 Effects of Node Mobility

The neighborhood set of a mobile node is always changing, so that the prime for any object will also change. Thus, the probability of finding an object in the neighborhood of a mobile node decreases with mobility, increasing average access delay. Discussions in [12, 11] have shown that users of a wireless LAN can be safely assumed to be stationary. As Figure 6(b) shows, the performance of our method remains surprisingly stable over a large range of node velocities.

5.6 Effect of Number of Objects

Increasing the total number of broadcasted objects increases the number of segments. Thus, broadcast cycle length increases with the number of objects. This effect is shown in Figure 7(a). However, the average access latency can be reduced by increasing the size of segment as Figure 6(a) shows.

5.7 Effect of Node Density

The number of nodes in each 1-hop neighborhood increases with node density. Consequently, each node is prime for fewer objects, and must cache fewer prime

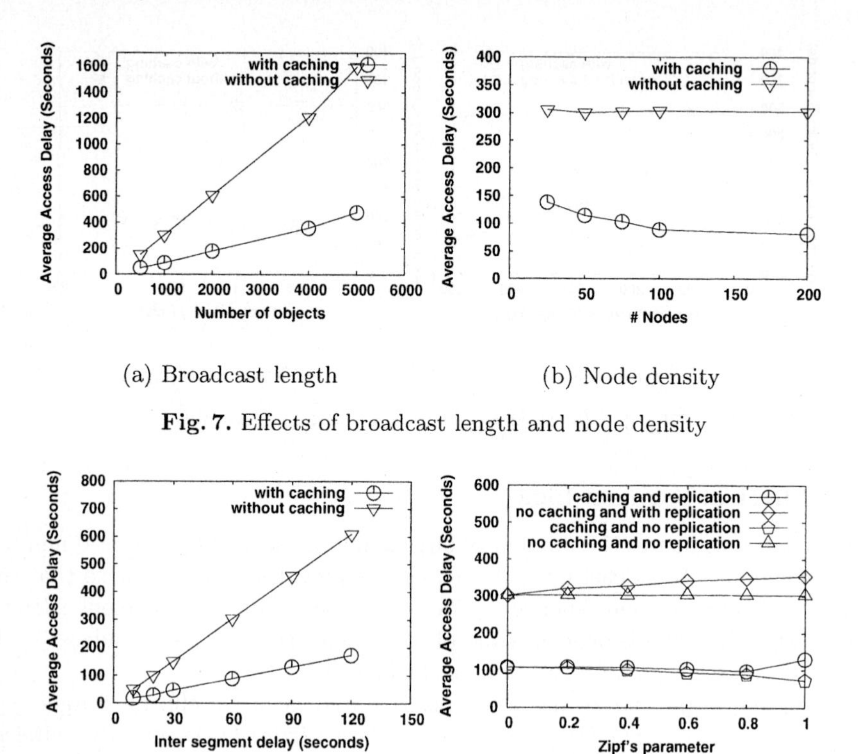

(a) Broadcast length (b) Node density

Fig. 7. Effects of broadcast length and node density

(a) Intersegment delay (b) Zipf parameter

Fig. 8. Effects of intersegment delay and object replication. With no caching, objects are obtained directly from the broadcast.

objects. It can use rest of the cache space to cache other popular non-prime objects. This leads to an improvement in performance as is evident from 7(b).

5.8 Effect of Inter-segment Delay

The broadcast cycle length increases as inter segment delay increases, so that average access latency increases. However, longer the inter segment delays allow mobile devices greater flexibility to communicate during the silences. Since clients can obtain objects from peers, performance degrades only moderately with our mechanism (see Figure 8(a)) as inter-segment delay increases.

5.9 Broadcast with Object Replication

Servers can reduce access latencies by broadcasting an object several times in each cycle, in proportion to its popularity [17]. We assume that the server knows object popularities, and creates the broadcast schedule as in [17]. We have compared the replicated broadcast scheme without caching and our caching scheme without replication. Figure 8(b) shows that our caching scheme outperforms object replication in broadcasts. We can achieve even better results by combining caching with replication.

6 Conclusion and Future Work

We have argued that the standard approach of continuous broadcasts is unsuitable for typical wireless environments. The server monopolizes the medium, preventing communication among clients. We have addressed this issue by advocating the use of segmented broadcast. Wireless devices can communicate among themselves in the silence period between two consecutive broadcasts.

We counteract increases in access times due to these silences using a novel cooperative caching scheme that allows clients to obtain objects from peers, rather than from server broadcasts alone. Our approach uses limited cache space at clients effectively, and uses a hash-based mapping function, making it completely decentralized. Earlier approaches have required inter-cache cooperation, which is an unreasonable requirement in wireless broadcast schemes. Our experimental results clearly demonstrate the efficiency of our approach.

A standard approach in broadcasting is to replicate each object in proportion to its popularity. Although this scheme reduces the average object access delay, we have been able to achieve similar performance by our caching approach, even without the use of replication. We have assumed homogeneous mobile clients with identical transmission and receiving ranges. The case of heterogeneous ranges will be the subject of future work.

References

1. Ns. http://www.isi.edu/nsnam/ns/.
2. S. Acharya, R. Alonso, M. Franklin, and S. Zdonik. Broadcast disks: data management for asymmetric communication environments. *SIGMOD Rec.*, 24(2):199–210, 1995.
3. J. Broch, D. A. Maltz, D. B. Johnson, Y.-C. Hu, and J. Jetcheva. A performance comparison of multi-hop wireless ad hoc network routing protocols. In *Mobile Computing and Networking*, pages 85–97, 1998.
4. J. Cai, K.-L. Tan, and B. C. Ooi. On incremental cache coherency schemes in mobile computing environments. In *Proceedings of the Thirteenth International Conference on Data Engineering*, pages 114–123. IEEE Computer Society, 1997.
5. G. Cao, L. Yin, and C. R. Das. Cooperative cache-based data access in ad hoc networks. *IEEE Computer*, 37:32–39, 2004.
6. S. Chen, B. Shen, S. Wee, and X. Zhang. Adaptive and Lazy Segmentation Based Proxy Caching for Streaming Media Delivery. In *Proceedings of the 13th international workshop on Networks and operating systems support for digital audio and video*, number 1-58113-694-3, pages 22–31, June 2003.
7. C.-Y. Chow, H. V. Leong, and A. Chan. Peer-to-peer cooperative caching in mobile environments. In *Proceedings of the 24th International Conference on Distributed Computing Systems Workshops*. IEEE Computer Society, 2004.

8. S. Hameed and N. H. Vaidya. Efficient algorithms for scheduling data broadcast. *Wirel. Netw.*, 5(3):183–193, 1999.
9. T. Hara. Cooperative caching by mobile clients in push-based information systems. In *Proceedings of the eleventh international conference on Information and knowledge management*, pages 186–193. ACM Press, 2002.
10. D. Karger, A. Sherman, A. Berkheimer, B. Bogstad, R. Dhanidina, K. Iwamoto, B. Kim, L. Matkins, and Y. Yerushalmi. Web Caching with Consistent Hashing. In *Proceedings of 8th International World Wide Web Conference*, May 1999.
11. D. Kotz and K. Essien. Analysis of a campus-wide wireless network. In *Proceedings of International Conference on Mobile Computing and Networking(MOBICOM)*, pages 107–118, 2002.
12. P. Nuggehalli, V. Srinivasan, and C.-F. Chiasserini. Energy-efficient caching strategies in ad hoc wireless networks. In *Proceedings of the 4th ACM international symposium on Mobile ad hoc networking & computing*, pages 25–34, 2003.
13. W.-C. Peng and M.-S. Chen. Allocation of Shared Data based on Mobile User Movement. In *Proceedings of the Third International Conference on Mobile Data Management (MDM'02)*, 2002.
14. C.-J. Su and L. Tassiulas. Joint Broadcast Scheduling and User's Cache Management for Efficient Information Delivery. In *Proceedings of International Conference on Mobile Computing and Networking(MOBICOM)*, 1998.
15. J. B. Tchakarov and N. H. Vaidya. Efficient content locations in wireless ad hoc networks. In *Proceedings of the IEEE International Conference on Mobile Data Management (MDM'04)*. IEEE Computer Society, 2004.
16. D. G. Thahler and C. V. Ravishankar. Using Name-Based Mapping to Increase Hit Rates. *IEEE/ACM Transactions on Networking*, 6:1–13, February 1998.
17. W. Wang and C. V. Ravishankar. Adaptive data broadcasting in asymmetric communication environments. In *Proceedings of 8th International Database Engineering and Applications Symposium (IDEAS)*, pages 27–36. IEEE Computer Society, 2004.
18. K. L. Wu, P. S. Yu, and J. L. Wolf. Segment-Based Procy Caching of Multimedia Streams. In *Proceedings of the tenth internation conference on World Wide Web*, number 1-58113-348-0, pages 36–44, May 2001.
19. B. Xu, O. Wolfson, and S. Chamberlain. Cost based data dissemination in broadcast networks with disconnection. In J. V. den Bussche and V. Vianu, editors, *Database Theory - ICDT 2001, 8th International Conference, London, UK, January 4-6, 2001, Proceedings*, volume 1973 of *Lecture Notes in Computer Science*, pages 114–128. Springer, 2001.
20. B. Xu, O. Wolfson, S. Chamberlain, and N. Rishe. Cost based data dissemination in satellite networks. *Mob. Netw. Appl.*, 7(1):49–66, 2002.
21. E. Yajima, T. Hara, M. Tsukamoto, and S. Nishio. Scheduling and Caching Strategies for Correlated Data in Push-based Information Systems. volume 9, pages 22–28. ACM Press, 2001.
22. W. G. Yee, S. B. Navathe, E. Omiecinski, and C. Jermaine. Efficient Data Allocation over Multiple Channels at Broadcast Servers. In *IEEE Transactions on Computers*, October 2002.
23. G. K. Zipf. *Human Behavior and the Principles of Least Effort.* Addison-Wesley, Cambridge, MA, 1949.

Range Adjustment for Broadcast Protocols with a Realistic Radio Transceiver Energy Model in Short-Range Wireless Networks*

Jialiang Lu[1], Fabrice Valois[1], and Dominique Barthel[2]

[1] INRIA ARES Project, CITI Lab., INSA Lyon, France
{firstname.lastname}@insa-lyon.fr
[2] France Telecom R&D TECH/ONE, Meylan, France
Dominique.barthel@rd.francetelecom.com

Abstract. Broadcasting in short-range wireless ad hoc networks is a key issue. Currently, these networks can be formed not only by computer hosts with full PC functionalities, but also by small communicating objects. Hence energy-efficiency becomes a big challenge for broadcasting protocols. To design an energy-efficient broadcasting protocol, an energy model should be defined to evaluate protocols performances. Nevertheless, all the performance evaluations are based on a one-to-one energy model without taking into account the wireless multicast advantage: such a model is not able to describe correctly the energy consumption in a broadcast. In this paper, we propose to use a more realistic one-to-all energy model to quantify the broadcast cost. By applying this model to the main broadcasting protocols, we will discuss how a range adjustment technique takes effect. We will also show that in a broadcast, a range adjustment provides a trade-off between the number of rebroadcasts and the number of nodes touched by each rebroadcast. It is through this trade-off that a broadcasting protocol can achieve energy-efficiency.

Keywords: energy model, broadcast protocols, performance evaluation, range adjustment.

1 Introduction

Broadcasting is a communication paradigm that allows users to send data packets from a source to multiple receivers. It is the basic mechanism to developp routing protocols, services discovery, etc. In a short-range wireless ad hoc network, a broadcast message transmitted by a host is only received by all its neighbors within the radio range: it is the wireless multicast advantage (WMA). By default the broadcast message is not relayed by its neighbors, hence the broadcast task is not achieved. The simplest idea is to use flooding: each node retransmits the broadcast message once it receives the first copy of the message.

* This work is financed and supported by France Telecom R&D under CRE No 46128746 with PACIFIC team.

X. Jia, J. Wu, and Y. He (Eds.): MSN 2005, LNCS 3794, pp. 197–206, 2005.

Nevertheless, this protocol often engenders unproductive and harmful networks, congestions and the inefficient use of node resources [15].

Recent broadcasting techniques aim to either minimize the number of retransmission or minimize the total energy cost on all the nodes [10, 8]. These techniques use range adjusting to save energy in the whole network. In these researches, path-loss model is usually used to compute the radiated power whereas reception energy consumptions are rarely taken into account. In this paper, we will highlight that the energy consumption on the receivers is even more important than that on the transmitter: in reception mode, in addition of energy consumption of the electronic circuits, decoding scheme and equalization are also applied. More, due to WMA, many receivers are implied. We propose in this article to utilize a more realistic one-to-all energy model for evaluating the cost of broadcasting protocols. It is commonly understood that a range adjustment saves energy on the transmitter side. We show that with our model a range adjustment has especially influence on reception consumption, which is a dominant factor to examine the energy-efficiency of a broadcasting protocol.

This paper is organized as follows. In Section 2, we present the common energy models used in major broadcasting protocols and an overview of energy-efficient broadcasting techniques based on these models. Section 3, we give a realistic energy model in which the cost of receptions is integrated. It describes better energy consumption in a broadcast. Simulation results and analysis of broadcasting protocols are presented in the Section 4 using our energy model. Finally, conclusion and future works are given.

2 Energy-Efficient Broadcasting with Common Energy Model

2.1 Communication Model and Energy Model

A common communication model is to represent a wireless ad hoc networks by a unit disk graph $UDG(G) = (V, E)$, where V is a set of nodes and $E \subseteq V^2$ the edge set. If we assume radio links are symmetric, then the graph is a non-orientation graph. Unless there are obstacles in the communication area, an edge between host pairs (u, v) belongs to E when distance between u and v is smaller than the transmission range, denoted by R. We also define the neighborhood set $N(u)$ of the node u as the subset of the nodes who has an edge with u. And the degree of a given vertex u is the number of nodes in $N(u)$. The density of the graph is defined as the average degree of all nodes.

In most common consideration, the energy consumption of a transmitter depends on the radius $r(u)$:

$$e(u) = E_{tm} * r(u)^{\alpha} \quad (1)$$

Where E_{tm} is a relative energy coefficient on (J/m^{α}), α is a constant greater or equal to 2 due to the radio environment and $r(u)$ is the transmission range of u. This model is used in some works: *e.g.* [14].

[13] points out that a constant should be added in order to take the overhead due to signal processing, minimum energy needed for successful reception into account. The general energy consumption model turns to:

$$e(u) = E_{tm} * r(u)^{\alpha} + C \quad (2)$$

2.2 Broadcasting with Fixed Range

Firstly, in a wireless ad hoc network where transmission range of each host is non-adjustable, the obvious idea is to reduce the number of rebroadcasts to optimize energy consumption. The proposed broadcast techniques can be classified into neighbor elimination based, coverage based, forwarding based and probabilistic.

NES (Neighbor Elimination Scheme) is a simple wait-and-see technique for broadcasting [17]. When receiving a new broadcast message, the node generates, for this broadcast, a list of neighbors that have not yet received this message. On receiving the copy of the message, the node eliminates the neighbors which are estimated to receive the copy, from its list. After a given timeout, if the list is not empty, the node rebroadcasts the message. This technique can reduce significantly the number of rebroadcasts in the network.

The coverage based broadcasting protocols concentrate on computing a small *CDS* (Connected Dominating Set) S for a set of nodes V. Each node of V is either in S or a neighbor of a node in S. Furthermore S should be connected. Many algorithms has been proposed to generate a small cardinality *CDS* in a distributed fashion: [3], [12], [19]. The rule of broadcasting is then defined as: when receiving a broadcast message for the first time, a dominator (i.e. in CDS) will forward the message while a non-dominant node won't perform forwarding action.

Opposite to coverage based broadcasting protocols of which the broadcast structure is source-independent, forwarding based approach is source-dependent. *MPR* (Multi-Point Relay) [1] is the most known forwarding based protocol. Each node computes a subset (*MPR* set) of its 1-hop neighbor set that covers its whole 2-hop neighborhood. When a node receives a broadcast message, it will forward it only if it is a MPR of the sender. Another kind of forwarding based broadcasting protocols are based on RNG, which is a geometric concept introduced by Toussaint [18]. *RRS* (RNG Relay Subset) [11] is in this family. It is also a variant of *NES*, in which nodes watch over only their RNG-neighborhood during the *NES*.

In the probabilistic approach, the messages are forwarded with a probability on each node. This probability can be fixed or computed locally by nodes with different parameters.

It is worth noting that the previous protocols assume the common energy model (1). Though all of them evaluate the number of retransmission in a broadcast as the unique criterion.

2.3 Broadcasting with Range Adjustment

While range adjustment is possible, it is natural to think that a node does not always need to transmit a message at full power for various reasons (for instance, some neighbors may have already received the message). According to the energy model in (1), shorter the transmission range is, less energy is consumed. So in the first category of range adjustment algorithms, it is assumed that the shortest radius is the optimal one.

[10] proposes *RBOP* (RNG Broadcast Oriented Protocol) and *LBOP* (LMST Broadcast Oriented Protocol), in which each node applies a range adjustment on its *RNG* (Relative Neighborhood Graph) and *LMST* (Local Minimum Spanning Tree) subgraph: a node u using *NES*, rebroadcasts the message with its range $r(u)$ determined by the appropriate subgraph (*RNG* or *LMST*). After a given timeout, if the list is not empty, the node rebroadcasts the message with a range allowing it to reach its furthest neighbor left in its list. In this protocol neighbors' relative position information is needed.

[8] proposes another algorithm to compute the needed range. Each node broadcasts at each power level to verify whether the subgraph of all its 1-hop neighbors is still connected. Then, each node chooses the minimal range, with which the connectivity of the subgraph is conserved, as its transmission radius. The advantage of this scheme is that the procedure is localized and no position information is needed. However the communication overhead that occurs when determining the transmission radius seems significant, and cannot achieve topology changes.

With the energy model in (2), the previous assumption that the shortest radius is the optimal one is no more validated. [5] proposes a protocol named *TR-LBOP* aiming to determine a target radius for a given broadcast, to reserve an optimal energy consumption. Considering a geometrical area S divided to a honeycomb mesh, 2 nodes are placed on each hexagon in the best case, so that each node receives once the message. Then the total energy consumption for a broadcast is a function of radius r: $PC(r) = (r^\alpha + C)\frac{k}{r^2}$ with $k = \frac{4S}{3\sqrt{3}}$, α is the path-loss model coefficient. Using this configuration, they can compute optimal radii for a given space with a given number of nodes, depending on coefficient α and C. The second idea in this protocol is to add this optimal radius in LBOP. Each node increases its radius up to the optimal one if possible, when the rebroadcast is needed.

3 A Energy Model for Short-Range Transmission

According to the previous solutions, we investigate a one-to-all energy model dedicated to short-range transmission using range adjustement. The common idea is to adjust the radius, taking effects of exponential coefficient α and constant C into account. The total energy consumption is represented by the sum of all transmission power. However in a short range wireless radio transmission, the energy consumed in electronic circuits during the transmission and reception is much more important than radiated energy cost (in antenna). For example, the power values in Table 1 are for Freescale MC13192 chipset [16], used as a 802.15.4 transceiver. It is clearly shown that the electronic circuits' consumptions represent the major part in wireless transmission and a reception is as costly as a transmission.

Table 1. Radio power characteristic

Radiated power (mW)	Transmit mode (mW)	Receive mode (mW)
1	81	100

As we mentioned earlier, some works assume that the energy consumption in electronic circuits can be integrated in the constant C of Equation (2) [4]. However when a node rebroadcasts the message, all its neighbors receive a copy of it. Hence an amount of energy for reception is spent on each neighbor. It signifies that the reception energy consumption is along with the number of neighbors within the transmission range of the rebroadcast node, and can not be simply represented by a constant.

If we note E_{tx} and E_{rx} as the unit energy consumption per message during a transmission and a reception respectively, then the energy consumption E_u for a node u rebroadcasting the message is:

$$E_u = E_{tx} + n_u E_{rx},\ where\ n_u = |N(u)| \tag{3}$$

Assuming that the number of the forwarding nodes in a broadcasting protocol is m, then the total power consumption for a broadcast is:

$$E_{total} = \sum_{i=1}^{m} \{E_{tx} + n_i E_{rx}\}. \tag{4}$$

Further, E_{tx} and E_{rx} have their expressions along with the radio communications and decoding techniques used. Typically a transmission cost is composed of energy needed by transmitter electronics E_{te}, energy needed for transmitting a packet over a distance $E_{tm}(d)^\alpha$ (according to path loss model) and E_{st} which is the startup energy to activate a transmitter electronics from idle (or reception...) mode to transmit mode [20]. For reception, in addition of energy needed by receiver electronics E_{re} and startup energy E_{sr}, a decoding cost E_{dec} has to be taken into account.

$$\begin{cases} E_{tx} = E_{te} + E_{tm}(d)^\alpha + E_{st} \\ E_{rx} = E_{re} + E_{dec} + E_{sr} \end{cases} \tag{5}$$

It is generally accepted, the transmission radius which achieves the minimum value of $E_{tm}(d)^\alpha$ is optimal. However it is obvious in our model that transmission radius has effects on not only radiated power but also the number of transmission and the number of receptions occurring in each rebroadcast. Reducing transmission radius is useful to reduce the number of neighbors touched by a broadcast. On the other hand, it leads to increase of rebroadcast nodes. Furthermore, the total energy consumption is dominated by the number of transmissions and the number of receptions. It is primary through a trade-off between these two parameters that a broadcasting protocol can achieve its energy-efficiency.

It is worth highlighting that we assume an ideal MAC layer which activates the transceiver from idle to reception model upon detecting a message. A receiver which is able to receive messages at any time by randomly sensing the medium spends as much energy as a receiver always in reception mode.

Taking reception energy consumption into account has already been introduced in [6, 20]. But as far as our knowledge, only the path-loss model (1) and its variant (2) are used as the energy model for broadcasting protocol performance evaluation and design consideration. In the next section, we analyze some well-known broadcasting protocols: *CDS*, *MPR* and *NES*, in order to show the influence of range adjustment on these protocols.

4 Simulations and Analysis

4.1 Simulation Guideline

We use a simulator developed in C language. It generates random unit disk graphs to model ad hoc networks. We adopt certain assumptions to define the area of our studies: 1) an ideal MAC protocol is used, which provides a collision-free broadcast; 2) Nodes are static while the broadcasting is in progress; 3) the channel is time-slotted and each transmission takes one slot. A random UDG is generated in an area of 1*1 with 200 nodes. An edge is added if the distance between two nodes is shorter than radius R in the interval [0, 1]. In order to make sure that each node participates in the broadcast, we ignored the disconnected graphs and considered the connected networks only.

In this simulation, the study is carried out on *CDS* (algorithm of Wu and Li [19]), *MPR* ([1]), *NES* ([9]), *MPR* with additional *NES* and *NES* with range adjustment. In order to add range adjustment on *NES*, after the backoff timer, each node transmits the message with the power which is enough to cover the furthest neighbor in the listening list. We collect the number of rebroadcasts, the number of receptions and the cost function given in (4) as statistics along with transmission radius.

4.2 Results

The ratio of the nodes which forward the message is measured (Fig 1). Since each node rebroadcasts once at most, this value is the number of rebroadcasts per node. It is obvious that the number of rebroadcasts decreases when the radius increases. It is easy to understand this phenomenon because a rebroadcast with a big radius touches more nodes in the network. Though, the number of rebroadcast nodes is smaller. When the radius is larger than 0.4, all the broadcast protocols exhib the same behavior. In Table 2, the relation between network density and transmission radius is given. We note that the network density increases when radius becomes bigger. So the number of rebroadcast nodes decreases when the network density increases. This result can also been found in a recent work [7]. Our simulation results show that the Wu and Li's algorithm (*CDS*) achieves a small number of rebroadcasts, because their algorithm favors the nodes who have many neighbors to be dominators. We find also that adding *NES* to *MPR* can reduce significantly the

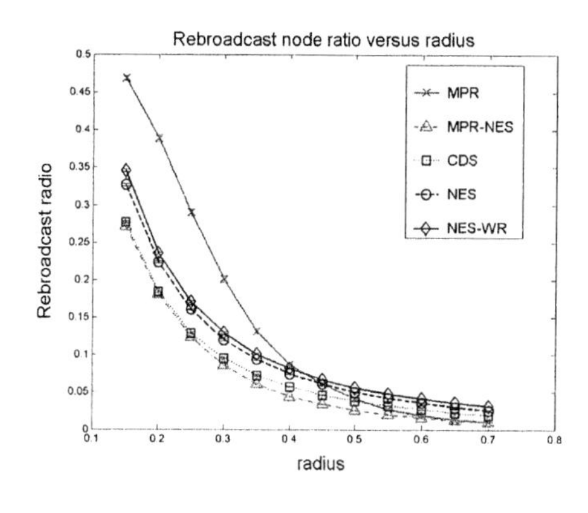

Fig. 1. Rebroadcast nodes

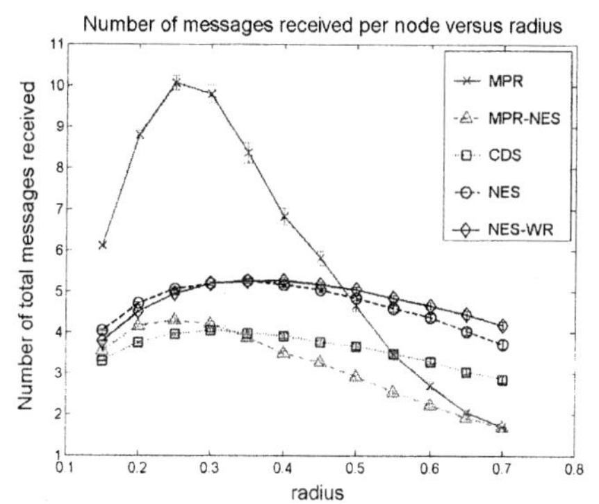

Fig. 2. The number of receptions

Table 2. Network degree versus radius

Radius	0.15	0.20	0.25	0.30	0.35	0.40	0.45	0.50	0.55	0.60	0.65	0.70
Density	12.3	20.9	31.1	42.7	55.3	68.7	82.3	96.1	109.8	123.3	136.1	148.2

number of rebroadcasts. However adding range adjustment to *NES* does not make any improves on the number of forwarding nodes.

We also measure the average number of times that the copies of a message are received by a node, given in Fig 2. Here, our simulation results are very interesting. None of the curves is a monotone function and the maximums of these curves do not occur on a same radius value. This observation differs from the results in [7], where the average receptions per node increases with respect to network density. In fact, their results are with a network degree up to 20, which are in the waxing part of our curves.

One clarification has been given in [2] in which author points out that the ratio between the number of receptions and the number of rebroadcasts is a factor of r^2. The reason is that the average number of receptions per transmission is linear to the area covered by the transmission, which is itself linear to r^2. In our opinion, this stochastic relation explained above is correct only if almost all the nodes transmit. However we think this simple relation cannot describe the reception and forwarding relation in a broadcasting protocol. In order to figure out clearly, we compared the average number of reception copies per node obtained in our simulation to the value estimated by multiplying r^2 and the number of rebroadcasts. The curves carried out on *CDS* and *MPR* are presented in Fig 3 and Fig 4. We remark that the factor r^2 can be almost identified with MPR in Fig 4 in which many rebroadcasts occur. But neither in CDS in Fig 3 nor in MPR with NES in Fig 4, the curves show this relation. It is because that the nodes chosen to rebroadcast have some special proprieties depending on the algorithms used. We can conclude that the relation between the number of receptions and the number of forwarding nodes depends on the way in which rebroadcast nodes are chosen, and it is not simply a factor of r^2.

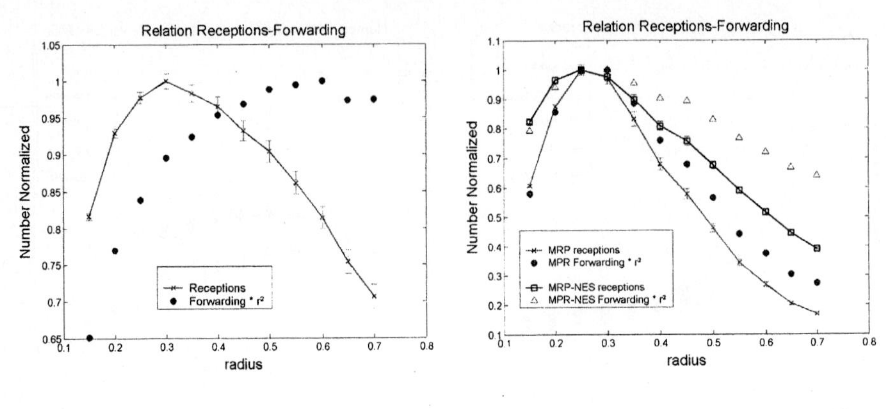

Fig. 3. CDS

Fig. 4. MPR and MPR with NES

We remark also that *CDS* achieves a relatively low reception redundancy and a low sensibility of densities. Once more there is no significant different between simple *NES* and *NES* with range adjustment. Combining Fig 1 and Fig 2, we can point out that adding traditional range adjustment on *NES* is useless to saving energy consumption.

We use the radio characteristics of Freescale MC13192 (Fig 5) in this simulation to compute the total energy consumption in Fig 6. Since there is few difference between the energy consumed on transmitter and that on receiver, and the number of receptions is much more important than the number of re-broadcasts due to the WMA, energy consumption curves has an appearance very close to the number of receptions in Fig 2. This proves that the energy consumption is dominated by receptions in a broadcast. In order to determine in which way we could save energy (i.e. by increasing or by decreasing radius), the energy consumption curves should be examined on the possible radius interval. For instance, in the waxing part of the curve, less energy is spent when using small

Packet size	50 Bytes (400b)
Rate	250Kb/s
Packet duration	1.6ms per packet
Transmission power	81mW
Reception power	100mW
Startup power	50mW
Startup time	144μs

Fig. 5. Simulation parameters based on Freescale MC1392

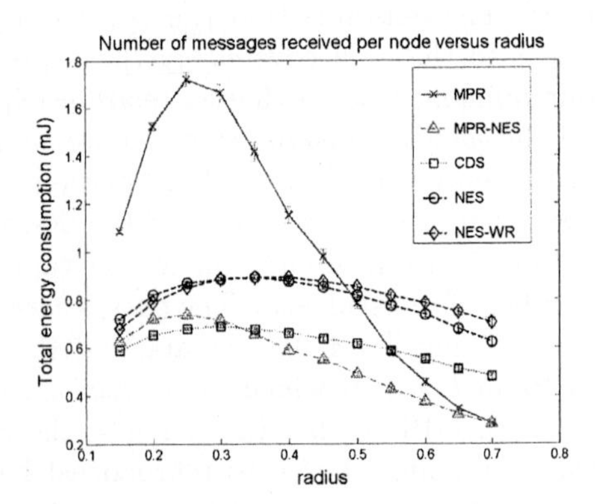

Fig. 6. Total energy consumption for one broadcast

radius. While in the waning part of the curve, it's better to transmit as strong as possible. However, this is only validated for short-range communication.

5 Conclusion and Future Research Directions

Energy saving is a big challenge in broadcasting in short-range wireless ad hoc networks. We highlight in this paper that the common one-to-one energy models used in this field of research are too simple to describe the energy consumption in a broadcast. Some of them even lead to incorrect assumptions on range adjustment. The main contribution of this paper is that we give a more realistic one-to-all energy model for broadcast taking into account the wireless multicast advantage and re-examine the assumptions on energy-efficient broadcasting. Range adjustment techniques effect primarily on the number of transmission and the number of neighbors in each transmission, but very few on the radiated energy. Furthermore, in some short range communication networks such as sensor networks, minimizing the total energy consumption in a broadcast is not enough. In order to achieve a reasonable network life time for given applications, the residual power on each node has to be also taken into account in protocols. The solutions like static CDS are not the suitable approach for this kind of networks, because retransmission always occurs on the *dominators* which spend much more energy than others. Forwarding nodes and no forwarding nodes must switch their roles quite often. Total and individual energy consumption in a same broadcasting protocol will be considered in our future research.

Acknowledgments

The authors want to thank their colleagues in FT R&D PACIFIC team and CITI Laboratory for those discussions which have improved this work.

References

1. L. Viennot A. Qayyum and A. Laouiti. Multipoint relaying for flooding broadcast messages in mobile wireless networks. In *Hawaii International Conference on System Sciences (HICSS02)*, 2002.
2. G. Chelius. *Architectures et communications dans les reseaux spontanes sans fil.* PhD thesis, l'INSA de Lyon, 2004.
3. Y.P. Chen and A.L. Liestman. Approximating minimum size weakly-connected dominating sets for clustering mobile ad hoc networks. In *Proc. of Internatinal Symposium on Mobile Ad Hoc Networking and Computing*, 2002.
4. D. Simplot F. Ingelrest and I. Stojmenovic. Energy-efficient broadcasting in wireless mobile ad hoc networks. Chartiper of Resource Management in Wireless Networking, 2004.
5. D. Simplot-Ryl F. Ingelrest and I. Stojmenovic. Target transmission radius over LMST for energy-efficient broadcast protocol in ad hoc networks. *The 39th IEEE International Conference on Communications (ICC'04) Paris, France*, 2004.

6. L. Feeney. An energy-consumption model for performance analysis of routing protocols for mobile ad hoc networks. *ACM MONET*, 6:239–249, 2001.
7. H. Guo, F. Ingelrest, D. Simplot-Ryl, and I. Stojmenovic. Performance evaluation of broadcasting protocols for ad hoc and sensor networks. *The Fourth Annual Mediterranean Ad Hoc Networking Workshop (Med-Hoc-Net 2005), France*, 2005.
8. Z. Huang and June 2003. C.-C. Shen. Distributed topology control mechanism for mobile ad hoc networks with swarm intelligence. In *ACM MOBIHOC03, Annapolis, MD, USA*, 2003.
9. M. Seddigh I. Stojmenovic and J. Zunic. Dominating sets and neighbor elimination based broadcasting algorithms in wireless networks. *IEEE Transactions on Parallel and Distributed Systems*, 13(1):14–25, January 2002.
10. D. Simplot-Ryl J. Cartigny, F. Ingelrest and I. Stojmenovic. Localized LMST and RNG based minimum energy broadcast protocols in ad hoc networks. In *IEEE INFOCOM, San Francisco, CA, USA, April 1-3*, 2003.
11. F.Ingelrest J. Cartigny and D.Simplot. RNG relay subset flooding protocols in mobile ad hoc networks. *International Journal of Fondations of Computer Science (IJFCS)*, 14(2):253–265, April 2003.
12. P. Wang K.M. Alzoubi and O.Frieder. Message-optimal connected dominating sets in mobile ad hoc networks. *Proceedings of the 3rd ACM international symposium on Mobile ad hoc networking & computing*, 2002.
13. V. Rodoplu and T. Meng. Minimum energy mobile wireless networks. *IEEE Journal on Selected Areas in Communication*, 17(8), August 1999.
14. A. Misra S. Banerjee. Minimum energy paths for reliable communication in multi-hop wireless networks. In *ACM MobiHoc 2002, Switzerland*, 2002.
15. Y. Chen S. Ni, Y. Tseng and J. Sheu. The broadcast storm problem in a mobile ad hoc network. In *International Conference on Mobile Computing and Networking, USA*, 1999.
16. Freescale Semiconductor. MC13192 / MC13193 2.4 GHz Low Power Transceiver for the IEEE 802.15.4 Standard: Technical Data, April 2005.
17. I. Stojmenovic and M. Seddigh. Broadcasting algorithms in wireless networks. In *International Conf Advances in Infrastructure for Electronic Business, Science, and Education on the Internet SSGRR*, 2000.
18. G. Toussaint. The relative neighborhood graph of finite planar set. *Pattern Recognition*, 1980.
19. J. Wu and H. Li. On calculating Connected Dominating Set for efficient routing in ad hoc wireless networks. In *Third Int' Workshop on Discrete Algorithms and Methods for Mobile Computing and Communications*, 1999.
20. Carlos Pomalaza-Rez Zach Shelby and Jussi Haapola. Energy optimization in multihop wireless embedded and sensor networks. *IEEE PIMRC2004, Spain*, 2004.

Reliable Gossip-Based Broadcast Protocol in Mobile Ad Hoc Networks

Guojun Wang[1,3], Dingzhu Lu[1], Weijia Jia[2], and Jiannong Cao[3]

[1] School of Information Science and Engineering, Central South University, Changsha, China, 410083
csgjwang@mail.csu.edu.cn, ldz214cn@163.com
[2] Department of Computer Engineering and Information Technology, City University of Hong Kong, Kowloon, Hong Kong
itjia@cityu.edu.hk
[3] Department of Computing, Hong Kong Polytechnic University, Kowloon, Hong Kong
csjcao@comp.polyu.edu.hk

Abstract. Based on existing reliable broadcast protocols in MANETs, we propose a novel reliable broadcast protocol that uses clustering technique and gossip methodology. We combine local retransmission and gossip mechanisms to provide reliability in MANETs. The proposed protocol can dynamically change system parameters for reliable broadcast communication in order to improve the adaptability in the rapidly changing network environment. In our proposed protocol, the adaptive gossip probability is explored to make the protocol insensitive to changing environment. In sparse or boundary areas, a large gossip probability is adopted in order to improve the reliability; while in dense or inner areas, a small gossip probability is used to alleviate the contention and collision. Theoretical analysis shows that the proposed protocol has high delivery ratio and low end-to-end delay for broadcasting.

1 Introduction

Mobile Ad hoc NETworks (MANETs) are multi-hop wireless networks, which consist of mobile nodes and require no fixed infrastructure. MANETs can be potentially used for military operations, rescue missions, virtual class rooms, etc.

Broadcasting is a communication primitive in MANETs. Because radio signals are likely to overlap with each other in a geographical area, a straightforward broadcasting by flooding is usually very costly and will result in serious redundancy, contention and collision, to which is referred as the broadcast storm problem [9].

Besides blind flooding [6], these broadcast protocols can be classified into three categories: probability-based methods [9], area-based methods [9] and self-pruning methods [11] [12] [13] [15]. Blind flooding requires each node to rebroadcast a broadcast packet once. Probability-based methods use some basic understanding of the network topology to assign a probability to a node to rebroadcast. Area-based methods assume nodes have common transmission distances; and a node rebroadcasts a packet only if the rebroadcast will reach sufficient additional coverage area [1]. Self-pruning

X. Jia, J. Wu, and Y. He (Eds.): MSN 2005, LNCS 3794, pp. 207 – 218, 2005.

methods maintain state on the neighborhood of each node, with which each node decides its role (forward node or non-forward node) in a specific broadcasting, where the forward nodes relay the broadcast packet, while the non-forward nodes do nothing. In these broadcast protocols, blind flooding and self-pruning based protocols behave more reliably than others in sparse networks [1]. For probability-based protocol, its unreliability in sparse area results from the low probability adopted by such protocol.

Since MANETs suffer from transmission contention and congestion as a consequence of the broadcasting nature of radio transmission, it is challenging to provide reliable broadcasting in such a dynamic environment. We propose a reliable gossip-based broadcast algorithm that uses the clustering technique, which decreases data redundancy by selecting only a part of nodes to rebroadcast without sacrificing broadcast delivery ratio.

The rest of the paper is organized as follows. Section 2 gives a brief review of reliable broadcast protocols in MANETs. In section 3, we present our Reliable Gossip-based Broadcast protocol (RGB). Section 4 gives the mathematical analysis of the RGB protocol. Section 5 concludes the paper and discusses our future work.

2 Related Works

As pointed out by many researchers, providing deterministic reliability for broadcasting in MANETs is impractical and unnecessary when the physical communication channels are prone to errors. Usually, acknowledgments (ACKs) are used to ensure reliable broadcast delivery. However, the requirement of sending ACKs in response to the receipt of a packet for all the receivers may cause channel congestion and packet collision, which is called the ACK implosion problem.

The reliable broadcast protocol proposed in [4] is flooding-based. The node only forwards the fresh message and sends an ACK back to the sender. If the sender does not receive an ACK from any of its neighbors for a predefined period, it resends the message. The flooding based protocols are subject to the broadcast storm problem. Furthermore, the ACK implosion may worsen the broadcast storm problem.

A cluster-based reliable broadcast protocol for MANETs proposed in [10] constructs a forwarding tree, which is rooted from the cluster head of source to each cluster head in order to forward the broadcast packet. The packet is forwarded down the tree from the root to the leaf nodes; the ACKs are first collected by each cluster head in each cluster and then forwarded up the tree from the leaves to the root. The algorithm switches to flooding when the topology change of the network becomes too high. The protocol is sensitive to mobility for the destruction of its tree structure.

The reliable broadcast protocols presented in [2] [8] adopt self-pruning methods, where only the selected nodes (the relayer in [2] and the forward node in [8]) relay the broadcast packet. Although ACK implosion is alleviated by the self-pruning protocols, they suffer from low delivery ratio in highly mobile networks [3].

In this paper, we present a Reliable Gossip-based Broadcast protocol (RGB) in MANETs. In sparse or boundary areas, a large gossip probability is adopted in order to improve the reliability; while in dense or inner areas, a small gossip probability is used to alleviate the contention and collision. And the light overhead of gossip mechanism avoids ACK implosion.

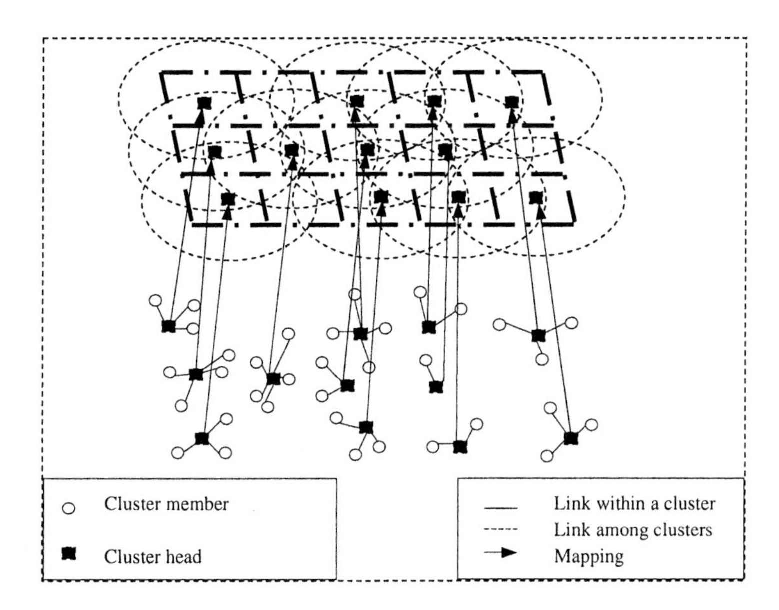

Fig. 1. The System Model of RGB

3 Proposed Reliable Gossip-Based Broadcast Protocol (RGB)

RGB is a reliable gossip-based broadcast protocol using a hierarchical structure. To deal with the rapidly changing environment in MANETs, RGB uses an adaptive mechanism in the sense that different gossip probabilities are employed by the cluster heads in different network environments. This mechanism improves the data-delivery reliability and the adaptive capability of the protocol by the local retransmission, which consists of two levels: the retransmission among cluster heads within a pre-defined maximum hops; and the retransmission within each cluster in MANETs.

3.1 System Model

Figure 1 shows the system model with clusters constituting the fundamental network architecture. In this paper, we assume to use our proposed location-based clustering technique presented in [14]. In each cluster, one node is selected as cluster head. Besides functioning as normal mobile nodes, the cluster head is also responsible for collecting information about neighbor cluster heads and exchanging messages with them. Messages including the broadcast packets and the retransmitted messages spread among cluster heads by way of gossip. All the other nodes in a cluster called cluster members send and receive broadcast packet via their cluster head. Each node is given a node *ID* which identifies a node uniquely. Each broadcast packet is given a sequence number (*SEQ*) by the sender. The *ID* of the sender and a *flag bit* (see below) are included in the packet. If the *flag bit* in the packet is set to 1, this packet is a normal broadcast packet; otherwise, it is a LRG packet defined in subsection 3.2. The triple, [*ID*, *SEQ*, *flag bit* (1)], named *Data ID*(*DID*), tags a broadcast packet uniquely.

Cluster head broadcasts beacon messages to its neighbors periodically, through which the cluster head gets its neighborhood information. The cluster members ignore these beacons. When a cluster head leaves the cluster where it resides, a new cluster head should be elected, but the cluster members can leave the cluster freely.

3.2 Major Operations of RGB

The operations of RGB include two major parts: operations among clusters and operations within each cluster. And duplicate data packets are discarded in both cases. AODV is used as the underlying unicast routing protocol. We assume the cluster heads have powerful radio capability for them to communicate directly with neighbor cluster heads [16]. Multiple channels are exploited in RGB, one of which is for normal broadcast packet delivery, and the other one of which is for local gossip-based retransmission. In this way, communication interferences between the nodes in different levels of the hierarchical structure can be avoided.

A. Operations among Clusters

Operations among the clusters include normal broadcast packet delivery and local gossip-based retransmission. In this paper, we borrow the basic idea of GBAR [5] for operations among clusters. The nodes in this part are supposed to be cluster heads and the distance between two neighbor clusters is characterized as one hop.

(1) Normal Broadcast Packet Delivery

In the wireless network, messages transmitted by a node could be received by all of its neighbors due to the broadcast nature of radio communication. In GOSSIP1(p) [5], the source sends the route request with probability 1.0. If a node first receives a fresh route request, with probability p it broadcasts the request to its neighbors, and with probability 1.0-p it discards the request.

In a random network, a node may have very few neighbors. In this case, the probability that none of the node's neighbors propagates the gossip is high. GOSSIP2($p1$, k, $p2$, n) and GOSSIP3(p, k, m) in [5] solve this problem partly to make route request fairly reliable. We borrow their basic idea in this paper for broadcast. In GOSSIP2($p1$, k, $p2$, n), $p1$ is the typical gossip probability as p in GOSSIP1 and k is the number of hops within which packets are gossiped with probability 1.0. If the neighbors of a node is fewer than n, it gossips with probability $p2$, where $1.0 \geq p2 > p1 > 0$. (p, k) in GOSSIP3(p, k, m) is just like ($p1$, k) in GOSSIP2($p1$, k, $p2$, n). A node that originally did not broadcast a received message (because it happened to fall into 1.0-p), but then did not get the message from at least m other nodes within the timeout period, broadcasts the message immediately after the timeout period. GOSSIP2 and GOSSIP3 do better in refraining gossip from dying before its spreading into the network. In [5], gossip is used to look for routing knowledge; however, we extend it to propagate broadcast packets.

There is another probability in the network that a node has too many neighbors which may result in many collisions. In this case, if packets are gossiped with small probability, the collisions can be decreased. Then we introduce a new parameter $n1$, an upper limit of the number of neighbors, into gossip. If a node's number of neighbors is larger than $n1$, the gossip probability should be decreased to $p3$, which is less than average level $p1$. We call this gossip mechanism with upper limit UGOSSIP($p1$, k, $p2$, n, $p3$, $n1$), where $1.0 > p2 \geq p1 > p3 > 0$.

In RGB, the basic data structures kept by cluster heads include:

DID: Identity of the broadcast packet received; *numnbrs*: Number of neighbors; *sendflag*: A flag to show whether the broadcast packet *DID* is sent out or not;

receivetimes: Times that the broadcast packet was received by a node; *timer*: Timer for *m* in GOSSIP3(*p*, *k*, *m*).When the broadcast packet's *receivetimes* of a node in $T_{timeout}$ is smaller than *m* and this packet was never forwarded by the node, it will be broadcast.

Normal broadcast packet delivery includes three major procedures: Timeout(*DID*), ReceiveB(*DID*) and Gossip(*p*, *DID*). Gossip(*p*, *DID*) is called by the former two procedures. The pseudo code of the procedures is presented in the following:

```
Procedure Gossip(p, DID)
   /*A broadcast packet DID is gossiped with probability p. If it is sent out, the
    sendflag of DID is set to true.*/
   {Gossip DID with probability p;
    If it is sent out Then
       sendflag[DID]=true;
   }
Procedure ReceiveB(DID)
   /*When a node receives a fresh broadcast packet, this procedure calls Gossip(p,
    DID) where p is given different value according to k , n and n1. If it is not sent
    out, timer starts working and receivetimes is set to 1. If the packet is a
    duplication never broadcasted and its timer does not timeout, we increase its
    receivetimes by 1 until reaching m when sendflag is set to true, otherwise it is
    discarded.*/
   {If it is not a duplication Then //UGOSSIP
       {If hop < k Then
           {hop++; Gossip(1,DID)}
     Elseif numnbrs < n Then
           Gossip(p2,DID);
       Elseif numnbrs ≥ n1 Then
           Gossip(p3,DID);
       Else Gossip(p1,DID); //UGOSSIP
     If sendflag[DID] = false Then
           {receivetimes[DID] = 1; timer[DID] = 0;}
       }
    Elseif (sendflag[DID] = false and timer[DID]< T_timeout) Then
       {receivetimes[DID]++;
        If receivetimes[DID] ≥ m Then
          sendflag[DID] = true;
       }
    Else discard the packet;
   }
Procedure Timeout(DID)
   /*If a broadcast packet was never forwarded and its receivetimes is smaller than
    m, let it broadcast with probability 1.0.*/
   {If (timer[DID] ≥ T_timeout and receivetimes[DID] < m) Then
       Gossip(1.0, DID);
   }
```

Although GOSSIP2, GOSSIP3 and UGOSSIP improve data delivery reliability, they do not provide retransmission mechanism. We present a local retransmission mechanism based on gossip called Local Retransmission-based Gossip (LRG). Accordingly, we use LRG packets for local retransmission. The cluster heads send LRG packets by periodically exchanging messages about the broadcast packets they received and lost. In this way, a node knows that which packets should be sent with unicast to the node who loses them and that which packets could be requested from the node who holds them.

The LRG packet includes the following information:

1. *ID* of the source node; *SEQ*: Sequence number of the packet given by the source node; *flag bit*: It is set to 0 to make it different from that of normal broadcast packet. The triple, [*ID*, *SEQ*, *flag bit* (0)], named *DID*, identifies the LRG packet uniquely.
2. *RID*: An array consisting of the received broadcast packets' *DIDs*. The lost packets could be detected by observing gaps in broadcast packet ID sequence. *RID* of each node is renewed periodically.
3. (*LRGP*1, *LRGP*2): Gossip probabilities.
4. H_m: Maximum hops that the LRG packet can travel to.
5. H_t: Hops that LRG packet has traveled through.

The LRG packet is forwarded locally with H_m defined in the packet. When a node receives the packet, it compares its own *RID* with that in the packet. If it holds the broadcast packets the source node lost, it would unicast them back with probability *LRGP*1 and would not forward this LRG packet any longer. If it does not hold them and $H_t<H_m$, H_t in the LRG packet is increased by 1 and the changed LRG packet is forwarded with the probability *LRGP*2. Of course, it requests for the packets listed in *RID* which it lost. H_t could also help cluster heads to get local member information and unicast route information.

(*LRGP*1, *LRGP*2) and H_m are initialized by the source node. If the number of its neighbors is smaller than threshold *n* or the times of the last broadcast packet it received is smaller than *m* in $T_{timeout}$, (*LRGP*1, *LRGP*2) and H_m will be increased accordingly. A large H_m and a large *LRGP*2 help the LRG packet to travel remotely in case of boundary effect. On the other hand, if the node's number of neighbors is larger than *n*1, (*LRGP*1, *LRGP*2) should be decreased to alleviate collisions. Furthermore, (*LRGP*1, *LRGP*2) could be changed by intermediate nodes with dynamic situation of network environment. The ReceiveLRG() procedure is presented in the following:

Procedure ReceiveLRG(*DID*)
/*If the LRG packet is a fresh one, the node compares its own *RID* with that in the packet, then decides what to do, broadcasting and responding the LRG packet. The parameters (*LRGP*1, *LRGP*2) will be changed according to parameters *numnbrs* and *receivetimes*. Duplicate LRG packet is discarded.*/
{**If** it's a duplication **Then**
discard the packet;
Else
{**If** *numnbrs*<*n* or *receivetimes* of last broadcast packet in $T_{timeout}<m$ **Then**
increase the value of *LRGP*1 and *LRGP*2;

```
      Elseif numnbrs > n1 Then
     decrease the value of LRGP1 and LRGP2;
      If holding the packets the source lost Then
         retransmit them to the source with probability LRGP1;
         Else {hop++; Gossip(DID) with LRGP2;}
      If lost the packet listed in source's RID Then
         unicast to the source for retransmission;
     }
  }
```

Besides n, $n1$ and m, some other system parameters can also be used as references to adjust gossip probability $LRGP1$, $LRGP2$ and the number of hops H_m.

B. Operations within a Cluster

Cluster heads have more powerful capability than cluster members in the sense that packets can travel the cluster heads to their neighbor cluster heads directly. Thus, when broadcast packets are forwarded among cluster heads, they are also received by cluster members simultaneously. In a cluster, cluster members send requests of retransmission to and receive retransmissions from their cluster head, which are limited within the cluster. And the cluster head receives its members' requests and responds to them accordingly. Two major procedures are presented in the following:

```
(1)Pseudo Code Operated by Cluster Members
Procedure LocalReceiveB (DID)
   /*If the broadcast packet DID is duplicate, it is discarded; otherwise it is
     accepted. The node could judge whether it has lost packets or not by the SEQ
     of packet DID, then makes decision whether sending out the retransmission
     request or not.*/
   {If it is a duplication Then
       discard the packet;
    Else
     {accept the packet;
      If lost some broadcast packets Then
            send the retransmission request;
     }
   }
(2) Pseudo Code Operated by Cluster Heads
Procedure Resend()
   {On receiving a retransmission request from a cluster member:
      Resend the broadcast packet being requested;
   }
```

4 Theoretical Analysis

This section provides a mathematical analysis on the reliability of the proposed RGB protocol, and shows that our proposed local retransmission-based gossip mechanism does not incur too much overhead. In fact, it balances the overhead of nodes in the whole network.

4.1 Reliability Analysis

Since the gossip mechanism has a strong similarity with epidemic spreading of disease, we analyze the proposed RGB protocol using the simple epidemic model called the SI model in [7]. Since only cluster heads forward packets with gossip, here this model is used to analyze the reliability of packet delivery among cluster heads. The SI model we used is somewhat different from the original one in [7]. Firstly, they have different analytical objects, where [7] aims at normal data delivery, and we aim to analyze the local retransmission. Furthermore, their analytical parameters are different from each other. In [7], the density of nodes in the network is analyzed as a key factor affecting the data delivery, while in RGB, we study how the product of $p*x$ affects the reliability.

Some parameters mentioned in the SI model are defined in the following:

CH: Number of cluster heads in the network; $S(t)$: Number of cluster heads that do not receive the broadcast packet *md* until time t; $I(t)$: Number of cluster heads that have received the packet *md* until time t; x: Number of a cluster head's neighbors in unit time; p: Probability with which a cluster head gossips to its neighbors; $a= p*x/CH$: Broadcast packet transmission rate; $\alpha=a*I(t)$: Probability per unit time for a cluster head to receive the packet *md*; $i(t)=I(t)/CH$: Spreading ratio of the packet *md*.

The balance equations are presented as follows:

$$\begin{cases} \dfrac{dS(t)}{dt} = -\alpha * S(t) \\ \dfrac{dI(t)}{dt} = \alpha * S(t) \end{cases} \tag{1}$$

Since $\alpha=a*I(t)$ and $S(t)=CH-I(t)$, thus

$$\frac{dI(t)}{dt} = a*I(t)*[CH - I(t)] = a*CH*I(t) - a*I^2(t) \tag{2}$$

This first order ordinary differential equation has the following general solution:

$$I(t) = \frac{CH}{1 + C*CH*e^{-a*CH*t}} \tag{3}$$

Where C is a constant value that only depends on the initial conditions. Here C is computed as follows. At the beginning (t=0), we assume that there are $r(0<r\leq CH)$ nodes that carry such information. So $I(t)$ should satisfy $I(0)= r$.

$$I(0) = r = \frac{CH}{1+C*CH*e^{-p*x*0}} \Rightarrow C = \frac{CH - r}{CH*r} \tag{4}$$

In order to study the relationship between $i(t)$ and parameters p and x, we replace a with $p*x/CH$, then we get:

$$i(t) = I(t)/CH = \frac{1}{1 + (CH - r) * e^{-p*x*t} / r} \tag{5}$$

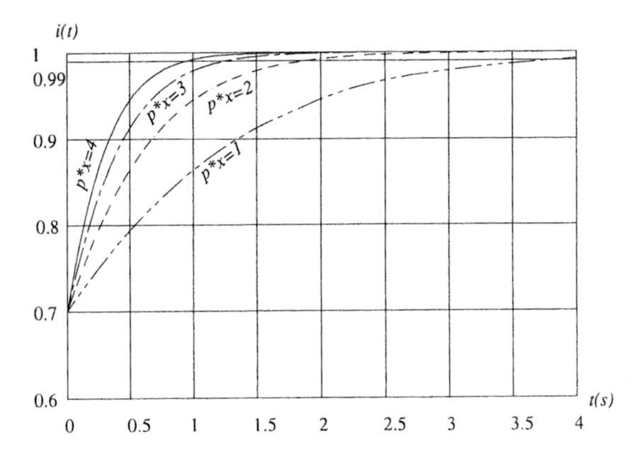

Fig. 2. $i(t)$ versus t with r =0.7C

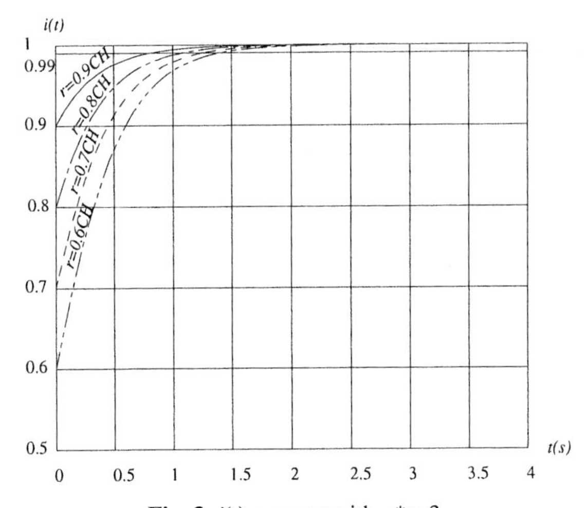

Fig. 3. $i(t)$ versus t with $p*x$=3

If initial condition r is constant, the value of $i(t)$ depends on $p*x$ and t. It is obvious that the time t taken to get similar spreading ratio $i(t)$ decreases with the increase of $p*x$. In Figure 2, r =0.7CH, when the values of $p*x$ are 1, 2, 3, 4 separately, and $i(t)$=99%, the values of time t are 3.8s, 1.8s, 1.3s, 0.9s, respectively.

If $p*x$ is kept constant, the retransmission is in a better initial condition, i.e., at the beginning, more of cluster heads have got the packet *md*, the time to get similar spreading ratio $i(t)$ becomes less. The curves of the comparison in different conditions are shown in Figure 3 and Figure 4, respectively. In Figure 3, $p*x$=3, r=0.6CH, 0.7CH, 0.8CH, 0.9CH and if $i(t)$ comes at 99%, the time t is separately 1.4s, 1.3s, 1.1s, 0.8s. Likewise, in Figure 4, $p*x$=1, r=0.6CH, 0.7CH, 0.8CH, 0.9CH and if $i(t)$ comes at 99%, the time t is separately >4s, 3.8s, 3.2s, 2.4s.

The value of $p*x$ in the SI model reflects the communication intensity between a cluster head and its neighbor cluster heads in unit time. It causes more impact on spreading ratio $i(t)$ than initial condition r. If the initial condition becomes worse, an applicable value of $p*x$ could still obtain a larger spreading ratio $i(t)$ in shorter time. In Figure 3, r=0.6CH, that means only 60% of cluster heads in the network have received the packet *md* before operation of retransmission, $p*x$=3, $i(t)$=99%, the needed time t is only 1.4s. However, less $p*x$ could not produce an ideal $i(t)$ in a short period of time, even if the initial condition is good. The result is shown in Figure 4, where $p*x$=1, r=0.6CH, $i(t)$=99%, the needed time t is 2.4s, which is longer than the needed time 1.4s in Figure 3.

In fact, when the cluster head has few neighbors, even if a large gossip probability is adopted, we still could not get an ideal spreading ratio $i(t)$ in a short period of time. We explain this point in the following. In the SI model, x means the number of a cluster head's neighbors, that is to say, the relationship with cluster heads more than one hop away is not taken into consideration. Thus, if the cluster head has few neighbors, we can not attain a larger $p*x$ through increasing the gossip probability. For example, in case of x=1, even though p is assigned maximal value 1.0, $p*x$ gains its maximal value, which still results in worse spreading ratio in shorter time t as shown in Figure 4.

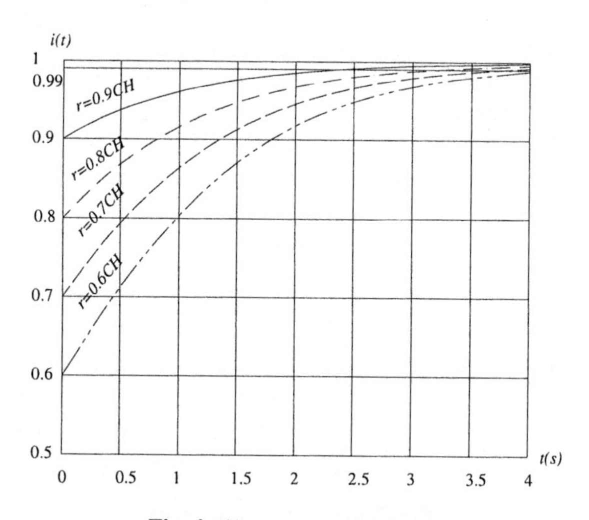

Fig. 4. $i(t)$ versus t with $p*x$=1

As a matter of fact, if the cluster head has fewer neighbors, spreading ratio could be improved by intensifying communication with cluster heads more than one hop away. Thus, it is reasonable for us to define x as the number of nodes with which cluster head communicates in unit time. This extended SI model allows us to assign different value to x in dynamic network environment in order to improve reliability. It shows that, if a cluster head has fewer neighbors, then it is necessary to extend communication scope of the cluster head. The RGB protocol works just like this.

In a large scale network, r is much less than CH. The above analysis shows that a suitable value of $p*x$ could lead to a satisfactory spreading ratio in a short period of time, i.e., RGB could provide higher data delivery reliability in a large scale network.

4.2 Analysis on Nodes in Boundary or Sparse Areas

Normal broadcast packets are forwarded only by cluster heads among clusters, meanwhile these packets are received by cluster members within each cluster. This mechanism greatly reduces communication overhead. Multi-channels are adopted in RGB, where gossip-based retransmission uses different channel from that of normal data delivery, so the communication among cluster members is not affected.

In sparse or boundary areas, the nodes' local retransmission with high probability may increase the communication overhead. However, it does improve the performance of RGB, and we explain this point as follows:

1. For nodes in the sparse or boundary areas, the explicit reason that they may not get broadcast packet is that they communicate rarely with nodes far away. Simply stated, rare communication makes them idler than other nodes, i.e., their overhead is relatively lighter, so the overhead caused by high gossip probability will not degrade their performance.
2. For nodes that respond to gossip for local retransmission, their burden will be aggravated actually. But there are a large number of nodes close to the inner parts of the network or in the dense areas, so the average overhead burdened by them is light.

Based on the above two points, we conclude that this discrimination for nodes in different network environments does not deteriorate the performance of the network, conversely this mechanism balances the overhead of nodes, and improves the capability of the network.

5 Conclusions

Our main contribution in this paper is that we present a reliable broadcast mechanism based on gossip and local retransmission including local gossip-based retransmission among clusters and local retransmission within each cluster. The dynamic adjustment of system parameters makes the proposed protocol outperform self-pruning based protocols in adapting to the dynamic network situation, thus improving the broadcast reliability and increasing the network throughput. In our future work, we will study how to fine tune the gossip parameters by extensive simulations.

Acknowledgments

This work is supported in part by the Hunan Provincial Natural Science Foundation of China No. 05JJ30118 (Secure group communications in large-scale mobile ad-hoc networks), in part by the City University of Hong Kong Strategic Research Grant No. 7001709, and in part by the Hong Kong PolyU research fund A-PF77.

References

[1] T. Camp and B. Williams, "Comparison of broadcasting techniques for mobile ad hoc networks," *Proc. ACM/IEEE MOBIHOC*, pp. 194-205, June 2002.
[2] S.Y. Cho and J.H. Sin, "Reliable broadcast scheme initiated by receiver in ad hoc networks," *Proc. IEEE LCN*, pp. 281-282, 2003.
[3] F. Dai and J. Wu, "Performance analysis of broadcast protocols in ad hoc networks based on self-pruning," *IEEE Transactions on Parallel and Distributed Systems*, Vol. 15, Issue 11, pp. 1027-1040, Nov. 2004.
[4] J.J. Garcia-Luna-Aceves and Y.X. Zhang, "Reliable broadcasting in dynamic network," *Proc. IEEE ICC*, pp.1630-1634, 1996.
[5] Z.J. Haas, J.Y. Halpern, and L. Li, "Gossip-based ad hoc routing," *Proc. INFOCOM*, pp.1707-1716, June 2002.
[6] C. Ho, K. Obraczka, G. Tsudik, and K. Viswanath, "Flooding for reliable multicast in multi-hop ad hoc networks," *Proc. DIALM*, pp. 64-71, 1999.
[7] A. Khelil, C. Becker, J. Tian, and K. Rothermel, "An epidemic model for information diffusion in MANETs," *Proc. MSWiM*, pp. 54-60, Sept. 2002.
[8] W. Lou and J. Wu, "A reliable broadcast algorithm with selected acknowledgements in mobile ad hoc networks," *Proc. IEEE GLOBECOM*, Vol. 22, No.1, pp. 3536-3541, Dec 2003.
[9] S. Ni, Y. Tseng, Y. Chen, and J. Sheu, "The broadcast storm problem in a mobile ad hoc network," *Proc. ACM/IEEE MOBICOM*, pp. 151-162, 1999.
[10] E. Pagani and G.P. Rossi, "Providing reliable and fault tolerant broadcast delivery in mobile ad hoc networks," *ACM/Baltzer Mobile Networks and Applications*, Vol.4, No. 3, pp. 175-192, Aug. 1999.
[11] W. Peng and X. Lu, "AHBP: An efficient broadcast protocol for mobile ad hoc networks," *Journal of Science and Technology (Beijing, China)*, 2002.
[12] M.Q. Rieck, S. Pai, and S. Dhar, "Distributed routing algorithms for wireless ad hoc networks using *d*-hop connected dominating sets," *Proc. HPC-ASIA*, pp. 443-450, Dec. 2002.
[13] I. Stojmenovic, M. Seddigh, and J. Zunic, "Dominating sets and neighbor elimination based broadcasting algorithms in wireless networks," *IEEE Transactions on Parallel and Distributed Systems*, Vol. 13, No. 1, pp. 14-25, Jan. 2002.
[14] G. Wang, L. Zhang, and J. Cao, "A virtual circle-based clustering algorithm with mobility prediction in large-scale MANETs," *Proc. ICCNMC*, pp. 364-374, Aug. 2005.
[15] J. Wu and F. Dai, "Broadcasting in ad hoc networks based on self-pruning," *Proc. INFOCOM*, pp. 2240-2250, Mar./Apr. 2003.
[16] K. Xu, X. Hong, and M. Gerla, "Landmark routing in ad hoc networks with mobile backbones," *Journal of Parallel and Distributed Computing*, Vol. 63, Issue 2, pp.110-122, 2003.

An Energy Consumption Estimation Model for Disseminating Query in Sensor Networks*

Guilin Li, Jianzhong Li, and Longjiang Guo

Harbin Institute of Technology, 150001, Harbin, China
{liguilin, lijzh, guolongjiang}@hit.edu.cn

Abstract. There are three approaches to disseminate queries into the sensor network, which are the multi-hop unicast, full flood and geocast. Since the energy is very limited resource of the sensor network, query optimizer must select an optimal approach consuming minimum energy to disseminate the query into the sensor network. As the selection relies on energy consumption estimation, the accuracy of the energy consumption estimation is very critical for query processing in sensor network. In this paper, a general energy consumption estimation model is proposed. Compared with other models, our model takes both routing protocol and MAC protocol into consideration. In experiments we compared our energy model's estimation for the multi-hop unicast, full flood and geocast with simulation results. Results showed that our model's accuracy is very high.

1 Introduction

With the development of technology, wireless sensor networks have been widely used in various fields. For instance, they are used to extract environmental data from the physical world. Many researchers look upon sensor networks as a new kind of database [1,2] and carry out various algorithms to let the sensor network support query processing. As the energy of each sensor is very limited and it is inconvenient to change its battery, the energy consumption is the most important factor to think about when designing query processing algorithms.

Sensor networks are composed of ad hoc wireless sensor nodes. Sink is a special node, which has more energy and stronger processing capability than other ordinary nodes. One of the most important work for sink is to send the user's queries into the sensor network. There are three types of query dissemination approaches, multi-hop unicast, full flood and geocast. In Mulit-hop unicast, sink sends the queries directly to the destinations by some routing protocol. Paper [3] discusses the idea of data centric storage, which mainly relies on the multi-hop unicast to send queries. In Full flood approach, sink broadcasts the queries to all nodes in the sensor network and the direct diffusion [4] relies on this approach to send queries. Geocast, the queries sent by the sink are only broadcasted in a certain area of the sensor network, which has been used in [5].

* Supported by the National Natural Science Foundation of China under Grant No.60473075; the Natural Science Foundation of Heilongjiang Province of China under Grant No.ZJG03-05.

X. Jia, J. Wu, and Y. He (Eds.): MSN 2005, LNCS 3794, pp. 219 – 228, 2005.

Given a query, the query processor can find a lot of destinations to send the query. According to the set of destinations, the query optimizer can develop a lot of dissemination plans to send the query. The optimizer must use some cost model to compare the energy consumption for finding a plan consuming minimum energy. So a precise energy consumption estimation model is a critical problem for query processing in sensor network.

But the proposed cost model can't accurately estimate the energy consumption of different dissemination plans. Paper [3] gives an abstract cost model from the message complexity aspect. This model assumes that there are N nodes in the sensor networks then multi-hop unicast will cost $O(\sqrt{N})$ messages and full flood will cost $O(N)$ messages. As the model just gives the asymptotic bound of the messages, it can't be used to estimate the cost of a particular query plan rather than compare energy consumption among different query dissemination plans. Paper [6] tries to estimate the energy consumption for the spatio-temporal query, but it ignores the effects the MAC protocol brings to the cost model. From our experiments in section 4 we can see that the estimated energy consumption of this model has a big gap with the simulation in some conditions.

In this paper we present that the cost model is related with both routing protocol and MAC protocol used by the query dissemination approach and propose a new cost model taking the two factors into consideration. First we look on the hop number experienced by the query, which is determined by the multi-hop routing protocol, and it shows the cost model is related with routing protocol. Second, the multi-hop routing protocol is composed of hop by hop communication and the hop by hop communication is controlled by the MAC protocol, which means the cost model is also related with MAC protocol. So we conclude that the cost model is related with both routing protocol and MAC protocol used by the query dissemination approach. Experimental results showed that the accuracy of our cost model is very high.

The remainder of this paper is organized as follows. In section 2, we present a general cost model which takes both routing protocol and MAC protocol into consideration. In section 3, we analyze the energy consumption for multi-hop unicast, full flood and geocast according to our general cost model. In section 4, we present the experiments to show the accuracy of our cost model. We draw the conclusion in section 5.

2 Energy Consumption Model for Disseminating Queries

2.1 Differences Between Cost Models Considering MAC Protocol or Not

Compared with other cost models neglecting the MAC protocol, there are three different aspects between these two models. First, the packet times sent by the sender are different. The cost model neglecting the MAC protocol thinks the sender sends a packet only one time. But in our model we think the time is a parameter determined by the MAC protocol used by the sensor. Because we think that a routing packet must be transformed into a MAC packet first and the MAC packet may be sent several times determined by the MAC protocol. For example, if the sensor uses a TDMA MAC protocol, a broadcast packet may be sent several times, one time for a particular neighbor of the sender. But if the sensor uses a CSMA MAC protocol, a broadcast packet will be sent only once and all neighbors receive the packet. We can see that the times of sending a packet are different for different MAC protocols so the simple model neglecting the MAC layer is inaccurate. Second, the number of nodes receiving the packet is different. In unicast, the cost model neglecting the MAC layer considers that only the receiver will receive the packet. But the nature of wireless media determines that even a MAC packet is sent to a particular neighbor, the other neighbors of the sender may also hear the packet and consume energy for receiving the packet so the number of neighbors receiving the MAC packet is also determined by the MAC protocol. Third, the cost model neglecting the MAC protocol ignores all the control packets. But the MAC protocol may rely on some control packets and the energy consumed by these control packets is another energy consumption source we should consider. So the MAC protocol is very important to energy consumption estimation and can't be ignored.

2.2 General Energy Cost Model

In this section we give a general cost model considering both routing and MAC protocol. Our energy cost model is determined by two factors: the number of packets sent by the routing layer and the energy consumed by the MAC layer to send the MAC packet transformed from the routing packet, so the general energy model is:

$$E = (\sum_{i=1}^{N} I_i) * E_{MAC} \quad I_i=0,1,2\ldots m \tag{1}$$

Assume that there are N nodes in the sensor networks. I_i stands for the packet number sent by the routing layer of the ith node in a particular query dissemination approach and I_i may be any number from 0 to m. $\sum_{i=1}^{N} I_i$ stands for the total number of messages sent by all nodes' routing layer. And E_{MAC}, defined by formula (2), is the energy consumed by a sensor's MAC layer to send a routing packet. By multiplying these two factors we get the general energy estimation model.

$$E_{MAC} = M_d*[S_{sd}*(P_s+M_{sdr}*P_r)+(\Sigma S_{sc})*(P_s+M_{scr}*P_r)+(\Sigma S_{rc})*(P_s+M_{rcr}*P_r)] \tag{2}$$

In formula (2), P_s and P_r are the energy a sensor consumes to send and receive a bit. By considering the first aspect ignored by the cost model neglecting the MAC protocol, we use M_d to represent the times that a MAC packet will be sent by a particular MAC protocol. Each time a MAC packet is sent, the energy consumption estimation is $[S_{sd}*(P_s+M_{sdr}*P_r)+(\Sigma S_{sc})*(P_s+M_{scr}*P_r)+(\Sigma S_{rc})*(P_s+M_{rcr}*P_r)]$ which is composed by three parts and we consider the other two aspects ignored by other cost models. $S_{sd}*(P_s+M_{sdr}*P_r)$ is the energy consumed by the networks when the sender sends a MAC packet. S_{sd} is the size of the MAC packet, so $S_{sd}*P_s$ means the energy consumed by the sender to transmit a MAC packet. M_{sdr} is the number of nodes receiving the MAC packet, so $S_{sd}*M_{sdr}*P_r$ means the total energy consumed by the other nodes of the networks to receive a MAC packet. The MAC protocol may rely on two kinds of control packets when sending a MAC packet, which are control packets sent by the sender and the receiver. $(\Sigma S_{sc})*(P_s+M_{scr}*P_r)$ is the energy consumed by the networks when the sender sends control packets. ΣS_{sc} is the total size of control packets sent by the sender when the sender wants to send a MAC packet, so

$(\Sigma S_{sc})*P_s$ stands for the energy consumed by the sender to send the control packets. M_{scr} is the number of nodes receiving these control packets, so $(\Sigma S_{sc})*M_{scr}*P_r$ stands for the energy consumed by the other nodes receiving the control packets sent by the sender. $(\Sigma S_{rc})*(P_s+M_{rcr}*P_r)$ is the energy consumed by the networks when the receiver sends control packets. ΣS_{rc} is the total size of control packets the receiver replies to the sender when the receiver receives the MAC packets from the sender. M_{rcr} is the node number receiving the control packets sent by the receiver. $(\Sigma S_{rc})*P_s$ and $(\Sigma S_{rc})*M_{rcr}*P_r$ have the same explanation with that of $(\Sigma S_{sc})*P_s$ and $(\Sigma S_{sc})*M_{scr}*P_r$.

2.3 Estimating the Average Number of Neighbors for a Node

In this section we analyze the average number of neighbors N_n for a node. Assume that N is the total number of sensors and A is the distribution area of the sensor networks. These N nodes are uniformly distributed in the area A. Each node has a circle communication range whose radius is r. Fig.1 illustrates the three kinds of nodes with different average number of neighbors. The first kind of nodes lie in rectangleⅠand their average number of neighbors N_{n1} can be estimated by $\pi*r^2*(N/A)-1$. The second kind of nodes lie in rectangleⅡ,Ⅲ.Their average number of neighbors N_{n2} can be estimated by $[(\int_0^r V_1(x)dx)/r]*(N/A)-1$, where $V_1(x) = \pi r^2 - r^2\cos^{-1}(x/r) + x\sqrt{r^2-x^2}$ is the intersection between a node's communication range and the whole rectangle. The third kind of nodes lie in rectangle Ⅳ, the average number of neighbors N_{n3} can be calculated by $[(\int_0^r\int_0^r V_2(x,y)dxdy)/r^2]*(N/A)-1$, where $V_2(x,y) = \pi r^2 - r^2\cos^{-1}(x/r) + (x/2)\sqrt{r^2-x^2} - r^2\cos^{-1}(y/r) + (3y/2)\sqrt{r^2-y^2}$ is also the intersection between the node's communication range and the whole rectangle. The final average number of neighbors N_n is calculated by formula (3), in which A_i (i=1…4) is the area of the four kinds of rectangles illustrated in Fig. 1.

$$N_n=(A_1/A)*N_{n1}+[2*(A_2+A_3)/A]*N_{n2}+(4*A_4/A)*N_{n3}. \tag{3}$$

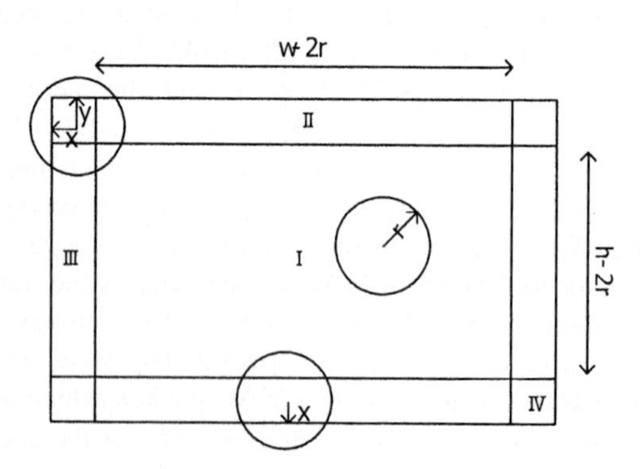

Fig. 1. Estimating the average number of neighbors for a node

3 Cost Estimations for Query Dissemination Approaches

3.1 Energy Analysis for Hop by Hop Protocol

We use the 802.11 MAC protocol [7] as an example to analyze its energy consumption according to formula (2). There are two methods for 802.11 protocol to transmit a packet which are broadcast and unicast.

A broadcast packet will be sent only one time, which means M_d equals to 1. And this message will be received by all the neighbors of the sender, which means M_{sdr} equals to N_n given by formula (3). As broadcast does not rely on any control packet, all energy consumed by control packets can be removed from formula (2). The cost to broadcast a MAC packet is:

$$S_{sd} * (P_s + N_n * P_r) \tag{4}$$

In order to guarantee the reliability and avoid hidden [8] and exposed terminal phenomena [9], unicast in 802.11 protocol uses the RTS/CTS/DATA/ACK handshake mechanism to transmit a DATA packet where RTS/CTS/ACK are control packets. If the receiver fails to receive the MAC packet the sender will retry for at most x times, which is a system defined parameter. It means M_d equals to x. For 802.11 protocol, even though the MAC packet is for a particular node, the nature of wireless media determines that all of the neighbors of the source will also hear the packet and consume energy for receiving this packet but only the destination will send the packet to the routing layer. This means M_{sdr} equals to N_n. The source will send a RTS message per MAC packet and the destination will reply a CTS and an ACK packet to the source. The RTS packet will be heard by all neighbors of the source and CTS/ACK packet will be heard by all neighbors of the destination, which means M_{scr} and M_{rcr} equal to N_n. According to formula (2) we get the maximum cost estimation for 802.11 protocol to send a MAC packet:

$$\begin{aligned} & x * [S_{sd} * (P_s + N_n * P_r) + S_{rts}(P_s + N_n * P_r) + (S_{cts} + S_{ack}) * (P_s + N_n * P_r)] \\ & = x * (S_{sd} + S_{rts} + S_{cts} + S_{ack}) * (P_s + N_n * P_r) \end{aligned} \tag{5}$$

where S_{rts}, S_{cts} and S_{ack} stands for the packet size of RTS, CTS and ACK packets. Given the hop by hop cost analysis, we can analyze the cost for multi-hop unicast approach.

3.2 Cost Estimation for Multi-hop Unicast

The geographical greedy routing protocol [10] is used as the routing protocol of the multi-hop unicast approach. This kind of routing protocol is based on the following idea: among the neighbors y of node z closer to target t than z, z picks the one closest to t. As multi-hop is composed of single hops, we must accurately estimate the distance l that a sensor can transmit in a single hop. If the distance between the target and the source is d, after getting l, the hop number of multi-hop unicast can be estimated by $\lceil d/l \rceil$.

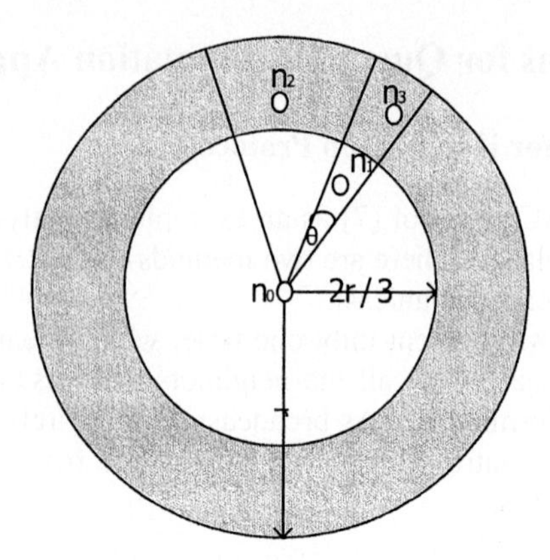

Fig. 2. Average distance of a single hop

When the angle between two successive neighbors n_1 and n_2 of node n_0 is larger than a threshold angleθ, illustrated in Fig.2, we use the average distances between the node n_0 and its neighbors to estimate one hop distance l, which is $\iint_{A_r} \sqrt{x^2+y^2}\,dxdy / A_r = 2r/3$, where A_r is the area of the circle whose radius is r.

When the angle between two successive neighbors n_1 and n_3 of node n_0 is smaller than the threshold angleθ, shown in Fig.2, we can think of the node n_1 and n_3 are almost in the same direction and the geographical greedy routing algorithm will choose the further one as the next hop node. And we use $\iint_{A_r'} \sqrt{x^2+y^2}\,dxdy / A_r' = 38r/45$ to estimate one hop distance l, where A_r' is the area between the inner circle and the outer circle.

For multi-hop unicast protocol based on geographical greedy routing protocol, there are $\lceil d/l \rceil$ nodes involved in the dissemination and the routing layer of each node will sends one messages, which means the I_i of these $\lceil d/l \rceil$ nodes equals to 1. The energy consumption for sending a routing packet is given by formula (5). Putting these parameters into formula (1) we get the cost estimation for multi-hop unicast:

$$\sum_{i=1}^{\lceil d/l \rceil} x^*(S_{sd}+S_{rts}+S_{cts}+S_{ack})^*(P_s+N_n{}^*P_r) \tag{6}$$

3.3 Cost Estimation for Full Flood and Geocast

When full flood is used to disseminate query, the sink broadcasts the query to all its neighbors, the neighbors will broadcast this query to all their neighbors and so on. Finally all the nodes in the networks will receive the query sent by the sink. Every node may receive the same query from its neighbors several times but it only broadcasts the first query received and drops the others. From the description we can see that in full flood protocol, the routing layer of every node will send a message, which means all nodes take part in the full flood and I_i for all nodes equals 1. The energy consumption for sending a routing packet using broadcast can be estimated by (4). Putting these into formula (1) we get the cost estimation for full flood:

$$\sum_{i=1}^{N} S_{sd} * (P_s + N_n * P_r) \tag{7}$$

Finally we give the cost estimation for geocast. The geocast refers to a message flooded in a certain area in the sensor networks. Geocast is composed of two phases: in the first phase, the sink sends the message to a node in the flood area by multi-hop unicast protocol; in the second phase, the node receiving the message will flood the message in the area. In the first phase, there are $\lceil d/l \rceil$ nodes taking part in the transmission and each node's routing layer sends a packet. The energy consumption for sending a routing protocol is given by formula (6). In the second phase, the message is flooded within the area and there are $(N/A)A_d$ nodes in that area, where N is the total number of sensors, A is the distribution area of all sensors and A_d is the destination area. The routing layer of each node in the area will broadcast a message. As same as the analysis of formula (7) for full flood, we get the second part of formula (8).

$$\sum_{i=1}^{\lceil d/l \rceil} x * (S_{sd} + S_{rts} + S_{cts} + S_{ack}) * (P_s + N_n * P_r) + \sum_{j=1}^{(N/A)A_d} S_{sd} * (P_s + N_n * P_r) \tag{8}$$

4 Experiments and Analysis

We use ns2 [11] as the simulation platform for wireless sensor networks. 200 nodes are uniformly distributed in a 200*200 rectangle. We set the transmitting power of the sensor to 0.660W, the receiving power to 0.395W. The transmission range of a sensor is 40 meters and the sensor's bandwidth is 200Kbps. The data packet size is 180 bytes. We use 802.11 as MAC protocol and GPSR as the routing protocol. The sensors are assumed to be stationary and the sink node lies in the left up corner of the rectangle.

We do three groups of experiments to show the accuracy of our cost estimation for multi-hop unicast, full flood and geocast by comparing the estimation calculated by our cost models with the simulation results gotten by ns2. Furthermore we compare the estimation between our cost model and the one neglecting the effect of MAC layer to show our cost model is more accurate.

4.1 Accuracy of the Cost Estimation for Multi-hop Unicast

The sink sends messages to randomly selected 10~50 destinations by multi-hop unicast approach. The energy consumption of multi-hop unicast using the 802.11 hop by hop protocol is estimated according to formula (6). We set x to 1 because in query dissemination application, the network is in the unsaturated state and most MAC packet can be sent successful in one time. The comparison between the estimation and the simulation result is shown in Fig.3. From this figure we can see that our estimation (estimation1) and the simulation result are very similar which means our cost model can accurately estimate the energy consumption. For comparison, we illustrate

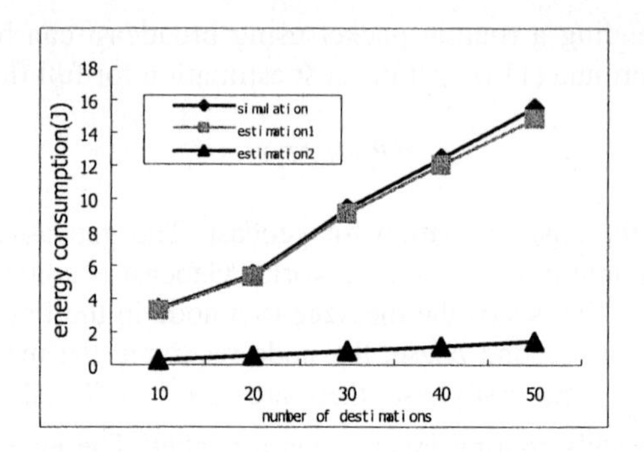

Fig. 3. Comparison between estimation of multi-hop unicast cost model with simulation result

the energy consumption (estimation2) estimated by the cost model neglecting the effect of MAC protocol. We can see that the gap between the estimation cost (estimation2) and the simulation result becomes bigger and bigger with the increasing of destination numbers.

4.2 Accuracy of the Cost Estimation for Full Flood

We produce five topologies with 200 nodes uniformly distributed in a 200*200 rectangle area. The sink floods a query throughout the networks. We compare the energy consumption estimation calculated by formula (7) with the simulation results. The comparison result is shown in Fig.4 (a). Then we vary the node number from 100 to 500 and keep the node density constant and repeat the same experiment. The comparison result is shown in Fig.4 (b). In the two experiments, the cost model neglecting the MAC protocol and our model give the same cost estimation and the experiments show that the cost estimation for full flood is accurate.

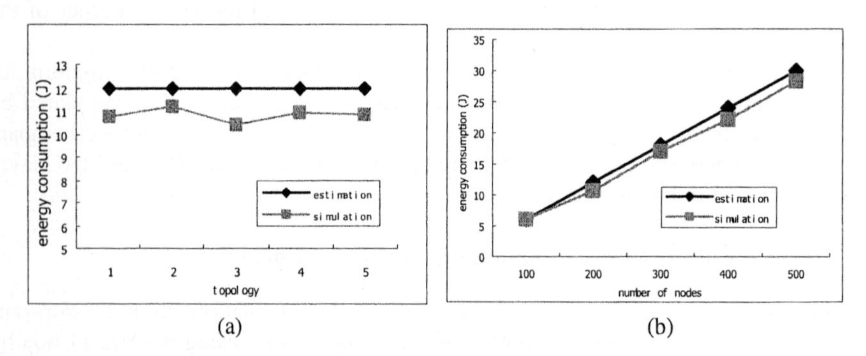

Fig. 4. Comparison between estimation of full flood cost model with simulation result

4.3 Accuracy of the Cost Estimation for Geocast

The shape of the geocast area is a circle. We set the circle's radius to 40 meters and send a query to 9 random circle areas. We repeat the experiment for 5 times and get

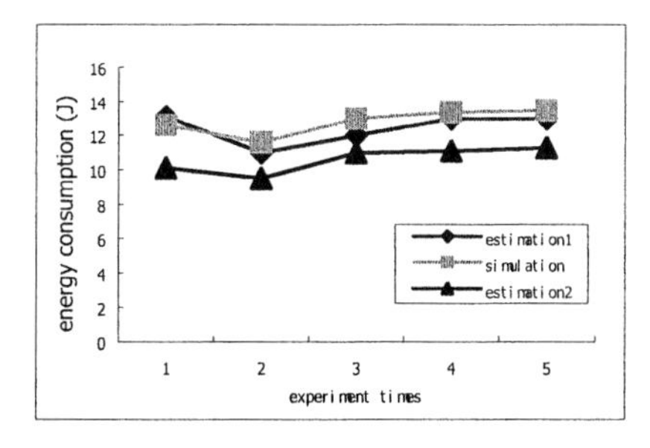

Fig. 5. Comparison between estimation of geocast cost model with simulation result

the comparison result in Fig.5. Our cost estimation is calculated according to formula (8) and the x is set to 1. The experimental results showed that our model (estimation1) gives more accurate cost estimation than the cost (estimation2) given by the model neglecting the effect of MAC layer.

5 Conclusion

In this paper we present that the energy cost model is related with both routing protocol and MAC protocol used by the query dissemination approaches. We propose a general model that takes these two factors into consideration. Then we use geographical greedy based routing protocol and 802.11 MAC protocol to implement different query dissemination approaches and analyze the cost estimation for these query dissemination approaches according to our general cost model. Finally extensive experiments showed that the accuracy of our model is very high.

References

1. S. Madden and H. Kung. The design of an acquisitional query processor for sensor networks. In Proc. of ACM SIGMOD, pages 491-502, 2003.
2. P. Bonnet, J. Gehrke, P. Seshadri. Towards sensor database systems. In Proc. 2nd International Conference on Mobile Data Management, Hong Kong, pages 3-14, 2001
3. S. Shenker, S. Ratnasamy, B. Karp, R. Govindan, and D. Estrin. Data-Centric Storage in Sensornets. ACM SIGCOMM, Computer Communications Review, Vol. 33, Num. 1, January 2003
4. C. Intanagonwiwat, R. Govindan, and D. Estrin. Directed Diffusion: A Scalable and Robust Communication Paradigm for Sensor Networks. In Proceedings of the Sixth Annual International Conference on Mobile Computing and Networks (MobiCOM 2000), Boston, Massachusetts, August 2000.
5. A. Coman, M.A. Nascimento J. Sander. A Framework for Spatio-Temporal Query Processing Over Wireless Sensor Networks. Proc. of the Intl. Workshop on Data Management for Sensor Networks pages. 104-110, Aug. 2004.
6. A. Coman, J.Sander, M.A. Nascimento, An analysis of Spatio-Temporal Query Processing in Sensor Networks, in Proceedings of the 1st International Workshop on Networking Meets Databases in conjunction with 21st ICDE, pages 45-50, April 2005

7. Wireless LAN Medium Access Control (MAC) and Physical Layer (PHY) Specifications, IEEE Standard 802.11-1999, IEEE Computer Society LAN MAN Standards Committee, Ed., 1999.
8. D. Allen. Hidden terminal problems in Wireless LAN's, IEEE 802.11 Working Group Papers, 1993.
9. V. Bharghavan, A. Demers, S. Shenker, and L. Zhang, "MACAW: A Media Access Protocol for Wireless LAN's," ACM SIGCOMM, 1994
10. B. Karp and H.T. Kung. GPSR: Greedy Perimeter stateless Routing for Wireless Networks. MobiCom2000, Boston, Massachusetts. Aug 6-11, 2000.
11. UCB/LBNL/VINT Network Simulator – ns2. http://www-mash.cs.berkeley.edu/ns/, 1998

EasiSOC: Towards Cheaper and Smaller

Xi Huang[1,2], Ze Zhao[1], and Li Cui[1]

[1] Institute of Computing Technology, Chinese Academy of Sciences, Beijing 100080 P.R. China
[2] Graduate School of the Chinese Academy of Sciences, Beijing 100039 P.R. China
lcui@ict.ac.cn

Abstract. With the goal to make the wireless sensor network nodes small in size, light in weight, cheap in cost, as well as low in power consumption, projects have been carried out to study the hardware and software co-design and to develop sensor node SOC technology. We have proposed a general "sensor node on a chip" approach, namely EasiSOC, with two typical SOC architectures for different application areas. The first architecture of sensor node supports basic functionalities which performs relatively simple tasks with fixed routines. The second architecture of sensor node favors complex functionalities and advanced jobs. Current research progresses on the development of the first SOC structures are also introduced.

Keywords: SOC (System on Chip), embedded system, sensor node, wireless sensor network.

1 Introduction

The next generation wireless sensor networks are composed of small, ultra low-power, ultra low-cost sensor nodes which are deployed densely within or near the areas of interests. Those sensor nodes collect information on entities and transfer the data wirelessly to the remote end user in an Adhoc network fashion. While sensor networks for various applications may be quite different, they share common technical issues. One of the most challenging issues is how to make the sensor nodes small in size, light in weight and inexpensive in cost. Besides, sensor nodes should be made operating at a very low power consumption level to increase the overall lifetime of the network.

Nowadays, most of the wireless sensor network nodes are made using SCM (Single Chip Micyoco) or other embedded technology. Some of the representative sensor nodes include Berkeley Motes [1,2], Sensoria WINS[3], MIT μAMPs [4], Intel iMote [5], Intel XScale nodes [6], SmartMesh Dust mote [7] and ICTCAS EasiNet Node [8-10]. There are some common features for those embedded systems. They are consisted of a number of individual components integrated on a PCB board. Those components include a CPU, memory unit, input-output modules and data bus. These nodes are all controlled by supporting embedded software. Depends on the design of the specific software, these all-purpose sensor nodes can be adjusted to meet various application requirements. Thus the design flexibility is one of the main advantages using embedded

X. Jia, J. Wu, and Y. He (Eds.): MSN 2005, LNCS 3794, pp. 229 – 238, 2005.

technology. However, the disadvantages of the system are revealed as big in size and relatively low in operation efficiency especially when there are complicated application software or protocols running on it, e.g. software for calculating the realtime location of the node. Besides, the hardware manufacture cost is confined by the total cost of the individual components and this may not be managed very low easily. The power consumption of the system is also determined by the overall power to run all the individule components and this may not be kept at an ultra-low level easily.

Nevertheless, with the development of advanced SOC technology, it is now possible to design and make the next generation sensor nodes on a single chip to solve all of the above mentioned problems at the same time. In this work, we proposed a general "sensor node on a chip" approach, namely EasiSOC, with two typical SOC architectures based on SOC FPGA/ ASIC technology for different application areas. The first architecture of sensor node supports basic functionalities which performs relatively simple tasks with fixed routines. The second architecture of sensor node favors complex functionalities and advanced jobs. Compared with the all-purpose embedded system, the "sensor node on a chip" approach is based on hardware and software co-design SOC technology with additional integrated network protocol modules special algorithms hardware modules and various interface modules. The SOC nodes may be made very small in size, light in weight as well as low in cost for a massive production. More importantly, the operation efficiency of SOC sensor node to process information or to run other computing tasks may be improved greatly due to the support from functional hardware modules. As a consequence, the power consumption level may be lowered by less occupancy of limited system resources.

2 EasiSOC Node Architecture

Nowadays with the advanced semiconductor technology, the conversion from FPGA to ASIC is relatively mature. Hence the proposed EasiSOC project is mainly focused on the design of system architecture and the FPGA verification of the functionalities. It is well known that the appearance of modular IP Cores makes the IC design procedure much easier. Thus we propose to design sensor network nodes in a similar modulated manner. The idea is that we divide the functions of sensor nodes into a series of functional modules with standardized interchangeable interfaces. The sensor network developer may choose to use different modular IP cores to build SOC sensor nodes for their specific applications. In this work we classified sensor nodes into two major groups and proposed two system architectures accordingly.

The first architecture of sensor node supports basic functionalities which perform relatively simple tasks with fixed routines. Figure 1 shows the schematic diagram of the first architecture.

Here the functional modules are divided into the following parts:

- **Sensor Interface Module:** It couples external sensors with the MCU (Micro Control Unit) module. It collects data from sensors and transfers the data to the MCU. The module also receives commands from MCU to control the sensors reversely. Communication protocols have been established between the Sensor Interface Module and MCU as a standard IP core.

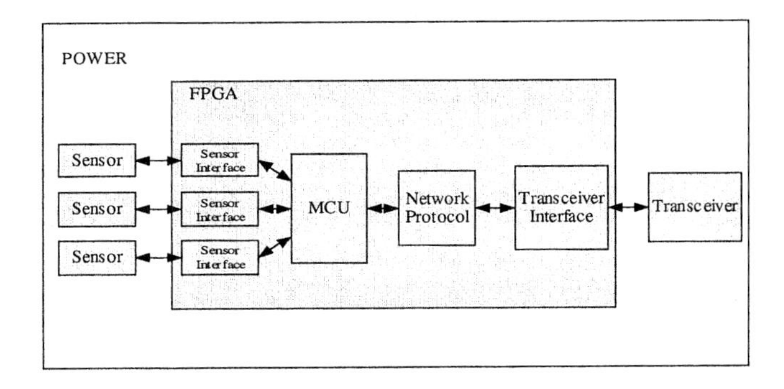

Fig. 1. Schematic architecture diagram of a SOC sensor node of basic functionality

- **MCU Module:** It controls the node's working state by sending commands to the Sensor Interface Module and the Network Protocol Module, respectively. It transfers sensor data from the Sensor Interface Module to the Network Protocol Module. This module also receives external commands from the Network Protocol Module and adjusts the node's working state accordingly.
- **Network Protocol Module:** It is in responsible for building up the network. It receives commands from MCU to control the Transceiver Interface Module, compresses and packs data according to the network protocol and transfers packets to the Transceiver Interface Module for wireless transmission. The module also receives the external packet from the Transceiver, unpacks and decompresses the packets, and transfers it to the MCU.
- **Transceiver Interface Module:** It couples the external transceiver. It receives commands from the Network Protocol Module to control the transceiver's performance, transfers data packages between the Network Protocol Module and the Transceiver, and transfers data between them.

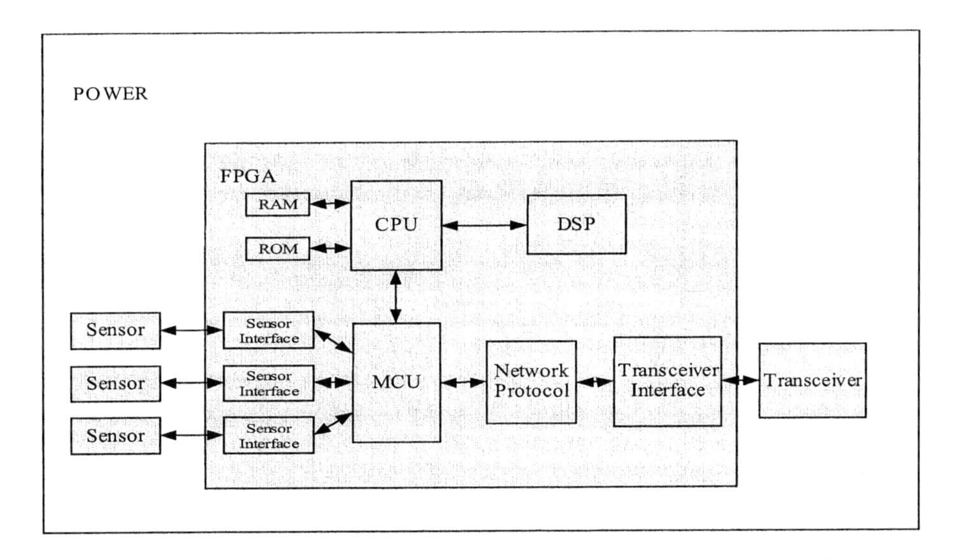

Fig. 2. Schematic architecture diagram of a SOC node of complex functionalities

The second architecture of sensor node favors complex functionalities and advanced jobs. Figure 2 shows the schematic diagram of the second architecture. It is suitable for applications where the sensor node have to process complex control procedures or run complex protocols. In this case, additional modules such as CPU, ROM, RAM and DSP are combined.

In this design, we use DSP module to run complex algorithms. We use CPU and RAM modules in conjunction with the system software for the system control. The software is stored in the ROM module.

It can be seen that a SOC sensor node can be composed of different functional hardware modules. These modules are linked with standard interchangeable communication protocols. Thus our main task is to optimize the design of these modules using hardware and software co-design method. For a specific application, what we need to do is to integrate these module's IP cores properly.

3 Design Details

There are a lot of works need to be carried out for the realization of a SOC sensor node. In this paper, we developed the Sensor Interface Module, MCU Module, partial of Network Protocol Module and the Transceiver Interface Module for the first architecture of basic functionality.

3.1 Design Platform

The platform we used is Virtex-II™ V2MB1000 FPGA developing system (from Memec). The sensors we used are SHT-71 temperature and humidity sensor (from SENSIRION) and tsl-2550 light sensor (from TAOS). The transceiver we used is CC1000 RF chip (from Chipcon).

3.2 Design of Modules

3.2.1 Sensor Interface Module

Figure 3 shows the design of the Sensor Interface Modules for SHT-71 temperature and humidity sensor and tsl-2550 light sensor, respectively.

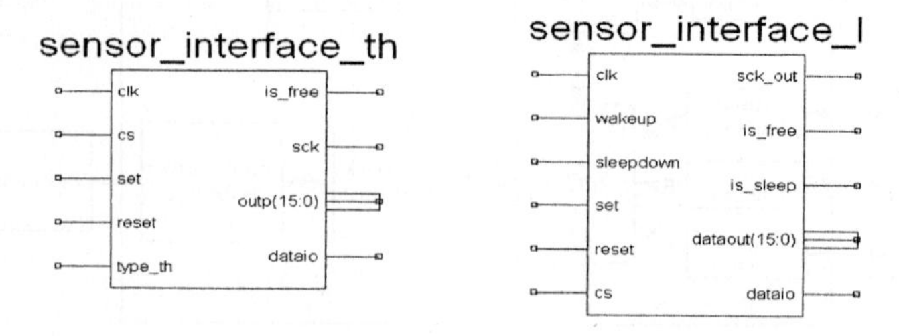

Fig. 3. Sensor Interface Modules for temperature, humidity and light sensors

Sensor_interface_th is the interface for SHT-71 temperature and humidity sensor. It collects temperature or humidity data from the sensors depending on the input signal of the type_th port. When the cs input port is high, the module sends the data to MCU through the outp output port.

Sensor_interface_l is the interface for tsl-2550 light sensor. It can enable or disable the sensor to work according to the input signal on wakeup and sleepdown ports. If both of the inputs are low, the module collects data from the sensor. When the input on the cs input port is high, it sends the data to MCU through the outp ports.

3.2.2 MCU Module

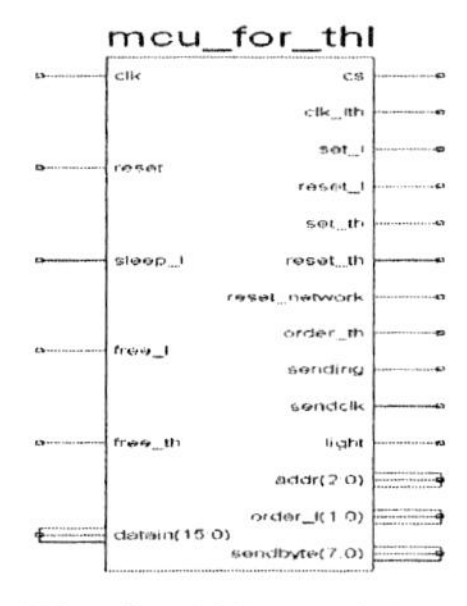

Fig. 4. MCU Module

The design of MCU module is shown as in Figure 4. It collects real-time temperature, humidity and light data form the Sensor Interface Modules once per second, and sends the data to the Network Protocol Module through the sending, sendclk and sendbyte output ports. Once the data sent, the status of the light output port turns to its opposite, e.g. from high to low or from low to high. So that we can tell if the MCU is working well by the shining of indicating ccd lights on the platform.

3.2.3 Network Protocol Module

protocol
clk
sending
reset
sendclk
reading
reset_out
readclk
sendbyte(7:0)
readbyte(7:0)

Fig. 5. Network Protocol Module

The Network Protocol Module is designed as in Figure 5. It receives the data from MCU through the reading, readclk and readbyte input ports, packets the data, then transfers the packet to the Transceiver Interface Module through the sending, sendclk and sendbyte output ports.

The proposed SOC Network Protocol Module is built on top of optimized network protocols. Once the protocols software are optimized and fixed, it is possible to transform the complete or part of the software design into a hardware module to improve the computation and power efficiency. In order to achieve the conversion, it is important to study the software and hardware co-design in the future work. To illustrate the idea, we have designed a simple network protocol in this work and studied its FPGA implementation.

The protocol designed is listed as below:

Start Byte	Node ID	Packet number	Data	End Byte
1 Byte	1 Byte	4 Bytes	6 Bytes	1 Byte

Where the start byte and the end byte are 0x7e. When the Node ID or Packet number or Data contains byte 0x7e, 0x7e will be expanded into two bytes 0x7d5e. Similarly, when the Node ID or Packet number or Data contains byte 0x7d, 0x7d will be expanded into two bytes 0x7d5d.

3.2.4 Transceiver Interface Module

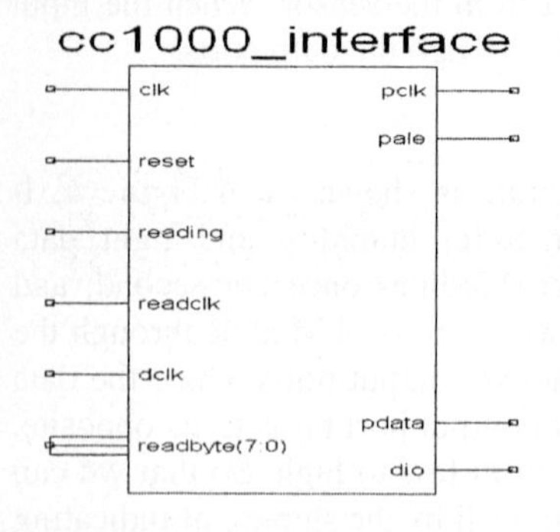

Fig. 6. Transceiver Interface Module for CC1000

Figure 6 shows the design of the Transceiver Interface Module for CC1000.

The CC1000 RF chip was set initially working at a frequency of 915.9988 MHz, in Manchester code and power off mode. The baud rate was 19.2 kbps. When data is received from the Network Protocol Module, the Transceiver Interface Module turns CC1000 RF chip into transmission mode. Data is then sent out wirelessly by CC1000 to complete the whole procedure.

4 Simulation

Simulations have been carried out to verify the performance of the Sensor Interface Modules, MCU Module, and the Network Protocol Module, respectively.

4.1 Simulation on the Sensor Interface Modules and MCU Module

We designed a sensor_th module and a sensor_l module to simulate the performance of the SHT-71 temperature and humidity sensor and the tsl-2550 light

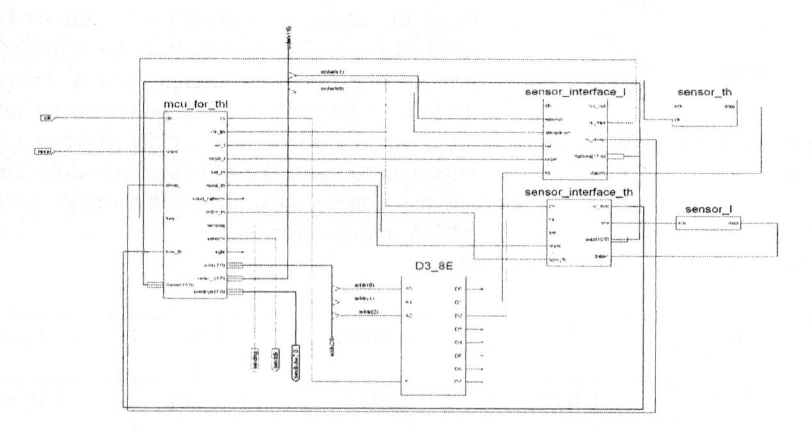

Fig. 7. Diagram to show the system connection of the simulation on the Sensor Interface Modules and MCU Module

sensor, respectively. The only difference between the simulated sensor and the real sensor is that the data provided by the stimulated sensor is set and fixed by the designer. Here the fixed data is 0x00FF for sensor_th and 0x0303 for sensor_l, respectively. The D3-8E module in Figure 3 was a 3 to 8 Decoder module used to decode the 3 bits address bus. The system connection of the sensor modules with the Sensor Interface Modules and MCU Module is shown as in Figure 7.

Then we used simulation software tool to simulate the performance of the Sensor Interfaces and MCU Modules. Figure 8 shows the waveforms of the Sensor Interface and MCU Modules.

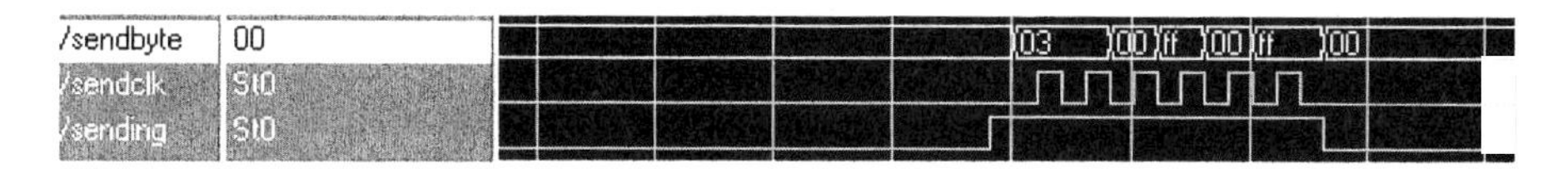

Fig. 8. Simulation waveforms of the Sensor Interface and MCU Modules

We set 6 bytes data into the Sensor Simulation Modules sensor_l and sensor_th first. It can be seen from Figure 8 that the MCU Module collected those data and the then sent them out (/sendbyte). The transferred data was 0x030300ff00ff, where the first two bytes was the data form sensor_l and the rest was from sensor_th, which proved that the Sensor Interface and MCU Modules worked correctly.

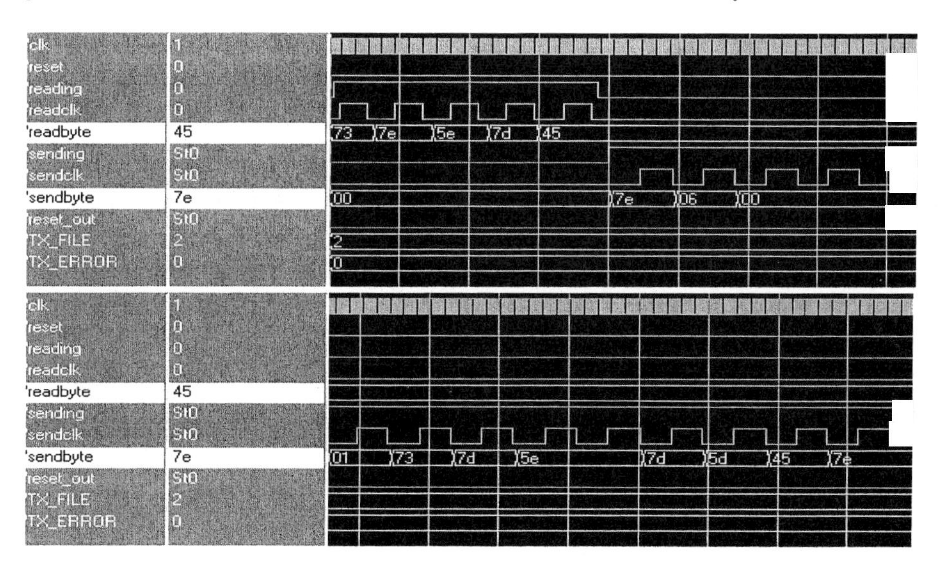

Fig. 9. Simulation waveform for the Network Protocol Modules

4.2 Simulation on the Network Protocol Module

The performance of the Network Protocol Module was also simulated. Figure 9 shows the result. First, we input 5 bytes data 0x73, 0x7e, 0x5e, 0x7d and 0x45 into the waveform table manually. From Figure 9 we can see that the sendclk and sendbyte output ports sent out 14 bytes data: 0x7e, 0x06, 0x00, 0x00, 0x00, 0x01, 0x73, 0x7d, 0x5e, 0x5e, 0x7d, 0x5d, 0x45 and 0x7e. Where the first one was the start byte 0x7e, the second one 0x06 was the node ID, the next four bytes 0x00000001 were packet number, and the next seven bytes 0x73, 0x7d, 0x5e, 0x5e, 0x7d, 0x5d and 0x45 were packed data. whilst 0x7e was indeed expanded into two bytes 0x7d and 0x5d and 0x7d was expanded into two bytes 0x7d and 0x5d, the other bytes were kept unchanged. The last byte was 0x7e for the end. This result has proved that both the design and performance of the Network Protocol Module are correct.

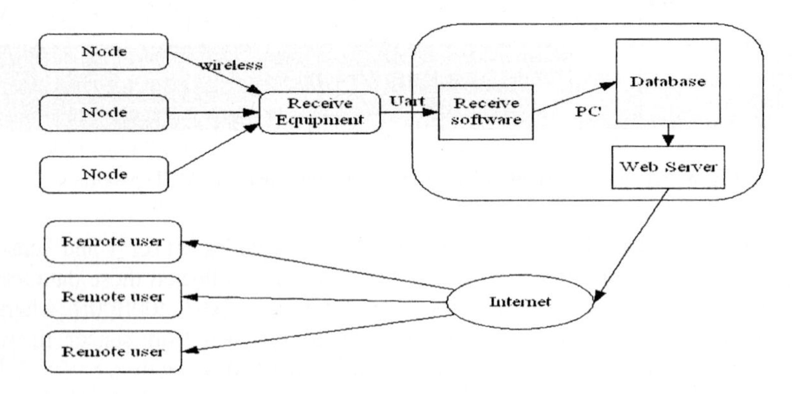

Fig. 10. Experimental verification system for the real time data collection, wireless transmission and result display

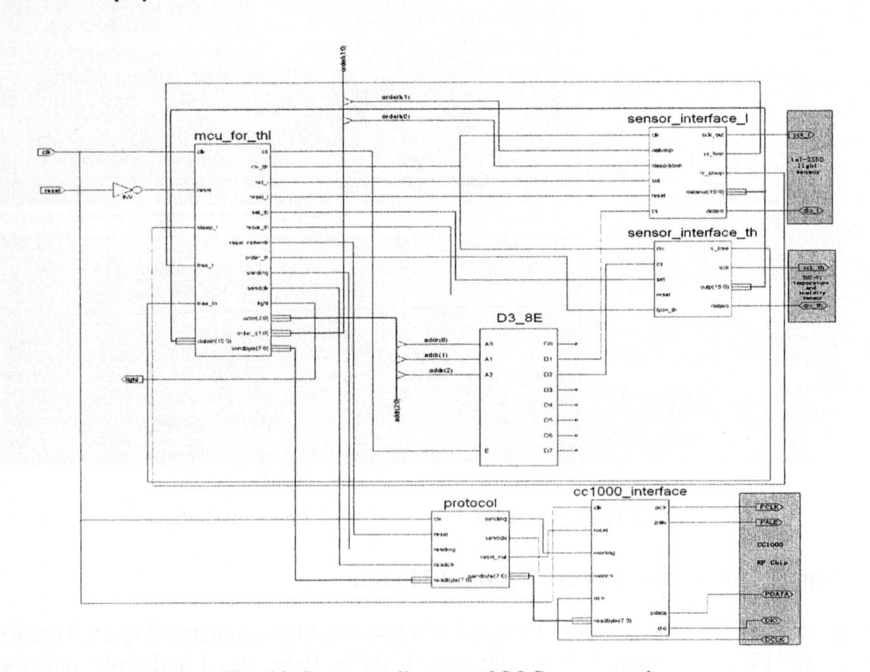

Fig. 11. Structure diagram of SOC sensor node

5 Experimental Verification of Real Time Data Collection and Transmission

Experimental verification of the performance of the designed SOC node was based on the system shown in Figure 10. Where the SOC node was constructed from temperature, humidity and light sensors, functional Modules and CC1000 RF transceiver. Detailed node structure is shown in Figure 11. We have developed in-house a data receive equipment and supporting software, database and data display interface, e.g. to display real time data on a webpage.

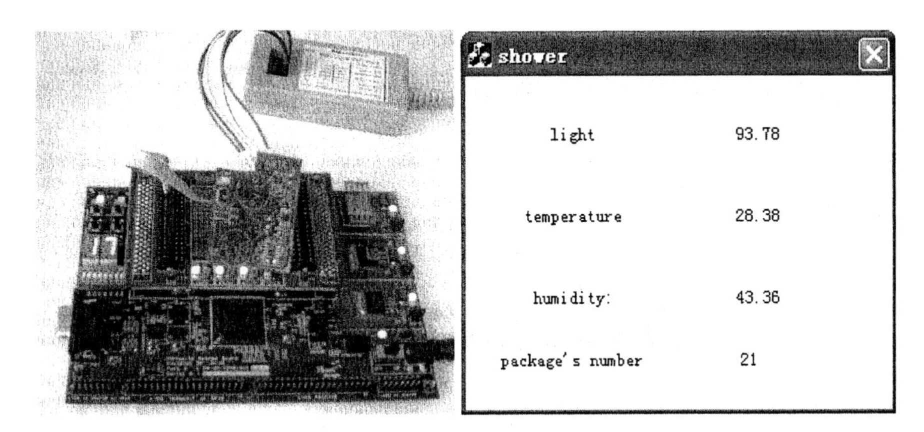

Fig. 12. Real-time data collection and wireless transmission by a completed SOC FPGA node of basic functions.

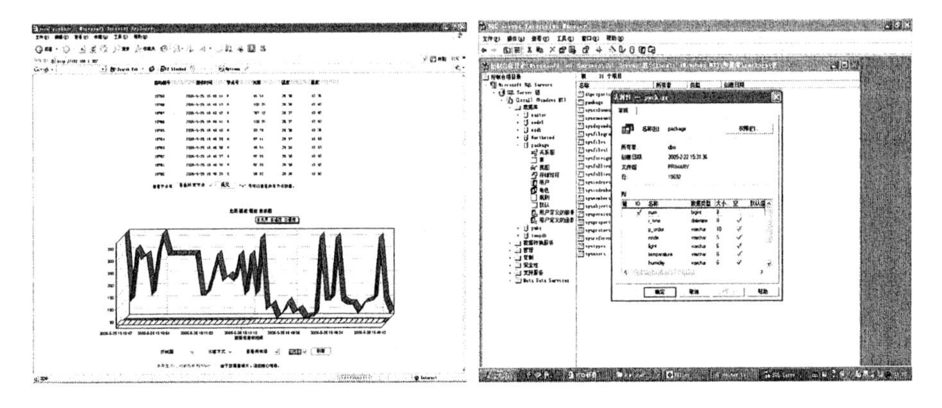

Fig. 13. Database management interface and data display interface on webpage

First the node transmitted the collected data from the temperature, humidity and light sensors to the receive equipment wirelessly. Then the receive equipment transferred the data to a PC through uart port. The processed information was stored into the database. The sensing information can be displayed either on the local PC for local users or on an internet linked webpage for the remote users. Figure 12 shows experimental results indicating real time data collection (from temperature, humidity and light sensors) and wireless data transmission by a completed SOC FPGA node of basic functions. Figure 13 shows the database management interface and the data display interface on internet webpage, respectively. The experimental results showed that the completed SOC FPGA node of basic functions worked successfully.

6 Conclusions and Future Work

In this work, a general "sensor node on a chip" approach, namely EasiSOC, has been presented with two proposed SOC architectures targeting on different application areas. The successful design and performance of the first type SOC FPGA node of basic functions have been verified by device simulation and tested experimentally. However there is still a lot of work need to be carried out before the SOC nodes are available as a massive product because the further development of SOC sensor node depends intensively on the progresses in other related research areas, e.g. optimized interface modules, specific algorithm and optimized network protocols, etc. Nevertheless the design concept technical progress and experience gained from this work can be applied directly to the further development of SOC node for various wireless sensor network applications.

Acknowledgements

The authors wish to acknowledge CAS "Plan of 100 Scientist" and ICTCAS innovation funds for financial supports.

References

[1] http://www.xbow.com/Products/Wireless_Sensor_Networks.htm
[2] J. M. Kahn, R. H. Katz, and K. S. J. Pister, "Next Century Challenges: Mobile Networking for Smart Dust", In Proceedings of the 5th Annual ACM/IEEE International Conference on Mobile Computing and Networking, 271-278, ACM Press, 1999.
[3] G.J. Pottie and W.J. Kaiser, "Wireless integrated network sensors", ACM Communications, 43(5), 51-58, 2000.
[4] A. Chandrakasan, R. Min, M. Bharwaj, S.-H. Cho, and A. Wang, "Power Aware Wireless Microsensor Systems", In Proceedings of ESSCIRC, Florence, Italy, September 2002.
[5] D. Cullar, D. Estrin, and M. Strvastava, "Overview of sensor network", Computer, 37(8), 41-49, 2004.
[6] R. Venugopalan, P. Ganesan, P. Peddabachagari, et al, "Encryption overhead in embedded systems and sensor network nodes: modeling and analysis", In Proceedings of the 2003 international conference on Compilers, architecture and synthesis for embedded systems, San Jose, California, USA , 188 -197, 2003.
[7] http://www.dust-inc.com/
[8] L.M. Ni, L. Cui, Q. Luo, et al, "Status of CAS-HKUST joint project BLOSSOMS", IEEE RTCSA 2005, invited
[9] W. Gao, L.M. Ni, Z.W. Xu, S.C. Cheung, L. Cui, Q. Luo, "BLOSSOMS: Building lightweight Optimized Sensor Systems on a Massive Scale", JCST, 20(1), 105-117, 2005.
[10] L. Cui, F. Wang, H.Y. Luo, H.L. Ju and T.P. Li, "A Pervasive Sensor Node Architecture", NPC2004, LNCS3222, 565-567, 2004.

Deployment Issues in Wireless Sensor Networks

Liping Liu*, Feng Xia, Zhi Wang*, Jiming Chen*, and Youxian Sun

National Laboratory of Industrial Control Technology,
Institute of Modern Control Engineering,
Zhejiang University, Hangzhou 310027, China
{lpliu, wangzhi, jmchen}@iipc.zju.edu.cn

Abstract. The performance of wireless sensor networks depends largely on the deployment of sensor nodes as well as their lifetime mainly determined by the energy consumption. Most current attention, however, has been paid to energy-efficient deployment. With the goal of facilitating further evolution of wireless sensor networks, recently proposed deployment schemes for wireless sensor networks are surveyed. The focus is on coverage and connectivity, which are regarded as the most important respects of network performance and energy-efficiency. Depending on the application and different actions in the network, coverage issues are classified into static and dynamic ones, while connectivity issues into pure connectivity and routing algorithm based connectivity. An overview of each of these areas is presented, and the performance of existing methodologies is discussed. In order to spark new interests and developments in this field, some crucial open issues are pointed out.

1 Introduction

The emergence of wireless sensor networks as one of the dominant technologies in the coming decades has posed numerous challenges to researchers. These networks are generally composed of hundreds, and potentially thousands of tiny sensor nodes, functioning autonomously, and in many cases, without access to renewable energy resources. A comprehensive overview of different aspects of research in sensor networks is provided by [1]. Wireless sensor networking challenges are proposed in four broad categories in detail, including: limiting radio operation, data management, geographic routing, and system monitoring and maintenance [2]. Foremost among these is development of long-life sensor networks in spite of energy-constraints of individual nodes. Sensor nodes are expected to be battery equipped, and deployed in a variety of terrains. Most of the applications are unattended or barren environment. Minimizing energy consumption is an important challenge in wireless sensor networks. There are two methods to prolong the life of the networks: 1) low-power device design; 2) coordination among sensor nodes. Though there are some progresses in low-power hardware design [3-5], most of the lower-power devices still have limited battery life and replacing batteries on tens of thousands of these devices is infeasible. As a consequence, it is well acknowledged that a sensor network should be deployed

* Corresponding author.

X. Jia, J. Wu, and Y. He (Eds.): MSN 2005, LNCS 3794, pp. 239 – 248, 2005.

with high density and coordinate in the inner networks in order to prolong the network lifetime.

Deployment is an important issue in wireless sensor networks, and affects the performance of the networks directly. The large number of nodes expected in sensor network deployments and unpredictable nature of deployment conditions introduce significant scalability and reliability concerns as well. According to the function of the networks, we discuss deployment in two fractions: 1) Coverage: any point on the desired region can be monitored; 2) Connectivity: the network of active sensors remains connective. We classify coverage into static and dynamic categories, while dividing connectivity into P-connectivity and RAB-connectivity. Due to the limited power supply used on board batteries, an important issue in wireless sensor networks is minimizing power consumption. Prolonging the networks lifetime must be deliberated in the whole deployment algorithm design.

In this paper, we give an overview of deployment issues in wireless sensor network, with special focus on coverage and connectivity. The remainder of the paper is organized as follows. In section 2 we discuss coverage in terms of static coverage and dynamic coverage. The current works related to connectivity that includes P-connectivity and RAB-connectivity are surveyed in Section 3. In the last section, we point out some crucial open issues in this field.

2 Coverage

In wireless sensor networks, each node has a detecting range, and the monitored area must be sensed through the networks. In order to make the resources more efficient and prolong the networks lifetime, algorithms that ensures the monitored area can be fully covered are needed. This is what we call the coverage problem. Under the assumption that an (acoustic or light) signal can be detected with certain minimum signal to noise ratio by a sensor node if the sensor is within a certain range of the signal source, and that each node can monitor a disk (the radius of which is called the detecting range of the sensor node) centered at the node on a two dimensional surface, the coverage problem can be described as follows: What is the minimum set of nodes that should be put in the active mode in order to cover the entire area? What is the proper distribution with the smallest energy cost? Regarding different requirements of different applications, we classify coverage issues into the following categories.

1) Static coverage problem: assuming that the detecting range is a disk with the radius r in two dimensions, what is the minimum set of nodes that should be put in the active mode in order to cover the entire area? Considering the overlap among the sensor nodes, what is the coverage algorithm that can ensure the overlaps among the sensors be the smallest while covering the whole monitored area.
2) Dynamic accommodation: since demands on performance are different in different areas, the number of active nodes in a local area should satisfy the corresponding requirements. That is to say, the nodes density of local areas can dynamically change according to the actual demands. We can shut off some sensor nodes when the current data precision exceeds the requirement. Vice versa, the precision can be improved while open more sensor nodes in the area. On the other hand, shutting off some nodes can reduce energy consumption. Another way to save energy is to change the duty

time of each sensor node through shifting the node's working mode between sleep and active. Note that the ratio of active time to sleep time is called duty time.

2.1 Static Coverage

The motivation of investigation on the coverage of sensor networks is to guarantee that the monitored area can be detected by the networks. In most cases, sensors collect data from the environment without any variations. Under the assumption that the architecture of wireless sensor network does not change with the network development, a number of coverage algorithms have been proposed according to the initial task assignment and requirements. This coverage problem is a static one. The difficulty in this issue is the definition of the detecting range and its influence on the entire network. The interest is to fully cover the monitored area with the least number of sensor nodes. It can also be stated as developing a sensor distribution algorithm to make the exposure area (where the sensor node can not detect) be the least or to minimize the overlapped sensing area of sensor nodes, as shown in Fig.1.

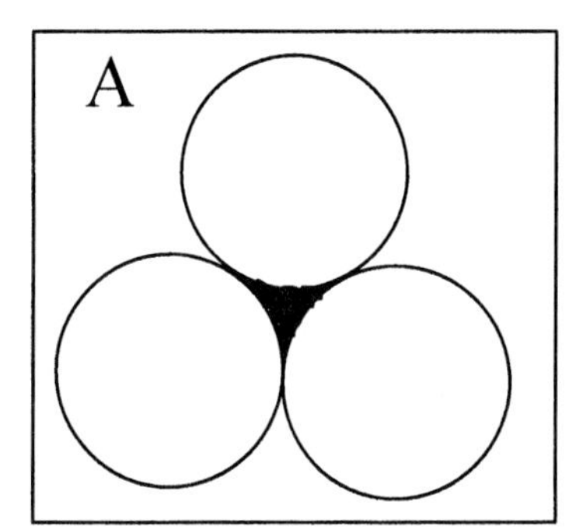

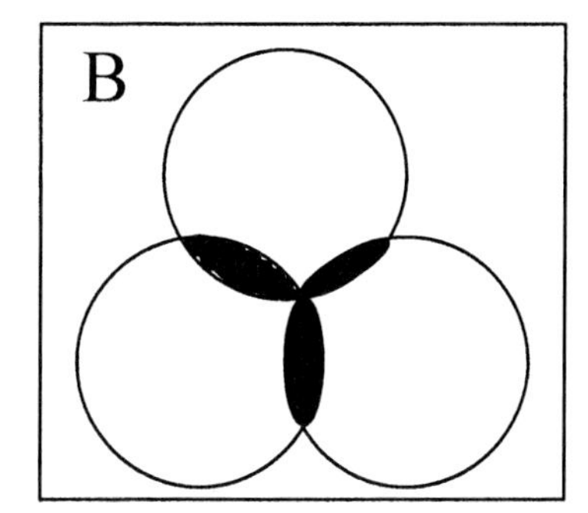

Fig. 1. Illustration of the static coverage problem (black represents exposure area in A and overlapped area in B)

There exists a close relationship between the static coverage and the art gallery problem (AGP) [6], which is about how to determine the minimum number of guards required to cover the interior of an art gallery (represented by a polygon). Introducing the AGP problem into the sensor coverage, the algorithm can be solved optimally in two-dimensional while shown to be NP-hard in the three-dimensional case. However, accurate coverage solutions can only be used in the regular small monitoring area. To address this problem, new local optimization schemes are presented to solve coverage problem in large irregular areas. A linear programming approach was induced in [7,8], and a least active node set for full coverage is suggested. Furthermore, the coverage uniformity is considered in the network to obtain more robust coverage.

On the exposure-based coverage problem, [9] established an optimal polynomial time worst and best case algorithm for coverage calculation by combining the Voronoi diagram and graph search algorithms. An algorithm is proposed to find the shortest exposure path in [10]. Under the assumption that each sensor node has equal distance to its neighbor nodes, Slijepcevic [11] presented a heuristic coverage scheme.

Virtual force algorithm (VFA) was also introduced into sensor coverage problem [12]. Each sensor behaves as a *source of force* for all other sensors. This force can be either positive (attractive) or negative (repulsive) according to the distance between each other. It is found that this method can fully cover the monitored area.

2.2 Dynamic Coverage

In the wireless sensor networks, coverage must adapt well to the application and network variations, such as tracking mobile object, node disabled, sensing block, etc. Dynamic coverage with dual-space approach has been introduced in event tracking and sensor management [13]. The dual-space transformation maps a non-local phenomenon, e.g. the edge of a half-plane shadow, to a single point in the dual space, and maps locations of distributed sensor nodes to a set of lines that partitions the dual space. Then nodes can decide to turn on or off according to the transformed space. Alberto [14] divided nodes into many small groups, and each group has a header. The header is on the duty to detect the moving object. Once finding the object, the leader node will wake up the nodes in the detecting range. According to the message transmitted through the nodes, the object moving direction, velocity and location can be estimated. When the object moves out of the detecting range, the node will turn into sleep mode. Also targeting at turning off some redundant nodes, [8] proposed a probe-based density control algorithm to put some nodes in a sensor-dense area to a doze mode to ensure a long-lived, robust sensing coverage. A coverage preserving node scheduling scheme is presented in [15] to determine when a node can be turned off and when it should be rescheduled to become active again. Some turning off mechanisms are used to change the nodes' working mode. However, turning off mechanism will debase the coverage degree, resulting in some blind sensing area.

Coverage methodologies should not only consider the full coverage of the area, but also satisfy the coverage degree request variations. A polynomial-time algorithm [16] is presented in terms of the number of sensors to determine whether every point in the service area of sensor networks is covered by at least K sensors, where K is a predefined value. With this algorithm, network works well in situations that require stronger environmental monitoring and impose more stringent fault-tolerant capability. Some works are targeted at particular applications, while the basic idea is still related to the coverage issue.

In the literature, mobile agents or robots have also been introduced into coverage algorithms. Mobile agent can exchange messages through the static nodes that have some difficulty in contacting with the networks, thereby optimizing the current coverage. For networks lack of connectivity, Winfield [17] proposed a self-deployment method. Mobile nodes are utilized to enhance network coverage and system lifetime is extended via configuration of uniformly distributed node topologies from random node distributions. Coverage process itself is very energy consuming due to the locomotive action as well as computation and communications associated with it. Not only minimizing average moving distance, but also reducing difference of the remaining energy among sensor nodes is essential for a longer lifetime. Taking into account the dynamic and distributed nature of coverage, [18] proposed an intelligent energy-efficient deployment algorithm for cluster-based wireless sensor network by a synergistic combination of cluster structuring and a peer-to-peer deployment scheme.

Simulation results show that the algorithm can successfully obtain a uniform distribution from initial uneven distributions in an energy-efficient manner.

3 Connectivity

Another key issue in wireless sensor network is to keep the network connective so that certain network performance can be achieved. In the network, each node has a radio range. Out of their ranges, the nodes cannot reach each other directly, though they can connect in a multi-hop manner. For a network with certain node density, increasing the radio range can increase the connectivity, while leading to more transmission power requirements and shorter network lifetime. It is essential for the networks that deployment should ensure that any pair of nodes in the networks can reach each other. We call it pure-connectivity (P-connectivity). P-connectivity is independent of the employed routing algorithm. Considering the energy consumption and the system performance, some specific routing algorithms can be used to make connection between an arbitrary pair. This is called routing algorithm based connectivity (RAB-connectivity). RAB-connectivity is closely related to the employed routing algorithm. It values quite different for different routing algorithms. The former requires that any pair of nodes must be connective either directly or in a multi-hop manner. The latter emphasizes making the connectivity more efficient through some special algorithms.

3.1 Pure Connectivity (P-Connectivity)

P-connectivity of a random network is defined as the average probability of being able to make connection between an arbitrary node pair in the network. As long as network is fully connected in terms of graphic theory, its P-connectivity is equal to one.

The P-connectivity of a randomly distributed network as a function of radio range can be described in several ways. The most direct way is similar to coverage problem. For instance, what is the certain distance between arbitrary pair of sensor nodes to make connection in the whole networks? What is the appropriate nodes density with nodes connecting with each other? How many neighbor nodes should the node have to make network connective? Considering the energy cost, it is valuable to find the smallest radio range in a connective network. All current schemes are based on the assumption that nodes are static. When there are mobile nodes in the network, novel techniques will then be required. Furthermore, to the best of our knowledge, most theoretic analysis on P-connectivity in a multi-hop manner has been carried out based on some communication models under the assumption that all nodes are distributed uniformly in spatio-temporal direction and there is no intervention among the nodes.

The research on optimal number of neighbor nodes is started in the 1970s. Based on the analysis of the slotted ALOHA protocol, Kleinrock and Silvester [19] have found that there exists a *magic number* when the transmission powers are the same for all modes. When the network is fully connective, each node should have 6 neighbor nodes. Later, some researchers have made further study on *magic number*. As a result, it is certain that the *magic number* ensuring the network connective is between 5 and 8. Philips et al [20] study the connectivity properties of randomly distributed radio network theoretically. They prove that if the average number of neighbors of each station is a constant, then for sufficiently large area the network will almost surely be disconnected. This implies that from network connectivity point of view, no *magic number* could ensure the randomly distributed network over a sufficiently large area being fully connected. However, this does not mean that the *magic number* is useless.

There exists valuable critical number in the connectivity problem. Gibert[21] has shown that there is a critical number N_0. If the average number of neighbor nodes exceeds N_0, the random plane network contains an infinite connected component with non-zero probability. He has found N_0 to be bounded between 1.64 and 17.94. Later, Li and Halpern [22] give a more precise deduction, i.e. $2.195<N_0<10.526$. It is also confirmed that if the average number of neighbor nodes exceed N_0, the network is connective in the finite range. Gilbert run simulations to verify that the vast majority of node can be contained in a single giant component in practice. In addition, he estimate the true value of N_0 to be about 3.2. Xue and Kumar [23] argue again that the network with *magic number* neighbor nodes change to be not connective as the number of nodes increases in networks with finite nodes. In order to maintain the connectivity, the number of neighbor should be proportional to $\log n$ (n is the number of the nodes in the network). It is also mentioned that the network cannot connect when the average number of neighbor nodes is less than $0.074\log n$ or larger than 5.1774.

With the help of zero-one law, [24] found the critical density of path connectivity. For networks with higher density, path connectivity gets better with larger networks. It is proved that connectivity increases quickly from 0.1 to 0.9 with the critical density between 1.6 and 1.7. This is a wonderful discovery in network connectivity research.

The radio range relies on the characters of nodes, energy, and application circumstance. The characteristics of the wireless sensor network and residual energy may vary over time, so we should find the appropriate radio range that can adapt with respect to the variations. One natural approach to solving this problem is the polynomial-time algorithm in one dimension. Unfortunately, it is a NP-hard problem in 2-D or 3-D case [25, 26]. Santi et al [27] investigated the smallest radio range problem in the cube circumstance, and proposed the relationship between the radio range and length of edge in 2-D and 3-D.

3.2 Routing Algorithm Based Connectivity (RAB-Connectivity)

Since RAB-Connectivity is closely related to the employed routing algorithm and it values quite different for different routing algorithms, we will focus on the routing algorithm in the subsection. Routing in wireless sensor networks is very challenging due to several characteristics that distinguish them from contemporary communication and wireless ad hoc networks. First of all, it is not possible to build a global addressing scheme for the deployment of sheer number of sensor nodes. Secondly, almost all applications of sensor networks require the flow of sensed data from multiple regions (sources) to a particular sink. Thirdly, generated data traffic has significant redundancy. Finally, sensor nodes are tightly constrained in terms of transmission power, on-board energy, processing capacity and storage [28].

For these reasons, many new algorithms have been proposed in wireless sensor networks, and can be classified into data-centric, hierarchical, location-based, network-flow based and QoS aware ones. We next survey and evaluate the four routing mechanisms for wireless sensor networks developed in recent years.

(1) Data-centric Protocols

Data-centric protocols are query-based and depend on the naming of desired data, which help in eliminating many redundant transmissions. In data-centric routing, the sink sends queries to certain regions and waits for data from the sensors located in the selected regions. SPIN [29] is the first data-centric protocol, which considers data negotiation between nodes in order to eliminate redundant data and save energy. Later, Directed Diffusion [30] has been developed and has become a breakthrough in data-centric routing. Then, many other protocols have been proposed following a similar concept, such as CADR, COUGAR, ACQUIRE, and so on.

Data-çentric routing protocols are quite energy efficient since the query is performed only when it is needed and global topology needn't be maintained. While there are also some disadvantages such as not performing well in condition monitoring applications, the naming depends on different applications and must be done firstly, and the process of data querying and matching introduces extra communication load and will cause broadcast storm.

(2) Hierarchical Protocols

Hierarchical protocols aim at clustering the nodes so that cluster headers can do some aggregation and reduction of data to save energy. Hierarchical routing protocols can efficiently maintain the energy consumption of sensor nodes by involving them in multi-hop communication within a particular cluster and by performing data aggregation and fusion in order to decrease the number of transmitted messages to the sink. LEACH [31] is one of the first hierarchical routing approaches for sensors networks. The idea proposed in LEACH has been an inspiration for many hierarchical routing protocols such as PEGASIS, Hierarchical-PEGASIS, TEEN, APTEEN, EARCSN, etc.

Dynamic clustering in Hierarchical routing protocols can strengthen the connectivity and prolong lifetime of the network, while the main disadvantage is using single hop communication and unsuitable in large area applications.

(3) Location-based protocols

Location-based protocols utilize the position information to relay data to the desired regions rather than the whole network. In most cases, location information is needed to estimate energy consumption between two particular nodes. Since, there is no addressing scheme for sensor networks like IP-addresses and they are spatially deployed on a region, location information can be utilized in routing data in an energy efficient way. In order to stay with the theme, we limit the scope of coverage to only energy-aware location-based protocols, for instance, MECN, SMECN, GAF, GEAR, etc.

Location based routing protocols are energy efficient when the sensor nodes are deployed densely. That is, when the density of sensor nodes is small, these protocols may not keep the connectivity efficiently.

(4) Network-flow Based and QoS aware Protocols

The last category includes routing approaches based on general network-flow modeling and protocols that strive for meeting some QoS requirements along with the routing function. The routing protocols include, e.g. Maximum Lifetime Energy Routing, Maximum Lifetime Data Gathering, Minimum Cost Forwarding, SAR, SPEED etc.

Network flow based and QoS aware routing protocols can significantly improve the performance of QoS such as delay of end-to-end and rate of packets losing, while the energy efficiency must be considered ulteriorly in maintaining the connectivity of the network.

4 Discussion and Open Issues

Deployment issue in wireless sensor networks has attracted a lot of attention in recent years and introduces unique challenges compared to traditional deployment in networks. In this paper, we have summarized and discussed recent research results on node deployment in wireless sensor networks. In particular, coverage and connectivity are of our main interest. Though a considerable amount of approaches to node deployment has been suggested recently, much more issues still remain open today.

In the context of coverage, most of current works are based on two assumptions: 1) sensor nodes are immobile and uniform distributed; 2) the number of sensor nodes is huge, so that networks can coverage the monitored area with any algorithm. Little attention is paid to the optimization of the topology and sensor numbers. In fact, varying topology gives rise to a challenging issue in nodes deployment. Further open issues related to coverage may include regulation of coverage degree, modulation on sensing frequency, determination of neighbor node coverage boundary and so on. The smallest sensor number and distribution for a given application are also among main spots in wireless sensor network. In addition, optimizing energy consumption and prolonging the network lifetime are still necessary to be effectively addressed.

As regarding P-connectivity, special attention has been paid to theoretic analysis. Besides the problems mentioned above, other open issues can be characterized, for instance, given the radio range and distribution of nodes, find the greatest coverage area, or given the radio range and the density of nodes, find proper distribution for given properties. In addition, the energy consumption should also be considered in P-Connectivity. Although great progress has been made in routing algorithms for connectivity, there still lack of new approaches that are adaptable with respect to the application of wireless sensor networks. Possible directions, which are mostly driven by practical applications, include mobile sensor node routing algorithm, uniting wireless sensor network, wired network routing protocol, etc.

Furthermore, the relationship between coverage and connectivity is also interesting for further study. Detecting range is different from radio range. Coverage ensures that nodes can sense the monitored area, while connectivity ensures that nodes can communicate with each other. There are comparability between coverage and connectivity. Coverage implies connectivity when radio range is larger than twice of detecting range [32, 33]. But the relationship between them is rarely investigated in case of radio range less than twice of detecting range, because radio interference exists. This also remains to be a valuable open issue in the deployment of sensor nodes.

Acknowledgement. This work is supported by National Natural Science Foundation of China (NSFC 60434030), National Basic Research Program of China (2002CB312200) and Advanced Research Program of France-China (PRA SI03-02).

References

1. Akyildiz I. F., Su W., Sankarasubramaniam Y., et al.: Wireless sensor networks: a survey. Computer Networks, 2002, 38(4): 393-422.
2. Ganesan.D, Cerpa A, Ye,W, et al. Networking issues in wireless sensor networks. Journal of Parallel and Distributed Computing Volume: 64, Issue: 7, July, 2004, pp. 799-814
3. Perkins M., Corrreal N. and B. O'Dea: Emergent Wireless Sensor Network Limitations: A Plea for Advancement in Core Technologies. The first IEEE international conference on sensors (Sensors02). Orlando, Florida, USA. June 2002.
4. Sibley G.T., Rahimi M.H., and Sukhatme G.S. Robomote: a tiny mobile robot platform for large-scale ad-hoc sensor networks. Robotics and Automation, 2002. Proceedings. ICRA '02. IEEE International Conference on, May 2002.(2)11-15.
5. Abrach H., Bhatis S., Carlson J. et al: MANTIS: System Support for MultimodAl NeTworks of In-situ Sensors. WSNA03, San Diego, CA, USA. September 2003.
6. O'Rourke, J., Art Gallery Theorem and Algorithms. New York: Oxford University Press, 1987.
7. K. Chakrabarty, S. S. Iyengar, H. Qi, et al. Grid coverage for surveillance and target location in distributed sensor networks, IEEE Transactions on Computers, December 2002.51(12): 1448-1453.
8. Nirupama Bulusu, John Heidemann, and Deborah Estrin: Adaptive Beacon Placement, in Proceedings of the 21th International Conference on Distributed Computing Systems, Phoenix, AZ, Apr. 2001. 489-498.
9. S. Meguerdichian, F. Koushanfar, M. Potkonjak, et.al. Coverage Problems in Wireless Ad-Hoc Sensor Networks. INFOCOM'01, April 2001(3), 1380-1387.
10. S. Meguerdichian, F. Koushanfar, G. Qu et al. Exposure in Wireless Ad Hoc Sensor Networks. Procs. Of 7th Annual International Conference on Mobile Computing and Networking (MobiCom'01). July 2001. 139-150.
11. Slijepcevic, S.; Potkonjak, Power efficient organization of wireless sensor networks. Communication, IEEE International Conference on, 11-14 June 2001, 2:472 - 476
12. Y.Zou and K.Chakrabarty Sensor deployment and target localization based on virtual forces, in Proc.of the IEEE INFOCOM Conference, 2003.
13. Jie LIU, Patrick Cheung, et al. A Dual-Space Approach to Tracking and Sensor Management in Wireless Sensor Networks. WSNA'02, September 28, 2002, Atlanta, Georgia, USA. 131-139.
14. Alberto Cerpa, Jeremy Elson. et al. Habitat monitoring: application driver for wireless communications technology. UCLA Computer Science Technical Report 200023, December 2000.
15. Di Tian, Georganas ND. A node scheduling scheme for energy conservation in large wireless sensor networks. Wireless Communications and Mobile Computing, Wiley, UK March 2003, 3(2) 271-290.
16. C. Huang and Y. Tseng. The coverage Problem in a Wireless Sensor Network. *WSNA03*, September 19, 2003, San Diego, CA. 2003

17. A.F.T. Winfield, Distributed sensing and data collection via broken ad hoc wireless connected networks of mobile robots, in Distributed Autonomous Robotic Systems 4, eds. LE Parker, G Bekey & J Barhen, Springer-Verlag, 273-282, 2000.
18. D. Tian and N. D. Georganas. A coverage-preserving node scheduling scheme for large wireless sensor networks. In ACM Int'l Workshop on Wireless Sensor Networks and Applications (WSNA), 2002.
19. Kleinrock.L. and Silvester. J.A. Optimum transmission radii for packet radio networks or why six is a magic number, in Proc. of IEEE Nat.Telecommun. Conf. December 1978:431-435.
20. Philips.T.K, Panwar. S.S. and Tantawi.A.N. Connectivity properties of a packet radio network model, IEEE Transactions on Information Theory 35 September 1989:1044-1047.
21. Gilbert.E.N. Random plane networks, SIAM, vol 9,533-543,1961.
22. Li L, Halpern J Y. Minimum energy mobile wireless networks revisited. Proceedings of IEEE International Conference on Communications (ICC01), Helsinki,Finland, June 2001.
23. Xue F and Kumar P.R.The number of neighbors needed for connectivity of wireless networks. wireless Networks vol. 10 169-181 2004.
24. Yu D, Li H. On the definition of ad hoc network connectivity. Communication Technology Proceedings, 2003. ICCT 2003. International Conference on, Vol2, 990 - 994, 9-11 April 2003.
25. Clementi A. E. F, Penna P, and Silvestri R, Hardness Results for the Power Range Assignment Problem in Packet Radio Networks, Proc. 2nd International Workshop on Approximation Algorithms for Combinatorial Optimization Problems (RANDOM/APPROX 99), LNCS (1671), 197 - 208, 1999.
26. Kirousis L, M, Krabakis E, Krizanc D. et.al, Power Consumption in Packet Radio Networks, Theoretical Computer Science, vol. 243, 289 - 305, July 2000.
27. Santi P, Blough D. M, Vainstein F. A probabilistic analysis for the range assignment problem in ad hoc networks. Proceedings of the 2nd ACM international symposium on Mobile ad hoc networking & computing, 212-220, October 2001.
28. Akkaya K, Younis M, A Survey on Routing Protocols for Wireless Sensor Networks. Ad Hoc Networks. 2005,3(3):325-349.
29. Heinzelman W, Kulik J, Balakrishnan H. Adaptive protocols for information dissemination in wireless sensor networks. Proceedings of the 5th Annual ACM/IEEE International Conference on Mobile Computing and Networking (MobiCom99), Seattle, WA, Aug.1999.
30. Intanagonwiwat C, Govindan R, Estrin D. Directed diffusion: a scalable and robust communication paradigm for sensor networks. Proceedings of the 6th Annual ACM/IEEE International Conference on Mobile Computing and Networking (MobiCom00), Boston, MA, August 2000.
31. Heinzelman W, Chandrakasan A, Balakrishan H. Energy-efficient communication protocol for wireless sensor networks. Proceeding of the Hawaii International Conference System Sciences, Hawaii, January 2000.
32. Di.T, Georganas N.D. Connectivity maintenance and coverage preservation in wireless sensor networks. Electrical and Computer Engineering, 2004. Canadian Conference on, 2-5 May 2004, 2, 1097 - 1100.
33. Ye F, Zhong G, Lu S et.al. PEAS: A Robust Energy Conserving Protocol for Long-lived Sensor Networks. The 23rd International Conference on Distributed Computing Systems (ICDCS'03), May 2003.

A Geographical Cellular-Like Architecture for Wireless Sensor Networks*

Xiao Chen and Mingwei Xu

Department of Computer Science and Technology,
Tsinghua University, Beijing 100084, P. R. China
{csun, xmw}@csnet1.cs.tsinghua.edu.cn

Abstract. The mobility of sink and stimulus brings a huge challenge to WSN. It wishes that the location update packets be continuously transmitted through the sensor network to keep the data path between the sink and the source node sensing the stimulus. Unfortunately, frequent location update packets would not only result in excessive energy consumption of sensor node with highly energy constraint, but also increase the communication collision in wireless propagation. In this paper we propose a novel Geographical Cellular-like Architecture (GCA) for WSN, which can efficiently manage the mobility of both sink and stimulus. GCA proactively builds a cellular-like structure assisted by the geographical location of each node, where each cell includes a header and several members. The header acts as the role of base station in cellular network, receiving the information from sink and forwarding the data report; the members just take charge of watch-ing the target. Meanwhile, the hierarchical design of GCA can also save energy remarkably. The performance of GCA is evaluated through analysis and the simulation in ns2, which demonstrates that GCA is an efficient solution to mobility in WSN.

1 Introduction

The advancements of Micro-Electro-Mechanical System (MEMS) technology, wireless communication and embedded processing have enabled the development of small-size, low-cost, low-power sensor nodes [1, 2]. These sensor nodes, with functions of wireless communication, data sensing, processing and collaborative effort, are constructed to Wireless Sensor Networks (WSN) by self-organization.

Although most of the sensor nodes are stationary, the sink and stimulus are often mobile in many scenarios [3]. It hopes that the location update be continuously propagated across the sensor network for keeping the data path connective from source to sink. But the excessive energy consumption and communication collision resulting from frequent location update does not adapt to the WSN that characterized as energy constraint. Therefore, how to solve the mobility of the sink and stimulus is a new challenge the WSN faces.

* This work was supported by the Natural Science Foundation of China (No60373010), National 973 Project Fund of China (No 2003CB314801).

X. Jia, J. Wu, and Y. He (Eds.): MSN 2005, LNCS 3794, pp. 249–258, 2005.

In this paper we propose a Geographical Cellular-like Architecture (GCA) for sensor networks. GCA proactively constructs a two-tiered cellular-like architecture in terms of the location information of each node, in order to efficiently manage the mobility of both sink and stimulus. The architecture is divided into many cells, each of which includes a *header* and several *members*. The header acts as the role of base station in cellular network, receiving the information from sink and forwarding the data report; the members just take charge of watching the target. The source node got the stimulus would propagate data packet to its header, which would further forward the data among headers according to the Greedy Energy-aware Alternative Forwarding Algorithm (GEAFA), until the data reached the last header to the sink node. Eventually, the header would broadcast the data packet to the sink.

The design of GCA adopts the hierarchical structure, which is demonstrated as an energy-efficient solution for WSN in many papers [4,5,6,7]. The hierarchy and node classification can avoid the collision in data transmission and carry out efficient data fusion, both of which could reduce the energy cost. Furthermore, we exploit the advantage of hierarchy in the mobility management of WSN in this paper. GCA can benefit from some mature techniques in cellular networks [8], such as cell splitting, to solve the problems appeared in WSN.

The remainder of the paper is organized as follows. Section 2 discusses the related work of data dissemination and routing; Section 3 describes the details of GCA, including the construction, the forwarding algorithm and the mobility management; Section 4 simulates the GCA and evaluates the performance of the architecture in ns2 [9]; and Section 5 concludes the paper.

2 Related Work

Though most of the sensor nodes are considered stationary, the mobility in WSN is still existed. The sink mobility has been studied in many research works [3] [10]. In TTDD [3], each data source builds a grid structure that enables mobile sinks to continuously receive data on the move by flooding queries within a local cell only, which reduces the overhead of frequent location update packet of sink. But the per-source grid construction is a new overhead; especially in the scenario that stimulus is mobile. Moreover, the square is not optimal shape of the cell. [10] considers the sink mobility as a solution to balance the energy cost of nodes near the sink. The data collection protocol can then be optimized by taking sink mobility and multi-hop routing into account.

The geographical routing is a kind of routing methods in wireless networks, which bases on the assumption that each node can obtain its relative coordinate. GPSR [11] makes greedy forwarding decisions using only information about a node's immediate neighbors. When a packet reaches a region where greedy path is invalid, the algorithm recovers by routing around the perimeter of the region. GEAR [12] uses energy aware neighbor selection to route a packet towards the target region and Recursive Geographic Forwarding or Restricted Flooding algorithm to disseminate the packet inside the destination region.

3 Geographical Cellular-Like Architecture

In order to efficiently manage the mobility of both sink and stimulus in WSN, we develop the GCA which is not only a good solution to mobility but also an energy efficient forwarding scheme. This section will present the details of GCA.

In GCA, each sensor node is aware of its own location through receiving GPS signals or some localization algorithm [13] [14]. And through the periodical interaction, nodes can also learn the location of its neighbors. Another important assumption in GCA is that the transmission distance d and the radius of cell r must satisfy the condition: $d \geq \sqrt{3}r$. Under this consumption, the header could directly select the neighbor headers and communicate with them.

In GCA, once the source produces a data report, it will propagate the packet to its header. Then the header would further forward the packet hop-by-hop at header layer according to GEAFA algorithm, until the packet reached the last header to the sink node. Finally, the header would broadcast the data packet to sink. The above procedure can be described as Fig.1.

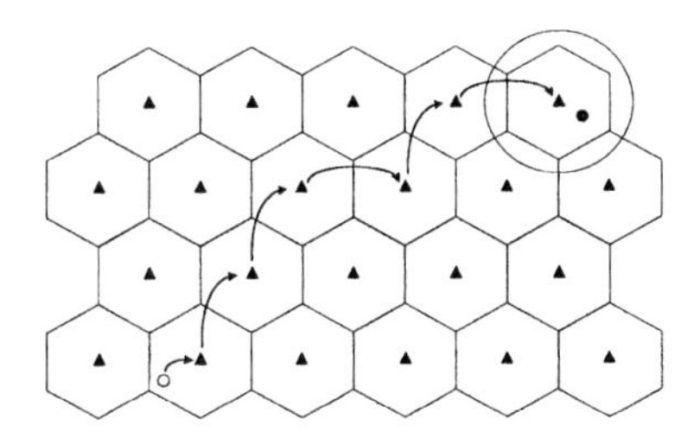

Fig. 1. the Geographical Cellular-like Architecture and forwarding procedure The white point is the source node and the black point is the sink. The black triangle is logical header of each cell.

3.1 Construction

The construction of the cellular-like architecture is initiated by the control node out of the sensor network. Some sensor node is selected as the base point, and then constructs the architecture through control packets step by step. Firstly, the base point becomes header and calculates the *standard coordinate* of six neighbor headers. For some header at location (x_0, y_0), the neighbor headers are located at $(x_0 \pm \sqrt{3}r, y_0)$ and $(x_0 \pm \sqrt{3}r/2, y_0 \pm 3r/2)$, shown as Fig.2. Then the base point selects the nearest node to each neighbor header, and informs them to become header serving as the base station located at the standard coordinate. Afterwards, the new headers should return a packet for acknowledgement. Other nodes will become members, each of which could choose its header by comparing the signal strength of the control packets received from different headers.

After the new header is elected, it can select the five neighbor headers except the upstream header from which it receives the control packet. But if the distance between the nearest node and corresponding standard coordinate is bigger than

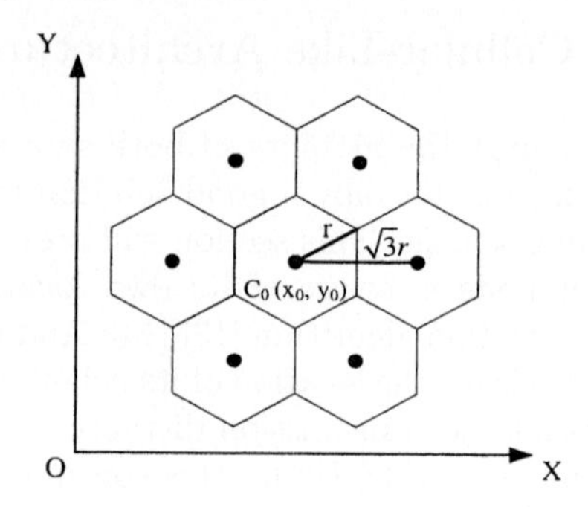

Fig. 2. the hexagon cell

r, it would not select the node as header, because the packet may reach the network border or a "hole". Through the recursive construct-ing procedure, the two-tiered cellular-like architecture is accomplished.

3.2 Forwarding Algorithm

There are four kinds of packet in GCA: control packet, query packet, inform packet and data packet. The control packet is used to construct the architecture and control the behavior and status of node, described as above section. The query packet is sent by sink, and then is received at the header that the sink belongs to. Afterwards, the packet is flooded to the network at the header layer. Each header maintains a task table, in which every entry specifies the task and the location of sink. In fact the location of sink is the location of the sink's header, which would be used as the destination location in the following data packet forwarding. The inform packet is issued by the node captured target, and would received at corresponding header. If there is matching entry in the task table of the header, then data will be forwarded to the sink location; else the packet could be simply dropped.

The forwarding algorithm of data packet is the keystone. Known the sink location, the packet could be routed to the destination header by greedy forwarding algorithm [12] among headers, and finally transmit to the sink node. However, as the greedy path between source and sink is unique, it would lead to excessive energy cost of headers on the path.

The energy-awareness can be introduced to forwarding algorithm for load balance. We develop the Greedy Energy-aware Alternative Forwarding Algorithm (GEAFA) that can be de-scribed as follows:

1. If the sink locates between the two radials from current header to two adjacent neighboring headers, the neighbor header with more energy will be selected as next hop.
2. If the sink locates on the radial from current header to some neighbor header,
   (a) the neighbor header will be selected as next hop (GEAFA-1). or

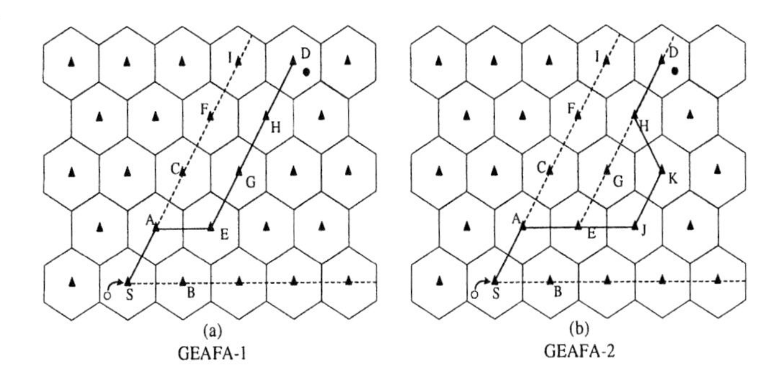

Fig. 3. the two variants of GEAFA

(b) it could arbitrarily choose another neighboring header as a candidate, and then the neighbor header with more energy will be selected as next hop (GEAFA-2).

There are two variants of GEAFA, whose difference is the decision when the sink locates on the radial. Shown as Fig.3.(a), after the header S receives the data, it judges that the header D that sink belongs to locates between $\boldsymbol{SA}$ and $\boldsymbol{SB}$. Supposed the energy of A is more than B, the packet is forwarded to A. As the same, the packet could be forwarded to E from A. Now the location of D is on the radial , so header E would send the packet to D along the path EGHD according to GEAFA-1. However, the choice of header E is different in GEAFA-2. E could arbitrarily choose a neighbor header adjacent to D. Assuming that J is chosen and its energy is more than G, the packet would be sent to J not G. Thus, the forwarding path is different, as Fig.3.(b).

There are different characteristics of the two variants. If selecting the unique neighbor header as next hop (GEAFA-1), the headers on the radial would be charged with much energy cost. If choosing another candidate (GEAFA-2), the load balance would be better but the forwarding path maybe not optimal.

3.3 Mobility

The mobility of WSN in which most sensor nodes are stationary can be divided into the stimulus mobility and the sink mobility. In GCA, the impact of stimulus mobility to the network and forwarding is very little. Because the source node captured the target just sends the data to the header it belongs to.

The mobility of sink can be separated to two situations. If sink just moves in the cell size, it does not need any other process as the transmission range of header can cover the whole cell. But if sink moves to another cell, it needs the new header register to the old header called *home agent*. When no data report returns to sink, the simple register is enough. Once data is produced, the packet still propagates to the home agent, which further forwards the packet to the new header. All of nodes except home agent do not know the motion of sink, and the data will firstly be sent to home agent. The new sink location is informed through keep-alive flood-ing initiated by sink, whose period is relative long, in order to reduce the affect and cost of mobility.

3.4 Maintenance

The main problem of network maintenance is the header failure resulted from excessive energy cost, which could consequently lead to network collapse. So the header dynamic rotation is introduced to maintain the architecture of network. The header needs to select neighboring members as *successors*, which should meet the requirement that it locates in the overlap area of all the neighbor headers' communication range so that it can communication with them directly. Thus, the successor could take charge of the old header's role to collect and forward data.

Another aspect of network maintenance is periodically updates the query task. The query packet sent by sink would setup a task entry on every node and maintain it dynamically. When sink node wishes carry out the task for a long time, it needs periodical keep-alive flooding to refresh the timer of task entry. If no keep-alive flooding packet is received before timer expires, the entry would be deleted, which means the query task is out of date. Generally, the period is relative long so that it can reduce the flooding traffic and save energy.

3.5 Compared with TTDD

TTDD is a good solution to large-scale WSN with multiple mobile sinks. But it still can be improved in several aspects. GCA is a novel scheme to solve the mobility of sink and stimulus, which is essentially different from the design of TTDD, and it can overcome the defects of TTDD.

The stimulus mobility is ignored in TTDD. If the stimulus is mobile, the per-source grid construction would be a great overhead to network. However, GCA proactively constructs a unified two-tiered architecture, which can efficiently manage the mobility of both stimulus and sink. Meanwhile, the GCA construction overhead is as the same order of magnitude as the construction of per-source in TTDD.

Secondly, the square is not the optimal shape of cell. GCA adapts the hexagon as cell shape, which is similar with current cellular network. In the worst case, the ratio of forwarding path length to straight distance in GCA is $2/\sqrt{3}$; but the ratio in TTDD is $\sqrt{2}$. So the GCA's forwarding overhead is less than TTDD's. Furthermore, the hexagon is better than square in mobility management.

4 Performance Evaluation

In this section, we evaluate the performance of GCA through simulation. We firstly describe our implementation and metrics in simulation environment. Then we evaluate how exterior factors and control parameters affect the performance of GCA. The results demonstrate the efficiency of data forwarding and mobility management in GCA.

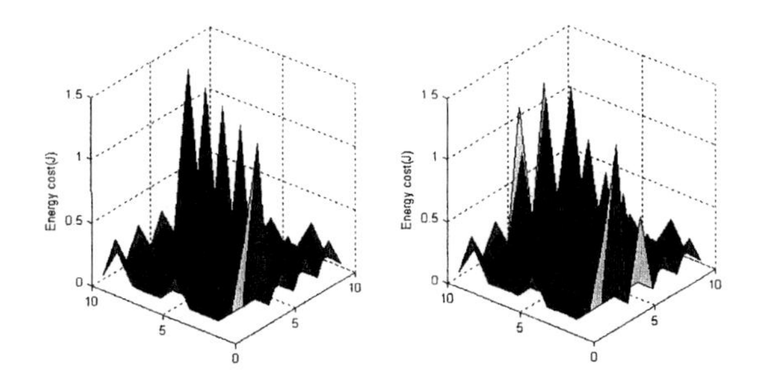

Fig. 4. energy cost of GEAFA-1 and GEAFA-2. (a) GEAFA-1, (b) GEAFA-2.

4.1 Experiment

We simulate GCA in ns2 environment [9]. The energy consumption is the main metric to evaluate the efficiency of GCA, which is the most important factor considered in WSN. Meanwhile, the load balance and the amount of data packets are used to evaluation.

The default scenario is that one sink and 100 sensor nodes are uniformly distributed in 400m $*$ 400m field with one stimulus. The bandwidth of wireless channel is 2 Mbps. The communication range is 100m and the initial energy is 2J. The power of transmitting and receiving are 0.6W and 0.3W. The radio propagation model is two-ray ground reflection, of which the energy cost is proportional to the fourth power of distance. A source generates 20 data packets per-second. The size of data packet is set to 500 Bytes, and control packet is 25 Bytes. The cell radius is 50m.

4.2 Two Variants of GEAFA

As analyzed in section 3.2, the two variants of GEAFA have each feature, demonstrated in Fig.4. Because the header of sink is nearly located on the radial from source's header to its neighbor header, the data packet would be forwarded along the unique path under GEAFA-1. The path is optimal, but the energy cost is unbalanced, of which the total energy consumption is 17.18J, as Fig.4.(a). If GEAFA-2 is used as forwarding algorithm, the total cost is larger (20.82J) as the path is not optimal, but the load balance is better, as Fig.4.(b). In both forward-ing algorithms, all of the data packets are received by sink.

4.3 Mobility

The design of GCA guarantees that stimulus mobility would not affect the network, so we focus on the sink mobility. The speed of sink is 20m/s. Fig.5 shows that the moving sink can be efficiently managed in GCA, and the moving can make the load balance better than static situation. As the sink motion is limited in top left region, the header cost of this region is relative large. The sink

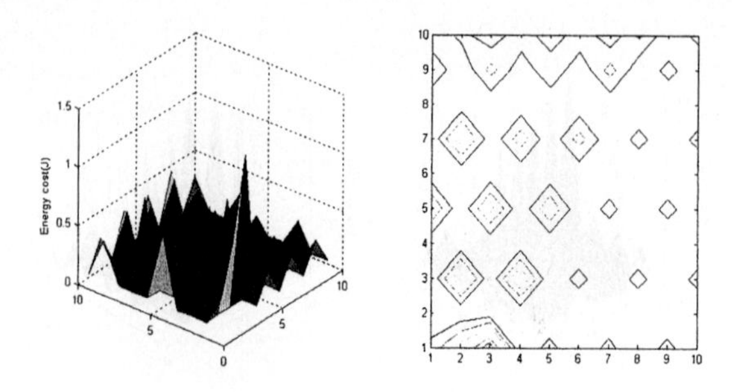

Fig. 5. Energy cost when sink is moving

receives all the packets sent by source. In order to clearly describe the energy consumption of each node, the contour figure is also given.

4.4 Data Rate

The data rate of source will influence the energy cost distribution among sensor nodes. The default value of simulation is 20 packets per-second, but the value is generally much lower in practical sensing application. So the large deviation of energy cost among nodes would be smoothed when data rate is low.

In probability and statistics, the standard deviation is the most commonly used measure of statistical dispersion. The standard deviation is defined as the square root of the variance, in order to give us a measure of dispersion that is a non-negative number and has the same units as the data.

The relation between standard deviation of header's energy cost and data rate is shown as Fig.6. The metric of X axis is data cycle, which is the reciprocal of data rate. When the data cycle is 0.05s (namely, data rate is 20 pkt/s), the standard deviation is large, which means the load balance among headers is not good. But with the cycle becoming longer, the deviation is gradually trending smaller. And the figure also demonstrates that the GEAFA-2 is better than GEAFA-1 on load balance.

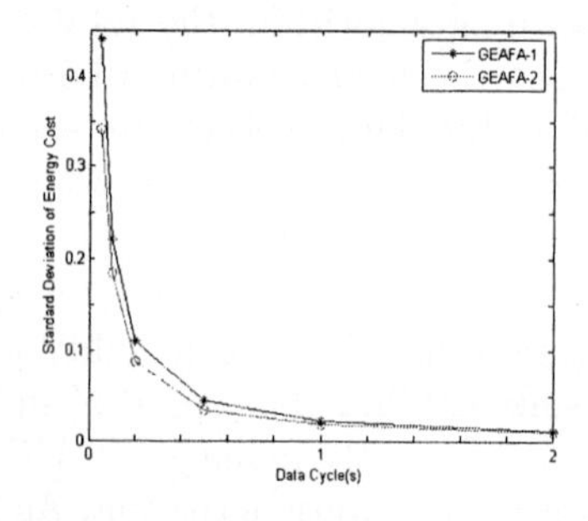

Fig. 6. The standard deviation with various data cycle

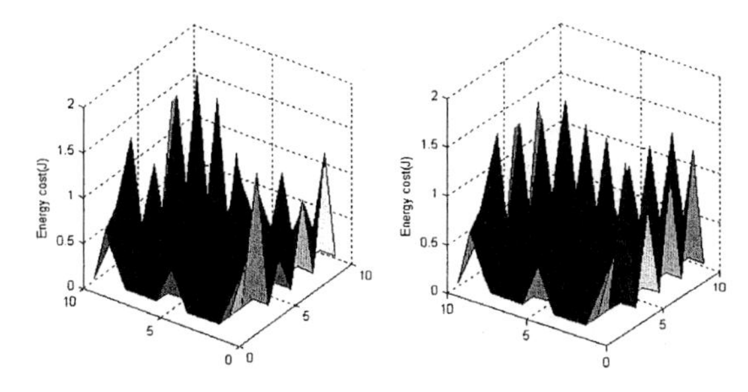

Fig. 7. energy cost of GEAFA-1 and GEAFA-2 when there are two pairs of sink and stimulus. (a) GEAFA-1, (b) GEAFA-2.

4.5 Multi-sinks and Multi-stimuli

The multiple sinks and multiple stimuli can be efficiently managed in GCA. Fig.6 shows the energy consumption situation when there are two pairs of sink and stimulus in the network. The distribution of energy consumption largely depends on the data stream from source to sink. Fig.7.(a) is the energy cost of network under GEAFA-1, whose total energy consumption is 29.38J. But the node at the cross point of two data stream fails as the energy is exhausted, so some packets are dropped after the forwarding path is broken. The load balance and robustness of GEAFA-2 is better, shown as Fig.7.(b), although the total energy cost is larger (35.18J) than GEAFA-1.

5 Conclusion

In order to efficiently manage the mobility of both sink and stimulus, we design a Geographical Cellular-like Architecture (GCA) for WSN, which proactively builds up a two-tired architecture according to location information of each node. The network is divided into many hexagon cells, each of which includes a header and several members. The header plays the role of base station in cellular network, receiving the data from sink and forwarding it cell by cell. Further more, we develop Greedy Energy-aware Alternative Forwarding Algorithm (GEAFA), and describe the two variants of it, which have different characteristics. The mobility management and network maintenance are contained in GCA. Through the simulating experiment in ns2 environment, we verify the correctness and effectiveness of GCA, and evaluate two variants of GEAFA and several important parameters, such as sink speed, data rate and the number of sinks and stimuli.

References

1. G. Pottie and W. Kaiser: Wireless Integrated Network Sensors. Communication ACM. **43** (2000) 51–58
2. Ian F. Akyildiz, Weilian Su, Y Sankarasubramaniam, E. Cayirci: A Survey on Sensor Networks. IEEE Communication Magazine. (2002)

3. Fan Ye, Haiyun Luo, Jerry Cheng, Songwu Lu, Lixia Zhang: A Two-Tier Data Dissemination Model for Large-scale Wireless Sensor Networks. In Proc. 8th Annu. ACM Int. Conf. Mobile Computing and Networking(MobiCom). (2002)
4. W. Heinzelman, A. Chandrakasan, and H. Balakrishnan: Energy-Efficient Routing Protocols for Wireless Microsensor Networks. In Proc. 33rd Hawaii Int. Conf. System Sciences (HICSS), Maui, HI. (2000)
5. W. Heinzelman, A. Chandrakasan, H. Balakrishnan: An Application-specific Protocol Architecture for Wireless Microsensor Networks. IEEE Transactions on Wireless Communications. **vol. 1, No. 4** (2002) 660–670
6. A. Manjeshwar, D. Agrawal: TEEN: A Routing Protocol for Enhanced Efficiency in Wireless Sensor Networks. Int. Proc. 15th Parallel and Distributed Processing Symposium. (2001)
7. S. Lindsey, C. Raghavendra: PEGASIS: Power-Efficient Gathering in Sensor Information Systems. Int. Conf. on Communications. (2001)
8. William Stallings: Wireless Communications and Networks. Prentice Hall, Inc. (2003)
9. UCB/LBNL/VINT Network Simulator - ns2. Available: `http://www.isi.edu/nsnam/ns/`
10. Jun Luo, Jean-Pierre Hubaux: Joint Mobility and Routing for Lifetime Elongation in Wireless Sensor Networks. IEEE InfoCom. (2005)
11. Brad Karp, H. T. Kung: GPSR: Greedy Perimeter Stateless Routing for Wireless Networks. In Proc. 6th Annu. ACM Int. Conf. Mobile Computing and Networking(MobiCom). (2000)
12. Yan Yu, Ramesh Govindan, Deborah Estrin: Geographical and Energy Aware Routing: a recursive data dissemination protocol for wireless sensor networks. Technical Report, UCLA/CSD-TR-01-0023. (2001)
13. Xiuzhen Cheng, Andrew Thaeler, Guoliang Xue, Dechang Chen. TPS: A Time-Based Positioning Scheme for Outdoor Wireless Sensor Networks. IEEE InfoCom. (2004)
14. Yi Shang, Wheeler Ruml. Improved MDS-Based Localization. IEEE InfoCom. (2004)

EAAR: An Approach to Environment Adaptive Application Reconfiguration in Sensor Network*

Dongmei Zhang, Huadong Ma, Liang Liu, and Dan Tao

Beijing Key Laboratory of Intelligent Telecommunications Software and Multimedia,
School of Computer Science and Technology,
Beijing University of Posts and Telecommunications, Beijing 100876, China
{zhangdm, mhd}@bupt.edu.cn

Abstract. Application reconfiguration provides a powerful mechanism to adapt component-based distributed applications for changed environmental conditions. In this paper, we propose an effective environment adaptive application reconfiguration (EAAR) mechanism based on the feedback-loop concept of the control theory. Based on this mechanism, a pull-based paradigm is introduced to represent the sensing-judging-acting process in sensor networks. We utilize rule-based knowledge to analyze the change of environment, thus perform self-adaptive application reconfiguration efficiently. To demonstrate how EAAR mechanism works, we simulated a scenario of reconfiguring applications in sensor networks.

1 Introduction

Sensor networks consist of tiny sensors deeply embedded within environment [1]. Advances in sensors and wireless technologies enable sensor networks to be deployed for a wide range of monitoring applications. Typical applications include environment monitoring, habitat monitoring, intelligent alarms, vital sign monitoring, etc [8,9,10].

Sensor networks focus on monitoring physical environment [2]. As the objects in environment are dynamic and unpredictable, the applications of sensor networks must be dynamically configurable and adaptive [11].

Flexibility is a key issue of application in sensor networks. The existing software for sensor networks is not flexible enough to meet the demands of many applications. The main method of improving flexibility is reconfiguring applications. Application reconfiguration provides powerful mechanism to adapt component-based distributed applications to the dynamic environment. But it is very challenging to implement application reconfiguration, because sensor nodes are memory-constrained, they can't store all possible applications in their local memories.

Adaptability is another key issue of application. It's difficult to predict all possible applications in the initial deployment of the sensor networks. The reasons are

* The work reported in this paper is partly supported by the National Natural Science Foundation of China under Grant 60242002 and the NCET Program of MOE, China.

X. Jia, J. Wu, and Y. He (Eds.): MSN 2005, LNCS 3794, pp. 259 – 268, 2005.

two-fold. First, the sensing data coming from dynamic environment can be influenced by various factors. Second, users' requirements to the applications are varied. Even if know all the possible states of environment, we wouldn't predict its exact state at a given time. For example, in a building environment monitoring system most nodes only collect temperature data under normal situation. But under certain situation, such as abnormal high temperature, some sensor nodes need to collect smoke data adaptively. In order to perform the sensing application efficiently, a sensor network should be aware of the variations of environment condition dynamically at runtime and reconfigure the applications adaptively.

Reprogramming technology [14] is a popular approach to application reconfiguration. SensorWare [3], Mate [12] and Agilla [5] are exemplary frameworks to provide reprogramming mechanisms that enable users to inject new instructions into a sensor network. However, they have some limitations. First, they cannot automatically cope with unexpected changes in the sensing environment. Second, their approaches mainly focus on mobile code techniques for sensor nodes. Our mechanism is a systemic level approach to application reconfiguration in sensor networks.

Meanwhile, a variety of techniques that allow software to adapt to environment dynamically have been proposed in traditional wired networks [16, 17, 18]. The typical methods have been based on the control theory and fuzzy logic, which allows for control and deduction of the future state under consideration [15].

The notion of bringing adaptability to the software system in sensor networks has been addressed in [6]. However, [6] only presents a repairing approach to node failures, it cannot dynamically deploy the applications into sensor networks. Moreover, periodically information exchanging between sensor nodes and the central control node brings expensive communication consumption and long response delay.

Our work adopts the mobile code idea of Mate to implement application script delivery, and introduces the method of adaptive software founded on the mathematics of the control theory in traditional networks. We propose an efficient environment adaptive application reconfiguration (EAAR) mechanism, which executes application reconfiguration according to sensing-judging-acting process. We denote environmental information as rule-based knowledge, and utilize knowledge to analyze the environment data, thus determine which application should be supported in the current environment. In this way, users deploy all the related knowledge into the sensor networks once, and the sensor networks will self-adaptively reconfigure applications according to the environmental changes.

The rest of the paper is organized as follows. Section 2 presents the approach for dynamic application reconfiguration based on environmental changes. In Section 3, we introduce a pull-based paradigm to represent the application script delivery process in sensor networks and describe the operations of EAAR approach for the paradigm. A scenario to demonstrate how the environment adaptive application reconfiguration works in sensor networks is described in Section 4. Finally, Section 5 concludes the paper and outlines the future works.

2 Environment Adaptive Application Reconfiguration Approach

A sensor network is a distributed system, in which the software on each sensor node consists of multiple components organized hierarchically into layers. These layers range from the operating system and networking subsystem at the bottom to various middleware components to the application at the top.

Before detail our EAAR mechanism, we first provide some preliminaries. EAAR introduces two new functional entities: Script Node (SN) and Reconfiguration Node (RN). SN is defined as a server that stores and manages scripts. Users may upload new application scripts to a SN by traditional networks. RN is a sensor node equipped with executing-engine and application-supporting components. RN may execute new application scripts from SN, thus reconfigure local applications. Application tasks encapsulate the know-how to reconfigure applications. These tasks are described by scripts, which can move from one RN to another.

2.1 Model of EAAR Mechanism

Fig.1 illustrates the model of EAAR mechanism composed of three functional modules: *Decision making*, *Script providing* and *Scripts executing*. The *Decision making* is responsible for detecting changes and determining whether they are significant enough to warrant reconfiguration. The *Script providing* generates, stores, manages and sends scripts to appointed RNs. The *Script executing* is responsible for executing application scripts transmitted from *Script providing*.

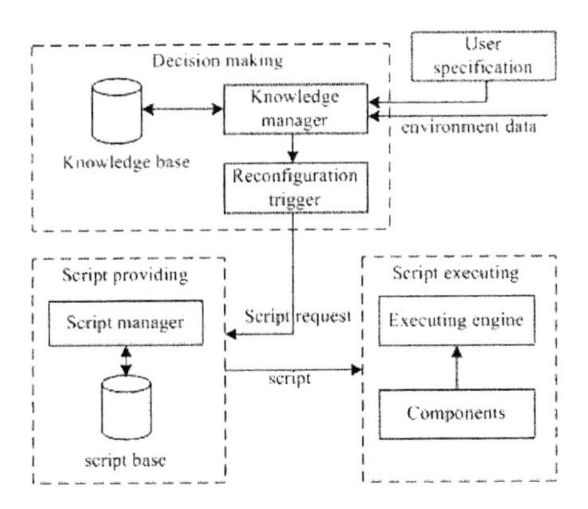

Fig. 1. EAAR Model

Decision making includes three components: Knowledge base, Knowledge manager, and Reconfiguration trigger. Knowledge base stores knowledge defined by users. The relationship between environment state and corresponding applications can be described using rule-based knowledge. We denote knowledge as the following formula: P->Q (or IF P THEN Q), where

(1) P is a prediction in the formula, which represents the environment condition that matches the knowledge.

(2) Q is the action, which denotes the applications that should be supported by SN in current environmental state

Knowledge manager is responsible for managing knowledge. Users can define domain knowledge and submit them to Knowledge base. Utilizing the knowledge and the input from hardware sensors, Knowledge manager can deduce some results associated with the current environment, and send the action to Reconfiguration trigger. Reconfiguration trigger combines the action from Knowledge manager with the task's running state of current node, thus determine how, when and where to reconfigure, and send request to *Script providing*.

Script providing includes two components: Script manager and Script base. Script base stores scripts defined by users or generated by Script manager. Scripts can migrate from SN to RN via multi-hop communication. A light-weigh script language is needed in our model to meet the constraints of sensor nodes. Script manager is responsible for generating and managing scripts. Utilizing it, scripts can be extracted from Script base according to the requirements. Moreover, if needed, Script manager can generate scripts according to the requirements.

The key component of *Script executing* is Executing engine. It is responsible for executing scripts deployed on the node. It works as a lightweight script interpreter and map scripts to lower-layer components. The main factors affecting the choice of script interpreter are hardware platform of RN, operation system and the resource constraints. One kind of script interpreter, which is well suited for EAAR framework, is Mate [12].

2.2 Process of EAAR Mechanism

Based on the above model and concepts, the overall adaptive reconfiguration process of EAAR mechanism can be divided into three phases: reconfiguration detection, script acquirement, and reconfiguration action.

Reconfiguration detection phase. The first phase involves detecting the variations of environment and determining whether it would be necessary to reconfigure. Knowledge manager utilizes the knowledge to analyze environment data, judges whether environment has changed. Thus, Reconfiguration trigger determines whether a reconfiguration action should be taken referring to the result. If it is necessary to reconfigure, a reconfiguration request message indicating the type and key parameters of the scripts will be sent to script manager.

Script acquirement phase. The second phase is an acquirement and distribution process of scripts. In this phase, Script manager receives script request message from Reconfiguration trigger and query corresponding scripts in Script base. The Script manager extracts scripts from Script base or generates scripts according to reconfiguration request message from Reconfiguration trigger and transmits scripts to RN.

Reconfiguration action phase. The final phase is scripts executing. Given the scripts from Script manager, Executing engine maps scripts to lower-layer components and implements application reconfiguration.

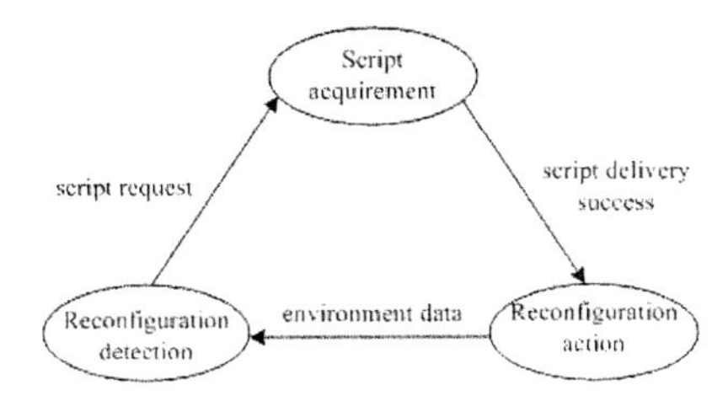

Fig. 2. State transition diagram of EAAR mechanism

Fig.2 illustrates the state transition diagram of EAAR mechanism in sensor networks. In general, a sensor network system can be in one of the three states: Reconfiguration detection, Script acquirement and Reconfiguration action. A sensor network starts at the reconfiguration detection state. It enters the script acquirement state if a scripts request message is sent which indicates some nodes need to be reconfigured. A sensor network enters reconfiguration action state if it is in the script acquirement state and the scripts delivery is successful, then it returns to reconfiguration detection state once new scripts have been performed on RNs and new environment data have been collected.

2.3 Pull-Based Paradigm for EAAR

Based on the feedback control concept of the control theory, we present a pull-based paradigm for EAAR mechanism. Under this paradigm, SN sends scripts in response to explicit requests of RNs.

As far as RN is concerned, pull-based paradigm is a kind of active approach to application reconfiguration, which is sponsored by RN. In this case, RN performs *Decision making* and *Script executing*, and SN performs *Script providing*. Fig.3 illustrates the control flows of EAAR mechanism between SN and RNs. Our approach of pull-based adaptive paradigm for EAAR supports a closed-loop reconfiguration, which includes a primary sensing-judging-acting process. We detail the operations as follows:

Step1: RN senses the current environment data, and analyzes these environment data based on knowledge, thus judges whether there are some variations in environment or not.

Step2: If environment has changed, Reconfiguration trigger of RN will analyze the running states of tasks on current node, thus determines whether it is necessary to reconfigure or not.

Step3: When reconfiguration is needed, Reconfiguration trigger will send script request message, which describes how, when and where to reconfigure, to Forwarding manager of RN. Forwarding manager is responsible for sending/receiving messages to/from neighbor nodes according to some routing protocols.

Step4: Having received the script request message from Reconfiguration trigger, Forwarding manager of RN would send the message to SN.

Step5: Forwarding manager of SN receives script request message and judges whether the message is correct and legal or not. If the script request message is correct and legal, it will be forwarded to the Script manager of SN, otherwise, it will be dropped.

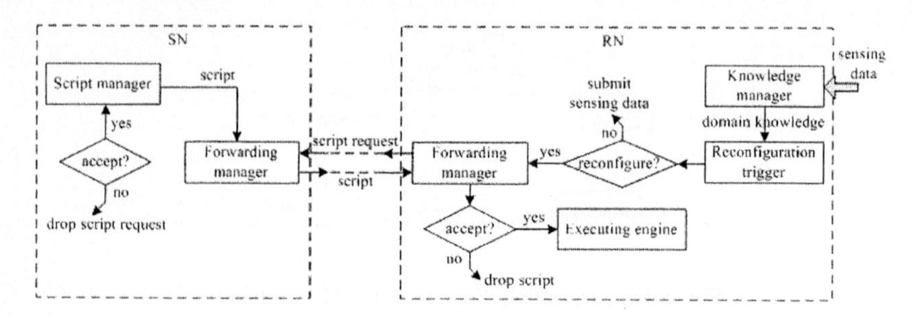

Fig. 3. Flows of EAAR in pull-based paradigm

Step6: Script manager of SN extracts scripts from Script base or generates scripts according to the script request message, then sends scripts to corresponding RN.

Step7: Executing engine of RN deploys scripts, then executes the special application task and senses the environment data.

3 Application Example

Environment monitoring is one of the most key application fields of sensor networks. Because the environment data is various and complicated, this kind of application takes more advantages of EAAR. We have utilized sensor networks to establish an environment monitoring system for smart building, with one sensor per room to monitor the environment data, such as temperature, light, smoke. With the introduction of EAAR, the application running on sensor networks can automatically adjust with the changes of the environment, and reconfigure applications.

3.1 Hardware Platform and Software Architecture

Our system has been implemented and tested on MicaZ motes [13], each node is configured with three basic sensors such as light, temperature and smoke. Refering to Mate's architecture [12], we have designed the software architecture based on TinyOS [4], which is illustrated in Fig.4. The architecture of RN can satisfy the need of adaptive reconfiguration.

Operations layer performs the system's basic commands, mainly including operating commands to hardware, simple calculations and control commands. Middle layer provides a set of more complicate commands composed of several simple ones. Execute engine will read out the command sequence in the script capsule and then execute. There are mainly two types of capsules. One is the entity to implement the application, which is triggered to execute by events. The other is the knowledge capsule in the knowledge base, which can be called by other running tasks, but not be triggered by events. Capsule manager will have two main functions: one is to act as concurrency manager and scheduler, the other is to act as the reconfiguration trigger which can decide how to reconfigure according to the current state of tasks and the knowledge.

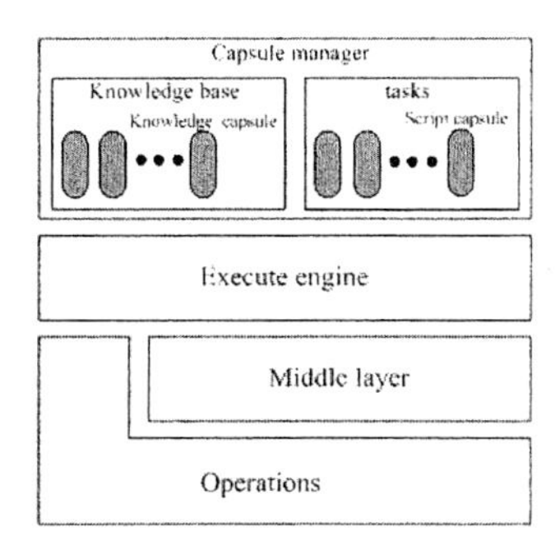

Fig. 4. RN architecture

3.2 A Scene Implementing

Here's an application scenario in environment monitoring of intelligent building, which gives a detail implementation of EAAR. For instance, in fire monitoring the RNs is in temperature monitoring state, each node will judge the temperature by its knowledge. When the temperature is higher than a threshold, the node will request SN, which then generate a script capsule for smoke sensor, and deploys the capsule on this node in order to monitor both smoke and temperature. This will be more useful to judge if fire has occurred around this node. In this case, the system can analyze the environment temperature, and autocontrol smoke sensor's running state to satisfy the need of application reconfiguration. We describe the operations as following steps.

```
pushc 0    #push 0 on the operand stack
sense      # read sensor 0 (temperature)
setvar 0   # set current
pushc 1    # push 1 on the operand stack
call       # call knowledge 1
halt
```

Fig. 5. Temperature sensing

Step1: SN deploys the initial temperature monitoring program which is described in Fig. 5 onto RNs.

Step2: RN will analyze the temperature data with knowledge, and output the result. When temperature data is beyond the threshold, RN will send a script request message with the format illustrated in Fig.6, which describes the demanded sensor type and the sampling frequency. Reconfiguration trigger can judge whether the requested sensor is running or not, if not it will send a message to the SN. The process of knowledge deducing and reconfiguration judging is showed in Fig.7.

Step3: SN analyzes the message from RN, and generates new scripts.

Step4: SN deploys the new scripts onto the corresponding RN. Then RN will collect temperature data and smoke data synchronously, thus complete the application reconfiguration.

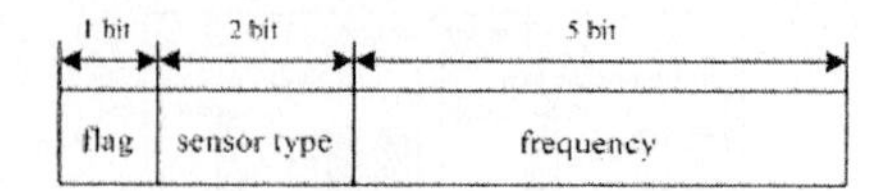

Fig. 6. Reconfiguration request message format

```
          getvar 0    #get the value of temperature
          pushc 40    # push 40 on the operand stack
          push
          gt          # is temperature greater than 40 ?
          not
          jumps label0  # jump to halt
          pushc 161   # push 161 on the operand stack
          setlocal 0    # set a message
          getvar 1    # get the value of state
          pushc 4     # push 4 on the operand stack
          land        # take logic and, and push the result
          pushc 0
          eq          # is the result equal to 0?
          not
          jumps label1  # jump to halt
          getlocal 0   # get the value of message
          send        # send the message
lable 1:
lable 0:  halt
```

Fig. 7. Knowledge deducing and reconfiguration judging

4 Conclusions and Future Work

Unpredictable variations of surveillance environment demand an efficient reconfiguration method for increasing flexibility and adaptability of applications in sensor networks. Motivated by this, this paper proposes an approach for environment adaptive application reconfiguration (EAAR) in sensor networks based on the control theory. First, we detail the model of EAAR mechanism and functional modules. According to the model, we present a pull-base paradigm to perform the sensing-judging-acting process. Finally, an application example is performed to demonstrate how the EAAR mechanism works for sensor networks.

The main advantages of our approach are listed as follows:

(1) Adaptability: The reconfiguration triggering is self-adaptive. We denote reconfiguration strategies using rule-based knowledge. Once users deploy reconfiguration strategies in sensor networks, RN can trigger the reconfiguration according to environment state automatically. Based on the feedback-loop concept of the control theory, EAAR mechanism implements application reconfiguration utilizing the process of sensing-judging-acting.

(2) Flexibility: There are two aspects of flexibility. One aspect is that all kinds of scripts have been stored in SN, and RNs will download these scripts from SN only if necessary. Thus RNs can realize flexible applications even with limited resource. The

other aspect is that users can easily upload required application to SN through traditional networks. Consequently, SN can provide all kinds of applications for RNs flexibly.

Our method still has its limitation. Although EAAR mechanism has self-adaptability, it is only a closed-adaptive [7] one. That is, the system is self-contained and not able to support the addition of new knowledge dynamically. Our future work will consider the addition of new knowledge, in order to improve the adaptability and flexibility of application reconfiguration further.

References

[1] I. F. Akyildiz, W. Su, Y.Sankarasubramaniam, E. Cayirci. "Wireless sensor networks: a survey". Computer Networks, vol.38(4), 2002, pp.393-422.

[2] D.Culler, D.Estrin, and M.Srivastava. "Overview of sensor networks", IEEE Computer, vol.37(8), 2004, pp.41-49.

[3] A. Boulis and M. B. Srivastava. "A framework for efficient and programmable sensor networks", Open Architecture and Network Programming Proceedings, 2002, pp.117-128.

[4] J.Hill, R.Szewczyk, A.Woo, S.Hollar, D.E.Culler, and K.S.J.Pister. "System architecture directions for networked sensors". Architectural Support for Programming Languages and operating Systems, 2000, vol.35(11), pp.93-104.

[5] C.-L. Fok, G.-C. Roman, and C. Lu, "Rapid development and flexible deployment of adaptive wireless sensor network applications", Proceedings of the 25th International Conference on Distributed Computing Systems (ICDCS'05), 2005, pp.653-662.

[6] H.C. Kim, H.J. Choi, and I.Y.Ko. "An architectural model to support adaptive software systems for sensor networks", Proceedings of the 11th Asia-Pacific Software Engineering Conference (APSEC'04), 2004, pp.670-677.

[7] P.Oreizy et al. "An architecture-based approach to self-adaptive software", IEEE Intelligent Systems, vol.14(3), 1999, pp.54-62.

[8] Huadong Ma, Yonghe Liu. "Correlation based video processing in video sensor networks", IEEE WirelessCom05, Hawaii, USA, June 2005.

[9] Dan Tao, Huadong Ma and Yonghe Liu, "Energy-efficient Cooperative Image Processing in Video Sensor Network", 2005 Pacific-Rim Conference on Multimedia, Lecture Notes in Computer Science, Springer, Nov.13-16, 2005, Jeju. Korea.

[10] Huadong Ma and Yonghe Liu. "On Coverage Problems of Directional Sensor Networks", International Conference on Mobile Ad-hoc and Sensor Networks (MSN05), Lecture Notes in Computer Science, Springer, 13-15 Dec 2005, Wuhan, China

[11] Z. Liu and Y. Wang, "A secure agent architecture for sensor networks", Proceedings of The 2003 International Conference on Artificial Intelligence--Intelligent Pervasive Computing Workshop (IC-AI'03) June 23-26, 2003, pp.10-16.

[12] P. Levis and D. Culler. "Mat´e: a tiny virtual machine for sensor networks", Proceedings of the 10th International conference on Architectural Support for Programming Languages and Operating Systems (ASPLOS X), Oct. 2002, pp.85-95.

[13] Micaz http://www.xbow.com/Products/ productsdetails.aspx?sid=3

[14] A.T.Campbell et al., "A Survey of Programmable Networks", ACM SIGCOMM Computer Communication Review, Vol. 29(2), 1999, pp. 7-23.

[15] J. Cangussu, K. Cooper, E. Wong, and X. Ma, "A run-time adaptable persistency service using the SMART framework", Proceedings of the 38th Annual Hawaii International Conference on System Sciences, 2005, pp. 276a-276a.
[16] P.K. McKinley et al., "Composing adaptive software". IEEE Computer, vol.37(7), 2004, pp.56-64.
[17] Junrong Shen, Qianxiang Wang, and Hong Mei. "Self-adaptive software: Cybernetic perspective and an application server supported framework". Proceedings of the 28th Annual International Computer Software and Applications Conference (COMPSAC'04), 2004, vol.2, pp.92-95.
[18] P.Robertson, and R.Laddaga. "The GRAVA self-adaptive architecture: history; design; applications; and challenges". Proceedings of the 24th International Conference on Distributed Computing Systems Workshops (ICDCSW'04), 2004, pp.298-303.

A Secure Routing Protocol SDSR for Mobile Ad Hoc Networks

Huang Chuanhe, Li Jiangwei, and Jia Xiaohua

Computer School, Wuhan University,
National Engineering Technology Research Center for Multimedia Software,
Wuhan University
huangch@whu.edu.cn

Abstract. Mobile Ad Hoc networks (MANET) are composed of a group of wireless mobile nodes that can communicate with each other. Different from normal networks, MANET is easy to be attacked. The attacking to the protocol can paralyze the network, so the security of routing protocol is an important part of the Ad Hoc networks security. This paper proposes a secure on-demand routing protocol SDSR. It uses digital signature and node monitoring to ensure security. It can protect route information from many kinds of attacks, and has good network performance as well.

Keywords: Mobile Ad Hoc Networks, Secure Routing Protocol, Digital Signature, Node Monitoring.

1 Introduction

A mobile Ad hoc network is a network composed of a group of wireless mobile nodes, in which nodes cooperate by forwarding packets for each other to allow them to communicate with other nodes beyond direct wireless transmission range. In Ad hoc networks, there is no wired infrastructure or centralized administration. They can be used in scenarios where no infrastructure exists or the existing infrastructure does not meet application requirements for reasons such as security or cost. Applications such as military, rescue, temporary and urgent working place may benefit from MANET. Secure and reliable communication is a necessary prerequisite for such applications.

MANET may not be secure due to many factors, such as insecure protocol, open and broadcast radio signal, non-perfect network security support. Similar to wired networks, the communication protocol is the source of insecurity.

Dynamic Source Routing protocol (DSR) [1,2] is an adaptive on-demand protocol for Ad hoc networks. But it does not provide security support. This paper proposes a secure on-demand routing protocol SDSR based on DSR. SDSR uses digital signature and node monitoring to ensure security. SDSR can protect route information from all kinds of attacks. SDSR also focus on network performance and rooting overhead. Because a good secure routing protocols needs not only powerful security capability but also good network performance.

X. Jia, J. Wu, and Y. He (Eds.): MSN 2005, LNCS 3794, pp. 269 – 277, 2005.

Threatens in Ad hoc networks mainly come from two sources: interior attack and exterior attack. The exterior attackers do not invade into the Ad hoc networks. The attackers always simulate a node and forge or send the messages. Interior attackers are those nodes who have invaded into the networks and knew the key system about the networks. The interior attackers are more dangerous than the exterior attackers, because we can not use key system to prevent the interior attackers to participate in the communication of the networks and can not use signature to identify those malicious nodes.

Attacks can be classified into the following four types: attack using modification, attack using impersonation, attack using fabrication and Energy consumption.

Attack using modification: Malicious nodes can cause redirection of network traffic and DOS attacks by altering control message fields or by forwarding routing messages with falsified values.

Attack using impersonation: Spoofing occurs when a node misrepresents its identity in the network, such as by altering its MAC or IP address in outgoing packets.

Attack using fabrication: The generation of false routing messages can be classified as fabrication attacks. Such attacks can be difficult to verify, especially the case of fabricated error messages that claim a neighbor cannot reach.

Energy consumption: Energy is a critical parameter in the MANET. Battery-powered devices try to conserve energy by transmitting message only when absolutely necessary. An attacker can attempt to consume batteries by requesting routes or forwarding unnecessary packets to a node continuously.

All these attacks can be done either from interior or from exterior.

The purpose of this paper is to design a protocol that is able to resist all exterior attacks and avoid some interior attacks.

The rest of this paper is organized as follows. Section 2 introduces some related research work on ad hoc routing protocols. Section 3 presents the secure ad hoc routing protocol SDSR. Section 4 analyzes the capability of SDSR to resist security threat and simulate the network performance of SDSR. Section 5 summarizes the paper.

2 Related Work

Security issues in Ad hoc networks have attracted considerable attention over the past few years and many solutions have been proposed to provide a security environment for Ad hoc network. Encryption, digital signature, reputation system and HAMC are four main ways to ensure the routing security.

Dahill et al proposed ARAN[3], a routing protocol for Ad hoc networks that uses authentication and requires the use of a trusted certificate server. In ARAN, every node that forwards a route discovery or a route reply message must also sign it. This method consumes computing power and causes the size of the routing messages to increase at each hop. In addition, it is prone to reply attacks using error messages unless the nodes have time synchronization.

Papadimitratos and Haas proposed a protocol SRP[4] that can be applied to several existing routing protocols (in particular DSR [1,2] and IERP [5]).

Ariadne[6] is a DSR-based and on-demand secure routing protocol. S. Yi proposed a secure aware routing protocol SAR[7], which uses safety rate to ensure security. Resnick presented a special way called Reputation Systems[8] to provided security. The method creates a trust network and assigns each node a mark. Each node judges its actions according to assigned mark. Bo Zhu et al proposed a protocol ASR[9] which can be used to protect the network framework information. B.Lu et al proposed a protocol CSER[10], by using node cooperation to ensure security. V.Sumathy proposed a protocol GLSSR[11], which uses grid location service to ensure security.

Although much research work has been done, there are no available practical secure protocols.

3 Secure Ad Hoc Network Routing Protocol SDSR

The proposed secure Ad hoc network routing protocol SDSR bases on DSR. It is an on-demand routing protocol composed of two parts: Route Discovery and Route Maintenance. In each part, node monitoring is used to reduce transmitted messages, and authentication is used to ensure security.

3.1 Assumptions

Network links are bidirectional. That is, if node A is able to transmit to some node B, then B is able to transmit to A.

Each node has an equivalent computational ability and storage.

Each node in Ad hoc network has an exclusive private key and a public key. When a node wants to send a message, it uses its private key to sign it. The receiver uses the sender's public key to validate the message. Each node in the Ad hoc network knows other nodes' public keys.

Table 1. Variables and notations

IP_A	A's IP address
Sig_{S-}	Private-key of S
Sig_{S+}	Public-key of S
$[d]Sig_{S-}$	Data d digitally signed by private-key of S
$Sig_{S+}[d]$	Validate data d use public-key of S
t	Timestamp
$RREQid$	Exclusive request Id
$A \rightarrow brdcst[d]$	Node A broadcast message d
$A \rightarrow B[d]$	Node A send message d to node B
$RREP$	Route reply
ERR	Error packet identifier

Table 1 shows the notations to describe security of protocols and cryptographic operations.

3.2 Route Discovery in SDSR

SDSR is an on-demand protocol. Route discovery of SDSR allows any node in the Ad hoc network to dynamically discover a route to any other node, whether directly reachable within wireless transmission range or reachable through one or more hops through other nodes. A node initiating a route discovery broadcasts a route request packet which may be received by those nodes within its wireless transmission range. Before broadcasting the request packet, the node should use its private key to sign it.

$$S \rightarrow brdcst : [\mathrm{RREQid}, IP_X, \{IP_S\}, t]Sig_{S-} \tag{1}$$

The request packet includes a request ID, the destination IP address, the route information about the source IP address, timestamp of the packet. The request is signed with the S's private key.

If the route discovery is successful the initiating node receives a route reply packet listing a sequence of network nodes through which the request packet may reach the target.

In addition to the address of the original initiator of the request and the target of the request, each route request packet contains a route record, which accumulates a record of the sequence of hops taken by the route request packet as the request packet is propagated through the Ad hoc network during this route discovery. Each route request packet also contains a unique request ID, set by the initiator from a locally-maintained sequence number. In order to detect duplicate route requests received, each node in the Ad hoc network maintains a list of the initiator address, request ID pairs that it has recently received in any route request.

When a node receives a route request packet, it processes the request according to algorithm_rcv.

```
Algorithm_rcv()
{
    receive a Request;
    if ValidateSig(Request) then
         {
              drop(Request);
              Broadcast(Err)
         }
         else if (request in team) then
              { drop(Request) }
                else if (ValidateSelfAddr(Request)) then
                     {
                          AddrSelfAddr(Request);
                          AddSig(Reply);
                          SendReply(Reply);
                     }
```

```
                    else if (ValidateAddrNum(Request)) then
                              {
                                    AddrSelfAddr(Request);
                                    AddSig(Request);
                                    broadcast Request (Request);
                              }
                        else  {
                                    RemoveperSig(Request);
                                    AddSig(Request);
                                    broadcast Request (Request);
                              }
}
```

The route request propagates through the Ad hoc network until it reaches the target node, which then replies to the initiator. A node sends the reply according to algorithm_reply.

```
Algorithm_reply()
{  If the signature is NOT valid
   { drop(Request);
     broadcast(error information);
   }
   else if the target of the request matches this node's own address
          finish the process;
      else if the node's sequence number is 1
          { sign it by using its own private key ;
             send the reply;
      }
          else
              { remove pre-node's signature;
                 sign it by using its own private key;
                 re-broadcast the request ;
                  send the reply;
          }
}
```

3.3 Route Maintenance in SDSR

Conventional routing protocols integrate route discovery with route maintenance by periodically sending routing updates. If the status of a link or router changes, the periodic update will eventually reflect the changes to all other routers, presumably resulting in the computation of new routes. However, using route discovery, there are no periodic messages of any kind from any of the mobile nodes. Instead, while a route is in use, the route maintenance procedure monitors the operation of the route and informs the sender of any routing errors.

The usual way to be used to detect the link of the route is: A node forwarding a packet to the next hop along the source route returns a ROUTE ERROR to the original sender of the packet if it is unable to deliver the packet to the next hop after a limited number of retransmission attempts.

Fig. 1 illustrates the method.

The above method may suffer some attacks and waste network resource. Our method works as follows: A node monitors its neighbor nodes, if it has not gotten any messages from the neighbor node in a given particular period, a ROUTE ERROR would be returned to the sender of the packet. The way need not send any packet and will not judge the route link from the reply packet.

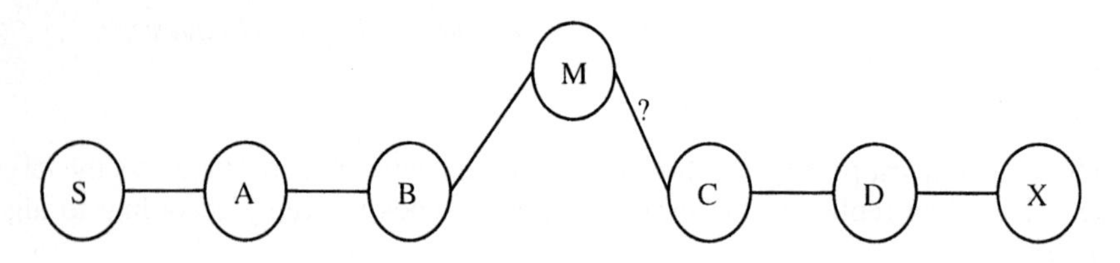

Fig. 1. Route link failure

By using the usual way, node M sends a packet d to C, and if C replies a packet to M, then the link between M and C is in good condition. Otherwise M sends a route error to S to report the error.

By using our way, M needs not send any packet to C. When the link between M and C is in good condition, M is in the wireless transmission range of C. According to the assumption about network, network links are bidirectional, C is also in the wireless transmission range of M. That means when C sends any packet to D, M can receive that packet. M does not care about the content of the packet. It only pays attention to whether it receives that packet. This method has two advantages: one is to reduce the consumption of the resource, the other is to improve the security of the network.

Sending a route error also need sign signature. For example, a route such as {A-B-C-X-D}, when the link between C and X is broken, C need send a route error message to B. The route error format is

$$C \to B : [ERR, IP_X, t]Sig_{C-}$$

ERR is the identifier of route error. IPx is the broken router. t is the timestamp. All of these are signed with C's private key. It can prevent malicious nodes from forging the route error message.

3.4 Node Monitoring

Node monitoring ensures that when network links are bidirectional, the node can receive its neighbor's message, and the node can estimate its neighbor by analyzing these messages.

In SDSR, each node need maintain an action table to save and analyze the messages which are sent by its neighbor. The action table saves node id, real actions, theory actions and scores. Each node uses "real actions" to save its neighbor's real action. Each node compares it with its neighbor's theory action and gives a score. By using those scores, the node can judge whether its neighbor is a malicious node. When a malicious node has been detected, its neighbor should broadcast its id to the network. For example after T detected malicious node A, it broadcasts the message as follow:

$$T \rightarrow brdcst : [IP_A, ERR, t]Sig_{t-}$$

Action table described as table 2.

Table 2. Action table

Node id	Real actions	Theory actions	Scores
A	....	....	....
B	....	....	....
C	....	....	....

There are too many attacking actions, such as dropping messages, modifying messages, sending vicious error messages etc. It is difficult to solve these problems only by using node monitoring. We need unite other secure technology to solve these problems.

4 Analysis and Simulation

4.1 Security Analysis

In this section, we provide a security analysis of SDSR by evaluating its ability to resist the attacks introduced in Section 3. We also compare the performance of SDSR with DSR and Ariaden routing protocol through simulation.

Ability to resist attack using modification: Each node needs a private key and a public key, when the Malicious nodes want to modify the information, they should know the other node's private key. So it is difficult to modify the information. And if information has been modified, the good node can detect it by checking its signature.

Ability to resist attack using impersonation: In SDSR, spoofing occurs when a node knows the other node's private key. So if the key is safe, the spoofing attack will not succeed in the networks.

Ability to resist attack using fabrication: Most false routing messages can be detected by the node monitoring. Maybe some fabrication can not be detected, however, because messages are signed, malicious nodes cannot generate messages for other nodes. It decreases the possibility of the attack from the malicious node.

4.2 Simulation

We implement evaluations using the NS2. We used 802.11 MAC layer and CBR traffic over UDP. We simulate the performance and compare the overhead with DSR, Ariadne.

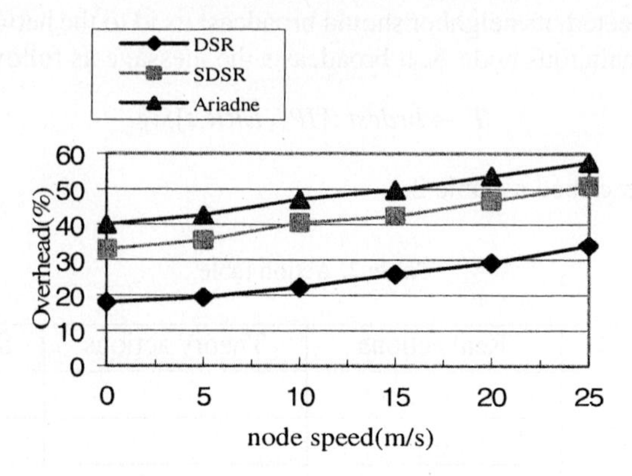

Fig. 2. Overhead evaluation results

Overhead is defined as $overhead=\dfrac{overhead\ data}{total\ data}$

The overhead is illustrated in Fig. 2.

Figure 2 shows that: (1)When the network structure remains unchanged while the node moving speed changes, the SDSR's overhead is smaller than Ariadne's. (2) When the node speed increases, the overhead of the protocol increases too.

The relationship between overhead and average number of hops is illustrated in fig. 3.

Fig. 3 shows that: (1) When the node speed does not change, and the average number of hops increases, the overhead of the protocol increases. (2) When the average number of hops is smaller than 2 hops, the SDSR's overhead is equal with the Ariadne's, when the average number of hops is greater than 2, the SDSR's overhead is smaller than the Ariadne's.

Simulation result shows that the change of node's speed influences the overhead of protocol. When speed is increased, the overhead is also increased. When the average

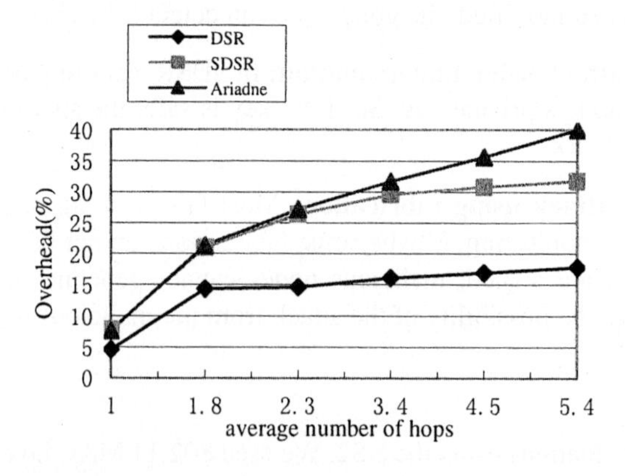

Fig. 3. The relationship between overhead and average number of hops

number of hops is longer, the SDSR is more advantageous. The overhead of the SDSR is smaller than the Ariadne's, and the SDSR has better network performance than Araidne.

5 Conclusion

This paper analyses the security of exiting protocols, presents a secure on-demand routing protocol SDSR based on DSR. SDSR uses digital signature and node monitoring to ensure security. The analysis and simulation results show that SDSR not only can protect route information from many kinds of attacks but also has a good network performance. The disadvantages of SDSR are that SDSR can not resist some interior attacks and can not be used in the high risk environment.

The node monitoring can improve the security of Ad hoc networks, but how to resist complicated attacks is a open problem.

How to resist attack by Energy consumption is another open problem. It is very difficult to avoid energy consummation only by using secure protocol.

The tasks will be studied in the future.

Reference

1. D.B. Johnson, D.A. Maltz, Y.C. Hu, "The Dynamic Source Routing Protocol for Mobile Ad hoc Networks (DSR)", IETF Mobile Ad hoc Networks Working Group, Internet Draft, Apr 2003.
2. Johnson D B, Maltz D A, "Mobile Computing". Boston: Kluwer Academic Publishers, 1996.
3. Kimaya Sanzgiri ,Bridget Dahill, "A Secure Routing Protocol for Ad Hoc Networks". IEEE International Conference on Network Protocols (IC2 NP)., Paris ,France ,2002.
4. Papadimitratos , Z J Haas, "Secure Routing for Mobile Ad Hoc Networks [C]" . Proceed-ings of SCS Communication Networks and Distributed Systems Modeling and Simulation Conference, San Antonio ,USA ,2002.
5. E M Royer ,C K Toh, "A Review of Current Routing Protocols for Ad Hoc Mobile Wireless Networks". IEEE Personal Communications., 1999, (4): pp.46 - 55.
6. Y. Hu, A. Perrig, and D. Johnson, "Ariadne: A Secure On-demand Routing Protocol for Ad Hoc Networks," ACM MOBICOM02., 2002.
7. S Yi, P Naldurg, R Kravets, "Security Aware Ad Hoc Routing for Wireless Networks". UIUCDCS-R-2001-2241 Technical Report., 2001.pp.299-302
8. Resnick, Paul, Zeckhauser et al, "Reputation Systems". Communications of the ACM., 2000, 43(12): 45-48.
9. Bo Zhu, Zhiguo Wan, "Anonymous Secure Routing in Mobile Ad-Hoc Networks", Pro-ceedings of the 29th Annual IEEE International Conference on Local Computer Net-works(LCB'04), 2004, pp.102-108.
10. 10.B.Lu, U.W.Pooch, "Cooperative Security-Enforcement Routing in Mobile Ad Hoc Networks", Proc. the 4th International Workshop on Mobile and Wireless Communications Network, Feb 2002, pp.157-161.
11. 11.V.Sumathy, Dr.P.Narayanasmy, "GLS With Secure Routing in Ad Hoc Networks", Proc. Conference on Convergent Technologies for Asia-Pacific Region, Feb 2003, pp.1072-1076.

Secure Localization and Location Verification in Sensor Networks*

Yann-Hang Lee[1], Vikram Phadke[2], Jin Wook Lee[3], and Amit Deshmukh[4]

[1] Computer Science and Engineering Dept.,
Arizona State University, Tempe, AZ 85287, USA
[2] Qualcomm USA, San Diego,CA 92121, USA
[3] Samsung Advanced Institute of Technology(SAIT), South Korea
[4] Siemens USA, San Diego, CA 92126, USA

Abstract. Evolving networks of wireless sensing nodes have many attractive applications. Localization, that is to determine the locations of sensor nodes, is a key enabler for many applications of sensor networks. As sensor networks move closer to extensive deployments, security becomes a major concern, resulting in a new demand for secure localization and location proving systems. In this paper, we propose a secure localization system based on a secure hop algorithm. A location verification system that allows the base station to verify the locations of sensor nodes is also presented. It prevents a node from falsifying reported location information. The evaluation of the secure localization system shows that the proposed scheme performs well under a range of scenarios tested and is resilient against an attacker creating incorrect location information in other nodes.

1 Introduction

Distributed sensor nodes can employ communications to form wireless ad hoc networks capable of collaborative processing and collecting valuable information. The physical location of these sensor nodes can prove to be useful in geographic routing, node authentication and location critical applications, such as target tracking. To determine the locations of sensor nodes, either centralized approach [1] or distributed algorithms [2] [3] have been proposed. Niculescu et al. [3] insisted that a positioning algorithm has to be distributed because in a very large network of low memory and low bandwidth nodes, even shuttling the entire topology to a server in a hop by hop manner would put too high a strain on the nodes close to the observer(also called base station).

When sensor networks are deployed in remote and often hostile environments, security properties of these networks shall be critical. Attacks on such nodes can disrupt a distributed location sensing algorithm resulting in nodes storing and propagating erroneous location information. Hence, it is imperative to look into secure localization for sensor networks with the consideration of compromised nodes.

Most of the existing work in localization does not consider any security issues. To best of our knowledge, there have been three works proposed for location verification

* This work was supported by DARPA (Defense Advanced Research Projects Agency, USA) IXO NEST (Network Embedded Software Technology) program.

X. Jia, J. Wu, and Y. He (Eds.): MSN 2005, LNCS 3794, pp. 278–287, 2005.

for wireless networks so far. Sastry et al. in [4] introduced the secure in-region verification problem to address false location claims. The disadvantage of this approach is that it requires ultrasound and time-of-flight techniques and their focus is on one-hop verification with an echo back protocol. The authors of paper [5] designed a location proving system that offers integrity and privacy. The round-trip latency of wireless communication between a node and location manager was chosen to determine the proximity of the node to the location manager. This idea requires a quite precise time synchronization. Recently, Capkun et al. in [6] proposed two mechanisms for the position verification, based on multilateration and Time Difference Of Arrival(TDOA) techniques. They also proposed mechanisms for verifying nodes' mutual distances. However, the mechanisms are based on GPS and radar technologies for all nodes which may not be available to sensor nodes of limited resources.

In this paper, we show the use of one-way hash functions and existing trilateration techniques to create a Secure Localization System (SLS). The proposed system is resilient against various types of attacks. We also propose a Location Verification System (LVS) based on the secure localization system. Location verification is the process of verifying location claims from sensor nodes. We believe that location verification has an important role in various applications especially those, where resources are granted or computation is performed based on location information.

In section 2, the threat model we attempt to address is provided. Sections 3 present our Secure Localization System. Then we propose our Location Verification System in section 4. Section 5 gives the results from simulation experiments. Section 6 presents the conclusions drawn from this work.

2 System and Threat Models

The localization system focused in the paper is based on hop count and trilateration. This is similar to the APS [3] where a distance vector exchange is taken place so that all nodes get distances in hops to all landmarks. Hop count information gets propagated from landmarks following the minimum hop spanning tree. The system is assumed with the following characteristic:

- Individual nodes are not trusted: Nodes are subject to compromise from capture and reprogramming attacks. We do not assume presence of tamper resistance hardware.
- Base station and landmarks are trustworthy: We assume that either some kind of tamper resistance is provided or other methods are used to prevent base station and landmarks against compromise.
- Landmarks have a priori knowledge of their own locations with respect to some global coordinate system.
- Sensor nodes are not mobile during the localization phase.
- Radio links are vulnerable: we assume that these links are insecure. They are vulnerable to eavesdrop, replay and injection attacks.

Almost all of the distributed localization systems rely on intensive communication and exchange of information between nodes. Below are a list of some generic attacks for hop-based localization systems: systems.

- *Modification Attack*: The most direct attack against a localization system is to target the localization information exchanged between nodes. Malicious nodes may adversely affect the working of the system by falsifying/modifying hop count information.
- *Replay Attack*: In case of hop-based localization systems, hop count information should be incremented to reflect the number of hops to a Landmark. A malicious may just replay the hop count it received, thus create a false hop count information.
- *Spoofing Attack*: A malicious node may spoof localization information. In the example of hop based localization systems hop counts may be spoofed.
- *Sinkhole Attack*: In a sinkhole attack, the attacker's goal is to attract nearby nodes creating a metaphorical sinkhole. This can be achieved by advertising a low hop count and infecting neighboring nodes. Although the source of the attack is a bad node, the attacks gets propagated via good nodes who forward false localization information.
- *Impersonation Attack*: Most distributed localization systems assume the presence of some Landmarks. Impersonation attack involves malicious nodes impersonating/pretending to be these Landmarks. Attackers can propagate incorrect information while impersonating Landmarks and thus cause nodes to localize to wrong locations.

3 Secure Localization System

In this section, we present a secure localization system based on hop by hop trilateration. Using the proposed secure hop (SecHop) algorithm, every node can securely determine the number of hops it is from each of the Landmarks. Similar to APS [3], every node then estimate range between node and each Landmark, and then compute its location by solving trilateration equations.

3.1 Algorithm

The SecHop localization algorithm has four phases of operations, which we now describe in detail.

Commitment Distribution - Phase I. Each Landmark will start with a seed $s(0,1)^n$ and compute a set of *Links* to form a one-way hash chain. *n* number of links are created, where n is the estimated network diameter. In our notation scheme, the *Links* created by Landmark *i* will be shown as below.

$$Link_1^i,\ Link_2^i,\ Link_3^i,\ \ldots,\ Link_{n-1}^i,\ Link_n^i \tag{1}$$

However a Landmark needs not store all these links. Of particular interest is $Link_n^i$ which is called 'commitment ' to the chain. At the start of the localization process each Landmark will broadcast its nth link, i.e., $Link_n^i$. It is also possible to bootstrap the commitment corresponding to the Landmarks into the nodes prior to deployment. This mechanism prevents impersonation attack under which a malicious node may pretend to be a Landmark.

Algorithm 1. SecHop Localization

when a localization packet is received
  h_{recv} = received hop count and $Link_{recv}$ = received Link
  if new Landmark i **then**
    if $F_{netDiam-h_{recv}}(Link_{recv}) == Link_n^i$ **then**
      store hop count $h_i = h_{recv}$ for this Landmark i
      store $Link_h^i$ = $Link_{recv}$ and compute $Link_{h+1}^i$
      broadcast new packet for this Landmark
  else if lower hop count **then**
    if $F_{h_i-h_{recv}}(Link_{recv}) == Link_{h_i}^i$ **then**
      store hop count $h_i = h_{recv}$ for this Landmark i
      store $Link_h^i$ = $Link_{recv}$ and compute $Link_{h+1}^i$
      broadcast new packet for this Landmark
when a calibration packet received
  if first packet **then** store *avgDistPerHop*
if calibrated **and** $number\ of\ Landmarks \geq dimensions\ of\ space$ **then** triangulate

Secure Hop By Hop Propagation - Phase II. In the classical distance vector exchange approach to get distances in hops to all Landmarks, malicious nodes may spoof, alter or replay distance vector updates resulting in corruption of distance vector at nodes, and eventually causing them to localize incorrectly. We propose the use of SecHop algorithm (as illustrated in Algorithm 1.) based on one-way hash functions to prevent against these attacks. The proposed SecHop algorithm is described below.

Each node maintains a table whose attributes are (X_i, Y_i, h_i) and exchanges distance vector updates only with nodes it can communicate via a local broadcast. The localization algorithm begins with each Landmark broadcasting its ID, hop count and the first element of the hash chain. The Landmark initiates the localization process by sending the following packet, called 'positioning' packet as shown below:

$$Landmark\ i \rightarrow Broadcast(i, HC = 0, Link_1^i) \tag{2}$$

Essentially a Landmark broadcasts to the nodes around it that its hop count is 0 corresponding to itself, and it provides the Link corresponding to hop count 1, i.e., $Link_1^i$. If a normal node gets the positioning packet, it checks whether it has already has such a positioning packet corresponding to the same Landmark in question. If so, the hop count is validated by checking the authenticity of the corresponding Link. If the hop count and Link are verified and found to be authentic, the information corresponding to the smallest hop count is retained and the other discarded. The hop count is incremented by 1 while the Link of the chain corresponding to the next hop count is calculated and is forwarded to the current node's neighbors. If Link authentication fails, the hop count and corresponding Link are discarded.

Correction Information Propagation - Phase III. This phase is similar to that proposed originally in APS [3]. We call 'correction' for the average distance travelled per hop. At the end of SecHop phase, each Landmark gets all positioning packets from all other Landmarks and knows the number of hops between itself and other Landmarks.

With the known Landmark positions, it can compute the average distance per hop. All Landmarks now broadcast the average hop distance (embedded in correction packet) to the network.

Location Estimation - Phase IV. Once a node with unknown location gets has estimates to a number of ($\geq$ 3) Landmarks, it can compute its own location using trilateration [3]. Each node uses the correction packet and uses that as the average distance travelled per hop for calculating its distance to all Landmarks it is connected to. The trilateration is performed using the Landmark coordinates and the corresponding distances to each of them from this node. This distance to each Landmark is calculated as a product of the average hop distance of the nearest Landmark (in terms of number of hops) and the number of hops to that Landmark.

3.2 Security Analysis

Comparing with APS, the *Link* information is added to the hop count broadcast packets in SecHop localization. Thus, the hop count can be checked. In this subsection we discuss security properties of the proposed secure localization system.

Landmark Impersonation Attack. In subsection 3.1, we mentioned the need for countermeasures against Landmark impersonation attack. We propose the use of public key cryptography for authentication to prevent against Landmark impersonation attack wherein a node pretends to be a Landmark and floods fraudulent information into the network causing nodes to localize incorrectly. Authentication for a node is done just once in a network session so public key mechanism is not much overhead.

Modification Attack. Since we take a hop-distance vector based approach we can view the hop by hop propagation phase as equivalent to advertising routes to Landmarks. In SecHop algorithm, given that a malicious node receives a localization packet with a metric of hop count h hops corresponding to Landmark i, three conditions as described below hold.

- The malicious node can generate localization packets for h hops or longer routes to the Landmark i. This is because a node is able to use a one-way hash function to determine *next Link* of the one it received (forward Links of the one-way hash chain). However, this behavior may not be very harmful in most cases because most nodes will choose the lowest hop count available. Sending out the same Link that it received is replay attack. The replay attack can also be viewed as a one node wormhole. Although we do not propose any countermeasures against this attack, systems such as TIK [7] can handle this attack.
- A malicious node cannot generate a *Link* less than hop count h for Landmark i.
- A malicious node cannot generate a *Link* for Landmark j, where $i \neq j$.

An adversary, that has not compromised any node and therefore does not possess the crypto-keys, cannot produce any localization packets as it will be rejected by its neighbors.

Spoofing Attack. A node that has either been compromised or been introduced by an adversary cannot spoof localization information because even though it can claim to have an arbitrary hop count value, it will not be able to generate a Link corresponding to the hop count. Thus, even if a good node that contained the commitment corresponding to Landmark i is turned into a malicious node by an adversary, the adversary cannot spoof a link from the commitment.

4 Location Verification System

Location verification is the process of verifying location claims from nodes in the base station [4]. We present a scheme that shall allow a centralized entity like a base station to be *Verifier (V)* and verify location claims from nodes, called *Provers (P)*, in the network. We envision that the location verification scheme will be used in conjunction with the secure localization system proposed in this work. It is possible however, in principle, to combine the proposed secure location verification system with other location determination schemes.

4.1 Design

A location claim from a node will be of the form:

$$(node's\ ID,\ Claimed\ Location,\ \{Verification\ Tokens\})$$

Verifier V returns 'success' if can verify the authenticity of the claim else it returns 'failure'. The crux of our scheme is that each Prover P needs to prove to the Verifier that the hop count h it claims to have corresponding to each *Landmark i* is indeed the actual hoop count and the hop count has not been spoofed or altered. Each node has a hashed *Link* for every reachable Landmark. It can use the *Links* to prove the authenticity of the its hop count claim. For P to prove that it has hop count h corresponding to Landmark i, it should provide the V with $Link^i_h$.

In order to verify location claims we require that nodes not be able to modify the *Links*, i.e., given a $Link^i_h$ that corresponds to hop count h the node should not be able to compute forward $Link_{h+1}$ or a backward $Link^i_{h-1}$. A malicious node cannot create a *backward Link* because of the one-wayness of hash functions, however a malicious node that wants to claim that it is in a different location from its actual location can create *forward Links* by using the one-way hash function thus pretending to have a higher hop count than it actually has and eventually fooling the verifier into thinking that the prover is in a different location than it actually is.

We propose some modifications to the secure localization system in order to facilitate the verification process. The aim is two-fold.

1. Even though it is difficult to prevent a malicious node from generating a higher hop count and corresponding forward Link we propose a scheme to prevent the node from being able to report this higher hop count to the verifier V.
2. Prevent framing or blackmail attacks. These are attacks where in a malicious node successfully frames another node. In the context of location verification it could mean causing the legitimate location claim from a good node to be rejected by the verifier.

Neighbor Authentication and Non-repudiation. Each node maintains a one-way hash chain for every reachable Landmark in the system. The chain is used to authenticate the source of *Link* that a node broadcast.

For notational convenience we represent the one-way hash chain element as *ALink* (being different from the *Link* chain is the SecHop algorithm). For example, the *ALink* chain corresponding to Landmark 3 for node 45 is shown below.

$$ALink^3_{1,45},\ ALink^3_{2,45},\ ALink^3_{3,45}, \ldots,\ ALink^3_{n-1,45},\ ALink^3_{n,45}$$

We assume that the *n th ALink* values or *commitments* corresponding to every node are known to all neighboring nodes and the Verifier V. When a node propagates hop count information corresponding to a certain Landmark it also attaches a *ALink* from its one-way hash chain corresponding to the Landmark. Every node reveals an *ALink* from the end towards beginning, i.e., in a direction opposite to that of computation. For example node 45 reveals $Link^3_{n-1,45}$, to authenticate hop count *n-1* corresponding to Landmark 3. In general to authenticate a hop count h, for Landmark i, node N reveals $Link^i_{n-h,N}$. A neighboring node that receives a hop count and an *ALink* from a neighboring node can authenticate the hop count using the commitment information corresponding to the neighboring node. The added hash chains at each node provide the following security features.

- ***Authentication***: The one-way hash chain at every node allows a node to authenticate localization information from neighboring nodes.
- ***Non-Repudiation***: At each node commitments corresponding to each Landmark are revealed to the verifier and the neighboring nodes. This provides for non repudiation and prevents a malicious node from framing other nodes.

During location verification a prover node *P* sends a verification token that has the following format, $Link^i_h$, $ALink^i_{n-h,N}$ where i indicates the Landmark. 'N' is some neighbor of *P* and is the source of both *links* and $P \neq N$.

4.2 Security Analysis

Our main aim in the location verification scheme was to prevent an attacker(malicious node *M*) from claiming a higher or lower hop count *h´* than its actual hop count *h*. We explain these concepts below.

- ***Link Alteration-Backward Links***: Consider a malicious node *M* that is the Prover P. If the actual hop count of *M* is h, the lowest hop count that it can claim is h. This is because this node will receive a $Link_h$ from one of the neighboring nodes. Due to the one-way property of the one-way hash chain it cannot derive the link corresponding to hop count *h-1*, i.e., $Link_{h-1}$.
- ***Link Alteration - Forward Links***: If we only use one-way hash chains corresponding to Landmarks the *M* will be able to claim a hop count $(h + \Delta)$, because it is able to calculate forward links. However in our location verification scheme the verification token also contains a link of the form $ALink^i_{n-h,N}$. To fake a higher hop count of *h*, M is expected to send $ALink^i_{n-(h+\Delta),N}(M \neq N)$. Given that M will have received a link of the form $ALink^i_{n-h,N}$ corresponding to hop count h, it is infeasible to compute $ALink^i_{n-(h+\Delta),N}$.

– ***Link Spoofing***: A malicious node does not know the *seed* of any other node in the network and so it cannot spoof a *Link* or an *ALink*.
– ***Framing***: A malicious node *M* may still attempt to frame other nodes. It may send a garbage link and node *P* will report a corrupt link and its claim may be rejected by *V*. This attack is thwarted because one-way hash chains at each node facilitate neighbor authentication and non repudiation.

5 Simulation Results

We chose the *NS-2* simulator to evaluate the performance of SLS and to realistically model physical radio propagation effects such as signal strength, and interference capture effect. We also used an implementation of the S-MAC protocol, which has been specifically designed for sensor networks. The one-way hash function being used in our simulations is MD5. We generate random topologies of 100 nodes and place these nodes in a 400m by 400m square grid with uniform distribution. The radius of radio range is assumed to be 70m. We place 3 Landmarks such that they are on the edges of the network topology. This is because distributed localization algorithms generally perform better when the Landmarks surround most of the nodes.

5.1 Location Error

We define 'Location Error' as the difference between the actual location of the node and the calculated location (as determined by the localization algorithm). The average location error is the location error averaged over all nodes. Figure 1 shows the location error in percentage of the radio range for different scenarios:

– Hop-Based-No Attacks: There are no attacks to the localization system. This scenario is simulated to determine a baseline location error so that we can evaluate other attack scenarios against it.
– Hop-Based-Sink-Hole: In this scenario the malicious nodes in the network attack the localization system by broadcasting a very low hop count, i.e., 1. to infect as many nodes as possible.

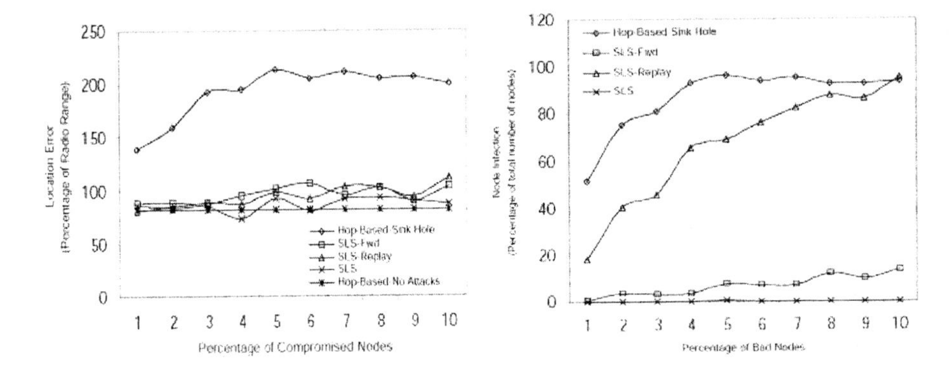

Fig. 1. Location Error

Fig. 2. Node Infection

- SLS-Sink Hole: This represents the case in which the proposed SLS is in use and the localization system is under sink hole attack.
- SLS-Replay: This scenario represents the case of replay attack or one node wormhole attack on SLS.
- SLS-Fwd: The malicious nodes attempt to thwart SLS using a forward attack. Forward attack is most effective in C-shaped networks, i.e., anisotropic topology.

As can be seen in the Figure 1, SLS performs much better in the face of different attacks on the hop propagation based localization systems. This is because the one-way hash chain of *Links* allows a node to verify the authenticity of the hop counts in localization packets that it receives. As can be seen from SLS-Replay and SLS-Fwd curves, even more subtle attacks like the replay attacks do not have a major impact on the location error.

5.2 Node Infection

With respect to the hop by hop propagation based localization systems, an infected node is defined as one that includes a malicious node as part of its route path to any Landmark. In Figure 2, we show the the percentage of infected nodes in several attack scenarios similar to the ones described in the context of the location error. The results show again that the SLS is *SLS* robust against Alteration/Modification/Spoofing attacks as well as Sink Hole attacks etc. As can be seen from the SLS-Replay curve more subtle attacks like the replay attacks result in more nodes being negatively impacted by malicious nodes. However as the location error curves in Figure 2 clearly illustrates the impact in terms of location error is rather limited.

5.3 Network Diameter Estimation

The network diameter is estimated so that a commitment can be generated at every Landmark which then forms the basis of SLS. In this experiment, we study the effect of the difference between estimated network diameter, n_{est}, and actual network diameter,

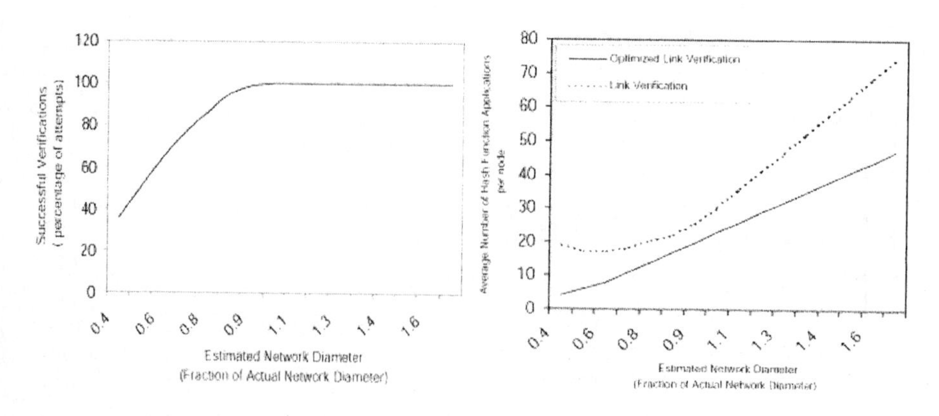

Fig. 3. Effective Link Verification

Fig. 4. Computational Overhead

n_{act}, on proposed solution. Results are shown in Figure 3. The X axis represents n_{est} as a fraction of n_{act}. The Y axis represents the percentage of feasible verification operations. The results show that successful verification can be more than 80% when the estimated diameter is 0.8 of the real network diameter.

5.4 Computational Overhead

The overhead resulting from the computation of Link Authentication is shown in Figure 4 where the average number of hash function operations per node corresponding to three Landmarks is plotted. Two scenario are included in the figure. The dotted curves represents that in every *Link* is authenticated by iterative application of hash function to obtain $Link^n$ and then matching it against the *Commitment*. The solid curve represents the results from a simple yet efficient optimization in which the first *Link* a node receives is authenticated with the same approach. However for subsequent verifications of any link $Link^j$ where $j < i$ iterative application of hash function to obtain $Link^i$ is sufficient. This results in reduction of computational overhead.

6 Conclusion

This work presents the SLS as a secure distance vector based localization system which is robust against multiple uncoordinated attacks of compromised and malicious nodes. SLS is based on a computational security model and relies on one-way hash functions. A simulation-based evaluation and a security analysis are included in the paper. In addition, we explain the location verification problem in the context of sensor networks and describe how SLS can be extended to facilitate location verification. This work does not take into consideration the mobility of sensor nodes nor possible collusion attacks. Such scenarios will entail an extension of the proposed secure localization system to address the variation in operational parameters that result form the mobility of nodes and to eliminate falsified information from propagation through the networks.

References

1. L. Doherty, K. S. J. Pister, L. E. Ghaoui, *Convex Position Estimation in Wireless Sensor Networks*, Proceedings of IEEE infocom 2001. pp. 1655-1663.
2. C. Savarese, K. Langendoen, J. Rabaey: Robust Positioning Algorithms for Distributed Ad-Hooc Wireless Sensor Networks, USENIX Tech. Annual Conference, 2002. pp. 317-328.
3. D. Niculescu, B. Nath, *Ad Hoc Positioning System(APS)*, IEEE GlobeCom 2001. pp. 2926-2931.
4. N. Sastry, U. Shankar, D. Wagner, *Secure Verification of Location Claims*, Proceedings of the 2003 ACM workshop on Wireless Security, 2003. pp. 1-10.
5. B. Waters and E. Felten, *Secure, Private Proofs of Location*, Technical report TR-667-03, Princeton University, 2003.
6. S. Capkun and J. P. Hubaux, *Securing position and distance verification in wireless networks*, Technical report EPFL/IC/200443, May 2004.
7. A. Perrig, Y. C. Hu and D. B. Johnson, *Wormhole Protection in Wireless Ad Hoc Networks*, Technical Report TR-01-384, Department of Computer Science, Rice University, 2001.

Secure AODV Routing Protocol Using One-Time Signature

Shidi Xu, Yi Mu, and Willy Susilo

School of Information Technology and Computer Science,
University of Wollongong
{sdx86, ymu, wsusilo}@uow.edu.au

Abstract. Mobile ad hoc network (MANET) has been generally regarded as an ideal network model for group communications because of its specialty of instant establishment. However, the security of MANET is still a challenge issue. Although there are some existing security schemes such as ARAN (Athenticated Routing for Ad hoc Networks) protocol that makes use of cryptographic certificate to provide end-to-end authentication during routing phases, the overhead of security computation is still a serious hurdle for real application. In this paper, we propose a comparatively efficient scheme to perform ARAN protocol, based on AODV, by using one-time signature in place of conventional signature, aiming at achieving the same level of security but improved efficiency.

1 Introduction

The Mobile Ad hoc Networks (MANET) are a specific type of network. Just as its name implies, it is formed by mobile nodes, such as laptops and PDAs. The construction of the networks is generally impromptu, therefore, networks can be formed whenever required and topology is changing from time to time. Ideally, any nodes satisfy general entering conditions will be accepted as a legitimate member of the network. These properties make MANET very suitable for group communications, in which, a number of people get together, forming a network to share documents and exchange conversations.

On the other hand, the wide-open environment makes this network super vulnerable to inside and outside attacks [1]. Especially in the case of routing [2], since the absence of central control, it is extremely difficult to prevent nodes from behaving improperly. Although there exist a large number of MANET routing protocols [3,4,5, 7,10], most of them were designed without any security considerations (generally it is assumed that all nodes are friendly). Besides, the resource constraints (both computation and bandwidth) of MANET put up great difficulties over the deployment of security. Two widely known reactive routing protocols are AODV (Ad hoc On-Demand Distance Vector Routing) [7] and DSR (Dynamic Source Routing) [5], which are both very efficient but are subject to a variety of attacks.

To reinforce the security of routing, ARAN [10] makes use of cryptographic techniques to offer security in an open-manage environment. Since the security is based on public key cryptography, the efficiency of ARAN is under suspicion. In this paper, we pursue the advantages of one-time signature, which is more efficient in signing

X. Jia, J. Wu, and Y. He (Eds.): MSN 2005, LNCS 3794, pp. 288 – 297, 2005.

and verification, to replace conventional digital signature in protecting routing packets, though, at the same time, maintaining the same level authentication.

The rest of the paper is organized as below. Section 2 takes a look at the ARAN routing scheme. Section 3 briefly introduces AODV routing mode and its security requirements, in addition to the one-time signature model used in our proposal. Section 4 describes the scheme used to secure AODV, called authenticated AODV. Section 5 discusses the security of our proposal. The last section concludes the paper.

2 Backgrounds

ARAN was proposed by Sanzgiri et al in 2002, targeting to combat attacks including unauthorized participation, spoofed route signaling, alteration of routing messages, replay attacks, etc. Similarly to other secure routing protocols, ARAN is also a security adds-on over on-demand routing protocols. It provides authentication, message integrity and non-repudiation as part of minimal security policy for ad hoc environment.

ARAN consists of three stages: a preliminary certification process, a mandatory end-to-end authentication stage and an optional stage providing secure shortest path. To deploy these three stages, ARAN requires the use of a trusted certificate server T and public key cryptography. Each node, before entering the network, must request a certificate from T, and will receive exactly one certificate after securely authenticating their identities to T.

The route discovery phase of ARAN is performed similarly to AODV, but with security supplements. When a node initializes a route discovery, it broadcasts a RDP (route discovery packet) consisting of RDP identifier, destination IP address, the certificate of source, a nonce and a timestamp. The RDP is signed by the source node. Upon the first hop the node receives this RDP, it signs over the whole RDP and rebroadcast it along with its own certificate. Then the next hop nodes will validate all the signatures with given certificates, and replace previous signature with its own along with its own certificate. The same activities continue until the RDP reaches the destination. Accordingly, the destination node will unicast a REP (route reply packet) back to the source, following the same activities as in sending RDP.

3 Models

In this section, we simply introduce the AODV routing process and the one-time signature scheme that is about to be used.

3.1 AODV Routing

AODV is a simple and efficient on-demand ad hoc routing protocol. Basically, it uses RREQ (route request), RREP (route reply) and RRER (route error) messages to accomplish route discovery and maintenance operations. It also utilizes sequential numbers to prevent routing loops. Routing decision making is based on sequence numbers and routes maintained in each node's routing table.

From the security point of view, AODV requires at least two security attributes: sender authentication at each receiving node and routing message integrity. Message integrity is of the most concern in AODV routing. In the route request broadcasting phase, each node has to check the originator sequential number in the RREQ packet with the one recorded in its routing table, and updates its routing table to the newest one; in route reply phase, instead of checking originator sequential number, each node check the destination sequential number and keeps it up-to-date. Any exploits of changing sequential number will result in routing loops.

Besides message alteration, spoofing is also a serious attack. A node forwarding RREP might claim itself to be someone else, misleading the receiving nodes falsely recording the fake identity as the next hop towards destination. This is another way of disrupting topology by creating route loops.

3.2 HORS One-Time Signature Scheme

As we observed, since ARAN uses public key cryptography to protect routing process, the time delay of signature generation and verification is significant. In general, significant time delay at each hop causes unacceptable route acquisition latency. Thus, we are looking for some digital signature schemes that maintain all the traits of conventional DSS, but are efficient enough in signature generation and verification.

The very first one-time signature scheme was introduced by Lamport in 1979 [6], to sign just 1 bit information. In 2002, Reyzin et al [9] proposed a one-time signature scheme, which is both efficient in signing and verification, and generating short signatures. This resulting scheme is called HORS, which stands for Hash to Obtain Random Subset. The major operation in signature generation is using a hashed message to obtain a random subset to form the signature.

- **HORS Key Generation**

On constructing this scheme, several security parameters are predefined. To sign b-bit messages, we firstly pick t and k such that $\binom{t}{k} \geq 2^b$ and then choose a security parameter l, and a one-way hash function f that operates on l-bit strings. To generate public key, randomly generate l-bit string $(s_1, s_2, \ldots, s_t)$. Let $v_i = f(s_i)$ for $1 \leq i \leq t$. The resulting public key is PK = $(k, v_1, v_2, \ldots, v_t)$, private key is SK = $(k, s_1, s_2, \ldots, s_t)$.

- **HORS Signature Generation**

To sign a message m, with secret key SK = $(k, s_1, s_2, \ldots, s_t)$, firstly let $h = hash(m)$; then split h into k substrings $h_1, h_2, \ldots, h_k$, of length $\log_2 t$ bits each; finally, interpret each h_j as an integer i_j for $1 \leq j \leq k$. The resulting signature is $\sigma = (s_{i1}, s_{i2}, \ldots, s_{ik})$.

- **HORS Signature Verification**

The verification is the same as the signature generation. Suppose the verifier has the message m, signature $\sigma = (s'_{i1}, s'_{i2}, \ldots, s'_{ik})$, and public key PK = $(k, v_1, v_2, \ldots, v_t)$. Firstly, let $h = hash(m)$; then split h into k substrings $h_1, h_2, \ldots, h_k$, of length $\log_2 t$ bits each and interpret each h_j as an integer i_j for $1 \leq j \leq k$. If for each j, $1 \leq j \leq k$, $f(s'_j) == v_{ij}$, accept the signature; otherwise, reject the signature.

3.3 One-Time Key Generation for Routing

Here, we describe the HORS one-time key generation process.

- **Notations:**

  $h()$, h, $h^i()$ – one way function
  $Sign_{K_n}$ – conventional digital signature generated by node n
  $<>K_n^{-1}$ – one-time signature generated by node n

- **Key Chain generation:**

Suppose that the decision has been made regarding security parameters l, k and t according to message length b.

1. Each node chooses t secret key components x_j (j=1,…,t) at random.
2. Each node creates a n hash chain of length t (see **Figure 1**):
3. Public key components are obtained through a one-way function h, namely $v_i = h(x_i)$. We assume that h is a hash function for simplicity.
4. Public key components are disclosed periodically.

Generating a set of one-time keys to sign routing messages has been discussed by Zhang in 1998 [15]. Two schemes called chained one-time signature scheme (COSP) and independent one-time signature scheme (IOSP) were proposed. These two schemes activate us to generate our novel scheme.

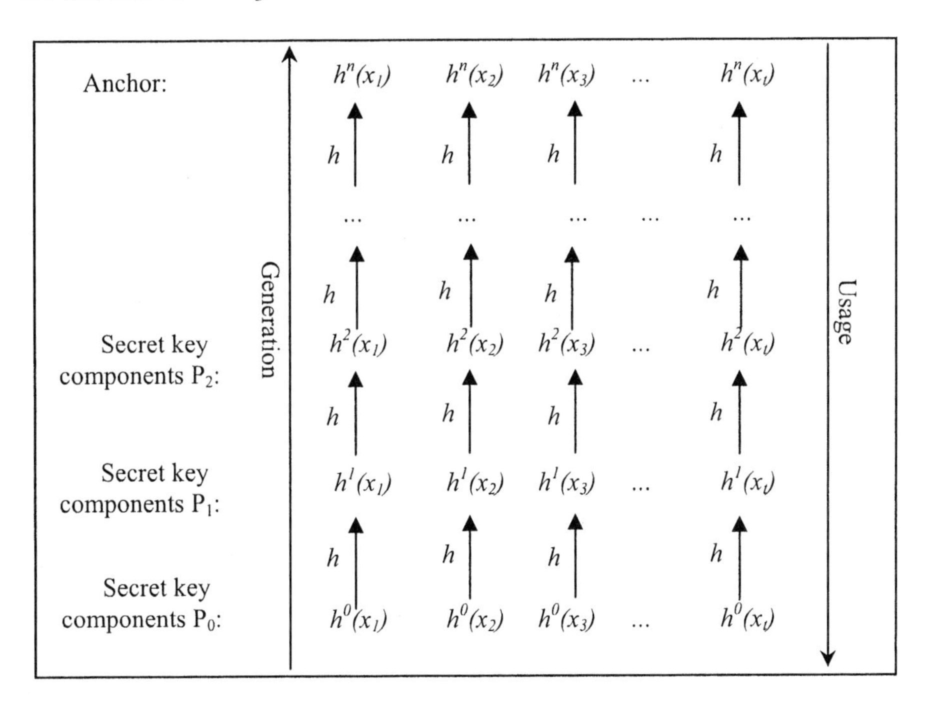

Fig. 1. Secret key components hash chain

4 Authenticated AODV Routing Protocol

Based on the one-time signature scheme described above, we propose a security adds-on for AODV, which containing ARAN's authenticated routing features. This proposed protocol will provide following security properties:

1. The target node can authenticate the originator;
2. Each receiving node can authenticate its previous hop from which the routing message coming;
3. Each intermediate node can authenticate the sender for updating its routing table entry;
4. The hop count value is protected using hash chain. It cannot be reduced by a malicious node, but could be increased more than one or retained unchanged, as in SAODV [12].

To achieve security features listed above, we firstly assume the existence of an offline CA, which issues certificate for each node when entering the network. Thus, each node possesses a public key and private key pair. The conventional digital signature will still be used to provide sender authentication, whereas the one-time signature will offer end-to-end authentication.

4.1 Public Key Handling

The public key in our proposed protocol is disseminated in two different ways. One aims at providing keys for authentication among neighbors. Another one tries to enable sender authentication during message transmission.

End-to-end authentication is achieved through neighbor authentication. Each node will generate a set of one-time keys as described in section 3.1. The one-time public key components are distributed locally among neighbors. Since one-time keys can only be used once or limited times, nodes need to update their one-time public keys periodically. To guarantee that each neighboring node has an authentic copy of node's public key, the very first public key, *anchor*, is distributed safely during system setup. When a node enters the network, it signs its anchor and broadcasts to its neighbors, along with its certificate. Thus, successive one-time public keys can be distributed in a more efficient manner by using *Hello* message, which is broadcasted periodically. The verification of updates is straightforward.

For example, the first secret key SK_1 is $(k, h^n(x_1), h^n(x_2), h^n(x_3), \ldots, h^n(x_t))$. The corresponding public key PK_1 is $(k, h^{n+1}(x_1), h^{n+1}(x_2), h^{n+1}(x_3), \ldots, h^{n+1}(x_t))$. The second secret key SK_2 is $(k, h^{n-1}(x_1), h^{n-1}(x_2), h^{n-1}(x_3), \ldots, h^{n-1}(x_t))$, thus the corresponding public key PK_2 is $(k, h^n(x_1), h^n(x_2), h^n(x_3), \ldots, h^n(x_t))$, which can be verified by hashing once and comparing to PK_1.

On the other hand, sender authentication is achieved through conventional digital signature. The sender's public key is contained in its certificate which is obtained when entering the network.

4.2 System Setup

This phase is used for initial key distribution (see **Figure 2**). Suppose when a mobile node enters the network, it is soon informed about the security parameters agreed in this network. It then chooses its secret key components and generates a hash chain according to section 3.2. Then it performs as follows:

N: Choose secret key component *SK*
Construct hash chain
The first public key component PK_1 is the *Anchor*
N→*: $Sign_{KN}$< N, PK_1, timestamp>, $Cert_N$

Fig. 2. Initial key distribution and authentication (in System Setup)

4.3 Route Discovery

Route Discovery is performed as in **Figure 3**. When the originator (*S*) initiates a route discovery to a certain destination, it simply generates a signature over the RREQ, using conventional digital signature.

Upon the first hop node (*A*) receives the RREQ, it firstly verifies the signature of the originator. If the signature is fine, the neighboring node hashes the received

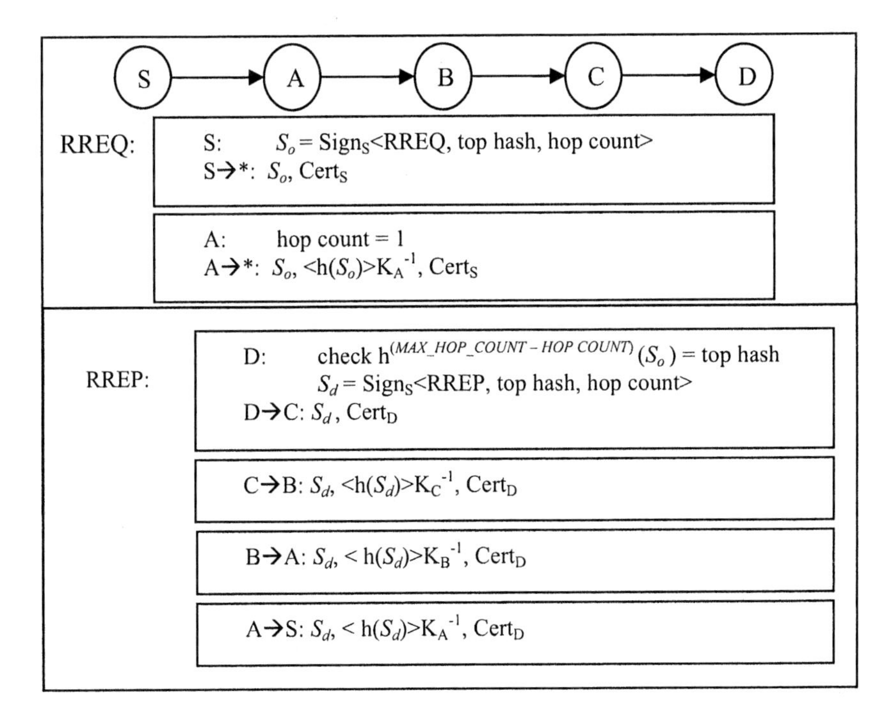

Fig. 3. Route Request and Route Reply

message S_o again and generates its own signature over it. This time, the signature is generated using HORS one-time signature scheme. Then the whole message is re-transmitted to second hop. From now on, there are two signatures. One is over S_o, another is over the hash of S_o.

Once the second hop node (B) receives this double signed RREQ, it firstly verifies the pervious hop (A) using public key of A (which might receive through *Hello* messages). If the one-time signature is fine, B hashes S_o one more time and creates a signature over the hash to replace the signature of A. Then this new message is broadcasted to next hop neighbors. Notice that the verification of conventional signature could be delayed. Only if both conventional signature and one-time signature are fine, does B update its routing table entry according to RREQ. These operations repeated until RREQ reaches the destination.

When RREQ reaches the destination, the destination node performs verifications the same as each intermediate node. Then a RREP is generated and signed the same as RREQ. Each intermediate node will transmit it back to the originator through the reverse route and same operations are performed along the route.

4.4 Gratuitous Route Reply

In AODV, gratuitous route reply enables an intermediate node to reply RREQs which it has an active route towards the destination. This feature is optional in AODV, though turning on this feature will highly enhance the efficiency of routing discovery. However, to enable this feature, additional technique is needed. The conceptual idea is that since we used digital signature to protect each routing message at each hop, for an intermediate node to reply RREQs instead of the destination, the intermediate node should be able to sign the RREQ properly on behalf of the destination.

To solve this problem, we borrow the idea from proxy signature proposed by Varadharajan et al. [11], in which delegation is enabled by using a warrant. The warrant appears as a delegation token, containing the identities of primary signer and proxy signer, the privilege (Pr_a) given to proxy signer, an identifier (r_a) used by primary signer, and a timestamp (t_a). This delegation token is signed by the primary signer.

In AODV, the delegation token can be further simplified, because the token does not need to be designated to certain nodes. Any node that has received the token is automatically proved to be having an active route towards that node. Otherwise, it would not be able to obtain this token. Thus, we only require three fields in token: the destination's identity, an identifier and a timestamp, and is also signed by the creator using conventional digital signature. We assume that each node is able to store the received token for future gratuitous route reply.

If the gratuitous route reply option is turned on, nodes broadcasting RREQs must create tokens for gratuitous route reply delegation. The whole message including the token will be signed again, using the same public key as signing the token. Then, the originator broadcast the RREQ as usual.

Upon receiving the RREQ, the node processes the authentication as normal. Then it checks the timestamp to see if the token has expired. If the token is valid, the nodes will store the token for future use.

The originator firstly checks if this RREP was created by destination or by intermediate node. If it is a gratuitous route reply, the originator checks the timestamp to determine if the route is still active. Then the token and the RREP will be authenticated as described before.

5 Discussion and Improvement

The most outstanding point of this scheme is the efficiency of one-time signature generation and verification at each hop. It is the same as HORS [9], each time, key generation requires t evaluation of one-way function. The secret key size is lt bits, and the public key size is $f_l t$ bits, where f_l is the length of the one-way function output on input of length l. The signature is kl bits long. There is a tradeoff between t and k, since the public key size will be linear in t, and the signature size and verification time will be linear in k.

The security of this scheme stems from the system setup phase. In this phase, a conventional digital signature is used to guarantee the authenticity of the first public key component. This can be achieved through using a public key certificate issued by an offline CA, namely, each node must present a creditable identity when entering the network. The signature verification and generation may be inefficient, but since this message is broadcasted locally, it should be practical for each node.

The update of public key component is done along with *Hello* message, which is broadcasted periodically. Since the public key component comes from a hash chain, the verification is straightforward – the previous public key component is used to authenticate the new one. The trustworthiness of the new public component depends totally on the security of one-way hash function and the digital signature over *anchor*. The anchor is used only once. It is replaced by newly coming public key component after the first *Hello* message is broadcasted. In this way, nodes only need to do one hash to authenticate new public key component each time, which is much more efficient than hashing repeatedly back to anchor.

Sender authentication is performed with some compromise of efficiency, using conventional DSS. This method is much more secure than in SAODV, because in SAODV, the originator simply signs on its own public key without the support of PKI. Attackers can easily forge RREQ and RREP packets during transmission. On the other hand, the efficiency can be enhanced to some degree through the way that each node verifies conventional digital signature after broadcasting routing packets. Therefore, these will be no verification delay. Only both conventional signature and one-time signature is fine, will the routing table entry be updated.

Double signing over the received message does not provide more security than single signature from cryptographic point of view. Nevertheless, it provides non-repudiation hop-by-hop, which can be sued as an evidence for future intrusion detection. This thought comes from ARAN. It is considered as impractical because the use of conventional signature schemes. If there is a technique to produce even shorter signature in more efficient manner, this scheme can be extended to allow each node to sign on the received messages.

One significant drawback of one-time signature is that it can sign only predefined number of messages, which, in our scheme, is limited by the size of hash chain n. We generally consider it is not a serious problem, because nodes in MANET are mobile devices which are leaving and entering the network frequently. Consequently, the hash chain will be refreshed. In this sense, we can set n to a proper value according to network scale and average active time of nodes.

6 Conclusion

This paper presented a novel scheme to implement ARAN protocol based on AODV routing protocol. However, it is more efficient than original ARAN in signature generation and verification since we used HORS one-time signature in place of conventional digital signatures. We also enable the protections over gratuitous route reply feature, under the concept of proxy signature's delegation by warrant. The warrant here is represented as a token, which contains creator's identity and public key, and is signed by the creator.

The security of our scheme needs to be enforced by performing conventional digital signature. With the help of asymmetric cryptography or public key certificate, we can ensure the authenticity of mobile nodes and the secure distribution of initial keys. Hence, the security of sub-sequential keys can be guaranteed by one way hash chain.

References

1. A. Burg. Ad Hoc Network Specific Attacks (pdf). In *Seminar. Ad hoc networking: concepts, applications, and security*, Technische Universität München, Nov. 2003.
2. Y. C. Hu, D. Johnson, and A. Perrig. SEAD: Secure Efficient Distance Vector Routing for Mobile Wireless Ad Hoc Networks (pdf). In *4th IEEE Workshop on Mobile Computing Systems and Applications (WMCSA' 02), June 2002*, pages 3-13, June 2002.
3. Y. C. Hu, A. Perrig, and D. Johnson. Ariadne: A Secure On-demand Routing Protocol for Ad Hoc Networks (pdf). In *Proc. ACM MOBICOM*, Sep, 2002.
4. Y. C. Hu, A. Perrig and D. B. Johnson. Rushing Attacks and Defense in Wireless Ad Hoc Network Routing Protocols (pdf). In *Proc. the 2003 ACM workshop on Wireless*, Sep. 2003.
5. D. B. Johnson, D. A. Maltz and Y. C. Hu. The Dynamic Source Routing Protocol for Mobile Ad Hoc Networks (DSR). IETF INTERNET DRAFT, MANET working group, July. 2004. draft-ietf-manet-dsr-10.txt.
6. L. Lamport. Constructing digital signature from a one way function. Technical Report CSL-98, SRI International, October 1979.
7. C. E. Perkins, E. M. Royer, and S. R. Das. Ad Hoc On-Demand Distance Vector (AODV) Routing. IETF INTERNET DRAFT, MANET working group. Feb. 2003. Draft-ietf-manet-aodv-13.txt.
8. A. Perrig. The BiBa one-time signature and broadcast authentication protocol. In *8th ACM Conference on Computer and Communication Security, page 28-37. ACM, November 508, 2001.*
9. L. Reyzin and N. Reyzin. Better Than BIBA: Short One-Time Signatures With Fast Signing and Verifying (pdf). In *Proc. 7th Australasian Conference on Information Security and Privacy, LNCS 2384*, Apr. 2002.

10. K. Sanzgiri, B. Dahill, B. N. Levine, C. Shields, and E. M. Royer. A Secure Routing Protocol for Ad Hoc Networks (pdf). Technical Report: UM-CS-2002-032, 2002.
11. V. Varadharajan, P. Allen, and S. Black. An Analysis of the Proxy Problem in Distributed Systems (pdf). In *Proceedings of the IEEE Symposium on Security and Privacy, 1991*, pages 255-275, May 1991.
12. M. G. Zapata. Secure Ad hoc On-Demand Distance Vector (SAODV) Routing. IETF INTERNET DRAFT, MANET working group, Nov. 2004. draft-guerrero-manet-saodv-02.txt.
13. K. Zhang. Efficient Protocols for Signing Routing Messages (pdf). In *Symposium on Network and Distributed Systems Security (NDSS '98)*, 1998.

A Security Enhanced AODV Routing Protocol

Li Zhe, Liu Jun, Lin Dan, and Liu Ye

Institute of Telecommunications and Information Systems,
Faculty of Information Science & Engineering,
Northeastern University, Shenyang 110004, China
lizhe@mail.neu.edu.cn

Abstract. Ad-hoc networks are characterized by open medium, dynamic topology, distributed cooperation and constrained capability. These characteristics set more challenges for security. If the routing protocol is attacked, the whole network would have been paralyzed. As a result, routing security is the most important factor in the security of the entire network. However, few of current routing protocols have the consideration about the security problems. The potential insecure factors in the AODV protocol are analyzed. Furthermore, a security routing protocol based on the credence model is proposed, which can react quickly when detecting some malicious behaviors in the network and effectively protect the network from kinds of attacks.

1 Introduction

MANET (Mobile Ad-hoc Networks) [1], comprised of a collection of mobile nodes connected with wireless link, is a multi-hop and self-organized system. Ad-hoc networks are autonomic, provisional, infrastructure less and easily-constructed, which is widely used in military information system of battle field, civil emergency search-and-rescue operations and other occasion. The Ad-hoc networks are gradually adopted for commercial and civilian purpose since the technology of the Internet and mobile telecommunication is maturing and being broadly applied. However, the security of ad hoc networks is also broadly concerned [2] [3]. According to continuous random motion of the nodes, the topology of the network keeps changing and varying. In addition, the features of Ad-hoc networks are the same as that of normal wireless system, all of which make security problems of Ad-hoc networks more complex, especially the security of routing protocols. Some research about security issues of AODV protocol, which is one of the most popular routing protocols in Ad-hoc networks, is done. Moreover a new routing security enhanced protocol with two methods, attack detection and credence mechanism, is proposed in this paper.

2 Rationale of AODV Routing Protocol

AODV (Ad-hoc On Demand Distance Vector) [4] has been considered as one of the most popular and promising on-demand (route will be built just when one node wants

X. Jia, J. Wu, and Y. He (Eds.): MSN 2005, LNCS 3794, pp. 298 – 307, 2005.

to communicate with some other) routing protocols, which has been standardized by IETF and attracts many concerns because of its lower network overhead and algorithm complexity.

The routing procedure of AODV is described as follows.

2.1 Route Creating

1. The source node broadcasts Route Request (RREQ) message.
2. Once the intermediate node receives the RREQ message, a reverse route towards the upstream node that sends the RREQ message is built. If the node has a fresh route to the destination, it will send Route Reply (RREP) message along the reverse route to the source node, else the RREQ message will be forwarded one by one.
3. The destination node sends RREP message to the source node through reverse route after it receives RREQ message.
4. All nodes on the reverse route update their routing tables, in which a route to the destination node will be built.
5. Once RREP reaches the source node, the route searching process is terminated. A new route is built in its routing table by which the transmission can be done.

2.2 Route Maintaining

If a node detects a link break for the next hop of an active route in its routing table, it deletes all route entries including this broken link, and broadcasts Route Errors (RERR) message to notify those upstream nodes to delete the corresponding entries in their routing tables.

A node may detect connectivity with others by broadcasting local Hello messages periodically and broadcast RERR message when a link break is detected.

3 Security Problems in AODV Protocol

The conclusion from the above description of the working procedure is that AODV doesn't need to maintain all routing information of each node in one's routing table, and the intermediate node can reply with RREP message if it has a fresh route to the destination. This method can reduce route searching and building time, but giving an opportunity to malicious nodes. The main security threats [5][6][7] of AODV just take advantage of this feature.

Several main attacks have been listed as follows.

1. Black Hole Attack. The attacker broadcasts some fraudulent messages to make others believe that data can be transmitted through itself with the shortest path or the least cost, while this trickster never forwards these data packets, which forms a "black hole", that is, absorbing in everything but never giving out. The main attack methods are fabricating distance vector and fabricating sequence number.
2. Routing Table Overflow Attack (RTO). A malicious node keeps sending a large number of RREQ messages for some node that does not exist, which consumes lots of computation and network bandwidth, even the paralysis of entire network.

3. Network Segmentation Attack.
(1) Fabricating RERR Packet Attack. Malicious nodes broadcast fabricated RERR packets to destroy the route table, which causes network segmentation.
(2) Interrupting Routing Attack. Selfish node drops the received routing messages from its neighbors for limited battery energy and computation ability.

4 A Security Mechanism Built on AODV Protocol

In this paper, some necessary improvements are made to AODV protocol, adding attack detection and building up credence mechanism, which would collaboratively guarantee the network security.

When a malicious node is judged as an attacker by the credence mechanism, the security routing protocol will implement routing reconstruction to isolate the attacker from the network.

In order to provide secure and reliable data forwarding services, nodes should firstly consider the route with high credence value when routing packets. In figure 1, as S wants to send packets to D, the number of hops is 2 when B as the next hop, and is 3 when C as the next hop. S will then compare the credence value of B and C. If C has higher credence value, S may choose C as the next hop ignoring its more hops.

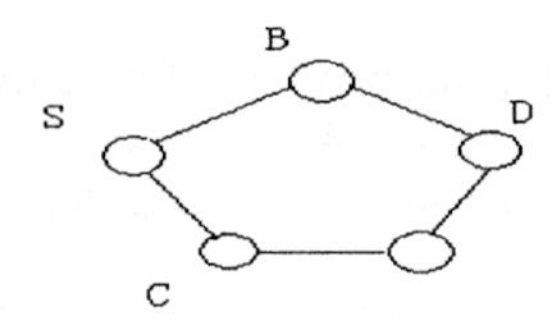

Fig. 1. Data forwarding based on the credence value

4.1 AODV-AD Protocol (AODV with Attack Detection)

Some necessary modifications are made to AODV protocol. Nodes in the network are working under the promiscuous mode and monitoring the communication behaviors of their neighbors. The intermediate node adds the next hop information in the RREP packet, through which every node store part of its neighbor's routing information as the reference of the attack detection. Detections for usual attacks are as follows.

(1) Fabricating Distance Vector Attack Detection
In figure 2, node S wants to communicate with node D. X is supposed to be malicious node and sends a RREP indicating Y as the next hop.

a) Y is not the neighbor of X. X's neighbor receives the RREP sent by X and knows that Y is the next hop. It then looks up Y in the X's neighbor list. If Y is not in the list, X will be considered as malicious node.
b) Y is the neighbor of X. Y also receives the RREP and looks up its own routing table. If there is no route to D, X will be considered as malicious node.

Otherwise, Y verifies whether the hops in the RREP equals the hops that from Y to D plus 1. If not equals, X will also be considered as malicious node.

c) Some other node, like A in the figure, is the common neighbor of X and Y. After receiving the RREP, it judges X's behavior through the stored local information

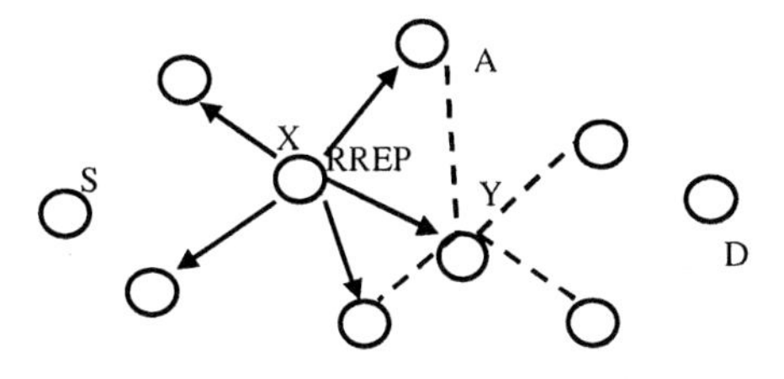

Fig. 2. Fabricating Distance Vector Black hole attack detection

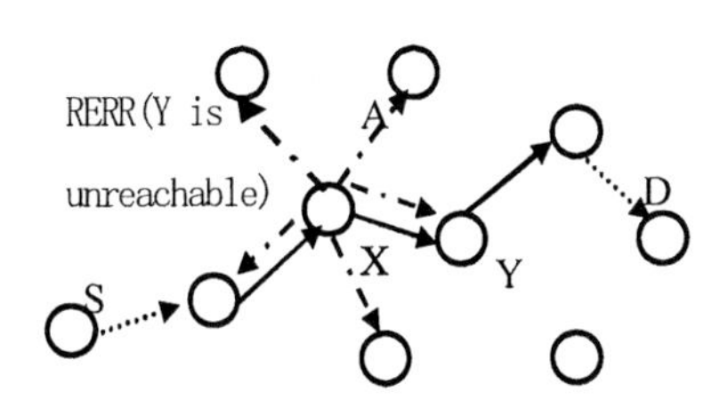

Fig. 3. Fabricating RERR attack detection

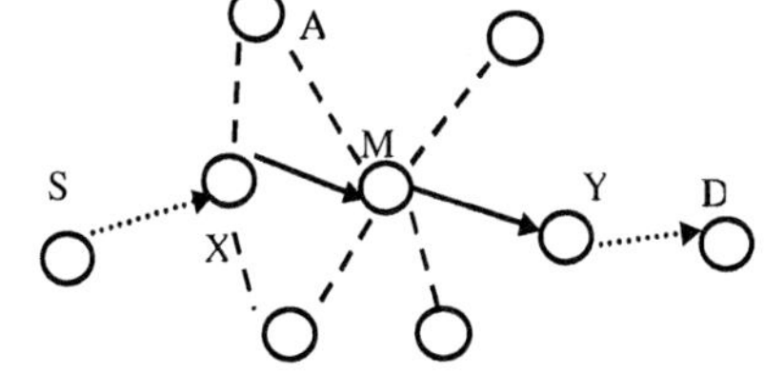

Fig. 4. False package forwarding detection

(2) Fabricating Sequence Number Attack Detection

This attack usually uses a big enough sequence number in the RREP packet in order to avoid the AODV sequence number increment mechanism disabling the attack. Thus, nodes can check the destination's sequence number in the routing information. If it is large dramatically, there must be a sequence number attack.

(3) Routing Table Overflow Attack (DDOS) Detection

The detecting solution is creating a table recorded the number of RREQ packets from other nodes and a timer. During a period of time, if the number of RREQ is bigger than the threshold, the sender would be suspicious.

(4) Fabricating RERR Packet Attack Detection

In figure 3, X sends RERR to its neighbor to inform that Y is unreachable, which makes network segmentation happen. For RERR is broadcasting packet, Y is X's neighbor and also receives this packet. Thus, Y can deduce whether this RERR is correct through the communication between itself and X. When Y judges that the RERR from X is fabricated, Y will alarm the whole network; other nodes are neighbors of both X and Y, like A in the figure. When they receive RERR, they will estimate X's behavior based on their local information.

(5) Interrupt Routing Attack (Selfish Node) Detection

Node A in figure 4 is the neighbor of both M and M's upstream neighbor X. Therefore, A can monitor that X sends packets to M and M's forwarding behavior. If A senses that

X sends data packets to M but doesn't hear M forwarding them, A will consider that M has dropped these data packets maliciously.

In addition, the credence mechanism can detect this selfish behavior.

4.2 Credence Mechanism Built on AODV Protocol

Definition. Credit is a belief that the entities will behave without malicious intent and a belief that the whole system will resist malicious manipulation. Credence is just a credit measurement of entities in the network.

The credence mechanism is mainly used to prevent the security threats brought by malicious nodes, especially selfish nodes. The main goals are as follows.

1. Offering reliable information to decide whether a node is trustful.
2. Encouraging cooperation among nodes.
3. Preventing the cooperation service protected by the mechanism from being accessed by malicious nodes.

4.2.1 Credence Establishment

The establishment of credence is much close to monitoring behavior of nodes. The nodes' behavior can be classified into three kinds.

(1) Routing packet processing credence subcategory, which evaluates the nodes' processing behaviors to routing packet.
(2) Data packet forwarding credence subcategory which evaluates the nodes' forwarding behaviors to data packet.
(3) Malicious behaviors credence subcategory which evaluates the attack behaviors on AODV protocol, such as the black hole attack.

This credence classification can make more accurate evaluation on nodes' behaviors, which will benefits making better use of network resources and finding malicious nodes more quickly.

For the dynamic topology of Ad Hoc networks makes credence value changing all the time, the quantified representation of credence value is the number that changes from -1 to 1. "-1" means distrust at all; smaller than 0 means distrust; bigger than 0 means trust; "+1" means completely trust. Nodes monitor communication behaviors between neighbors and exchange the information with others to obtain credence values, and store them in the credence table, each node one entry.

4.2.2 Credence Computation

4.2.2.1 Computation of Routing Credence Category Routing Credence Computation. R_r denotes the value of routing credence category, depending on two parameters. R_{rs} is the number of routing packets that are forwarded successfully; R_{rf} is the number of routing packets that are failing to forward. When forwarding routing packet successfully, the credence value is increased, and then the R_r value changing range should be [0, +1]. Moreover, as R_{rs} increases, R_r becomes closer to 1. Hence, it can be prompt for the new nodes to join in the network when they forward routing packet successfully. And the appropriate curve plotted against R_{rs} should like figure 5. The formula used is (1).

$$R_r = 1 - \frac{2 \times R_{rf}}{R_{rs} + R_{rf}} \quad \text{in which } R_{rs} + R_{rf} \neq 0\text{; otherwise } R_r = 0 \tag{1}$$

When failing to forward routing packet, the credence value is decreased, and then the R_r value changing range should be [-1, 0]. And as R_{rf} increases, R_r becomes closer to -1. So it should be quick to decrease the credence value of the stranger who doesn't forward routing packet and finally isolates the node from network. And the appropriate curve plotted against R_{rf} should like figure 6.

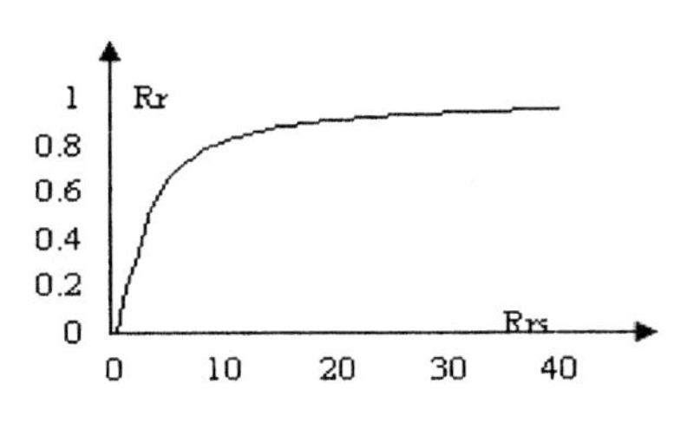

Fig. 5. R_{rs} versus R_r

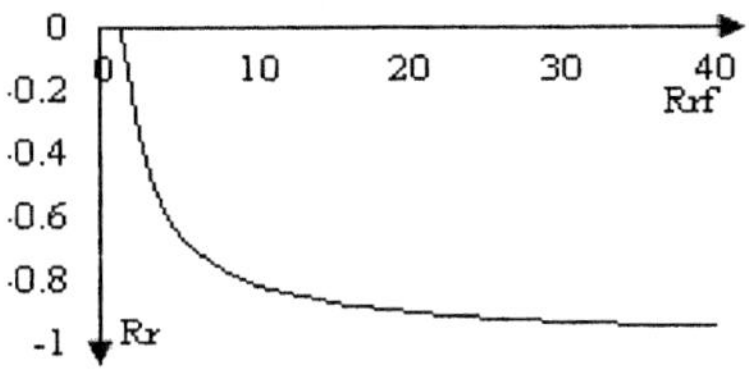

Fig. 6. R_{rf} versus R_r

The formula used is (2)

$$R_r = \frac{2 \times R_{rs}}{R_{rs} + R_{rf}} - 1 \text{ in which } R_{rs} + R_{rf} \neq 0\text{; otherwise } R_r = 0 \tag{2}$$

According to (1) (2), we can generalize the formula of routing credence category:

$$R_r = \frac{R_{rs} - R_{rf}}{R_{rs} + R_{rf}} \text{ in which } R_{rs} + R_{rf} \neq 0\text{; otherwise } R_r = 0 \tag{3}$$

However, when the values of R_{rs} and R_{rf} become too high, their impact on R_r in the formula is not so apparent. So they should be scaled down periodically, which would keep them at an appropriate level.

4.2.2.2 Forwarding Credence Computation. In a similar way, the formula of forwarding credence category is as follows.

$$R_f = \frac{R_{fs} - R_{ff}}{R_{fs} + R_{ff}} \text{ in which } R_{fs} + R_{ff} \neq 0\text{; otherwise } R_f = 0 \tag{4}$$

R_f denotes the value of forwarding credence category; R_{fs} denotes the number of data packets that are forwarded successfully; R_{ff} denotes the number of data packets that are failing to forward.

4.2.2.3 Malicious Behavior Credence Computation. R_m denotes the value of malicious behavior credence, in which m means the number of node behaviors

a). Increasing credence value for legitimate behavior.

b). When illegitimate behavior happens, this credence category evaluates whether an entity has attack behavior. The credence value should be decreased largely when attack happens. If this entity's previous works are normal ($R_m>0$), its credence will be cut into a half. Meanwhile, if R_m is close to zero, the value will be further decreased by ΔR besides halving; if the entity has had abnormal performance before ($R_m<0$), its credence will be decreased greatly according to linearity strategy. Thereby the formula of attack behavior credence is as follows.

$$R_{m+1}=\begin{cases} R_m+\Delta R & \text{when entity has legitimate behavior} \\ R_m/2-\Delta R \quad R_m>0 & \text{when entity has illegitimate behavior} \\ R_m-2\times\Delta R \quad R_m<0 & \text{when entity has illegitimate behavior} \end{cases} \tag{5}$$

When the behavior of the entity (node in the network) is legitimate, ΔR denotes the increment of credence for each normal behavior. The formula, $R_m= m\times\Delta R+ R_0$, can be derived from $R_{m+1}=R_m+\Delta R$. This is a linear function about m, in which R_m is increasing along with the incremental of m. Moreover, the speed of credence changing depends on the value of ΔR.

For example, when R_m is bigger than 0, according to $R_{m+1}=R_m/2-\Delta R$, we can get

$$Rm = 2\times R_0 - 2\times\Delta R\times(1-2^{-m}) \tag{6}$$

Make $R_m=0$, then

$$m=\log_2(\frac{R_0}{2\times\Delta R}+1) \tag{7}$$

If the entity is completely trusty ($R_0=1$) and the entity is considered trusty after just six legitimate behaviors ($\Delta R=1/6$), m equals 2. That is, if the entity suddenly makes an attack after a series of legitimate behaviors and gets complete trust, just malicious behaviors for twice could make the credence returned to zero. Thus the malicious node can be found promptly.

When $R_m<0$, R_m is reduced in term of linear rule. Speeding up the reduction make it fast to reach the alarm threshold.

4.2.2.4 The Whole Credence Computation. Under different conditions, the expectation of one entity to the service provided by others is also different. Therefore, the whole credence is the weight sum of all credence categories. The ideal method is dynamically computing the whole credence and increasing the weight of any credence category whenever needed. However, this dynamic method is very hard to implement. In the paper, the weight of each credence category is configured manually according as network using. The formula is:

$$R_0 = W_f\times R_f + W_r\times R_r + W_m\times R_m \tag{8}$$

In which, R_o denotes the whole credence of entity in network; R_f denotes forwarding credence category; R_r denotes routing credence category; R_m denotes malicious behavior credence category; W_f denotes the weight of forwarding credence; W_r denotes the weight of routing credence; W_m denotes the weight of malicious behavior credence.

4.2.3 Credence Purge

When a node has a malicious behavior, its credence will be accordingly decreased. If the credence value drops to a certain threshold, the detecting node may send alarm information. After ascertaining it as malicious node, it is recorded in the black sheet of other nodes. What's more, all other nodes won't communicate with it and delete its credence value information, which complete the credence purge.

5 Performance Analysis of the Algorithm

NS2.26 with open source code is adopted as the simulation platform of this experiment. The scenario is defined with a set of parameters as follows.

Number of nodes: 30 Packet size: 512 byte
Topology range: 1000 m×1000 m Transmission range: 250 m
Simulation time: 500s High-level flow: CBR (Constant Bytes Rate Stream)
Motion of nodes: random motion with 20 m/s as the highest speed

5.1 Impact of these Attacks on the Network Performance

Packet delivery ratio means the ratio between the number of packets successfully received by the destination node at application layer (the network throughput) and the number of packets originated by the source node at application layer in the network, which reflects the network processing ability and data transferring ability, and is the main symbols of reliability, integrity, effectiveness and correctness of the protocol.

Routing overhead equals to the ratio between the number of routing control packets transferred during the whole simulation process and the number of data packets. That is how many routing control packets are needed for one data packet transmission. It is an important index that compares the performance among different routing protocols; moreover it can evaluate the scalability of routing protocol, the network performance and the energy consumption efficiency under lower bandwidth or congestion.

Both indexes are respectively simulated when the network is under kinds of attacks. And the routing table overflow attack and fabricating sequence number attack are taken as examples to analyze the performance of the improved AODV protocol.

5.1.1 Routing Table Overflow Attack

There is a malicious node existing in the network. After 10 seconds it starts to attack, sending 10 RREQ packets every second for the nodes not existing in the network at all, which will make the request packets broadcast unceasingly in the network and destroy the normal work of the network.

Figure 7 is the routing overhead curve when under the attack. The massive RREQ packets make the routing overhead of the AODV protocol much higher, but the improved AODV protocol can find the attack quickly and isolate the malicious node from the network to maintain the routing overhead at the normal level.

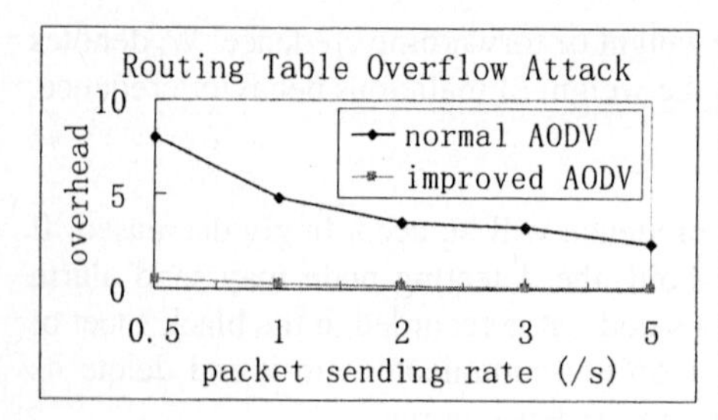

Fig. 7. The overhead curve under RTO attack

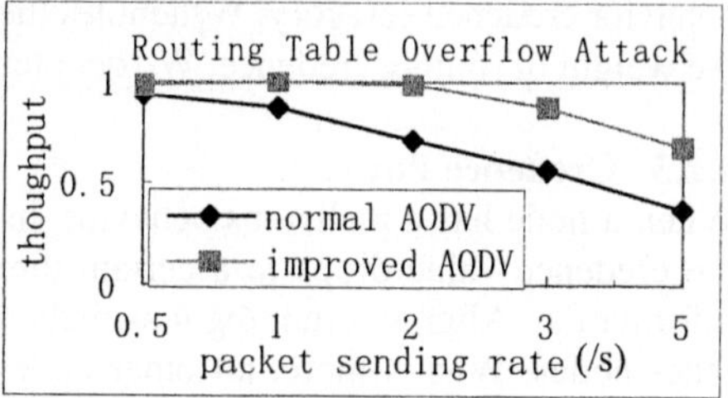

Fig. 8. The throughput curve under RTO attack

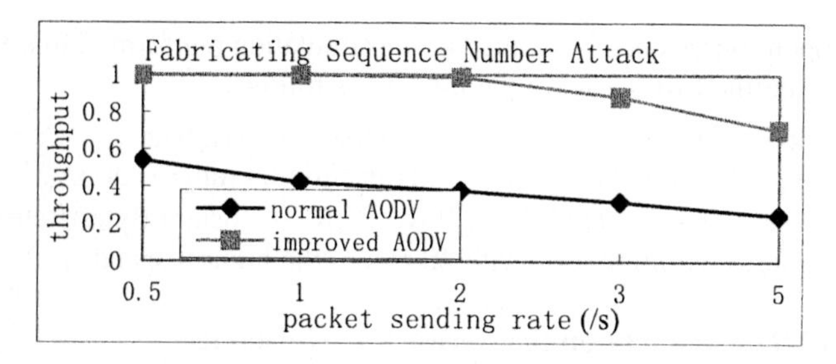

Fig. 9. The throughput curve under fabricate sequence number attack

Figure 8 is the throughput curve when under the routing table overflow attack. The massive RREQ packets maliciously occupy great bandwidth and affect normal use of the network resource of other nodes, which makes the throughput become low. But the improved AODV protocol can maintain the normal work of network at a high throughput level.

5.1.2 Fabricating Sequence Number Attack

Because the improved algorithm detects sequence number attack by analyzing the routing packets in the network and does not produce extra routing overhead, we just simulate the average throughput of the network. Figure 9 is the throughput curve under the attack. The figure illustrates that the fabricated routing information produces black hole attack; thus the data packets from the source is not able to reach the destination, which affects the network throughput seriously. While the improved AODV protocol can find the malicious behavior promptly and isolate the attacker from the network to maintain the network working normally at high throughput level.

5.2 Precision of the Detection Algorithm

The two parameters representing detecting effect are: 1. attack-detecting accuracy, the ratio of the number of detected attackers to that of actual attackers; 2. attack-detecting inaccuracy—the ratio of the number of error-detected nodes to that of normal ones.

The performance indexes of each attack detecting algorithm obtained from simulation are listed in table 1. And the condition is that credence alarm threshold is -0.1; the variation of credence ΔR is 0.1667.

Table 1. Attack detection algorithm performance

Index / Attack	Accuracy	Inaccuracy	Reaction time (s)
Sequence number attack	81.2%	14.8%	12.7
Fabricating distance vector attack	73.6%	22.6%	12.5
Routing table overflow attack	74.8%	9.2%	13.2
Fabricating RERR attack	74.5%	27.6%	16.3
Interrupt routing attack	71.5%	13.8%	21.4

6 Conclusion

Some routing security issues have been researched, and the security hidden trouble of AODV protocol has been analyzed in this paper. Moreover, an attack detection function, which can detect several normal attacks, such as black hole attack, routing table overflow attack, and interruption routing attack, etc, is added to the AODV protocol, and a credence mechanism is also built up. Both of them are working interactively to provide rapid detection and reaction to malicious behaviors inside or outside the network and guard the network security

The simulation analysis resulting from the NS2 platform can prove that the performance of the modified AODV protocol is better than the normal one, which guarantees high-level routing security and availability of Ad-hoc networks.

Reference

1. C.-K. Toh. Ad hoc Mobile Wireless Networks: Protocols and Systems[M]. Prentice Hall PTR. 2002:55-77
2. Yang H. Security in Mobile Ad Hoc Networks: Challenges and Solutions[J]. IEEE Wireless Communications, 2004,11(1)38-47
3. Papadimitos P, Hass Z. Secure Routing for Mobile Ad Hoc Networks[A]. CNDS[C], 2002. http://wnl.ece.cornell.edu/Publications/cnds02.pdf
4. Perkins CE, Royer EM, Das S R. Ad-hoc On-Demand Distance Vector Routing(AODV)[EB/OL].http://www.ietf.org/internet-drafts/draft-ietf-manet-aodv-12.txt, Nov. 2002
5. Yih-Chun Hu, Adrian Perrig, David B. Johnson. Ariadne: A Secure On-Demand Routing Protocol for Ad hoc Networks[C], in Proceedings of the MobiCom 2002, September 23-28, 2002, Atlanta, Georgia, USA
6. A. Habib, M. H. Hafeeda, B. Bhargava, Detecting Service Violation and DoS Attacks [C], in Proceedings of Network and Distributed System Security Symposium (NDSS), 2003.
7. Yih-Chun Hu, David B. Johnson, and Adrian Perrig, SEAD: Secure Efficient Distance Vector Routing for Mobile Wireless Ad Hoc Networks [C], in Proceedings of the 4th IEEE Workshop on Mobile Computing Systems & Applications(WMCSA 2002), pp.3-13, IEEE, Calicoon, NY, June 2002

A Constant Time Optimal Routing Algorithm for Undirected Double-Loop Networks*

Bao-Xing Chen[1,2], Ji-Xiang Meng[1], and Wen-Jun Xiao[3]

[1] College of Mathematics & System Science, XinJiang University, P.R. China
mjx@xju.edu.cn

[2] Dept. of Computer Science, Zhangzhou Teacher's College, P.R. China
cbaoxing@hotmail.com

[3] Dept. of Computer Science, South China University of Technology, P.R. China
wjxiao@scut.edu.cn

Abstract. Undirected double-loop networks $G(n; \pm s_1, \pm s_2)$, where n is the number of its nodes, s_1 and s_2 are its steps, $1 \leq s_1 < s_2 < \frac{n}{2}$ and $\gcd(n, s_1, s_2)$=1, are important interconnection networks. In this paper, by using the four parameters of the L-shape tile and a solution $(\overline{x}, \overline{y})$ of a congruence equation $s_1 x + s_2 y \equiv 1 \pmod n$, we give a constant time optimal routing algorithm for $G(n; \pm s_1, \pm s_2)$.

Keywords: Distributed systems; undirected double-loop networks; optimal routing; shortest path; algorithm.

1 Introduction

Let n, s_1 and s_2 be positive integers such that $1 \leq s_1 < s_2 < n$. A directed double-loop network $G(n; s_1, s_2)$ is a directed graph (V, E), where $V = \mathbb{Z}_n = \{0, 1, 2, \cdots, n-1\}$, and E=$\{i \to i + s_1 \pmod n, i \to i + s_2 \pmod n \mid i = 0, 1, 2, \cdots, n-1\}$.

Let n, s_1 and s_2 be positive integers such that $1 \leq s_1 < s_2 < \frac{n}{2}$. An undirected double-loop network $G(n; \pm s_1, \pm s_2)$ (see Fig. 1) is an undirected graph (V, E), where V=$\mathbb{Z}_n$=$\{0, 1, 2, \cdots, n-1\}$, and E=$\{i \to i + s_1 \pmod n, i \to i - s_1 \pmod n, i \to i + s_2 \pmod n, i \to i - s_2 \pmod n \mid i = 0, 1, 2, \cdots, n-1\}$. It is known that $G(n; \pm s_1, \pm s_2)$ is strongly connected if and only if n, s_1 and s_2 are relatively prime, i.e., $\gcd(n, s_1, s_2)$=1. In the following we always suppose that $\gcd(n, s_1, s_2)$=1.

Double-loop networks have become popular in designs of computer networks and distributed memory multiprocessor systems. Many researchers are interested in the case of undirected double-loop networks [2, 3, 12, 15], while others are interested in the case of directed ones [1, 4-9, 11, 13, 14]. They mainly focus on routing [3, 4, 8, 12, 15], diameters [2, 13, 14] and optimal double-loop networks [1, 2, 5, 7, 11]. For more details we refer readers to [9, 10] and their references.

* This work was supported by the Natural Science Foundation of Fujian Province(No. Z0511035)and the Scientific Research Foundation of Fujian Provincial Education Department(No. JA04249).

X. Jia, J. Wu, and Y. He (Eds.): MSN 2005, LNCS 3794, pp. 308–316, 2005.

The primary function of a computer network is routing of messages between pairs of computers. It is desirable to route each message along a shortest path from its source to its destination. It is called optimal routing if each message is sent along a shortest path from its source to its destination. Some optimal routing algorithms for undirected double-loop networks have been presented, for example, the algorithms in [3, 12, 15].

The algorithm in [12] computes a shortest path from node 0 to any other nodes for undirected double-loop networks $G(n; \pm 1, \pm h)$, its time complexity is $O(\sqrt{n})$. Chalamaiah and Ramamurty [3] give an algorithm to obtain the shortest path in $O(h/g + \log(h))$ time for $G(n; \pm 1, \pm h)$, where $g = \gcd(n, h)$.

In [4], by using the four parameters of the L-shape tile and a solution $(\overline{x}, \overline{y})$ of a congruence equation $s_1 x + s_2 y \equiv 1 \pmod{n}$, we give a constant time optimal routing algorithm for directed double-loop networks $G(n; s_1, s_2)$. We can use the algorithm in [4] to compute a shortest path between any two nodes i, j for undirected double-loop networks $G(n; \pm s_1, \pm s_2)$. For example, if we want to find a shortest path from 0 to i in $G(n; \pm s_1, \pm s_2)$, what we need to do is to find a shortest path P_1 from 0 to i in $G(n; s_1, s_2)$, a shortest path P_2 from 0 to i in $G(n; n-s_1, n-s_2)$, a shortest path P_3 from 0 to i in $G(n; n-s_1, s_2)$, and a shortest path P_4 from 0 to i in $G(n; s_1, n - s_2)$ respectively. The shortest path among P_1, P_2, P_3 and P_4 is the one we wanted. But it is indirect and its computation time is about four times as much as that of the one we will propose. In this paper, by using the four parameters of the L-shape tile determined by $G(n; \pm s_1, \pm s_2)$ and a solution $(\overline{x}, \overline{y})$ of the congruence equation $s_1 x + s_2 y \equiv 1 \pmod{n}$, we will propose a constant time optimal routing algorithm for $G(n; \pm s_1, \pm s_2)$.

2 Preliminary Observations

Let $\mathbb{Z}$ be the set of integers and $\mathbb{Z}^+$ the set of nonnegative integers. For $G(n; \pm s_1, \pm s_2)$, an edge from i to $(i + s_1) \pmod{n}$ (or $(i - s_1) \pmod{n}$) is called $[+s_1]$ edge (or $[-s_1]$ edge) and an edge from i to $(i + s_2) \pmod{n}$ (or $(i - s_2) \pmod{n}$) is called $[+s_2]$ edge(or $[-s_2]$ edge). Consider a path from i to j involving w, x, y, and z (all non-negative integers) number of $[+s_1]$, $[-s_1]$, $[+s_2]$, $[-s_2]$ edges. Then $j = (i + w*s_1 - x*s_1 + y*s_2 - z*s_2) \pmod{n}$ holds irrespective of orders the edges appear in the path. Since we are interested only in the length of the path, we shall denote such a path by $w[+s_1] + x[-s_1] + y[+s_2] + z[-s_2]$. If a path $w[+s_1] + x[-s_1] + y[+s_2] + z[-s_2]$ from i to j is the shortest one, then at most one of w and x is nonzero, and at most one of y and z is nonzero.

For convenience, sometimes we represent a path $x_1[\pm s_1] + x_2[\pm s_2]$ $(x_1, x_2 \in \mathbb{Z}^+)$ as $(\pm x_1)[+s_1] + (\pm x_2)[+s_2]$. For example, a path $6[+2] + 3[-7]$ can be represented as $6[+2] + (-3)[+7]$.

Since $G(n; \pm s_1, \pm s_2)$ is vertex symmetric, to compute a path from node i to node j, we only need to compute a path from node 0 to node $j - i \pmod{n}$. For simplicity in the following we always suppose that node 0 is the source node.

We obtain the following construction of a pattern for $G(n; \pm s_1, \pm s_2)$ as Wong and Coppersmith did in [13].

In the Euclidean plane, all lattice points (x, y) (i.e., x, y being integers) of the first quadrant are arranged in the following order: $(0,0), (1,0), (0,1), (2,0), (1,1), (0,2), \cdots, (j,0), (j-1,1), (j-2,2), \cdots, (j-i,i), \cdots, (1, j-1), (0,j), \cdots$. We start from $(0,0)$, then $(1,0), (0,1)$, and then $(2,0), (1,1), (0,2), \cdots$. At each lattice point (x, y) if the value k which is determined by the equation $xs_1 + ys_2 \equiv k \pmod{n}$ has not appeared so far, we write it down, otherwise we just leave a blank. We stop when all values of k, i.e., k=0, 1, 2, $\cdots$, $n-1$, have been accounted for.

From [9, 13] we know that the above construction of the pattern is an L-shape tile and an L-shape tile always tessellates the plane.

Definition 1. *The L-shape area determined by $G(n; \pm s_1, \pm s_2)$ is called an L-shape tile (see Fig. 2) with parameters a, b, p, q, where a, b, p, q are all integers such that $a, b \geq 2, 0 \leq p < a, 0 \leq q < b$. This L-shape tile is written as $L(n; a, b, p, q)$, where $n = ab - pq$.*

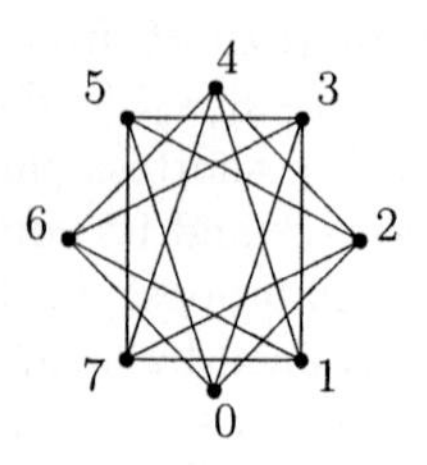

Fig. 1. $G(8; \pm 2, \pm 3)$

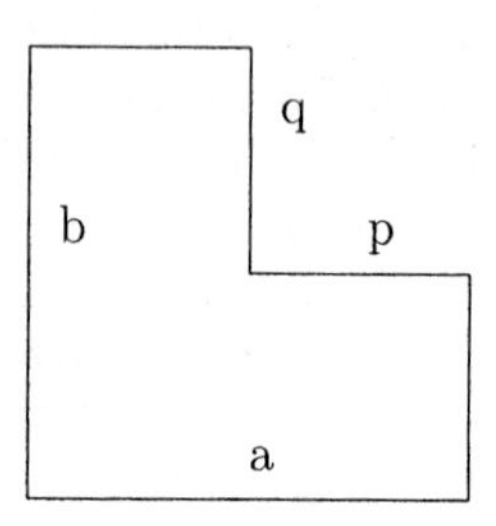

Fig. 2. L(n; a, b, p, q)

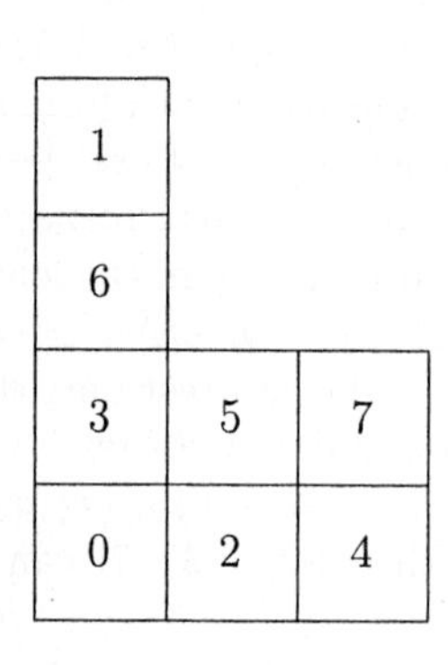

Fig. 3. $L(8; 3, 4, 2, 2)$ determined by $G(8; \pm 2, \pm 3)$

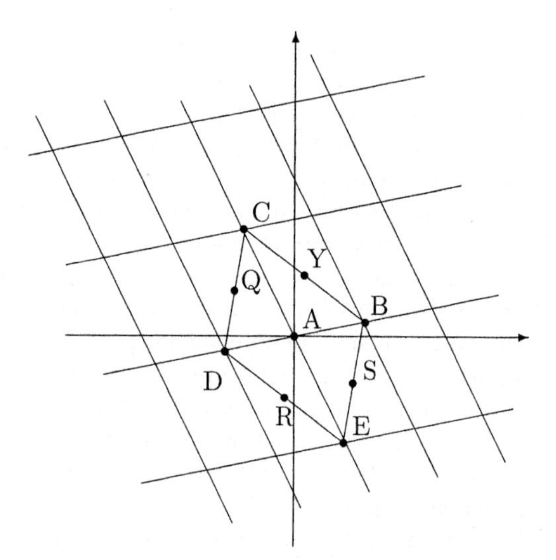

Fig. 4. A, B, C, D, E are zero lattice points

The L-shape tile determined by $G(8; \pm 2, \pm 3)$ is $L(8; 3, 4, 2, 2)$ (see Fig. 3).

Lemma 1 ([6]). *If the L-shape tile $L(n;a,b,p,q)$ is determined by $G(n;s_1,s_2)$, then $a>p\geq 0, b>q\geq 0, b\geq p, a>q$ and $ab-pq=n$.*

Lemma 2 ([7]). *If the L-shape tile $L(n;a,b,p,q)$ is determined by $G(n;s_1,s_2)$, then $-as_1+qs_2\equiv 0 \pmod n$, $-ps_1+bs_2\equiv 0 \pmod n$ and $(a-p)s_1+(b-q)s_2\equiv 0 \pmod n$.*

The above two lemmas hold for the case of undirected double-loop network $G(n;\pm s_1,\pm s_2)$.

Definition 2. *Two lattice points (x_1,y_1) and (x_2,y_2) are called equivalent if $x_1s_1+y_1s_2\equiv x_2s_1+y_2s_2 \pmod n$. A lattice point (x_1,y_1) is called a zero lattice point if $x_1s_1+y_1s_2\equiv 0 \pmod n$.*

For example, lattice points $(4,0)$ and $(1,2)$ are equivalent, and $(2,2)$ is a zero lattice point in $G(10;\pm 2,\pm 3)$.

Definition 3. *The distance $dist((x_1,y_1),(x_2,y_2))$ between any two lattice points (x_1,y_1) and (x_2,y_2) is defined to be $|x_1-x_2|+|y_1-y_2|$.*

Let $\lfloor x\rfloor$ denote the largest integer not larger than x. Let round(x) be $\lfloor x+0.5\rfloor$ and round$((x,y))$ denote (round(x), round(y)).

Definition 4. *The distance $d(u_1,u_2)$ between any two nodes u_1 and u_2 in $G(n;\pm s_1,\pm s_2)$ is defined to be the length of a shortest path from the node u_1 to the node u_2.*

In order to obtain a new routing algorithm, we give the following two lemmas first.

Lemma 3. *Given $G(n;\pm s_1,\pm s_2)$. Let u,v,l,h be four nonnegative integers such that $u\geq v>0, u>l, h\geq l, h\geq v$. Let lattice points A,B,C,D,E be $(0,0),(u,v)$, $(-l,h)$, $(-u,-v)$, $(l,-h)$ respectively (see Fig. 4). Suppose that B,C are zero lattice points. Let $P=x_1(u,v)+x_2(-l,h)$, where x_1,x_2 are two real numbers, $|x_1|\leq 0.5, |x_2|\leq 0.5$, and $x_1u-x_2l\in \mathbb{Z}$, $x_1v+x_2h\in\mathbb{Z}$. If $uh+vl=n$, then $d((x_1u-x_2l)s_1+(x_1v+x_2h)s_2,0)=\min\{dist(P,A),dist(P,B),dist(P,C), dist(P,D),dist(P,E)\}$.*

Proof. Since B,C are zero lattice points, we know that A,D,E are also zero lattice points. Let Y,Q,R,S be the middle points of edges BC,CD,DE,EB respectively. Let d_0,f,g denote $\min\{dist(P,A),dist(P,B),dist(P,C),dist(P,D),dist(P,E)\}$, x_1u-x_2l and x_1v+x_2h respectively. We know that $P=(f,g)$, where $f,g\in\mathbb{Z}$, and P is in the parallelogram $YQRS$ (see Fig. 4). As $d(fs_1+gs_2,0)\leq dist(P,X)$, where $X=A,B,C,D,E$, we have $d(fs_1+gs_2,0)\leq d_0$.

Suppose that a shortest path from fs_1+gs_2 to 0 is $t_1[+s_1]+t_2[+s_2]$, where $t_1,t_2\in\mathbb{Z}$. We know that (t_1+f,t_2+g) is a zero lattice point. Let (i,j) denote (t_1+f,t_2+g). As (i,j), $(-l,h)$, (u,v) are zero lattice points, we have $is_1+js_2\equiv 0 \pmod n$, $-ls_1+hs_2\equiv 0 \pmod n$ and $us_1+vs_2\equiv 0 \pmod n$. Thus $hi+lj\equiv 0 \pmod n$, $-vi+uj\equiv 0 \pmod n$. So there exist two integers k_1,k_2, such that $hi+lj=k_1n$ and $-vi+uj=k_2n$.

As $uh + vl{=}n$, we have $(i, j){=}\ k_1(u, v) + k_2(-l, h)$. That is, $(t_1 + f, t_2 + g){=}k_1(u, v) + k_2(-l, h)$. So $t_1{=}k_1u - k_2l - f$, $t_2{=}k_1v + k_2h - g$.

There are four cases should be considered: (1) $x_1 \geq 0, x_2 \geq 0$, (2) $x_1 < 0, x_2 \geq 0$, (3) $x_1 < 0, x_2 < 0$, (4) $x_1 \geq 0, x_2 < 0$. In the following, we only prove the cases (1) and (2). As the cases (3) and (4) can be similarly proved, their proofs are omitted.

Case (1): When $x_1 \geq 0, x_2 \geq 0$, one can easily get that $dist(P, A){=}\ |f| + g$, $dist(P, B){=}u - f + |v - g|$, $dist(P, C){=}l + f + h - g$, $dist(P, D){=}u + f + v + g$ and $dist(P, E){=}|l - f| + h + g$.

In the following five subcases of k_1, k_2 should be considered: (a) $k_1 \geq 1, k_2 \geq 1$; (b) $k_1 < 0$, $k_2 \leq 0$; (c) $k_1 \leq 0$, $k_2 > 0$; (d) $k_1 > 0$, $k_2 \leq 0$; (e) $k_1 = 0$, $k_2 \leq 0$.

Subcase (a): when $k_1 \geq 1$, $k_2 \geq 1$, we have $|t_1| + |t_2| \geq t_1 + t_2 = k_1(u + v) + k_2(h - l) - f - g \geq u + v + h - l - f - g = (1 - x_1)(u + v) + (1 - x_2)(h - l)$. If $f \geq 0$, then $|t_1| + |t_2| \geq (1 - x_1)(u + v) + (1 - x_2)(h - l) \geq 0.5(u + v) + 0.5(h - l) \geq x_1u - x_2l + x_1v + x_2h = f + g = dist(P, A)$. If $f < 0$, then $|t_1| + |t_2| \geq u{+}v{+}h{-}l{-}f{-}g = -f{+}u{-}l{+}v{+}h{-}g \geq -f{+}u{-}l{+}g \geq -f{+}g = dist(P, A)$.

Subcase (b): when $k_1 < 0$, $k_2 \leq 0$, we have $|t_1| + |t_2| \geq -t_1 - t_2 = -k_1(u + v) - k_2(h - l) + f + g \geq u + v + f + g = dist(P, D)$.

Subcase (c): when $k_1 \leq 0$, $k_2 > 0$, we have $|t_1| + |t_2| \geq -t_1 + t_2 = -k_1(u - v) + k_2(l + h) + f - g \geq l + h + f - g{=}dist(P, C)$.

Subcase (d): when $k_1 > 0$, $k_2 \leq 0$ and $v \leq g$, we have $|t_1| + |t_2| \geq t_1 - t_2 = k_1(u - v) - k_2(l + h) - f + g \geq u - v - f + g{=}dist(P, B)$.

When $k_1 > 0$, $k_2 = 0$ and $v > g$, we have $|t_1| + |t_2| \geq t_1 + t_2 = k_1(u + v) - f - g \geq u + v - f - g{=}dist(P, B)$.

When $k_1 > 0$, $k_2 < 0$ and $v > g$, we have $|t_1| + |t_2| \geq t_1 - t_2 = k_1(u - v) - k_2(l + h) - f + g \geq u - v + l + h - f + g = (1 - x_1)u + (1 + x_2)l + g + h - v > x_1u + x_2l + g \geq |f| + g{=}dist(P, A)$.

Subcase (e): when $k_1 = 0$, $k_2 = 0$, we have $|t_1| + |t_2| = |f| + |g|{=}\ dist(P, A)$.

When $k_1 = 0$, $k_2 < 0$ and $f \geq l$, we have $|t_1| + |t_2| \geq -t_1 - t_2 = f + k_2l + g - k_2h = f + g - k_2(h - l) \geq f + g + h - l = dist(P, E)$.

When $k_1 = 0$, $k_2 < 0$ and $l > f$, we have $|t_1| + |t_2| \geq t_1 - t_2 = -f - k_2l + g - k_2h = -f + g - k_2(h + l) \geq -f + g + h + l = dist(P, E)$.

Case (2): When $x_1 < 0, x_2 \geq 0$, we have $dist(P, A){=}\ -f + |g|$, $dist(P, B){=}u - f + |v - g|$, $dist(P, C){=}|l + f| + h - g$, $dist(P, D){=}u + f + v + g$, and $dist(P, E){=}l - f + h + g$.

In the following five subcases of k_1, k_2 should be considered: (a) $k_1 \geq 1$, $k_2 \geq 1$; (b) $k_1 < 0$, $k_2 \leq 0$; (c) $k_1 \leq 0$, $k_2 > 0$; (d) $k_1 > 0$, $k_2 \leq 0$; (e) $k_1 = 0$, $k_2 \leq 0$.

Subcase (a): when $k_1 \geq 1$, $k_2 \geq 1$, we have $|t_1| + |t_2| \geq t_1 + t_2 = k_1(u + v) + k_2(h - l) - f - g \geq u + v + h - l - f - g = (1 - x_1)v + (1 - x_2)h - f + u - l \geq |g| - f = dist(P, A)$.

Subcase (b): when $k_1 < 0$, $k_2 \leq 0$, we have $|t_1| + |t_2| \geq -t_1 - t_2 = -k_1(u + v) - k_2(h - l) + f + g \geq u + v + f + g = dist(P, D)$.

Subcase (c): when $k_1 \leq 0$, $k_2 > 0$ and $l \geq -f$, we have $|t_1|+|t_2| \geq -t_1+t_2 = -k_1(u-v)+k_2(l+h)+f-g \geq l+h+f-g = dist(P,C)$.

When $k_1 = 0$, $k_2 > 0$ and $l < -f$, we have $|t_1|+|t_2| \geq t_1+t_2 = -k_2l - f + k_2h - g \geq -l - f + h - g = dist(P,C)$.

When $k_1 < 0$, $k_2 > 0$ and $l < -f$, we have $|t_1|+|t_2| \geq -t_1+t_2 = -k_1(u-v)+k_2(l+h)+f-g \geq u-v+l+h+f-g$. If $g \leq 0$, then $|t_1|+|t_2| \geq u-v+l+h+f-g = (u+l+f)+(h-v)-g \geq -f-g = dist(P,A)$. If $g > 0$, then $|t_1|+|t_2| \geq t_1-t_2 = u-v+l+h+f-g = u-v+l+h+x_1(u-v)-x_2(l+h) = (1+x_1)(u-v)+(1-x_2)(l+h) \geq -x_1(u-v)+x_2(l+h) = -f+g = dist(P,A)$.

Subcase (d): when $k_1 > 0$, $k_2 = 0$ and $g \geq 0$, we have $|t_1|+|t_2| \geq t_1-t_2 = k_1(u-v)-f+g \geq u-v-f+g \geq dist(P,A)$.

When $k_1 > 0$, $k_2 = 0$ and $g < 0$, we have $|t_1|+|t_2| \geq t_1+t_2 = k_1(u+v)-f-g \geq u+v-f-g > dist(P,A)$.

When $k_1 > 0$, $k_2 < 0$, we have $|t_1|+|t_2| \geq t_1-t_2 = k_1(u-v)-k_2(l+h)-f+g \geq u-v+l+h-f+g \geq dist(P,E)$.

Subcase (e): when $k_1 = 0$, $k_2 = 0$, we have $|t_1|+|t_2| = |f|+|g| = dist(P,A)$.

When $k_1 = 0$, $k_2 < 0$, we have $|t_1|+|t_2| \geq t_1-t_2 = -f-k_2l+g-k_2h = -f+g-k_2(h+l) \geq -f+g+h+l = dist(P,E)$.

From above we have $|t_1|+|t_2| \geq d_0$. So $d(fs_1+gs_2,0)=d_0$. □

As the following Lemma 4 can be similarly proved just like Lemma 3, its proof is omitted.

Lemma 4. *Given $G(n;\pm s_1,\pm s_2)$. Let u,v,l,h be four nonnegative integers such that $0 < u < v, l \geq u, l > h, v > h$. Let lattice points A,B,C,D,E be $(0,0)$, (u,v), $(-l,h)$, $(-u,-v)$, $(l,-h)$ respectively and B,C are zero lattice points. Let $P = x_1(u,v)+x_2(-l,h)$, where x_1,x_2 are two real numbers and $|x_1| \leq 0.5, |x_2| \leq 0.5$ and $x_1u - x_2l \in \mathbb{Z}$, $x_1v+x_2h \in \mathbb{Z}$. If $uh+vl=n$, then $d((x_1u-x_2l)s_1+(x_1v+x_2h)s_2,0)=\min\{dist(P,A), dist(P,B), dist(P,C), dist(P,D), dist(P,E)\}$.*

3 Optimal Routing Algorithm

Given an undirected double-loop network $G(n;\pm s_1,\pm s_2)$, as the L-shape tile determined by $G(n;\pm s_1,\pm s_2)$ is the same as that by $G(n;s_1,s_2)$, we can compute four parameters a,b,p,q of the L-shape tile determined by $G(n;\pm s_1,\pm s_2)$ by using an $O(\log\ n)$ algorithm given in [9, 13]. Since $\gcd(n,s_1,s_2)=1$, by using the Euclid's *GCD* algorithm whose time complexity is $O(\log\ n)$, we can get a solution $(\overline{x},\overline{y})$ of an equation $s_1x+s_2y \equiv 1 \pmod{n}$. By using four parameters a,b,p,q of the L-shape tile determined by $G(n;\pm s_1,\pm s_2)$ and a solution $(\overline{x},\overline{y})$ of the equation $s_1x+s_2y \equiv 1 \pmod{n}$ computed in advance, we can give a constant time optimal routing algorithm for $G(n;\pm s_1,\pm s_2)$ in the following.

Let $u=a-p$, $v=b-q$, where a,b,p,q are four parameters of the L-shape tile determined by $G(n;\pm s_1,\pm s_2)$. From Lemma 1 and Lemma 2, we know that $u > 0, v > 0$ and $us_1+vs_2 \equiv 0 \pmod{n}$. In the following we consider two cases: (1) $u \geq v$, (2) $u < v$ separately.

3.1 When $u \geq v$

Routing Algorithm 1. Given four parameters a, b, p, q of the L-shape tile determined by $G(n; \pm s_1, \pm s_2)$ and a solution $(\overline{x}, \overline{y})$ of the congruence equation $s_1 x + s_2 y \equiv 1 \pmod{n}$. Let u=$a-p$, v=$b-q$. Compute a shortest path from node 0 to node s when $u \geq v$.

Step 1: $t := \lfloor \frac{p}{u} \rfloor; a_0 := p - tu; b_0 := b + tv;$
Step 2: $(a_1, b_1) :=(s\overline{x}, s\overline{y})$- round $(\frac{1}{n}\ (s\overline{x}, s\overline{y}) \begin{pmatrix} b_0 & -v \\ a_0 & u \end{pmatrix}) \begin{pmatrix} u & v \\ -a_0 & b_0 \end{pmatrix}$;
Step 3: $P_1 = a_1[+s_1] + b_1[+s_2]$; $P_2 = (a_1 - u)[+s_1] + (b_1 - v)[+s_2]$;
$P_3 = (a_1 + a_0)[+s_1] + (b_1 - b_0)[+s_2]$; $P_4 = (a_1 + u)[+s_1] + (b_1 + v)[+s_2]$;
$P_5 = (a_1 - a_0)[+s_1] + (b_1 + b_0)[+s_2]$;

The shortest path P' among P_1, P_2, P_3, P_4, P_5 is a shortest one from node 0 to node s.

Theorem 1. *When $u \geq v$, the path P' got from Routing Algorithm 1 is a shortest one from node 0 to node s.*

Proof. As $a_0 = p - tu, b_0 = b + tv$, where $t = \lfloor \frac{p}{u} \rfloor$, we have $0 \leq a_0 < u, 0 < v \leq b_0$. Since $b \geq p$, we have $b_0 \geq a_0$. It can be easily proved that $(-a_0, b_0)$ is a zero lattice point. That is, $-a_0 s_1 + b_0 s_2 \equiv 0 \pmod{n}$. As $(-a_0, b_0)$ and (u, v) are zero lattice points, we know that (a_1, b_1) and $(s\overline{x}, s\overline{y})$ are equivalent.

Let lattice points A, B, C, D, E be $(0,0), (u,v), (-a_0, b_0), (-u,-v), (a_0, -b_0)$ respectively. Let M=$\begin{pmatrix} u & v \\ -a_0 & b_0 \end{pmatrix}$. As $ub_0 + va_0 = u(b+tv) + v(p-tu) = ub + vp = (a-p)b + (b-q)p = ab - pq = n$, where $t = \lfloor \frac{p}{u} \rfloor$, we have M^{-1}=$\frac{1}{n}\begin{pmatrix} b_0 & -v \\ a_0 & u \end{pmatrix}$. From Step 2 of Routing Algorithm 1, we have (a_1, b_1)=$(s\overline{x}, s\overline{y})$- round $((s\overline{x}, s\overline{y})M^{-1})M$ $=t_1(u,v) + t_2(-a_0, b_0)$, where $|t_1| \leq 0.5, |t_2| \leq 0.5$. Let $X = (a_1, b_1)$. From Lemma 3 we have $d(0,s)$=$d(s,0) = d(a_1 s_1 + b_1 s_2, 0)$=$\min\{dist(X,A), dist(X,B), dist(X,C), dist(X,D), dist(X,E)\}$. As lengths of paths P_1, P_2, P_3, P_4 and P_5 are $dist(X,A)$, $dist(X,B)$, $dist(X,C), dist(X,D), dist(X,E)$ respectively, we know that the shortest path P' among P_1, P_2, P_3, P_4, P_5 is a shortest one from node 0 to node s. □

Example 1. Compute a shortest path from node 0 to node 291 in an undirected double-loop network $G(500; \pm 1, \pm 82)$.

By using the algorithm given in [9, 13], we can get four parameters a, b, p, q of the L-shape tile determined by $G(500; \pm 1, \pm 82)$. We have a=42, b=37, p=34, q=31. Let u=$a-p$=8, v=$b-q$=6. As $u > v$, we use Routing Algorithm 1 to compute a shortest path from 0 to 291. We know that $(1,0)$ is a solution of the congruence equation $x + 82y \equiv 1 \pmod{500}$.

In Step 1, t=$\lfloor \frac{34}{8} \rfloor$=4; a_0=$34 - 4*8$=2; b_0=$37 + 4*6$=61. In Step 2, $(a_1, b_1) = (291, 0)$ -round $(\frac{1}{500}\ (291, 0) \begin{pmatrix} 61 & -6 \\ 2 & 8 \end{pmatrix}) \begin{pmatrix} 8 & 6 \\ -2 & 61 \end{pmatrix}$=$(-3, -33)$.

In step 3, $P_1 = (-3)[+1] + (-33)[+82]$, $P_2 = (-11)[+1] + (-39)[+82]$, $P_3 = (-1)[+1] + (-94)[+82]$, $P_4 = (+5)[+1] + (-27)[+82]$, $P_5 = (-5)[+1] + (+28)[+82]$. As the shortest path among P_1, P_2, P_3, P_4, P_5 is P_4, we know that P_4 is a shortest path from node 0 to node 291.

3.2 When $u < v$

Routing Algorithm 2. Given four parameters a, b, p, q of the L-shape tile determined by $G(n; \pm s_1, \pm s_2)$ and a solution $(\overline{x}, \overline{y})$ of the congruence equation $s_1 x + s_2 y \equiv 1 \pmod{n}$. Let u=$a - p$, v=$b - q$. Compute a shortest path from node 0 to node s when $u < v$.

Step 1: $t := \lfloor \frac{q}{v} \rfloor; a_0 := a + tu; b_0 := q - tv;$
Step 2: $(a_1, b_1) :=(s\overline{x}, s\overline{y})$- round$(\frac{1}{n}\ (s\overline{x}, s\overline{y}) \begin{pmatrix} b_0 & -v \\ a_0 & u \end{pmatrix}) \begin{pmatrix} u & v \\ -a_0 & b_0 \end{pmatrix}$;
Step 3: $P_1 = a_1[+s_1] + b_1[+s_2]$; $P_2 = (a_1 - u)[+s_1] + (b_1 - v)[+s_2]$;
$P_3 = (a_1 + a_0)[+s_1] + (b_1 - b_0)[+s_2]$; $P_4 = (a_1 + u)[+s_1] + (b_1 + v)[+s_2]$;
$P_5 = (a_1 - a_0)[+s_1] + (b_1 + b_0)[+s_2]$;

The shortest path P' among P_1, P_2, P_3, P_4, P_5 is a shortest one from node 0 to node s.

Theorem 2. *When $u < v$, the path P' got from Routing Algorithm 2 is a shortest path from node 0 to node s.*

Since Theorem 2 can be similarly proved just like Theorem 1 by using Lemma 4, its proof is omitted.

Example 2. Compute a shortest path from node 0 to node 227 in an undirected double-loop network $G(450; \pm 2, \pm 7)$.

By using the algorithm given in [9, 13], we can get four parameters a, b, p, q of the L-shape tile determined by $G(450; \pm 2, \pm 7)$. We have a=7, b=66, p=6, q=2. Let u=$a - p$=1, v=$b - q$=64. As $u < v$, we use Routing Algorithm 2 to compute a shortest path from 0 to 227. We know that $(-3, 1)$ is a solution of the congruence equation $2x + 7y \equiv 1 \pmod{450}$. In Step 1, t=$\lfloor \frac{2}{64} \rfloor$=0; a_0=$7 + 0 * 1$=7; b_0=$2 - 0 * 64$=2. In Step 2, (a_1, b_1)=$(227 * (-3), 227)$ -round$(\frac{1}{450}\ (-681, 227) \begin{pmatrix} 2 & -64 \\ 7 & 1 \end{pmatrix}) \begin{pmatrix} 1 & 64 \\ -7 & 2 \end{pmatrix}$=$(-3, -31)$.

In step 3, $P_1 = (-3)[+2] + (-31)[+7]$, $P_2 = (-4)[+2] + (-95)[+7]$, $P_3 = 4[+2] + (-33)[+7]$, $P_4 = (-2)[+2] + 33[+7]$, $P_5 = (-10)[+2] + (-29)[+7]$. As the shortest path among P_1, P_2, P_3, P_4, P_5 is P_1, we know that $P_1 = (-3)[+2] + (-31)[+7]$ is a shortest path from node 0 to node 227.

4 Conclusion

Many optimal routing algorithms for undirected double-loop networks have been presented, but their time complexities are not constant. By using the four parameters a, b, p, q of the L-shape tile and a solution $(\overline{x}, \overline{y})$ of a congruence equation $s_1 x + s_2 y \equiv 1 \pmod{n}$, we propose a constant time optimal routing algorithm for undirected double-loop networks $G(n; \pm s_1, \pm s_2)$ in this paper. To the best of our knowledge, fast optimal routing algorithms for triple-loop networks $G(n; \pm s_1, \pm s_2, \pm s_3)$ or $G(n; s_1, s_2, s_3)$ are not available now. They are deserving further research.

Acknowledgement. The authors are grateful to referees for many helpful comments and suggestions.

References

1. Aguilo, F., Fiol, M. A.: An efficient algorithm to find optimal double loop networks. Discrete Mathematics **138** (1995) 15-29
2. Boesch, F. T., Wang, J. F.: Reliable circulant networks with minimum transmission delay. IEEE Trans. Circuits Syst. **CAS-32** (1985) 1286-1291
3. Chalamaiah, N., Ramamurty, B.: Finding shortest paths in distributed loop networks. Information Processing Letters **67** (1998) 157-161
4. Chen, B. X., Xiao, W. J.: A constant time optimal routing algorithm for directed double loop networks $G(n; s_1, s_2)$. In the proceeding of 5th International Conference on Software Engineering, Artificial Intelligence, Networking, and Parallel/Distributed Computing(SNPD 2004) 1-5
5. Chen, B. X., Xiao, W. J.: Optimal designs of directed double-loop networks. International Symposium on Computational and Information Sciences (CIS'04), Lecture Notes in Computer Science (LNCS), Springer Verlag, 2004, 19-24
6. Chen, C., Hwang, F. K.: The minimum distance diagram of double-loop networks. IEEE Trans. Comput. **49**(2000)9:977-979
7. Fiol, M. A., Yebra, J. L. A., Alegre, I., Valero, M.: A discrete optimization problem in local networks and data alignment. IEEE Trans. Comput. **36** (1987)6:702-713
8. Guan, D. J.: An optimal message routing algorithm for double-loop networks. Information Processing Letters **65**(1998) 255-260
9. Hwang, F. K.: A complementary survey on double-loop networks. Thereotical Computer Science **263** (2001) 211-229
10. Hwang, F. K.: A survey on multi-loop networks. Theoretical Computer Science **299** (2003) 107-121
11. Li, Q., Xu, J. M., Zhang, Z. L.: Infinite families of optimal double loop networks. Science in China, Ser A **23** (1993) 979-992
12. Mukhopadhyaya, K., Sinha, B. P.: Fault-tolerant routing in distributed loop networks. IEEE Trans. Comput. **44** (1995)12:1452-1456
13. Wong, C. K., Coppersmith, D.: A combinatorial problem related to multimodule memory organizations. J. ACM **21** (1974) 392-402
14. Ying, C., Hwang, F. K.: Diameters of weighted double loop networks. Journal of Algorithm **9** (1988) 401-410
15. Ying, C., Hwang, F. K.: Routing algorithms for double loop networks. International Journal of Foundations of Computer Science **3**(1992) 3:323-331

A Directional Antenna Based Path Optimization Scheme for Wireless Ad Hoc Networks*

Sung-Ho Kim and Young-Bae Ko

College of Information and Communication,
Ajou University, South Korea
{likeleon, youngko}@ajou.ac.kr

Abstract. Using directional antennas in wireless ad hoc networks is an attractive issue due to its potential advantages such as spatial reuse, low power consumption and low chance of interference. To acquire these advantages, we propose a new routing scheme called DAPOS (Directional Antenna based Path Optimization Scheme). The proposed scheme enables us to acquire an efficient path by considering the characteristics of directional antennas, resulting a routing performance improvement. DAPOS focuses on shortening the length of routing path gradually by considering the higher gain of directional antennas at a receiver side. Simulation results show that DAPOS significantly reduces the number of hops in route, thereby it improves the overall network performance.

1 Introduction

A wireless ad hoc network is an infrastructure-less network consisting of mobile nodes that are typically equipped with an omnidirectional antenna for communicating with each other through wireless links. An omnidirectional antenna has an uniform radiation pattern from the view of top so it emits the signal over a whole sphere and receives the signal from all directions. On the contrary, a directional antenna has a preferential direction which has a more powerful communication capability. Due to this property, we can expect that using directional antennas instead of omnidirectional antennas gives us several benefits such as spatial reuse, low energy consumption and increased signal to interference and noise ratio (SINR).

In this paper, we propose a new routing protocol, DAPOS (Directional Antenna Based Path Optimization Scheme), by which each node tries to optimize the route by considering the high *reception gain* of directional antennas. Fig. 1 shows the characteristic of a directional antenna which our DAPOS protocol utilizes. In Fig. 1(a), node A equipping a directional antenna transmits

* This research was supported by the MIC (Ministry of Information and Communication), Korea, under the ITRC (Information Technology Research Center) program, and the Ubiquitous Computing and Network project (part of the MIC 21st Century Frontier R&D program). It was also partially supported by grant No. R05-2003-000-10607-02004 from the Korea Research Foundation Grant funded by the Korean Government.

X. Jia, J. Wu, and Y. He (Eds.): MSN 2005, LNCS 3794, pp. 317–326, 2005.

its signal intended to node B having an omnidirectional antenna, but B can't do a successful reception because the receiving signal is too weak. However, in Fig. 1(b), the reception process by node B is successfully completed even though the transmission profile of node A is same as Fig. 1(a). This is possible because the reception capability of B (now being equipped with a directional antenna) can be improved by setting its direction of main lobe to A. This implies that if a transmission condition (such as a transmission gain, a transmission power, a direction of transmission) doesn't become different, the range of possible communication between two nodes is dependent on a reception condition (such as a reception gain, an antenna type). Motivated by this, DAPOS shortens the length of route by including the link shown in Fig. 1(b) to the route.

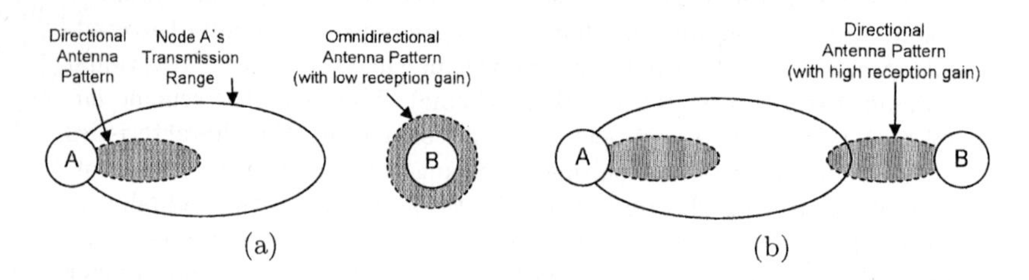

Fig. 1. An example showing the effect of the receiver's different antenna pattern on the success of communication

This paper is organized as follows. In Section 2, we discuss related work on routing schemes using directional antennas and explain our directional antenna model. The problems occured when past routing protocols designed for omnidirectional antennas is used with directional antennas are discussed in Section 3, and our new routing protocol DAPOS solving the problem is proposed in Section 4. We evaluate the performance of DAPOS with simulation in Section 5. Finally, Section 6 is for the conclusion.

2 Preliminaries

2.1 Related Works

To expolit the capabilities of directional antennas at the network(routing) layer, not a little researches have been proposed. In [1], the authors have investigated the impact of directional antennas over the dynamic source routing (DSR) [2] protocol and suggested some efficient routing schemes by considering the properties of directional antennas. The authors of [3] have proposed MAC and routing protocols suited to electronically steerable passive array radiator (ESPAR) antenna which allows an arbitrary radiating structure. In case of [4], they proposed routing scheme finding mutually disjoint routes which minimizes the effect of route coupling. It is observed that the proposed routing schemes above have a tendency to focus on a secondary point such as improving network throughput or network connectivity, rather than finding more efficient route. As mentioned earlier, this observation motivates our work here.

2.2 Antenna Model

We assume a switched beamforming antenna system as in [1, 7]. Fig. 2 shows the simplified pattern of directional antenna. Antenna system supports two mode, omnidirectional mode and directional mode, and a node must exist in only one mode at a time. A node stays in omnidirectional mode while in idle to receive the signal from all directions because it is impossible to know where the signals may come from. Also omnidirectional mode is used when a node wants to transmit the signal to all directions. Whereas, a directional mode is used when it can know the direction of coming signal and when it tranmsits the signal directionally. The angle of arrived signal can be acquired by the assumption of the ability to detect the beam-of-arrival of the signal. Each node receiving the signal caches the angle and uses it when the direction is needed to know. We assume that a node reduces the transmit power when transmitting in directional mode such that communication range becomes equal in omnidirectional mode and directional mode.

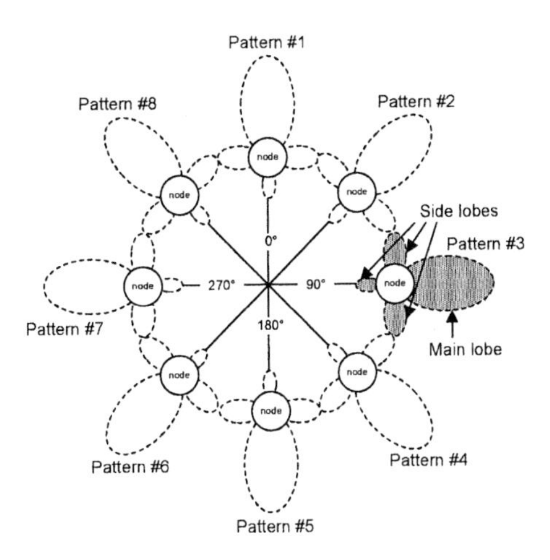

Fig. 2. The simplified pattern of directional antenna

3 Problems of Existing Ad Hoc Routing Protocols

As investigated in [1], most of omnidirectional antenna-assumed routing protocols for ad hoc networks such as AODV [5] and DSR also may work in the environment of directional antennas. However they experience the effect of using directional antenna and miss an opportunity of obtaining the advantages. They also provide a bad performance in acquiring shortest path, as seen later.

In case of typical reactive routing protocols, the flooding of route request (RREQ) packet is performed to elect the nodes that need to be included in route. Each node receives flooded RREQ in omnidirectional mode because nodes can't know when it comes and where it comes from. So the signal strength of received RREQ is proportional to a multiplication of the transmitter's gain in omnidirectional mode (G^o) and the receiver's gain in omnidirectional mode (G^o). The friss equation explains the relationship between transmission, reception gain (G_T, G_R) and transmittion power, reception power (P_T, P_R) [6].

$$P_R = \frac{P_T \times G_T \times G_R}{d^\alpha \times L} \tag{1}$$

As a result, two nodes composing a link of acquired route communicate with each other by $G^d \times G^o$ if a sender can be in directional mode. However, if sender and receiver are all in directional mode, communication can be accomplished by $G^d \times G^d$. In this case, communication range is expanded because reception strength is enlarged by Eq. 1. This implies that two nodes which are far away to communicate properly can communicate directly if they adjust their patterns properly as shown in Fig. 1.

Most of existing reactive routing protocols find the route comprised of links bounded by $G^d \times G^o$, although $G^d \times G^d$ links exist. Fig. 3 shows an example of this situation. We assume that each node equips a directional antenna and node S and D can communicate directly by beamforming to each other. That is to say, the link between S and D is a link bounded by only $G^d \times G^d$, not $G^d \times G^o$. At first, node S which wants to tranfer data to D initiates the flooding of RREQ to acquire the route. This RREQ is relayed by node A, B and C, and finally arrives in destination node D. As a result, acquired path is path 1, {S, A, B, C, D}, although shorter path {S,D} (path 2) can also be used.

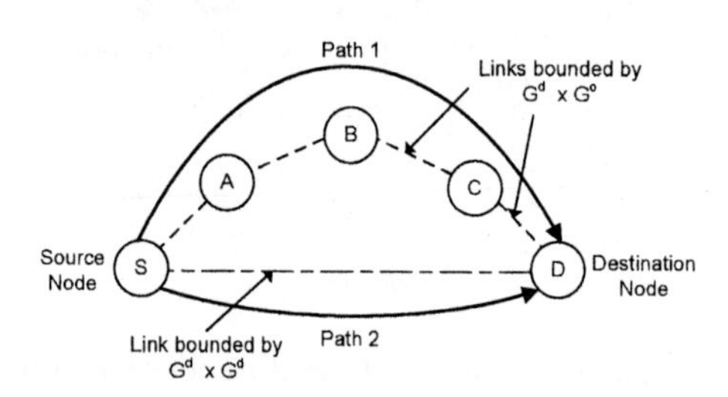

Fig. 3. Two possible routes between S and D

In this paper, we propose a new routing scheme to find more efficient route (i.e., path 2 in Fig. 3) by including $G^d \times G^d$ and $G^o \times G^d$ links into the route. For the simplicity of explnanation, we will refer the link requiring receiver node to be in directional mode ($G^d \times G^d$ and $G^o \times G^d$ links) as directional link and the link which can be used in omnidirectional mode ($G^d \times G^o$ and $G^o \times G^o$ links) as omnidirectional link.

4 DAPOS: Directional Antenna Based Path Optimization Scheme

4.1 Basic Mechanism of DAPOS

DAPOS performs following procedures to use directional links. While in the period of flooding RREQ, some node which received RREQ packet determines the link status between itself and the previous intermediate nodes relayed this RREQ before. If the link with a node which is more than 1 hop away is a directional link, it means that they can communicate directly without intermediate nodes. Then a node can exclude intermediate nodes from the path to shorten the length of path. In example of Fig. 3, node B received RREQ from A may know that the link between itself and S is a directional link and modifies the path {S,A,B} to {S,B}. This procedure is performed by each relay node until the RREQ packet is received by destination node D.

The question is how to determine the link status whether it is a directional link or not. We suggest utilizing wireless channel propagation model to estimate the communication range. Eq. 2 represents one of well-known propagation model, two-ray ground path loss model. By using this equation, we can estimate received power(P_R) if we know transmit power(P_T), the distance between two nodes(d), antenna gain(G_T, G_R) and antenna height(H_T, H_R).

$$P_R = \frac{P_T \times G_T \times G_R \times ({H_T}^2 \times {H_R}^2)}{d^4 \times L} \qquad (2)$$

To simply provide distance information d, we assume that each node is able to know its own location information by location detection device like GPS. Each node relaying RREQ packet puts its own location information and transmit power level into RREQ. A node receiving RREQ gets distance d from location in RREQ and its own location, then calculates the received power P_R by using directional antenna gain G^d as G_R. If P_R is greater than receving threshold RX_Thresh, the node determines the link as an directional link.

4.2 DSR Variation with the Proposed Scheme

Flooding of Route Request Packet

Some extra information must be offered to enable the determination, so we modified the packet structure of RREQ. Observe that the information of node deleted from path are not needed anymore after optimization procedure. Therefore, the number of information newly attached to RREQ is limited to two sets, each consisting of location information and transmit power information. One set is for the node X which is the target of directional link determination and the other set is for the next potential target node Y. The information of Y is used when optimization can't be performed anymore. The optimized path created by optimization procedure is also attached to RREQ. The followings are additional information must be attached to RREQ.

– Transmit power 1 ($TxPower_1$): Transmit power of node X
– Transmit power 2 ($TxPower_2$): Transmit power of node Y
– Location information 1 ($Location_1$): Location information of node X
– Location information 2 ($Location_2$): Location information of node Y
– Optimized Path($OptPath$): Path which is optimized by DAPOS

Fig. 4 shows the flooding of RREQ with DAPOS. Initially, source node S initiating RREQ stores its location information (L_S) and transmit power (T_S) in $TxPower_1$ and $Location_1$ fields of genereated RREQ packet. $TxPower_2$ and $Location_2$ are set to zero. $OptPath$ is same as DSR route request option header, {S}.

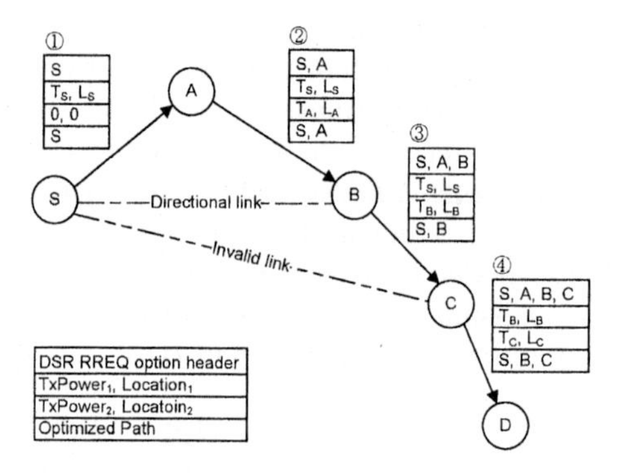

Fig. 4. Flooding of RREQ with DAPOS

Node A receiving RREQ from source node S doesn't need to optimize the path because it is a neighbor node of source node. A fills $TxPower_2$ and $Location_2$ with its own information T_S and L_S, and attaches its id to $OptPath$. Please note that $OptPath$ is also same as DSR route request option header, {S, A}. A relay the RREQ by broadcasting.

When node B receives RREQ from A, it performs the directional link detection algorithm. By algorithm, B knows that the link between S and itself B is a directional link so does optimization by deleting A from the path. With this optimization, $OptPath$ {S, A, B} becomdes {S, B}. Observe that the informations of deleted node A doesn't need anymore. $TxPower_2$ and $Location_2$ are set by the information of node B, T_C and L_C and these informations are needed in future when S cannot be an one end of directional link anymore. At that time, node B becomes a new target node for directional link detection.

Next, RREQ packet relayed by B is received by C and C knows that it can't communicate with S directly. This implies that the path {S, B} can't be optimized anymore. Therefore, $TxPower_1$ and $Location_1$ are replaced by the value of $TxPower_2$ and $Location_2$ which are the information of the next target node B. C puts its information T_C and L_C into $TxPower_2$ and $Location_2$, then rebroadcast the RREQ.

Finally, RREQ arrives at the detination node D. D determines that it can't construct a directional link with B and no optimization is performed here. D knows the passed sequence {S, A, B, C} by original DSR RREQ option header and also gets the optimized path {S, B, C} by *OptPath*. With DAPOS, destination node gets the opportunity of providing optimized path to source node.

Return of Route Reply Packet

For use of directional link, each node using directional link must know each other's direction. It is possible by caching the location informations in relayed RREQ and by providing the location information of previous hop in RREP packet. Therefore one location information field is needed. One another field for optimized path, *OptPath* is also attached to original DSR RREP packet. The followings are the additional field of RREP packet.

- Location information (*Location*): Location information of one end node using directional link
- Optimized Path (*OptPath*): Path which is optimized by DAPOS

In Fig. 5, we can know how RREP packet return to source node. While in back of RREP, found directional links can not be used because a node relayed RREQ doesn't know whether it woulde be included in path at the conclusion and also doesn't know the direction of corresponding node of directional link. Therefore, RREP is sent back by sequence {D, C, B, A, S}. One important thing is that node A doesn't update *Location* to give location information of B to S.

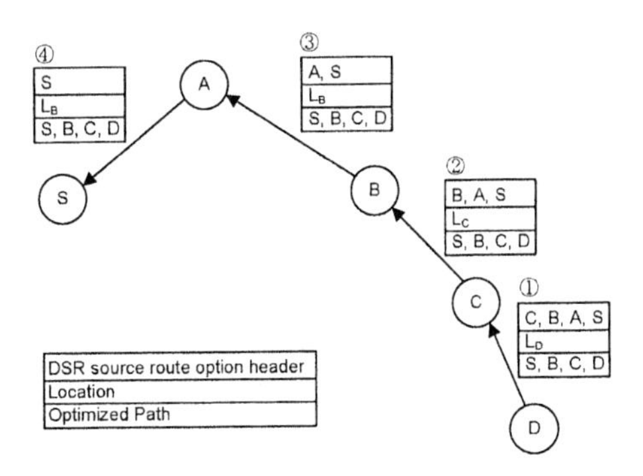

Fig. 5. Return of RREP with DAPOS

5 Performance Evaluation

We use the Qualnet simulator [8] version 3.8 for simulating our proposed protocol DAPOS and comparing it with DSR which is one of the most famous routing protocol. In all considered scenarios, constant bit rate (CBR) traffic is used at each source node. Transmission range is set to 250m in $G^d \times G^o$ and $G^o \times G^o$

condition. Note that power control is used to equal the transmission range in directional mode and omnidirectional mode. The gain in omnidirectional mode G^o is 0.0dBi (no gain) and the peak gain in directional mode G^d is 15.0dBi. Of course, the gain of side lobes are weaker than that of main lobe.

5.1 Grid Topology

We simulated grid topology illustrated in Fig. 6 to investigate the routing control overhead in order to evaluate the performance in more simplified scenario. The distance between adjacent node is 200m and three CBR traffics are placed in centered 3 rows. Source nodes are the leftmost three nodes (12, 23, 34) except top and bottom, destination nodes are selected by varying the distance between source and destination node from 200m to 2000m. In Fig. 6, black solid lines indicate the path acquired by DSR and dotted lines indicate DAPOS, in distance 1000m. Gradual optimization by each relaying node enables local optimization and this leads to acquire more efficient path compared to DSR as seen Fig. 6.

The effect of packet's increased size can be found in Fig. 7(a), route discovery latency (the time for route discovery) of DAPOS is slightly higher than that of DSR. However, route discovery is successfully complished without any packet drops and average end-to-end delay of DAPOS is much lower than DSR, shown in Fig. 7(b).

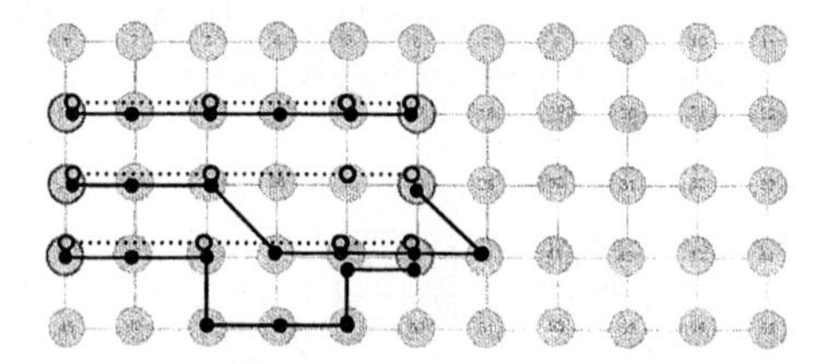

Fig. 6. Grid topology

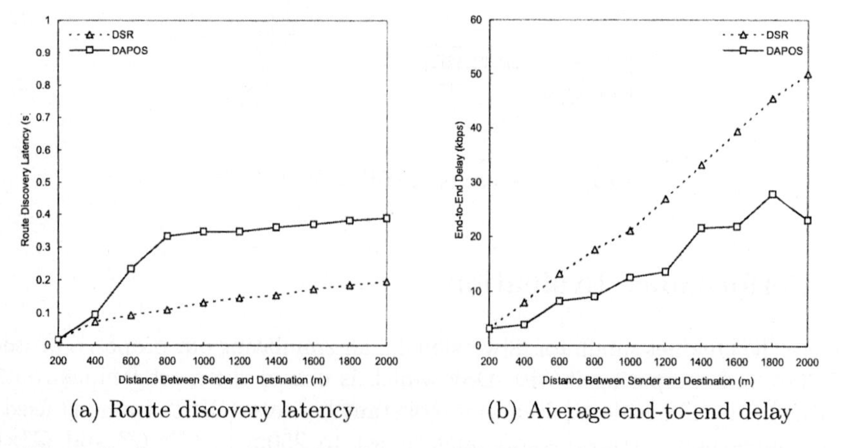

(a) Route discovery latency (b) Average end-to-end delay

Fig. 7. Performance comparison between DAPOS and DSR in grid topology

5.2 Simple Linear Topology

Finally, simulation is performed in a simple linear topology to evaluate the effect of traffic load and the level of interference. Five nodes called A, B, C, D, E (from left to right) are located in linear position separated by 200m. One CBR traffic from leftmost node A to rightmost node E is used during 900 seconds simulation time.

Fig. 8 shows the performance of DAPOS and DSR in various aspect. Fig. 8(a) shows throughput by varying traffic load. From 2ms to 8ms of traffic generation interval, throughput of DAPOS is about 450kbps, the maximum throughput this network can provide. From 8ms to 28ms, DAPOS fully utilize the network capacity. However, in case of DSR, througput is under 50kbps from 2ms to 16ms and increases slowly. Performance barely becomes same with DAPOS in 28ms. In all cases, DSR can't provide full capacity so packet delivery ratio is relatively low.

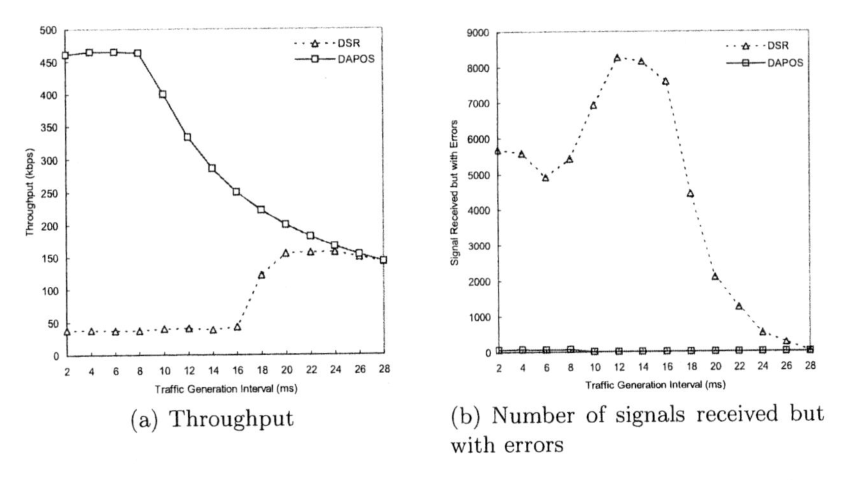

(a) Throughput

(b) Number of signals received but with errors

Fig. 8. Performance comparison between DAPOS and DSR in linear topology

This is because of the interference between adjacent nodes. After node B receives data from A, it communicates with next node C. While in this communication, node C beamforms to the direction of B to increase the SINR. At that time, A may want to initiate communication with B for delivery of next data packet, thus RTS frame transmitted from A can interfere C. This is because directional link between A and C becomes available temporarily if C is in directional mode toward A. As traffic load increases, the probability of interference also increases. Observe that route may be consisted of directional link like {A, C} in case of DAPOS, so the probability is very low. Fig. 8(b) shows the number of collision. As you can see, collision of DSR is much higher than DAPOS. This is why the performances are very different from each other in this scenario.

6 Conclusion

In this paper, we investigated the problem and limitation of existing routing protocols and proposed a new routing protocol, directional antenna based path optimization scheme (DAPOS), to overcome the problem. DAPOS aims to shorten the length of path by including directional link to the route at the first route discovery phase. By simulation, we show that overall network performance can be improved with DAPOS. Future work would include more simulations with mobility scenarios and more realistic environment with a random topology.

References

1. R. Roy Choudhury and N. H. Vaidya, "Impact of directional antennas on ad hoc routing," in *Proceedings of Personal and Wireless Communication (PWC)*, September 2003.
2. D. Johnson and D. Maltz, "Dynamic source routing in ad hoc wireless networks," in *Mobile Computing, Kluwer Academic Publishers*, 1996.
3. S. Bandyopadhyay, et. al., "An adaptive MAC and directional routing protocol for ad hoc wireless networks using ESPAR antenna," in *Proceedings of ACM MobiHoc*, October 2001.
4. S. Roy, et al., "A network-aware MAC and routing protocol for effective load balancing in ad hoc wireless networks with directional antenna," in *Proceedings of ACM MobiHoc*, June 2003.
5. S. Das, C. Perkins, and E. Royer, "Ad hoc on demand distance vector (AODV) routing," Internet-Draft Version 4, IETF, October 1999.
6. T. Rappaport, "Wireless communications principles and practice," Prentice Hall, 2002.
7. R. Roy Choudhury and N. H. Vaidya, "Deafness: A MAC problem in ad hoc networks when using directional antennas," in *Proceedings of IEEE International Conference on Network Protocols (ICNP)*, October 2004.
8. Scalable Network Technologies, "Qualnet simulator version 3.8," www.scalable-networks.com.

Optimized Path Registration with Prefix Delegation in Nested Mobile Networks*

Hyemee Park, Tae-Jin Lee, and Hyunseung Choo**

School of Information and Communication Engineering,
Sungkyunkwan University 440-746, Suwon, Korea +82-31-290-7145
{hyemee, tjlee, choo}@ece.skku.ac.kr

Abstract. As the need for a solution to support mobile network is consistently increasing, NEtwork MObility (NEMO) Basic Support protocol is proposed by using a bi-directional tunnel between a Mobile Router (MR) and its Home Agent (HA). However, the multiple levels of MR-HA tunnels in nested mobile networks lead to significant suboptimal routing, called pinball routing. In this paper we propose a Route Optimization (RO) scheme based on improved Prefix Delegation in order to pass ingress filtering. In an effort to provide the RO based on the improved procedure, it performs an address translation mechanism and a registration procedure to inform other nodes of an optimal path. The proposed scheme supports nested mobile networks without requiring additional tunneling, thus reducing packet overhead and latency with regard to the NEMO basic solution. We evaluate the proposed scheme with previous schemes by analytical models. According to the results, our newly proposed scheme shows 20% and 50% of performance improvements for the worst case and the best case, respectively.

1 Introduction

As convenient devices and services on wireless networking proliferate, users expect to be connected to the Internet from *anywhere* at *anytime*. In particular, the demand for the Internet access in mobile platforms such as trains, buses, and ships is consistently increasing. In order to satisfy such demands, the protocols for supporting mobility need to be extended from an individual mobile device to an entire mobile network. As a result, a working group (WG) called NEtwork MObility (NEMO) has been organized within the IETF to extend host mobility support protocol such as Mobile IPv4 [1] and Mobile IPv6 [2]. The ultimate objective of a network mobility solution is to allow all nodes in the mobile network to be reachable via their permanent IP addresses, as well as to maintain ongoing sessions when the Mobile Router (MR) changes its point of attachment within the Internet.

* This work was supported in parts by Brain Korea 21 and the Ministry of Information and Communication in Republic of Korea.

** Corresponding author.

X. Jia, J. Wu, and Y. He (Eds.): MSN 2005, LNCS 3794, pp. 327–336, 2005.

The NEMO WG has been standardizing a protocol for the network mobility and the NEMO Basic Support [3] is recently proposed. It ensures session continuity for all nodes in the mobile network, even as the MR changes its point of attachment to the Internet. It also supports transparent mobility to every node by using a bi-directional tunnel between the MR and the Home Agent (HA) [3]. As the number of nested levels increases, it suffers from the overhead due to the increased number of packets and transmission delay as the packets are repeatedly encapsulated by all the HAs of the MRs. Also, it results in pinball routing. Therefore, we need a Route Optimization (RO) solution for NEMO to allow packets between CN and MNN to be routed along the optimal path.

In this paper, we propose a novel route optimization scheme based on Prefix Delegation. The proposed scheme is a solution that the entire mobile network forms a logical multilink subnet in a nested mobile network, and thus eliminates any nesting. We improve the Prefix Delegation procedure to preserve the MNN operation unchanged. As the scheme performs an address translation and registration for RO, it allows direct communications between MNN and CN reducing the packet overhead and delay. Moreover, our proposed scheme supports cases that CNs and MNNs are either standard IPv6 nodes or MIPv6 nodes.

The rest of this paper is organized as follows. In Section 2, we introduce the related works with some discussion. Section 3 presents the proposed scheme and deploys various cases for the scheme. After that, we evaluate the performances of the proposed scheme in Section 4. Finally, we conclude in Section 5.

2 Related Works

The NEMO WG has proposed various route optimization schemes. In this section, we describe two schemes recently proposed in the literature.

Reverse Routing Header (RRH) [4] has proposed a new routing header called RRH and RH2 extension to record addresses of intermediate MRs into the packet header with pre-allocated slots and to avoid packet delivery through all HAs of nested MRs. This scheme avoids the multiple encapsulations of the traffic but maintains only one bi-directional tunnel between the first MR and its HA. The first MR encapsulates the packet over its reverse tunnel using the RRH. The upstream MR simply swaps their CoA and the source of the packet to save the original source in the RRH. Therefore, when HA of the first MR receives the packet, it obtains the optimal path information to first MR from the packet's source field and RRH. Then, if the HA receives the packet from CN to MNN, it inserts CoAs of all intermediate MRs into an extended type-2 routing header based on the path information in the RRH of the packets sent from the MR.

Even though RRH approach supports route optimization for nested mobility, it still suffers from an increasing overhead as the depth of the nested mobile networks increases. The more MRs are nested, the more slots need to be made in the RRH. Remark that this approach raises security issues. The RRH process is performed by every nested MR and it does not protect the malicious MR intentionally supplying false upstream information to its HA. So, the issue for consideration is whether the HA should trust the upstream information supplied by the nested MR.

Neighbor Discovery Proxy (ND-Proxy) [5] approach achieves route optimization by having mobile routers to act as neighbor discovery proxy. MR will configure a CoA from the network prefix advertised by its Access Router (AR), and also relay this prefix to its subnets. Before MR advertises the prefix information through Router Advertisement (RA) message, it must set O (route optimization) flag indicating that this prefix can be used for RO of mobile nodes. If a mobile node receives the new prefix information option through RA message, it makes its new CoA with access network prefix and performs the binding procedure with its CN. By performing BU with the prefix of the access network, the mobile node can optimize the routes between its CNs and itself.

This mechanism proposes the proxying function of MR to overcome weaknesses of previous work - Prefix Delegation [6]. *i.e.* it solves the limitation that previous scheme needs the additional delay for the prefix delegation in route optimization. However a simple change in the point of attachment for the root MR will require every nested MR and MNN to change their CoAs and delegated prefixes. These will significantly cause a burst of BUs to their HAs and CNs whenever the MR changes its point of attachment. As a result, they become aware of network mobility. This is against the basic rule for NEMO. Also, it requires a new protocol to perform extended MNN operation since each MNN has two CoAs based on its Mobile Network Prefix (MNP) and Delegated Prefix (DP). We cannot expect MNNs to run a complex protocol to the their physically limited capabilities as typical mobile devices.

3 Proposed Scheme

3.1 Improved Prefix Delegation

The proposed scheme is based on Prefix Delegation method by MR performing ND-Proxy described in previous chapter. Before this method is applied to our scheme, we improved it to provide transparency for network mobility. Like previous method, MR relays the DP advertised by AR through RA message. But, when it relays the DP into the whole mobile network, it multicasts the RA message to only routers on the link. It uses all-routers multicast address (FF02::2) defined in IPv6 addressing architecture. That is, MRs configure its CoA using the DP and MNNs configure its CoA using the MNP advertised by default MR. Figure 1 shows how improved scheme is operated in the nested NEMO.

In Figure 1, the MR1 configures its own CoA (3ffe::1) based on the DP and the DP is allocated to its ingress interface link. So, the MR1 manages the MNP (1ffe:1::/20) and the DP (3ffe:1::/20). It can forward the packet with the source address set to the addresses configured by MNP and DP without ingress filtering. Then, it multicasts the RA message using all-router multicast address to its subnet (LFN5, LMN6, and MR2). Then, only MR2 receives this RA message

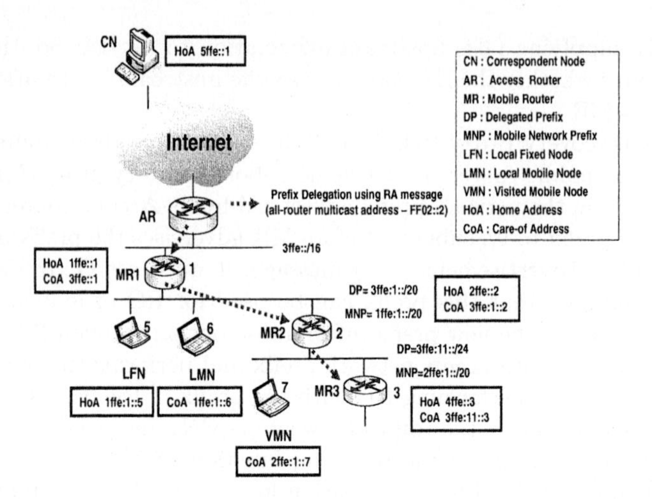

Fig. 1. The proposed Improved Prefix Delegation mechanism

and performs as in MR1. As we know, LFN5 and LMN6 have the addresses configured by the MNP (1ffe:1::/20).

Whenever a mobile network changes its point of attachment, a DP of the link is replaced old DP with new DP advertised by the AR. But, each MR maintains its MNP for the mobile network link regardless of network mobility. As the MNN configures its CoA using the MNP, the proposed scheme solves the problem on bursty binding update messages of MNN whenever a mobile network changes its point of attachment. Moreover, MNN does not need a new operation for RO.

3.2 Delegated Prefix Translation

According to Prefix Delegation procedure, nested MRs and links located within a mobile network are allocated the DP. However, MNNs within the mobile network have their CoAs configured by the MNP. If an MNN sends packets directly to the CN using its CoA as a source address, ingress filtering causes the MNN's packets to be dropped by the default router, since the source address is not derived from the DP. In order to solve this situation, we need an address translation mechanism that is very similar to Network Address Translation (NAT) [7]. The need for IP address translation arises since the network's internal IP addresses cannot be used to provide optimized routes between hosts and those addresses configured by MNP.

In this paper, we define a mechanism called Delegated Prefix Translation (DPT) that replaces an original address with a new delegated address. DPT translates the source address of outbound packets and the destination address of inbound packets. This mechanism utilizes one-to-one mapping between the original address and the new delegated address and replaces a prefix of address with a new DP to avoid ingress filtering problem. The MRs can perform this operation without searching the table or address pools because they know two prefix, as MNP and DP.

The MR acting as a gateway of a subnet checks the source address of an outbound packet sent by MNNs. If the source address is configured by MNP, the MR replaces with the address configured by a DP, and forwards the modified packet to the destination address. Then, this packet is directly forwarded to the CN. On the contrary, if the MR receives the packet from CNs, it checks the destination address of an inbound packet. After that, the MR conducts the Reverse DPT (RDPT) procedure, replacing the destination address with the original MNN's address. That is, this operation can be performed by processing a simple operation without searching the table.

Also, the MR has a the DPT table, which is similar to the NAT table. This table has four fields. The flag field is used to record the state of MNNs. We proposed this field for the mechanism introduced in Section 3.3. After finishing the operation by MR, the MR records the source and the destination address of the packet sent by a MNN in the inside source address and the outside destination address field, respectively. And the outside source address field contains the new delegated address created by the MR. Whenever an MR changes its point of attachment, the MR resets this table because it receives a new DP from the new AR. The DPT table is represented in Figure 3.

3.3 Optimized Path Registration (OPR)

In this paper, we propose Optimized Path Registration (OPR) procedure other than binding update procedure of MIPv6 scheme. The binding update, the procedure to support host mobility, is a limit to support an entire nested network that changes its point of attachment. If we solve the RO problem using the binding procedure, the solution needs the protocol complexity and the additional delay as the number of nested levels increases. So, we define a new procedure, OPR.

OPR is a procedure that registers a new delegated address created by DPT at HA and CN. This address represents an optimized path between the end nodes. So, if nodes use this address to communicate with other nodes, the packets will be directly delivered, following an optimized path. We define a new option for registration. Home Address destination Option (HAO) is supported by only MIPv6-enabled nodes, so we consider that proposed option can be used to optimize the routes for all nodes regardless of mobility support. The OPR option format is shown in Figure 2.

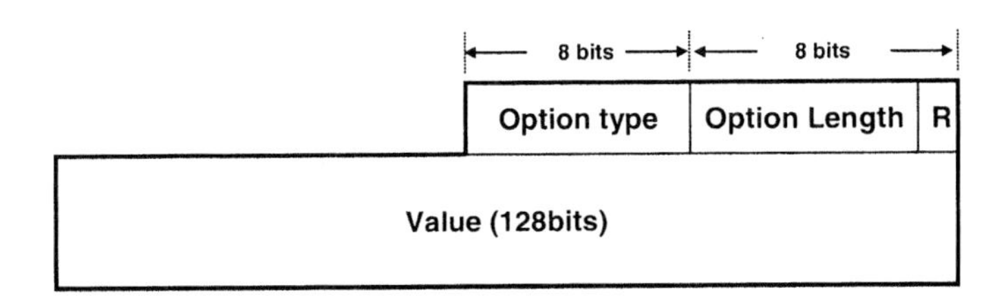

Fig. 2. OPR option format

This option is carried by the Destination Option extension header. The value of Option Type is 202. Length field of the option, in octets, excluding the Option Type and Option Length fields must be set to 16. Value field contains MNN's original HoA or CoA. R flag is set to indicate to the HA and CN that they must register the address of the OPR option. If the flag is set to 0, the HA and CN assume that registration is already finished and validate OPR option using an existing registration table. The reason why we add R flag is that it provides a transparency to all MNNs when the MR changes its point of attachment. If the MR detects its movement and receives a new DP, R flag is set to by the MR to notify HA or CN that an optimized path is changed. The HA and CN update a new address then the packets will flow through this changed path. Therefore, whenever a mobile network changes its point of attachment, MNNs cannot be aware of network mobility and have no need to perform the binding update.

MR operation. The OPR procedure is performed by the default MR acting as a gateway. MR operation is divided into two types according to packet format. If HAO is included the packet sent by MNNs, MR recognizes that MNN is a MIPv6-enabled node. So it replaces HAO with OPR option that carries the address of HAO, *i.e.* VMN's HoA. On the contrary, if HAO is not included, MR knows that MNN is a non-MIPv6 node. So MR adds the destination option header containing the OPR option to its original packet. The option carries the source address of the original packet. There is no corresponding DPT table entry, MR assumes that the registration is not performed and sends the packet with R flag of OPR option set to 1. After that, the former case is recorded MIPv6-enabled (M) flag and the latter one is Non-MIPv6 (N) flag in DPT table. When the MR receives the packet from CNs, it searches its table to find the corresponding entry with the destination address of the packet and confirms the flag. If it is M flag, MR forwards this packet containing type-2 routing header to MNN without any further processing. But it confirms N flag, the MR processes the packet, *i.e.* eliminating the type-2 routing header if existed, or forwarding the packet without any further processing.

CN and HA operation. CN and HA need a data structure to register the Optimal Path Address (OPA). If they have the Binding Cache, a new field is added to this cache so that can be used for the binding and OPR procedure. They also record R (Route optimization) in flag field if R flag of OPR option is set. If CN is non-MIPv6 node and do not have the cache, it has a new RO table.

CN and HA register the address if R flag of OPR option is set to 1. Before they send the packet to MNNs, they search their cache with the MNN's HoA and the destination address of the packet is set to OPA if R flag is existed. Both the extended Binding Cache and RO table are represented in Figure 3.

Figure 3 (a) and (b) show how the proposed scheme is operated. (a) represents the packet transmission from VMN8 to CN1. 1) VMN sends the packet; 2) MR3 performs DPT and replace with OPR option; 3) MR3 records the information in DPT table; 4) MR3 forwards the packet to CN; 5) The receiving CN registers the OPA in its extended Binding Cache. On the contrary, (b) describes the packet

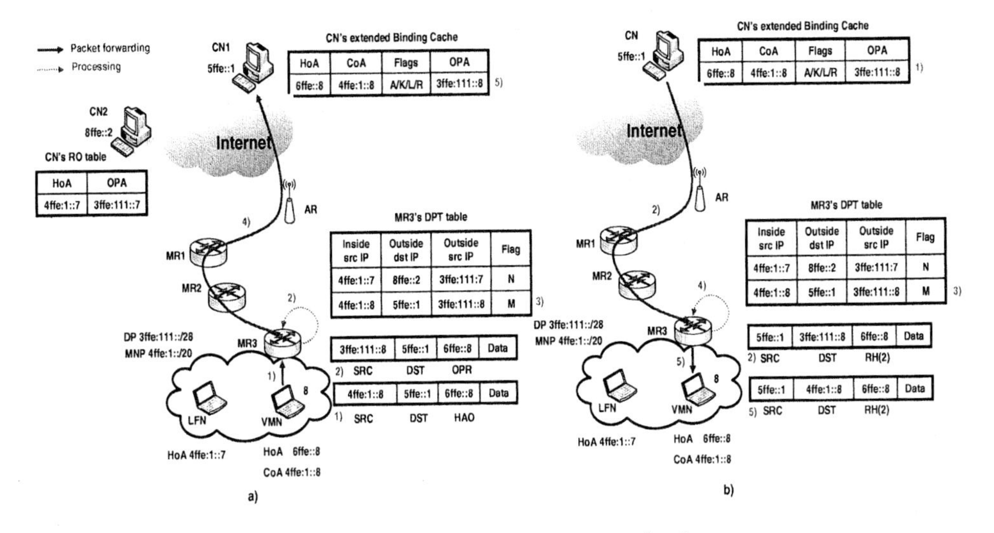

Fig. 3. The operation of the proposed scheme

transmission from CN1 to VMN8. 1) Before the packet sends, CN searches its cache and confirms the R flag; 2) CN sends the packet with the destination address set to the OPA; 3) MR3 finds the corresponding entry and confirms the M flag; 4) MR3 performs RDPT; 5) The packet is forwarded to MNN; The communication between different nodes such as LFN, IPv6 CN could be derived using the same method.

4 Performance Evaluation

In this section we will evaluate the performances of the proposed scheme in comparison with the NEMO Basic Support [3] and RRH [4].

4.1 Analytical Model

Before we start the analysis for a new performance measure, we describe some assumptions to compare the route optimization schemes. The details are as follows.

- The processing capacities for all nodes are identical.
- The wired propagation speed is the same as the wireless one.
- The transmission speed of wired routers is 10 times faster than that of wireless ones.
- HAs and CNs are uniformly distributed in the Internet.

Table 1 shows the parameters needed in the analysis of the total delay for data transmission between MNN and CN.

With these parameters, we calculate the total delay presented as D_{TOT} and evaluate the proposed scheme with others. Total delay between two nodes is represented as follows.

$$D_{TOT} = H(D_{TRANS} + D_{PROC} + D_{PROP}) \tag{1}$$

Table 1. Parameters for performance analysis

N	Nesting Level	B_{wl}	Wireless transmission Speed (bit/sec)
S_{pkt}	Packet size (bit)	B_{wd}	Wired transmission Speed (bit/sec)
S_{slt}	Size of slots (bit)	T_{ch}	Time to change source address(sec)
S_{tu}	Tunnel header size (bit)	T_{tu}	Tunnel processing time(sec)
T_{sch}	Cache searching time (sec)	T_{dpt}	DPT processing time (sec)
P	Propagation speed (m/sec)	T_{opr}	OPR processing time (sec)

Where D_{TOT} : total delay, D_{TRANS} : transmission delay, D_{PROC} : processing delay, D_{PROP} : propagation delay and H : the number of hops

At first, in order to get the total delay, we calculate the transmission time, processing time, and propagation time in each hop between two end nodes. The following shows the procedure for the calculation through the example and the generalized formula of three schemes one by one.

The number of nesting levels for the example is 3 as shown in Figure 4. The dotted lines denote wireless connections in the meantime straight lines are used for wired communications. We formulate each step of the proposed scheme and they are as follows.

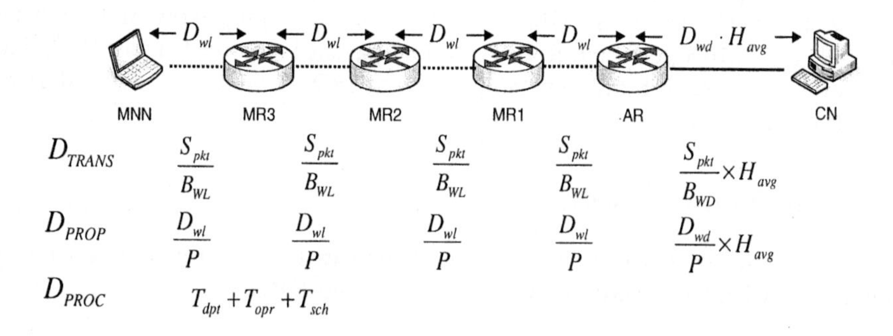

Fig. 4. The procedure for evaluation in the proposed scheme

For the generalization of the total delay in terms of the number of nesting level N, we have the following equation.

$$D_{TOT} = \frac{S_{pkt} + S_{pkt} \cdot N}{B_{wl}} + \frac{S_{pkt} \times H_{avg}}{B_{wd}} + \frac{(N+1) \cdot D_{wl} + D_{wd} \cdot H_{avg}}{P} + T_{dpt} + T_{opr} + T_{sch} \qquad (2)$$

The generalized formula of the NEMO Basic Support and RRH could be derived using the same method as return in preceding columns. Equation (3) and (4) show the generalized form of the total delay in terms of the number of nesting level N for two schemes, respectively.

$$D_{TOT} = \sum_{k=0}^{N} \frac{S_{pkt} + k \cdot S_{tu}}{B_{wl}} + \sum_{k=0}^{N} \frac{(S_{pkt} + k \cdot S_{tu}) \times H_{avg}}{B_{wd}} + \frac{(N+1)(D_{wl} + D_{wd} \cdot H_{avg})}{P} + 2 \cdot N \cdot T_{tu} \quad (3)$$

$$D_{TOT} = \frac{S_{pkt} + (S_{pkt} + S_{tu} + S_{slt} \cdot N) \cdot N}{B_{wl}} + \frac{(2 \cdot S_{pkt} + 2 \cdot S_{tu} + S_{slt} \cdot N) \cdot H_{avg}}{B_{wd}} + \frac{(N+1) \cdot D_{wl} + 2 \cdot D_{wd} \cdot H_{avg}}{P} + 2 \cdot T_{tu} + (N-1) \cdot T_{ch} \quad (4)$$

4.2 Numerical Results

Table 2 shows the parameters values for our performance analysis. We assume that all the processing time is identical. Based on those values, we calculate the total delay for the proposed scheme along with two previous schemes. Figure 5 (a) and (b) show the result of evaluation.

Table 2. Parameters value for analysis

S_{pkt}	S_{slt}	D_{wl}	D_{wd}	H_{avg}
1480×6	128	30	10000	10
S_{tu}	B_{wd}	B_{wl}	P	T_{all}
320	10^8	10^7	2×10^8	0.00001

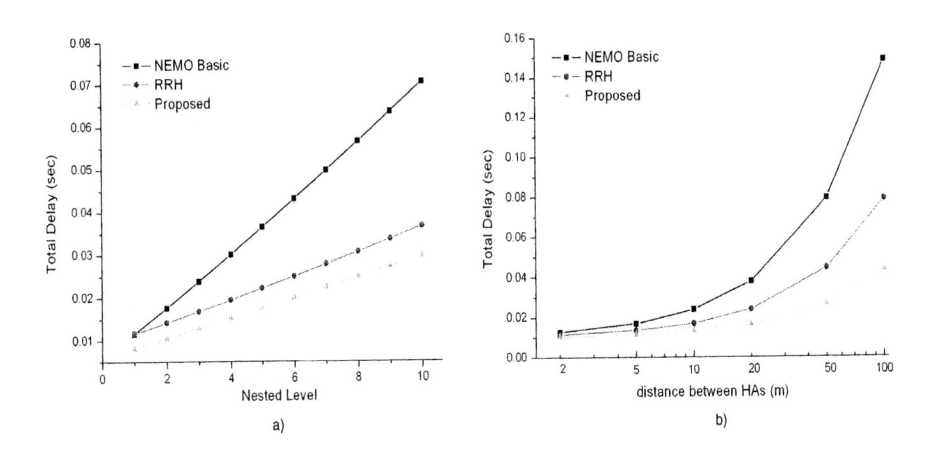

Fig. 5. The total delay for three schemes

The results represent that our proposed scheme shows better performance than the previous two. Because the proposed scheme has a small packet size in comparison with others, in addition, it remains constant irrespective of the number of nesting levels. As the packet is delivered via the optimal route, the transmission and propagation delays are smaller than others. Furthermore, our scheme is operated only by one MR acting as a gateway while two existing schemes must be operated by all MRs. As a result, the processing time is significantly reduced as the nesting level increases.

5 Conclusion

In this paper, we have proposed a new mechanism for route optimization in nested mobile networks. The scheme supports the RO solution between the various end nodes regardless of mobility support. Therefore, it provides the optimized communication as well as scalability in heterogeneous network environments and backward compatibility with existing mechanisms.

Our evaluation represents advantages of the proposed scheme compared to two earlier proposed scheme - NEMO Basic Support and RRH. According to the results, our newly proposed scheme shows 20% and 50% of performance improvements for the worst case and the best case, respectively. Thus our solution allows all Internet nodes to communicate with mobile network node with minimum routing overhead and delay. We conclude that the proposed scheme can be used to handle RO problems reported in literature.

References

1. C. Perkins, *"IP Mobility support for IPv4"*, IETF, RFC 3220, January 2002.
2. D. Johnson, C. Perkins, and J. Arkko, *"Mobility Support in IPv6"*, IETF, RFC 3775, June 2004.
3. V. Devarapalli, R. Wakikawa, A. Petrescu, P. Thubert, *"Network Mobility (NEMO) Basic Support Protocol"*, IETF, January 2005, RFC 3963.
4. P. Thubert, M. Molteni, *"IPv6 Reverse Routing Header and its application to Mobile Networks"*, draft-thubert-nemo-reverse-routing-header-05.txt, March 4, 2005,
5. L. KyeongJin et al., *"Route Optimization based on ND-Proxy for Mobile Nodes in IPv6 Mobile Networks"*, IEEE 59th Vehicular Technology Conference 17-19 May, 2004.
6. L. KyeongJin et al., *"Route Optimization for Mobile Nodes in Mobile Network based on Prefix Delegation"*, IEEE 58th Vehicular Technology Conference 6-9 October, 2003.
7. P. Srisuresh, M. Holdrege, *"IP Network Address Translator (NAT) Terminology and Considerations"*, IETF, August 1999, RFC 2663.

BGP-GCR+: An IPv6-Based Routing Architecture for MANETs as Transit Networks of the Internet*

Quan Le Trung and Gabriele Kotsis

Dept. of Telecooperation, Johannes Kepler University,
Altenberger Str. 69, A-4040 Linz, Austria
{quanle, gk}@tk.uni-linz.ac.at
http://www.tk.uni-linz.ac.at

Abstract. Internetworking mobile ad-hoc networks (MANETs) with the Internet has been a hot issue for many years. However, most researches have been concentrated on the use of MANETs as the access networks for the Internet. This paper introduces another use of MANETs: backup or load-balancing transit networks for the Internet. To achieve this goal, a scalable, stable, low-overhead, QoS-support ad-hoc routing architecture with the address auto-configuration is required. Moreover, how an Internet gateway selects an external route via MANETs to another autonomous system (AS) also needs to be solved. In this paper, BGP-GCR+, a combination of the border gateway protocol (BGP), the gravitational cluster routing (GCR), and the passive/weak IPv6-based address stateless auto-configuration, is developed towards the standards to achieve the required functions.

1 Introduction

Advances on the medium access control (MAC) and physical layers of MANETs such as the ultra-wideband (UWB) technology [11] have led to a proliferation of MANETs applications, esp. on the field of internetworking MANETs with the Internet. This paper concentrates on the use of MANETs as backup or load-balancing transit networks of the Internet, considering the following features: (1) a unicast, scalable, stable, low-overhead, QoS-support ad-hoc routing architecture, (2) an IPv6-based address auto-configuration solution, (3) metrics and procedures at Internet gateways to select external routes via MANETs to other autonomous systems (ASs).

Numerous ad-hoc routing architectures have been proposed. Gravitational cluster routing (GCR) [1] belongs to the scalable, QoS-support class, designing specially to improve the end-to-end perceived quality of the application by increasing the stability of the active connection. In this paper, GCR has been extended in the following aspects: (1) GCR assumes that each node is assigned a unique ID before, and a unique cluster ID for each created cluster, (2) cluster maintenance: when a link is broken, there is no selection procedure for the downstream node to choose another stable neighbor to continue maintaining the repairable feature of GCR with the highest probability, (3) source routing is needed to send data to the downstream nodes.

* This work has been fully supported by Austrian Exchange Service (OÄD) under the PhD research program in Department of Telecooperation - Johannes Kepler University.

X. Jia, J. Wu, and Y. He (Eds.): MSN 2005, LNCS 3794, pp. 337 – 350, 2005.

BGP [2] is used by exit points (Internet gateways) of an AS to choose the best external routes connecting to other ASs. Fig. 1 shows scenarios in which MANETs are used as backup or load-balancing transit networks for the Internet.

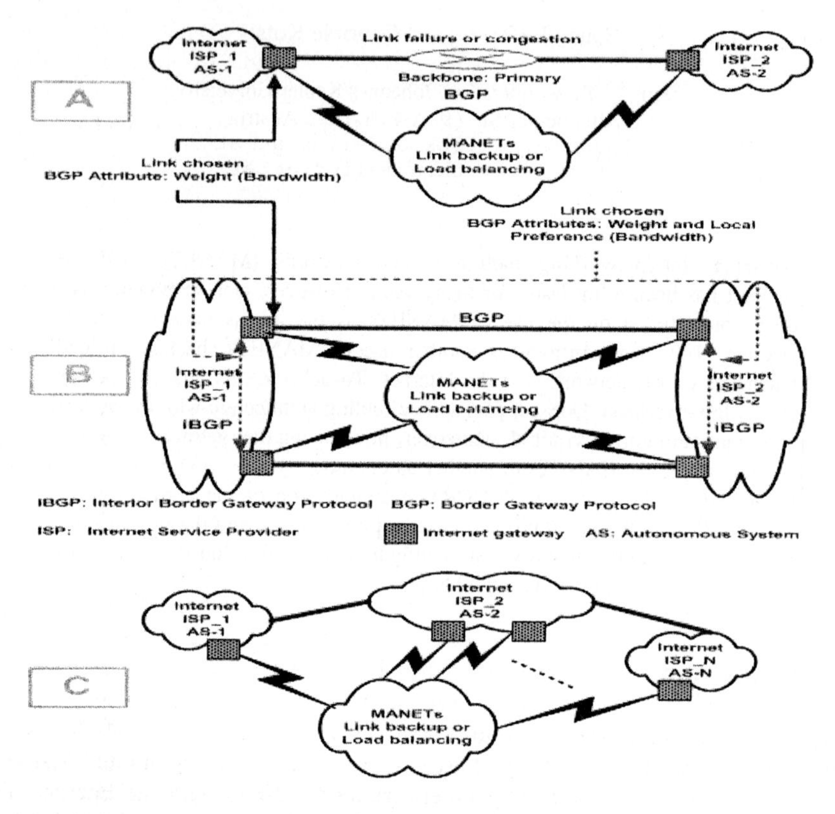

Fig. 1. Scenarios of MANETs as transit networks for the Internet: *(A)* One bacbone link between two AS systems, *(B)* Multiple backbone links, *(C)* Connection of multiple ASs

Two attributes of BGP have been chosen to select the best external routes: (1) weight: each exit point chooses the external route with the highest weight, (2) local preference (LOCAL_PREF): exit points exchange their weights for their best external routes to other ASs. Address auto-configuration is another needed feature to insure the self-organization of the MANETs. In this paper, it is integrated into BGP-GCR+, including the use of Weak DAD [4] for intra-cluster detection and a combination of the weak DAD [4] and passive DAD [3] for inter-cluster detection.

This paper is organized as follows. Sec. 2 reviews the features of GCR. The operations of BGP-GCR+ and its prototypes towards the standards are presented in Sec. 3 and 4, respectively. Sec. 5 reviews the related works of MANETs routing, address auto-configuration, and internetworking with the Internet. Sec. 6 ends this paper with conclusions and our future works.

2 A Review on Features of GCR

2.1 The Operations of GCR

The GCR [1] is a two-level cluster routing protocol. In each cluster with the scope of R_{max} hops, a node with the local maximum degree is elected to be the cluster head and initiate its non-overlapped cluster construction. Each cluster head is responsible for maintaining the newest topological status of its cluster. Proactive routing based on the unicast Tree-Links maintained by the cluster heads has been adopted inside the clusters and reactive routing based on AODV [6] outside the clusters.

When a node switches on, it detects its neighbors and waits for a chance to join in one of the constructed clusters. Otherwise, it periodically broadcasts its degree, i.e. the number of its neighbors, within the scope of R_{max} hops for electing a new cluster head to construct the new cluster.

To construct a new cluster, an elected cluster head generates a unique cluster ID (Cid). It attaches this cluster ID and a level initialized to zero to a control packet called the Gravitational Packet with a hop count of R_{max} hops and broadcasts it to its neighbors. The neighbors of the cluster head attach their IDs (called Sub-Tree ID: STid) to this packet, increase the level and continue re-broadcasting this packet. Each downstream node receiving this packet will store its (Cid, STid, level) values from this packet, increasing the level and rebroadcast this packet provided that the hop count is larger than zero. A leaf node, i.e. an end node of the cluster having at least two paths to the cluster head (called joinless), then sends the Join Packet to its upstream neighbors towards its cluster head. Each node receives this packet will add its link states and continue forwarding this packet to its cluster head. The cluster head receives this packet and updates its intra-routing table, i.e. the unicast Tree-Links.

The constructed clusters is repairable (one-node fault-tolerant) since there are at least two paths from any node to the cluster head. Fig. 2 shows an example where the link (u,v) is failed, then node v can maintain its connection to the cluster head via nodes w/x/y/.../H.

The cluster maintenance procedure is carried out by nodes recognizing the changes of link states and sending updates to the cluster head via Link Change Packets.

2.2 The Open Problems of GCR

Addressing: independent or dependent routing protocol. GCR assumes that each node is assigned a unique ID and the cluster ID is also unique among clusters. Thus an address auto-configuration protocol is required in self-organizing the MANETs for their fast deployments as the backup or load-balancing transit networks of the Internet. BGP-GCR+ uses the address protocol with the following features:

- GCR-dependent, weak/passive DAD [3, 4].
- Address space: Part of IPv6 site-local address space (FEC0:0:XX:XX/64) [9]. The XX:XX is the 32-bit network prefix used for the cluster ID.

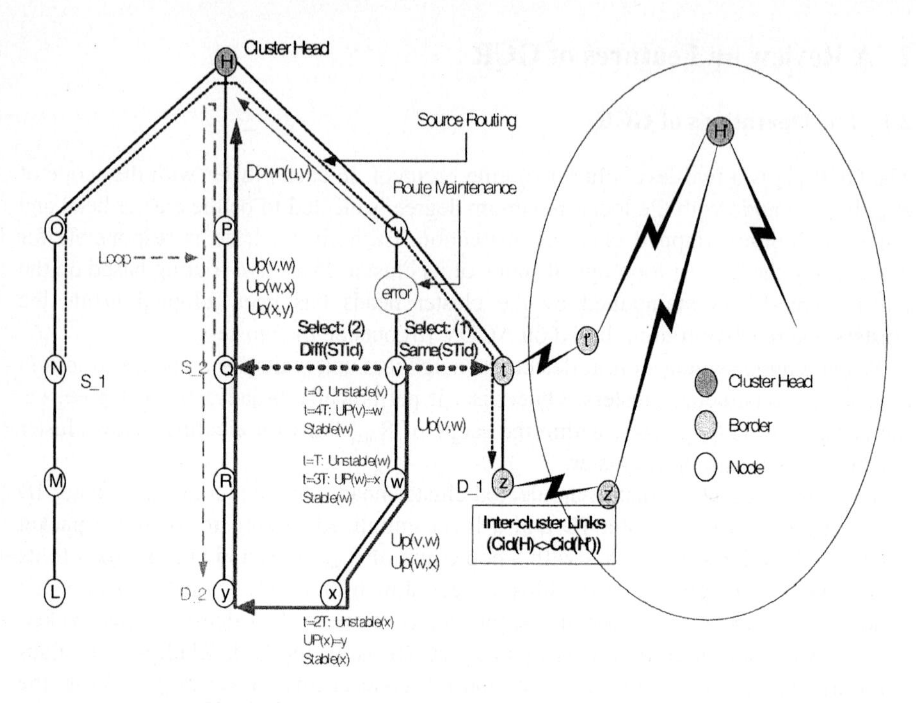

Fig. 2. Operations of gravitational cluster routing *(GCR)*

Neighbor selection for cluster construction and maintenance. Minimum level is the metric for nodes to select neighbors to join in a cluster or to repair a broken link.

The advantage of this approach is to shorten the route length to its cluster head, but creating an un-repairable cluster.

We suggest another metric, the STid of neighbor, to continue maintaining the repairable cluster with higher probability. The rules of BGP-GCR+ are: (1) neighbors with the same Cid is preferred than those with different Cids, (2) for neighbors with the same Cid, those with the same STid is preferred than those with the different Stids, (3) for neighbors with the same Cid and STid, select the minimum level one.

Source routing. Fig. 2 shows an example when the source S_1 (node N) sends data packet to the destination D_1 (node z). Data packet is first forwarded to the cluster head based on its upstream nodes (N→O→H). The cluster head finds a route based on its intra-cluster routing table (H→u→t→z) and forwards the data packet to the down stream node (u). However, node (u) does not know which downstream node (v or t) to forward the data packet since its does not have any routing information.

Although this problem is not recognized in GCR, a possible solution is to use the source routing technique as in DSR [7] or the routing header in IPv6 [13]. BGP-GCR+ uses the following rules:

- The source first checks if the destination is in its neighbor cache. If yes, send data packet directly. If not, send data packet to the cluster head via its upstream node.
- The cluster head adds the route to the downstream destination into the data packet using IPv6 routing header, then forwards to the corresponding downstream node.

3 The Operations of BGP-GCR+

The main operations of BGP-GCR+ is similar to those of GCR described in Sec. 2. This section will continue discussing (1) how to integrate the IPv6 address stateless auto-configuration protocol into GCR, (2) how BGP is extended to select an external route at an Internet gateway (exit point) to another autonomous system via the MANETs, and (3) main procedures and algorithms of the BGP-GCR+.

3.1 How to Integrate the IPv6 Addressing Protocol into BGP-GCR+?

We have to distinguish between intra-cluster and inter-cluster routing. For inter cluster, Each cluster is constructed by the leaf nodes sending the Join Packet containing their link states to the cluster head. Upstream nodes receive this packet, adding their link states into this packet and continue forwarding it towards the cluster head. Each cluster is maintained by any nodes detecting the link state changes, generating the Link Change Packet upstream towards the cluster head for updating. The Route from the cluster head to its downstream nodes are attached to the header of data packet (source routing).

Rule 1. *Link state information in the Join Packet and Link Change Packet (IPv6 link/site local, R-ID)*[1] *is used by BGP-GCR+ to insure the detection of any duplicated addresses within a cluster, but not for the cluster merging.*

For inter-cluster, each cluster is assigned the 32-bit IPv6 network prefix for its cluster ID. Inter-cluster routing in GCR is the reactive AODV-based algorithm, which uses the sequence number to detect the stale routes and avoid the routing loops.

Rule 2. *The AODV-based route discovery based on the Weak DAD (Cid, R-CID)*[2] *is used by BGP-GCR+ to insure the detection of duplicated cluster IDs in the whole MANETs, supporting for the cluster merging detection. Two PDAD techniques: PDAD-RNS (RREQ-Never-Sent) and PDAD-SN (Sequence Number) can be used to increase the reliability, i.e. both (Cid,R-CID) are duplicated.*

3.2 An Extension of BGP for BGP-GCR+

The following BGP attributes are extended to determine when the exit point of an autonomous system chooses external routes via MANETs to other ASs: (1) weight: if each exit point of an AS has many external routes to another AS, it selects the one with highest weight. We use the bandwidth to calculate the weight, (2) local preference (LOCAL_PREF): if an AS has many exit points connecting to another AS, it select the one with highest weight (largest total bandwidth).

Equations 1 to 5 are used to calculate these two above parameters for the use of MANETs as the load-balancing transit networks for the Internet.

---

[1] R-ID is a 32-bit unsigned random number generated once by each node.

[2] R-CID is a 32-bit unsigned random number generated once by each cluster head.

$$B_{Ad-hoc-static} = \lambda_{\max}[packet/\sec]\, sizeof(packet)[bits] \quad (1)$$

$$B_{Temp} = \frac{[sizeof(RREQ) + sizeof(RREP)][bits]}{RTT[\sec]} \quad (2)$$

$$B_{Ad-hoc-dynamic} = \alpha.B_{Ad-hoc-dynamic} + (1-\alpha).B_{Temp} \quad (3)$$

$$W_{Ad-hoc} = \frac{B_{Ad-hoc-static/dynamic}}{B_{Ad-hoc-static/dynamic} + \sum_{j=1}^{n} B_j} \quad (4)$$

$$\beta = \frac{\left[B_{Ad-hoc-static/dynamic} + \sum_{j=1}^{n} B_j\right]_{Current-Exit-Point}}{\left[B_{Ad-hoc-static/dynamic} + \sum_{j=1}^{n} B_j\right]_{Received-Exit-Point}} \quad (5)$$

The weight of each external route from an exit point of an AS to another AS is the ratio of its bandwidth over the total bandwidth of all external routes connecting the two ASs. For selecting an external route via MANETs, we suggest two approaches for determining the bandwidth of MANETs:

- Static (Equations 1, 4): an analysis on the capacity MANETs is needed. The works in [5] shows an upper bound of arrival packet rate (λ_{max}) for 802.11 MANETs is $O(1/n^{0.5})$. Then the exit point will send packets over an external route via MANETs with the probability equal to the weight ($W_{Ad\text{-}hoc}$) of this route.
- Dynamic (Equations 2, 3, 4): the exit point measures the bandwidth of the MANETs by either periodically sending the RREQ and wait for a RREP, or integrating with MANETs route discovery. To reduce the dynamic feature of MANETs, Equation 3 is used with $\alpha \in [0.8, 1.0]$.

If an AS has many exit points with many external routes connecting to a different AS, the selected exit point is the one with highest total bandwidth. Equation 5 is used by exit point to compare its total bandwidth to that of other ones.

Another point is the use of AS numbers of exit points in inter-cluster routing of BGP-GCR+. Since we use IPv6 site-local address space for MANETs, an exit point using an external routes via MANETs does not know the IPv6 site-local addresses of the exit points in the other sides except for their global IPv6 unicast address or their AS numbers.

The last point is the reduction of congestion inside MANETs in the neighboring area of the exit point. When the MANETs is used as the backup or load-balancing transit networks for the Internet, almost all RREQs are generated by the exit points. Therefore, we suggest that all exit points joining in the MANETs are the cluster heads. This is achieved by setting the degree of each exit point to a maximum value. Here we see the advantages of chosen exit points to be the cluster heads: (1) load-balancing for MANETs neighbor nodes of the exit points, (2) the cluster head is the exit point, so it has its own AS number. The use of AS number instead of IPv6 site-local address requires no additional overhead, (3) the use of AS number makes the DAD process transparent among exit points.

3.3 Main Procedures and Algorithms of BGP-GCR+

When a node switches on, it automatically assigns itself the IPv6 link-local address and generates a random number R-ID. It periodically sends Hello packet to inform its existence to its neighbors, and checks its neighbor cache to determine its neighbor status and its degree. If all its neighbor status is not-joined or joined with level equal to R_{max}, it broadcasts its degree to its neighbors, joining in the new cluster head election and construction. Otherwise, it selects and joins in an already constructed cluster using our modified selection rules:

- Cluster having two or more stable neighbors in its neighbor cache with different STids is the first choice, i.e. a new node joins in the constructed cluster while continue maintaining the one-node fault-tolerant of BGP-GCR+ is the first choice.
- Cluster having minimum level, stable neighbors in its cache is the second choice.

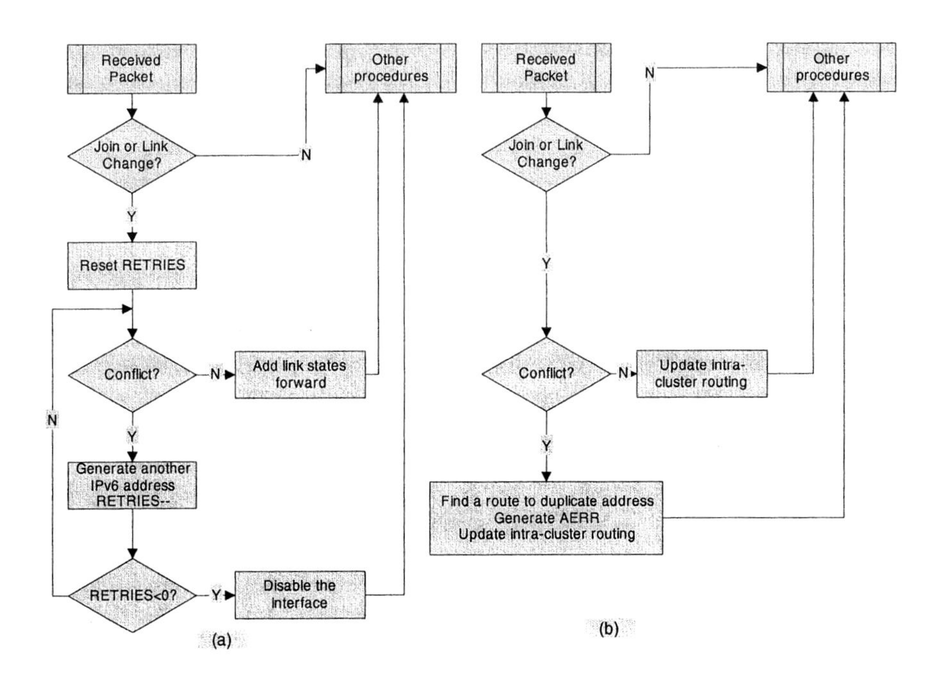

Fig. 3. Intra-cluster addressing protocol: *(a)* node detection, *(b)* cluster head detection

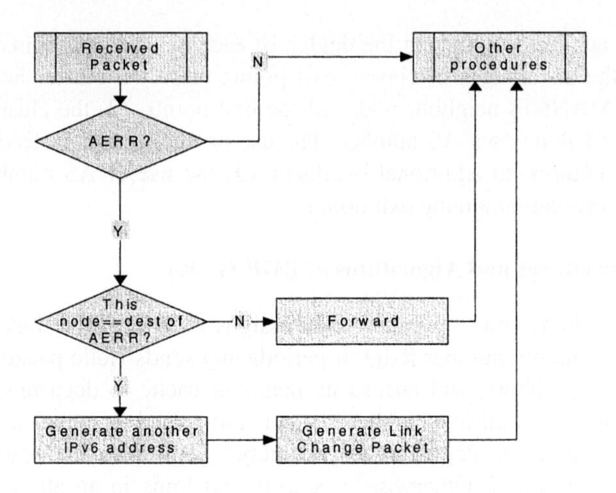

Fig. 4. Intra-cluster addressing protocol: processing

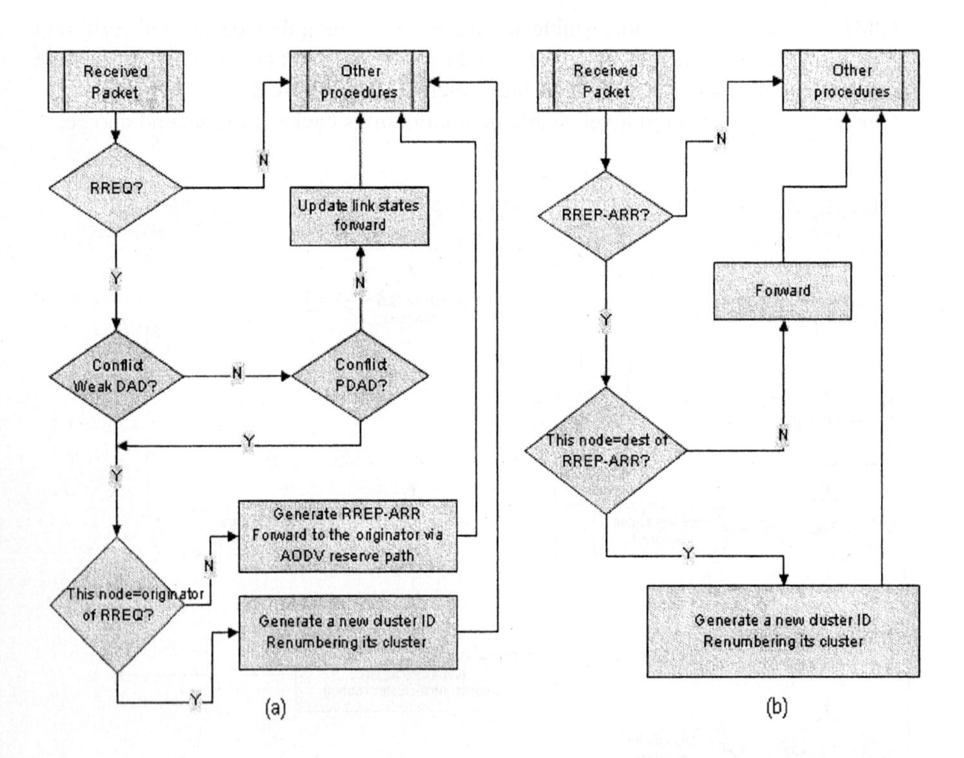

Fig. 5. Inter-cluster addressing protocol: *(a)* detection, *(b)* processing

Nodes have already joined in the constructed cluster, but later do not find any stable neighbors of the same cluster, the selection procedure is modified from GCR, see Sec. 2.2. When a node detects the duplicated addresses, the conflict is solved by using the algorithms in Fig. 3-4 and 5 for intra-cluster and inter-cluster, respectively. Inter-cluster routing is AODV-based. We have shown that the setting of Internet gateways to be the cluster heads results in many advantages, esp. the use of AS numbers as the destination. Intra-cluster routing in BGP-GCR+ is similar to GCR. Nodes send data packets to the cluster head via their upstream nodes, the cluster head calculates the routes to the corresponding destination nodes and forwarding the data packets downstream using IPv6 routing header.

4 The Prototype of BGP-GCR+

Each node has one of the following states during its operations in BGP-GCR+:

- State 1: (no-J, no-A), not joined and not assigned[3] address.
- State 2: (J , A, no-H), joined and assigned address, but not the cluster head.
- State 3: (J, A, H), joined and assigned address, the cluster head.

Details in the transitions among these states are shown in the corresponding state transition tables in Fig. 7-9.

The following open alternative approaches needs to be considered in future versions of BGP-GCR+:

- In this prototype of BGP-GCR+, a node joined in the cluster stops broadcasting its degree. This approach reduces the broadcast traffic, but the cluster head later can not be the highest degree, which also reduces the stability of its cluster. Another approach is for each node to broadcast its degree periodically and elect the new cluster head with the highest degree in its cluster even though it has joined in the cluster. This approach takes much overhead, esp. in re-constructed the cluster, but the more stable unicast Tree-Links is maintained at the cluster head, which is useful for inter-cluster routing and QoS support.
- In this prototype of BGP-GCR+, a cluster head re-assigns (Cid=NULL) if its Cid is duplicated. Another approach is for the cluster head send the Gravitational packet containing its new generated Cid (network prefix) to other nodes in its cluster, a process called "re-numbering cluster". However, there are also some other problems when this approach is used such as the selection procedure if one node receives multiple Gravitational packets, or the cluster head sends the Gravitational packet once when it is needed or periodically.

Finally, the prototype of BGP-GCR+ is a step towards the standards. For this purpose, all packet formats in BGP-GCR+ are developed based on IETF Internet drafts and rfcs [8, 9, 13]. Fig. 6 shows a summary on types of packet formats used in BGP-GCR+ and its corresponding IETF documents used to develop.

---

[3] When a node switches on, it has already initialized its own IPv6 link-local address and a random number R-ID. A node is considered "not assigned" if it has not received the Gravational packet to get the network prefix to configure its IPv6 site-local address from the pair (IPv6 link-local, network prefix).

Packet Formats used in BGP-GCR+				
Index	Type	Modification of	Towards Standards (IETF Drafts, RFCs)	Meanings
1	Hello	Neighbor Solicitation (NS)	Neighbor Discovery Protocol (IPv6 NDP)	Neighbor discovery
2	Degree	Neighbor Solicitation (NS)	Neighbor Discovery Protocol (IPv6 NDP)	Broadcast degree of node for cluster head election
3	Gravitational	Router Advertisement (RA)	Neighbor Discovery Protocol (IPv6 NDP)	The elected cluster head broadcasts it for cluster construction
4	Join	DSR RREQ	Dynamic Source Routing (IPv6 DSR)	Nodes send their link states to cluster head for building the intra-cluster routing cache Attach information for intra-DAD
5	Link Change	DSR RREQ	Dynamic Source Routing (IPv6 DSR)	Nodes send their link states to cluster head for building the intra-cluster routing cache Attach information for intra-DAD
6	AERR	Neighbor Advertisement (NA)	Neighbor Discovery Protocol (IPv6 NDP)	Intra-cluster duplicate address detection
7	RREQ	AODV RREQ	Ad-hoc On-demand Distance Vector Routing Protocol (IPv6 AODV)	Inter-cluster route discovery
8	RREP	AODV RREP	IPv6 AODV	Inter-cluster route discovery
9	RERR	AODV RERR	IPv6 AODV	Inter-cluster route maintenance
10	RREP-ARR	AODV RREP	IPv6 AODV	Inter-cluster duplicate address detection
11	Data	IPv6 Data	IPv6	Data packet
12	Source routing	IPv6 Routing header	IPv6	All packets (downstream)

Fig. 6. Packet formats used in BGP-GCR+

I. Timer handlers				
Index	Incoming Event	Current State	Action	Next State
I.1	Hello interval expired	X∈{1,2,3}	Broadcast Hello packet to neighbors	X∈{1,2,3}
I.2	Neighbor cache update cycle expired	1	Purge stale neighbors in Neighbor cache Detect link state changes: +either determine its degree for electing the new cluster head, +or join in the constructed cluster, send JOIN Packet to cluster head	X∈{1,2,3}
		2	Purge stale neighbors in Neighbor cache Detect link state changes: +either send LINK CHANGE Packet to its cluster head, +or leave (assign its Cid=NULL) the current cluster	Y∈{1,2}
		3	Purge stale neighbors in Neighbor cache Detect link state changes: +either update its intra-cluster and inter-cluster routing caches +or leave (assign its Cid=NULL) the current cluster (isolated node)	Z∈{3,1}
I.3	Degree interval expired	1	Broadcast its DEGREE within its scope R_{max} if its status (Cid=NULL, i.e. not joined in any cluster) and the status of all its neighbors: +either NOT JOIN (Cid=NULL) +or JOIN (Cid≠NULL), but level=R_{max}	1
I.4	Cluster head election cycle expired	1	Compare its degree with those from received DEGREE packet: +either its degree is MAX, be the cluster head +or its degree is less than some, wait to join in the cluster	X∈{1,3}
I.5	Broadcast ID cache cycle expired	X∈{1,2,3}	Purge stale broadcast ID entries in the Broadcast ID cache	X∈{1,2,3}
I.6	Inter-cluster routing cache cycle expired	3	Purge stale route entries in the inter-cluster routing cache	3

Fig. 7. State transition table of BGP-GCR+ *(timer handlers)*

II. Cluster construction, maintenance, and intra-cluster DAD handlers				
Index	Incoming Event	Current State	Action	Next State
II.1	Send Gravitational Packet	3	New elected cluster head constructs its cluster: +Generate its Cid and R-CID +Attach Cid to Gravitational Packet and broadcast it (Hop-Count = R_{max})	3
II.2	Received Gravitational Packet	X∈{2,3}	Drop the Packet	X∈{2,3}
		1	Save its upstream node to the cluster head Join in the cluster, save its (Cid, STid, level) Attach (level++, Hop-Count--) and forward downstream if Hop-Count>0	2
II.3	Send JOIN Packet	2	Attach its (IPv6 site-local, R-ID) for intra-DAD Attach its link states for updating intra-cluster routing at its cluster head Send JOIN Packet to its cluster head via its upstream node	2
II.4	Received JOIN Packet	1	Drop the Packet	1
		2	Check address conflict (using intra-DAD), if duplicate: +Generate another IPv6 site-local, continue check RETRIES times +Still duplicated, disable interface, drop JOIN Packet If no duplicate: add its link states, forward JOIN Packet to its cluster head	2
		3	Check address conflict (using intra-DAD), if duplicate: +Find a route and send AERR Packet to the duplicated node Update its intra-cluster and inter-cluster routing caches	3
II.5	Send AERR Packet	3	Add routing header to AERR, send it to corresponding downstream node	3
II.6	Received AERR Packet	X∈{1,3}	Drop the Packet	X∈{1,3}
		2	Generate another IPv6, send Link Change Packet to its cluster head	2
II.7	Send Link Change Packet	2	Add pair (IPv6 site-local, R-ID) for intra-DAD Add its link states and send the packet to its cluster head	2
II.8	Received Link Change Packet	1	Drop the Packet	1
		2	Check address conflict (using intra-DAD), if duplicate: +Generate another IPv6 site-local, continue check RETRIES times +Still duplicated, disable interface, drop JOIN Packet If no duplicate: add its link states, forward JOIN Packet to its cluster head	2
		3	Check address conflict (using intra-DAD), if duplicate: +Find a route and send AERR Packet to the duplicated node Update its intra-cluster and inter-cluster routing caches	3

Fig. 8. State transition table of BGP-GCR+ *(cluster construction and intra-cluster DAD)*

5 Related Work

Although there are different types of internetworking MANETs with the Internet, most researches have only focused on the use of MANETs as the access networks for the Internet and mobility management [10]. Further classifications are differed in the way how the MANETs nodes detect their Internet gateways for Internet access, how to select an Internet gateway if multiple ones exist for different objectives such as

III. Inter-cluster routing and DAD handlers				
Index	Incoming Event	Current State	Action	Next State
III.1	Send RREQ	3	Attach pair (Source IPv6, R-CID) instead of Source IPv6 for inter-DAD If it is internet gateway, used destination AS-# instead of IPv6 Increase its sequence number, attach to RREQ for PDAD-SN Border-cast this RREQ to its borders using its unicast Tree-Links	3
III.2	Received RREQ	1	Drop the Packet	1
		2	Upstream direction: forward to its cluster head via its upstream node Downstream direction: forward to its downstream node based on routing header	2
		3	Check address conflict (using inter-DAD), if duplicate: +If it is duplicated, **re-assigned cluster (its Cid=NULL)** +If another is duplicated, send RREP-ARR to it via reversed AODV route If no duplicate: +Update its inter-cluster routing cache towards originator of RREQ +If it is destination of RREQ, or having a route to destination of RREQ, send RREP to the originator of RREQ +If not, reborder-casting this RREQ Packet	X∈{3,1}
III.3	Send RREP-ARR	3	Send to the duplicated node via reversed AODV route	3
III.4	Received RREP-ARR	1	Drop the Packet	1
		2	Upstream direction: forward to its cluster head via its upstream node Downstream direction: forward to its downstream node based on routing header	2
		3	Check if it is the destination in the RREP-ARR Packet +If not, forward this RREP-ARR Packet via AODV reversed route +If yes, **re-assigned cluster (its Cid=NULL)**	X∈{3,1}
III.5	Send RREP	3	Send RREP to the originator of RREQ via AODV reversed route	3
III.6	Received RREP	1	Drop the Packet	1
		2	Upstream direction: forward to its cluster head via its upstream node Downstream direction: forward to its downstream node based on routing header	2
		3	Update its inter-cluster routing cache If it is the destination in RREP Packet, start transmiting Data Packet If not, forward this RREP to the originator of RREQ via AODV reversed route	3
III.7	Send RERR	3	Send RERR to the originator of RREQ via AODV reversed route to inform the route error	3
III.8	Received RERR	1	Drop the Packet	1
		2	Upstream direction: forward to its cluster head via its upstream node Downstream direction: forward to its downstream node based on routing header	2
		3	Check if it is the destination in the RERR Packet +If not, forward this RERR Packet via the AODV reversed route +If yes, re-generate RREQ for a RREQ_RETRIES	3
III.9	Send Hello	X∈{1,2,3}	Broadcast Hello to all its neighbors (Cid, Stid, level,Hop-Count=1,...)	X∈{1,2,3}
III.10	Received Hello	X∈{1,2,3}	Update its corresponding neighbor status in its neighbor cache	X∈{1,2,3}
III.11	Send Data	X∈{2,3}	Send to its cluster head via its upstream node	X∈{2,3}
III.12	Received Data	1	Drop the Packet	1
		X∈{2,3}	Check if it is the destination in the Data Packet If not: +Upstream direction: forward to its cluster head via its upstream node +Downstream direction: forward to its downstream node based on routing header If yes, send data to its upper layer	X∈{2,3}

Fig. 9. State transition table of BGP-GCR+ *(Inter-cluster routing and DAD handlers)*

load-balancing, minimum hop-count,...[10]. To our best knowledge, there are currently no works concentrating on the use of MANETs as the transit networks for the Internet.

Numerous ad hoc routing protocols for the stand-alone MANETs have also been proposed. They differ in the way how a route is established (table-driven or on-demand or hybrid) [6, 7]; how the network topology is organized (flat or hierarchical) [1, 6, 7]; how the service is supported (best-effort or QoS) [1]; whether or not they are scalable, considered cross-layer relation,...[1, 10]. Extensions on the stand-alone ad hoc routing protocols have been carried out to support for the internetworking MANETs with the Internet [6, 10].

Address auto-configuration protocols for MANETs can be classified as either stateful/stateless or independent/dependent routing protocols. Further classifications are specified in [12].

6 Conclusions and Future Works

This paper introduces a new use of MANETs: backup or load-balancing transit networks for the Internet. To achieve the required functions, we have proposed and prototyped towards standards the BGP-GCR+ architecture, a combination of BGP, GCR, and the passive/weak IPv6 address stateless auto-configuration protocol.

The current version of BGP-GCR+ uses the GCR with an enhancement on the neighbor and the cluster selection. Our objective is to continue maintaining the stable cluster with the highest probability. We also use the link states in Join and Link Change Packets to integrate the weak duplicate address detection within the cluster. For the correctness of the weak DAD, each node in the cluster generates a unique random number (R-ID) and we also assume that the MAC addresses of nodes are unique within two-hop neighbor.

Intra-cluster routing is maintained by the cluster head based on its intra-cluster routing cache. Inter-cluster routing is AODV-based. The weak DAD and passive DAD schemes have been integrated into the route discovery for detecting the duplicated cluster. When a cluster ID is duplicated, it re-assigns (Cid=NULL) the cluster. Renumbering cluster is another approach we will consider in future works.

To find an external route to another AS via MANETs, an exit point needs a selection procedure. We have developed the weight and local preference attributes of BGP for this purpose. An exit point will select an external route with the highest weight, and an AS will select an external route with the highest total bandwidth among its exit points. In BGP-GCR+, the bandwidth is used to calculate the weight. We also suggest two approaches for determining MANETs capacity, either static or dynamic.

When a route between two exit points is established via MANETs, the destination AS number is used instead of MANETs IPv6 site-local address. This is because an exit point only know either the AS number or the global IPv6 unicast address of the destination exit point in another AS. To further make the DAD transparent to the exit points, and increase the load-balancing capacity of MANETs in the neighboring area of exit points, we have suggested setup the exit point to be the cluster head.

Our future works will be the implementation of our proposed BGP-GCR+ in ns-2 [14] for performance evaluation. We also consider alternative approaches to open problems of GCR and compared results with those of BGP-GCR+.

References

1. C. Y. Chiu and C. Gen-Huey, "A stability aware cluster routing protocol for mobile ad hoc networks," *Wireless Communications and Mobile Computing 2003*; 3:503-515.
2. Y. Rekhter and T. Li, "A Border Gateway Protocol 4 (BGP-4)," *Request for Comments (RFC1771)*, Mar.1995.
3. Killian Weniger, "PACMAN: Passive Autoconfiguration for Mobile Ad hoc Networks," *IEEE Journal on Selected Areas in Communication,* March 2005, Vol.23 No.3.
4. Nitin H. Vaidya, "Weak Duplicate Address Detection in Mobile Ad Hoc Networks," *MOBIHOC'02*, June.9-11 2002, EPLF Lausanne, Switzerland.
5. J. Li, C. Blake, D. S. J. De Couto, H. I. Lee, and R. Morris, "Capacity of Ad Hoc Wireless Networks," *MOBICOM'01*, July 2001, Rome, Italy.
6. C. Perkins, E. Belding-Royer and S. Das, "Ad hoc On-Demand Distance Vector (AODV) Routing for IP version 6," *draft-perkins-manet-aodv6-01.txt*, July 2003.
7. D. B. Johnson, D. A. Maltz, and Yih-Chun Hu, "The Dynamic Source Routing Prototol for Mobile Ad Hoc Networks (DSR)," *draft-ietf-manet-dsr-10.txt*, July 2004.
8. S. Thomson, T. Narten, and T. Jinmei, "IPv6 Stateless Address Autoconfiguration," *draft-ietf-ipv6-rfc2462bis-08.txt*, May 2005.
9. R. Hidden and S. Deering, "IP Version 6 Addressing Architecture," *draft-ietf-ipv6-addr-arch-v4-04.txt*, May 2005.
10. C. E. Perkins, J. T. Malinen, R. Wakikawa, A. Nilsson, and A. J. Tuominen, "Internet connectivity for mobile ad hoc networks," *Wireless Communications and Mobile Computing,* 2002; 2:465-482.
11. D. Porcino and W. Hirt, "Ultra-Wideband Radio Technology: Potential and Challenges Ahead," *IEEE Communications Magazine*, July 2003.
12. K. Weniger and M. Zitterbart, "Address Autoconfiguration in Mobile Ad Hoc Networks: Current Approaches and Future Directions," *IEEE Network* ,July/August 2004.
13. S. Deering and H. Hidden, "Internet Protocol, Version 6 (IPv6) Specification ," *rfc2460.txt*, December 1998.
14. The Network Simulator ns-2. *http://www.isi.edu/nsnam/ns/.*

A Local Repair Scheme with Adaptive Promiscuous Mode in Mobile Ad Hoc Networks*

Doo-Hyun Sung[1], Joo-Sang Youn[1], Ji-Hoon Lee[2], and Chul-Hee Kang[1]

[1] Broadband Communication Lab., Dep. Of Electronics Engineering, Korea University
Sungbuk-gu, Anam-dong 5-1ga, Seoul, 136-701 Korea
{ssungdoo, ssrman, chkang}@widecomm.korea.ac.kr
[2] Comm. & Network Lab., Samsung Advanced Institute of Technology,
San 14-1, Nongseo-Ri, Kiheung-Eup, Yongin, Kyungki-Do, 449-712 Korea
vincent.lee@samsung.com

Abstract. In current routing protocols in ad hoc networks, a source node unnecessarily re-discovers the whole path when just one node moves, even if the rest of path needs not to be re-arranged. The time for re-discovery of the whole path may often take too long and affects the network efficiency adversely. In this paper, a local repair scheme has been proposed, where the source node recovers the route breakage caused by a shifting of node with an aid of adjacent node instead of re-routing the whole path. Adjacent nodes are under operation in adaptive promiscuous mode. Consequently, the proposed scheme has advantages in shortening the recovery time of the route breakage and in minimizing the energy consumption under promiscuous mode. The ns-2.27 simulator has been utilized for the evaluation of the proposed scheme.

1 Introduction

The MANET (Mobile Ad-hoc NETwork) is a part of network system consisted of a group of mobile nodes communicating with each other over wireless channel under circumstances where no regular network infrastructure is established. Due to the limitation of radio propagation range, nodes usually communicate through several intermediate nodes. Each node is a host and also serves as a router. The topology of network changes frequently and randomly by nodes' mobility. For this reason, proactive routing protocols may not function effectively due to their periodical updates of routing information. Instead, reactive routing protocols have been mostly researched. They instantly construct a route to the destination and maintain it during data transmissions. By doing this, routes are not sensitive to the topology changes. Under circumstances where route breakages occur frequently and randomly due to shifting of nodes, the recovery of route breakage will be as important as route establishment in ad hoc routing protocols. When an existing route is disconnected by

* This paper was supported by the Samsung Advanced Institute of Technology, under the project "Research on Core Technologies ant its Implementation for Service Mobility under Ubiquitous Computing Environment".

X. Jia, J. Wu, and Y. He (Eds.): MSN 2005, LNCS 3794, pp. 351 – 361, 2005.

node's mobility, a source node has to re-establish a new route. The route recovery time can be a critical factor affecting the network performance. Because route discovery under reactive routing algorithm is based on broadcasting Route Request messages from a sender node, it takes considerable time to recover the whole route. The main point of local repair is how fast the route can be restored in a seamless manner. To solve the inefficiency problem of route restoration, researchers have proposed local repair schemes. In [1], "Localized Route Repair (LRR)" technique has been devised. The LRR defines "neighbor node" as a node which is on the route and is in the immediate vicinity of the moved node. When the route is disconnected, the adjacent node detecting breakage automatically, broadcasts the "route-repair" packet with TTL limit 2. When the route-repair packet reaches the member of existing route, the route recovery is complete. Instead of network-wide flooding, the recovery message is shared among the limited adjacent nodes surrounding the route breakage. If there is no reply from any node within a limited time, it activates the route maintenance mechanism of the underlying protocol to find a new route. LRR suggests local repair algorithm based on broadcasting twice. To restore the route breakage more promptly, the neighbor nodes around the active route have to know the situation around them. Witness Aided Routing (WAR)[2], AODV-Backup Routing (AODV-BR)[3] adopt the promiscuous mode[4]. WAR uses witness hosts to overcome transient problems, which greatly reduces the overall packet delivery time and the network traffic generated by route discovery messages. The witness hosts forward packets instead of broken node. AODV-BR establishes backup routes when Route Reply (RREP) comes to source node. When a node has broken, backup routes around the node activated and recover the breakage. Both protocols recover the breakage locally utilizing promiscuous mode, where an additional overall communication burden is unavoidable as long as nodes keep the mode. Continuous promiscuous mode operation may be a large overhead to the network.

In this paper, our work focuses on recovering the route breakage quickly and maintaining end-to-end hop counts by help of neighborhood nodes. It is the local repair scheme with adaptive promiscuous mode. Nodes adaptively convert promiscuous mode and non-promiscuous mode. Each node while receiving packets in a promiscuous manner, maintain automatically its data being updated by neighbor nodes. When a route has broken, a node detecting the breakage transmits a help message to neighbor nodes. When any node which can restore the route among those adjacent nodes, responds to the help message, the recovery procedure is completed. Adaptive promiscuous mode reduces the restoration time and its required energy and thus increases network efficiency through a quick restoration of route breakage.

The rest of the paper is organized as follows: Section 2 describes the proposed scheme in details. In Section 3 we show the numerical results and discuss conclusions in Section 4.

2 A Local Repair Scheme with Active Promiscuous Mode

This section illustrates the design of proposed scheme in more details. The main algorithm, "Adaptive Promiscuous Mode" is proposed, in which a node decides independently whether to keep the promiscuous mode. It depends upon the number of neighbor nodes, participating in a specific routing path. With adaptive promiscuous mode, the proposed scheme can recover route breakage caused by node's mobility locally and quickly.

2.1 Adaptive Promiscuous Mode

To achieve the goal of quick route repair, our scheme adopts promiscuous mode, in which each node keep processing automatically the overheard packets. By processing the overheard packets, each node obtains the routing information about the adjacent nodes. A continuous operation of the whole network system in a promiscuous mode causes an excessive energy overhead of the system and reduces the efficiency. The proposed scheme provides with a system to improve the network efficiency by making each node adapt itself to the changing environments and decides autonomically whether to maintain promiscuous mode or not.

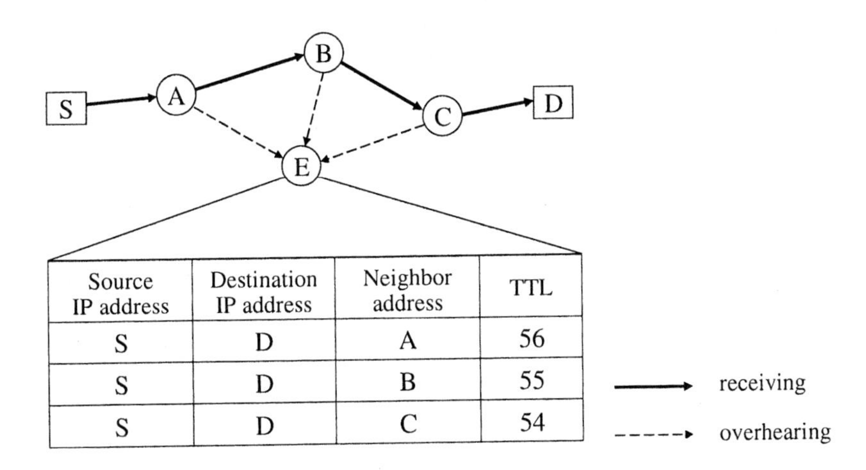

Source IP address	Destination IP address	Neighbor address	TTL
S	D	A	56
S	D	B	55
S	D	C	54

Fig. 1. Note E overhears packets from node A, B, and C, then creates a table

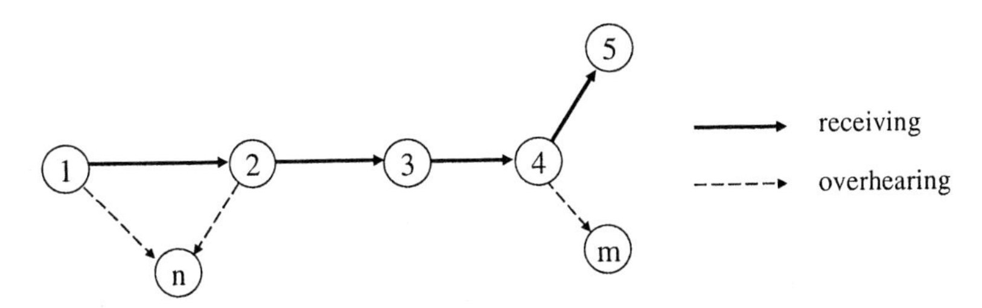

Fig. 2. Node n and m overhear their neighbor nodes, however, they have no ability to recover neighbor nodes' breakage (node 1, 2 and 4)

When nodes operate in promiscuous mode, they read the packet header of overheard packet. From packet header, they acquire information like where the packet comes from, where it goes to, and how many hops it can be forwarded. Based on these, an entry of table is created. Consider that source node S communicates with destination node D in Fig. 1. Node E existing around the active route overhears packets of node A, B, and C. From the overheard packets, node E extracts the path information and creates entries for each neighbor node A, B, and C. The entry contains neighbor's address, source (SRC) IP address, destination (DST) IP address, and TTL (Time To Live) value. For example, the entry is [SRC S, DST D, A, 56] in case of neighbor node A. When packets are transmitted from node B and C, the entries are added with neighbor address as B and C. Each entry has an expiration timer, which means the life time of entry. If the path information of overheard packet already exists in the table, the entry needs not to be updated and simply the expiration timer is reset.

The reason for adopting promiscuous mode is to recover the route breakage in a quick manner. When a node loses its ability to recover the connection, it has no need to keep the promiscuous mode. Thus, an adaptive promiscuous mode has been devised with above consideration. Consider the two cases in which nodes exist in the middle of a routing path and its neighbor as shown in Fig. 2. A packet is forwarded from node 1 to node 5. The nodes n, m are receiving packets promiscuously. Node n processes overheard packets and creates a table containing 2 entries (node 1, 2). The table of node m has just 1 entry (node 4). When node 1 moves away, node n has nothing to do any process. It happens due to the lack of information about the neighborhood nodes. Similarly, node m is not in a position to recover any route breakage due to the node's mobility, either. To enable the recovery of route breakage caused by mobility, the nodes staying within the active route (i.e. node n, m in Fig. 2) need continuously the respective information about the pre- and post-node of the shifted node when any nodes have no further information to their requirements, those nodes may discontinue promiscuous mode. The nodes of n, m are in the same situation of losing promiscuous mode.

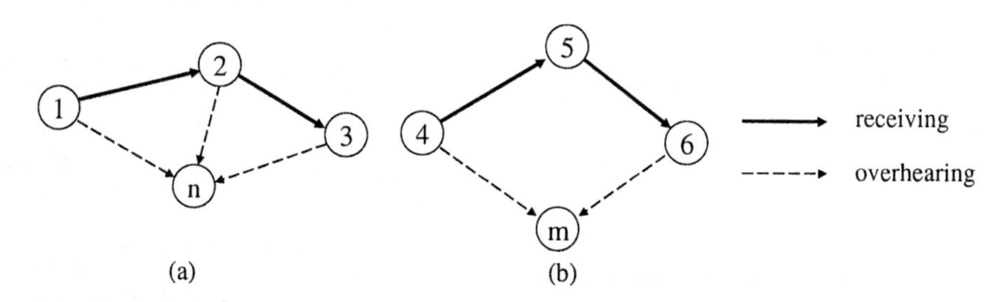

Fig. 3. Two requirements for continuous promiscuous mode: (a) 3 neighbor nodes, (b) 2 neighbor nodes

For route restoration, a node needs to keep contact with two immediate adjacent nodes of the disconnected node, i.e. both the former and latter node of breakage point. Any node satisfying either one of the following two requirements maintains the promiscuous mode. The requirements are that (1) nodes to maintain more than three neighbors for SRC-DST IP addresses pair, and (2) the difference of TTL value has to exceed 2 when there exist only two neighbors for SRC-DST IP addresses pair. The

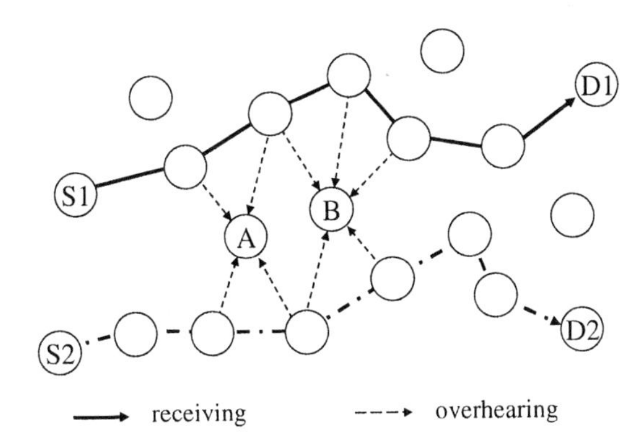

Fig. 4. Both node A and B begin to operate in promiscuous mode, but node A stops the mode because it receives only two unlike node B

criterion of selecting promiscuous mode is the number of nodes that actively participate in one routing path. Fig. 3 shows the example as mentioned above. In Fig. 3(a), node n recognizes that there exist three nodes consisting of one routing path. When the path breaks by mobility of node 2, node n replaces node 2 as the route without any additional changes like end-to-end hop count. After recovery, the previous route [1-2-3] will change to [1-n-3]. Consequently, a node which has 3 neighbors for one routing path maintains promiscuous mode. The Fig. 3(b) displays another requirement. Node m overhears packets of 2 neighbor nodes, node 4, 6. The difference of TTL between node 4, 6 is two. Although node m does not recognize existence of node 5, it may become a new intermediate node of the route when node 4 signals node m to recover the route vacancy caused by shift of node 5. Thus the node of having two neighbors with TTL value difference of 2, will keep its promiscuous mode.

Fig. 4 shows the difference between total number of neighbors and number of neighbors per routing path. Node A overhears packets from total 4 neighbor nodes. Node B has total 5 neighbors. Although both nodes have more than 3, only node B maintains promiscuous mode. Because node A has actually 2 neighbors per routing path and difference of TTL is 1, it does not satisfy the requirements. Accordingly node A does not meet the qualification. Node B, however, has 3 neighbors for S1-D1 route. In this way, it will maintain the promiscuous mode.

The operational procedure of adaptive promiscuous mode is described as follows. A node operates in promiscuous mode when it enters into the network. It overhears forwarded packets and checks neighbor address, SRC-DST IP addresses and TTL value. If the table doesn't have been created or the information is a new one, the node creates or updates an entry based on overheard information. When the node finds that the information has already existed, it does not update anything. Every table has a timer and the node maintain the promiscuous mode during at least ‘t’ seconds. Within ‘t’ seconds, the node waits any additional entry updates. If there is any update, the timer resets. If not, the node checks the contents of table. When the contents do not meet the requirements, it operates in non-promiscuous mode for next ‘t’ seconds. In

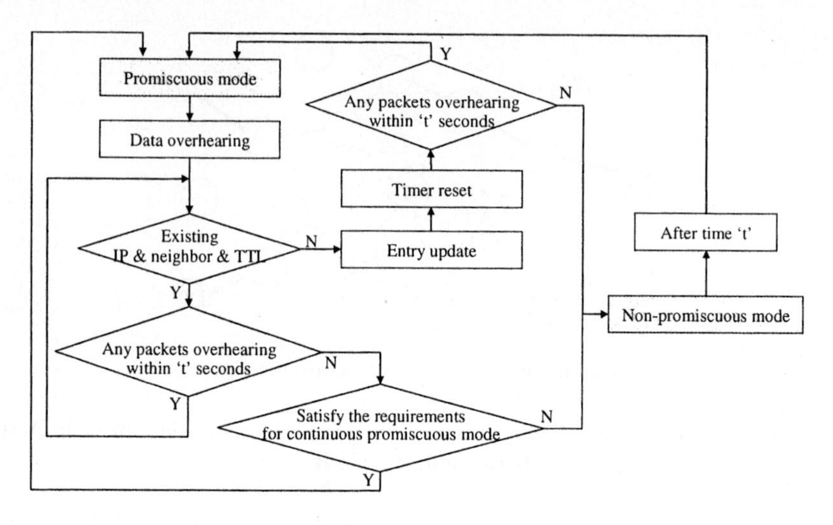

Fig. 5. The flow chart of adaptive promiscuous mode

Variables
acvt : whether promiscuous mode ON or OFF
src_ip_addr : source IP address
dst_ip_addr : destination IP address
nei_addr : neighbor address
ttl : ttl of IP packet

```
initially promiscuous mode ON
acvt = TRUE
for every overheard packet
{
    if new SRC-DST IP addresses && neighbor address && TTL?
    {
        entry creation
        [SRC address, DST address, neighbor address, TTL]
    }
    entry timer reset                //determine entry invalidation
}
for every expiration timer           //determine promiscuous mode
{
    if (3 neighbors / IP address pair)
        | | (2 neighbors / IP address pair && TTL difference ==2)
    {
        keep promiscuous mode
    }
    convert to non-promiscuous mode
}
```

Fig. 6. The pseudo-code of adaptive promiscuous mode

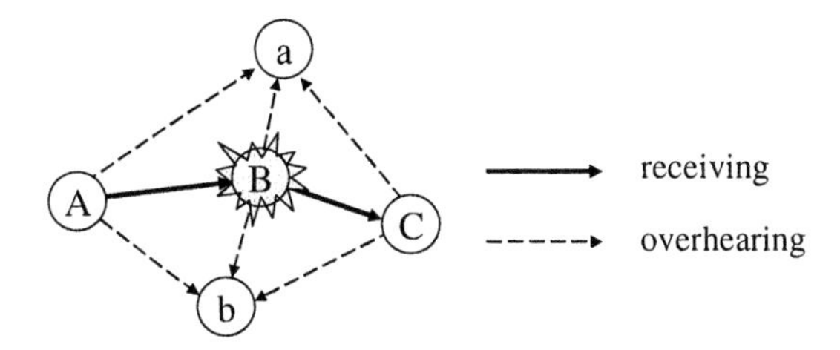

Fig. 7. When node B moves away, node A activates local repair mechanism, and then node a or b may recover the route breakage

this paper, the time 't' is fixed as 0.5ms. After that, it resumes the promiscuous mode for 't' seconds again. The node reiterates these processes. The whole process is shown in Fig. 5. Fig. 6 shows the pseudo code of adaptive promiscuous mode.

2.2 Quick Local Repair

Unlike the existing route recovery processes, proposed scheme can recover the route breakage as soon as it is detected by means of overhearing nodes. Also it does not change end-to-end hop count. As a brief summary, a node detecting route breakage asks its neighbor nodes route recovery. Then the neighbor nodes which have information about local area replace the break point as a route quickly.

The operation of proposed local repair scheme proceeds through 4 steps as described below (refer to Fig. 7).

Step 1. Failure detection: In on-demand routing protocols, a node participating in a routing path periodically checks the local connectivity with its next hop node. There are link layer notification and passive acknowledgment, etc. A node is able to confirm the connectivity to its next hops by link layer ACK or RTS & CTS in link layer notification. If link layer notification is not available, passive acknowledgment can be used by receiving any packet including "Hello" message from next hop nodes in AODV. In case of Fig. 7, when node B moves away, node A detects that, after local connectivity procedures.

Step 2. Asking for help: If a node notices that its next hop node does not exist any longer, it broadcasts "HELP" message toward any node which has the ability to repair the break point quickly. Node A transmits HELP message after detection and the message reaches at node a, b as shown in Fig. 7 above. The HELP message contains its own (node A) and next hop node's (node B) IP address.

Step 3. Reply to HELP message: When an overhearing node receive "HELP" message, it verifies the possibility. Because it will replace the error node as a routing path with unchanged hop count, it should have knowledge of nodes existing right before and after the break point. A node which satisfies requirements responds to HELP message. If not, it does nothing. Node A and B deserve to reply.

Step 4. Completion of local repair: When the node which requests for help receives more than 2 reply packets, it selects the previous one and changes next hop of own routing table. If both node a and b send reply packets, node A change its routing table according to a packet which most previously arrives. But in case of no replies, node A

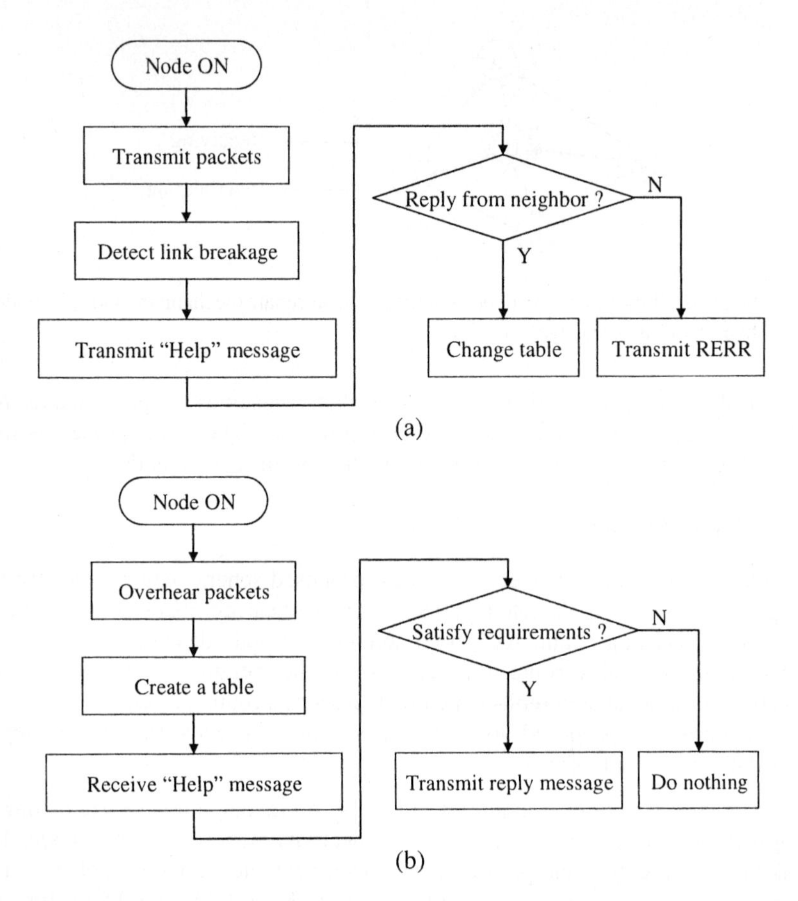

Fig. 8. The flow charts of quick local repair: (a) the flow chart of node A in Fig 7, (b) the flow chart of node a, b in Fig 7

transmits route error message back to the source node. The flow chart of local repair process is described in Fig. 8. Fig. 8(a) is the flow chart of node A. Fig. 8(b) is the one of node a and b.

3 Simulation Studies

In this section, simulation environment is created and we show the results to evalute performance improvements. We adopt AODV[5] protocol a routing protocol, and implement our scheme within AODV. In addition, we experiment promiscuous mode to verify efficiency of proposed scheme. The simulation is implemented within ns-2.27[9].

The simulation model has the number of mobile hosts varied from 10 to 50 with increment of 10. These hosts are randomly deployed within a 1500m x 500m field except the case of 10 hosts. 1000m x 500m field is used for 10 hosts. The host mobility follows random waypoint mobility model. The velocity is kept at a uniform

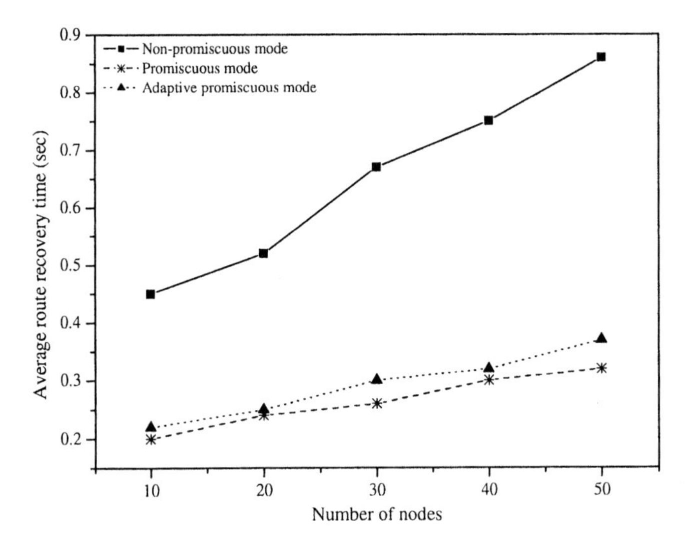

Fig. 9. The average time for route recovery

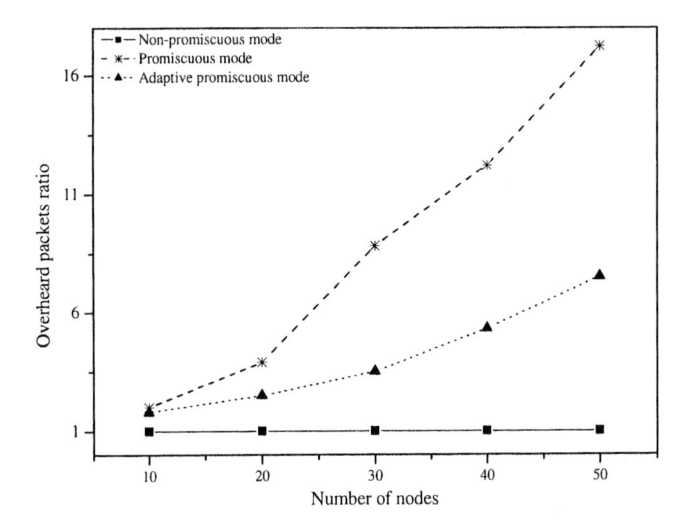

Fig. 10. The average ratio of total (overheard & forwarded) packets to forwarded packets

rate of 0-20 m/s. The transmission range of each node is 250m. The underlying MAC is 802.11 DCF. And the channel capacity is 2 Mbps. The type of traffic generated each node is Constant Bit Rate (CBR). Each simulation lasts for 300 seconds. The results are shown in Fig. 9 and Fig. 10.

The route recovery time is showed in Fig. 9. As the number of nodes grows, recovery time of non-promiscuous mode (original AODV) sharply increses more than that of promiscuous mode and adaptive promiscuous mode. The difference between promiscuous mode and adaptive promiscuous mode is about 0.03s. The non-promiscuous period in adaptive promiscuous mode leads to this result. Fig. 10 shows the overhead caused by using promiscuous mode. The overhead means the average ratio of received packets to received and overheard packets for one connection. When nodes operate in non-promiscuous mode, the ratio value equals 1. As nodes are densely deployed, they overhear and process much more packets.

From the above results, the local repair scheme considerably decreases the route recovery time, and nodes reduce overhead by discarding unneccesarily overheard packets when they operate in adaptive promiscuous mode.

4 Conclusions

In this paper, we proposed the adaptive promiscuous mode operation and the local repair scheme using the adaptive promiscuous mode. Nodes adaptively convert their mode according to the number of neighbors in a specific route. With these nodes, the route can be quickly recovered with less overhead. Through the proposed scheme, we achieved quick & local route repair by aids of overhearing nodes. Besides the overhead owing to the promiscuous receiving has been considerably reduced by adopting the promiscuous mode adaptively. The simulation results showed that our scheme not only takes similar time to recover the route compared with promiscuous mode, but also reduces considerable overhead. As a result, our scheme is an efficient route recovery solution suitable for mobile ad-hoc networks.

References

[1] R. Duggirala, R. Gupta, Q. A. Zeng, and D. P. Agrawal.: Performance Enhancements of Ad Hoc Networks with Localized Route Repair. *IEEE Transactions on Computers*, Vol. 52, no. 7, pp.854-861, Jul 2003

[2] I. Aron and S. Gupta.: A Witness-Aided Routing Protocol for Mobile Ad-Hoc Networks with Unidirectional Links. In *Proceedings of First International Conference on Mobile Data Access (MDA '99)*, pp.24-33, Hong-Kong, Dec 1999.

[3] S. J. Lee and M. Gerla.: AODV-BR: Backup Routing in Ad hoc Networks. *IEEE Wireless Communications and Networking Conference (WCNC)*, Vol. 3, pp.1311-1316, Sep 2000

[4] D. Johnson, D. A. Maltz, and Y. Chu.: The Dynamic Source Routing Protocol for Mobile Ad Hoc Networks (DSR) (Internet Draft). Mobile Ad-hoc Network (MANET) Working Group, IETF, Jul 2004

[5] C. Perkins, E. Belding-Royer, and S. Das.: Ad hoc On-Demand Distance Vector (AODV) Routing. IETF RFC 3561, Jul 2003

[6] G. Liu, K. J. Wong, B. S. Lee, B. C. Seet, C. H. Foh, L. Zhu.: PATCH: a novel local recovery mechanism for mobile ad-hoc networks. *IEEE Vehicular Technology Conference*, Vol 5, pp.2995-2999, Oct 2003

[7] M.Spohn and J. J. Garcia-Luna-Aceves.: Neighborhood-aware Source Routing. In *Proceedings of the ACM Symposium on Mobile Ad Hoc Networking and Computing (MobiHOC'01)*. pp.11-21, Oct 2001

[8] R. Castaneda and S. Das.: Query Localization Techniques for On-Demand Routing Protocols in Ad Hoc Networks. In *Proceeding of Mobile Computing and Communication Conference*. Vol. 3, pp.113-120, Aug 1999
[9] The ns-2 Network Simulator. *http://www.isi.edu/nsnam/ns/*
[10] The CMU Monarch Project. *http://www.monarch.cs.emu.edu*

PSO-Based Energy Efficient Gathering in Sensor Networks*

Ying Liang[1,2] and Haibin Yu[1]

[1] Shenyang Institute of Automation,
Chinese Academy of Sciences, 100016, China
{liangying, yhb}@sia.cn
[2] Graduate School of Chinese Academy of Sciences, 100039, China

Abstract. Sensor networks consisting of nodes with limited battery power and wireless communications is among the fastest growing technologies. There are many challenges in implementation of such systems: energy dissipation and data gathering being one of them. Gathering sensed information in an energy efficient manner is critical to operate the sensor network for a long period of time. While cluster-based data collection is efficient at energy and bandwidth, it's difficult to ensure balanced energy depletion on all cluster-members after a longtime run. In this paper we propose a new distributed cluster-based data gathering algorithm using PSO to optimize clustering process. The election of cluster-heads need synthetically consider the state information including location and energy reserved about candidates and their neighbors. Our preliminary simulation results show that the proposed algorithm balances the energy dissipation over the whole network thus prolongs the network lifetime.

1 Introduction

Wireless sensor networks are a prime example of a second generation distributed system. Such dynamic, adaptive and distributive systems will enable the reliable monitoring of a variety of environments for application that include home security, machine failure diagnosis, chemical/biology detection, medical monitoring, and a variety of military applications[1], [2]. There are many fundamental problems that sensor networks research will have to address in order to ensure a reasonable degree of cost and system quality. Some of these problems include energy dissipation and data gathering. Since sensor nodes are severely constrained by the amount of battery power available, the key challenge of gathering and routing data in wireless sensor network is conserving the sensor's energies, so as to maximize their lifetime[3].

In most of the applications, sensors are required to collect audio, seismic, and other types of data and communicate with one another to transfer the collected data to base stations. They also need to collaborate to route the control information from the base

* This paper is supported by Natural Science Foundation of China under contract 60434030 and 60374072.

X. Jia, J. Wu, and Y. He (Eds.): MSN 2005, LNCS 3794, pp. 362–369, 2005.

stations to a specific sensor. Communication is usually the main source of energy dissipation in sensors, which greatly depends on the distance between the source and destination of a communication link. Previous researches[4], [5], [6] have shown that energy dissipation e to transmit a message to a receiver at distance d can be estimated by the following formula:

$$e = kd^c \tag{1}$$

where k and c are constants for a specific wireless system (usually $2 < c < 4$).

Since message transmission energy consumption is proportional to d^c where $c > 2$, significant amount of energy savings can be made by partitioning the sensor nodes into clusters and transmitting the information in a hierarchical fashion. Moreover data aggregation at cluster-heads can drastically reduce the amount of data transmitted and tolerate node failures. Hence, clustering, specifically cluster-heads selecting, is one of the pivotal problems in sensor network applications and can drastically affect the network's communication energy dissipation.

There are already a lot of works related to cluster for data collection. Several heuristics have been proposed to choose cluster-heads in ad hoc networks. They include ID-based, connectivity-based and weight-based. Clustering algorithm for ad hoc network is to maintain network topology, while the main goal of sensor network' clustering is to balance energy consumption over the whole network. The LEACH protocol presented in [6] is an elegant solution where a small number of clusters are formed in a self-organized manner. A designated node in each cluster collects and fuses data from nodes in its cluster and transmits the result to the BS. LEACH uses randomization to rotate the cluster heads and achieves a factor of 8 improvement compared to the direct approach, before the first node dies.

While cluster-based data collection is efficient at energy and bandwidth, it's difficult to ensure balanced energy depletion on all cluster-members after a longtime run. In LEACH, the decision of cluster-heads is simply based on the suggested percentage of them for the network and the number of times the node has been a cluster-head so far. The method ensures that none of the sensors are overloaded because of the added responsibility of being a cluster-head, however, it neglects the distance between cluster-heads and their cluster-members. As mentioned above, message transmission energy consumption is proportional to d^c, that is, sensor nodes further from their cluster-heads will spend more energy for longer distance transition. As a result, the remaining energy level for each node in the network could be unbalanced if running the LEACH protocol for several rounds. If there exists several nodes in clusters having little residual energy and longer distance from their cluster-heads, we could imagine that those would quickly exhaust all their rest energy for data transmission and become dead nodes. The emergence of dead nodes with high frequency would reduce the network lifetime of sensor networks and lead to inefficiency of the protocol.

Fig.1 demonstrates an example of cluster-head affect on communication energy consumption. In this figure, we assume the distance between node A and cluster-head is R1 and the remaining energy of it is EA, the distance between node B and cluster-head is R2 and the remaining energy of it is EB. We have EB>>EA and R1> R2. Since the energy ET consumed in transmitting to destination is proportion to the distance, it is obvious to have the result EB-ETB>>EA-ETA, which declare node A could have high probability to become a dead node. If cluster-head selection in clustering enables to consider the location and the residual energy of its cluster-members, it would efficiently lower the probability of dead nodes occurrence and further improve the lifetime of sensor network through evenly energy dissipation in clusters.

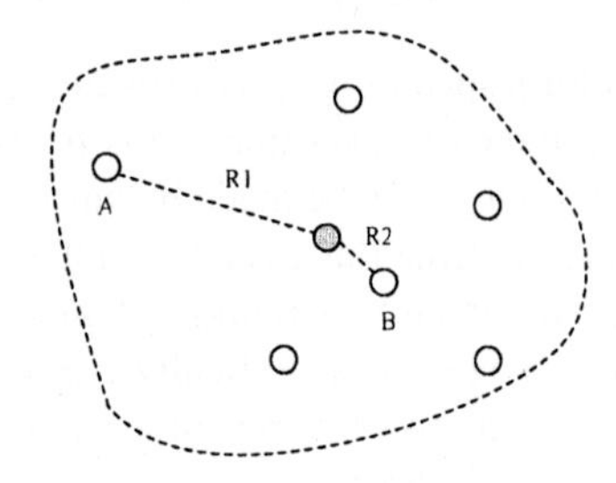

Fig. 1. Cluster of ignoring the energy and location of cluster-members

Therefore, an important optimization question to be answered is: to which cluster-head each sensor node has to communicate such that the total energy dissipation of the sensors is balanced and minimized. To addressing the issue, we propose a novel distributed cluster-based data gathering algorithm using PSO to optimize clustering process. In this work, we assume that sensors are distributed in a spatial region and have location information about all other nodes. Our model sensor network has the following properties:

The BS is fixed at a far distance from the sensor nodes.
The sensor nodes are homogeneous and energy constrained with uniform energy.
No mobility of sensor nodes.

The rest of the paper is organized as follows. Section 2 describes the algorithm of Particle Swarm Optimization. Section 3 provides a basis description of the optimal PSO-based clustering mechanism. Section 4 presents simulation results showing the effectiveness of our solution. Finally, we present our conclusion in section 5.

2 Particle Swarm Optimization

PSO is a kind of evolutionary computation proposed by Eberhart and kennedy. The algorithm is inspired by the research of the social behavior of bird flocking. PSO simulates the behaviors of bird flocking. Suppose the following scenario: a group of birds are randomly searching food in an area. There is only one piece of food in the area being searched. All the birds do not know where the food is. But they know how far the food is during each of search iterations. So what's the best strategy to find the food? The effective one is to follow the bird which is nearest to the food. PSO learned from the scenario and used it to solve the optimization problems. In PSO, each single solution is a "bird" in the search space. We call it "particle". All of particles have fitness values which are evaluated by the fitness function to be optimized, and have velocities which direct the flying of the particles. The particles are "flown" through the problem space by following the current optimum particles[7].

PSO is an iteration-based optimal algorithm. The system is initialized with a population of random solutions and searches for optima by updating generations. In the search process, PSO combine local information with global information, that is, a particle adjusts its current position not only according to its own historical information but also according to the related information of neighboring particles, and finds optimal solution through iteration. In every iteration, each particle is updated by following two "best" values. The first one is the best solution (fitness) it has achieved so far. This value is called individual best. Another "best" value that is tracked by the particle swarm optimizer is the best value, obtained so far by any particle in the population. This best value is a global best. When a particle takes part of the population as its topological neighbors, the best value is a local best.

After finding the two best values, the particle updates its velocity and positions with following formulas:

$$V_{id} = WV_{id} + c_1\text{rand}(\,)(P_{id} - X_{id}) + c_2\text{Rand}(\,)(P_{gd} - X_{id}) \tag{2}$$

$$X_{id} = X_{id} + V_{id} \tag{3}$$

Where V_{id}, called the velocity for particle i, is the distance to be traveled by this particle from its current position, X_{id} is the particle position, P_{id} is its best previous position, and P_{gd} is the best position among all particles in the population. Rand and rand are two random functions with a range [0,1]. c_1 and c_2 are positive constant parameters called acceleration coefficients (which control the maximum step size the particle can do). The inertia weigh, W is a user-specified parameter that controls, with c_1 and c_2, the impact of previous historical values of particle velocities on its current one. A large inertia weigh pressures toward global exploration (searching new area) while a smaller inertia weight pressures toward fine-tuning the current search area[8].

3 Optimal PSO-Based Clustering

The unbalance of energy depletion caused different distance from the cluster-heads is prone to lead cluster-members to die easily and thus reduce the network lifetime. To ensure fair energy dissipation in the network, additional parameters could be considered to balance the transmission energy consumption of nodes that are both low-energy and long-distance from cluster-heads. We address the problem by using PSO-based optimal clustering which is motivated from the LEACH. Our method extend LEACH's stochastic cluster-head selection by synthetically consider two essential parameters including the energy reserved and location of nodes-in-cluster. With PSO iteration iterative computation, optimal cluster-heads would be selected. The description of the algorithm is as follows:

Step1: Phase of the initial cluster-head formulation

In the phase of cluster formulation we can use LEACH to initially organize the sensor nodes into clusters and then determine the cluster-head of each cluster. We call the cluster-heads formed in the stage as initial cluster-heads. As stated above, Clusters formed in the stage are difficult to ensure balanced energy depletion on all cluster-members after a longtime run.

Step 2: Phase of initial cluster-heads collecting the information of in-cluster nodes

In the stage, nodes of each cluster would transmit their state information including location and rest energy to the initial cluster-head, in which the state information of cluster-members stored could be represented as $P\{p_1,p_2...p_n\}$ and $E\{e_1,e_2...e_n\}$, where n is the amount of nodes in each of cluster, p_i and e_i represents the location and the rest energy of node i.

Step 3: Phase of using PSO to optimize and determine optimal cluster-heads

The stage is the core of the PSO-based clustering protocol. In order to suit PSO to our problem domain, it is necessary to alternate Eq. (2) and Eq. (3) mentioned in PSO and give the according fitness function f(x).

Since searching is proceeded in flat network, the velocity is a vector which has magnitude and director, that is, the velocity includes x-velocity component and y-velocity component. From this, we have Eq.(4) and Eq.(5).

$$V_{xid} = WV_{xid} + c_1\text{rand}(\,)(P_{id} - X_{xid}) + c_2\text{Rand}(\,)(P_{gd} - X_{xid}) \tag{4}$$

$$V_{yid} = WV_{yid} + c_1\text{rand}(\,)(P_{id} - X_{yid}) + c_2\text{Rand}(\,)(P_{gd} - X_{yid}) \tag{5}$$

For the same reason, there exists x-location component and y-location component. We have Eq.(6) and Eq.(7).

$$X_{xid} = X_{xid} + V_{xid} \tag{6}$$

$$X_{yid} = X_{yid} + V_{yid} \tag{7}$$

Since node distribution in sensor network is discrete, the value calculated by formula (6) and (7) could not map the according nodes. It is necessary to adjust the calculated value of location to satisfy $X_{xid} \in \{p_{x1},p_{x2}...p_{xn}\}$ and $X_{yid} \in \{p_{y1},p_{y2}...p_{yn}\}$, where p_{xi} is the component of x direction and p_{yi} is the component of y direction.

Let $\triangle p_{xj} = | X_{xid} - p_{xj} |$, $\triangle p_{yj} = | X_{yid} - p_{yj} |$, $\triangle p_j = (\triangle p_{xj})^2 + (\triangle p_{yj})^2$, where $\triangle p_{xj}$ is the absolute value of the difference between X_{xid} and p_{xj} and $\triangle p_{yj}$ is the absolute value of the difference between X_{yid} and p_{xj}. Assume that $\triangle p_k = \min\{\triangle p_1 , \triangle p_2 ...\triangle p_n \}$, which means node k is the nearest node to X_{id}. After that, the adjusted value is $X_{xid} \approx p_{xk}$ and $X_{yid} \approx p_{yk.}$, that is, the searching node is on the location of node k.

The determination of the fitness function is closely related to the characteristic of special field. We should determine the fitness value of a node by considering not only the energy reserved in itself but also the energy distribution of its neighbors. The farther is the distance between the node and its neighbors, the greater is the energy reserved in the neighbors. By contraries, the nearer is the distance between the node and its neighbors, the fewer is the energy reserved in the neighbors. So, according to the characteristic talked above, the fitness function is constructed as Eq.(8).

$$f(k)=\eta e_k+\frac{\lambda}{n-1}\sum_{\substack{i=1\\ i\neq k}}^{n}\frac{e_i\cdot r_i}{r_i+1} \tag{8}$$

where $\eta+\lambda=1$ and $\eta,\lambda\in[0,1]$. In Eq.(8), k is the current number of the nodes, n is the amount of nodes, r_i is the distance between the current and node i, e_k and e_i are the value of energy reserved individually in the current node and node i, η and λ are the impact factor of the energy reserved individually in the current node and the neighbors. Properly adjusting the impact factor η and λ would demand the proportion of which neighbors devoted to the fitness value. The proposed PSO-based clustering algorithm summarized as follows.

a (Initialization) : Let n be the size of the PSO population, i.e., the amount of nodes in the cluster

b (Initial operation): For each partial i in the population

1 Initialize X_{xid}, X_{yid} of $PSO[i]$ randomly and adjust them according to the method above;

2 Initialize V_{xid}, V_{yid} of $PSO[i]$ randomly;

3 Evaluate f_i ;

4 Let P_{gd} be the best fitness among the population;

5 Let $P_{id}=f_i$.

c Repeat until the prescribed number of the cycle or a special stopping condition is satisfied.

1 Find the best fitness P_{gd} among the population;

2 For each particle i: evaluate P_{id} according to Eq.(8);

3 For each particle i: update V_{xid}, V_{yid}, X_{xid}, X_{yid} according to Eq.(4), Eq.(5), Eq.(6), Eq.(7) and adjust X_{xid}, X_{yid} according to the method above;

4 Evaluate the fitness value of particle i.

d Select nodes whose mapping particles have the bigger fitness value as the optimal cluster-head.

Step 4: Phase of optimal cluster formulation

In the phase of optimal cluster formation, each of initial cluster-heads has to decide the nodes acting as optimal cluster-head and then distribute the information to all in-cluster nodes. Since the cluster-heads have been optimized in clustering process, the algorithm we proposed efficiently balances the energy dissipations in cluster-members and avoids the untimely occurrence of blind nodes.

4 Simulations

In our simulations, we use the same radio model as discussed in [6] which is the first order radio model. To evaluate the performance of our proposed algorithm, we simulated it and LEACH using ns-2 with the 100-node network in a play field of size 100m x 100m. The base station is located at position (50,300). According to PSO algorithm, five different parameters values were chosen as follows: η=0.6, λ=0.4, c_1=c_2=2, W=0.9. The value of random function rand( ) and Rand( ) would be given dynamically. We assume the program stop running when the number of iteration is 100.

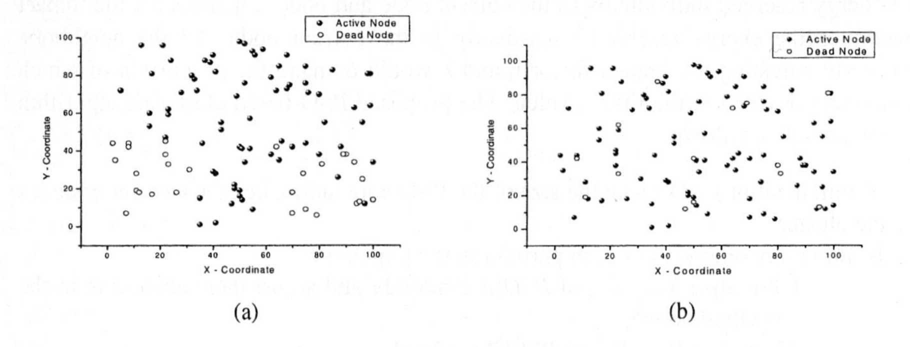

Fig. 2. Topology of sensors that remain alive (circles) and those that are dead (dots) after 1200 rounds with 0.5 J/node for (a) LEACH and (b) optimal PSO-based clustering

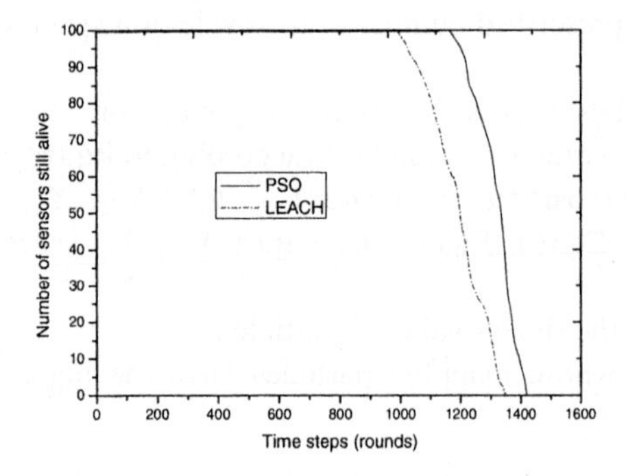

Fig. 3. Network lifetime using LEACH and optimal PSO-based clustering

We ran the simulation to determine the network lifetime and had each sensor send a 2000-bit data packet to the base station during each round of the simulation. After the energy dissipated in a given node reached a set threshold, that node was considered dead for the remainder of the simulation.

Fig.2 shows the topology of sensors that remain alive and those that dead after 1200 rounds with each node initially given 0.5 J of energy for the two protocols. This plot shows the distribution of dead nodes in the network using our proposed algorithm is more even than that using LEACH. This is as expected, since LEACH using randomization to select the cluster heads while neglecting the state information of cluster-members, the unbalance of energy depletion could be occurred. In addition, the uneven distribution of dead nodes would lead to information vacuum in a certain region, which decrease the network quality and thus shorten the network lifetime.

Fig.3 shows a comparison of network lifetime using our algorithm versus LEACH. This plot shows that our algorithm improves more than 10% of the useful network lifetime compared with LEACH.

5 Conclusions

In this paper, we describe a novel distributed cluster-based data gathering algorithm based on PSO to optimize clustering process. Optimal PSO-based clustering extend LEACH's stochastic cluster-head selection algorithm by synthetically consider two additional parameters including the energy reserved and location of nodes-in-cluster. Optimal cluster-heads could evenly distribute the energy dissipation among the whole network and save a lot of communication energy of sensor nodes, thus increase the lifetime and quality of the system. Our simulation shows that the distribution of dead nodes using our proposed algorithm is more even than that using LEACH. In addition, the optimal PSO-based clustering show 10% improvement of the useful network lifetime compared with LEACH. On base of the results we have achieved, our future research will focus on the interplay of choosing different parameters in PSO optimal algorithm and will give the contrast data in detail.

References

1. Estrin D., Govindan R., Heidemann J. and Kumar S.: Next Century Challenges: Scalable Coordination in Sensor Networks. Proceedings of the ACM/IEEE International Conference on Mobile Computing and Networking, Seattle, Washington, USA, pp. 263-270, 1999.
2. Kahn J., Katz R., and Pister K.: Mobile Networking for Smart Dust. In Proc. Mobicom, Aug. 1999, pp. 271-278.
3. Kohonen T.: The self-organizing map. Proceedings of the IEEE, Sept. 1990. pp.1464-1480.
4. Chang, J., Tassiulas, L.: Energy Conserving Routing in Wireless Ad hoc Networks. Proceedings of INFOCOM '00, 2000, vol. 1, pp. 22--31.
5. Li, Q., Aslam, J., Rus, D: Hierarchical Power aware Routing in Sensor Networks. Proceedings of the DIMACS Workshop on Pervasive Networking 2001.
6. Heinzelman, W. R., Chandrakasan A.: Balakrishnan, H. Energy efficient Communication Protocols for Wireless Microsensor Networks. Proc. Hawaaian Int'l Conf. on Systems Science, 2000.
7. Kennedy J., Eberhart R. C.: Particle swarm optimization. Proceedings of the IEEE International Conferrence on neural Networks , 1995, pp. 1942-1948.
8. Salman A., Ahmad I.: Particle swarm optimization for task assignment problem. Microprocessors and Microsystems, 2002, vol. 26, pp. 363-371.

Efficient Data Gathering Schemes for Wireless Sensor Networks

Po-Jen Chuang, Bo-Yi Li, and Tun-Hao Chao

Department of Electrical Engineering, Tamkang University,
Tamsui, Taipei County, Taiwan 25137, R.O.C
pjchuang@ee.tku.edu.tw

Abstract. Microsensor nodes, charged with battery power and capable of wireless communications, are distributed around to survey the environments and send the needed data to the base station periodically. The limited energy resources are consumed in both communication and computation -- especially in communication. To reduce such power consumption, this paper presents a new data gathering scheme based on the hypercube topology for microsensor networks. By gathering data from all nodes to the base station through the communication tree in the constructed hypercube, the new scheme is able to shorten communication delay by parallel transmissions and to replace dead nodes through reconfiguration. Simulation results show that compared with other data gathering schemes, the proposed hypercube-based scheme brings up favorable results, including reduced transmission delay, balanced energy loads, satisfying system scalability and as a result prolonged system lifetime. To complete our investigation, the distributed approach for constructing a hypercube and a binary tree for data gathering are also proposed.

1 Introduction

With advances in microelectromechanical (MEM) technology [1], cheap and wireless microsensor nodes are now available. Microsensor nodes, charged with battery power and capable of wireless communications, are distributed around to survey the environments and send the needed data to the base station. A microsensor system can be employed in both military and civil applications, such as detecting landmines, moving troops, monitoring building structures, observing animal behaviors, and so on.

In practice, microsensor nodes carry limited energy resources and need to consume energy in both communication and computation -- especially in communication [2]. If the energy consumption problem can be effectively addressed, the nodes will be more widely utilized. A number of schemes have been proposed to reduce such power consumption. For example, some use data-centric algorithms to collect and transmit data to the base station; others adopt power control mechanics to reach the intended recipients with minimum required energy.

A data gathering scheme which gathers sensed data from all sensor nodes to the base station (BS) efficiently is desirable for utilization of microsensor networks. This

X. Jia, J. Wu, and Y. He (Eds.): MSN 2005, LNCS 3794, pp. 370 – 379, 2005.

paper presents a new data gathering scheme for microsensor networks. The new scheme is built on the constructed hypercube topology whose communication tree transmission is able to shorten communication delay by parallel transmission and to replace dead nodes through reconfiguration. Simulation results show that compared with the other data gathering schemes, such as the Low-Energy Adaptive Clustering Hierarchy (LEACH) [3] and the Power-Efficient GAthering in Sensor Information Systems (PEGASIS) [4], the proposed scheme yields more favorable performance in terms of reduced transmission delay, balanced energy loads, satisfying system scalability and as a result prolonged system lifetime. To complete our investigation, the distributed approach for constructing a hypercube and a binary tree for data gathering are also proposed.

2 The Proposed Hypercube-Based Data Gathering Scheme

The proposed hypercube scheme forms the total sensor system into a (log N)-cube where N is the number of nodes. A node will be the root by turns at each round. After data are transmitted to the root through the communication tree embedded in the hypercube, the root will transmit the gathered data to the BS. Our proposed data gathering scheme constructs hypercubes by the bottom-up approach. That is, we form every two separate nodes into a 1-cube, two 1-cubes into a 2-cube and two 2-cubes into a 3-cube until all nodes are included in the constructed hypercube.

Assume that all nodes transmit their locations to the BS after deployment. The BS then takes the following steps to build links between two nodes:

1. Compute all links between any two nodes in the network.
2. Recurrently delete the longest of these links until a node, say a, retains only one link to another node, say b. Connect nodes a and b.
3. Delete all other links of node b until there remains only a link between the two nodes.

Construction of 1-cubes is thus completed; the connected node pairs are dimension 0 neighbors of each other. Constructing dimension-1 links involves a similar process except that the square of distances between two nodes is now computed to form two 1-cubes into a 2-cube. (This is because transmission energy grows according to the square of transmission distances.) Two values (sums of squares) will result from two possible connections; take the connection with the smaller value as the dimension-1 link.

To facilitate further discussion, some notations are provided below.

C_i : a set of i-cubes.

E_i : a set of connections between all pairs of i-cubes. Each element of the set represents the selected connection (including necessary dimension-i links) of a pair of i-cubes. The selected connection is the one with the minimum weight (the minimum sum of squares of the link lengths) of all possible connections between the pair of i-cubes.

d(c) : the degree of cube c (the number of connections relevant to c in E_i).

An i-cube has 2^i nodes, denoted by $X_0X_1X_2......X_{2^{i-1}}$ in which each X_i is a node labeled as i. Two nodes will be connected if and only if the binary representation of their labels differs in one bit position.

2.1 The Algorithm for Constructing the Hypercube

To form all nodes into 1-cubes, first put all nodes in C_0 and all links between any two nodes in E_0. Each link between any two nodes has the square of distance between the two nodes, i.e., the square of the link length, as the weight (reflecting the required transmission energy). Repeatedly delete the maximum weight of links until a node, say a, with d(a)=1 is found. Remove node a and the node connected with it, say b. Combine the two 0-cubes into a 1-cube and put it in C_1. The link between nodes a and b is marked as the dimension 0 link. Delete all other links relevant to the two nodes. In case two or more node pairs have degrees of 1, choose the node pair with the least weight. Repeat the above process until C_0 is null, i.e., until the construction of 1-cubes is completed. Similarly, higher dimensional cubes (say (i+1)-cubes) can be constructed from lower dimensional cubes (say i-cubes) using the same minimum weight criteria until a hypercube including all the nodes is constructed. The algorithm for constructing the hypercube is listed below.

```
START
i = 0
while (the number of cubes in Ci is greater than 1)
{
    for each pair of cubes in Ci select the connection with the minimum
    weight (i.e., sum of squares) put the selected connection into Ei.
    while (Ci is not empty)
    {
        if (there is only one cube in Ci)
            increase the dimension number of the cube's links and let
            the dimension 0 neighbor of each node in the cube to be itself.
            Remove it from Ci add it to Ci+1
        if (a cube has no related connections in Ei)
            re-compute connection between it and each cube in Ci and put
            all such connections into Ei
        while (a cube c in Ci has only one related connection in Ei with
                another cube c' in Ci)
            connect c and c' to form an (i+1)-cube by the related connection
            delete all the other related connections of c'. Remove c and c' from
            Ci and add the formed (i+1)-cube to Ci+1.
        Delete the connection with the maximum weight in Ei
    }
    i ++
}
END
```

After the hypercube is constructed, data can be transmitted through the communication tree embedded in it. The transmission scheme [5,6] is illustrated in Fig. 1 for a 4-cube, where the node at the starting end of an arrow line is a sending node and the one at the arrow head is a receiving node.

When transmission starts, first choose a head node by turns. Each node then compares its own label with the head's label from MSB to LSB. Among the different bits, the smallest bit number (say k) indicates the stage in which the node has to transmit to its dimension-k neighbor. For example, if the head node is 0000, node 0101 will transmit data at stage 0 and the receiver will be its dimension 0 neighbor, i.e., node 0100. Likewise, if the head node is 1001, node 0101 will transmit data at stage 2 to node 0001. Transmission completes at the log N stage, where N is the number of nodes. As the probability for a link to be used is $1/2^k$ (k being the dimension of the link), we decide to find the lowest dimension link first when constructing the cube.

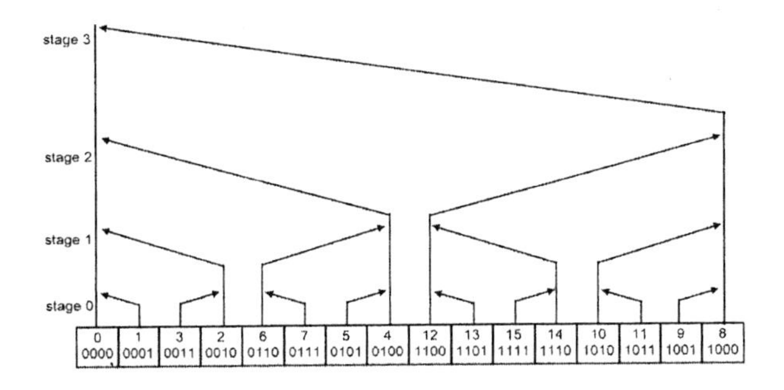

Fig. 1. Communication-tree transmission for a 4-cube

2.2 Hypercube Maintenance

In our data gathering scheme, the BS performs the task of constructing the hypercube. Whenever a node is found dead, the hypercube needs to be reconstructed. To reduce the probability of re-constructing the cube (or to cut down the complexity), it will be desirable if we can find another node to replace the dead one as described in the following.

Each node will broadcast a packet to all of its neighbors at each round to notify it is alive. If a node does not receive such a packet from a neighbor node, say *i*, for a period of time, it will send out the information that node *i* is dead. Each of *i*'s neighbor nodes then broadcasts a message including its location and the connection bit number *k* between itself and node *i* to the network. After receiving the information, each node *n* computes $C(n)$, which is the increased average cost per round when *n* gets the label of the dead node *i*

$$C(n) = \sum_{i \in N} \frac{d(i,n)^2}{2^k} \tag{1}$$

where N is the set of neighbors of node i, and $d(i, n)$ is the distance between nodes i and n. The cost is computed by the square of distance $d(i, n)$ times the probability of the link to be used ($1/2^k$) in a round. After all neighbor nodes of i broadcast the computed values to the network, the node with the minimum value will take the label of the dead node.

3 The Proposed Distributed Approaches

The proposed hypercube transmission scheme constructs the cube by the bottom-up approach which is conducted in a centralized way by the BS. In the following, we propose to construct the cube by the top-down approach carried out in a distributed way to show the difference.

3.1 The Distributed Hypercube Approach

To construct the communication tree of a hypercube, we choose the node closest to the BS to be the head node and start constructing the cube in log N steps (N = the number of nodes). The head node first chooses a nearest neighbor node as its downstream. The head and the chosen downstream node again take their nearest neighbor nodes as their further downstream. Repeat the same process until all nodes are connected into a communication tree of a hypercube. Fig. 2 shows an example of constructing a hypercube in a distributed way.

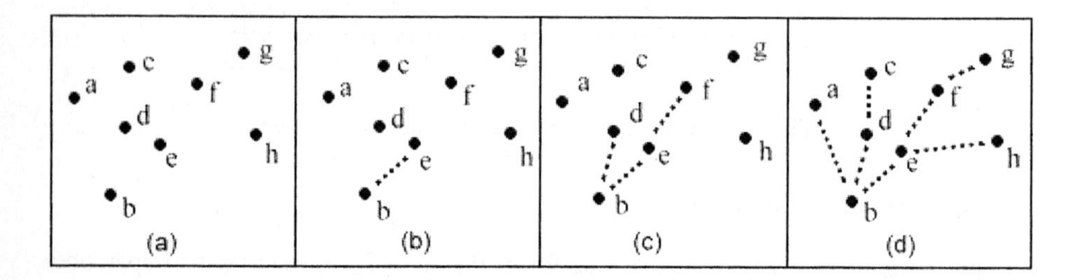

Fig. 2. Constructing a hypercube in a distributed way

1. Fig. 2(a): Choose node b as the head.
2. Fig. 2(b): Node b takes node e as its downstream node.
3. Fig. 2(c): Nodes b and e respectively take nodes d and f, the nearest nodes, as their downstream.
4. Fig. 2(d): Nodes b, e, d and f again choose their own downstream nodes in the same way until all nodes are taken. The communication tree of a hypercube is thus constructed.

Transmission for such a distributed hypercube is simple. A node will transmit data upstream after receiving data from all of its downstream nodes. When receiving data from all downstream nodes, the head node then transmits the fused data to the BS. The approach takes log N steps to gather the data. If a node dies, the topology needs to be re-constructed.

3.2 The Binary Tree Approach

When gathering data by the distributed hypercube approach, each node needs to have a fixed number of immediate downstream nodes to practice complete communication tree transmission. Constructed also in a distributed way, the binary tree approach proposed below is more flexible on the allowed number of immediate downstream nodes: Each node in the binary tree may have at most two immediate downstream nodes. Such flexibility will save a significant amount of energy in gathering data.

The binary tree approach first selects the node closest to the BS as the root for constructing the tree. At each round, data are collected through the binary tree to the root which then transmits the fused data to the BS. When a node, possibly the root, dies, the topology needs to be re-constructed. By this way, the dead nodes will always appear in the neighboring area of the BS.

As mentioned, the BS first selects a nearest node as the root to construct the binary tree. The root then chooses two closest nodes to be its children and asks the two children to find their own children. The offspring of all nodes will be located in the same way. To avoid choosing a child with long distance, we set up a maximum distance constraint: A node will not be the child of another node if the distance between them exceeds the constraint.

In case a node fails to find any child or receives tables from all of its children, it will return an integrated table (including its connection status) to the parent. After receiving tables from all children, the root will broadcast a query to see if any node is left out of the tree. If a node responds (that it is not included in the tree), the root then passes a command along the tree to spread the information. Obtaining the information, those nodes with less than two children now get the chance to re-find their children with a larger distance constraint.

Tree construction will be complete when all nodes are included in the tree. Repeatedly widening the distance constraint to construct a complete tree may be time consuming, but it is not a critical situation because after a tree is constructed, it will not be re-constructed unless a dead node appears.

4 Performance Evaluation

4.1 Data Gathering Delay Time

Table 1 lists the required transmission stages, or the data gathering delay time, at each round for different schemes. Among the schemes, we can see the cube transmission scheme takes only O(log n) stages to gather data, less than the O(n) stages of both PEGASIS and LEACH, where n is the number of nodes. Such an advantage makes the proposed scheme desirable for time critical applications.

Table 1. Transmission steps for various schemes

	Data gathering delay
Cube	O(log n)
Binary tree	O(log n)
Leach	O(n)
Pegasis	O(n)

4.2 The Simulation Model

Our simulation is conducted using NS-2 and adopts the same radio model and network topology as PEGASIS [4]. The system has 100 nodes, initially charged with 0.25J energy and randomly scattered to a 100m*100m network. The BS is located at (50, 250). Every node has a packet to send to the BS at each round. The packet size is 2000 bits. The simulation is carried out to evaluate the performance of various data gathering schemes, including our proposed schemes -- the Hypercube, DH (distributed hypercube) and Binary tree, and other schemes, such as the PEGASIS, PEGBin (the binary approach of PEGASIS), LEACH and Direct. All data gathering schemes will reconstruct their topologies when a node dies.

Our binary tree approach sets a 10m basic distance constraint between any two nodes. When such a distance constraint fails to include all nodes in the tree, increase 10m at each construction attempt until the tree is completed. LEACH is set to have only one head at each round to save energy (as the BS is located far from the sensor nodes). In PEGASIS and Hypercube, the BS performs the task of constructing topologies. The cost for constructing topologies is left out because it is negligible. Simulation results collected under a number of parameters are provided below to exhibit performance comparison between different schemes.

4.3 Simulation Results

Fig. 3 displays **the rounds vs. the number of living nodes**. It is assumed in our simulation that the BS is located 200m away from the center of the topology. Due to such a long distance, the Hypercube scheme which transmits data in parallel instead of by shorter paths fails to retain as many living nodes as PEGASIS. The Binary tree approach yields the best performance because it is more flexible with the number of downstream nodes and because it selects the node that uses the least energy to transmit data to the BS.

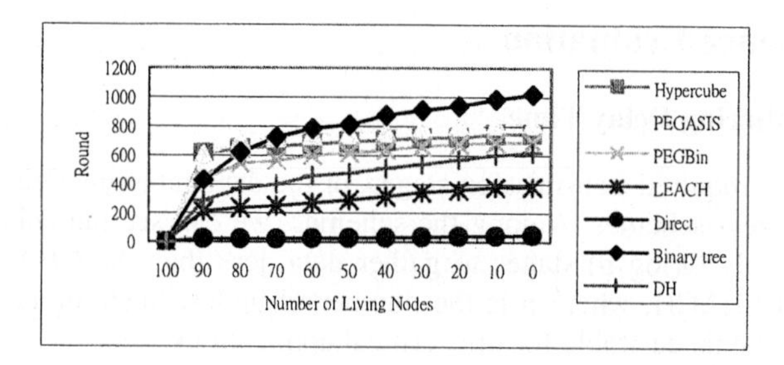

Fig. 3. Rounds vs. the number of living nodes

Fig, 4 gives **the average delay of total inter-node transmissions per round** before all nodes turn dead. As it shows, the DH, the binary approach of PEGASIS and the Hypercube are the three schemes with the lowest average delay, while the Binary tree yields twice as much delay because it takes double time of log n, where n is the number of living nodes in the network.

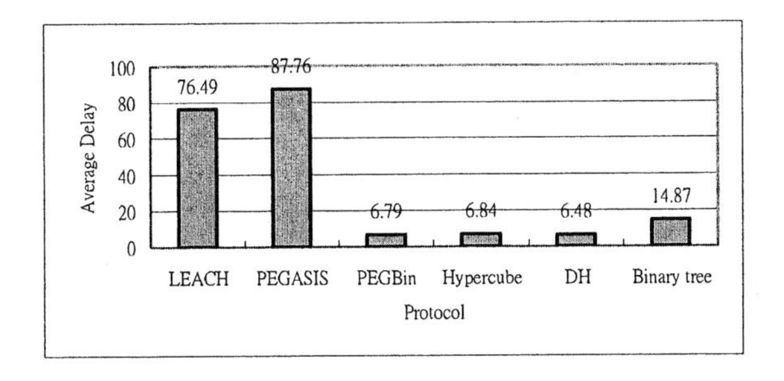

Fig. 4. The average delay of total inter-node transmissions per round

Fig. 5 demonstrates **the average energy dissipation for inter-node transmissions per round**. (To obtain clear comparison, the large amount of energy the head node consumes when sending the gathered data to the BS is not counted in.) As the result shows, the Binary tree, PEGASIS and Hypercube consume much less energy per round than the other schemes. The binary approach of PEGASIS fails to perform as favorably because it optimizes only the first transmission step of each round.

The average energy*delay between nodes in Fig. 6 is the required energy times the needed delay before data are gathered to the head in a round. It can be taken as the

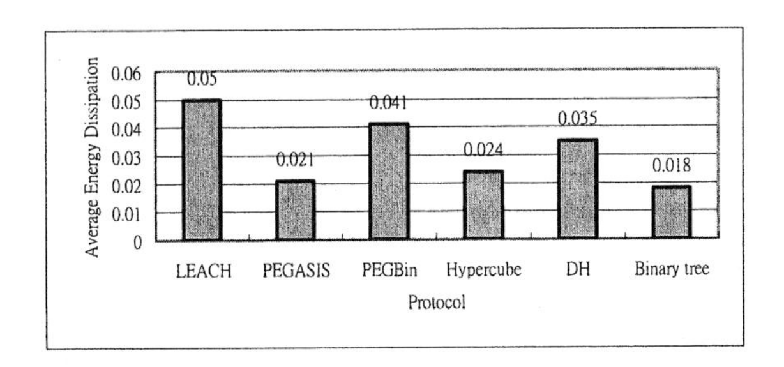

Fig. 5. The average energy dissipation for inter-node transmissions per round

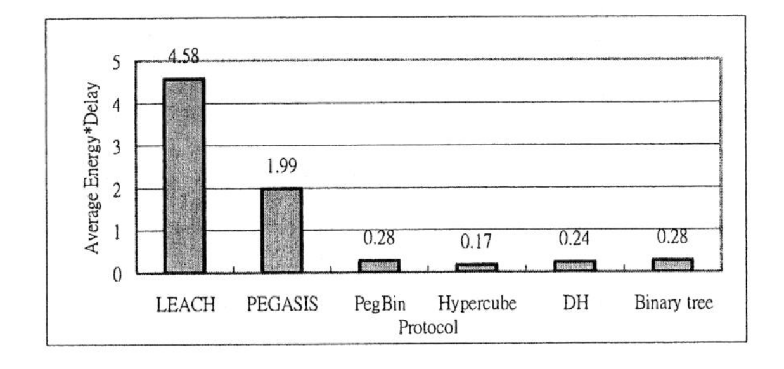

Fig. 6. The average energy*delay between nodes

combined evaluation of energy dissipation and transmission delay. The figure shows that the proposed Hypercube gives the best performance at this parameter. This is because the scheme takes both time and energy into consideration when conducting transmission. The Binary tree is found to perform slightly behind as it puts more concern on energy than on time.

Fig. 7 depicts **the rounds vs. the number of living nodes** for the Hypercube scheme undergoing **re-construction** or **reconfiguration** when a node dies. "Reconfiguration" indicates the hypercube goes through a reconfiguring process by replacing the dead node, instead of re-constructing the hypercube itself. The result shows that the performance of the "reconfiguration" approach, which saves a lot of work for the BS, comes close to that of the re-construction approach. Thus when it is undesirable for the BS to reconstruct a faulty hypercube, system reconfiguration can be another choice.

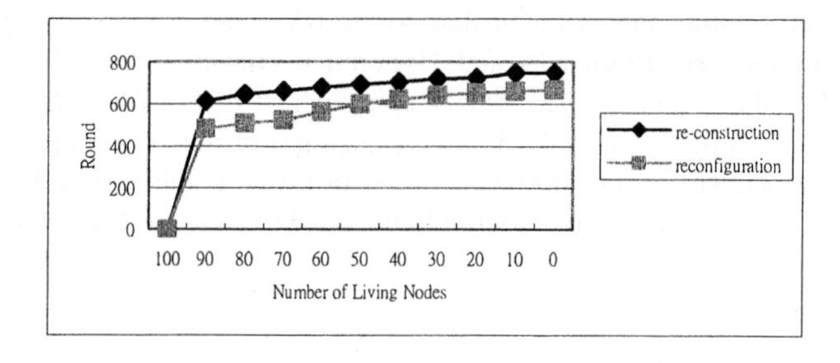

Fig. 7. The round vs. the number of living nodes for the Hypercube undergoing re-construction or reconfiguration

5 Conclusions

A new data gathering scheme for the microsensor network is proposed in this paper to reduce energy consumption during data transmission. The new scheme is built based on the tree communication of the hypercube topology. In the proposed hypercube-based data gathering scheme, data are gathered from all microsensor nodes to the BS in parallel through the communication tree of the hypercube. Gathering data by parallel transmissions shortens the communication delay, and a dead node can be replaced through reconfiguration of the hypercube topology. The construction of hypercubes and the operation of communication-tree transmission are specified in the paper. Performance of the new scheme and other related schemes are evaluated and simulated for comparison. The results show that compared with other data gathering schemes, the hypercube-based scheme turns over favorable results, such as reduced data gathering time, more balanced energy loads, satisfying system scalability, and as a result prolonged system lifetime. To complete our investigation, constructing the hypercube and the binary tree in a distributed way for data gathering in microsensor networks are also proposed and compared.

References

1. M. Dong, K. Yung, and W. Kaiser., "Low Power Signal Processing Architectures for Network Microsensors," *Proc. 1997 International Symposium on Low Power Electronics and Design*, Aug. 1997, pp. 173-177.
2. I. F. Akyildiz, W. Su, Y. Sankarasubramaniam, and E. Cayirci, "A Survey on Sensor Networks," *IEEE Communications Magazine*, Vol. 40, No. 8, pp.102-114, Aug..2002.
3. W. Heinzelman, A. Chandrakasan, and H. Balakrishnan, "Energy-Efficient Communication Protocol for Wireless Microsensor Networks," *Proc. 33rd Hawaii Int'l Conf. on System Sciences*, Jan. 2000.
4. S. Lindsey, C. Raghavendra, and K. M. Sivalingam, "Data Gathering Algorithms in Sensor Networks Using Energy Metrics," *IEEE Trans. on Parallel and Distributed Systems*, Vol. 13, No. 9, pp. 924-935, Sept. 2002.
5. Y.-R. Leu and S.-Y. Kuo, "A Fault-Tolerant Tree Communication Scheme for Hypercube Systems, " *IEEE Trans. on Computers*, Vol. 45, No. 6, pp. 641-650, June 1996.
6. P.-J. Chuang, S.-Y. Chen, and J.-T. Chen, "Constructing Fault-Tolerant Communication Trees in Hypercubes," *Journal of Information Science and Engineering*, Vol. 20, No. 1, pp. 39-55, Jan. 2004.

Delay Efficient Data Gathering in Sensor Networks

Xianjin Zhu, Bin Tang, and Himanshu Gupta

Department of Computer Science,
State University of New York at Stony Brook, Stony Brook, NY 11794
{xjzhu, bintang, hgupta}@cs.sunysb.edu

Abstract. Data gathering is a very important functionality in sensor networks. Most of current data gathering researches have been emphasized on issues such as energy efficiency and network lifetime maximization; and the technique of data aggregation is usually used to reduce the number of radio transmissions. However, there are many emerging sensor network applications with different requirements and constraints. Rather, they are time critical, i.e., delivering sensed information of each individual sensor node back to a central base station quickly becomes most important. In this paper, we consider collision-free delay efficient data gathering problem in sensor networks, assuming that no data aggregation happens in intermediate nodes. We formally formulate this problem and propose optimal and near-optimal algorithms for different topologies. Particularly, in general topology, we present two approximation algorithms with performance ratio of 2 and $1+1/(k+1)$, respectively.

1 Introduction

1.1 Motivation

Recent advances in miniaturization of computing devices with advent of efficient short-range radios have given rise to strong interest in sensor networks [1],[2]. Sensor networks are ad hoc multi-hop wireless networks formed by a large number of low-cost sensor nodes with limited battery power and processing capacity. Wireless sensor networks are expected to be used in a wide range of applications, such as military surveillance, environmental monitoring, target tracking, etc. One of the most important communication primitives that has to be provided by sensor networks is data gathering, i.e., information collected at sensors has to be transmitted back to a sink node, which is responsible for further processing for end-user queries.

Since sensor nodes are usually deployed in adverse environments, it is often not feasible to replace or recharge their batteries. In order to prolong network lifetime, how to maintain energy efficient data gathering has attracted a lot of attention. Most of current data gathering researches have been conducted towards this goal [3],[4],[5],[6]. Some works also use data aggregation techniques [7],[8],[9] to reduce the number of radio transmissions, which is the main drain of the

X. Jia, J. Wu, and Y. He (Eds.): MSN 2005, LNCS 3794, pp. 380–389, 2005.

battery of sensor nodes. Data aggregation is especially appropriate when sensor networks have high density and as a result, data sensed at neighboring nodes are either highly correlated or simply redundant.

However, there are many emerging sensor network applications with different requirements and constraints. Rather, they are time critical, i.e., delivering sensed information of each individual sensor node back to a central base station as fast as possible becomes most important. Since the data is usually time varying in sensor networks, it is essential to guarantee that data can be successfully received by the sink the first time instead of being retransmitted due to collisions. In such scenarios, efficient data gathering scheduling is needed, which speciftes how the data collected at each individual node is transmitted to the sink or base station without any collisions, within the shortest period of time. In this paper, we restrict our focus on collision-free delay efficient data gathering problem in sensor networks, assuming that no data aggregation happens in intermediate nodes. We first use a concrete example to motivate our research.

Motivating Example. For patient monitoring in hospital, a sensor network may consist of many sensor instruments, which are attached to patients to monitor their physical conditions. Medical data sensed at each sensor is transmitted back to a central platform for diagnosis. Due to the short radio transmission range, sensed data is transmitted in multi-hop manner. Such application has several characteristics. First, energy is not a concern as the battery of each sensor can be easily replaced. Secondly, the data collected from each individual patient is independent and not correlated in anyway. All the illness information of each patient has to be sent back to the central platform separately and no aggregation on the intermediate nodes is desired. Thirdly, the time criticality of such application is obvious without further explanation.

The delay efficient data gathering problem is not limited in sensor networks. It is suitable for any real time data gathering applications. For example, in network monitoring, wherein each individual computer in the network is required to send back to the server of its own source or traffic information. This monitoring happens in rounds and all the nodes are synchronized to send the data back. This way server can gather all the information and generate a complete picture of the network; and react quickly to coordinate for resource allocation and traffic congestion control, etc. In this scenario, data aggregation in intermediate nodes is obvious unwanted. Rather, the delay window (the period within which messages from all the nodes are received by the base station) is more critical since the quicker the central server gets all the necessary information, the more timely decisions it can make as to the actions each node has to take to prevent undesirable situations from happening.

1.2 Related Work

The work which is most related to the problem we consider in this paper is [10]. It studies the problems of data distribution and data collection in wireless sensor networks via simple discrete mathematical models. Our approach differs with it in two aspects. First, instead of solving a converse distribution problem wherein a base station transmits data packets to nodes as proposed in [10], we design optimal or near optimal algorithms to directly solve the data gathering problem. Second, to further reduce delay while avoiding collisions, we introduce multiple channels in our work. In [10], a transmission from node N_i to node N_j is successful only if none of the neighbor nodes of receiver N_j is transmitting simultaneously. With this model, they present a strategy for general graph networks within a factor of 3 of the optimal performance. In our model, each node is equipped with two channels. This way its two non-adjacent neighbors can transmit simultaneously as long as they choose different channels. As we will show later, such model will result in an approximation algorithm with performance ratio 2 in general graph networks. Multiple channels were shown to be more efficient for multi-hop wireless networks [11],[12].

Another closely related work is [13]. It studies the problem of minimum latency broadcast in ad hoc networks. It presents a collision-free broadcast algorithm which simultaneously produces provably good solutions in terms of latency and the number of retransmissions. Our proposed problem is not a simple reverse engineering of above problem. Unlike [13], in which one single message is broadcast to all the nodes, our work studies the condition whereby different messages are gathered and sent back to one node. In this sense, our problem is conjectured to be even harder than that in [13].

There are some research work [14],[15],[6] also realizing the importance to consider the delay incurred in gathering sensed data, in addition to minimizing energy. They capture this with the $engergy \times delay$ metric and present schemes that attempt to balance the energy and delay cost for data gathering from sensor networks. Our goal in this paper is solely delay minimization.

Paper Organization. The remainder of this paper is organized as follows. In section 2, we specify our network model and present the problem statement. In section 3, we design and analyze our data gathering algorithms for different special topologies. Two heuristics for general topology are then presented in section 4. We conclude our paper in section 5.

2 Network Model and Problem Statement

Network Model and Assumptions. We model the sensor network as a disk graph $G = (V, E)$, where $V = \{1, 2, \ldots, N\}$ is the set of sensor nodes and E is the set of undirected edges. The edge exists between two nodes if their Euclidean distance is within the transmission range of each other (for simplicity, we assume all the nodes have the same transmission range). There is only one sink node, which is used to collect and analyze data for interest queries. All nodes are sensing data and trying to send/relay data back to the sink node. A *collision* happens at a node if it hears a message from more than two transmitters at the same time. Our goal is to intelligently schedule each node's receiving/transmitting time in order to guarantee collision-free data gathering with minimum delay.

We assume that each node has half duplex interface and is equipped with two channels. Therefore, it can not receive/transmit at the same time, but its non-adjacent neighboring nodes can transmit simultaneously as long as they choose different channels. Each node has one message waiting to be transmitted back to the sink node, but it has limited buffer size such that relay nodes only perform simple receive and forward type operations, i.e., packets received must be forwarded in the next time slot following its arrival. Such constraint is also used in [10]. Furthermore, in this work, we do not consider data aggregations, which may be explored in future research.

Problem Statement. Given a disk graph $G = (V, E)$ and a sink node S, node $S_i \in V$ is the source of data item i, our data gathering problem is to schedule for each data item j and each node S_i, the time slot at which node S_i transmits/receives data item j collision-free, such that the time when the last data item received by S is minimized.

3 Algorithms for Special Topologies

In this section, we address the delay efficient data gathering problem in some special topologies. We start with the simplest linear topology and prove the optimality of our solution. Then we propose the approach to solve our problem in star and tree topologies by exploring the relationship between these two topologies and the linear topology.

3.1 Linear Topology

In linear topology, all nodes are arranged in a line and the sink node S is at one end of the line. The case wherein S is at an arbitrary position instead of two ends of the line can be considered as a special case of star topology, which will be discussed later in Section 3.2. We label the nodes in sequence from 1 to N, with nodes closer to the sink assigned lower IDs (see Fig. 1). Before explaining the details of our algorithm, we present a method to assign channels for each node at each time slot, such that nodes two hops away from each other can send packets simultaneously without causing collisions.

Channel Assignment. Suppose every node can use two channels, viz. channel 1 and channel 2. The sink always listens to and receives packets at channel 1. For any other node i ($1 \leq i \leq N$), the channel that it should use at time slot t is chosen based on the values of i and t. Basically, nodes with even IDs switch between two channels alternatively while nodes with odd IDs always stay at one assigned channel, so that a node is always assigned a different channel with its 2-hop neighbors and a pair of transmitter and receiver stay on a same channel. Such assignment guarantees that node i and node $i+2$ can send packets collision free at the same time.

Algorithm for Data Gathering. As in channel assignment algorithm, we divide nodes into two sets based on their IDs, viz. odd set and even set. Similarly,

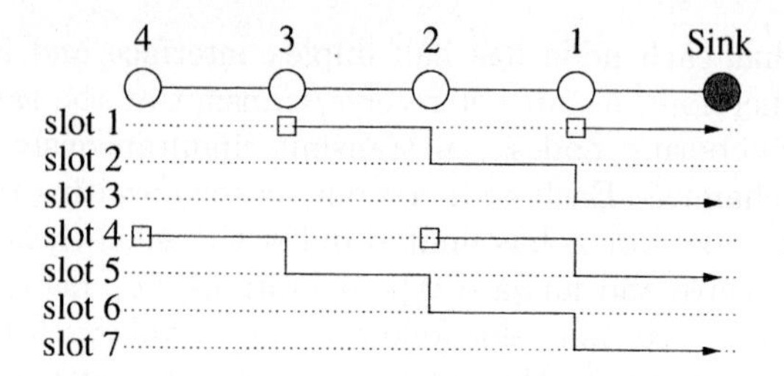

Fig. 1. Scheduling in linear topology: nodes are divided into odd set and even set; nodes in the same set transmit packets simultaneously

packets are also divided into two sets, viz. packets from nodes in odd set P_{odd} and from nodes in even set P_{even}. With the channel assignment presented previously, for any $i > 0$, node i and node $i+2$ use different channels to send packets at any time slot. Therefore, nodes in the same set can transmit packets collision-free simultaneously. Since in our network model, every node has a packet to be sent back to the sink, we let all packets in P_{odd} be transmitted first. During this process, nodes both in odd set and in even set are involved in relaying those packets. After the sink gets all packets in P_{odd}, nodes in the even set then start to transmit their own packets in following time slots. Such scheduling guarantees that every sender-receiver pair uses the same channel so that each packet can be forwarded one hop per slot once the transmission starts.

A simple example is illustrated in Fig. 1. Initially, node 1 and 3 simultaneously send their packets to the sink. At the end of time slot 3, the sink has received packets both from 1 and 3. Then node 2 and 4 start their transmissions at time slot 4. The total delay in this example is 7.

Proof of Optimality. In the following, we show the above algorithm is optimal.

Lemma 1. *$2N - 1$ is the lower bound for delay of data gathering in linear topology, i.e., for any algorithm, the resulting delay is at least $2N - 1$, where N is the number of nodes in the network excluding the sink.*

Proof. In the linear topology, before packets are received by the sink, they must pass through node 1 (the closest node to the sink). For packets from node $2, 3, \ldots, N$, each of them needs two slots to be received and forwarded by node 1. For the packet of node 1 itself, it only needs one slot to be sent to the sink. Thus, the total delay is at least $2(N - 1) + 1 = 2N - 1$.

Theorem 1. *The above data gathering algorithm in linear topology is optimal.*

Proof. First, we prove the delay produced by our algorithm is exactly $2N - 1$. The total delay equals to the time taken by packets in both P_{odd} and P_{even} to be received by the sink. For each set of packets, since packets are forwarded one hop per slot in our scheduling, the delay is bounded by the largest hop count away from the sink. Thus, the delay is N for one set and $N - 1$ for the other set. Straightforwardly, total delay is $2N - 1$. Since the result delay of our algorithm exactly matches with the lower bound, our algorithm is optimal.

3.2 Star Topology

In star topology, the sink S has m linear branches. A branch i has arbitrary number of nodes N_i and the total number of nodes is $N = \sum_{i=1}^{m} N_i$. Each branch can be assigned channel separately with the channel assignment algorithm for linear topology[1].

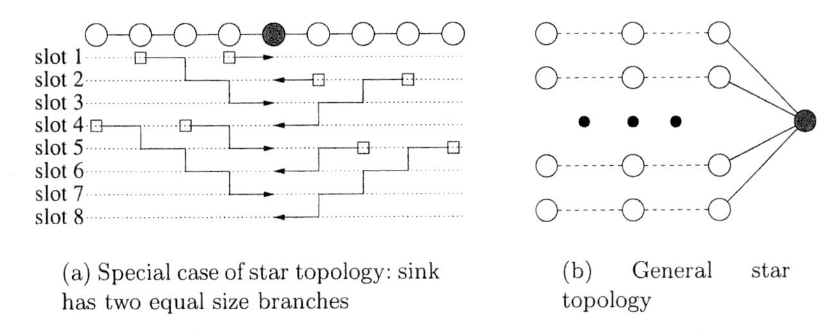

(a) Special case of star topology: sink has two equal size branches

(b) General star topology

Fig. 2. Data Gathering in Star Topology

Obviously, the data gathering algorithm for linear topology can also be applied directly on each branch, and the result delay is $2N - m$, since the sink can receive one packet every two slots on average. However, an ideal scheduling to get minimum delay in star topology must fill every idle slot so that the sink can get one packet one slot. In such ideal case, the total delay is N. In a special case of star topology, wherein the sink only has two equal size branches (see Fig. 2(a)), transmissions of each branch can be pipelined in a proper way to meet the target of one packet one slot. Now we generalize the scenario to m branches (see Fig. 2(b)). If we can divide all branches into two groups which contain same number of nodes, then transmissions of these two groups can be pipelined in a similar way to minimize the delay. Thus, the key becomes how to make such division.

NP-Completeness of Group Division Problem. We prove the NP-completeness of our group division problem by reduction from the well-known Integer Partition problem. We first give the decision version of these two problems separately, and then show the proof.

Definition 1. *Group Division problem is defined as: given a star topology with m branches, is it possible to divide m branches into two groups so that each group has equal number of nodes?*

[1] This channel assignment algorithm can also be easily generalized to tree and general topology. For general topology which can be decomposed into several independent components, each component can be assigned channel separately. Otherwise, we may need 3 channels.

Definition 2. *Integer Partition problem $IP = (X, y)$ is defined as: given a set of integers $X = \{x_1, x_2, \ldots, x_n\}$ and a target number y. Is there a subset $X' \subseteq X$, such that the sum of all the elements in X' is equal to y?*

Theorem 2. *Group division problem is NP-complete.*

Proof. Given a group division, since there are only polynomial number of nodes in total, we can decide whether this division is valid or not in polynomial time. Thus, the group division problem is NP-hard.

To prove it is NP-complete, we show a polynomial reduction from Integer Partition problem $IP = (X, y)$. For each $x_i \in X$, we create a branch with x_i nodes. After doing this, we also create an extra branch with $(\sum_{x_i \in X} x_i - 2y)$ nodes. All these branches are connected to the sink. If $IP = (X, y)$ has a solution, i.e., we can find a subset X' such that $\sum_{x_i \in X'} x_i = y$, there is also a solution to the group division problem. Since $\sum_{x_i \in X'} x_i + (\sum_{x_i \in X} x_i - 2y) = (\sum_{x_i \in X} x_i + (\sum_{x_i \in X} x_i - 2y))/2$, the corresponding branches of all $x_i \in X'$ plus the extra branch contain exact half of the total number of nodes. On the other hand, if we can divide all branches into two equal sets, the extra branch with $(\sum_{x_i \in X} x_i - 2y)$ nodes must be in one set. Then, the integers corresponding to the rest branches in this set make up of the solution to the integer partition problem, since $(\sum_{x_i \in X} x_i + (\sum_{x_i \in X} x_i - 2y))/2 - (\sum_{x_i \in X} x_i - 2y) = y$.

3.3 Tree Topology

In a tree topology with the sink S as the root, all N other nodes are arranged as an arbitrary tree. First, let us look at a simple situation when the root has only one child. The basic idea here is to view it as conjunction of multiple lines, which can be processed one by one.

Algorithm for Single Child Tree. For the topology in Fig. 3, it can be viewed as conjunction of two lines, viz. the top line (Fig. 3(a)) and the bottom line (Fig. 3(b)), which are jointed at node J. Suppose there are n nodes in the top line, and there are m nodes in the bottom line except the joint part. In this case, $N = n + m$. The first step is to process the top line, where the data gathering algorithm for linear topology can be applied. Supposing that node J is k hops away from the sink S, it is easy to know that the last packet of the top line will leave node J at time slot $2n - 1 - (k - 1)$. Otherwise, all the packets of the top line cannot be collected by the sink in $2n - 1$ time slots. Until this time, all the

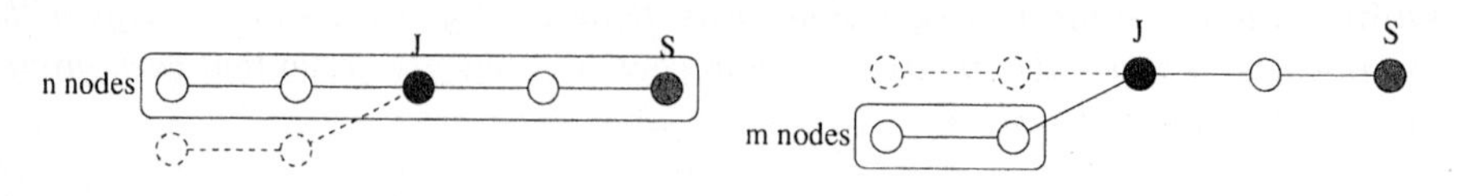

(a) Processing of the top line. (b) Processing of the bottom line.

Fig. 3. Data gathering for the single child tree topology

nodes in the top line after node J have been processed, and we can continue to the bottom line. For the bottom line without the joint part, it can be handled as if node J is the sink. Beginning from slot $2n - k + 1$, the nodes in the bottom line can begin to send one packet to node J every two time slots. As soon as node J receives the first packet from the bottom line, it immediately forwards the packet to the sink S. As a result, the first packet from the bottom line arrives at S at slot $2n - k + 1 + k = 2n + 1$. Thereafter, S can continue to receive one packet from J every two time slots until all the remaining $m - 1$ packets have been collected. Then, the total gathering delay is $2n + 1 + 2(m - 1) = 2N - 1$.

Generalizing the above algorithm, for any given single child tree, it can be eventually decomposed as a bunch of lines, which can be processed one-by-one in a similar way according to certain order.

Lemma 2. *For a single child tree whose root has only one child, the minimum gathering delay is $2N - 1$. N is the total number of nodes except the root.*

Proof. Similar as analysis in linear topology, for the single direct child of the sink, it takes two time slots to transmit a packet from any of the rest $N - 1$ nodes and one slot to send its own packet to the sink. Thus, the total gathering delay is at least $2(N - 1) + 1 = 2N - 1$.

Theorem 3. *The proposed data gathering algorithm for the single child tree topology achieves the optimal delay.*

After finding the optimal solution for the single child tree, we continue to the general tree topology, as illustrated in Fig. 4. Based on the above discussion, it can be seen that, from the point view of the sink, a single child subtree is identical to a line. Hence, a general tree can be processed as a star with multiple branches. We define a direct subtree as a subtree whose root is the direct child of the sink. Then, if all direct subtrees can be divided into two equal size groups, the two groups can be fully pipelined so that the sink can receive one packet each time slot alternately from the two groups, and an optimal delay of N can be achieved. As discussed earlier, the group division problem is NP-complete, and no efficient polynomial solution exists. However, in the worst case, the direct subtrees can be processed one by one, and the upper bound of the delay is $2N-1$.

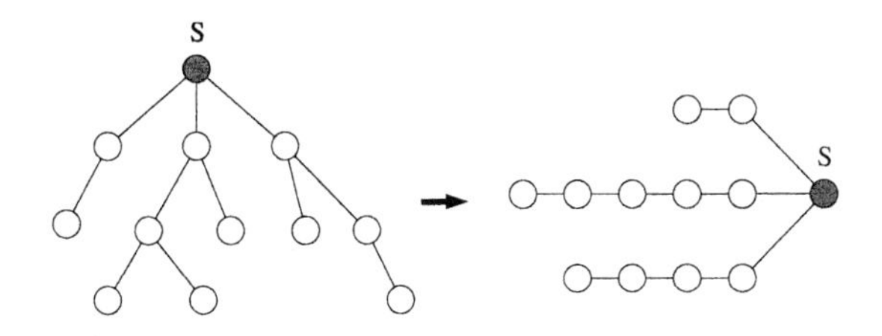

Fig. 4. A tree whose root has multiple children can be viewed as a star

4 Heuristics for General Topology

In this section, we propose two heuristics for general topology based on the techniques designed in the previous section. The first heuristic is simple. We first generate a BFS tree for the original graph [16]. Then, the data gathering algorithm for tree topology can be applied. We have known that the upper bound of the gathering delay for any tree is $2N - 1$, and the lower bound of the gathering delay is N. Thus, the tree based solution for general topology guarantees 2-approximation.

In the following, we further propose a more efficient heuristic to divide the set of direct subtrees into two equal size groups and then pipeline transmissions of these two groups to fill as many idle slots as possible at the sink.

Group-based Heuristic Algorithm. For Integer Partition problem, there exists a $(1+1/k)$ approximation algorithm [17], for a constant integer k. Such algorithm can be used to get an approximate group division. The basic idea is as follows. We consider each of the subsets with at most k direct subtrees for a given constant k. If the number of nodes contained in the subset has not exceeded half of the total number, we add the remaining direct subtrees as many as possible. A subset with maximal number of nodes is selected at last. Obviously, the algorithm will run more times and get more accurate answer with a larger k. After dividing all direct subtrees into two groups, we pipeline the transmissions of these two groups so that the sink can receive packets from each group alternatively.

Theoretical Analysis. With the group-based heuristic, all direct subtrees can be divided into two groups B' and $B - B'$. The size of B' and $B - B'$ must satisfies that $kN/2(k+1) \leq |B'| \leq N/2$ and $N/2 \leq |B - B'| \leq N - kN/2(k+1)$. Thus, in the worst case, the difference of the number of nodes in these two sets is equal to $N - kN/2(k+1) - kN/2(k+1) = N/(k+1)$. From the observations we get in the previous section, the total delay is at most $2kN/2(k+1) + 2N/(k+1) - 1 \leq N + N/(k+1)$. Therefore, with any $1 + 1/k$ approximation algorithm for group division, we can get a $1+1/(k+1)$ approximation algorithm for data gathering.

Theorem 4. *For any $1+1/k$ approximation algorithm for group division, there exists a corresponding $1+1/(k+1)$ approximation algorithm for data gathering.*

5 Conclusions

In this paper, we have studied the collision-free delay efficient data gathering problem in sensor networks by tackling with different topologies. In linear topology, we prove that an optimal delay of $2N-1$ can be achieved with two channels for each node, where N is the number of nodes except the sink in the graph. In star topology, we conclude that if all branches can be divided into two equal size groups, an optimal delay of N can be achieved. This group division problem is proved to be NP-complete. In tree topology, we show that a general tree can be processed as a star topology and a 2-approximation algorithm also applies. Finally, for the general topology, we propose the group based heuristic algorithm, which is able to achieve $1 + 1/(k + 1)$ approximation.

References

1. Estrin, D., Govindan, R., Heidemann, J., eds.: Special Issue on Embedding the Internet, Communications of the ACM. Volume 43. (2000)
2. Badrinath, B., Srivastava, M., Mills, K., Scholtz, J., Sollins, K., eds.: Special Issue on Smart Spaces and Environments, IEEE Personal Communications. (2000)
3. Intanagonwiwat, C., Govindan, R., Estrin, D.: Directed diffusion: A scalable and robust communication paradigm for sensor networks. In: Proceedings of the International Conference on Mobile Computing and Networking (MobiCom). (2000)
4. Heinzelman, W., Kulik, J., Balakrishnan, H.: Adaptive protocols for information dissemination in wireless sensor networks. In: Proceedings of the International Conference on Mobile Computing and Networking (MobiCom). (1999)
5. Chang, J., Tassiulas, L.: Energy conserving routing in wireless ad hoc networks. In: Proceedings of the IEEE INFOCOM. (2000)
6. Lindsey, S., Raghavendra, C., Sivalingam, K.M.: Data gathering algorithms in sensor networks using energy metrics. IEEE Transactions on parallel and distributed systems (2002)
7. Solis, I., Obraczka, K.: The impact of timing in data aggregation for sensor networks. In: Proceedings of the International Conference on Communications (ICC). (2004)
8. von Rickenbach, P., Wattenhofer, R.: Gathering correlated data in sensor networks. In: Proceedings of the 2004 joint workshop on foundations of mobile computing. (2004)
9. Cristescu, R., Beferull-Lozano, B., Vetterli, M.: On network correlated data gathering. In: Proceedings of the IEEE INFOCOM. (2004)
10. Florens, C., McEliece, R.: Packets distribution algorithms for sensor networks. In: Proceedings of the IEEE INFOCOM. (2003)
11. Nasipuri, A., Zhuang, J., Das, S.: A multichannel csma mac protocol for mobile multihop neworks. In: Proceedings IEEE Wireless Communications and Networking Conference. (1999)
12. So, J., Vaidya, N.: Multi-channel mac for ad hoc networks: Handling multi-channel hidden terminals using a single transciver. In: Proceedings of the International Symposium on Mobile Ad Hoc Networking and Computing (MobiHoc). (2004)
13. Gandhi, R., Parthasarathy, S., Mishra, A.: Minimizing broadcast latency and redundancy in ad hoc networks. In: Proceedings of the International Symposium on Mobile Ad Hoc Networking and Computing (MobiHoc). (2003)
14. Yu, Y., Krishnamachari, B., Prasanna, V.: Energy-latency tradeoffs for data gathering in wireless sensor networks. In: Proceedings of the IEEE INFOCOM. (2004)
15. Raghavendra, C., Sivalingam, K., Lindsey, S.: Data gathering in sensor networks using the energy*delay metric. In: Proceedings of IPDPS Workshop on Issues in Wireless Networks and Mobile Computing. (2001)
16. Skiena, S.S., Revilla, M.: Programming Challenges. (2003)
17. Baase, S., Gelder, A.V.: Computer Algorithms: Introduction to Design and Analysis. (2001)

Localized Recursive Estimation in Wireless Sensor Networks

Bang Wang, Kee Chaing Chua, Vikram Srinivasan, and Wei Wang

Department of Electrical and Computer Engineering (ECE),
National University of Singapore (NUS), Singapore
{elewb, eleckc, elevs, g0402587}@nus.edu.sg

Abstract. This paper proposes a localized recursive estimation scheme for parameter estimation in wireless sensor networks. Given any parameter occurred at some location and time, a number of sensors recursively estimates the parameter by using their local measurements of the parameter that is attenuated with the distance between a sensor and the target location and corrupted by noises. Compared with centralized estimation schemes that transmit all measurements to a sink (or a fusion center), the recursive scheme needs only to transmit the final estimate to a sink. When a sink is faraway from the sensors and multihop communications have to be used, using localized recursive estimation can help to reduce energy consumption and reduce network traffic load. Furthermore, the most efficient sequence of sensors for estimation is defined and the necessary condition for such a sequence is determined. Some numerical examples are also provided. By using some typical industrial sensor parameter values, it is shown that recursive scheme consumes much less energy when the sink is three hops or more faraway from the local sensors.

1 Introduction

Recently, *wireless sensor networks* (WSNs) that consist of a great number of sensors each capable of sensing, processing and transmitting environmental information have attracted a lot of research attentions [1]. Sensor networks can be used for example in environmental monitoring, military surveillance, space exploration etc. One of the important characteristics of sensor networks is that sensors are cooperative to perform some functionalities. For example, sensors can cooperate to track an intruder or estimate a signal. This paper studies the estimation problem in a WSN: how to efficiently estimate a parameter of a signal (e.g., signal amplitude) occurred on a particular location by cooperation of sensors.

In WSNs, parameter estimation can be realized by processing K local measurements of the same parameter measured by K sensors. Signal might be attenuated with distances and local measurement might be corrupted by noises. Since sensors are geographically distributed, their measurements of a same parameter thus have some spatial correlations. Furthermore, since each sensor is

X. Jia, J. Wu, and Y. He (Eds.): MSN 2005, LNCS 3794, pp. 390–399, 2005.

normally installed with a limited energy battery, energy consumption is a critical issue in sensor networks. Exploiting correlations in sensor networks can lead to significant potential advantages for the development of efficient communication protocols well-suited for the paradigm of WSNs [2]. The estimation efficiency here mainly refers to reducing energy consumption from the computations and transmissions when performing an estimation.

In WSNs, one way to reduce energy consumption is to use least bits to encode measurements while still meeting certain performance requirements [3]. Some schemes have been proposed for energy efficient estimation along this line [4][5][6]. In these schemes, the basic idea is to reduce the bits needed to encode the local measurements. An universal distributed estimation scheme with one-bit message encoding functions has been proposed in [4] for a homogeneous WSN where sensors have observations of the same quality. In [5][6], an inhomogeneous sensing environment was considered, and the length of encoded message is decided by the local signal to noise ratio. However, these schemes require that all K sensors transmit their encoded measurements (messages) to a central fusion center where an estimate is obtained via batch processing of the K messages. When the sink is far away from the K sensors and multi-hop communications have to be used, the transmission of these messages not only consumes much energy but also increases network traffic load even if the messages are short. Furthermore, it is generally believed that in sensor network the energy required for local computation is much less than that used for communications [1][7]. This motivates us to propose a localized recursive estimation scheme that performs the estimation locally and recursively by the sensors measuring the parameter and only transmits the final estimate to the sink. Hence communications overhead can be reduced significantly. Unlike the measurement model used in [4][5][6], a more realistic distance based measurement model is used where signals decay with distance when a wavefront expands [7]. Based on the used measurement model, a most efficient sequence of sensors is determined for estimation by which the variance of estimation error of an estimator reduces faster than others. Some numerical examples are also provided to illustrate the performance improvement of the proposed scheme.

The rest of the paper is organized as follows. Section 2 presents the problem formulation for parameter estimation in sensor networks. A localized recursive parameter estimation scheme is provided and its property is analyzed in Section 3. Section 4 gives some numerical examples and some concluding remarks are provided in Section 5.

2 Problem Formulation

Consider a set of K geographically distributed sensors, each making a measurement on an unknown parameter θ at some location and time. The scenario can be the sensing an acoustic or seismic signal of amplitude θ. Let $d_k, k = 1, 2, ..., K$ denote the distance between a sensor k and the parameter θ (d_k can be obtained via, e.g., GPS system). The parameter θ is assumed to decay with the distance,

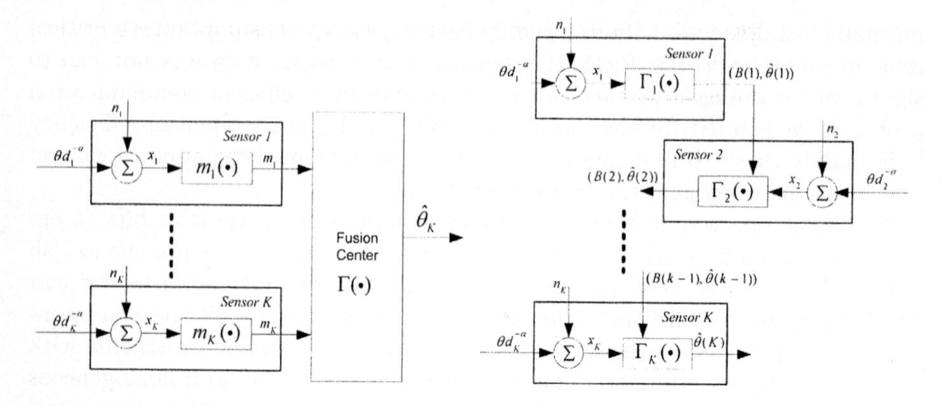

Fig. 1. Illustration of a centralized batch estimation scheme (CBES)

Fig. 2. Illustration of a localized recursive estimation scheme (LRES)

and at distance d it is θ/d^α, where $\alpha > 0$ is the decay component. The above sensing model has been used in [8] for determining path exposure. Furthermore, the measurement of the parameter at a sensor is corrupted by an additive noise. Let x_k denote the measurement of the parameter θ at a sensor k, and is given by

$$x_k = \frac{\theta}{d_k^\alpha} + n_k, k = 1, 2, ..., K. \tag{1}$$

The objective of a parameter estimator is to estimate θ based on the corrupted measurements. Let $\hat{\theta}$ and $\tilde{\theta} = \theta - \hat{\theta}$ denote the estimate and the estimate error, respectively. A most common used criterion is to minimize the *mean squared error* (MSE) of an estimator, i.e., to minimize $\mathbb{E}[\tilde{\theta}^2]$. Since sensors are geographically distributed, they need to encode the measurements to messages and then these messages are transmitted to a fusion center which generate an estimate according to a fusion function. The design of these message functions ($\{m_k : k = 1, 2, ..., K\}$) and the fusion function ($\Gamma(m_1, m_2, ..., m_K)$) has been discussed in [4][6], where an estimate is made based on a batch process of all the measurements. We note that these schemes are referred to as distributed estimation schemes in [4][6] as the measurements are distributed. However, the estimation processing (the fusion function) is a centralized way and is based on the availability of the batch measurements, which needs all these K measurements are transmitted to the center. Hence, their estimation scheme is called *centralized batch estimation scheme* (CBES) in this paper, and is illustrated by Fig. 1.

It is generally believed that in sensor networks the energy required for local computation is much less than that used for communications. In CBES, transmitting all measurements to the fusion center might cause high traffic load and consume many energy in a WSN, especially when the fusion center is far away where multi-hop communications are needed. This motivates us to propose to use a *localized recursive estimation scheme* (LRES) to address this problem. In

LRES, K sensors are considered to be tandem connected (via radio) for parameter estimation. Without loss of generality, assume sensor k is a predecessor of sensor $k+1$. This means sensor $k+1$ makes an estimate based on its own measurements x_k and its predecessor's output. Finally, the sensor K makes the last estimate and sends it to the sink. Suppose that each sensor is i hops away from the sink while any two successive sensors are only one hop away, a simple calculation shows a total of a number of $(i-1) \times (K-1)$ communications can be saved. Furthermore, if an estimate is suitably transmitted, a sensor no longer needs a message function to encode x_k to a message m_k, which further reduces energy consumption. An LRES is illustrated by Fig. 2. We note that to estimate $\hat{\theta}(k)$, another parameter $B(k-1)$ together with the estimate $\hat{\theta}(k-1)$ from its predecessor sensor need to be transmitted. The definition and computation of $B(k)$ is given in next section. Compared with the CBES illustrated by Fig. 1, the LRES needs not to encode the measurements, however, the parameter $B(k)$ and the estimate $\hat{\theta}(k)$ might need be suitably encoded and transmitted. The next section presents a general recursive estimation algorithm to construct a local fusion function based on a best linear unbiased estimator.

Some notation remarks are summarized here. We use $\hat{\theta}_k$ to denote the estimate from a CBES when k measurements are used for estimation, and $\hat{\theta}(k)$ to denote the estimate produced by the kth sensor in LRES after the estimates from its previous $k-1$ sensors. Boldface is used to denote vectors and matrixes; and $\mathbf{D}^T$ denotes the transpose of matrix $\mathbf{D}$.

3 A Localized Recursive Estimation Scheme

3.1 A Recursive Estimation Algorithm

The measurement given by (1) can be written in matrix format for K sensors as

$$\mathbf{X} = \mathbf{D}\theta + \mathbf{N}, \tag{2}$$

where $\mathbf{X} = (x_1, x_2, ..., x_K)^T$, $\mathbf{D} = (d_1^{-\alpha}, d_2^{-\alpha}, ..., d_K^{-\alpha})^T$, and $\mathbf{N} = (n_1, n_2, ..., n_K)^T$. The additive noises are assumed to be spatially uncorrelated white noise with zero mean and σ_k^2 variance, but otherwise unknown. The covariance matrix of noises $\{n_k : 1, 2, ..., K\}$ is given by

$$\mathbf{R} = \mathbb{E}[\mathbf{N}\mathbf{N}^T] = \text{diag}[\sigma_1^2, \sigma_2^2, ..., \sigma_K^2]. \tag{3}$$

Note that $\mathbf{R}$ is symmetric and positive definite. A well-known *best linear unbiased estimator* (BLUE) [9] can be applied to estimate $\hat{\theta}_K$ and to achieve a minimum MSE. According to BLUE, when K measurements are available, the estimate $\hat{\theta}_K$ of the original signal θ is given as

$$\hat{\theta}_K = \left[\mathbf{D}^T\mathbf{R}^{-1}\mathbf{D}\right]^{-1}\mathbf{D}^T\mathbf{R}^{-1}\mathbf{X}. \tag{4}$$

The MSE of BLUE is given as

$$\mathbb{E}[(\theta - \hat{\theta}_K)^2] = (\mathbf{D}^T\mathbf{R}^{-1}\mathbf{D})^{-1}, \tag{5}$$

and the estimation error $\tilde{\theta}$ is given as

$$\tilde{\theta}_K = \theta - \hat{\theta}_K = -[\mathbf{D}^T\mathbf{R}^{-1}\mathbf{D}]^{-1}\mathbf{D}^T\mathbf{R}^{-1}\mathbf{N}. \tag{6}$$

Note that Eq.(4) also gives the fusion function Γ_K when K measurements are available. The following theorem provides a recursive BLUE for our problem.

Theorem 1. *The* k*th sensor can make an estimate $\hat{\theta}(k)$ by using the following recursive structure*

$$\hat{\theta}(k) = \hat{\theta}(k-1) + \frac{B(k)}{d_k^\alpha \sigma_k^2}\left(x_k - \frac{\hat{\theta}(k-1)}{d_k^\alpha}\right), k = 1, 2, ..., K, \tag{7}$$

where

$$B(k) = \left(\frac{1}{B(k-1)} + \frac{1}{d_k^\alpha \sigma_k^2}\right)^{-1} \tag{8}$$

These equations are initialized by $\hat{\theta}(0) = 0$ and $B(0)$ equal to a very large number.

Proof. With a little abuse of notation, we use subscript k to indicate the dimension for vectors and matrix. For example, $\mathbf{X}_k = (x_1, x_2, ..., x_k)^T$ denotes a $k \times 1$ vector, and $\mathbf{R}_k$ denotes a $k \times k$ matrix. The objective of our recursive estimator is to produce the same estimate as that of the batch estimator. This implies $\hat{\theta}_k = \hat{\theta}(k)$, $k = 1, 2, ..., K$. Hence we rewrite (4) as

$$\hat{\theta}(k) = \hat{\theta}_k = [\mathbf{D}_k^T\mathbf{R}_k^{-1}\mathbf{D}_k]^{-1}\mathbf{D}_k^T\mathbf{R}_k^{-1}\mathbf{X}_k. = B(k)A(k) \tag{9}$$

where $A(k) = \mathbf{D}_k^T\mathbf{R}_k^{-1}\mathbf{X}_k$ and $B(k) = [\mathbf{D}_k^T\mathbf{R}_k^{-1}\mathbf{D}_k]^{-1}$. With some algebra, we have

$$\begin{aligned} A(k) = \mathbf{D}_k^T\mathbf{R}_k^{-1}\mathbf{X}_k &= \sum_{i=1}^{k} \frac{x_i}{d_i^\alpha \sigma_i^{-2}} = \sum_{i=1}^{k-1} \frac{x_i}{d_i^\alpha \sigma_i^2} + \frac{x_k}{d_k^\alpha \sigma_k^2} \\ &= A(k-1) + \frac{x_k}{d_k^\alpha \sigma_k^2}. \end{aligned} \tag{10}$$

and

$$\begin{aligned} B^{-1}(k) = \mathbf{D}_k^T\mathbf{R}_k^{-1}\mathbf{D}_k &= \sum_{i=1}^{k} \frac{1}{d_i^{2\alpha} \sigma_i^2} = \sum_{i=1}^{k-1} \frac{1}{d_i^{2\alpha} \sigma_i^2} + \frac{1}{d_k^{2\alpha} \sigma_k^2} \\ &= B^{-1}(k-1) + \frac{1}{d_k^{2\alpha} \sigma_k^2} \end{aligned} \tag{11}$$

Rewrite (11) as

$$B(k) = \left(\frac{1}{B(k-1)} + \frac{1}{d_k^{2\alpha} \sigma_k^2}\right)^{-1} \tag{12}$$

Now we come to derive the relationship between $\hat{\theta}(k)$ and $\hat{\theta}(k-1)$. By using (10), (11) and (12), and the fact $A(k-1) = B^{-1}(k-1)\hat{\theta}(k-1)$, we obtain

$$\begin{aligned}\hat{\theta}(k) &= B(k)A(k) = B(k)\left[A(k-1) + \frac{x_k}{d_k^{\alpha}\sigma_k^2}\right] \\ &= B(k)\left[(B^{-1}(k) - \frac{1}{d_k^{2\alpha}\sigma_k^2})\hat{\theta}(k-1) + \frac{x_k}{d_k^{\alpha}\sigma_k^2}\right] \\ &= \hat{\theta}(k-1) + \frac{B(k)}{d_k^{\alpha}\sigma_k^2}\left(x_k - \frac{\hat{\theta}(k-1)}{d_k^{\alpha}}\right) \end{aligned} \tag{13}$$

Note that when choosing $\hat{\theta}(0) = 0$ and $B(0)$ equal to a very large number, $B(1) \approx d_1^{2\alpha}\sigma_k^2$ and hence

$$\hat{\theta}(1) = d_1^{\alpha}x_1 = \theta + d_1^{\alpha}n_1 \tag{14}$$

which is the same as $\hat{\theta}_1$, i.e., set $K = 1$ in (4). ■

The first sensor only needs its local measurement x_1, its distance information d_1^{α} and its local noise variance σ_1^2 to compute $\hat{\theta}(1)$. For kth sensor ($k = 2, 3, ...K$), the computation of $\hat{\theta}(k)$ needs the estimate and $B(k-1)$ from its predecessor sensor $\hat{\theta}(k-1)$ as well as its local measurement, its distance information and its noise variance. Since $B(k-1)$ is a scalar, the transmission overhead for $B(k-1)$ is small. In (7), the term $\hat{\theta}(k-1)/d_k^{\alpha}$ can be considered as a prediction of the actual measurement of x_k, and hence $[x_k - \hat{\theta}(k-1)]/d_k^{\alpha}$ can be considered as the prediction error. Consequently, $\hat{\theta}(k)$ can be considered as a combination of the just computed $\hat{\theta}(k-1)$ with a linear transformation of the prediction error.

Since $\hat{\theta}(k) = \hat{\theta}_k$ for $k = 1, 2, ..., K$, some properties of the batch estimator (4) are also applicable to the recursive estimator (7). Recall that the estimator given by (4) is an unbiased estimator since $\mathbb{E}[\hat{\theta}_K] = \theta$. Accordingly, the following unbiasedness definition is used for the recursive estimator. Furthermore, the recursive estimator (7) is also an unbiased estimator.

Definition 1. *A recursive estimator $\hat{\theta}(k)$ is an unbiased recursive estimator of a deterministic parameter θ if*

$$\mathbb{E}[\hat{\theta}(k)] = \theta, \textit{for all } k = 1, 2, ..., K. \tag{15}$$

Corollary 1. *The recursive estimator given by (7) is an unbiased recursive estimator.*

The proof can be based on the fact that $\hat{\theta}(k) = \hat{\theta}_k$ or by a simple induction argument. From Corollary 1, the variance of the kth estimate $\mathbb{V}[\hat{\theta}(k)] = \mathbb{E}[(\hat{\theta}(k) - \theta)^2]$ and can be obtained via (5). Furthermore, $\mathbb{V}[\hat{\theta}(k)] = B(k)$ and (8) can be considered as a recursive structure of $\mathbb{V}[\hat{\theta}(k)]$.

Corollary 2. *The variance of the kth estimate $\mathbb{V}[\hat{\theta}(k)]$ of the recursive estimator satisfies*

$$\mathbb{V}[\hat{\theta}(k)] = \mathbb{E}[(\hat{\theta}(k) - \theta)^2] = \left(\frac{1}{\mathbb{V}[\hat{\theta}(k-1)]} + \frac{1}{d_k^{2\alpha}\sigma_k^2} \right)^{-1} \tag{16}$$

and

$$\mathbb{V}[\hat{\theta}(k)] < \mathbb{V}[\hat{\theta}(k-1)] \tag{17}$$

for $k = 1, 2, ..., K$, where $1/\mathbb{V}[\hat{\theta}(0)] \approx 0$ by appropriately choosing $\mathbb{V}[\hat{\theta}(0)]$.

Since the recursive estimator computes the variance recursively, the recursive estimator is also said as the *variance form* of recursive BLUE. Furthermore, (17) indicates that the more sensors are used for parameter estimation, the better the estimation performance in terms of mean squared estimation error.

3.2 Selection and Termination of Estimation Sequence

The choice of the order of the sensors to perform parameter estimation impacts on the efficiency of the recursive estimator. In some cases, we might not use all the K sensors when we already obtain an estimate with required error performance. Corollary 1 states that the recursive estimator is an unbiased one and the mean of the estimation error $\mathbb{E}[(\theta - \hat{\theta}(k))] = 0$ for all k. However, the variance of the estimation error of the recursive estimator might be dependent on the choice of the sequence of sensors as well as the number of measurements. This motivates us to optimize the estimator in two ways. One is to select sensors judiciously and the other is to terminate estimation procedure wisely.

Recall that BLUE is designed for minimizing MSE. This motivates us to use the variation of the estimation error to measure the estimation efficiency. From Corollary 2, this equals to use $\mathbb{V}[\hat{\theta}(k)]$ to measure the estimation efficiency. Let $\overline{K} = (k_1, k_2, ..., k_K)$ denote a permutation of the K sensors. The following definition gives efficiency measurement for choices of the order of sensors, and Theorem 2 provides necessary conditions for a most efficient sequence.

Definition 2. *A sequence of sensors $\overline{K} = (k_1, k_2, ..., k_K)$ is said to be more efficient than another sequence $\overline{K}' = (k'_1, k'_2, ..., k'_K)$ if*

$$\mathbb{V}[\hat{\theta}(k_i)] \leq \mathbb{V}[\hat{\theta}(k'_i)], \textit{ for all } i = 1, 2, ..., K. \tag{18}$$

Theorem 2. *The most efficient sequence of sensors $\overline{K} = (k_1, k_2, ..., k_K)$ should satisfy*

$$d_{k_1}^{2\alpha}\sigma_{k_1}^2 \leq d_{k_2}^{2\alpha}\sigma_{k_2}^2 \leq \cdots \leq d_{k_K}^{2\alpha}\sigma_{k_K}^2. \tag{19}$$

Proof. The proof proceeds by induction on the construction of $\overline{K}$. For $i = 1$, we have $\hat{\theta}(1) = \theta + d_1^{\alpha} n_1$ and hence

$$\mathbb{V}[\hat{\theta}(k_1)] = \mathbb{E}[(\hat{\theta}(k_1) - \theta)^2] = d_{k_1}^{2\alpha}\sigma_{k_1}^2. \tag{20}$$

Obviously the smallest $\mathbb{V}[\hat{\theta}(k_1)]$ is achieved when choosing the sensor k_1 such that $d_{k_1}^{2\alpha}\sigma_{k_1}^2$ is the smallest among all sensors. Now assume the construction of $\overline{K}$ is the most efficient for i sensors, i.e., the sequence of selected sensors is as $(k_1, k_2, ..., k_i)$ such that $d_{k_1}^{2\alpha}\sigma_{k_1}^2 \leq d_{k_2}^{2\alpha}\sigma_{k_2}^2 \leq ... \leq d_{k_i}^{2\alpha}\sigma_{k_i}^2$ and $\mathbb{V}[\hat{\theta}(k_i)]$ is the most efficient for $1, ..., i$. From (16), we have

$$\mathbb{V}[\hat{\theta}(k_{i+1})] = \frac{1}{\frac{1}{\mathbb{V}[\hat{\theta}(k_i)]} + \frac{1}{d_{k_{i+1}}^{2\alpha}\sigma_{k_{i+1}}^2}} \tag{21}$$

Consider another sequence construction $\overline{K}'$ which is also the most efficient sequence for the first i sensors, however, the $(i+1)$th sensor is different from $\overline{K}$. From (21), it is easily to see that $\mathbb{V}[\hat{\theta}(k_{i+1})] \leq V[\hat{\theta}(k_{i+1})]$ implies that $d_{k_{i+1}}^{2\alpha}\sigma_{k_{i+1}}^2 \leq d_{k'_{i+1}}^{2\alpha}\sigma_{k'_{i+1}}^2$, that is, the selection of the k_{i+1} sensor should also satisfy (19). Hence the desired result is obtained from the induction. ∎

When all sensors have the same noise variance, we have the following corollary.

Corollary 3. *When all sensors have the same noise variance, i.e.,* $\sigma_1^2 = \sigma_2^2 = ... = \sigma_K^2$*, the most efficient sequence of sensors* $\overline{K} = (k_1, k_2, ..., k_K)$ *satisfies*

$$d_{k_1} \leq d_{k_2} \leq \cdots \leq d_{k_K}. \tag{22}$$

The recursive estimator can stop the estimation when the error performance is achieved or when the improvement of the estimate is small. Let ϵ denote the required error variation. When $B(k) \leq \epsilon$, the kth sensor stops forwarding the estimate $\hat{\theta}(k)$ and $B(k)$ to its successor but sends to the sink. Another criterion is that when $B(k) - B(k-1) \leq \epsilon'$, the recursive estimation stops.

4 Numerical Examples

We present some numerical examples in this section. For simplicity, we assume that all noises have the same unit variance. The unit distance is set to achieve

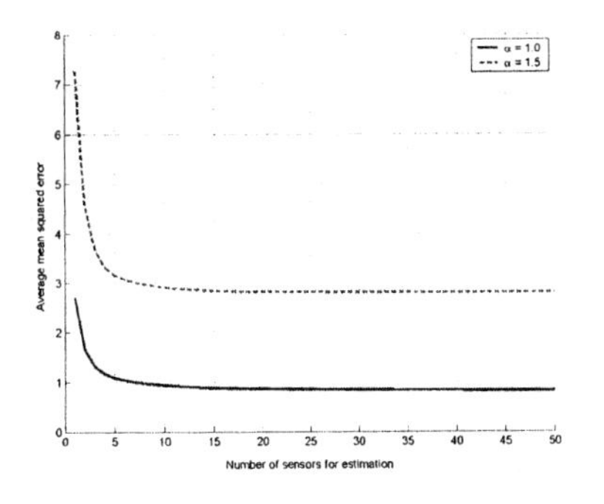

Fig. 3. Average MSE of the most efficient selection for a sensor sequence

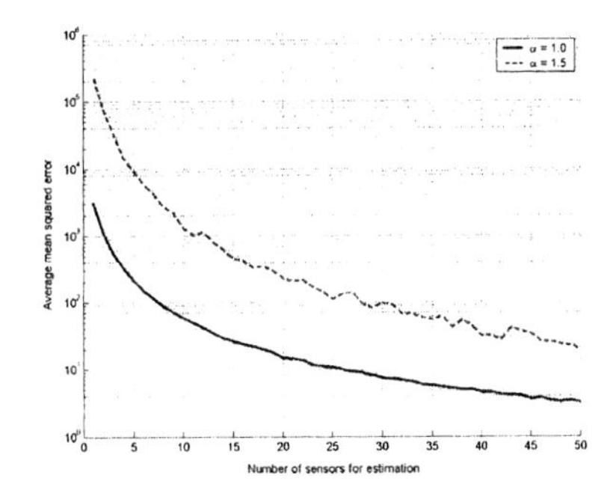

Fig. 4. Average MSE of the random selection for a sensor sequence

a unit mean squared error when only one sensor is used. 100 sensors are used for estimation with their distances to the target location uniformly random distributed from 0.1 to 100 distance units. A most efficient selection scheme is to select K sensors according to Corollary 3. A random selection scheme is to randomly select K sensors from the 100 sensors, and the obtained MSE is averaged over 25 selections. The MSE of the two schemes are then average over 20 simulation runs. The results are shown in Fig. 3 and Fig. 4 for the most efficient and random schemes, respectively. It is first observed that the average MSE is lower for a smaller distance decay component α since low α introduce low distance attenuation. We also observe that the two schemes have the similar average MSE when K is larger. However, it is not unexpectedly to observe that the average MSE of the most selection scheme (10 sensors) converges fast compared with that of the random selection scheme (50 sensors). Finally, the long smooth tail of the average MSE in Fig. 3 indicates that only a few number of sensors, e.g., 10 sensors in the figure, is enough to achieve a target estimation error.

We next compare power consumption for CBES and LRES. However, since we do not have exact power consumption values for sensor computations and communications, some approximations are used in the comparison. Let a_1 denote the precessing power for computing $B(k)$ and $\hat{\theta}(k)$. Let a_2 and a_3 denote the receive and transmit power, respectively. Though the transmit/receive power might be different for different lengths of packets, we assume that the transmit/receive power is the same in CBES and LRES. This is a reasonable assumption as the encoded $B(k)$ and $\hat{\theta}(k)$ might only be a few bytes long. When peak transmit power is used, the transmit distance is also fixed to achieve a target transmission error (though it might be different for different modulation and encoding schemes). Hence we simply use *average hops* to measure the distance between local sensors to a sink (or a fusion center). Furthermore, we assume that sensors for estimation are within one hop away from each other. This is also reasonable since we only choose sensors close to the target location. Suppose there are K sensors and their average distance to the sink is i hops. Then the total power consumption for CBES is: $K \times i \times (a_2+a_3)$; and for LRES is: $K \times (a_1+a_2+a_3)+i \times (a_2+a_3)$.

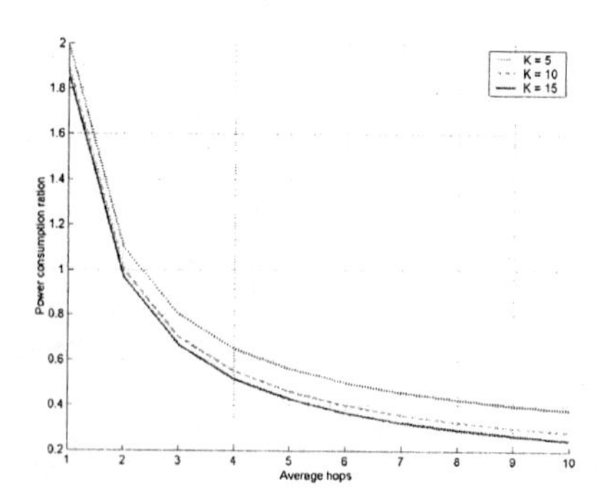

Fig. 5. Power consumption ratio: power consumption of LRES divided by power consumption of CBES

Some typical values of a mote sensor by Crossbow Inc., are used for computation, and they are $a_1 = 16mA$, $a_2 = 8mA$ and $a_3 = 12mA$ [10]. The power consumption ratio defined by the power consumption of LRES divided by the power consumption of CBES is plotted against the number of average hops in Fig. 5. It is observed that when the sink is only one hop away from the sensors, the CBES is more energy efficient; however, when the sink is three or more hops away, the LRES is more energy efficient.

5 Concluding Remarks

This paper has proposed a localized recursive estimation scheme for parameter estimation in sensor networks. Some properties of the scheme have been analyzed and the performances of the scheme have been illustrated via numerical examples.

It is worth noting that even in a local recursive estimation scheme the quantization, modulation and encoding, and transmission (in noisy channels) of the intermediate estimates and error variance are also indispensable. Their impacts on the performance of an estimator are one of our future work.

References

1. Akyildiz, I.F., Su, W., Sankarasubramaniam, Y. and Cayirci, E.: Wireless Sensor Networks: A Survey. *Computer Networks, Elsevier Publishers* (2002) vol. 39, no. 4, 393–422
2. Vuran, M.C., Akan, Ö.B., and Akyildiz, I.F.: Spatio-Temporal Correlation: Theory and Applications for Wireless Sensor Networks. *Computer Networks, Elsevier Publishers* (2004) vol. 45, 245–259
3. Goldsmith, A.J., Wicker, S.B.: Design Challenges for Energy Constrined Ad hoc Wireless Networks. *IEEE Wireless Communications* (2002), vol. 9, no. 4, 8–27
4. Luo, Z.Q., and Xiao J.J.: Universal Decentralized Estimation in A Bandwidth Constrained Sensor Network. *accepted for publication in IEEE Transactions on Information Theory* (2004)
5. Luo, Z.Q., and Xiao J.J.: Decentralized Estimation in An Inhomogeneous Environment. *IEEE International Syposium on Information Theory* (2004)
6. Luo, Z.Q., and Xiao J.J.: Decentralized Estimation in An Inhomogeneous Sensing Environment. *submitted to IEEE Transactions on Information Theory* (2004)
7. Pottie, G.J. and Kaiser, W.J.: Wireelss Integrated Network sensors. *Communications of the ACM* (2000) vol. 43, no. 5, 51–58
8. Meguerdichian, S., Koushanfar, F., Qu G., and Potkonjak M.: Exposure in Wireless Ad hoc Sensor Networks. *ACM International Conference on Mobile Computing and Networking (MobiCom)* (2001) 139–150
9. Mendel, J.M.: *Lessons in Estimation Theory for Signal Processing, Communications and Control*, Prentice Hall, Inc, (1995)
10. Crossbow Technology Inc., *MPR-Mote Processor Radio Borad, MIB-Mote Interface/Programming Board User's Manual (Document 7340-0021-05)*, Crossbow Technology, Inc., (2003)

Asynchronous Power-Saving Event-Delivery Protocols in Mobile USN

Young Man Kim

Kookmin University, School of Computer Science,
Seoul 136-702, South Korea
ymkim@kookmin.ac.kr
http://cclab.kookmin.ac.kr/

Abstract. Recently, Radio Frequency IDentification(RFID) technology has come to maturity such that numerous applications in many industries have been devised and implemented in the real world. In the conventional RFID application, the reader is located at some place without motion, periodically polling the nearby RFID tags. However, the emerging *Mobile RFID* technology[1] provides the user with the mobile RFID reader function by installing it into the portable wireless device like cellular phone or PDA so that he can access the spontaneous RFID-attached object information by approaching the device to RFID tag and triggering the information acquisition operation. Since the information service is immediately offered at any time and place, the whole service domain of mobile RFID is broader than that of conventional RFID. Currently, many commercial applications and their service senarios are developed and standardized in the mobile RFID Forum[1].

In this paper, we propose a new concept called as *mobile USN (Ubiquitous Senser Network)*, the next-generation version of the newly emerging ubiquitous technologies like mobile RFID and NFC(Near Field Communication)[2]. The effective working range of mobile RFID(NFC) is limited within $1m(10cm)$ from the mobile(NFC) phone. This problem can be relieved by adopting RF transmission chip like CC2420 ZigBee RF Tranceiver[3] into the phone. By this upgrade, the effective working range is enlarged to 30-100m so as to invite much more diverse ubiquitous services into realization.

Then, we study energy saving issue in the data delivery from USN tag to phone. In mobile USN, USN phone is in charge of collecting the sensing and event data occuring at the nearby USN tags(that are USN version of RFID tags). However, since USN tag must be equipped with battery to support the active data acquisition, power-saving becomes the fundamental problem in achieving the commercial success of mobile USN. In particular, we introduce three asynchronous power-saving communication protocols. Finally, we study the energy efficiency and QoS aspects of three protocols.

1 Introduction

Recently, RFID technology has come to maturity such that numerous applications in many industries have been devised and implemented in the real world.

X. Jia, J. Wu, and Y. He (Eds.): MSN 2005, LNCS 3794, pp. 400–412, 2005.

The basic communication paradigm in the RFID facility is the polling of the RFID tag-attached object by the RFID reader. In response to the polling request of the reader, the RFID tag answers with its own ID information that is employed to dig the whole object information(object description, delivery history, current status, etc.) from the RFID server. Traditionally, the RFID technology has been utilized in the various industries to improve the quality of services.

In the conventional RFID application, the reader is located at some place without motion. However, the emerging new technology called *Mobile RFID* provides the user with the mobile RFID reader function by installing it into the portable wireless device like cellular phone or PDA so that he can access the spontaneous RFID-attached object information by approaching the device to RFID tag and triggering the information acquisition operation. Since the information service is immediately offered at any time and place by utilizing the wireless phone communication channel, the whole service domain of mobile RFID is broader than that of conventional RFID. Currently, many commercial applications and their service senarios are developed and standardized in the mobile RFID Forum[1].

In this paper, we propose another new concept called *mobile USN(Ubiquitous Senser Network)*, the advanced version of mobile RFID. The effective working range of the mobile RFID reader installed in a mobile device(let's say phone) is practically limited within the circle of $1m$ radius centered the mobile phone. This condition can be overcome by adopting RF wireless transmission chip(for example, ZigBee chip) instead of RFID reader chip. Then, the effective working data acquisition range is enlarged to 30-$100m$. Furthermore, the USN tag, another USN device inheriting the RFID tag function, may collect the dynamic sensor data, detect any event(for example, fire, medical emergency, breakdown, etc.), and inform users of the sensed data and/or the event in real time via USN phone. Once again, the transmission range enhancement and the additional physical/real-time information will invoke much more diverse applications and services for mobile USN.

In the mobile USN configuration, one of the major problem to be seriously tackled is *power-saving issue* since the active function of the USN tag is realized due to the energy supply from the battery. Since the electric power is the most precious among all USN resources, we should devise the USN tag power management scheme that guarantees the optimal energy consumption and thus, the longest operating lifetime. On the other hand, the energy consumption rate in USN phone is not so critical as that in the USN tag, since, in any way, the user should repeat the battery recharge operation several times each week. In fact, the energy consumption rate for the phone operation in voice communication is known to be one order higher than that for USN communication.

In this paper, we introduce the concept of mobile USN. Then, we propose three energy-efficient, asynchronous event-reporting protocols adopted between mobile USN device and USN tag. The performance evaluation is then presented so as to be applied in the practical configuration of mobile USN.

The rest of this paper is organized as follows. In Section 2, the previous researches related to the energy saving problem in the wireless mobile networks are reviewed. Following that, Section 3 presents the general USN phone-tag communication model that will be used in the remaining sections. Then, we introduce three energy-efficient, asynchronous event-reporting communication protocols: *Cyclic Monitoring vs Continuous Reporting(CMCR)*, *Cyclic Beaconing vs. Continuous Monitoring(CBCM)*, *Cyclic Monitoring vs Continuous Jamming(CMCJ)* protocols. We present the performance evaluation for three protocols in Section 5. Finally, we conclude in Section 6.

2 Related Works

Power-saving is a fundamental problem inherent in any mobile devices supported by the battery . Without power, any mobile device will become useless. Furthermore, battery technology is not likely to progress as fast as computing and communication technologies do. Hence, how to lengthen the lifetime of batteries is an important issue for mobile node.

More and more wireless devices become to support low-power sleep modes. IEEE 802.11[4] has a power-saving mode in which a radio only needs to be awake periodically. HyperLAN allows a mobile host in power-saving mode to define its own active period. An active host may save powers by turning off its equalizer according to the transmission bit rate. Comparisons are presented in [5] to study the power-saving mechanisms of IEEE 802.11 and HIPERLAN in ad hoc networks. Bluetooth[6] provides three different low-power modes: *sniff*, *hold*, and *park*.

Although a lot of power-saving methods[7][8][9] in wireless mobile networks are proposed, the network configuration between USN phone and tag has several unique properties that make the power-saving problem of the mobile USN different from that of the other wireless networks like 802.11, Bluetooth and conventional sensor network(SN).

Since mobile USN is a particular branch of SN, we will specifically enumerate these mobile USN characteristics in comparison with SN. First, the communication between USN phone and tag is asymmetric, contrary to that between the identical parties in SN. While USN tag collects data and detects event, USN phone waits up indefinitely until the neighboring tags are ready to send them. Second, the sink node of SN is usually fixed and supplied with ample power so that power-saving of the sink node is trivial. On the other hand, USN phone has its own energy saving problem so as to minimize the USN burden over the original telecommunication lifetime, although it is recharged frequently. Thus, it is necessary to provide a reasonable, asymmetric energy-expense sharing rule, on which an efficient message delivery protocol should be designed.

In the next section, an example energy-expense sharing rule is proposed. Following that, in Section 4, we propose three efficient data communication protocols.

3 Mobile USN

3.1 Mobile USN Architecture

The communication in the typical RFID application is initiated with polling RFID tag by RFID reader. In response to the polling request from the reader, RFID tag answers with its own ID information that is used to get the whole object information(e.g., object description, delivery history, current status, etc.) from RFID server. Traditionally, RFID technology has been utilized in the various industries to identify the object location in a restricted work area such as storage yard.

In the conventional RFID application, the reader is located at some place without motion. However, *Mobile RFID* technology provides the user with the mobile RFID reader function by installing it into the portable wireless device like cellular phone or PDA so that he can access the spontaneous RFID-attached object information by approaching the device to RFID tag and triggering the information acquisition operation. Since the information service is immediately offered at any time and place, the whole service domain of mobile RFID is broader than that of conventional RFID. For example, as an ordinary person strolls in the street corner or shopping center, he may encounter with some interesting object(say, a movie poster). Once he decides to invest his time to access more information about the movie, he approaches his mobile phone toward the RFID tag attached on the surface of the poster. Then, the ID information written in the tag is transfered over wireless medium to the phone that, in its turn, makes a connection with the poster server to get the detailed movie information. After reviewing that information, he will probably decide to purchase the ticket on-line.

Mobile USN(Ubiquitous Senser Network) is the result of another technological leap from mobile RFID. Fig. 1 denotes technological relation among RFID, Mobile RFID, Mobile USN and AN(Access Network)-CN(Core Network) backbone network. The operational working range of mobile RFID is enlarged from $1m$ to 30-$100m$ by employing an active wireless transmission module like ZigBee chip. The range enhancement in mobile USN invites much more diverse services into realization. For example, suppose that a lot of RFID tags are attached along the street sideways in some ubiquitous city. When a visitor becomes to lose his way to the hotel, he has to find out the nearby RFID tag and move his RFID phone close to the tag location explicitly. In contrast to this, he does not have to find the USN tag location nor move his USN phone to get the current location information since the location data from several RFID tags are gathered automatically from the user's physical area and the current location is predicted in a good precision by using some location data processing algorithm.

In mobile USN, USN tag, that is the counterpart of RFID tag in mobile USN, may collect the dynamic sensor data, detect any event(for example, fire, medical emergency, breakdown, etc.) and inform the user the sensing data and/or the event in real time by sending it to the neighboring USN phone. In case the user would be the owner of the phone, the work is done and he has the information

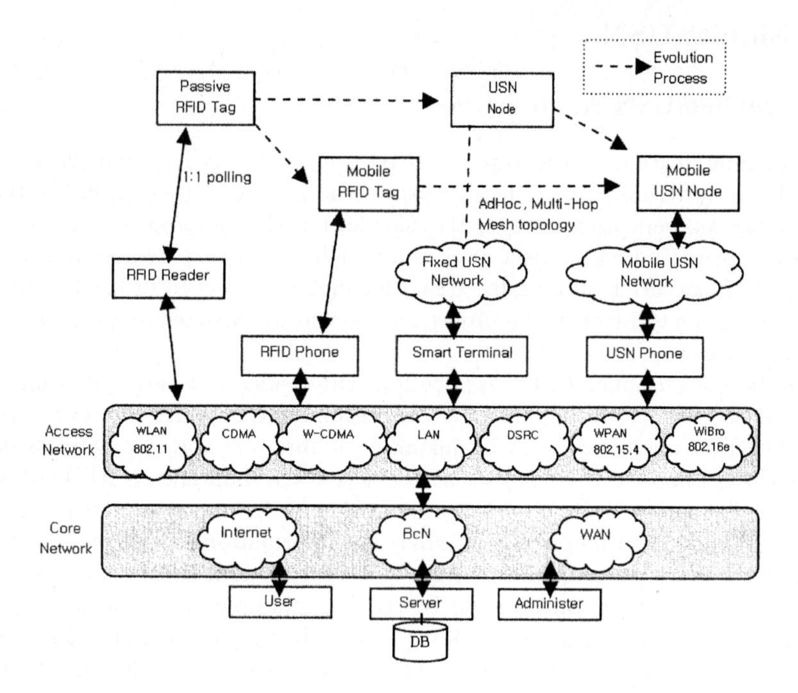

Fig. 1. Technological Relation among RFID, Mobile RFID, Mobile USN and AN-CN backbone network

at hand. Otherwise, it will be routed through AN-CN global backbone network and, sooner or later, delivered to the remote user's terminal device.

Formally, mobile USN consists of back-end global data networks, called AN (Access Network consisting of CDMA phone network, WLAN, WiBro, DSRC, etc.) and CN (Core Network consisting of PSTN, Internet, Sattellite Network, etc.) and three kinds of devices; *USN end-user terminal*, *USN phone*, and *USN tag*. USN end-user terminal is a data monitoring device that displays, to the user, the USN data gathered at USN tag, delivered to the USN phone, and routed via AN-CN internetwork up to itself. Thus, USN end-user terminal may be PDA, laptop, PC, mobile phone or any other handheld wireless device. The USN phone is in charge of data delivery gateway between USN tag and AN-CN internetwork. Finally, USN tag is an intelligent data source node producing any valuable sensor or ID or pure abstract data for the user. Typically, USN tag consists of sensor, memory, RF transmission module, and processing unit.

3.2 USN Phone-Tag Communication Model

There are two categories of information classified in terms of data formation method; *scheduled* data and *event* data. Scheduled data denotes any data that are acquisited from sensor according to the predetermined measurement schedule. Since the schedule is fixed, the power-saving rule for USN tag under such measurement schedule is relatively simple by setting the tag activation and data delivery schedule identical to the measurement schedule. Furthermore, if the local USN phone neighboring to the tag has the scheduling information of the latter, the phone's USN module may sleep and wake up following the same schedule to minimize the energy consumption in USN data communication.

On the other hand, event data is inherently unpredictable and asynchronous, being generated randomly. Therefore, developing an power-saving data communication scheme between USN tag(data source) and USN phone(data sink) is nontrivial. In the following section, we propose three efficient power-saving delivery protocols for event data. All proposed protocols are based on two rules; (i) USN tag tries to make a contact with the neighboring USN phone only if the event actually occurs at that time, otherwise, the RF communication module remains in sleep mode minimizing the energy consumption, (ii) USN phone wakes up periodically and stays in active mode for a while, making it possible for a tag to send the event. Since the mobile phone should offer the basic communication function without significant change in battery usage, we suggest an energy-expense sharing rule between voice communication and USN functions such that the total energy consumption due to USN should not exceed 10% of total energy. We believe that 10% reduction in the phone operation time may not bother users too much.

4 Three Asynchronous Event-Reporting Communication Protocols

USN phone is relatively free from the energy problem since the phone user may recharge the phone battery each day. The energy consumption rate of mobile phone is known to be one order higher than that of USN communication. For example, it is found that an example Samsung mobile phone spends $300mA$ of current during the voice communication interval. However, low-power, short-range(30-100m), RF transmission chip spends 20-30mA depending on the current mode type(listening or sending) and the signal strength. This results from the fact that the typical communication distance between the phone and the nearby base station is much longer, up to one thousand meter. By rule of thumb as explained in the previous section, it may be acceptable to user that the energy consumption ratio between USN and the other phone operations is less than 10%.

The USN event, that would be detected in USN sensor tag, is unpredictable in the time of occurance. Once it happens, the event message should be delivered as soon as possible since the event information(e.g., patient's heart attack, fire, burglary, etc.) is usually of real-time characteristics. Therefore, we should employ an asynchronous event-reporting protocol between USN phone and tag that compromise three conflicting trade-off requirements: (i) reducing the delivery time of the event message, (ii) improving the power-saving in USN tag operation, and (iii) minimizing USN phone user's annoyance due to frequent battery recharge.

It is well known that both RF transmission and listening modes have high power consumption rate with the same order. Thus, the effective method to reduce the energy consumed in RF module is to stay in the sleeping mode as long as possible, since the power consumption in the sleeping mode is three to five order less than that in active(transmission or listening) modes. In this section, three asynchronous message-delivering protocols are introduced. Basically, USN phone follows the periodic short-wakeup-and-long-sleep schedule to save the phone battery power. The cyclic wakeup is unavoidable because of the unpredictability and real-time property of the event. On the other hand, USN tag should transmit the event message quickly and with little energy consumption. However, the existence of sleep period in USN phone prevents the immediate message delivery, resulting in the message delay, in addition to more power consumption due to active-sleep mode asynchronism.

The period of the phone wakeup-sleep cycle is denoted as T_C. In the beginning of each period, the phone USN module remains in active listening mode to catch up the possible event message transmission. Let T_L be the wakeup interval remaining in listening state. The event message transfered from tag to phone takes the transmission time T_D. After the successful event message transmission, the tag waits ACK message from USN phone until time T_A, the ACK transmission time, passes over.

If ACK message is received during ACK waiting interval, the complete event delivery process is over. Otherwise, the event transmission and ACK waiting process repeats itself until the transmission succeeds or the delivery process failure is declared after a predefined number of retransmissions. To make the protocol description simple, the message transmission is supposed to succeed whenever both USN tag and phone are in active state at the time the event or ACK message is transmitted. Furthermore, only the communication between a pair of USN phone and tag is considered without any interrupt from the other devices in the field.

4.1 Cyclic Monitoring vs Continuous Reporting(CMCR) Protocol

The first event delivery protocol introduced in this subsection is the simplest among all such that, whenever the tag detects some event, it keeps transmitting event message until receiving the corresponding ACK message. On the other hand, USN phone repeats the basic listening-sleep cycle, as depicted in Fig. 2. This protocol is called as *Cyclic Monitoring vs Continuous Reporting(CMCR)*, since the phone enters the cyclic monitoring state at the beginning of each cycle while the tag transmits the event message continuously until the phone replies with ACK message.

4.2 Cyclic Beaconing vs. Continuous Monitoring(CBCM) Protocol

In *Cyclic Beaconing vs. Continuous Monitoring(CBCM)* protocol, as shown in Fig. 3, the initial transmission in event message delivery sequence occurs at USN phone instead of tag. USN phone broadcasts a beacon periodically. Since the

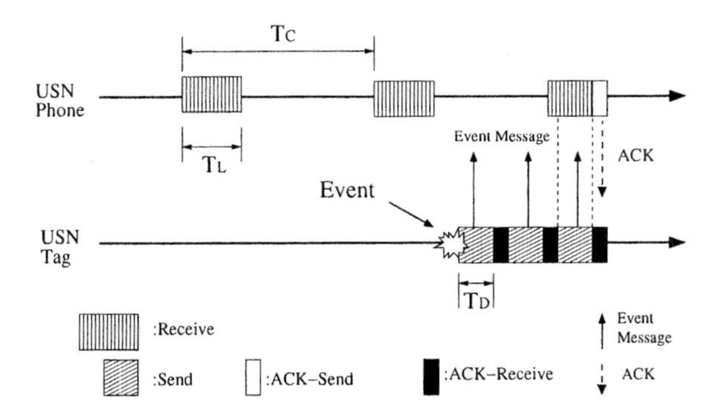

Fig. 2. CMCR event-delivery protocol message sequence diagram

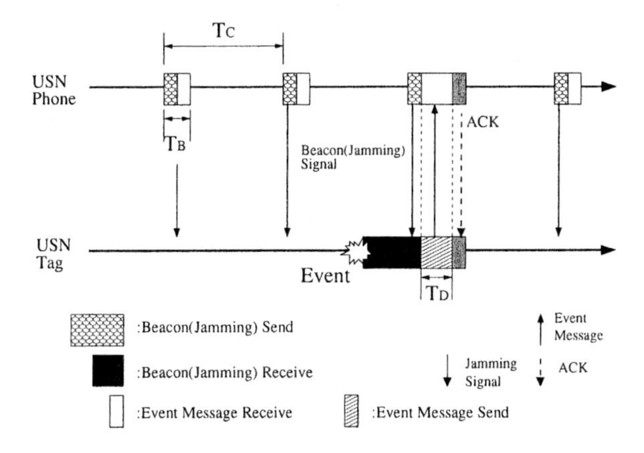

Fig. 3. CBCM event-delivery protocol message sequence diagram

purpose of the beacon is only to inform the neighboring tag(s) having the event message of the start of the delivery sequence, the size of beacon message is much shorter than that of event message. Let T_B and T_E be the beacon transmission and extra waiting time, respectively. Extra waiting time is used to make sure no existence of event message in the neighboring tag. Otherwise, when a tag replies with the event message, the phone answers with ACK message in its turn, so that the phone wakeup time extends from $T_B + T_E$ to $T_B + T_D + T_A$. Usually, the USN phone wakeup time with no event, $T_B + T_E$, is much shorter than the event message transmission time T_D, since the triggering operation of beacon requires no data byte in the beacon message.

4.3 Cyclic Monitoring vs Continuous Jamming(CMCJ) Protocol

In *Cyclic Monitoring vs Continuous Jamming(CMCJ)* protocol, as depicted in Fig. 4, the protocol sequence follows the similar procedure to CBCM except that the role of two parties, phone and tag, is swapped. USN tag, instead of USN

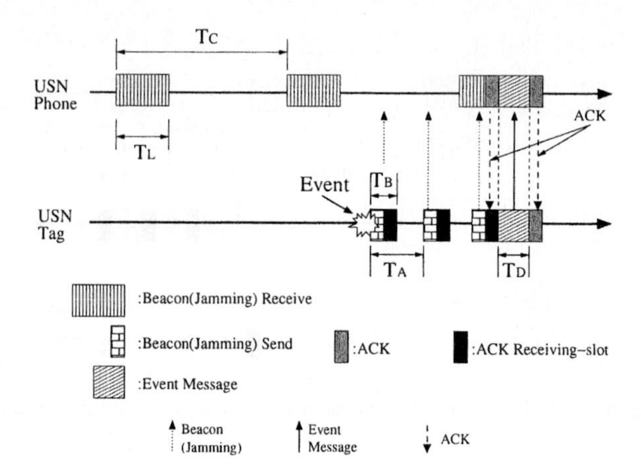

Fig. 4. CMCJ event-delivery protocol message sequence diagram

phone, broadcasts a beacon periodically as it holds an event to be reported. Since the purpose of the beacon is to inform the neighboring phone to receive the event message, the size of beacon message is again much shorter than that of event message. Let T_B and T_E be the time of transmitting beacon and extra waiting time (in case of no ACK message following the beacon), respectively. Extra waiting time T_E is used to make sure no existence of USN phone in the wakeup state. Otherwise, when a phone replies with ACK message, the tag sends the event message immediately to the phone. Following that, the phone then sends ACK message, informing the normal receipt of event message.

Likewise to CBCM, in CMCJ, the USN tag wakeup time with no active phone in the neighborhood, $T_B + T_E$, is much shorter than the event message transmission time, T_D, since the triggering operation of beacon to send the event requires no data byte. Otherwise, the USN tag wakeup time extends to $T_B + T_D + 2 \cdot T_A$.

5 Performance Evaluation

In this section, we study the protocol performance quantitatively. The performance measures used in the analysis are related to power saving and transmission delay. The specific parameter values and assumptions employed in the performance evaluation are presented as follows. The event message transmission time T_D is set to $20ms$, that corresponds to the transmission time of 50 bytes via 20Kbps ZigBee specification. Also, T_L is set to $2 \cdot T_D$, since asynchronism between phone wakeup period and tag event transmission period needs double size of waiting interval for the perfact transmission from tag to phone during this overlap time. For simplicity, extra waiting time T_E is ignored due to the fact that the fast transmission signal detection will make the waiting interval so small. The ACK transmission time T_A is also ignored. Beacon transmission time T_B is set to $4ms$, one fifth of T_D.

The typical electrical capacity of AAA battery like Duracell is $1.1Ah$. Moreover, the low-power rechargeable battery used for mobile phone also has the similar capacity, so that we assume both phone and tag batteries have the same capacity of $1.1Ah$. To make a simplified power-saving analysis, we suppose that both phone and tag spend all energy for the USN message transmission function so as to clearly catch the energy characteristics of three protocols. We assume that USN RF transmission module spends $20mA$ either in transmission or listening mode. During the sleeping state, the energy consumption is supposed to be zero.

Phone energy saving ratio(PESR) is defined as the ratio between the exhaustive energy consumption, in the case that the phone preserves the wakeup state all the time, and the actual energy consumption employed in each proposed protocol. PESR, for two protocols CMCR and CMCJ, is denoted as $\frac{T_C}{T_L}$. On the other hand, PESR for CBCM is set as $\frac{T_C}{5 \cdot T_B}$, since the beacon period in CBCM is, in the role, equivalent to the wakeup time of $T_L (= 5 \cdot T_B)$ in the other protocols. Then, two performance measures indicating the power-saving property of USN phone and tag are *phone operation lifetime(POL)*, total time span in which the phone can remain in normal operating state, and *event transmission number(ETN)*, the number of events that can be successfully delivered from the tag until the tag battery becomes empty.

Figures 5 and 6 depict the energy performance aspects, POL and ETN, of three protocols vs PESR, respectively. CBCM shows the best performance as the phone energy efficiency is concerned. For example, at $PESR = 10$, the phone lifetime of CBCM is about 7 times longer than that of the other protocols. On the other hand, CMCJ has the highest power-saving efficiency in USN tag. With $PESR = 10$, the event transmission number of CMCJ is about three times larger than that of the others.

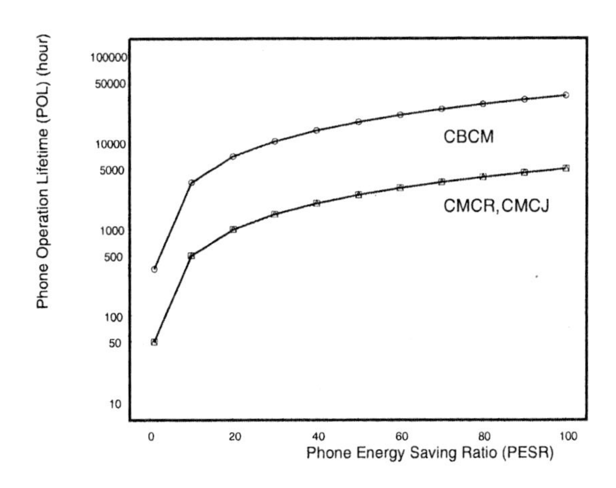

Fig. 5. Phone operation lifetime(POL) (*hours*) vs phone energy saving ratio(PESR)

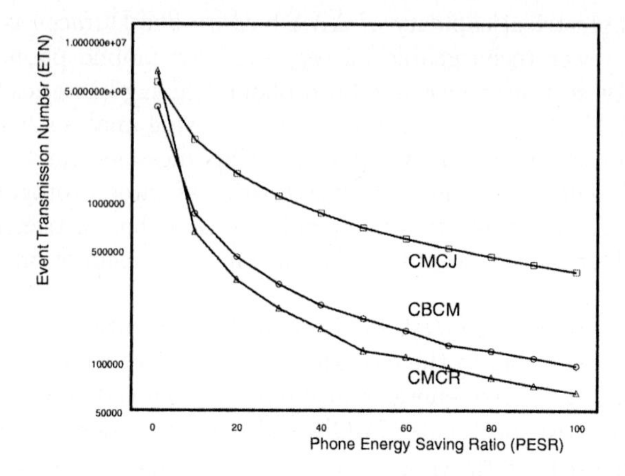

Fig. 6. Event transmission number(ETN) vs phone energy saving ratio(PESR)

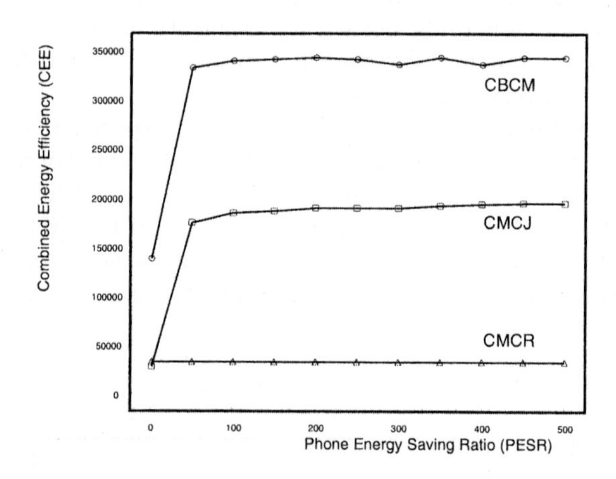

Fig. 7. Combined energy efficiency(CEE) vs phone energy saving ratio(PESR)

Combined energy efficiency(CEE) is defined as the multiplication of POL and ETN, revealing the total power-saving efficiency of both USN phone and tag. Therefore, the higher CEE denotes the better USN overall energy-effieiency performance such that the phone operational lifetime and/or the event transmission number of USN tag increase. Fig. 7 depicts CEE values of three protocols vs PESR. Note that CEE arrives at the upper boundary as PESR increases up to 100. Among the three protocols, CBCM shows the highest performance, since the major energy consumption source, that is the duty ratio of CBCM at the phone, is the smallest among all.

Fig. 8 depicts the last performance measure, the *delay time* between the event occurance at the tag and the event message arrival at the phone. The delay

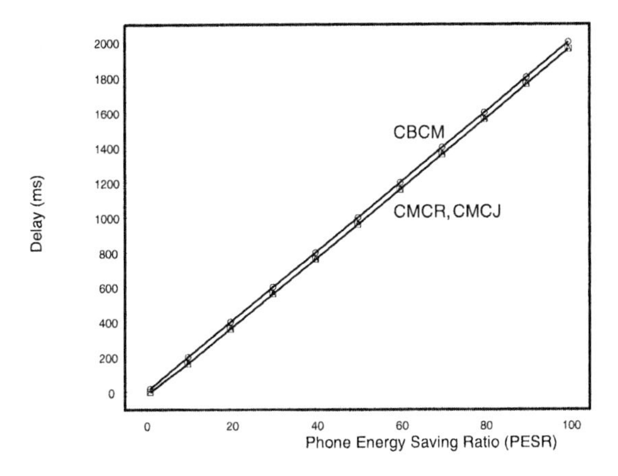

Fig. 8. Tag-to-phone average event-reporting delay(ms) vs phone energy saving ratio(PESR)

increases proportional to PESR regardless of the protocol differences among three. In addition, the delay curves for three protocols are also almost identical to each other.

All in all, we recommend protocol CBCM as the candidate mobile USN protocol in the event message delivery. Applying 10% rule of thumb in phone energy consumption for USN, CBCM yields the highest ETN and the delay close to that of the other protocols.

6 Conclusion

In this paper, we proposed a new concept called as *mobile USN(Ubiquitous Senser Network)*, the next-generation version of the newly emerging ubiquitous technologies like mobile RFID and NFC(Near Field Communication). The effective working range of mobile RFID(NFC) is limited within $1m(10cm)$ from the mobile(NFC) phone. This problem can be relieved by adopting RF transmission chip like CC2420 ZigBee RF Tranceiver, instead of RFID reader chip, into the phone. By this upgrade, the effective working range is enlarged to 30-100m so as to invite much more diverse ubiquitous services into realization.

Then, we studied energy saving issue in the data delivery from USN tag to phone. In mobile USN, USN phone is in charge of collecting the sensing and event data occuring at the nearby USN tags(that are USN version of RFID tags). However, since USN tag must be equipped with battery to support the active data acquisition, power-saving becomes the fundamental problem in achieving the commercial success of mobile USN.

Next, we showed two general USN phone-tag communication models: scheduled and asynchronous event communication. Following that, we introduced three asynchronous event-reporting communication protocols: *CMCR*, *CBCM*, and *CMCJ*. Then, we analyzed the energy efficiency and QoS aspects of three protocols. We recommend protocol CBCM as the candidate mobile USN protocol used in the event message delivery. Applying 10% rule of thumb in phone energy consumption for USN, CBCM yields the highest ETN and the shortest delay.

References

1. Mobile RFID Forum, http://www.mrf.or.kr/.
2. NFC Forum, http://www.nfc-forum.org/.
3. ChipCon AS, http://www.chipcon.com/.
4. LAN MAN Standards Committee of the IEEE Computer Society, IEEE Std 802.11-1999, Wireless LAN Medium Access Control (MAC) and Physical Layer (PHY) specifications, IEEE, 1999.
5. H.Woesner, J. P. Ebert,M. Schlager, and A.Wolisz, Power-Saving Mechanisms in Emerging Standards for Wireless LANs: The MAC Level Perspective, *IEEE Personal Communications*, pp. 40-48, June 1998.
6. J. C. Haartsen, The Bluetooth Radio System, *IEEE Persinal Communications*, pp. 28-36, Feb. 2000.
7. Y.C.Tseng, C.S.Hsu and T.Y.Hsieh, Power-Saving Protocols for IEEE 802.11-Based Multi-Hop Ad Hoc Networks, *IEEE INFOCOM*, 2002.
8. Wei Ye, John Heidemann, and Deborah Estrin, "An Energy-Efficient MAC Protocol for Wireless Sensor Networks," *Proceedings of the IEEE INFOCOM*, pp. 1567-1576, New York, NY, USA, June 2002.
9. Alec Woo and David Culler, A transmission control scheme for media access in sensor networks, *Proceedings of the ACM/IEEE International Conference on Mobile Computing and Networking*, Rome, Italy, July 2001.

Low-Complexity Authentication Scheme Based on Cellular Automata in Wireless Network*

Jun-Cheol Jeon, Kee-Won Kim, and Kee-Young Yoo**

Department of Computer Engineeing,
Kyungpook National University, Daegu, 702-701 Korea
{jcjeon33, nirvana}@infosec.knu.ac.kr,
yook@knu.ac.kr

Abstract. This paper proposes an efficient authentication protocol for wireless networks based on group and non-group cellular automata (CA). Most previous protocols employed typical cryptosystems such as symmetric and asymmetric encryptions in spite of their high computation and communication cost. In addition, previous protocols are problematic in that the mobile user copies his/her anonymous ticket and shares it with friends. We propose an efficient and secure authentication protocol which provides a mobile user's anonymity and which has a low computation cost to withstand the above weakness without employing any typical cryptosystems.

1 Introduction

In wireless communication networks, when a mobile station (MS) wants to roam freely in a visiting network (VN), it has to send its temporal mobile subscriber identity to the VN. Then the VN charges some private information from the MS's home network (HN) for identifying the MS. However, the HN and the VN should not know the MS's location or private information for the applications of the anonymous channel such as the anonymous conference key distribution, e-cash payment, and voting scheme [1-4].

Various protocols have been proposed to ensure the privacy of the MS, but most of them have employed asymmetric encryptions to protect the anonymity of the MS. Lin and Jan (2001) proposed an authentication protocol for an anonymous channel of the wireless communications based on asymmetric encryptions [5]. Barbancho and Peinado (2003) pointed out that Lin and Jan's protocol was vulnerable to the forgery attack [6]. Yang et al. (2005) have proposed a secure and efficient authentication protocol for anonymous channel of the wireless communications to overcome the above problems [7]. However their protocols still employ many exponentiation operations and symmetric encryptions.

Meanwhile, a CA introduced by John Von Neumann [8] has been accepted as a good computational model for simulating complex physical systems, and has been

* This work was supported by the Brain Korea 21 Project in 2005.
** Corresponding author.

X. Jia, J. Wu, and Y. He (Eds.): MSN 2005, LNCS 3794, pp. 413 – 421, 2005.

used in evolutionary computation for over a decade. It can be used to simulate complex growth patterns as well as various applications, such as parallel processing computations and number theory [9]. Recently, the CA has been employed to design pseudo-random number generators and cryptographic protocols [10, 11].

Various studies have discussed the reversibility and non-reversibility of CA based on group and non-group CA [12, 13]. A group property provides that the previous state is reachable from the present state of the CA, while a non-group property is not. We employ a group CA based on the reversibility to conceal the private information of the MS and VN, and a non-group CA based on the non-reversibility to guarantee the anonymity of the MS. The computation in our scheme is only composed of the logical operations such as XOR, AND, OR, and NOT, while the typical cryptographic schemes have complex and heavy operations such as hash functions, inversions, and exponentiations.

In section 2, the background and properties of CA are illustrated. Section 3 presents the proposed authentication scheme based on group and non-group CAs, and section 4 presents a security analysis of our scheme. Finally, section 5 presents our concluding remarks.

2 Cellular Automata

A CA is a collection of simple cells arranged in a regular fashion. CAs can be characterized based on four properties: cellular geometry, neighborhood specification, the number of states per cell, and the rules to compute to a successor state. The next state of a CA depends on the current state and rules [14]. A CA can also be classified as linear or non-linear. If the neighborhood is only dependent on an XOR operation, the CA is linear, whereas if it is dependent on another operation, the CA is non-linear. If the neighborhood is only dependent on an XOR or XNOR operation, then the CA can also be referred to as an additive CA.

Among additive CAs, a CA whose dependency on neighbors is shown only in terms of XOR is called a non-complemented CA, and the corresponding rule is called the non-complemented rule. If the dependency on neighbors is shown only in terms of XNOR, the CA is called a complemented CA, and the corresponding rule is called the complemented rule. A hybrid CA can be subject to either the complemented or non-complemented rule. Additionally, there are 1-dimensional, 2-dimensional, and 3-dimensional CAs according to the structure of their cells.

If the same rule applies to all the cells in a CA, the CA is called a uniform or regular CA, whereas if different rules apply to different cells, it is called a hybrid CA. In the structure of CAs, the boundary conditions should be taken into consideration, where the boundary conditions incur since there exist no left neighbors in the leftmost cell or right neighbors in the rightmost cell among the cells composing CA. According to the conditions, they are divided into three types: null boundary CA, periodic boundary CA, and intermediate boundary CA [9].

1) Null Boundary CA (NBCA): CA of which left neighbor of the leftmost cell and right neighbor of the rightmost cell are regarded to be ‘0’.
2) Periodic Boundary CA (PBCA): CA of which leftmost cell and rightmost cell are regarded to be adjacent to each other, i.e., the left neighbor of the leftmost cell becomes the rightmost cell, and the right neighbor of the rightmost cell becomes the leftmost cell.
3) Intermediate Boundary CA (IBCA): The left neighbor of the leftmost cell is regarded to be the second right neighbor, and right neighbor of the rightmost cell is regarded to be the second left neighbor.

A one-dimensional CA consists of a linearly connected array of n cells, each of which takes the value of 0 or 1, and an evolutionary function $F(s)$ on the state configuration, s, with q variables. The value of cell state s_i is updated in parallel using this function in discrete time steps as $s_i(t+1)= F(s_{i+j}(t))$ where $-r \leq j \leq r$ [12]. The parameter q is usually an odd integer, i.e. $q = 2r+1$, where r is often named the radius of the function $F(s)$; the possible configuration and the total number of rules for radius r neighborhood are 2^q and 2^n, where $n = 2^q$.

The new value of the i-th cell is calculated using the value of the i-th cell itself and the values of r neighboring cells to the right and left of the i-th cell. If a group rule is applied to a CA, then the CA is called a group CA; otherwise, the CA is a non-group CA [15]. In a two-state 3-neighborhood CA, there are 256 rules. Table 1 specifies three particular sets of transition using the rules 90, 171, and 129 from a neighborhood configuration to the next state. The notations ‘⊕’, ‘∨’, and ‘¬’ indicate the bitwise XOR operation, OR operation, and NOT operation respectively.

Table 1. State transaction for rule 90, 171, and 129 in 2-state 3-neigborhoods

rules	111	110	101	100	011	010	001	000
90	0	1	0	1	1	0	1	0
171	1	0	1	0	1	0	1	1
129	1	0	0	0	0	0	0	1

rule 90 : $s_i(t+1) = s_{i-1}(t) \oplus s_{i+1}(t)$
rule 171: $s_i(t+1) = (\neg(s_{i-1}(t) \vee s_i(t))) \oplus s_{i+1}(t)$
rule 129: $s_i(t+1) = \neg((s_{i-1}(t) \oplus s_i(t)) \vee (s_{i-1}(t) \oplus s_{i+1}(t)))$

A CA provides both reversible and non-reversible properties according to $F()$. In a group CA, the previous state can be easily found by computing the inverse of a group rule, but it is computationally infeasible to find the inverse of a non-group rule in a non-group CA [14].

3 Proposed Authentication Scheme

The security of our scheme is based on reversibility and non-reversibility. A group rule function must be tractable to compute in both directions, while a non-group function must be tractable to compute in the forward direction, but computationally infeasible to invert. In addition, the evolved states must be distinctive. To make sure that a state is used only once, a system should choose and compute the rule and the length according to the given initial state. A length represents the number of the unique states in a CA.

Table 2 shows the evolution configurations and lengths when the initial state is 4-bits vector (1011). As shown in Table 2, the group CAs applying rules 90 and 150 possess a cyclic property in which the initial state appears after certain evolutions, and the non-group CAs applying rules 171 and 129 have the property such that one of the previous states appears indefinitely after a certain period which is called a length.

Table 2. The evolution configurations and length according to the given initial state (1011) by the group or non-group rule {90, 150, 171, 129}

	group CAs		non-group CAs	
rule #	rule 90	rule 150	rule 171	rule 129
expression	$s_{i-1} \oplus s_{i+1}$	$s_{i-1} \oplus s_i \oplus s_{i+1}$	$(\neg(s_{i-1} \vee s_i)) \oplus s_{i+1}$	$\neg((s_{i-1} \oplus s_i) \vee (s_{i-1} \oplus s_{i+1}))$
evolution configuration	1011→0011→0111→1101→1100→1110→1011	1011→1000→1100→0010→0111→1010→1011	1011→0110→1100→1001→0010→1100	1011→0000→1111→0110→0000
length	6	6	5	4

3.1 Notation

- ID_{MS}, ID_{VN}, and ID_{HN} denote the identity of the MS, VN, and the HN, respectively.
- X → Y: Z denotes the sender X sends a message Z to the receiver Y.
- ${}_G\#_{h,m}$ and ${}_G\#_{h,v}$ denote the secret group rules that is shared between the HN and the MS, and between the HN and the VN, respectively.
- ${}_N\#_m$ and ${}_N\#_v$ denote the non-group rule or the combination of rules of the MS and the VN.
- ${}_GF()$ and ${}_NF()$ denote the evolutionary function operated by a group rule and non-group rule, respectively.
- ${}_NF^L()$ denotes the evolutionary function performed by a non-group rule L times, where L is the length of the non-group CA.
- C_{MS} and C_{VN} denote the counter for the MS and the VN.

3.2 Proposed protocol

The proposed anonymous channel protocol consists of two phases: the ticket-issuing phase and the ticket-authentication phase. The MS has to purchase an anonymous ticket from the HN before roaming anonymously in the VN. The following illustrates the ticket issuing phase in which the MS purchases an anonymous ticket from the HN via the VN.

I-1. MS → VN: ID_{HN}, $ID_{MS} \oplus {}_{G}\#_{h,m}$, ${}_{G}F_{h,m}(ID_{MS}, T_1, {}_{G}\#_{h,m})$

The MS computes ${}_{G}F_{h,m}(ID_{MS}, T_1, {}_{G}\#_{h,m})$ using the shared secret ${}_{G}\#_{h,m}$. The MS sends $ID_{MS} \oplus {}_{G}\#_{h,m}$ and ${}_{G}F_{h,m}(ID_{MS}, T_1, {}_{G}\#_{h,m})$ with the HN's identity to the VN, where T_1 is a timestamp.

I-2. VN → HN: ID_{VN}, $ID_{MS} \oplus {}_{G}\#_{h,m}$, ${}_{G}F_{h,m}(ID_{MS}, T_1, {}_{G}\#_{h,m})$, ${}_{G}F_{h,v}(ID_{VN}, T_2, {}_{G}\#_{h,v})$

The VN computes ${}_{G}F_{h,v}(ID_{VN}, T_2, {}_{G}\#_{h,v})$ using the shared secret ${}_{G}\#_{h,v}$, then forwards the received message and ${}_{G}F_{h,v}(ID_{VN}, T_2, {}_{G}\#_{h,v})$ with his identity to the HN, where T_2 is a timestamp.

I-3. HN → VN: ${}_{G}\#_{h,m} \oplus (T_1, {}_{N}F_v^{Lv}(ID_{VN}), {}_{N}\#_m, {}_{N}\#_v, L_m, L_v, N_{expire})$, ${}_{G}\#_{h,v} \oplus (T_2, {}_{N}F_m^{Lm}(ID_{MS}), {}_{N}\#_v, {}_{N}\#_m, L_v, L_m, N_{expire})$

The HN firstly records a received time T_3. Then, the HN finds the MS's ID_{MS} and their session key ${}_{G}\#_{h,m}$ from $ID_{MS} \oplus {}_{G}\#_{h,m}$. Then, the HN computes the inverse of the group rule, ${}_{G}\#_{h,m}^{-1}$ and extracts $(ID_{MS}, T_1, {}_{G}\#_{h,m})$ from the received message. Similarly, the HN finds ID_{VN} and the VN's session key ${}_{G}\#_{h,v}$. If $(T_3 - T_2 \leq \Delta T)$ and the ID_{VN} is correct, the HN authenticates the identity of the VN, where ΔT denotes a valid time interval. Similarly, if $(T_2 - T_1 \leq \Delta T)$ and the ID_{MS} is correct, the HN authenticates the identity of the MS. Finally, the HN chooses ${}_{N}\#_m$ and ${}_{N}\#_v$, and finds L_m and L_v, computes ${}_{N}F_v^{Lv}(ID_{VN})$ and ${}_{N}F_m^{Lm}(ID_{MS})$ using the chosen non-group rules, and sends ${}_{G}\#_{h,m} \oplus (T_1, {}_{N}F_v^{Lv}(ID_{VN}), {}_{N}\#_m, {}_{N}\#_v, L_m, L_v, N_{expire})$ and ${}_{G}\#_{h,v} \oplus (T_2, {}_{N}F_m^{Lm}(ID_{MS}), {}_{N}\#_v, {}_{N}\#_m, L_v, L_m, N_{expire})$ to the VN where N_{expire} denotes the login numbers of the MS.

I-4. VN → MS: ${}_{G}\#_{h,m} \oplus (T_1, {}_{N}F_v^{Lv}(ID_{VN}), {}_{N}\#_m, {}_{N}\#_v, L_m, L_v, N_{expire})$

The VN uses ${}_{G}\#_{h,v}$ to get $(T_2, {}_{N}F_m^{Lm}(ID_{MS}), {}_{N}\#_v, {}_{N}\#_m, L_v, L_m, N_{expire})$. If the T_2 of the received message is correct, the VN believes the message is sent from the HN, stores ${}_{N}F_m^{Lm}(ID_{MS})$, ${}_{N}\#_v$, ${}_{N}\#_m$, L_v, L_m, N_{expire} and $C_{MS} = 1$ in the database, and forwards ${}_{G}\#_{h,m} \oplus (T_1, {}_{N}F_v^{Lv}(ID_{VN}), {}_{N}\#_m, {}_{N}\#_v, L_m, L_v, N_{expire})$ to the MS. The MS uses ${}_{G}\#_{h,m}$ to get $(T_1, {}_{N}F_v^{Lv}(ID_{VN}), {}_{N}\#_m, {}_{N}\#_v, L_m, L_v, N_{expire})$. Then the MS checks whether the T_1 is correct. If it is correct, the MS believes the message is sent from the HN, and stores ${}_{N}F_v^{Lv}(ID_{VN})$, ${}_{N}\#_m$, ${}_{N}\#_v$, L_m, L_v, N_{expire} and $C_{VN} = 1$ in the database.

When the MS wants to use his anonymous ticket, he has to make a request to the VN. While the i-th request is asked, they perform the following steps.

A-1. MS → VN: ${}_{N}F_m^{Lm-i}(ID_{MS})$, N_{expire}, C_{VN}, L_v

The MS computes ${}_{N}F_m^{Lm-i}(ID_{MS})$ by applying the evolutionary function ${}_{N}F()$ with ${}_{N}\#_m$ to his identity L_m-i times, and sends ${}_{N}F_m^{Lm-i}(ID_{MS})$, N_{expire}, C_{VN}, and L_v.

A-2. VN → MS: ${}_{N}F_v^{Lv-i}(ID_{VN})$, N_{expire}, C_{MS}, L_m

According to C_{VN} and L_v, the VN can find ${}_{N}F_m^{Lm}(ID_{MS})$, ${}_{N}\#_v$, ${}_{N}\#_m$, L_v, L_m, N_{expire} and $C_{MS} = i$ in the database. Then the VN checks if the $N_{expire} > 0$ and $C_{VN} = C_{MS}$, and checks on the MS's authentication by applying the evolutionary function

${}_N F()$ with ${}_N\#_m$ to the received ${}_N F_m^{Lm-i}(\text{ID}_{\text{MS}})$ once. Then the VN sets $\text{C}_{\text{MS}} \leftarrow \text{C}_{\text{MS}} + 1$, and saves ${}_N F_m^{Lm-i}(\text{ID}_{\text{MS}})$ for the next session verification. Similarly, the MS can also verify the VN's authentication.

Figure 1 shows the configuration of the proposed authentication scheme, composed of a ticket-issuing phase and ticket-authentication phase.

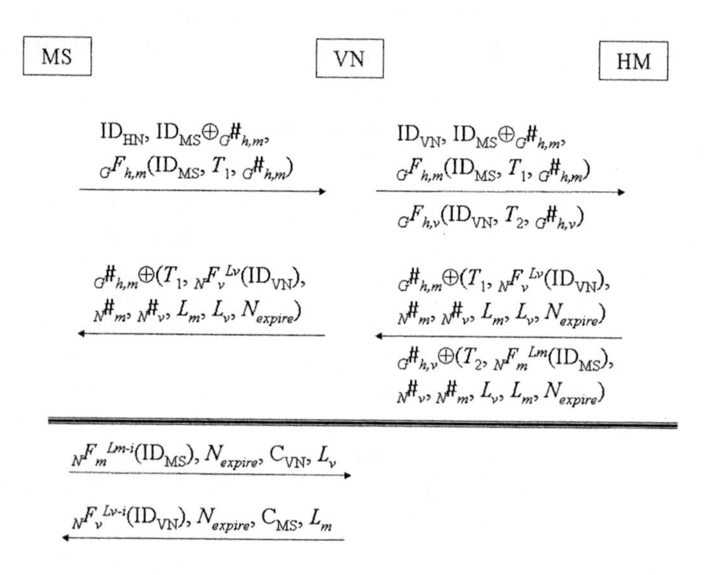

Fig. 1. Proposed authentication scheme based on a cellular automata composed of two layer (upper layer illustrates the ticket-issuing phase and lower layer illustrates the ticket-authentication phase)

In the ticket-issuing phase, a group rule can be applied only once or more times according to demanded security level. ${}_N F()$ can be achieved by just one non-group rule or the combination of rules, e.g., <107, 230>. That is, the rules are applied to the CA alternately. An evolutionary non-group rule function is used to define the authenticating value's sequence decreasingly from ${}_N F^{L-1}(\text{ID})$. The authenticating value for the i-th identification session, $1 \leq i \leq L$, is defined to be ${}_N F_m^{Lm-i}(\text{ID}_{\text{MS}})$ and ${}_N F_v^{Lv-i}(\text{ID}_{\text{VN}})$. In the ticket-authentication phase, for i-th session, the MS and VN already has the other side's length and counter along with the verifier, ${}_N F^{L-i}(\text{ID})$, so that they only need to simply check the other side's authentication mutually by applying the evolutionary function ${}_N F$ once.

4 Discussion and Analysis

In spite of the growth of wireless authentication schemes, previous schemes still have problems such as high computation, an additional communication path, the time or sequence synchronization, and ticket copy. To solve the high computation problem, our scheme takes advantage of the characteristics of a CA in which computation only consists of logical bitwise operations such as XOR, AND, OR, and NOT, while typical cryptosystems need heavy computations such as hash functions, inversions, and exponentiations.

The proposed scheme has a minimal path to check mutual authentication in the ticket-authentication phase, and does not require synchronization of time. If the MS and the VN have gotten out of the sequence synchronization because of an unstable network – the MS sending ${}_{N}F_{m}{}^{Lm-i}(\mathrm{ID_{MS}})$ and the VN using ${}_{N}F({}_{N}F_{m}{}^{Lm-k}(\mathrm{ID_{MS}}))$ to authenticate it, with $i \neq k$ – this can be detected by repeatedly applying ${}_{N}F$ to the VN's authenticating value until a match is obtained.

Meanwhile, L should be a reasonable length. If the L of a CA is too short or too long, it requires frequent renewals of the rule or guessing the verifier from the attacker. Thus we recommend that L is several hundreds to several thousands in length. We can determine the reasonable length by a combination of the group rules and non-group rules.

Table 3. The length, L corresponding to the non-group rule and the combination of group rule and non-group rule

	non-group rule		combination of group rule and non-group rule	
rule #	193	230	90, 129	171,90
length	770	127	2,524	2,471

Table 3 shows the results of a test when the rules are applied to the initial value, 'wirelessnetwork'. The CAs applied the non-group rules only are relatively shorter length, 770 and 127, which correspond to rules 193 and 230. Meanwhile, the CAs applied to the combinations of group rule and non-group rule <90, 129>, <171, 90> have a long enough length, 2,524 and 2,471. There are a number of combination rules which have the proper length.

The following criteria are crucial for authentication schemes to show the robustness of security.

- *Confidentialities of the MS* : In the ticket-issuing phase, the $\mathrm{ID_{MS}} \oplus {}_{G}\#_{h,m}$ and the session key ${}_{G}\#_{h,m}$ are only known between the MS and HN. Therefore, when the MS requests an anonymous ticket, the VN cannot know the identification of the MS according to XORed message. Because the VN does not know the identity of the MS, the VN has to broadcast the evolved anonymous ticket to the MS via the wireless systems. Therefore, the HN cannot know the location of the MS according to its anonymous ticket.
- *Mutual authentication*: In the ticket-issuing phase, the session key ${}_{G}\#_{h,m}$ is only known between the MS and HN. Hence they can use it to authenticate each other. Similarly, ${}_{N}F_{m}{}^{Lm-i}(\mathrm{ID_{MS}})$ and ${}_{N}F_{v}{}^{Lv-i}(\mathrm{ID_{VN}})$ can only be computed by the MS and the VN. They can use them to authenticate each other.
- *Forgery attack*: In the ticket-issuing phase, the session key ${}_{G}\#_{h,m}$ is only known between the MS and the HN; hence, no one can send the forged message to the

HN and the MS in I-1 and I-4. Similarly, the ${}_{G}\#_{h,v}$ is only shared between the HN and the VN; hence, no one can send the forged message to the HN and the VN in I-2 and I-3. In the ticket-authentication phase, ${}_{N}F_{m}^{Lm-i}(ID_{MS})$ and ${}_{N}F_{v}^{Lv-i}(ID_{VN})$ are secretly computed by the MS and the VN; therefore, no one can send the forged message to them.

- *Replay attack*: In I-1 of the ticket-issuing phase, if the attacker replays $ID_{MS} \oplus {}_{G}\#_{h,m}$, ${}_{G}F_{h,m}(ID_{MS}, T_1, {}_{G}\#_{h,m})$ to the VN, and the VN sends the ${}_{G}F_{h,v}(ID_{VN}, T_2', {}_{G}\#_{h,v})$ to the HN. The request will be rejected by checking the invalid time interval, $(T_2' - T_1 > \Delta T)$. Similarly, if the attacker replays ${}_{N}F_{m}^{Lm-i}(ID_{MS})$, N_{expire}, C_{VN}, L_v to the VN, the VN will reject this login request since the next authentication value is ${}_{N}F_{m}^{Lm-(i+1)}(ID_{MS})$, which can only be computed by the MS.
- *Ticket copy*: If the valid user copies his anonymous ticket and shares it with his/her friends, it does not effect the rights of the VN since the VN computes N_{expire} -1 in each time until $N_{expire} = 0$.
- *Impersonating attack*: It is infeasible that an attacker can acquire the shared secret ${}_{G}\#_{h,m}$ or ${}_{G}\#_{h,m}$ since he/she cannot extract it from any information he can obtain.
- *Stolen verifier attack*: Although the attacker stoles the verifier, ${}_{N}F^{L-i}(ID)$, in *i*-th session, ${}_{N}F^{L-(i+1)}(ID)$ cannot be computed by any method since it is impossible to find the inverse of a rule in a non-group CA.

5 Conclusions

In a wireless communication environment, the most important element is the computation and communication cost of the mobile station. Most previous protocols have employed typical cryptosystems such as symmetric and asymmetric encryptions in spite of their high computation and communication cost, and failed to provide the user's anonymity and protect the ticket copy. Our scheme provides not only low computational cost, but also the safety and secure authentication to withstand the above weakness without employing any typical cryptosystems. We have shown that our scheme only has the logical bitwise operations, and that it can resist the typical attacks. We believe that our authentication scheme is well-suited and practical for wireless network applications.

References

1. ETSI/TC Recommendation GSM 03.20, Security related network function, Tech. Rep., (1991).
2. C.C. Yang, T.Y. Chang, M.S. Hwang, A new anonymous conference key distribution system based on the elliptic curve discrete logarithm problem, Computer Standards and Interfaces, vol.25, pp.141-145, (2003)
3. 3.C.C. Chang, Y.P. Lai, A flexible date attachment scheme on e-cash, Computers and Security vol.22, no.2, pp.160-166, (2003)
4. J.K. Jan, C.C. Tai, A secure electronic voting protocol with ic cards, Journal of Systems and Software, vol.39, pp.93-101, (1997)

5. W.D. Lin and J.K. Jan, A wireless-based authentication and anonymous schannels for large scale area, iin: Proc. 6th IEEE Symposium on Computers and Communications (ISCC'01), pp. 36-41, (2001)
6. A.M. Barbancho, and A. Peinado, Cryptanalysis of anonymous channel protocol for large-scale area in wireless communications, Computer Networks, vol.43, pp.777-785, (2003)
7. C.C. Yang, Y.L. Tang, R.C. Wang, and H.W. Yang, A secure and efficient authentication protocol for anonymous channel in wireless communications, Applied Mathematics and Computation, article in press, (2005)
8. J. Von Neumann, The theory of self-reproducing automata, University of Illinois Press, Urbana and London, (1966)
9. J.C. Jeon and K.Y. Yoo, Design of Montgomery Multiplication Architecture based on Programmable Cellular Automata, Computational intelligence, vol. 20, pp. 495-502, (2004)
10. F. Seredynski, P. Bourry, and A.Y. Zomaya, Cellular automata computations and secret key cryptography, Parallel Computing, vol. 30, pp. 753-766, (2004)
11. R. Alonso-Sanz and M. Martin, One-dimensional cellular automata with memory: patterns from a single site seed, Int. J. Bifur. Chaos Appl. Sci. Engrg., vol.12, pp.205-226, (2002)
12. M. Seredynski, K. Pienkosz, and P. Bouvry, Reversible Cellular Automata Based Encryption, NPC 2004, LNCS 3222, pp. 411-418, (2004)
13. S. Das, B. K. Sikdar, and P. Pal Chaudhuri, Charaterization of Reachable/Nonreachable Cellular Automata States, ACRI 2004, LNCS 3305, pp. 813-822, (2004)
14. A.K. Das, A. Ganguly, A. Dasgupta, S. Bhawmik, and P.P. Chaudhuri, Efficient characterization of cellular automata, IEE proceedings, vol.137, no.1, pp.81-87, (1990)
15. C.K. Koc, A.M. Apohan, Inversion of cellular automata iterations, IEE Proc. Comput. Digit. Tech., vol. 144, pp. 279-284, (1997)

SeGrid: A Secure Grid Infrastructure for Sensor Networks

Fengguang An[1], Xiuzhen Cheng[2,*], Qing Xia[3], Fang Liu[2], and Liran Ma[2]

[1] Institute of Computing Technology,
Chinese Academy of Sciences, Beijing 100080, P.R. China
[2] Department of Computer Science,
The George Washington University, Washington DC 20052, U.S.A
[3] Kellar Institute, George Mason University, Fairfax, VA 22030, U.S.A

Abstract. In this paper, we propose SeGrid, a secure grid infrastructure for large scale sensor networks. The basic idea relies on the availability of a low-cost public cryptsystem (e.g. Blom's key management scheme [4]) that can be used for shared key computation between the source and the destination, as long as the public shares are known to each other. In SeGrid, each sensor resides in a grid computed from its physical location. Within a grid, one or a few number of sensors, with one of them the grid head, are active at any instant of time and all other sensors fall asleep for energy conservation. We intend to compute a shared key for two grids instead of two nodes, such that the grid heads can securely communicate with each other. The public shares of a grid are stored at designated locations based on our public share management protocol such that the closer two grids, the shorter distance to obtain each other's public shares. We instantiate SeGrid based on Blom's key management scheme [4] to illustrate the computation of a grid key. To our best knowledge, this is the first work that simultaneously considers both key management and network lifetime extension, which explores along the dimension of network density.

Keywords: Blom's key management scheme, secure grid infrastructure.

1 Introduction

Security provisioning is a critical service for many sensor network applications. However, the severely-constrained resources (memory, processor, battery, etc.) within a sensor render many of the very popular security primitives inapplicable. Therefore, much research effort [1, 5, 9, 10, 12, 14, 16] has been placed on how to establish a shared key between two sensors such that their communications can be secured with low-cost symmetric encryption techniques.

Most existing schemes [9, 10, 12, 16] for distributed key agreement in sensor networks intend to design light weight (in computational complexity) algorithms

* The research of Dr. Xiuzhen Cheng is supported by the NSF CAREER Award No. CNS-0347674.

X. Jia, J. Wu, and Y. He (Eds.): MSN 2005, LNCS 3794, pp. 422–432, 2005.

to compute pairwise keys for communicating nodes. The induced key graph containing only edges incident at two sensors sharing common keys should be globally connected in order for the network to function well. Another constraint considered by these techniques is the memory budget allocated for pre-deployment key information storage. The tradeoff between memory cost and security has been well-studied in most of these works.

As understood by the research society, the utmost problem in a sensor network is its operation time elongation. Even though the above-mentioned works do take resource (especially energy) consumption into consideration, none of them explores the *density* dimension for further energy conservation. In this paper, we propose an idea of establishing a secure grid infrastructure (termed SeGrid) for sensor networks. We envision that all sensors within a grid are equivalent in routing and thus a secret key is needed between two grids (instead of two nodes) that demand communication. In this secure grid infrastructure, only one or a few number of sensors (for fault-tolerance) within a grid are active at any instant of time and all other sensors fall asleep to conserve energy. This design explores the fact that sensors are low-cost and are densely deployed in a typical network. When a new sensor becomes active, or an active sensor dies due to energy depletion, the shared grid keys should be recomputed. We instantiate this idea by applying Blom's key management scheme [4] to demonstrate the grid key computation. Note that putting redundant sensors to sleep for energy conservation is a popular method in all layers of the protocol [19, 20, 17] design for sensor networks. However, to the best of our knowledge, this work is the first to combine it with key management.

The basic idea of SeGrid is outlined in the following. We assume that there exists a public cryptosystem with low computation overhead (e.g. Blom's key management scheme [4]) such that each sensor can be preloaded with a crypto pair containing a public share and a private share before deployment. In SeGrid, sensors compute the grids they are residing in and choose to sleep or wake-up based on some schedule (e.g. the wake-up schedule proposed in [20]). Each grid has a *grid head*, an active node for message transmission. The grid head stores the public shares of all active nodes within its grid at designated locations and queries the nearest grid that stores the public shares of another grid based on our public share management protocol. After obtaining the public shares of the destination grid, source grid computes a key k_s that will be used to secure all transmissions between these two grids. The destination grid can follow the same procedure to compute the grid key k_s. The public share management protocol ensures that the closer two grids, the shorter the query distance to obtain each other's public shares. This protocol involves only simple algebraic (shift and addition) operations, thus has very low computation overhead. We finally instantiate SeGrid based on Blom's key management scheme [4] to demonstrate how the grid key can be computed based on the underlying public cryptosystem.

This paper is organized as follows. We elaborate the network model and the underlying assumptions in Section 2, then propose our secure grid infrastructure for sensor networks (SeGrid) in Section 3. An example instantiation of SeGrid is outlined in Section 4. We conclude this paper with a discussion in Section 5.

2 Network Model and Assumptions

We are considering a large-scale sensor network deployed in outdoor environments. Each sensor is able to position itself through any of the techniques proposed in literature (eg. [6, 15, 18]). A virtual grid will be computed based on position information and each sensor resides in one grid. The id of a grid is denoted by (X, Y). At any instant of time, one or t number of sensors, where t is a small integer, are active and all other sensors fall asleep for energy conservation. A sleeping sensor wakes up periodically in order to replace a sensor with depleted energy. An active sensor is in full operation and all active sensors collaborate together to guarantee the functioning of the network. Sensors within neighboring grids can communicate directly. The wakeup/sleep schedule, the active/inactive status transition, and the underlying routing protocol for message dissemination, are out of the scope of this paper. We just simply assume that they are available for us to employ. Existing works related to these topics can be found in [3, 20].

We will explore a public cryptosystem that contains public and private crypto pairs. The public share can be disseminated as plain text while its corresponding private share must be kept secret. By exchanging their public shares, two nodes can compute a shared secret key based on their private shares and the exchanged public share. Example cryptosystems satisfying these conditions include the Diffie-Hellman key exchange protocol [11] and Blom's key management scheme [4]. In Section 4, we are going to instantiate a secure grid sensor network infrastructure based on Blom's method.

We assume each sensor is preloaded with a crypto pair before deployment. The operation of the sensor network is unattended after the initial bootstrap procedure for sensor localization and key management is done. Each grid may have more than one public shares, if it has more than one active sensors. An update message will be directed to all locations storing the public shares of the grid such that the public shares of newly introduced active (old inactive) sensors can be inserted (removed). A grid demanding the public shares of another grid can just query the nearest grid storing the corresponding information. We will propose a simple protocol for public share management in Subsection 3.2.

We envision that in a sensor network all nodes within a grid are equivalent. Therefore we only consider the secure communication between two grids. The computation of the shared key k_s between the two grids depends on the underlying public cryptosystem. We will show how to compute k_s based on Blom's key management scheme in Section 4. Note that intra-grid secure communication may be needed when more than one sensors are active simultaneously within a grid. The shared keys between these active nodes can be computed based on the underlying public cryptosystem too.

Note that even though $t > 1$ number of sensors may be active at any instant of time, we assume that only one sensor within a grid is in charge of transmission.

This sensor is the *grid head*. All active nodes other than the grid head listen to the messages directed from neighboring grids. This assumption is realistic since in a sensor network, a measurement from one sensor may not be attractive due to dynamics. Usually sensor readings in close neighborhood need to be combined for fault-tolerance [7] and an aggregated summary will be reported to the base stations [7, 8].

3 SeGrid: The Secure Grid Infrastructure

In this section, we propose the basic idea of our secure grid infrastructure for outdoor sensor networks. Note that this elaboration does not depend on any public cryptosystem. We will instantiate this idea in Section 4 based on Blom's key management scheme [4].

We will first describe a simple algorithm for each sensor to locally and independently compute the id of the grid it resides in. Then we give a novel protocol for each grid to determine where to store its public shares. In the last, we propose how to apply the secure grid infrastructure for protecting the unicast communications between two grids.

3.1 Grid Determination

In GAF [19], the size of a grid is determined based on node equivalence for routing. In other words, any node within a grid can communicate directly with any other node in any neighboring grid. This constraint specifies that the size of a grid, denoted by r, can be at most $\frac{R}{\sqrt{5}}$, where R is the nominal transmission range. In our study, we adopt this idea since we also intend to turn off most of the sensors within a grid for energy conservation in order to extend network lifetime. GAF specifies the length of the grid edge but does not specify how to determine the grid a node resides. In the following, we propose a very simple algorithm to allow each node independently and locally determine its grid.

Let (x, y) be the location of any sensor residing at grid (X, Y). Then we have $X = \lfloor x \div 2^{\lfloor \log_2 r \rfloor} \rfloor$, $Y = \lfloor y \div 2^{\lfloor \log_2 r \rfloor} \rfloor$. Note that X and Y can be computed through shift operations only, as long as $2^{\lfloor \log_2 r \rfloor}$ is computed off-line and uploaded into each sensor before deployment. This is a reasonable assumption since r depends only on the nominal transmission range R, which can be made available before deployment. Therefore we can simply shift the binary representations of x and y to the right for k positions, where $k = \lfloor \log_2 r \rfloor$, to obtain X and Y.

Remark: The procedure of computing the grid id of a sensor based on its physical location does not require precise location information. We may treat k as an error bound. In other words, as long as the position error in both x and y directions are upper-bounded by 2^k, we can still determine the grid id. In this case, the constraint that two nodes within neighboring grids should communicate directly may be violated.

3.2 Public Share Management

In this subsection, we propose a simple protocol for storing and querying the public shares of a grid. We need to answer two questions. First, for any grid (X_0, Y_0), where shall we store its public shares? Second, if grid (X_1, Y_1) would like to securely communicate with (X_0, Y_0), where to find out the latter's public shares? Our protocol is based on the following assumption: the closer two grids, the higher the probability they may communicate. Therefore, the public shares of a grid will be stored at designated locations such that the closer to the grid, the shorter the query distance involved in public share acquisition. In our protocol, the density of the grids storing the public shares of a grid drops logarithmically as the distance to the grid increases. Fig. 1 gives a simple example to illustrate the storage locations of the public shares for the grid (4, 4).

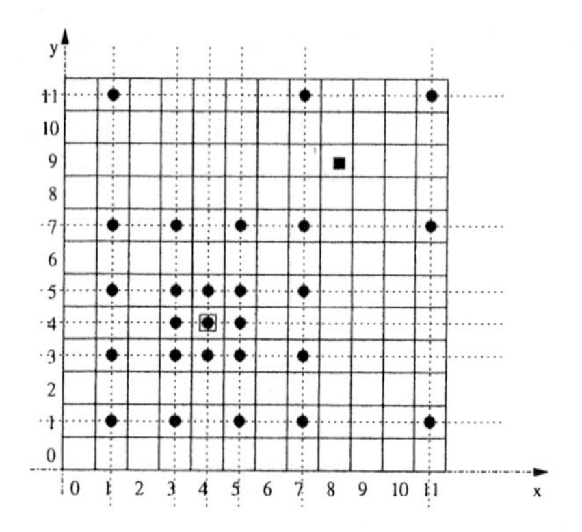

Fig. 1. The public shares of the grid (4, 4) are stored at the grids denoted as "." in the figure. If the grid (8, 9) needs the public shares of (4, 4), it can query either (7, 7) or (7, 11) since they are closer.

The answer to the first question is very simple. The public shares of the grid (X_0, Y_0) will be stored at (x, y) if $x = X_0 \pm (2^i - 1)$ or $x = X_0 \pm (2^{i+1} - 1)$, and $y = Y_0 \pm (2^i - 1)$ or $y = Y_0 \pm (2^{i+1} - 1)$, where $i = 0, 1, 2, \cdots$. To identify the nearest grid that stores the public shares of (X_0, Y_0), the grid (X_1, Y_1) will compute

$$\Delta X_L = 2^{\lfloor \log_2 |X_1 - X_0| \rfloor} - 1\,, \quad \Delta Y_L = 2^{\lfloor \log_2 |Y_1 - Y_0| \rfloor} - 1,$$
$$\Delta X_H = 2^{\lceil \log_2 |X_1 - X_0| \rceil} - 1\,, \quad \Delta Y_H = 2^{\lceil \log_2 |Y_1 - Y_0| \rceil} - 1,$$

to figure out the following four grids that contain the public shares of (X_0, Y_0): $(X_1 - \Delta X_L, Y_1 - \Delta Y_L)$, $(X_1 - \Delta X_L, Y_1 + \Delta Y_H)$, $(X_1 + \Delta X_H, Y_1 - \Delta Y_L)$, $(X_1 + \Delta X_H, Y_1 + \Delta Y_H)$, and then choose the nearest one to query.

Note that if Manhattan distance instead of Euclidean distance is used as a routing metric for public share queries and updates, the computation overhead is very low since only simple addition and subtraction operations are involved. For example in Fig. 1, $\Delta X_L = 1$, $\Delta X_H = 3$, $\Delta Y_L = 2$, and $\Delta Y_H = 2$. The

Manhattan distance from $(8, 9)$ to $(7, 11)$ is $\Delta X_L + \Delta Y_H = 3$. Similarly we can compute the Manhattan distance to the other three points. We choose either $(7, 7)$ or $(7, 11)$ to query since they are the closest among the four grids that are close to $(8, 9)$.

Remarks:

- The computation of the storage locations for a grid contains only shift and addition operations. However, the identification of the nearest grid for public share query involves the complicated log functions. Nevertheless, this can be done easily through a lookup table.
- This protocol guarantees that closer grids obtain the public shares within shorter distance. Therefore, the farther away a grid, the higher the communication overhead for public share query. In reality, closer grids intend to communicate securely with higher probability.
- The update of the public shares for a grid always take the same number of messages, as long as the routing protocol remains unchanged.

3.3 Secure Grid Communication

Now we are ready to propose our secure grid communication framework. We assume there exists a routing protocol, either geography-based (e.g. [13]) or topology-based (e.g. [3]), such that we can employ directly.

Let (X_A, Y_A) and (X_B, Y_B) be the two grids that require a secure communication. The following procedure will be conducted at both ends:

1. Query the nearest grid that contains the public shares of the other party based on the procedure proposed in Subsection 3.2 to obtain the public shares.
2. Compute a secret key K_s shared by these two grids and secure the future communication with this key.

Note that Step 2 depends on the underlying public cryptosystems. In the following section, we will show how to apply Blom's key management scheme [4] to compute the shared key between two grids.

4 A Simple Realization

In this section, we implement a secure grid infrastructure for sensor networks based on Blom's key management scheme [4]. For completeness, we give a brief overview on Blom's scheme first. Then we describe how to compute a grid key based on Blom's scheme.

4.1 Preliminary: Blom's Key Management Scheme

Blom's λ-secure key management scheme [4] has been well-tailored for light-weight sensor networks by [9]. In the following, we will give an overview on Blom's scheme based on [9].

Let G be a $(\lambda+1) \times M$ matrix over a finite field $GF(q)$, where q is a large prime. The connotation of M will become clear latter. G is public, with each column called a public share. Let D be any random $(\lambda+1) \times (\lambda+1)$ symmetric matrix. D must be kept private, which is known to the network service provider only. The transpose of $D \cdot G$ is denoted by A. That is, $A = (D \cdot G)^T$. A is private too, with each row called a private share. Since D is symmetric, $A \cdot G$ is symmetric too. If we let $K = (k_{ij}) = A \cdot G$, we have $k_{ij} = k_{ji}$, where k_{ij} is the element at the ith row and the jth column of matrix K, $i, j = 1, 2, \cdots, M$.

The basic idea of Blom's scheme is to use k_{ij} as the secret key shared by node i and node j. D and G jointly define a *key space* (D, G). Any public share in G has a unique private share in A, which form a so-called *crypto pair*. For example, the ith column of G, and the ith row of A form a crypto pair and the unique private share of the ith column of G, a public share, is the ith row of A. Two sensors whose crypto pairs are obtained from the same key space can compute a shared key after exchanging their public shares. From this analysis, it is clear that M is the number of sensors that can compute their pairwise keys based on the same key space.

In summary, Blom's scheme states the following protocol for nodes i and j to compute k_{ij} and k_{ji}, based on the same key space:

- Each node stores a unique crypto pair. Without loss of generality, we assume node i gets the ith column of G and the ith row of A, denoted by g_{ki} and a_{ik}, where $k = 1, 2, \cdots, \lambda+1$, respectively. Similarly, node j gets the jth column of G and the jth row of A, denoted by g_{kj} and a_{jk}, where $k = 1, 2, \cdots, \lambda+1$, respectively.
- Node i and node j exchange their stored public shairs drawn from their crypto pairs as plain texts.
- Node i computes k_{ij} as follows: $k_{ij} = \sum_{k=1}^{\lambda+1} a_{ik} \cdot g_{kj}$; Similarly, node j computes k_{ji} by $k_{ji} = \sum_{k=1}^{\lambda+1} a_{jk} \cdot g_{ki}$.

Blom's key management scheme ensures the so-called λ-secure property, which means the network should be perfectly secure as long as no more than λ nodes are compromised. This requires that any $\lambda+1$ columns of G must be linearly independent. An interesting method of computing G is proposed by Du *et. al* in [9]. This idea is sketched as following. Let len be the number of bits in the symmetric key to be computed. Choose q as the smallest prime that is larger than 2^{len}. Let s be a primitive element of $GF(q)$ and $M < q$. Then

$$G = \begin{bmatrix} 1 & 1 & 1 & \cdots & 1 \\ s & s^2 & s^3 & \cdots & s^M \\ s^2 & (s^2)^2 & (s^3)^2 & \cdots & (s^M)^2 \\ & & & \vdots & \\ s^\lambda & (s^2)^\lambda & (s^3)^\lambda & \cdots & (s^M)^\lambda \end{bmatrix}$$

Note that G is a Vandermonde matrix. Each column of G represents the public share of some sensor node storing that column. In Blom's key management scheme, public shares need to be exchanged between sensors that require secure peer-to-peer communication. Based on the structure of G, we observe that only the second element of each column, the *seed* of the column, needs to be stored and exchanged. Thus both storage and communication overheads can be greatly decreased.

4.2 Shared Key Computation

Assume a large key space (D, G) following Blom's key management scheme has been computed off-line. Before deployment, each sensor receives a crypto pair from the key space. Note that the crypto pairs to different sensors may not be unique, as the key shared by two grids will be computed based on multiple public shares. But we require that all active sensors within one grid have different crypto pairs.

Let t_A (t_B) be the number of active sensors in a grid (X_A, Y_A) $((X_B, Y_B))$. Following the public share management protocol proposed in Subsection 3.2, all these t_A (t_B) public shares will be stored at designated grids, and are available to other grids. If (X_A, Y_A) and (X_B, Y_B) need secure communication, they acquire the public shares of the other party first in order to compute a shared key.

In Blom's key management scheme, two sensors can compute a shared key as long as they know each other's public share. We can derive a shared key k_s between two grids from the keys shared by all pairs of sensors within the two gride, as shown in Fig. 2.

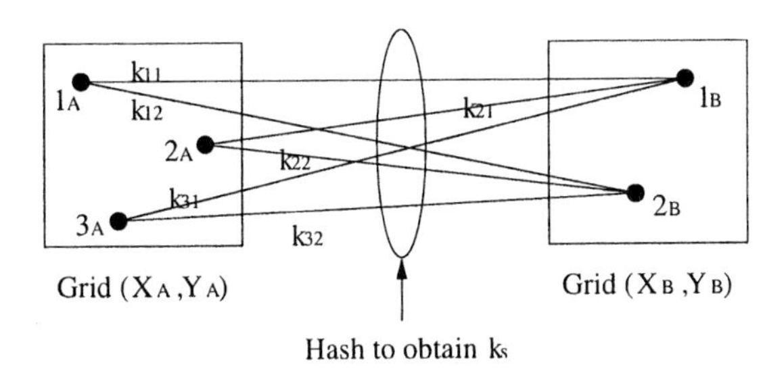

Fig. 2. The grid $(X_A, Y_A)((X_B, Y_B))$ contains three (two) active sensors with node 1_A (1_B) as the grid head. After obtaining the public shares of (X_B, Y_B), each node i_A in grid (X_A, Y_A) first computes k_{i1} and k_{i2}, the shared keys with the two nodes in grid (X_B, Y_B), then transmits securely to node 1_A the value $k_{i_A} = Hash(k_{i1}, k_{i2})$. Finally, node 1_A computes k_s as $k_s = Hash(k_{1_A}, k_{2_A}, k_{3_A})$. Similarly, node 1_B computes k_s based on the public shares of (X_A, Y_A).

Let's use grid (X_A, Y_A) as an example to demonstrate the procedure of computing a shared key with the grid (X_B, Y_B). After obtaining the public shares of grid (X_B, Y_B), each node i in (X_A, Y_A) computes a shared key with each node j in grid (X_B, Y_B). These pairwise keys are denoted by k_{ij}, where $i = 1, 2, \cdots, t_A$ and $j = 1, 2, \cdots, t_B$. Then i computes $k_i = Hash(k_{i1}, \cdots, k_{it_B})$. This value will be securely transmitted to the grid head h of (X_A, Y_A). After receiving all k_t's, where $t \neq i$, h derives the grid key k_s by computing $Hash(k_1, k_2, \cdots, k_{t_A})$.

Remarks:

- The private shares of each sensor must be kept secret. Therefore if one node within a grid is compromised, grid keys can still be computed. Further, the grid key remains secure if an active node other than the grid head is compromised.
- When the active nodes within a grid changes, the public shares of the grid can be modified easily since only the public share of the node with role change needs to be updated.
- The hash function exploited must be linear, and must be able to take any number of inputs. The XOR function is a simple example.
- Two nodes within a grid can compute a shared key based on Blom's method easily after exchanging their public shares as plain texts. This key can be used to secure the intra-grid communication.
- The security of the grid key computation protocol based on Blom's key management scheme [4] is determined by the λ-secure property of the key space (D, G). Therefore if the crypto pairs of more than λ sensors are exposed to the adversary, the security of the whole network is compromised. This is the major drawback of applying Blom's key management scheme for grid key computation since the memory budget within a sensor for security information storage is limited.
- The space consumed for storing the crypto pairs within a sensor is exclusively determined by λ. The larger the λ, the higher the security level, and the larger the storage space.
- The computation overhead of a grid key is determined by λ too. Each shared key computation between two active nodes takes $\lambda + 1$ number of modular multiplications.

5 Conclusion and Future Research

In this paper, we have proposed SeGrid, a secure grid infrastructure based on public cryptosystems for large scale sensor networks. We have instantiated SeGrid based on Blom's key management scheme to demonstrate how to compute a grid key shared by two grids. To our knowledge, SeGrid is the first work that targets key management in a level that is above the physical sensors. This is a more practical consideration since sensors may stay in sleep mode most of the time for network lifetime extension.

As a future work we will explore the applicability of ID-Based Cryptosystems [2] in SeGrid. In an ID-based encryption system, the public key can be any string (e.g. an email address), and the private key needs to be computed from the public key and other system parameters. The idea of using the grid ID as a public key in SeGrid is very attractive since public key management can be totally avoided.

References

1. F. An, X. Cheng, M. Rivera, J. Li, and Z. Cheng, PKM: A Pairwise Key Management Scheme for Wireless Sensor Networks, to appear in *2005 International Conference on Computer Networks and Mobile Computing (ICCNMC'05)*, Zhangjiajie, Hunan, China, August 2-4, 2005.
2. D. Boneh and M. Franklin, Identity-Based Encryption from the Weil Pairing, *Proceedings of CRYPTO'01*, Spinger-Verlag, 2001.
3. A. Boukerche, X. Cheng, and J. Linus, Energy-Aware Data-Centric Routing in Microsensor Networks, *MSWiM 2003*, pp. 42-49, 2003.
4. R. Blom, An optimal class of symmetric key generation systems, in *Proc. of the EUROCRYPT 84 workshop on Advances in cryptology: theory and application of cryptographic techniques.* Springer-Verlag New York, Inc., pp. 335-338, 1985.
5. H. Chan, A. Perrig, and D. Song, Random Key Predistribution Schemes for Sensor Networks, *IEEE SP 2003.*
6. X. Cheng, A. Thaeler, G. Xue, and D. Chen, TPS: a time-based positioning scheme for outdoor sensor networks, *INFOCOM 2004*, Vol. 4, pp.2685-2696, HongKong, March 7-11, 2004.
7. M. Ding, D. Chen, K. Xing, and X. Cheng, Localized Fault-Tolerant Event Boundary Detection in Sensor Networks , *IEEE INFOCOM 2005*, 13-17 March 2005.
8. M. Ding, D. Chen, A. Thaeler, and X. Cheng, Fault-Tolerant Target Detection in Sensor Net works, *IEEE WCNC 2005*, Vol. 4, pp. 2362-2368, 13-17 March 2005.
9. W. Du, J. Deng, Y. S. Han, and P. K. Varshney, A pairwise key pre-distribution scheme for wireless sensor networks, in *CCS '03: Proceedings of the 10th ACM conference on Computer and communications security.* ACM Press, pp. 42–51, 2003.
10. W. Du, J. Deng, Y.S. Han, S. Chen, and P.K. Varshney, A Key Management Scheme for Wireless Sensor Networks Using Deployment Knowledge, *IEEE INFOCOM 2004.*
11. W. Diffie and M.E. Hellman, New Directions in Cryptography, *IEEE Transactions on Information Theory*, Vol. 22, No. 6, pp. 544-654, 1976.
12. L. Eschenauer and V.D. Gligor, A Key-Management Scheme for Distributed Sensor Networks, *CCS'02*, pp.41-47, November 18-22, 2002, Washington DC, USA.
13. B. Karp and H.T. Kung, GPSR: Greedy Perimeter Stateless Routing for Wireless Networks, *ACM MOBICOM 2000*, pp. 243-254.
14. D. Liu and P. Ning, Establishing Pairwise Keys in Distributed Sensor Networks, *ACM CCS'03*, pp. 52-60, 2003.
15. F. Liu, X. Cheng, D. Hua, and D. Chen, TPSS: A Time-based Positioning Scheme for Sensor Networks with Short Range Beacons, to appear in *2005 International Conference on Computer Networks and Mobile Computing (ICCNMC'05)*, Zhangjiajie, China, August 2-4, 2005.
16. F. Liu, L. Ma, X. Cheng, D. Hua and J. Li, S-KMS: A Self-configured Key Management Scheme for Sensor Networks, submitted to *IEEE INFOCOM 2006.*
17. L. Ma, Q. Zhang, and X. Cheng, A Power Controlled Interference Aware Routing Protocol for Dense Multi-Hop Wireless Networks, *submitted.*

18. A. Thaeler, M. Ding, and X. Cheng, iTPS: An Improved Location Discovery Scheme for Sensor Networks with Long Range Beacons, Special Issue on *Theoretical and Algorithmic Aspects of Sensor, Ad Hoc Wireless, and Peer-to-Peer Networks* of *Journal of Parallel and Distributed Computing*, Vol. 65, No. 2, pp.98-106, Feburary 2005.
19. Y. Xu, J. Heidemann, and D. Estrin, Geography-Informed Energy Conservation for Ad Hoc Routing, *Proceedings of the 7th annual international conference on Mobile computing and networking*, pp. 70-84, Rome, Italy, 2001.
20. W. Ye, J. Heidemann and D. Estrin, An energy-efficient MAC protocol for wireless sensor networks, *Proceedings of IEEE INFOCOM*, pp. 1567-1576, 2002.

Handling Sensed Data in Hostile Environments

Oren Ben-Zwi[1] and Shlomit S. Pinter[2]

[1] Computer Science Haifa University, Haifa 31905, Israel
nbenzv03@cs.haifa.ac.il
[2] IBM, Haifa Research Lab., Haifa 31905, Israel
shlomit@il.ibm.com
http://cs.haifa.ac.il/~shlomit/

Abstract. Systems that track sensed data trigger alerts based on the evaluation of some condition. In the presence of loss data a conservative condition may not generate a necessary alert and an aggressive condition may generate an alert that could have never happened. We observe that some lost values can be predicted and suggest new classes of conditions that provide more accurate alerts. We motivate the use of such conditions, provide a method for comparing two condition systems, and investigate the systems' properties in both replicated and non replicated architectures. In addition, we propose a weak completeness property, discuss its merit and show a motivation for its use. Our main result shows that a triggering algorithm, used in one of our condition systems, strictly dominates another algorithm for conservative system, yet, both algorithms satisfy the same set of properties; thus, with some simple observations, we have a strong evidence for its optimality.

1 Introduction

Sensing data is essential in a variety of application areas. The environment and the requirements dictate the type of sensors used, the filtering and conditions applied to control the data, the communication network used - be it fixed or mobile, the timing and accuracy of the delivered alerts.

In this work we consider a system comprised of a collection of low capability transmitting (possible wireless) sensor devices locating close to the sensed sources, a computing device (more than one in the replicated case) that checks whether a certain condition is satisfied on the sensed data and a collection of alert display devices. The sensed data is sent to the computing device that may trigger an alert based on its monitoring condition. The alert is then being sent to the alert displayers.

A common problem in such monitoring systems is the loss of a value sent from the data monitoring device (sensor). In Ad-hoc network this can happened, for example, when out of reach. The condition that triggers the alert, based on the values received, pertains to the lost values and may use more than a single update from a sensor (historical condition). A conservative historical condition is formulated with respect to the values sent from the sensors, thus

X. Jia, J. Wu, and Y. He (Eds.): MSN 2005, LNCS 3794, pp. 433–442, 2005.

when a value is missing no alert is triggered. This may be critical in some areas. In contrast, an aggressive historical condition relates only to the values received at the computing device and thus may trigger an alert that could never happen.

We observe that in many cases missing values can be predicted and used to improve the accuracy of the computed results. Based on this observation we suggest alternative classes of conditions, motivate their use, and explore their properties. A method for comparing different condition systems is provided.

In the presence of communication losses or computing device failures it is desire to increase reliability and availability in the network. This is supported by replication of the computing devices that are each receiving data from the sensors and triggering alerts based on the same condition. The replication may introduce surplus alerts, and the alerts may get to the displayer out of order. Thus, the alert device is responsible for displaying only and all alerts that could be generated in none replicated system; specifically, forbid duplicate alerts and preserve the arrival order between received alerts. Huang and Garcia-Molina introduced in [7] three properties (orderedness, consistency, and completeness) such a system might satisfy, and then present several algorithms and examine which of the properties each algorithm satisfies.

We investigate our new classes of conditions with respect to their algorithms. Our main result, which is counter intuitive, is that an algorithm that preserves consistency (delivers only correct alerts) and removes duplications, produces more alerts that happened when used with one of our new conditions, compared with an algorithm that only removes duplications but uses a *similar* condition that is conservative. We also propose a weaker completeness property, discuss its merit and show a motivation for its use.

The rest of this document is organized as follows. We briefly review the system specification and introduce our notation in Section 2. Then, in Section 3 we present and investigate different classes of conditions. In Section 4 we suggest a different completeness property and explore its use. Finally, we discuss related work in Section 5.

2 Replicated Condition Monitoring

A *Condition Monitoring System* tracks real world variables and tries to apply a (predefined) condition on them. In this section we follow the modeling and terminology of [7]. We view a system with one or more *Data Monitors (DM)* (also called event sources) that send updates values to a *Condition Evaluator (CE)* (or event broker). The CE applies the condition, generates an alert and sends it to one or more *Alert Displayers (AD)* (event displayers). The DM is usually a sensor device located near the data source that is monitored and its link to the CE is called the front link. The AD is a functional unit that resides on the user's alert device and listens to alerts sent from the CE over the back links. Wireless 'links' can be assumed both for the front and for the back links. The CE typically resides on a separate computer in a network (possible a mobile network).

A failure on the CE or on a link to or from it (e.g. loss of a message when an alert displayer is shut down), might causes an alert miss. In addition, there may be a need to increase availability of the CE for the ADs. Therefore, *replication* is used, i.e., multiple CEs are used to monitor the same condition. Replication may cause conflicts, for instance, the system can create alerts that have never happened. Next, we define the system formally and address the various problems that arise as a result of possible communication failures and replication.

2.1 System Specification and Notations

A *data update* is a tuple *(varname, seqno, value)* where *varname* is a name of monitored variable, and *value* is its sampled value during an update whose serial number is *seqno*. The sequence numbers of updates sent from the same variable are consecutive. We sometime use *seq* and *val* to denote *seqno* and *value* respectively. A *condition* is an expression on real world values that can only be evaluated to true or false. The set of variables for a condition is denoted by V. The set of all updates (*history*) used for a condition is denoted by H. The *degree* of a condition on a variable is the number of updates (of that variable) the condition has to check in order to evaluate. A condition is defined to be *non-historical* if its degree with respect to all its variables is at most 1. On historical conditions we define *conservative triggered* to be those conditions that never evaluate to true if an update was lost (missing for the evaluation of the condition). *Aggressive triggered* is an historical condition without this property. In Section 3 we explore different systems for historical conditions. We denote by $H_x[0]$ the last update of x received at the CE, $H_x[-1]$ the one before the last, and so on. When a condition evaluates to true an alert is sent to the AD. The alert is of the form $a(condname, history)$ where *history* is the set of sequence numbers of updates used. The AD uses its algorithm to decide which alerts to display to the user.

Without loss of generality, or for simplicity reasons the following standard assumptions are used for discussing replications: a replicated system has two CEs; only a single condition is being evaluated (on both CEs); the system contains one AD and one DM; both *front* and *back* links (links from DM to CEs and from CEs to AD, respectively) maintain order; and the front links may lose messages while the back links are lossless.

Given a sequence S let $\Phi(S)$ denote the set of the elements of S. Given two sequences S_1, S_2 we say that S_1 is a subsequence of S_2, denoted by $S_1 \sqsubseteq S_2$ if S_1 can be obtained from S_2 by removing zero or more elements. If both S_1 and S_2 are ordered (non decreasing) sequences, their ordered union, $S_1 \bigsqcup S_2$ is the ordered sequence that satisfies $\Phi(S_1 \bigsqcup S_2) = \Phi(S_1) \cup \Phi(S_2)$. A *SpanningSet* (S) is the set of consecutive integers between the smallest and the largest in the set S, inclusive. For example, $\text{SpanningSet}(\{1,2,5\}) = \{1,2,3,4,5\}$.

The sequence of updates produced by the DM is denoted by U and the two sequences received by CE1 and CE2 are denoted by U_1 and U_2, respectively. CE1 and CE2 produce the sequences A_1 and A_2, respectively. These sequences arrive to the AD. We use $T(\cdot)$ to denote the action preformed by the CE (for instance $T(U_1) = A_1$). Finally, the AD receives A_1 and A_2 and outputs A. We focus on the way the AD works, and explore properties of algorithms on the AD.

3 A New Triggering System

In this section we investigate the tradeoff between conservative and aggressive triggering conditions and propose new classes of conditions. In order to compare the different conditions we define similarity between two conditions.

Definition 1. *(similar) We call two conditions similar if both produce exactly the same sequence of alerts on every update sequence when no updates were lost.*

For example [7], for a condition c_1 = '*reactor temperature has risen for more than* 200^o_c' the aggressive variant is c_2 = '*reactor temperature has risen for more than* 200^o_c *since last reading received at the CE*' and the *similar* conservative variant of the condition is c_3 = '*reactor temperature has risen for more than* 200^0_c *since last reading at the DM*'. Formally: $c_2(H) : (H_x[0].val - H_x[-1].val > 200)$ and $c_3(H) : (H_x[0].val - H_x[-1].val > 200) \wedge (H_x[0].seq = H_x[-1].seq + 1)$.

While a conservative triggering condition (system) is likely to satisfy more properties than an aggressive one, an aggressive triggering condition has more alerts delivered to the user. A more careful look reveals that both systems have major disadvantages. The conservative system tends to loose alerts even though we can be sure the condition, when evaluates on the SpanningSet of the updates.seqno, triggers an alert. For example, the conservative condition (c_3) on the following:

Example 1. $(x, 1, 3000^o_c)$ and $(x, 3, 3402^o_c)$

does not trigger an alert if the updates received are consecutive, yet, any value of the lost update would have caused an alert.

The aggressive system, however, tends to trigger alerts even though we can almost be sure the condition (evaluates on the SpanningSet of the updates) should not trigger an alert. For example, the aggressive condition (c_2) on the following:

Example 2. $(x, 1, 3000^o_c)$ and $(x, 210, 3201^o_c)$

triggers an alert if the updates are consecutive but it is hard to find values of the 208 lost updates such that the system will be evaluated to true. This is true if we assume that the function values are within the parallelogram whose endpoints are: (1,2900^o_c), (210,3000^o_c), (1,3100^o_c), (210,3200^o_c).

3.1 Choosing a Different Triggering System

The major disadvantages described above, lead us to seek a better triggering system. One that triggers an alert on Example 1 but does not trigger on Example 2. The triggering system has to produce alerts at least as many as the conservative system does and has to have at least the properties the aggressive system has. For that we should suggest a system that triggers all the alerts the conservative system does and some more alerts we suspect the SpanningSet system (a system with any of the conditions evaluated on the SpanningSet of updates) would trigger as well.

A natural question raised is which condition system triggers more alerts. The next definition is used to compare the amount of triggered alerts of two systems.

Definition 2. ***(Triggering Dominates)*** *Given two different triggering systems with a similar condition, we say that one system* Triggering Dominates *the other system if it produces more (or the same) alerts on every identical sequence of updates. We use* dominates *when the context is clear.*

Note that when two conditions are *similar* we can compare systems that may satisfy different properties. Naturally, there is a tradeoff between domination and properties satisfactions. I.e., a filtering algorithm that forces more properties is likely to be dominated by an algorithm that forces fewer properties.

Next we suggest two new classes of conditions (systems). In the same way many more systems can be derived. The first one is called *Liberal Triggering.*

Definition 3. ***(Liberal triggering condition)*** *A condition is called* liberal triggering *if it can evaluate to true only if no updates were lost or some updates were lost but for all existing values to the lost updates the condition evaluates to true as well.*

The following is an example of a *similar* condition to the previous ones, only a liberal triggering one:

$$c_4(H) : H_x[0].val - H_x[-1].val > 200(H_x[0].seq - H_x[-1].seq).$$

A liberal triggering system is the closest we can think of to the conservative triggering one. Nevertheless, it should trigger more alerts than (dominate) the conservative one, namely the obvious ones. For example, the system under liberal condition c_4, with the same set of updates as in Example 1, triggers an alert (unlike the conservative one), but does not trigger an alert on Example 2 (or any other example 'in between').

We think of all the triggering condition systems as lying on an interval from the conservative to the aggressive. On this interval we want to define two different systems besides the conservative and aggressive, one close to the conservative and one to the aggressive. Then use these two to define by a convex combination many more classes in between these two. We can not use the aggressive system for that as a portion of infinite number of possible updates is of-course infinite. Next we assume that lost updates are within the limits of the not lost ones, and include no repetitions. Note that if the DM only produces updates when there is a new update, the sequence will be with no repetitions.

The system close to the aggressive triggering system is the *Ultra-Liberal* system defined below.

Definition 4. *(An Ultra-Liberal triggering condition) A condition is called* Ultra-Liberal *triggering if it can evaluate to true only if no updates were lost or some updates were lost but there are some possible values to the lost updates, within the limits of the not lost ones, and without repetitions, such that the condition evaluates to true on them as well.*

The following is an example of a condition *similar* to the previous ones only an ultra-liberal triggering one:

$$c_5(H) : H_x[0].val - H_x[-1].val > 200 \wedge H_x[0].val - H_x[-1].val > H_x[0].seq - H_x[-1].seq$$

An ultra-liberal triggering system triggers many more alerts than the conservative and the liberal ones, but less than the aggressive one. It omits only those alerts we are quite sure the system on the SpanningSet will not produce as well. The system under the ultra-liberal triggering condition c_5, with the same set of updates as in Example 2 does not trigger an alert (unlike the aggressive one), yet, it triggers an alert on Example 1. However, it also triggers an alert on any example 'in between'.

These two new definitions can be extended to many more, varying the quantifier on the definition into something between the universal one ('for all') in the liberal condition and the existential one ('there are') in the ultra-liberal condition. The formall definition of those systems is left for the full paper.

3.2 Liberal Triggering Merits

The following three properties were defined and motivated for a proper operation of a replicated system [7]. A replicated system is said to have a property if every alert sequence A it produces satisfies this property.

- **Orderedness**: A is ordered.
- **Completeness**: $\Phi(A) = \Phi(T(U_1 \bigsqcup U_2))$.
- **Consistency**: $\exists U'$ s.t. $\Phi(A) \subseteq \Phi(T(U')) \wedge U' \sqsubseteq (U_1 \bigsqcup U_2)$.

Consistency means that no alert, that never occurred, is being displayed. A system is complete if it generates all the alerts that occurred and only them. Completeness is a stronger property than consistency, meaning every complete system is consistent.

None of the algorithms that utilize historical conditions and proposed in [7] satisfies completeness. Thus, in this section we discuss the other properties. In Section 4 we explore a different definition for completeness and discuss its merit.

We next examine the properties for systems that use our new conditions. The proof of the following proposition is derived directly from the definitions.

Proposition 1. *Given two different systems such that one's alerts sequence is a supersequence of the other's for every input sequence. If the first system satisfies orderedness, and consistency properties then so does the second system.*

We illustrate the proposition with an example for c_4.

Example 3. For liberal condition c_4, let
$U =< 1(3000), 2(3400), 3(3410) >$, $U_1 =< 1(3000), 3(3410) >$ and $U_2 = U$. Consequently $A_1 = T(U_1) =< 3 >$ and $A_2 = T(U_2) =< 2 >$.

An algorithm that only discards duplicate alerts might produce A as $< 3, 2 >$ which is of course not ordered. A is also inconsistent since in any U' we choose, alerts 2 and 3 cannot both be included. Now a conservative triggering system on a *similar* condition produces A as $< 2 >$, it is of course ordered and consistent (and even complete).

Theorem 1. *A replicated monitoring system with lossy front links and a liberal triggering historical condition is neither ordered nor consistent under an algorithm that only discards duplicate alerts.*

Proof. The proof follows Example 3. □

Corollary 1. *A replicated monitoring system with lossy front links and any one of our predefined triggering historical conditions is neither ordered nor consistent, under an algorithm that only discards duplicate alerts.*

The proof follows directly from Theorem 1 and Proposition 1. The same condition, under a conservative system that discards duplicates is consistent. Therefore a liberal system, under an algorithm that only discards duplicate alerts, is inferior to a conservative one from the properties perspective. Furthermore, an aggressive system strictly dominates a liberal one and has exactly the same properties (none) as a liberal system has.

However, one can state that although liberal system is inferior to the aggressive system in 'domination' and to the conservative in 'properties' aspects, it still produces only alerts that we are sure that 'had happened', and any other system that produces only those alerts will produce less than it does. The next definition tries to capture this intuitive claim of quality:

Definition 5. ***(Quality)*** *A system which produces an alert only when an alert should have been triggered on the spanningSet of the updates is better than a system that can produce an alert when there is not any in the system.*

We consider algorithms AD-2,3,4 of [7] (see below) and state three theorems on the properties of liberal systems under those algorithms.

Algorithm AD-2 preserves orderedness by removing every alert that was preceded by a later alert (discard alerts that arrived late - with respect to their order) and removes duplicates. Algorithm AD-3 guaranties consistency by collecting the indices in its delivered alerts and use this growing set to discard alerts. An alert is discarded, based on these indices, only if there was a delivered alert whose history is not equal to the SpanningSet of its history, or if it is a duplicate alert. AD-3 is not order preserving. Algorithm AD-4 combines both AD-2 and AD-3 to enforce both orderness and consistency. Algorithm AD-1 removes duplicate alerts.

Theorem 2. *A replicated monitoring system with lossy front links and a liberal triggering historical condition is ordered but not consistent under AD-2.*

Proof. It is ordered trivially from the correctness of AD-2. We can show it is not consistent by Example 3. □

Theorem 3. *A replicated monitoring system with lossy front links and a liberal triggering historical condition is consistent but not ordered under AD-3.*

Proof. It is consistent of-course. We give a counter-example illustrating the lack of orderedness. Assume the condition is c_4 $([H_x[0].val - H_x[-1].val] > 200(H_x[0].seq - H_x[-1].seq))$. Let $U =< 1(3000), 2(3205), 3(3410) >$, $U_1 =< 2, 3 >$ and $U_2 =< 1, 2 >$. Consequently, $A_1 = T(U_1) =< 3 >$ and $A_2 = T(U_2) =< 2 >$. Algorithm AD-3 might produce A as $< 3, 2 >$ which is of course not ordered. □

Theorem 4. *A replicated monitoring system with lossy front links and a liberal triggering historical condition is ordered and consistent under AD-4.*

3.3 Triggering Systems Conclusions

We conclude this section by a serious of observations on the differences between the systems in every aspect, and a proof of an optimality theorem. For two conditions c_1, c_2 we denote by $c_1 <_T c_2$, $c_1 <_Q c_2$, $c_1 <_{SP} c_2$, when c_1 is triggering dominated by c_2, c_1 does not have the *quality* property while c_2 does, and c_1 has a subset of c_2's system properties, respectively.

Observation 5 ***(triggering dominates)***
Under all algorithms
conservative $<_T$ ***liberal*** $<_T \ldots <_T$ *ultra-liberal* $<_T$ *aggressive.*

Observation 6 *(quality)*
Under all algorithms
conservative $=_Q$ ***liberal*** $>_Q \ldots =_Q$ *ultra-liberal* $=_Q$ *aggressive.*

Observation 7 ***(system properties)***
Under AD-1,2
conservative $>_{SP}$ ***liberal*** $=_{SP} \ldots =_{SP}$ *ultra-liberal* $=_{SP}$ *aggressive.*
Under AD-3,4
conservative $=_{SP}$ ***liberal*** $=_{SP} \ldots =_{SP}$ *ultra-liberal* $=_{SP}$ *aggressive.*

Although AD-1 may have more alerts (on any input) than AD-3 when the same condition is considered the concept of similarity between conditions gives us the possibility to compare algorithms on different triggering systems, and to conclude almost the opposite. The next theorem shows that in a liberal condition system, algorithm AD-3, strictly dominates AD-1 in a conservative system, and both preserve the same set of properties.

We denote by G(c) the operation of algorithm G on condition c.

Theorem 8. ***(optimality of liberal condition)*** *Given a conservative condition cc, and a similar liberal condition lc. Algorithm AD-3 on lc strictly dominates AD-1 on cc (AD-1(cc) $<_T$ AD-3(lc))*

Proof. AD-3 filters out an alert, a, only if a was already displayed (and of course AD-1 filters it as well) or if a.history $\neq$ spanningSet(a.history). A conservative system will not trigger an alert if it does not have all updates, therefore AD-3 can only filter out an alert which is already not in AD-1's output. Therefore we conclude that AD-1(cc) $\leq_T$ AD-3(lc). Now together with Example 1 the proof is complete. □

The last theorem and the previous observations lead us to conclude that a liberal system under AD-3 is optimal, in the sense that it has the *quality* property, the maximum set of properties, and it triggering dominates the conservative system (which is the only one that has the two former requirements).

4 Different System Properties

In this section we suggest an alternative definition of the completeness property. Instead of the predefined definition we suggest:

Definition 6. ***(alternative-completeness)***
A system is alternative-complete *if every alert sequence A it produces satisfies*

$$\Phi(T(U1 \bigsqcup U2)) \subseteq \Phi(A).$$

If a system is alternative-complete it produces all the alerts that happened, and maybe some more. From the definitions it is clear that a system is complete iff it is consistent and alternative-complete. We next investigate the three properties: orderedness, consistency and alternative-completeness.

With this different definition an alternative-complete system is no longer consistent by definition. In other words, we have more possibilities than before: a system can be inconsistent but alternative-complete. We next argue that an alternative-complete system that is not consistent is an interesting system. An inconsistent system together with the original completeness has no properties at all (except orderedness in trivial cases), in other words, it has exactly the same properties as a system that produces a random set of alerts. However, our alternative-complete system above, without consistency, is a system that produces all the alerts that happened, and in some cases this is very important. Consider for example the nuclear reactor temperature reading with the condition: reactor temperature is over 3000^o_c. A system that guarantees all the alerts that happened, namely all the time instances in which the reactor temperature has gone above 3000^o_c, even if it produces some mistaken alerts is much better than a system which guarantees nothing at all.

An algorithm which achieves alternative-completeness exists, for example the trivial algorithm that produces an alert for every update (regardless of what the CEs evaluate) is trivially alternative-complete.

The next theorem shows that a system with any triggering condition and any one of the existing algorithms (AD-1,...,4) is not alternative-complete.

Theorem 9. *Every replicated monitoring system with lossy front links, any of the predefined triggering historical conditions, and any AD-i algorithm* ($i \in \{1, 2, 3, 4\}$) *is not alternative-complete.*

The proof is ommited and will apear in the full paper.

The new property provides more possibilties, and is motivated, namely there are scenarios for which it is highly required, whereas the original properties say nothing on system which fulfill such requirements. In addition, we introduced an algorithm that achieves the new requirement, namely it is not an empty property.

5 Related Work

The monitoring system discussed in this work can be viewed as a publish/ subscribe system that collects events and delivers alerts to subscribed users in a distributed environment [8, 5]. In [1] the authors suggest a way to embed such system in mobile networks.

Publish/subscribe systems [3, 9] are being used for a variety of application areas including financial analysis, and alert systems. Contents can be used to specify subscription as in content-based publish/subscribe systems [4].

More discussion on replications and the problems raised can be found in [2, 6].

References

[1] E. Anceaume, A. K. Datta, M. Gradinariu, and G. Simon. Publish/subscribe scheme for mobile networks. Technical report, POMC 2002 Workshop on Principles of Mobile Computing, 2002. October 30-31, Toulouse, France.

[2] P. A. Bernstein and E. Newcomer. *Principles of Transactions Processing.* Morgan Kaufmann, 1997.

[3] K. P. Birman. The process group approach to reliable distributed computing. *Communications of the ACM*, 36(12):37–53, 1993.

[4] A. Carzaniga, E. Nitto, D. Rosenblum, and L. Wolf. Issues in supporting event-based architectural styles. In *Proceedings of the third international workshop on Software architecture*, ACM SIGPLAN Notices, pages 17–20, Orlando, Florida, 1998. ACM Press.

[5] G. Cugola and H.-Arno Jacobsen. Using publish/subscribe middleware for mobile systems. *ACM SIGMOBILE Mobile Computing and Communications Review*, 6(4):25–23, 2002.

[6] J. Gray, P. Helland, P. O'Neil, and D. Shasha. The dangers of replication and a solution. In *ACM SOGMOD conference*, pages 173–182. ACM Press, 1996.

[7] Y. Huang and H. Garcia-Molina. Replicated condition monitoring. In *Proc. 20th ACM Symp. on Principles of Distributed Computing*, ACM SIGPLAN Notices, pages 229–237, Newport, Rhode Island, 2001. ACM Press.

[8] Y. Huang and H. Garcia-Molina. Publish/subscribe in a mobile environment. *Wireless Networks*, 10:643–652, 2004.

[9] Sun Microsystems. Jini technology core platform spec - distributed events. Technical report, Sun Microsystems, 2000. `http://www.sun.com/software/jini/specs/core2_0.pdf`.

Detecting SYN Flooding Attacks Near Innocent Side*

Yanxiang He[1], Wei Chen[1], and Bin Xiao[2]

[1] Computer School, The State Key Lab of Software Engineering,
Wuhan University, Wuhan 430072, Hubei, China
{yxhe, chenwei}@whu.edu.cn

[2] Department of Computing,
The Hong Kong Polytechnic University,
Hung Hom, Kowloon, Hong Kong
csbxiao@comp.polyu.edu.hk

Abstract. Distributed Denial-of-Service (DDoS) attacks seriously threat the servers in the Internet. Most of current research is focused on the detection and prevention methods at the victim side or the source side. However, defense at the innocent side, whose IP is used as the spoofed IP by the attacker, is always ignored. In this paper, a novel method at the innocent side has been proposed. Our detection scheme gives accurate detection results using little storage and computation resource. From the result of experiments, the approach presented in this paper yields accurate DDoS.

1 Introduction

Distributed Denial-of-Service (DDoS) attacks are a large-scale cooperative attack, launched from a large number of compromised hosts. DDoS attacks have posed a major threat to internet since 1990's and they have caused some popular web sites on the world, such as Yahoo, eBay, Amazon, become inaccessible to customers, which caused huge financial losses. Current events have shown that DDoS attacks continue to bring increasing threats to the internet. While many methods have been proposed, there still is a lack of efficient defense.

Most DDoS attacks exploit Transmission Control Protocol(TCP)[1]. It has been shown that more than 90% of the DoS attacks use TCP[2]. The most efficient and commonly-used SYN flooding attacks[3, 4] exploit the standard TCP three-way handshake in which the server receives a client's SYN (synchronization) request and replies with a SYN/ACK (synchronization/acknowledge) packet. The server then waits for the client to send the ACK (acknowledge) to complete the handshake. While waiting for the final ACK, the server maintains a half-open connection. As more and more half-open connections are maintained on a victim server, DDoS attacks will deplete the server's resources.

* This work is supported by the National Natural Science Foundation of China under Grant No. 90104005 and partially by HK Polyu ICRG A-PF86 and CERG Polyu 5196/04E.

X. Jia, J. Wu, and Y. He (Eds.): MSN 2005, LNCS 3794, pp. 443–452, 2005.

Lots of research work has been done to detect and prevent the TCP based DDoS attack. According to the deployment location of defense systems, the current DDoS detection and prevention methods can be classified into three categories: defense at the source-end, at victim-end or at intermediate-network. However, the information at the innocent host, whose IP is utilized as the spoofed IP, is totally ignored. Each kind of mechanisms has its limitations, for example detection at the side of a victim server can hardly produce an alarm at the early stage because abnormal deviation can only be easily found until the DDoS attack turns to the final stage. Furthermore, it is difficult for victim side to take efficient response after DDoS is detected due to numerous malicious packets aggregating at this side. Providing an early DDoS alarm near the source is a difficult task because the attack signature is not easy to capture at this side. The information at the innocent host side will be used in our approach because the innocent side will receive abnormal TCP control packets during a DDoS attack. Compared to other detection mechanisms, defense at the innocent host has two main advantages:

- Detection at innocent side is more hidden since it is deployed apart from attacking path. To avoid being detected, attackers usually sniffer and deceive defense systems deployed around victims and attacking source before launching attacks. It is difficult for attackers to be aware of the existence of detection mechanism at the innocent side.
- It has little vulnerability to DDoS attack. The burden of monitoring numerous attacking packets congesting at the victim side makes the defense system itself vulnerable to DDoS attack. Defense at innocent side will avoid this problem due to limited attack streams near innocent side. This enables the defense system itself to have little risk of becoming potential target of DDoS attacks.

Detection at innocent side has several challenges. On the one hand, accurate detection is not easy to achieve since the abnormal signature is distributed in a backscatter way [2]. To capture the small quantity of attack signatures, more packets should be monitored and recorded for analysis, which implies expensive storage cost. On the other hand, defense at innocent side is deployed far from the victim side and detection effect depends on the number of participant of Internet Service Providers(ISPs). To attract more ISPs to participate detection, the detection scheme should use limited resource in defense, which will not bring evident degradation to ISPs' service. In this paper, we provide a detection scheme which gives satisfying result with little storage and computation requirement.

In this paper we propose a novel detection method against SYN flooding attack near innocent side. We summarize our contributions as follows:

- The detector performs detection at the side of innocent hosts because this side can provide valuable information. Another benefit of such deployment is to make the detection system itself invulnerable to DDoS attacks.
- A Bloom filter based detection scheme is proposed. We apply Bloom filter to DDoS detection, which can record huge traffic information on high speed network.

– With this space-efficient data structure, detection scheme monitors abnormal handshake with little computation overhead. The detection scheme only requires simple hash function operation, addition operation and subtraction operation, which bring little overhead to present computers.

The remainder of the paper is organized as follows: Section 2 will introduce some related works in spoofed DDoS detection. In Section 3, the TCP-based DDoS attack will be discussed and analyzed. The techniques for DDoS detection at innocent side will be proposed in section 4. Some experiment results will be given in Section 5 to evaluate the performance of the proposed method. In Section 6 we will conclude our work and discuss future work.

2 The Related Work

Hash table is a high performance data structure used for quick data look up and is versatilely applicable in network packet processing. The Bloom filter is a kind of space-efficient hash data structure, which is first described by Burton Bloom [5]. It is originally used to reduce the disk access times to different files and other applications and now it has been extended to network packet processing. Song present a hash table data structure and lookup algorithm using extended Bloom filter, which can support better throughput for router applications based on hash tables. NetFlow [6] maintains a hash table of connection record in DRAM and monitors network traffic. The concept of multiple hashing, which is similar to Bloom filter, was used to track large flows in network traffic.

Hash table is also used to defend DDoS attack. Snoeren [7] present a hash-based technique for IP traceback that generates audit trails for traffic within the network, and can trace the origin of a single IP packet delivered by the network in the recent past. Hash table is employed to look for an imbalance between the incoming and outgoing traffic flows to or from each IP address [8]. IDR [9] is a router equipped with DDoS protection mechanism, which uses Bloom filter to detect DDoS attack.

3 The TCP-Based DDoS Attack

In this section, we analyze the difference between normal TCP handshakes and spoofed one. During spoofed DDoS attack, the source IP address of attacking packet is usually modified, which is not the attacker's IP address anymore. The normal three-way handshake to build a connection would be changed consequently.

The normal three-way handshake is shown in Figure 1(a). First the client C sends a $Syn(k)$ request to the server S_1. After receiving such request, server S_1 replies with a packet, which contains both the acknowledgement $Ack(k+1)$ and the synchronization request $Syn(j)$. Then client C sends $Ack(j+1)$ back to finish the building up of the connection. k and j are sequence numbers produced randomly by the server and client respectively during the three-way handshake.

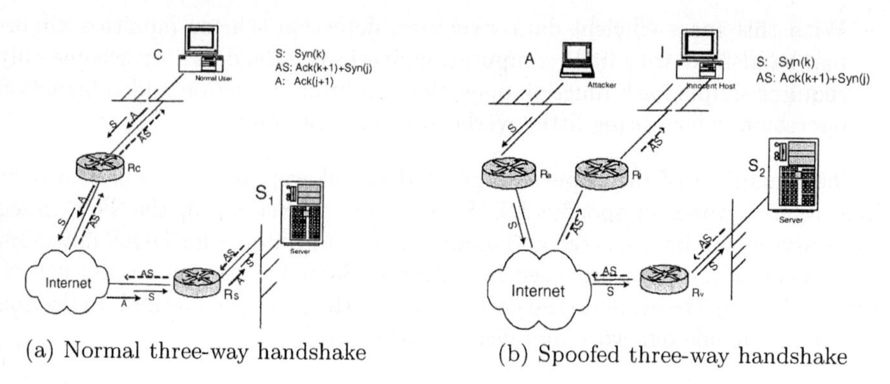

(a) Normal three-way handshake (b) Spoofed three-way handshake

Fig. 1. The process of the TCP three-way handshake

In the remainder of the paper, SYN shows a message sent to server S inside the TCP control packet during the first round of the three-way handshake protocol. A packet containing both $Ack(k+1)$ and $Syn(j)$ (denoted as ACK/SYN in the following sections) is delivered back from server S in the second round. A control package with $Ack(j+1)$(denoted as ACK) involves in the third round. During the normal three-way handshake procedure, SYN, ACK/SYN and ACK can be observed at the edge router(R_c in Figure 1(a)) near the client.

If the packet at the first round is a malicious one with a spoofed IP address, a valid authentication process is modified. As Figure 1(b) shown, the innocent host I, whose IP is used as spoofed source IP, is usually not in the same domain with the attacker host A. In other words, the edge router working for attacker host A does not route for the innocent host I. In fact, to avoid being traced back, the attacker usually uses the IP address belonging to other domains to make a spoofed packet. Under this assumption, there exists difference between the normal TCP three-way handshake and the spoofed one. In Figure 1(b) the edge router R_a in the attacker domain forwards the SYN packet with the spoofed address P_I, the IP address of the innocent host I, to the server S_2. The sever S_2 replies with an ACK/SYN packet. This ACK/SYN will be sent to the innocent host I because the server S thinks the SYN packet is from I according to the spoofed source IP P_I in it. So the edge router R_I at the innocent host side will receive the ACK/SYN packet from victim server S_2. But there is no previous SYN request forwarded by the client detector at R_I. This scenario is different from the normal one. Our approach is proposed on the base of this difference.

4 The Bloom Filter Based Detection Scheme

In order to detect DDoS, the TCP control packets for handshakes are monitored and analyzed at the edge router of innocent side. For example, the detector will be installed on the router R_I in Figure 1(b). To save storage cost, a Bloom filter based method is applied to monitor two-way traffic between innocent side and the rest of Internet. It checks the TCP control packets flowing through the edge router. When it captures suspicious handshakes, the alarm about the potential DDoS attack will be launched.

4.1 The Bloom Filter Based Hash Table

The basic idea of innocent-side detection is to monitor two-way TCP control packets and capture abnormal handshakes. A TCP connection may hold for several seconds or even for several minutes but most three-way handshake can be finished in a very short period(e.g., less than 1 seconds) at the beginning phrase of the connection. However, it is expensive to keep a record for each handshake considering numerous traffic volume on the high speed link network. To record useful information with limited storage, Bloom filter, a kind of space-efficient hash data structure, is applied in our method.

Original Bloom Filter. Bloom filter is first described by Burton Bloom [5] and originally used to reduce the disk access to differential files and other applications, e.g. spell checkers. Now it has been extended to defend against DDoS attack [7, 8, 9]. The idea of Bloom filter is to allocate a vector v of m bits, initially all set to 0, and then choose k independent hash functions, $h_1, h_2, \ldots, h_k$, each with range $\{1, \ldots, m\}$. For each element $a \in A$, the bits at positions $h_1(a), h_2(a), \ldots, h_k(a)$ in v are set to 1(Figure 2). Note that a particular bit might be set to 1 multiple times which may cause potential false result. Given a query for b we check the bits at positions $h_1(b), h_2(b), \ldots, h_k(b)$. If any of them is 0, then certainly b is not in the set A. Otherwise we conjecture that b is in the set. However there is a certain probability that Bloom filter give false result, which is called a "false positive". The parameters k and m should be chosen such that the probability of a false positive is acceptable.

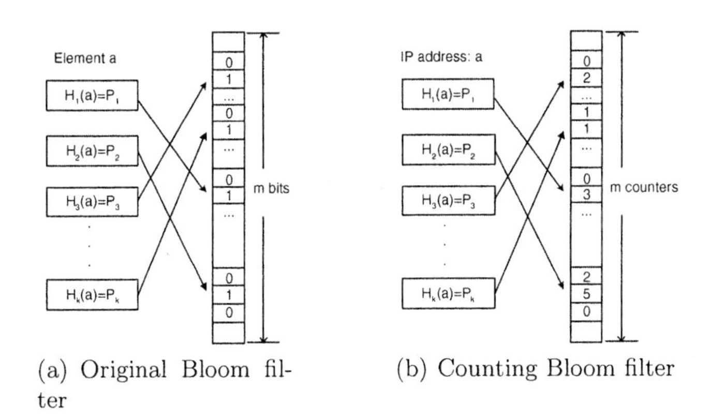

(a) Original Bloom filter

(b) Counting Bloom filter

Fig. 2. Bloom filter uses independent hash functions to map input into corresponding bits

Counting Bloom Filter. The original Bloom filter is not suitable to monitor handshakes. We will use a variant of a Bloom filter called *Counting Bloom filter* which substitutes m bits with countable integers table as shown in Figure 2(b).

After using counters table to replace m bit array, all the counts are initialized to 0. When a key is inserted or deleted, the value of count is incremented or decremented by 1 accordingly. When a count changes from 0 to 1, the corresponding bit is turned on. When a count changes from 1 to 0 the corresponding bit is turned off. The value in the count indicates the current statistic results of traffic.

4.2 Detection Scheme

To detect attacking traffic with spoofed source IP, the destination IP(the server's IP) is recorded in the hash tables. When a SYN packet, the TCP control packet for the first round handshake, is captured from the outgoing traffic, the destination IP(the server's IP) is hashed into the hash table by k independent hash functions. For the output of each hash function, if the corresponding counter is 0, the corresponding counter is turned on. If the counter is already turned on, the counter is incremented by 1 accordingly. If corresponding ACK/SYN packet for the second round of handshake is soon captured in the incoming traffic. The source IP(the server's IP) is hashed into the hash table again. But this time the corresponding counter is decremented by 1 for each independent hash function. When a count changes from 1 to 0, the corresponding counter is turned off. For a normal TCP handshake, ACK/SYN packet can be hashed to a turn-on counter by each independent hash function because the corresponding counter has already been turned on by previous SYN packet. The counter will keep unchanged if the first two rounds of three-way handshake are completely captured at the ingress and egress router at the innocent side. The detection scheme is depicted in Figure 3. These counts are reset to 0 for every period t.

On the contrary, in the scenario of the spoofed handshake, suspicious ACK/SYN packet should meet at least turn-off counter since previous SYN

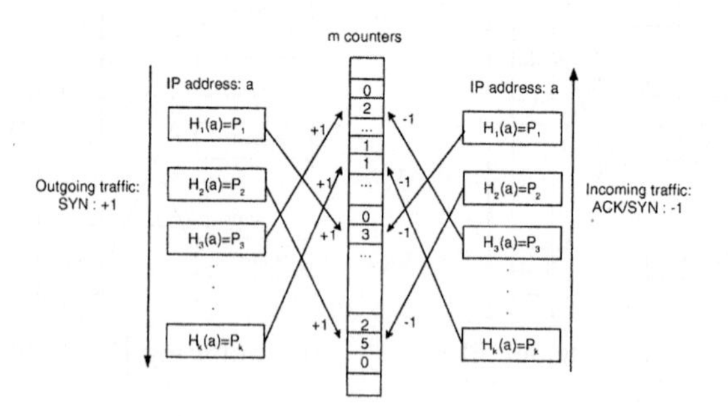

Fig. 3. The detection scheme increases or decreases the value of the count according to the three-way handshake

```
Detection_Scheme (INPUT: P) {
if P is a SYN packet then
  for i = 0; i < k; i++ do
    j=Hash_i(P)
    Counter_j++
  end for
else if P is a ACK/SYN packet then
  for i = 0; i < k; i++ do
    j=Hash_i(P)
    //Check whether exists turn-off counter
    if Counter_j == 0 then
      Report Suspicious Alarm(SA)
      RETURN
    end if
  end for
  //If no turn-off counter, do subtraction
  for i = 0; i < k; i++ do
    j=Hash_i(P)
    Counter_j- -
  end for
end if
RETURN }
```

Fig. 4. The Bloom filter based Detection Scheme

packet has turned on the counter. During a DDoS attack, the second round handshake packets, ACK/SYN, are backscattered according to the spoofed source IP of attacking packets. The innocent side will receive part of these ACK/SYN packets and try to hash them with k independent hash functions. It is possible for the abnormal ACK/SYN to hit one turn-on counter since there may exist collision for one of hash functions. However, the possibility of hitting all k turn-on counters is rather low since it is not likely that all k independent hash functions have collisions at the same time. If ACK/SYN packet hits one turn-off counter, the detailed information of this packet is recorded for further analysis. The detection scheme only requires addition and subtraction operations. These operations bring little overhead to system considering today's computation ability.

When a new Suspicious Alarm(SA) is reported, the detector will analyze the source IP distribution of SAs in database. Assumed a DDoS attack takes place, the detector will find asymmetric ACK/SYN packets sent from victim server with its IP P_{victim} as the source IP. When SAs with the same source IP P_{victim} are reported in a short period, there probably exists a DDoS attack targeting the host P_{victim}. But if each SA has a different source IP, it is most likely caused by some reasons other than a DDoS attack. To evaluate the distribution of the source IP of the alarms, an expression is presented below:

$$score = \sum_{s \in IPList} (|X_s| - 1)^2$$

Where X_s stands for a subset of the total SA set. All the elements in X_s are SAs that have the same IP value s in a certain period. The score will increase dynamically when the number of SAs with the same source IP increases. On the other hand if each of the SAs has a different source IP, the score will reach minimum.

5 Experiment

Experiments are designed to evaluate the detection method. In the experiment, 10 zombies are simulated to perform SYN flooding attacks toward the server. The rate of the attack packets rises from 10 packets/sec to 1000 packets/sec in 10 seconds and 100 seconds separately. The maximum rate is set to 1000 packets/sec because it is enough to shut down some services as Chang reported [10]. In simulation only 1% of ACK/SYN packets replied by the victim server are designed to arrive detectors near innocent side. These packets will trigger detectors to generate SAs. The number of SAs generated by the client detectors is shown in the Figure 5.

Although only 1% spoofed attack packets can be received by detector, detector still give accurate SAs. The number of SAs is enough for detector to give a further potential DDoS alarm at the early stage of the DDoS attack. The detection results are satisfying even when the DDoS attacker increases the attack packets slowly. From experiment results, the SAs number raises stably in the Figure 5(a) because the DDoS attack is launched in a short time. In the Figure 5(b) the number of SAs fluctuates a little because the attacking packets rise up at a much slower rate.

The storage and computation cost of our scheme is compared with another hash method used in Snort [11]. Snort uses a hash method to monitor network traffic and each connection is recorded as a 5-tuple entry. In simulation, 1000000 IP addresses are randomly generated and inserted into hash tables using our hash

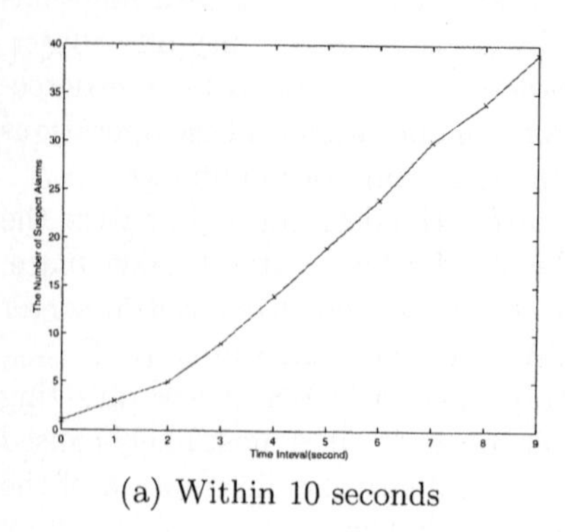

(a) Within 10 seconds

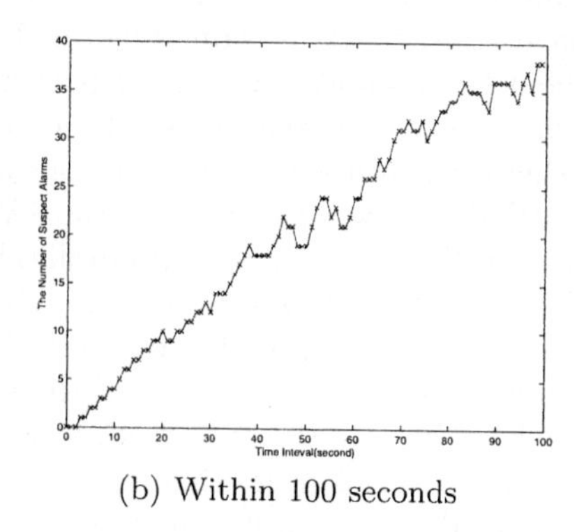

(b) Within 100 seconds

Fig. 5. The number of SA generated by the detector after receiving 1% of ACK/SYN packets from the protected server. The attacking packets rate reaches to 1000 packets/sec within (a)10s (b)100s.

Table 1. Comparison of time and storage consumption between 5-tuple hash and our hash method

Hash method	Time Consumption(sec)	Storage Consumption(byte)
5-tuple hash	0.328	14336
Our hash method	0.187	4096

method and the 5-tuple hash method. The simulation platform uses a Pentium 4 1.7G processor and 256M memory. The time and storage consumption are compared and list in Table 1. It shows that our method uses less memory and spends less time than 5-tuple hash method.

6 Conclusion and Future Work

In this paper a novel detection method against the DDoS attack is presented. The detection system is deployed at the innocent side, which is quite different from current methods which are often deployed at the sides of the victim server or the attacking source. Detection at innocent side makes defense more hidden to attackers. The basic idea in the detection is to differentiate between the normal TCP three-way handshake and the spoofed one. The proposed Bloom filter based detection scheme can give accurate detection with little storage cost. The experiment results are given in section 5 and the proposed approach is effective to detect DDoS.

In the future research work we will apply method to real environment to test the memory and computing cost for actual running. The optimization of k, m will also be studied.

References

1. Postel, J.: Transmission Control Protocol : DARPA internet program protocol specification,RFC 793 (1981)
2. Moore, D., Voelker, G., Savage, S.: Inferring internet Denial of Service activity. In: Proceedings of USENIX Security Symposium, Washington, D.C, USA. (2001) 9–22
3. Wang, H., Zhang, D., Shin, K.G.: Detecting SYN flooding attacks. In: Proceedings of Annual Joint Conference of the IEEE Computer and Communications Societies(INFOCOM). Volume 3. (2002) 1530–1539
4. Schuba, C.L., Krsul, I.V., Kuhn, M.G., Spafford, E.H., Sundaram, A., Zamboni, D.: Analysis of a denial of service attack on TCP. In: Proceedings of the 1997 IEEE Symposium on Security and Privacy, IEEE Computer Society, IEEE Computer Society Press (1997) 208–223
5. Bloom, B.H.: Space/time trade-offs in hash coding with allowable errors. Communications of the ACM **13** (1970) 422–426
6. Estan, C., Keys, K., Moore, D., Vargese, G.: Building a better NetFlow. In: ACM SIGCOMM. (2004) 39–42

7. Snoeren, A.C.: Hash-based IP traceback. In: Proceedings of the ACM SIGCOMM Conference, ACM Press (2001) 3–14
8. Abdelsayed, S., Glimsholt, D., Leckie, C., Ryan, S., Shami, S.: An efficient filter for denial-of-service bandwidth attacks. In: IEEE Global Telecommunications Conference(GLOBECOM'03). Volume 3. (2003) 1353–1357
9. Chan, E., Chan, H., Chan, K., Chan, V., Chanson, S., etc.: IDR: an intrusion detection router for defending against distributed denial-of-service(DDoS) attacks. In: Proceedings of the 7th International Symposium on Parallel Architectures, Algorithms and Networks 2004(ISPAN'04). (2004) 581–586
10. Rocky.K.Chang: Defending against flooding-based distributed denial-of-service attacks: a tutorial. Communications Magazine, IEEE **40** (2002) 42–51
11. Snort: (Open source network intrusion detection system, http://www.snort.org)

Network Capacity of Wireless Ad Hoc Networks with Delay Constraint

Jingyong Liu[1], Lemin Li[1], and Bo Li[2]

[1] Key laboratory of broadband optical
fiber transmission and communication networks,
University of Electronic Science and Technology of China,
Chengdu, Sichuan, China, Postfach 61 00 54
liu_jingyong@163.com, lml@uestc.edu.cn
[2] Computer Science Dept.,
Hong Kong University of Science and Technology,
Hong Kong, China
bli@cs.ust.hk

Abstract. The capacity of wireless ad hoc networks has been widely studied since the pioneer works by Gupta and Kumar [3], and Grossglauser and Tse [4]. Various asymptotic results on capacity and capacity-delay tradeoffs have been obtained, whereas most of those investigated the asymptotic performance in large networks and addressed the delay caused by node mobility. In this paper the tradeoff between throughput capacity and average transmission delay of arbitrary scale wireless ad hoc networks is investigated. We consider a network with N nodes randomly distributed in a disk area of radius R. A new criterion of *transmission concurrency* is introduced to describe the concurrent transmitting capability of the network. And by using a constructive scheme, where a balanced scheduling strategy and a minimal length route selection mechanism are proposed, an analytical expression of the relationship between the throughput capacity T and the average transmission delay D is derived. A case study is also provided to give numerical results.

1 Introduction

Wireless ad hoc networks or multihop radio networks are typically composed of equal nodes that communicate over wireless links without any central control or infrastructure, wherein nodes may cooperate in routing each other's data packets. Wireless ad hoc networks inherit the traditional problems of wireless and mobile communications, such as bandwidth optimization, power control, and transmission quality enhancement. A key aspect of wireless communications is the radio propagation channel, which introduces cochannel and adjacent channel interference among users. For ad hoc networks where nodes share the same spectrum, there is still fundamental work to be done in this area.

The capacity of wireless ad hoc networks means to calculate theoretical upper bounds on the capabilities of networks. Consider a complete network model and

X. Jia, J. Wu, and Y. He (Eds.): MSN 2005, LNCS 3794, pp. 453–465, 2005.

assume perfect coordination, network capacity is the traffic carrying capability of the network under this condition. The capacity of wireless ad hoc networks has been widely studied since Gupta and Kumar, they presented asymptotic results for static networks based on an information-theoretic framework in [3], where it is shown that per-user network capacity is $O(1/\sqrt{N})$, and hence vanishes as the number of users N increases. The effect of mobility on the capacity of ad hoc wireless networks was first explicitly developed in [4], where a 2-hop relay algorithm was developed and shown to support constant per-user throughput. These works did not consider the associated network delay, and analysis of the fundamental queueing delay bounds for general networks remains an important open question.

Tradeoff between capacity and delay has thereafter been the focus on studying the capacity of ad hoc wireless networks. Many literatures [12, 13, 14, 15, 16, 17, 18, 19, 20, 21, 22] has come forth since [3] and [4] in recent years. In [12], the authors considered a network with stationary nodes and mobile relays, and showed that the delay can be improved by exploiting the velocity information and selectively relaying packets to those nodes which are moving in the direction of the destination. In [14], the authors determined the delay limited capacity of ad hoc networks using the diversity coding approach given in [15]. Delay improvement via redundant packet transfers is considered in [16]. In [17] the authors considered a cell partitioned network with an infinite mobility model and in [18] the authors considered a more realistic random walk mobility model with a cellular TDMA scheme. They characterized the delay and determined the throughput-delay tradeoff aline the models of Gupta-kumar and Grossglauser-Tse. In [19], the authors considered two canonical random mobility models which are used very often in the literature on MANETs. They estimated the packet delay under the distributed version of the 2-hop relaying protocol proposed in [4], and considered two alternative protocols for delay improvement. The idea behind these protocols is to reduce the packet delay by trading off the throughput capacity.

For the multihop and shared media nature of ad hoc networks, there are some tradeoffs on analyzing the performance of throughput capacity and transmission delay. The main factors that involved are transmission range and connectivity. Some definitions and explanations are given in the following.

Definition 1. *A throughput $\lambda > 0$ is feasible if every node in the network can send at a rate of λ bits per second to its chosen destination. And the* ***throughput capacity*** *of the network is the maximum feasible throughput under a given transmission delay constraint and denoted by T.*

When there is no delay constraint, the throughput capacity get the maximum value and is the same as in [3] and [4].

Definition 2. *The transmission delay of a packet is the time it takes the packet to reach the destination after it has been generated at the source and it equals the sum of the time spent at each relay and the source. The* ***average transmission delay*** *for a network is the mean value of all packets over all source-destination pairs and is denoted by D.*

As to the transmission range and the connectivity there are not explicit definitions. In an interference-constrained wireless network, the transmission range of an individual node is dependant on the transmission power of the node and constrained by interference caused by other nodes transmitting data simultaneously. As stated in [3], a smaller transmission range helps to obtain a higher throughput capacity, to maximize throughput capacity one should try to find the minimum transmission range. Whereas the minimum value of transmission range is restricted by network connectivity since too small a transmission range may cause the network to be disconnected. In [5] Gupta and Kumar gave an intuitive explanation by the requirement of the network to be connected: in a wireless ad hoc network, a critical requirement is that each node can find a way to every other node. In the literature the number of neighbors of a node is usually associated with the connectivity level of the network and various results are proposed based on different approaches and criteria [5]–[7]. In addition, the route length is a major factor that affects the transmission delay and the minimum route length is mainly determined by transmission range.

In this paper the capacity-delay tradeoff of an arbitrary scale wireless ad hoc network is investigated under a constructive scheme. In the scheme we assume the network is ideally scheduled and propose a minimum length routing selection mechanism, the objective is to maximize the throughput capacity while minimize the transmission delay. The transmission delay of a common traffic is obtained by analyzing the queuing delay at each node of the route and the critical transmission range is derived from the mean hop length. Finally an analytical expression of the relationship between the throughput capacity T and the average end to end transmission delay D is obtained. The minimum transmission range that satisfies connectivity of the network and the capacity-delay tradeoff with the minimum transmission range is also derived. moreover, the maximum throughput capacity without delay constraint is also provided and the result is agreed with that of [3].

The rest of the paper is organized as follows. In section 2 the network model and some hypothesis are proposed. In section 3 we discuss the relationship between *transmission range* and *transmission concurrency*, which is an indicator of the concurrent transmitting capability of the network introduced in this paper and a major factor that affect the network capacity. In section 4 we analyze the average transmission delay and In section 5 the analytical expression of tradeoff between throughput capacity and average transmission delay is derived and a case study is also proposed to give numerical results. Finally we conclude in section 6.

2 Model and Hypothesis

2.1 Network Model

The ad hoc network consists of N nodes randomly located in a disk of radius R on planar surface $\Re^2$, whose area is $A = \pi R^2$. All nodes are fixed and uniformly distributed in the disk area, this means each node is equally likely to be anywhere in this area, and the location of different nodes are independent and identically distributed (i.i.d.). We assume that each of the N nodes is a source node of one session and a destination node of another session. Moreover, each node may act as a relay node of another session as well. We also assume that each source has an infinite packets flow to transmit to the destination. And the source-destination pair will not change since they have been set up. Each node sends packets at uniform power P using omni-direction antenna and through the same radio channel. We do not consider the case that the channel is divided into several subchannels. This does not loss generality for it is shown in [3] that it is immaterial to the capacity result.

2.2 Transmission Model

Let $P_i(t)$ denote the transmission power of node i at time t, and $\gamma_{ij}(t)$ denote the channel gain from node i to node j , then the received power at node j is $P_i(t)\gamma_{ij}(t)$. At time t, node i can transmit to node j at W packets/s if the SINR at node j satisfy

$$\frac{P_i(t)\gamma_{ij}(t)}{N_0 + \sum\limits_{k \neq i} P_k(t)\gamma_{kj}(t)} > \beta \tag{1}$$

where β is the minimum signal-to-interference ratio (SINR) necessary for successful receptions, N_0 is the background noise power. In this paper, we only consider large scale path loss characteristics in the fading channel model. The channel gain is given by

$$\gamma_{ij}(t) \triangleq \frac{1}{|X_i(t) - X_j(t)|^\alpha} \tag{2}$$

where α is a parameter greater than 2, $X_i(t)$ and $X_j(t)$ are the positions of node i and node j respectively.

3 Transmission Range and Concurrent Transmission

It is obvious that all the nodes in the network can not send packets simultaneously, at any time instant only a subset of all nodes can transmit. in this section we introduce a new criterion *transmission concurrency* as an indicator of this concurrent transmitting capability of the network. Given an certain transmission range, when nodes in the network are ideally scheduled such that the network works in a balanced state and at any time there are as much as possible nodes transmitting data, the network can obtain the maximum concurrent transmitting capability. The scheduling strategy is depicted in section 4.4. In this section we investigate the relationship between transmission range and the transmission concurrency. Some definitions used in this section are given in the following.

Definition 3. *The **transmission range** of a sender is the area that receivers within it can receive data from the sender successfully while those outside it can not. When the network is load-balanced this area is a disk that centered at the sender and it is denoted by its radius r.*

Definition 4. *Define the subset of nodes simultaneously transmitting data at some time instant in a network as the* ***concurrent transmitting nodes*** *of the network. When ideally scheduled the network have the maximum number of concurrent transmitting nodes and denote this maximum number by C.*

Definition 5. *Define the ratio of the maximum number of concurrent transmitting nodes to the number of total nodes in a network as* ***transmission concurrency*** *of the network and denoted by ξ, hence $\xi = C/N$.*

3.1 Transmission Range and Transmission Concurrency

Since every transmitting node causes interference to other nodes, it can be imagined that there is no other transmitting nodes in an area around each transmitting node. We call this area *interference range* and denote it by its radius r_{i}. In this paper we neglect the edge effect of the nodes that located at the boundary of the network and treat each sender-receiver pair as the same. Since all nodes use omni-direction antenna and we only consider large scale path loss characteristics in the fading channel model, it is reasonable to assume that the interference range is round in shape.

When the network is load-balanced there are C transmitting nodes which are located uniformly in the network region and each corresponding interference range is a circumcircle of other surrounding one. Around each interference range there are other six one, as can be demonstrated by Fig. 1. It can be proved that this distribution consumes minimum area and thus the network has the maximum number of concurrent transmitting nodes. Moreover, by replacing each interference range with its circumscribed regular hexagon, we get a hexagonal tessellation of $\Re^2$.

Suppose the radius of interference range in Fig. 1 is $r_{\mathrm{i}} = \Delta r$, where $\Delta > 1$ is an unknown constant to be calculated, if we can find a critical value Δ_{th} such that inequation (1) is satisfied for every sender-receiver pair when $\Delta \geq \Delta_{\mathrm{th}}$

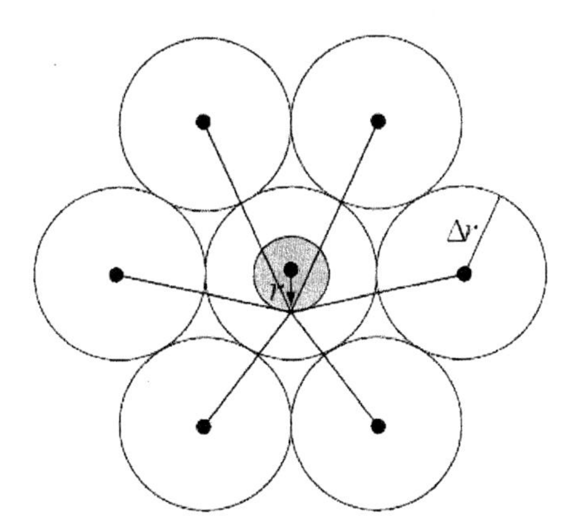

Fig. 1. Distribution of concurrent transmitting nodes when load-balanced

we get the transmission concurrency of the network. Let X_i and X_j be the location of a sender and its receiver respectively and let $|X_i - X_j| = r$, denote the distance between two transmitting nodes by $D_{\rm i} = 2r_{\rm i}$. We only consider the nearest tier of interference nodes, this situation resembles the cells of cellular networks, see [23] for reference. The approximate distances from the receiver to the nearest two transmitting nodes are both $D_{\rm i} - r$, those to the two fastest ones are approximately $D_{\rm i} + r$, and those to the other two nodes are approximately $D_{\rm i}$. Therefore we get the SINR at a receiver:

$$\begin{aligned}\mathrm{SINR} &= \frac{Pr^{-\alpha}}{N_0 + P(2(D_{\rm i}-r)^{-\alpha} + 2(D_{\rm i}+r)^{-\alpha} + 2D_{\rm i}^{-\alpha})} \\ &= \frac{r^{-\alpha}}{\gamma + (2(D_{\rm i}-r)^{-\alpha} + 2(D_{\rm i}+r)^{-\alpha} + 2D_{\rm i}^{-\alpha})}\end{aligned}$$

where $\gamma = N_0/P$. When the condition

$$\frac{r^{-\alpha}}{2(D_{\rm i}-r)^{-\alpha} + 2(D_{\rm i}+r)^{-\alpha} + 2D_{\rm i}^{-\alpha}} > \beta \tag{3}$$

is satisfied, there exists a $P_{\min}$ so as to when $P \geq P_{\min}$ inequation (1) is satisfied. Let $Q = D_{\rm i}/r = 2\Delta$ and substitute it into (3) we get an inequation:

$$\frac{1}{2(Q-1)^{-\alpha} + 2(Q+1)^{-\alpha} + 2Q^{-\alpha}} > \beta\,. \tag{4}$$

As it is very hard to find the analytical resolution of this inequation, we give a numerical method to acquire the value of Q corresponding to any given value of β. Let

$$\beta = F(Q_{\rm th}) = \frac{1}{2(Q_{\rm th}-1)^{-\alpha} + 2(Q_{\rm th}+1)^{-\alpha} + 2Q_{\rm th}^{-\alpha}}\,, \qquad 1 < Q_{\rm th} < \infty\,.$$

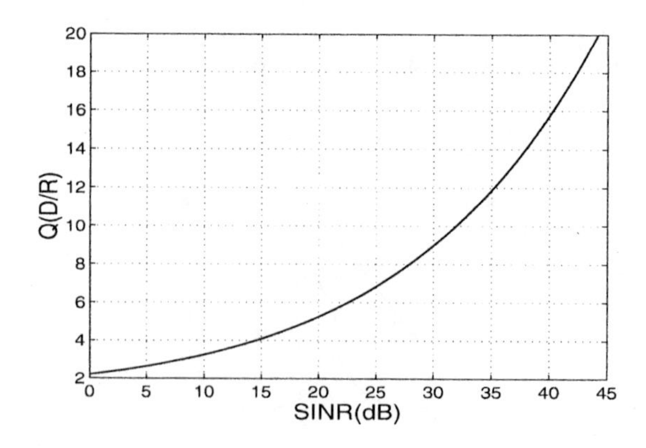

Fig. 2. The relationship between Q and SINR

It is a monotone increase function of Q_{th} and its inverse function exists and $Q_{\mathrm{th}} = F^{-1}(\beta)$. Let $\alpha = 4$, we plot the inverse function in Fig. 2. Recall $Q = 2\Delta$, to a given value of β, we get $\Delta_{\mathrm{th}} = Q_{\mathrm{th}}/2 = F^{-1}(\beta)/2$. When $Q > Q_{\mathrm{th}}$, (4) is satisfied. And since the circumscribed regular hexagons of the interference ranges make up a hexagonal tessellation of $\Re^2$, the area occupied by each transmitting node is the area of the hexagon inclosing it. Denote this area by S_{i}, we have $S_{\mathrm{i}} = 2\sqrt{3}r_{\mathrm{i}}^2 = (\sqrt{3}/2)Q_{\mathrm{th}}^2 r^2$. Thus we get the relationship between transmission range and transmission concurrency:

$$\xi = C/N = \frac{2\sqrt{3}}{3}\frac{1}{Q_{\mathrm{th}}^2 r^2 \rho_{\mathrm{s}}} \tag{5}$$

where $\rho_{\mathrm{s}} = N/A$ is the node spatial density of the network.

4 Average Transmission Delay

4.1 Minimal Length Route Selection Mechanism

According to different routing protocol, there are various criteria on route selection, such as minimum hops and minimum power consumption. In this paper our objective is to maximize the network capacity. Since smaller transmission range helps to achieve more capacity, we introduce two strategies to achieve this objective: to minimize the total route length and to minimize the mean hop length. And the following mechanism is adopted to implement these strategies.

Consider an arbitrary node n_i as the source and it randomly selects another node $Dst(n_i)$ as its destination. Denote by L_{i} the straight-line segment connecting n_i and $Dst(n_i)$, we call it the reference route. An actual route will deviate from it. In order that the actual route can be selected as close as possible to the reference route and not deviate from the destination, we make two parallel lines beside L_{i} with distance d to L_{i}, each node that selected as an intermediate node choose the nearest node in the forward direction from in the band region enclosed in the two lines as next-hop node, see Fig. 3. The distance d determines the departure of the actual route to the reference route and the hop length distribution.

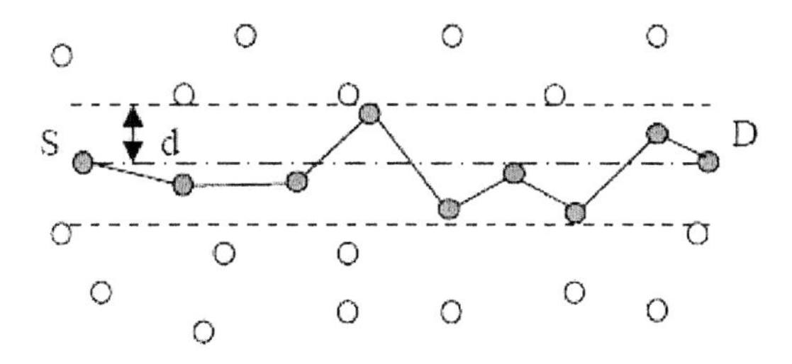

Fig. 3. The sketch map of the minimal length route selection mechanism

4.2 Mean Route Length (Number of Hops)

Let L denote the distance between the source and its corresponding destination. Without loss of generality, we assume that the source is placed at the origin of the network and let X and Y be the coordinates of the destination. Since X and Y are independent and the location of the destination is uniformly distributed in the network area, the CDF (cumulative distribution function) of L can be written as

$$F_L(l) = \iint_{\sqrt{x^2+y^2} \le l} \frac{1}{\pi R^2} dxdy = \frac{l^2}{R^2}\,, \qquad 0 \le l \le R\,. \tag{6}$$

Taking the derivative of $F_L(l)$ with respect to l yields the pdf of L:

$$f_L(l) = 2l/R^2\,, \qquad 0 \le l \le R\,. \tag{7}$$

Denote by random variable W the hop length, by W' the length of projection of W on the reference route L_i. First consider the probability that W' is greater than some distance w', which is equal to the probability that there is no node in the rectangle of length w' and width $2d$. Since the positions of nodes in the network are independent and uniformly distributed, the probability of a node lying in this area is $p = 2dw'/A$. And since the total number of nodes N is large enough it can be shown that the number of nodes in this area has a Poisson distribution with parameter $\lambda = Np = 2d\rho_\text{s}w'$. Hence the probability that there is no node in a rectangle of length w' and width $2d$ is $P(W' > w') = e^{-2d\rho_\text{s}w'}$. The CDF of W' can then be written as:

$$F_{W'}(w') = 1 - e^{-2d\rho_\text{s}w'}\,. \tag{8}$$

And the corresponding probability density function (pdf) can be obtained by taking the derivative of $F_{W'}(w')$ with respect to w', is

$$f_{W'}(w') = 2d\rho_\text{s}e^{-2d\rho_\text{s}w'}\,. \tag{9}$$

Intuitively the number of hops of a route is in direct proportion to the distance between the source and the destination and in inverse proportion to the hop length. The distance between a source and its destination is the length L of the reference route L_i, and the hop length is replaced by its projection on the reference route. Thus the mean number of hops from the source to the destination can be approximately computed by

$$\bar{n}_\text{h} = \bar{L}/\bar{W'} = E(L)/E(W') \tag{10}$$

where $E(L)$ and $E(W')$ can be derived from (7) and (9) respectively:

$$E(L) = \int_0^R lf(l)dl = \frac{2}{3}R\,; \quad E(W') = \int_0^\infty w'f(w')dw' = \frac{1}{2\rho_\text{s}d}\,. \tag{11}$$

Substituting (11) into (10) yields the mean route length:

$$\bar{n}_\text{h} = \frac{4}{3}R\rho_\text{s}d\,. \tag{12}$$

4.3 Mean Number of Routes Traversing Each Node

Denote by s the mean number of routes that traverse each node, since the total number of nodes is N and mean route length is $\bar{n}_{\mathrm{h}}$, we get:

$$s = N(\bar{n}_{\mathrm{h}} - 1)/N = \bar{n}_{\mathrm{h}} - 1 . \tag{13}$$

4.4 Average Transmission Delay

Suppose arrival of data packets at each node is a Poisson process with arrival rate λ and all data packets have the same length L, service time is a fixed-length distribution and the mean service time is $1/\mu = L/W$. We also assume the network is under a scheduling strategy as follows.

Balanced scheduling strategy: Assume that the transmitting time is divided into slots and each time slot is $1/\mu$, nodes are divided into groups and each group is a group of concurrent transmitting nodes called *concurring group*. At each timeslot a concurring group is assigned to transmit data. As shown in definition 4 and 5, each group has C nodes and all the nodes in the network can be divided into N/C groups. Hence each node can transmit once every $1/\xi$ timeslots.

Under above scheduling strategy, data packets traversing an mean length route can be regarded as a tandem $M/D/1/\infty$ queueing system. Within this system the arrival of data packets at each node is a Poisson process with parameter $(s+1)\lambda$, service time is also a fixed-length distribution and the mean service time equals $1/\mu\xi$.

It has been deduced that for a $M/D/1/\infty$ queueing system, where the arrival of customers is a Poisson process with arrival ate of λ and the service time series are independent and obey the same fixed-length distribution with mean service time $1/\mu$, the mean queue length $\bar{N}$ and the mean stay time $\bar{W}$ can be expressed as:

$$\begin{cases} \bar{N} = \frac{\rho(2-\rho)}{2(1-\rho)} \\ \bar{W} = \frac{2-\rho}{2\mu(1-\rho)} \end{cases}, \qquad \rho < 1 \tag{14}$$

where $\rho = \lambda/\mu$ is the traffic intensity of the system [27].

In our model the queue length means the buffer length and the stay time is the transmission delay of packets at a node. Replace arrival rate $(s+1)\lambda$ and service time $1/\mu\xi$ into (14) and also recall $s = \bar{n}_{\mathrm{h}} - 1$, we get the mean buffer length and the mean transmission delay at a node:

$$\begin{cases} \bar{N} = \frac{2\bar{n}_{\mathrm{h}}\lambda\mu\xi - \bar{n}_{\mathrm{h}}^2\lambda^2}{2\mu\xi(\mu\xi - \bar{n}_{\mathrm{h}}\lambda)} \\ \bar{W} = \frac{2\mu\xi - \bar{n}_{\mathrm{h}}\lambda}{2\mu\xi(\mu\xi - \bar{n}_{\mathrm{h}}\lambda)} \end{cases}, \qquad \frac{\bar{n}_{\mathrm{h}}\lambda}{\mu\xi} < 1 . \tag{15}$$

Since the mean route length is $\bar{n}_{\mathrm{h}}$, the average transmission delay of all the packets in the network is:

$$D = \bar{n}_{\mathrm{h}}\bar{W} = \frac{2\mu\bar{n}_{\mathrm{h}}\xi - \bar{n}_{\mathrm{h}}^2\lambda}{2\mu\xi(\mu\xi - \bar{n}_{\mathrm{h}}\lambda)}, \qquad \lambda \in \left(0, \frac{\mu\xi}{\bar{n}_{\mathrm{h}}}\right) . \tag{16}$$

5 Capacity with Delay Constraint

Expression (16) is an important result of this paper. From (16) one can see that under ideal network state the average transmission delay of the network is determined by the arrival rate of the packets and the mean service time L/W at each node. On the other hand when the delay constraint is given the arrival rate of packets i.e. the throughput capacity is also determined. Thus we get the delay constraint capacity of the network:

$$T = \frac{2L\mu\xi(D\mu\xi - \bar{n}_\mathrm{h})}{(2D\mu\xi - n_\mathrm{h})\bar{n}_\mathrm{h}}, \qquad D \in \left(\frac{\bar{n}_\mathrm{h}}{\mu\xi}, \infty\right). \tag{17}$$

Given an adequate SINR β the value of Q_{th} is determined. And by choosing a proper value of d the transmission concurrency ξ is also determined and hence the value of throughput capacity can be acquired accordingly. Since W' is the length of the projection of W on the reference path, let $\bar{W}_{\mathrm{max}}$ denote the maximum hop length that corresponding to $\bar{W}'$, to assure the network is connected the condition $\bar{W}_{\mathrm{max}} \leq r$ should be satisfied. Since $\bar{W}'$ is a function of d, from (11) $\bar{W}_{\mathrm{max}}$ can be written as

$$\bar{W}_{\mathrm{max}} = \sqrt{4d^2 + E^2(W')} = \sqrt{4d^2 + 1/4d^2\rho_\mathrm{s}^2} \tag{18}$$

and it is plotted in Fig. 4. When $d = 1/2\sqrt{\rho_\mathrm{s}}$, $\bar{W}_{\mathrm{max}}$ acquire its minimum value $\min(\bar{W}_{\mathrm{max}}) = \sqrt{2/\rho_\mathrm{s}}$. Thus we get the minimum transmission range $r_{\mathrm{min}} = \sqrt{2/\rho_\mathrm{s}}$. Substituting it into (5) and (12) we have

$$\begin{cases} \xi = 1/\sqrt{3}Q_{\mathrm{th}}^2 \\ \bar{n}_\mathrm{h} = 2\sqrt{N}\Big/3\sqrt{\pi} \end{cases} . \tag{19}$$

And substituting (19) into (17) yields the throughput-delay tradeoff corresponding to r_{min}:

$$T = \frac{2L\sqrt{3\pi N}Q_{\mathrm{th}}^2\mu - 3D\pi\mu^2}{2NQ_{\mathrm{th}}^4 - 2D\sqrt{3\pi N}Q_{\mathrm{th}}^2\mu}, \qquad D > \frac{2\sqrt{N}Q_{\mathrm{th}}^2}{\sqrt{3\pi}\mu}. \tag{20}$$

The maximum throughput capacity T_{max} is the throughput that is feasible without delay constraint and it can be drawn from (20):

$$T_{\mathrm{max}} = \lim_{D\to\infty} T(D) = \frac{\sqrt{3\pi}\mu}{2Q_{\mathrm{th}}^2\sqrt{N}}. \tag{21}$$

This result agrees with that of [3].

A case study is given in the following to show numerical results of the relationship between the average transmission delay and the throughput capacity in an ad hoc wireless network. The results are obtained based on the assumptions that the total number of nodes is $N = 1000$, the radius of the network area is

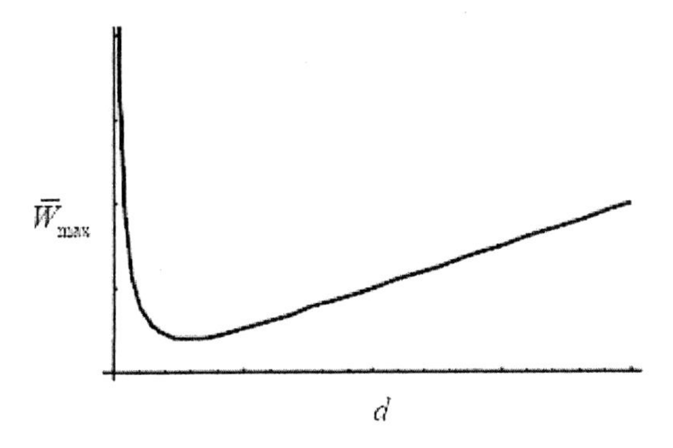

Fig. 4. The relationship between d and $\bar{W}$max

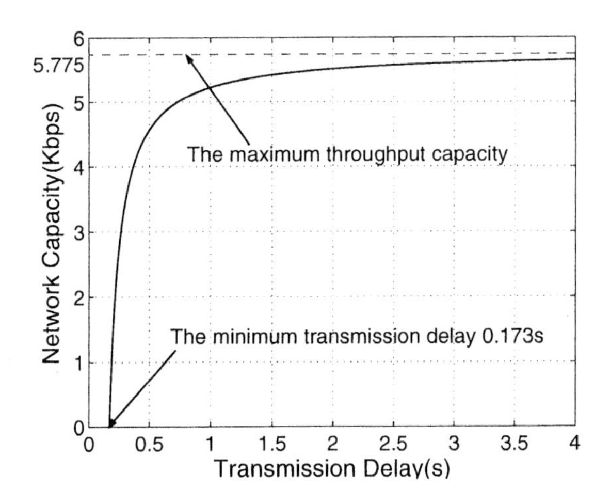

Fig. 5. Transmission delay and throughput capacity tradeoff

$R = 100$ m, the minimum SINR is $\beta = 15$ dB, the data rate is $W = 2$ Mb/s, and the packet length is $L = 1000$ bits. These Parameters are typical for a low power sensor network. Fig. 5 shows the maximum tradeoff between throughput capacity and average transmission delay.

6 Conclusion

The capacity and tradeoff between capacity and delay of wireless ad hoc networks have been widely studied and different results have been derived. However most of the former works investigated the asymptotic performance of large networks and concern delay caused by node mobility. The matter of transmission delay caused by queueing and multihop transmission is of paramount effect to the performance of network capacity for general networks. In this paper the relationship between network capacity and average transmission delay in an arbitrary scale network is investigated. And based on a constructive scheme, where a balanced scheduling strategy and a minimal length route selection mechanism are proposed, we managed to find deterministic analytical results. The relationship between the throughput capacity T and the average transmission delay D is $T = \frac{2L\mu\xi(D\mu\xi-\bar{n}_h)}{(2D\mu\xi-n_h)\bar{n}_h}$, the minimum transmission range that satisfies connectivity of the network is $r_{\min} = \sqrt{2/\rho_s}$ and the capacity-delay tradeoff with the minimum transmission range is $T = \frac{2L\sqrt{3\pi N}Q_{th}^2\mu-3D\pi\mu^2}{2NQ_{th}^4-2D\sqrt{3\pi N}Q_{th}^2\mu}$. The maximum throughput capacity without delay constraint is $T_{\max} = \frac{\sqrt{3\pi\mu}}{2Q_{th}^2\sqrt{N}}$ and it is agreed with that of [3].

Acknowledgment

The research was supported in part by a grant from NSF China and RGC HK under the contract 60218002, grants from RGC under the contract HKUST6204/03E and the contract HKUST6104/04E, a grant from NSF China under the contract 60429202, a NSFC/RGC joint grant under the contract N_HKUST605/02, and a grant from Microsoft Research under the contract MCCL02/03.EG01.

References

1. Ivan Stojmenovic, Handbook of Wireless Networks & Mobile Computing, John Wiley & Sons, Inc., 2002, 605 Third Avenue, New York, NY.
2. Mohammad llyas, The handbook of Ad hoc wireless networks, CRC press LLC, 2003
3. Piyush Gupta and P. R. Kumar, The Capacity of Wireless Networks, IEEE Transactions on Information Theory, vol. IT-46, no. 2, pp. 388-404, March 2000
4. M. T. Grossglauser and D. Tse, Mobility increases the capacity of ad hoc wireless networks, Networking, IEEE/ACM Transactions on, Vol. 10, Iss. 4, Page(s): 477 -486, 2002.
5. Piyush Gupta and P. R. Kumar, Critical Power for Asymptotic Connectivity in Wireless Networks, pp. 547-566, in Stochastic Analysis, Control, Optimization and Applications: A Volume in Honor of W.H. Fleming. Edited by W.M. McEneany, G. Yin, and Q. Zhang, Birkhauser, Boston, 1998. ISBN 0-8176-4078-9.
6. Feng Xue and P. R. Kumar, The number of neighbors needed for connectivity of wireless networks, Wireless Networks, pp. 169–181, vol.10, no. 2, March 2004.
7. G. Ferrari and O.K. Tonguz, Minimum Number of Neighbors for Fully Connected Uniform Ad Hoc Wireless Networks, IEEE International Conference on Communications (ICC'04), accepted, Paris, France, May 2004.
8. S. Panichpapiboon, G. Ferrari, and O.K. Tonguz, Impact of Interference in Uniform and Random Ad Hoc Wireless Networks, IEEE Vehicular Technology Conference (VTC 2004), Milan, Italy, accepted.
9. S. Toumpis and A.J. Goldsmith, Capacity regions for wireless ad hoc networks, IEEE Trans. on Wireless Communications., Vol. 2, No. 4, pp. 736-748, July 2003.
10. M. V. Gastpar, M, On the capacity of wireless networks: the relay case, presented at INFOCOM 2002. Twenty-First Annual Joint Conference of the IEEE Computer and Communications Societies. Proceedings.IEEE, 2002.

11. S. Toumpis and A. J. Goldsmith, Ad Hoc Network Capacity, Asilomar Conference on Signals, Systems and Computers, 2000.
12. N. Bansal and Z. Liu. Capacity, delay and mobility in wireless ad-hoc networks. IEEE Proceedings of INFOCOM, April 2003.
13. M. Grossglauser and M. Vetterli. Locating nodes with ease: Last encounter routing in ad hoc networks through mobility diffusion. IEEE Proceedings of INFOCOM, April 2003.
14. E. Perevalov and R. Blum. Delay limited capacity of ad hoc networks: Asymptotically optimal transmission and relaying strategy. IEEE Proceedings of INFOCOM, April 2003.
15. A. Tsirigos and Z. J. Haas. Multipath routing in the presence of frequent topological changes. IEEE Communications Magazine, Nov. 2001.
16. M. J. Neely and E. Modiano. Improving delay in ad-hoc mobile networks via redundant packet transfers. Proceedings of the Conference on Information Sciences and Systems, Johns Hopkins University: March 2003.
17. Michael Neely and Eytan Modiano, Capacity and Delay Tradeoffs in Mobile Ad Hoc Networks, (Invited Paper) IEEE BroadNets 2004, San Jose, CA, October, 2004.
18. A. El Gamal, J. Mammen and B. Prabhakar and Devavrat Shah, Throughput Delay Trade-off in Wireless Networks, presented in Proceedings of IEEE INFOCOM, Hong Kong, March 2004.
19. Sharma, G. and Mazumdar, R.; On Achievable Delay/Capacity Trade-offs in Mobile Ad Hoc Networks, WiOpt 2004, Conference on Wireless Networks and Optimization, Cambridge, UK, March 2004.
20. H. Wu, S. De, C. Qiao, E. Yanmaz, O.K. Tonguz, Queuing Delay Performance of the Integrated Cellular and Ad hoc Relaying System, in Proc. of the IEEE International Conference on Communications(ICC'03), vol. 2, pp. 923-927, Anchorage, Alaska, May 2003.
21. S. Toumpis and A. J. Goldsmith, Large wireless networks under fading, mobility, and delay constraints, in Proc. IEEE INFOCOM, Hong Kong, China, Mar. 2004.
22. G. Sharma and R. Mazumbdar, Scaling laws for capacity and delay in wireless ad hoc networks, in Proc. IEEE ICC, Paris, France, June 2004.
23. M. Bhatt, R. Chokshi, S. Desai, S. Panichpapiboon, N. Wisitpongphan, and O.K. Tonguz, Impact of Mobility on the Performance of Ad Hoc Wireless Networks, IEEE Vehicular Technology Conference (VTC 2003), Orlando, FL, October 2003.
24. S. Toumpis, Capacity bounds for three types of wireless networks: Asymmetric, cluster and hybrid," in Proc. ACM MobiHoc, Roppongi, Japan, May 2004.
25. T. S. Rappaport, Wireless Communications. Principles & Practice. Upper Saddle River, NJ, U.S.A.: Prentice-Hall, 2002, second edition.
26. Leonard Kleinrock, Queueing Systems, volume I: Theory, 1st ed. New York: John Wiley & Sons, Inc., Jan. 1975.
27. Yinghui Tang and Xiaowo Tang, Queueing theory—fundamental and applications, University of Science and technology of China Press, 2000.5 (in Chinese).

Load-Based Dynamic Backoff Algorithm for QoS Support in Wireless Ad Hoc Networks

Chang-Keun Seo[1], Weidong Wang[1], and Sang-Jo Yoo[1]

[1] Graduate School of Information Technology & Telecommunications, Inha University, 253 YongHyun-dong, Nam-ku, Incheon 402-751, Korea
{seochangkeun, wangwd}@gmail.com, sjyoo@inha.ac.kr
http://multinet.inha.ac.kr

Abstract. The stations that use a shared medium like IEEE 802.11 wireless LAN have transmission opportunities by contention in the contention period. If there are collisions in the contention period, a medium access control protocol may solve collisions by using a backoff algorithm. The backoff algorithm is an important part of the medium access control protocol, but a legacy backoff method used in IEEE 802.11 standard is not efficiently adjusted depending on the load condition and priorities. In this paper, we propose a new load-based dynamic backoff algorithm to improve throughput of medium and to reduce the number of collisions. The proposed backoff algorithm can increase network utilization about 20% than that of the binary exponential backoff algorithm[1].

1 Introduction

IEEE 802.11 Wireless LAN standard [1] has been the most successful standard in the internet-based wireless device market. IEEE 802.11 uses CSMA/CA (Carrier Sense Multiple Access with Collision Avoidance) as a medium access method. IEEE 802.11 has two modes: DCF (Distributed Coordination Function) which has opportunities to transmit by the contention of each station and PCF (Point Coordination Function) which is controlled by a base station. IEEE 802.11 can also use RTS (Request To Send)/CTS (Clear To Send) to resolve a hidden node problem and exposed node problem. Recently, with the growth of wireless market, there are many demands for QoS (Quality of Service) support. So especially, IEEE suggests IEEE 802.11e [2] for QoS support that extends the legacy medium access control layer of IEEE 802.11. IEEE 802.11e divides each priority frame by interframe space, and the differentiated size of the contention window during a backoff procedure when a collision occurs. But in the contention-based period, a BEB (Binary Exponential Backoff) algorithm which is used in IEEE 802.11 and IEEE 802.11e has a fairness problem, is suffered from more collisions and dramatically decreased throughput with the increasing load, because the BEB algorithm declares same size of the contention window when a collision occurs.

In this paper, we propose a new method of the backoff procedure that is named a load-based dynamic backoff algorithm. Proposed algorithm uses dynamic contention

[1] This research was supported by University IT Research Center Project (INHA UWB-ITRC), Korea.

X. Jia, J. Wu, and Y. He (Eds.): MSN 2005, LNCS 3794, pp. 466 – 477, 2005.

windows according to expected collision rate as a long-term value and slot utilization as a short-term value. By using LDB algorithm, we can increase throughput of the medium.

The rest of this paper is organized as follow: Section 2 introduces IEEE 802.11e MAC and its backoff procedure. After that, we propose our new backoff algorithm, LDB in Section 3. In Section 4, we give the simulation results. Finally in Section 5, we conclude this paper.

2 IEEE 802.11e Structure for QoS Support

IEEE 802.11e provides four AC (Access Category) to support QoS by each UP (User Priority) [3-4]. Each AC has different parameters per priority, AIFS (Arbitration Interframe Space) has different access rights, minimum contention window CW_{min}[AC]s, maximum contention window CW_{max}[AC]s and TXOP[AC]s (Transmission Opportunity) to support a different transmission interval. In EDCA (Enhanced Distributed Channel Access) mode, each station has a transmission opportunity by the contention. In HCCA (HCF Controlled Channel Access) mode, HC (Hybrid Coordinator) that is included in QBSS (QoS Basic Service Set) uses polling to control frame transmission orders. From 8 UPs of IEEE 802.1D to 4 multiple queue is shown in Figure 1.

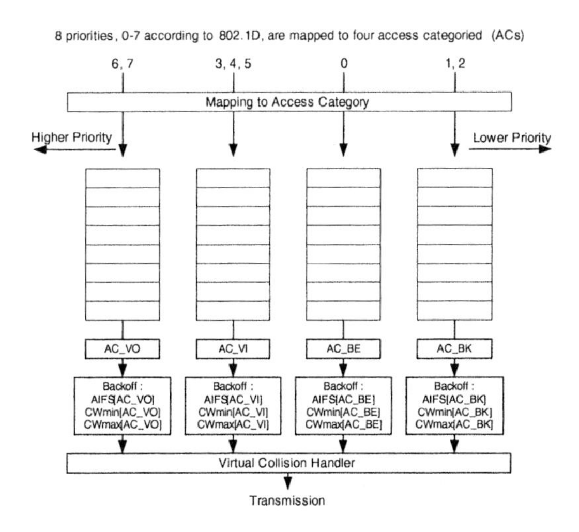

Fig. 1. Multiple queue of station in IEEE 802.11e

When stations sense the medium in idle state, each station selects time slot randomly within their set size of the contention window after AIFS[TC] which is access delay time for each priority. After reduce backoff counter, if the counter becomes zero, station can transmit frame. But in this process, when one station transmits frame because their time slot becomes zero, the others must stop their process until transmission is ended. Expected time is offered by NAV (Network Allocation Vector)

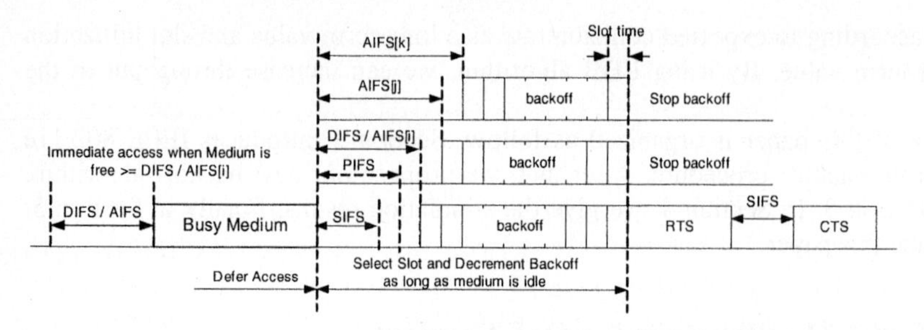

Fig. 2. EDCA mode of IEEE 802.11e

factor which is included in header. If the number of backoff counter becomes zero, station can transmit frame (data frame or RTS frame). But if some stations become zero at the same time, there is collision. In IEEE 802.11e, virtual scheduler manages queues to prevent collision in a station because IEEE 802.11e has four queues. Figure 2 shows access right of each priority queue in IEEE 802.11e.

Although IEEE 802.11e can support differentiated service in contention period by modified legacy method, its performance is dramatically decreased when the number of stations increases or the traffic load increases in QBSS. To solve this problem, HCCA can be used to improve the utilization of medium by using a polling method by HC as a central controller is introduced. But this mode can be only used in the fixed network topology, it can restrict mobility of stations. As a result, HCCA mode is not acceptable in ad-hoc networks that a network topology is variable.

3 Backoff Methods for QoS Support

There are many researches about backoff methods used in the contention medium [5-6]. A backoff method defines how to treat the contention window when there is a successful or unsuccessful transmission. Especially stations in ad-hoc networks (especially using the contention-based shared medium) must communicate by themselves without any central controller and finding an efficient backoff method to increase throughput is very important.

The backoff method in EDCA uses a binary exponential backoff (BEB) algorithm as follows.

1) When the transmission of a frame is successful,

$$CW_{new}[AC]=CW_{\min}[AC] \quad (1)$$

2) When the transmission of a frame is failed,

$$CW_{new}[AC]=\min(CW_{\max}[AC],\ (CW_{old}[AC]+1)\times 2-1) \quad (2)$$

The BEB algorithm is selected by IEEE 802.11 standard because its definition is simple, but its performance is dramatically decreased when load is increased. Also BEB has a ‘fairness problem’ that there is a special station that can transmit frames with high probability than other stations. It is happened because the station successfully transmitting a frame sets CW as CW_{min}. So the other stations with bigger contention window have low probability of the transmission. To overcome these problems, there are many researches, their scopes are defining functions how to set contention windows effectively.

AEDCF (Adaptive Enhanced Distributed Coordination Function) [7] introduces an average collision rate as a factor to control the contention window. AEDCF reduces the collision window by multiplying MF[AC] (Multiplicator Factor) when the transmission is success. If the transmission of a frame is failed, the station increases the contention window by multiplying PF[AC] (Persistence Factor). The advantage of AEDCF is that it can control the contention window according to the state of network by using the average collision rate. It can adjust the window size increasing and decreasing parameters depending on the collision status. In AEDCF, the average collision rate at period *j* is updated as in (3).

$$f_{avg}^{j} = (1-\alpha)\times f_{curr}^{j} + \alpha\times f_{avg}^{j-1} \ . \tag{3}$$

Where, f^{i}_{avg} is the average collision rate up to *j*th period and f^{i}_{curr} is the measured the collision rate at *j*th period. After calculating the f^{i}_{avg} that becomes an index of the state of networks, MF[AC] is computed to determine the next size of the contention window by using f^{i}_{avg}. The final decision of the contention window is shown in (5) and (6).

$$MF[AC] = \min((1+(level\times 2)\times f_{avg}^{j}, 0.8) \ . \tag{4}$$

The *level* is the weight per priority.

1) When the transmission of a frame is successful,

$$CW_{new}[AC] = \max(CW_{\min}[AC], CW_{old}[AC]\times MF[AC]) \ . \tag{5}$$

2) When the transmission of a frame is failed,

$$CW_{new}[AC] = \min(CW_{\max}[AC], CW_{old}[AC]\times PF[AC]) \ . \tag{6}$$

4 Proposed Load-Based Dynamic Backoff Algorithm

In this section, we propose a new backoff mechanism named LDB (Load-based Dynamic Backoff algorithm) to improve the throughput of medium and reduce collisions. LDB uses two factors to determine the variable contention window. Basic concept is from that we can find easily a long-term and short-term factor as an index of status of the load in the communication procedure. LDB uses a collision rate as a long-term factor and slot utilization as a short-term factor to dynamically determine the contention window. We have defined a new function with the short-term and long-term factors.

4.1 Long-Term Decision Factor (LDF)

First, we define LDF (Long-term Decision Factor) as an indication of the collision status of networks. A collision rate is calculated by dividing the collisions with the number of transmission attempts at each station p during time period j [7]. When a station has frames to transmit in its queue, RTS frame is sent first, ACK frame is replied by the receiver when the transmission is successful. The number of attempts is simply calculated by the number of RTS frames.

$$f_j = \frac{E(collisions_j[p])}{E(frame\_sent_j[p])} . \tag{7}$$

f_j of (7) that is currently observed a collision rate is used to decide the contention window for the next time period. In this paper, we propose a new method to predict a collision rate for the next time period. We use a LMS (Least Mean Square) predictor to determine the next contention window, we define the predicted collision rate as the LDF factor. Fig. 3 shows the concept of LMS predictor [8].

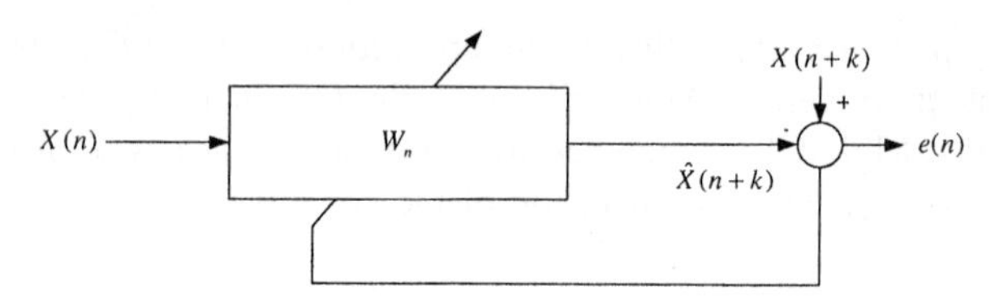

Fig. 3. LMS prediction

The LMS prediction scheme is a useful method for real-time prediction. The concept of LMS scheme is illustrated in Fig. 3 and equation 8 and 9. The LMS predictor is derived from error *e(n)* which is the gap between the predicted value and real value, a time varying coefficient vector W_n which minimizes the mean square error, matrix *F(n)* with the number of real collisions p. the step size $\mu$is fixed during the entire prediction. If $0 < \mu < 2$, then the LMS will converge on the mean.

Next predicted collision rate (case of k=1) is calculated with p numbers of previous real collision rate as in (8). LMS predictor can be used for the medium access control because it can dynamically predict the network conditions in long-term time scale.

$$\hat{f}(n+k) = \sum_{l=0}^{p-1} w_n(l) f(n-l) = W_n^T F(n).$$

$$\begin{cases} W_n = [w_n(0), w_n(1), \ldots, w_n(p-1)]^T . \\ W_{n+1} = W_n + \mu \times e(n) \dfrac{F(n)}{\|F(n)\|^2} . \\ e(n) = f(n+k) - \hat{f}(n+k) . \\ F(n) = [f(n), f(n-1), \ldots, f(n-p+1)]^T . \end{cases} \tag{8}$$

$$LDF_{j+1} = \hat{f}_{j+1} = \sum_{l=0}^{p-1} W_n(l) f(j-l) \cdot \tag{9}$$

4.2 Short-Term Decision Factor (SDF)

The second factor for the contention window deciding function is slot utilization. Slot utilization can be expressed as a ratio of used time slots and total time slots. In this paper, we use slot utilization as a SDF (Short-term Decision Factor). It is apparent that the utilization of time slot indicates whether the medium is busy or not. Stations must execute backoff procedure when stations have frames to transmit in its queues. Fig. 4 shows use of the time slot related to the backoff procedure. If a backoff counter of a station reaches zero, then it transmits a frame first and other stations must stop their backoff counter. Each station calculates slot utilization, the number of stopped backoff cause by busy medium and divided by the total number of backoff time slots in a certain period *i*. When we calculate slot utilization, we don't consider whether there are collisions or not but only sense the time slots used or not. Figure 4 explains how to calculate slot utilization as SDF in LDB.

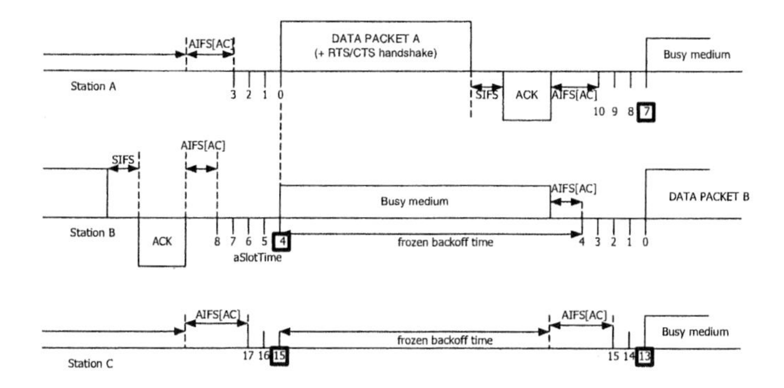

Fig. 4. Behavior of backoff procedure

SDF is updated at every period *i*. For period *i*, slot utilization is calculated as (10).

$$SDF_i = U^i_{slot} = \frac{\sum \text{Used time slots by other station}}{\sum \text{Backoff time slots for transmitting frames}} . \tag{10}$$

Generally the time period for SDF is much shorter than that of LDF. The range of LDF and SDF is [0, 1]. These SDF and LDF values are used to determine a medium decision coefficient in section 3.3.

4.3 Medium Decision Coefficient (D_{medium}[AC])

After calculating LDF and SDF with the expected collision rate and slot utilization, D_{medium}[AC] is derived to decide the value of contention window. D_{medium}[AC] is defined as follow.

$$D_{medium} = (1-\gamma)\times LDF + \gamma\times SDF . \tag{11}$$

$$D_{medium}[AC] = (D_{medium})^{\frac{3+AC\_level}{2}} . \tag{12}$$

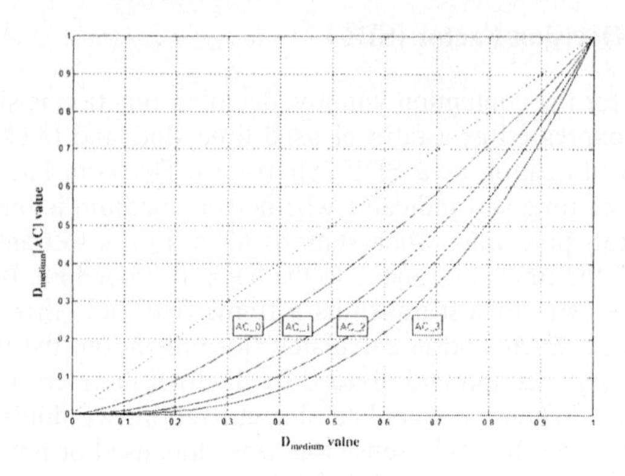

Fig. 5. D_{medium} mapping curve per priorities

AC_level is a priority value. The highest priority is 0, and the lowest priority is 4. γ is for balance with LDF and SDF. Fig. 5 is shows a mapping curve of $D_{medium}[AC]$ in accordance with each AC value.

4.4 Contention Window Decision Method

In the proposed method, the value of contention window is determined as follow with $D_{medium}[AC]$.

1) If the transmission of a frame is successful,
The new contention window for the next time period is determined as (13).

$$CW_{new}[AC] = CW_{old}[AC] \times \min\{(Weight + 2 \times AC\_level) \times (D_{medium}[AC]),\ 1\}\ . \tag{13}$$

The *weight* is used to select the efficient contention window, we set 3 to *weight*. The reason that $D_{medium}[AC]$ is limited with value 1 when the transmission of a frame is successful is that the new contention window should be not bigger than the size of current contention window.

2) If the transmission of a frame is failed,
When the transmission of a frame is failed, we use a new factor named scale vector V[AC] and finally calculate $V_{scale}[AC]$. $V_{scale}[AC]$ is limited range of [0.9, 1.1]. When the transmission is failed, the former and the current $D_{medium}[AC]$ will affect on the next size of contention window.

$$V[AC] = \frac{D^{i}_{medium}[AC]}{D^{i-1}_{medium}[AC]}\ . \tag{14}$$

$$V_{scale}[AC] = \begin{cases} 0.9, & V[AC] < 0.9 \\ V[AC], & 0.9 \le V[AC] \le 1.1 \\ 1.1, & V[AC] > 1.1 \end{cases}\ . \tag{15}$$

The new contention window for the next time period is determined as (16) with the V_{scale}[AC] and PF[AC].

$$CW_{new}[AC] = \min(CW_{\max}[AC],\ CW_{old}[AC] \times PF[AC] \times V_{scale}[AC]) \ . \tag{16}$$

5 Simulation Results

In this paper, we used ns2 [9-10] in order to verify the performance of the proposed LDB. EDCA, AEDCF were implemented for the performance comparison with the proposed LDB algorithm. Table 1 and 2 show the parameters used in this simulation. For physical layer, IEEE 802.11b was used [11].

Each host has three types of traffics with different priorities. The host which has a data to send tries to transmit the data to other nodes. The number of sending hosts is increased from two to forty. This paper supposes that all hosts are not moved and are located in the communication range and hosts transmit the data after exchanging RTS/CTS frames. Table 3 remarks the characters of priority in experiment. IEEE 802.11e defines four ACs as AC_VO (voice), AC_VI (video), AC_BE (best effort), and AC_BK (background). In this experiment, we use three ACs as AC_VO, AC_VI, AC_BE. The size of data frames of each priority is same so that we can compare the characteristic of processing in the variable priority data. The rate of transmitting data is 480Kbps. As we increase the number of hosts from 2 to 40, the total transmitting data rate changes from 8% to 174% of the link speed.

Table 1. Environment of simulation

Channel	WirelessChannel
Propagation type	TwoRayGround
Interface queue	PriQ(priority queue)
Antenna model	OmniAntenna
Interface Queue length	50

Table 2. Parameters of IEEE 802.11b PHY in ns2

SIFS	10 (us)
RTS	48 (bytes)
CTS	44 (bytes)
ACK	44 (bytes)
Data rate	11 (Mbps)
aSlotTime	20 (us)
CCA time	15 (us)
MAC header	30 (bytes)
Preamble length	144 (us)
RxTxTurnaround time	5 (us)
PLCP header length	48 (bits)

Table 3. Simulation parameters

Parameter	AC_VO	AC_VI	AC_BE
Level(Priority)	0	1	2
CWmin	7	15	31
CWmax	255	511	1023
AIFSN	2	3	4
PF	2	3	4
Packet size(bytes)	300	300	300
Packet interval(ms)	10	10	10
Sending rate(Kbit/s)	160	160	160
AIFS[AC] (us) = SIFS + AIFSN × aSlotTime			

As the first experiment, we try to find suitable values of LDF, SDF and γ. By comparing 10 hosts communicate simultaneous with different parameter sets. We can see in Figure 6, the number of successful transmission is constant or decreasing when the value of γ is larger than 0.6. So we select 15,000 time slots for LDF, 3000 time slots for SDF, and 0.6 for γ.

For the second experiment, we evaluate the performance of proposed LMS predictor that is used to predict the long-term collision frequencies in this paper. In this experiment, the value of μ of the LMS predictor is set by 0.05. The difference of the actual collisions and the predicted collisions are defined by error. Table 4 shows the mean square errors of the EWMA (AEDCF) and the LMS (LDB). We can conclude that LDB can predict the actual collisions the more approximately than AEDCF.

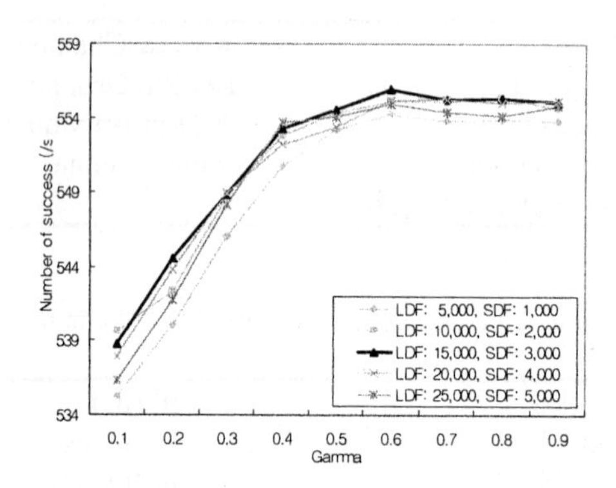

Fig. 6. Decision of LDF, SDF, and γ value

Table 4. LMS performance

Parameter	Mean square error
EWMA(AEDCF)	0.0033
LMS(LDB)	0.0026

For the third experiment, we evaluate the performance of the proposed LDB when the number of hosts is increased. Figure 7 shows the number of successfully transmitted frames per second with increasing the number of stations (the total load of network is increased). The proposed LDB provides differentiated services depending on the each class priority.

The Figure 8 shows the average number of total successful transmissions per second with increasing the number of stations. The proposed LDB method can archive the highest throughput, and others are sorted by AEDCF, EDCA in order. With this experiment, we know that LDB increases the utilization of medium because LDB by increasing the number of hosts according to the status of networks. We can see that the number of successful transmissions in EDCA is decreased by increasing the number of hosts because of suddenly increased collisions.

The last simulation is to compare the number of collisions in the medium. It is very important performance parameter in the wireless LAN because the collisions affect the efficiency of the medium in terms of throughput, the number of retransmissions, and delay. Figure 9 shows the number of collisions per second with increasing the number of hosts. LDB shows the smallest collisions.

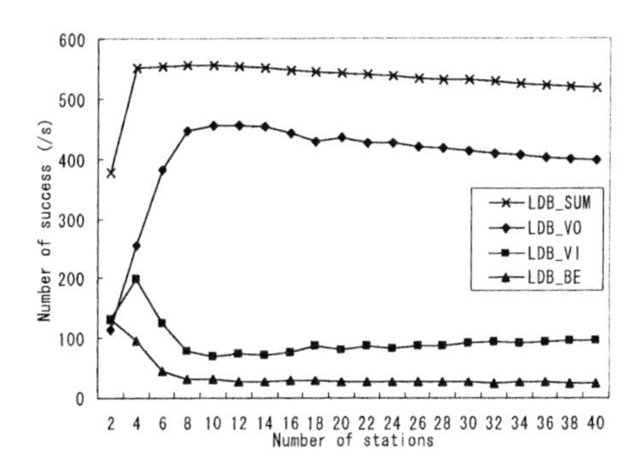

Fig. 7. Performance of LDB each priorities

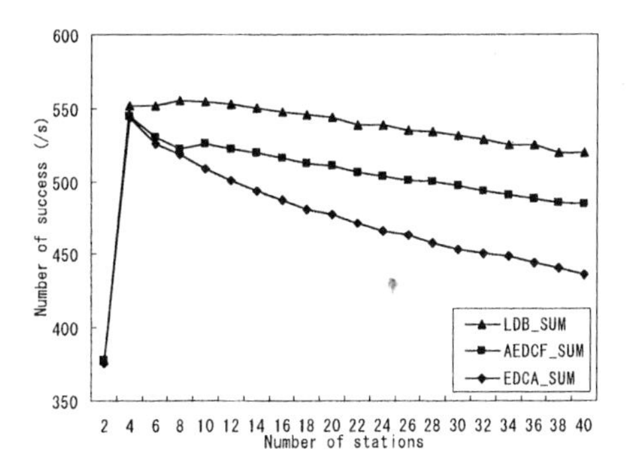

Fig. 8. The average number of total successful transmission frames per second

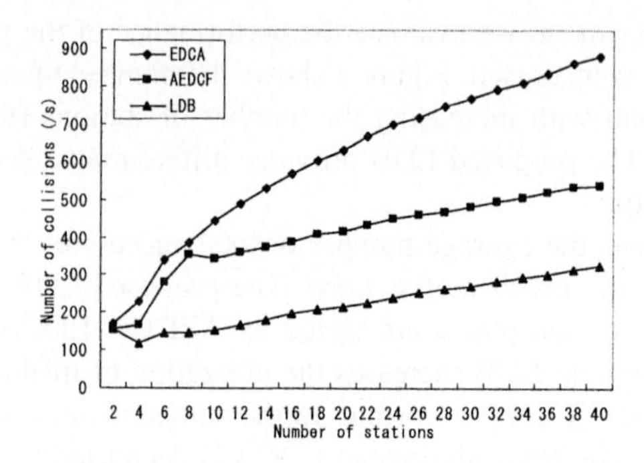

Fig. 9. The number of collisions per second

6 Conclusion

This paper proposes a new load-based dynamic backoff algorithm and it can increase the throughput of medium in the contention-based shared medium. Proposed LDB uses dynamic contention windows changed by the expected collision rate as a long-term factor and slot utilization as a short-term factor. These factors can be computed easily. Proposed method doesn't need to change the structure of the existing wireless LAN standard. It only changes the backoff method to improve the performance of medium access control. Simulation results show that the proposed LDB method can provide differentiated QoS per each class and increase the network utilization by effectively decreasing collisions.

References

1. IEEE Std 802.11-1997, Information technology – Telecommunications and information exchange between systems – Local and metropolitan area networks – Specific requirements – Part 11 : Wireless LAN Medium Access Control (MAC) and Physical Layer (PHY) specifications.
2. IEEE 802.11 WG, IEEE 802.11e/D6.0: Draft Amendment to Standard for Information Technology – Telecommunications and Information Exchange Between Systems – LAN/MAN Specific Requirements – Part 11 : Wireless Medium Access Control (MAC) and Physical Layer (PHY) Specifications : Medium Access Control (MAC) Quality of Service (QoS) Enhancements (2003)
3. Choi, S.H.: Overview of emerging IEEE 802.11 protocols for MAC and above, Telecommunications Review, Special Edition (2003) 102-127
4. Xiao, Y.: IEEE 802.11e: QoS provisioning at the MAC layer, IEEE Transactions on Wireless Communications, Vol.11, No. 3 (2004) 72-79
5. Kang, S.S., Mutka, M.W.: Provisioning service differentiation in ad-hoc networks by modification of the backoff algorithm, Proc. of IEEE Computer Communications and Networks (2001) 577-580
6. Haas, Z.J., Deng, J.: On optimizing the backoff interval for random access schemes, IEEE Transactions on Communications, Vol. 51, No. 12 (2003) 2081-2090

7. Romdhani, L., Ni, Q., Turletti, T.: Adaptive EDCF: enhanced service differentiation for IEEE 802.11 wireless ad-hoc networks", Proc. of IEEE Wireless and Communications and Networking Conference (2003) 1373-1378
8. Yoo, S.J.: Efficient traffic prediction scheme for real-time VBR MPEG video transmission over high-speed networks, IEEE Transactions on Broadcasting, Vol. 48, No. 1 (2002) 10-18
9. The Network Simulator – ns2, http://www.isi.edu/nsnam/ns/
10. TKN: Design and Verification of an IEEE 802.11e EDCF simulation model in ns-2.26, Berlin (2003)
11. Mangold, S., Choi, S.H., Hiertz, G.R., Klein, O., Walke, B.: Analysis of IEEE 802.11e for QoS support in wireless LANs, IEEE Wireless Communications, Vol. 10, No. 6 (2003) 40-50

Efficient Multiplexing Protocol for Low Bit Rate Multi-point Video Conferencing*

Haohuan Fu, Xiaowen Li, Ji Shen, and Weijia Jia

Department of Computer Science, City University of Hong Kong,
83 Tat Chee Ave., Kowloon, Hong Kong
wjia@cs.cityu.edu.hk

Abstract. This paper discusses an efficient implementation of the multiplexing protocol H.223, which is an important part of 3G-324M protocol stack required for 3G mobile multimedia communications. Our implementation of the protocol aims to support the multi-point video conferencing with the capability of transmitting/receiving multiple video/audio streams simultaneously. Conference managements such as admission and audio channel scheduling are also discussed. As a result, the implementation improves efficiency and makes the conference more convenient to set up and operate. Our prototype system is stable and its performance is satisfactory.

1 Introduction

Currently, multimedia streams of 3G video calls is transferred using 3G-324M protocol [1] derived from ITU-T H.324 [2] standard by International Telecommunications Union (ITU) to enable multimedia communication over low-bit rate terminals (in the following, "ITU-T" is dropped for simplicity). H.324 is an umbrella protocol, referencing other important standards such as H.223 [4] that specifying data multiplexing/demultiplexing, and H.245 [3] that specifying the control messages and procedures. H.324 and its several mobile specific annexes are usually referred to as H.324M (M stands for *mobile*). The 3rd Generation Partnership Project (3GPP) [12] has adopted the H.324M with some modifications in codec and error handling requirements to create 3G-324M standard for 3G wireless networks [14].

This paper describes our prototype implementation called *anyConference*, which is a prototype multi-point video conferencing system over the mobile terminals, based on our 3G-324M protocol stack implementation. We mainly focus on the enhanced implementations of the multiplexing protocol H.223, which is designed to be capable to coordinate the sending and receiving of multiple media streams from different participating entities so as to support a multi-point conference rather than a simple video call.

The rest of the paper is organized as follows. A brief introduction of the 3G concerned standards and concepts used are given in section 2. In section 3, we describe our prototype video conference system first, and then we illustrate our

* This work is supported by CityU strategic grant nos: 70001587, 70001709 and 70001777.

X. Jia, J. Wu, and Y. He (Eds.): MSN 2005, LNCS 3794, pp. 478–487, 2005.

efficient implementation of the H.223 protocol with support for multi-point conversation; conference management such as admission control and audio channel scheduling will also be discussed. Section 4 gives evaluation of our implementation in the prototype system. We conclude the paper and give the future research direction in Section 5.

2 Background

2.1 Overview of 3G-324M/H.324M Standard

The architecture of 3G-324 protocol stacks is shown in Fig. 1 and the components protocols are illustrated as follows:

(1) H.324 — the Base Protocol, which consists of main protocol components.

(2) H.245 — specifies the Call Control Protocol which provides the end-to-end signaling for proper operations of the H.324 terminal. These signaling operations include audio and video capabilities exchange, opening and closing of logical channels, exchange of the multiplex tables of each party, and etc.

(3) H.223 —provides a multiplexing/demultiplexing service for the upper-layer applications. More details will be described in Section 2.1.

(4) Standards for video/audio codecs, such as H.263 [8], H.261 [9], G.723.1 [10], AMR (Adaptive Multi Rate) [13], and etc.

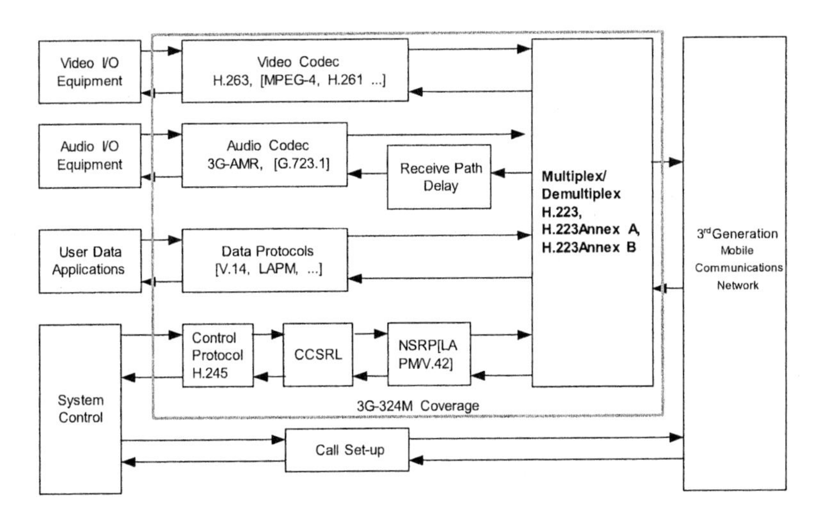

Fig. 1. Architecture of the 3G-324M protocol stack

2.2 Multiplex Protocol H.223

H.223 is usually used between two multimedia terminals, or between a terminal and a gateway adapter. As the interface in the middle of the above application layer (video/audio codecs and system control) and the below physical layer (WCDMA or other 3G physical link), H.223 provides low delay/overhead by using segmentation, reassembly, and multiplexing information from different logical channels into one single packet. The entire function of this recommendation is divided into two layers:

(1) The first layer is Adaptation Layer (AL), which is mainly responsible for error detection/correction and optional retransmission for lost or corrupt packets. It is actually an interface for upper-layer applications, and it still deals with different source separately. This layer could be further divided into three more specific sub-layers (1) AL1 for control information and data; (2) AL2 for audio stream; (3) AL3 for video stream.

(2) The second layer is Multiplex Layer (MUX), which performs the actual multiplexing. In this layer, data traffics from different sources are regarded as streams from different logical channels, which are identified by a unique Logical Channel Number (LCN), in the range from 0 to 65535. LCN 0 shall be permanently assigned to H.245 control channel. Other channels can be assigned to other streams such as video/audio. Different logical channels would be multiplexed into one packet according to some rules which are negotiated by two peers at the beginning of the communication. These rules are described in the forms of multiplex table. A multiplex table has maximum 16 different table entries. And each of them would define a specific multiplexing pattern. For each packet, the terminal will choose one of these patterns to do the multiplexing.

2.3 Multi-point Conference for 3G

Based on the existing point-to-point 3G video calls, multi-point conference becomes increasingly important as it permits a large number of mobile terminals to concurrently participate in a meeting. Multi-point conferences involve three or more terminals each time, and call control and media handling in multipoint conferences are more complicated than that in point-to-point calls.

Multipoint Control Unit (MCU) [11] supports multi-conferencing between three or more terminals and gateways. There are generally two types of multi-point conferences controlled by MCU as shown in Fig. 2 below:

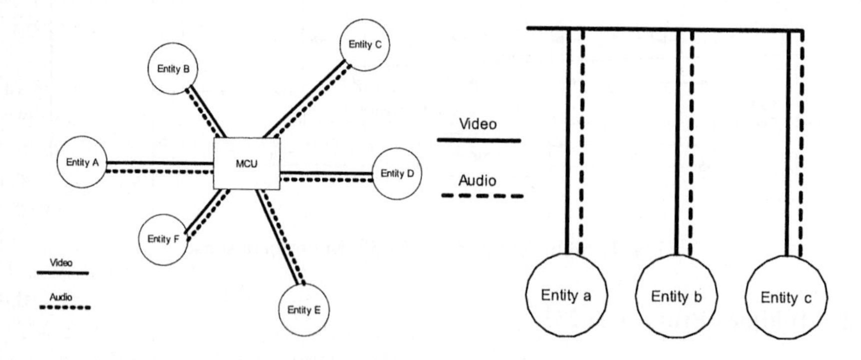

Fig. 2. Centralized and decentralized multi-point conferencing

(1) Centralized multipoint conference: All participating terminals are point-to-point connected to a MCU. MCU is regarded as the centre of the conference. Its functions include audio mixing, data distribution and video switching/mixing and sending back the resulting media streams to the participating parties. In this model, the endpoints that participate in the conference are not required as powerful as in the decentralized model. Each endpoint only has to encode its locally produced media streams and decode the set sent by the MCU. The MCU's capability to provide custom-mixed media streams allow otherwise constrained endpoints to participate in conferences.

(2) Decentralized multipoint conference: The MCU is not involved in this operation and the terminals communicate directly with each other. If necessary, the terminals assume the responsibility for summing the received audio streams and selecting the received video signals for display.

Thus, differing from the participants in point-to-point video calls, the MCU in centralized conferences and terminals in decentralized conferences are required to have the capability of handling multiple media streams from multiple end points and the original H.223 protocol needs to be enhanced for such multi-point situations.

3 Design and Implementation of Enhanced H.223

In this section, we will firstly give an introduction to our prototype video conference system, which includes a full implementation of 3G-324M protocol stack. Then we will focus on the H.223 part of 3G-324M. Enhanced capabilities to this recommendation, such as concurrent handling of multiple media streams and conference management, will be discussed.

3.1 Prototype Video Conference System

A prototype video conference system called *anyConference* has been implemented to test our 3G-324M protocol stack [5]. The system includes four general modules and each module is designed to be supported by other modules through various functional calls:

(1) H.324 Module -- This is a main module of the system which is responsible for cooperating with its umbrella modules such as H.245 Control Module and H.223 Multiplexing Module, to perform necessary functions such as call set-up etc.

(2) Video Capture/Display Module -- This module is designed for encoding and displaying the video captured from the local camera. For video codec, an open implementation of H.263 is used to encode video frames collected from the camera and decode the video frames from remote terminal. For video display, basic Windows Multimedia Software Development Kit (SDK) such as 'DrawDibDraw' is applied to handle the video as the sequence of Bitmap images.

(3) Audio Recorder/Player Module -- The Recorder module is responsible for capturing the audio data from the microphone devices. These audio data is then passed to the corresponding modules by Call-back functions once the buffer is full. The Player module is responsible for playback of the audio data received from the remote terminal.

(4)Socket Interface -- This is a network processing module provides a channel for the system to interact with air interface.

3.2 Handling of Multiplex Table

According to H.223 standard, there should be four different kinds of streams from the upper layer: control information, data information, audio and video streams. Every information stream is identified as a logical channel with a unique LCN. And the multiplex table entry specifying a LCN and the corresponding data length can describe how a data packet is multiplexed.

The data structure of multiplex table entry, which is called multiplex descriptor, takes the form of an element list. Each element in the list represents a slot of data from a specific information source. A typical example of element list with two elements is shown below

{LCN1, RC 24}, {LCN2, RC UCF}

where LCN is the logical channel number; RC is the repeat count and UCF is until closing flag. In the first element, RC 24 means that the first 24 bytes of the packet will be filled with data from logical channel 1. In the second element, RC UCF means that after the 24 bytes from logical channel 1, bytes from channel 2 will be filled in the packet until the closing flag (end of the packet). More specific description about the multiplex table entry could be found from [3] and [4].

For the simple multiplex patterns, this kind of multiplex descriptor is easy to handle. However, when the descriptor becomes complex, it will not be so easy to manipulate. The problem is that the multiplex descriptor uses a nested form for complicated multiplex patterns. Each element in the element list can be extended to a sub-element-list, which also contains other elements. Thus, when the structure is complicated with many sub-lists in the descriptor, the processing overhead may be high as recursive function calls will be needed to handle the nested lists.

In order to tackle the problem stated above, we apply a serialization approach [6] to the multiplex table, which transforms the 'nested' structure of a multiplex descriptor by flattening it into two serialized linear lists. The key point for the approach is that RC UCF will appear only once in the multiplex descriptor, i.e., only one part would be repeated until the end of the packet. Thus the whole descriptor could be divided into two parts: the RC part with finite data length; the UCF part, which is repeated until the closing flag. The whole serialization process could be divided into two steps, as shown in Fig. 3 below. At the first step, we find the point where the UCF part starts and divide the whole descriptor into the RC part and the UCF part. Then at the second step, we serialize the two parts into two separate lists of Descriptor Atoms.

Descriptor Atom is used to distinguish that our concept is different from the element we have described above. The Descriptor Atom is a very simple data structure as shown in the Fig. 3. The logical channel number specifies the information source. The repeat count is a finite number that specifies how many bytes from that source would be filled in. And it also contains a pointer which links up the following atom. As it is called an 'Atom', it is indivisible, and can not be extended to be a sub-list. Using these two serialized lists can save much processing time during the multiplexing process and introduces less overhead for the modification of the multiplex descriptors [6].

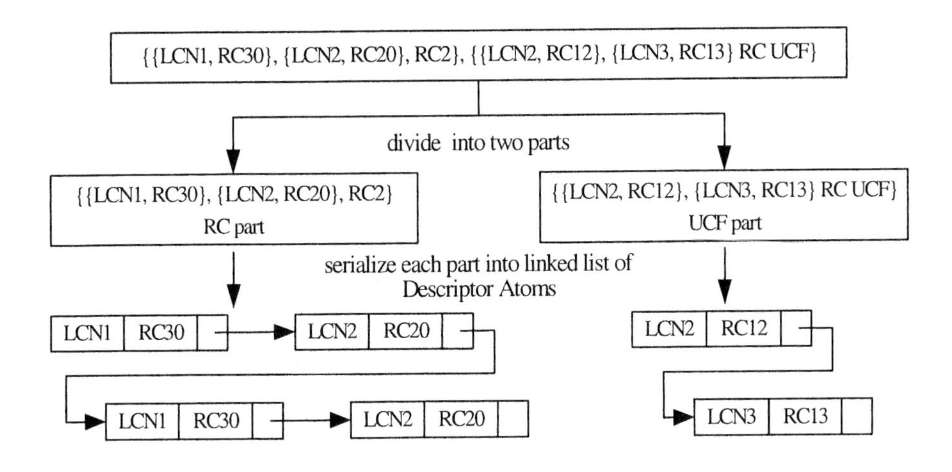

Fig. 3. Example of serialization

3.3 Structure Design of the Enhanced H.223 Module

Original H223 Design. Due to the convenience and efficiency of object-oriented approach, the major components of 3G-324M protocol stack, including H.324, H.245 and H.223, are designed and implemented as classes. H.324 class performs the negotiation about video/audio settings, etc., with the remote terminal through H.245 class, and handles the video/audio data exchange through H.223 class. The H.223 and H.245 classes also keep a pointer to each other so that the H.245 class can send out or receive control messages through H.223. The H.223 can also retrieve the information about various settings from H.245.

Thus, simply speaking, H.223 provides the sending/receiving interface for the above layers. In the point-to-point conversations, a H.223 object with a send() function, a recv() function and a buffer set for multiplexing/demultiplexing and sending/receiving, is already capable to handle all the tasks. However, for multi-point conference, H.223 might need to handle multiple incoming streams and a number of outcoming streams concurrently. Therefore, a more complex structure design might be needed.

Improved MultiH223 Design. Firstly, decomposition is applied to the original H223 design. The original H223 class is divided into two separate parts: H223_Sender assumes the responsibilities of multiplexing and sending out the outcoming streams; meanwhile, H223_Receiver handles the task of receiving the incoming streams and demultiplexing. Both H223_Sender and H223_Receiver provide convenient sending or receiving interfaces and keep their own buffer set for concerned operations.

After that, in order to handle the multiple media streams in a multi-point conference, the new designed MultiH223 module will keep two separate linked lists of H223_Senders and H223_Receivers. Thus in a real-time multi-point conference, the MultiH223 module will initialize a new sender/receiver at any time to handle a new incoming/outcoming stream. On the other hand, the sender/receiver can be destroyed at any time during the conference to release the resources.

3.4 Simple Management of the Multi-point Conference

Although H.245 has already defined many messages and procedures for the set-up of a point-to-point video call, there is no information defined for a multi-point conference. Thus, in order to make the set-up and management of the multi-point conference more convenient, some simple conference management functions have also been added into our H.223 module.

As noted in section 2, maximum 16 different multiplex descriptors can be defined to describe 16 different multiplex patterns. For each H.223 data packet, there is an MC (Multiplex Code) field which identifies the multiplex descriptor used for this packet. However, in practical experiments, we have found that current 3G mobile terminals only use 3 different multiplex descriptors (MC = 0, 1, 2) at most. Thus, we can use MC = 15 to identify that the content in this packet is information used for conference management. The following are the major control messages defined to accomplish the most common management functions:

Table 1. Major Control Messages Defined for Conference Management

Command	Description
CONNECTDEFAULT	Request to establish a connection to the remote terminal.
CONNECTED	Accept the connection request from the remote terminal.
DISCONNECT	Leave the current video conference.
KICKOFF	Expel the specified terminal from the conference.
TALK_O	Request a token to speak out in the group
TALK_X	Return token to the group
CONNECTION_REJECTED_F	Notify that Max. Number of Clients is Reached
CONNECTION_REJECTED_I	Notify that illegal Connection! Please Connect to Host
CONNECTION_REJECTED_B	Notify that No Permission to Join this Conference

In our management mechanism, one terminal will act as the host of the conference. The host is the creator of the conference and keeps the real-time information of the conference, such as the list of clients and their IDs. It is also responsible to broadcast the conference information to participating clients timely. A terminal can send the CONNECTDEFAULT message to the host to indicate its intention to join in the conference. The host has to make the admission control decisions according to terminal information, such as ID, or the constraints of the conference, such as maximum number of participants. After that, the host sends back CONNECTED or CONNECTION_REJECTED_F/I/B message to admit or decline the requesting terminal.

For the voice channel scheduling, our conference system allows only one terminal to speak at any time, and the voice is generally broadcast to all participants of the conference. Similar to the mechanism used in token-ring network, a terminal is granted a permission to broadcast its voice data over the network as long as having the "token". However, the way of token passing is different from the token-ring network. The grant of token is governed by the host of the conference. A terminal has to send a request `TALK_O` to the host in order to speak out in the conference. After finished speaking, the terminal has to send an indication `TALK_X` to the host for return of the token.

4 Performance Evaluation

4.1 System Interface

Our prototype system is implemented on PC but not a mobile handset. The PC can access 3G mobile network through 3G modem card. However, as the current 64 kbps video call bandwidth is not large enough to support multi-point video conferences, we test the performance of our system on the physical channel of 802.11 WLAN.

Fig. 4 shows the user interface of our system. Currently, the terminal supports the display of four different video channels; while there can be more than 4 clients in the conference. For audio channels, as noted in section 3.4, only one terminal is permitted to speak at any time.

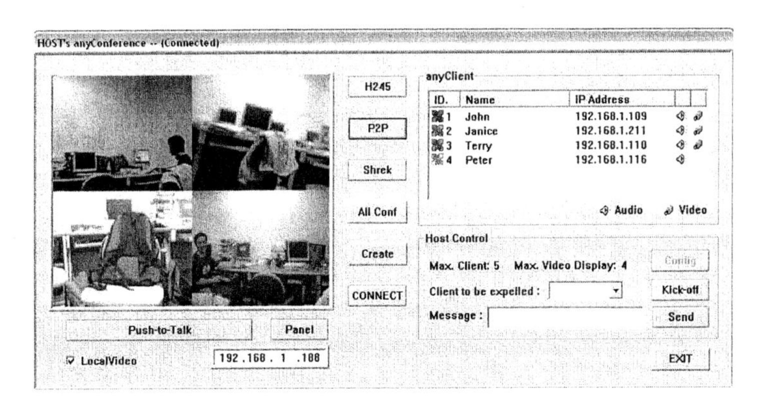

Fig. 4. GUI of the prototype system - anyConference

4.2 Performance of H.223 Implementation

Multiplex Descriptor Serialization. The performances of original and serialized multiplex descriptors are compared. With the original descriptor, the processing cost of multiplexing one packet can be divided into two parts: C_m is the cost of filling bytes into a packet, i.e. the cost of memory operations; C_n is the cost of handling the nested descriptor structure, i.e. the cost of recursive function calls.

For multiplexing of k packets, the total cost is $C(k)=k*(C_m+C_n)$. With the serialized multiplex descriptors, the processing cost of multiplexing one packet is C_m only. However, there is an additional cost to serialize the original descriptors at the initialization stage, which is defined as C_i.

For the initialization stage, maximum 16 nested structures need to be handled. Thus it is reasonable to assume that $C_i \approx 16*C_n$. For multiplexing of k packets, the total cost can be calculated as: $C(k)' = C_i + k*C_m \approx 16*C_n + k*C_m$.

A number of experiments have been done to evaluate the parameters C_m and C_n. The result is as follows: time taken for handling a nested descriptor structure (C_n) for one million times is 0.0472s and the time taken for filling bytes into a packet (C_m) for one million times is 0.353s. From the measurement result, we can estimate that: $\frac{C_n}{C_m} \approx \frac{0.0472}{0.353} = 13.4\%$. For the multiplexing of k packets,

$$\frac{C(k)'}{C(k)} = \frac{16*C_n + k*C_m}{k*C_n + k*C_m} = \frac{16*0.134 + k}{0.134k + k} \xrightarrow{k \to \infty} \frac{k}{0.134k + k} = 88.2\% .$$

Thus, the overall execution cost can be optimized by 11.8% when serialization is used as compared with recursive approach.

MultiH223 Module and Simple Conference Management. With the MultiH223 module and the simple conference management functions added into H.223, a video conference could be set up and managed conveniently and efficiently. Based on our conference testing over 802.11 WLAN, the performance of video and audio communication is satisfactory and stable.

5 Conclusions and Future Work

We have presented an efficient implementation of the data multiplexing protocol H.223, targeted for the case of multi-point video conferences. The design and implementation of the protocol is based on our prototype system '*anyConference*'. With serialization approach to the multiplex descriptors, support for multiple media streams and simple conference management functions, our H.223 implementation makes the set-up and performing of video conferences much more convenient and efficient. With tests on the prototype system, the performance of our protocol implementation is satisfactory and stable.

For the future work, we intend to do some work for the transcoding and compatibility between 3G-324M terminals and H.323 terminals. We'll try to make our protocol stack and conference system compatible with both 3G-324M and H.323. Thus, the video conference can be performed over both 3G mobile and internet networks, with support of MCUs and gateways.

References

1. ITU-R Rec. PDNR WP8F, Vision, Framework and Overall Objectives of the Future Development of IMT-2000 and Systems beyond IMT-2000, 2002.
2. ITU-T Rec. H.324, Terminal for low bit rate multimedia communication, March 2002.
3. ITU-T Rec. H.245, Control protocol for multimedia communication, July 2003.
4. ITU-T Rec. H.223, Multiplexing protocol for low bit rate mobile multimedia communication, July 2001.

5. B. Han, H. Fu, J. Shen, P. O. Au and W. Jia, Design and Implementation of 3G-324M - An Event-Driven Approach, Proc. IEEE VTC'04 Fall.
6. H. Fu, B. Han, P. Au and W. Jia, Efficient Data Transmission Multiplexing in 3G Mobile Systems, Proc. Globe Mobile Congress 2004.
7. ITU-T Rec. T.120, Data protocols for multimedia data conferencing, 1996.
8. ITU-T Rec. H.263, Video coding for low bit-rate communication, 1998.
9. ITU-T Rec. H.261, Video codec for audiovisual services at p × 64 kbit/s, 1993.
10. ITU-T Rec. G.723.1, Speech coders: Dual rate speech coder for multimedia communications transmitting at 5.3 and 6.3 kbit/s, 1996.
11. ITU-T Rec. H.243, Procedures for establishing communication between three or more audiovisual terminals using digital channels up to 1920 kbit/s, 2000.
12. The 3rd Generation Partnership Project (3GPP): http://www.3gpp.org.
13. 3GPP TS 26.071 V4.0.0, AMR Speech Codec; General Description, 2001.
14. 3GPP TS 26.111 V5.1.0, Codec for circuit switched multimedia telephony service: Modifications to H.324, June, 2003.

A New Backoff Algorithm to Improve the Performance of IEEE 802.11 DCF*

Li Yun, Wei-Liang Zhao, Ke-Ping Long, and Qian-bin Chen

Special Research Centre for Optical Internet & Wireless Information Networks,
ChongQing University of Posts & Telecommunications, ChongQing 400065, China
{liyun, longkp, zhaowl}@cqupt.edu.cn

Abstract. For IEEE 802.11 DCF, Backoff Timers of all stations in the wireless LAN are decreased by the same step if the state of channel is idle, and are paused and resumed at the same time. In this paper, we defined this as the "synchronization" feature of 802.11 DCF. The synchronization feature makes the packet collision probability of 802.11 DCF high. To break up the synchronization, this paper proposes a novel asynchronous backoff algorithm for 802.11 DCF, named *asyn*-DCF. A Markov model is built to analyze the performance of *asyn*-DCF. The simulation results indicate that *asyn*-DCF can decrease the packet collision probability significantly and utilize the channel more efficiently comparing to 802.11 DCF.

1 Introduction

IEEE 802.11 covers the MAC sub-layer and the physical (PHY) layer of wireless LAN (WLAN). The MAC sub-layer defines two medium access coordination functions, the basic Distributed Coordination Function (DCF) and the optional Point Coordination Function (PCF).

Much research has been conducted on the performance of IEEE802.11 DCF. In [1] and [2], the author gave a Markov chain model for the backoff procedure of 802.11 DCF. Wang C. et al. [3] proposed a new efficient collision resolution mechanism to reduce the collision probability. [4] analyzed the throughput of 802.11 DCF under finite load traffic. [5] proposed an improved analytical model that calculates IEEE 802.11 DCF performance taking into account both packet retry limits and transmission errors for the IEEE 802.11a protocol.

In recent years, there are growing researches on the QoS supporting for IEEE 802.11 WLAN. In [6-9], various schemes have been proposed to modify the backoff stage to support different priority source. In [10], an Advance Access (AA) MAC protocol was proposed to support differentiated service, power control, and radio efficiency. In [11], Wang, W. et al. proposed a voice multiplex–multicast (M–M)

* Supported by the National Science Foundation of China (90304004); the Science and Technology Research Project of Chongqing Municipal Education Commission of China (050310, KJ050503), the Research Grants by the Science & Tech. Commission of Chongqing(8817), the Research grants by the Ministry of Personnel of China.

X. Jia, J. Wu, and Y. He (Eds.): MSN 2005, LNCS 3794, pp. 488 – 497, 2005.

scheme for overcoming the large overhead effect of VoIP over WLAN. In [12], IQD, a distributed coordination function with integrated quality of QoS differentiation was proposed.

Although much work evaluated the 802.11 DCF in term of saturation throughput, fairness, spatial reuse, and QoS supporting, few of them studied the packet collision probability thoroughly. In this paper, we disclose the "synchronization" feature of 802.11 DCF, which causes high packet collision probability, and proposes a novel asynchronous backoff algorithm for 802.11 DCF, named *asyn*-DCF, to reduce the packet collision probability of 802.11 DCF.

2 Synchronization Feature of 802.11 DCF

In 802.11 DCF, there are two rules that all stations in an 802.11 WALN must obey, which are given as follows.

R1. If no medium activity is indicated for the duration of a particular backoff slot, then the Backoff Timer was decreased by *aSlotTime.*

R2. If the medium is sensed busy during a backoff slot, then the Backoff Timer is suspended until the medium is idle for the duration of DIFS period again, after which the Backoff Timer is resumed.

R1 means that Backoff Timers of all stations in an 802.11 WALN are decreased by the same step, *aSlotTime*. ***R2*** means that Backoff Timers of all stations in an 802.11 WALN are paused and resumed simultaneously. ***R1*** and ***R2*** mean that Backoff Timers of all stations in an 802.11 WALN are decreased, paused and resumed synchronously. We define this as the "synchronization" feature of 802.11 DCF.

Since the synchronization feature of 802.11 DCF, if two or more stations randomly set their Backoff Timers to the same value, then collision happens because these Backoff Timers will reach to zero at the same time.

The synchronization feature makes the packet collision probability of 802.11 DCF high. To break up the synchronization, we will propose an asynchronous backoff algorithm for 802.11 DCF in Section 3.

3 Asyn-DCF and Its Performance

3.1 Asyn-DCF

To avoid stations synchronously decreasing their Backoff Timers, the basic idea of *asyn*-DCF is that the Backoff Timer is decreased by *aSlotTime* with a probability p_w, named walking probability, if no medium activity is indicated for the duration of a particular backoff slot.

The *asyn*-DCF algorithm is as follows:

After transmitting, the station sets its Backoff Timer to $uniform(0, CW_{\min} - 1) \times aSlotTime$ if the transmission is successful, and sets its Backoff Timer to $uniform(0, 2^i \times CW_{\min} - 1) \times aSlotTime$ otherwise. If no medium activity is indicated for the duration of a particular backoff slot, then the Backoff Timer was decremented by *aSlotTime* with a probability of p_w. If the medium is sensed busy during a backoff slot, the Backoff Timer is suspended until the medium is idle for the duration of DIFS period again, and then the Backoff Timer resumes.

In *asyn*-DCF, since stations randomly decrease their Backoff Timer with probability p_w, the probability that two or more Backoff Timers reach to zero at the same time is very small even if these Backoff Timers are set to the same value initially. This feature of *asyn*-DCF makes the packet collision probability reduce significantly, which will be shown in the simulation.

3.2 Performance of Asyn-DCF

We assume that each packet has a constant and independent collision probability p regardless of the number of retransmissions, and channel is saturated. Let $b(t)$ and $s(t)$ be the stochastic processes of the backoff timer and the retransmission stage, respectively. The two-dimensional discrete time process, $\{b(t),s(t)\}$, can be modeled as a Markov chain shown in Figure 1, with the non-null one-step transition probabilities given by

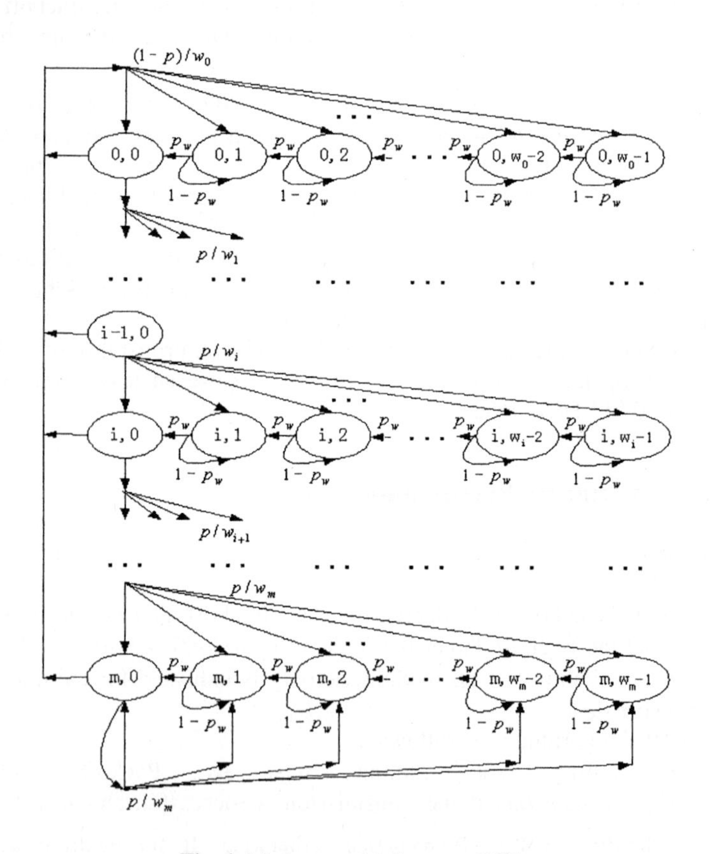

Fig. 1. Markov model for *asyn*-DCF

$$\begin{cases} P\{i,k \mid i,k+1\} = p_w, k \in [0, w_i - 2], i \in [0,m] \\ P\{i,k \mid i,k\} = 1 - p_w, k \in [1, w_i - 1], i \in [0,m] \\ P\{0,k \mid i,0\} = (1-p)/w_0, k \in [0, w_0 - 1], i \in [0,m] \\ P\{i,k \mid i-1,0\} = p/w_i, k \in [0, w_i - 1], i \in [1,m] \\ P\{m,k \mid m,0\} = p/w_{m,} k \in [0, w_m - 1] \end{cases} \tag{1}$$

where, $P\{i_1, k_1 \mid i_0, k_0\} = P\{s(t+1) = i_1, b(t+1) = k_1 \mid s(t) = i_0, b(t) = k_0\}$, and $w_i = 2^i \times CW_{\min}$.

Let $b_{i,k} = \lim_{t \to \infty} P\{s(t) = i, b(t) = k\}, k \in [0, W_i - 1], i \in [0,m]$, be the stationary distribution of $\{b(t), s(t)\}$, owing to the chain regularities, we obtain

$$b_{i,0} = p \cdot b_{i-1,0} \Rightarrow b_{i,0} = p^i \cdot b_{0,0}, 0 \le i < m \tag{2}$$

$$b_{m,0} = p \cdot b_{m-1,0} + p \cdot b_{m,0} \Rightarrow b_{m,0} = \frac{p^m}{1-p} \cdot b_{0,0} \tag{3}$$

$$\sum_{i=0}^{m} b_{i,0} = \sum_{i=0}^{m-1} p^i \cdot b_{0,0} + \frac{p^m}{1-p} \cdot b_{0,0} = b_{0,0}/(1-p) \tag{4}$$

and

$$b_{0,w_0-1} = \sum_{i=0}^{m} b_{i,0} \cdot (1-p)/w_0 + (1-p_w) \cdot b_{0,w_0-1} \Rightarrow b_{0,w_0-1} = \frac{1-p}{p_w \cdot w_0} \cdot \sum_{i=0}^{m} b_{i,0} = \frac{b_{0,0}}{p_w \cdot w_0} \tag{5}$$

$$b_{m,w_m-1} = p \cdot b_{m,0}/w_m + p \cdot b_{m-1,0}/w_m + (1-p_w) b_{m,w_m-1} \Rightarrow b_{m,w_m-1} = \frac{p^m b_{0,0}}{p_w (1-p) w_m} \tag{6}$$

$$b_{i,w_i-1} = p \cdot b_{i-1,0}/w_i + (1-p_w) \cdot b_{i,w_i-1} \Rightarrow b_{i,w_i-1} = \frac{p^i}{p_w \cdot w_i} \cdot b_{0,0}, 0 < i < m \tag{7}$$

When $0 < i < m, 0 < k < w_i - 1$,

$$b_{i,k} = (1-p_w) \cdot b_{i,k} + p_w \cdot b_{i,k+1} + b_{i-1,0} \cdot p/w_i \Rightarrow b_{i,k} = b_{i,k+1} + \frac{p^i \cdot b_{0,0}}{p_w \cdot w_i} \tag{8}$$

By relations (7) and (8), we obtain

$$b_{i,k} = \sum_{j=k}^{w_i-1} \frac{p^i \cdot b_{0,0}}{p_w \cdot w_i} = \frac{w_i - k}{w_i} \cdot \frac{p^i}{p_w} \cdot b_{0,0}, 0 < i < m, 0 < k \le w_i - 1 \tag{9}$$

Note that

$$\begin{aligned} b_{m,k} &= (1-p_w) \cdot b_{m,k} + p_w \cdot b_{m,k+1} + b_{m-1,0} \cdot p/w_m + b_{m,0} \cdot p/w_m \\ &\Rightarrow b_{m,k} = b_{m,k+1} + \frac{p^m \cdot b_{0,0}}{p_w \cdot (1-p) \cdot w_m}, 0 < k < w_m - 1 \end{aligned} \tag{10}$$

By relations (6) and (10), we obtain

$$b_{m,k} = \sum_{j=k}^{w_m-1} \frac{p^m \cdot b_{0,0}}{p_w(1-p)w_m} = \frac{w_m - k}{w_m} \cdot \frac{p^m b_{0,0}}{p_w(1-p)}, 0 < k \le w_m - 1 \tag{11}$$

When $i = 0, 0 < k < w_0 - 1$,

$$b_{0,k} = (1-p_w) \cdot b_{0,k} + p_w \cdot b_{0,k+1} + \sum_{i=0}^{m} b_{i,0} \cdot (1-p)/w_0 \Rightarrow b_{0,k} = b_{0,k+1} + \frac{b_{0,0}}{p_w \cdot w_0} \tag{12}$$

By relations (5) and (12), we obtain

$$b_{0,k} = \sum_{j=k}^{w_0-1} \frac{b_{0,0}}{p_w \cdot w_0} = \frac{w_0 - k}{w_0} \cdot \frac{b_{0,0}}{p_w}, 0 < k \le w_0 - 1 \tag{13}$$

Integrate relations (2),(3),(9) ,(11)and (13), it is easy to obtain that

$$b_{i,k} = \begin{cases} \frac{w_i - k}{w_i} \cdot \frac{p^i}{p_w} \cdot b_{0,0}, 0 \le i < m, 0 < k \le w_i - 1 \\ p^i \cdot b_{0,0}, 0 \le i < m, k = 0 \\ \frac{w_m - k}{w_m} \cdot \frac{p^m}{p_w \cdot (1-p)} \cdot b_{0,0}, i = m, 0 < k \le w_m - 1 \\ \frac{p^m}{1-p} \cdot b_{0,0}, i = m, k = 0 \end{cases} \tag{14}$$

Let τ be the probability that a station transmits a packet in a randomly chosen slot, which can be expressed as a function of $b_{0,0}$ and p given by

$$\tau = \sum_{i=0}^{m} b_{i,0} = \sum_{i=0}^{m-1} p^i \cdot b_{0,0} + \frac{p^m}{1-p} \cdot b_{0,0} = b_{0,0} / (1-p) \tag{15}$$

with $b_{0,0}$ to be finally determined by imposing the normalization condition as follows:

$$\sum_{i=0}^{m} \sum_{k=0}^{w_i-1} b_{i,k} = 1 \tag{16}$$

Inserting (14) into equation (16), we can obtain

$$\sum_{i=0}^{m} \sum_{k=0}^{w_i-1} b_{i,k} = \sum_{i=0}^{m-1} \sum_{k=0}^{w_i-1} b_{i,k} + \sum_{k=0}^{w_m-1} b_{m,k} = \sum_{i=0}^{m-1} \left(\sum_{k=1}^{w_i-1} \frac{w_i - k}{w_i} \cdot \frac{p^i}{p_w} \cdot b_{0,0} + p^i \cdot b_{0,0} \right)$$

$$+ \sum_{k=1}^{w_m-1} \frac{w_m - k}{w_m} \cdot \frac{p^m b_{0,0}}{p_w(1-p)} + \frac{p^m b_{0,0}}{1-p} = \frac{\left[1 - p - p(2p)^m\right]w - (1-2p)(1-2p_w)}{2p_w \cdot (1-p) \cdot (1-2p)} b_{0,0} = 1$$

$$\Rightarrow b_{0,0} = \frac{2p_w \cdot (1-p) \cdot (1-2p)}{\left[1 - p - p \cdot (2p)^m\right] \cdot w - (1-2p)(1-2p_w)} \tag{17}$$

For simplicity, we replace CW_{min} with w in expression (17), and $w = CW_{min}$.

Inserting (17) into (15), we can express τ as

$$\tau = \frac{2p_w \cdot (1-2p)}{\left[1-p-p\cdot(2p)^m\right]\cdot w-(1-2p)\cdot(1-2p_w)} \tag{18}$$

Note that p, the probability that a transmitted packet encounters a collision, is equal to the probability that, in a slot, at least one of the n-1 remaining station transmit, therefore,

$$p = 1-(1-\tau)^{n-1} \tag{19}$$

We can obtain the value τ by solving the equations (18) and (19) using numerical techniques.

As in [2], the saturation throughput, S, can be expressed as:

$$S = \frac{P_s \cdot E[P]}{(1-P_{tr})\cdot\sigma + P_s \cdot T_s + (P_{tr}-P_s)\cdot T_c} \tag{20}$$

where, $E[P]$ is the average packet payload size, T_s is the average time the channel is sensed busy because of a successful transmission, T_c is the average time the channel is sensed busy during a collision, σ is the duration of an empty slot time, P_{tr} is the probability that there is at least one transmission in the considered slot time, P_s is the probability that a transmission occurring on the channel is successful, and

$$P_{tr} = 1-(1-\tau)^n \tag{21}$$

$$P_s = n\tau\cdot(1-\tau)^{n-1} \tag{22}$$

4 Validation

We validate the correctness of analytical model by comparing the simulated saturation throughput to the analytical saturation throughput.

The simulation platform is NS-2 [13]. Stations are randomly placed in an area of 200×200 meters and have no mobility, constituting an ad-hoc network, in which a station can transmit packets to any other station with the packet size fixed to 1024 bytes at the same CBR rate. Each CBR flow runs 200 seconds. The physical layer is DSSS. The stations transmit packets by means of RTS/CTS mechanism, and the simulation parameters are shown in Table 1.

For simplicity, we set the minimum contention windows to 32, and the walking probability to 0.75 for all stations in the networks. The analytical and simulated saturation throughputs of *asyn*-DCF are drawn in Fig. 2 under different stations.

Fig.2 shows that the analytical and simulated results are very closed. The analytical saturation throughput is just about 2% higher than the simulated saturation throughput, as is attributed to that we ignore the frame retry limits in the analytical model.

Table 1. Simulation Parameters

Channel Bit Rate	2Mbit/s
Slot Time	20μs
SIFS	10μs
DIFS	50μs
PHYHeader	192bits
MACHeader	144bits
RTS Length	160bits
CTS Length	112bits
CW_{min}	32
CW_{max}	1024
CBR Packet Size	1024Bytes

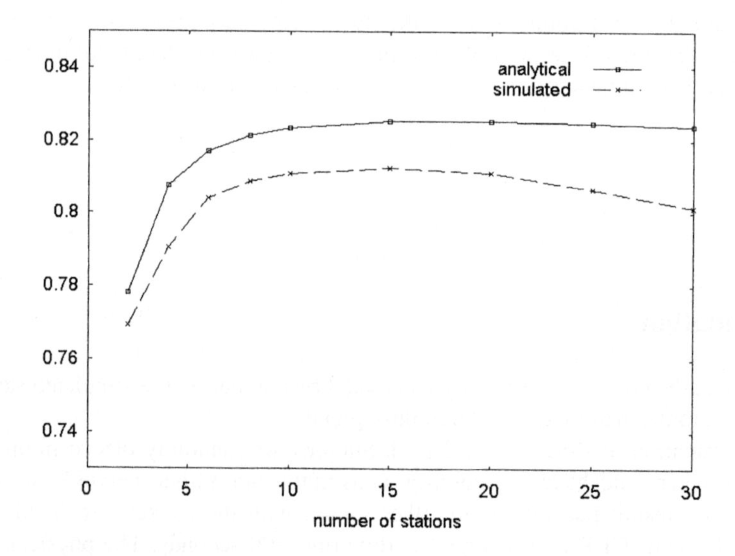

Fig. 2. Saturation throughput of *asyn*-DCF: analytical and simulated

5 Performance Evaluation

In this section, we firstly compare the performance of *asyn*-DCF to 802.11 DCF in terms of saturation throughput, packet collision probability and packet delay. Then, we evaluate the influence of p_w on the saturation throughput and packet collision probability of *asyn*-DCF, respectively. The simulation scenes and parameters are same as that in Section 4.

Fig.3 shows the packet collision probability of *asyn*-DCF and 802.11 DCF when the total number of contending stations is different. Fig.3 demonstrates that the packet collision probability of *asyn*-DCF is much lower than that of 802.11 DCF, which is consistent to the analysis in section 3.

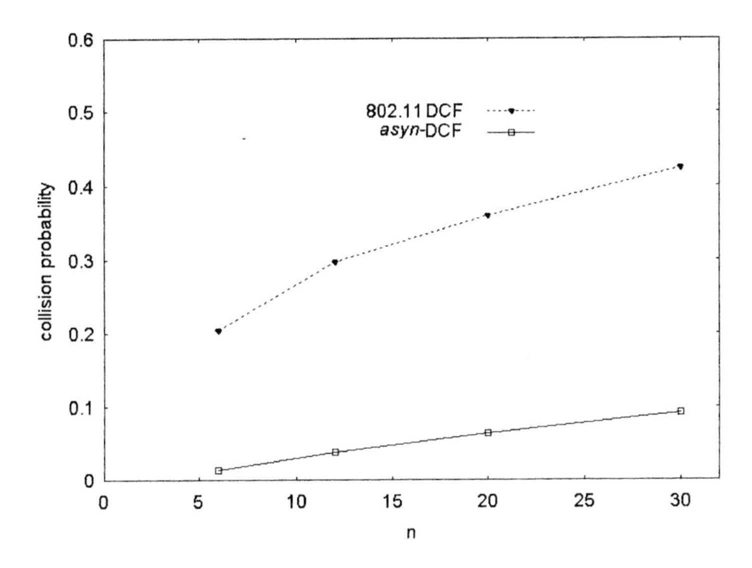

Fig. 3. Collision probability, CW_{min}=32, CW_{max}=1024; p_w=0.75

The saturation throughput of *asyn*-DCF and 802.11 DCF are shown in Table 2. From simulation results of Table 2, we can obtain that saturation throughput of *asyn*-DCF is higher than that of 802.11 DCF.

The packet delay for a given flow of *asyn*-DCF and 802.11 DCF are shown in Fig.4, when the total number of contending stations is 20, and the traffic load is 200%. Fig. 4 demonstrates the packet average delay of *asyn*-DCF (which is 0.09436 s) is a bit higher than that of 802.11 DCF (which is 0.07747s), but the delay jitter of *asyn*-DCF (which is 0.01810 s) is much lower than 802.11 DCF (which is 0.16698 s). This feature is favorable to real-time services.

Table 2. Saturation throughput, CW_{min}=32, CW_{max}=1024; p_w=0.75

Number of stations	802.11 DCF (Kbps)	*asyn*-DCF (Kbps)
6	1538.05	1586.70
12	1574.76	1678.75
20	1479.17	1624.87
30	1427.45	1594.00

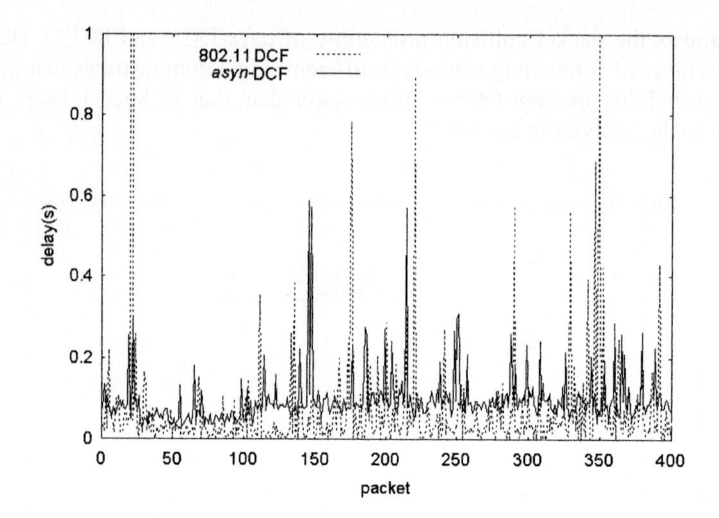

Fig. 4. Packet delay, CW_{min}=32, CW_{max}=1024; p_w=0.75

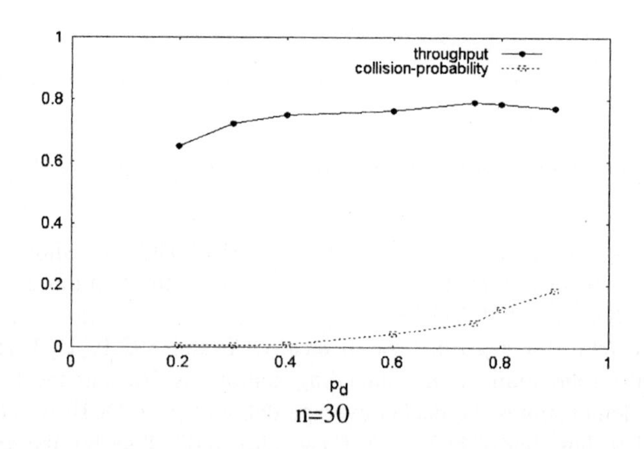

Fig. 5. Saturation throughput collision probability vs. p_w, CW_{min}=32, CW_{max}=1024

The saturation throughput and packet collision probability vs. p_w of *asyn*-DCF are shown in Fig.5, when the number of stations is 30 and the traffic load is 200%.

The saturation throughput normalized by the channel raw bandwidth is given in Fig.5. Fig. 5 illustrates that, increasing p_w increases collision probability also, because more stations transmit at a given slot. But the collision probability is lower than 0.1, even if p_w increase to 0.9. The throughput enhances when increasing the p_w because the transmitting probability of station increases in a given slot, but at some point, increasing p_w continuously can decrease the throughput because the collision probability is too large. In the range of [0.6, 0.8] of p_w, the throughput changes little and the collision probability keeps low (under 0.1), which suggest that we can have a wide configurable range of p_w.

6 Conclusion

Because of the "synchronization" feature, the packet collision probability of 802.11 DCF is high. In this paper, we propose a novel asynchronous backoff algorithm for 802.11 DCF, named *asyn*-DCF, to break up the synchronization. The simulation results indicate that *asyn*-DCF can decrease the packet collision probability significantly and utilize the channel more efficiently comparing to 802.11 DCF.

References

1. G.Bianchi. IEEE802.11 – Saturation throughput analysis. IEEE Communication Letters, 1998, Vol.2: 318 – 320.
2. G.Binachi. Performance analysis of the IEEE 802.11 distributed coordination function. IEEE Journal on Selected Areas in Commu. , 2000, 18(3): 535-547.
3. Chonggang Wang, Bo Li, Lemin Li. A new collision resolution mechanism to enhance the performance of IEEE 802.11 DCF. IEEE Transactions on Vehicular Technology. 2004, 53(4): 1235-1246.
4. Zaki, A.N., El-Hadidi, M.T..Control. Throughput analysis of IEEE 802.11 DCF under finite load traffic. In First International Symposium on Communications and Signal Processing, 2004, 535 - 538.
5. Chatzimisios, P., Boucouvalas, A.C.; Vitsas, V.. Performance analysis of IEEE 802.11 DCF in presence of transmission errors. In IEEE International Conference on Communications, 2004, Vol. 7: 3854 - 3858.
6. Michael Barry and Andrew T. Campbell and Andras Veres. Distributed Control Algorithms for Service Differentiation in Wireless Packet Networks. IEEE INFOCOM, April 2001
7. Imad Aad and Claude Castelluccia. Differentiation mechanisms for IEEE 802.11. IEEE INFOCOM, April 2001.
8. V.Kanodia and C.Li and A.Sabharwal and B.Sadeghi and E.Knightly. Distributed Multi-Hop Scheduling and Medium Access with Delay and Throughput Constraints. MOBICOM, August 2001.
9. S. Mangold, S. Choi, P. May, O. Klein, G. Hiertz, L. Stibor, IEEE 802.11e wireless LAN for quality of service (invited paper). Proceedings of the European Wireless, Florence, Italy, February 2002, Vol. 1, pp. 32-39
10. Robinson J.W.and Randhawa T.S..Saturation throughput analysis of IEEE 802.11e enhanced distributed coordination function. IEEE Journal on Selected Areas in Communications, 2004, 22(5):917 - 928
11. Wang, W., Liew, S.C., Li, V.O.K.. Solutions to Performance Problems in VoIP Over a 802.11 Wireless LAN. IEEE Transactions on Vehicular Technology, 2005, 54(1): 366 – 384
12. Liqiang Zhao; Changxin Fan. Enhancement of QoS differentiation over IEEE 802.11 WLAN. IEEE Communications Letters, 2004, 8(8): 494 – 496
13. The networks simulator ns-2, http:// www.isi.edu/nsnam/

Enhanced Power Saving for IEEE 802.11 WLAN with Dynamic Slot Allocation*

Changsu Suh, Young-Bae Ko, and Jai-Hoon Kim

Graduate School of Information and Communication,
Ajou University, Republic of Korea
{scs, youngko, jaikim}@ajou.ac.kr

Abstract. In the area of wireless mobile communication, minimizing energy consumption as well as maximizing data throughput in medium access control (MAC) layer is a very important research issue. The Distributed Coordination Function (DCF) of IEEE 802.11 provides the power saving mechanism (PSM) that allows nodes to remain silent in 'doze mode' for reducing energy consumption. However, IEEE 802.11 PSM is known to cause unnecessary energy consumption due to the problems of an overhearing, a back-off time delay and possible packet collisions. To overcome these problems, we present a new MAC protocol called 'Slotted-PSM' for IEEE 802.11 WLANs. In our proposed scheme, the Beyond-ATIM window is divided into a number of time slots, and any node that participates in communication wakes up in its slot times only and other nodes can enter to doze mode during the slot times. Slotted-PSM can reduce unnecessary energy consumption and idle listening problem in the Beyond-ATIM window.

1 Introduction

Recently, there have been lots of attentions on developing energy efficient MAC protocols in wireless networks. These protocols are similar in the sense that they try to turn off their radio transceivers for reducing energy consumption whenever they are not involved in communication. Such a state of turning off is known as 'sleep' or 'doze' state. Neither a transmission nor a reception is allowed when a node stays in this inactive state, resulting in little energy consumption. The IEEE 802.11 DCF (Distributed Coordination Function) standard [9], the dominating WLAN technology today, also incorporates such a power saving mechanism (PSM) that uses awake and doze states. In 802.11 PSM, time is divided into beacon intervals and each node tries to synchronize with its neighbors to ensure that all nodes wake up at the same time. Any node can announce its

* This research was supported by the MIC (Ministry of Information and Communication), Korea, under the ITRC(Information Technology Research Center) program, and the Ubiquitous Computing and Network Project (part of the MIC 21st Century Frontier R&D Program). It was also partially supported by grant No. R01-2003-000-10794-0 from Basic Research Program of Korea Science & Engineering Foundation.

X. Jia, J. Wu, and Y. He (Eds.): MSN 2005, LNCS 3794, pp. 498–507, 2005.

pending data information during the period called the announcement traffic indication map (ATIM) window. During the period following ATIM window (we call this period as *Beyond-ATIM window*), nodes exchanging ATIM and ACK frames in the ATIM window period must remain in an awake state and perform data communication based on Carrier Sense Multiple Access with Coliision Avoidance (CSMA/CA) mechanism, whereas all other nodes go to 'doze state'. We are now arguing that, Beyond-ATIM window in the current 802.11 standard causes energy consumption with the following reasons:

Overhearing: A node may receive packets that are not destined for it. Unnecessarily receiving such packets will cause energy consumption.
Back-off time: A node in CSMA/CA is required to sense the medium in randomly chosen back-off time duration before sending any packets. During this time, node remains in awake state.
Packet collision: If two nodes transmit packets simultaneously, the packets are corrupted and the follow-on retransmission causes more energy consumption.

With respect to these problems, we propose a new energy efficient MAC protocol named "Slotted Power Saving Mechanism (i.e., Slotted-PSM)". Slotted-PSM divides the time of the Beyond-ATIM window period into equal slots of length L. When a node wants to send data packets, it first selects free slots which are not reserved by other nodes during the ATIM window, and then waits unless the slot time is not started in the Beyond-ATIM window. While waiting for its slot time, a node can enter to the doze mode for saving energy. Here, slots are allocated dynamically, and the allocated slot information will be piggybacked into legacy ATIM frame.

2 Related Works

Fig. 1 shows the basic PSM in DCF mode. As mentioned earlier, time in DCF is divided into beacon intervals. All nodes try to synchronize with each other by sending beacon frames and wake up at the beginning of every beacon interval. A beacon interval consists of two parts: ATIM window and Beyond-ATIM window. Generally, one ATIM window size is long enough to allow the transmission and retransmission of several ATIM packets [1,5]. In Fig. 1, nodes A, B and C intend to send data packet to node B, A and B respectively. So, during ATIM window, sender nodes announce the traffic information by sending an ATIM packet based on CSMA/CA mechanism (In Fig. 1, 'CS' means the carrier sensing after back-off time). A receiver node replies by sending an ATIM-ACK packet. A successful exchange of ATIM and ATIM-ACK packets between two nodes implies that both nodes continuously maintain awake mode for the data transmission after finishing ATIM window. However, other nodes (e.g., node D) are allowed to enter a doze mode until the next beacon interval. In the normal PSM in IEEE 802.11, ATIM window size as well as a beacon interval is fixed for synchronization between neighbor nodes. However, the fixed ATIM window critically

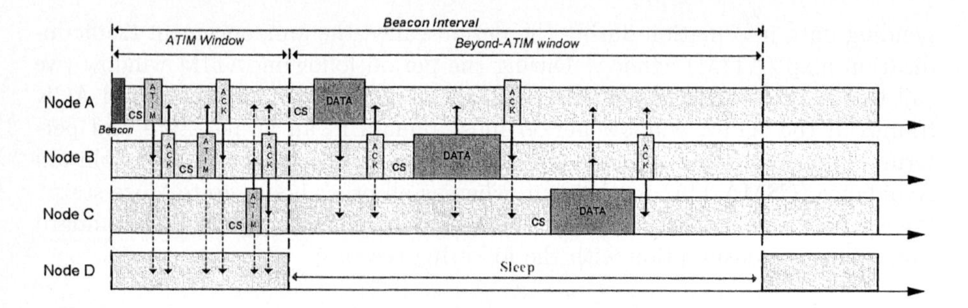

Fig. 1. The operation of basic PSM in IEEE 802.11 DCF

affects on throughput and energy consumption [1]. One of the previous research issues of the PSM IEEE 802.11 is to dynamically adjust the size of ATIM window. [2] proposes a mechanism to dynamically choose an ATIM window size for improving data throughput based on three indications. Generally, most of PSM schemes need the synchronization between neighbors, but it is difficult to achieve in a multi-hop, mobile ad hoc networking environment. However, [3] investigates three power management protocols based on IEEE 802.11 PSM scheme in unsynchronized multi-hop environments. [4] suggests three schemes for reducing the energy consumption in listening (1) by carrier sensing, (2) by dynamically re-size the ATIM window and (3) by scheduling wake-up times of sender and receiver based on past traffic patterns. [5] suggests scheme to enter sleep state early without being awake during the whole Beyond-ATIM window as well as to adjust the ATIM window size. [5] is able to reduce more energy consumption than normal PSM of IEEE 802.11 DCF, but it has some weakness that several nodes have to wait until other nodes finish data transmission despite of not participating in the communication and it has the hidden terminal problem in multi-hop environments.

3 Proposed Scheme

3.1 The Operation of Slotted-PSM

In our proposed scheme, we extend the Beyond-ATIM window to be divided into fixed time slots such that nodes can effectively use them for conserving energy. As each node bearing traffic reserves the time slot during ATIM-window period, it can exactly know when it needs to wake up for data communication. One whose slot has not yet arrived remains in the sleep state conserving additional energy. The operation of the Slotted-PSM mode is explained in detail by the following steps:

– At the starting of a new beacon interval, a node having data packets to send make a slot reservation. We perform this slot reservation as soon as back-off time is over and a node verifies that the wireless medium is free. After this,

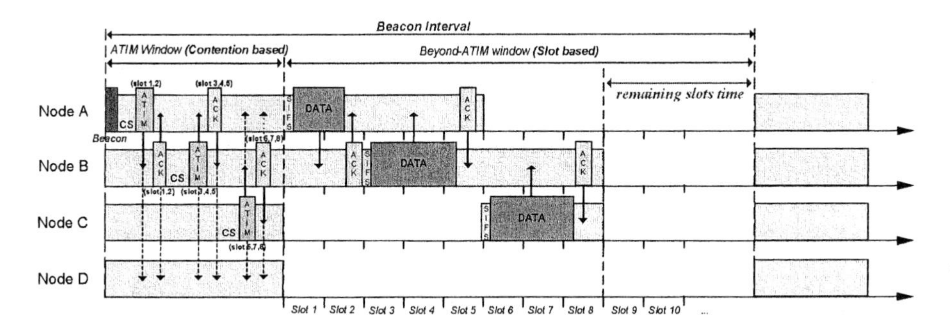

Fig. 2. The operation of Slotted-PSM

node transmits the ATIM packet piggybacked with the proposed list of slots required for the data transmission during the Beyond-ATIM window. To avoid overlapping during slot reservation, all nodes maintain the slot table that is updated every time the ATIM or ATIM-ACK is received. Slot table provides the information to the node about whether the slot is free or not. Note that, sender selects the slot based on amount of data traffic and the available slot table information.

- Actual slot reservation is performed after sender transmits the ATIM packet with proposed slot list to the target node as mentioned in the previous step. Here, we explain two different methods for reserving slots: (1) sender-initiated and (2) receiver-initiated. In (1), the slots for future data transmission are reserved by the sender side, which are same slots in the proposed slot list. However, it is possible that the same slots are already occupied by another sender(s) in the receiver. In this case, the receiver replies with ATIM-RE packet with free slot list to the sender. After receiving ATIM-RE packet, sender tries to re-selects the slots from the receiver's free slots and immediately sends a new ATIM packet. In (2), slots are reserved at the receiver side. Sender transmits ATIM packet with free slot list (proposed slot list) to the receiver. The receiver then chooses the appropriate slot by considering its slot table and sender's free slot list. After this, it replies the ATIM-ACK with the selected slots back to sender. In this case, nodes around the sender may not receive the ATIM-ACK and may remain unaware of the used slots. Therefore, after receiving ATIM-ACK, senders need to send another packet for announcing the occupied slots. Since this can cause more overhead, we have used the first method for slot reservation in our implementation.
- After the ATIM window phase is over, each sender transmits data according to the reserved slots. Other nodes maintain doze state and only wake up at the beginning of their slots. With the starting of the new beacon interval, all slots are reset and each node tries to reserve new slots for further data transmission.

Fig. 2 illustrates our scheme ((1) sender-initiated). Each node selects its slots for future data transmission based on its empty slot table. Slot 1 and 2 are reserved by node A for sending its data to node B. Slot 3, 4 and 5 are reserved by node B for its data transmission. So both nodes will have to maintain awake state from slot 1 to 5. After slot 5, node A enters to doze state but node B still maintain awake state to receive data from node C at slot 6, 7 and 8. Hence, we see that any node without a reserved slot or after completing its data communication may go into sleep state conserving energy. Moreover, the possibility of collision is reduced as the packets are sent only in the reserved slot. Reduction in collision leads to increase in data throughput. Therefore, using our scheme, we expect to not only save energy but also increase throughput.

3.2 Collision, Hidden Terminal Problem and Synchronization in Slotted-PSM

In order to avoid a data packet collision by newly joined nodes, a sender in slotted-PSM waits for the SIFS time interval before transmitting its data. Nevertheless, the data collision problem in the Beyond-ATIM window continuously remains due to long range of carrier sensing [10] or interference from other devices. If the nodes do not finish the communication in their reserved slot time, they uses remaining slot time (refer to Fig.2). The remaining slot time is some extra time in the Beyond-ATIM window after all reserved slots become over. Any data communication that occurs in this time is based on CSMA/CA mechanism. Additionally, our scheme also prohibits the hidden terminal problem during data communication in the Beyond-ATIM window phase. In IEEE 802.11 and other related protocols this problem is handled by RTS/CTS mechanism. In Slotted-PSM, since sender knows the reserved slot information of other senders through ATIM/ATIM-ACK packets and simultaneous data transmission never occurs, it avoids the hidden terminal problem without the overhead of RTS/CTS exchanging. Finally we have assumed that the synchronization is achieved through beacon packet or out-of-band solutions such as GPS [12]. All nodes wake up at the same time for exchanging the ATIM and ATIM-ACK packets. If out-of-band solution is used, synchronization is perfect and there will be no other overhead. However, by using beacon packet solution, a perfect synchronization is not guaranteed in multi-hop environments. To improve the performance, our scheme has to be combined with some synchronization mechanism. However, developing a scheme for synchronization is outside the scope of this paper.

3.3 Energy Saving Analysis

Since the primary goal of our scheme is to reduce energy consumption as compared to the IEEE 802.11 PSM, we present the mathematical analysis on saving energy. We can express the total energy consumed by a node for a beacon interval T as equation (1).

$$E_{node} = E_{ATIM} + E_{BeyondATIM} \tag{1}$$

E_{ATIM} and $E_{BeyondATIM}$ mean the energy consumption during the ATIM window and Beyond-ATIM window period. Equation (2) shows the energy consumption during the ATIM window in more detail. In this equation, E_{beacon} is the energy consumed while transmitting or receiving beacon packet and $E_{ctrlpkt}$ is a combination of the transmitted and received power for exchanging ATIM and ATIM-ACK packets. During the remaining ATIM window (T_{ATIM} - T_{beacon} - $T_{ctrlpkt}$), a node stays in the idle mode and consumes P_{idle} energy. In the ATIM window, the operations for both Slotted-PSM and basic PSM are the same, and its equation is expressed to

$$E_{ATIM} = E_{beacon} + E_{ctrlpkt} + P_{idle} \cdot (T_{ATIM} - T_{beacon} - T_{ctrlpkt}) \tag{2}$$

Equation (3) is only valid for the nodes that do not perform data communication during the Beyond ATIM-window. Here, T_{Beyond} is the time interval for the Beyond-ATIM window.

$$E^{sleep\_node}_{Beyond} = P_{sleep} \cdot T_{Beyond} \tag{3}$$

In the Beyond-ATIM window, both schemes have different operations and accordingly different energy consumption. Basic PSM's energy consumption is given by (4). In this equation, P_{rx} and P_{tx} mean the energy for transmitting and receiving data packets. T_{rx_data} and T_{tx_data} are times required for receiving and transmitting. During the remaining Beyond-ATIM window (T_{Beyond} - T_{rx_data} - T_{tx_data}), a node stays in the idle mode and consumes P_{idle} energy.

$$\begin{aligned} E^{Basic}_{Beyond} = {} & P_{rx} \cdot T_{rx\_data} + P_{tx} \cdot T_{tx\_data} \\ & + P_{idle} \cdot (T_{Beyond} - T_{rx\_data} - T_{tx\_data}) \end{aligned} \tag{4}$$

Equation (5) shows the energy consumption of Slotted-PSM in the Beyond-ATIM window. For simplifying analysis, we omit the benefit of our scheme such as eliminating RTS/CTS overhead and back-off time in T_{rx_data} and T_{tx_data}. The main difference between (4) and (5) is that Slotted-PSM's node stays in the idle mode during only T_{idle_slot} and then it can stay in sleep mode during the rest time except in T_{idle_slot}, T_{rx_data} and T_{tx_data}.

$$\begin{aligned} E^{Slotted}_{Beyond} = {} & P_{rx} \cdot T_{rx\_data} + P_{tx} \cdot T_{tx\_data} + P_{idle} \cdot T_{idle\_slot} \\ & + P_{sleep} \cdot (T_{Beyond} - T_{rx\_data} - T_{tx\_data} - T_{idle\_slot}) \end{aligned} \tag{5}$$

T_{idle_slot} means the remained times except to T_{rx_data} or T_{tx_data} in *reserved slots*, which is expressed in (6). In this equation, T_{one_slot} is the length of a slot. In Slotted-ATIM, communication nodes try to reserve the enough slots for transmitting data packets. Therefore, generally, transmission time may be not fixed with reserved slot times. In this case, T_{idle_slot} is generated.

$$\begin{aligned} T_{idle\_slot} = {} & \left(\left\lceil \frac{T_{rx\_data}}{T_{one\_slot}} \right\rceil \cdot T_{one\_slot} - T_{rx\_data}\right) \\ & + \left(\left\lceil \frac{T_{tx\_data}}{T_{one\_slot}} \right\rceil \cdot T_{one\_slot} - T_{tx\_data}\right) \end{aligned} \tag{6}$$

The benefit in terms of energy obtained by using Slotted-PSM is the difference between (4) and (5). By ignoring the minute power of the sleep state because its energy consumption is much smaller than Tx state or Rx state, we can approximate the benefit of Slotted-PSM as (7):

$$\begin{aligned}\Delta E_{node} &= E^{Basic}_{node} - E^{Slotted}_{node} = E^{Basic}_{Beyond} - E^{Slotted}_{Beyond} \\ &= P_{idle} \cdot (T_{Beyond} - T_{rx\_data} - T_{tx\_data}) - P_{idle} \cdot T_{idle\_slot} \\ &\quad - P_{sleep} \cdot (T_{Beyond} - T_{rx\_data} - T_{tx\_data} - T_{idle\_slot}) \\ &\approx P_{idle} \cdot (T_{Beyond} - T_{rx\_data} - T_{tx\_data}) - P_{idle} \cdot T_{idle\_slot} \qquad (7)\end{aligned}$$

If T_{one_slot} can divide T_{rx_data} and T_{tx_data} without remainder, T_{idle_slot} becomes zero in (6). In this case, the performance of Slotted-PSM is highest and equation (7) can be described to (8).

$$Max(\Delta E_{node}) = P_{idle} \cdot (T_{Beyond} - T_{rx\_data} - T_{tx\_data}) \qquad (8)$$

Eventually, the maximum energy benefit of Slotted-PSM comparing to basic PSM can be defined as $P_{idle} \cdot (T_{Beyond} - T_{rx\_data} - T_{tx\_data})$. Hence, our scheme shows the significant energy conservation which is crucial for the wireless mobile devices that are operated using limited power sources.

4 Performance Evaluations

4.1 Simulation Environments

In our simulation, we modified the CMU extended version of ns-2 [11]. We have assumed the maximum bit rate of 1Mbps, the transmission range for each node to be approximately 250m and packet size fixed at 512 bytes. For calculating energy consumption of two protocols, we use 1.65W, 1.4W and 0.045W as value of power consumed by the MAC layer while transmitting, receiving and in doze state [8]. We have set beacon interval to 100ms and ATIM window to 20ms as in [2, 4]. We fix the ATIM window size and have not considered adjusting the size of ATIM window scheme. Each simulation is performed for 50 seconds in a wireless LAN environment, in which all nodes are within each other's transmission range and every source node can reach its destination in a single hop. Analysis of the performance evaluation is for the two scenarios. In the first scenario, data traffic is changed from high traffic to low traffic in 16 nodes. In the second scenario, we focus on throughput of both protocols in the high data traffic given 30 nodes. Both scenarios are for the wireless LAN environment [2,4]. Half of the nodes are sources and rests are destinations. One of the important issues in our scheme is to decide the optimal size of a slot. If the size of a slot is larger than the period for DATA/ACK transmission, it causes unnecessary time after data transmission is finished within the reserved slot. Otherwise, if the size of a slot is small, many number of slots causes the problems such as complexity of slot scheduling and large size of ATIM packet. Fig. 3 shows the average energy consumption of Slotted-PSM with the increasing slot size in various message inter-arrival times. The message inter-arrival time affects amount of data traffic in the simulation. As shown in Fig. 3, a reasonable slot size for Slotted-PSM is 2.6ms. Therefore, the comparison of our scheme with Basic-PSM is performed with slot size equal to 2.6ms.

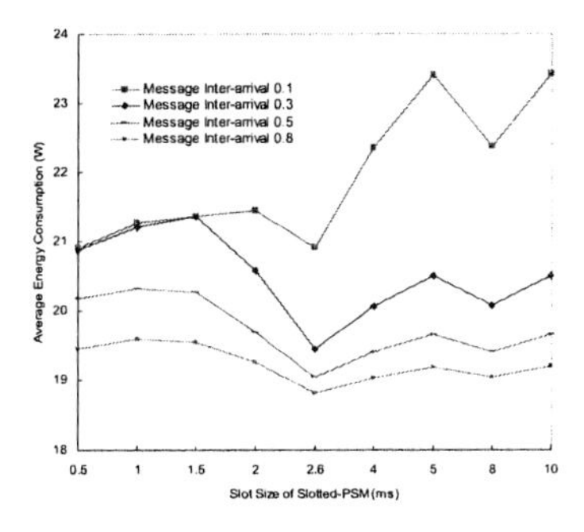

Fig. 3. Average energy consumption according to the slot size

4.2 Simulation Results

Fig. 4(a) shows the total energy consumption as the message inter-arrival period is increased. According to the graph, Slotted-PSM consumes less energy than other protocol as the network load is increased . The reason is that, Slotted-PSM does not spend the unnecessary energy consumption for data transmission, but the basic PSM consumes more energy due to overhearing, waiting for back-off time and packet collisions. Therefore, the energy saving performance of Slotted-PSM is better in situation where data packets are frequently transmitted.

Fig. 4(b) shows the total number of packets (i.e. Beacon, ATIM, ATIM-ACK, RTS, CTS, DATA and ACK) which are generated during simulation time. In our scheme, when data packet is transmitted in reserved slots, RTS and CTS packets are not necessary (we mentioned this fact in the section 3). So RTS or CTS packet is not sent and counted in Slotted-PSM. Therefore, total number of packets of our proposed scheme is smaller than that of Basic-PSM. Fig. 4(c) shows the total number of retransmission by packet collision and loss. Our scheme can reduce the data collision in Beyond-ATIM window because each node should try to transmit its data packet only in its reserved slot time. Therefore, total retransmissions in Slotted-PSM are counted smaller than Basic-PSM.

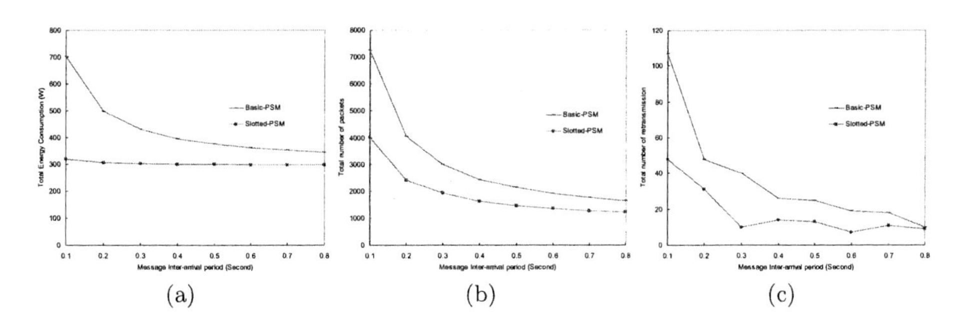

Fig. 4. Comparison of Slotted-PSM to Basic Basic-PSM with a variation of *traffic loads* :(a)Total Energy Consumption, (b)Total number of packets and (c)Total number of retransmission

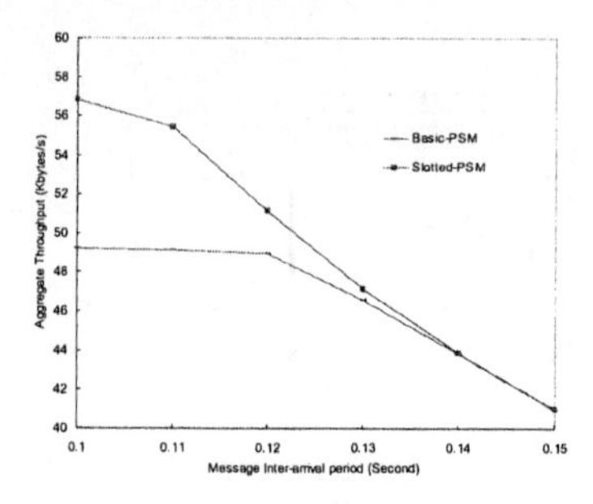

Fig. 5. Aggregate throughput (Kbytes/s) in the high data traffics

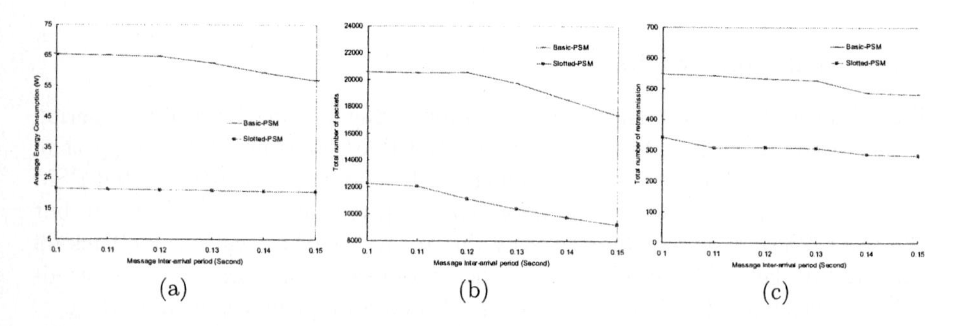

Fig. 6. Comparison of Slotted-PSM to Basic Basic-PSM in the high data traffics :(a)Total Energy Consumption, (b)Total number of packets and (c)Total number of retransmission

Fig. 5 and 6 are the results of the second scenario in the high data traffic environments (30 nodes). In this scenario, we focus more on the analyzing the aggregate throughput of both protocols. As mentioned above, our scheme can reduce the RTS/CTS overhead and the data collision problem. These characteristics may lead to increase of data throughput. Fig. 5 shows the aggregate throughput (Kbytes/s) of both protocols, and the result shows that the aggregate throughput of Slotted-PSM is larger than Basic-PSM. The throughput performance of Slotted-PSM is better in situation where data packets are frequently transmitted because high data traffic causes lots of RTS/CTS overhead and packet collision in Basic-PSM. Other Fig. 6 shows the results of energy consumption, total number of packets and total number of retransmission in the high traffics environment. Through these figures, our proposed scheme can expect to not only save energy but also increase throughput more than Basic-PSM.

5 Conclusions

In this paper, we proposed the energy efficient slot based MAC protocol for IEEE 802.11 wireless LANs, named the Slotted-PSM (Slotted Power Saving Mechanism). To solve the problem of unnecessary energy consumption due to

CSMA/CA scheme, Slotted-PSM divides Beyond-ATIM window into several slots, and allows the node which participates in communication wakes up only in its slot times. We believe that our scheme can effectively save the energy of communication as well as increase data throughput. We verified our scheme's effectiveness by performance evaluation, and made very positive results of saving 70% energy in our Slotted-PSM compared to the IEEE 802.11 PSM.

References

1. H. Woesner, J.-P., Ebert, M. Schlager, and A. Wolisz, "Power-Saving Mechanism in Emerging Standards for Wireless LANs: The MAC Level Perspective," *in IEEE Personal Communications*, June 1998.
2. E.-S. Jung and N. H. Vaidya, "An Energy Efficient MAC Protocol for Wireless LANs," *in IEEE INFOCOM'02*, June 2002.
3. Y.-C. Tseng, C.-S. Hsu, and T.-Y. Hsieh, "Power-Saving Protocols for IEEE 802.11-Based Multi-Hop Ad Hoc Networks," *in IEEE INFOCOM'02*, June 2002.
4. M. J. Miller and N. H. Vaidya, "Improving Power Save Protocols Using Carrier Sensing and Dynamic Advertisement Window," *Technical Report, University of Illinois at Urbana-Champaign*, Dec. 2004.
5. S.-L. Wu and P.-C. Tseng, "An Energy Efficient MAC Protocol for IEEE 802.11 WLANs," *in IEEE CNSR'04*, May 2004.
6. E.-S. Jung and N. H. Vaidya, "A Power Control MAC Protocol for Ad Hoc Networks," *in ACM MOBICOM'02*, June 2002.
7. J-. So and N. H. Vaidya, "Multi-Channel MAC for Ad Hoc Networks: Handling Multi-Channel Hidden Terminals Using A Single Transceiver," *in ACM MOBIHOC'04*, May 2004.
8. M. Stemm and R. H. Katz, "Measuring and Reducing Energy Consumption of Network Interfaces in Hand-Held Devices," *in IEICE Transactions on Communications, special Issue on Mobile Computing*, 1997.
9. IEEE Std 802.11-1999, Wireless LAN Medium Access Control (MAC) and Physical Layer (PHY) Specifications. LAN/MAN Standards Committee of the IEEE Computer Society, IEEE, November 1999.
10. J.-M. Choi, J-. So and Y.-B. Ko, "Numerical Analysis of IEEE 802.11 Broadcast scheme in Multihop Wireless Ad Hoc Networks," *in ICOIN'05*, Feb. 2005.
11. The CMU Monarch Project, "The CMU Monarch Project's Wireless and Mobility Extensions to NS."
12. I.A. Getting, "The Global Positioning System," *in IEEE Spectrum 30*, Dec. 1993.

DIAR: A Dynamic Interference Aware Routing Protocol for IEEE 802.11-Based Mobile Ad Hoc Networks

Liran Ma[1], Qian Zhang[2], Fengguang An[3], and Xiuzhen Cheng[1,*]

[1] Department of Computer Science,
The George Washington University, Washington D.C.20052, U.S.A
[2] Wireless and Networking Group,
Microsoft Research Asia, P.R.China
[3] Institute of Computing Technology,
Chinese Academy of Sciences, P.R. China

Abstract. A fundamental issue impacting the performance of mobile ad hoc networks is the wireless interference among neighboring nodes. In this paper, we derive an interference aware metric NAVC based on the information collected from the IEEE 802.11 Medium Access Control (MAC) layer. We then propose a novel Dynamic Interference Aware Routing protocol (DIAR) building on NAVC and AODV [3]. Both mathematical analysis and experimental study indicate that NAVC can effectively predict available bandwidth and delay. Simulation results indicate that the overall system performance can be improved by DIAR compared to AODV.

1 Introduction

A wireless mobile ad hoc network (MANET) is an autonomous system of mobile nodes that wish to communicate with each other. Usually, the mobile nodes are powered by battery. The communications among neighboring nodes suffer from limited bandwidth, co-channel and cross-channel interferences, unidirectional links, etc. Therefore, a MANET is featured by infrastructurelessness, strict resource constraints (such as power supply, CPU processing capability, storage budget, etc.), nodes mobility, multihop communications, just name a few. As a result of these characteristics, designing a high performance routing protocol for MANET is a very challenging problem.

Many recent routing protocols [1, 2, 3, 4] choose routes with a minimum hop-count, ignoring the possibility that a longer path might offer higher throughput. Further, in these protocols, an arbitrary path among those with the same hop-count may be selected, regardless of their large differences in performance. These protocols have been shown in simulation to work very well on small to medium networks [5]. However, they only achieve a small portion of the network capacity,

* The research of Dr. Xiuzhen Cheng is supported by the NSF CAREER Award No. CNS-0347674.

X. Jia, J. Wu, and Y. He (Eds.): MSN 2005, LNCS 3794, pp. 508–517, 2005.

as reported by Das *et al.* [6]. For example, in one of their network scenarios containing 100 nodes with 2Mbps links, the throughput of each node is in the order of a few kilobits per second.

The pure hop-count metric is misleading for a long time because it may choose routes that have significantly less capacity than the best one in the network [7], especially for dense networks in which many paths with the same minimum hop-count may exist. Therefore, there still exist significant challenges in finding and choosing better routes in MANET.

Recently, several routing protocols [8, 9] with new metrics have been proposed aiming at improving performance parameters such as power consumption, throughput, and the entire network life time. In PARO [8], a power-efficient route can be constructed via a cost function based on the low energy-consuming route between a pair of nodes. In [9], a high throughput path metric called Expected Transmission Count (ETX), which is obtained from link loss characteristics between the two directions of each link and the interference among the successive links of a path, has been developed by Douglas *et al.* ETX-driven routing protocols can substantially improve the system performance, though in a scenario involving a mobile sender, the minimum hop-count routing performs considerably better because the metric does not react sufficiently quickly [10].

However, none of these state-of-the-art routing protocols touches the key to the performance improvement of MANET. As pointed out by Gupta and Kumar [11], the fundamental reason leading to the degradation of the performance as the number of nodes increases is the fact that each node has to share the radio channel with its neighbors. Subsequently, finding practical wireless interference-aware metric for network layer to make routing decisions becomes critical. As reported by [12], there still exists an opportunity for achieving throughput gains by employing an interference-aware routing protocol.

To the best of our knowledge, [13] is the only work that intends to utilize the interference information obtained from the MAC layer to improve routing efficiency. However, the interference metric in this work is not the core part leading to the routing decisions but a small portion of the proposed metric. Further, this work only considers low mobility nodes such as stationary objects or pedestrians with low speed.

To maximize the system performance, we propose DIAR, a novel Dynamic Interference-Aware Routing protocol that extends the AODV [3] with an interference aware metric, which replaces the hop-count metric in AODV. The main contributions of this paper are the followings: (i) We propose a novel interference-aware metric named Network Allocation Vector Count (NAVC) for MANET. (ii) We derive a function to predicate the possible delay and the available bandwidth based upon NAVC. (iii) The design of DIAR with the interference-aware metric NAVC is described in detail. (iv) We show that in scenarios with moderately high traffic load, our DIAR outperforms the traditional AODV though the former uses longer paths.

The remaining of this paper is organized as follows. Section 2 details the derivation of our interference aware metric. Our dynamic interference aware routing protocol is proposed in Section 3. Simulation settings and results are reported in Sections 4 and 5, respectively. We conclude this paper with a future research discussion in Section 6.

2 Interference Analysis and Metric

In this paper, we adopt the point-to-point coding model of [11] in which at any given time, one receiver only decodes messages from one sender, considering all other simultaneous transmissions purely as noise. Similarly, at any given time, one sender transmits information only to one receiver. According to this model, each node will interfere with its neighbors when sending packets, while encounter interference from its neighborhood when they are transmitting. Any node-to-node transmission adds to the level of interference experienced by other nodes.

2.1 Mathematical Preliminary

Consider the following stochastic integral equation:

$$x(t) = x(0) + \int_0^t f(x(\tau), \tau)\, d\tau + \int_0^t g(x(\tau), \tau)\, dN_\tau. \tag{1}$$

Definition: $x(\cdot)$ is a solution to Eq. (1) in the Itô sense if on an interval where N is a constant and x satisfies $\dot{x} = f(x, t)$, and if when N jumps at t_1, x changes according to Eq. (2):

$$\lim_{t \to t_1^+} x(t) = g\left(\lim_{t \to t_1^-} x(t), t_1\right) + \lim_{t \to t_1^-} x(t). \tag{2}$$

Rewrite Eq.(1) as

$$dx(t) = f(x, t)dt + g(x)dN. \tag{3}$$

Eq. (3) is called a *Poisson* counter driven stochastic differential equation (SDE) [14] and [15].

2.2 Interference Analysis

To analyze the interference, we must first study the channel contending mechanism in IEEE 802.11 protocol. The primary MAC technique of IEEE 802.11 is called *distributed coordination function* (DCF). Specifically, DCF uses a random back-off procedure to resolve medium contention conflicts and a virtual carrier-sense mechanism that exchanges RTS/CTS frames to announce the impending use of the medium. Once a node hears other nodes transmission, its Network Allocation Vector (NAV) will be set to busy state and this node has to keep silence for a duration equal to the value in the *Duration-ID* field of the RTS header. The higher the traffic rate, the larger the accumulated NAV duration value, and vice versa.

The NAV in the 802.11 MAC protocol can be a good indicator of the neighboring traffic. Therefore, according to the definition given by Eq. (3), the channel situation (free or busy) sensed by one node can be regulated via Eq. (4):

$$dx\left(t\right) = \left(1 - x\left(t\right)\right) dN_1 - x\left(t\right) dN_2 \quad x\left(0\right) \in \{0, 1\} \tag{4}$$

Eq. (4) models the on-off Markov modulated sources, which refer to the neighboring traffic. Compared to Eq. (3), Eq. (4) does not contain the term $f(x,t)$, but contains two *Poisson* counters N_1 and N_2. Subsequently, $x(t)$ remains unchanged in an interval where N_1 and N_2 are constant. When N_1 or N_2 does jump, apply Eq. (2) and $x(t)$ will flip between 0 (channel free) and 1 (channel busy). Assume λ and μ are the rates of $N_1$and N_2, respectively. Then, we can easily get the expectation of $x(t)$ in steady state, denoted by εx

$$\varepsilon x = \frac{\lambda}{\lambda + \mu} \tag{5}$$

From Eq. (5), we can conclude that the interference (channel busy) encountered by one node from its neighborhood is determined by λ and μ .

For simplicity, we assume that at each node the packet processing rate C is a constant and the buffer size is reasonably large [16]. Thus, the increment of the queue length at one node can be described by Eq. (6):

$$dv\left(t\right) = -cI_v dt + hx\left(t\right) dt \tag{6}$$

where $hx(t)$ represents the rate of packets inserted into the queue caused by the interference. Here, the notation I_v is the indicator function for $v> 0$, while v is the queue length.

Now, consider the expectation of $dv(t)$, denoted by $d\varepsilon v$.

$$d\varepsilon v = -c\varepsilon I_v dt + \varepsilon\left[hx\right] dt \tag{7}$$

Dividing both sides of Eq. (7) by dt, we get

$$\frac{d}{dt}\varepsilon v = -c\varepsilon I_v + \varepsilon\left[hx\right] \tag{8}$$

Based on [15], we can deduce Eq. (9) and Eq. (10) based on the Itô formula, the Chain Rule, Eq. (5) and Eq. (8):

$$\frac{d}{dt}\varepsilon v^2 = -2c\varepsilon v + 2h\varepsilon xv \tag{9}$$

$$\frac{d}{dt}\varepsilon vx = \varepsilon\left(1 - x\right) v\lambda - \varepsilon xv\mu - c\varepsilon x + h\varepsilon I_v x \tag{10}$$

It's very important to observe that $\varepsilon\ I_v x = \varepsilon x$ because v is positive whenever x is positive. Through Eqs. (8), (9) and (10), we get Eq. (11) and Eq. (12)

$$\varepsilon v = \left(c - h\frac{\lambda}{\lambda + \mu}\right)^{-1} \frac{h - c}{\lambda + \mu}\varepsilon[hx] \tag{11}$$

$$\varepsilon v^2 = 2\left(c - h\frac{\lambda}{\lambda+\mu}\right)^{-1}\frac{h-c}{\lambda+\mu}c\varepsilon v \tag{12}$$

Remark: $h\lambda/(\lambda+\mu) > c$ always holds under normal situation, i.e., the average packet arrival rate ought to be lower than the node's packet processing rate. Otherwise the whole system can not work properly.

$1/(\lambda+\mu)$ represents the interference situation caused by neighboring nodes. From Eq. (11) and Eq. (12), we observe that the queue length monotonously increases when $1/(\lambda+\mu)$ increases and $h\lambda/(\lambda+\mu)$ keeps constant. Furthermore, it can be proved that a queue accumulating many packets may cause the Tandem Queue effect at the downstream nodes in a path [15].

2.3 Metric

We observe that the NAV in the 802.11 MAC protocol is a good indicator of surrounding traffic because it is set by the neighboring nodes' carriers. To measure the network interference condition parameter $1/(\lambda+\mu)$, we define the NAVC as follows:

$$NAVC = \frac{The\ total\ time\ that\ the\ NAV\ is\ set}{Observation\ period} \tag{13}$$

We notice that NAVC has the following features based on our extensive simulation study:

- NAVC is insensitive to the number of users and the traffic pattern.
- If NAVC $>$ 65%, system may overflow.
- If NAVC $<$ 20%, the delay is usually negligible. The relationship between the average delay and the NAVC is plotted in Fig. 1(a).
- With average packet size, we can easily estimate the residual bandwidth by using the NAVC. The relationship between the residual bandwidth and the NAVC is plotted in Fig. 1(b).

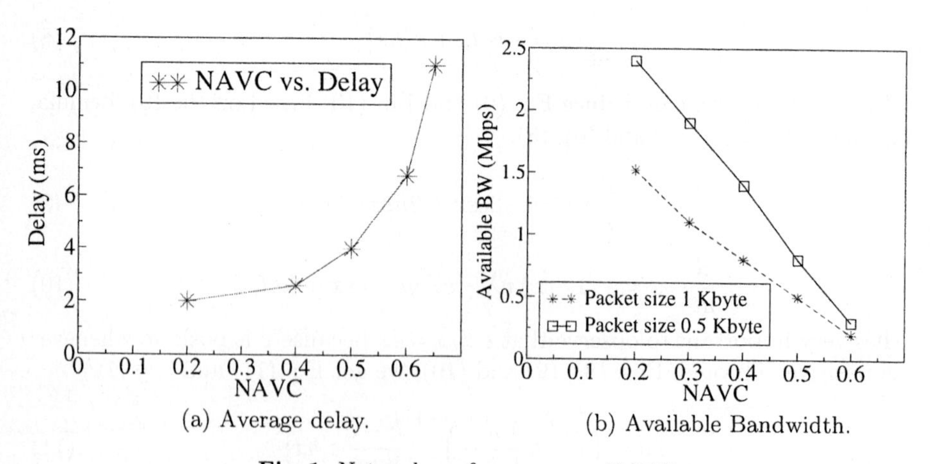

(a) Average delay. (b) Available Bandwidth.

Fig. 1. Network performance vs. NAVC

Therefore, as long as the NAV value can be sensed from the MAC layer, we can estimate the delay characteristics. If the average packet length is available, we can also estimate the residual bandwidth easily. We can approximate the delay and the available bandwidth with the following equations:

$$Delay(x) = \begin{cases} 2 \quad ms, & 0 \le x \le 0.2 \\ 2.0e^{7.9\ x - 0.2^2} ms, & 0.2 < x \le 0.65 \end{cases} \tag{14}$$

$$BW\,(L, x) = (0.0024L + 0.9) - (0.0036L + 1.4)\,x \tag{15}$$

where L is the average packet length and x refers to the NAVC value. For the detailed derivation of equations (14) and (15), we refer the readers to our previous paper [17].

Now we have introduced our interference aware metric NAVC. In the next, we will show how NAVC can be applied to make routing decisions.

3 The DIAR Scheme

As an interference aware metric, NAVC can be applied to any routing protocol proposed for 802.11-based networks. In this paper, we choose to build our dynamic interference aware routing protocol, i.e. the DIAR, by extending AODV with NAVC. AODV has been well-studied by the wireless ad hoc network society. It is a *reactive* routing protocol proposed for wireless ad hoc networks.

As an extension to AODV, DIAR is also an on-demand algorithm, building routes via a route request and route reply query cycle with the interference aware metric NAVC. The key idea of DIAR is to avoid selecting a route which has nodes that are severely interfered by their neighborhood such that the energy and delay caused by contention are minimized. In DIAR, each node computes its NAVC by periodically collecting the NAV values from its MAC layer. The operations of DIAR is sketched below.

Step 1: Route Request

When a source node desires a route to a destination for which it does not already have a path, it broadcasts a *route request* (RREQ) packet across the network. There are two extra fields namely *heavy_node_number* and *nav_sum* in the RREQ message compared to the AODV RREQ message. After receiving the RREQ packet, a node other than the destination has three options before continue broadcasting this message.

- If its NAVC is larger than 65%, increase the value of *heavy_node_number* by one and the value of *nav_sum* by the square of its NAVC.
- If its NAVC is somewhere between 20% and 65%, increase the value of *nav_sum* by the square of its NAVC only.
- In other cases, make no modification to these two fields.

Step 2: Route Reply

When the destination node receives any non-duplicated and valid RREQ message, it will send the *route reply* (RREP) message immediately. If an intermediate node already has a valid route to the destination, it can generate the RREP too.

Step 3: Route Update and maintenance

When any of the following conditions is satisfied, the route update process will be activated at the source, the destination or intermediate nodes:

- Receives a *route renew* message.
- Receives a *route failure* message.
- Receives another RREQ message indicating a better route: a new route with fewer *heavy_node_number*; or less *nav_sum* with the same *heavy_node_number*; or smaller hop-count with the same *heavy_node_number* and *nav_sum.*

Otherwise, keep the most recently used paths for forwarding packets. Via this approach, our DIAR may achieve the desired objective of improving the network throughput in a fully distributed manner.

4 Simulation Settings

In order to be consistent with previous works [5] and [6], we have exploited the following parameters in our simulations.

- There are various numbers of nodes in an area with length and width of 670 meters. The nodes are located according to a *Poisson* point process over the plane.
- The mobility model is the random *waypoint* model [5]. Each node moves from a random location to a random destination with a randomly chosen speed (uniformly distributed between 0 and 20 meter/sec).
- The traffic sources are UDP Constant Bit Rate (CBR) and the communication pairs are randomly chosen from the entire network.
- The number of source-destination pairs and the packet sending rate of each pair is varied to change the offered load in the network. Each pair of UDP session starts at a staggered time.
- Every single simulation run lasts for 300 simulated seconds.
- We fixed the packet size to 1024 byte.
- Radio link bandwidth is 2Mbps with 250 meter nominal range in maximum.

5 Simulation Results

We have implemented DIAR in *ns-2* [18] with its wireless extensions to evaluate the performance of DIAR. In the first scenario, the networks contain 30, 40 or 50 nodes with 6 to 10 CBR traffic communication pairs. The throughput (Kbps), which is defined over all data packets successfully received only at the destination nodes, is obtained by averaging over all the flows in all given runs. The average improvement of total throughput of DIAR compared with AODV is around 15%; For the best case, the improvement of the total throughput is about 30%. Details are shown in Figures 2, 3 and 4.

Next, in order to give a higher level view on the throughput improvements, we then compare the total throughput for different numbers of nodes ranging from 20 to 50 at a step of 5. The average improvement of the total throughput is 13%;

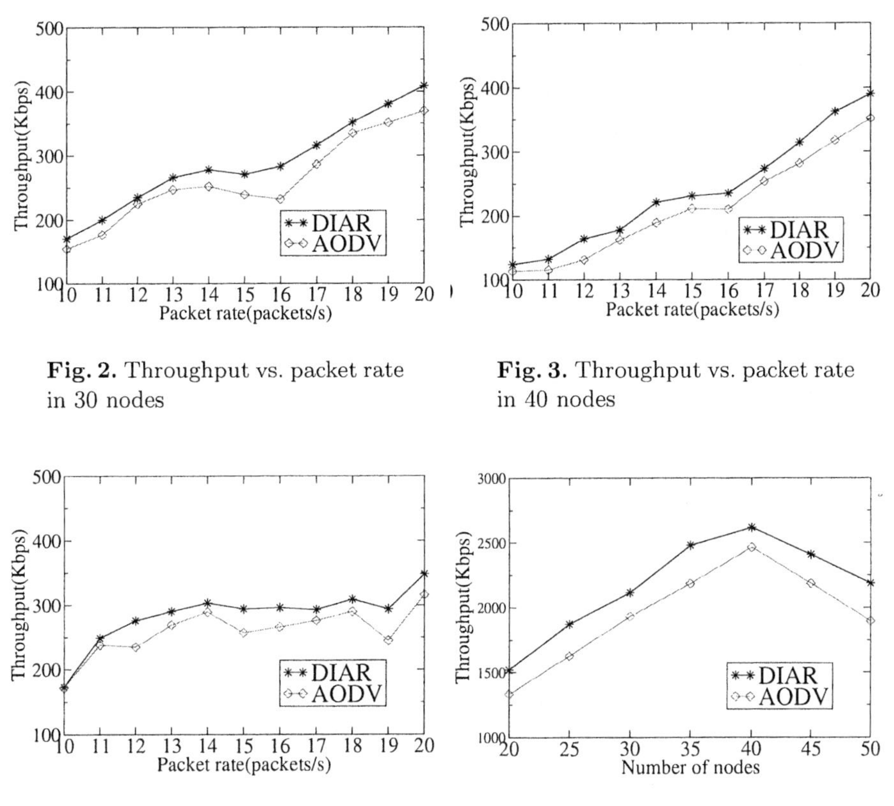

Fig. 2. Throughput vs. packet rate in 30 nodes

Fig. 3. Throughput vs. packet rate in 40 nodes

Fig. 4. Throughput vs. packet rate in 50 nodes

Fig. 5. Total throughput vs. the number of nodes

For the best case, the improvement of the total throughput is 29%. Detailed information is given in Figure 5.

From our simulation we draw the following conclusions: (i) The throughput improvement is the best at moderate traffic loads and is negligible at very low or very high loads; (ii) The total throughput improvement is not as high as expected because there does not exist many better routes, i.e. DIAR ends up selecting the same path as AODV. One potential solution to this problem is to apply power control mechanism to increase the spatial reuse level so that more paths could become available. (iii) NAVC is sensitive under highly dynamic situation, which leads to unnecessary rerouting. Solving this problem needs to investigate the appropriate mapping between NAVC and the hop-count.

6 Conclusions and Future Work

In this paper, we have derived an interference aware metric NAVC and proposed a dynamic interference aware routing protocol DIAR for MANET. Our protocol is fully distributed, which exploits the MAC layer information to make routing decisions. We have theoretically and experimentally shown that DIAR is interference-aware and it improves the network throughput. As a part of future works, we will investigate better equations regulating the mapping between NAVC and delay, and NAVC and available bandwidth. We also propose to apply our DIAR mechanism to wireless sensor networks.

References

1. D. B. Johnson. "Routing in ad hoc networks of mobile hosts," *In Proc. of the IEEE Workshop on Mobile Computing Systems and Applications*, pp. 158-163, December 1994.
2. C. E. Perkins and P. Bhagwat. "Highly dynamic Destination-Sequenced Distance-Vector routing (DSDV) for mobile computers," *In Proc. ACM SIGCOMM Conference*, pp. 234-244, August 1994.
3. C. E. Perkins, E. M. Belding-Royer, and S. R. Das, "Ad hoc ondemand distance vector (AODV) routing," http://www.ietf.org/internetdrafts/draft-ietf-manet-aodv-10.txt, January 2002, work in progress.
4. D. B. Johnson, D. A. Maltz, Y.-C. Hu, and J. G. Jetcheva, "The dynamic source routing protocol for mobile ad hoc networks (DSR)," http://www.ietf.org/internet-drafts/draft-ietf-manetdsr-07.txt, February 2002, work in progress.
5. J. Broch, D. A. Maltz, D. B. Johnson, Y.-C. Hu, and J. Jetcheva. "A performance comparison of multi-hop wireless ad hoc network routing protocols," In *Proc. ACM/IEEE MobiCom*, October 1998.
6. S. R. Das, C. E. Perkins, and E. M. Royer, "Performance Comparison of Two On-demand Routing Protocols for Ad Hoc Networks," *in Proceedings of Infocom 2000*, Tel-Aviv, Israel, March 2000.
7. D. S. J. De Couto, D. Aguayo, B. A. Chambers, and R. Morris, "Performance of multihop wireless networks: Shortest path is not enough," in *Proceedings of the First Workshop on Hot Topics in Networks (HotNets-I)*. Princeton, New Jersey: ACM SIGCOMM, October 2002.
8. J. Gomez, A. T. Campbell, M. Naghshineh, and C. Bisdikian, "PARO: Supporting Dynamic Power Controlled Routing in Wireless Ad Hoc Networks," *ACM/Kluwer Journal on Wireless Networks (WINET)*, 2003.
9. D. S. J. De Couto, D. Aguayo, J. Bicket, and R. Morris, "A high throughput path metric for multi-hop wireless routing," *in Proc. of the ACM MOBICOM*. ACM Press, 2003, pp. 134-146.
10. R. Draves, J. Padhye, and B. Zill, "Comparison of routing metrics for static multi-hop wireless networks," *ACM SIGCOMM' 2004*, Portland, OR, August 2004.
11. P. Gupta and P. R. Kumar, "The capacity of wireless networks," *IEEE Transactions on Information Theory*, vol. IT-46, no. 2, pp. 388-404, March 2000.
12. K. Jain, J. Padhye, V. N. Padmanabhan, and L. Qiu, "Impact of interference on multi-hop wireless network performance," in *Proc. of the ACM MOBICOM*, San Diego, CA, Sept. 2003.
13. W. H. Yuen, H. Lee, and T. D. Andersen, "A simple and effective cross layer networking system for mobile ad hoc networks," *The 13th IEEE International Symposium on Personal, Indoor and Mobile Radio Communications,* Vol. 4, pp. 1952 -1956, 2002.
14. R.W. Brockett*, Stochastic Control,* Lecture Note, Harvard University, Cambridge, MA. 1983.

15. R. Brockett, W. Gong, and Y. Guo, "Stochastic analysis for fluid queueing systems," in *IEEE Conference on Decision and Control (CDC) 99*, vol. 3, Dec 1999, pp. 3077-3082.
16. Z. Fu, P. Zerfos, H. Luo, S. Lu, L. Zhang and M. Gerla "The Impact of Multihop Wireless Channel on TCP Throughput and Loss," *Proceedings of IEEE INFOCOM'03, 2003,* San Francisco.
17. Q. Zhang, C. Guo, Z. Guo, and W. Zhu, "Efficient mobility management for vertical handoff between wwan and wlan," *IEEE Communications Magazine*, vol. 41, pp. 102-108, Nov 2003.
18. The Network Simulator - ns-2, http://www.isi.edu/nsnam/ns/.

A Low-Complexity Power Allocation Scheme for Distributed Wireless Links in Rayleigh Fading Channels with Capacity Optimization*

Dan Xu[1], Fangyu Hu[1], Qian Wang[2], and Zhisheng Niu[3]

[1] University of Science and Technology of China, Hefei 230027, P.R. China
xudan03@mail.ustc.edu.cn
[2] Chinese Academy of Sciences,
Institute of Microsystem and Information Technology, Shanghai 200050, P.R. China
[3] Tsinghua University, Department of Electronic Engineering,
Beijing 100084, P.R. China

Abstract. In this paper, a distributed power allocation scheme for wireless links in Rayleigh fast fading environments is proposed. An optimization model is formulated to minimize total transmit powers with constraint conditions that outage probability of each link should be satisfied. A closed-form near optimal solution to the model is presented based on the exploration of power ratio factor at the boundary of system capacity. In terms of the scheme, each link can determine its power ratio factor independently. The inequity problem of the power allocation algorithms in incremental fashion is avoided. Meanwhile the scheme is in low computational complexity due to closed-form solution and it is more time-saving than those schemes in iterative manner. By performance evaluation, we verify our analysis on the optimization model and validate a better result of our scheme in comparison with the counterpart. That is, a larger system capacity is admissible by our method.

1 Introduction

Power allocation or power control is very significant for wireless link. A wise power allocation scheme can prolong the battery life of each node, moreover, quality of service(QoS) of each link is guaranteed and system capacity can be enhanced. The difficulty to design efficient power allocation schemes for distributed wireless links mainly comes from the distributed attribute. Meanwhile the time-varying nature of wireless links inflicts another great challenge.

Much work [1][2][3][4][5][7][9] has investigated distributed power allocation for distributed wireless links or ad hoc networks. Among these schemes, there is a so-called incremental fashion such as PCDC[1], PCMA[2] and DRNP[3]. In the incremental method, no optimization objective, e.g., minimizing total transmit

* This work is done during the visit of the first author to Department of Electronic Engineering, Tsinghua University. This work has been partially supported by the Nature Science Foundation of China (No.60272021).

X. Jia, J. Wu, and Y. He (Eds.): MSN 2005, LNCS 3794, pp. 518–527, 2005.

powers is available. An existing link will restrict transmit power of an incoming link to guarantee its QoS requirement, i.e., the signal-to-interference-and-noise ratio(SINR) threshold. Thus inequity problem that former links have more privileges than the latter is unavoidable. Also improper power allocation to an existing link will block a new link, which is the defect of the lack of renegotiation of network resource for optimization. Different from the incremental scheme, [4]-[10] propose power allocation schemes including F-M, DCPC, DPC/ALP, DPCMAC etc., for ad hoc networks or distributed wireless links in the iterative manner. An optimization model to minimize total transmit powers is often formulated. In this kind of scheme, SINR requirement of each link is satisfied simultaneously. Each link steps to the optimal value of transmit power iteratively. The iterative method will inevitably be time-consuming. However, in real scenario, network topology of wireless links or ad hoc networks varies fast. Thus the optimal convergence point may not be reached.

The schemes discussed above only consider large-scale fading by assuming static channel gains between transmitters and receivers. However, fast fading makes channel gains fluctuate randomly. In the paper we will discuss power allocation under Rayleigh fast fading channels. In [11], the authors also investigate power allocation scheme for affected by Rayleigh fast fading. An optimization model to minimize total transmit powers is formulated as a geometric programming problem. However, no distributed power allocation scheme is proposed. Another limitation is that background interference at each receiver is ignored. These drawbacks make it not suitable for distributed wireless links. The rest of this paper is organized as follows. Section 2 presents the network model and a novel definition of system capacity. Section 3 formulates a global optimization model and gives the near optimal solution. Section 4 is performance evaluation and discussions. Finally, the conclusions are drawn in Section 5.

2 Network Model and Definition

Suppose the ith link in a scenario, the transmit power of the ith link is P_{ti}, the receive power is P_{ri}. Since the link is affected by distance dependent path loss, shadowing loss and fast fading, the relation between P_{ti} and P_{ri} is

$$P_{ri} = F_{ii} G_{ii} P_{ti}, \tag{1}$$

where G_{ii} represents path loss and shadowing loss, which can be deemed as a constant for a relatively long time scale. F_{ii} represents the effects by fast fading. In Rayleigh fast fading channels, receive power is deferring to negative exponential distribution. Then both F_{ii} and P_{ri} are negative exponential random variables. Also there is $E[P_{ri}]=G_{ii}P_{ti}$. Assume the kth link is adjacent to the ith link, whose transmitter will bring interference to the receiver of the ith link. The interference is denoted by I_{ki}. Here we assume interference also suffers fast fading. Then I_{ki} is also a random variable deferring to negative exponential distribution. There is $E[I_{ki}]=G_{ki}P_{tk}$, where G_{ki} represents path loss form the kth transmitter to the ith receiver. SINR at ith receiver is given by

$$\Gamma_i = \frac{P_{ri}}{\sum_{k \neq i} I_{ki} + I_i^0}, \tag{2}$$

here I_i^0 is a constant which represents background interference at the ith receiver. Γ_i is also a random variable. Let γ_i is the minimum SINR requirement of the ith link, the outage probability of the ith link is

$$O_i = \Pr\left\{ \frac{P_{ri}}{\sum\limits_{k \neq i} I_{ki} + I_i^0} < \gamma_i \right\} = \Pr\left\{ P_{ri} < \gamma_i \sum_{k \neq i} I_{ki} + \gamma_i I_i^0 \right\}. \tag{3}$$

Since P_{ri} and I_{ki} are deferring to negative exponential distribution. It can be deduced that O_i is represented by the following expression:

$$O_i = 1 - (1 + \gamma_i) e^{-\frac{I_i^0 \gamma_i}{G_{ii} P_{ti}}} \prod_k \left(\frac{G_{ii} P_{ti}}{G_{ii} P_{ti} + \gamma_i G_{ki} P_{tk}} \right), \tag{4}$$

here subscript k includes i. Outage probability should be smaller than a predefined threshold. The threshold is determined in terms of the traffic type of the ith link.

2.1 The Definition of System Capacity

For distributed wireless links with multimedia traffics carried, system capacity does not mean the number of active links only. QoS of each link should be the other factor of system capacity. In the paper, QoS of a link can be represented by (Γ_i, O_i), where Γ_i and O_i are SINR and outage probability respectively. We give the definition of system capacity as follows,

$$S = \{N, ((\Gamma_i, O_i), (i = 1, \ldots, N))\}. \tag{5}$$

Here N is the number of active links. System capacity S has boundary S_{max}, beyond which no any power allocation is feasible, i.e., the capacity can not be sustained. A wiser power allocation scheme makes capacity admissible closer to the boundary. Equivalently, with the same capacity, a better power allocation scheme will provide a lower transmit power for each link.

3 Optimization Model and Solution

Suppose there are N active links in the networks, each link is associated with specific QoS requirements. We explore the optimization model as

$$\begin{cases} \min \sum_{i=1}^{N} P_{ti} \\ s.t. \\ O_i = 1 - (1 + \gamma_i) e^{-\frac{I_i^0 \gamma_i}{G_{ii} P_{ti}}} \prod_{k=1}^{N} \left(\frac{G_{ii} P_{ti}}{G_{ii} P_{ti} + \gamma_i G_{ki} P_{tk}} \right) \leq \delta_i, \\ \qquad (i = 1, \ldots, N). \end{cases} \tag{6}$$

In the optimization model (6), the constraint conditions require outage probability of each link not exceed a threshold. Concerning the model, we give the following theorem.

Theorem 1. *If (6) has optimal solution, the solution exactly makes outage probability of each link equal to the threshold.*

Proof. The outage probability of the ith link is a mono-decreasing function of P_{ti}. While it is a mono-increasing function of P_{tj} $(j = 1, \ldots, N, j \neq i)$. Suppose the optimal solution to (6) is P_{ti}^0 $(i = 1, \ldots, N)$. If there is a link's outage probability is not equal to the upper limit, without loss of generality, let it be the kth link , i.e., $O_k < \delta_k$. Since O_k is a mono-deceasing function of P_{tk}, there must be a decrement ΔP_{tk}, by which P_{tk}^0 is reduced to $P_{tk}^0 - \Delta P_{tk}$. Meanwhile the resulted outage probability will increase but inequity $O_k < \delta_k$ is still met due to the continuous attribute of function O_k. The outage probability of other links, i.e., O_i $(i = 1, \ldots, N, i \neq k)$ will decease with the increment of P_{tk}^0, that is, the constraint condition in (6) of every link will still be satisfied. So we find a feasible solution, i.e., $(P_{t1}^0, \ldots, P_{tk}^0 - \Delta P_{tk}, \ldots, P_{tN}^0)$ for (6), which makes the constraint condition still met while the sum of the transmit powers is smaller than that of the optimal solution P_{ti}^0 $(i = 1, \ldots, N)$. Then the hypothesis of $O_k < \delta_k$ is invalid and all constraints in (6) must be met equity at optimal solution.

According to Theorem 1, the optimal solution to optimization model (6) is equivalent to that of the following equation set:

$$e^{-\frac{I_i^0 \gamma_i}{G_{ii} P_{ti}}} \prod_{k=1}^{N} \left(\frac{G_{ii} P_{ti}}{G_{ii} P_{ti} + \gamma_i G_{ki} P_{tk}} \right) = \frac{1 - \delta_i}{1 + \gamma_i}, \qquad (i = 1, \ldots, N). \qquad (7)$$

Equation (7) is a non-linear equation set, the solution is very hard to derive. Thus we should find a near-optimal solution with low computation complexity. Meanwhile, the distributed links require the solution can decide each transmit power as distributed as possible. We explore the follow theorem under the criterions.

Theorem 2. *With the constraint conditions of (6) are met equity, increasing each transmit power to infinite by the same scale makes system capacity boundary S_{max} admissible.*

Proof. By our definition of system capacity, it includes two factors: the number of links and QoS. Capacity boundary may mean the largest number of links, or the highest QoS, thus proof to this theorem should be divided into two parts. First, given a unique QoS requirements, we prove Theorem 2 by the angle of link numbers. After that, we will prove Theorem 2 from the layer of QoS.

Given the power allocation P_{ti} $(i = 1, \ldots, N)$ for each transmitter, the current active number of links is N. Outage probability of each link is a mono-increasing function of N, i.e., $\frac{\partial O_i}{\partial N}$ $(i = 1, \ldots, N)$. Suppose the maximum number of links admissible is N_{max}, meanwhile transmit power of the ith link is P_{ti}, $(i = 1, \ldots, N_{max})$. When the maximum number of links N_{max} is given, by Theorem 1, there is

$$O_i = 1 - (1 + \gamma_i) e^{-\frac{I_i^0 \gamma_i}{G_{ii} P_{ti}}} \prod_k \left(\frac{G_{ii} P_{ti}}{G_{ii} P_{ti} + \gamma_i G_{ki} P_{tk}} \right) = \delta_i, \quad (i = 1, \ldots, N_{max}). \tag{8}$$

We use the counterevidence method to prove that P_{ti} $(i = 1, \ldots, N_{max})$ must tend to be infinite simultaneously to accommodate the maximum number of links. Assuming P_{ti} $(i = 1, \ldots, N_{max})$ is finite, then we can set $P_{ti}{}' = M P_{ti}$ $(i = 1, \ldots, N_{max})$. Now outage probability of link the ith link is

$$O_i = 1 - (1 + \gamma_i) e^{-\frac{I_i^0 \gamma_i}{G_{ii} M P_{ti}}} \prod_k \left(\frac{G_{ii} M P_{ti}}{G_{ii} M P_{ti} + \gamma_i G_{ki} M P_{tk}} \right), \quad (i = 1, \ldots, N_{max}). \tag{9}$$

By (8) and (9) we can see scaling P_{ti} $(i = 1, \ldots, N_{max})$ up by M times is equivalent to scaling I_i^0 down by M times. Thus outage probability of each link is deceased by this adjustment, i.e.,

$$O_i < \delta_i \qquad (i = 1, \ldots, N_{max}). \tag{10}$$

Since O_i $(i = 1, \ldots, N)$ is a mono-increasing function of the number of links N, then we can find an increment ΔN, which makes the outage probability of each link increase yet (10) can still be met. Then we find a larger system capacity $N_{max} + \Delta N$, with transmit power $P_{ti}{}'$ $(i = 1, \ldots, N_{max} + \Delta N)$. Then it violate the fact that N_{max} is the maximum number of links. Hence the assumption that power P_{ti} $(i = 1, \ldots, N_{max})$ is finite is invalid.

The above has proved Theorem 2 in the number of links sense. Considering QoS, the theorem also follows.

Actually increasing transmit power of each link is equal to deceasing background interference at each receive node to zero. Then capacity loss caused by background interference is eliminated, thus system capacity boundary is approached. Here background interference is distributed at each receiver with different levels. Nevertheless, the distinction of background interference is negligible when transmit power of each link reaches infinite. We have mentioned in [12] that in wireless cellular networks, capacity loss caused by background noise at BS can be compensated by increasing each user's power. Actually, the circumstance that each transmit power reaches infinite should be a ideal condition, by which we can investigate the nature of the relations among each transmit power. We further give a theorem as below.

Theorem 3. *When transmit power of each link reaches infinite, the ratio among* P_{ti} $(i = 1, \ldots, N)$ *uniquely determine outage probability of each link.*

Proof. Let $P_{ti}=\eta_i\phi$, $(i=1,\ldots,N)$. η_i is power ratio factor of the ith link , ϕ is a constant. Then outage probability O_i can be written as

$$O_i = 1-(1+\gamma_i)e^{-\frac{I_i^0\gamma_i}{G_{ii}\eta_i\phi}}\prod_k\left(\frac{G_{ii}\eta_i}{G_{ii}\eta_i+\gamma_i G_{ki}\eta_k}\right), \quad (i=1,\ldots,N). \tag{11}$$

By (11), We see that increasing ϕ leads to a smaller O_i. When ϕ reaches infinite, $e^{-\frac{I_i^0\gamma_i}{G_{ii}\eta_i\phi}}$ is equal to one. Thus O_i is exclusively determined by η_i $(i=1,\ldots,N)$, and has no relation with ϕ. Theorem 3 follows.

By the above analysis, we first explore power ratio among each transmit power at capacity boundary. By Theorem 3, transmit power of the ith link can be represented by $P_{ti}=\eta_i\phi$, here ϕ is a power constant factor which is the same of each link. At system capacity boundary, ϕ reaches ∞. η_i is the power ratio factor of the ith transmitter. Then η_i $(i=1,\ldots,N)$ determines O_i $(i=1,\ldots,N)$ completely at capacity boundary. At the boundary S_{max}, (7) can be written as

$$1-(1+\gamma_i)\prod_{k=1}^{N}\left(\frac{G_{ii}\eta_i}{G_{ii}\eta_i+\gamma_i G_{ki}\eta_k}\right)=\delta_i, \quad (i=1,\ldots,N). \tag{12}$$

Equation (12) is equivalent to

$$\ln G_{ii}\eta_i = \frac{1}{N}\ln\frac{1-\delta_i}{1+\gamma_i}+\frac{1}{N}\sum_{k=1}^{N}\ln(G_{ii}\eta_i+\gamma_i G_{ki}\eta_k), \quad (i=1,\ldots,N). \tag{13}$$

Concerning the ith link and the jth one, in terms of (13) , there is

$$\ln\frac{G_{ii}\eta_i}{G_{jj}\eta_j}=\frac{1}{N}\ln\frac{(1-\delta_i)(1+\gamma_j)}{(1-\delta_j)(1+\gamma_i)}+\frac{1}{N}\sum_{k=1}^{N}\ln\frac{(G_{ii}\eta_i+\gamma_i G_{ki}\eta_k)}{(G_{jj}\eta_j+\gamma_j G_{kj}\eta_k)}, \quad (i,j=1,\ldots,N). \tag{14}$$

The second item on the left hand side depends on the power ratio factor of each link. However, we can make an approximation here. Notice the last item of the equation should multiply a factor $1/N$, we can suppose power ratio factor of each link is similar, this will not bring too much influence on the last item, especially when QoS requirements of each link is similar. Then (14) can be written as

$$\ln\frac{G_{ii}\eta_i}{G_{jj}\eta_j}=\frac{1}{N}\ln\frac{(1-\delta_i)(1+\gamma_j)}{(1-\delta_j)(1+\gamma_i)}+\frac{1}{N}\sum_{k=1}^{N}\ln\frac{(G_{ii}+\gamma_i G_{ki})}{(G_{jj}+\gamma_j G_{kj})}, \quad (i,j=1,\ldots,N). \tag{15}$$

By (15), we get power ratio between the ith link and the jth one as follows,

$$\frac{\eta_i}{\eta_j}=\frac{G_{jj}}{G_{ii}}\frac{[(1-\delta_i)(1+\gamma_j)\prod_{k=1}^{N}(G_{ii}+\gamma_i G_{ki})]^{\frac{1}{N}}}{[(1-\delta_j)(1+\gamma_i)\prod_{k=1}^{N}(G_{jj}+\gamma_j G_{kj})]^{\frac{1}{N}}}, \quad (i,j=1,\ldots,N). \tag{16}$$

Then we get power ratio factor of the ith link at system capacity boundary. That is

$$\eta_i = \frac{1}{G_{ii}} \left[\frac{(1-\delta_i)}{(1+\gamma_i)} \prod_{k=1}^{N} (G_{ii} + \gamma_i G_{ki}) \right]^{\frac{1}{N}}, \qquad (i = 1, \ldots, N). \tag{17}$$

In term of expression of (17), each link can determine its power ratio factor by its own information, which includes the link's QoS requirements and path loss between the receiver and other transmitters. However, the independency sacrifices a little capacity loss by the approximation. Another merit of power ratio factor is the inequity problem is avoided. Determination of transmit power has no relation with the sequence of affiliation of each link. By (17), transmit power of the ith link at system capacity boundary is

$$P_{ti} = \frac{\phi}{G_{ii}} \left[\frac{(1-\delta_i)}{(1+\gamma_i)} \prod_{k=1}^{N} (G_{ii} + \gamma_i G_{ki}) \right]^{\frac{1}{N}}, \phi \to \infty, \qquad (i = 1, \ldots, N). \tag{18}$$

Equation (18) is the solution to optimization model (6) at the capacity boundary S_{max}. We know power level of each transmitter is very limited, the infinite value of power constant factor ϕ can not be achieved. However, within capacity boundary, increasing transmit power of each link to infinite will make outage probability smaller than the threshold. Thus, we can scale ϕ down to the level that the outage probability, i.e., O_i $(i = 1, \ldots, N)$, is equal to δ_i exactly. Concerning the ith link, according to constraint condition in (6), there is

$$1 - (1+\gamma_i) e^{-\frac{I_i^0 \gamma_i}{G_{ii}\eta_i \phi}} \prod_{k=1}^{N} \left(\frac{G_{ii}\eta_i}{G_{ii}\eta_i + \gamma_i G_{ki}\eta_k} \right) \leq \delta_i. \tag{19}$$

By deduction, The minimum value of ϕ satisfying (19) is

$$\phi_i = \frac{I_i^0 \gamma_i}{G_{ii}\eta_i \ln \left(\frac{1+\gamma_i}{1-\delta_i} \prod_{k=1}^{N} \frac{G_{ii}\eta_i}{G_{ii}\eta_i + \gamma_i G_{ki}\eta_k} \right)}. \tag{20}$$

The maximum one of ϕ_i should be selected as the uniform measure for every link, i.e., $\phi_{max} = \max\{\phi_i\}$ $(i = 1, \ldots, N)$. Then we have given a near-optimal closed-form solution for (6), i.e., $P_{ti} = \eta_i \phi_{max}$ $(i = 1, \ldots, N)$.

The solution is a near-optimal power allocation scheme for distributed wireless links in fast fading channels, which is called NOPSF. We have mentioned the existed algorithms often regard channel states be constant. When the channels suffer fast fading, each link must enhance transmit power to overcome deep fading. We select a scheme form these methods which can provide the largest capacity as the optimal power allocation scheme for static channels (OPSSC), the OPSSC method can be formulated by the following model.

$$\begin{cases} \min \sum_{i=1}^{N} P_{ti} \\ s.t. \\ \Gamma_i = \frac{G_{ii} P_{ti}}{\Sigma_{k=1,k \neq i}^{N} G_{ki} P_{tk} + I_i^0} \geq \gamma_i, (i = 1, \ldots, N). \end{cases} \tag{21}$$

Similar to optimization model (6), the solution will make constraint condition in (21) met equity. Hence OPSSC method is converted to the solution to the following equation set.

$$\frac{G_{ii}P_{ti}}{\Sigma_{k=1,k\neq i}^{N} G_{ki}P_{tk} + I_i^0} = \gamma_i, \quad (i = 1, \ldots, N). \tag{22}$$

Equation (22) is a linear equation set whose solution is easy to derive. Here we do not give the expression of the solution. The OPSSC method make the largest capacity admissible among all the schemes which suppose channel gains are static. However, concerning OPSSC scheme, it must increase each transmit power when the channels suffer fast fading. That is, outage probability requirement of each link should also be satisfied. Hence we will compare our scheme with OPSSC to verify if our scheme can sustain a larger capacity.

4 Performance Evaluation

First, by Fig. 1, we compare the performance of the proposed NOPSF with that of OPSSC method in all fast fading channels. We simulate a series of power allocations for a specific time duration. In this procedure, a link with specific QoS requirements will affiliate in the network, or leave the network. QoS requirements of each link is randomly selected from (4, 0.09), (5, 0.1), (5, 0.08), and (6, 0.12). Background interference ranges from 0.1W to 0.15W randomly at each link. Under the same capacity, network topology and background interference, both NOPSF and OPSSC are implemented respectively. The y-axis denotes the sum of transmit powers of all the links, while x-axis is the power allocation cycle. Although the parameter studied is sum of transmit powers, we can also analyze which scheme makes a larger system capacity admissible. That is, under the same capacity, the scheme with higher powers or failed power allocation provides a smaller capacity admissible. It is observed in the figure, in each cycle, the proposed NOPSF always provides a lower sum of transmit powers. While

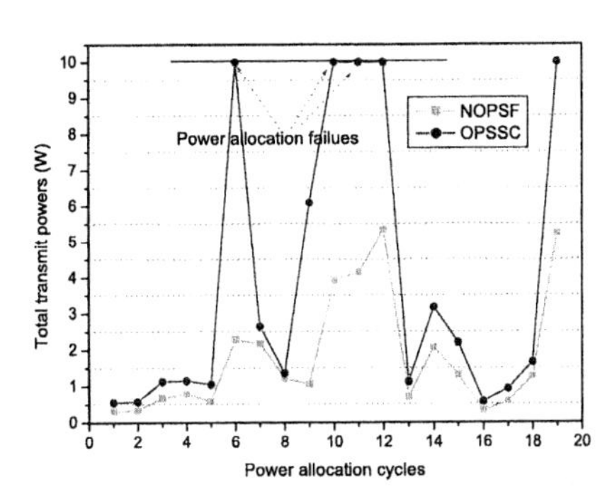

Fig. 1. Comparison between OPSSC and NOPSF in fast fading channels

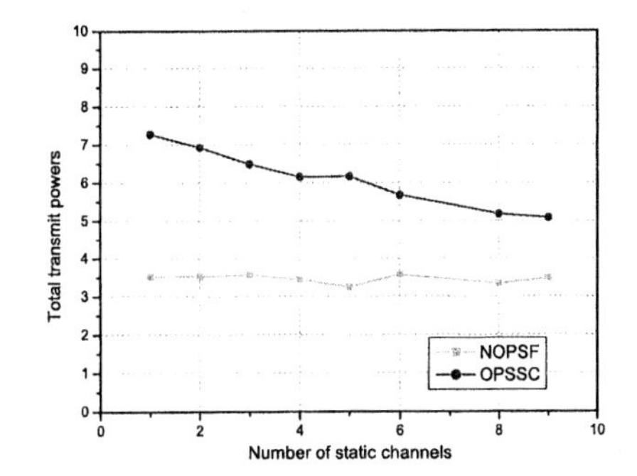

Fig. 2. Comparison between OPSSC and NOPSF in terms of the number of static channels

Table 1. Parameters of the active links

Parameter	Symbol	Link					
		1	2	3	4	5	6
SINR	γ_i	4	5	3	5	4	5
Outage probability	δ_i	0.122	0.134	0.097	0.118	0.135	0.126
Background interference	I_i^0(W)	0.13	0.15	0.16	0.17	0.16	0.13

for OPSSC, the sum of transmit powers is higher. Meanwhile, under relatively larger capacity(A large number of links, or strict QoS requirements), OPSSC method fails to provide a feasible power for each link. Here the points of failed power allocations are marked by a top solid line. Under the same circumstances, it is observed that NOPSF still make the capacity admissible by feasible power allocation. Hence NOPSF can sustains a capacity larger than OPSSC.

When the static channels coexist with fast fading channels in the networks. OPSSC method should provide a better performance comparing to the scenario where all the channels suffer fast fading. Under the circumstances, NOPSF may not be so excellent. Then we compare the two schemes in terms of the number of static channels by Fig. 2. Here we study six fixed links under specific network topology. The parameters are listed in Tab. 1. We set a scenario that static channels coexist with fast fading channels. We Stat. the average sum of transmit powers allocated by NOPSF and OPSSC under the same number of static channels. It is illustrated in the figure, when the static channels increase, the sum of transmit powers of OPSSC method decrease, this is reasonable since OPSSC is optimal for the scenario where there is no fast fading. Concerning NOPSF, there is no obvious decease of sum of transmit powers. However, it is observed NOPSF can still provide a larger capacity, or a low transmit power when static channels coexist with fast fading channels.

Actually the bad performance of OPSSC in fast fading channels comes from the improper determination of power ratio factor of each link. OPSSC method provides a optimal solution for static channels, which means power ratio among each link is optimal in this instance. However, it must enhance transmit power of each link when it suffers fast fading, then the optimal power ratio factors become improper, which will bring capacity loss. The proposed NOPSF first provides a wise power ratio factor for each link in the circumstances of fast fading, although it is not the optimal. Thus a better performance is brought forth by NOPSF.

5 Conclusions

In this paper, we propose a novel power allocation scheme called NOPSF for distributed wireless links in fast fading channels. We formulate an optimization model to minimize total transmit powers. The system capacity is defined as a combination of the link number and QoS of each link. A near optimal solution is presented with low computational complexity, which provides a basis for our distributed power allocation scheme. By performance evaluation, it is proved that NOPSF makes a larger system capacity admissible than the OPSSC scheme.

References

1. A. Muqattash and M. M. Krunz: A distributed transmission power control protocol for mobile ad hoc networks. IEEE Transactions on Mobile Computing, vol. 3, no. 2, Apr.-June. 2004.
2. J. Monks, V. Bharghavan, and W. Hwu: A Power Controlled Multiple Access Protocol for Wireless Packet Networks. Proc. IEEE INFOCOM'01, Apr. 2001.
3. S. Lal and E. S. Sousa: Distributed resource allocation for DS-CDMA-based multimedia ad hoc wireless LAN's. IEEE J. Select. Areas Commun., vol. 17, no. 5, May. 1999.
4. J. Sun, Y. Hu, W. Wang, and Y. Liu: Channel Access and Power Control Algorithms with QoS for CDMA based Wireless Ad-Hoc Networks. IEEE PIMRC'03, Sept. 2003.
5. T. ElBatt, A. Ephremides: Joint scheduling and power control for wireless ad hoc networks. IEEE Trans. on Wireless Communications, Vol. 3, pp. 74-85, Jan. 2004.
6. G. J. Foschini and Z. Miljanic: A simple distributed autonomous power control algorithm and its convergence. IEEE Trans. Veh. Tech., vol. 42, pp. 641-646, Apr. 1993.
7. S. A. Grandhi, J. Zander, and R. D. Yates: Constrained Power Control. Wireless Personal Commun., vol.1, no. 4, pp. 257-270, 1995.
8. N. Bambos, S. C. Chen, and G. J. Pottie: Channel access algorithms with active link protection for wireless communication networks with power control. IEEE/ACM Trans. Networking, vol. 46, no. 2, pp. 388-404, Mar. 2000.
9. A. Behzad, I. Rubin, A. Mojibi-Yazdi: Distributed power controlled medium access control for ad-hoc wireless networks. Proc. 2003 IEEE 18th Annual Workshop on Computer Communications, CCW 2003, 20-21 Oct. 2003.
10. A. Dua: Power Controlled Random Access. Proc. IEEE ICC'04, vol.6, pp. 3514-3518, Jun. 2004.
11. S. Kandukuri and S. Boyd: Optimal Power Control in Interference-Limited Fading wireless Channels with Outage-Probability Specifications. IEEE Transactions on Wireless Communications, vol. 1, no. 1, Jan. 2002.
12. Tao Shu and Zhisheng Niu: Uplink capacity optimization by power allocation for multimedia CDMA networks with imperfect power control. IEEE J. Select. Areas Commun., vol. 21, no. 10, Dec. 2003.

On Energy Efficient Wireless Data Access: Caching or Not?*

Mark Kai Ho Yeung and Yu-Kwong Kwok**

Department of Electrical and Electronic Engineering,
The University of Hong Kong, Pokfulam, Hong Kong
{khyeung, ykwok}@eee.hku.hk

Abstract. We consider a typical wireless data access scenario: a number of mobile clients are interested in a set of data items kept at a common server. A client sends a request to inform the server of its desired data item while the server replies in the common broadcast channel. To study the energy consumption characteristics in such a scenario, we first define a power aware utility function. Based on the utility function, we propose a novel wireless data access scheme, which is a non-cooperative game—*wireless data access (WDA) game*. Although it does not rely on client caching (without-cache), our theoretical analysis shows that it is not always necessary for clients to send requests to the server. Simulation results confirm that our proposed scheme, compared with a simple always-request one, increases both the utility and lifetime of *every client* while reducing the number of requests sent, at the cost of slightly larger average query delay.

1 Introduction

Nowadays, we are accustomed to using different types of mobile equipments such as laptop computers, hand-held devices, smart cellular phones, etc. With the popularity of personal area networks (PANs), such as 802.11x and Bluetooth, there has been increasingly large amount of information being delivered over the wireless medium. Furthermore, the world-wide deployment of third generation cellular systems (3G) is going to complement the coverage limitations in PANs. It is expected to have more and more data applications being deployed in the near future [10]. One fundamental support to many interesting data applications is to provide efficient *on-demand information access*.

Although 3G and PANs offer promising data rates, the aggregate bandwidth requirement of different services still makes bandwidth conservation a significant concern. Moreover, mobile clients are inevitably battery-operated. This makes power awareness become an indispensable issue in the protocol design of wireless networks [5]. Researchers have proposed two techniques to meet these challenges:

* This research was supported by a grant from the Research Grants Council of the HKSAR under project number HKU 7157/04E.

** Corresponding author.

X. Jia, J. Wu, and Y. He (Eds.): MSN 2005, LNCS 3794, pp. 528–537, 2005.

(1) data broadcasting; and (2) client caching. Previous research studies mainly focused on quantifying performance improvement from client caching. However, the effect of maintaining consistency on mobile clients is largely ignored. For example, the most popular class of consistency schemes mandates that each client retrieves periodic cache invalidation information *indefinitely* from the server. This is not always desirable since the extra energy cost incurred may not justify the amount of energy conserved from caching. Thus, we are interested in designing a power efficient on-demand data access scheme.

In this paper, we study the following wireless data access scenario: a number of clients send query requests to a common server for their desired data items; the server replies with the requested content in the common broadcast channel. Specifically, we consider two classes of data access schemes: *with-cache schemes* use client caching while *without-cache schemes* do not. In without-cache schemes, there is no need to maintain cache consistency but clients have to send an uplink query request for each new query. It is observed that the server may receive more than one request for some popular data items. These duplicated requests are a waste of (1) battery energy; and (2) uplink bandwidth. Ideally, only one request is enough to trigger the server to broadcast a desired data item. However, this would require *explicit* coordination among clients. Such extra communication overheads may not justify the energy conserved from uplink requests. Instead, we model the without-cache data access problem as a non-cooperative game—*wireless data access (WDA) game*—each *player* (client) maximizes its own *utility* (number of queries that can be completed) with *no* explicit communication with one another.

2 Proposed Power Aware Wireless Data Access Scheme

Figure 1 depicts the system model for wireless data access. It consists of a server and a set of clients, N. The clients are interested in a set of data items, D, which are kept at the server. To request a specific data item, d_a, client i is required to inform the server by sending an uplink request, represented by $q_i(d_a)$. Server then replies with the content of the requested data item, d_a, in the common broadcast channel. This allows the data item to be shared among different clients. As illustrated in Figure 1, both clients j and k request the same data item, d_c, in the second interval. However, server is required to broadcast the content of d_c only *once* in the next broadcast period, which reduces the bandwidth requirement. Recently, this system model has attracted much attention [1, 3, 4, 8, 9, 10, 11].

To successfully complete a query, a client expends its energy in two different parts: (1) informing the server of the desired data item, E_{UL}; and (2) download-

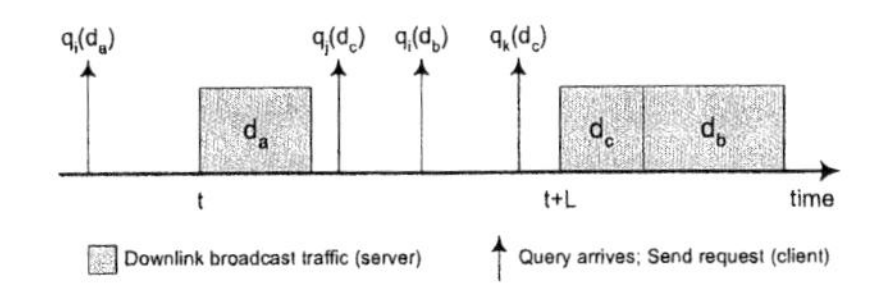

Fig. 1. System model for wireless data access

ing the content of the data item from the common broadcast channel, E_{DL}. The energy cost of sending a request to the server is represented by E_s. Note that E_{UL} does not necessarily equal E_s (see below).

Define E_{total} as the amount of energy available to a client. We use "number of queries that can be completed" to quantify the performance of a wireless data access scheme. Mathematically, the *utility* of client i's is given by:

$$U_i = \frac{E_{total}}{E^i_{UL} + E^i_{DL}} \tag{1}$$

It is assumed that E_{DL} is the same for all clients, but its value depends on the size of a data item. In practice, E_s is a function of various quantities [7], including spatial separation, speed, instantaneous channel quality, bit-error rate requirement, etc. For simplicity, however, E_s is also assumed to be a fixed quantity.

Our objective is to reduce the amount of energy consumed in the query process such that *every* client's utility (Equation (1)) is increased with bounded query delay. For a simple data access scheme, defined as **simple access**, the client is required to send an uplink request to inform the server whenever a new query arrives, i.e., $E_{UL} = E_s$. Thus, we have $U_i = \frac{E_{total}}{E_s + E_{DL}}$. Due to data locality, some data items are more popular than the others [2]. It is possible that more than one client independently requests the same piece of "hot" data item. Based on this observation, we model the data access scenario as a non-cooperative game—*wireless data access game* (**WDA game**)—each *player* (client) maximizes its own *utility* (Equation (1)) with *no* explicit communication with one another.

Denote $N = \{1, \ldots, n\}$ as the set of players (clients). Each player determines its *request probability*—the probability of sending an uplink request to the server. The strategy space of player i is given by, $S_i = \{s_i | 0 \leq s_i \leq 1\} \subset \Re^1$. The strategy combination is denoted as, $s = (s_1, \ldots, s_n) \in S$, where $S = \times_{j \in N} S_j \subset \Re^n$ is the Cartesian product of the n players' strategy spaces. Furthermore, define $s_{-i} = (s_1, \ldots, s_{i-1}, s_{i+1}, \ldots, s_n) \in S_{-i}$, where $S_{-i} = \times_{j \in N \setminus \{i\}} S_j \subset \Re^{n-1}$, as the strategy combination of all the players, except i. $U_i(s) \in \Re^1$ represents the utility of player i when the strategy combination is s. A strategy combination, s^*, is said to achieve the state of *Nash equilibrium* when:

$$U_i(s^*) \geq U_i(s^*_{-i}, s_i) \qquad \forall s_i \in S_i, i \in N \tag{2}$$

Notice that the utility function, U_i, depends on every player's strategy, s_i, which in turn affects the strategies of all the other players, s_{-i}. In other words, a strategy combination is said to achieve the state of *Nash equilibrium* when *no* player can improve its utility by *unilaterally* deviating from its own strategy. A salient feature is that there is *no* coordination among the set of players. In general, a game may have multiple equilibria or even none at all.

2.1 WDA Game—2-Person Version

To study the performance of the WDA game, we start by analyzing the 2-person version, i.e., $N = \{1, 2\}$. There are two clients in the system, each of which determines its request probability, s_i, independently. If client 1 chooses *not* to request, there are two possible consequences:

1. Client 2 sends a request for the same data item. Server then broadcasts the data item in the next scheduled period. Thus, the uplink cost for client 1 is zero, i.e., $E_{UL} = 0$;
2. Client 2 *does not* sends a request for the same data item. Client 1 cannot complete the query but dissipates energy in waiting.

In the second case, if client 1 is not patient enough to wait for another period, its next strategy is to send the request with probability one. The uplink costs for clients 1 and 2 are, $E_{UL}^1 = s_1 E_s + (1 - s_1)(1 - s_2)(E_s + E_w)$ and $E_{UL}^2 = s_2 E_s + (1 - s_1)(1 - s_2)(E_s + E_w)$, respectively, where E_w is the energy cost of waiting for a broadcast period.

Thus, the utility function (Equation (1)) takes power consumption in transmit, receive and idle modes, into consideration. From the expressions of E_{UL}^1 and E_{UL}^2, we observe that the utility functions are symmetrical. This motivates us to search for symmetric equilibrium strategies. Let $E_w = \alpha E_s$ and differentiate U_1 with respect to s_1 gives: $\frac{\partial U_1}{\partial s_1} = -\frac{E_{total} E_s}{E_{UL}^1 + E_{DL}} \{s_2(1 + \alpha) - \alpha\}$. Depending on the value of s_2, $\frac{\partial U_1}{\partial s_1}$ takes on different values:

1. $s_2 < \frac{\alpha}{1+\alpha} \Rightarrow \frac{\partial U_1}{\partial s_1} > 0$
   Client 1's best-reply strategy is, $s_1^* = 1$, which reduces to the original simple access scheme.
2. $s_2 > \frac{\alpha}{1+\alpha} \Rightarrow \frac{\partial U_1}{\partial s_1} < 0$
   Client 1's best-reply strategy is, $s_1^* = 0$, i.e., client 2's request probability is so large that client 1 is always advantageous to wait for a broadcast period.
3. $s_2 = \frac{\alpha}{1+\alpha} \Rightarrow \frac{\partial U_1}{\partial s_1} = 0$
   The best-reply for client 1 is any feasible strategy, i.e., player 1 is indifferent between request and wait. In particular, the strategy combination, $(s_1^*, s_2^*) = (\frac{\alpha}{1+\alpha}, \frac{\alpha}{1+\alpha})$, achieves a *weak Nash equilibrium* [6].

To avoid the weak equilibrium and the associated degenerated solutions ($s^* = 0$ or 1), consider the case that client 1 is patient enough to wait for one extra broadcast period. In other words, client 1 waits for a maximum of two server broadcasts before resolving to request with probability one. Thus, the uplink energy cost becomes: $E_{UL} = E_s\{s_1 + (1 - s_1)(1 - s_2)(s_1 + \alpha) + (1 - s_1)^2(1 - s_2)^2(1 + 2\alpha)\}$. From that, the best-reply strategy for client 1 is shown to be:

$$s_1 = \frac{2(1 + 2\alpha)(1 - s_2)^2 - (1 - \alpha)(1 - s_2) - 1}{2(1 + 2\alpha)(1 - s_2)^2 - 2(1 - s_2)} \tag{3}$$

If $\alpha = 0.5$, we can find from Equation (3) that the unique equilibrium strategy is $s_1^* = s_2^* = 0.4131$.

To determine the optimal symmetric strategy, we take $s_1 = s_2 = s$ in the E_{UL} expression and then substitute into Equation (1). With some manipulations, we get: $4(1+2\alpha)s^3 - 3(3+8\alpha)s^2 + 2(4+13\alpha)s - 2(1+5\alpha) = 0$. Solving this equation, we obtain the optimal strategy as: $s_1 = s_2 = 0.6567$. Thus, the equilibrium utility is *Pareto inefficient*, which is a common characteristic in non-cooperative game models. In the case of our WDA game, it is possible to leverage the presence of the central server such that each client plays the optimal strategy (see Section 2.3).

2.2 WDA Game—n-Person Version

Based on the analysis in Section 2.1, the 2-person WDA game is transformed to the n-person version as follows:

If client i has a query pending for the data item, d_x, the equilibrium strategy, s_i^*, is used. Otherwise, $s_i = 0$ for that data item, since client i is not interested in d_x. With reference to the 2-person version, client i's uplink energy cost for d_x is shown to be: $E_{UL}^i = E_s\{s_i + (s_i + \alpha)\prod_{j\in M}(1 - s_j) + (1 + 2\alpha)\prod_{j\in M}(1 - s_j)^2\}$, where $M \subseteq N$ is the set of clients interested in d_x. Thus, the symmetric equilibrium strategy is given by:

$$s^* = \frac{2(1+2\alpha)\chi^2 - (1-\alpha)\chi - 1}{2(1+2\alpha)\chi^2 - 2\chi} \tag{4}$$

where $\chi = (1 - s^*)^m$ with $m = |M|$.

Theorem 1. *There exists an equilibrium strategy for the n-person WDA game.*

Proof: First, each player's strategy space, $S_i \in \Re^1$, is nonempty, convex and compact. Second, the utility function, U_i , are continuous on S, $\forall i \in N$. Furthermore, the best-reply mapping is single-valued. Thus, we can conclude that there exists an equilibrium strategy for the n-person WDA game [6].

Although Theorem 1 does not rule out the possibility of more than one equilibrium, we observe, from simulations, that the equilibrium strategy, i.e., Equation (4), appears to be unique.

2.3 The Protocol

To achieve the stated equilibrium strategy, we design algorithms for server and clients, which are formalized as follows:

Each client executes Algorithm 1 to determine the equilibrium strategy. Client does not solve Equation (3) each time. Instead, it is implemented as table lookup to facilitate the calculation. If there is more than one query pending, the client plays with the joint request probability.

Using Algorithm 2, the server keeps track of the set of m and announces the values to each client via periodic broadcast. For $m = 1$, the request probability is 0.4132 (see Section 2.1). If the server intentionally announces $m = 0.4668$, each client would play 0.6567, which corresponds to the optimal strategy. Thus, clients plays with the optimal request probability as if it is an equilibrium strategy. This

Algorithm 1 Client—Query

```
1: INPUT: L; {m_j}
2: OUTPUT: Q (query list); s (request probability)
3:
4: for each new query, q_i(d_j) do
5:    Use m_j to determine s* from Equation (4);
6:    s = 1 − (1 − s)(1 − s*);
7:    Q = Q ∪ d_j;
8:    Send(s);
9: end for
10: for each server broadcast period, L do
11:    if Q ≠ ∅ then
12:       Check the index list, I;
13:       Download the desired data item(s);
14:    end if
15:    if Q ≠ ∅ then
16:       if a query has missed two server broadcast
          then
17:          s = 1;
18:       end if
19:       Send(s);
20:    end if
21: end for
22: Function Send(s)
23: if rand < s then
24:    Send a request;
25:    Q = ∅; s = 0;
26: end if
```

Algorithm 2 Server—Reply

```
OUTPUT: I (index list); Query replies

for each received query q_i(d_x) do
   I = I ∪ d_x;
end for
for each server broadcast period, L do
   Broadcast I;
   Broadcast the content of data items in I;
end for
```

Algorithm 3 Server—m-values

```
INPUT: θ; L; T_q^i, i ∈ N' ⊆ N (alive clients)
OUTPUT: {m_j}, j ∈ D

for j = 1 to |D| do
   m'_j = p_j Σ_{i∈N'} L / T_q^i;
   m'_j → m_j, using actual-to-optimal m table;
end for
```

requires the server to perform the actual-to-optimal m mapping. To improve efficiency, the mapping is also done via table lookup. It is assumed that the server knows (1)the mean query generation time of a typical client, e.g., from previous usage statistics; and (2) the number of alive clients, e.g., from the data link layer, from which the true values of m can be estimated.

Furthermore, the server runs Algorithm 3 to reply clients' queries. In addition to the data content, the server also broadcasts an index list, similar to the one used in [3, 8]. Using an index list, clients can tune to their desired data item(s) and ignore the others. This saves energy in monitoring the broadcast channel, which is also a costly operation [5].

3 Performance Evaluation

We evaluate our proposed game theoretic power aware wireless data access scheme using MATLAB simulation. In particular, the performance is quantified in terms of (1) *utility*—number of queries completed; (2) *lifetime*—time taken for a client to deplete its available energy; (3) *uplink traffic*—number of query requests sent; and (4) *average query delay.* We also compare our proposed approach with two popular with-cache schemes: (1)IR; and (2)IR+UIR, described in [3, 9]. "WDA game (Optimal)" represents that the server performs the actual-to-optimal m mapping while "WDA game (Nash)" means that the server announces the true values of m.

Table 1 shows the values of major simulation parameters, most of which are similar to those used in [3, 10]. Each client generates a stream of exponentially distributed queries with mean generation time, T_q, drawn uniformly between $T_q^{\min}$ and $T_q^{\max}$. This models different levels of interest of the database among the set of clients. We use the *Zipf-like* distribution [5], with $\theta = 0.9$, to model

[1] For illustration purposes, the energy available, E_{total}, is set to $100J$. In fact, different values would only change the scale of the performance graphs.

Table 1. Simulation parameters—wireless data access

Parameter	Value
Database size ($\|D\|$)	1,000
Data item size	1KB–10KB
Broadcast interval (L)	20s
Client population size ($\|N\|$)	200
Transmit power/Receive power	0.5W/0.1W
Relative cost of waiting (α)	0.5
Bandwidth	384Kbps
Energy available[1] (E_{total})	100J
Minimum query generation time ($T_q^{\min}$)	5s
Maximum query generation time ($T_q^{\max}$)	50s

the non-uniform access pattern. Each client queries the jth data item with probability given by $p_j(\theta) = \frac{1}{j^\theta \sum_{k=1}^{|D|} \frac{1}{k^\theta}}$, where $0 \leq \theta \leq 1$. Server replies clients' queries every L seconds and announces m-values whenever the set of *alive* clients changes. It is assumed that both the uplink and downlink bandwidth are used exclusively for data access. Simulation results are presented as follows:

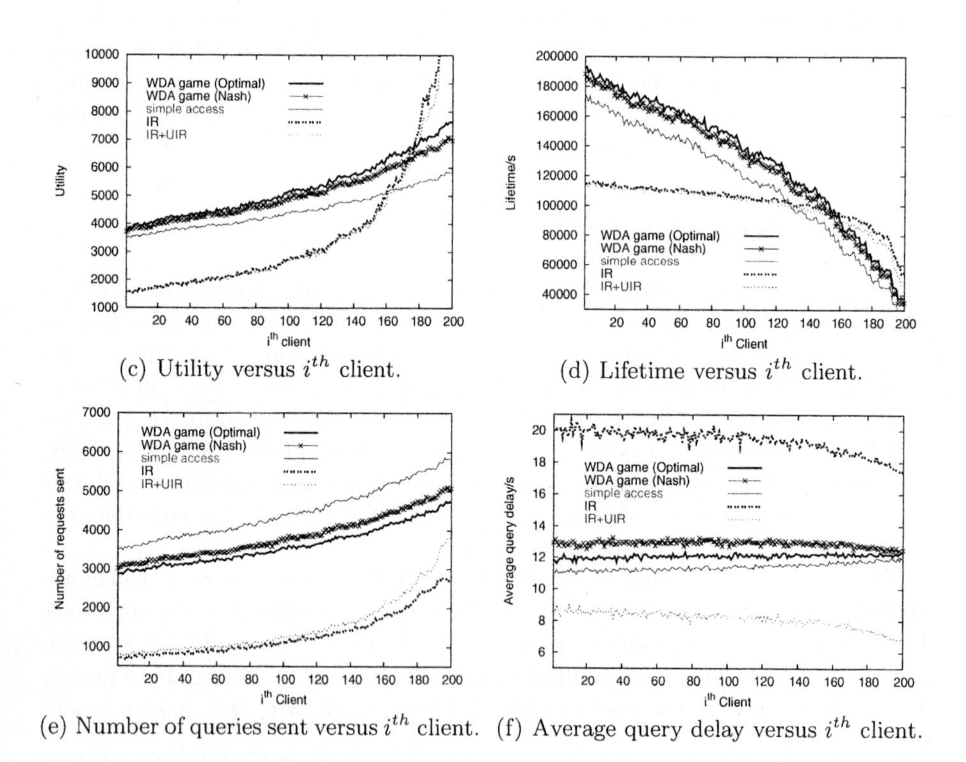

(c) Utility versus i^{th} client.

(d) Lifetime versus i^{th} client.

(e) Number of queries sent versus i^{th} client.

(f) Average query delay versus i^{th} client.

Fig. 2. Variations of performance metrics with clients

Utility: Figure 2(c) shows the utility versus i^{th} client. Notice that client indices are in descending order of their mean query arrival time, i.e., the 1st client has the lowest query rate. We observe that each client's utility is increased in the WDA game compared to the simple access scheme, which confirms our analysis in Section 2. This is because each client individually and independently optimizes its utility function (Equation (1)) to determine its equilibrium request probability. It is also observed that the performance of optimal and Nash strategies are similar, which indicates the original Nash equilibrium (without server to perform actual-to-optimal m mapping) is quite efficient. In with-cache schemes (IR and IR+UIR), clients with high query rates (those with large index) perform much better than the without-cache counterparts. On the other hand, clients with low query rates are more advantageous to adopt a without-cache scheme. Based on the above observation, we can conclude that client caching and the use of invalidation reports (IRs and UIRs) are more suitable for clients with frequent queries. Another observation is that the change occurs at about the 150th client, i.e., $T_q = 25s$, which is fairly demanding given the mean update time to be $5,000s$. However, more than half of the clients achieve a higher utility value (number of queries completed) in a without-cache scheme. This suggests that clients should not always adopt caching for energy efficient operations.

Lifetime: Figure 2(d) shows the lifetime versus i^{th} client. The lifetime of a client is defined as the time taken to deplete its available energy, E_{total}. First, the lifetime of *every client* is increased in the WDA game compared to the simple access scheme. Second, we observe that high-utility clients have relatively shorter lifetimes because these clients have high query generation rates and energy is dissipated more rapidly. Third, low-utility clients achieve larger improvements in lifetime compared to the high-utility ones. This is because the energy conserved from uplink requests has a more significant effect in the low-utility group, which is mainly in the idle state with low power consumption. In with-cache schemes, the lifetime of most clients are between $100,000s$ and $120,000s$ since *every* client is required to (1) download invalidation reports; and (2) replace invalidated caches with updated ones. Both factors consume considerable proportions of energy (see Section 3). Therefore, low-query-rate clients expend most of their energy in the cache consistency process, leaving little for actual query operations.

Uplink Traffic: Figure 2(e) shows the uplink traffic (number of requests sent) versus i^{th} client. With-cache schemes *always* result in much lower amount of uplink traffic than without-cache schemes do. This is due to the aggregate effect in with-cache schemes: each client has to wait for the next invalidation report (IR or UIR) before sending a request. The result is that there may have more than one query pending at the client. Therefore, a single request can inform the server of multiple desired data items. For the simple access scheme, each client sends a request upon every new query. This results in the largest number of requests. WDA game exploits the possibility of duplicated requests, which reduces the amount of requests sent. Similarly, the optimal and Nash strategies result in comparable lifetime. This confirms that the original Nash equilibrium is fairly efficient.

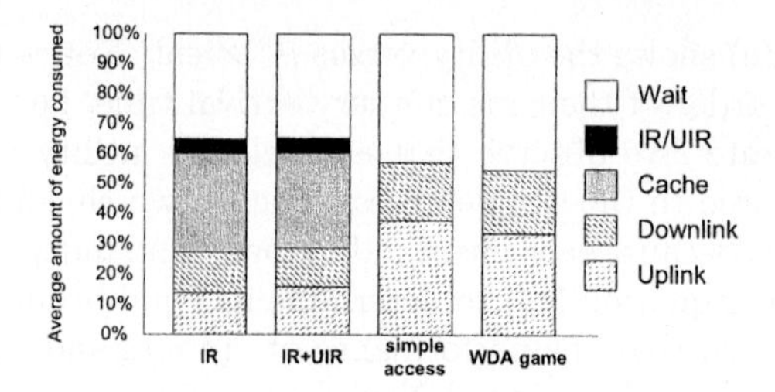

Fig. 3. Comparison—energy consumption characteristics

Average Query Delay: The performance improvements of the WDA game come with a cost: increase in average query delay, which is defined as the time between arrival of a query and complete reception of the reply from the server. Figure 2(f) shows the clients' average query delay versus i^{th} client. As expected, clients playing the WDA game experience slightly longer average query delay. This is inherently in the design of the game: wait for some one else to make the request. However, the delay is bounded by the additional requirement: if the desired data item does not appear in the immediately *two* server's broadcast, the client is forced to send a request with probability one. Indeed, if a client can tolerate a larger worst-case delay or energy conservation becomes very significant (e.g., in low-energy state), we can relax the above condition. We also conducted simulations with the strategy: check n server's broadcast before forcing to request. Both the utility and lifetime improves slightly but the average query delay increases significantly. However, due to space limitations, the results are not shown here. On the other hand, clients experience a much smaller average query delay in IR+UIR than those in IR. This is consistent with the goal of IR+UIR—to reduce the large query delay in IR. It is also observed that high-query-rate clients have shorter query delay for both IR and IR+UIR. This is due to the increase in the number of cache hits.

Energy Consumption Characteristics: We have compared the performance of with-cache schemes and without-cache schemes in terms of a number of metrics. In this section, we take a closer look at the difference between the two approaches. Specifically, we look at the energy characteristics. In a without-cache scheme (simple access and WDA game), a client consumes its energy in three different ways: (1) sending query requests (Uplink); (2) receiving replies from server; and (3) waiting (energy cost of being "alive"). For a with-cache scheme (IR and IR+UIR), a client is also required to (4) receive IRs; (5) receive UIRs (IR+UIR only); (6) replace invalidated caches with the updated copies (cache maintenance).

Figure 3 shows the average amount of energy consumed in different ways among the 4 schemes. In the with-cache schemes, receiving invalidation reports (IRs and UIRs) takes up only about 5% of the available energy, which is insignificant. Although, client caching reduces both the uplink and downlink cost via cache hits, the energy cost of keeping the cache up-to-date, is considerable. As discussed previously, the extra energy cost outweighs the benefit from cache hits for most clients, except those with high query rates. In without cache schemes, client does not dissipate energy to update the caches but consumes significant amount of energy in sending query request.

References

1. D. Barbará and T. Imieliński, "Sleepers and Workaholics: Caching Strategies in Mobile Environments," *Proc. of the 1994 SIGMOD Int'l Conf. on Management of Data*, pp. 1–12, May 1994.
2. L. Breslau, P. Cao, L. Fan, G. Phillips and S. Shenker, "Web Caching and Zipf-like Distributions: Evidence and Implications," *Proc. of the* 18th *Int'l Conf. on Computer Communications*, vol. 1, pp. 126–134, Mar. 1999.
3. Guohong Cao, "A Scalable Low-Latency Cache Invalidation Strategy for Mobile Environments," *IEEE Trans. on Knowledge and Data Engineering*, vol. 15, no. 5, pp. 1251–1265, Sept.–Oct. 2003.
4. A. Datta, D. E. VanderMeer, A. Celik and V. Kumar, "Broadcast Protocols to Support Efficient Retrieval from Databases by Mobile Users," *ACM Trans. on Database Systems*, vol. 24, no. 1, pp. 1–79, Mar. 1999.
5. L. M. Feeney and M. Nilsson, "Investigating the Energy Consumption of a Wireless Network Interface in an Ad Hoc Networking Environment," *Proc. of the* 20th *Int'l Conf. on Computer Communications*, vol. 3, pp. 1548–1557, Apr. 2001.
6. D. Fudenberg and J. Tirole, *Game Theory*, MIT Press, 603 pages, ISBN: 0262061414, Aug. 1991.
7. J. D. Parsons, *The Mobile Radio Propagation Channel*, Second Edition, John Wiley & Sons, 436 pages, ISBN: 047198857X November 2000.
8. K.-L. Tian, J. Cai and B. C. Ooi, "An Evaluation of Cache Invalidation Strategies in Wireless Environments," *IEEE Trans. on Parallel and Distributed Systems*, vol. 12, no. 8, pp. 789–807, Aug. 2001.
9. M. K. H. Yeung and Y.-K. Kwok, "Wireless Cache Invalidation Schemes with Link Adaptation and Downlink Traffic," *IEEE Trans. on Mobile Computing*, vol. 4, no. 1, pp. 68–83, Jan.–Feb. 2005.
10. L. Yin and G. Cao, "Supporting Cooperative Caching in Ad Hoc Networks," *Proc. of the* 23rd *Int'l Conf. on Computer Communications*, Mar. 2004.
11. B. Zheng, J. Xu and D. L. Lee, "Cache Invalidation and Replacement Strategies for Location-Dependent Data in Mobile Environments," *IEEE Trans. on Computers*, vol. 51, no. 10, pp. 1141–1153, Oct. 2002.

An Efficient Power Allocation Scheme for Ad Hoc Networks in Shadowing Fading Channels*

Dan Xu[1], Fangyu Hu[1], and Zhisheng Niu[2]

[1] University of Science and Technology of China, Hefei 230027, P.R. China
xudan03@mail.ustc.edu.cn
[2] Tsinghua University, Department of Electronic Engineering,
Beijing 100084, P.R. China

Abstract. In this paper, a distributed power allocation scheme for wireless ad hoc networks in shadowing fading environments is proposed. Considering fluctuated SINR during the reception of long packets, especially in multimedia communications, outage probability is introduced as a QoS parameter. An optimization model is formulated to minimize total transmit powers with constraint conditions that outage probability requirement of each link is satisfied. We explore a near-optimal closed-form solution by exploration of power ratio factor at system capacity. The power allocation scheme is based on the solution. In the scheme, power ratio factor of each link can be determined independently. The inequity problem of the power allocation algorithms in incremental fashion is avoided. Meanwhile the scheme is more time-saving than those schemes in iterative manner. The scheme can be implemented on a time scale longer than that of shadowing fading. Thus network overload is alleviated. By numerical example, we verify a better performance of our power allocation scheme than its counterpart.

1 Introduction

Power allocation for mobile ad hoc networks is a very significant topic and much work has investigated efficient schemes with various application contexts. The traditional power allocation schemes [1][2][5][6][7] often assume channel gain from a transmitter to receiver be constant for a specific period. These schemes should be inefficient when the channel incurs fast fading which makes the channel state change within millisecond[3]. In this circumstance, to make these power allocation schemes valid, high speed channel gain estimation and algorithm execution are necessary. However, these should be impractical for ad hoc network since very frequent information exchange and collection will make the network break down.

In [4], the authors analyze power allocation for interference-limited wireless links considering Rayleigh fast fading of the channels. They propose a scheme

* This work is done during the visit of the first author to Department of Electronic Engineering, Tsinghua University. This work has been partially supported by the Nature Science Foundation of China (No.60272021).

X. Jia, J. Wu, and Y. He (Eds.): MSN 2005, LNCS 3794, pp. 538–547, 2005.

that power allocation can be implemented at a time scale far larger than fast fading. In the scheme, the effects of fast fading can be compensated by mounting a high expectation of SINR. However, during the period of a packet reception, SINR also varies randomly due to lognormal shadowing fading which changes not fast. Thus a packet maybe improperly received if SINR excessively goes below a predefined threshold. If we define the state SINR goes below the threshold as outage, then there is a mapping between outage probability and the failure of packet reception. Hence we will use outage probability as a QoS parameter. In ad hoc networks, the distances between the nodes are often not too far, especially in the indoor environment. With relatively slower motion speed, path loss for a specific time can be negligible, while the change of location still bring shadowing fading due to the complicated obstacles. Thus a static distance dependent path loss during the period of a packet reception is assumed here.

In this paper, an optimization model which minimizes total receive powers is formulated. Due to distributed nature of ad hoc networks, we explore a near optimal closed-form solution that each link can determine transmit power to the best of its own information. The paper is organized as follows, Section 2 presents the network model and the definition of system capacity. Section 3 formulates a global optimization model and gives a near optimal solution. This is followed by a distributed power allocation scheme in Section 4. Section 5 is a numerical example and discussions. Finally, the conclusions are drawn in Section 6.

2 Network Model and Definition

2.1 Signal Model

Considering Fig. 1, we analyze the node pair with an active link denoted by L_i. The source node S_i of the link transmits data to node D_i with power P_{ti}. The receive power at node D_i is P_{ri}. The path loss of link L_i is g_{ii}. Then $P_{ri} = g_{ii}\xi_{ii}P_{ti}$, where ξ_{ii} represents shadowing fading deferring to lognormal distribution. Receive power P_{ri} is also a lognormal random variable with mean m_i and variance v_i^2. Then there is $m_i = E[P_{ri}] = g_{ii}P_{ti}$. Let B_{ri} represent the decibel format of

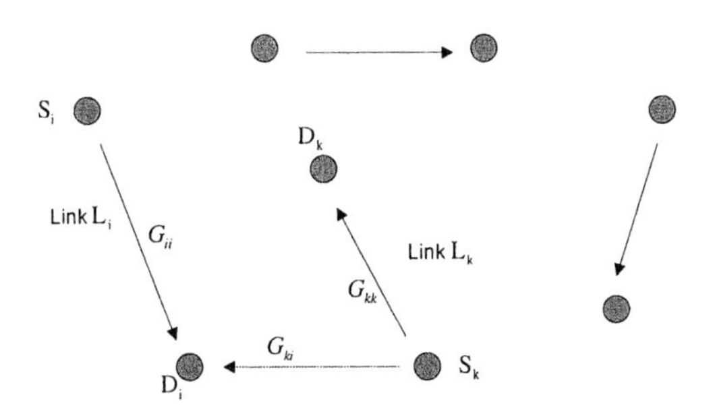

Fig. 1. A scenario of ad hoc networks

P_{ri} , $B_{ri} = 10\lg P_{ri}$. Then B_{ri} is a Gaussian random variable with mean μ_i and variance σ_i^2. There exists the following relations

$$m_i = e^{c\mu_i + \frac{1}{2}c^2\sigma_i^2}, \tag{1}$$

$$v_i^2 = e^{2c\mu_i + c^2\sigma_i^2}(e^{c^2\sigma_i^2} - 1), \tag{2}$$

where $c = \frac{\ln 10}{10}$. The parameter σ_i^2 reflects the receive signal quality at the receiver D_i. It is mainly determined by the shadowing fading rate and the receiver structure. Here we assume the nodes have slow and similar motion speed, as well as the same receiver structures. Then we replaced σ_i^2 by σ^2 for all the links discussed below. Receive signal variance σ^2 has much influence on system capacity as discussed later.

The dotted lines in Fig. 1 represent interference from adjacent nodes. Suppose there is another active link L_k whose transmitter is S_k. The transmit power of S_k is P_{tk}. Path gain from S_k to D_i is g_{ki}. Then the interference by S_k at node D_i is $I_{ki} = g_{ki}\xi_{ki}P_{tk}$, where I_{ki} is also deferring to lognormal distribution. Then total interference plus receive power of link L_i at node D_i is

$$I_i = P_{ri} + \sum_{k \neq i} I_{ki} + I_i^0, \tag{3}$$

here I_i^0 is processed as a constant which represents background interference at node D_i. Since I_i is the sum of a set of independent lognormal random variables, it can also be approximated by lognormal distribution[8], whose mean and variance are given by

$$E[I_i] = m_i + \sum_{k \neq i} m_{ki} + I_i^0 = g_{ii}P_{ti} + \sum_{k \neq i} g_{ki}P_{tk} + I_i^0, \tag{4}$$

$$D[I_i] = v_i^2 + \sum_{k \neq i} v_{ki}^2. \tag{5}$$

Let $E_i = 10\lg I_i$. E_i is also a Gaussian variable with mean λ_i and variance τ_i^2. Ref. [8] has given the following relations

$$\lambda_i = \frac{1}{c}\ln E\left[I_i\right] - \frac{1}{2c}\ln\left[1 + \frac{D[I_i]}{(E[I_i])^2}\right], \tag{6}$$

$$\tau_i^2 = \frac{1}{c^2}\ln\left[1 + \frac{D[I_i]}{(E[I_i])^2}\right]. \tag{7}$$

2.2 Description of Outage Probability

Assume CDMA technology is used in data channels for the ad hoc networks with system bandwidth denoted by W, transmission rate of link L_i is represented by R_i. Then SINR denoted by Γ_i of link L_i is given by

$$\Gamma_i = \frac{W}{R_i}\frac{P_{ri}}{I_i - P_{ri}}. \tag{8}$$

If γ_i is the minimum SINR requirement of link $\mathrm{L_i}$, the outage probability of link $\mathrm{L_i}$ can be written as

$$O_i = \Pr\left\{\frac{W}{R_i}\frac{P_{ri}}{I_i - P_{ri}} < \gamma_i\right\} = \Pr\left\{P_{ri} < \frac{I_i}{1 + \frac{W}{R_i \gamma_i}}\right\}. \tag{9}$$

Let $F_i{=}-10\lg\left(1+\frac{W}{R_i\gamma_i}\right)$. Then by using the decibel representation of P_{ri} and I_i, (9) can be rewritten as

$$O_i = \Pr\left\{B_{ri} < E_i + F_i\right\}. \tag{10}$$

Since B_{ri} and E_i are both Gaussian random variable with mean μ_i and λ_i and variance σ^2 and τ_i^2 respectively, then the outage probability of link $\mathrm{L_i}$ is given by

$$O_i = Q\left(\frac{\mu_i - \lambda_i - F_i}{\sqrt{\sigma^2 + \tau_i^2}}\right), \tag{11}$$

where $Q(x) = \frac{1}{\sqrt{2\pi}}\int_x^{\infty} e^{-\frac{t^2}{2}\,dt}$ is a mono-decreasing function of x.

2.3 The Definition of System Capacity

For wireless networks, system capacity does not mean the number of active links only. QoS of each link should be the other factor of system capacity. In the paper, QoS of a link can be represented by (R_i, Γ_i, O_i), where R_i, Γ_i and O_i are data rate, SINR and outage probability respectively. We give the definition of system capacity as follows,

$$S = \{N, ((R_i, \Gamma_i, O_i), (i = 1, \ldots, N))\}. \tag{12}$$

Here N is the number of active links.

3 Analysis of the Optimization Model for Power Allocation

Suppose there are N active links in the network which is an interference limited CDMA system, minimizing total transmit powers is equivalent to system capacity optimization, we explore capacity optimization model as

$$\begin{cases} \min \sum_{i=1}^{N} P_{ti} \\ s.t. \\ O_i = Q\left(\frac{\mu_i - \lambda_i - F_i}{\sqrt{\sigma^2 + \tau_i^2}}\right) \le \delta_i \quad (i = 1, \ldots, N). \end{cases} \tag{13}$$

It can be proved the closed-form accurate solution to (13) is equivalent to the following non-linear equation set

$$\frac{\mu_i - \lambda_i - F_i}{\sqrt{\sigma^2 + \tau_i^2}} = Q^{-1}(\delta_i), \quad (i = 1, \ldots, N). \tag{14}$$

This is a non-linear equation set and the solution is very hard to drive. Thus we should also present a near optimal solution with lower complexity and less parameters needed.

We have proved in [9] that for CDMA cellular networks, uplink capacity loss in a cell resulting from background interference at the BS can be eliminated completely by enhancing transmit power of each user simultaneously to infinite. This conforms to the factor that scaling all transmit powers up is equivalent to decreasing background interference. In ad hoc networks, we draw the same conclusion that background interference at each node will induce capacity loss which can be compensated by increasing transmit power of each link to infinite. However in ad hoc networks, background interference is inflicted at each node with different power level in the distributed manner. Nevertheless, it can be proved when transmit power of each link reach infinite, the distinction of background interference at each node is eliminated, no effects will be made by them. This reveals if we can scale all transmit powers up to infinite, system capacity boundary S_{max} can be sustained. Meanwhile the ratio among transmit powers uniquely determines the outage probability of each link. First, we will determine power ratio factor of each link when all transmit powers reach infinite.

Let $P_{ti} = \eta_i \phi$, here ϕ is a power constant factor which is the same of each link. At system capacity boundary, ϕ reaches ∞. η_i is the power ratio factor of link $\mathrm{L_i}$. Then η_i $(i = 1, \ldots, N)$ will determine O_i $(i = 1, \ldots, N)$ completely. The following steps will determine the representation of power ratio factor.

At system capacity boundary, by (14), there is

$$\mu_i = \lambda_i + F_i + Q^{-1}(\delta_i)\sqrt{\sigma^2 + \tau_i^2}. \tag{15}$$

In terms of (1), we get

$$\mu_i = 10 \lg m_i - \frac{1}{2} c\sigma^2 = 10 \lg g_{ii} P_{ti} - \frac{1}{2} c\sigma^2. \tag{16}$$

Then between the two transmitters associated with link $\mathrm{L_i}$ and $\mathrm{L_j}$ respectively, there is the following relation:

$$\begin{aligned} 10 \lg \frac{g_{ii}\eta_i}{g_{jj}\eta_j} &= \mu_i - \mu_j \\ &= \lambda_i - \lambda_j + F_i - F_j + Q^{-1}(\delta_i)\sqrt{\sigma^2 + \tau_i^2} - Q^{-1}(\delta_j)\sqrt{\sigma^2 + \tau_j^2} \\ &\qquad i, j \in \{1, 2, \ldots, N\}. \end{aligned} \tag{17}$$

By (2) (6) (7), there is

$$\begin{aligned} \lambda_i &= \frac{1}{c} \ln \left(\sum_{k=1}^{N} g_{ki}\eta_k\phi + I_i^0 \right) - \frac{1}{2c} \ln \left[1 + \frac{\alpha \sum_{k=1}^{N} (g_{ki}\eta_k\phi)^2}{\left(\sum_{k=1}^{N} g_{ki}\eta_k\phi + I_i^0 \right)^2} \right] \\ &= \frac{1}{c} \ln \phi + \frac{1}{c} \ln \left(\sum_{k=1}^{N} g_{ki}\eta_k + \frac{I_i^0}{\phi} \right) - \frac{1}{2c} \ln \left[1 + \frac{\alpha \sum_{k=1}^{N} (g_{ki}\eta_k)^2}{\left(\sum_{k=1}^{N} g_{ki}\eta_k + \frac{I_i^0}{\phi} \right)^2} \right], \end{aligned} \tag{18}$$

$$\tau_i^2 = \frac{1}{c^2} \ln \left[1 + \frac{\alpha \sum_{k=1}^{N} (g_{ki}\eta_k\phi)^2}{\left(\sum_{k=1}^{N} g_{ki}\eta_k\phi + I_i^0\right)^2} \right], \tag{19}$$

where $\alpha = e^{c^2\sigma^2} - 1$. By (18), we know when ϕ approaches infinite, the background interference item I_i^0 is removed. Then the difference between λ_i and λ_j lies in the different path gains from the other nodes. However, we can first estimate the ratio between η_i and η_j in the circumstances that the path gains from all the transmitters to node D_i are the same as node D_j. Then we will make rectification considering the different path gains of the two nodes with the transmitters. When g_{ki}=g_{kj} (i=1,...,N), there is λ_i=λ_j. Concerning (19), when ϕ approaches infinite, it is equal to

$$\tau_i^2 = \frac{1}{c^2} \ln \left[1 + \frac{\alpha \sum_{k=1}^{N} (g_{ki}\eta_k)^2}{\left(\sum_{k=1}^{N} g_{ki}\eta_k\right)^2} \right]. \tag{20}$$

In (20), it is evident when $g_{ki}\eta_k$ (k=1,...,N) is equal, τ_i^2 reaches the lowest value, which is denoted by τ_{min}^2, $\tau_{min}^2 = \ln \left[1 + \frac{(e^{c^2\sigma^2}-1)}{N} \right]$. We use τ_{min}^2 as approximation to substitute τ_i^2 for the determination of the ratio factor. The approximation will make the ratio factor η_i $(i = 1, \ldots, N)$ a bit closer. However, the impacts should be trivial. By (20), the ratio between η_i and η_j can be represented in the form of

$$\frac{\eta_i}{\eta_j} = \frac{g_{jj} \frac{1}{1+\frac{W}{R_i\gamma_i}} 10^{\frac{1}{10}\sqrt{\sigma^2+\tau_{min}^2}Q^{-1}(\delta_i)}}{g_{ii} \frac{1}{1+\frac{W}{R_j\gamma_j}} 10^{\frac{1}{10}\sqrt{\sigma^2+\tau_{min}^2}Q^{-1}(\delta_j)}}. \tag{21}$$

Eq. (21) is deduced on the premise of node D_i and D_j has the same path gains with the other nodes. However, this is impossible in an asymmetrical network. Thus we should rectify (21) to make it more accurate. Notice the last item of (18) is smaller than $\frac{1}{2c}\ln(1+\alpha) = \frac{1}{2}c\sigma^2$. This is a very small value and will be made trivial by $\lambda_i - \lambda_j$. Then in terms of (18), there is

$$\lambda_i - \lambda_j = \frac{1}{c} \ln \left[\frac{\sum_{k=1}^{N} (g_{ki}\eta_k + \frac{I_i^0}{\phi})}{\sum_{k=1}^{N} (g_{kj}\eta_k + \frac{I_j^0}{\phi})} \right]. \tag{22}$$

Since the rectification for (21) is in terms of the different path gains between D_i or D_j with the other nodes, we can ignore the impact of power ratio factors here. Then $\lambda_i - \lambda_j$=$\frac{1}{c}\ln\frac{g_i'}{g_j'}$, where $g_i' = \sum_{k=1}^{N} g_{ki}$ represents the sum of path gains from the transmitters to node D_i. Substituting the value of $\lambda_i - \lambda_j$ to (17), then we get the rectified ratio of η_i to η_j as

$$\frac{\eta_i}{\eta_j} = \frac{g_{jj} g_i^{'} \frac{1}{1+\frac{W}{R_i \gamma_i}} 10^{\frac{1}{10}\sqrt{\sigma^2+\tau_{min}^2} Q^{-1}(\delta_i)}}{g_{ii} g_j^{'} \frac{1}{1+\frac{W}{R_j \gamma_j}} 10^{\frac{1}{10}\sqrt{\sigma^2+\tau_{min}^2} Q^{-1}(\delta_j)}}. \tag{23}$$

Thus power ratio factor of link L_i is

$$\eta_i = \frac{g_i^{'}}{g_{ii}} \frac{1}{1+\frac{W}{R_i \gamma_i}} 10^{\frac{1}{10}\sqrt{\sigma^2+\tau_{min}^2} Q^{-1}(\delta_i)}. \tag{24}$$

The representation of P_{ti} (i=1,...,N) is

$$P_{ti} = \frac{g_i^{'}}{g_{ii}} \frac{10^{\frac{1}{10}\sqrt{\sigma^2+\tau_{min}^2} Q^{-1}(\delta_i)}}{1+\frac{W}{R_i \gamma_i}} \phi, \qquad \phi \to +\infty. \tag{25}$$

Eq. (25) is the closed-form representation of the optimal transmit power at system capacity S_{max}. However the infinite value of transmit power can't be achieved. Since if system capacity S is within the boundary S_{max}, the infinite transmit power of each link make outage probability far smaller than the predefined threshold. Thus we can scale ϕ down to the level that the outage probability, i.e., O_i ($i = 1, \ldots, N$), is equal to δ_i exactly. By deduction, we derive the minimum power constant factor ϕ of link L_i that rightly satisfy constraint condition of (13) as

$$\phi_i = I_{ri}^0 \Big/ \left[\sqrt{Y / \left(\sqrt{2XY+1} - 1 \right)} - Z \right], \tag{26}$$

where there is $X = 10^{\frac{1}{5}\left[\sqrt{\sigma^2+\tau_{max}^2} Q^{-1}(\delta_i) + \frac{1}{2} c\sigma^2 - 10 \lg g_{ii}\eta_i + F_i\right]}$, $Y = 2\alpha \sum_{k=1}^{N} (g_{ki}\eta_k)^2$, and $Z = \sum_{k=1}^{N} g_{ki}\eta_k$. Since ϕ_i are different for each link, the maximum one must be selected as the uniform measure for every link, i.e., $\phi_{max} = \max\{\phi_i\}$ ($i = 1, \ldots, N$). Then we have given a near optimal closed-form solution for (14), i.e., $P_{ti} = \eta_i \phi_{max}$ ($i = 1, \ldots, N$).

A merit of the solution is that each link can determine its power ratio factor independently. The confirmation of power ratio factor will overcome the inequity and blockage problems in the incremental methods[1] [2], where transmit power assigned to a new link will be restricted by the existing links with QoS requirements in the networks.

4 A Distributed Power Allocation Scheme

We design a distributed power allocation scheme as follows.

Step1: Power allocation is performed in a cycle manner. An implementation cycle is divided into a number of slots. In a initial slot, a transmitter S_i broadcast a Hello_Msg via the control channel. The Hello_Msg includes the fields as below. The first field records its own ID, the second is the ID of the destination node.

ID(S$_i$)	ID(D$_i$)	PN	P_i	P_{ti}	η_i^c	ACTIVE	QOS

If ACTIVE field is set TRUE, then S$_i$ and D$_i$ is communicating with each other in the data channel. If ACTIVE is FALSE, it means S$_i$ will reserve data channel with node D$_i$, the QoS requirements of the expected link which include data rate, SINR and outage probability items are kept in the QOS field. In this case, the next slot should be reserved for node D$_i$ to return a feedback packet. P_i is the maximum power level of S$_i$. Hello_Msg is broadcast with this power level. η_i^c represents the current value of power ratio factor if S$_i$ is active. If not, it records the value of the last link. P_{ti} is the transmit power level of link L$_i$ if it is active.

Step2: Each node receives the Hello_Msg will average the receive power. Suppose a node D$_j$, the average receive power is P_{ij}, then path gain $g_{ij}=\frac{P_{ij}}{P_i}$. Path gain is recorded and updated in each cycle. When path gains have been estimated, node D$_i$ will estimate background interference I_i^0 by the equation $I_j^0=P_{rj}^{total}-\Sigma g_{ij}P_{ti}$, here P_{ri}^{total} is the total receive power of D$_i$ which can be measured. After this, node D$_j$ will check if ID(D$_j$)=ID(D$_i$). If it is, and in the data channel, node D$_j$ is idle, it will return a ACK packet to node S$_i$. If not, it will send a NAK packet to tell the expiration time of its current link.

Step3: After several slots for the Hello_Msg exchanges, each receiver associated with an active link or a reserved link will calculate power ratio factor of the link by (24). After that the value will be publicized. When each receive node gathers power ratio factors of the other links, it will calculate ϕ_i according to (26). A maximum value of ϕ_i, $(i = 1, \ldots, N)$ is selected at each receive node. Then node D$_i$ will calculate the expectation of P_{ti}^{exp} and send it to S$_i$ via a NOTIF packet. The format is as follows.

ID(D$_i$)	ID(S$_i$)	g_{ii}	P_{ti}^{exp}	η_i

Step4: If P_{ti}^{exp} exceeds P_i, meanwhile node S$_i$ has not set up the link with D$_i$ yet, S$_i$ will delay sending data to node D$_i$ and wait until next power allocation cycle. If link L$_i$ is already active, but P_{ti}^{exp} exceeds P_i, S$_i$ will broadcast a FORBID packet to suspend the incoming link.

5 Numerical Analysis

First, we analyze the system capacity influenced by the background interference at each node by Fig. 2. The system spread spectrum bandwidth is W=5MHz. Here we assume a unique voice traffic with QoS requirements as (32kbps, 5, 0.01), thus system capacity S can be represented by the number of the active links N. The background interference at each node is set identical. It is observed that to maintain the same system capacity, it requires each transmitter to increase the power level when interference at each node is raised. This best illustrates the point that the system capacity loss caused by the background interference can

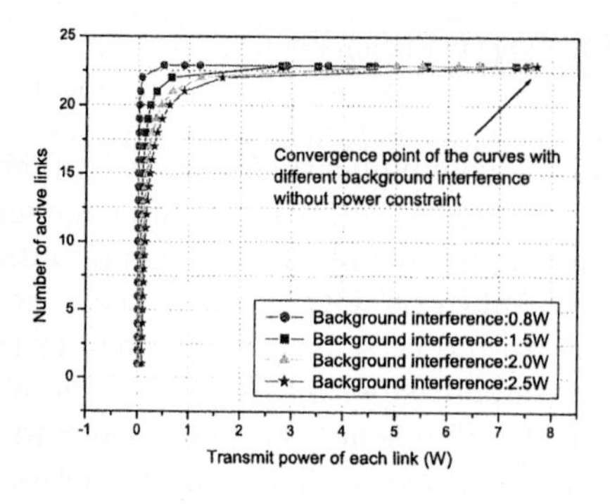

Fig. 2. The impacts on system capacity by background interference at each node

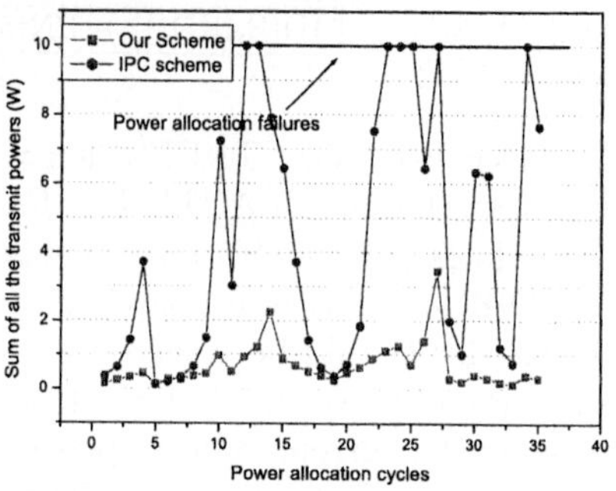

Fig. 3. Performance comparison between our scheme and insufficient power allocation

be compensated by scaling transmit power up. If each transmit power reaches infinite, capacity boundary S_{max} is the same under different background interference, i.e., the four curves will converge to the same point. This explains no matter what background interference is, it will be eliminated by the infinite value of transmit power.

By Fig. 3, we evaluate the performance of our power allocation scheme. A scheme named insufficient power control (IPC) is compared with the proposed scheme. The IPC scheme does not determine power ratio first, it just let each transmit power keep equal. However, outage probability of each link should also be satisfied. We simulate a series of affiliation and left of four types of links with QoS requirements represented by (32kbps, 4, 0.01), (64kbps, 6, 0.1), (64kbps, 5, 0.01) and (128kbps, 6, 0.01) respectively. The system spread spectrum bandwidth is W=5MHz. The value of σ^2 which represents signal quality is set equal to 0.45. The distance dependent path loss is randomly created at each power allocation cycle. It is observed at each cycle, under the same system capacity, our scheme always provides a lower total transmit powers than IPC scheme. This means the proposed scheme makes a larger capacity admissible. Moreover, under specific system capacity , IPC fails to provide feasible power allocations. In Fig. 4, the points of failed power allocations of IPC scheme are marked at the top line. Under the same capacity, it is observed our scheme still provides feasible power allocations. However, the value of them are larger than those of smaller system capacity. Actually the better performance of the proposed scheme comes from a wise power ratio factor for each link, although it is not the optimal.

6 Conclusions

In this paper, we propose a distributed power allocation scheme for mobile ad hoc networks. We fully consider shadowing fading and introduce outage probability as a QoS parameter. Under the constraint condition of outage probability requirement, an optimization model is formulated to maximize system capacity for ad hoc networks. The system capacity is defined as a combination of the link number and QoS. Based on the propositions in our previous work, a near optimal solution is presented with low computational complexity. Our distributed power allocation is based on the solution. By numerical analysis, we verify a better performance of the proposed scheme.

References

1. S. Lal and E. S. Sousa: Distributed resource allocation for DS-CDMAbased multimedia ad hoc wireless LANs. IEEE J. Select. Areas Commun., vol. 17, no. 5, May. 1999.
2. A. Muqattash and M. M. Krunz: A distributed transmission power control protocol for mobile ad hoc networkks. Mobile Computing, IEEE Transactions on, vol. 3, no. 2, Apr.-June. 2004.
3. T. Rappaport: Wireless Communications: Principles and Practice. Prentice Hall, 1996.
4. S. Kandukuri and S. Boyd: Optimal Power Control in Interference- Limited Fading wireless Channels with Outage-Probability Specifications. Wireless Communications, IEEE Transactions on, vol. 1, no. 1, Jan. 2002.
5. T. Elbatt and A. Ephremides: Joint Scheduling and power control for wireless ad hoc networks. Wireless communications, IEEE Transactions on, vol. 3, no. 1, Jan. 2004.
6. S. Agarwal, R.H. Katz, S.V. Krishnamurthy, and S.K. Dao: Distributed Power Control in Ad-Hoc Wireless Networks. Proc. IEEE Intl Symp. Personal, Indoor and Mobile Radio Comm., vol. 2, pp. 59-66, Oct. 2001.
7. C. Comaniciu and H. V. Poor: QoS provisioning for wireless ad hoc data networks. in Decision and Control, 2003. Proceedings. 42nd IEEE Conference on, vol. 1, 9-12 Dec. 2003
8. N. C. Beaulieu, A. A. A. Dayya and P. J. Mclane: Estimating the distribution of a sum of independent lognormal random variables. Communications, IEEE Transactions on, vol. 43, no. 12, pp. 2869-2873, Dec. 1995.
9. Tao Shu and Zhisheng Niu: Uplink capacity optimization by power allocation for multimedia CDMA networks with imperfect power control. IEEE J. Select. Areas Commun., vol. 21, no. 10, Dec. 2003.

A Soft Bandwidth Constrained QoS Routing Protocol for Ad Hoc Networks

Xiongwei Ren and Hongyuan Wang

Department of Electronics and Information Engineering,
Huazhong University of Science and Technology, Wuhan, Hubei 430074, P.R. China
xwren@vip.sina.com, wythywl@public.wh.hb.cn

Abstract. This paper proposes an on demand source routing protocol with multiple disjoint paths for ad hoc networks to support soft bandwidth constrained QoS requirement, which is termed as the Active Multipath QoS Routing (AMQR) protocol. A distributed route discovery algorithm is proposed to find multiple disjoint paths with associated path stability and network resource information. In addition, an actively dynamic route maintenance algorithm based on periodic state update combined with gratuitous QoS_ACK replies is introduced to refresh network topology and resource information, which helps the source dynamically adjust the traffic load on the desired route for data dispersion. Simulation results compared with DSR show that AMQR provides excellent end-to-end QoS.

1 Introduction

An ad hoc network is a multi-hop wireless network temporarily and dynamically formed by a collection of mobile nodes without the use of any preexisting network infrastructure or centralized administration. Owing to its self-organization, rapid deployment and absence of any fixed infrastructure, ad hoc networks are gaining popularity as a significant and promising research domain. In recent years, the need to provide support for multimedia services has raised certain research issues concerning Quality of Service (QoS) routing. The effects of node movement, signal interference, and power outages, however, make the available link state and network topology information inherently imprecise. On the other hand, heavy traffic, frequent link failure and network partition will incur QoS disruptions, causing packets to be delayed and dropped. Therefore, no hard QoS but soft QoS is supported in ad hoc networks.

There exist three main challenges for QoS routing in ad hoc networks: (1) to find a feasible path with sufficient network resources to satisfy the QoS constraints of a requested connection, (2) to avoid QoS disruptions, and (3) to optimize network resource utilization. Many research efforts have been done to solve these problems. For example, a ticket-based QoS routing protocol was proposed in [1] under a quite ideal model that the bandwidth of a link can be determined independently of its neighboring links. A CDMA-over-TDMA channel model was assumed in [2] to develop a QoS Routing protocol in a MANET, where the use of time slot on a link is only dependent of the status of its one-hop neighboring links. A QoS multi-path routing approach was developed in [3], which does not consider the radio interference problem, and load

X. Jia, J. Wu, and Y. He (Eds.): MSN 2005, LNCS 3794, pp. 548 – 559, 2005.

balancing scheme for multi-path routing protocols is also proposed in [4]. The number of packets waiting in the queue proposed in [5] is used to dynamically adjust the traffic flow on different paths, only resulting in a best-effort delay-sensitive routing. In order to prevent network from being overloaded with two much traffic, an admission control scheme and a feedback scheme are incorporated to meet the QoS requirements of real-time applications and an approximate bandwidth estimation to react to network traffic is used in [6].

QoS supporting in ad hoc networks involves physical channel, MAC layer, network layer, and application layer. The aim of these protocols described above is to find feasible paths that guarantee a hard end-to-end QoS requirement in a specific layer. Instead of providing hard QoS support, this paper proposes an on demand active source routing protocol, which utilizes disjoint paths based on path stability and used bandwidth for data dispersion and periodic state update mechanism to implement soft resource reservation and network resource optimization.

The remainder of this paper is organized as follows. Section II describes the bandwidth constrained QoS problem statement and Section III introduces the basic AMQR protocol. After presenting the simulation environment and performance analysis compared with DSR in Section IV, the conclusion follows in the last section.

2 The Problem Statement

2.1 Network Model

An ad hoc network is modeled as a graph G = (V, E), where V is a finite set of nodes and E is a set of bidirectional links. A radius R defines a coverage area within which each node can communicate with each other directly and form links between neighbors. A path from source node s to destination node d, p(s,d) is defined as a sequence of intermediate nodes, such that p(s,d)={s,i,j,…,m,d} without loops. V_p is the set of nodes belong to p(s,d) and L_p is the length of p(s,d). P(s,d) is defined as the set of acquirable paths from s to d, such that P(s,d)={p(s,d)}, P'(s,d) is defined as the set of all acquirable disjoint paths from s to d, $P'(s,d) \subset P(s,d)$. $\forall p_1(s,d), p_2(s,d) \in P(s,d)$, the paths p_1 and p_2 are disjoint paths if $V_{p1} \cap V_{p2} = \{s,d\}$; otherwise, the paths p_1 and p_2 are joint paths where $\{s,d\} \subset V_{p1} \cap V_{p2}$. For joint paths, if $L_{p1}=L_{p2}$, p_1 is an equal-length path of p_2, if $L_{p1}<L_{p2}$, p_1 is a shorter-length path of p_2 and p_2 is a longer-length path of p_1.

2.2 QoS Metrics

It is too difficult to calculate the accurate bandwidth of a given path because the bandwidth calculation is influenced by many factors including the channel load, congestion, packet collision, etc. Even when the sufficient bandwidth is reserved in the route discovery phase, it will become uncertainty over time. So this paper does not propose a formula to evaluate the path bandwidth, whereas it gives an assumption that the required bandwidth of a given connection will be satisfied if the used bandwidth of the selected path is less than a bandwidth threshold.

For simplicity, assumes that (1) a node cannot receive and transmit at the same time, and (2) when a node transmits a data packet, only the addressed neighbor can receive and the others (i.e. overhearing neighbors) cannot transmit and receive. For an intermediate node i of a given path p(s,d), such as p(s,d)={s,i,j,…,m,d}, let P_{Ni}, L_{ki} ($1 \le k_i \le N_i$), and $B_a(p_{ki})$ be the set of active routes joined by node i, the length of current route p_{ki}, and the allocated payload on route p_{ki}, respectively. The used bandwidth of a given path p(s,d) is defined as

$$B_u(p(s,d))=\max(\sum B_a(p_{ki})*L_{ki}) \quad (1)$$

where $p_{ki} \in P_{Ni}$, $i \in V_p$, $i \neq s$, $i \neq d$.

2.3 Path Stability

Path stability is a crucial problem in designing protocols for ad hoc networks. Many metrics, such as associativity [7], signal strength [8], link life [9], and link expiration time [10], have been used to evaluate the stability of a given path. This paper selects the link expiration time based on mobility prediction as stability criterion, and the calculation equation taken from [10] follows.

Assume that two nodes i and j are within transmission range R of each other. Let (x_i,y_i,v_i,θ_i) and (x_j,y_j,v_j,θ_j) be the coordinates, the speed, the moving directions of nodes i and j, respectively, then the link expiration time is predicted by

$$T_e(l_{i,j})=[-(ab+cd)+((a^2+c^2)R^2-(ad-bc)^2)^{1/2}]/ (a^2+c^2) \quad (2)$$

where $a=v_i\cos\theta_i-v_j\cos\theta_j$, $b=x_i-x_j$, $c=v_i\sin\theta_i-v_j\sin\theta_j$, $d=y_i-y_j$.

Note that when $v_i=v_j$, $\theta_i=\theta_j$, or $v_i=v_j=0$, $T_e(l_{i,j})$ is set to ∞ without applying the above equation. Given T_s is the link stability threshold, this paper divides links into three classes. If $T_e(l_{i,j})<T_s$, $l_{i,j}$ is an invalid link, if $T_s \le T_e(l_{i,j})<2T_s$, $l_{i,j}$ is an unstable link, if $2T_s<T_e(l_{i,j})$, $l_{i,j}$ is a stable link. The path expiration time is defined as $T_e(p(s,d))=\min(T_e(l_{i,j}))$, where p(s,d)={s,i,j,…,m,d}, and the paths can be divided into invalid path, unstable path, and stable path correspondingly. Note that invalid links are not considered in the route discovery and in the route maintenance.

2.4 The Routing Problem

Path bandwidth and path stability are key problems in QoS routing. The criteria for path selection are described in the followings.

1. If $T_e(p_2)<2T_s$, $T_e(p_1)<2T_s$, and $B_u(p_1)<B_u(p_2)$, then p_1 is superior to p_2;
2. If $T_e(p_2)<2T_s$, $T_e(p_1)<2T_s$, $B_u(p_1)=B_u(p_2)$, $T_e(p_1)>T_e(p_2)$, then p_1 is superior to p_2;
3. If $T_e(p_2)<2T_s$ and $T_e(p_1)\ge 2T_s$, then p_1 is superior to p_2;
4. If $T_e(p_2)\ge 2T_s$, $T_e(p_1)\ge 2T_s$, and $B_u(p_1)<B_u(p_2)$, then p_1 is superior to p_2;
5. If $T_e(p_2)\ge 2T_s$, $T_e(p_1)\ge 2T_s$, $B_u(p_1)=B_u(p_2)$, $T_e(p_1)>T_e(p_2)$, then p_1 is superior to p_2.

Ordered by the quality, the set of all acquirable disjoint paths from source node s to destination node d is $P'(s,d)=\{p_1,p_2,\ldots,p_n\}$ but only the paths p_1 and p_2 are reserved. The following three steps are used to select two disjoint optimal paths.

Step1: select the best path from all joint paths and remove the rest;
Step2: select two best paths from all disjoint paths and remove the rest;
Step3: allocate traffic load on the desired paths.

Given a bandwidth constraint B_r, an allocated bandwidth B_a, and a bandwidth threshold B_t, where B_r is the minimum bandwidth that a connection need, B_a is the actually allocated bandwidth on a desired path, and B_t is the limit that denotes the maximum available bandwidth of a path and the value is changed with varying environmental conditions. When there are two disjoint routes to the destination, the traffic allocation algorithm used in this paper is described as follows:

1. If B_t-$B_u(p_1)\geq B_r$, then the optimal path p_1 is an active route, $B_a(p_1)=B_r$, and the suboptimal path p_2 is a backup route, $B_a(p_2)=0$; if the first condition does not meet, then
2. If $2B_t$-$B_u(p_1)$-$B_u(p_2)\geq B_r$, then the optimal path p_1 and the suboptimal path p_2 are active routes, $B_a(p_1)=B_t$-$B_u(p_1)$, $B_a(p_2)=B_r$-$B_a(p_1)$; if the second condition dose not meet, then
3. The optimal path p_1 and the suboptimal path p_2 are active routes, $B_a(p_1)=B_r*\max(B_u(p_1), B_u(p_2))/(B_u(p_1)+B_u(p_2))$, $B_a(p_2)=B_r$-$B_a(p_1)$.
   If only one path is available, traffic allocation degrades as a best-effort algorithm.

3 Active Multipath QoS Routing (AMQR) Protocol

The proposed Active Multipath QoS Routing (AMQR) protocol is an on-demand source routing protocol consisting of route discovery phase and route maintenance phase, which selects two optimal disjoint paths based on link expiration time and used bandwidth for data dispersion. The outstanding difference between any other on-demand routing protocols and AMQR exists in three areas: (1) a distributed route discovery algorithm is proposed to find multiple disjoint paths with network resource information and path stability, (2) a periodic state update mechanism combined with gratuitous QoS_ACK replies is introduced to refresh network topology and resource information, which helps the source dynamically adjust the traffic load on the desired route, and (3) the active routes involved, the length of each route, and the allocated bandwidth on each route are utilized to calculate the used bandwidth and to provide soft resource reservation.

3.1 Information Stored in Each Node

- Route cache: Each node stores disjoint paths it has learned in its route cache while processing RREP and QoS_ACK messages. When a RERR message is received, any route using the broken link specified in the RERR packet are removed from its route cache.
- Request cache: Nodes broadcasting a RREQ packet store information in its request cache, which will be used to prevent a node from rebroadcasting the same RREQ message and to record the optimal path reverse to the source.
- Active route table: Each node maintains an active route table to calculate the used bandwidth. When receiving QoS_Update or QoS_Reserve messages, the addressed node realizes that it is in an active route and then inserts or updates an active route entry. On the other side, when it receives a RREQ to the same connection, receives a RERR packet notifying the broken route, or does not receive a QoS_Update message

during a period of time, all nodes except for the source will remove the corresponding entry from its active route table.

- Neighbor table: Overhearing nodes and downstream nodes of an active route maintain a neighbor table that records its link expiration time with surrounding neighbors and relative resource information of the current route. When these nodes find they can construct a better or stable path via promiscuously listening QoS_Update and QoS_Ack messages, they will send a gratuitous QoS_Ack reply to the original source.

3.2 Route Discovery Protocol

3.2.1 Route Request

The route discovery mechanism is based on request-reply cycles.

```
// destination node of the path
IF (destAddr of the RREQ packet = its own address)
  IF (the first RREQ packet)
    -Delete all stale routes from route cache
    -Insert route from the source to the destination into its route cache
    -Insert the relative information into request cache
    -Initiate route reply operation
  ELSE (the pair of <srcAddr, seqNo> of the RREQ has been received already)
    IF (path of the RREQ is disjoint from the other paths found earlier)
      -Insert route information into route cache
      -Initiate route reply operation
    ELSE IF (the path is joint path but is superior to the cached one)
      -Delete the old joint path
      -Insert route information into route cache
      -Initiate route reply operation
    ELSE
      -Discard the RREQ packet
// intermediate node of the path
ELSE
  IF (its own address is in the path of the RREQ packet)
    -Discard the RREQ packet
  ELSE IF (the pair of <srcAddr, seqNo> of the RREQ has been received already)
    IF (the route of RREQ is superior to that of the request cache)
      -Set bUpdate to true and update the information of request cache entry
    ELSE
      -Discard the duplicate RREQ packet
  ELSE
    -Append its own address to the path of the RREQ packet
    -Update the PET, upNodeInfo, usedBW, and hopCount of the RREQ packet
    -Insert the relative information into request cache
    -Rebroadcast the RREQ packet to its neighbors
```

When a source needs a bandwidth constrained route to the destination but no route is available, it broadcasts a route request (RREQ) message to its neighbors. Each intermediate node forwards the first RREQ packet and records the optimal path reverse to the source via utilizing duplicate RREQ packets traversed through different routes. The destination may receive multiple RREQ packets but only these packets containing disjoint paths or joint paths with better quality are replied along the reverse path to the original source, and the others are discarded.

When a node receives a RREQ packet, the above procedures are invoked.

A RREQ packet carries *<srcAddr, destAddr, upNodeInfo, seqNo, hopCount, path, PET, usedBW>* information, where *srcAddr* and *destAddr* refer to the source and the destination address, respectively, *seqNo* is a sequence number assigned by the source, *hopCount* and *path* record the source route information, *upNodeInfo* includes the coordinates, the velocity, and the moving direction of the upstream node, which is used to calculate path expiration time (*PET*), and *usedBW* is calculated by equation (1).

Note that the above *bUpdate* flag in request cache denotes whether the intermediate node receives a duplicate RREQ packet with a better route than that of the first one.

3.2.2 Route Reply

Each intermediate node of the selected path forwards the route reply (RREP) message, and replaces the unprocessed partial of the path with the cached one only if its *bUpdate* flag is true. The source, according to the path selection criteria, selects two optimal disjoint paths from all routes carried by RREP packets, whose format is *<targetAddr, srcAddr, seqNo, hopCount, segLeft, path, upNodeInfo, PET, usedBW>*.

```
-Delete all unstable path from route cache
// source node of the path
IF (targetAddr of the RREP packet = its own address)
  IF (no route is available in route cache)
    -Insert route information into route cache
  ELSE IF (This route has already exist in its route cache)
    -Update route information existed in its route cache
  ELSE
    IF (path of the RREP is disjoint from other paths received earlier)
      -Insert route information into route cache
    ELSE IF (route of the RREP packet is superior to old ones)
      -Update route information in route cache with new one
    ELSE
      -Discard the RREP packet
// intermediate node of the path
ELSE
  IF (its own address is not in the processed path of the RREP packet)
    IF (no route to the destination is available in route cache)
      -Insert route information into route cache
    ELSE
      -Update route information with new one
    IF (bUpdate of the request cache entry with the same seqNo is true)
      -substitute path in request cache for the unprocessed partial path of the RREP packet
    -Forward the RREP packet along path of the RREP packet
  ELSE
    -Discard the RREP packet
```

When a node receives a RREP packet, the above procedures are invoked.

The source sets up a timer once it receives the first RREP packet, and performs the selection operation and allocates the traffic load until this period is expired. Note that if the resulting paths do not satisfy the required bandwidth, AMQR acts as a best-effort protocol.

If no reply is received within a timeout period, another route request is flooded throughout the entire network and the waiting time is double to limit the frequency of route discovery flooding.

3.2.3 Soft Resource Reservation

Generally, QoS routing is combined with some form of resource reservation mechanism to guarantee that data are transmitted along the same path with guaranteed network resources as long as the connection is available. Different from any other on-demand routing protocol, AMQR does not require the support of resource management protocols but only utilizes soft state to estimate the used bandwidth.

After finding feasible paths satisfied QoS constraints and allocating traffic on the desired paths, the source unicasts a QoS_Reserve packet along the selected path, which carries <*srcAddr, destAddr, seqNo, hopCount, segLeft, allocatedBW*> information. Each node receiving this QoS_Reserve packet inserts the state information into its active route table, which will be used to calculate the used bandwidth (this procedure equal to the soft resource reservation). When a node receives a QoS_Update packet periodically unicasted by the source, it will update the soft state. If it is the intended destination, it will reply a QoS_Ack packet immediately. On the other side, if no QoS_Update packet is received within a certain update period, or RERR and RREQ packet for the same connection is learned, the record will be deleted automatically (this procedure equal to the soft resource release). Although this method does not explicitly reserve bandwidth for the current route, allocating traffic on the light load routes will result in optimal resource utilization. That is to say, AMQR is an efficient congestion avoidance mechanism and degrades as a best-effort routing protocol under heavy traffic.

3.3 Route Maintenance Protocol

Due to highly frequent topology and resource changes in ad hoc networks, route maintenance is more important than route discovery, especially for QoS constraints. Unlike existing approaches, AMQR actively utilizes periodic state update mechanism to maintain up-to-date network topology and soft resource information. A QoS_Update message is unicasted along the optimal path when the path update period reaches. If the resulting path does not meet the bandwidth requirement, a new QoS_Update message is initiated along the suboptimal path.

Each intermediate node of the current route updates its active route table entry when receiving QoS_Update packet, which carries <*srcAddr, destAddr, upNodeInfo, seqNo, hopCount, segLeft, path, PET, allocatedBW, usedBW*> information. Once receiving this QoS_Update packet, the destination immediately returns a QoS_Ack packet back to the source, which includes <*srcAddr, destAddr, upNodeInfo, seqNo, hopCount, segLeft, path, PET, usedBW*> information. If an overhearing node or a downstream node working in promiscuous mode (a node can listen to neighbors' packets even though the packet is not addressed to it) of the current route finds it can construct a new path with less used bandwidth or exclusion of unstable link via processing these QoS control packets, it will send a gratuitous QoS_Ack packet back to the source. With these QoS_Ack packets, the source can update the topology and resource information, reselect the optimal route, and reallocate the traffic if the resulting route satisfies the QoS constraint. Otherwise, a new RREQ is issued in the cases that there is no stable route or only one route with insufficient bandwidth is available. Periodic state update and gratuitous QoS_Ack reply make it possible to adapt to network topology and resource changes and to avoid QoS disruptions. Note that the above calculation of *usedBW* in route maintenance excludes the current route.

Overhearing nodes and downstream nodes actively take part in the dynamic route maintenance only in five cases depicted in Fig. 1 and in Table 1.

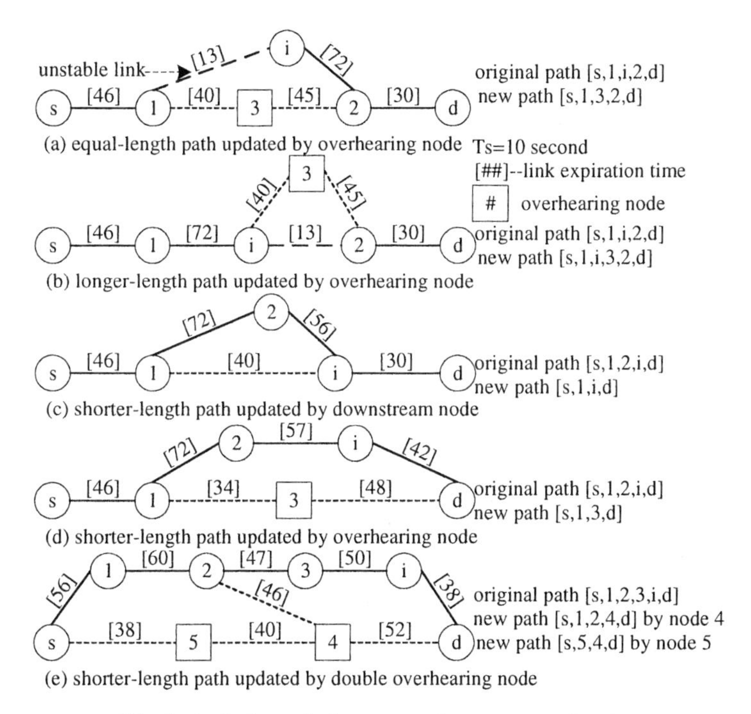

Fig. 1. Actively periodic route maintenance mechanism

Table 1. Typically dynamic route maintenance

Trigger conditions	Action
1,2,5	Equal-length path update by an overhearing node
1,2,6	Longer-length path update by an overhearing node
1,3,4	Shorter-length path update by a downstream node
1,3,4	Shorter-length path update by an overhearing node
1,3,4	Shorter-length path update by overhearing nodes

Let p_1 and p_2 refer to the original path and the new path, respectively, trigger condition of replying gratuitous QoS_Ack can be summarized as follows: (1) $T_e(p_2)\geq 2T_s$, (2) $T_e(p_1)<2T_s$, (3) $B_u(p_2)<B_u(p_1)$, (4) p_2 is the shorter-length path of p_1, (5) p_2 is the equal-length path of p_1, and (6) p_2 is the longer-length path of p_1.

In Fig. 1a, for example, when an overhearing node 3 overhears a QoS_Update with *PET* equal to 46 transmitted by node 1, a QoS_Update with *PET* equal to 13 transmitted by node i, and a QoS_ACK with *PET* equal to 30 transmitted by node 2, it realizes that it can construct a new stable path [s,1,3,2,d] and the original path [s,1,i,2,d] includes an unstable link [1,i], which meets the condition 1, 2, and 5, so it returns a gratuitous QoS_ACK to the original source.

Note that these cases may occur simultaneously, so the process of QoS_Ack packets is similar to that of RREP packets.

3.4 Route Error

A link of an active route can be disconnected because of mobility, congestion, packet collision, or a combination of the above. How to deal with link failure will significantly affect the performance of routing protocols. Instead of initiating a route error (RERR) packet whenever a unicast packet (data packet or control packet) fails to be delivered to the next node, AMQR sends RERR reverse to the source only when the data packet is undeliverable and has not been salvaged. On the other hand, if there is an alternate route in its cache, the immediate upstream node of this broken link will use it to salvage data. When receiving the RERR packet, the source changes the backup route into the active route, or it initiates route discovery if no feasible route is available.

4 Performance Simulation

Simulation study was evaluated using the library-based GloMoSim [11] simulator and ten runs with different random seed numbers were conducted for each scenario and collected data during 1000 seconds was averaged over these runs. Simulation parameters are given in Table 2.

In the conventional wisdom, it is impossible to support QoS guarantee with feasible control overhead under heavy load and in the cases of highly frequent topology and resource changes. Therefore, these severe network environments, such as very high speed, too heavy traffic, and sparse node density, are not discussed in this paper.

Table 2. Simulation parameters

Parameter	Value
Physical area (X×Y)	1000 m×1000 m
Propagation model	free space propagation model
Transmission range	250 m
Mobility model	random waypoint model [12]
Mobility rate (m/s)	0-5, 0-10, 0-15, 0-20
Pause time (s)	5, 10, 15, 20
Number of nodes	48, 64, 80, 96
Traffic sources	10 CBR
Packet rate	2.5, 5, 7.5, 10 (packets/s)
Packet size	512 bytes
Channel capacity	2 Mbps
MAC protocol	IEEE 802.11 DCF
Path update period	5 s
Link stability threshold	10 s

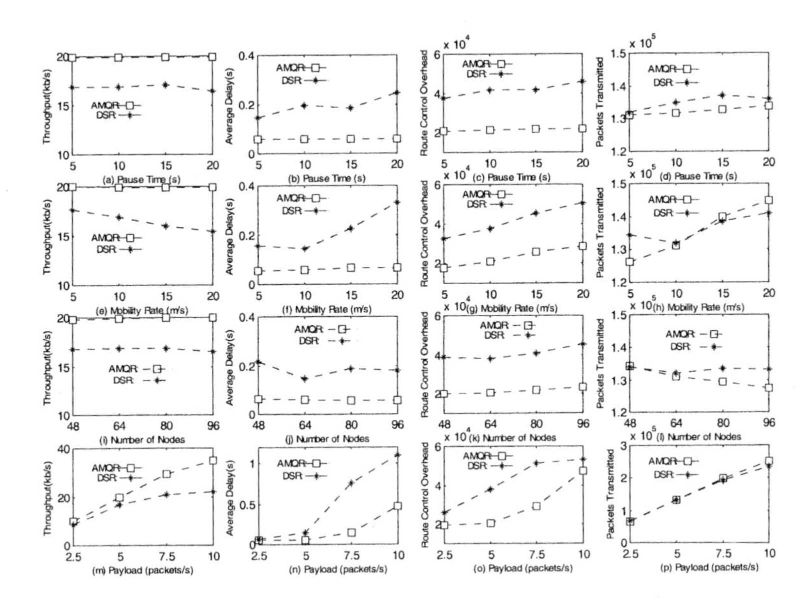

Fig. 2. Performance comparison between DSR and AMQR

In each run, one input parameter, such as pause time, mobility rate, number of nodes, and traffic load, was varied while the other parameters were kept constant. Average received throughput per connection, average end-to-end delay, route control overhead, and the total number of data packets transmitted successfully are selected as performance metrics to compare the performance between AMQR and DSR [13] with respect to varying environmental conditions.

The first set of experiments varies pause time to show the effect of pause time with a moderate packet rate of 5 packets/s, a maximum mobility rate of 10 m/s, and a network size of 64. Experiments done in [12] indicate that the mobile network is quite stable for all pause times over 20 s, so the pause time of this paper is not over 20 s.

As shown in Fig. 2a-d, the performance of throughput, delay, and control overhead of AMQR are nearly constant with respect to the pause time. AMQR provides about 17 percent higher throughput, 60 percent lower delay, 45 percent lower routing overhead, and less 2 percent lower packets transmitted than DSR.

The second set of experiments illustrates the effect of mobility when packet rate, network size, and pause time is set to 5 packets/s, 64, and 5 s, respectively. Fig. 2e-h reveals that mobility rate has a significant influence on the overall performance of DSR while AMQR is slightly affected by the mobility. When the mobility rate increases from 5 m/s to 20 m/s, AMQR reduces throughput only by 1 percent and prolongs delay by 23 percent while DSR does by 12 percent and 112 percent, respectively. This is because that when a link of the current route becomes unstable, an alternate link will substitute for it by utilizing gratuitous QoS_ACK message. Comparing with DSR in the case of higher mobility (20 m/s), AMQR gives 29 percent higher throughput and 80 percent lower delay.

The effect of network size is depicted by the third set of experiments (Fig. 2i-l) with a moderate payload of 5 packets/s, a pause time of 5 s, and a maximum mobility rate of 10 m/s. The periodic route update-based scheme would intuitively have higher control overhead with the increased network size, which is proved by Fig. 2k. When the number of nodes changes from 48 to 96, routing overhead of AMQR increases by 14 percent while the number of packets transmitted decreases by 5 percent because AMQR can get a greater amount of shorter paths by making aggressive use of gratuitous QoS_ACK messages in the denser network. For the size of control packet is by far less than that of data packet, the performance of throughput and average delay of AMQR has been improved roughly by 1 percent and by 16 percent respectively at a network size of 96 compared with the case of network size of 48.

The fourth set of experiments has been conducted to explore the effect of offered load by varying packet rate with a moderate mobility rate of 10 m/s, a pause time of 5 s, and a network size of 64. The results of Fig. 2m-p demonstrate that the overall performance of both AMQR and DSR becomes poor under heavy load. Varying offered load per connection from 10 kb/s to 40 kb/s (i.e. 2.5 packets/s to 10 packets/s), AMQR yields about 143 percent higher control overhead and 281 percent higher data overhead, respectively. DSR becomes saturated when the offered load reaches 30 kb/s and AMQR starts saturating only at an offered load of about 40 kb/s since AMQR take advantage of multiple disjoint paths for data dispersion.

As illustrated above, AMQR outperforms DSR by giving significantly higher throughput and lower delay in all cases, especially in the case of high mobility. The overall performance of AMQR is hardly changed with respect to the pause time, slightly affected by the mobility rate, degraded under heavy load while improved as the increased network size.

The key advantages of AMQR protocol are contributed to the adoption of actively periodic state update mechanism combined with gratuitous QoS_ACK messages, which helps AMQR obtain a significantly greater amount of more stable and better routes than DSR as long as the current route is available. Path expiration time and used bandwidth are used to prevent from link failure and congestion. Multiple disjoint paths may be used to dynamically adjust the traffic load on the desired route for data dispersion and to avoid QoS disruptions.

5 Conclusions

This paper presents a novel distributed algorithm based on path expiration time and used bandwidth to find multiple disjoint paths satisfying soft bandwidth constrained QoS requirement. The key distinguishing feature of AMQR is the use of periodic state update and active route maintenance. In that the source can obtain the network topology and resource information in a timely manner, it is possible to allocate the traffic on the light load paths and to substitute stable route for the old one with unstable link, which results in avoiding QoS disruptions and implementation of optimal resource utilization.

References

1. S. Chen and K. Nahrstedt, "Distributed Quality-of-Service Routing in Ad Hoc Networks," IEEE Journal on Selected Areas in Communications, vol. 17, no. 8, Aug. 1999, pp. 1488-1505.
2. C.-R. Lin and J.-S. Liu. "QoS Routing in Ad Hoc Wireless networks," IEEE Journal on Selected Areas in Communications, vol. 17, no. 8, Aug.1999, pp. 1426-1438.

3. W.-H. Liao, Y.-C. Tseng, J.-P. shen, et al, "A Multi-Path QoS Routing protocol in a Wireless Mobile Ad Hoc Network," IEEE International Conference on Networking, Part II, pages 158-167, July, 2001.
4. Linifang Zhang, Zenghua Zhao, Yantai Shu, et al., "Load Balancing of Multipath Source Routing in Ad Hoc Networks," IEEE International Conference on Communications, ICC 2002 , vol. 5, Apr. 2002, pp. 3197-3201.
5. Min Sheng, Jiandong Li, and Yan Shi, "Routing protocol with QoS guarantees for ad-hoc network," Electronics Letters, vol. 39, no. 1, Jan. 2003, pp. 143-145.
6. Lei Chen and W. B. Heinzelman, "QoS-Aware Routing Based on Bandwidth Estimation for Mobile Ad Hoc Networks," IEEE Journal on Selected Areas in Communications, vol. 23, no. 3, Mar. 2005, pp. 561-572.
7. C. K. Toh, "Associativity-Based Routing For Ad Hoc Mobile Networks," Wireless Personal Communication, Special Issue on Mobile Networking and Computing Systems, vol. 4, no. 2, Mar. 1997, pp. 103-139.
8. Y. Hwang and P. Varshney, "An Adaptive QoS Routing Protocol with Dispersity for Ad-hoc Networks," Proc. HICSS'03, Hawaii, 2002, pp. 302-312.
9. B. S. Manoj and R. Ananthapadmanabha, R. M. C. Siva, "Link life Based Routing Protocol for Ad hoc Wireless Networks," Proc. IEEE IC3N, Phoenix, Arizona, Oct. 2001, pp. 573-576.
10. S. H. Bae, S. J. Lee, and W. Su, et al, "The Design, Implementation, and Performance Evaluation of the On-Demand Multicast Routing Protocol in Multihop Wireless Networks," IEEE Network, vol. 14, no. 1, Jan. 2000, pp. 70-77.
11. L. Bajaj, M. Takai, and R. Ahuja R, et al, "GloMoSim: a scalable network simulation environment," UCLA Computer Science Department, Technical Report-990027, May 1999.
12. T. Camp, J. Boleng, and V. Davies, "A Survey of Mobility Models for Ad Hoc Network Research," Wireless Communication and Mobile Computing (WCMC): Special Issue on Mobile Ad Hoc Networking: Research, Trends, and Applications, vol. 2, no. 5, 2002, pp. 483-502.
13. D. A. Maltz, J. Broch, and J. Jetcheva, et al, "The Effects of On-Demand Behavior in Routing Protocols for Multihop Wireless Ad Hoc Networks," IEEE Journal on Selected Areas in Communications, vol. 17, no. 8, Aug. 1999, pp. 1439-1453.

Optimal QoS Mechanism: Integrating Multipath Routing, DiffServ and Distributed Traffic Control in Mobile Ad Hoc Networks

Xuefei Li and Laurie Cuthbert

Department of Electronic Engineering, Queen Mary, University of London,
Mile End Road, London, E1 4NS, United Kingdom
{xuefei.li, laurie.cuthbert}@elec.qmul.ac.uk
http://www.elec.qmul.ac.uk

Abstract. Future mobile Ad hoc networks (MANETs) are expected to be based on all-IP architecture and be capable of carrying multitude real-time multimedia applications such as voice, video and data. It is very necessary for MANETs to have an optimal routing and quality of service (QoS) mechanism to support diverse applications. Providing multipath routing is beneficial to avoid traffic congestion and frequent breaks in communication due to mobility in MANETs. Differentiated Services (DiffServ) can be used to classify network traffic into different priority levels and apply priority scheduling and queuing management mechanisms to obtain QoS guarantees. Dynamically distributed traffic control can effectively take advantage of multiple routing paths to avoid congestion and decrease delay. In this paper, we propose an optimal QoS mechanism: Integrating Multipath routing, DiffServ and distributed Traffic control in mobile ad hoc networks (IMDT). Simulation results show that IMDT achieves better performance in terms of packet delivery ratio and average delay.

1 Introduction

A Mobile ad hoc network is a collection of mobile nodes that can communicate with each other using multihop wireless links without utilizing any fixed based-station infrastructure and centralized management. Each mobile node in the network acts as both a host generating flows or being destination of flows and a router forwarding flows directed to other nodes. With the rising popularity and development of wireless network based on all-IP architecture and multimedia applications, potential uses of MANETs in military and civilian life are attracting more and more researchers' attentions on QoS support. QoS provisioning in MANETs faces a number of technical challenges because of the network restrictions such as dynamically and unpredictably variable topology resulting from nodal mobility and bandwidth constraints caused by the shared wireless medium.

For the current Internet there are two different models to obtain a QoS guarantee: the Integrated Services (IntServ) [1] and Differentiated Services (DiffServ) [2]. IntServ uses the RSVP protocol [3] to carry the QoS parameters from the sender to the receiver to make resource reservations along the path. IntServ/RSVP provides for a

X. Jia, J. Wu, and Y. He (Eds.): MSN 2005, LNCS 3794, pp. 560 – 569, 2005.

rich end-to-end QoS solution, by way of end-to-end signalling, state-maintenance (for each RSVP-flow and reservation), and admission control at each network element. DiffServ on the other hand, does not have any end-to-end signalling mechanism and works on a service level agreement between the provider and the user. All packets from a user are marked to specify the service level and are treated accordingly. Multiple flows in DiffServ model are mapped to a single service level and state information about every flow need not be maintained along the path.

IntServ-based model on per-flow resource reservation is not particularly suitable for MANETs because of the frequently changing topology and limited resources in MANETs, resulting in more signalling overhead and unaffordable storage and computing process for mobile nodes. DiffServ-based is a lightweight model using a relative-priority scheme to soften the hard requirements of hard QoS models like IntServ. The service differentiation is based on per-hop behaviours (PHBs) [4], so no flow states need to be maintained within the core of the network. Thus the model could be a potential QoS model in MANETs.

The current existing solutions for QoS provisioning in MANETs are mainly based on the IntServ or DiffServ model. AQOR [5] uses a reservation-oriented method to decide admission control and allocate bandwidth for each flow. INSIGNIA [6] employs an in-band signalling protocol rather than out-of-band signalling protocol as RSVP to decrease reservation overhead. FQMM [7] is designed to provide QoS in ad hoc networks by mixing the IntServ and DiffServ mechanisms. High priority applications are provided by IntServ per-flow QoS guarantee, while lower priority applications are provided with per-class differentiation based on DiffServ. SWAN [8] is based on reservation-less approach. By avoiding signalling, it simplifies the whole architecture and provides a differentiation between real-time and best effort in spite of not being able to guarantee the QoS needs of each flow for the whole session due to frequently changing topology and limited wireless bandwidth restriction.

Multipath routing allows the establishment of multiple paths between a single source and single destination node during a single route discovery. Some multipath routing protocols [9-12] in MANETs have been proposed to provide load balancing, fault-tolerance and higher aggregate bandwidth as well as eliminate route discovery latency after a link break by making use of the availability of multiple route paths. However, these multipath routing protocols lack QoS support in the process of transmission of data packets.

In this paper, we propose an optimal QoS mechanism: Integrating node-disjoint Multipath routing, DiffServ and distributed Traffic control in mobile ad hoc networks (IMDT). IMDT can distribute dynamically data traffic into multiple routing paths to reduce congestion and delay.

The remainder of this paper is organized as follows. In section 2, an overview of NDMR is presented. Section 3 presents IMDT. In section 4 a simulation model based on OPNET is proposed. Performance evaluation and comparison of NDMR and IMDT are presented in Section 5 and concluding remarks are made in Section 6.

2 NDMR

Ad hoc On-Demand Distance Vector (AODV) [13] and Dynamic Source Routing (DSR) [14] are the two most widely studied on-demand ad hoc routing protocols. Previous work [15] has shown limitations of the two protocols. The main reason is that both of them build and rely on a unipath route for each data session. Whenever there is a link break on the active route, each of the two routing protocols has to invoke a route discovery process. Each route discovery flood is associated with significant latency and routing overhead.

NDMR [9] modifies and extends AODV to enable path accumulation feature of DSR in route request/reply packets and discover multiple node-disjoint routing paths with a low routing overhead. NDMR routing computation has three key components: Path Accumulation, Decreasing Routing Overhead and Selecting Node-Disjoint Paths.

2.1 Path Accumulation

The main goal of NDMR is to build multiple node-disjoint paths with a low routing overhead during a route discovery. We modify AODV to include path accumulation in RREQ packets. When the RREQ packets are generated or forwarded by the nodes in the network, each node appends its own address to the routing request packets. When a RREQ packet arrives at its destination, the destination is responsible for judging whether or not the routing path is a node-disjoint path. As an example, consider five nodes A, B, C, D and E as shown in Fig. 1. Node A wants to send data to node E. Since A does not have a route for E in its routing table, it broadcasts a route request. Node B receives the route request, appends its own address to the request, and forwards the request since it also has no route to E. Similarly, when node C and node D receive the RREQ, they append their address to the request and forward it.

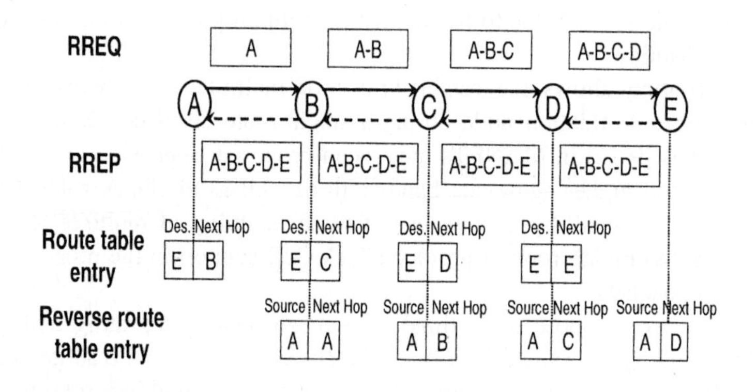

Fig. 1. Path Accumulation in NDMR

2.2 Decreasing Routing Overhead

In this section a novel approach is introduced to decrease broadcast overhead. When a node receives a RREQ packet for the first time, it checks the path accumulation list from the packet and calculates the number of hops from the source to itself and records the number as the shortest number of hops in its reverse route table entry. If the node receives the RREQ duplicate again, it computes the number of hops from the source to itself and compares it to the number of the shortest hops recorded in its reverse route table entry. If the number of hops is larger than the shortest number of hops, the node drops the RREQ packet. Otherwise (less than or equal to), the node appends its own address to the route path list of the RREQ packet and broadcasts the RREQ packet to its neighbouring nodes.

Fig. 2 illustrates the route request process with low routing overhead in the entire network. Source S broadcasts a route request packet. Each intermediate node uses the approach with low routing overhead to propagate and discard packets. Therefore, only seven packets (S-c-f-D, S-a-d-g-D, S-b-e-h-D, S-c-d-g-D, S-c-e-h-D, S-c-f-g-D, S-c-f-h-D) can reach the destination D. However, not all of paths packets that arrive in destination are node-disjoint. In next section we discuss how to choose node-disjoint paths.

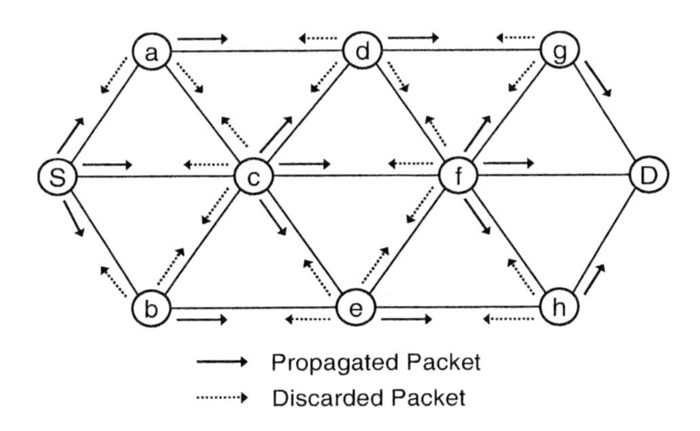

Fig. 2. Route Request Process with Low Overhead

2.3 Select Node-Disjoint Paths

In NDMR, the destination is responsible for selecting and recording multiple node-disjoint route paths. In Fig. 2, its three node-disjoint route paths are: S-a-d-g-D, S-c-f-D, S-b-e-h-D. When receiving the first RREQ packet (the shortest route path: S-c-f-D), the destination records the list of node IDs for the entire route path in its reverse route table and sends a RREP that includes the route path towards the source along the reverse route. When the destination receives a duplicate RREQ, it will compare the whole route path in the RREQ to all of the existing node-disjoint route paths in its route table entry. If there is not a common node (except source and destination) between the route path from the current received RREQ and any node-disjoint route path recorded in the destination's reverse route table entry, the route path of the current RREQ (such as S-a-d-g-D or S-b-e-h-D) satisfies the requirement of node-disjointness and is recorded in the reverse route table of the destination. Otherwise, the route path (such as paths: S-c-d-g-D, S-c-e-h-D, S-c-f-g-D, S-c-f-h-D) and the current received RREQ are discarded.

3 Integrating Multipath Routing, DiffServ and Distributed Traffic Control (IMDT)

Although NDMR provides node-disjoint multipath routing with low route overhead in MANETs, it is only a best-effort routing approach, which is not enough to support QoS. DiffServ is an approach for a more scalable way to achieve QoS in an IP network. It could be a potential QoS model in MANETs because it acts on aggregated flows and minimises the need for signalling. Distributing dynamically data traffic into multiple routing paths can reduce the probability of network congestion and average delay. IMDT integrates multipath routing, DiffServ and distributed traffic control to provide QoS guarantee. In the section, we introduce some important characteristics of IMDT.

3.1 Resource Management of IMDT

An effective QoS mechanism can be used to provide better service to certain flows in the environments of limited wireless bandwidth. In IMDT this is done by either raising the priority of a flow or limiting the priority of another flow. In order to support service differentiation, scheduling and queue management are thought to be two important aspects of resource management. The former is done by the scheduler which decides the opportunities of flows for link access and the latter holds the valid packets when necessary drops some packets from the buffer in case of network congestion.

Priority Scheduling. In MANETs, when a mobile node is receiving traffic faster than it can transmit, the node may buffer the extra traffic until bandwidth is available. This buffering process is called queuing. Using a queuing algorithm to sort the traffic has been deployed to handle an overflow of arriving traffic in wired networks. In IMDT, priority queuing is used to build a priority scheduler. The priority scheduler includes two queues: a high-priority queue and a low-priority queue. The high-priority queue must be emptied before packets are emptied from low-priority queues.

Although DiffServ has a lot of classes defined, the most essential use of DiffServ is to provide support for the two most common applications:

(A) Voice and Video traffic. (B) Best Effort data.

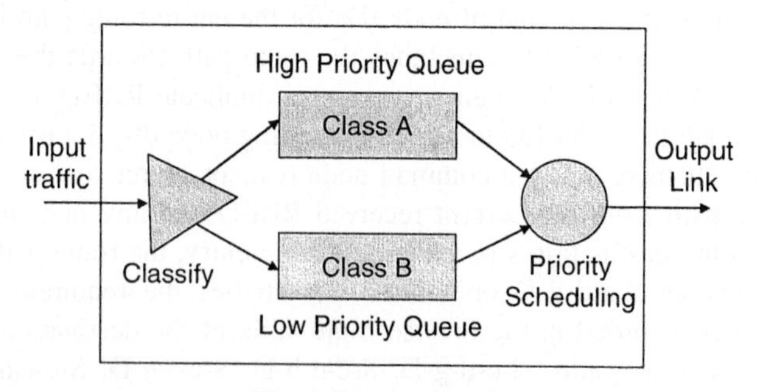

Fig. 3. Priority Scheduler

Let us denote the two classes as A and B. Class A applications require generally low loss, low latency and assured bandwidth service, so packets of class A are classified as Expedited Forward (EF) traffic. Class B is classified as Best Effort (BE) traffic which offers a lower priority service. Our priority scheduler (see Fig. 3) is designed to transmit any available Class A packets ahead of Class B packets. On the other hand, Class B packets are not sensitive to delay, as the application which they service are primarily HTTP and FTP sessions.

Queue Management. While the scheduling does play a big role in the QoS provided by the network, it is only effective if there are sufficient to hold incoming packets. Because queues are not of infinite size, they can fill and overflow. When a queue is full, any additional packets cannot get into the queue and will be dropped. This is a tail drop. The issue with tail drops is that the router cannot prevent this packet from being dropped (even if it is a high-priority packet). So, the purpose of queue management is to make sure the queue does not fill up so that there is room for high-priority packets.

The random early detection (RED) algorithm [18] is implemented to avoid congestion before it becomes a problem. The minimum threshold specifies the number of packets in a queue before the queue considers discarding packets. The probability of discard increases until the queue depth reaches the maximum threshold. After a queue depth exceeds the maximum threshold, all other packets that attempt to enter the queue are discarded.

3.2 Recording Maximum Queue Usage Rate of Intermediate Nodes

In order to decrease transmission delay, an effective strategy in multiple routing paths is to choose dynamically a routing path in which there is minimum queue usage rate to shorten waiting time of data packets in queues.

How to collect some information about queue usage rate of intermediate nodes without increasing extra overhead is an important problem. In IMDT, when a destination node receives a data packet, it needs return an acknowledgement message to the source node along reverse route path to confirm the data packet has been received safely. A new field: <maximum queue usage rate> is added to acknowledgement message to record the maximum queue usage rate of nodes along the routing path. When any intermediate node receives the acknowledgement message, it reads the queue usage rate from the message and compares it with its own queue usage rate. If the compared result is less than zero, the node updates the maximum queue usage rate of the message by using its own queue usage rate. Then the node forwards the acknowledge message to next intermediate node. When the acknowledge message arrives in the source node, the source node update maximum queue usage rate of the routing path in its routing table.

3.3 Distributed Traffic Control

By recording maximum queue usage rate of intermediate nodes in every routing path into the routing table of sources, source nodes can choose a right routing path to transmit data packets. Before sending a data packet, source node compares all of queue usage rates of multiple routing paths to the destination and chooses a routing path whose queue usage rate is the smallest to send data packets. By the approach to opt a routing path, data traffic can be distributed dynamically into multiple routing paths. This actually results in the increase of network bandwidth and the decrease of transmission delay. Distributing dynamically data traffic is also a good solution to provide excellent load balancing in multiple paths.

4 Simulation Model

OPNET 8.1 Modeler [19] was used to create a simulation environment to develop and analyze the proposed IMDT and compare performances with NDMR.

4.1 Mobility and Traffic Model

The random waypoint model [20] is used to model mobility. Each node starts its journey from a random location to a random destination with a specific speed. Once the destination is reached, another random destination is targeted after a pause. Field configuration of 1000m x 1000m field with 50 nodes is used and each node uses the IEEE 802.11 [21] with a 250m transmission radius. The pause time is kept constant at 30 seconds for all our simulation experiments. Traffic sources with 512 byte data packets are CBR (constant bit rate). The source-destination pairs are spread randomly over the network and the number of sources is varied to change the offered load in the network. When varying the number of sources, velocity is kept at a uniform rate of 0-20 m/s.

In order to investigate the usage of network ability, the number of EF (Expedited forwarding) sources with 80kbps (20pkt/s) bandwidth requirement is varied from 5 to 20 in intervals of 5. Twenty other nodes are randomly chosen to send background BE (Best Effort) traffic with 2pkt/s. Simulations are run for 800 simulated seconds. Each data point represents an average of five runs with identical traffic models, but different randomly generated mobility scenarios by using different seeds.

4.2 Performance Metrics

The following metrics are used in varying scenarios to evaluate the three different protocols:

- *Packet delivery ratio:* The ratio of the data packets delivered to the destinations to those generated by the CBR sources.
- *Average delay of data packets:* This includes all possible delays from the moment the packet is generated to the moment it is received by the destination node.

5 Simulation Results

Fig. 4 shows that the packet delivery ratio for IMDT has better performance than that of NDMR with the increase in the number of FE service sources. The reason is that IMDT can distribute dynamically data traffic into multiple routing paths to avoid congestion. When the number of sources increases, NDMR drops a larger fraction of the packets than that of IMDT. Although the delivery ratio of NDMR is more than 75%, it decreases more quickly with larger numbers of FE sources. The reason is that there are more collisions in the air and queue congestion in mobile node buffers when the number of sources increases.

It can be seen from Fig. 5 that IMDT has a lower average delay than NDMR. With the increase of the number of EF sources average delay of IMDT is a little increased. The reason is that an increase in the number of EF sources leads to higher network load traffic in the ad hoc networks. Because of the limitation of a constrained wireless bandwidth, packets that will be sent or forwarded have to stay in buffers and wait for a longer time to get a radio channel available in order to avoid traffic congestions and collisions in the air. Because IMDT distributes dynamically traffic into multiple routing paths, it can decrease network congestion and average delay.

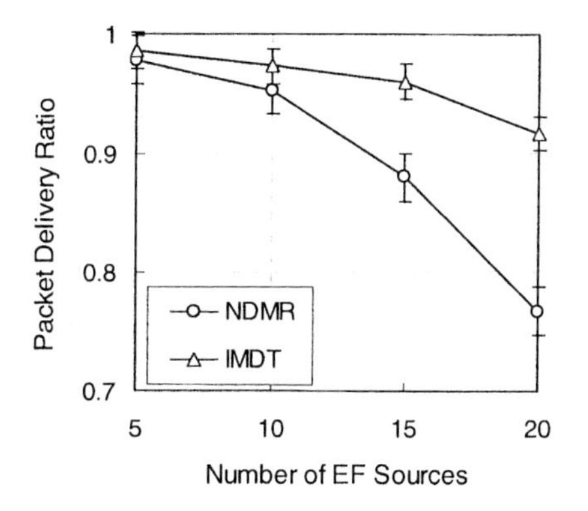

Fig. 4. Packet Delivery Ratio

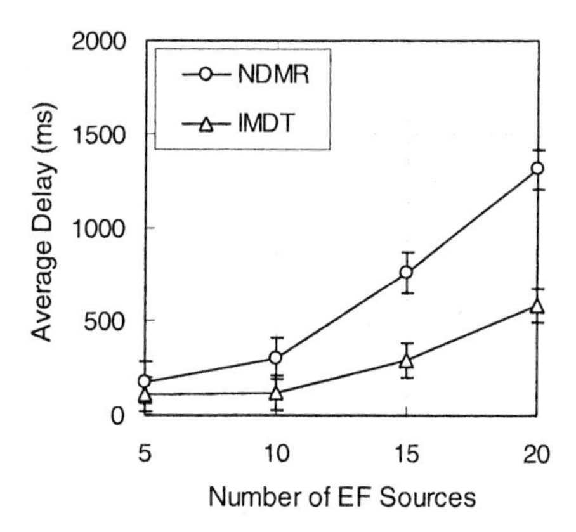

Fig. 5. Average Delay

6 Conclusions

In this paper, we have introduced an optimal QoS provisioning mechanism which integrates multipath routing, DiffServ and distributed traffic control in MANETs to overcome the shortcomings of the NDMR. The performance evaluation and comparison are studied by extensive simulations using OPNET Modeler. Simulation results show that IMDT achieves better performance than NDMR in MANETs. We can conclude that IMRD has a good potential to serve as a QoS model to provide real-time multimedia applications under the dynamically changing environment of ad hoc networks.

References

1. R. Braden, D. Clark, and S. Shenker, "Integrated Services in the Internet Architecture – an Overview", IETF RFC1663, June 1994.
2. S. Blake, "An Architecture for Differentiated Services", IETF RFC2475, December 1998.
3. R. Braden, L. Zhang, S.Berson, S. Herzog, and S. Jamin, "Resource reservation Protocol (RSVP) – Version 1 Functional Specification", RFC2205, Sept. 1997.
4. D. Black, S. Brim, B. Carpenter, "Per Hop Behavior Identification Codes", June 2001
5. Q. Xue and A.Ganz. Ad hoc QoS on-demand routing (AQOR) in mobile ad hoc networks. *Journal of parallel and Distributed Computing*, 154-165, 20003
6. X. Zhang S.B. Lee, A. Gahng-Seop and A.T. Campbell, INSIGNIA: An IP-Based Quality of Service Framework for Mobile Ad hoc Networks. *Journal of Parallel and distributed Computing*, 60(4), 374-406, April 2000
7. A. Lo H.Xiao, W.K.G..Seah and KC Chua. A Flexible Quality of Service Model for Mobile Ad hoc networks. In *IEEE Vehicular Technology Conference Fall 2000*, pages 445-449, May 2000
8. A. Veres G.Ahn, A.T. Campbell and L.Sun. SWAN: Service Differentiation in Stateless Wirelass Ad hoc network. In *Conference on Computer Communications(IEEE infocom)*, Jun 2002
9. X. Li and L. Cuthbert, Stable Node-Disjoint Multipath Routing with Low Overhead in Mobile Ad Hoc Networks, In *Proceedings of the The IEEE MASCOTS'04*, Netherland, Oct. 2004
10. S.J.Lee and M.Gerla, Split Multipath Routing with Maximally Disjoint Paths in Ad Hoc Networks, In *Proceedings of the IEEE ICC*, pages 3201-3205, 2001.
11. Marina, M.K., Das, S.R.: On-demand Multipath Distance Vector Routing in Ad Hoc Networks, *Proceedings of the International Conference for Network Procotols (2001).*
12. Ye, Z., Krishnamurthy, S. V., Tripathi, S.K., A Framework for Reliable Routing in Mobile Ad Hoc Networks. *IEEE INFOCOM (2003).*
13. Charles E. Perkings, Elizabeth M. Belding-Royer, Samir R.Das, Ad Hoc On-Demand Distance Vector (AODV) Routing, http://www.ietf.org/internet-drafts/draft-ietf-manet -aodv-13.txt, IETF Internet draft, Feb 2003
14. J.Broch, D.Johnson, and D. Maltz, "The Dynamic Source Protocol for Mobile Ad hoc Networks, http://www.ietf.org/internetdrafts/draft-ieft-manet-dsr-03.txt, IETF Internet draft (work in progress), Oct.1999.
15. Charles E. Perkings, Elizabeth M.Royer and Samir R.Das, Performance Comparison of Two On-Demand Routing Protocols for Ad Hoc Networks, *IEEE Personal Communications*, Feb 2001.

16. V. Jacobson and K. Nichols, An Expedited Forwarding PHB, RFC2598, June 1999
17. J. Heinanen, F. Baker, W. Weiss, and J. Wroclawski, Assured Forwarding PHB Group, RFC2597, June 1999
18. S. Floyd and V. Jacobson, Random Early Detection Gateways for Congestion Avoidance
19. OPNET Technologies, Inc. http://www.opnet.com/
20. J. Broch, D.A. Maltz, D.B. Johnson, Y.C. Hu, and J.Jetcheva, A Performance Comparison of Multi-Hop Wireless Ad Hoc Network Routing Protocols, In *Proceedings of the Fourth Annual ACM/IEEE MobiCom'98,* October 25-30, 1998, Dallas, Texas, USA.
21. IEEE Standards Department, Wireless LAN medium access control (MAC) and physical layer (PHY) specifications, *IEEE standard 802.11*, 1997

A New Backoff Algorithm to Support Service Differentiation in Ad Hoc Networks*

Li Yun[1], Ke-Ping Long[1], Wei-Liang Zhao[1], Chonggang Wang[2], and Kazem Sohraby[2]

[1] Special Research Centre for Optical Internet & Wireless Information Networks, ChongQing University of Posts & Telecommunications, ChongQing 400065, China
{liyun, longkp, zhaowl}@cqupt.edu.cn
[2] Department of Electrical Engineering, University of Arkansas, Fayetteville, AR 72701, USA
cgwang@uark.edu

Abstract. It is essential to design QoS supported MAC mechanism for supporting QoS in WLAN. In [1], we have proposed a new backkoff algorithm, named RWBO+BEB, to decrease the packet collision probability significantly. In this paper, we explore how to make RWBO+BEB support service differentiation in WLAN, and propose a novel proportional service differentiation algorithm, named *p*-RWBO, to allocate the wireless bandwidth according to the bandwidth ratio of each station. The basic idea of *p*-RWBO is that different priority stations use different walking probability, p_w, which is a key parameter in RWBO+BEB. An analytical model is proposed to analyze how to choose p_w according to the bandwidth ratios of stations. The simulation results indicate that *p*-RWBO can allocate the wireless bandwidth according to the bandwidth ratio of each station.

1 Introduction

IEEE 802.11 wireless LAN is the most widely used WLAN standard today. It covers the MAC sub-layer and the physical (PHY) layer. The MAC sub-layer defines two medium access coordination functions, the basic Distributed Coordination Function (DCF) and the optional Point Coordination Function (PCF).

Much research has been conducted on the performance of IEEE802.11 DCF. In [2] and [3], the author gives a Markov chain model for the backoff procedure of 802.11 DCF and studies its saturation throughput. In [4], the authors evaluate the performance of 802.11 DCF in terms of the spatial reuse and propose an improved virtual carrier-sensing scheme to enhance the spatial reuse. Wang C. et al. [5] propose a new efficient collision resolution mechanism to reduce the collision probability. [6] proposes an improved analytical model that calculates IEEE 802.11 DCF performance taking into account both packet retry limits and transmission errors for the IEEE 802.11a protocol.

---

* Supported by the National Science Foundation of China (90304004); the Science and Technology Research Project of Chongqing Municipal Education Commission of China (050310, KJ050503), the Research Grants by the Science & Tech. Commission of Chongqing(8817), the Research grants by the Ministry of Personnel of China.

X. Jia, J. Wu, and Y. He (Eds.): MSN 2005, LNCS 3794, pp. 570 – 579, 2005.

In recent years, there are growing researches on the QoS supporting for IEEE 802.11 WLAN. In [7-11], various schemes have been proposed to modify the backoff stage so that different priority source stations use different "Contention Window" generation functions. In [12], an Advance Access (AA) MAC protocol was proposed to support differentiated service, power control, and radio efficiency. Mohammad M. et al [13] proposed an adaptive fair EDCF that extends EDCF, by increasing the contention window during deferring periods when the channel is busy, and by using an adaptive fast backoff mechanism when the channel is idle. In [14], Wang, W. et al proposed a voice multiplex–multicast (M–M) scheme for overcoming the large overhead effect of VoIP over WLAN. In [15], IQD, a distributed coordination function (DCF) with integrated quality of service (QoS) differentiation, is proposed in this letter to enhance QoS over IEEE 802.11 WLAN.

In [16], we have quantitatively analyzed the Probability Distribution of Slot Selection (PDoSS) of the 802.11 DCF and proved that it is extremely uneven, as makes the packet collision probability very high. In [1] we have proposed a new backkoff algorithm, named Random Walk Backoff with Binary Exponential Backoff (RWBO+BEB), to make the stations select the slots uniformly and dramatically decrease the packet collision probability. In this paper, we explore how to make RWBO+BEB support proportional service differentiation. An analytical model is given to analyze how to choose the walking probability for each station according to its bandwidth ratio. Based on the analysis, a novel proportional service differentiated backoff algorithm for IEEE 802.11 WLAN, named *p*-RWBO, is proposed.

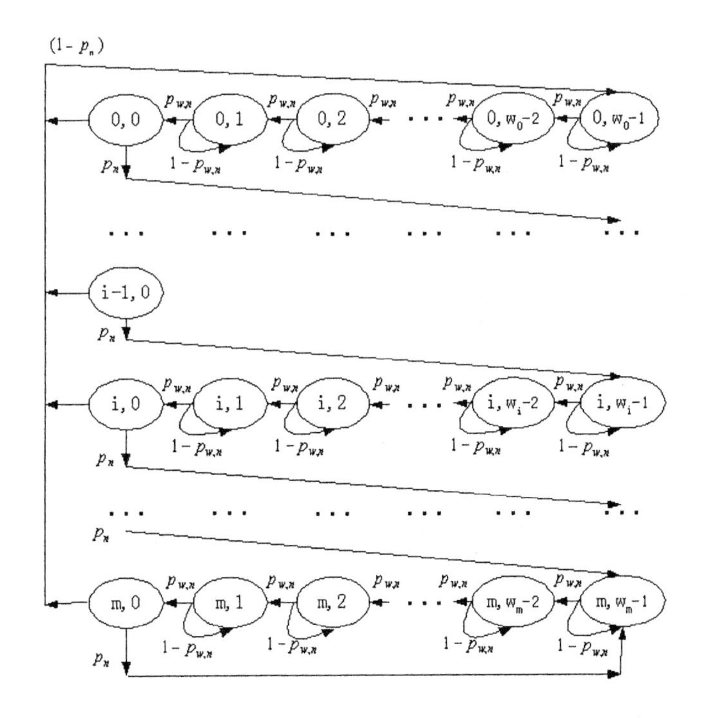

Fig. 1. New Markov model for RWBO+BEB

2 RWBO+BEB and the Basic Idea of *p*-RWBO

In [16], we have quantitatively analyzed the PDoSS of the 802.11 DCF and proved that it is extremely uneven. In 802.11 DCF, after transmitting a packet, the station uniformly selects a slot from 0 to CW_{min} as its Backoff Timer if the transmission is successful, and uniformly selects a slot from from 0 to $2^i \times CW_{\min} - 1$ otherwise. The Backoff Timer was decremented by aSlotTime if no medium activity is indicated for the duration of a particular backoff slot. This backoff algorithm makes the PDoSS decrease linearly as the slot number increases in the range of $[0, CW_{\min} - 1]$ and $[2^{i-1} \times CW_{min}, 2^i \times CW_{\min} - 1], i \in [1, m]$, respectively [16]. Here, m is the maximum backoff stage, whose default value is 5 in 802.11 DCF.

In order to achieve a uniform PDoSS, we have given a novel backoff algorithm for 802.11 DCF in the precious work [1], named RWBO+BEB, by which the packet collision probability can be extremely decreased.

2.1 RWBO+BEB

The RWBO+BEB is as follows:

After transmitting, the station selects slot w_0-*1* if the transmission is successful, and selects slot $2^i \times CW_{\min} - 1$ otherwise. If no medium activity is indicated for the duration of a particular backoff slot, then the Backoff Timer was decremented by aSlotTime with a probability of p_w. If the medium is sensed busy during a backoff slot, the Backoff Timer is suspended until the medium is idle for the duration of DIFS period again, and then the Backoff Timer resumes.

For simplicity, we let the contention windows, w, equal CW_{min}, and denote p_w as the *walking probability* in the following. The combination of w and p_w determines the performance of the new algorithm.

The performance analysis in [16] has denoted that RWBO+BEB can decrease the packet collision probability to a large extent, utilize the channel more efficiently, and make the packet delay jitter much lower comparing to 802.11 DCF. In the following, we improve RWBO+BEB to support proportional service differentiation.

2.2 Basic Idea of *p*-RWBO

The basic idea of *p*-RWBO is that different priority stations use different walking probability, p_w.

Given the normalized throughput, *S*, and the bandwidth ratio, $r = [r_1, r_2, \ldots, r_N]$, for *N* stations, the walking probability, $[p_{w,1}, p_{w,2}, \ldots, p_{w,N}]$, for all *N* can be determined such that the bandwidth obtained by each station equals its own ratio.

The key issue of *p*-RWBO is how to choose $[p_{w,1}, p_{w,2}, \ldots, p_{w,N}]$ so that the bandwidth ratios of *N* stations are equal to $[r_1, r_2, \ldots, r_N]$. In section 3, we propose an analytical model to solve this issue.

3 Analytical Model

As in [11], it is assumed that each station has a constant and independent packet collision probability p_n and transmission probability τ_n, and the channel is saturated. Let $b(t)$ and $s(t)$ be the stochastic processes of the backoff timer and the retransmission stage for a given station n, respectively. The two-dimensional discrete time process, {b(t),s(t)}, can be modeled as a Markov chain shown in Figure 1, with the non-null one-step transition probabilities given by

$$\begin{cases} P\{i,k \mid i,k+1\} = p_{w,n}, k \in [0, w_i - 2], i \in [0,m] \\ P\{i,k \mid i,k\} = 1 - p_{w,n}, k \in [1, w_i - 1], i \in [0,m] \\ P\{0, w_0 - 1 \mid i,0\} = 1 - p_n, i \in [0,m] \\ P\{i, w_i - 1 \mid i-1,0\} = p_n, i \in [1,m] \\ P\{m, w_m - 1 \mid m,0\} = p_n \end{cases} \tag{1}$$

where

$P\{i_1, k_1 \mid i_0, k_0\} = P\{s(t+1) = i_1, b(t+1) = k_1 \mid s(t) = i_0, b(t) = k_0\}$, and $w_i = 2^i \times CW_{\min}$.

Let $b_{i,k} = \lim_{t \to \infty} P\{s(t) = i, b(t) = k\}, k \in [0, W_i - 1], i \in [0,m]$, be the stationary distribution of {b(t),s(t)}, owing to the chain regularities, we obtain

$$b_{i,0} = p_n \cdot b_{i-1,0} \Rightarrow b_{i,0} = p_n^i \cdot b_{0,0}, 0 \le i < m \tag{2}$$

$$b_{m,0} = p_n \cdot b_{m-1,0} + p_n \cdot b_{m,0} \Rightarrow b_{m,0} = \frac{p_n^m}{1 - p_n} \cdot b_{i,0} \tag{3}$$

$$\sum_{i=0}^{m} b_{i,0} = \sum_{i=0}^{m-1} p_n^i \cdot b_{0,0} + \frac{p_n^m}{1 - p_n} \cdot b_{0,0} = b_{0,0} / (1 - p_n) \tag{4}$$

and

$$b_{0,w_0-1} = (1 - p_n) \cdot \sum_{i=0}^{m} b_{i,0} + (1 - p_{w,n}) \cdot b_{0,w_0-1} \Rightarrow b_{0,w_0-1} = \frac{1 - p_n}{p_{w,n}} \cdot \sum_{i=0}^{m} b_{i,0} = b_{0,0} / p_{w,n} \tag{5}$$

$$b_{m,w_m-1} = p_n \cdot b_{m,0} + p_n \cdot b_{m-1,0} + (1 - p_{w,n}) \cdot b_{m,w_m-1} \Rightarrow b_{m,w_m-1} = \frac{p_n^m}{p_{w,n} \cdot (1 - p_n)} \cdot b_{0,0}) \tag{6}$$

$$b_{i,w_i-1} = p_n \cdot b_{i-1,0} + (1 - p_{w,n}) \cdot b_{i,w_i-1} \Rightarrow b_{i,w_i-1} = \frac{p_n}{p_{w,n}} \cdot b_{i-1,0} = \frac{p_n^i}{p_{w,n}} \cdot b_{0,0}, 0 < i < m \tag{7}$$

When $0 < i < m, 0 < k < w_i - 1$,

$$b_{i,k} = (1 - p_{w,n}) \cdot b_{i,k} + p_{w,n} \cdot b_{i,k+1} \Rightarrow b_{i,k} = b_{i,k+1} \tag{8}$$

By relations (7) and (8), we obtain

$$b_{i,k} = b_{i,k+1} = \frac{p_n}{p_{w,n}} \cdot b_{i-1,0} = \frac{p_n^i}{p_{w,n}} \cdot b_{0,0}, 0 < i < m, 0 < k < w_i - 1 \tag{9}$$

Note that

$$b_{m,k} = (1 - p_{w,n}) \cdot b_{m,k} + p_{w,n} \cdot b_{m,k+1} \Rightarrow b_{m,k} = b_{m,k+1} = \frac{p_n^m}{p_{w,n} \cdot (1 - p_n)} \cdot b_{0,0}, 0 < k < w_m - 1 \tag{10}$$

and by relations (2),(3),(5) and (9), it is easily to obtain that

$$b_{i,k} = \begin{cases} \frac{p_n^i}{p_{w,n}} \cdot b_{0,0}, 0 \le i < m, 0 < k \le w_i - 1 \\ p_n^i \cdot b_{0,0}, 0 \le i < m, k = 0 \\ \frac{p_n^m}{p_{w,n} \cdot (1 - p_n)} \cdot b_{0,0}, i = m, 0 < k \le w_m - 1 \\ \frac{p_n^m}{1 - p_n} \cdot b_{0,0}, i = m, k = 0 \end{cases} \tag{11}$$

Let τ_n be the probability that station n transmits in a randomly chosen slot, τ_n can be expressed:

$$\tau_n = \sum_{i=0}^{m} b_{i,0} = \sum_{i=0}^{m-1} p_n^i \cdot b_{0,0} + \frac{p_n^m}{1 - p_n} \cdot b_{0,0} = b_{0,0} / (1 - p_n) \tag{12}$$

By relation (12), τ_n is expressed as function of $b_{0,0}$ and p_n. $b_{0,0}$ is finally determined by imposing the normalization condition as follows:

$$\sum_{i=0}^{m} \sum_{k=0}^{w_i - 1} b_{i,k} = 1 \tag{13}$$

Plugging (11) into equation (13), we can obtain

$$\sum_{i=0}^{m} \sum_{k=0}^{w_i-1} b_{i,k} = \sum_{i=0}^{m-1} \sum_{k=0}^{w_i-1} b_{i,k} + \sum_{k=0}^{w_m-1} b_{m,k} = \sum_{i=0}^{m-1} \left(\sum_{k=1}^{w_i-1} \frac{p_n^i}{p_{w,n}} \cdot b_{0,0} + p_n^i \cdot b_{0,0} \right) + \sum_{k=1}^{w_m-1} b_{m,k} + \frac{p_n^m}{1 - p_n} \cdot b_{0,0}$$

$$= \frac{(1 - p_n) \cdot \left[1 - (2p_n)^m\right] \cdot w - (1 - 2p_n) \cdot (1 - p_n^m) + p_{w,n} \cdot (1 - 2p_n) + (1 - 2p_n) \cdot p_n^m \cdot (2^m \cdot w - 1)}{p_{w,n} \cdot (1 - p_n) \cdot (1 - 2p_n)} \cdot b_{0,0} = 1$$

$$\Rightarrow b_{0,0} = \frac{p_d \cdot (1 - p) \cdot (1 - 2p)}{(1 - p) \cdot \left[1 - (2p)^m\right] \cdot w - (1 - 2p) \cdot (1 - p^m) + p_d \cdot (1 - 2p) + (1 - 2p) \cdot p^m \cdot (2^m \cdot w - 1)} \tag{14}$$

Plugging (14) into (12), we can express the τ as follows:

$$\tau_n = \frac{p_{w,n} \cdot (1 - 2p_n)}{(1 - p_n) \cdot \left[1 - (2p_n)^m\right] \cdot w - (1 - 2p_n) \cdot (1 - p_{w,n}) + (1 - 2p_n) \cdot p_n^m \cdot 2^m \cdot w} \tag{15}$$

Note that p_n, the probability that a packet transmitted by station n encounters a collision, is equal to the probability that, in a slot, at least one of the N-1 remaining station transmit, therefore,

$$p_n = 1 - \prod_{j=1, j \ne n}^{N} (1 - \tau_j) \tag{16}$$

The probability of successful transmission of station n, P_s^n, can be express as:

$$P_s^n = \tau_n \cdot \prod_{j=1, j \ne n}^{N} (1 - \tau_j) \tag{17}$$

Adding together the probabilities of all N stations leads to the probability of total successful transmission, P_s :

$$P_s = \sum_{s=1}^{N} P_s^n \tag{18}$$

Let S_n be the saturation throughput of station n. Then the system saturation throughput and the allocated bandwidth ratio can be written as

$$S=\sum_{n=1}^{N}S_n=\frac{P_s\cdot E[P]}{(1-P_{tr})\cdot\sigma+P_s\cdot T_s+(P_{tr}-P_s)\cdot T_c} \tag{19}$$

and,

$$S_1:S_2:\cdots:S_N=P_s^1:P_s^2:\cdots:P_s^N=\frac{\tau_1}{1-\tau_1}:\frac{\tau_2}{1-\tau_2}:\cdots:\frac{\tau_N}{1-\tau_N}=r_1:r_2:\cdots:r_N \tag{20}$$

respectively, where

$$P_{tr}=1-\prod_{n=1}^{N}(1-\tau_n) \tag{21}$$

Relation (19) expresses the saturation throughput, S, as in [3], where

$$S=\frac{P_s\cdot E[P]}{(1-P_{tr})\cdot\sigma+P_s\cdot T_s+(P_{tr}-P_s)\cdot T_c} \tag{22}$$

$E[P]$ is the average packet payload size, T_s is the average time the channel is sensed busy because of a successful transmission, T_c is the average time the channel is sensed busy during a collision, σ is the duration of an empty slot time, P_{tr} is the probability that there is at least one transmission in the considered slot time, P_s is the probability that a transmission occurring on the channel is successful. P_s and P_{tr} are expressed in relations (18) and (21), respectively.

Solving equations (17), (18) and (21), we can obtain:

$$P_s=(1-P_{tr})\cdot\sum_{n=1}^{N}\frac{\tau_n}{1-\tau_n} \tag{23}$$

Expression (20) allows us to rewrite the sum portion in expression (23) as

$$\sum_{n=1}^{N}\frac{\tau_n}{1-\tau_n}=\frac{\tau_n}{1-\tau_n}\cdot\frac{\sum_{n=1}^{N}r_n}{r_n} \tag{24}$$

Inserting (24) into (23), and reorganizing the equation, we can obtain the expression of τ_n as follows:

$$\tau_n=\frac{P_s\cdot r_n}{P_s\cdot r_n+(1-P_{tr})\cdot\sum_{n=1}^{N}r_n} \tag{25}$$

where P_s can be solved from Equation (19), that is,

$$P_s=\frac{(1-P_{tr})\cdot\sigma+P_{tr}\cdot T_c}{\frac{E[P]}{S}-T_s+T_c} \tag{26}$$

Equations (21), (25) and (26) constitute a set of n+2 dimension nonlinear equations, and solving these equations can yield the results of P_{tr}, P_s and τ_n.

Moreover, the walking probability of station n can be derived from equation (15):

$$p_{w,n}=\frac{\tau_n\cdot\left[1-p_n-p_n\cdot(2p_n)^m\right]\cdot w-\tau_n\cdot(1-2p_n)}{(1-\tau_n)\cdot(1-2p_n)} \tag{27}$$

where p_n can be derives from equation (16).

Equation (27) suggests that: given the normalized throughput S, and the bandwidth ratio r= [r_1, r_2, …, r_N], we can obtain the walking probability [$p_{w,1}$, $p_{w,2}$,…, $p_{w,N}$] analytically to make the bandwidth obtained by each station equal to its own ratio.

4 Model Validation

In this section, we validate the correctness of analytical model by comparing the simulated saturation throughput to the analytical saturation throughput. To obtain the analytical saturation throughput, we can firstly yield p_n and τ_n by solving equations (15) and (16), and obtain P_{tr} and P_s by plugging τ_n into expressions (17), (18) and (21). Then, the analytical saturation throughput can be obtained by equation (22).

The simulation platform is NS-2 [17]. Stations are randomly placed in an area of 200×200 meters and have no mobility, constituting an ad-hoc network, in which a station can transmit to any other station packets with the packet size fixed to 1024 bytes at the same CBR rate. Each CBR flowe runs 200 seconds. The physical layer is DSSS. The stations transmit packets by means of RTS/CTS mechanism, and the simulation parameters are shown in Table 1.

Table 1. Simulation Parameters

Parameter	Value
Channel Bit Rate	2Mbit/s
Slot Time	20μs
SIFS	10μs
DIFS	50μs
PHYHeader	192bit
MACHeader	144bit
RTS Length	160bit
CTS Length	112bit
CW_{min}	32
CW_{max}	1024
CBR Packet Size	1024Byte

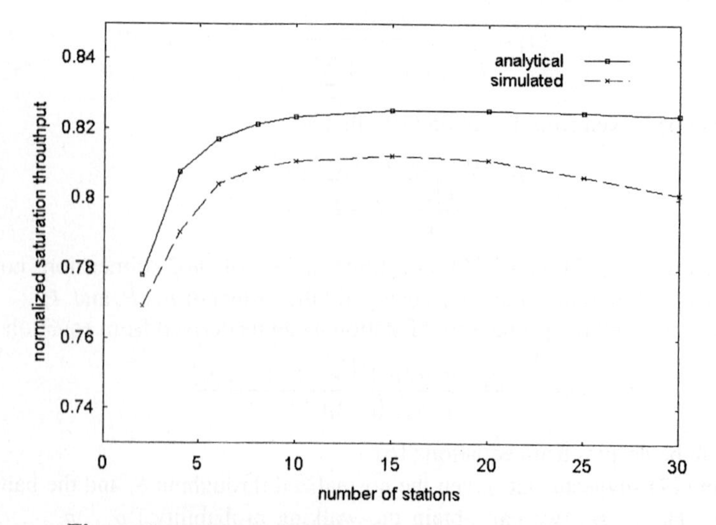

Fig. 2. Saturation throughput of p-RWBO: analytical and simulated

For simplicity, we set the minimum contention windows to 32, and the walking probability to 0.75 for all stations in the networks. The analytical and simulated saturation throughputs of p-RWBO are drawn in Fig. 2 under different stations.

Fig.2 shows that the analytical and simulated results are very closed. The analytical saturation throughput is just about 2% higher than the simulated saturation throughput, as is attributed to that we ignore the frame retry limits in the analytical model.

5 Performance Evaluation

We place 12 stations on 250m×250m flat space, and establish a CBR flow on each pair of stations. The normalized saturation throughput S=0.8, the bandwidth ratio $\boldsymbol{r}$=[1.0,0.5,0.5,0.25,0.25,0.125]. We set the minimum contention windows to 32. Through equation (27), we calculate numerically the value of walking probability of every station, which are [0.845588,0.428992,0.428992,0.216110,0.216110,0.108467], and set the walking probability of each station to its calculated value in simulation. The good throughputs of 6 CBR flows are drawn in Fig. 3, respectively. The average good throughput of each CBR flow is shown in Table 2.

Fig. 3 shows that DS-RWBO can differentiate each CBR flow according to its bandwidth ratio, and Table 2 indicates that DS-RWBO can allocate bandwidth to each flow according to expected ratio.

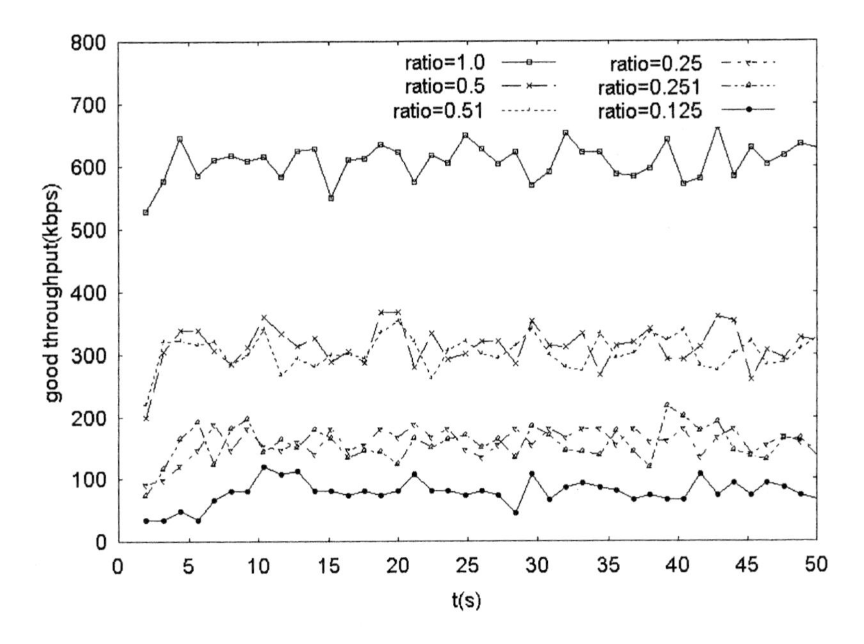

Fig. 3. The good throughput for different CBR flows

Table 2. Actual bandwidth ratio vs. expected bandwidth ratio

Flow#	1#	2#	3#	4#	5#	6#
Expected bandwidth ratio	1.0	0.5	0.5	0.25	0.25	0.125
Actual bandwidth ratio	1.0	0.4964	0.4966	0.2543	0.2576	0.1275

6 Conclusion

It is essential to design QoS supported MAC mechanism for supporting QoS in WLAN. We have explored how to make RWBO+BEB support proportional service differentiation. An analytical model is given to analyze how to choose the walking probability for each station according to its bandwidth ratio. Based on the analysis, a novel proportional service differentiated backoff algorithm for WLAN, named *p*-RWBO, is proposed. *p*-RWBO can allocate bandwidth to each station according to expected ratio.

References

1. Li Yun, Long Ke-ping, Zhao Wei-liang. RWBO+BEB: A Novel Backoff Algorithm for IEEE 802.11 DCF. Accepted to appear in Chinese Journal of Electronics.
2. G.Bianchi. IEEE802.11 – Saturation throughput analysis. IEEE Communication Letters, 1998, Vol.2: 318 – 320.
3. G.Binachi. Performance analysis of the IEEE 802.11 distributed coordination function. IEEE Journal on Selected Areas in Commu. , 2000, 18(3): 535-547..
4. Fengji Ye, Su Yi, Sikdar, B.. Improving spatial reuse of IEEE 802.11 based ad hoc networks. IEEE GLOBECOM '03. 2003, Vol. 2: 1013 - 1017.
5. Chonggang Wang, Bo Li, Lemin Li. A new collision resolution mechanism to enhance the performance of IEEE 802.11 DCF. IEEE Transactions on Vehicular Technology. 2004, 53(4): 1235-1246.
6. Chatzimisios, P., Boucouvalas, A.C.; Vitsas, V.. Performance analysis of IEEE 802.11 DCF in presence of transmission errors. In IEEE International Conference on Communications, 2004, Vol. 7: 3854 - 3858.
7. Michael Barry and Andrew T. Campbell and Andras Veres. Distributed Control Algorithms for Service Differentiation in Wireless Packet Networks. IEEE INFOCOM, April 2001
8. Imad Aad and Claude Castelluccia. Differentiation mechanisms for IEEE 802.11. IEEE INFOCOM, April 2001.
9. V.Kanodia and C.Li and A.Sabharwal and B.Sadeghi and E.Knightly. Distributed Multi-Hop Scheduling and Medium Access with Delay and Throughput Constraints. MOBICOM, August 2001.

10. S. Mangold, S. Choi, P. May, O. Klein, G. Hiertz, L. Stibor, IEEE 802.11e wireless LAN for quality of service (invited paper). Proceedings of the European Wireless, Florence, Italy, February 2002, Vol. 1, pp. 32-39
11. Jie Hui, Devetsikiotis, M.. Designing improved MAC packet schedulers for 802.11e WLAN. IEEE GLOBECOM '03, 2003 Vol.1: 184 – 189
12. Chi-Hsiang Yeh. The advance access mechanism for differentiated service, power control, and radio efficiency in ad hoc MAC protocols. Proceeding of Vehicular Technology Conference, 2003, Vol.3: 1652 – 1657.
13. Mohammad Malli, Qiang Ni, Thierry Turletti, and Chadi Barakat. Adaptive fair channel allocation for QoS enhancement in IEEE 802.11 wireless LANs. IEEE International Conference on Communications (ICC 2004), Paris, 2004, Vol.6: 3470 - 3475
14. Wang W., Liew S.C., Li V.O.K.. Solutions to Performance Problems in VoIP Over a 802.11 Wireless LAN. IEEE Transactions on Vehicular Technology, 2005, 54(1): 366 – 384
15. Zhao Liqiang, Changxin Fan. Enhancement of QoS differentiation over IEEE 802.11 WLAN. IEEE Communications Letters, 2004, 8(8): 494 – 496
16. Li Yun, Long Ke-ping, Wu Shi-qi, Chen Qina-bin. Performance analysis and improvement of IEEE802.11 DCF. Acta Electronica Sinica, 2003, 31(10): 1446 – 1451
17. The networks simulator ns-2, http:// www.isi.edu/nsnam/

Power Aware Multi-hop Packet Relay MAC Protocol in UWB Based WPANs*

Weidong Wang, Chang-Keun Seo, and Sang-Jo Yoo

[1] Graduate School of Information Technology & Telecommunications, Inha University, 253 YongHyun-dong, Nam-ku, Incheon 402-751, Korea
{wangwd, seochangkeun}@gmail.com, sjyoo@inha.ac.kr
http://multinet.inha.ac.kr

Abstract. Ultra wide band (UWB) technology will be applied in the high rate wireless personal area networks (WPANs) for its high rate, low power, and innate immunity to multipath fading. In this paper, a power aware multi-hop packet relay MAC protocol in UWB based WPANs is proposed and a power aware path status factor (PAPSF), which is derived from SINR and power resource condition of each device, is used to select a suitable relay node. Compared with relaying by piconet coordinator (PNC), which is easily chosen by other ad hoc routing protocol, the new scheme can achieve higher throughput, decrease the time required for transmitting high power signal and we can easily distribute the battery power consumption from PNC to other devices in the piconet to prevent the PNC device using up its battery too fast and finally avoid PNC handover too frequently.

1 Introduction

The next generation communication will try to achieve the connectivity for "everybody and everything at any place and any time". This ambitious view assumes that the new wireless world will be the result of a comprehensive integration of existing and future wireless system [1]. By now, the cellular phone system, wide area networks (WAN) and wireless local area networks (WLAN) services can only satisfy people basic connection requirement such as voice, instant message or relative low rate data exchange. However, to achieve the high rate data exchange, a new technology for high speed short range wireless communication is required and will play an important role in the future. Due to the high rate data exchange will be applied frequently in the future, a new wireless personal access networks (WPANs) based on ultra-wide band (UWB) is expected for its characteristics such as high data rate, low power consumption, low complexity in terms of transmitter/receiver.

Limited by the power of signal and required high channel speed, the coverage of WPAN is quite small, about a circle area with 10 meter radius. Even within 10 meters, the link speed may change from 480 Mbps to 110 Mbps and the difference makes it quite attractive to improve the throughput by selecting high rate link for data

---

* This research was supported by University IT Research Center Project (INHA UWB-ITRC), Korea.

X. Jia, J. Wu, and Y. He (Eds.): MSN 2005, LNCS 3794, pp. 580 – 592, 2005.

relay. Some researches utilize UWB characteristic distance awareness to select the shortest relay path [2]. But the distance can not represent all. Sometimes, the shortest path may be combined with highest noise or interference. In this paper, channel conditions and power issue are taken into account and power aware path status factor (PAPSF) is used as a criterion to select a relay node.

This paper is organized as follows: Overview of UWB and MAC protocol for WPANs are shown in Section 2. Section 3 presents the proposed power aware MAC layer packet relay protocol and the simulation results will be shown in Section 4. Finally, the conclusion is given in the Section 5.

2 Overview of UWB and WPAN MAC Protocol

FCC (Federal Communication Commission) defines UWB that its fractional bandwidth is bigger than 0.2 or the bandwidth is bigger than 500MHz. Compared with conventional narrowband communications and spread spectrum wideband communications, much higher bandwidth is applied in UWB technologies. The fraction is defined as $(f_H-f_L)/f_C$, where f_H is the upper frequency, f_L is the lower frequency at the -10dB emission points and the f_C is the center frequency of the spectrum.

Shannon's equation for channel capacity is a basic one in the communication as in (1).

$$C = W \log_2\left(1+\frac{P_0}{N_0}\right) bps \; . \tag{1}$$

Where C is the channel capacity, W is the bandwidth in Hz, P_0 is the signal power in watts/Hz and $N_0/2$ is the noise spectral density in watts/Hz. Two methods can be used to increase the channel capacity: increasing the bandwidth W and increasing the transmitting power of signal P_0. However, C increases linearly with W but increases only as the log arithmetic with power of signal P_0. So, the bandwidth of UWB system is much higher than the required data rate, which makes UWB can operate on a very low signal to noise ratio (SNR) and makes UWB a power efficient technology. FCC also approved the bandwidth for UWB from 3.1GHz – 10.6GHz, and the spectral masks for UWB indoor and outdoor system are shown in Fig. 1. UWB system operates at a very low transmitting power with the effective isotropic radiated power (EIRP) lower than -41dBm/MHz, which allow UWB systems to overlay already available services.

There are two different UWB technologies emerged in these years. One is the DS-UWB (Direct Sequence-UWB) which is proposed by UWB Forum and another is the Multi-Band OFDM proposed by Multi-Band OFDM Alliance (MBOA). DS-UWB uses the whole available spectrum defined by FCC with the Impulse Radio (IR) technique, and it is sometimes termed as carrierless radio because it needs not a higher carrier frequency [3]. Unlike the DS-UWB, Multi-Band OFDM divides the available spectrum into several bands, each band with about 500 MHz bandwidth [4].

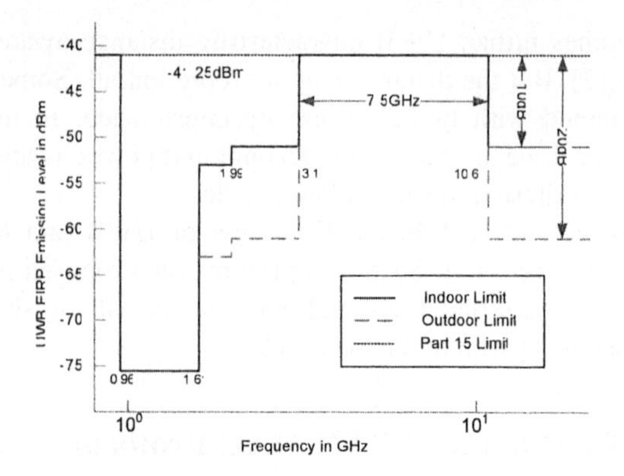

Fig. 1. FCC mask for indoor and outdoor communications

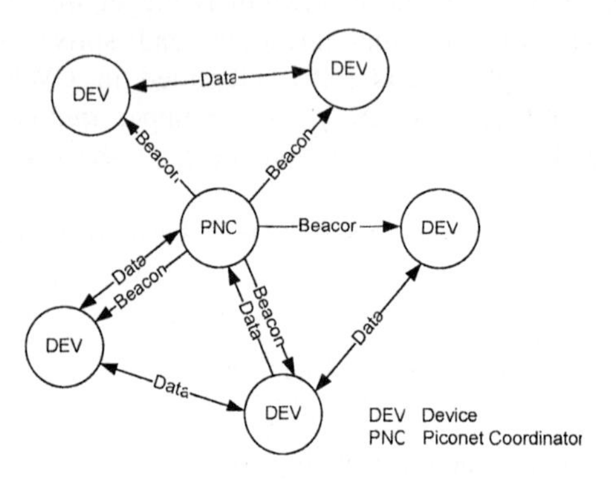

Fig. 2. IEEE 802.15.3 Piconet elements and messages/data exchange

The IEEE 802.15.3a group is working on the standard for high speed WPAN with the UWB as its physical layer. The MAC protocol will mostly adopt the original MAC protocol in IEEE 802.15.3. Piconet is a basic concept in IEEE 802.15.3 MAC protocol. A piconet is a wireless ad hoc data communications system which allows a number of independent data devices to communicate with each other [5]. A piconet is composed of several parts, devices (DEVs) and piconet coordinator (PNC). The PNC controls its piconet by its beacon, including timing, quality of service (QoS) requirements, power saving and access control. Fig. 2 shows the piconet elements in the standard. A DEV in a piconet hears the beacon sent by PNC and the data can be transmitted between devices and PNC. Management message can be transmitted between PNC and DEV.

The difference between PNC and devices is that PNC sends beacon information periodically. Any DEV can be elected as the PNC if it's satisfied with criteria of PNC such as preference for become a PNC and ability of receiving power from the mains.

Timing in piconet is based on the superframe as shown in Fig. 3. A typical superframe is composed of three parts, beacon, contention access period (CAP) and the channel time allocation period (CTAP). A beacon is sent by PNC to control all

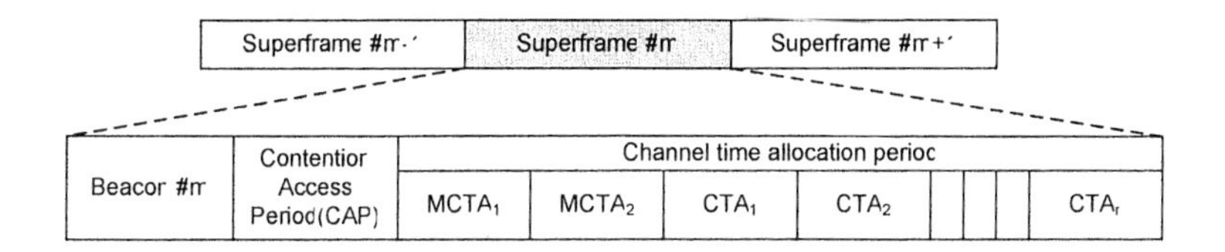

Fig. 3. Structure of IEEE 802.15.3 piconet superframe

devices' behaves and set parameters of this piconet. If a CAP is expressed in a beacon, devices can use this period to send commands or asynchronous data using the carrier sense multiple access with collision avoidance (CSMA/CA) as the medium access mechanism. The channel access in the CTAP is based on the TDMA mechanism, CTAP is used for data or commands. Management CTAs (MCTAs) may be allocated to a device or accessed by devices using slotted aloha mechanism.

WPAN is confined to a small area around person or object that typically covers at least 10m in all directions and envelops the person or a thing whether stationary or in motion [5]. Child piconet is proposed to extend the area of coverage of a piconet or the parent PNC can shift some computing requirement to another PNC capable device which is the child piconet PNC. Similar to a child piconet, a neighbor piconet is also dependent on a parent piconet and gets the private CTA from its parent PNC. But the purpose of setting up a neighbor piconet is to share the frequency between different piconets when there is no vacant PHY layer. The application of neighbor and child piconet enlarges the piconet coverage and shares the channel resource.

3 Proposed MAC Layer Packet Relay Protocol for WPANs

WPAN as a short range personal communication system, devices can exchange data within about 10 meters. Because all devices need to associate to PNC, listen to beacon sent by PNC and exchange management information with PNC, we can only guarantee that devices can communicate with PNC directly. Some of them must use the multi-hop schemes to relay data when two devices can not communicate directly.

Because the IEEE 802.15.3 standard does not define a relay scheme [5], ad hoc routing protocols will be adopted to relay data. These ad hoc routing protocols such as AODV [6] use the hop account as criterion to select a route. It prefers to use PNC as intermediate node rather than other devices, because the hop account via PNC is 2 from any source to any destination. Once a route via PNC is selected, it can not be changed until PNC handover procedure happens, because of its reactive characteristic. As the result, the traffic concentrating problem will be generated and PNC will be deprived of all its battery much faster than other devices. Also, lowest link rate and highest transmitting signal power is required by source device and PNC, because PNC is usually located in the center of a piconet. Higher signal power conflicts the demanding of decreasing signal power; lower link speed conflicts the requirement of increasing system throughput.

A multi-hop scheme is also meaningful even two devices can communicate directly. Assume that the channel status is not good enough between two devices and the lowest link speed is adopted by them. If there is a device close to these two

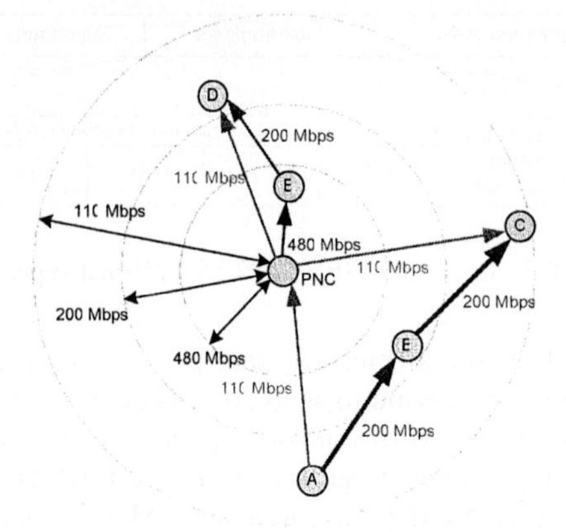

Fig. 4. Frame relay concept in MAC layer

devices, the device may operate as a relay node and increase the total transmitting speed between the source and destination. However, conventional ad hoc routing protocols only consider the hop account and can not support this situation at all. Even some researches exploit the signal stability in [7] or power consumption in [8] as factors to select intermediate node for data relay, the mechanism is too complex and require heavy overload. More important, they only pay attention to one issue. Two situations are shown in Fig. 4. Node A tries to send data to node C, which is out of its coverage. Conventional routing protocols prefer to use PNC for relaying the data that means node A first sends data to PNC then PNC sends data to the destination C. We can see that if node A select node B as relay node, total link speed will be higher than that of using PNC for relay. In the second situation, a node (e.g. PNC) wants to send data to node D which is in its coverage. Node E can be used to relay data and we can get much better performance than one hop transmission.

So, a MAC layer relay protocol is required for improving the performance of the system and the MAC layer relay protocol will be more efficient than other ad hoc routing protocols because routing protocols requires high overhead and complexity for routing discovering and routing maintenance. Also, a MAC layer relay protocol is transparent to routing protocol, which makes routing protocol only select intermediate node between piconets (parent piconet and its children piconets) but not within a piconet.

Many researches exploit the distance as the parameter to select the relay path for the location awareness capability of the UWB systems. Based on the location information, the suitable device is selected as a relay node [2]. However, as mentioned above, shortest distance is not always the best choice and an extended distance awareness module increases the complexity and cost of system. The routing scheme based on SINR named Angle-SINR Table Routing (ATR) is proposed in [9]. This routing is based on the Angle-SINR Table (AT) and each terminal gets AT through the 360-degree scanning of an adaptive antenna and measures the Signal-to-Interference and Noise Ratio (SINR). By using this AT, a route from the source to the destination is created. This scheme can reduce the routing traffic and power

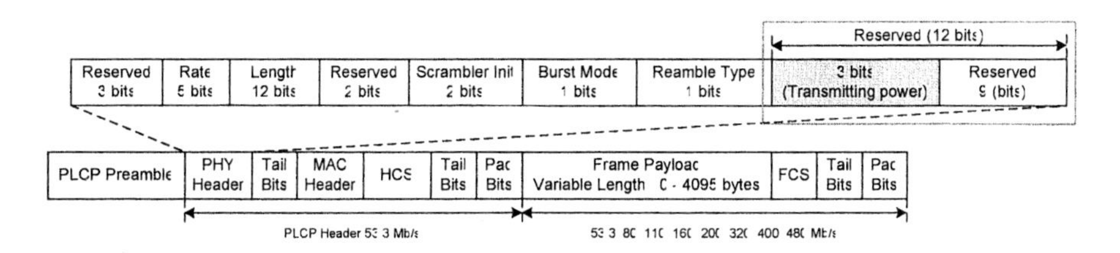

Fig. 5. UWB(Multi-band OFDM) PLCP frame format

consumption of the terminals. Obviously, SINR is better than distance on representing the channel status and SINR can be easily gotten by transceivers. But SINR is not only related with noise, interference and signal decay, but also related with the emitted signal power. In a piconet, as the emitted signal power from devices may be different and same device may emit different power of signal from moment to moment, quite different SINRs may be gotten by devices when they have same channel conditions. In the proposed protocol, a parameter named channel status factor (CSF) is used to help to select relay node. CSF is defined as the ratio of SINR at a receiver to transmitting power P_t from the transmitter and shown in (2).

$$CSF = \frac{SINR}{P_t} = \frac{S}{(I+N)P_t} \ . \tag{2}$$

Where S is received signal power at the receiver, I is the interference to the receiver, N is the noise of the system and P_t is the transmitting power from the sender device. In order to make each device know the signal power from transmitter, sending device needs to send packets with information of transmitting power in the PHY header.

In [10], the PHY service primitive TXPWR_LEVEL ranging from 1 to 8 is used by MAC layer to indicate which of the available TxPowerLevel attributes defined in the management information base (MIB) shall be used for the current transmission. This value (3 bits) can be inserted to the last reserved 12 bits field in PLCP header (PHY header). This modification is shown in Fig. 5.

When a device is sending a packet, other devices in its coverage should listen to it, then calculate the CSF and report it to PNC periodically. In order to select the relay path, PNC also needs to build a matrix named channel status relay matrix (CSRM). CSRM is an $(N+1)\times(N+1)$ matrix; N is the number of devices and one PNC in the piconet. According to the standard of IEEE 802.15.3, N+1 will not bigger than 256. CSRM is shown in (3). In the equation, CSF_{ij} indicates the channel status factor calculated by node j via detecting the signals sent by node i. The CSF_{ij} does not exist when $i = j$.

$$CSRM = \left\{ \begin{array}{ccccc} - & CSF_{01} & \cdots & CSF_{0(N-1)} & CSF_{0N} \\ CSF_{10} & - & \cdots & CSF_{1(N-1)} & CSF_{1N} \\ \vdots & \vdots & \vdots & \vdots & \vdots \\ CSF_{(N-1)0} & CSF_{(N-1)1} & \cdots & - & CSF_{(N-1)N} \\ CSF_{N0} & CSF_{N1} & \cdots & CSF_{N(N-1)} & - \end{array} \right\} . \tag{3}$$

Based on CSRM, three kinds of possible path status factors (PSF) can be derived by PNC. The first one is the direct path from source node s to destination node d as shown in (4), the PSF^{sd} is equal to the CSF_{sd} .The second one is the path via PNC, PSF^{sd} is the sum of CSF from source node s to PNC (node ID 0) and from PNC to destination node d, as in (5). The last one is the PSF via any possible relay node. It's also composed of two parts, CSF from source nod s to relay node k and CSF from relay node k to destination node d as in (6).

$$PSF^{sd} = CSF_{sd}\,. \tag{4}$$

$$PSF_0^{sd} = CSF_{s0} + CSF_{0d}\,. \tag{5}$$

$$PSF_k^{sd} = CSF_{sk} + CSF_{kd}\,. \tag{6}$$

Where k is not unique because there are maximum (N-3) possible relay candidates in the piconet (exclude PNC, source and destination). In the proposed protocol, increasing the throughput of system is considered mostly. Since PNC knows the CSF of each path and PNC provides the maximum limitation of signal power broadcasted by beacon, it is not difficult for PNC to estimate maximum possible link speed of each relay path according to the link budget. We assume that PNC evaluates the maximum possible link speed by using the maximum power limitation of its piconet. In this paper, only those paths that total link speed is equal to or bigger than the link speed of direct path will be kept and evaluated in the latter steps. Otherwise, two devices will communicate directly.

Next, we will consider about battery issue for selecting relay node. If a device is using the AC as power, it should be selected more as a relay node because AC power seemed to be infinite energy to this device. However, those devices using batteries are always limited by their battery and can not be selected as relay nodes frequently. We evaluate the power weight factor (PWF) of each device and ranging from 0 to 1. A device using AC power, higher battery capacity, or more energy remains in its battery is given a bigger PWF. PWF can be easily reported to PNC with piggyback information. After PNC selects several candidates as relay nodes and each PSF of a relay path multiples the PWF of possible relay node as (7), we will get the power aware path status factor (PAPSF) and the device with biggest PAPSF will be chosen as the selected relay device (SRD^{sd}) as shown in (8).

$$PAPSF_k^{sd} = PSF_k^{sd} \times PWF_k\,. \tag{7}$$

$$SRD^{sd} = k \mid_{\max\{PAPSF_k^{sd}\},\ \forall k}\,. \tag{8}$$

We defined a new channel time allocation for the proposed power aware multi-hop packet relay MAC protocol and named relay channel time allocation (RCTA). If PNC selects a relay node, a RCTA will be signed to sender and relay node, relay node and receiver node at the same time. Such information will be indicated in the beacon information element (IE) CTA block. The comparison of original CTA and RCTA is given in Fig. 6. In Fig. 6 (a), we can see that the original MAC protocol of WPANs does not support the relay scheme and all relay function is achieved by routing

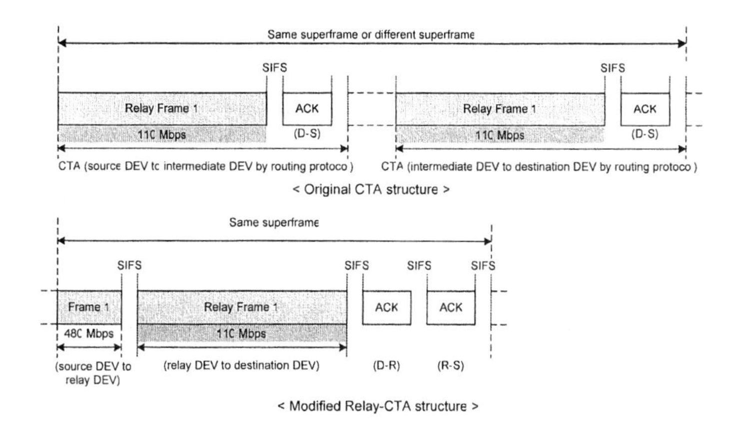

Fig. 6. Modified relay-CTA structure

protocols. Both sender and intermediate node should request the channel resource form PNC independently and PNC is preferred as the intermediate node. However, in (b), only source node request the channel resource one time and PNC will allocate all possible required channel resource to source node and relay node to finish data transmission. After selecting the SRD^{sd} as stated above, we may get a highest total link speed compared with other routing protocols. In the figure, D means destination, S means source and R means relay node.

4 Simulation Results

In this paper, the simulation scenario is set as shown in the Fig. 7. There are 72 devices in three piconet systems, one is the parent piconet and others are child piconets. Each piconet has 24 random moving devices and one PNC. PNC doesn't have data to transmit but control its piconet and relay the data. Assume that half of these devices have data to send, others have no data in a relative long time.

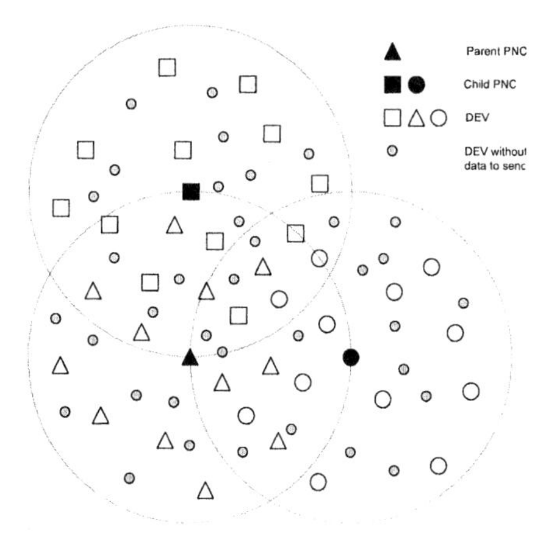

Fig. 7. Simulation topology

Table 1. Simulation parameters

Duration of Superframe (fixed)	20ms
Length of MPDU	4024 bytes
Power consumption (110Mbps)	$P_t = 93$mW, $P_r = 155$mW
Power consumption (200Mbps)	$P_t = 93$mW, $P_r = 169$mW
Power consumption (480Mbps)	$P_t = 145$mW, $P_r = 236$mW
Power consumption (Idle)	$P_{idle} = 0$mW
Path loss at 1 meter	$L_1 = 44.2$dB
Path loss at d meter	$L_2 = 20log_{10}$(d)
Average noise power per bit (110Mbps, 200Mbps, 480Mbps)	-87.0dBm, -84.4dBm, -80.6dBm
Reference average power	-10.3dBm

Since the IEEE 802.15.3 MAC protocol doesn't define the scheme for data transfer between child piconet DEV and parent piconet DEV. The child PNC is a member of both parent piconet and child piconet so that we regard child PNC as a gateway in the simulation. Except getting private CTA from parent PNC and exchange data with parent piconet and another child piconet, the data exchanging within a child piconet is quite independent with parent piconet. So in the simulation, data within parent piconet, data sent to or from parent piconet and data via the parent piconet is considered. Data exchanged within a child piconet will not be calculated. Simulation time is 20 seconds. Because all devices are random moving fast and within relative short time, we can regard that in the simulation period, all devices use PNC as relay node and traffic concentration problem is happened and we compare this situation with proposed protocol.

The PWF is set to 0 or 1 in the simulation. By increasing the number of devices with PWF=1, we can evaluate both the performance of increasing number of possible relay nodes and the performance of increasing device with better power supplying. According to the [10], channel mode CM1 and CM2 can achieve about 480 Mbps within 2 meters distance, 200 Mbps between 2 and 6 meters and 110 Mbps bigger than 6 meters. In this simulation, we suppose that all devices can use the maximum possible channel speed to transmit data without negotiation procedure. Relay through only PNC, which is usually chosen by typical ad hoc routing protocols, and relay through any devices including PNC are compared. Simulation is based on NS-2 environment [11] and the simulation parameters are set as in Table 1. These parameters are from [10].

The Fig. 8 shows the throughput of the system as increasing more devices with PWF=1. Zero means that PWF of all devices is set to 0. Any device will not be used as relay node, only PNC itself can relay data for devices. The dashed line is a reference line of using PNC to relay data, and it will appear in the following figures if needed. We can also get the conclusion easily from the figure that the throughput is increasing with the number of devices with good power supplying increased. Equivalently, more relay capable devices will increase the throughput of piconet system. The reason is obviously, when there are more competent relay nodes in the system, there are more possible candidates that can achieve high link capacity

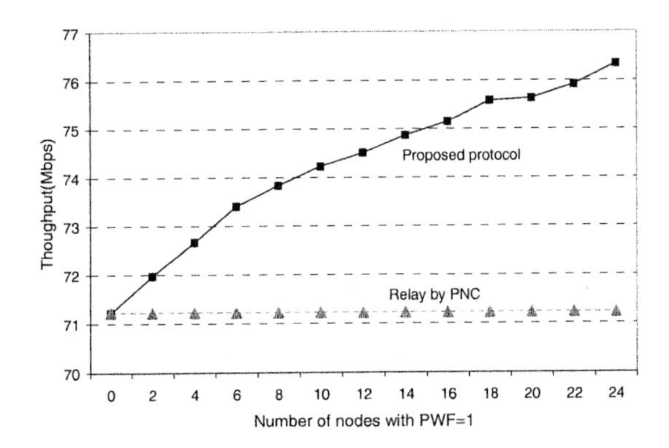

Fig. 8. Throughput as number of possible relay nodes increasing

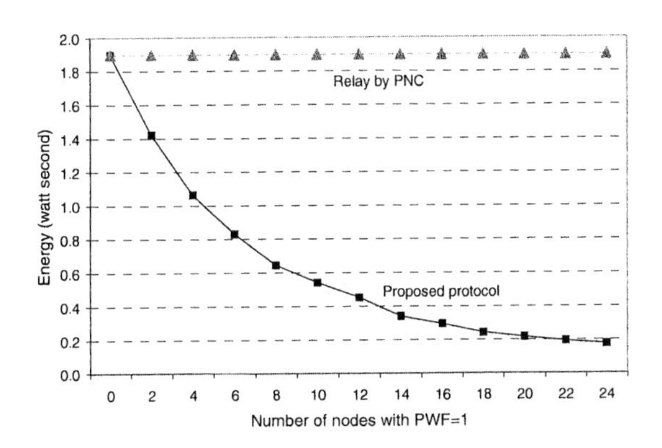

Fig. 9. Energy consumption of PNC as number of possible relay nodes increasing

between source and destination. PNC will have more opportunity to select an optimal relay node. When the number of relay nodes is bigger than 12, it means that those devices, which have no data to transmit in a relative long time, participate in the relay function if they are required by PNC. The throughput will not always increase and we can expect a maximum point, however, we need quite a lot relay devices to reach it and it's not practical in a real network.

The Fig. 9 shows the battery energy consumption of PNC under two situations. 1) only PNC is used to relay data. 2) any device including PNC can be used to relay data based on (8). We can easily get the conclusion that the consumption of PNC battery energy under the proposed situation 2 is 24% of situation 1, when PWF of 12 devices are set to 1 (or we can say 12 devices are relay capable device) and decreases to 9% when PWF of 24 devices is set to 1.

It is reasonable that the total system power consumption is increased with total throughput of the piconet system increased because devices need more power to transmit additional amount of data. However, if total energy consumption amount compares with

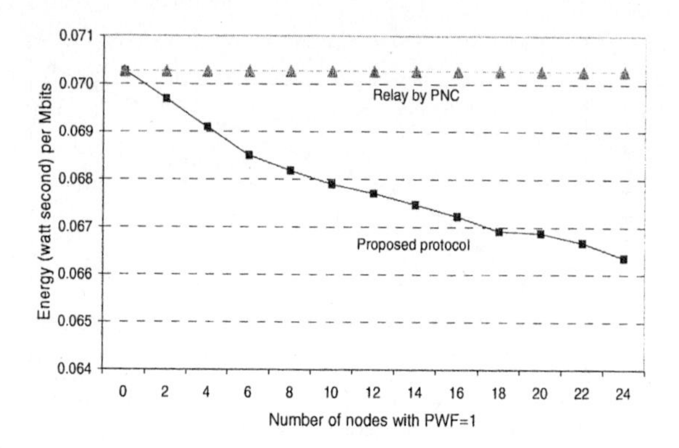

Fig. 10. Energy consumption for 1Mbits as number of possible relay nodes increasing

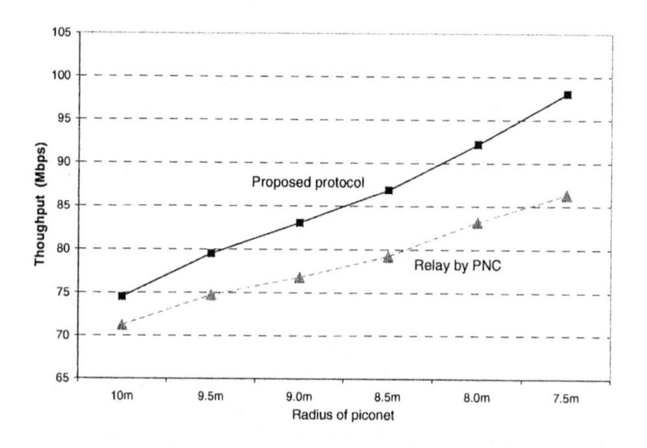

Fig. 11. Throughput comparison between PNC relay and node relay

total throughput amount, the energy consumption for transmitting 1Mbits will decrease as shown in Fig. 10. On the other word, the energy efficiency for transmitting data will be increased when there are more devices as relay nodes in the system.

In the Fig. 11, it shows the throughput as reducing the coverage of piconet. The x axis is the radius of the piconet covered area, and y axis is the throughput. We can understand from the figure that the throughput is increased by decrease radius of piconet. When the radius is decreased, the probability of short distance between two devices will be increased. Even we don't use the MAC relay scheme, the throughput will be increased. But we can see clearly that throughput of using more devices as relay nodes increases much faster than that of only using PNC as relay node. We must notice that the throughput can not increase continually by decreasing the piconet coverage. We can expect that the throughputs of two situations will reach the same maximum point when all devices are close enough and all devices can communicate directly with the maximum channel speed.

At last, we will exam the issue about emitted signal power. In this simulation, we use -10.3dBm watt as reference maximum signal power in the piconet. We assume that each device can only transmit signal not higher than -10.3dBm. As stated above, PNC should evaluate the channel link speed by using the reference maximum signal power and allocate the relay path. However, to achieve the link speed evaluated by PNC, devices do not need to use the reference power every time. Because the path status factor is considered in the proposed protocol to select channel with better channel conditions, there are spaces for devices decreasing the transmitting signal power even after achieving the higher link speed. When more devices operate on relay function, the distance between two nodes will be greatly decreased and the channel conditions will be improved. It results the total transmitting time of high level signal will be decreased. The Fig. 12 shows the result of simulation. We can see that the time of transmitting high power level signal is distributed to the time of low power level signal and the time for transmitting the highest power signal is decreased 42.7%.

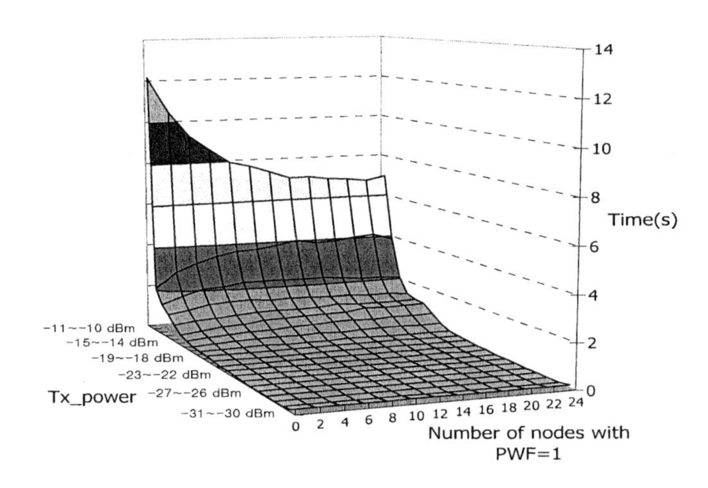

Fig. 12. Time of signal power used vs. number of possible relay nodes

5 Conclusion

In this paper, we proposed a power aware multi-hop packet relay MAC protocol in UWB based WPANs. In the new protocol, each device in the piconet should evaluate the channel status factor named CSF which is calculated from SINR and transmitted signal power sent with the PHY header. The value of CSF is reported to PNC periodically. If two devices can not communicate directly or there is a device can increase the total rate between sender and receiver, PNC will select the device as relay node. If there are more than one devices can be relay nodes, the one with maximum power aware path status factor (PAPSF) will be chosen. After applied the new protocol, the throughput is increased; the battery consumption of PNC is distributed to other devices; the efficiency of power for data transmitting is increased; also the total transmitting time of high level signal is decreased.

References

1. Porcino, D., Hirt, W.: Ultra-wideband radio technology: Potential and challenges ahead, IEEE Communications Magazine, vol. 41, No. 7 (2003) 66-74
2. Horie, W., Sandara, Y.: Novel Packet Routing Scheme based on Location Information for UWB Ad-hoc Network, Proc. of IEEE Conference on Ultra Wideband Systems and Technologies (2003) 185-189
3. Win, M.Z., Scholtz, R.A.: Ultra-Wide Bandwidth Time-Hopping Spread-Spectrum Impulse Radio for Wireless Multiple-Access Communications, IEEE Transactions on Communications, vol. 48 (2001) 679 - 689
4. Balakrishnan, J., Batra, A., Dakak, A.: A multi-band OFDM system for UWB communication, IEEE Ultra Wideband Systems and Technologies Conference (2003) 354-358
5. IEEE std. 802.15.3, "Information technology - Telecommunications and information exchange between systems - Local and metropolitan area networks - Specific requirements - Part 15.3: Wireless Medium Access Control (MAC) and Physical Layer (PHY) Specifications for High Rate Wireless Personal Area Networks (WPANs)" (2003)
6. Perkins, C.E., Royer, E.M.: Ad hoc on-demand distance vector routing, Proc. of IEEE Workshop on Mobile Computing Systems and Applications (1999)
7. Dube, R., Rais, C.D., Wang, K.Y., Tripathi, S.K.: Signal stability-based adaptive routing for ad hoc mobile networks, IEEE Personal Communications Magazine (1997) 36-45
8. Toh, C.K.: Maximum battery life routing to support Ubiquitous mobile computing in wireless ad hoc networks, IEEE Communications Magazine, vol. 39, no. 6 (2001) 138-147
9. Gyoda, K., Kado, Y., Ohno, Y., Hasuike, K., Ohira, T.: WACNet wireless ad-hoc community network, IEEE International Symposium, Volume 4 (2001) 862-865
10. IEEE P802.15 Working Group for WPANs, "Multi-band OFDM Physical Layer Proposal for IEEE 802.15 Task Group 3a" (2004)
11. The Network Simulator – ns2, http://www.isi.edu/nsnam/ns/

Traffic-Adaptive Energy Efficient Medium Access Control for Wireless Sensor Networks*

Sungrae Cho[1], Jin-Woong Cho[2], Jang-Yeon Lee[2],
Hyun-Seok Lee[2], and We-Duke Cho[3]

[1] Computer Sciences,
Georgia Southern University,
Statesboro, P.O. Box 7997, GA 30460 USA
srcho@georgiasouthern.edu
[2] Wireless Network Research Center,
Korea Electronics Technology Institute (KETI),
463-816 South Korea
[3] Center of Excellence for Ubiquitous Computing and Networking,
Ajou University, Suwon, Kyunggi,
443-749, South Korea

Abstract. Design of medium access control (MAC) in wireless sensor networks (WSNs) poses several key factors such as energy conservation and latency. This paper proposes a new MAC protocol referred to as TEEMAC which reduces energy consumption by making the idle nodes sleep to reduce idle listening. TEEMAC is a cluster-based MAC protocol where each cluster is dynamically formed based on cluster-head. Numerical analysis is provided and it shows that the proposed MAC protocol outperforms other existing MAC protocols in terms of energy consumption.

1 Introduction

In wireless sensor networks (WSNs), small sensor nodes with sensing, computation, and wireless communications capabilities, form and coordinate among themselves to produce high-quality information about the physical environment. The power of wireless sensor networks lies in the ability to deploy a large number of tiny nodes that assemble and configure themselves to detect sensory information. Usage scenarios for these devices range from real-time tracking, to monitoring of environmental conditions, to in situation monitoring of the health of structures or equipment, and to almost everywhere.

In order for sensor node to collaborate by themselves, sensor nodes need a medium access control (MAC) protocol. When designing MAC protocols, there are several design spaces which have been conventionally considered. These include: bandwidth utilization, fairness, latency, scalability, and adaptability. Yet,

* This research is supported by the Ubiquitous Autonomic Computing and Network Project, the Ministry of Information and Communication (MIC) 21st Century Frontier R&D Program in South Korea.

X. Jia, J. Wu, and Y. He (Eds.): MSN 2005, LNCS 3794, pp. 593–602, 2005.

in wireless sensor networks, nodes tend to rely on battery and therefore energy consumption is considered as a primary design factor [1].

There have been huge efforts [2, 3, 4] to develop a medium access control (MAC) protocol in wireless sensor networks. Most of them considered energy-efficiency as a paramount design factor. They identified major causes of energy waste as follows [4]. Collision is the first source of energy waste. MAC protocols try to avoid collisions one way or another. Collision is a major problem in contention-based protocols, but is generally not a problem in schedule-based protocols.

The second source is idle listening. It happens when the radio keeps listening to the channel to receive possible data even if many of the sensor network applications are expected to have a low duty cycle. Many MAC protocols (such as CSMA and CDMA protocols) always listen to the channel when active, assuming that the user would power off a device if there is no data to send. The exact cost of idle listening depends on radio hardware and mode of operation.

The third source is overhearing, which occurs when a node receives packets that are destined to other nodes. Overhearing unnecessary traffic can be a dominant factor of energy waste when traffic load is heavy and node density is high. The last major source that we consider is control packet overhead. Sending, receiving, and listening for control packets consume energy. Since control packets do not directly convey data, they also reduce the effective goodput.

Bearing the above in mind, we propose a traffic-adaptive energy efficient MAC (TEEMAC) protocol for wireless sensor networks. TEEMAC protocol deals with aforementioned four major sources of energy waste. Using a schedule, TEEMAC prevents collisions in data transmission, idle listening, overhearing, unnecessary control overhead such as request to send (RTS) and clear to send (CTS) packet, and thereby conserving energy. To share the schedule with other nodes, TEEMAC protocol uses a cluster, and the cluster is dynamically formed based on remaining power. To reduce the energy consumption, the proposed protocol uses traffic-adaptive cluster formation mechanism where most of the time the nodes turn off their radios.

The remaining part of this paper is organized as follows: in section 2, the proposed MAC protocol referred to as TEEMAC is presented. In section 3, performance is evaulated and followed by conclusions in section 4.

2 TEEMAC Protocol Description

2.1 Network Assumptions

In designing medium access control, we have assumed the following sensor application. Traffic rate in sensor networks is assumed to be very low. Typical communication frequency is at minutes or hours level. The location of sensor nodes is assumed to be not fixed, but generally stationary after their deployment. Sensor nodes are assumed to be battery- powered and recharging is usually not available so energy is an extremely expensive resource. Sensor nodes in the network coordinate with each other to implement a certain function so traffic is not randomly generated as those in mobile ad hoc networks. But rather traffic is bursty when a certain event is detected.

2.2 Operation

Contention-based channel access schemes such as CSMA protocols are clearly not suitable for sensor networks, due to their requirement for radio transceivers to monitor the channel at all times. This is a particularly expensive proposition for the low radio ranges of interest for sensor networks, where transmission and reception have almost the same energy cost. We would like to turn off the radios when no information is to be sent or received.

Use of schedule-based MAC schemes is viewed as a natural choice for sensor networks because radios can be turned off during idle period in order to conserve energy. Transmission schedule allows to send or receive traffic in the network and makes nodes sleep when they are not scheduled to send or receive. In addition, clustering is a promising distributed technique used in large-scale WSNs.

Clustering solutions can be combined with schedule-based schemes to reduce the cost of idle listening and overhearing problem. Such an approach lends itself well to sensor network, when compared to ad hoc networks in general, because a relatively static topology is expected and traffic patterns may be more regular (e.g., periodically sending updates to sink). The key research challenge is determining how slots can be assigned.

When nodes in a sensor field detect events or a sink node sends its interest, traffic is generated to be transmitted. This traffic is not frequently nor periodically generated, but rather generated as a bursty source. Therefore, most of the time, sensor nodes are in sleep mode to turn off their radios and turn only their sensors. When traffic is generated, TEEMAC forms a cluster and begins its operation. The operation of TEEMAC is divided into two phases: clustering phase and data transmission phase.

The clustering phase is a time duration for nodes to form a cluster as in Fig. 1. In clustering phase, nodes try to form a cluster by scanning process. During scanning process, each node scans less noisy medium (or channel). We assume that there are ranges of frequency bands available at physical layer so that they can be used as a transmission medium. This approach is similar to

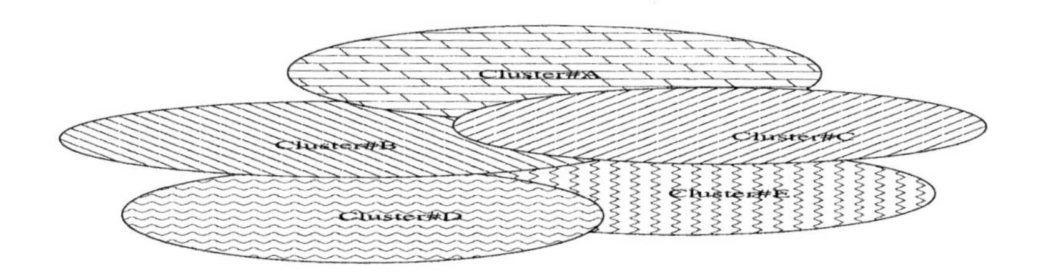

Fig. 1. TEEMAC Clustering

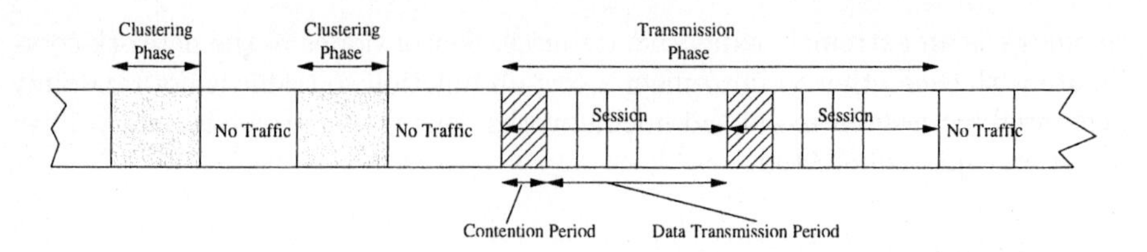

Fig. 2. Timing in TEEMAC

cellular architecture in terrestrial wireless systems to avoid interference problem. As in Fig. 1, cluster#A need to have different frequency channel from the other clusters. This clustering phase is done when nodes are deployed, and periodically done after deployment to reflect the recent topology changes. By forming cluster before the actual transmission all the time, we can reduce the cluster formation latency.

The transmission phase is a time duration for nodes to access a channel to transmit their data. The transmission phase is performed when the actual traffic is generated. Clustering and transmission phases are described in detail in sections 2.3 and 2.4 and the complete timing is depicted in Fig. 2.

2.3 Clustering

During the clustering phase [1] in Fig. 2, each node must decide whether it could become a cluster-head, based on its remaining power-level. There is a specific radius called *sensing radius*, which can be predefined and built in the sensor nodes before deployment. In that specific sensing radius, a cluster is formed. If an application needs more accurate sensing, then the sensing radius can be smaller. This sensing radius is thus an application-specific parameter.

Each node which wants to become a cluster-head broadcasts an advertisement message with its remaining power-level to all other nodes claiming to be a new cluster-head by using non-persistent CSMA with random delay T. Each node has its power-level threshold in decision-making, i.e., if its power-level is greater than the threshold, it will broadcast its advertisement message. If no cluster-head is chosen during T due to lack of nodes wishing to be a cluster-head, etc., another T is used to select a cluster-head. In this case, the likelihood of broadcasting advertisement should be increased by adjusting the power-level threshold.

Based on the power-level advertised, a node with the highest power-level will be elected as a cluster-head simply because it is expected to play a good role with enough power. Next each non cluster-head node (or *member node*) joins the cluster by sending its information e.g., identification number to the cluster-head. When information from all member nodes are gathered, the cluster-head node broadcasts transmission schedule for each member node. This schedule tells ith

[1] Aforementioned the clustering phase is repeated periodically in order to reflect the current topology change.

node to send its data at jth time slot [2]. Once the clusters are built, the system is ready to transmit any traffic generated, but the transmission does not take place until real traffic is generated.

If non cluster-head (or member) node wants to join or leave a cluster, it can be done in either other clustering phases or transmission phase. Joining or leaving in the transmission phase can be done by sending join message during contention period in the transmission phase. So, the clusters can be formed more dynamically by allowing join and leave operation in the clustering phase and transmission phase. If the current cluster-head dies under some unexpected reason, the node with the next highest power-level will take over the previous cluster-head's role. Using periodic clustering phase, another cluster will be formed with different group of nodes based on current situation, e.g., node's power level and mobility.

Assuming Poisson arrival to a cluster, the probability that $N+1$ nodes including the cluster-head forms a cluster during T is $(\lambda T)^N e^{-\lambda T}/N!$ where λ is the join rate to the cluster. Based on this probability, T can be determined.

2.4 Data Transmission

Schemes in the previous section allow each node to know which time slot it can use. However, without synchronization, this information is useless. Therefore, cluster-head node periodically broadcasts a special packet called *SYNC packet* so that the other member nodes to synchronize to it. Based on this synchronization, a transmission phase starts.

The transmission phase is divided into a number of sessions which is a time required to convey a number of fragments from the above layer, i.e., network layer. This phase is synchronized by transmitting a SYNC packet by the cluster-head. Each session consists of a contention period and a data transmission period as shown in Fig. 2. During the contention period, the nodes can send their *interest to send* (IS) packets to the cluster-head. The IS packet indicates that the node wish to send its data. Also, the IS packet conveys information such as leave or join message as described in the clustering phase. During each contention period, all nodes keep their radios on.

After the contention period, the cluster-head knows all the nodes that have data to transmit and their intended references and the system enters into the data transmission period, which is divided into several number of time slots. The number of time slots is equal to the number of members in the cluster. The cluster-head builds a TDMA schedule and broadcasts it to all nodes within the cluster. There is one data slot allocated to each node in each session. Each node that has data to send or receive is awake; all other nodes keep their radios off.

The first time slots are reserved for the cluster-head which sets up and broadcasts a transmission schedule for the cluster members. Based on this transmission schedule, each member turns on or off its radio. If a member has no data to send, the radios of all other members can be turned off. If the schedule indicates that a member node is a recipient in a certain time slot, its radio needs to be turned on.

[2] Details are described in section 2.4.

Slot#1	Slot#2	Slot#3	Slot#4	Slot#5
Data from cluster–head is broadcasted Following schedule is broadcasted – node1–> node2 in slot#2 – node2 broadcasts in slot#3 – node3–> node2 in slot#4 – node4 –> node3 in slot#5	– node1 –> on – node2 –> on – node3 –> off – node4 –> off	– node1 –> on – node2 –> on – node3 –> on – node4 –> on	– node1–> off – node2 –> on – node3 –> on – node4 –> off	– node1–> off – node2–> off – node3 –> on – node4 –> on

Fig. 3. An example of actions performed in data transmission period

Fig. 3 show an example in the data transmission period. Suppose that there are 5 nodes in a cluster including a cluster-head. As shown in the figure, the cluster-head broadcasts transmission schedule at slot #1. During this time period, all nodes turn on their radios. Node 1 sends its data to node 2 at slot #2 during which nodes 1 and 2 turns on and nodes 3 and 4 turns off the radios. Similarly, on or off of radios are scheduled in the next time slots.

3 Performance Evaluation

3.1 Power Consumption Analysis of TEEMAC

We assume that a clustered network has already been formed and there are N member nodes within a cluster. Each transmission consists of k sessions. Suppose n nodes out of N want to transmit in the ith session. The event whether a node has data to transmit or not can be viewed as a Bernoulli trial. The probability that a node has data to transmit is p. Therefore n_i, the number of nodes at the ith session to transmit is a Binomial random variable and

$$E[n_i] = N \times p = n, \text{ for } i = 1, 2, \ldots, k. \tag{1}$$

Since the number of nodes to transmit is independent from frame to frame, the expectation of the total number of nodes to transmit in a transmission round is

$$E\left[\sum_{i=1}^{k} n_i\right] = \sum_{i=1}^{k} E[n_i] = kn \tag{2}$$

Let the power consumption during the transmit mode, the receive mode, and the sleep mode, are denoted by P_t, P_r , and P_s respectively. When a node spends τ seconds transmitting a packet, the radio dissipates the energy equivalent to

$$E_{\text{tx}}(\tau) = P_t \tau, \tag{3}$$

and during τ in receiving a packet, the radio consumes

$$E_{\text{rx}}(\tau) = P_r \tau. \tag{4}$$

Similarly, the energy dissipated by the radio during an sleep period of τ is

$$E_S(\tau) = P_s\tau. \quad (5)$$

Let the time required to transmit or receive a data packet in the data transmission period is T_d. Let denote the time required to receive a schedule from a cluster head as T_{ch} and the time spent in contention period as T_c.

All nodes keep their radios on during the whole contention period. Each node that wants to transmit a packet uses one time slot and remains sleep or in receiving during $N-2$ slots. So, each node that has transmitted during a single session dissipates

$$E_{sn} = P_tT_c + P_rT_{ch} + P_tT_d + q(N-2)P_rT_d + (1-q)(N-2)P_sT_d \quad (6)$$

where q is the probability that a node receives a data which will be the same as p on average if the network is in equilibrium state. By applying $q = p$,

$$\begin{aligned} E_{sn} &= P_tT_c + P_rT_{ch} + P_tT_d + p(N-2)P_rT_d + (1-p)(N-2)P_sT_d \\ &= P_tT_c + P_rT_{ch} + \{P_t + p(N-2)P_r + (1-p)(N-2)P_s\}\, T_d \end{aligned} \quad (7)$$

During a single session, the node that does not have a slot allocated to it will consume the following energy:

$$\begin{aligned} E_{in} &= P_sT_c + P_rT_{ch} + p(N-1)P_rT_d + (1-p)(N-1)P_sT_d \\ &= P_sT_c + P_rT_{ch} + \{p(N-1)P_r + (1-p)(N-1)P_s\}\, T_d. \end{aligned} \quad (8)$$

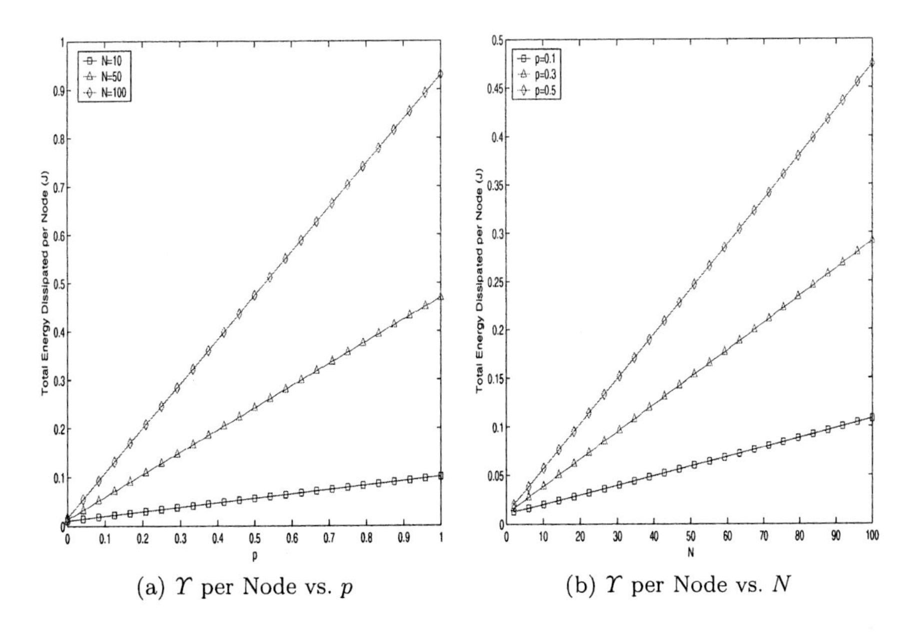

(a) Υ per Node vs. p (b) Υ per Node vs. N

Fig. 4. Average Energy Dissipated per Node ($k = 10$, $T_d = 1452$ bytes/2 Mbps, $T_{ch} = 2 \times T_d$, $T_c = N \times T_{ch}$, $P_t = 0.462$ mW, $P_r = 0.346$ mW, $P_s = 0.003$ mW)

Likewise, energy dissipated in the cluster-head during one session is

$$E_{ch} = P_r T_c + P_t T_{ch} + \{p(N-1)P_r + (1-p)(N-1)P_s\} T_d. \quad (9)$$

Since each transmission round has k sessions and therefore the average total energy consumed is

$$\Upsilon = k[nE_{sn} + (N-1-n)E_{in} + E_{ch}]. \quad (10)$$

Consequently,

$$\Upsilon = k\left[\{pNP_t + (N-1-pN)P_s + P_r\} T_c + \{(N-1)P_r + P_t\} T_{ch} + N\{pP_t + p(N-p-1)P_r + (N-1-p^2-pN)P_s\} T_d\right]. \quad (11)$$

Fig. 4(a) shows the average energy dissipation per node (Υ/N) versus p when $N = 10$, $N = 50$, and $N = 100$. We assume $P_t = 0.462$ mW, $P_r = 0.346$ mW, $P_i = P_r$, $P_s = 0.003$ mW from [6]. As shown in the figure, the average energy linearly increases with p, but at different slopes with N. We observed that if traffic load (p) increases, more energy will be dissipated.

Fig. 4(b) also shows the average energy dissipation per node (Υ/N) versus N when $p = 0.1$, $p = 0.3$, and $p = 0.5$. Similar result can be observed in that the more energy is dissipated when the larger number of nodes form a cluster.

In order to see the benefit of the scheduled sleep mode of the TEEMAC protocol, we compare two cases: one is the TEEMAC protocol and the other is the TEEMAC protocol but without sleep mode. For the latter case, we use idle

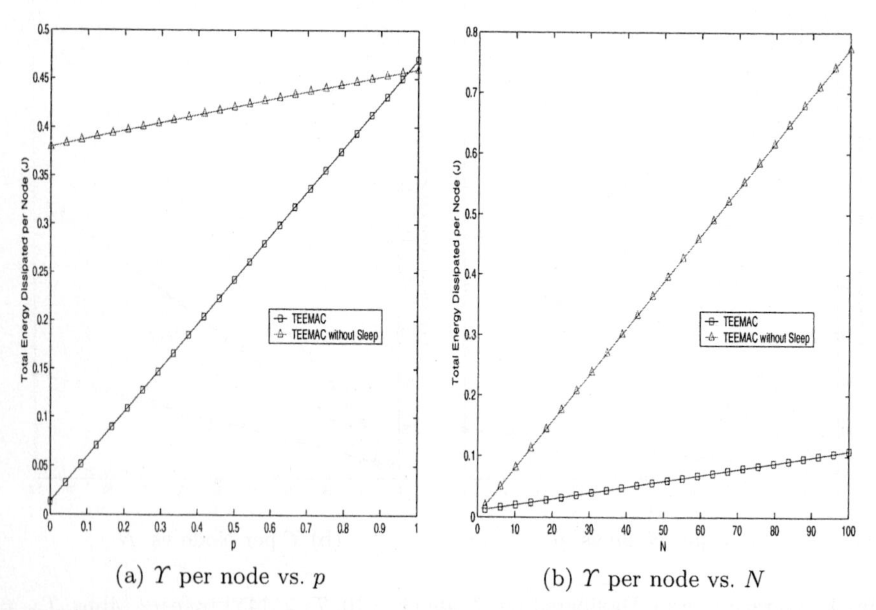

(a) Υ per node vs. p (b) Υ per node vs. N

Fig. 5. Average Energy Dissipated per Node ($k = 10$, $T_d = 1452$ bytes/2 Mbps, $T_{ch} = 2 \times T_d$, $T_c = N \times T_{ch}$, $P_t = 0.462$ mW, $P_r = 0.346$ mW, $P_i = P_r$, $P_s = 0.003$ mW)

listening with power consumption of P_i instead of sleep mode. From [6], we found that $P_i \approx P_r$. Applying $P_s = P_i$, we observe the average energy dissipation per node (Υ/N) versus p as in Fig. 5(a). We observe that at lower range of p the TEEMAC protocol has a significant gain against the protocol without scheduled sleep. However, when p approaches to 1 the gain becomes negligible, and even TEEMAC dissipates more energy than the protocol without scheduled sleep when $p \approx 1$. It is because when traffic is high most of the time slots are used for transmission and reception, and the time slots are never used for sleep mode. Also, more collision is expected during the contention period.

Another interesting result is observed in Fig. 5(b). When N increases, we are benefited with more energy conservation using the scheduled sleep scheme of the TEEMAC protocol.

3.2 Simulation

We evaluated the performance of TEEMAC and SMAC [4] through simulation using ns-2 network simulator [5]. The schemes are compared in terms of energy consumption. The topology consists of a cluster with N sensor nodes and one cluster head. The nodes are deployed randomly through the 10 m x 10 m area. The node location patterns are generated using CMU's movement generator. We use an omni-directional antenna for sensor nodes and a two ray ground propagation model. We implement TEEMAC and SMAC schemes under ns-2, and use the energy model contained in ns-2. Each node begins with 100 J of energy. The power consumption for transmitting is 462 mW, 346 mW for receiving, 346 mW for idle listening, and 0.003 mW for sleep mode. The bandwidth is set to 2 Mbps. The data packet size is 1452 bytes, which includes 1400 bytes payload and 52 bytes header. The control packet size is set to 152 bytes, which consists of 100 bytes payload and 52 bytes header. For TEEMAC, the source to cluster head control packet size is set 72 bytes, which contains 20 bytes payload and 52 bytes header.

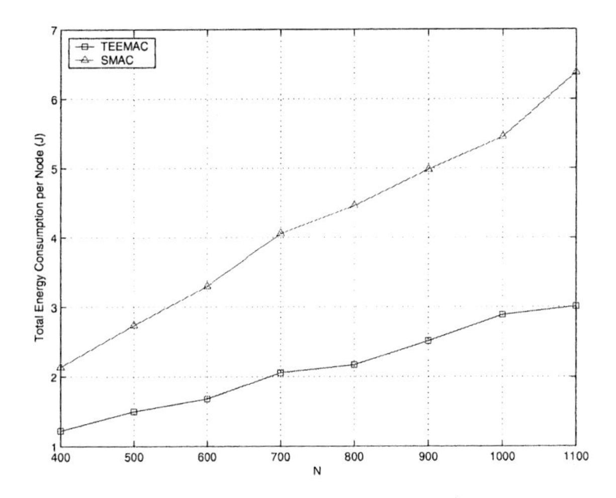

Fig. 6. Average Energy Dissipated per Node vs. N

Fig. 6 provides the simulation results of the two MAC protocols in terms of the average total energy consumption per node as a function of N for the case of traffic load of $p = 0.3$ and 4 transmission rounds. As can be seen, TEEMAC protocol provides less energy consumption against SMAC protocol.

4 Conclusions

In this paper, we proposed a cluster based energy efficient technique referred to as TEEMAC, which can potentially improve the efficiency of intra-cluster MACs. TEEMAC is intended for event-driven applications, where sensor nodes transmit data to the cluster-head only if significant events are observed. We compared TEEMAC with SMAC through numerical analysis. We have shown that TEEMAC outperforms SMAC significantly in terms of energy consumption.

References

1. I. F. Akyildiz, W. Su, Y. Sankarasubramaniam, and E. Cayirci, "A survey on sensor networks," *IEEE Communications Magazine* , August 2002.
2. K. Sohrabi, J. Gao, V. Ailawadhi, and G. J. Pottie, "Protocols for Self-Organization of a Wireless Sensor Network," *IEEE Personal Communications*, October 2000.
3. B. Tavli and W. B. Heinzelman, "MH-TRACE: Multihop Time Reservation Using Adaptive Control for Energy Efficiency," *IEEE Journal on Selected Areas in Communications,* vol. 22, no. 5, June 2004.
4. W. Ye, J. Heidemann, and D. Estrin, "Medium access control with coordinated adaptive sleeping for wireless sensor networks," *IEEE/ACM Transactions on Networking,* vol. 12, no.3, June 2004.
5. http://www.isi.edu/nsnam/ns/, Network Simulator 2.
6. http://www.xbow.com/ MICA2 Mote Datasheet

An Energy-Conserving and Collision-Free MAC Protocol Based on TDMA for Wireless Sensor Networks*

Biao Ren[1,2], Junfeng Xiao[1], Jian Ma[2], and Shiduan Cheng[1]

[1] State Key Laboratory of Networking and Switching,
Beijing Univ. of Posts & Telecommunications, Beijing 100876, China
{renb, xiaojf, chsd}@bupt.edu.cn
[2] Nokia Research Center, Beijing 100013, China
jian.j.ma@nokia.com

Abstract. This paper proposes TDMA-EC, a novel energy-conserving and collision-free media access control protocol based on TDMA principle. Through assigning separate time slots for neighboring nodes and resolving the "hidden terminal" problem, TDMA-EC performs well with collision avoidance and achieves better energy-efficient performance compared to contention-based protocols. Meanwhile, TDMA-EC schedules receiving and transmitting periods according to node's position in the spanning tree of network, which also guarantees end-to-end latency with less jitter. Simulation results show that TDMA-EC achieves both energy conservation and lower latency compared to SMAC and DMAC.

1 Introduction

Wireless Sensor networks (WSN) [1,2] are integrated information systems combined with message-sampling, message-processing and message-transmitting capabilities. Such distributed networks consist of large number of nodes and typically these nodes coordinate to perform a common task.

Medium access control (MAC) is an important technique in all shared-medium networks. One fundamental task of the MAC protocol is to avoid collisions so that two interfering nodes do not transmit at the same time. Furthermore, a well-designed MAC protocol should reduce cost and optimize overall network performance.

This paper proposes an Energy-conserving, Collision-free and TDMA-based MAC (TDMA-EC) explicitly designed for wireless sensor networks.

In TDMA-EC, shared-channel is composed of fixed-length frames; each frame is divided into a certain amount of slots. Since each node wakeups only in its predetermined slots to receive or transmit data while keeps sleep in other slots with the least energy consumption, TDMA-EC not only achieves a good collision-free capability, but also conserves more energy than CSMA-based MAC protocols. Furthermore, TDMA-EC uses a data-gathering tree as its network topology. Through scheduling work status of each node according to node's position in the spanning tree,

* Supported by NSFC Grant No. 90204003, 60402012; Beijing Municipal Key Disciplines XK100130438; National 973 No. 2003cb304806; EU project: FP6-2004-IST-3-015774.

X. Jia, J. Wu, and Y. He (Eds.): MSN 2005, LNCS 3794, pp. 603 – 612, 2005.

TDMA-EC guarantees lower end-to-end latency with less jitter. TDMA-EC reduces the potential energy wastes from the following aspects. First, it is collision avoidance benefiting from TDMA-based mechanism. Each node within Neighborhood is assigned unique time slot for transmitting. No two nodes within two-hop distance can share the same transmitting slot. This is similar to the 2-distance graph-coloring [3] problem. After establishing spanning tree of the network, TDMA-EC assigns slot to node according to the position associated with that node in the spanning tree. Second, each node doesn't care about other activities irrelevant to itself over shared-channel, so it needn't much overhearing. Thirdly, TDMA-EC reduces significantly the amount of worthless idle listening due to the knowledge of the time to wakeup.

TDMA-EC also reduces the end-to-end latency through dividing one TDMA frame into two periods, one for receiving and the other for transmitting. Once received a data packet from its children nodes in the spanning tree, node can immediately forward it up to its parent node in the same frame, so the data forwarding interruption [4] is reduced significantly.

2 Related Works

Low-power MAC protocols for wireless sensor networks have being researched many years. These protocols can be broadly divided into two categories: one is contention-based and the other is TDMA-based.

SMAC [5,6] is a classical contention-based protocol. Locally managed synchronizations and periodic sleep listen schedules based on these synchronizations are its basic idea. Neighboring cells form virtual clusters to set up a common sleep schedule. But periodic sleep may result in high latency especially for multi-hop routing algorithms. Adaptive listening technique is also proposed to improve the sleep delay in SMAC. WiseMAC [7] protocol uses non-persistent CSMA with preamble sampling to decrease idle listening. All nodes in a network sample the medium with a common period but their relative schedule offsets are independent. WiseMAC dynamically determines the length of the preamble. According to neighbors' sleep schedule table, WiseMAC schedules transmissions so that the destination node's sampling time corresponds to the middle of the sender's preamble. DMAC [4] is based on convergecast communication pattern. Low latency is achieved by assigning subsequent slots to the nodes that are successive in the data transmission path. Once received a data packet, current node would forward it to its next node as soon as possible. So DMAC can achieves very good latency compared to SMAC. TMAC [8] is proposed to enhance the poor results of SMAC protocol under variable traffic load. A node keeps listening and potentially transmitting as long as it is in an active period. An active period ends when no activation event has occurred for a certain time. The activation time events include reception of any data, the sensing of communication on the radio, etc. TDMA-W [11] proposed by Zhihui Chen etc. is based on TDMA principle, in which a sensor utilizes its assigned slot only when it is sending or receiving information, otherwise its receiver and transmitter are turned off to avoid unnecessary neighbor listening.

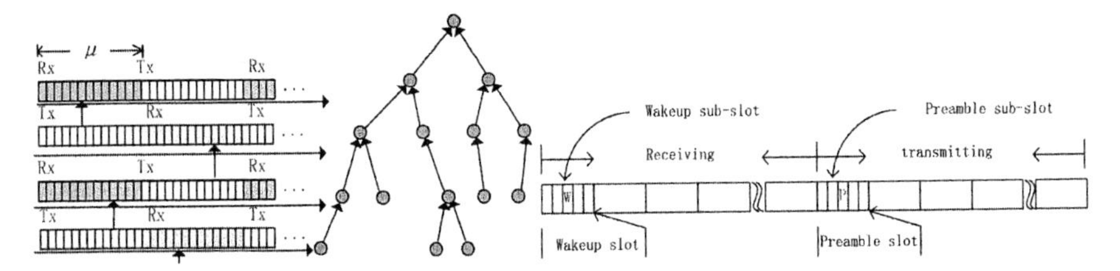

Fig. 1. TDMA-EC in a data-gathering tree **Fig. 2.** Slot Assignments of TDMA-EC frame

3 TDMA-EC Protocol Design Details

3.1 Traffic Pattern and Staggered Schedule

Generally, in sensor networks, the significant traffic pattern is data gathering from sensor nodes to sink. For a sensor network application with multiple sources and one sink, the data delivery paths from sources to sink consist of a tree structure, that is, a spanning tree or a data-gathering tree [9]. We assume that all the nodes are fixed and that a route to the sink is fairly durable, so that the spanning tree remains stable for a reasonable length of time. All nodes except the sink will forward any packets they received to the next hop. This traffic pattern mentioned here is similar to that in DMAC to some extent. But there are two differences. Firstly the data flows in the data-gathering tree are bidirectional, that is, sink can dispatch control messages to any nodes down to the tree at any time, whereas DMAC is unidirectional from sources to sink. Second, DMAC staggers activity schedule of nodes on the multihop path to wake up sequentially like a chain reaction. A node with depth d skews its wake-up scheme $d\mu$ ahead from the schedule of the sink, where $\mu$is the unit offset of schedule. While in TDMA-EC, only two types of schedules are adopted. One is the same as the sink, the other has a offset difference μ with the first one. Any pair of parent and its child node will be assigned different schedules so that the two schedules are just staggered on the multihop path. Figure 2 shows a data gathering-tree and its staggered scheduler. One frame consists of two periods: *Rx* for receiving and *Tx* for transmitting, and both are equal in length.

This scheme brings some advantages to WSN. Firstly, the traffic can be bidirectional so that control packets can also reach any sensor nodes in the tree. In most of applications this function is necessary. Secondly, when its transmitting data, a node parent is just in the state of receiving, and vice versa. These staggered schedules can avoid transmitting collision of any pair of parent and children. Thirdly, interruption latency at each node can be reduced about *1/2* of the length of one frame since received packets can be sent out within the same frame if no queuing needed. So the end-to-end latency of packets is decreased dramatically compared with traditional TDMA schemes.

A significant characteristic of TDMA-EC is that nodes can operate well without transmitting collision through slot assignment in advance, while staggered schedule mentioned above only resolves the collision problem partially. Furthermore, TDMA-EC lets each node randomly choose their transmitting slot and then negotiates if two nodes within two hops range have selected the same slot. Decentralized slot assignment scheme is outlined as following:

1) Assume schedule has been synchronized with sink or the offset to sink is equal to μ.
2) New node randomly selects a slot among all the transmitting slots.
3) During its preamble slot (interpreting in the following subsection), new node broadcasts its ID, its transmitting slot(t-slot) number, its one-hop neighbors' Ids and their t-slot assignments in preamble packet. It also broadcasts the slot number of any t-slot during which this node has identified a collision.
4) New node then listens to the traffic from neighbors. The node should record all the information being broadcast by all its neighbors, i.e. their t-slot number and their node IDs, and the slot number of any slot being broadcast as a collision-prone slot.
5) If a node determines that it is involved in a collision or finds out that one of its two-hop neighbors has the same transmitting slot, it then randomly selects an unused slot and go to step 3.
6) Each node broadcasts its and neighbors' information periodically through preamble packet in its preamble slot. It is necessary for topology maintaining and neighboring discovery.

Above idea originates from [10] and make some improvements. A small preamble packet is used to carry control information. The preamble packet would be sent in the preamble sub-slot relative to that node. In [11], each node has its own wakeup slot and this slot number should be aware by its all neighbors through broadcast. A TDMA-EC frame maintains one wakeup slot and divides it into sub-slots corresponding to different nodes. The function of preamble slot and preamble packet will be described later.

3.2 Frame Structure and Packet Delivery

A TDMA-EC frame consists of receiving period and transmitting period. Both have the same length and are composed of the same numbers of time slots. In receiving period, the first slot is named as Wakeup slot composed of sub-slots. And the numbers of sub-slots are equal to the numbers of the receiving slots. In transmitting period, a preamble slot is used in the first place, and it is also divided into sub-slots with the same numbers as wakeup sub-slots. Figure 2 shows the details of one TDMA-EC frame. For space constraints, we only show the fixed number of sub-slots as 5.

When forwarding data, each node has got the knowledge of its parent's schedule. It is guaranteed through staggered schedule scheme as mentioned before. With this assumption, the preamble slot of child node is exactly corresponding to the wakeup slot of its parent. The node firstly sends a short preamble packet to its parent to notify that a data packet will be sent later, and this action should be accomplished in the ith sub-slot where i is determined by the t-slot number of the child node. Normally the parent node would receive this preamble packet right during the ith sub-slot of its wakeup slot. Based on a presumption that an incoming packet will arrive in the receiving slot i, the parent node will wake up to listen channel and receive data at that moment. Herein the exchange is finished successfully. All the other nodes would sleep if there were no data packet to receive or transmit except that they have to listen in wakeup slot. So TDMA-EC obtains much lower duty-cycle and the energy-conserving is also achieved significantly.

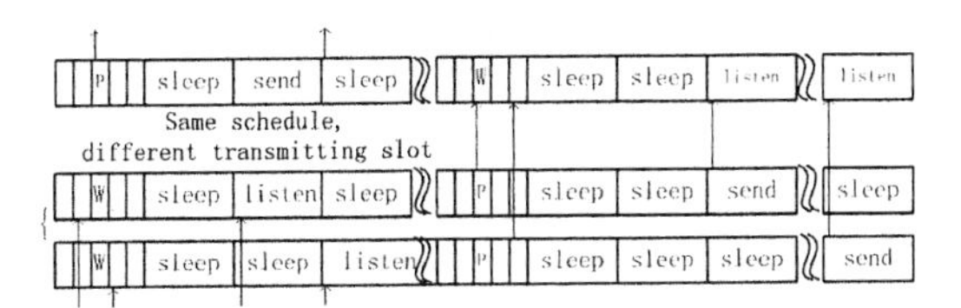

Fig. 3. The process of data delivery

Figure 3 illustrates the process of data delivery involving one parent node and its two child nodes. One should notice that in DMAC, if two child nodes want to send packets to their parent at the same time, a contention would occur inevitably because of shared media channel. The node that failed to hold the channel has to wait for the next work interval after sleep. As to TDMA-EC, when the above scenario happens, the two child nodes would firstly send preamble packets in different sub-slots and then send their data packets in different transmitting slots respectively. So the two child nodes can finish their transmitting in one frame. The parent node also wakes up and then obtains data in the corresponding receiving slots. So TDMA-EC avoids the channel contention and potential collision and so its energy wasting is decreased accordingly.

4 Latency Analyses

Assume that the duration of a slot is one unit of time. So a frame length T_{frame} of TDMA-EC means one frame consists of T_{frame} slots. Since there is one wakeup slot and one preamble slot in each frame, the number of both transmitting or receiving slots is equal to $(T_{frame}-2)/2$.

4.1 Average Delay at One Node

Before analyzing the delay performance of TDMA-EC, we assume that the traffic is so low that there is no additional queuing delay. Each node chooses one slot as its own transmitting slot uniformly at random, thus, one data packet would arrive at random in the receiving period. It is reasonable despite that two nodes with two-hop distance can't have the same slot to send data. Let T_w denotes time interval between the arrival of a packet and departure from this node, and obviously it is also the length of delay at this node (neglecting transmitting delay).

T_w is distributed from 2 to $T_{frame}-2$. When $T_w <= (T_{frame}-2)/2$, the possibility density distribution $p(T_w)$ of T_w is:

$$p(T_w) = \frac{T_w - 1}{((Tframe-2)/2)^2} . \quad (1)$$

when $T_{frame}-2 >= T_w > (T_{frame}-2)/2$,

$$p(T_w) = \frac{(T_{frame} - 2) - T_w + 1}{((Tframe-2)/2)^2} . \quad (2)$$

So the average delay at one node can be computed as:

$$E(p(T_w)) = \sum_{T_w=2}^{(T_{frame}-2)/2} \frac{T_w - 1}{((T_{frame}-2)/2)^2} T_w + \sum_{T_w=((T_{frame}-2)/2)+1}^{T_{frame}-2} \frac{(T_{frame}-2) - T_w + 1}{((T_{frame}-2)/2)^2} T_w \quad (3)$$

$$= \frac{1}{((T_{frame}-2)/2)^2} \left\{ \sum_{T_w=1}^{(T_{frame}-2)/2} (T_w - 1)T_w + \sum_{T_w=1}^{(T_{frame}-2)/2} \left(\frac{T_{frame}-2}{2} + T_w\right)\left(\frac{T_{frame}-2}{2} - T_w + 1\right)T_w \right\}$$

$$= \frac{T_{frame}}{2} \quad (4)$$

Apparently, the average delay for a packet is approximately equal to half of the length of one frame when no queuing delay needed.

4.2 Packet End-to-End Latency

Giving a source node and a sink node, they are N-hop distance between them, let $f(i)$ denotes the number of transmitting slot of the ith node on the path from source to sink. Assume that packet arrives at source node at t_0. So we can deduce the end-to-end latency L:

$$L = (f(0) + T_{frame}/2 - t_0) + (f(1) + T_{frame}/2 - f(0)) + ... + (f(N-1) + T_{frame}/2 - f(N-2)).$$

$$= N\frac{T_{frame}}{2} + f(N-1) - t_0 \quad (5)$$

Formula (5) verifies the correctness of formula (4) and implies that packet end-to-end latency is linearly proportional to the number of hops on the path from source to sink. Figure 4 shows the principle of packet delivery through predetermined slots at each node.

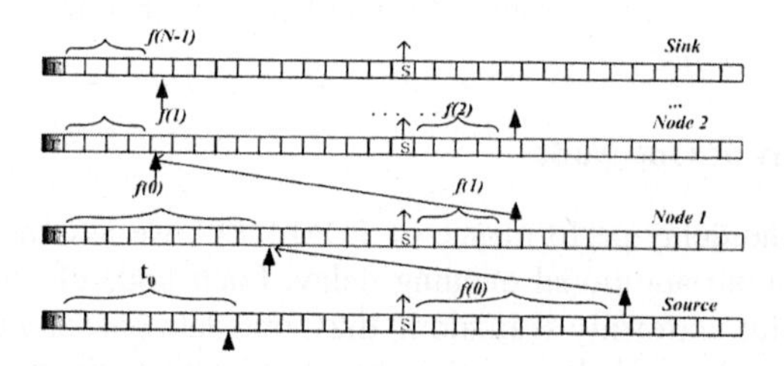

Fig. 4. The process of packet delivery through slots

5 Performance Evaluation

We implement our prototype in the ns-2 network with the CMU wireless extension [12]. Comparisons are performed by TDMA-EC, DMAC and adaptive SMAC. In simulation scenarios, the radio bandwidth is set to 100Kbps and transmission range is 250m for all nodes. Assume that the size of one data packet is 100 bytes and one time slot can only transmit one data packet each time, so a slot length is equal to 10 milliseconds. The length of TDMA-EC frame is set to 200 milliseconds. The relative energy costs of the Tx:Rx:Idle radio modes are assumed to be about 1.67:1:0.88. The sleeping power consumption is set to 0 (i.e. considered negligible). The active period is set to 20ms for SMAC with adaptive listening. Both S-MAC and D-MAC have the basic duty cycle of 10%.

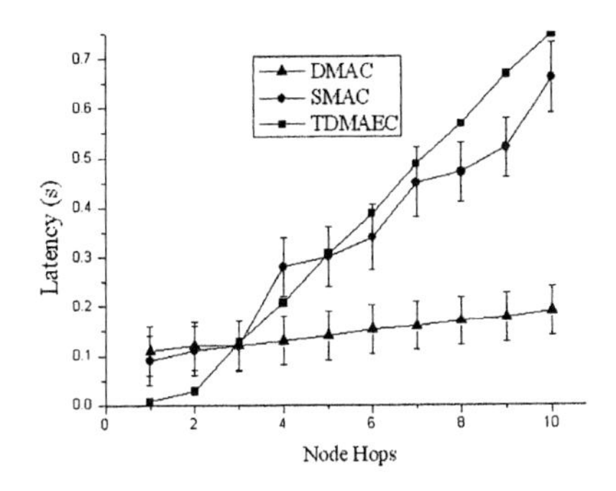

Fig. 5. Comparison of latency on linear multi-hop chain

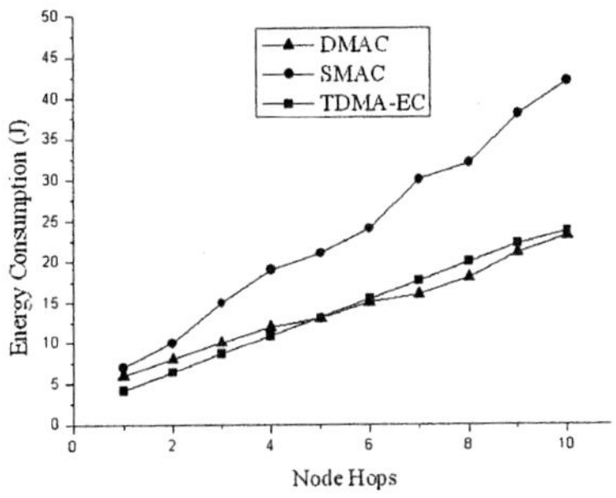

Fig. 6. Comparison of energy consumption on linear multihop chain

In the following comparisons, End-to-end latency, Energy cost and Delivery ratio are used to evaluate the performance of TDMA-EC. Delivery ratio means the ratio of received data packets successfully to the total packets originating from all source nodes. Energy cost is total energy cost to deliver a certain number of packets. The metric of end-to-end latency will be measured at each hop.

5.1 Linear Multihop Chain Traffic Model

For comparison with D-MAC and S-MAC, we firstly performed a test on a simple multihop chain with 11 nodes. To eliminate the impact of queuing delay on end-to-end latency, a light traffic rate of source report interval 0.5s is used. Following figures show the end-to-end latency and relevant energy consumption under increasing hop lengths.

In figure 5, packet latency is about $0.1ms$ at each hop and it is linearly proportional to the number of hops. Generally, compared to contention-based MAC protocol, packet delay achieved by TDMA-based MAC protocol is larger due to fixed length of frame. But TDMA-EC can obtain approximate performance as adaptive SMAC in slight traffic. It is adequate for latency-insensitive WSN applications. Moreover, with the increase of traffic load, the latency performance of TDMA-EC would be more predominant. On the other hand, from figure 6 we can see TDMA-EC consume some amount of energy similar to DMAC and much less than SMAC.

5.2 Random Spanning Tree Traffic Model

In order to reduce power consumption for routing operations we first construct a spanning tree in the network. The construction of the spanning tree in sensor network is an independent problem and will be addressed in a different work. The data is communicated only along the edge of the spanning tree. In the first topology, 50 nodes are employed randomly in a 1000m×500m area. The sink node is at the right bottom corner. Five nodes at the margin are chosen as sources with the same reporting rate.

Comparison results are shown in figure 7 regarding to average end-to-end latency. TDMA-EC presents similar performance to DMAC and both achieve lower latency

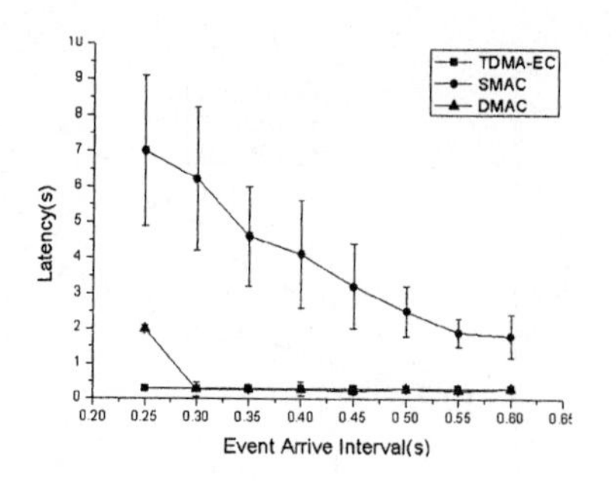

Fig. 7. different latency comparison under different event arrive rate on spanning tree

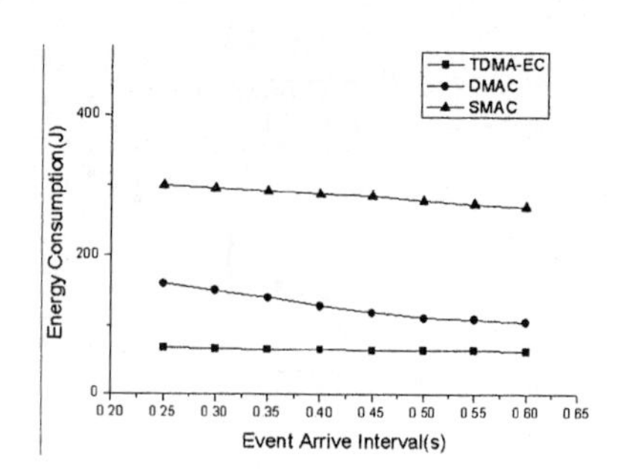

Fig. 8. Comparison of energy consumption under different event arrive rate on spanning

than SMAC. But more importantly, TDMA-EC remains stable low-latency with the increase of traffic load. Since SMAC and DMAC are not based on pre-schedule, higher traffic load would results them into the channel contention frequently, so the end-to-end latency would increase inevitably. Figure 8 shows the results of energy consumption. Due to inherent low duty cycle of TDMA, TDMA-EC achieves more energy conservation than SMAC and DMAC. In figure 8, the average energy consumed by TDMA-EC is about 65J, which yields 2 and 4 folds enhancement in battery lifetime compared to SMAC and DMAC.

Figure 9 compares the delivery ratios of the number of received packets to the number of transmitted packets under SMAC, DMAC and TDMA-EC. The delivery ratio of SMAC decreases when traffic load is heavy whereas TDMA-EC and DMAC achieve better and stable delivery ratios. So TDMA-EC can also help to improve the reliability of network in this regard.

Dense network means high probability of transmitting interference. Following evaluations are destined for a dense network, in which 100 nodes are randomly deployed in a 1000m×500m area. Source nodes generate one data packet per 3 seconds and they are chosen randomly from the margin nodes in the network. The number of source nodes is added from 6 up to 36.

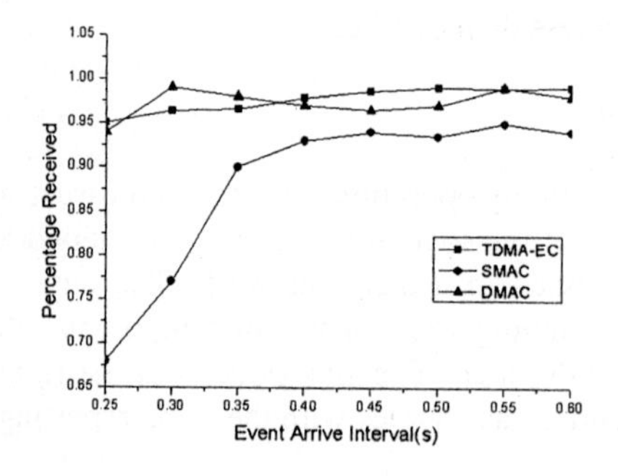

Fig. 9. Comparison of delivery ratio under different event arrive rate on spanning tree

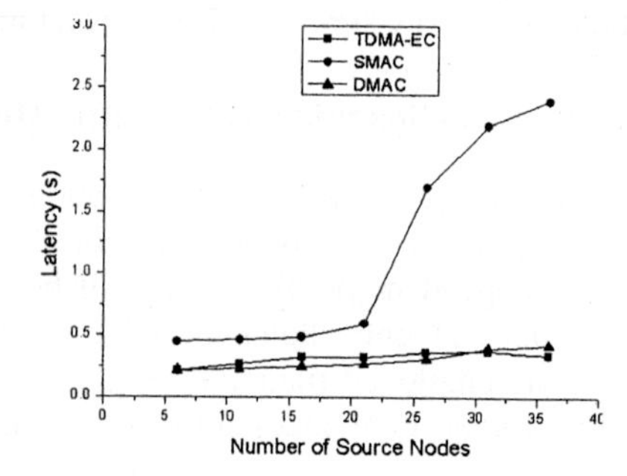

Fig. 10. Latency comparisons in dense network on spanning tree

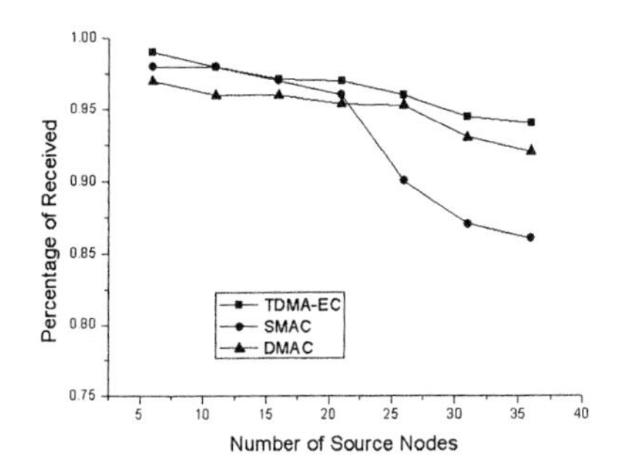

Fig. 11. Comparison of delivery ratio in dense network on spanning tree

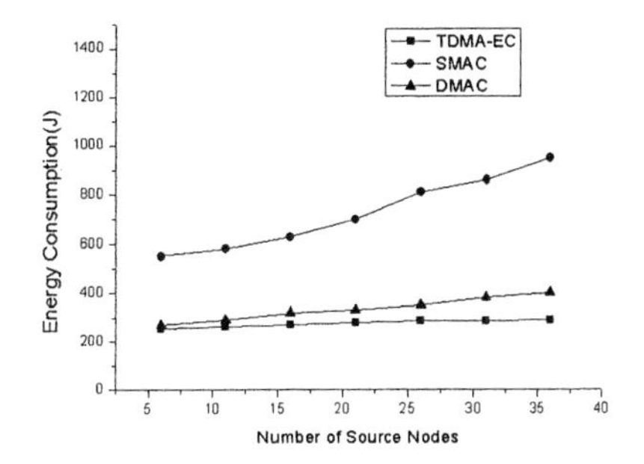

Fig. 12. Energy consumptions in dense network on spanning tree

Fig 10 shows the end-to-end latency under different number of source nodes. TDMA-EC brings better and stable low latency performance as DMAC while SMAC show the deteriorative effect with the increase of number of source nodes. Similar results also shown in figure 11, where TDMA-EC can maintain high delivery ratio compared to DMAC and SMAC.

Both DMAC and TDMA-EC show close performance as shown above. Moreover, due to the property of contention-avoidance, TDMA-EC can achieve much better energy-efficiency than DMAC. Fig. 12 shows the energy comparisons by TDMA-EC, DMAC and SMAC. We can see that the predominance of TDMA-EC is obvious and there is about 50J difference on average.

6 Conclusions

In this paper, we have proposed TDMA-EC, an energy efficient and collision avoidable protocol in wireless sensor networks. TDMA-EC employs the well-known TDMA principle so that it is energy efficient for the nature of low duty cycle. TDMA-EC protocol also achieves collision avoidance posed by simultaneous data transmits and "hidden terminal" problem. The collision avoidance is beneficial to both decreasing latency and conserving energy. On the other hand, based on the data gathering tree, TDMA-EC staggers the receiving and transmitting periods of nodes in the tree according to the position, that is, parent and its child nodes are scheduled with a difference of half of the length of one frame. This allows continuous packet forwarding during one frame in each node. By this means TDMA-EC also obtains low packet delivery latency. Simulation results have shown that TDMA-EC achieves better energy conservation and lower latency compared to SMAC and DMAC. Further work will emphasize on Designing and optimizing a distributed, efficient, and low latency-ensured slot assignment algorithm which is crucial to the performance of TDMA-EC.

References

1 Akyildiz I F, Su W, Sankarasubramaniam Y, and Cayirci E. A survey on sensor networks. IEEE Communications Magazine, 2002, 40(8): 102~114.

2 Ren FY , Huang HN , Lin C. Wireless Sensor Networks. Journal of Software, 2003, 14(7) : 1282~1291

3 Assefaw Hadish Gebremedhin, Fredrik Manne and Alex Potheny. Graph Coloring in Optimization Revisited. http://www.cse.psu.edu/~zha/CSE555/color.pdf

4 Gang Lu, Bhaskar Krishnamachari, Cauligi Raghavendra, "An Adaptive Energy-Efficient and Low-Latency MAC for Data Gathering in Sensor Networks", in *4th IEEE WMAN*, April 2004.

5 Wei Ye, John Heidemann, and Deborah Estrin, "An Energy-Efficient MAC protocol for Wireless Sensor Networks," in *IEEE Infocom*, 2002.

6 W. Ye, J. Heidemann, and D. Estrin, "Medium Access Control with Coordinated, Adaptive Sleeping for Wireless Sensor Networks", Technical Report USC ISI-TR-567, January, 2003.

7 C. C. Enz, A. El-Hoiydi, J-D. Decotignie, V. Peiris, "WiseNET: An Ultralow-Power Wireless Sensor Network Solution", *IEEE Computer,* 2004.

8 T. van Dam and K. Langendoen. An adaptive energyefficient MAC protocol for wireless sensor networks. In *SenSys 2003*, pages 171–180, Los Angeles, CA, November 2003.

9 B. Krishnamachari, D. Estrin and S.Wicker, "The impact of data aggregation in wireless sensor networks", in International Workshop on Distributed Event-based Systems, 2002

10 L. C. Pond and V. O. K. Li, "A distributed time-slot assignment protocol for mobile multi-hop broadcast packet radio networks", *MILCOM*, pp. 70-74, 1989.

11 Zhihui Chen, Ashfaq Khokhar,, "Self Organization and Energy Efficient TDMA MAC Protocol by Wake Up For Wireless Sensor Networks" in Secon2004

12 UCB/LBNL/VINT Network Simulator – ns (version 2). http://www.isi.edu/nsnam/ns/

Experiments Study on a Dynamic Priority Scheduling for Wireless Sensor Networks

Jiming Chen and Youxian Sun

National Laboratory of Industrial Control Technology,
Zhejiang Univ., Hangzhou 310027, P.R. China
{jmchen, yxsun}@iipc.zju.edu.cn

Abstract. Wireless Sensors Networks (WSNs) represent a new generation of embedded systems, which deploy wireless ad hoc communication mechanisms to route a sensory data from the originator sensor node to the base station. One of the most important issues in wireless sensor networks is to support real-time QoS, which is fundamentally a challenging problem. In this paper, we introduce EDBP scheduling algorithm for sensor network along with a MAC protocol based on the dynamic priority assignment EDBP. Two different platforms based experiment projects are designed to test our proposal.

1 Introduction

The wireless Sensor Networks (WSNs) technology is one of the dominant research trends and has been the subject of numerous studies with recent advances in micro-electro- mechanical system (MEMS) in the last few years. Sensor networks represent a significant improvement over traditional sensors. The potential applications of WSNs are highly varied: e.g., military applications; environmental monitoring; health applications; home application.

WSNs have arisen diverse challenges due to its wireless nature, node density, limited resources, low reliability of the nodes, distributed architecture and frequent mobility and whose issues are different from those of classical wireless ad hoc networks[1]. Many researches have been done concerning system architecture and design. Energy consideration has dominated most of the researches in WSNs, because unattended sensors are energy constrained and their batteries cannot be recharged. QoS supporting is not the primary concern in most of the published work on WSNs and a still largely unexplored research field. It is actually a big challenge for sensor networks to provide real-time QoS.

WSNs may consist of many different types of sensors such as seismic, thermal, which are able to monitor a wide variety of ambient conditions including temperature, humidity, lightning condition, pressing etc. However, the increasing interest in real-time applications along with the introduction of audio/video sensors has posed additional challenges. Consider the following scenario: in a battle environment to identify a target, imaging sensors are used. After detecting and locating a target using contemporary types of sensors, e.g. acoustic, imaging sensors can be turned on to capture a picture of such a target periodically

X. Jia, J. Wu, and Y. He (Eds.): MSN 2005, LNCS 3794, pp. 613–622, 2005.

and then send to the gateway. This requires a real-time data exchange between sensors and controller in order to take the proper actions, since it is a critical situation[2]. Delivering such time-constrained data requires certain bandwidth with minimum possible delay and thus a service differentiation mechanism will be needed in order to guarantee timeliness. Therefore, collecting critical real-time data like alerting signal and sensed imaging requires both energy and QoS aware network protocols in order to ensure efficient usage of the sensors and effective access to the gathered measurements.

As far as we know, little attention has been paid by researchers to addressing QoS supporting in WSNs. Especially for loss-tolerant applications like audio/video streaming, consecutive losses of a certain fraction of packet, will result in noticeable degradation in QoS[3], even if the loss-rate is below given expected value. A suitable performance metric in this case is a window-based loss-rate, i.e. loss-rate constrained over a finite range, or window, of any consecutive packets. More precisely, an application might tolerate at most k-m packet losses for every k arrivals at the various service points across a network. Any service discipline attempting to meet these requirements must ensure that the number of violations to the loss-tolerant specification is minimized (if not zero) across the whole stream. In another way, it is same meaning that at least m packets must be serviced before their deadline in any consecutive k packets. We refer to such QoS requirement as (m,k)-firm constraint[4].

In WSNs, the packet losses are not inevitable because of wireless nature and limited bandwidth. Furthermore, loss-tolerant real-time applications permit occasional packet losses. Such behavior can be described appropriately by concept of (m,k)-firm. An improved scheduling algorithms based on (m,k)-firm EDBP has been proposed[5]. Two experiments projects based on different WSNs hardware platform are designed to validated that our proposal EDBP is more effective to provide better QoS than others.

The rest of paper is structured as follows. Section2 reviews some related work. In section3, EDBP scheduling is described in brief. In section4, two experiments are designed to evaluate QoS in WSNs. Finally, we make some concluding remarks.

2 Related Work

In wireless sensor networks, QoS and real-time mechanisms (scheduling, routing, medium access control) are typically defined at the MAC and Network Layers. Each layer may support QoS for real-time communications in sensor networks. A major problem is that, so far, there have been no standards for MAC and Network Layer. Some solutions were proposed for these layers to support QoS and real-time requirements and seem to be promising.

In MAC level, several typical protocols were proposed for wireless networks based on contention and carrier sense[6][7]. Such protocols try to maximize the throughput, but do not provide any real-time guarantee. While some energy-efficient MAC ptotocols have been put forward for WSNs[8][9]. In order to get QoS guarantee such as dynamic failure, bounded delay, many specific real-time packet scheduling algorithms have been introduced to protocol design[10][11][12]. But these protocols can not be applied to WSNs because a amount of control packet overhead is a burden for the energy-constraint sensor nodes. Recently, A MAC protocol based on Earliest Deadline First (EDF) Scheduling is presented in one cluster of WSNs by Caccamo et al in order to ensure timeliness for real-time traffic [13]. Because of periodic nature of sensor date, The idea is to create a scheduler rather than using control packets to channel reservation and collision avoidance. Simulation experiment shows that the protocol gets better performance in terms of throughput and average dealy in heavy load conditions comparing to CSMA/CA.

Another scheduling algorithm, Rate-Monotonic is proposed in RAP project for MAC protocol, which is in order to minimize deadline miss ratio[14].

These scheduling based MAC protocols do not explore the characteristics of loss-tolerant real-time applications. They schedule each packet to meet its deadline and retransmit colliding packets that results in extra energy cost.

In fact, many real-time applications like audio/image is loss-tolerant. Packet missing deadline can be dropped and did not occupy the bandwidth.

In this paper, EDBP will be applied to schedule packets in a cluster of WSNs, which is proposed by us in [5]. Two different experiments projects are designed to test the performance of EDBP in terms of loss-rate and dynamic failure rate on the platform of wireless sensor networks.

3 EDBP Scheduling in WSNs

In order to describe loss-tolerant real-time applications better, a window based loss-rate called (m,k)-firm constraint model was firstly proposed by Hamdaoui and Ramanathan[15]. A real-time audio/video stream is said to have (m,k)-firm constraint if at least m out of any k consecutive packets issued from the same streaming meet their respective deadlines. When a stream violates (m,k)-firm constraint, a condition known as *dynamic failure* occurs. For example, if a few consecutive audio packets miss their deadlines, a vital portion of the talkspurt may be missing and the quality of the reconstructed audio signal may not be satisfactory. The *dynamic failure rate* is then used as a dynamic performance metric of the QoS perceived by a (m,k)-firm real-time stream.

Especially, resource including bandwidth and energy is limited in WSNs, it is much important to design an effective scheduling based on MAC protocol to provide real-time QoS guarantee. EDBP is a dynamic priority scheduling algorithm, which is validated much more effectively by simulation than previous scheduling algorithms to assure better QoS in terms of dynamic failure and delay.

In order to encapsulate the (m,k)-firm model with scheduling in WSNs, EDBP was applied to WSNs. The key idea to provide EDBP at the MAC layer is to replicate contention free schedule at each sensor node for packet transmission. Assuming that each node has knowledge about the periodic in a given cluster, a local EDBP scheduler can run at each node synchronously. As a result, each sensor is aware of the next packet that is scheduled for transmission according to EDBP.

In the EDBP, a streaming is specified by 5-tuple $(C_i, T_i, D_i, m_i, k_i)$ and its history state information is maintained. This state represents the history of the most recently last k packets transmitted and a value given by EDBP scheduling algorithm is associated with each state. In successful state, the EDBP value of a stream is the number of consecutive packets deadline misses, which makes the current state reach a failure state. The packet in the stream with lower EDBP value Ω will get higher priority.

Formally, given a real-time stream τ_j with (m_j,k_j)-firm, priority is evaluated as follows. Let $s_j = (\delta^j_{i-k_j+1}, \ldots, \delta^j_{i-1}, \delta^j_i)$ denotes the state of the previous k consecutive packets of τ_j, $l_j(n, s)$ denotes the position (from the right) of the nth meet (or 1) in the s_j, then the priority of the $(i+1)^{th}$ packet of τ_j is given by equation (1):

$$\Omega^j_{i+1} = k_j - l_j(m_j, s_j) + 1 \quad (1)$$

Example1:
A stream τ_1 with (3,5)-firm constraint, current state is 11011, we can get $l_1(3, s_1)$ =4 and Ω^1_{i+1}=5-4+1=2.

If there are less than n 1s in s, we should consider the distance to exit a failure state. The distance to exit a failure state is thus the number of consecutive 1s adding to the right side of the current state. Formally, given a stream τ_j with constraint parameter m_j and k_j in a failure state, and let $s_j = (\delta^j_{i-k_j+1}, \ldots, \delta^j_{i-1}, \delta^j_i)$ be its current k-sequence. Define $\tilde{l}_j(n, s)$ as the position (from the right side) of the nth miss in the state of s_j , so the distance to exit a failure state of stream is given by equation (2):

$$\Phi^j_{i+1} = k_j - \tilde{l}_j(k_j - m_j + 1, s_j) + 1 \quad (2)$$

Example2:
A stream τ_1 with (4,6)-firm constraint, current state is 100011, we can get Φ^1_{i+1}=2. A stream τ_2 with (3,5)-firm constraint, current state is 00011, we can get Φ^2_{i+1}=1.

In a successful state, priority is assigned according to Value Ω, while in a failure state priority is assigned by equation (2), the packet in the each node with small Φ will be assigned higher priority. In case of priority equality, EDF is adopted, which is referred as EDBP.
EDBP precedence among all being selected packets

- If all packet streams are in successful states, the smaller Ω, the higher priority. If Ω is the same, EDF is adopted.
- If just only packet stream τ_j is in a failure state, others are in successful states, the packet in the stream τ_j, gets higher priority.
- If many streams are in failure states at the same time, the smaller Φ, the higher priority. If Φ value is the same, EDF is adopted.
- For all cases, if the same deadline, then FIFO.

Table 1. Experiment Results(ms)

Times Steps	1	2	3	4
A-B	1678.7	1678.6	1678.2	1678.3
C-D	700.8	700.8	700.8	700.8
D-E	18.9	18.9	18.9	18.9
E-F	10776.2	10666.7	10602.1	10822.5

4 Experiment Design

4.1 Project1 Based on Motes Platform

In order to test our proposed scheduling based MAC protocol in WSNs, the Motes platform is considered, which is developed by Crossbow. Acoustic sensor in MICA2 node is used to acquire audio signal.

ADC. According to Windows SDK, the least frequency of replay audio signal is 5kHz. ADC clock frequency is 115.20kHz in MICA2 node, which can be gotten from prescaler. Based on the performance of Atmegal128 ADC, the highest sampling frequency is 8.26kHz, which means the sampling period 121μs.

In fact, sampling period includes the time of ADC conversion and data saving. So it is necessary to increase the ADC clock frequency, resulting in decrease of ADC conversion time. But it causes side effect on precision of ADC conversion, which will lead to serious distortion and low quality of audio.

Program Design. The programs of the experiment are divided into two parts: Searching MICA2 node in the cluster, then audio signal sampling and packets transmission.

Experiment Results. In the experiment, sampling period is set $\frac{140}{921.6kHz} =$ 152μs, 921.6kHz is system clock frequency, 140 is the number of system clock period in Micro-Timer. This sampling period is larger than ADC conversion time 121μs when sampling frequency is 8.26kHz.

For each sensor node, five thousands sampling is done for one time, the total data size is 5000×2 bytes and the time is 5000×152μs=760ms. Data rate of Motes is 19.2kHz and length of each packet is 36 bytes. Effective sound data only 28 bytes in the packets, so the time for sending is $\frac{5000\times2\times8\times36}{28\times19.2Kbps} = 5356.8$ms.

During the process of experiment, the time point of A Ready, B Erase Flash Finish, C Sampling Start, D Sampling Finish, E Transmit Start and F Transmit Finish, is recorded. The original data was gotten from experiment as shown in Table 1.

The experiment results indicate that:

- Actual transmission time (≈10770ms) is more than theoretical value (5356.8ms);
- Time interval of A-B plus E-F is 16 times sampling period;
- There is a difference of E-F interval, because a mount of data need to transmit.

Fig. 1. Experiment Platform based on WLAN

The above experiment results shows that the processor capability and bandwidth of WSNs based on Motes is limited. It is found that this experiment test-bed is not appropriate for the audio signal acquiring and transmission. So, the project2 is designed.

4.2 Project2 Based on WLAN

Project2 based on WLAN includes several Pocket PCs (HP iPAQ), Desktop PC, and Laptop PC, which tries to overcome the drawbacks in Project1 and can be thought as enhance WSNs platform.

Pocket PC can be used as a powerful sensor to acquire audio signal, Desktop is considered as a base station. Laptop PC was used to adjust the bandwidth by FTP. The experiment platform is shown as in Fig.1.

The program for Desktop PC and Pocket PC is developed by C♯ and eVC.

In this project, sampling parameters: frequency 11.025kHz, 8bit, ADC conversion time is 300ms. Experiment time is 90s. The parameters of average transmission time, delay time, loss-rate, dynamic failure rate are recorded.

Average transmission time of packet is illustrated in Fig.2 that performance of wireless networks is stable and reliable. The transmission time changes from 98 to 107ms, the difference can be accepted. EDBP (3,4)-firm means each streaming has (3,4)-firm constraint and its scheduling algorithm is EDBP.

The delay of packet includes transmission time and queue delay. Fig.3 reveals that the largest delay value is the case that EDF scheduling is applied. It is evident that EDF always assigns higher prority to the packet that has the earliest deadline, which makes the long delay. When deadline is smaller (D=450ms), the difference of delay is not obvious because many packets are dropped and not serviced when deadline is missed.

In two experiments, each streaming is set (3,4)-firm, (6,8)-firm respectively. Performance of these scheduling algorithms are compared in terms of loss-rate and dynamic failure rate. Loss-rate is corresponding to the utilization of network bandwidth. Dynamic failure rate denotes the percentage that real-time

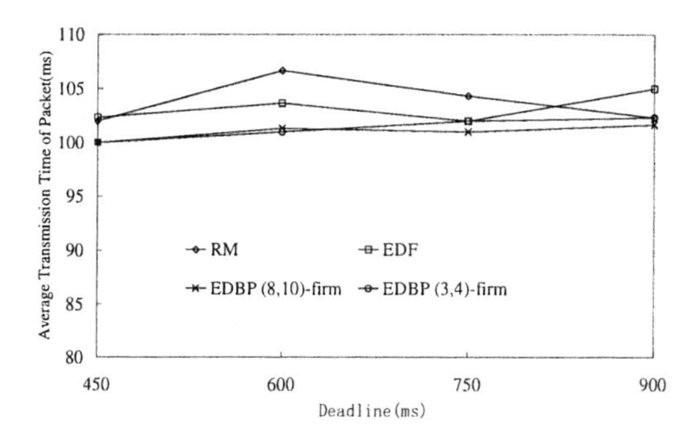

Fig. 2. Average Transmission Time with Variable Deadline

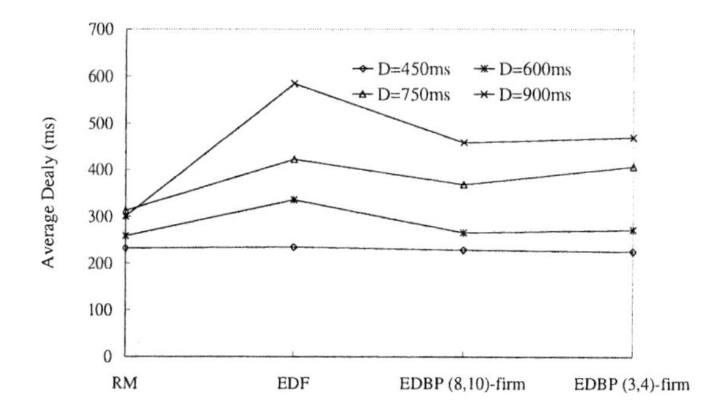

Fig. 3. Delay with Variable Deadline

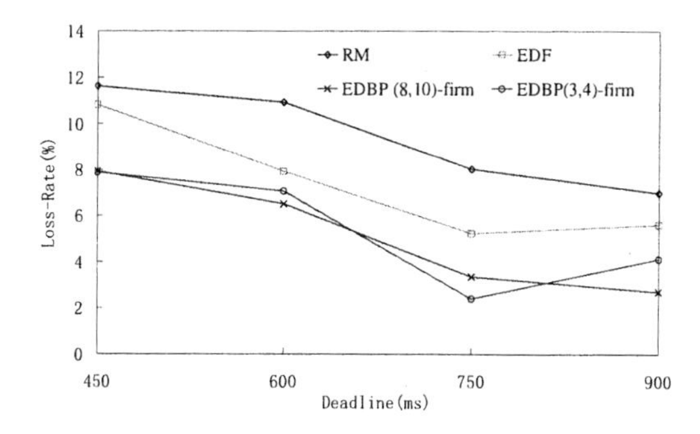

Fig. 4. Loss Rate with Variable Deadline

application gets expected QoS requirement. Table.2 shows the loss-rate data in the experiments. It is clear that it is descending according to RM, EDF, EDBP.

In order to find effect of bandwidth on dynamic failure, FTP is setup by Laptop PC and bandwidth is adjusted by the download velocity. Table.2 presents the experiment result of streams with (8,10)-firm constraint. It is demonstrated that EDBP get better dynamic performance. The maximum decreasing percentage of EDBP to RM, EDF is 53.62%, 34.84% respectively.

Table 2. D=600ms, (8,10)-Firm, Dynamic Failure Rate compare under Variable Bandwidth(%)

adjustment load(KBps) scheduling algorithms	0	25	50	75	100
RM	19.62	21.11	46.98	51.35	67
EDF	12.51	18.48	45.12	49.62	66.22
EDBP	9.1	12.04	36.01	41.42	60.12

Table 3. D=600ms, (3,4)-Firm, Dynamic Failure Rate Compare Under Variable Bandwidth(%)

adjustment load(KBps) scheduling algorithm	0	25	50	75	100
FP	12.4	17.15	34.45	38.44	50.61
EDF	8.28	15.41	32.32	38.85	48.22
EDBP	7.04	13.2	26.8	31.22	41.48

The experiment data in Table.3 for streams with (3,4)-firm has the similar trend that EDBP can provide better QoS in terms of dynamic failure than EDF and RM.

In the experiment on EDBP scheduling, there is a fraction of consecutive losses in each audio signal streaming. It maybe caused by the unstable nature of wireless network. Fig.5 shows the replay audio signal that is transmitted from Pocket PC2 to based station under different scheduling algorithm. It is obviously discovered that EDBP scheduling is a more effective proposal and can provide better QoS for loss-tolerance real-time audio streaming than others.

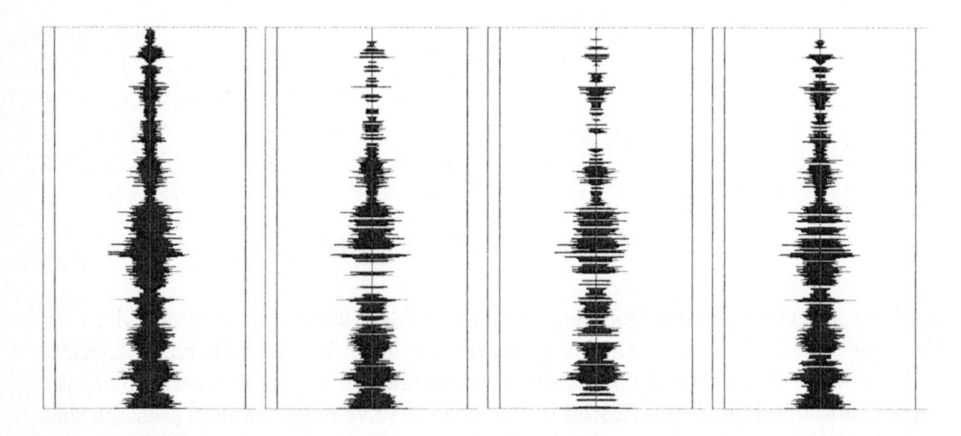

Fig. 5. D=600ms; FTP=50KBps; From Left to Right, Replay Audio Wave in Pocket PC2 is original Wave and generated by Scheduling Algorithm is RM, EDF, EDBP respectively

5 Conclusions

In this paper, we presented a novel MAC protocol based on a dynamic priority assignment EDBP. The key idea consisted of exploiting the characteristic of loss-tolerant real-time application.

The performance of proposed technique has been proved in our experiment. Our protocols have been implemented in a real experiment test-bed. PDA is designed as a powerful acoustic sensor and WLAN is extended to enhance the bandwidth of communication. The experiment shows the effectiveness of EDBP providing lower dynamic failure and loss-rate under different bandwidth conditions. As a future work, we plan to consider the energy constraint in our proposed algorithms and investigate the routing and mobility problem.

Acknowledgements

The authors thank the financial support of Key Project of National Natural Science Foundation of China: Distributed Coordination Theory and Key Techniques for Autonomous Wireless Sensor Networks (No. 60434030).

References

1. Akyildiz, I., Su, W., Sankarasubramaniam, Y., Cayirci, E.: Wireless Sensor Networks: A Survey. Computer Networks **38** (2002) 393–422
2. Younis, M., Akkaya, K., Eltoweissy, M., Wadda, A.: On Handling QoS Traffic in Wireless Sensor Networks. In: Proceedings of the 37th Annual Hawaii International Conference on System Sciences (HICSS'04). (2004) 1–10
3. Mittal, A., Manimaran, G., Siva Ram Murthy, C.: Integrated Dynamic Scheduling of Hard and QoS Degradable Real-Time Tasks in Multiprocessor Systems. Journal of Systems Architecture. **46** (2000) 793–807
4. Bernat, G., Burns, A., Llamosi, A.: Weakly Hard Real-Time Systems. IEEE Transaction on Computers. **50** (2001) 308–321
5. Chen, J.M., Wang, Z., Song, Y.Q., Sun, Y.X.: Extended DBP for (m,k)-firm based QoS. IFIP Network and Parallel Conference, Lecture Notes in Computer Science **3-540-23388-1** (2004) 357–365
6. Bharghavan, V., Demers, A., Shneker, S., Zhang, L.: Macaw: A Media Access Protocol for Wireless LAN. In: Proceedings of ACM SIGCOMM'94 Conference. (1994) 212–225
7. Kleinrock, L., Tobagi, F.: Packet Switching in Radio Channels: Part 1-Carrier Sense Multiple Access Modes and Their Throughput Characteristics. IEEE Transaction on Communication **23** (1975) 1400–1416
8. Woo, A., Culler, D.: A Transmission Control Scheme for Media Access in Sensor Networks. In: Proceedings of AMC/IEEE Conference on Mobile Computing and Networks, Rome (2001) 221–235
9. Ye, W., Heidemann, J., Estrin, D.: An Energy-Efficient MAC Protocol for Wireless Sensor Networks. In: Proceedings of the IEEE INFOCOM, New Yorkw (2002) 1567–1576

10. Garcia-Luna-Aceves, J., Fullmer, C.: Floor Acquisition Multiple Acess (FAMA) in Single-Channel Wireless Networks. In: ACM Mobile Networks and Applications Journal, Special Issue on Ad-Hoc Networks. (1999) 157–174
11. Adamou, M., Khanna, S., Lee, I., Shin, I., Zhou, S.: Fair real-time traffic scheduling over a wireless lan. In: Proceedings of the 22th IEEE Real-Time Systems Symposium, London (2001) 279–288
12. Sobrinho, J., Krishnakumar, A.: Quality of service for ad hoc carrier sense multiple access networks. IEEE Journal on Selected Areas in Communications **17** (1999) 1353–1368
13. Caccamo, m., Zhang, Y.L., Sha, L., Buttazzo, G.: An implicit prioritized access protocol for wireless sensor networks. In: Proceedings of 23rd IEEE Real-Time Systems Symposium, Texas (2002) 39–48
14. Lu, C., Blum, B., Abdelzaher, T., Stankovic, J., He, T.: Rap:a real-time communication architecture for large-scale wireless sensor networks. In: Proceedings of the IEEE Real-Time and Embedded Technology and Applicants Symposium, San Jose (2002) 55–66
15. Hamdaoui, M., Ramanathan, P.: A dynamic priority assignment technique for streams with (m,k)-firm deadlines. IEEE Transactions on Computers **44** (1995) 1443–1451

A BPP-Based Scheduling Algorithm in Bluetooth Systems*

Junfeng Xiao, Biao Ren, and Shiduan Cheng

State Key Laboratory of Networking and Switching Technology,
Beijing University of Posts and Telecommunications, 100876, Beijing, China
{xiaojf, renb, chsd}@bupt.edu.cn

Abstract. In this paper we view the scheduling problem in Bluetooth as a bin packing problem (BPP) with *k*-preview items. We analytically demonstrate that even though the *First Fit Decreasing* with *k*-preview Items (*FFD-k*I) algorithm has the same asymptotic worst case performance ratio as the Look Ahead Round Robin (LARR) algorithm, it achieves a competitive asymptotic average case performance ratio which is superior to LARR and Round Robin (RR). We present both worst case and average case results. Extensive simulation results show that in a Bluetooth piconet, the *FFD-k*I algorithm achieves significantly better throughput and channel utilization performances over LARR and RR.

1 Introduction

In Bluetooth, a piconet is formed by at most eight Bluetooth enabled devices, one of which acts as the Master and the remaining ones as Slaves. Bluetooth systems are scheduled by the Master-driven Time Division Duplex (TDD) scheme. When a Master sends poll or data packets to a Slave, the Slave can send packets to the Master immediately after receiving packets from the Master [1]. In other words, the transmission always occurs in pairs. Two types of Radio Frequency links have been defined: Asynchronous Connectionless Link (ACL) and Synchronous Connection Oriented (SCO) Link. SCO link is applied for time-bound services such as voice, with slots reserved at regular intervals. ACL links are used for data communication. In Fig. 1, the schedule of SCO and ACL slots is shown.

Unspecified in the Bluetooth standard, the algorithm of scheduling ACL slots may significantly influence the throughput performance of a piconet. When SCO links existing in a piconet, there has an important constraint, which is that a data packet transmission from a Master or Slave cannot span across different SCO slots. Due to the constraint, the scheduling algorithm of ACL slots becomes different from current ones. Considering the effect of SCO slots, Kalia et al. proposed scheduling policies that utilized information about the size of the *Head-of-Line* (HOL) packet at the

---

* This work was supported by the National Nature Science Foundation of China (Grant No.9020-4003, 60402012), the National Basic Research Program of China (Grant No.2003CB314806, 2006CB701306) and Beijing Municipal Key Disciplines (XK100130438).

X. Jia, J. Wu, and Y. He (Eds.): MSN 2005, LNCS 3794, pp. 623 – 632, 2005.

Master and Slave queues to schedule the TDD slots [2]. However, the Round Robin (RR) algorithm, adopted when the Master-Slave pairs are the same class, will result in a lot wastage of slots. Yang Chun-Chuan et al. suggested a bandwidth-based polling scheme for QoS support in Bluetooth [3]. Although the channel utilization of ACL links is promoted, the performance of SCO links is sacrificed. Yang Daqing et al. [4] concluded the scheduling of ACL slots as an online bin packing problem (BPP). They proposed a new scheduling policy, named Look Ahead Round Robin (LARR), which performs significantly better than the RR algorithm.

In this paper, we assume that there is only one SCO link in a piconet as in [2-4], and that the transmission buffers of Master and Slaves are always loaded for the next transmission so that we can only see the effect of algorithms. The queue statuses at the slaves can be obtained through feedback as in [2, 4]. We distinguish the Master-Slave pairs based on the size of the HOL packets at the Master and Slaves queues. Master-Slave pairs (M-S pairs) are referred to on the basis of the size of HOL packet at both queues. To analyze the scheduling problem, we model it as a problem of ***Bin Packing*** with ***k***-preview ***Items*** (*BP-k*I) and show that the two are strongly related. The *k* denotes the number of enabled Slaves in a piconet.

The remainder of the paper is organized as follows. In Section 2 we formally define the problem. Section 3 presents worst case analysis of scheduling algorithms. And the average case analysis of them is also illustrated. In Section 4, simulations and numerical results are presented. Finally, we conclude the paper in Section 5.

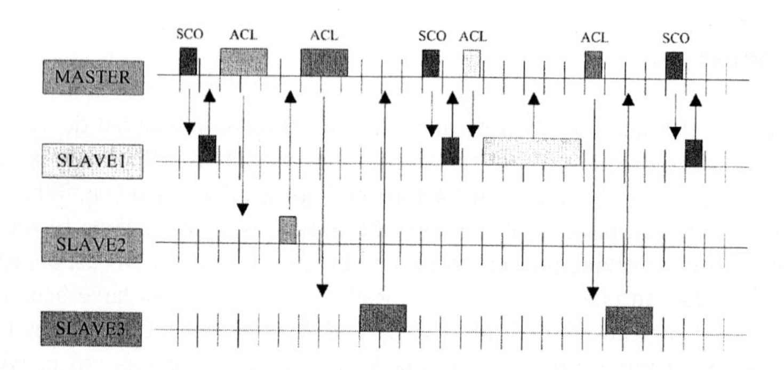

Fig. 1. Interleaved SCO and ACL slots for communication between the Master and the Slaves

2 Problem Statement and Definitions

In this section we formally define the problem of bin packing with *k*-preview items. Meanwhile, the relation between the scheduling problem and the problem of bin packing with *k*-preview items is shown.

2.1 Bin Packing Problem

In the classical one-dimensional bin packing problem, we are given a sequence $L=\{a_1, a_2, \ldots, a_n\}$ of items, each with a size $s(a_i)\in(0,1]$ and are asked to pack them into a minimum number of unit-capacity bins [5].

The classical one-dimensional bin packing problem, as many of its derivatives, is NP-hard [6]. The limitation preview of items renders the bin packing problem non-trivial, so *BP-k*I is also NP-hard. Many approximation algorithms have been developed for it (see [5]). The *First-Fit (FF)*, the *Best-Fit (BF)* and the *Next-Fit (NF)* are best known [5]. The *FF* algorithm is an improvement of *NF*. It can be explained in two ways [7]: One explanation uses an *on-line* model. Another one uses an *off-line* model. In this case, the list of items must be given in advance. There is only one open bin at a time, but we check the items left in the list one by one and put the first item which fits in the open bin and repeat this process. Only if no item fits in the open bin, the bin is closed and a new bin is opened. We will use the second explanation for our suggested algorithm in this paper. It is well known that if the items in the set are arranged into a *non-increasing* order of the item size, the performance bound can be further improved. The corresponding versions of the *FF* and *NF* techniques are known as *First Fit Decreasing* (*FFD*) and *Next Fit Decreasing* (*NFD*) respectively.

The analysis of bin packing algorithms is traditionally divided into worst case analysis and average case analysis [5]. In worst case analysis we are usually interested in the asymptotic worst case performance ratio. For a given list of items L and algorithm A, let $A(L)$ be the number of bins used when algorithm A is applied to list L, let $OPT(L)$ denote the optimum number of bins for a packing of L, and let $R_A(L) \equiv A(L)/OPT(L)$ [5]. To measure worst case behavior, the asymptotic worst case performance ratio R_A^∞ of algorithm A is defined by [8]

$$R_A^\infty \equiv \lim_{m \to \infty} \sup R_A^m \tag{1}$$

where $R_A^\infty \equiv \sup_L \{A(L)/OPT(L) \mid OPT(L) = m\}$. A different approach for estimating the performance of an algorithm is an average case analysis. In this case we assume the items are taken from a given distribution H and we try to estimate the performance ratio of an algorithm, when it is applied to a list taken from that distribution. For a given algorithm A and a list of n items L_n, generated according to distribution H, the ***asymptotic expected performance ratio*** is defined as follows [5]:

$$\overline{R}_A^\infty(H) \equiv \lim_{n \to \infty} E[R_A(L_n)] = \lim_{n \to \infty} E\left[\frac{A(L_n)}{OPT(L_n)}\right] \tag{2}$$

2.2 Bluetooth Scheduling Problem and *FFD-k*I Algorithm

As the description in section 1, when one SCO link exists in a Bluetooth piconet the gap between two SCO slot pairs comes into being a bin. The key problem of scheduling is how to efficiently arrange M-S pairs in the gap to obtain the best throughput performance. The scheduling problem becomes a bin packing problem. Due to the lower asymptotic worst case performance ratio of *FFD*, we adopt it for our scheduling. *FFD* is an *off-line* algorithm. But there has only a limited knowledge of all items. Because if there are k Slaves presented in the piconet, the scheduling algorithm can examine the k M-S pairs and decide which pair to place in the current bin (or gap). At the meantime, a closed bin cannot be opened again in Bluetooth. As a result, the packing algorithm turns to a new version of *FFD*, named *FFD-k*I. The *FFD-k*I algorithm for Bluetooth can be viewed as *semi-offline* bin packing algorithm with a limited amount of future knowledge. Its sorting overhead is in $O(\log N)$ amount of computation, where N is the number of enabled Slaves in a piconet. This amount of computational overhead is high acceptable in Bluetooth. As the algorithm always attempts to schedule the M-S pair with the largest HOL, it may lead to starvation for some of M-S pairs in the extreme case. To avoid it, we assume that a M-S pair, which hasn't been scheduled in k successive bins, will be served first in the $(k+1)^{th}$ bin. This algorithm schedules M-S pairs in the following way.

Step 1. The Master obtains k M-S pairs' HOL based on RR.
Step 2. The Master sorts k M-S pairs with non-increasing.
Step 3. Judge whether space of current bin is not less than the minimum M-S pair. If 'yes', go to Step 4. Otherwise, go to Step 7.
Step 4. Judge whether there is any M-S pair, which isn't scheduled in successive k bins. If 'yes', go to Step 5. Otherwise, go to Step 8.
Step 5. Pack the maximum M-S pair, which fits the current bin, from these M-S pairs.
Step 6. Obtain a new HOL from the scheduled M-S pair. Then go to Step 2.
Step 7. Wait next bin. Then go to Step 3.
Step 8. Pack the maximum M-S pair, which fits the current bin, from all M-S pairs. Then go to Step 6.

3 Analysis of Scheduling Algorithms

In this section we analyze the performance of the *FFD-k*I algorithm in the worst case and the average case. At the meantime, LARR and RR are compared with *FFD-k*I. Because the ACL link packets can occupy only 1, 3 or 5 slots, the size of one M-S pair will be 2, 4, 6, 8 or 10 slots. That is the sizes of items are 2, 4, 6, 8 or 10. In order to guarantee the correct transmission of packets, we assume that the gap between every two SCO slot pairs is at least 10 slots (i.e., the bin size $C \geq 10$). Let A denote any of algorithms which we will analyze. Let n denote the number of all items. We use n_1, n_2, n_3, n_4 and n_5 denote the number of items whose sizes are 2, 4, 6, 8 and 10 respectively. So, $n=n_1+n_2+n_3+n_4+n_5$.

3.1 Worst Case Analysis

The performance metric of choice for bin packing algorithms is the worst case asymptotic performance ratio [8]. It presents an upper bound of the performance ratio of algorithms.

Theorem 1: For the *FFD-k*I scheduling algorithm in Bluetooth, $R_A^\infty \leq 1.5$.

Proof: As the above assumption, the bin size is larger than or equal to 10. First, we analyze the R_A^∞ in the case of the bin size being 10. We have

Lemma 1: For the RR scheduling algorithm in Bluetooth when the bin size is 10, $R_A^\infty = 5/3$. [9]

Lemma 2: For the LARR scheduling algorithm in Bluetooth when the bin size is 10, $R_A^\infty = 1.5$ [9].

Proof: The proof of the LARR algorithm is similar to the proof of following lemma 3. Due to size constraints the detailed proof of the LARR algorithm appears in [9].

Lemma 3: For the *FFD-k*I scheduling algorithm in Bluetooth when the bin size is 10, $R_A^\infty = 1.5$.

Proof: To find the asymptotic worst case performance ratio of an algorithm, we must get a list of items L to make $A(L)$ much more and $OPT(L)$ much less. The detailed proof is shown in Appendix A.

Now, we analyze the R_A^∞ in the case of the bin size C>10. Item sizes are still only 2, 4, 6, 8 and 10. The larger bin will result in smaller normalized item sizes. It is obvious that smaller normalized items will bring out less wasted space. So, combining Lemma 3, the **Theorem 1** is proved. That is the asymptotic worst case performance ratio of the *FFD-k*I algorithm will be not more than 1.5. ∎

3.2 Average Case Analysis

The worst case analysis presented above provides an upper bound on the performance ratio of the *FFD-k*I algorithm. However, one drawback of relying on worst case analysis is that in many applications the worst case never seems to occur [5]. To learn about the typical behavior of the algorithm we present an average case analysis on experiments. Since the results of an average case analysis depend on the item-size distribution, it is desirable to calculate results for any given distribution. But, the asymptotic expected performance ratio acting as one performance metric of an algorithm, its closed form solution is hard to derive. We can resort to a large number of experiments.

We simulated the algorithms of *FFD-k*I, LARR and RR to get an idea how they behavior. We used sequences of items whose sizes are drawn from the combination of any two packet sizes. The packet size is drawn from a discrete uniform distribution. We used the Matlab tool to produce the input sequence for the program and to simulate the behaviors of algorithms. In Fig. 2, we present the asymptotic expected

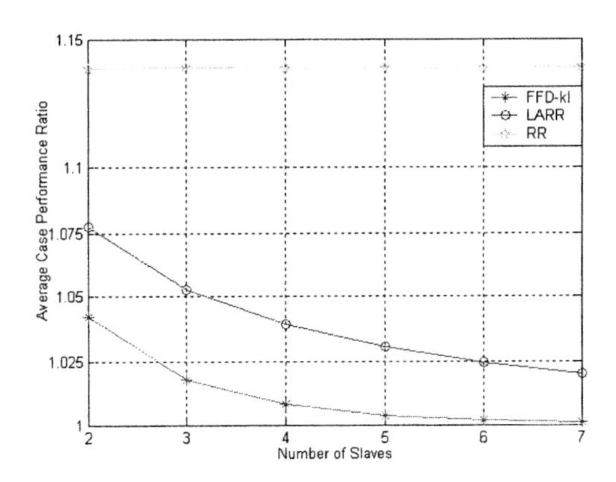

Fig. 2. Expected case performance ratios of algorithms

performance ratios of various algorithms when the number of enabled Slaves varies from 2 to 7. From this figure, we observe that the asymptotic average case ratios of *FFD-k*I and LARR go down with the increase of Slaves. When the number of Slaves is 7, the average case performance ratio of *FFD-k*I is very close to 1. It is rational that the ratio of *FFD-k*I is equal to 1 when k becomes infinite. Because the *FFD-k*I algorithm with infinite k transforms to the pure *FFD* algorithm and the asymptotic expected performance ratio of *FFD* is 1. At the meantime, it is obviously found that the asymptotic average case ratio of *FFD-k*I is better than that of LARR and RR. The average case results are not as bad as the worst case analysis indicates. They reflect the typical behaviors of various algorithms. In addition, the essence of RR decides that its average case performance ratio keeps constant, about 1.1385.

4 Simulations and Numerical Results

In the previous sections we analytically and experimentally showed the worst case and the average case performances of the *FFD-k*I algorithm. Now, we apply this algorithm into Bluetooth systems to evaluate the channel utilization, throughput and fairness performances. We simulate a single piconet consisting of a Master and k active Slaves. The Master has a corresponding queue for each Slave. The packet sizes are drawn from a discrete uniform distribution as in [2, 4]. Table 1 presents the different packet types used in our simulations.

Fig. 3 and Fig. 4 show the channel utilization, i.e. the percentage of time that the channel is busy in carrying data packets, and the throughput respectively. From these two figures, we can find that the *FFD-k*I algorithm has higher channel utilization and

Table 1. ACL data packet types used in simulations

Type	Payload Header (Byte)	User Payload (Byte)	FEC	CRC
DH1	1	0-27	No	Yes
DH3	2	0-183	No	Yes
DH5	2	0-339	No	Yes

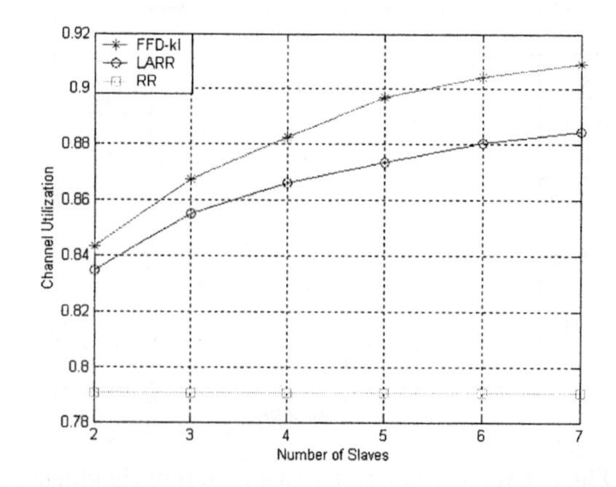

Fig. 3. Channel utilization with different number of Slaves

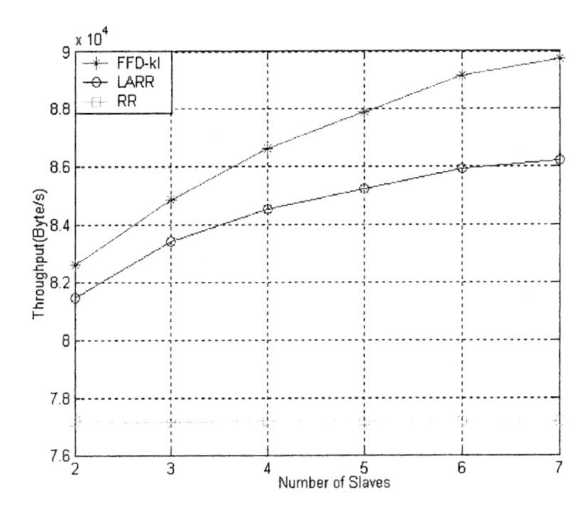

Fig. 4. Throughput with different number of Slaves

throughput than other two. The RR scheme has a lower throughput than other schemes, because it acts as the *NF* algorithm. The *NF* algorithm performs poorly. Both *FFD-k*I and LARR improved channel utilizations and throughputs as the number of enabled Slaves. However, the proposed algorithm has much higher throughput and channel utilization. The difference is increased at high *k*. This is quite expected because in this situation the *FFD-k*I algorithm has a larger number of queues to examine to fill the current bin.

In Fig. 5, fairness performance is presented. The Master-Slave pairs' turns of polling are the number of service chances received from the scheduler. Fairness index is measured by equation (19) to show the fairness performance among the Master-Slave pairs' turns of polling.

$$f = \left(\sum_{i=1}^{k} x_i\right)^2 \Bigg/ \left(k\sum_{i=1}^{k} x_i^2\right) \tag{19}$$

The M-S pair's turns of polling are denoted by x_i, $2 \le i \le k$. The fairness index always lies between 0 and 1. The scheduling algorithm is fairer with the fairness index

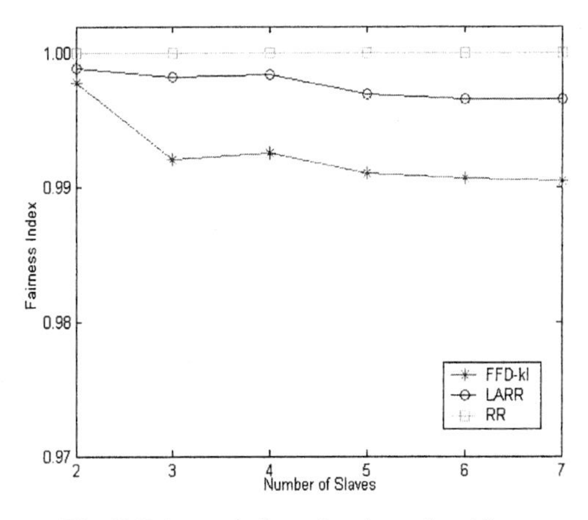

Fig. 5. Fairness index of various algorithms

approaching 1. From Fig. 5, it can be found that although the *FFD-k*I algorithm has a lower fairness index than the LARR algorithm and the RR algorithm, its fairness index also approaches 1. That is the *FFD-k*I algorithm achieves the competitive channel utilization and throughput performances at the expensive of a trivial fairness performance.

5 Conclusion

In this paper, we presented a scheduling problem of ACL links when there is one SCO link in a piconet. To analyze the scheduling problem we introduced a new variant of bin packing problem that the packing algorithm has a limited knowledge of all items, and showed that the two problems are strongly related. We defined the *FFD-k*I algorithm and analyzed both worst case and average case to evaluate the schedule efficiency of the algorithm. We found that *FFD-k*I achieved a competitive asymptotic expected performance ratio, which is superior to LARR and RR. In addition, we observed that a limited knowledge of all items can considerably improve the schedule efficiency. An important characteristic is that schedule efficiency increases with k. Simulation results demonstrated that in a Master-driven TDD Bluetooth piconet, the *FFD-k*I algorithm achieved significantly better throughput and channel utilization performances over LARR and RR. Meanwhile, the *FFD-k*I algorithm has a similar fairness performance to LARR.

References

1. Bluetooth Special Interest Group: Specification of the Bluetooth System 1.2, (2003)
2. M. Kalia, D. Bansal and R. Shorey: Data Scheduling and SAR for Bluetooth MAC, in Proc. of IEEE VTC, (2000), 716-720
3. Chun-Chuan Yang and Chin-Fu Liu: A bandwidth-based polling scheme for QoS support in Bluetooth, Computer Communications, (2004)
4. Daqing Yang, G. Nair and B. Sivaramakrishnan: Round Robin with Look Ahead: A New Scheduling Algorithm for Bluetooth, Proceedings of the International Conference on Parallel Processing Workshops, (2002), 45-50
5. E.G. Coffman, Jr., M.R. Garey and D.S. Johnson: Bin Packing Approximation Algorithms: A Survey, In D. Hochbaum (ed.), PWS Publishing, Boston. Approximation Algorithms for NP-Hard Problems, (1996), 46-93
6. M.R. Garey and D.S. Johnson: Computers and Intractability: A Guide to the Theory of NP-Completeness, W.H. Freeman and Co., San Francisco, (1979)
7. Guochuan Zhang: A new version of on-line variable-sized bin packing, Disc. Appl. Math., (1997), 193-197
8. E.G. Coffman, Jr., J. Csirik and G. Woeginger: Bin Packing Theory, Handbook of Applied Optimization, P. Pardalos and M. Resende, Eds., Oxford University Press, New York, (2002)
9. Junfeng Xiao, Biao Ren and Shiduan Cheng: Bin packing problem with *k*-preview items. Technical report, (2005)

Appendix A: Proof of Lemma 3

We will prove this lemma in two aspects.

1. $n_2 \geq n_3$

1.1 $n_1 \geq n_4 + (n_2 - n_3)/2$

We assume $n_1 \geq n_5$. When $2 \leq k \leq 5$, the asymptotic worst case performance ratio of *FFD-k*I is

$$R_A^\infty = \lim_{n\to\infty} \frac{A(L)}{OPT(L)} \leq \frac{n_5 + n_4 + n_3 + n_5/k + (n_1 - n_5)/5 + n_2/2}{n_5 + n_4 + n_3 + (n_2 - n_3)/2 + (n_1 - n_4 - (n_2 - n_3)/2)/5} \quad (3)$$

When $k \geq 6$, the asymptotic worst case performance ratio of *FFD-k*I is

$$R_A^\infty = \lim_{n\to\infty} \frac{A(L)}{OPT(L)} \leq \frac{n_5 + n_4 + n_3 + n_1/5 + n_2/2}{n_5 + n_4 + n_3 + (n_2 - n_3)/2 + (n_1 - n_4 - (n_2 - n_3)/2)/5} \quad (4)$$

It's obvious that the right of *ineq (4)* is less than that of (3). We can have larger R_A^∞ with less k. So, k is 2 in the following proof. At the meantime, when $n_1 \leq n_5$ the R_A^∞ is less than the right of *ineq* (3). So we can only consider $n_1 \geq n_5$ in the following proof. It is obtained that

$$R_A^\infty \leq \frac{1.3n_5 + n_4 + n_3 + 0.5n_2 + 0.2n_1}{n_5 + 0.8n_4 + 0.6n_3 + 0.4n_2 + 0.2n_1} \quad (5)$$

Comparing the coefficients of n_5, n_4, n_3, n_2 and n_1, we can find that the right of *ineq (5)* has the maximum when $n_1 = Max[n_4 + 2(n_3 - n_2), n_5]$.

1.1.1 When $n_4 + (n_2 - n_3)/2 \geq n_5$, *ineq (5)* can be

$$R_A^\infty \leq \frac{1.3n_5 + 1.2n_4 + 0.9n_3 + 0.6n_2}{n_5 + n_4 + 0.5n_3 + 0.5n_2} \leq \frac{1.3n_5 + 1.2n_4 + 1.5n_3}{n_5 + n_4 + n_3} \leq 1.5 \quad (6)$$

1.1.2 When $n_4 + (n_2 - n_3)/2 \leq n_5$, *ineq (5)* can be

$$R_A^\infty \leq \frac{1.5n_5 + n_4 + n_3 + 0.5n_2}{1.2n_5 + 0.8n_4 + 0.6n_3 + 0.4n_2} \leq \frac{1.5n_5 + n_4 + 1.5n_3}{1.2n_5 + 0.8n_4 + n_3} \leq 1.5 \quad (7)$$

1.2 $n_1 \leq n_4 + (n_2 - n_3)/2$

The asymptotic worst case performance ratio of *FFD-k*I is

$$R_A^\infty = \lim_{n\to\infty} \frac{A(L)}{OPT(L)} \leq \frac{n_5 + n_4 + n_3 + n_5/k + (n_1 - n_5)/5 + n_2/2}{n_5 + n_4 + n_3 + (n_2 - n_3)/2} \quad (8)$$

When $k=2$, we have

$$R_A^\infty \leq \frac{1.3n_5 + n_4 + n_3 + 0.5n_2 + 0.2n_1}{n_5 + n_4 + 0.5n_3 + 0.5n_2} \leq \frac{1.3n_5 + 1.2n_4 + 0.9n_3 + 0.6n_2}{n_5 + n_4 + 0.5n_3 + 0.5n_2} \quad (9)$$

$$\Rightarrow \quad R_A^\infty \leq \frac{1.3n_5 + 1.2n_4 + 1.5n_3}{n_5 + n_4 + n_3} \leq 1.5 \quad (10)$$

2. $n_2 \leq n_3$

2.1 $n_1 \geq n_4 + 2(n_3 - n_2)$

As the above explanation, when $2 \leq k \leq 5$, the asymptotic worst case performance ratio of *FFD-k*I has a maximum.

$$R_A^\infty = \lim_{n\to\infty} \frac{A(L)}{OPT(L)} \leq \frac{n_5 + n_4 + n_3 + n_5/k + (n_1 - n_5)/5 + n_2/2}{n_5 + n_4 + n_3 + (n_2 - n_3)/2 + (n_1 - n_4 - (n_2 - n_3)/2)/5} \quad (11)$$

When k=2, we have

$$R_A^\infty \le \frac{1.3n_5 + n_4 + n_3 + 0.5n_2 + 0.2n_1}{n_5 + 0.8n_4 + 0.6n_3 + 0.4n_2 + 0.2n_1} \quad (12)$$

Comparing the coefficients of n_5, n_4, n_3, n_2 and n_1, we can find that the right of *ineq (12)* has the maximum when $n_1 = Max[n_4 + 2(n_3 - n_2), n_5]$. So

2.1.1 When $n_5 \ge n_4 + 2(n_3 - n_2)$, *ineq (12)* can be

$$R_A^\infty \le \frac{1.5n_5 + n_4 + n_3 + 0.5n_2}{1.2n_5 + 0.8n_4 + 0.6n_3 + 0.4n_2} \le \frac{1.5n_5 + n_4 + n_3}{1.2n_5 + 0.8n_4 + 0.6n_3} \le \frac{2.5n_5 + 4n_4}{2n_5 + 3n_4} < 1.5 \quad (13)$$

2.1.2 When $n_5 \le n_4 + 2(n_3 - n_2)$, *ineq (12)* can be

$$R_A^\infty \le \frac{1.3n_5 + 1.2n_4 + 1.4n_3 + 0.1n_2}{n_5 + n_4 + n_3} \le \frac{1.3n_5 + 1.2n_4 + 1.5n_3}{n_5 + n_4 + n_3} \le 1.5 \quad (14)$$

2.2 $n_1 \le n_4 + 2(n_3 - n_2)$

As the above explanation, when $2 \le k \le 5$, the asymptotic worst case performance ratio of *FFD-k*I has a maximum.

$$R_A^\infty = \lim_{n\to\infty} \frac{A(L)}{OPT(L)} \le \frac{n_5 + n_4 + n_3 + n_5/k + (n_1 - n_5)/5 + n_2/2}{n_5 + n_4 + n_3} \quad (15)$$

It's obvious that when k=2, above equation has a maximum. So, we have

$$R_A^\infty \le \frac{1.3n_5 + n_4 + n_3 + 0.5n_2 + 0.2n_1}{n_5 + n_4 + n_3} \le \frac{1.3n_5 + 1.2n_4 + 1.4n_3 + 0.1n_2}{n_5 + n_4 + n_3} \quad (16)$$

$$\Rightarrow \quad R_A^\infty \le \frac{1.3n_5 + 1.2n_4 + 1.5n_3}{n_5 + n_4 + n_3} \le 1.5 \quad (17)$$

Comparing *inequations* (6), (7), (10), (13), (14) and (17), we can deduce that the asymptotic worst case performance ratio of the *FFD-k*I algorithm, $R_A^\infty \le 1.5$. Now, we need to find an example to demonstrate that 1.5 can be achieved. Consider the following example with two M-S pair queues. Each of them has n items of size 4 and n items of size 6.

$$\text{M-S 1: } \{\underbrace{6,\ldots,6}_{n}, \underbrace{4,\ldots,4}_{n}\} \qquad \text{M-S 2: } \{\underbrace{6,\ldots,6}_{n}, \underbrace{4,\ldots,4}_{n}\}$$

According to *FFD-k*I, the $A(L)=2n+(n-1)=3n-1$. In this case, we get

$$R_A^\infty = \lim_{n\to\infty}\sup \frac{A(L)}{OPT(L)} = \lim_{n\to\infty} \frac{3n-1}{2n} = 1.5 \quad (18)$$

Therefore, the asymptotic worst case performance ratio of the *FFD-k*I algorithm is $R_A^\infty = 1.5$. ■

On the Problem of Channel Assignment for Multi-NIC Multihop Wireless Networks

Leiming Xu, Yong Xiang, and Meilin Shi

Department of Computer Science and Technology,
Tsinghua University, Beijing, China, 100084
xlming@csnet4.cs.tsinghua.edu.cn

Abstract. Multihop wireless networks in which each node is equipped with multiple wireless NICs can utilize multiple wireless channels to improve performance. But how to assign wireless channels to interfaces to avoid collisions extremely while the network keeps a good topology is a problem not well solved for such a multi-NIC multihop wireless network. We reexamine the general requirements of this problem and propose an algorithm to solve it effectively and efficiently. We regard it as a particular edge coloring problem, and introduce a step of topology simplification to degrade the complexity of the problem. Novel policies are proposed to select edges, group edges and color edges. Some results demonstrate the effectiveness of our algorithms and performance gains of network by channel assignment.

1 Introduction

Multihop wireless network structure is receiving more and more interests, and regarded as a promising technology for next generation networks. However, interference between nodes is a big problem for multihop wireless networks. In the case of single channel, many MAC protocols are developed to solve the "hidden/exposed terminal" problem. But multihop wireless networks still suffer from scalability issues. Capacity and performance will degrade significantly when the size of network increases [1].

Since multiple channels are usually available, for instances, frequency points, channel codes, and hopping patterns, they can be used to increase the capacity. One method to use multiple channels is using only a single transceiver on one node and switching its working channel on the time axis. But the robustness and usability of this method is a problem. Performance of network will degrade significantly due to the "node-offline problem", which is explained in [2].

Another method to use multiple channels is using multiple wireless NICs (network interface cards) on one node and tuning these NICs to different channels. Each NIC has its own MAC layer and PHY layer, and works independently. For example, IEEE 802.11a NIC can select channel from 12 non-overlapped frequency channels. With commercial wireless network interfaces getting cheaper continuously, it is now acceptable to equip multiple wireless NICs on one node. Recently, some researchers

X. Jia, J. Wu, and Y. He (Eds.): MSN 2005, LNCS 3794, pp. 633 – 642, 2005.

have started working on multi-channel protocols using multiple wireless interfaces. For instance, [3] presented a clever policy to bind neighbors to interfaces and then bind channels to interfaces by constructing a tree. But, in scenarios other than an access network, the tree structure is not suitable for general traffic patterns. In addition, tree structure is vulnerable, so, they need complicated steps to handle network partitions. In [4], they divided multiple interfaces on each node into receiving interfaces and sending interfaces. Every receiving interface has a distinct fixed channel, and a node switches its sending interface to the channel of the receiving interface on the other end to do transmission. Although this policy can avoid collisions extremely, it needs a very big number of channels. It is also not efficient to use an interface only sending or receiving. More critically, it is hard to support broadcasting on it, thus, routing protocols can not be supported. And, in both of these two proposals, it is required at least two interfaces equipped on each node.

Although it seems that multi-NIC provide nodes with the ability of utilizing multiple channels, how to assign these channels to interfaces so that channels and interfaces are utilized efficiently and collisions are avoided extremely while the network keeps a good topology, is still a big problem. In this paper, we reexamine this problem and propose an algorithm to solve it effectively and efficiently.

The rest of the paper is organized as follows. Section 2 reviews the requirements of channel assignment. Setion 3 presents the formal description of the problem. In section 4, we analyze the problem and describe our algorithms. Section 5 gives some results to show the effectiveness and performance of channel assignment. Section 6 concludes the paper and gives some discussion about future works.

2 Multi-NIC Wireless Network

Figure 1 illustrates a diagram of such a multi-NIC multihop wireless network. In this network, interfaces on each node are tuned to different channels, and its neighbors are distributed to different interfaces. The network is still connected and interfaces do not need to switch their channels. Which channel is selected for each interface, is determined by a channel assignment algorithm. Each interface communicates independently and a virtual data link layer is required on top of these wireless NICs to make them transparent to upper layers. Thus, from the viewpoint of upper layer, it will perform as if there is only one NIC. Routing protocols and other upper layer protocols can be applied directly.

After analyzing the network diagram and existing methods for channel assignment, we propose some objectives for our algorithm designing as follows.

- The resulting network should have a mesh topology. This is suitable for general applications.
- On each node, different number of interfaces should be supported, even only single interface. And the algorithm should be adaptive to different numbers of total available channels.
- Collisions in the network are decreased extremely.
- Some redundant links can be omitted. Since high density of nodes will cause more collisions, omitting some links moderately will increase the capacity of network.

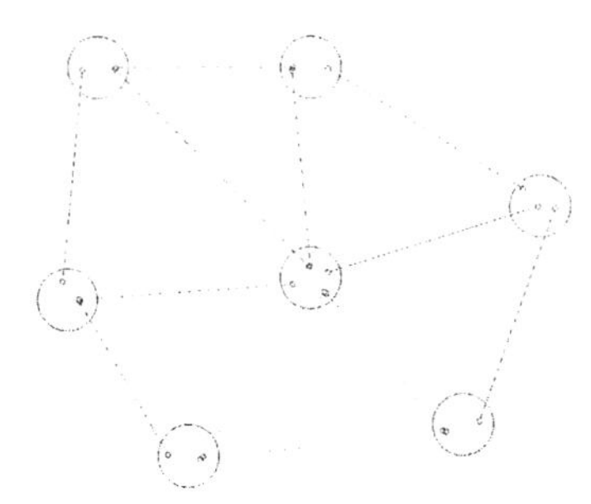

Fig. 1. A multi-NIC wireless network. Each circle represents a wireless node, and dots in each circle represent interfaces equipped on this node. Different line styles represent wireless links on different channels.

3 Problem Statement

In this paper, we model a multihop wireless network as a simple undirected graph, and use some knowledge of graph theory to help us to achieve our algorithm of channel assignment. Concepts of graph theory are used in this paper in their standard way. Readers can refer to [5] for terminologies and notations if they are not explained here. If not specified, all graphs in this paper are simple undirected.

Let $G = (V, E)$ be a simple undirected graph representing a multihop wireless network, where every vertex $v \in V=\{v_1,v_2,\ldots,v_n\}$ represents a wireless node, and every edge $e \in E$ represents that the two corresponding wireless nodes has a bidirectional link between them, i.e., they are in the transmitting range of each other. And, if we represent different channels as different colors, a channel assignment to wireless links can be represented as an edge coloring of the graph. Here, we do not utilize unidirectional links, and we assume that wireless channels are all orthogonal.

However, not every edge coloring corresponds to a feasible channel assignment because the number of interfaces on a node is limited. The number of distinct channels that can be assigned to a wireless node is bounded by the number of interfaces it has. Below, we give the definition of a feasible edge coloring.

Definition 1. Let $K = \{1, 2, \ldots, k\}$ be the set of colors. We call function $C: E' \rightarrow K$ a k-edge-coloring of G on E', where E' is a subset of E and forms a spanning subgraph of G. Then, for every $v \in V$, the set of colors occurred on its incident edges (called its incident colors) is denoted as $C(v)$, and $|C(v)|$ is called its incident color number. Given $f : V \rightarrow \mathbb{N}, f(v) \leq d(v)$, C is called feasible iff.

i) $(\forall v \in V)(|C(v)| \leq f(v))$, and

ii) $(\forall (u,v) \in E \setminus E')(C(u) \cap C(v) = \emptyset)$.

In other words, only those k-edge-colorings that the incident color number of each node v is not greater than its interface number $f(v)$, and edges between those nodes that share common incident colors have to be colored, are actually feasible assignments of k channels to wireless interfaces in the network.

Then, the problem is, what an optimal feasible coloring is like. For simplicity, we can define interference range as a Euclidean distance or a number of hops. Below, we give a simplified definition of interference range and total collision number of a network.

Definition 2. For any two vertices u and v, let $D(u,v)$ denote the distance between them. Then, for any two edges $e_1=(u,u')$ and $e_2=(v,v')$, distance between these two edges is defined as $D(e_1,e_2)=Min(D(u,v),D(u,v'),D(u',v),D(u',v'))$. Given collision distance D_{col}, total collision number of coloring C is defined as

$$Col = \sum_{e\in E} \left|\{e'\in E \mid e' \neq e \wedge D(e,e') < D_{col} \wedge C(e) = C(e')\}\right|$$

So, according to objectives of channel assignment described in section 2, we can give the problem statement as below.

Definition 3. Problem P: Given graph G, maximum incident color numbers $f(v)$, collision distance D_{col}, and k colors, find a feasible k-edge-coloring C on a connected spanning subgraph (denoted as G_{sol}) of G, such that for every vertex v of G, the incident color number of v is maximized and total collision number of coloring C is minimized.

Different from those extensively researched coloring problems, which are generally NP-Complete or harder [6], this particular coloring problem requires incident color numbers to be limited, and thus is much harder to solve. So, we do not seek the exactly optimum solution of the problem. We only try to develop an approximation algorithm to give a suboptimum solution.

4 Theory of Our Algorithms

Actually, the problem P can be divided into three subproblems:

(1) Deleting some redundant edges. First, we must determine which edges should be deleted to reduce collision possibility while keeping the connectivity of the graph.
(2) Neighbor grouping. Since a vertex's neighbors may be still more than its maximum incident color number, we must divide its neighbors into groups. Its incident edges to the neighbors in a same group will share a same color.
(3) Coloring. Then, we assign colors to every edge group of every vertex to give a solution of the problem.

However, each one of these three problems is a combinatorial optimization problem and hard to solve. But if the graph has the characteristic that $(\forall v \in V(G))(d(v) \leq f(v))$, then the subproblem (1) and (2) are solved immediately. So, if we can delete some redundant edges and reduce the graph to a connected spanning subgraph that $(\forall v \in V(G))(d(v) \leq f(v))$, then the problem can be extremely simplified and only subproblem (3) needs to solve. Fortunately, the problem to extract a spanning subgraph with given vertex degrees, which is called an f-factor, has been solved by Tutte[5]. We can use his method to help us solve the subproblem (1) and (2).

In the following parts of this section, first, we describe how to get a connected simplified graph from the original graph based on *f*-factor theory, and then, we develop a heuristic algorithm to solve the subproblem (3). We describe the whole algorithm as three algorithms, each for a subproblem.

4.1 Algorithm 1: *f*-Limit

First, we extend the concept of *f*-factor to a concept of *f*-limited spanning subgraph.

Definition 4. A spanning subgraph *F* of *G* is called *f*-limited if $(\forall v \in V(G))(d_F(v) \le f(v))$. For such an *F*, we say its size is the number of its edges, and say that a vertex *v* is filled or unfilled by *F* according as $d_F(v)$ is equal to or less than $f(v)$. *F* with the largest size is called the maximum *f*-limited spanning subgraph (called the *f*-limit for short). Particularly, an *f*-limit with all vertices filled, is called an *f*-factor of *G*.

In order to get an *f*-limit of *G*, we can construct a related graph *H* and transform the *f*-limit problem of *G* to the matching problem of *H*, which has been solved previously.

Definition 5. The related graph *H* of *G* is constructed as follows: Let $e(v) = d(v) - f(v) \ge 0$ be the "excess" degree at *v*. Replace each vertex *v* by a complete bipartite graph $K_{d(v),e(v)}$, with partite sets $A(v)$ of size $d(v)$ and $B(v)$ of size $e(v)$. For each edge (u,v) in *G*, join one vertex of $A(u)$ to one vertex of $A(v)$ so that each *A*-vertex participates in one such edge.

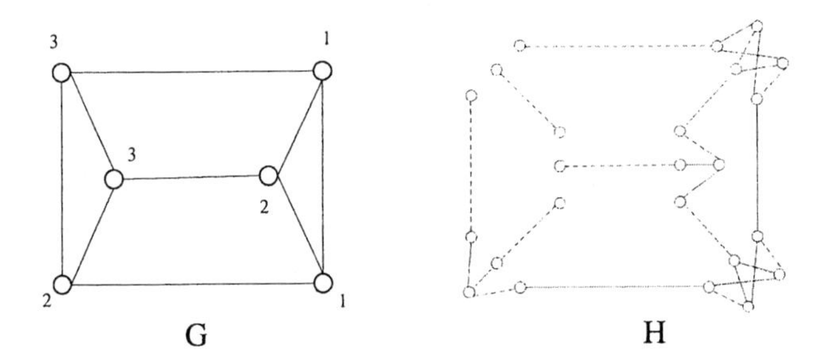

Fig. 2. Example for construction of the related graph H. Numbers beside vertices are $f(v)$. Dashed edges in *H* form a perfect matching of *H* that translates back into an *f*-factor of *G*.

Figure 2 presents an example of *G* and *H*. As illustrated in figure 2, *f*-limited spanning subgraphs of *G* can be translated from and into matchings of the related graph *H*. It is easy to see that for any matching of *H*, we can always augment or substitute some edges to make all *B*-vertices saturated. So, the following two lemmas are immediately proved.

Lemma 1. An f-limited spanning subgraph of size s in graph G corresponds to a matching of size $s+\sum_{v\in V(G)} e(v)$ in H such that all B-vertices are saturated. ∎

Lemma 2. There exists a maximum matching of H such that all B-vertices are saturated. ∎

Then, we can get the following theorem.

Theorem 1. A maximum matching of H with all B-vertices saturated corresponds to an f-limit of G.

Proof. Denote the given maximum matching as M. For each $A(v)$ in H, since $e(v)$ vertices in it are matched by vertices in $B(v)$, there are at most $d(v)$-$e(v)$=$f(v)$ vertices matched by other A-vertices. Hence M can be translated into an f-limited spanning subgraph F of G. Then, we prove that F is the maximum. Suppose the size of F is s and the size of M is $s+\sum_{v\in V(G)} e(v)$. If F is not maximum, and there is another f-limited spanning subgraph F' of size $s'>s$, then according to lemma 1, there is a matching M' in H of size $s+\sum_{v\in V(G)} e(v)$. It is contrary to the presumption that M is maximum. ∎

Theorem 1 described the relationship between the f-limit of G and the maximum matching of H. At the same time, it presents a method to obtain the f-limit. Since some efficient algorithms to obtaining maximum matching of a graph have been well discussed in literatures [8], we can immediately construct a f-limit of G for any f that $f(v)\le d(v)$.

4.2 Algorithm 2: Connecting and Grouping

At this time, the f-limit F may not be connected. We must add some edges to make the subgraph to be connected. Let J={J_1,J_2,…,J_s} be the connected components of F. We only need to add edges between these components to connect them together.

But after adding these augmenting edges, degrees of some vertices will probably exceed the limitation of f. Thus, at these vertices, the new added edges will have to share same colors with some existing edges if we color the edges of the f-limit priorly. So, at this step, we group a new added edge with some of its adjacent edges that already exist when it is selected.

At first, every edge of F is a group. Let $X=\{\{u,v\}|\ (u,v)\in E(F)\}$ represent initial groups. While a new edge (v_1,v_2) is added, we select two groups, $x(v_1)$ and $x(v_2)$, from X containing v_1 and v_2 respectively, and unite them together, i.e., $X = X\setminus\{x(v_1), x(v_2)\}\cup\{x(v_1)\cup x(v_2)\}$. In order to minimize potential collisions introduced by edge grouping, we always select the edge having the form $[J_i, J_j](i\neq j)$ such that $x(v_1)\cup x(v_2)$ is minimal.

After adding all augmenting edges, suppose X={x_1,x_2,…,x_t}, where every x_i is a vertex set. Then, we get a connected spanning subgraph of G, $G_{sol}=\bigcup G[x_i]$, and an edge grouping of G_{sol}, {$E(G[x_i])$|i=1,2,…,t}. It is obvious that, for any vertex, its occurrence time in X does not change when uniting two groups together. Since the original X represents the f-limit of G, we can immediately get the following theorem.

Theorem 2. For any v in $V(G)$, its occurrence time in X does not exceed $f(v)$.

4.3 Algorithm 3: Coloring

In this section, we develop a heuristic algorithm to color these edge groups using given k colors. Our method is transforming this edge coloring problem into a vertex coloring problem. First, we develop a concept of super line graph.

Definition 6. Given a sequence of vertex sets X, which represents an edge grouping of graph G, the super line graph G' is constructed as follows: for every vertex set x in X, put a vertex $v(x)$ in G', and if for any two sets x_1, x_2 in X, there exist some edge of $G[x_1]$ and some edge of $G[x_2]$ such that distance between them is less than D_{col}, then join $v(x_1)$ and $v(x_2)$ with an edge in G'.

Then, obviously, a vertex coloring of the super line graph G' can be translated into an edge coloring of G. Furthermore, according to theorem 2, the translated edge coloring is feasible. And, according to the definition of the super line graph, collisions between edge groups in a edge coloring for G (with edge grouping X) is translated into pairs of adjacent vertices with a same color (called colored improperly) in a vertex coloring for G'. So, it is generally true that the more pairs of vertices colored improperly exist in the vertex coloring of G', the more collisions in the edge coloring translated. Thus, the objective of problem P can be stated as: color vertices of G' using k colors, to make the total number of vertex pairs colored improperly minimized.

Although this problem has also not been well discussed in literatures, there have been many researches about the approximation algorithms for its related chromatic number problem, i.e., what is the minimal k such that all vertices of graph G can be colored properly. In [7], Brelaz presented a heuristic algorithm for the chromatic number problem. Here we present a similar policy to color vertices of the super line graph using k colors:

1) First, color the vertex of largest degree with color 1;
2) Then, repeatedly select the vertex with the highest color degree and color it with the smallest one of possible colors that makes the introduced collisions minimized. Color degree of a vertex is the number of different colors that have already been used on its adjacent vertices.

After translating the vertex coloring of G' into an edge coloring of G, we get a suboptimum solution of problem P. It is easy to see that algorithm 1 is the critical step in the whole algorithm. In [8], Gabow presented a efficient algorithm to obtain a maximum matching, which have the complexity of $O(n^3)$. So, we can estimate that the computational complexity of the whole algorithm does not exceed $O(|E(G)|^3)$ level.

5 Results and Analysis

We have implemented these algorithms and combined them with ns-2 simulator [9]. In this section, we show some results to demonstrate the effectiveness of these algorithms and study the performance gains of multi-channel assignment via simulations.

5.1 Effectiveness of Algorithm 1 and 2

First, we show some examples of the *f*-limit and edge grouping. Suppose a completely connected wireless network with 8 nodes as figure 3 shows. The interface number of each node is labeled by the side. We use algorithm 1 and 2 to delete redundant links and give a neighbor grouping. The results are showed in figure 3(a) and 3(b). It should be pointed out that the algorithms do not require every node to have multiple interfaces and nodes' interface numbers can be very different.

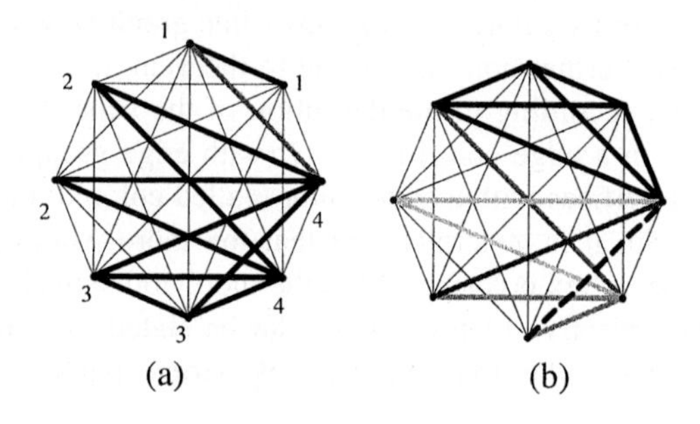

Fig. 3. Results of algorithm 1 and 2. (a) shows the f-limit (bold edges) and a augmenting edge (red edge) to make it connected. (b) shows an edge grouping, where different line styles and line colors represent different edge groups.

5.2 Effectiveness of Algorithm 3

Figure 4 shows a wireless network of 8*8 nodes that each of them has 3 interfaces. Figure 4(a) is its topology in the case of single channel. Algorithm 1 gives a 3-factor of it illustrated in figure 4(b), which is already connected.

Presume that two transmission can not exist simultaneously within any two hops range, i.e., regard number of hops as the distance function and D_{col}=2. By algorithm 3, we color these selected edges using different number of colors. Table 1 shows total collisions and edges exist in the result graph when k is from 1 to 14. We can see from this table that collisions are decreasing and less new edges are introduced by coloring while the color number increases. And, when color number reaches 14 (the result graph is showed in figure 4(c)), no new edges are introduced by coloring and no collision exists in the graph. This result is equal to the result of Brelaz's algorithm [7] computing the chromatic number of the super line graph.

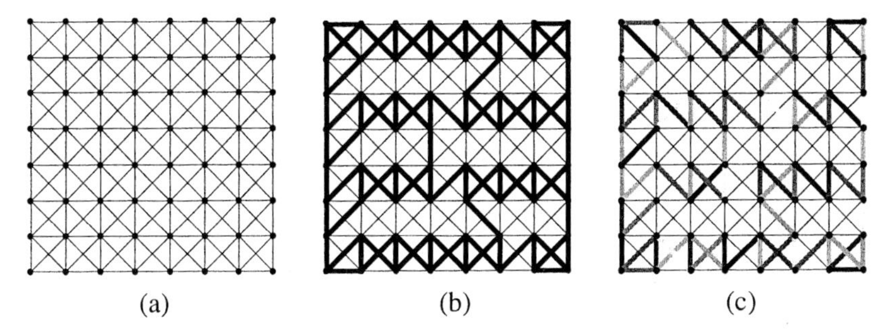

Fig. 4. 8*8 grid network and its results. (a) Topology. (b) *f*-limit. (c) Coloring of *k*=14.

Table 1. Total collisions and edges in result graphs

k	1	2	3	4	5	6	7	8	9	10	11	12	13	14
Col	10794	3018	1754	1124	642	420	236	162	114	84	34	20	2	0
|E|	210	166	161	150	133	126	113	107	106	104	97	98	96	96

5.3 Performance Evaluation Using ns-2

We extended ns-2 to support different number of wireless interfaces on each node. We simulated the network described in the previous subsection, and translated results of coloring with different *k* into channel assignments in the simulation. In the simulation model, Mac layer is 802.11, and Physical layer has been configured so that the collision range is of 2 hops. Other settings are default values in ns-2.

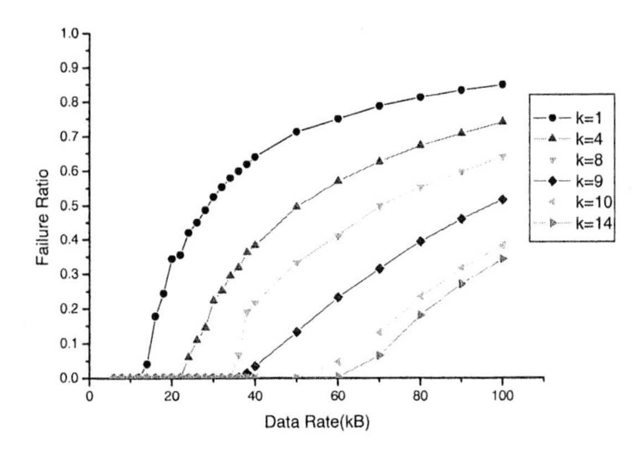

Fig. 5. Ratio of failed delivery when different *k* (number of channels) is used

To see the performance of channel assignments, we introduce a CBR data stream into the node at one corner of the square and the traffic floods throughout the network. Every node simply copies and forwards packets to each of his interfaces when he received nonduplicated packets. We monitor packets received at the other corner node on the diagonal. Changing data rate of the source and number of channels, we check the ratio of packets that can not be successfully delivered to the monitored node from the source node. Figure 5 shows some results.

From figure 5, we can see that, by distributing tranmissions onto different channels, our algorithm can effectively improve the network performance when more channels are used.

6 Conclusions and Future Works

Our algorithm presented a general policy to solve this kind of channel assignment problems for multi-NIC multihop wireless networks: First simplify the network and group the edges, then assign colors to edge groups. Simulation results have showed that this policy is feasible and effective. As the first step, this paper only proposes a static algorithm. For the future works, dynamic algorithms and online algorithms should be studied. And, results from areas of graph theory and topology control can be used to improve this algorithm.

References

1. S. Xu and T. Saadawi, "Does the IEEE 802.11 MAC protocol work well in multihop ad hoc networks," IEEE Commun. Mag., June 2001.
2. Leiming XU, Bo PANG, Yong XIANG and Meilin SHI, "A Novel Multi-channel MAC Protocol with Online-Prediction for Wireless Ad-hoc Networks", in the Proceedings of International Conference on Computer, Communication and Control Technologies (CCCT), Orlando, Florida, July 2003
3. A. Raniwala, T. Chiueh, "Architecture and Algorithms for an IEEE 802.11-based Multi-channel Wireless Mesh Network," in proceedings of IEEE Infocom 2005.
4. Pradeep Kyasanur and Nitin Vaidya, "Routing and interface assignment in multi-channel multi-interface wireless networks," in IEEE WCNC, 2005.
5. Tutte, W. T., Graph theory, Menlo Park, Calif. : Addison-Wesley Pub. Co., Advanced Book Program, 1984.
6. T. R. Jensen, B. Toft; "Graph Coloring Problems"; Wiley Interscience, New York, '95.
7. Brelaz, D. "New Methods to Color the Vertices of a Graph." Comm. ACM 22, 251-256, 1979.
8. H. Gabow, "An efficient implementation of Edmonds' algorithm for maximum weight matching on graphs," Tech. Rept. CU-CS-075-75, University of Colorado, Boulder, Colorado, 1975.
9. The Network Simulator: NS-2. http://www.isi.edu/nsnam/ns.

Validity of Predicting Connectivity in Wireless Ad Hoc Networks

Henry Larkin, Zheng da Wu, and Warren Toomey

Faculty of IT, Bond University, Australia
{hlarkin, zwu, wtoomey}@staff.bond.edu.au

Abstract. Wireless ad hoc networks face many challenges in routing, power management, and basic connectivity. The ability to predict the future communication connectivity between two or more autonomous wireless nodes is greatly beneficial to wireless ad hoc networks. Existing research has looked into using predicted node movement as a means to improve connectivity. While past research has focused on assuming wireless signals propagate in clear free loss space, our previous research has focused on using signal loss maps to improve predictions. This paper presents novel testing of predicting connectivity based on our previous concept designs. We test the performance of predicting future node connectivity by analysing various test cases and detailing results produced using custom simulator tools we have created.

Keywords: Future neighbours, predicting connectivity, wireless ad hoc networks.

1 Introduction

Many routing algorithms have already been designed to date to efficiently implement routing in wireless ad hoc networks ([13], [14], [15], [16], [17], [7], [18], etc). These algorithms have been designed to operate in various scenarios and take into consideration various information sources. However, none of these algorithms nor any wireless routing prediction algorithms found during this research utilise future node mobility based on a signal loss map. This information has the potential to improve connectivity in a number of routing protocols.

Signal loss maps represent the logical signal propagation topology over a physical area. They describe how signals are likely to propagate in various directions over various distances. Due to the constantly changing nature of the wireless environment, a perfect signal loss map is not possible to create with current technologies. However, various estimates may be developed to provide, with appropriate safety margins, predictions on whether two nodes at two locations will have connectivity in the future.

Our previous work [3] designed a novel signal loss map solution, dubbed the Communication Map (CM), which is tailored to be built in real time using only wireless ad hoc nodes. The map is created using signal strength information provided with each packet as it is received from any node. To provide a physical reference system, some form of location-providing device is required for each node. In this

X. Jia, J. Wu, and Y. He (Eds.): MSN 2005, LNCS 3794, pp. 643 – 653, 2005.

research, a system such as GPS [4] is assumed to be available to provide the coordinates of each node. From these two external sources the CM is constructed.

The CM is made up of cells, defined areas which are square in shape and represent an average signal loss modifier. The signal loss modifier is a value which represents how a signal's loss increases over distance, relative to free space loss. This approach of using cells describes to users of the CM the same information that vendors of wireless cards use to describe range and signal strength capabilities. Vendors of wireless cards often include the maximum range and signal strength of their product in a variety of general scenarios, for example outdoors, home environment, cluttered office, etc. In a similar fashion, the CM of this research generates such scenarios in real time, and delineates where on a map such areas exist.

Each cell represents the average signal loss of signals passing through that cell. The value stored for each cell is the modifier that a signal applies to the *physical distance* of a signal as it passes through that cell, which when multiplied together with the physical distance creates a *logical distance*. The minimum modifier value is 1.0, in other words a *logical distance* is identical to the *physical distance*, and thus represents perfect free-space loss. The modifier of each cell is used to extend the distance of a signal to the distance it would need to travel in perfect free space to achieve the same loss.

For an example of this process, consider Fig. 1 below. A signal will travel from Node A to Node B over the given CM. The direct line between the two nodes is formed, and a list of cells over which the signal will pass is created. Each of these cells multiply their modifier by the *physical distance* that the signal travels through their cell. The resulting *logical distance* can be used to find the signal loss, which can consequentially determine whether two nodes are predicted to be neighbours. This formula thus uses the modifier of each cell to *extend* the distance of a signal to the distance it would need to travel in perfect free space to achieve the same loss.

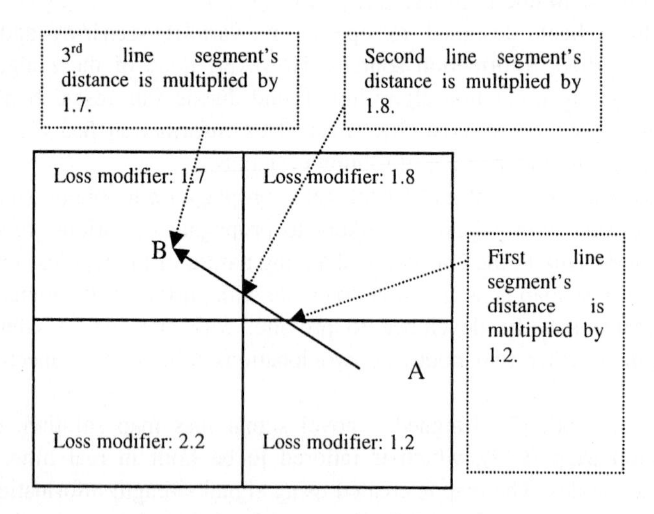

Fig. 1. Signal divided over multiple cells

2 Future Neighbours Table

Existing routing protocols have already been designed and tested to show their overall effectiveness. In order to improve existing routing protocols as effortlessly as possible, our research aimed to provide simple access to prediction information. More of our previous work [1] presented the Future Neighbours Table (FNT). The FNT concept provides a *table of future neighbours*. This table lists the times when nodes in the network will become neighbours with each node, and how long they will remain neighbours for. This solution can be immediately applied to existing routing protocols to improve connectivity in wireless ad hoc networks.

Time	Neighbours
0:00	true
1:42	false

Fig. 2. Example Future Neighbours Table Between Two Nodes

The FNT is based on the idea that in both autonomous wireless networks (e.g. search and rescue robots) and in human-controlled networks (e.g. a human riding a train) that future node movements may be available to routing protocols. This information can be combined with a signal loss map (such as the CM) to predict future node connectivity. The FNT is created by finding the points in time where a change in the *average signal strength* occurs between predicted future node movements. In the example below (Fig. 3), nodes A and B start with some value of signal loss between them. Because they are in a cell with an average signal loss modifier applied to distance, if the nodes move further apart at a linear rate, the signal loss will also increase at a linear rate. By finding the point in time where this signal loss is at a maximum, with regards to this cell alone, it is trivial to determine if both nodes will be communication neighbours during that time, or at what point in time they would become or lose the quality of being neighbours.

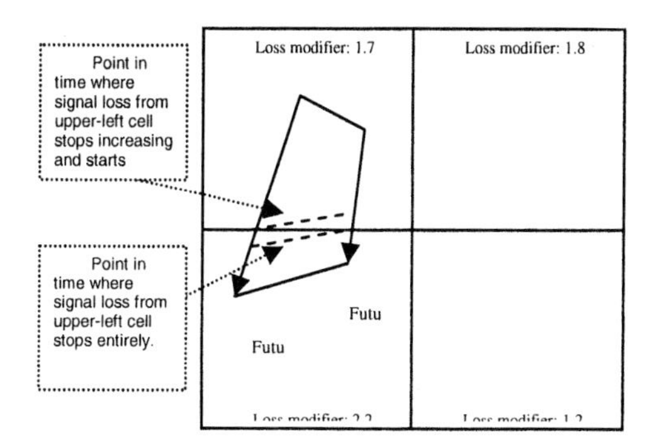

Fig. 3. Finding Points in Time

From this point on until a further point is reached, the signal loss will decrease linearly. As another cell's signal loss average comes into play, its starting minimum and finishing maximum signal loss can also be calculated, and added to the equation to produce a list of times and maximum / minimum signal losses. From this information it is computationally trivial to determine when nodes will and won't be neighbours over time, based on the maximum acceptable signal loss for all nodes.

3 Test Architecture

Existing simulators [9][10][11][12] already exist for basic network simulation. While many of these simulators are extensible, none specifically address the issues of predicting connectivity and signal loss map testing. Because of this, a custom simulator was created. The simulator was created in response to fulfil the need for an appropriate testing bed for wireless ad hoc protocols that relied specifically on prediction and signal loss maps. The design goals of this simulator are primarily to:

- Visually create networking environments.
- Run simulations in real or simulated time.
- Visually inspect the accuracy of the network and its performance.
- Obtain a variety of statistics with ease.
- Save and load simulations for later re-use.

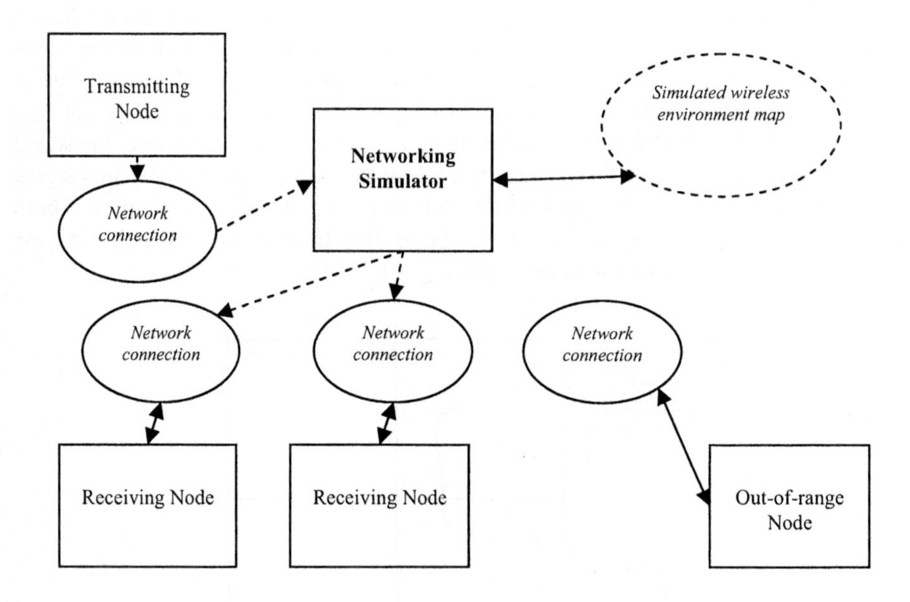

Fig. 4. Example Packet Broadcast

The simulator routes packets between nodes based on a simulated wireless environment map, which details user-created scenarios of how signals will propagate over real-world simulated objects. As each packet is successfully transmitted or lost (based on signal propagation), the simulator records the signal loss values, along with the predicted signal loss based on the FNT. Fig. 4 below shows how a packet would be routed from a transmitting node to nodes within range, based on the simulated wireless environment map for the given scenario.

4 Scenarios

Several scenarios have been created to validate the concept of the FNT, while also identifying its weaknesses. The field of wireless communication has an unlimited number of practical scenarios. This presents a significant challenge in representing a broad spectrum of possible scenarios to gauge the overall effectiveness of the FNT's ability to predict signal loss.

A total of five scenarios have been designed to both represent realistic environments and test the various aspects of the algorithms presented in this paper and in our previous work [3]. Some scenarios have been tested with a varying number of nodes to further analyse scenarios while also gaining an insight into the affects of network population. Two scenarios are explored in detail, Scenario 3 and Scenario 5. These scenarios have been designed to specifically test the FNT in artificial (Scenario 3) and realistic (Scenario 5) circumstances. Few nodes were used due to hardware requirements.

4.1 Scenario 3

Scenario 3 was designed to gain an insight into prediction algorithms. Eight nodes are placed within or outside a simulated area with a very high signal loss modifier of 8.0. Two of the nodes (C and E) are moving clockwise around the building for the purpose of generating a CM with varying levels of connectivity prediction, followed by node F's later move east across the map. The scenario diagram is included in Fig. 5.

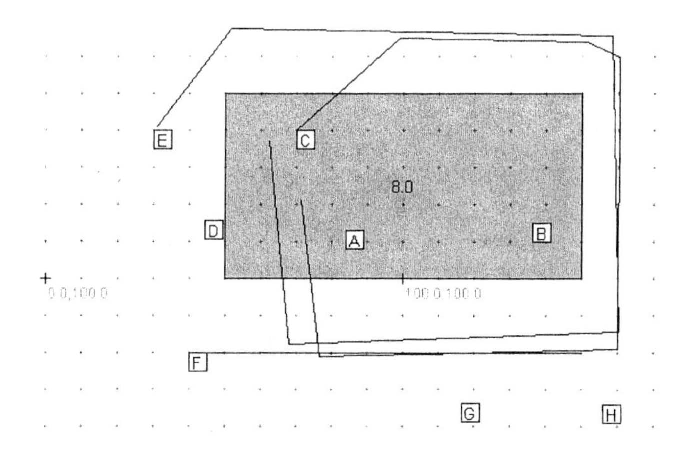

Fig. 5. Scenario 3 Overview

4.2 Scenario 5: Street - Moving

Scenario 5 is based on an actual neighbourhood street. It is based on signal loss readings measured using 802.11b equipment. Six nodes are based around a T-intersection of a street, with eleven houses of importance. Each house has walls having a logical distance of 50 meters, and an interior signal loss modifier of 2.0. Three of the nodes, A, B, and F, travel along the streets. The scenario diagram is included in Fig. 6.

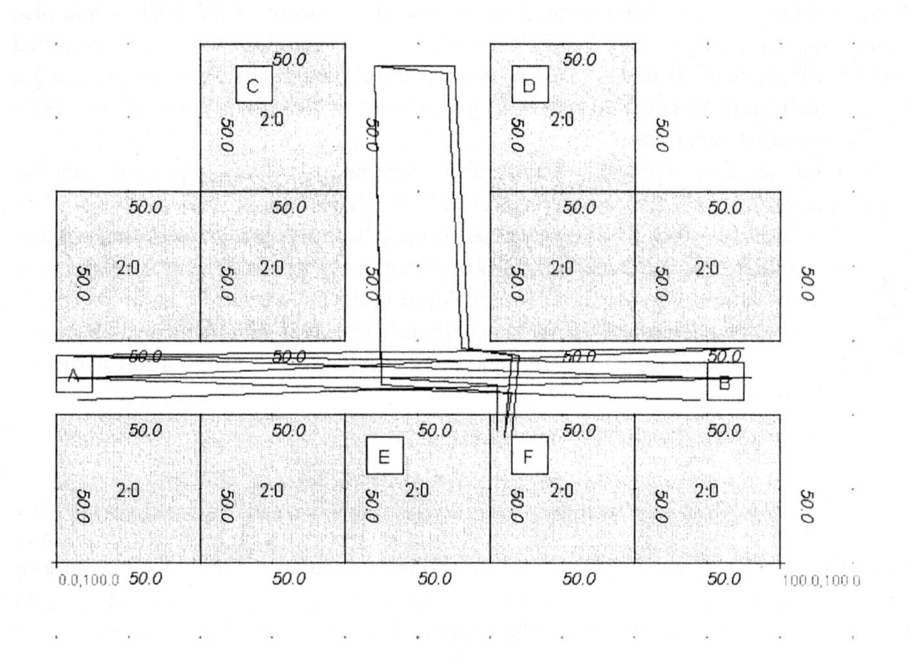

Fig. 6. Scenario 5 Overview

5 Results

Each of the five scenarios were executed for each variation in settings and for each variation in node population where the scenario allowed multiple node populations. The results were then processed and summarised across different combinations of settings so that well-informed analysis could be conducted.

5.1 Scenario 3

Scenario 3 was specifically designed to test the FNT concept. In this example (refer to Fig. 5 earlier) 8 nodes form a network over an area with a large building in the centre. This building has a very high signal loss modifier of 8.0 so that wireless connectivity between nodes will be broken often. The maximum acceptable signal loss for all nodes is set to 85 dBm. Two of the nodes, C and E, circumnavigate the building a single time. This is an important step for the Communication Map (CM) algorithms to map the signal loss appropriately. Due to the high signal loss of the building, connectivity between several nodes is broken often.

After these two nodes have finished their travel, an accurate Communication Map delineating the bounds of the signal loss has been created. The CM created by Nnode A at 600 seconds is shown in Fig. 7. Note that due to node map sharing, all nodes will have identical maps. From 600 seconds, Node F traverses the map from west to east, where connectivity between neighbouring nodes will change. The basis of the FNT concept is in predicting these connectivity changes to assist routing protocols.

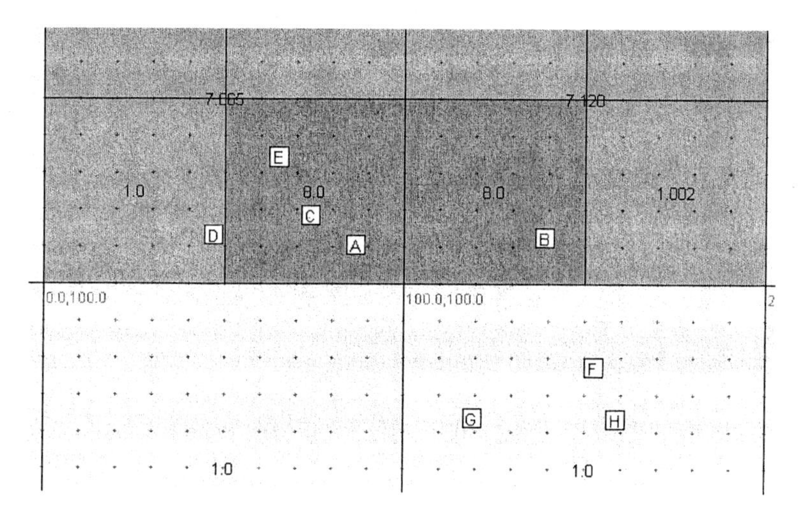

Fig. 7. CM of Scenario 3 after 600 seconds

Currently Neighbours:	false
Current loss:	86.31dBm
Time	**Connectivity Changes**
(12:36) Neighbours:	true
(14:51) Neighbours:	false

Fig. 8. Node A's FNT of Node F

Time	Connectivity
(12:36)	ALIVE: 84.95 dBm
(14:50)	[UNREACHABLE]

Fig. 9. Node A's Neighbour History of Node F

Each node in the network recalculates a FNT for each other node in the network that it could be communication neighbours with. This recalculation is done periodically, at 1 minute intervals in the case of Scenario 3. As the maximum acceptable signal loss is set to 85dBm, the FNT will list predicted changes in network connectivity based on when signal loss is predicted to fall below or rise above this value. Node A resides in the centre of the map, and thus initially has no connectivity with Node F. Node F will later travel and pass close enough to Node A that connectivity will be established, and then lost as Node F moves further away. The FNT generated during this scenario by Node A is shown in Fig. 8. The actual history of connectivity between nodes A and F is listed in **Fig. 9.**, which as can be seen closely matches that of the predicted connections in Node A's FNT.

The overall accuracy of all node predictions based on predicted signal loss (of any value, not only those lower than 85dBm) compared with actual signal loss values is shown in Fig. 10. This accuracy was graphed across three simulations. In each set, three Default Cell Size (DCS) settings are implemented, 25 meters, 50 meters, and 100 meters (the default). This is done to determine how reducing the DCS affects accuracy. DCS is the initial size of each cell when the CM is generated. In theory, the smaller the DCS, the more detail can be represented and thus the more accurately signal loss can be mapped. However, a smaller DCS increases bandwidth during periodic map sharing. With larger DCS settings, signal loss of smaller interferences is averaged into the larger cells. The results from these three experiments show that once the CM has been created with reasonable accuracy (that is, the signal loss has been mapped correctly), the error margin between predicted and actual signal losses is less than 20%. This is true for all DCS settings used. The default DCS value for this scenario produces marginally better results than lower DCS values. The reason behind this is that with the largest DCS value, the large building is mapped in entirety by default, whereas smaller DCS values take time before relevant cells are between nodes in order to map signal loss to them. It is important to note that a signal loss map cannot represent areas unless those areas have been communicated through. In this scenario the smaller DCS values take nodes marginally more time to discover and attribute loss to each cell as it is communicated through.

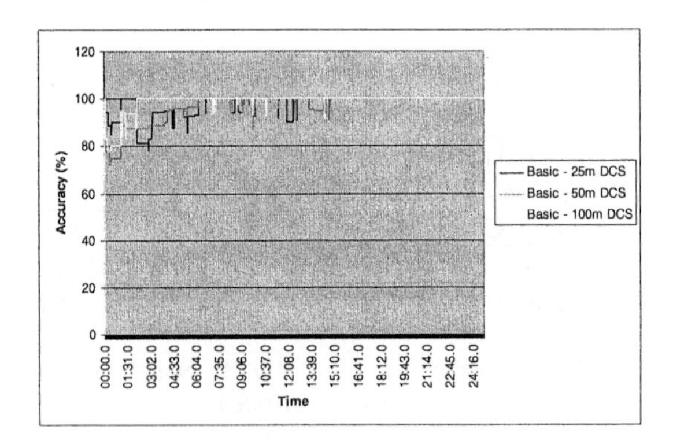

Fig. 10. Prediction Accuracy of Scenario 3

5.2 Scenario 5

Scenario 5 is based on a real world scenario, and thus the results from it hold particular significance. The predicted signal loss compared with the actual signal loss is shown in Fig. 11. Despite the DCS value used, the error in prediction is less than 40% for the majority of the simulation. As stated earlier, it is impossible to create a perfectly accurate signal loss map on which FNT is based. Without a comprehensive grid of well-positioned nodes it is not possible to fully understand and map the signal propagation environment.

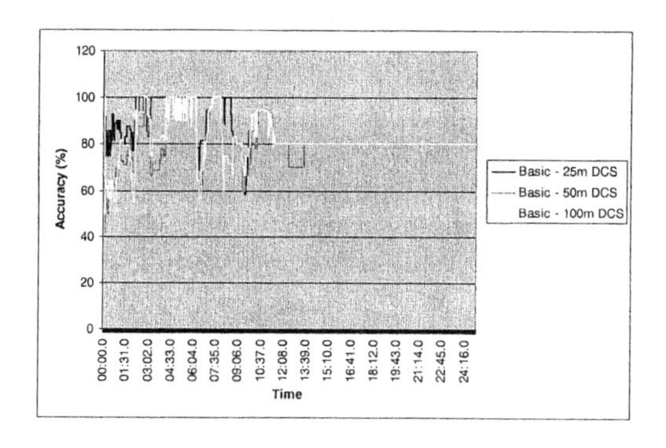

Fig. 11. Prediction Accuracy of Scenario 5

Accuracy across all scenarios for all node counts is shown in Fig. 12. It is seen from these results that overall the average prediction error is less than 20% across scenarios. With appropriate margins in place these predictions can be used by existing routing protocols to aid in connectivity and routing. With maximum "actual" acceptable signal loss at 85dBm, setting the "allowable" acceptable signal loss to a lower value of maybe 83dBm would allow false connectivity predictions to be largely avoided.

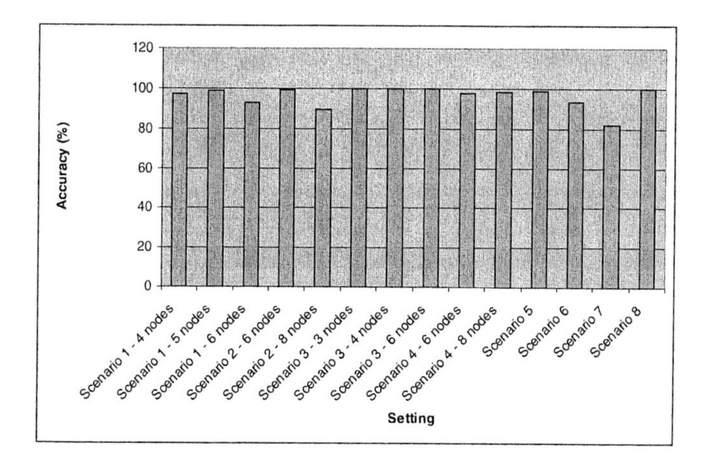

Fig. 12. Prediction Accuracy for all Scenarios

6 Conclusion

Predicting connectivity between nodes based on location information has the potential of significantly improving connectivity in wireless ad hoc networks. The Future Neighbours Table concept we have designed produces information that can be immediately applied to existing routing protocols. This paper has presented accuracy analysis of our solution under a number of custom-created scenarios, as well as the various settings which may be used to improve accuracy. The results of these tests conclude that the FNT can produce accurate results under a variety of situations. Further testing will be conducted after applying the FNT algorithms to existing routing protocols, and performance investigated.

References

1. Larkin, H., Wu, Z., Toomey, W. Predicting Network Topology for Autonomous Wireless Nodes. Proceedings of European Wireless 2005, pp. 413-417, VDE Verlag, Nicosia, Cyprus. 2005.
2. Howard, A., Siddiqi, S., Sukatme, G. S., An Experimental Study of Localization Using Wireless Ethernet, in 4th International Conference on Field and Service Robotics, July 2003.
3. Larkin, H., Wireless Signal Strength Topology Maps in Mobile Adhoc Networks, Embedded and Ubiquitous Computing, International Conference EUC2004, pp. 538-547, Japan. 2004.
4. Enge, P., Misra, P. Special issue on GPS: The Global Positioning System. Proceedings of the IEEE, pages 3–172, January 1999.
5. Shankar, P. M., Introduction to Wireless Systems. Wiley, USA. 2001.
6. Hu, Y., Johnson, D. B., Implicit source routes for on-demand ad hoc network routing, Proceedings of the 2nd ACM international symposium on Mobile ad hoc networking & computing, 2001.
7. Lee, S., Kim, C., Neighbor supporting ad hoc multicast routing protocol, Proceedings of the 1st ACM international symposium on Mobile ad hoc networking & computing, 2000.
8. Youssef, M. A., Agrawala, A., Shankar, A. U., WLAN Location Determination via Clustering and Probability Distributions, IEEE PerCom, 2003.
9. Fall, K. Network Emulation in the Vint/NS Simulator. Proceedings of ISCC'99, Egypt, 1999.
10. McDonald, C.S., A Network Specification Language and Execution Environment for Undergraduate Teaching. Proceedings of the ACM Computer Science Education Technical Symposium '91, San Antonio, pp25-34, Texas, Mar 1991.
11. Unger, B., Arlitt, M., et. al., ATM-TN System Design. WurcNet Inc. Technical Report, September 1994.
12. Keshav, S. REAL: A Network Simulator, tech. report 88/472, University of California, Berkeley, 1988.
13. Perkins, C. E., Bhagwat, P., Highly dynamic Destination-Sequenced Distance-Vector routing (DSDV) for mobile computers. In Proceedings of the SIGCOMM '94 Conference on Communications Architectures, Protocols and Applications, pp. 234-244, August 1994. London, UK.

14. Johnson, D. B., Maltz, D. A., Dynamic Source Routing in Ad-Hoc Wireless Networks. In Mobile Computing, pp. 153-181, 1996.
15. Park, V. D., Corson, M. S., A highly adaptive distributed routing algorithm for mobile wireless networks. In Proceedings of INFOCOM'97, pages 1405-1413, April 1997. University of Maryland, USA.
16. Ko, Y-B., Vaidya, N. H., Location-Aided Routing (LAR) in Mobile Ad Hoc Networks. Texas A&M University. College Station, TX, 1998.
17. Perkins, C., E., Royer, E., M., Ad-Hoc On-Demand Distance Vector Routing. Proceedings of 2nd IEEE Workshop on Mobile Computing Systems and Applications, February 1999.
18. Hu, Y., Johnson, D. B., Implicit source routes for on-demand ad hoc network routing. Proceedings of the 2nd ACM international symposium on Mobile ad hoc networking & computing, 2001.

A Novel Environment-Aware Mobility Model for Mobile Ad Hoc Networks

Gang Lu, Demetrios Belis, and Gordon Manson

Department of Computer Science, University of Sheffield,
Regent Court, 211 Portobello Street,
Sheffield, S1 4DP, United Kingdom
{ganglu, dimi, g.manson}@dcs.shef.ac.uk

Abstract. Simulation is the most important and widely used method in the research of Mobile Ad hoc NETworks (MANET). The topology of MANET and the mobility of mobile nodes are the key factors that have impact on the performance of protocols. However, most of the existing works are based on random movement, and the fact that the network topology is highly related to the environment of MANET is overlooked. In this paper, we propose a novel Environment-Aware Mobility (EAM) model which models a more realistic movement of mobile nodes. Environment objects such as Route and Hotspot, etc are introduced to represent the environment components by Scalable Vector Graphics (SVG). This model is considered to be a complex model with a combination of existing conventional mobility models and network environments. The results show that the intrinsic characteristics and properties of the environments have a significant influence on the performance of MANET protocols.

1 Introduction

Without any central administration, MANET is considered to be the dominating form of wireless network and has drawn more and more attraction from the industry. Using self-organised mobile nodes, MANET can be deployed in any environment, such as a gallery, a theatre, a shopping mall and even a street. Many researchers have shown interest in the fields of MANET and all sorts of protocols aiming at different issues have been proposed in recent years. The performance of these protocols needs to be carefully evaluated before they go to the commercial market. Therefore, network simulation plays a very important role in this area, and is widely used by researchers as a key method to better understand the overall performance of MANET.

Self-organization is the essential characteristics of this mobile wireless network. The mobile nodes within the network can be any moving or fix-position objects equipped with antennas. They can be either humans moving in the mall or gallery, or insects and animals in nature environments. The movement behaviors are highly dependent on their own mobility factors and the areas they are located. The mobility model is therefore regarded as an important component in ad hoc network simulation.

X. Jia, J. Wu, and Y. He (Eds.): MSN 2005, LNCS 3794, pp. 654–665, 2005.

In this paper, we propose a novel realistic model which is incorporated with environments where a MANET is deployed. Environments are presented by introducing some environment objects, such as a Route, a Junction, an Accessible Area, and a Hotspot etc. Signal-blocking issues are also covered in this model by introducing Closed Areas whose boundaries are considered to be obstacles with no radio penetration. In addition, with the use of Scalable Vector Graphics (SVG) [5], environments can be easily designed and the movements in the environment can be observed visually.

The remainder of this paper is organised as follows. Section 2 details some related work in the research of mobility models. Section 3 describes the detailed information about how the environment-aware mobility model works. The evaluation of this model is presented in Section 4, and the last section presents the conclusions.

2 Related Work

There exists many mobility models proposed for mobile wireless networks. Mobility models are used to describe the movement pattern of mobile nodes. In this section, some of the popular models used in MANET are briefly described. The relevancy between EAM and those models is clarified afterwards.

The category of conventional models is illustrated in Fig. 1. Entity Models are used to model the movement behaviour of an individual mobile node. The Random Waypoint model [1] is widely implemented by network simulators such as ns2 [4]. In each movement epoch, the mobile node picks a position within the simulation area and moves towards it with a speed distributed in the range $[v_{min}, v_{max}]$. Instead of moving to the next destination immediately as soon as it arrives at the current destination, the mobile node pauses for a specified time then repeats the procedure. The major drawback is that the nodes tend to move around the centre area so that they are not really distributed into the entire simulation area. The Random Direction [1] model is proposed to avoid this distribution problem. It forces the mobile node to move until it reaches the border of area before it starts its next epoch.

The models introduced previously are all considered to be memory-less models since the next movement segment has no dependency on the previous movement

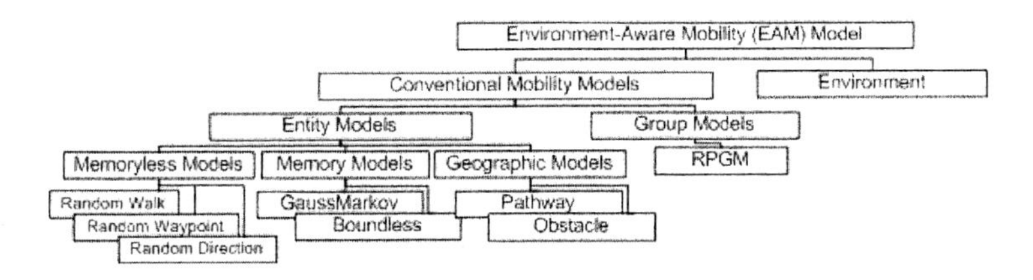

Fig. 1. Category of Mobility Models

regarding either speed or direction. This memory-less feature causes frequent sharp change in speed and direction of movement which is obviously not applicable on a realistic world. Some memory models are proposed. The Gauss-Markov model was first introduced in [6]. In this model, the velocity of the mobile node is assumed to be correlated over time and modelled as a Gauss-Markov stochastic process. The Boundless model [1] is another example of a memory model but with no geographical restriction. Mobile nodes are allowed to cross the boundaries and appear at the other side the area. The resulting effect is that the simulation area is modelled as a torus instead of a flat area.

In ad hoc networks, there are many situations where the movements of mobile nodes have some correlation with each other, i.e. mobile nodes have some group behaviour in common. To present this characteristic, Group Models have been proposed. One typical example of them is the Reference Point Group Mobility model (RPGM) [3]. In RPGM, a logical centre of a group is defined and its movement is used to direct the group-wide movement. Individual members of the group move not purely on a random basis, their movements are also effected by the group movement. RPGM is popularly used in research to depict some scenarios with group behaviours such as avalanche rescue. Other group models can also be found in [1].

The previous models all assume that the simulation area is a free space area where mobile nodes can move anywhere inside. They expose the self-organisation feature of mobile nodes, but they are not applicable to the situation geographic factors have to be considered. The Pathway models [7] and the Obstacle models [8] partially overcome this disadvantage. The Pathway Model forces each mobile node to move along the shortest path towards its destination. Similar behaviour is also modelled in the Freeway mobility model and in the Manhattan mobility model in [9]. The Obstacle mobility model was first introduced in [12]. Unlike the Pathway model, the Obstacle model defines some obstacles in simulation area. These obstacles have a signal- blocking effect on the communication of mobile nodes. This model also allows the nodes to change this movement trajectory when obstacles are encountered. Geographic models support more realistic scenarios than Entity and Group models.

Several works have brought the idea of real-world simulation into MANET simulation. The realistic environment in which the network exists needs to be constructed before running the simulation. [10] provides a way to achieve this by using Auto-CAD, and the realistic world is constructed by implementing the knowledge of the Voronoi diagram in [16]. The movements of mobile nodes are restricted by the pre-defined arbitrary obstacles and pathways. However, this real-world environment model does not provide a flexible approach to model the mobility variations. The movement trajectory of a mobile node might change over time. For example, the mobile nodes move along the pathway then may enter a particular terrain, such as a park where their movements become less deterministic, therefore they might walk in any direction, i.e. the pathway restraints are no longer applied.

It can be concluded that every model mentioned above has some abilities to model the mobility of mobile node in MANET. They can be applied to particular situations, but none of them is flexible and suitable enough for modelling more realistic scenarios. As a fact, Entity Models and Group Models can co-exist in some scenarios with obstacles. The Obstacle model seems to give a good solution for signal-blocking but it leaves entity and group mobility properties unconcerned. Obstacles and Paths are detected but the trajectories to deal with obstacles are simple and not general. The fact that environments factors can effect the movement is also overlooked by all of them. By taking the environment into account as illustrated in Fig. 1, the EAM model proposed in this paper is aimed to provide a more general approach to model more realistic ad hoc networks.

3 Environment-Aware Mobility Model

The EAM model proposed in this paper is designed to model the movement behaviour of mobile nodes in the environments of realistic ad hoc networks. By studying the possible environment where MANET is located, different sub-areas within the entire simulation area are abstracted to several environment objects, such as a Route, Junction, Hotspot, etc. The movement trajectory of the mobile node is correlated with the sub-area that it is located and is allowed to be changed during the simulation. The obstruction of radio propagation is also implemented in this model, and some of the conventional models are modified to take into account the environmental effects and to cope with the obstacles.

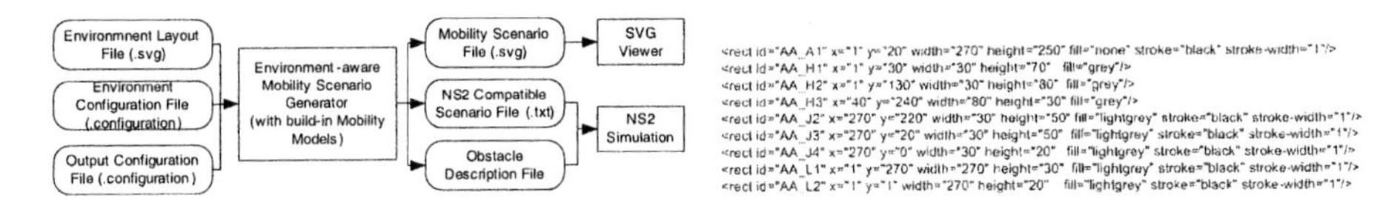

Fig. 2. Work flow for generating the Environment-Aware Model

Fig. 3. An example snippet of an Environment Layout File

In our proposal, the environment including all the environment objects is designed with SVG. The relationships between sub-areas and various properties of each sub-area are given by a text-formatted configuration file. A SVG file is one of the outputs after the simulation done. This animation-supported SVG file can be very useful for researchers to explicitly observe the movements in the simulated scenario. Some details shown by this SVG file can be configured by users by modifying the text-formatted configuration file. When the simulation starts, this model reads all three files to become aware of where the environment objects are and their properties. The mobile nodes are distributed into the accessible sub-areas. The mobility factors such as speed, direction and the mobility model attached to each node are determined by the properties of their locating area.

All the movements are recorded and converted to NS2 compatible movement scenario file at the end of the simulation. NS2 is also modified in order to be able to read an obstacle boundaries description file to block radio signal. Fig.2 illustrates the work flow of this model.

3.1 Environment Layout Design

Scalable Vector Graphic (SVG) is the key approach used by the Environment-Aware model to produce the simulation environment. SVG uses XML to describe 2-dimentions graphics. SVG contains a set of basic shape elements, such as the rectangle, circle, line, and polygon. These elements are used to represent the environment objects with arbitrary shapes. SVG also supports the abilities to change the vector graphic over time which provides the capability of visualising the movements of mobile nodes. A typical snippet of an SVG Environment Layout File (ELF) is attached in Fig.3. Note that the environment objects are indicated by the ID of the SVG shape elements. A Simple convention is set: the ID must start with AA (for Accessible Area) or NAA (for Non-accessible Area) then followed by an underscore then the objects type plus index. For example, if an element has an ID with AA_L1, which means the sub-area represented by this element is an accessible Lane with index 1.

3.2 Environment Objects

Different types of sub-areas are abstracted by Environment Objects (EOs). The EOs can be classified into two categories: Non-Accessible Area (NAA) and Accessible Area (AA). NAA represents the restrict area where no movement is allowed. AA represents some areas where mobile nodes can move inside and might move in and out. An AA can include Lane, Path, Route, Junction, NORmal Accessible-Area (NORAA) and Hotspot objects.

Some hierarchical relationships are also given to EOs. For instance, a Route is composed of Paths, Junctions and even NORAAs, A Path can be composed of Lanes, and NORAAs can be a container area for Hotspots. Fig.4 illustrates the structure of EOs.

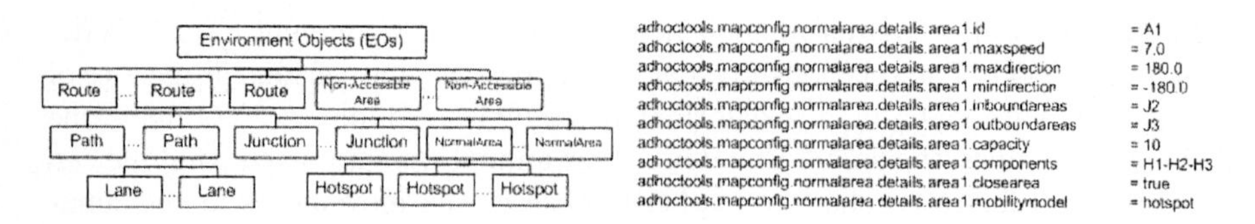

Fig. 4. Structures of Environment Object (EOs)

Fig. 5. Typical configurations for a Normal-Accessible Area

With the introduction of EOs, complex environments can be simplified and easily constructed. Each EO is given some intrinsic characteriscs which have influence on the mobility of the mobile nodes. Typically, the mobile node located in the Lane must follow the Lane to the exit with a maximum speed limit; if a Junction is encountered, a mobile node is forced to choose one of the exits to enter the next adjacent EO. NORAA is a very flexible EO because it can be used as a container area. If it is a free space area, a mobile node can move using its conventional mobility model. If it contains Hotspots, mobile nodes will be forced to commute among the Hotspots. If it contains a NAA, the accessible space inside it for mobile node is reduced.

3.3 Environment Configuration

The SVG layout file provides the model with locations of each EOs and helps to identify the type of each of them. The relationships among EOs mentioned in subsection 3.2 along with their properties are configured in a text-formatted Environment Configuration File (ECF). It is also assumed in EAM that some of the EOs might have some properties in common. They may have restrictions on speed, direction and capacity. All these properties are configurable in ECF file. As an example, a typical configuration for a NORAA is shown in Fig. 5.

3.4 Environment-Aware Movement

The EAM can integrate many Entity Models such as the Random Waypoint model, the Boundless model and the Group Models like RPGM with some modifications. In order to cope with the complex environments composed by the EOs, two sub-models are proposed: the Route model and the Hotspot model. These two models are described in detail and intensively studied in [13]. Boundary Effects are implemented to cope with the situation when NAAs are encountered.

Movement in the Route model. The Route model can model some complex environments, such as a city area or an urban road traffic network. The mobile nodes moving in a Route travel through all its components in a specified order given in the ECF file. When they reach the end of a Route, the mobile nodes will appear at the entry of a new Route in order to keep the number of nodes constant during the simulation. The Route Mobility model can be used to model the movement described in the Pathway model, the Manhattan model, the Virtual Track model, and the Obstacle model. Nevertheless here the movement in a route can use the Random Walk model as well since the Route model allows the existence of a NORAA to bridge two Paths. Movements can also be restrained by the route. For instance, the speed may not exceed the maximum allowed speed of a lane.

Movement in the Hotspot model. A hotspot is a place where most of the mobile nodes may visit. If hotspots exist in an area, the strategies with which the mobile nodes move between the hotspots can be varied because a hotspot might have a capacity to accommodate a maximum number of visitors and a maximum duration that the mobile nodes are allowed to stay. A number of different strategies are introduced in [17]. In this paper, a simple strategy is applied: the mobile nodes always adjust their speed to make sure that the hotspot can allow at least one more nodes to be added to when it arrives at the hotspot.

Boundary Effects. As mentioned before, various models are modified and given environment-aware capability. For example, different behaviors are designed for the Random Walk model and the Random Waypoint model when a NAA is encountered. Bouncing behaviour is applied to the Random Walk model. Fig. 6(a) illustrates the effect: the node bounces to another random direction at boundary $e1$. Compared with the Random Walk model, the Random Waypoint model features a target-driven behaviour so that Bouncing is not applicable. Thus, a different boundary effect called Surrounding is proposed. With Surrounding behaviour, once the next destination is determined, the mobile node will try to get there even if there are obstacles blocking the expected path. Fig. 6(b) illustrates this effect: the node starting at S moves towards D, it actually travels along the boundaries $e3 - e4 - e1$, and then gets to D.

3.5 Transmission Obstruction

Radio propagation could be one of the major issues that influence the performance of MANET. The received signal is very vulnerable to suffer significant power drop-down due to attenuation, multipath fading, etc. The Two-Ray Pathloss propagation model is used to calculate the power of the received signal. In EAM, it is assumed that if there is a boundary of an obstacle geographically located between the LOS range of the pair of mobile nodes, then the signal transmitted from either of them will not reach the other node.

Open Areas and Closed Areas are defined to help the model identify which EO should be considered as an Obstacle. A text-formatted Obstacle Description File (ODF) is generated for the environment. If one EO is configured to be a Closed Area, Its boundaries locations are added into the ODF which will be processed in the physical layer of NS2. Fig.7 gives an example ODF relating to the environment showed in Fig.8.

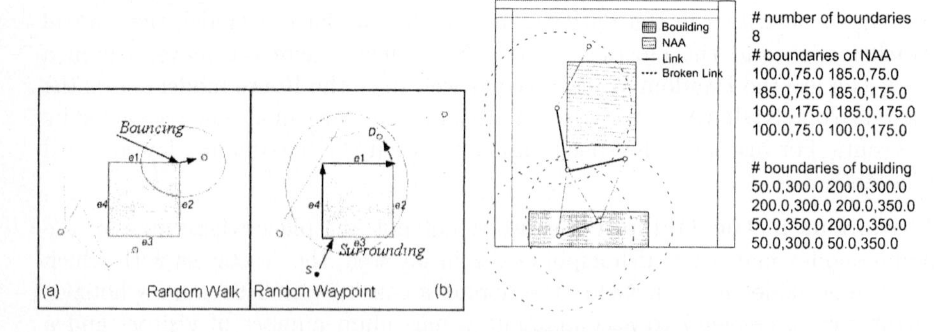

Fig. 6. Boundary Effects applied to different models

Fig. 7. Example of Obstacles Description File

4 Simulations

The Environment-Aware Mobility model is very flexible and can be applied to various environments. The main purpose of the simulation is to show that the performance of the ad hoc networks protocols is not only influenced by mobility models, but also significantly affected by the Environment Objects which are the fundamental sub-environments abstracted from the real environments.

With the presence of EOs, the movement behaviours of mobile nodes modelled by the mobility model are also restricted by the properties of the EOs. Moreover, by introducing Obstacles, the network connectivity is also highly influenced. The resulting topology change is evaluated by the measurement of the average numbers of connection changes at different speeds.

One popular routing protocol AODV [15] is simulated. Three typical metrics, Control Packet Overhead, End-to-End Delay and Data Packet Delivery Rate are used to evaluate the routing performance. For comparison, all the metrics are measured repeatly with increasing speed from 0m/s to 9m/s.

In order to show the impact caused by the environment, two ECF files are used and simulated individually. Both the Obstacle and Obstacles-less environment are simulated and compared. The SVG Layout is shown in Fig. 8.

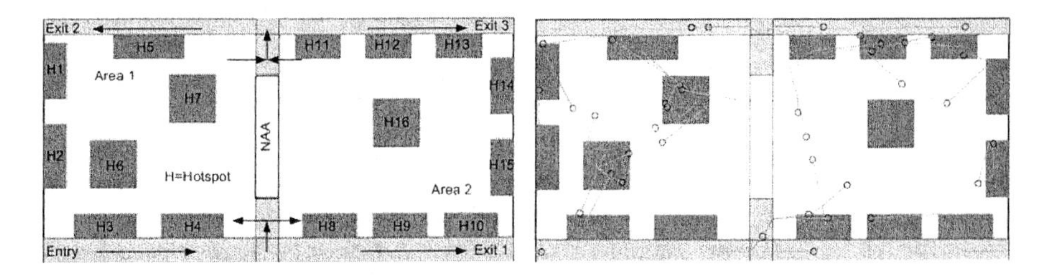

Fig. 8. Environment Layout of the simulation

Fig. 9. Snapshot of the simulation at the initial time

4.1 Environment Description

A 600x300m area is used to represent the simulation area. Two different MANET environments are used.

Art Gallery: In this environment, Area 1 and Area 2 can be considered to be two rooms separated by a wall (NAA). All the arts exhibited in the rooms are presented by Hotspots. Visitors enter this area from the entry. They may continue moving to the next exhibition section through Exit 1 or enter one room (Area 1 or Area 2). Once the visitors enter the room, they will move towards the art object that they find most attractive then the next and so on. If a visitor notices that his art object has too many visitors (up to the capacity of Hotspot), they will go to the next most interesting object. Once they have visited all the art objects or if they have been in the room too long (up to maximum duration allowed to stay in the area), they will leave the room and continue the visit through Exit 2 or Exit 3 as shown in Fig. 8. In this case, both areas are Closed Area because the signal is very likely to be blocked by the walls.

Festival Event: Compared to the Art Gallery environment, this environment exists in some public places, such as a park so that the two areas can be assumed to be Open Areas. In this case, when mobile nodes move into Area 1 or Area 2, they may just walk around freely for a while then leave. Note that if some entertainments or shops are distributed inside the areas as hotspots, the behaviours of nodes could be similar to the previous one.

4.2 Simulation Description

All the simulations are done by ns2. 40 mobile nodes with 100m omni-directional antenna transmission range are distributed over the simulation area. The Hotspot model is used for the modelling the Art Gallery environment and the random movements in the Festival Event are modelled by the Random Waypoint model. The simulation is repeated with different average nodes speeds ranging from 0m/s to 9m/s and the final result is an average of the simulation results from 40 runs.

To evaluate the impact of Obstacles on different mobility models, both environments are simulated with Obstacles and without Obstacles. Area 1 and Area 2 are assumed to contain minimum of 15 mobile nodes inside, and the other 10 nodes are distributed randomly along the routes. If any of the 10 nodes are located inside the two sub-areas, they will stay there for a certain time then leave. If they are located in Lanes, they will randomly choose a route. The initial snapshot of the simulation is shown in Fig. 9. For the Random Waypoint model, a 10 seconds pause time is spent by each node when it gets to the current destination and the pause time in the Hotspot model is depend on the property of each Hotspot. In the Art Gallery environment, nodes move about with a maximum speed of 7m/s inside the two rooms, but there is no limitation for the Festival Event because it is considered to be an out-door event. In order to keep the number of mobile nodes constant during the simulation, the nodes that arrive at any of the Exits will appear at the Entry to start a new journey. Each scenario is simulated for 500 seconds.

The measurement is started after 50 seconds to ensure all the nodes have been distributed throughout the areas. The number of connectivity between neighbours is calculated every second. The average changes of connectivity at each speed are measured as the metric to evaluate the influence on the network topology caused by the mobility models and the environments.

Environment Layout AODV routing protocol is used to evaluate the impact on the routing performance. The CBR data packet size is set to 512 bytes and the sending rate is set to 4packets per second. A maximum of 20 data connection is allowed at one time. A node will act as either a sender or receiver at any time during the simulation.

5 Results

5.1 Network Topology

The average number of connectivity changes with increasing speed is shown in Fig. 10. With the existence of Hotspots and NAA, the effective space in the simulation area is highly reduced. In the route environment, the movements of mobile nodes are also restricted. There are 16 hotspots located in the environment. Nodes gathered in one of the hotspots, generates very high connectivity, but this established connectivity can only be maintained for a short time because later they will head for another hotspot. The Obstacles partition the area into several sub-areas so that the possibility of connectivity loss is increased as the nodes move in and out of the sub-areas. It is shown that the Obstacles cause higher connectivity change and the Hotspot model normally have much higher connectivity change than the Random Waypoint mode in which nodes are more distributed throughout the area. An interesting point can be pointed out here is that for the Hotspot sub-model the connectivity change without Obstacles is increased sharply when Obstacles are introduced. At higher speed without Obstacles, the frequency of moving in and out is increased and the nodes moving along the routes are acting more likely as communication bridges to connect the nodes of two sub-areas. On the other hand, the connectivity created by those bridge nodes can not be kept for a long time at higher speed, which increases the change. This explanation is also applied to the Random Waypoint case. It can be seen that the two curves are jointed at 9m/s and the without-Obstacles situation has a faster increase. Most of the hotspots are located close to the boundaries which are adjacent to the route, therefore more nodes in the two sub-areas can be connected and also disconnected by movement of the bridge nodes. This explains why this effect is more significant in the Hotspots environment.

5.2 Routing Performance

The data delivery rate measured in different environments is shown in Fig. 11. Since the Obstacles obstruct the transmission, the mobile nodes located in different sub-areas are not communicable if a boundary of Closed Area detected, which obviously deteriorates the delivery rate. The hotspots generate very high density which also causes much higher possibility of collision and packets drop due to the mechanism of the MAC layer of MANET. The delivery rate in both Random Waypoint environments is much more stable than the Hotspot environments. There is a Faster drop-down in the Hotspots model without Obstacles (10% dropdown) than with-Obstacles (5% dropdown). This is due to the network topology change as explained in the previous subsection.

The number of control packets sent for routing is plotted in Fig. 12. This result is highly correlated to the packet reception showed before. Low delivery rate in the Hotspot environments causes the nodes to send more control packets to discover the routes as the requirement of AODV.

The high density of nodes in the Hotspot environment cause frequent transmission collision and contention as shown in Fig. 13. Consequently, the packets

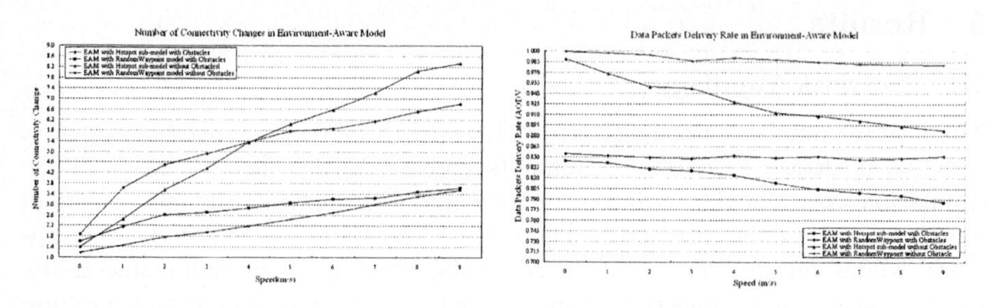

Fig. 10. Average number of connectivity changes

Fig. 11. Data packets delivery rate

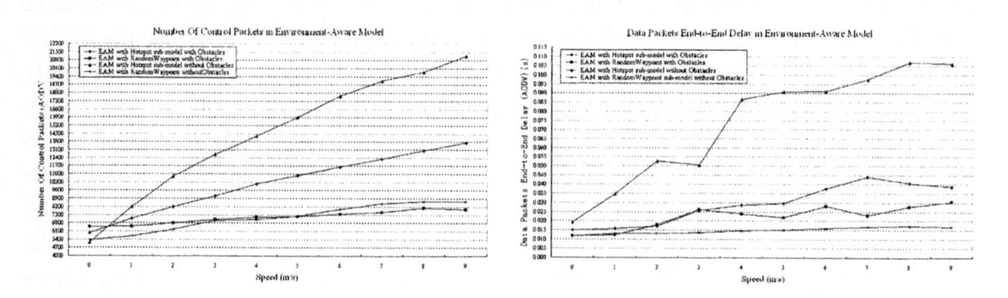

Fig. 12. Number of control packets sent

Fig. 13. Data packets end-to-end delay

are very likely to be delayed until an idle radio channel is found. Due to the higher topology change in the Hotspot without-Obstacles environment, the time spent in route discovery is much longer so that the latency of packets transmission is higher. The random distribution and lower topology change guarantees that the data packets can be sent and arrive at their destinations quickly in the Random Waypoint environment when no obstacles exist.

6 Conclusions

This paper proposes a novel mobility model for generating realistic scenarios of MANET. This model is an integration of current existing models with the environment where the network is deployed. With the aid of SVG, the layout of the environment can be easily designed. Some abstract Environment Objects are introduced to represent the various areas existing in the networks. It has been showed that the movements are highly influenced by both intrinsic characteristics and some properties of EOs. The behaviours of the mobile nodes are not only dependent on which mobility models they use but also on the area in which they are located. The signal-obstruction effect is also incorporated in this model with the introduction of the Open or Closed properties of EOs. The signal is assumed to be blocked if the transmission needs to go through any boundary of the Closed EOs. Two sub-models, Route model and Hotspot model are also proposed to model the mobility observed in some particular environments.

The simulations are done with different mobility models given the same geographical layout of network but with different environmental configurations. The average connectivity changes are measured and the throughput, end-to-end delay and delivery rate of AODV routing protocol are evaluated. The results prove that both the network topology and the performance of protocols are significantly influenced by the mobility models and environmental factors.

With this Environment-Aware model, more realistic environments can be easily handled and better understanding of MANET can be achieved.

References

1. T. Camp, J. Boleng, and V. Davies, A Survey of Mobility Models for Ad Hoc Network Research, in Wireless Communication and Mobile Computing (WCMC): Special issue on Mobile Ad Hoc Networking: Research, Trends and Applications, vol. 2, no. 5, pp. 483-502, 2002.
2. V. Tolety. Load reduction in ad hoc networks using mobile servers. Masters thesis, Colorado School of Mines, 1999.
3. X. Hong, M. Gerla, G. Pei, and C. Chiang. A group mobility model for ad hoc wireless networks, In Proceedings of the ACM International Workshop on Modeling and Simulation of Wireless and Mobile Systems (MSWiM), August 1999.
4. The Network Simulator 2 (ns-2). http://www.isi.edu/nsnam/ns.
5. Scalable Vector Graphics. http://www.w3.org/TR/SVG/
6. B.Liang, Z. J. Haas, Predictive Distance-Based Mobility Management for PCS Networks, in Proceedings of IEEE Information Communications Conference (INFOCOM 1999), Apr. 1999.
7. J.Tian, J. Hahner, C. Becker, I. Stepanov and K. Rothermel, Graph-based Mobility Model for Mobile Ad Hoc Network Simulation, in the Proceedings of 35th Annual Simulation Symposium, in cooperation with the IEEE Computer Society and ACM. San Diego, California. April 2002.
8. P. Johansson, T. Larsson, N. Hedman, B. Mielczarek, and M. Degermark, Scenario-based performance analysis of routing protocols for mobile ad-hoc networks, in International Conference on Mobile Computing and Networking (MobiCom'99), 1999, pp. 195-206.
9. F. Bai, N.Sadagopan, and A. Helmy, Important: a framework to systematically analyze the impact of mobility on performance of routing protocols for ad hoc networks, in Proceedings of IEEE Information Communications Conference (INFOCOM 2003), San Francisco, Apr. 2003.
10. Subodh Shah, Edwin Hernandez, Abdelsalam Helal, CAD-HOC: A CAD-Like Tool for Generating Mobility Benchmarks in Ad-Hoc Networks, SAINT 2002: 270-280.
11. C. E. Perkins, E. Royer, and S. R. Das, Ad hoc On Demand Distance Vector (AODV) Routing, In 2nd IEEE WorkShop on Mobile Computing Systems and Applications (WMCSA'99), pages 90-100, February 1999.
12. A.P. Jardosh, E.M.Belding-Royer, K.C. Almeroth, and S. Suri, Towards Realistic Mobility Models for Mobile Ad hoc Networks, In Proceedings of ACM MOBICOM, pages 217-229, San Diego, CA, September 2003.
13. Gang Lu,Demetrios Belis, Gordon Manson, "Study on Environment Mobility Models for Mobile Ad Hoc Network: Hotspot Mobility Model and Route Mobility Model", WirelessCom2005, Hawaii, USA.

A Low Overhead Ad Hoc Routing Protocol with Route Recovery

Chang Wu Yu[1], Tung-Kuang Wu[2], Rei Heng Cheng[3], and Po Tsang Chen[1]

[1] Department of Computer Science and Information Engineering,
Chung Hua University, Hsin-Chu, Taiwan, R.O.C.
cwyu@chu.edu.tw

[2] Department of Information Management,
National Changhua University of Education, Changhua, Taiwan, R.O.C.
tkwu@mail.tkwu.net

[3] Department of Information Management,
Hsuan Chuang University, Hsin-Chu, Taiwan, R.O.C.
rhc@hcu.edu.tw

Abstract. Many routing protocols have been designed for Ad Hoc networks. However, most of these kinds of protocols are not able to react fast enough to maintain routing. In the paper, we propose a new protocol that repairs the broken route by using information provided by nodes overhearing the main route communication. When links go down, our protocol intelligently replaces these failed links or nodes with backup ones that are adjacent to the main route. Experimental results show that our protocol finds a backup route around 50% of cases and achieve better (or as good) in term of the packet delivery rate than the major Ad Hoc routing protocols, but with much less overhead.

1 Introduction

Ad Hoc networks are wireless networks with no fixed infrastructure. Each mobile node in the network functions as a router that discovers and maintains routes for other nodes. These nodes may move arbitrarily, therefore network topology changes frequently and unpredictably. Other limitations of Ad Hoc networks include high power consumption, low bandwidth, and high error rates [17]. Applications of ad hoc networks are emergency search-and-rescue operations, meetings or conventions in which persons wish to quickly share information, data acquisition operations in inhospitable terrain, and automated battlefield [17].

For years numerous routing protocols have been developed for ad hoc networks including Destination-Sequenced Distance-Vector Routing protocol (DSDV) [15], Clusterhead Gateway Switch Routing protocol (CGSR) [3], Wireless Routing Protocol (WRP) [12], Ad Hoc On-Demand Distance Vector (AODV) [14, 16], Dynamic Source Routing (DSR) [1, 10], Temporally Ordered Routing Algorithm (TORA) [5, 13], Associativity-Based Routing (ABR) [18], and Zone Routing Protocol (ZRP) [8].

In general, the existing routing protocols either need to maintain routing information from each node to every other node in the network or try to create routes on an

X. Jia, J. Wu, and Y. He (Eds.): MSN 2005, LNCS 3794, pp. 666 – 675, 2005.

on-demand basis. Maintaining routing information takes a large portion of network capacity, even though most of the information is never used. On the other hand, the on-demand methods usually incur quite a significant latency in order to determine a route. When the rate of topological changes in the network is sufficiently high, most of the existing protocols may not be able to react fast enough to maintain necessary routing. In this paper, we propose a new routing algorithm, which overcomes the drawback associated with the conventional routing algorithms. Simulations show that the proposed algorithm achieves better in most aspects than most of these notable protocols, and with much less overhead.

The rest of paper is organized as follow. In Section 2, we give a survey of how relevant routing protocols react to link failure. The main ideas of our algorithm are described in Section 3, with experimental results of our proposed protocol presented in Section 4. We conclude the paper in Section 5.

2 Related Work

Ad Hoc routing protocols can generally be categorized as (1) table-driven routing and (2) source-initiated on-demand routing [17]. Table-driven routing protocols require constant propagation of routing information, which incurs extra communication overhead and power consumption. As a result, these become the limiting factors of their applications in Ad Hoc network environment since both bandwidth and battery power are scare resources in mobile devices [17]. On the other hand, on-demand routing protocols establish a route only when a source requires to send messages to some destination without requiring periodic update of routing information. However, various on-demand routing protocols differ in how they handle route maintenance and how they react to link failure, these will be the focus of the following survey.

AODV (Ad Hoc On-Demand Distance-Vector) [14, 16] Routing is an improvement to the table-driven and distance-vector based DSDV algorithm. With DSDV (Destination-Sequenced Distance-Vector) Routing [15], every mobile node maintains a routing table recording all the possible destinations and number of hops to each destination. In order to maintain routing table consistency, it requires nodes to periodically broadcast routing updates throughout the network. AODV minimizes the number of broadcast messages associated with DSDV by building routes on a demand basis. In case a broken link notification is received, source nodes in AODV would restart the route discovery process. But before sending the link failure notification to source, AODV allows the upstream node of the break to try to repair a recently used route by sending a Route Request (RREQ) message [16]. However, if the route repairing attempts were unsuccessful, more data packets would be lost.

DSR (Dynamic Source Routing) [1, 10] uses source routing, with each packet to be routed carrying in its header the complete, ordered list of nodes through which the packet have to go through. If a link fails, the upstream of this failed link sends a route error packet to the source node. When a route error is received, the hop in error is removed from this host's route cache, and all routes that contain this hop must be truncated at that point. A new route discovery process must be initiated by the source. A salvaging technique proposed by Maltz, Broch, Jetcheva, and Johnson [11] uses alternate route from caches when a data packet meets a failed link on its source route.

The Multiple Next Hops (MNH) routing protocol proposed by Jiang and Jan [9], applies the concepts of forward link and reverse link used in AODV. For each destination, each mobile node in MNH routing protocol maintains multiple next hops in its routing table. Hence the MNH may provide multiple routing paths for a source-destination pair. As link failure occurs, the upstream node will detect that and try to reconstruct a new route. However, the MNH constructs multiple routing paths on the first initiation, but does not update the information to reflect the current network status. Therefore, as the topology of network changes, there is little chance of using these backup paths.

Chung, Wang and Chuang [4] proposed the Ad Hoc Backup Node Setup Routing Protocol (ABRP) that is similar to the DSR. ABRP saves backup route information in certain on-the-route node. When a link fails, data messages are sent back to a backup node. The backup node checks its backup route cache, and pick one path (if there exists one) to replace the current broken one. Similarly, ABRP does not update its backup route information to reflect the network topology change.

TORA (Temporally Ordered Routing Algorithm) [5, 13], also a source-initiated routing algorithm, maintains multiple routes for any desired source-destination pair. The key idea of TORA is that control messages are restricted to a small set of nodes in case of topological change. However, to achieve this, it needs to build a directed acyclic graph (DAG) rooted at the destination. As links fail, only local routes are re-established to a destination-oriented DAG within a finite time. For maintaining a list of a node's neighbors, each node periodically transmits a BEACON packet, which is answered by each node hearing it with a Hello packet. Furthermore, in order to maintain the order, TORA needs a global timer to record the time of link failure.

As we have seen in the above review, the two major on-demand protocols (AODV and DSR) do not respond quickly enough to link failure. And they usually suffer from the risk of flooding the whole network for new route discovery. Variations of the two protocols try to cope with these issues by storing extra backup routes for use upon link failure. However, the backup routes are usually created during initial routes construction stage, no effort is done to modify these routes in order to reflect the changing network topology. On the other hand, TORA replies on periodic HELLO messages to monitor the status of network topology and tries to fix broken routes that may not be used later on, which considerably increases its protocol overhead.

3 The Proposed New Routing Protocol

Here we present a new fully distributed and on-demand based Ad Hoc routing protocol with nearly real time repairable route that handles the broken-link recovery in more efficient way. Before moving on to the details, we present the intuitive ideas behind our protocol in the following.

The proposed routing protocol begins by finding a route from a source S to a destination D, which we call the *main route*. All data packets are then sent along this main route to the destination. As the data packets proceed to move along the main route, nodes that are close enough to the path will overhear the messages. In other word, nodes that are able to overhear the messages should be close to the main route and are potentially good candidates for substituting the failed node. By piggybacking appropriate information within packets and applying proper procedure, our algorithm makes nodes that overhearing the packets the backup nodes for future route reconstruction in case of broken main route. Other than adding an additional field (for storing height value, as explain later) to a node's route table and header of message, there is no need for frequent message flooding and huge table maintenance. Our protocol consists of route construction and maintenance stages. Detailed description is given as follows.

3.1 Route Construction

In our proposed routing protocol, a routing path is constructed only when a node needs to communicate with another node. Assume that a source node S wants to send a packet to some destination node D. If the destination node D is a neighbor of source node S, the packet is sent directly to node D. Otherwise, the source node will first check if node D is in its main route table (MRT). If it is, packets will then be sent directly to the next-hop node as specified by the entry. On the other hand, a path (the main route) from source node S to destination node D need to be constructed before source node S can start the data transmission. The process of finding such a routing path is called the *main route construction*, which begins with the source node S sending a main route request (MREQ) to all its neighbors. Every host that receives the MREQ acts exactly the same as the source node does. MREQ is thus flooded over the network, and would eventually arrive at node D if a routing path exists between them. When node D receives a MERQ, it sends back main route reply (MRRP) and a value H, representing the hop number from this node (D) to the destination (which is zero in this case), to the host from which MREQ was received. Once a node receives MREQ, it adds a main route entry for node D to its MRT. It then propagates the MRRP, together with value $H+1$, to the host from which it receives the MREQ. Every other host receiving MRRP behaves similarly until node S receives MRRP and updates its MRT accordingly. A routing path from node S to D, referred to as the main route, is thus established.

3.2 Route Maintenance

The route maintenance process consists of two parts: the main route messages sniffering and the main route repairing. In main route messages sniffering stage, our concern is how to manage the messages go through the main route that are overheard by the neighboring nodes. With main route repairing, we need to take care of a broken main route using the messages collected and avoid the flooding of repair query (REPQ) packets as long as possible.

The Main Route Messages Sniffering. The main route messages sniffering stage begins as packets start delivering through the main route. However, we need to modify the packets delivery slightly to accomplish this task. First of all, we need to insert an H field into the header of data packet and require the nodes to maintain an extra height table for storing the H value. Secondly, with the broadcast nature of wireless communication, a node promiscuously "overhears" packets transmitted by nodes that are within the radio range. In case a node, which is not part of the main route, overhears a data packet transmitted by a neighbor on the main route, it records the H value within the packets header into its height table. If more than one such packet is received, the average of the received H values are computed and then recorded in its height table. The recorded H value can later be used to assist repairing of the route and to restrain flooding of control packets.

The Main Route Repairing. When a link on the main route is broken, the upstream node of this link would find out in a period of time and then initiate the main route repairing process by broadcasting a *repair query* (REPQ) packet. The REPQ packet contains the height value of the initiating node. Each node that receives the REPQ packet would first check if it has a route to the destination. If it does, a *repair reply* (REPR) packet is sent back to the node from which the *repair query* (REPQ) packet was received and the route repairing process is done. Otherwise, it compares the height value (denotes $Hreq$) contained in the REPQ packet to its own height value (denotes Ht) in the height table to determine what the next step will be. In case $Hreq$ is greater than (or equal to) Ht, which means very likely the route repairing request was sent from a node that is closer (or as close) to the source node, the receiving node would then rebroadcast the REPQ and hope the REPQ packet would propagate to the nodes that are closer to the destination. However, if $Hreq$ is smaller than Ht or the node has no Ht in its height table, the REPQ packet will be dropped.

4 Performance of the Proposed Protocol

Our proposed routing protocol described in previous section is simple, straightforward and should be viable to be included in any Ad Hoc networks without incurring much overhead. In this section, we will demonstrate how it performs as comparing to some major routing protocols.

We use NS-2 [7], which was adopted in numerous researches to evaluate the performance of existing Ad Hoc routing protocols [2, 6], as the simulation tools. The link layer of our simulator is IEEE 802.11 Distributed Coordination Function. Physical and data link layer models are devised and described in [2].

The simulation environment is a 1500 by 1500 meters rectangular region with 100 mobile nodes moving around. Each of the 100 nodes is placed randomly inside the region. Once the simulation begins, each node moves toward a randomly selected direction with a random speed ranges from 0 to 50 meters per second. Upon reaching the boundary of the rectangular region, the node pauses for a few seconds, then selects another direction and proceeds again as described earlier. The simulation period is set to 500 seconds for each simulation.

The radio coverage region of each mobile node is assumed to be a circular area with 250 meters in diameter. The transmission time for a hop takes 0.002 seconds, and the beacon period is one second. We assume the source node sends a data packet every half second (CBR: Constant Bit Rate), and use UDP as the transport protocol. Each data packet contains 512 bytes, and totally 920 data packets are sent. Each node has a queue, provided by network interface, for packets awaiting transmission that holds up to 50 packets and managed in a drop-tail fashion. One source-destination pair is selected arbitrarily from 100 nodes in each simulation. Each simulation scenario represents an average of at least 150 runs.

Three on-demand protocols including TORA, DSR, and AODV are used as the basis of comparison to our protocol. These three models and protocols described above are all supported natively by NS-2. Four aspects in evaluating how these algorithms perform are simulated and given in the following four subsections, include (1) data delivery rate, (2) routing overhead, (3) communication latency and (4) average number of hops a message needs to traverse to reach its destination.

4.1 Effectiveness of the Proposed Protocol in Repairing Broken Route

The first thing we would like to know is how the proposed protocol performs in term of broken route repairing. The experimental simulation result shows that in around 50% of cases of the simulations, the broken routes can be repaired, as shown in Table 1. As can be expected, the successful repair rate increases in accordance with the pause time since longer pause time implies more stable nodes.

Table 1. Percentage of Successful Route Repair

Pause Time (in seconds)	Percentage of Successful Repair
10	45.95 %
20	47.26 %
30	50.32 %
40	51.60 %
50	55.56 %

4.2 Data Delivery Ratio

Figure 1 shows the simulation results in packet delivery ratio. It can be easily seen that our protocol performs approximately as good as (or better than) AODV in term of packet delivery rate. In fact, in all the simulation scenarios, the margins are all around 1% range, with AODV slightly ahead in 4 of five cases. On the other hand, both of our protocol and AODV perform much better that DSR and TORA do, especially when nodes are moving fast (corresponding to higher node speed and short pause time). However, when nodes mobility reduces, the performance of DSR and TORA catches up a little since the pre-established main routes become more stable under such circumstance.

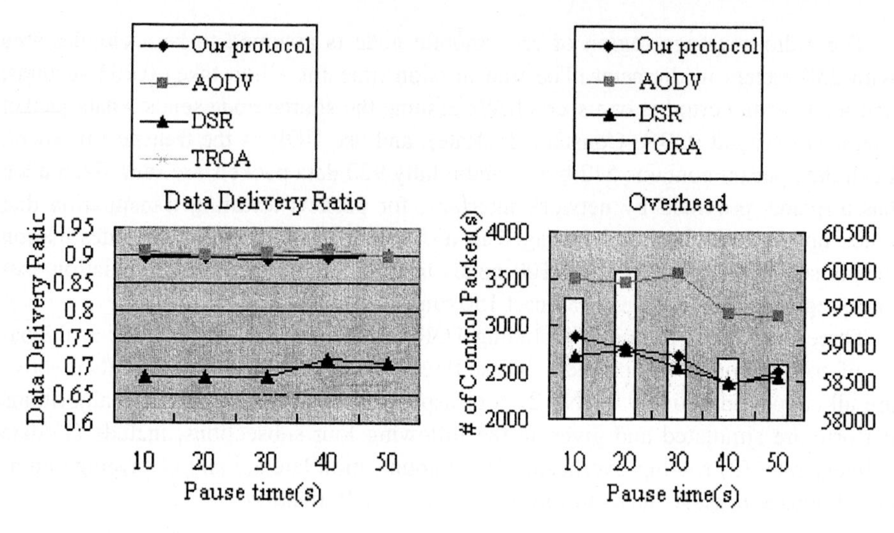

Fig. 1. Data Delivery Ratio

Fig. 2. Routing Overhead

4.3 Routing Overhead

Routing overhead, in term of number of control packets sent, is presented in Figure 2. We can see that our proposed protocol has much lower overhead than that of TORA. This is attributed to the fact that our protocol does not require to construct a backup route in advance, and thus save a large number of control packets broadcast. In addition, comparing to AODV, our protocol also performs better in this aspect. This is not surprising since AODV, instead of repairing the partially broken main route, always reconstructs a new main route from scratch when one is broken. As a result, every time a main route is broken, the main route request (MREQ) broadcast packets are flooded to the network. On the other hand, our protocol would try to repair the main route by taking advantage of the neighboring nodes that are close to the main route. The scope of repair query (REPQ) packet broadcast is thus effectively restricted within two-hops from the original main route. According to the results shown in Figure 1 and 2, we can see that AODV may have a slightly better chance in successfully delivering data packets, but it does that with substantially higher cost (approximately 700~1000 additional packets more than ours in each case). Finally, the gap in term of overhead between our protocol and DSR is only marginally.

4.4 Communication Latency and Variance

The communication latency is measured by the average data packet delivery delay per route in our simulation. In Figure 3 and Figure 4, the simulated average delay time and its variance are illustrated. It is expected that our protocol would perform better than the AODV and TORA do in this aspect. The reason for such speculation is based on the fact that instead of starting from scratch every time a route is broken, our protocol would try to repair the broken main route with some substitute route around the main link. As a result, our protocol can recover from link failure more quickly in half of cases (see table 1), which results to the low communication latency. As we can see, the results correspond exactly to what we expected.

On the other hand, it seems that our protocol does not perform as well as DSR does. However, the fact that the communication latency is calculated only according to data packets that are successfully received by the destination nodes and DSR has a very low delivery ratio make the difference insignificant.

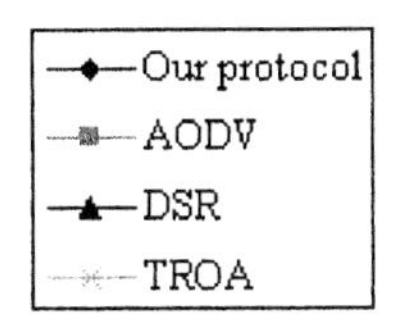

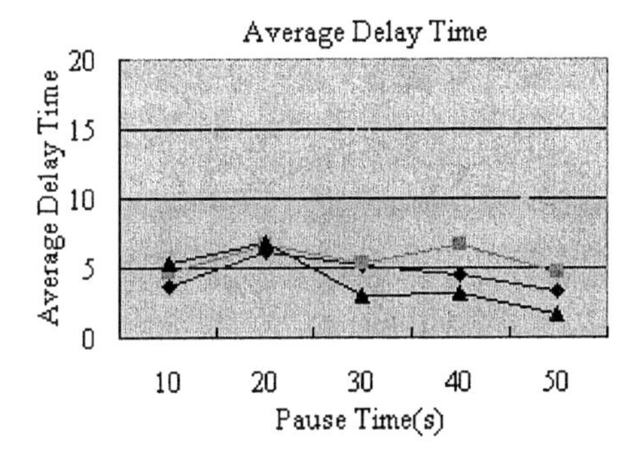

Fig. 3. Communication Latency

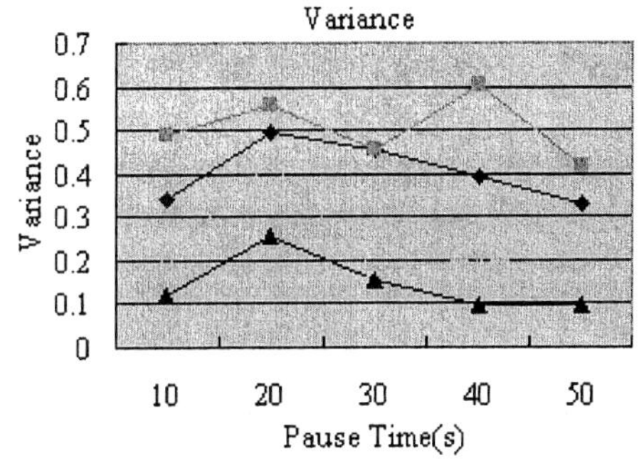

Fig. 4. Variance of communication latency

4.5 Average Number of Hops

Figure 5 shows the average number of hops that packets need to go through in order to reach the destination. The result indicates that our protocol in average requires up to 1 hop more than the AODV and DSR take. The reason is also resulted from our protocol's attempt to repair the main route. A successful repair action implies that packets now travel via an alternative route that is not as direct or straight as the original one, which contribute to the increase of the average hop count that a route traverses. On the other hand, both the AODV and DSR protocols always try to find a new main route, which tends to be a shorter one. In the case of TORA, the upstream node will search its routing table for another outgoing entry to the upstream node when the main route is broken. If there exist one, the upstream node does nothing. Otherwise, it sends Failure Query (FQ) to one of its upstream neighbors. However, upon receiving the FQ message, instead of rediscovering a brand-new route, TORA also tries to fix the existing ones, which could establish a potentially longer link as in our case.

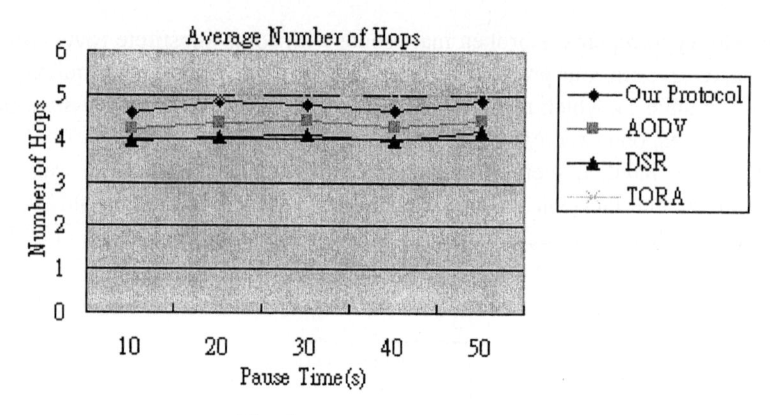

Fig. 5. Average number of hops

5 Conclusion and Future Work

In this paper, we present a new on-demand routing protocol that is able to quickly repair a link or node failure with less communication overhead. Comparing to the other notable on-demand routing protocols, our protocol has a much higher successful data delivery rate than those of TORA and DSR, while performs nearly as good as AODV does, but with approximately 25% less in communication overhead. Our protocol incurs much less overhead with respect to TORA, too. In term of communication latency, our protocol also performs very well, especially when node mobility is higher.

However, there remain issues that deserve further investigation. First of all, it would be interesting to explore ways in evaluating among various potential alternate routes so that a more robust or reliable backup route can be chosen under different network conditions or environments. Secondly, there may also be conditions or threshold that performing a new route discovery would be more economic than repairing the existing main route. As a result, in the future, we will be working on increasing the data delivery rate through a route maintenance mechanism that can dynamically determine when to initiate the route repairing process and how to select a more reliable alternative route. In other word, the alternative route construction process could be initiated at any time, not just when a route is broken. The dynamically constructed alternative routes information can be passed to the upstream nodes, which then determine by themselves when to direct their packets to the "optimal" alternative route. Finally, we will also be exploring the possibilities in applying similar idea to the sensor network environment.

References

1. J. Broch, D.B. Johnson, and D. Malt, "The Dynamic Source Routing Protocol for Mobile Ad Hoc Networks," IETF, Internet Draft, draft-ietf-manet-dsr-00.txt, Mar. 1998.
2. J. Broch, D.A. Malt, D.B. Johnson, Y.C. Hu, and J. Jetcheva, "A Performance Comparison of Multi-Hop Wireless Ad Hoc Network Routing Protocols," MOBICOM'98, pp. 85-97.

3. C.C. Chiang, "Routing in Clustered Multi-Hop, Mobile Wireless Networks with Fading Channel," Proc. IEEE SICON, 1997, pp. 197-211.
4. C.M. Chung, Y.H. Wang, and C.C. Chuang, "Ad hoc on-demand backup node setup routing protocol," Proceedings of 15th IEEE International Conference on Information Networking, pp. 933-937, 2001.
5. M.S. Corson and V.D. Park, "Temporally Ordered Routing Algorithm (TORA) version 1: Functional specification," Internet-Draft, draft-ietf-manet-tora-spec-00.txt, Nov.1997.
6. S.R. Das, C.E. Perkins, and E.M. Royer, "Performance Comparison of Two On-demand Routing Protocols for Ad Hoc Networks," Proceeding INFOCOM, 2000, pp. 3-12.
7. Kevin Fall and Kannan Varadhan, editors. ns notes and documentation, The VINT Project, UC Berkeley, LBL, USC/ISI, and Xerox PARC, November 1999. Available from http://www-mash.cs.berkeley.edu/ns/.
8. Z.J. Haas "The Zone Routing Protocol (ZRP) for Ad-Hoc Networks," IETF Internet Draft, draft-zone-routing-protocol-00.txt, Nov. 1997.
9. M.H. Jiang and R.H. Jan, "An efficient multiple paths routing protocol for ad-hoc networks," Proceedings of 15th IEEE International Conference on Information Networking, pp. 544-549, 2001.
10. D.B. Johnson and D. A. Maltz, "Dynamic source routing in Ad-Hoc wireless networks, " Mobile Computing, T. Imielinske and H. Korth, Eds., Kluwer, 1996, pp. 153-181.
11. D. Maltz, J. Broch, J. Jetcheva, and D. Johnson, "The effects of on-demand behavior in routing protocols for multi-hop wireless ad hoc networks," IEEE Journal on Selected Areas in Communication, 1999, to appear.
12. S. Murthy and J.J. Garcia-Luna-Aceves, "A routing protocol for packet radio networks, " ACM MOBICOM, 1995, pp. 86-94.
13. V.D. Park and M.S. Corson, "A Highly Adaptive Distributed Routing Protocol for Mobile Ad Hoc Networks," Proceeding INFOCOM '97, 1997, pp. 1405-1413.
14. C.E. Perkins, "Ad Hoc On Demand Distance Vector (AODV) Routing, " IETF, Internet-Draft, draft-ietf-manet-aodv-00.txt, Nov. 1997.
15. C.E. Perkins and P. Bhagwat, "Highly Dynamic Destination-Sequenced Distance-Vector Routing (DSDV) for Mobile Computers, " ACM SIGCOMM '94, pp. 234-244.
16. C.E. Perkins and E.M. Royer, "Ad-hoc On-demand Distance Vector routing, " Proceedings of the 2nd IEEE Workshop on Mobile Computing Systems and Applications, pp. 99-100, 1999.
17. E.M. Royer and C-K Toh, "A Review of Current Routing Protocols for Ad Hoc Mobile Wireless Networks," IEEE Personal Communication, pp. 46-55, 1999.
18. C.K. Toh, "Associativity Based Routing for Ad-Hoc Mobile Networks," Wireless Personal Communication, vol. 4, no. 2, pp. 103-139, 1997.

Recovering Extra Routes with the Path from Loop Recovery Protocol

Po-Wah Yau

Mobile VCE Research Group,
Information Security Group,
Royal Holloway, University of London,
Egham, Surrey TW20 0EX, UK
p.yau@rhul.ac.uk

Abstract. Reactive routing protocols for ad hoc networks typically flood the network with request packets, but only discover one route from the network flood. Multipath routing protocols have been proposed to increase the efficiency of request flooding by discovering several routes from one request flood. A novel routing protocol, Path From Loop Recovery (PFLR), is proposed which recovers additional link-disjoint routes from topological loops. PFLR operates as a sub-layer beneath either a single or multipath reactive route discovery protocol. Simulation studies have revealed that the PFLR protocol is at its most efficient when recovering additional paths from small topological loops.

1 Introduction

Reactive routing protocols for ad hoc networks [2, 6] typically discover routes by flooding a request throughout a large proportion of the network, each node creating a reverse route to the originator. The requested destination node also unicasts a reply which travels along the reverse route, allowing a forward route to be discovered. The majority of the nodes will not receive the reply, and thus will not form part of any useful route. Multipath reactive routing protocols [3, 5, 7, 8] are more efficient as, where possible, more than one route is maintained between pairs of network nodes. Multiple routes are also regarded as desirable by those seeking Quality of Service (QoS) guarantees, since they allow load balancing and they enhance the availability of a communication path between the originator and destination (which is, moreover, appealing from a security viewpoint [9]).

However multipath protocols may never find the full quota of possible routes due to topological loops, where part of the loop will form a subsection of a route between two end nodes. A request will propagate around the loop in both directions, but may converge on a node in the loop which cannot propagate the request any further. Hence, only one route is discovered instead of potentially two link-disjoint routes.

X. Jia, J. Wu, and Y. He (Eds.): MSN 2005, LNCS 3794, pp. 676–688, 2005.

In this paper we introduce the Path From Loop Recovery (PFLR) protocol, which can be used, together with any single or multipath discovery protocol, to find extra routes in addition to those found by the original routing protocol. As far as the author is aware, the PFLR protocol is the only protocol which always discovers link-disjoint routes from topological loops. While the additional routes may be sub-optimal, they will add redundancy and reliability to any optimal routes discovered by the reactive routing protocol.

This paper identifies and describes how current reactive routing protocols respond to topological loops in section 2, before giving an overview of the Path From Loop Recovery (PFLR) protocol in section 3. Section 4 provides details of the simulations conducted to study the protocol and, finally, brief conclusions are provided in section 5.

2 The Effect of Topological Loops

In ideal conditions, a topological loop will be discovered as two link-disjoint paths in an ad hoc network. However, there are situations where this will not be the case. We suppose that node A needs a route to node Z, and hence broadcasts a request. Consider the situation in Figure 1, where only the route $\{A, B, E, F, Z\}$ is discovered.

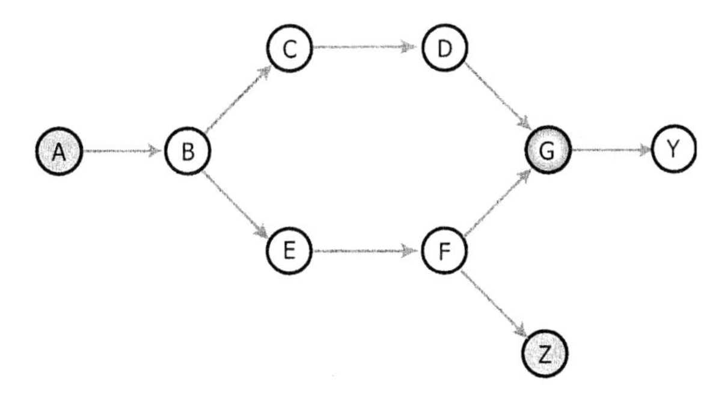

Fig. 1. Node A sends a request for node Z, where the arrows indicate the the propagation of a request packet. When all link latencies are equal, only one of the two possible routes is discovered $\{A, B, E, F, Z\}$ as a reply from Z is unicast along the reverse route to the originator A. Should node E leave the ad hoc network, an alternative route, $\{A, B, C, D, G, F, Z\}$, exists but will have to be discovered with another route discovery cycle.

Non-uniform link latencies will exacerbate the problem, by increasing the likelihood of topological loops occurring. In Figure 1, different link latencies could result in the request converging on nodes C, D or E, where again, only one route will be discovered.

To simplify subsequent discussions, we now introduce some terminology relating to the handling of a route request by a particular loop in a network. The loop entry node is the first node in a loop which receives the request, and thus this node rebroadcasts it in both directions round the loop[1]. The true loop exit node is the node which, in the ideal scenario, will receive more than one request packet and will eventually form part of two routes. In Figure 1, node B is the loop entry node and node F is the true loop exit node. A false loop exit node is one which receives more than one request packet but will not form part of the eventual route. In the example in figure 1, node G is the false exit loop node.

3 The Path from Loop Recovery Protocol

The Path From Loop Recovery (PFLR) protocol is a novel means of recovering link-disjoint paths, from topological loops, for reactive distance vector routing protocols. The PFLR protocol itself is divided into three stages — monitoring the route discovery cycle, loop discovery and path recovery. The PFLR protocol cannot be used alone; it must be used in conjunction with a reactive route discovery protocol. PFLR operates as a sub-layer between the link layer and the network layer routing protocol. Thus, no modifications to the routing protocol are required as PFLR can process the routing messages before passing them to the appropriate layer. However, PFLR must be allowed read access to the routing table, either directly or through some defined interface.

Table 1. Node sets for PFLR

$F =$	{For any particular node, the neighbour nodes which are next hops to the request destination, as discovered before use of PFLR}.
$R =$	{For any particular node, the neighbour nodes which are next hops to the request originator, as discovered before use of PFLR}.
$B =$	{For any particular node, the neighbour nodes from which a reply broadcast was received}.
$LC_{lco} =$	{For any particular node, the neighbour nodes which are next hops to the loop check originator lco}.

After briefly describing the PFLR protocol data structures, we present the monitoring operation of PFLR, which uses reply broadcasts and processes any received requests. Following this, a detailed description of loop discovery and path recovery is provided. Note that in this section we consider only packets from a single route discovery cycle. The mechanisms designed to ensure freshness are still applicable but they do not contribute to this discussion.

[1] Note that there may be more than one loop entry node. If this happens, then the loop that those loop entry nodes belong to is actually part of a larger, outer loop. There are situations where either the inner loop or the outer loop will be recovered as two routes, but not both.

We first outline some necessary terminology. As there are several types of packet, each of them with different originator and destination nodes, the text will always refer to the packet type originator and destination. For example, a node creates a reverse route to the request originator during one cycle of the route discovery. The packet type previous hop is the neighbour node from which the packet was received, i.e. the node will use the request's previous hop as a next hop to the request originator. Additional terminology is defined in Table 1.

3.1 PFLR Parameters and Data Structures

The protocol has three variable parameters which are described in Table 2. Values of these parameters must be selected before use of the scheme. Here, we present an overview of the data structures that are needed for nodes to use PFLR.

Each node must keep a PFLR routing table, indicating routes which were discovered before and after the use of PFLR. If the routing protocol does not use source routes, then each node must also maintain a PFLR Request Table. The PFLR Request Table is used to record the time when the first received request was successfully rebroadcast. The use of source routes negates the need for this table, as described in section 3.2.

Table 2. The variable parameters of PFLR

Parameter	Description
repmaxdist	The maximum number of hops a reply packet can be broadcast away from the unicast route. See section 3.3 for more information.
PFLRDelay	The period of time that a node should wait before attempting to initiate the PLFR protocol. If the node meets the preconditions then it may proceed with sending Loop Check packets (see section 3.4).
lpchmaxdist	The Loop Check maximum hop value is used to control the size of the topological loops that the PLFR protocol operates with.

Note that the information in the two aforementioned tables are often duplicated in the routing protocol, so to avoid duplication of effort PFLR could maintain tables containing the fields particular to PFLR, with links to the corresponding entries in the routing protocol tables.

Each node must also maintain a Reply Broadcast Table to record the details from any reply broadcasts received. Reply broadcasts are described in section 3.3. Finally, the PFLR Table is used to record the details of any PFLR control packets received. The use of this table is explained in sections 3.4 and 3.5.

3.2 Processing Requests and Initiating Route Recovery

An event which is described by the following rules triggers the PFLR sub-layer in a node to initiate an attempt to recover routes:

1. Two requests are received from the same route discovery cycle,
2. The two requests are from different neighbours, and
3. When the node receives the second request, the node has not yet rebroadcast the first request.

If these rules are met then the PFLR sub-layer can initiate a timer, in order to initiate the loop discovery stage after a period of *PFLRdelay* expires (see Table 2). These rules exploit the fact that requests are rebroadcast after a jitter period, in order to minimise collisions with neighbours also rebroadcasting the request. Hence, the PFLR Request Table is needed as described in section 3.1.

Note that with source routing reactive protocols, rule 3 is not needed as the source route from the packet can be used to ensure the node is not processing the same request it has previously rebroadcast.

3.3 Reply Broadcasts

When the destination node receives a route request, it replies according to the normal protocol rules. When the PFLR sub-layer in that node receives the reply, it appends two additional fields as described in Table 3 to create a reply broadcast packet. Note that the 'Reply Destination' field is not required if source routes are used by the reactive routing protocol, for example, as in the Dynamic Source Routing (DSR) protocol [2]. The source route in the original reply will already contain the information in the 'Reply Destination', namely the destination.

Once the reply broadcast is ready it is passed down to the link layer to be broadcast.

Table 3. Additional fields of a request broadcast packet

Reply next hop	The neighbours which should process the packet as a normal reply. This field is necessary because the unicast destination address in the IP header will be replaced by the IP broadcast address.
Number of hops from route	The number of hops that the reply broadcast has travelled away from the discovered forward route. This field is initialised as zero.

If an intermediate node receives a reply broadcast which contains the node's address in the 'Reply next hop' field, i.e. the node would have received a unicast reply according to the reactive routing protocol being used, then the two PFLR fields are removed before the packet is passed to the routing protocol to be processed. As the reply is passed back down to be sent, the PFLR sub-layer will again append and initialise the two PFLR fields before the reply broadcast is rebroadcast.

All other nodes which receive a reply broadcast which are not recorded in the 'Next hops field', i.e. those nodes which would not have received a unicast reply according to the routing protocol, will increment the 'Number of hops from route' count. These nodes will record the details of the reply broadcast in their Reply Broadcast Table (see section 3.1). The reply broadcast is not passed to the routing protocol so no forward path setup is performed. Instead, these nodes will rebroadcast the reply broadcast packet, if the 'Number of hops from route' count is less than *repmaxdist* hops. Therefore, all nodes within *repmaxdist* hops of reverse routes discovered by the routing protocol will receive a reply broadcast.

3.4 The Loop Discovery Stage

Once a node's *PFLRdelay* period expires, as described in section 3.2, the node must meet the Loop Discovery preconditions before the node can commence use of the PFLR protocol.

Originating a Loop Check Packet. The loop check preconditions must be met before a node can originate a Loop Check packet for loop discovery. These conditions are as follows, where F, R and B are the sets described in Table 1:

$$|R| \geq 2 \tag{1}$$

$$F = \varnothing \tag{2}$$

$$|B| \geq 1 \tag{3}$$

Precondition 1 states that during the route discovery cycle the node should have received the same request from at least two different neighbours. If the node has a forward route then it must be a true loop exit node, so the loop has already been discovered as two routes, hence precondition 2. Without precondition 3, every node in the network which satisfies the first two preconditions will perform loop discovery, even those nodes which are so far away from the initially discovered route that they will never form any meaningful routes between the originator and destination nodes. PFLR will only be useful when at least two nodes in the loop form part of the initially discovered route (see section 2). If the loop discovery preconditions are true, then the node may proceed and originate a Loop Check packet. The Loop Check packet must contain the information described in Table 4.

Table 4. The contents of a Loop Check Packet

Originator address	The address of the node originating the Loop Check packet.
Destination address	The address of the originator of the request.
Freshness number	This number is taken from the request packet which caused the loop discovery, and may take the form of a nonce, sequence number, etc.

The Loop Check originator node must then unicast the Loop Check packet to all nodes in R, or broadcast if R is equal to N, where all the node's neighbours are next hops to the request originator. Note that a node must only unicast the Loop Check packet to nodes representing routes discovered by the routing protocol, and not those discovered by any previous executions of the PFLR protocol.

Receiving a Loop Check Packet. All nodes which receive the Loop Check must use it to create a reverse route to the Loop Check Originator. Note that while this creates extra routes, their usefulness outside of PFLR will depend on the network traffic. When a node receives a Loop Check it should record its details in the PFLR Table, described in section 3.1.

If the Loop Check is the first copy received by a node, then the 'Previous hops' list is updated. The Loop Check can then be forwarded only if two conditions are met:

1. The hop count has not exceeded the *lpchmaxdist* value (see Table 2).
2. If the node is scheduled to originate its own Loop Check packets, then the node must delay forwarding the Loop Check until it has sent its own Loop Check packets.

As previously, the Loop Check packet is unicast or broadcast to all nodes in R. If a duplicate Loop Check is subsequently received from the same originator, then the node must perform the following steps:

1. If the Loop Check was received from the same neighbour before, as recorded in 'Previous hop', then the Loop Check can be ignored and dropped.
2. If the Loop Check was received from a different neighbour recorded in the 'Previous hops' list, then the neighbour's address can be appended to the list.

The intermediate node can now proceed with the path recovery stage if it meets the path recovery pre-conditions described in section 3.5. If the pre-conditions are not met, then the node drops the Loop Check. Any subsequent copies of the same Loop Check packet received after the node has initiated the Path Recovery stage are ignored and dropped. The motivation for this step is that such routes will have too large a latency to be considered. Note that the Loop Check destination node (the request originator) does not forward the Loop Check once it has received it, but the node must still process the packet details as above.

3.5 The Path Recovery Stage

As mentioned above, the Path Recovery stage can only begin if the preconditions are met.

Path Recovery Preconditions. The Path Recovery preconditions are:

$$|L_{lco}| \geq 2 \tag{4}$$

$$L_{lco} \cap F \neq \emptyset \neq L_{lco} - F \tag{5}$$

These preconditions are only met by a loop entry node which is part of a loop that forms part of the route discovered in the base route discovery protocol. Therefore, a loop entry node which has received no replies from any of the Loop Check previous hops must be part of loops that are not involved in the base protocol discovered route, and thus there will be no gain from path recovery. If a loop entry node has received route replies from all Loop Check previous hops, then it is clear that the loops have already been divided into two routes each. In this case the Loop Check message should not have been originated in the first place, so this could be an indication of an active attack.

Originating a Path Recovery Packet. A loop entry node which meets the preconditions outlined above must originate a Path Recovery packet. The Path Recovery packet must contain the information described in Table 5.

Table 5. The contents of a Path Recovery packet

Originator address	The address of the originator of the Path Recovery packet.
Destination address	The address of the originator of the Loop Check packet.
Freshness number	This is taken from the Loop Check packet.

As some of the node's previous hops are a next hop to the request destination, and some of the previous hops are not, the node must unicast the Path Recovery packet to all previous hops which are not a next hop, i.e. the path recovery next hops. Also, each path recovery next hop must be recorded as a next hop to the request destination (i.e. the node must create forward routes).

Receiving a Path Recovery Packet. There are four types of node which can receive a Path Recovery packet. These four types of node should process it as follows:

1. Type 1 nodes: These are nodes on the loop situated between the loop entry node and the Loop Check originator node. All these nodes can identify themselves as Type 1 nodes, since they will have received the Path Recovery packet from a neighbour $q \notin LC_{lco}$. These nodes must unicast the Path Recovery packet to all $s \in LC_{lco}$, and use the next hops to create forward routes to the request destination. These nodes already have the correct route to the request originator, so there is no need to create any additional reverse routes.
2. Loop Check Originator: The Loop Check originator can identify itself since its own address will be present in the Path Recovery packet destination field. The Loop Check Originator forms the Path Recovery Neighbour Set P, where P is defined to be the set of nodes in R for which the number of

hops from the unicast route in minimal, excluding the node specified as the previous hop in the Path Recovery packet.

The Loop Check originator node must unicast the Path Recovery packet to all nodes in P, and the node must also use these neighbours as next hops to create forward routes to the request destination. Finally, the node must delete the routes to the request originator which use a node in P as a next hop.

3. Type 2 nodes: These are nodes in the loop that are located between the Loop Check originator node and the true loop exit node. They can identify themselves as Type 2 nodes when the following two conditions are both true. Firstly, they do not meet one of the conditions that a Type 1 node must meet, i.e. they have received the Path Recovery from a neighbour which is used as a next hop to the Loop Check originator. Secondly, a Type 2 node will have also received the corresponding reply broadcast message. These nodes will perform the same steps as the Loop Check originator node. In addition, Type 2 nodes must also create a route to the request originator, using the Path Recovery's previous hop as a next hop.
4. True Loop Exit node: This node can identify itself because it will already have a route to the request destination as a result of the original route discovery process. When this node receives the Path Recovery packet, it will use the previous hop to create an additional reverse route to the request originator. The node then drops the Path Recovery packet. This signals the end of the current PFLR cycle.

3.6 Using Routes Discovered with PFLR

When a node receives a packet for forwarding and it has different possible routes to the destination, the node must only forward the packet to a neighbour from which the packet was not received. This is a technique similar to the split horizon check; such a check is already present in routing protocols for conventional wired networks [1]. This check prevents the packet circulating in a 2-hop loop between two neighbours.

4 Simulation Study

This section contains details of simulations used to test the PFLR protocol. The experiments described below were used to discover the optimum values for two of the three variable parameters listed in Table 2 — the *repmaxdist* and *lpchmaxdist* values.

4.1 Simulation Setup

The PFLR protocol, together with the Ad Hoc On-demand Distance Vector (AODV) reactive routing protocol, was simulated in Microsoft Excel using Visual Basic for Applications[2]. A simplified simplex MAC protocol was used, where a

[2] A simplified non-standard simulator was used to conduct a quick feasibility study.

node only begins transmission if the link is free. Nodes check the link every 500 time units. Each set of tests involve the following initialisation procedure.

Each node was allocated a randomly generated (x, y) coordinate where both x and y are less than 2500, i.e. the nodes were randomly placed on a 2500×2500 metre grid. In these simulations there was no node mobility.

Connectivity was calculated as follows. The simulations used a radio propagation range of 270 metres. This is a value which has been used in previously published research, where the radio propagation range matches those of the Lucent WaveLAN card [4] running at a medium speed of 5.5 Mb/s. Hence if two nodes were within 270 metres of each other then they were recorded as neighbours.

Once the complete set of links were recorded, each link was allocated a randomly generated latency value. This value determines (linearly) the time a packet takes to travel over the link.

The variables tested are described in section 4.2, before presenting an analysis of PFLR in terms of both memory and communication overheads in section 4.3.

4.2 Simulation Tests

Six main variables were used to create the different environments with which the PFLR protocol was tested. These were:

- *Node density.* As the simulation area remained constant, node density was varied by changing the total number of nodes involved in a test. The values used were 100, 150, 200 and 300. Calculations showed that average network connectivity increased from 20% to 100% as the node density increased.
- *Network topology.* This refers to the placement of nodes and the links between them. As the network topology can have a significant impact on the performance of a routing protocol, each set of tests was run over five different network topologies, randomly chosen as part of the initialisation procedure, as described above.
- *Link latency.* Again, to ensure fair sampling, each network topology was used with five different sets of link latencies. This allowed a mean average of results to be taken over the five sets of latencies.
- *Originator/destionation pair.* Each test required one originator/destination node pair. This introduced another variable, namely the distance between the originator and destination node in terms of hop number. Hence, the simulator randomly chose node pairs which were from 2 to 10 hops apart.
- *Reply maximum hop.* This is the value of *repmaxdist*. The simulations were performed with *repmaxdist* values ranging from 1 to 10.
- *Loop Check maximum hop.* This is the value of *lpchmaxdist*. The simulations were performed with *lpchmaxdist* values ranging from 1 to 10.

Thus, 135 000 individual tests were made to ensure fair and complete testing of all the above variables.

4.3 Simulation Results and Analysis

If no route discoveries occur, then there is no memory overhead as the PFLR tables will be empty. Moreover, there is also no communication overhead.

At any given time, the memory overhead in PFLR is proportional to the number of route discoveries which are occurring in the ad hoc network, where the maximum size of both the PFLR Request Table and Reply Broadcast Table is equal to the total number of nodes participating the ad hoc network.

If there are no topological routes where the false exit node receives more than one request, then there is no overhead with respect to the PFLR Table, and also no communication overhead due to PFLR. However, when this is not the case and PFLR loop recovery is initiated, then the size of the PFLR Table will be proportional to the number of loop discoveries taking place. This is related to the communication overhead, where the simulation results will help to gain an insight into how PFLR is affected by the choice of parameters.

The mean averages of the simulation results, over the variables, were used to calculate the following two effectiveness ratios relating to the number of packet transmissions and the overall network bandwidth used:

$$\text{Ratio1} = \frac{\text{\% increase in useful routes discovered}}{\text{Ave. number of packets per new route}} \tag{6}$$

$$\text{Ratio2} = \frac{\text{\% increase in useful routes discovered}}{\text{Ave. number of bits per new route}} \tag{7}$$

We define a route r as useful if $r \in F \cup R$, i.e. routes to Loop Check originators were not considered useful in these simulations.

Since both ratios revealed similar results, we only depict the results of ratio1, in figure 2. It can be seen that PFLR is most effective when the values of the two variable parameters are low. We write (a, b) where a represents the *repmaxdist* parameter and b represents the *lpchmaxdist* parameter. Thus, PFLR is most effective when we have value pairs of either (1,2) and (1,3). The random distributions of nodes and latencies reveal that the most common loops are as depicted

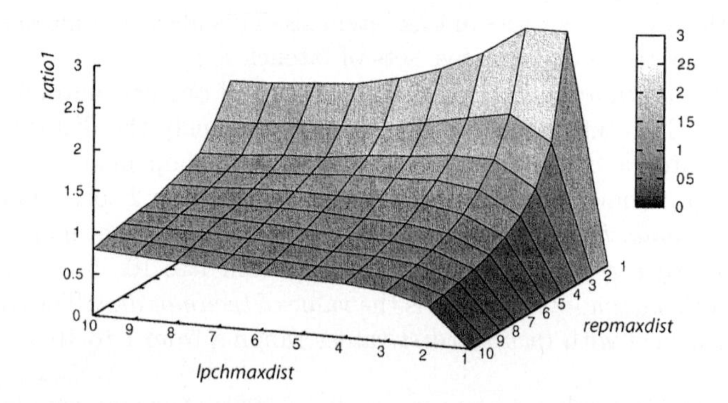

Fig. 2. The effectiveness of PFLR in terms of the number of packet transmissions

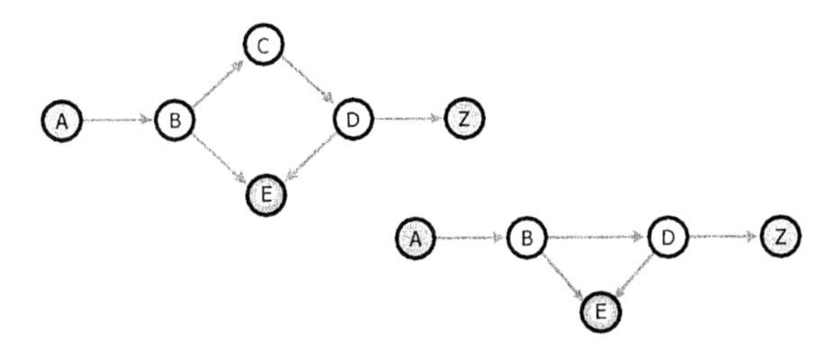

Fig. 3. Small loop topologies found in the simulations

in figure 3, i.e. PFLR is more effective when recovering paths from small loops rather than larger loops. The use of larger parameters will recover more routes from larger loops but will also incur a significant overhead.

This also helps to explain why statistical tests reveal that choosing larger values of the *repmaxdist* parameter has a negative influence on the effectiveness of PFLR. Allowing reply broadcast packets to travel further inevitably leads to more nodes initiating loop discovery. Increasing the *repmaxdist* parameter by a value of just 1 results in many nodes initiating loop discovery, which in turn leads to many more Loop Check packets, each being propagated up to a distance of *lpchmaxdist* hops.

The tests show that the size of the *lpchmaxdist* parameter and the effectiveness of PFLR are positively correlated, but the strength of this correlation is weak compared to the strength of the negative correlation between the protocol's effectiveness and the size of the *repmaxdist* parameter. Again, the conclusion is that the value of *repmaxdist* has to be kept to a minimum if PFLR is to be an effective method of finding extra routes.

5 Conclusion

Topological loops and differing link latencies in an ad hoc network can introduce discrepancies when using reactive protocols. These discrepancies are the result of packets being propagated around loops, where each loop could potentially be used as two link-disjoint paths. Both single route discovery protocols and multipath discovery protocols suffer from this problem, which may prevent the full quota of routes being discovered.

To recover link-disjoint routes from these loops, we propose the novel protocol Path From Loop Recovery (PFLR). PFLR acts as a sub-layer between the routing protocol and the link layer, in order to process or modify the routing protocol control packets before they are transmitted. The PFLR protocol itself works on the principle that if a node receives two request packets from the same route discovery cycle then it could be part of a loop. Thus such a node can use the reverse routes created to send Loop Check messages. If a node receives two copies of this Loop Check message from two different neighbours then a loop may have been discovered. Thus a Path Recovery packet may be sent to recover an additional route.

Initial simulation results have revealed that keeping the PFLR variable parameters to a minimum produces the most effective results in terms of the percentage increase in useful routes discovered compared to the number of packet transmissions and overall network bandwidth used. Further simulations are needed to examine the performance of PFLR in a mobile environment combined with a realistic traffic model.

Acknowledgments

The work reported in this paper has formed part of the Networks & Services area of the Core 3 Research Programme of the Virtual Centre of Excellence in Mobile & Personal Communications, Mobile VCE, www.mobilevce.com, whose funding support, including that of EPSRC, is gratefully acknowledged. Fully detailed technical reports on this research are available to Industrial Members of Mobile VCE. Thanks also go to Mr. E. Godolphin and Professor J. Essam.

References

1. D. E. Comer. *Internetworking with TCP/IP principles, protocols, and architectures.* Prentice Hall, 4th edition, 2000.
2. D. Johnson, D. Maltz, and J. Broch. DSR — The dynamic source routing protocol for multihop wireless ad hoc networks. In C. Perkins, editor, *Ad Hoc Networking*, chapter 5, pages 139–172. Addison-Wesley, 2001.
3. X. Li and L. Cuthbert. Node-disjointness-based multipath routing for mobile ad hoc networks. In Mohamed Ould-Khaoua and Franco Zambonelli, editors, *Proceedings of the 1st ACM International Workshop on Performance Evaluation of Wireless Ad Hoc, Sensor, and Ubiquitous Networks, Venezia, Italy, October 4, 2004*, pages 23–29. ACM Press, 2004.
4. Lucent Technologies. *Quick Installation Guide WaveLAN/IEEE Turbo 11Mb PC Card*, Sep 1999.
5. M. K. Marina and S. R. Das. On-demand multipath distance vector routing in ad hoc networks. In *Proceedings of the IEEE international conference on network protocols, Riverside, California, November 11-14, 2001*, pages 14–23. IEEE Press, Nov 2001.
6. C. Perkins, E. M. Belding-Royer and S. Das. Ad hoc On-Demand Distance Vector (AODV) routing. RFC 3561, Internet and Engineering Task Force, Jul 2003.
7. P. Sambasivan, A. Murthy, and E. M. Belding-Royer. Dynamically adaptive multipath routing based on AODV. In *Proceedings of the 3rd Annual Mediterranean Ad hoc Networking Workshop (MedHocNet), Bodrum, Turkey, June 27–30, 2004*, Jun 2004.
8. L. Wang, Y. T. Shu, O. W. W. Yang, M. Dong, and L. F. Zhang. Adaptive multipath source routing in wireless ad hoc networks. In *Proceedings of of IEEE International Conference on Communications, Helsinki, Finland, June 11-14, 2001*, pages 867–871. IEEE Press, Jun 2001.
9. P. Yau and C. J. Mitchell. Security vulnerabilities in ad hoc networks. In *The Seventh International Symposium on Communication Theory and Applications, July 13–18, 2003, Ambleside, Lake District, UK*, pages 99–104. HW Communications Ltd, July 2003.

Quality of Coverage (QoC) in Integrated Heterogeneous Wireless Systems

Hongyi Wu[1], Chunming Qiao[2], Swades De[3], Evsen Yanmaz[4], and Ozan Tonguz[4]

[1] University of Louisiana at Lafayette, Lafayette, LA 70504, USA
[2] State University of New York at Buffalo, Buffalo, NY 14260, USA
[3] New Jersey Institute of Technology, Newark, NJ 07102, USA
[4] Carnegie Mellon University, Pittsburgh, PA 15213, USA

Abstract. With the increased demand for high data rate wireless communication and the emergence of various wireless technologies, several integrated heterogeneous wireless systems, such as Multihop Cellular Network (MCN) and integrated Cellular and Ad hoc Relaying (iCAR) system, have been proposed recently. In this paper, we address the location management issue which is critical for the performance of heterogenous networks. We define a new performance metric called *Quality of Coverage (QoC)*. Using the iCAR system as an example, we compare various placement strategies in terms of their QoC values, and provide three rules of thumb as the guidelines for the placement of the ad hoc relay stations in iCAR. This QoC concept can be adopted in general, in deciding the optimum node locations (which need not be specialized ad hoc relay stations) for relaying traffic.

1 Introduction

The demand for high data rate wireless communication and the emergence of various wireless technologies have called for a ubiquitous network architecture integrating and taking advantage of various networking techniques, such as the bandwidth of the wired networks, the coverage of the cellular systems, and the flexibility of the infrastructureless mobile ad hoc networks.

In [1], Hsu and Lin have proposed a novel hybrid architecture called *Multihop Cellular Network (MCN)*, which preserve the benefits of the cellular infrastructure and incorporate the flexibility of ad hoc networks. Similar to the conventional cellular system, MCN relies on the base transceiver stations (BTS) to provide access service. However, instead of requiring all mobile hosts (MHs) to be within the transmission range of the BTSs, the MHs in MCN can connect with each other and communicate with the BTS through multiple hops. Accordingly, the required number of BTSs and/or the transmission range of BTSs and MHs can be reduced. In addition, the direct communication between the MHs can improve the throughput of intra-cell traffic. Similarly, the *opportunity driven multiple access (ODMA)* [2] proposal and the *MACA* [3] scheme also allow the MHs in the cellular system to communicate with each other using the ad hoc networking technique to improve system performance. In [4], the *integrated Cellular and Ad hoc Relaying (iCAR)* system is proposed to address the congestion problem due to limited bandwidth in a cellular system and provide interoperability in heterogeneous

X. Jia, J. Wu, and Y. He (Eds.): MSN 2005, LNCS 3794, pp. 689–700, 2005.

networks (e.g. connecting ad hoc networks and wireless LANs to the Internet). In iCAR, an MH is allowed to use the bandwidth available in a nearby cell (other than the cell it is located in) via relaying through *ad hoc relay stations (ARSs)* which are placed at strategic locations in a system.

In all of these integrated heterogeneous wireless networks involving ad hoc networking technologies, the location management of the ad hoc nodes is a critical design issue. For instance, in MCN, a number of MHs are needed at strategic locations to avoid the network partitioning and the "dead area". In iCAR, the ARSs need to be properly located to improve relaying efficiency. Similarly, the locations of the MHs in ODMA and MACA are also important to the system performance. In this paper, we discuss the location management of the mobile nodes in heterogeneous wireless systems. Instead of using the conventional performance metrics, such as the request blocking and dropping rate, which need either complex analysis or intensive simulations (as the classical models may be not applicable for the heterogeneous system), we define a new performance metric called *Quality of Coverage (QoC)* to evaluate various node placement strategies in a simple and straightforward way. In general, QoC indicates the system resource that can be acquired and utilized by a node to send/receive data or assist the communication of other nodes. The QoC value usually depends on the node connectivity (e.g., the number of nearby active nodes), their battery power, computation capacity, and mobility, the local traffic load, etc. For a particular system, one or several parameters listed above may be used to compute the QoC value.

The rest of this paper, we use iCAR as an representative system for illustrating the QoC concept, because the ARSs in iCAR are fully controlled by the system. Section 2 reviews the basic operations and key benefits of the iCAR system. Section 3 introduces the quality of coverage concept, describes various ARS placement strategies, and presents three rules of thumb as the guidelines to place ARSs. Section 4 discusses the simulation and the performance of the proposed rules of thumb in terms of request blocking and dropping rate. Finally, Section 5 concludes the paper.

2 An Overview of the iCAR System

To illustrate the QoC concept, we give an overview of the iCAR system in this section. An example of relaying is illustrated in Figure 1 where MH X in cell B (congested) communicates with the BTS (BTS A) in cell A (which is non-congested) through two ARSs, one called *proxy* (ARS 1), the other *gateway* (ARS 2). Note that a gateway ARS

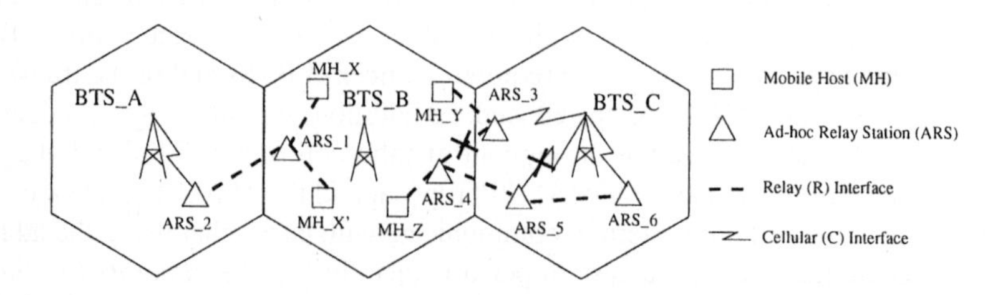

Fig. 1. Relaying examples

needs to have two air interfaces, the **C** (for cellular) interface for communicating with a BTS and the **R** (for relaying) interface for communicating with an MH or another ARS. As an example, the C interface may operate at or around 1900 MHz in the Personal Communication System (PCS), and consumes what we call *cellular bandwidth (CBW)*, and the R interface at or around 2.4 GHz (in the unlicensed Industrial, Scientific, and Medical (ISM) band) which consumes *relaying bandwidth (RBW)*. Also, MHs should have two air interfaces: the C interface for communicating with a BTS (without relaying), and the R interface for communicating with a proxy ARS. However, it is also possible to set aside a (subset of) cellular-band data channel(s) for relaying in order to avoid the need to equip an existing MH with the additional R-interface, while ensure no significant inference between R-interface and C-interface.

Figure 1 also shows that only one ARS, which acts as both the proxy and gateway, may be needed to establish a relaying route (e.g., from MH Y to BTS C). In addition, due to limited power and transmission range, as well as limited CBW and RBW, two or more ARSs may be needed in a cell to form a relaying path. In such a case, a number of intermediate nodes are needed between the proxy and the gateway ARSs. An example of relaying path from MH Z to BTS C is shown in Figure 1, where it is assumed that ARS 3 has no more RBW left (and hence it cannot relay data from ARS 4 to BTS C) and ARS 5 has no CBW left (and hence it cannot relay data to BTS C directly).

There are two basic relaying operations in iCAR, i.e., primary and secondary relaying.

Primary Relaying: In an existing cellular system, if MH X is involved in a new connection request[1] (as a sender or receiver) but it is in a congested cell (e.g., cell B), the new request will be rejected. In iCAR, the request may not have to be rejected. More specifically, MH X which is in the congested cell B, can use the R interface to communicate with an ARS nearby and ultimately with BTS A (see Figure 1 for an example) through what we call primary relaying. Hereafter, we will refer to the process of changing from the C interface to the R interface (or vice versa) at an MH as *switch-over*, which is similar to (but different from) the concept of frequency-hopping [5, 6]. It may be noted that the connection from MH X could also be relayed to another nearby non-congested cell other than cell A. In addition, a relaying route between MH X and its corresponding (i.e., sender or receiver) MH X′ may also be established (in which case, both MHs need to switch over from their C interfaces to their R interfaces), even though the probability that this occurs is typically low.

Secondary Relaying: If primary relaying is not possible, because, for example in Figure 1, ARS 1 is not close enough to MH X to be a proxy (and there are no other nearby ARSs), then one may resort to *secondary relaying* so as to *free up* some CBW from BTS B for use by MH X. Two basic cases are illustrated in Figure 2 (a) and (b), respectively, where MH Y denotes any MH in cell B which is currently involved in an active connection. More specifically, as shown in Figure 2 (a), one may establish a relaying route between MH Y and BTS A (or any other BTSs). In this way, after MH Y switches

[1] The relaying operations for hand-off connections are similar to those for the new connection requests.

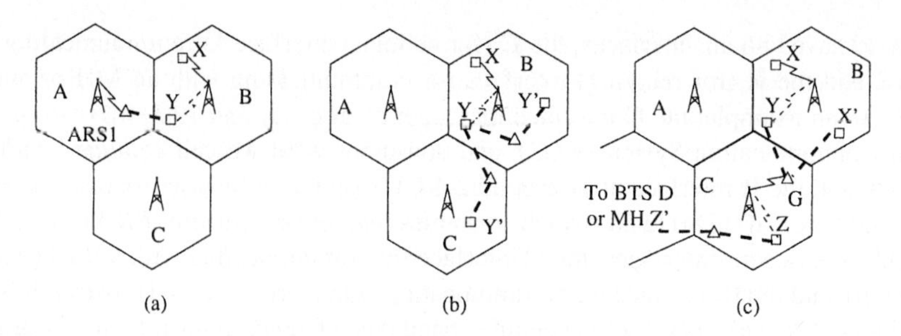

Fig. 2. Secondary relaying to free up bandwidth for MH X. (a) MH Y to BTS A, (b) MH Y to MH Y′.

over, the bandwidth used by MH Y can now be used by MH X. Similarly, as shown in Figure 2 (b), one may establish a relaying route between MH Y and its corresponding MH Y′ in cell B or in cell C, depending on whether MH Y is involved in an intra-cell connection or an inter-cell connection.

If neither primary relaying, nor basic secondary relaying works, the new call may still be supported by employing multiple primary and secondary relaying operations in a cascaded fashion (see Figure 2(c)).

3 Quality of Coverage (QoC)

The major motivation of introducing the concept of QoC is to quantitatively evaluate the mobile node placement via a simple and straightforward parameter, instead of the conventional performance metrics, such as the request blocking and dropping rate which may be obtained by either complex analysis or intensive simulations. But at the same time, a system with higher QoC should have a lower blocking/dropping probability and higher throughput. QoC indicates the system resource that can be acquired and utilized by a node for its communication or assisting the other nodes' communication. The value of QoC usually depends on the node connectivity (e.g., the number of nearby active nodes), their battery power, their computing power, their mobility, the local traffic load, etc. In this section, we use the iCAR system to illustrate the QoC concept. We define the QoC of ARSs and present the analytical results for the comparison between various ARS placement strategies. We also propose three rules of thumb as guidelines for cost-effective ARS placement in iCAR. For simplicity, we assume that there is no bandwidth shortage along any relaying routes in the following discussion (i.e., we have reserved *enough* bandwidth for relaying).

3.1 ARS Placement

The ARS placement is an important design issue in iCAR as iCAR's performance improvement over a conventional cellular system is largely due to its ability of relaying traffic from one cell to another.

The maximum number of relay stations needed for an ideal iCAR system which ensures a relaying route between any BTS and an MH located anywhere in any cell has been discussed in [7]. In case of limited availability of ARSs, an approach called *Seed Growing* may be used. The *seed ARSs* are placed on the boundary region of two cells, and additional ARSs may be placed around these seeds to form clusters to increase the ARS coverage and support the MHs that are far away from the cell boundary, which in turn provide service that is not available in a conventional cellular system. For example, although a hand-off connection involving an MH moving from cell A to cell B (which is congested), may be supported in the cellular system when the MH *just* moves into a congested cell (i.e., the MH is still around the cell boundary) due to the overlapped cell coverage or the soft hand-off (or/and cell breathing) in the CDMA system [8], the call will be eventually dropped while the MH moves farther away from BTS A. But in iCAR, because of the multi-hop relaying via ARSs, the relaying path can extend to any area inside a cell, thus further reduce request blocking or connection dropping probability.

When only partial coverage by ARSs can be provided for an iCAR system (i.e., only a limited number of ARSs are available), some locations are more important than others. For example, since an ARS provides a limited coverage (e.g., within a few hundreds of meters) compared to the size of a cell (e.g., a few kilometers in diameter), placing an ARS in the center of a cell without any nearby ARSs will be useless as it cannot relay any traffic between an MH in the cell and the BTS in another cell.

3.2 QoC of ARSs

We define the value (Q) of the quality of ARS coverage to be the *relayable traffic* in an iCAR system. For simplicity, we have ignored the battery power and the computing power of the mobile nodes, while they can be taken into account with minor modifications. If an ARS can directly or indirectly (i.e., via a multi-hop relaying path) relay traffic in one cell to a BTS in another cell, it has a non-zero Q value which is the amount of traffic covered by the ARS minus the part which will be blocked by the reachable BTSs. More formally, we have the following equation for the value of QoC.

$$Q(t) = \iint_C f(x,y,t) \times (1 - b(\overline{x,y},t))\, dx\, dy \tag{1}$$

where $f(x,y,t)$ is the location-dependent time varying traffic intensity at (x,y) at time t, C is the effective ARS coverage, and $b(\overline{x,y},t)$ is the time varying blocking probability of the cells reachable by the ARSs that cover the location (x,y), excluding the cell where (x,y) is located. When an ARS can reach more than one cells in addition to the cell where (x,y) is located, $b(\overline{x,y},t)$ is the product of their blocking probabilities, which means the probability that all of them are congested. The higher the Q value, the better the ARS placement strategy is. But note that, the effective ARS coverage (and in turn the Q value) is not always proportional to the ARS coverage. In particular, C equals to 0 if the area is covered by ARSs that can neither directly nor indirectly relay traffic in one cell to a BTS in another cell. For example, the Q values of ARS 1 and ARS 2 in Figure 3 are zero because they cannot relay any traffic between cells although they do cover a certain amount of traffic in a cell. In addition, if some area is covered by more than one ARSs, it is counted only once as the effective coverage. In other words, overlapping will not increase C. However, overlapping will help increase the Q value due to the decreased $b(\overline{x,y},t)$).

As shown in Equation 1, the Q value depends on the traffic intensity distribution, and the number, placement and coverage of ARSs. For a particular system, one may place the ARSs at the optimized positions by searching for locations that result in the globally highest Q value. However, this optimization probe may be intractable and in addition, requires a known static traffic intensity function at every location. Thus, in the rest of this section, we show several examples of computing the Q values and provide three rules of thumb which can be used as guidelines for placing ARSs in an iCAR system.

3.3 Various ARS Placement Strategies

In this subsection, we will discuss various ARS placement strategies, evaluate them according to their Q values, and provide several rules of thumb as the guidelines for placing ARSs. We will consider a cell A in Figure 3, which has six neighboring cells with the same traffic intensity. The cell coverage is modeled as hexagon. For simplicity, we assume that the traffic is uniformly distributed in each cell, and only consider the time-averaged values. We denote T_A and T_B to be the traffic intensity in cell A and each of its neighboring cells, respectively, and denote T_a and T_b to be the traffic intensity per unit area in cell A and cell B, respectively. Clearly, $T_A = T_a \times cell\ coverage$ and $T_B = T_b \times cell\ coverage$. In addition, we denote S to be the coverage of one ARS, and M to be the total amount of available cellular bandwidth (CBW) at each BTS, given that one unit of bandwidth is required by one connection. If the MHs' moving speeds and directions are assumed to be uniformly distributed (which is a typical assumption in the mobile computing analysis), the average traffic intensity (i.e., T_a and T_b) will not vary with the mobility of MHs, and consequently, the Q values will not be affected either (which is also verified by our simulation results to be shown later). Hence, the MH mobility will be ignored in the following analysis.

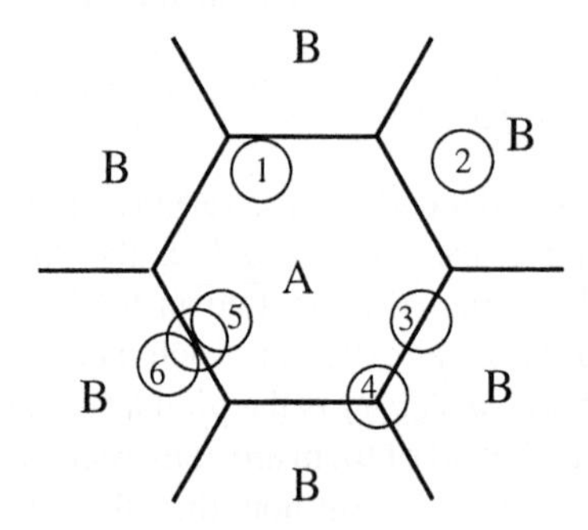

Fig. 3. The ARSs placed in cell A and its six neighbors

Seed ARS Placement. We first discuss where to put the seed ARS to achieve the best performance. Considering cell A in Figure 3, there are two approaches to place the seed ARS. One is to put it at the shared border of two cells (see ARS 3 in Figure 3), and the other is to put it at a vertex (see ARS 4 in Figure 3). An ARS in the two approaches has the same effective coverage. However, their Q values may not be the same. On one hand, assuming cell A is the hot spot, the ARS placed at the border will cover more high traffic intensity area than that covered by the ARS at a vertex. But on the other hand, the ARS at the vertex may relay traffic to two BTSs instead of one in the border approach, which in turn results in a lower blocking probability for the relaying requests. Applying Equation 1 in the above two approaches, we may have the following equations for their Q values.

$$Q_{Border} = \frac{S}{2} \cdot T_a \cdot (1 - b_B) + \frac{S}{2} \cdot T_b \cdot (1 - b_A) \quad (2)$$

$$Q_{Vertex} = \frac{S}{3} \cdot T_a \cdot (1 - b_B^2) + \frac{2S}{3} \cdot T_b \cdot (1 - b_A \cdot b_B) \quad (3)$$

in which b_A and b_B are the blocking probabilities without relaying in cell A and cell B. More specifically, $b_A = \frac{T_A^M/M!}{\sum_{i=0}^{M} T_A^i/i!}$ and $b_B = \frac{T_B^M/M!}{\sum_{i=0}^{M} T_B^i/i!}$, according to the Erlang-B model [9].

As shown in Equations 2 and 3, M will affect the Q values, because a different M may result in different b_A and b_B. However, although the results are not shown here, varying M within the normal range (e.g., choosing the values of M so that the blocking probability is around 2%), will not affect the performance comparison of various ARS placement approaches. Thus, in the following discussions, we will only consider the case where M is equal to the default value (50). In addition, although we have assumed that all neighboring cells of cell A have the same traffic intensity (T_B) and blocking probability (b_B), the analytical model can be easily extended to the case where the cell B's have different traffic intensities by replacing T_B and b_B with the corresponding values in Equations 2 and 3.

We have obtained Q_{Border} and Q_{Vertex} as shown in Figure 4 by varying T_A and T_B. The Q values of these two approaches depend on the traffic intensity in cell A and its neighboring cells. Approximately, when traffic intensity is high ($T_A, T_B > 50\ Erlangs$) or $T_A < T_B$, Q_{Vertex} is higher than Q_{Border}. But, when $T_B < T_A < 50\ Erlangs$, the border placement is better. Note that, $50\ Erlangs$ corresponds to about 10% blocking probability which is far beyond the normal network operation range. If we assume cell A is the hot spot, i.e., $T_A > T_B$, then the border placement approach is usually a good choice for seed ARSs. So, we may have the first rule of thumb as follows.

Rule of Thumb 1: *Place the seed ARSs at cell borders.*

In addition, it has been shown in [7] that, for an n-cell system, the maximum number of seed ARSs needed for each shared border of two cells is $3n - \lfloor 4\sqrt{n} - 4 \rfloor$.

Seed ARS vs. grown ARS. If additional ARSs are available, there are two approaches to place them. One is to place them as seeds according to what we discussed in Sec 3.3 without any overlap on the existing ARSs. This approach intends to maximize the total effective ARS coverage. The other way is to let them grow from the seeds which are already there (see ARS 5 shown in Figure 3, assuming border placement approach is adopted). The grown ARS is required to be within the coverage of at least one existing ARS so that they can relay traffic to each other. Thus, their coverage overlaps within some area, and not all of the area covered by the grown ARS will result in the increase

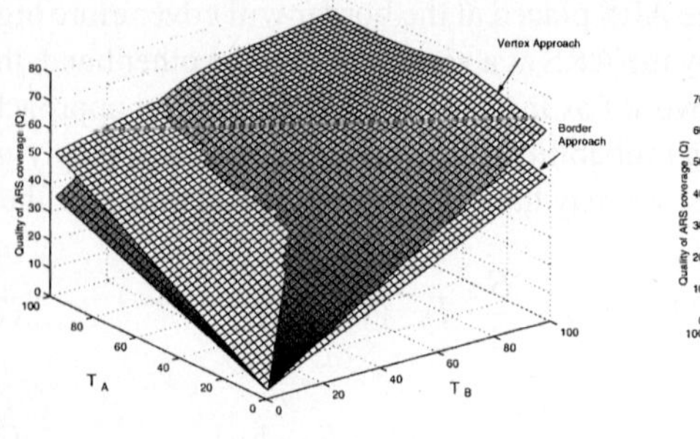

Fig. 4. Quality of ARS coverage: vertex placement vs. border placement

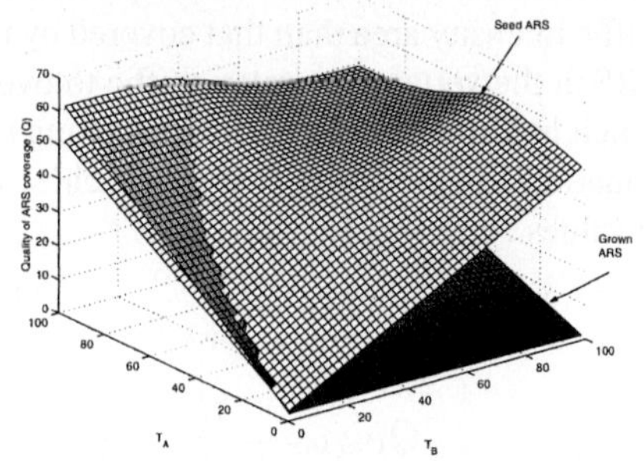

Fig. 5. Quality of ARS coverage: seed ARS vs. grown ARS

of the system's Q value. To minimize the overlapped area and maximize the effective coverage of the grown ARS, we place it just within the transmission range of the existing seed ARS (i.e., let the distance between the two ARSs as far as possible while they can still communicate with each other). Accordingly, we can compute the additional coverage of the grown ARS (i.e., its coverage minus the overlapped area), which is about $\frac{\pi r^2 - 2(\frac{1}{6}\pi r^2 - \frac{\sqrt{3}}{4}r^2)}{\pi r^2} \approx 0.61S$, where r is the radius of an ARS coverage area, and the increased Q value

$$Q_{Grow\_in} \approx 0.61 \cdot S \cdot T_a \cdot (1 - b_B) \tag{4}$$

assuming it grows inward cell A. Comparing it with Q_{Border} (i.e., placing a new seed ARS at cell border) in Figure 5, as we can see, only when T_B is very low and T_A is much higher than T_B, the grown ARS approach performs better than the seed ARS approach, and therefore we have the second rule of thumb.

Rule of Thumb 2: *Place an ARS as seed if it is possible.*

The direction of growing. If the additional ARSs can not be placed as seeds because there is no free space at the shared boundaries of cells, we have to let them grow from some of the seeds. An ARS can grow inward cell A (see ARS 5 in Figure 3) or outward cell A (see ARS 6 in Figure 3). Both of them have the same ARS coverage (S). But since the ARSs cover different cells with different traffic intensities, they may result in different Q values. When an ARS grow inward cell A, its Q value (Q_{Grow_in}) has been shown in Equation 4. When an ARS grows outward cell A, its Q value is

$$Q_{Grow\_out} \approx 0.61 \cdot S \cdot T_b \cdot (1 - b_A) \tag{5}$$

Similarly, we compute the Q values of these two approaches and conclude the third rule of thumb.

Rule of Thumb 3: *Grow an ARS toward the cell with high traffic intensity.*

The three rules of thumb may serve as the guidelines for ARS placement. More specifically, to optimize the system performance, the operators may first place ARSs at the shared borders of the cells. If there are additional ARSs, they may let them grow in the cell with higher traffic load. However, depending on the size of the cells and the coverage of the ARSs, there may be some exceptions. For example, when a number of seed ARSs have been deployed in a system, placing another seed ARS later may result in some overlap with the existing ARSs, and therefore result in a lower Q value. In this case, growing the additional ARSs may result in a better performance. Similarly, when there are already many ARSs growing in the cells with high traffic intensity, placing an ARS in the neighboring cells may be more beneficial.

4 Simulation and Discussions

To evaluate the performance of various ARS placement approaches in terms of the system-wide (i.e., weighted average) request blocking and dropping probability, we have developed a simulation model using the GloMoSim simulator [10] and the PAR-SEC language [11]. The simulated system includes a cell A and six neighboring cells (see Figure 6), which are modelled as hexagons with the center-to-vertex distance of 2 *Km*. We have assumed that 50 units of bandwidth are allocated for one cell, and for simplicity, each connection requires 1 unit bandwidth. In order to obtain converged statistical results, we have simulated $6,400$ MHs which are uniformly distributed in the system, and run the simulation for 100 hours for each traffic intensity before collecting the results. The traffic intensity is measured in Erlangs which is the product of the request arrival rate (Poisson distributed) and the holding time (exponentially distributed). In addition, we have used location dependent traffic pattern by default. More specifically, assuming cell A is the hot spot, the traffic intensity in cell B is about 80% of that in cell A.

Six ARSs with 500 m transmission range have been simulated in four scenarios (see Figure 6 (a)-(d)), which implement the different ARS placement approaches described in Section 3. Figures 6(a) and (b) show six seed ARSs placed according to the border and the vertex approaches, respectively, while in Figure 6(c) and (d), there are 3 seed ARSs placed at the borders and 3 additional ARSs growing from the seeds inward and outward cell A, respectively.

We have obtained Q values of all six ARSs for different placement approaches from the simulation, and compared them with the analytical results in Figures 7-8. As we can see, the analytical results (in Figure 7) and simulation results (in Figure 8) show a very

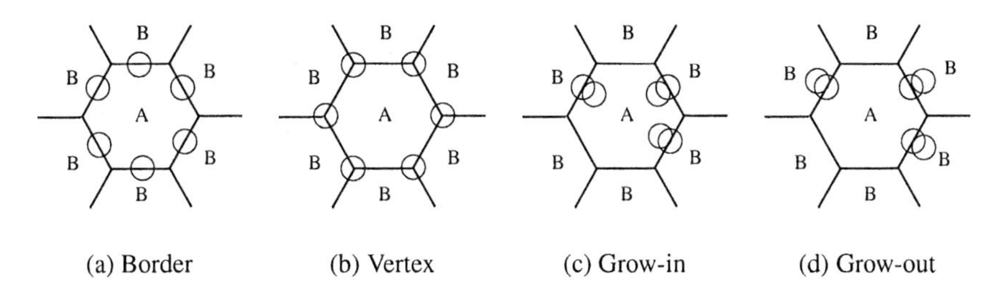

Fig. 6. Four scenarios of ARS placement in the simulated system

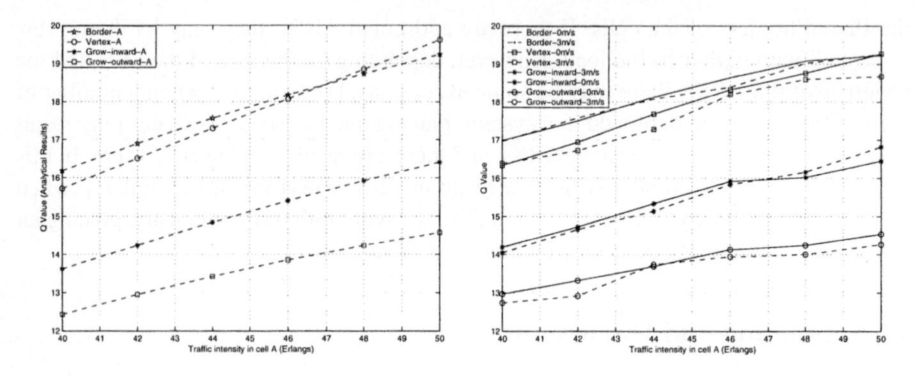

Fig. 7. Analytical Results; $T_b = 0.8T_a$

Fig. 8. Simulation Results; $T_b = 0.8T_a$

similar trend. The reason that the Q values obtained from the simulation are usually higher than those from the analysis is that the blocking probability without relaying is used in Equation 2 through 5, which is higher than the real blocking probability in iCAR (with relaying). In addition, as shown in Figure 8, the mobility of MHs has little affect on the Q values (although it does affect the connection dropping probability as to be shown later). In all cases within the normal operation range of an iCAR system (e.g., the traffic intensity of cell A is from 40 to 50 Erlangs), the grown ARSs yield lower Q values than that of the seed ARSs as we expected. However, the Q values of the border and the vertex approaches are very close, and when the traffic intensity is high, the vertex approach may result in higher Q values. This is because even though the ARSs in the border approach still cover more active connections in such a situation, a large fraction of the covered connections is *nonrelayable* because of the high blocking probability in the neighboring cells. On the other hand, the covered connections in the vertex approach may be relayed to either of the two neighboring cells, and therefore has higher Q values. As we discussed earlier, the real blocking probability of the cells in iCAR is lower than that we used in the analysis, thus the intersection point of the curves representing the Q values of the border and the vertex approaches in the simulation occurs at a higher traffic intensity than that in analysis (comparing Figures 7 and 8).

The request blocking rates of MHs in the systems with different ARS placement approaches are shown in Figure 9. As we can see, an iCAR system with a higher Q value usually has a lower blocking rate. The results have also verified the usefulness of the three rules of thumb established in Section 3. More specifically, the border ARS placement has the lowest blocking rate among all of these approaches, which may be kept below 2% (the acceptable level) even when the traffic intensity is as high as 50 $Erlangs$. As a comparison (though the results are not shown), if six ARSs are randomly placed in the seven cells of the system (with the Q value being close to 0), the request blocking rate is from about 2% to above 10% when the traffic intensity of cell A increases from 40 to 50 Erlangs.

Although the MHs mobility may affect the dynamics in relaying capability of an iCAR system due to switch-over, our results indicate that the blocking rates in all ARS

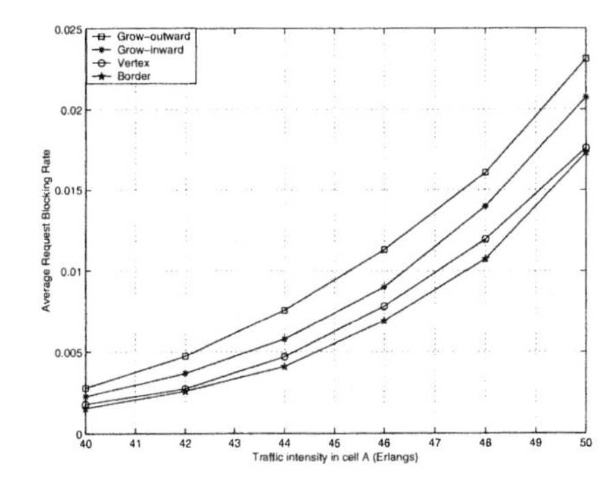

Fig. 9. Call Blocking rates for different ARS placement approaches, MH Speed=0m/s

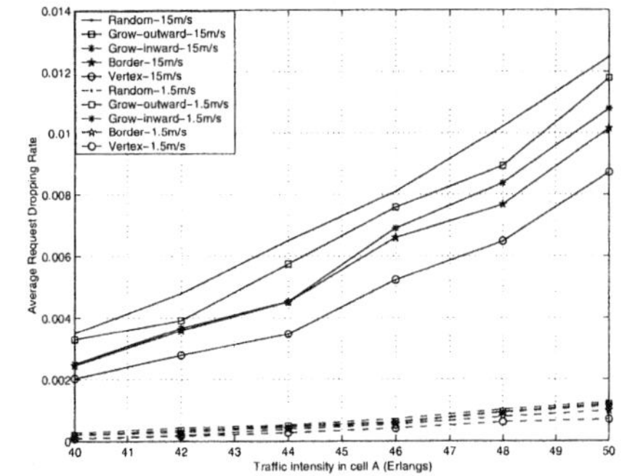

Fig. 10. Call Dropping rates for different ARS placement approaches

placement approaches increase very little with the MHs mobility. On the other hand, MH mobility affects the connection dropping probability significantly (see Figure 10[2]). More specifically, the dropping probability increases from 0 to the order of 10^{-3} and 10^{-2}, respectively, when the maximum MH moving speed increases from $0m/s$ to $1.5m/s$ and $15m/s$. Although the seed ARS placement approaches (i.e., the border and the vertex approaches) still perform better than the grown ARS placement approaches in terms of connection dropping rate, the difference among them is not as obvious as that in terms of the connection blocking rate. In addition, note that the vertex approach has a lower connection dropping rate than that of the border approach. This is because, when an active MH moves from one cell (i) to another cell (j), although there is the same probability that the MH is covered by ARSs at the moment crossing the shared border of the two cells, the ARS coverage is $\frac{2S}{3}$ in cell j in the vertex approach, which is larger than that in the border approach ($\frac{S}{2}$). The larger ARS coverage implies the longer time that the ARS can support the MH via relaying, and consequently results in a lower connection dropping rate.

We note that the proposed rules of thumb are based on the assumption that each ARS has an unlimited bandwidth at its R and C interface, that is, it can relay as many connections as needed to a BTS (provided that the BTS has free bandwidth). However, when there is only limited CBW, the grow-outward (cell A) approach may not be affected as much as the grow-inward approach, and hence the two approaches may perform equally well (or bad). This is because the amount of CBW determines the amount of traffic that can be relayed from cell A to its neighboring cells and consequently becomes the performance bottleneck, and placing an ARS outside cell A will increase the total amount of CBW (used to relay traffic from cell A to cell B) available to the ARS cluster, while placing inside cell A will not. Nevertheless, in a real system, only the connection requests that would be blocked without relaying, which is a small portion (e.g. about 5% of total requests if the initial blocking rate is 5%), will be supported by relaying although the relayable traffic (i.e., the Q value) may be much higher than that, and thus

[2] For simplicity, we assume that there is no priority given to the hand-off attempts over new connection attempts.

the assumption of having enough relaying bandwidth is valid in most situations, and the presented rules of thumb will be good guidelines for ARS placement.

5 Conclusion

In this paper, we have addressed the location management issue in heterogenous networks. In particular, we have defined a new performance metric called *Quality of Coverage (QoC)*, and used iCAR as an representative system. We have compared various placement strategies in terms of their QoC values, and provided three rules of thumb as the guidelines for the placement of ARSs in iCAR. The performance of the proposed ARS placement strategies have been evaluated via both analysis and simulations. We expect that the concept of QoC along with the results and guidelines will be useful not only for managing ARS placement (and their limited mobility) in iCAR, but also for planning of and routing in other integrated heterogeneous wireless systems including the wireless system with controllable mobile base stations or access points.

References

1. Y.D.Lin and Y.C.Hsu, "Multihop cellular: A new architecture for wireless communication," in *IEEE INFOCOM'2000*, pp. 1273–1282, 2000.
2. http://www.3gpp.org/.
3. X.-X. Wu, B. Mukerjee, and S.-H. G. Chan, "Maca – an efficient channel allocation scheme in cellular networks," in *IEEE Global Telecommunications Conference (Globecom'00)*, vol. 3, pp. 1385–1389, 2000.
4. H. Wu, C. Qiao, S. De, and O. Tonguz, "Integrated cellular and ad-hoc relay systems: iCAR," *IEEE Journal on Selected Areas in Communications special issue on Mobility and Resource Management in Next Generation Wireless System*, vol. 19, no. 10, pp. 2105–2115, Oct. 2001.
5. G. Stuber, *Principles of Mobile Communication.* Kluwer Academic Publishers, 1996.
6. R. Kohno, R. Meidan, and L. Milstein, "Spread spectrum access methods for wireless communications," *IEEE Communications Magazine*, vol. 33, no. 1, pp. 58–67, 1995.
7. C. Qiao and H. Wu, "iCAR : an integrated cellular and ad-hoc relay system," in *IEEE International Conference on Computer Communication and Network*, pp. 154–161, 2000.
8. A. Viterbi, *CDMA Principles of Spread Spectrum Communication.* Addison-Wesley, 1996.
9. R. L. Freeman, *Telecommunication System Engineering.* John Wiley & Sons Inc., 1996.
10. X. Zeng, R. Bagrodia, and M. Gerla, "GloMoSim: A library for parallel simulation of large-scale wireless networks," in *Proc. Workshop on Parallel and Distributed Simulation*, pp. 154–161, 1998.
11. R. Bagrodia, R. Meyer, M. Takai, Y. Chen, X. Zeng, J. Martin, B. Park, and H. Song, "Parsec: A parallel simulation environment for complex systems," *Computer*, pp. 77–85, Oct. 1998.

ACOS: A Precise Energy-Aware Coverage Control Protocol for Wireless Sensor Networks

Yanli Cai[1], Minglu Li[1], Wei Shu[2], and Min-You Wu[1,2]

[1] Department of Computer Science and Engineering,
Shanghai Jiao Tong University, Shanghai 200030, China
[2] Department of Electrical and Computer Engineering,
The University of New Mexico, Albuquerque, New Mexico, USA
{cai-yanli, li-ml,wu-my}@cs.sjtu.edu.cn,shu@ece.unm.edu

Abstract. A surveillance application requires sufficient coverage of the protected region while minimizing the energy consumption and extending the lifetime of sensor networks. This can be achieved by putting redundant sensor nodes to sleep. In this paper, we propose a precise and energy-aware coverage control protocol, named Area-based Collaborative Sleeping (ACOS). The ACOS protocol, based on the net sensing area of a sensor, controls the mode of sensors to maximize the coverage, minimize the energy consumption, and to extend the lifetime of the sensor network. The simulation shows that our protocol has better coverage of the surveillance area while waking fewer sensors than other state-of-the-art sleeping protocols.

1 Introduction

A wireless sensor network consists of a set of inexpensive sensors with wireless networking capability [1]. Applications of wireless sensor networks include battlefield surveillance, environment monitoring and so on [2].

As sensors may be distributed arbitrarily, one of the fundamental issues in wireless sensor networks is the coverage problem. The coverage of a sensor network, measured by the fraction of the region covered, represents how well a region of interest is monitored. On the other hand, a typical sensor node such as an individual mote, can only last 100-120 hours on a pair of AA batteries in the active mode [3]. Power sources of the sensor nodes are non-rechargeable in most cases. However, a sensor network is usually desired to last for months or years. Sleeping protocols to save energy are under intensive study, such as RIS [4, 5], PEAS [6] and PECAS [4]. These protocols presented different approaches to utilizing resources, but needs further improvement in coverage or efficient energy consumption.

Here we propose a sleeping protocol, named Area-based Collaborative Sleeping (ACOS). This protocol precisely controls the mode of sensors to maximize the coverage and minimize the energy consumption based on the *net sensing area* of a sensor. The net sensing area of a sensor is the area of the region exclusively covered by the sensor itself. If the net sensing area of a sensor is less than a given threshold, the sensor will go to sleep. Collaboration is introduced to the protocol to balance

X. Jia, J. Wu, and Y. He (Eds.): MSN 2005, LNCS 3794, pp. 701 – 710, 2005.

the energy consumption among sensors. Performance study shows that ACOS has better coverage and longer lifetime than other sleeping protocols.

The rest of the paper is organized as follows. Section 2 discusses previous research. Section 3 describes the basic design of the protocol. Section 4 improves the baseline ACOS for better performance. Section 5 provides a detailed performance evaluation and comparison. We conclude the paper in Section 6.

2 Related Work

Different coverage methods and models have been surveyed in [7, 8, 9]. Three coverage measures are defined [7], which are area coverage, node coverage, and detectability. Area coverage represents the fraction of the region covered by sensors and node coverage represents the number of sensors that can be removed without reducing the covered area, while detectability shows the capability of the sensor network to detect objects moving in the network. Centralized algorithms to find exposure paths within the covered field are presented in [8]. In [9], the authors investigate the problem of how well a target can be monitored over a time period while it moves along an arbitrary path with an arbitrary velocity in a sensor network.

Power conservation protocols such as GAF [10], SPAN [11] and ASCENT [12] have been proposed for ad hoc multi-hop wireless networks. They aim at reducing the unnecessary energy consumption during the packet delivery process. In [13], a heuristic is proposed to select mutually exclusive sets of sensors such that each set of sensors can provide a complete coverage. In [14], redundant sensors that are fully covered by other sensors are turned off to reduce power consumption, while the fraction of the area covered by sensors is preserved.

Sleeping protocols such as RIS [4, 5], PEAS [6] and PECAS [4] have been proposed to extend the lifetime of sensor networks. In RIS, each sensor independently follows its own sleep schedule which is set up during network initialization. In PEAS, a sensor sends a probe message within a certain probing range when it wakes up. The active sensor replies to any received probe message. The sensor goes back to sleep if it receives replies to its probes. In PEAS, an active node remains awake continuously until it dies. PECAS makes an extension to PEAS. Every sensor remains within active mode only for a duration and then goes to sleep.

3 Basic Protocol Design

In this section, we describe the basic design of ACOS protocol. This protocol precisely controls the mode of sensors so that the coverage of the sensor network can be maximized and the energy consumption minimized.

3.1 Notations and Assumptions

We adopt the following notations and assumptions throughout the paper.

- Consider a set of sensors $S = \{s_1, s_2, \ldots, s_n\}$, distributed in a two-dimensional Euclidean plane.

- Sensor s_j is referred as a *neighbor* of another sensor s_i, or vice versa, if the Euclidean distance between s_i and s_j is less than $2r$.
- Assume that each sensor knows its own location [15, 16, 17, 18]. As shown in Section 3.2, relative locations [19] are enough for our protocol.
- A sensor has two power consuming modes: *low-power* mode and *power-consuming* mode. Power-consuming mode is also called *active* mode.
- The *sensing region* of each sensor is a disk, centered at the sensor, with radius r, its *sensing range*. The *net sensing region* of sensor s_i is the region in the sensing range of s_i but not in the sensing range of any other active sensor. The *net sensing area* or *net area* of s_i is the area of the net sensing region. The *net area ratio*, denoted as a_i, is the ratio of s_i's net sensing area to s_i's maximal sensing area, πr^2.
- The *net area threshold*, denoted as φ, is a parameter between 0 and 1.

3.2 The Net Area Calculation

The shadowed region with bold boundary in Fig.1 shows an example of the net sensing region of sensor s_0. Before detailed description of ACOS, a solution to computing the net area is presented here.

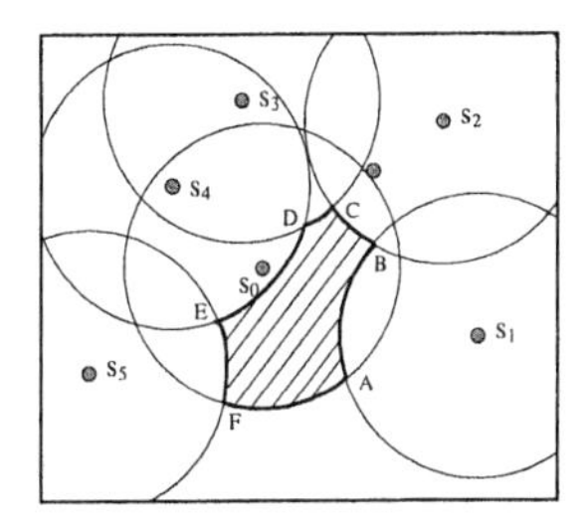

Fig. 1. The net sensing region of a sensor s_0

We use the algorithm in [20] to find boundaries of the net sensing regions inside a sensor, which is referred as *perimeter-coverage algorithm*. The perimeter-coverage algorithm costs polynomial time to find the coverage of the protected region by considering how the perimeter of each sensor's sensing region is covered. Consider sensor s_i, a segment of s_i's perimeter is *k-perimeter-covered* if all points on the segment are in the sensing range of at least k sensors other than s_i itself.

Consider s_0 as shown in Fig.1, we first find the 0-perimeter-covered segments of s_0's perimeter, the minor arc $\overset{\frown}{FA}$ in this example. Then in the sensing range of s_0, we find the 1-perimeter-covered segments for each of s_0's neighbors, the minor arcs $\overset{\frown}{AB}$, $\overset{\frown}{BC}$, $\overset{\frown}{CD}$, $\overset{\frown}{DE}$, $\overset{\frown}{EF}$ in this example. After all the segments are found, two segments are jointed together if they have common end points. The closed boundary of each net sensing region is determined by a segment sequence.

After finding the boundaries of each net sensing region, the area of a net sensing region can be computed by calculating the area of the polygon formed by its segment sequence, the polygon *ABCDEF* in this example, and the area of the region between each segment of arc and the corresponding chord. Here, each node only needs to know the relative locations of its neighbors. To determine the relative location among sensors is easier than determining each node's absolute location.

3.3 Basic Protocol Design

For simplicity of description, the basic design of ACOS protocol is presented in this section, which is called "baseline ACOS", leaving other intricate problems to be addressed in Section 4.

Each sensor node has four states: Sleep state, PreWakeUp state, Awake state, and Overdue state. The Sleep state corresponds to the low-power mode. The PreWakeUp, Awake, and Overdue states belong to the active mode. The PreWakeUp is a transit state and lasts for a short period of time, while the Awake and Overdue states may last for several minutes or hours. Every sensor remains in the Awake state for no more than $T_{wake\_Duration}$. The state transition diagram of the baseline ACOS is show in Fig.2.

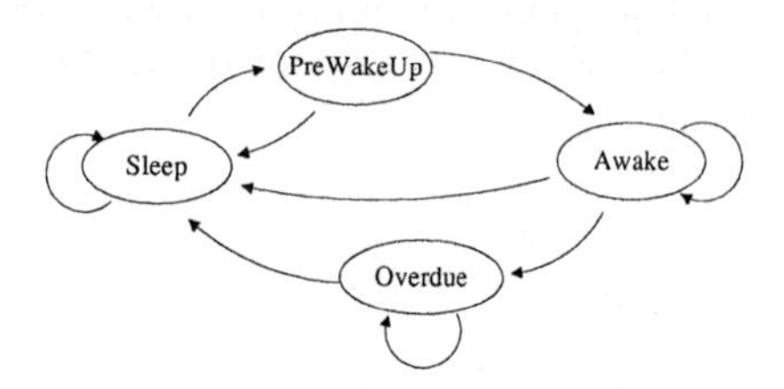

Fig. 2. State transition diagram of baseline ACOS

Consider any sensor s_i with a decreasing sleep timer $T^i_{sleep\_left}$ to represent the time left before s_i wakes up again. Sensor s_i also keeps a decreasing wake timer $T^i_{wake\_left}$, which is initialized as $T_{wake\_Duration}$ when the sensor turns from the low-power mode to the active mode. The value of $T_{wake\_Duration} - T^i_{wake\_left}$ indicates how long the sensor has been in the active mode since it turned from the low-power mode to the active mode. Sensor s_i also maintains an active neighbor list $nList^i$, collecting information from every message received. For any neighbor s_k in $nList^i$, s_k's location and $T^k_{wake\_left}$ are also stored in $nList^i$. All the above timers decrease as the time progresses, as shown in event *e0* in Fig.3.

When s_i wakes up, its state changes from Sleep to PreWakeUp. It broadcasts a message PreWakeUp_Msg to its neighbors within radius $2r$ and waits for T_w seconds. When any neighboring sensor s_j is in Awake state and receives this message, s_j sends back a Reply_PreWakeUp_Msg including its location and $T^j_{wake\_left}$. Upon receipt of a Reply_PreWakeUp_Msg from any neighbor s_j, s_i extracts the location of s_j and $T^j_{wake\_left}$ and stores them into its $nList^i$. At the end of T_w, s_i computes the net area ratio a_i.

If a_i is less than φ, it shows that s_i is not contributing enough coverage, and it is unnecessary for s_i to work at this moment. So it returns back to Sleep state and sleeps for a period time of the minimum value of all $T^k_{wake\_left}$ from $nList^i$. It is possible that several neighbors around an active sensor s_j get its $T^j_{wake\_left}$ and all wake up at the same time. The consequence is that not only they contend with communications, but also most of them may decide to start to work because of unawareness of each other. To avoid this situation, a random offset ε can be added to the sleep time.

The following is the event that occurs for any sensor s_i

Event e0: the clock of s_i ticks one time

if(s_i is in Sleep state){
  s_i's sleep timer $T^i_{sleep\_left} = T^i_{sleep\_left}$ -1;
}**else**{
  s_i's wake timer $T^i_{wake\_left} = T^i_{wake\_left}$ -1;
  Update the timers in local neighbor list $nList^i$, for any $s_k \in nList^i$, $T^k_{wake\_left} = T^k_{wake\_left}$ -1;
}

The following is the event that occurs when sensor s_i is in Sleep state

Event e1: the sensor s_i's sleep timer $T^i_{sleep\_left}$ has decreased to zero

Change to PreWakeUp state
Broadcast a PreWakeUp_Msg within radius $2r$;
Within T_w seconds, upon receipt of a Reply_PreWakeUp_Msg from neighbor s_j, *extract* the location of s_j and $T^j_{wake\_left}$ and *store* them into $nList^i$;
Compute the net area ratio a_i;
if ($a_i < \varphi$){
  Set $T^i_{sleep\_left} =$ **Min**{ $T^k_{wake\_left}$, for $s_k \in nList^i$ } + ε, ε is a random offset;
  Clear $nList^i$ and *change* back to Sleep state;
}**else**{
  Change to Awake state and *set* timer $T^i_{wake\_left} = T_{Wake\_Duration}$;
  Broadcast a Wake_Notification_Msg including its location and $T^i_{wake\_left}$ within radius $2r$;
}

The followings are events that occur only when sensor s_i is in Awake state

Event e2: sensor s_i receives a PreWakeUp_Msg from s_j

Reply s_j with Reply_PreWakeUp_Msg, including its location and $T^i_{wake\_left}$;

Event e3: the timer $T^i_{wake\_left}$ of the sensor s_i has decreased to zero

Change to Overdue state;

The followings are events that occur when sensor s_i is in Awake or Overdue state

Event e4: sensor s_i receives a Sleep_Notification_Msg from s_j

Remove s_j from $nList^i$;

Event e5: sensor s_i receives a Wake_Notification_Msg

Update $nList^i$ and *compute* the net area ratio a_i;
if ($a_i < \varphi$){
  Broadcast a Sleep_Notification_Msg within radius $2r$;
  Set $T^i_{sleep\_left} =$ **Min**{ $T^k_{wake\_left}$, for $s_k \in nList^i$ } + ε, ε is a random offset;
  Clear $nList^i$ and *change* to Sleep state;
}

Fig. 3. The events of baseline ACOS

If a_i is equal or greater than φ, s_i changes to Awake state, initialize its wake timer $T^i_{wake\_left}$ and broadcasts a Wake_Notification_Msg including its location and $T^i_{wake\_left}$ to its neighbors, as described by event *e1* in Fig.3.

When s_i is still in the Awake state and hears a PreWakeUp_Msg from s_j, it replies s_j with a Reply_PreWakeUp_Msg, including its $T^i_{wake\_left}$. Although sensor s_i in its Overdue state is also in the active mode, it does not reply to PreWakeUp_Msg, so that s_i is not counted by newly waked up sensors and is more likely able to go to sleep in a short time. This is how energy consumption balance among sensors is achieved. This procedure is described in event *e2* in Fig.3. When s_i is in the Awake state and its wake timer $T^i_{wake\_left}$ has decreased to zero, it changes from Awake to Overdue state, as shown in event *e3* in Fig.3.

When s_i is in the Awake or Overdue state and hears a Wake_Notification_Msg, it updates its list $nList^i$ and recalculates the net area ratio a_i first. If a_i is less than φ, this indicates that s_i can go to sleep safely, and therefore broadcasts a Sleep_Notification_Msg to its neighbors. Then it changes to the Sleep state and sleep for the minimum value of all $T^k_{wake\_left}$ from $nList^i$. Also, a random offset is added to the sleep time. This procedure is described in event *e5* in Fig.3.

If s_i is in the Awake or Overdue state and hears a Sleep_Notification_Msg from s_j, s_i removes s_j entry from $nList^i$, as shown in event *e4* in Fig.3.

Fig.3 demonstrates all events that occur in our protocol. The events drive a sensor to change from one state to another and precisely control its power consuming modes.

4 Optimizations

In the basic design of the ACOS protocol, two problems are not addressed. One problem is the unawareness of dead neighbors. When a sensor s_i receives a *Wake_Notification_Msg*, it computes its net area ratio a_i. The calculation of a_i depends on information stored in the local neighbor list $nList^i$. The information may be outdated, because some of neighbors may have died from physical failure or energy depletion without notification. The other problem is about sleep competition caused by waking up a sensor. Consider sensor s_i that decides to wake up after computing the net sensing area ratio a_i. It then broadcasts a *Wake_Notification_Msg* to its neighbors, and several neighbors may receive this message. Each of them computes their net area ratio without collaboration, and many of them may go to sleep. We call this situation *multiple sleeps*. In some cases, multiple sleeps are needed to reduce overlap, but in other cases, multiple sleeps should be avoided.

In this Section, we modify the baseline ACOS to solve dead neighbors problems and to reduce the effect of multiple sleeps problems by adding a new transit PreSleep state.

4.1 Dealing with Dead Neighbors

When sensor s_i receives a Wake_Notification_Msg and its net area ratio a_i is less than φ, it changes to PreSleep state and clears the current information in its $nList^i$. Then it broadcasts a PreSleep_Msg to its neighbors and waits for T_w seconds. When neighbor s_j is in its Awake or Overdue state and hears this message, s_j sends back a

Reply_PreSleep_Msg including its location and $T^j_{wake\_left}$. At the end of T_w, s_i recomputes the net area ratio a_i'. If a_i' is equal or greater than φ, this indicates that some neighbors died after the last time s_i woke up and that s_i should not go to sleep at the moment.

4.2 Dealing with Multiple Sleeps Caused by a Waking Up Sensor

The protocol is enhanced by making the neighbors that are ready to sleep collaborate with each other. When sensor s_i receives a Wake_Notification_Msg from s_j, it updates its net area ratio a_i' and broadcasts a SleepIntent_Msg to its neighbors, including a_i'. Within T'_w seconds, it receives SleepIntent_Msg from its neighbors, who are intent to sleep too. At the end of T'_w seconds, it chooses the sensor s_k who has the minimum value of net area among the neighbors from whom a SleepIntent_Msg had been received.

If a_i' is greater than s_k's net area ratio a_k', then s_i does not hold the minimum net area ratio and its neighbor s_k may go to sleep. Then s_i re-computes the net area ratio a_i'' by regarding s_k as a sleep node. If a_i' is less than a_k', it indicates that s_i does have the minimum net area ratio. If a_i'' is less than φ, it indicates the sleep of s_k does not largely increase s_i's net area. So s_i could go to sleep relatively safely in the case of a_i'< a_k' or a_i''< φ. We plan to find a more efficient strategy to select a set of s_j's neighbors and enable them to sleep, in order to achieve better coverage while making more s_j's neighbors sleep in future.

5 Performance Evaluation

In this section, we implemented ACOS and other three protocols RIS [4, 5], PEAS [6] and PECAS [7] for comparison. We evaluate the coverage with different node density using our protocol in Section 5.1, and compare the coverage achieved with an equal number of active nodes for all four protocols in Section 5.2.

In our simulation, the sensing range of each sensor is set as 20 meters, i.e. $r = 20m$, and the communication range is 40m. The sensors are uniformly distributed in a $400m \times 400m$ region, with bottom-left coordinate (0, 0) and top-right coordinate (400, 400). In order to evaluate the relations between coverage and different node densities, the numbers of distributed sensors are 400, 800, 1600 and 3200, respectively with density 1, 2, 4 and 8 per square of $r \times r$. From now on, we will abbreviate "square of $r \times r$" as "r-square."

In RIS, the time is divided into time slots of equal length T_{slot} at each sensor. Each T_{slot} is divided into two parts, the active period and the sleeping period. The duration of the active period is $p* T_{slot}$, where p depends on applications, and the sleeping period takes the rest part of a time slot. In PEAS, probing range *Rp* is given by the application depending on the degree of robustness it needs. In PEAS, a working node remains awake continuously until its physical failure or depletion of battery power. In PECAS, every sensor remains within the active mode for a duration indicated by parameter *Work_Time_Dur* each time it wakes up.

5.1 The Coverage of ACOS

In Fig.4(a), we can see the number of active nodes ascends sharply as the net area threshold goes smaller. When $\varphi = 0$, all nodes are active, however, even if φ is a small non-zero number, the number of active nodes is much smaller than the total number of nodes. Fig.4(b) shows that the maximal coverage may be approached by much fewer active sensors than the total number. For example, when the node density is 8 per r-square, i.e. 3200 sensors in total, ACOS wakes up only 361 nodes but covers 98.5% of the whole region. Fig.4(c) illustrates that the coverage percentage is approximately linear to the net area threshold φ.

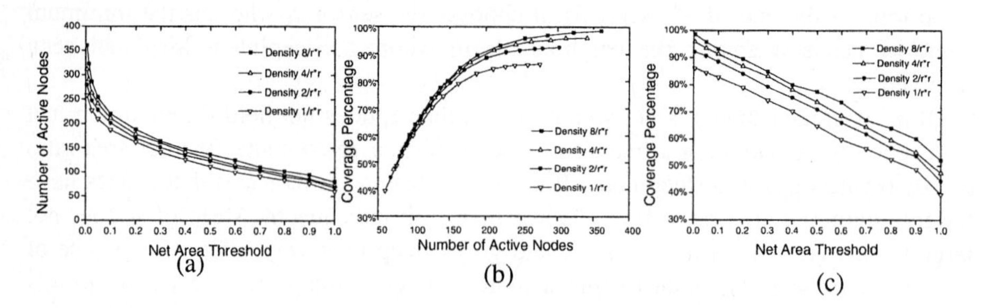

Fig. 4. The coverage over node density using ACOS protocol

5.2 Comparison of Coverage

We evaluate the coverage over the roughly equal number of active sensors in the case of 800 sensors deployed.

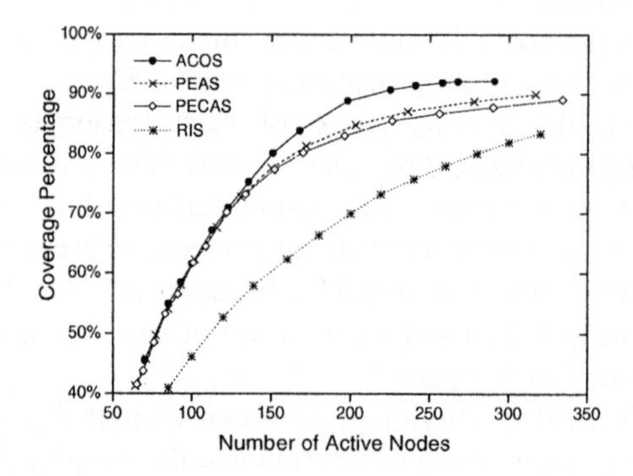

Fig. 5. Coverage over the number of active sensors with 800 sensors deployed

As shown in Fig.5, for the same number of active nodes, ACOS can achieve more coverage than others. And as the number of active nodes increases, ACOS approximates the maximal coverage that can be achieved quite sooner than the other three protocols.

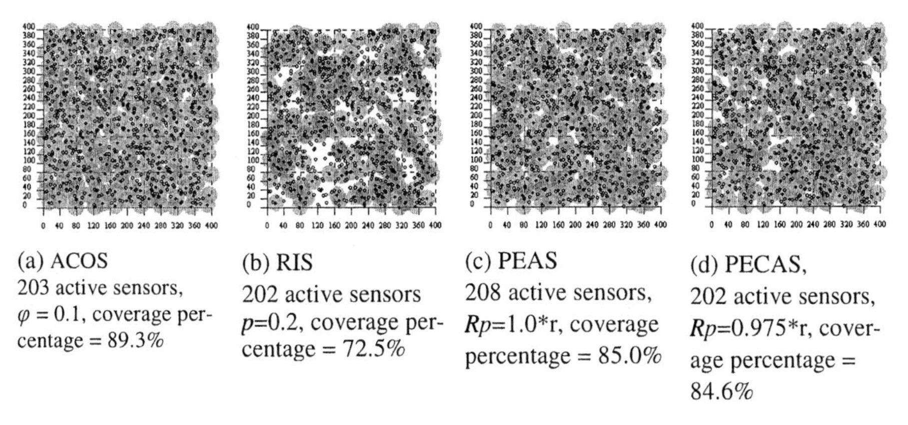

(a) ACOS
203 active sensors, $\varphi = 0.1$, coverage percentage = 89.3%

(b) RIS
202 active sensors p=0.2, coverage percentage = 72.5%

(c) PEAS
208 active sensors, Rp=1.0*r, coverage percentage = 85.0%

(d) PECAS,
202 active sensors, Rp=0.975*r, coverage percentage = 84.6%

Fig. 6. Spatial distribution of working sensors under different protocols

Fig.6 shows a typical scene of spatial distribution of working sensors under different protocols. The number of active nodes in each protocol is roughly 200. Fig.6(a) shows ACOS works quite well. There are no big holes and not much overlap. From the Fig.6(b) we can see that with the RIS protocol, there are many sensing holes and many sensors are densely clustered because of no collaboration. Fig.6(c) and Fig.6(d) show that PEAS and PECAS perform well but not as good as ACOS.

6 Conclusion

In this paper, we consider a fundamental problem of keeping sufficient coverage of the protected region while minimizing energy consumption and extending the lifetime of sensor networks. We have developed a sleeping protocol ACOS, which controls the mode of sensors to optimize the usage of energy as well as to maximize the coverage. We evaluate our protocol on a simulator and compare it with other sleeping protocols. The results demonstrate that our protocol has better coverage of the surveillance area while waking fewer sensors than other state-of-the-art sleeping protocols.

Acknowledgements. This research was supported partially by Natural Science Foundation of China grant #60442004.

References

1. G. Pottie and W. Kaiser. Wireless integrated network sensors, *Communications of the ACM*, 2000.
2. A. Mainwaring, J. Polastre, R. Szewczyk, D. Culler, and J. Anderson. Wireless sensor networks for habitat monitoring. In *First ACM International Workshop on Wireless Sensor Networks and Applications (WSNA)*, 2002.
3. Crossbow. Power management and batteries. Application Notes, available at *http://www.xbow.com/Support/appnotes.htm*, 2004.

4. Chao Gui and Prasant Mohapatra. Power Conservation and Quality of Surveillance in Target Tracking Sensor Networks. In *ACM MobiCom*, 2004.
5. Santosh Kumar, Ten H. Lai and J´ozsef Balogh. On k-Coverage in a Mostly Sleeping Sensor Network. In *ACM MobiCom*, 2004.
6. F. Ye, G. Zhong, J. Cheng, S.W. Lu and L.X. Zhang. PEAS: a robust energy conserving protocol for long-lived sensor networks. In *the 10th IEEE International Conference on Network Protocols (ICNP)*, 2002.
7. Benyuan Liu and Don Towsley. A Study of the Coverage of Large-scale Sensor Networks. In *IEEE International Conference on Mobile Ad-hoc and Sensor Systems (MASS)*, 2004.
8. S. Meguerdichian, F. Koushanfar, M. Potkonjak, and M. B.Srivastava. Coverage problems in wireless ad-hoc sensor networks. In *IEEE Infocom*, 2001.
9. S. Megerian, F. Koushanfar, G. Qu, and M. Potkonjak. Exposure in wireless sensor networks. In *ACM Mobicom*, 2001.
10. Y. Xu, J. Heidemann, and D. Estrin. Geography informed energy conservation for ad hoc routing. In *ACM Mobicom*, 2001.
11. B. Chen, K. Jamieson, and H. Balakrishnan. Span: An energy efficient coordination algorithm for topology maintenance in ad hoc wireless network. In *ACM Mobicom*, 2001.
12. A. Cerpa and D. Estrin. Ascent: Adaptive self-configuring sensor networks topologies. In *IEEE Infocom*, 2002.
13. S. Slijepcevic and M. Potkonjak. Power efficient organization of wireless sensor networks. In *IEEE Int'l Conf. on Communications (ICC)*, pages 472–476, 2001.
14. D. Tian and N. D. Georganas. A coverage-preserving node scheduling scheme for large wireless sensor networks. In *WSNA*, 2002.
15. P. Bahl and V. N. Padmanabhan. RADAR: An in-building RF-based user location and tracking system. In *IEEE Infocom*, 2000.
16. Koen Langendoen, Niels Reijers. Distributed localization in wireless sensor networks: a quantitative comparison. *Computer Networks*, pages 499–518, 2003.
17. R. L. Moses, D. Krishnamurthy and R. M. Patterson, "A self-localization method for wireless sensor networks", in *EURASIP J. Appl. Signal Process.*, pages 348-358, 2003.
18. Lingxuan Hu, David Evans. Localization for Mobile Sensor Networks. In *ACM MobiCom*, 2004.
19. Neal Patwari, Alfred O. Hero, III, Matt Perkins, Neiyer S. Correal and Robert J. O'Dea. Relative Location Estimation in Wireless Sensor Networks, *IEEE Trans. Signal Processing*, 2003.
20. C. Huang and Y. Tseng. The coverage problem in a wireless sensor network. In *WSNA*, 2003.

Coverage Analysis for Wireless Sensor Networks

Ming Liu [1,2], Jiannong Cao[1], Wei Lou[1], Li-jun Chen [2], and Xie Li[2]

[1] Department of Computing, Hong Kong Polytechnic University, Hung Hom, Kowloon, Hong Kong

[2] State Key Laboratory for Novel Software Technology, Nanjing University, Nanjing 210093, China

Abstract. The coverage problem in wireless sensor networks (WSNs) is to determine the number of active sensor nodes needed to cover the sensing area. The purpose is to extend the lifetime of the WSN by turning off redundant nodes. In this paper, we propose a mathematical model for coverage analysis of WSNs. Based on the model, given the ratio of the sensing range of a sensor node to the range of the entire deployment area, the number of the active nodes needed to reach the expected coverage can be derived. Different from most existing works, our approach does not require the knowledge about the locations of sensor nodes, thus can save considerably the cost of hardware and the energy consumption on sensor nodes needed for deriving and maintaining location information. We have also carried out an experimental study by simulations. The analytical results are very close to the simulations results. The proposed method can be widely applied to designing protocols for handling sensor deployment, topology control and other issues in WSNs.

1 Introduction

Technology advances in sensors, embedded systems, and low power-consumption wireless communications have made it possible to manufacture tiny wireless sensors nodes with sensing, processing, and wireless communication capabilities. The low-cost and low power-consumption sensor nodes can be deployed to work together to form a wireless sensor network. The sensor nodes in a sensor network are able to sense surrounding environment and carry out simple processing tasks, and communicate with the neighboring nodes within its transmission range. By means of the collaboration among sensor nodes, the sensed and monitored environment information (e.g. temperature, humidity) is transmitted to the base station for processing.

A large-scale wireless sensor network can consist of tens of thousands of tiny sensor nodes, with high density of sensors up to 20 nodes/m3. The high density of sensors may result in comparatively large energy consumption due to conflict in accessing the communication channels, maintaining information about neighboring nodes, and other factors. A widely-used strategy for reducing energy consumption while at the same time meeting the coverage requirement is to turn off redundant sensors by scheduling sensor nodes to work alternatively [1,2]. The coverage problem in wireless sensor networks (WSNs) is to determine the number of active sensor nodes needed to cover the sensing area.

X. Jia, J. Wu, and Y. He (Eds.): MSN 2005, LNCS 3794, pp. 711 – 720, 2005.

A broadly-used strategy [3][4][5][6][7][8] is to determine the active nodes by using the location information about the sensor nodes and their neighborhood. However, relying on complicated hardware equipment such as GPS (the Global Positioning System) or the directional antenna will greatly increase the hardware cost of sensor nodes and energy consumption; at the same time, the message transmission for and calculation of the locations and directions will also consume the energy of a node. Therefore, it is desirable for a solution to the coverage problem not to depend on any location information.

In this paper, we propose a mathematical model for coverage analysis of WSNs without requiring the use of location information. Based on the model, given the ratio of the sensing range of a sensor node to the range of the entire deployment area, the number of the active nodes needed to reach the expected coverage can be derived. The proposed analytical method is based on the random deployment strategy, which is the easiest and cheapest way for sensor deployment [10]. Comparing with similar work [13] using theoretical methods for coverage analysis without the use of location information, which is a special case of our work. It means our work is more general than [13].

Most applications may not require the maximal area coverage, and a small quantity of blind points generated at certain intervals can be accepted. If the working nodes in a sensor network can maintain a reasonable area coverage, most applications can realize. Coverage can be regarded as one quality of service of a sensor network to evaluate its monitoring capability [9]. If the coverage fraction is below certain threshold, the sensor network will be thought unable to work normally. So, it will be very significant to propose a simple method that can, in a statistical sense, evaluate the coverage fraction which meets the coverage requirement in application without depending on location information. This paper provides such a solution.

The rest of the paper is organized as follows. We introduce related work in section 2. In section 3, we present sensor network models and preliminary definitions. In section 4, we analyse the relationship between the ratio of the sensing range of a sensor node to the range of the entire deployment area and the coverage fraction. Numeric results and simulation results are provided in section 5.

2 Related Work

Coverage is one of the important issues in sensor networks. Because of different applications of sensor networks, maybe there are different definitions of coverage. We argue, in the case of K-cover, coverage in sensor networks can be simply described as: any point in the coverage area lies within the sensing range of at least K sensor nodes. Obviously, K is bigger than or equal to one. Wireless sensor networks are usually characterized by high density of sensor nodes and limited node energy. With the desired coverage fraction being guaranteed, working nodes density control algorithm and node scheduling mechanism are utilized to reduce energy cost and thus extend sensor network lifetime.

In [3] and [4], an approach is proposed to compute the maximal cover set: all the sensor nodes are divided into n cover sets which do not intersect one another, and the sensor nodes in each cover set can perform independently the task of monitoring the desired area; sensor nodes in all the cover sets take turns at performing the monitoring task. In [3], Slijepcevic et al. have proved that the calculation of the maximal cover set is as NP-complete problem. The two algorithms proposed in [3] and [4] are both centralized ones, so they are not suitable for the case in which there is a large quantity of sensor nodes. In addition, both the two algorithms have to rely on the location information of sensor nodes in reckoning cover sets.

In [1], Tian et al. propose a distributed coverage algorithm based on a node scheduling scheme. The off-duty eligibility rule proposed in this algorithm, relying on the geographical information of sensor nodes and AOA (Arrival of Angle) obtained through the directional antenna, can reckon the coverage relation between one node and its neighbors and then select working nodes. Obviously, sensor networks relying on GPS or AOA information are characterized by high cost and high consumption of energy. In addition, the off-duty eligibility rule fails to consider the problem that excessive overlap may be formed so that the number of working nodes selected becomes very large to cause extra energy consumption. In [11], it has been proved that this algorithm based on a node scheduling scheme is low-effective.

In [13], Gao et al. propose a mathematical method, which does not rely on location information, to describe the redundancy. According to this method, one sensor node can utilize the number of neighbors within its sensing range to calculate its own probability of becoming a redundant node. Since there is no need to be equipped with GPS or directional antenna, it is possible to get the cost of sensor nodes under control. In addition, it becomes unnecessary to derive location information through the exchange of message, and thus the energy consumption for communication in sensor networks is reduced. However, for most sensor nodes, the sensing hardware and the communicating hardware are two fully independent parts, and the communicating range is always not equal to the sensing range. Therefore, some specialized parts are needed to judge the number of neighbors within the sensing range.

As the above analysis suggests, most previously proposed coverage algorithms rely on outside equipment like GPS, directional antenna or positioning algorithm, etc. In this case, both the cost and the energy consumption are increased; in the mean time, some problems remain unsolved, e.g. GPS-based protocols have to correct some mistakes made in calculating location information; the work of GPS-based systems is not reliable in indoor environment, and thus some other positioning systems need to be deployed. For some positioning algorithms, each node needs to exchange a large quantity of information with the beacon node to calculate its location, and this will also result in high consumption of power. In [14], Stojemenvic makes a comprehensive analysis on location-based algorithms, and locations out that obtaining and maintaining location information will cause great consumption of energy.

In this paper, we provide an effective mathematical method to evaluate the number of nodes needed to reach the expected coverage fraction. In this method, only if the proportion of the node's sensing range to the range of the deployment area C is known, the relation between the number of sensor nodes in C and the expected coverage fraction can be derived by simple calculation. Therefore, our approach is applicable to many cases. It can be easily adopted in handling the problems of sensor deployment, topology control, etc.

3 Models and Assumptions

In this section, we first introduce two methods used in our research: the deployment method and the sensing method. Then we will give a few definitions to simplify the analytical process in Part 4.

3.1 Deployment Model

In [10], the commonly used deployment strategies are studied: random deployment, regular deployment, and planned deployment. In the random deployment strategy, sensor nodes are distributed with a uniformly distribution within the field. In the regular deployment strategy, sensors are placed in regular geometric topology such as a grid. In the planned deployment strategy, the sensors are placed with higher density in areas where the phenomenon is concentrated. In the planned deployment method, although sensors are deployed with a non-uniform density in the whole deployment area, however in a small range, sensors are approximately deployed randomly. In this sense, our analytical results of random deployment are also applicable to planned deployment.

The analysis in this paper is based on the random deployment strategy, which is reasonable in dealing with the application scenario in which priori knowledge of the field is not available. For convenience, we assume that sensor nodes are placed in a two-dimensional circular area C with a radius of R. Actually, we are not concerned about the shape of the deployment area, which can be circular or square, and the area C can represent a subset of the whole deployment area or represent the whole deployment area. Our research focuses on how to obtain the number of nodes required by C with the coverage of sensor network being guaranteed. We assume that sensor nodes are uniformly and independently distributed in the area C, and no two sensors can be deployed exactly at the same location.

3.2 Sensing Model

The analysis in this paper is based on Boolean sensing model, which is broadly adopted in the study of sensor networks [1][2][12]. In the Boolean sensing model, each sensor has a fixed sensing range. A sensor can only sense the environment and detect events within its sensing range. And in this paper all sensors are supposed to have the same sensing rang and the sensor's sensing range $r \leq R$. A point is covered if and only if it lies within at least one sensor's sensing range. So, the deployment area is partitioned into two regions, the covered region and the vacant region. An arbitrary point in the covered region is covered by at least one sensor node, while the vacant region is the complement of the covered region. Actually, some applications require a higher degree of accuracy in detecting objects, so an arbitrary point in the covered region has to lie within the sensing ranges of k nodes at the same time. The analytical results in this paper, however, can be easily extended into K-coverage.

3.3 Related Definitions

To facilitate later discussion, we introduce the following definitions:

Definition 1: Neighboring area. For an arbitrary point $(x, y) \in C$, its neighboring area is defined as

$$\aleph(x,y) = \left\{ (x',y') \in C \mid \forall \left((x'-x)^2 + (y'-y)^2 \leq r^2 \right) \right\}$$

Definition 2: The central area C'. For C', we have $C' \subset C$. And for an arbitrary point $(x, y) \in C'$, there is

$$x^2 + y^2 < (R - r)^2$$

Definition 3: Expected coverage fraction, denoted as q. Expected coverage fraction of a sensor network is defined as the expected proportion of the covered region to the whole deployment region. For example, an application requires the coverage fraction reach 85 percent of the whole region, and the expected coverage fraction equals 0.85. If the expected coverage fraction is known, it can be used to calculate the number of nodes needed to cover the deployment area.

As shown in Figure 1, for an arbitrary point, its neighboring area is actually the overlapped area of the circle centered at the point with a radius of r and the area *C*. The central area C', which and *C* are circles centered at the same point, has a radius of $R - r$. Obviously, the neighboring area of every point in C' is the same, and its value is πr^2. For any point in $C - C'$, its neighboring area is in inverse proportion to its distance away from the centered point of C, and is less than πr^2.

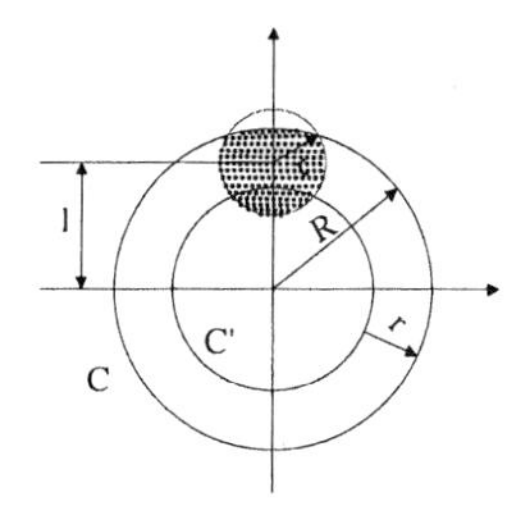

Fig. 1. Illustration of analysis

4 Analysis for Coverage

For an arbitrary point $(x, y) \in C$, if there exists at least one sensor node in its neighboring area, the point is covered. Since the sensors in *C* are distributed randomly and uniformly, the probability that an arbitrary node falls on the point's (x, y) neighboring area is $p = \aleph(x, y)_{area} / C_{area}$.

Assume that m sensor nodes are deployed randomly in *C*. In the case of single-cover, the probability that an arbitrary point is covered is equal to the one that at least one sensor node falls on its neighboring area, namely,

$$\begin{aligned} P_{\{(x,y)\in C\}} &= c_m^1 p(1-p)^{m-1} + C_m^2 p^2 (1-p)^{m-2} + \cdots\cdots + C_m^m p^m \\ &= \sum_{n=1}^{m} C_m^n p^n (1-p)^{m-n} \\ &= 1 - (1-p)^m \end{aligned} \tag{1}$$

Hence, for any two points in C, if the size of their neighboring area is the same, the probability of being covered is equal to each other. For $\forall(x,y)\in C'$, it's neighboring area $\aleph(x,y)_{area}=\pi r^2$, so the probability that an arbitrary node falls on certain point's neighboring area in C' is $p=\aleph(x,y)_{area}/C_{area}=\pi r^2/\pi R^2$. According to Formula 1, if there are m sensor nodes randomly deployed in C, for $\forall(x,y)\in C'$, the probability of being covered is

$$p_{((x,y)\in C')}=\sum_{n=1}^{m}C_m^n\left(\frac{r}{R}\right)^{2n}\left(1-\frac{r^2}{R^2}\right)^{m-n} \tag{2}$$

For each point that lies within the marginal region $C-C'$ of C, its neighboring area is less than πr^2; and especially for each point on the edge of C, its neighboring area is the smallest, and thus its probability of being covered is also the smallest. Assume the probability that each point on the edge of C is covered is $p_{\min}$, and then it is obvious that $p_{\min}\le p_{((x,y)\in C)}\le p_{((x,y)\in C')}$. When $R\gg r$, the area of $C-C'$ can be ignored in calculation. In this case, it can be approximately concluded that for an arbitrary point in C, the probability of being covered is the same: $p_{((x,y)\in C)}\approx 1-\left(1-\frac{r^2}{R^2}\right)^m$. Since the probability that each point in C is covered is $1-\left(1-\frac{r^2}{R^2}\right)^m$, the expected coverage fraction can be: $q=1-\left(1-\frac{r^2}{R^2}\right)^m$.

When the proportion of r to R is small enough that it cannot be ignored, to use Formula 2 to calculate the expected coverage fraction can lead to an error that is beyond tolerance. Therefore, in order to have an accurate evaluation of q, we have to compute the average probability of being covered for all the points in C. As shown in Figure ( ), there is one point (x_1,y_1), and l denotes the distance between this location and the center of Circle C, and $l>R-r$. Then the value of point's neighboring area $\aleph(x_1,y_1)_{area}$ is the area of the shadowed region:

$$\begin{aligned}\aleph(x_1,y_1)_{area}&=2\left(\int_{l-r}^{\frac{R^2-r^2+l^2}{2l}}\sqrt{r^2-(y_1-l)^2}\,dy_1+\int_{\frac{R^2-r^2+l^2}{2l}}^{R}\sqrt{R^2-y_1^2}\,dy_1\right)\\&=\frac{1}{2}\pi\left(r^2+R^2\right)+r^2\arcsin\frac{R^2-r^2-l^2}{2lr}+\frac{R^2-r^2-l^2}{2l}\sqrt{r^2-\frac{\left(R^2-r^2-l^2\right)^2}{4l^2}}\\&\quad-R^2\arcsin\frac{R^2-r^2+l^2}{2lR}-\frac{R^2-r^2+l^2}{2l}\sqrt{R^2-\frac{\left(R^2-r^2+l^2\right)^2}{4l^2}}\end{aligned} \tag{3}$$

From Formula 3, we can derive a common expression of the neighboring area of any point in $C-C'$. By the operation of integral calculus, we can calculate the average size of neighboring area of all the points in $C-C'$, denoted as $\overline{\aleph(x,y)_{C-C'}}$:

$$\overline{\aleph(x,y)}_{C-C'} = \iint_{C-C'} \aleph(x_1,y_1)_{area} d_\sigma \Big/ \pi\left[R^2-(R-r)^2\right]$$
$$= 2\pi \int_{R-r}^{R} l\aleph(x_1,y_1)_{area} d_l \Big/ \pi\left[R^2-(R-r)^2\right] \quad (4)$$

And the average neighboring area of all the points in C is:

$$\overline{\aleph(x,y)}_{area} = \left[2\int_{R-r}^{R} \aleph(x_1,y_1)_{area} d_l + r^2 \times \pi(R-r)^2\right] \Big/ R^2 \quad (5)$$

Hence, for all the points in C, the average probability of being covered, i.e. the expected coverage fraction in C is:

$$q = 1-(1-\frac{\overline{\aleph(x,y)}_{area}}{\pi R^2})^n \quad (6)$$

Our previous discussion only involves the single-cover. In the case of k-cover, there are at least k nodes in the neighboring area of an arbitrary point in the covered region. The probability of being covered is

$$P_{\{(x,y)\in C\}} = c_m^k p(1-p)^{m-k} + C_m^{k+1} p^2 (1-p)^{m-k+1} + \cdots\cdots + C_m^m p^m$$
$$= \sum_{n=k}^{m} C_m^n p^n (1-p)^{m-n} \quad (7)$$

Once the radius of C (denoted as R) and the node's sensing range (denoted as r) are determined, the probability that an arbitrary point in C is covered relates only to its neighboring area. Therefore, the above discussion based on the 1-cover is still applicable in dealing with multi-cover.

5 Analysis and Evaluation

5.1 Numerical Results

As Table 1 shows, the larger the proportion of r to R is, the smaller the size of the average neighboring area of all the points in C will be. When R = r, the proportion of the average neighboring area to the maximal neighboring area is only 0.5781. But when r<<R, e.g. r/R = 0.01, the proportion, as shown in Table 1, is approximately equal to 1. In this case, we can use Formula 2 to approximately compute the expected coverage fraction in C.

Table 1.

r_s/R	1	9/10	8/10	7/10	6/10	5/10	4/10	3/10	2/10	1/10	1/100
$\overline{\aleph}/\aleph_{max}$	0 5865	0 6259	0 6660	0 7066	0 7477	0 7891	0 8309	0 8730	0 9152	0 9576	0 9958

5.2 Simulation Methodology

Our simulation, based on MATLAB, gets started with the production of the deployment region C consisting of pixels. In order to make sure that the experimental

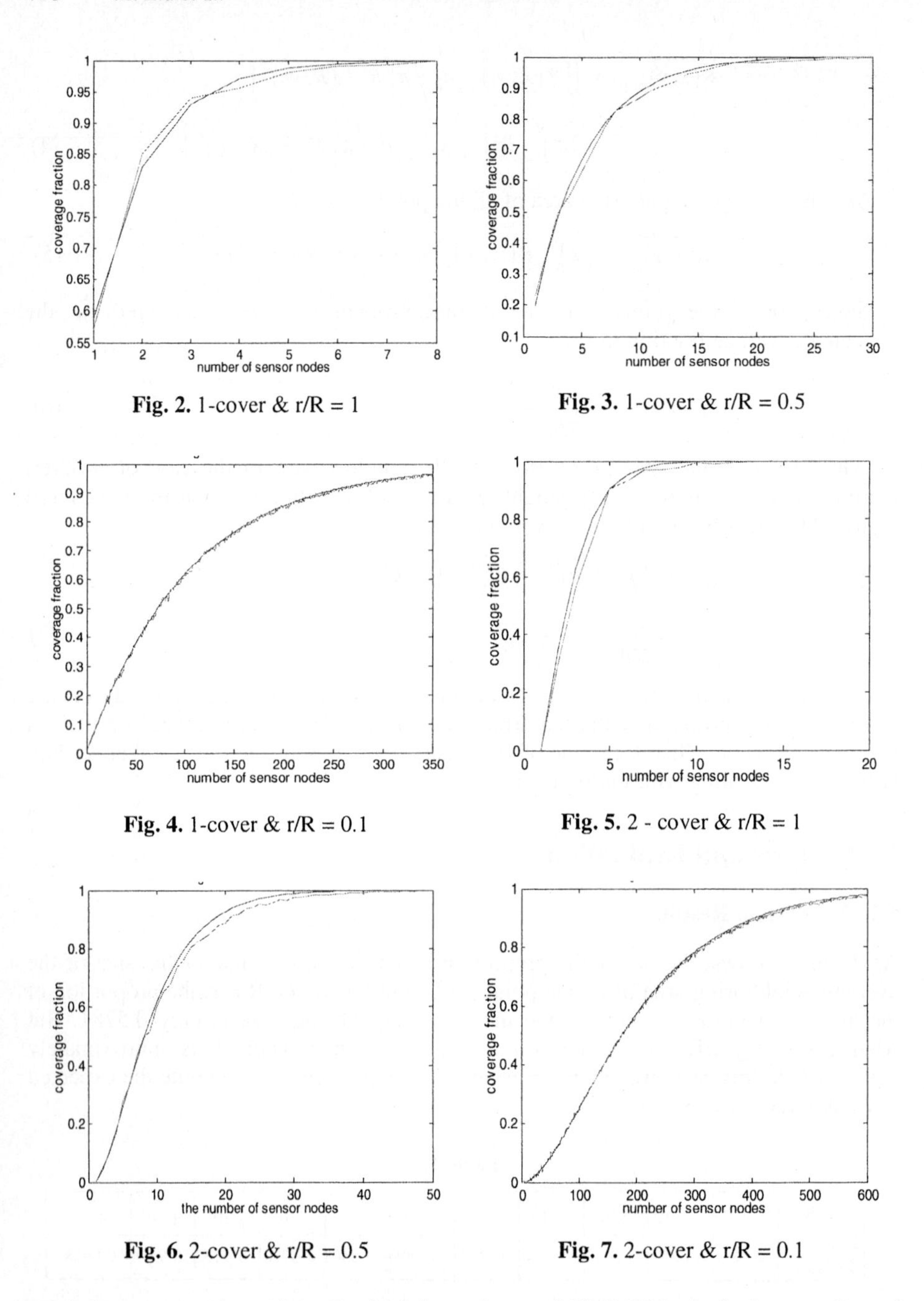

Fig. 2. 1-cover & r/R = 1

Fig. 3. 1-cover & r/R = 0.5

Fig. 4. 1-cover & r/R = 0.1

Fig. 5. 2 - cover & r/R = 1

Fig. 6. 2-cover & r/R = 0.5

Fig. 7. 2-cover & r/R = 0.1

result is accurate, the region *C*, which is made up of 19634803 pixels, has a radius of 5000 pixels. According to N, the number of the deployed sensor nodes, we randomly select N locations and let them distributed uniformly and independently in *C*. Each pixel is defined as a structure, and for each pixel we count the number covered by sensor nodes. It is obvious that the percentage of the coverage fraction is equal to the proportion of the number of the covered pixels to the number of the whole pixels.

5.3 Simulation Results

To fully test the accuracy of our analytical results, the experiment simulates 1-cover and 2-cover in the three cases in which the proportion of the sensing range r to the deployment range R is assumed to be 1, 0.5 and 0.1 respectively. All the results reported are averages of 50 simulation runs.

From figure 2 to figure 7 shows the simulation results of coverage fraction as a function of nodes number. As figure 2 shown, we observe that if there exist more than four nodes in deployment area C, the coverage fraction derived from simulation equals 0.95611 and the expected coverage fraction equals 0.97077. And when 6 nodes deployed in C area, the coverage fraction derived from simulation equal 0.99145 and the expected coverage equal 0.995. So it suggests that our analytical results are very similar to the simulation results.

As figure 2 to figure 7 shown, Under the random deployment, the deviation between the coverage fraction derived from our theoretical analysis and that obtained from the simulations is no larger than 5% of the analytical value; Given a coverage fraction, the deviation between the number of working nodes derived from analysis and that obtained from the simulation is less than 10% of the analytical value. This suggests that our results are identical to the experimental results.

6 Conclusion and Future work

In this paper, we proposed a mathematical method for coverage analysis of WSNs. Using the method, given the ratio of the sensing range of a sensor node to the range of the entire deployment area, the number of the active nodes needed to reach the expected coverage can be derived. The main contribution of this paper lies in a simple and effective approach to solving the coverage problem without the need of sensor nodes' location information and its potential application in developing heuristic algorithms for node scheduling, clustering, and other functions.

Our future work includes applying the proposed method to design energy efficient protocols for various WSN functions and extending our model to cover different node distribution and sensor network models.

Acknowledgement

This research is supported in part by Hong Kong Polytechnic University under the ICRG grant A-PF77 and the China National 973 Program Grant 2002CB312002.

References

1. D. Tian and N. Georganas. "A coverage-preserved node scheduling scheme for large wireless sensor networks," In Proceedings of First International Workshop on Wireless Sensor Networks and Applications (WSNA'02), pp 32–41. Atlanta, USA, September 2002.
2. F. Ye, G. Zhong, S. Lu, and L. Zhang, "PEAS: A robust energy conserving protocol for long-lived sensor networks," in Proceedings of the 23nd International Conference on Distributed Computing Systems (ICDCS), 2003

3. S. Slijepcevic and M. Potkonjak, "Power Efficient Organization of Wireless Sensor Networks," in Proceedings of IEEE Conference on Communications, Vol 2, pp 472-476, Helsinki, Finland, June 2001.
4. M. Cardei, D. MarCallum, X. Cheng, M. Min, X. Jia, D. Li, and D. Du, "Wireless Sensor Networks with Energy Efficient Organization," Journal of Interconnection Networks, Vol 3, No 3-4, pp 213-229, Dec 2002.
5. Y. Xu, J. Heidemann, and D. Estrin, "Geography-informed energy conservation for ad hoc routing," in Proceeding of ACM MOBICOM'01, pp 70-84, July 2001.
6. D. Tuan and N. D. Georganas, "A Coverage-preserving node scheduling scheme for large wireless sensor networks," in Proceedings of First ACM International Workshop on Wireless Sensor Networks and Applications, pp 32-41, 2002.
7. H. Zhang and J.C. Hou, "Maintaining scheme coverage and connectivity in large sensor networks," in Proceedings of NSF International Workshop on Theoretical and Algorithmic Aspects of Sensor, Ad Hoc wireless, and Peer-to-Peer Networks, 2004.
8. X. Wang, G. Xing, Y. Zhang, C. Lu, R. Pless, and C.D. Gill, "Integrated Coverage and Connectivity Configuration in Wireless Sensor Networks," in Proceedings of the First International Conference on Embedded Networked Sensor Systems, pp 28-39, ACM Press, 2003.
9. S. Meguerdichian, F. Koushanfar, M. Potkonjak, and M. Srivastava, "Coverage Problems in Wireless Ad-Hoc Sensor Networks," IEEE Infocom 2001, Vol 3, pp 1380-1387, April 2001.
10. S. Tilak, N.B. Abu-Ghazaleh, and W. Heinzelman, "Infrastructure tradeoffs for sensor networks," Proceedings of First InternationalWorkshop on Wireless Sensor Networks and Applications (WSNA'02), Atlanta, USA, September 2002, pp. 49-57.
11. O. Younis and S. Fahmy, "Distributed Clustering in Ad-hoc Sensor Networks: A Hybrid, Energy-Efficient Approach," in Proceedings of IEEE INFOCOM, March 2004.
12. S. Shakkottai, R. Srikant, and N. Shroff, "Unreliable sensor grids: Coverage, connectivity and diameter," in Proc. IEEE INFOCOMM, 2003.
13. Y. Gao, K. Wu, and F. Li, "Analysis on the redundancy of wireless sensor networks," Proceedings of the 2nd ACM international conference on Wireless sensor networks and applications (WSNA 03), September 2003, San Diego, CA.
14. I. Stojmenovic, "Position based routing in ad hoc networks," IEEE Communications Magazine, Vol. 40, No. 7, July 2002, pp. 128-134.

On Coverage Problems of Directional Sensor Networks*

Huadong Ma[1] and Yonghe Liu[2]

[1] School of Computer Science & Technology,
Beijing University of Posts and Telecommunications, Beijing 100876, China
mhd@bupt.edu.cn

[2] Dept. of Computer Science and Engineering,
The University of Texas at Arlington, Arlington, TX76019
yonghe@cse.uta.edu

Abstract. In conventional sensor networks, the sensors often are based on omni-sensing model. However, directional sensing range and sensors are great application chances, typically in video sensor networks. Thus, the directional sensor network also demands novel solutions, especially for deployment policy and sensor's scheduling. Toward this end, this paper evaluates the requirements of deploying directional sensors for a given coverage probability. Moreover, the paper proposes how to solve the connectivity problem for randomly deployed sensors under the directional communication model. The paper proposes a method for checking and repairing the connectivity of directional sensor networks for two typical cases. We design efficient protocols to implement our idea. A set of experiments are also performed to prove the effectivity of our solution. The results of this paper can be also used to solve the coverage problem of traditional sensor networks as a special case.

1 Introduction

Recently sensor networks have attracted tremendous research interests due to its vast potential applications [1, 2, 6, 7]. Conventional sensor networks often assume the omnidirectional sensing model. Actually, directional sensing range and sensors also have great application chances, typically in video sensor networks[14, 15]. Potential applications of video sensor networks span a wide spectrum from commercial to law enforcement, from civil to military. However, many methods for conventional sensor networks is not suitable for directional sensor networks. Thus, the directional sensor network also demands novel solutions, especially for deployment policy and sensor's scheduling.

In our best knowledge, no paper discussed the problems for directional sensor network, although a few papers have indeed studied the concept for video sensor networks [3, 4]. However, the work has mainly focused on low power hardware platform support while system level issues such as QoS capable networking and directional sensing features have been left unaddressed. Therefore, despite half decade of strong progress in sensor design and wireless communications, directional sensor networks remains an unanswered challenge.

* The work reported in this paper is partly supported by the NSFC under Grant 60242002 and the NCET Program of MOE, China.

X. Jia, J. Wu, and Y. He (Eds.): MSN 2005, LNCS 3794, pp. 721–731, 2005.

Fundamentally different from conventional sensor networks, directional sensor networks are characterized by its *directional sensing/communicating range*, typically camera's field of view. The feature affects the deployment of sensors, the capture of information and scheduling strategy. These differences are calling for novel approaches for sensor networking. In this paper we take the first step toward a solution for directional sensor networks. In particular, we propose a systematic method for deployment strategy, connectivity checking and repairing, sensor scheduling for randomly deployed directional sensor network.

The reminder of this paper is organized as follows. Related work is discussed in Section 2. In Section 3, we define the sensing model assumed and evaluate the coverage rate of directional sensor network. In Section 4 we discuss the connectivity checking and repairing problem in directional communication model. In Section 5 we describe the scheduling method of directional sensor networks. Experimental results are presented in Section 6 and we conclude the paper in Section 7.

2 Related Work

A lot of pioneer papers address the problem of coverage and connectivity maintenance faced by sensor networks. The common assumption of previous works are that the sensor is omnidirectional sensing/communicating [5, 9, 12, 13, 16, 18]. They can roughly be categorized into the following aspects:

Deployment decision: There are three kinds of deployment policies: regular deployment, planned deployment and random deployment [17]. In the regular deployment method, sensors are placed in regular geometric topology. An example of regular deployment is the grid-based sensor deployment where nodes are located on the intersection points of a grid. An example of planned deployment is the security sensor system used in museums. The most valuable exhibit objects are equipped with more sensors to maximize the coverage of the monitoring scheme. In many situations, deterministic deployment is neither feasible nor practical. The deployment policy often is to cover the sensor field with sensors randomly distributed in the environment. In these cases, the redundancy and density of sensor deployment are problems to focus on.

Sensor Scheduling: One of the main design challenges for sensor network is to obtain long system lifetime without sacrificing system original performance. Since sensors are arbitrarily distributed, sensors' on-duty time should be properly scheduled to conserve energy. Some node-scheduling schemes in [11, 16, 19] are proposed to conserve energy and thus extend the lifetime of the sensor network.

Coverage Completeness: A typical problems is k-coverage problem, whose goal is determine whether every point in the service area of sensor network is covered by at least k sensors. The authors of [8] proposed polynomial-time algorithm to determine k-coverage problem. However, k-coverage problem is formulated as a decision problem, which can only answer a yes/no question. A more general optimization problem is: how can we patch these insufficiently covered areas with the least number of extra sensors. This is still an open question and deserves further investigation. k-coverage is often used to find to the solutions to reliability or fault tolerance, object location, power savings for sensor networks.

Connectivity coverage: Its goal is determine whether every pair of sensors in the service area of sensor network are connected by at least k paths. The problem is how to determine whether the graph is k-connected, and if not, determine how to make the graph k-connected by inserting additional sensors to the network. Solving this problem with a minimum number of additional sensors is NP-hard [10]. The paper [18, 20] discussed how to combine coverage and connectivity maintenance in a single activity scheduling. However, random node deployment often makes initial sensing holes inside the deployed area inevitable even in an extremely high-density network. Therefore, the paper [16] discussed both connectivity maintenance and coverage preservation in wireless sensor networks.

Different from the previous works, this paper mainly focuses on the deployment and scheduling for randomly deployed directional sensor network.

3 Coverage Rate for Directional Sensor Network

3.1 Directional Sensing Models

From the concept of field of view in cameras, we employ a 2-D model where the sensing area of a sensor s is a sector denoted by 4-tuple $(L, r, \boldsymbol{V}, \alpha)$. Here L is the location of the sensor node, r is the sensing radius, $\boldsymbol{V}$ is the center line of sight of the camera's field of view which will be termed *sensing direction*, and α is the offset angle of the field of view on both sides of $\boldsymbol{V}$. Fig. 1 illustrates the directional sensing model. Note that the omni-sensing model is a special case of new model when α is π.

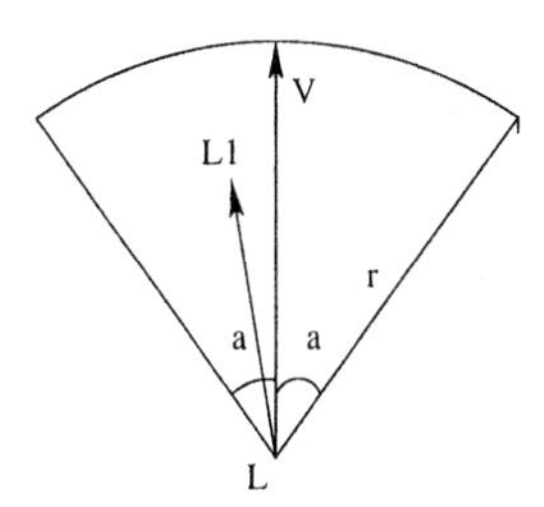

Fig. 1. Directional sensing model

A point L_1 is said to be covered by sensor s if and only if $d(L, L_1) \leq R$, and the angel between $\overrightarrow{LL_1}$ and $\boldsymbol{V}$ is within [-α, α]. An area A is covered by sensor s, if and only if for any point $L \in A$, L is covered by s.

3.2 Coverage Rate for Directional Sensing

In randomly deployed networks, it is hard or impossible to 100% guarantee complete coverage of the monitored region even if the node density is very high. So, it is a practical issue to study how to guarantee the given coverage rate of sensor network for a targeted region.

Assume that the area of the targeted region is S, and the locations of randomly deployed sensors obey uniform distribution. Therefore, after N directional sensors are deployed, the probability of covering the targeted region is represented in Equation (1):

$$p = 1 - (1 - \frac{\alpha r^2}{S})^N \tag{1}$$

For omni-sensing sensors with $\alpha = \pi$, the coverage rate for deploying N sensors is easily obtained by Equation (1).

Thus, if the coverage rate of the targeted region is at least p, the number of deployed directional sensors should be as follows:

$$N \geq \frac{\ln(1-p)}{\ln(S - \alpha r^2) - lnS} \tag{2}$$

For omni-sensing sensors with $\alpha = \pi$, the number of deployed sensors for the given coverage rate p is also obtained by Equation (2).

3.3 Adjustment of Sensing Coverage Rate

The sensing extensions of many sensors, including directional sensor, are adjustable. If the coverage rate must be a given value p in some cases, we can adjust the sensing radius to achieve the goal in Equation (3):

$$r = \sqrt{\frac{S}{\alpha}(1 - (1-p)^{\frac{1}{N}})} \tag{3}$$

In order to minimize the mean energy consumption thus extend the lifetime of sensor network, we can divide N sensors into n groups, and n groups of sensors are working alternatively. The another advantage of grouping working is that each group is activated in different time periods for different monitoring tasks.

According to the conventional model [1, 12], the energy consumption of sensor is in proportion to the k-power of its sensing radius, i.e., $E = Cr^k$, where C is the const and $k \geq 2$.

Theorem 1. When N sensors are partitioned into the groups with equal numbers of sensors, the deployment minimizes the energy consumption thus maximizes the lifetime of sensor network.

Proof: Assume that N sensors are deployed in the targeted region. We divide them into n groups to work alternatively. Their numbers of sensors are $m_1, m_2, \ldots, m_n$, respectively. The sensing radius for each group (denoted as $r_i, i = 1, \ldots, n$) is adopted to meet the coverage rate p according to Equation (3).

The energy consumption for a group can be represented as $E_i{=}n_i C r_i^k (i = 1, \ldots, n)$. Thus, finding the minimum energy consumption reduces to:

$$\min_{\{m_1, m_2, \ldots, m_n\}} \quad \Sigma_{i=1}^n m_i C r_i^k$$

subject to $0 \leq m_i \leq N$ $(i = 1, 2, \ldots, n)$, and $\Sigma_{i=1}^n m_i = N$.

The solution to this problem is $m_1 = m_2 = \cdots = m_n = N/n$. □

4 Connectivity Problem for Directional Sensor Network

4.1 Directional Communication Model

In some cases, the sensor only communicates with others in a specific direction, that is, the sensor is directional sending, and omni-receiving. Shown in Fig. 2, the communication area of a sensor s is a sector denoted by 4-tuple $(L, R, \boldsymbol{D}, \beta)$, where L is the location of the sensor node, R is the communication radius and generally greater than $2r$, $\boldsymbol{D}$ is the center line of the sending field which will be termed *sending direction*, and β is the offset angle of the sending field on both sides of $\boldsymbol{D}$. For the sensors of a directional sensor networks, the sensing direction and communication direction may be same, but communication direction is allowed to be different from sensing directions.

Assume that two sensors can directly communicate if their Euclidean distance is not larger than a communication range R and one node is in the communication area of the other node. We model a directional sensor network as follows:

Definition 1. The directional sensor network can be modelled as a directional communication graph $G(V, E)$ where V is a set of sensors, and E is the edge set. For a pair of node $s_1, s_2 \in V$, the edge $(s_1, s_2) \in E$ if $\|\overrightarrow{s_1 s_2}\| \leq R$ and $\overrightarrow{s_1 s_2} \cdot \boldsymbol{D} \geq \|\overrightarrow{s_1 s_2}\| \cos\beta$, where β is an offset angel of the sensor s_1.

Note that $(s_1, s_2) \in E$ means that s_2 can receive the message from s_1, but s_2 can not send message to s_1 if s_1 is not in the communication area of s_2.

Definition 2. The directional communication graph $G(V, E)$ of a sensor network is said to be *connected to s* if there is a path consisting of edges in E for a given node s and any other sensor s_i in V. The path P for the node s_i and s is represented as $P(s_i, s)$ $=\{(s_i, s_{i+1}), (s_{i+1}, s_{i+2}), \cdots, (s_{i+k}, s)\}$ where $(s_{i+h}, s_{i+h+1}) \in E$ $(0 \leq h \leq k$, s can be renumbered as s_{i+k+1})

Definition 3. Given that a directional communication sub-graph $G_1(V_1, E_1) \in G$ is connected to the node $s \in V_1$, if for any sensor $s_i \in V - V_1$, there is not path between s_i and s, then we call G_1 the maximal connected component to s.

Corollary 1. If $G_1(V_1, E_1) \in G$ is the maximal connected component to s, then for any sensor $s_i \in V - V_1$, there is not a sensor $s_1 \in V_1$ and $(s_i, s_1) \in E$.

Generally, we can find the maximal connected component G_1 to a node s for a directional communication graph G. If the node number of G_1 is equal to that of G, the graph G is connected to s, this is the expected deployment, but is often not true for randomly deployed directional sensor network.

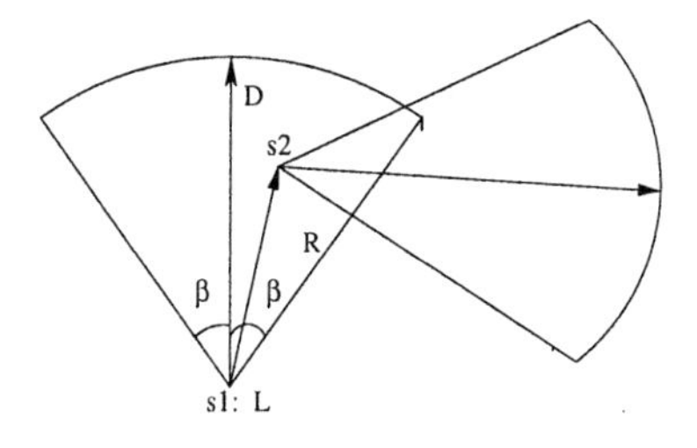

Fig. 2. Directional communication model

Definition 4. For a maximal connected component G_1 to the given sink node x, then we call G_1 *communicable sub-graph*. The other sensors can not communicate with x, they are called *incommunicable sensors*.

4.2 Connectivity Checking

The directional communication graph is a directed graph. First step of our method is to find a maximal connected component C to the given sink node x, and we can use the algorithm *MaxConComp* to find it. Assume that the graph G is represented as adjacency list. The algorithm takes use of depth-first search algorithm.

Procedure MaxConComp(x)

```
//use a visit flag array Visited
1. Visited[x]=TRUE; // the vertex x is visited
2. v:=*x.first; // take the first adjacent vertex
3. while (v is not NULL) do { // if there is adjacent vertex
3.1   if (!Visited[*v.vertex]) MaxConComp(*v.vertex);
3.2   v:=*v.next; //take the next node adjacent to v
3.3   }
```

After we found the maximal connected component C(x) for the sink node x, if the node number of C(x) is less than the node number of G then G is not completely connected to x. For randomly deployed directional sensor network, it is usual that some nodes can not connected to the given sink node x.

4.3 Connectivity Repairing

If there are incommunicable sensors, we need take some measures to make them connected to the sink. Thus, the connectivity repairing is very important for directional sensor network to achieve the monitoring task. Given that the locations of every sensors, we can find the nearest incommunicable node. For repairing connectivity, additional sensors should be added between the communicable component/sink and this incommunicable node. The algorithm is described as follows:

Procedure RepairConnectOne(G,x)

1. C:=MaxConComp(x); //find the maximal connected component
2. G1:=G-C; //take the remaining nodes;
3. while (G1 is not empty) do {

3.1 Find a node x_1 nearest to C in G1, the node nearest to x_1 in C is denoted as y, $d = dis(x_1, y)$;

3.2 Deploy $\lceil \frac{d}{R} \rceil$ sensors with communication radius R,these sensors with the sensing direction $\overrightarrow{x_1 y}$ are equally-spaced deployed between x_1 and y, the first sensor is located in the communicating area of x;

3.3 RepairConnectOne(G1,x_1); // repair the graph G1

3.4 } // end of while

In randomly deployed sensor networks, it is impossible to guarantee the connectivity of the communication graph in the first deployment. The re-deployment for achieving the connectivity is inevitable.

4.4 Grouping Connectivity for Directional Sensor Network

In order to prolong the lifetime of sensor network, we can deploy more than the expected number (N_0) of sensors for a given coverage rate. These sensors will be divided into some groups, each group with N_0 sensors is alternatively activated in turn, to maintain the coverage rate p of the target region.

In this case, we need to ensure each group of sensors are connected. A grouping and connectivity repairing method for directional sensor networks is as follows:

Procedure RepairConnectGroup(G,x)

//find n trees rooted by the node x, all nodes are initiated "not visited"
1. x is the shared root of trees ($T(x,i), i = 1, ..., n$), and marked as "visited";
2. ADJ= the set of nodes adjacent to x;
3. while (there is any not-visited node in G) {
/*Divide ADJ into n groups, and append them to n trees */
3.1 for each node s in ADJ do {
if s is only adjacent to one tree then append s into this tree
otherwise append it into one tree with the minimal number of adjacent nodes in ADJ.
The node s is marked as "visited";
}
3.2 for each tree without adding new node do {
3.2.1 if the node number of $T(x,i)$ is less than $\frac{\|G\|}{n}$ {
3.2.2 Find a pair of nodes with the nearest distance between $T(x,i)$ and
the not-visited nodes of G, denoted as (A,B) where $A \in T(x,i)$
and B is not visited;
3.2.3 B is marked as "visited", $d = dist(A,B)$;
3.2.4 Deploy $\lceil \frac{d}{R} \rceil$ sensors, these sensors with the sensing
direction $\overrightarrow{BA}$ are equally-spaced deployed between B and A,
the first sensor is located in the communicating area of B;
3.2.5 } //end of if
3.3 } //end of for
3.4 ADJ:= the set of not-visited nodes adjacent to the nodes of ADJ
3.5 } //end of while

5 Scheduling for Directional Sensor Network

5.1 Basic Idea

We propose a coverage and connectivity maintenance scheme for randomly deployed directional sensor networks, which works in three phases as follows.

Deployment: The directional sensors are randomly deployed for monitoring a targeted region. According to the analytical result of Section 3, we can calculate the required number of deploying sensors for the expected coverage rate.

Checking: During the checking phase, each sensor node find its own position and synchronizes time with neighboring nodes, and ready for sensing. The system will check the connectivity of deployed sensors, and if the sensor network is not connected then repair it for achieving the connectivity.

Sensing: The sensor network starts to execute sensing task according to the scheduling policy. Each sensor starts detecting the environmental events. Each node will establish a working schedule through our methods, which tell it when to sleep and when to work for each round. When a node goes to sleep, its sensing, computation and communication components can all be asleep and only a timer needs to work and wake up all components according to scheduling policy.

5.2 Scheduling Protocols

Assume that N_0 is the number of sensors for achieving the required coverage rate p under the sensing radius r and the communication radius R. We design two node scheduling schemes for directional sensor networks.

Simple Scheduling Protocol. We deploy N_0 sensors randomly in the targeted region; and call *RepairConnectOne* for checking and repairing the connectivity of N_0 sensors; then the sensor network begins working, all sensors are active until the sensor network is used up.

Grouping Scheduling Protocol. We take N sensors ($N = n * N_0$) to cover the targeted region. These sensors are divided into n groups in which there are N_0 sensors. These n groups are alternatively activated in turn, to maintain the coverage rate p of the target region for nT_0 time (T_0 is the lifetime of a group). Thus, the grouping scheduling protocol is presented as follows.

/* Deployment*/
1. Calculating the number of required sensors (N_0) for the coverage rate p;
2. Input the expected lifetime of the sensor network, i.e., T;
3. Take $n(=\lceil \frac{T}{T_0} \rceil)$ groups of sensors, each group consists of N_0 sensors;
4. Deploy N ($=n * N_0$) sensors randomly in the targeted region,
   get the directional communication graph G for this deployment;

/* Grouping and repairing */
5. Call RepairConnectGroup(G,x);// x is the sink node.
6. Set the starting time of sensors in i-th group is set to $(i-1)T_0$;

/* Scheduling sensor groups */
7. The sensor network begins working, the timer is counting from 0;
8. The first group of N_0 sensors is active at the time 0;
9. After T_0 time is passed, a new group (if there is) will be active
   until n groups of sensors have been used up.
10. End.

6 Experimental Results

6.1 Simulations

We want to verify our theoretic analysis for the relationships among the coverage rate p, the sensor number N, the offset angel α, and the sensor radius r.

We studied the coverage rate of the target region of 500*500 m^2 in our simulation. The number of randomly deployed sensors is varied from 0 to 1500. The offset angel of

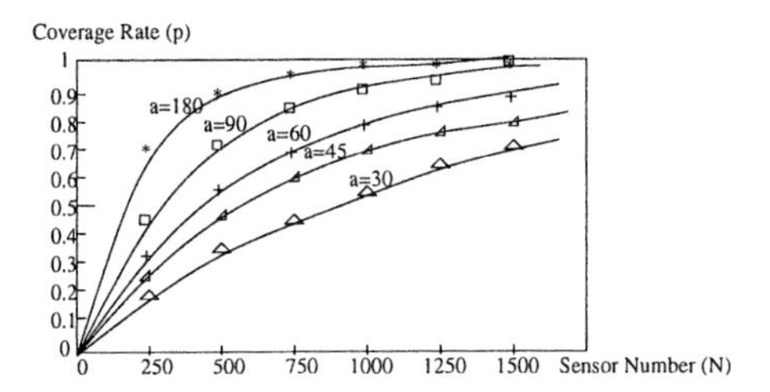

Fig. 3. The effect of the sensor number

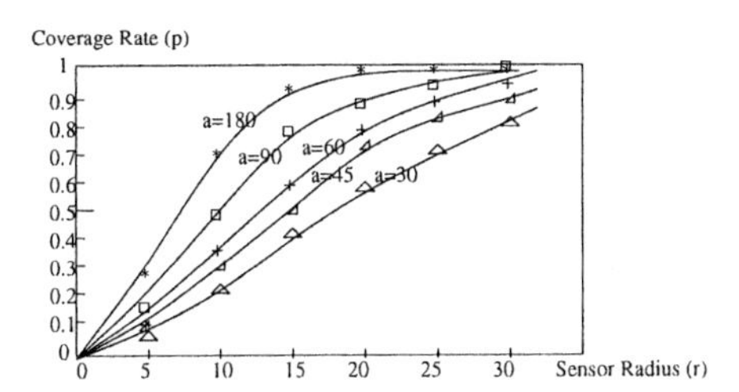

Fig. 4. The effect of the sensing radius

directional sensor (α) is varied from 0 to 180 (π), and the sensor radius is changed from 0 to 25m. The simulations was executed in OPNET platform. All simulation results are closely match with the theoretical analysis in Section 3.

We first consider the effect that sensor number make to the coverage rate. Fig. 3 shows that the greater the sensor number (N), the greater the coverage rate p becomes. In other words, the coverage rate will increase with the increasing of sensor number. We also examined the effect of the sensing radius for coverage rate. Fig. 4 shows the relationship between the coverage rate p and the sensing radius r. It indicates that the greater the sensing radius is, the greater the coverage rate p becomes.

6.2 Case Study

We use a case to illustrate the effectiveness of coverage and connectivity maintenance for randomly deployed directional sensor networks. In a 500*500 m^2 field, we want to deploy the video sensors with the sensing radius 50m and the offset angel $90(\frac{\pi}{2})$ for gathering the visual information. If the required coverage rate is at least 85%, we can calculate the sensor number to be deployed: N=87.

Assume that the lifetime of sensor is 50 hours. The expected lifetime of network is 150 hours, thus we can take 3*87=261 sensors to randomly deploy in the targeted region. A distribution of deployed sensors is generated.

We use grouping scheduling protocol to demonstrate the coverage and connectivity maintenance for this deployment. 261 sensors are divided into 3 groups in which there are at least 87 sensors with the sensing radius $50m$ and the communication radius $100m$, the additional 3 nodes and 5 pathes are re-deployed for repairing the connectivity of 3 groups. These 3 groups are alternatively activated in turn, to maintain the coverage rate 85% of the target region for 150 hours. The deployment and its distribution state of repaired grouping connectivity are not presented here because of the limit of page space.

7 Conclusions and Future Work

Different sensing model of directional sensor networks demands efficient methods for deployment policy and connectivity maintenance. Motivated by this, this paper proposes a systematic method for coverage and connectivity of randomly deployed directional sensor networks. By quantifying the requirements of deploying sensors for a given coverage rate, we can decide the deployment policy. Moreover, we model the directional sensor network as a directed communication graph to analyze its connectivity, and repairing the connectivity of a randomly deployed directional sensor network. Based on the theoretic works, we design the sensor deployment and scheduling protocols for the application of directional sensor networks. The methods are shown to be highly efficient and feasible for applications of directional sensor works.

References

1. I. Akyildiz, W. Su, Y. Sankarasubramaniam, and E. Cayirci, A survey on sensor networks, *IEEE Communications Magazine*, vol. 40, no. 8, Aug. 2002.
2. C. Chong and S. Kumar,"Sensor networks: evolution, opportunities, and challenges," *Proceedings of the IEEE*, vol. 91, no. 8, Aug. 2003.
3. W. Feng, B. Code, E. Kaiser, M. Shea, and W. Feng, "Panoptes: scalable low-power video sensor networking technologies," *Proc. the 11th ACM international conference on Multimedia*,pp.562-571, Berkeley, CA, Nov. 2003.
4. W. Feng, J. Walpole, W. Feng, and C. Pu, "Moving towards massively scalable video-based sensor networks, *Proceedings of the Workshop on New Visions for Large- Scale Networks: Research and Applications*, Washington, DC, Mar. 2001.
5. Yong Gao, Kui Wu, Fulu Li, "Analysis on the redundancy of wireless sensor networks", *ACM WSNA03*, pp.108-114, Sept.19, 2003, San Diego, CA.
6. J. Gehrke and S. Madden, "Query processing in sensor networks," *IEEE Pervasive Computing*, vol. 3, no.1, Jan. 2004.
7. W. Heinzelman, A. Murphy, H. Carvalho, and M. Perillo, "Middleware to support sensor network applications, *IEEE Network*, vol. 18, no. 1, Jan. 2004.
8. Chi-Fu Huang, Yu-Chee Tseng, "The coverage problem in a wireless sensor network", *ACM WSNA03*, pp.115-121, Sept.19, 2003, San Diego, CA.
9. S. Kumar, T. Lai, and J. Balogh, "On k-coverage in a mostly sleeping sensor network, *Proceeding of ACM Mobicom04*, Philadelphia, PA, Oct. 2004.
10. G.H. Lin and G. Xue, "Steiner tree problem with minimum number of Steiner points and bounded edge-length", *Inform. Process. Letter*, 69(2), pp.53-57, 1999.
11. S. Meguerdichian, M. Potkonjak, "Lower power 0/1 coverage and scheduling techniques in sensor networks", *UCLA Technical Report.*
12. S. Meguerdichian, F. Koushanfar, M. Potkonjak, M.B. Srivastava, "Coverage problems in wireless adhoc sensor networks, *Proceeding of IEEE Infocom01*, Anchorage, AK, Apr. 2001.

13. X. Li, P. Wan, Y. Wang, and O. Frieder, "Coverage in wireless ad-hoc sensor networks," *IEEE Transactions on Computers*, 52(6), pp.753-763, 2003.
14. Huadong Ma, Yonghe Liu, "Correlation based video processing in video sensor networks", *IEEE WirelessCom05*, Hawaii, USA, June 2005.
15. Dan Tao, Huadong Ma, Yonghe Liu, "Energy-efficient Cooperative Image Processing in Video Sensor Network", *2005 Pacific-Rim Conference on Multimedia, Lecture Notes in Computer Science*, Springer, Nov.13-16, 2005, Jeju. Korea.
16. Di Tian, Nicolas D. Georganas, "Location and calculation-free node-scheduling schemes in large wireless sensor networks", *Ad Hoc Networks*, 2004,2:65-85.
17. Sameer Tilak, et al., "Infrastructure Tradeoffs for Sensor Networks", *ACM WSNA02*, pp.49-58, Sept.28,2002, Atlanta, Georgia.
18. Wang et al., "Integrated coverage and connectivity configuration in wireless sensor networks", *Proceedings of the First ACM Conference on Embedded Networked Sensor Systems (SenSys 2003)*, Los Angeles, November 2003.
19. F. Ye, G. Zhong, S. Lu, L. Zhang, "Energy efficient robust sensing coverage in large sensor networks", *UCLA Technical Report*, http://www.cs.ucla.edu/ yefan, Oct. 2002.
20. Zhang and Hou. "Maintaining sensing coverage and connectivity in large sensor networks". *Technical report UIUCDCS-R-2003-2351*, June 2003.

Using MDS Codes for the Key Establishment of Wireless Sensor Networks*

Jing Deng[1] and Yunghsiang S. Han[2]

[1] Department of Computer Science,
University of New Orleans,
New Orleans, LA 70148, USA
jing@cs.uno.edu

[2] Graduate Institute of Communication Engineering,
National Taipei University, Sanhsia, Taipei, 237 Taiwan , R.O.C
yshan@mail.ntpu.edu.tw

Abstract. Key pre-distribution techniques for security provision of Wireless Sensor Networks (WSNs) have attracted significant interests recently. In these schemes, a relatively small number of keys are randomly chosen from a large key pool and loaded on the sensors *prior to* deployment. After being deployed, each sensor tries to find a common key shared by itself and each of its neighbors to establish a link key to protect the wireless communication between themselves. One intrinsic disadvantage of such techniques is that some neighboring sensors do not share any common key. In order to establish a link key among such neighbors, a multi-hop secure path may be used to deliver the secret. Unfortunately, the possibility of sensors being compromised on the path may render such establishment process insecure.

In this work, we propose and analyze an Incremental Redundancy Transmission (IRT) scheme that uses the powerful Maximum Distance Separable (MDS) codes to address the problem. In the IRT scheme, the encoded secret link key is transmitted through multiple multi-hop paths. To reduce the total information that needs to be transmitted, the redundant symbols of the MDS codes are transmitted only if the destination fails to decode the secret. One salient feature of the IRT scheme is the flexibility of trading transmission for lower information disclosure. Theoretical and simulation results are presented to support our claim.

1 Introduction

Wireless Sensor Networks (WSNs) have attracted significant interests from the research community due to their potentials in a wide range of applications such as environmental sensing, battlefield sensing, and hazard leak detection. The security problem of these WSNs are important as the sensors might be deployed

* This work was supported in part by grant number LBoR0078NR00C and by the National Science Council of Taiwan, R.O.C., under grants NSC 94-2213-E-305-001.

X. Jia, J. Wu, and Y. He (Eds.): MSN 2005, LNCS 3794, pp. 732–744, 2005.

to unfriendly areas. When any of the sensors is compromised or captured, the information on the sensor is disclosed to the adversary and its operation may become under the control of the adversary.

In order to secure the communication between a pair of sensors, a unique key is needed. Since public/private (asymmetric) keys consume large amount of system resource, the secret (symmetric) key technique is preferred in WSNs. However, distribution of a secret key for every possible communication link is non-trivial due to the large number of sensors and the limited on-board memory size. To this trend, key pre-distribution techniques have been proposed and studied [1, 2, 3, 4]. These techniques allow the sensors to randomly pick a relatively small number of keys from a large key pool and two neighboring nodes[1] then try to find a common key that is shared by themselves.

Due to the randomness of the key selection process in key pre-distribution, some communication links do not have any common key shared by the two neighboring nodes. In [2], a secret link key delivery technique using a multi-hop secure path was proposed: one of the two neighboring nodes finds a multi-hop secure path toward the other node.[2] Each pair of neighboring nodes on the secure path share at least a common key, which could be different throughout the path. Then a secret link key is generated from the source node and sent toward the destination through the secure path.

Such a multi-hop secure path scheme works quite well when all sensor nodes forward the secret key honestly and none of the nodes on the path is compromised. However, the scheme has security problems if any of the nodes is compromised or captured by the adversary. Such a compromise affects the multi-hop secure path scheme in the following way: 1) since the secret link key is decrypted and re-encrypted by each sensor on the path, it may be disclosed to the adversary; and 2) the adversary can modify or drop the information passing through.

In this work, we address the problem of compromised sensors modifying and eavesdropping the information passing through such multi-hop paths. We use the powerful Maximum Distance Separable (MDS) codes to develop the Incremental Redundancy Transmission (IRT) scheme to provide protection for information delivery. Our analysis and simulation results show that the proposed technique is highly efficient. Note that, in [2], Chan, Perrig, and Song proposed a multi-path reinforcement scheme that is similar to our scheme. However, the multi-path reinforcement scheme only concerns the information disclosure to the adversary but not information modification by the adversary.

The paper is organized as follows: in Section 2, we overview related work. The secret link key delivery problem is formulated in Section 3. In Sections 4 and 5, we present the IRT scheme and our analysis. The performance evaluation results are provided in Section 6. We summarize and conclude this work in Section 7, stating some possible future directions.

[1] In this work, we use "sensors", "nodes", and "sensor nodes" interchangeably.

[2] Note that the two communicating nodes are physical neighbors. Such a small geographical separation between the source and the destination enables prompt and efficient secret verifications.

2 Related Work

Reference [1] proposed a random key establishment technique for WSNs. In this technique, each sensor is pre-loaded a number of keys that are randomly selected from a large key pool. After deployment, two neighbors can establish a secure communication if they share a common key. Otherwise, they need to exchange a secret key via a multi-hop secure path. Reference [2] extended the technique into q-composite random key establishment technique which forces two neighbors to establish a secure communication only when they share q common keys, where $q \geq 2$. Based on [1], two similar random key pre-distribution techniques that used multi-space key pool to drastically improve network resilience and memory usage efficiency were developed independently [3, 4].

A multi-path key reinforcement technique was proposed in [2] to enable two nodes to establish secure communication even if they do not share enough common keys (with the use of the q-composite technique). These two neighbors first identify all secure paths between themselves. Then one node generates a set of random numbers (of the same size) for all the paths and send each number to the other node through each of the paths. After the destination receives all the numbers, it exclusive-ORs all of them to obtain the secret link key. The multi-path key reinforcement scheme significantly improves the protection of the secret link key from being disclosed to the adversary. However, the scheme fails if any of the paths is compromised by the adversary and the number is modified or dropped.

In [5, 6], combinatorial set was used to distribute keys to sensors *prior to* deployment. Such a deterministic combinatorial set technique allows each key in the key pool to be assigned to a constant number of sensor nodes. Therefore, the number of nodes each sensor shares a common key is fixed.

A Secure Routing Protocol (SRP) was proposed in [7] to send additional information to protect routing information being dropped. To combat the topology instability problem in wireless networks, a multi-path routing scheme was proposed and investigated in [8]. The scheme allows the sender to add extra overhead to each packet that is to be transmitted over multiple paths. The goal is to find the optimal way to fragment the packet into smaller blocks and deliver them over multiple paths. The focus of [7, 8] is on the problem of missing some of the messages but not modification of them.

In [9], an efficient information dispersal mechanism was developed to provide security, load balance, and fault tolerance for communication networks. Reed-Solomon codes were used to recover link faults and to provide security. However, all redundancy were sent along with the information symbols, increasing transmission overhead significantly.

MDS codes have been used in the Automatic-Repeat-reQuest (ARQ) protocols to reduce the transmission overhead on communication systems [10, 11]. In [10], a family of MDS codes, Reed-Solomon codes, was used in a type-II hybrid ARQ protocol. In the first transmission, a relatively high rate Reed-Solomon code with fewer redundancy is used. When an additional transmission is needed, only the redundant symbols are sent. With such a technique, the overall code rate is reduced. This scheme increases the system throughput by reducing the transmission overhead. In [11], punctured MDS codes were used for the type-II hybrid ARQ protocol and a modified version (with fewer decoding operations) of the scheme proposed in [10] was presented.

3 Problem Formulation

We explain the link key establishment problem in WSNs in more details in this section. The key pre-distribution schemes such as [1, 3, 4, 12] provide memory-efficient and resilient ways to establish secret link keys for a portion of potential communication links. The rest of the communication links need to establish their secret keys by other means such as multi-hop delivery.[3]

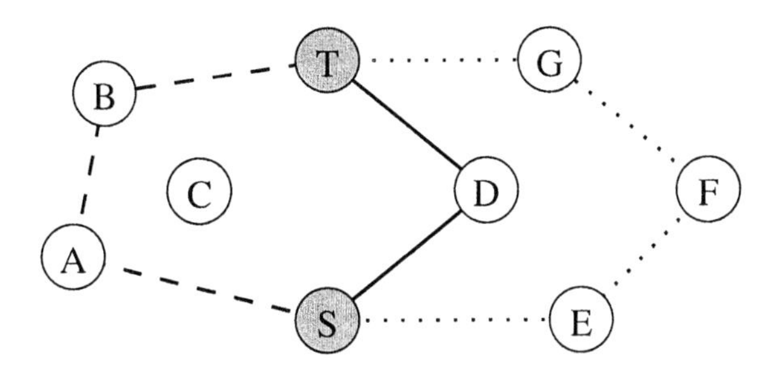

Fig. 1. Illustrations of multi-hop key establishment

A few sensor nodes are shown in Fig. 1. Each line segment connecting two nodes represents that these two nodes share at least a common key, e.g., nodes E and F share at least a key. Note that nodes S and T do not share any common key. Now assume that nodes S and T, which are physical neighbors, need to establish a secure communication that requires a secret link key. As suggested in [1, 2, 3], in order to establish a secret link key between nodes S and T, a multi-hop secure path may be used to deliver the secret key. For example, node D may be used to relay the secret key between nodes S and T. Since only one multi-hop path is used in the key delivery, we term it the Single Path (SP) scheme.

The SP scheme may be summarized as follows: when node S needs to establish a secret link key, K_{link}, with node T, node S finds a path $S - N_1 - N_2 - \cdots - N_h - T$, where S shares a key with N_1, N_1 shares a key (could be different) with N_2, etc.. Then node S encrypts K_{link} with the secret key shared by itself and node N_1 and sends the encrypted message to node N_1. Node N_1 decrypts K_{link} using the common secret key shared with node S. Then node N_1 sends K_{link} to

[3] We assume that some sensor nodes may be compromised even during the WSN initialization process and the adversary is able to decrypt the recorded information after it compromises some sensor nodes later on. An assumption of secure initialization process will make the key establishment issue trivial.

node N_2 using the same technique. This process continues until K_{link} reaches the destination, node T. Some examples of such multi-hop paths in Fig. 1 are $S-D-T$, $S-A-B-T$, and $S-E-F-G-T$.

Note that the secret link key K_{link} needs to be decrypted and then re-encrypted by each of the sensor nodes on the multi-hop path. This leads to the following problem:

Problem Statement: *In key pre-distribution schemes for WSNs, some neighboring sensors do not share any common key. Their secret link key needs to be established through multi-hop secure path. However, when any of the sensors on the multi-hop secure path is compromised or captured by the adversary, the secret link key is disclosed. A compromised sensor may also modify or drop the key information passing through itself. How do we provide a fault tolerant mechanism to send the secret link key between two physical neighbors efficiently and securely?*

In Section 4, we introduce the powerful Maximum Distance Separation (MDS) codes and use it in our Incremental Redundancy Transmission (IRT) scheme to solve this problem.

4 The Incremental Redundancy Transmission (IRT) Scheme

4.1 MDS Codes

We first review the MDS codes [10, 11] that will be used in the IRT scheme. Let Hamming distance between two vectors (codewords) be the number of distinct positions between two vectors. An (n, k, d_{min}) MDS code is a linear block code whose minimum Hamming distance d_{min} between any pair of distinct codewords must satisfy $d_{min} = n-k+1$, where n is the code length and k is the dimension of the code. Therefore, each codeword in the $(n, k, n-k+1)$ MDS code has exactly n symbols among which there are k information symbols. Usually, the extra $n-k$ symbols are called parity checks or redundancy of the code. Furthermore, an $(n, k, n-k+1)$ MDS code will be able to recover any t errors if

$$t \leq \left\lfloor \frac{n-k}{2} \right\rfloor .$$

MDS codes are optimal in the sense that they provide the largest possible Hamming distance between codewords and hence can correct most number of errors. The most famous family of MDS codes are Reed-Solomon codes. Efficient decoding algorithms for MDS codes have been studied extensively in [13]. The MDS codes have several nice properties that make them very useful. Two of such properties are given as follows without proof.

Property 1. Punctured (shortened) MDS codes are MDS.

A code is punctured by deleting parity symbols from each codeword in the code and a code is shortened by deleting information symbols from each codeword in the code. An $(n, k, n-k+1)$ MDS code can be punctured (shortened) to an $(n-j, k, n-j-k+1)$ $((n-j, k-j, n-k+1))$ MDS code by deleting j parity (information) symbols from each codeword.

Property 2. Any k coordinates of an MDS code can be used as information symbols.

According to this property, by knowing any k symbols of a codeword of an MDS code, one can recover other $n-k$ symbols for this codeword.

4.2 Multi-hop Paths

Before we present the IRT scheme, we discuss the path selection process and its effect on IRT. In this work, we assume that a source node identifies m multi-hop paths between itself and the destination. Such m paths could be chosen from the node-disjoint paths, where none of the paths has a common node besides the source and the destination [14]. Another option is to allow the source node to select randomly among the available paths. The result is that some paths may have common nodes and thus the security performance worsens. The benefit of such a selection technique is that it does not rely on the availability of node-disjoint multi-hop paths and eliminates the cost of identifying such paths.

As suggested in [2], it is always beneficial to choose short multi-hop paths instead of long multi-hop paths. As the length of a multi-hop path increases, the possibility of path compromise is higher. On the other hand, the number of short multi-hop paths is usually smaller than that of long multi-hop paths. We evaluate the values of m under various network conditions and the effect of such m paths on the security performance of our scheme in Section 6.

4.3 The Incremental Redundancy Transmission (IRT) Scheme

Let $\mathbf{c} = (c_0, c_1, \ldots, c_{k-1}, c_k, \ldots, c_{n-1})$ be a codeword of an (n, k) MDS code over $GF(q)$, where $c_i \in GF(q)$, $0 \leq i \leq n-1$, is a symbol. Here $c_0, \ldots, c_{k-1}$ are the information that the source needs to send to the destination. Assume that the secret link key between them is of length $\gamma < k$ and can be generated by a function f of these information if the destination decodes this codeword correctly. The design goal of our scheme is to tolerate up to e compromised paths. Since an (n, k) MDS code can recover up to $(n-k)/2$ errors, n should satisfy

$$n \geq k + 2e \ , \tag{1}$$

where we have implicitly assumed that $n-k$ is even. Since the identities of the compromised paths are unknown to the source node, the best strategy for the source is to split data evenly and send to all available paths.

A brief outline of the IRT scheme is given as follows: let $r_0 = k$ be the number of symbols sent by the source in the original transmission. If the destination receives all information correctly and re-generates the secret key sent by the source, our scheme terminates. Otherwise, the source sends r_1 symbols in its first additional transmission. If the destination succeeds in secret key re-generation, our scheme stops. Otherwise, the same process continues until the n symbols are exhausted.

Let $r_1, r_2, \ldots, r_e$ be the numbers of extra symbols sent out by the source in each additional transmission (the values of r_i, $1 \leq i \leq e$, will be determined in Section 5). Note that, in our scheme, the source only needs to send out up to $\sum_{i=0}^{t} r_i$ symbols when t paths are compromised. Let $\boldsymbol{y} = (y_0, y_1, \ldots, y_{\ell-1})$ be the corresponding received vector at the destination when the source has sent out ℓ symbols up to now.

The IRT scheme is given as follows:

1. The source first encodes k symbols, $\boldsymbol{c}_0 = (c_0, c_1, \ldots, c_{k-1})$, into a codeword with n symbols, $\boldsymbol{c} = (c_0, c_1, \ldots, c_{k-1}, c_k, \ldots, c_{n-1})$, where $n = \sum_{i=0}^{e} r_i = k + \sum_{i=1}^{e} r_i$ and $r_0 = k$. Initialize $i = 0$, $b = 0$, and $s = r_i - 1$.
2. The source transmits r_i symbols specified by b and s ($\boldsymbol{c}_i = (c_b, c_{b+1}, \ldots, c_s)$) evenly along the m paths. If r_i/m is not an integer, the last path will transmit less symbols.
3. Assume that the destination receives all symbols from the m paths as $\boldsymbol{y}_i = (y_b, y_{b+1}, \ldots, y_s)$.[4] The destination appends $\boldsymbol{y}_i$ to all the previously received symbols to form a longer codeword. Then it tries to decode this codeword in order to obtain the k symbols. If the decode process fails due to more than i errors occur, then go to Step 4 directly; otherwise, it verifies this result with the source through the challenge-response technique (recall that the source and the destination are physical neighbors). If the re-generated secret link key is verified, the transmission of the secret link key has succeeded; Otherwise, goes to Step 4.
4. If $i = e$, then the key establishment fails due to too many compromised paths. Otherwise, the destination asks for another round of additional transmission.
5. The source sets $i = i + 1$, $b = s + 1$, $s = s + r_i$, and repeats Step 2.

Therefore, compared with [10, 11], which have only one additional transmission, the IRT scheme sends multiple retransmissions when necessary.

5 Analysis

In this section, we derive the values of r_i, $0 \leq i \leq e$, for the IRT scheme. We will also discuss the security performance of the IRT scheme and the SP scheme, which sends the secret link key through a single multi-hop path.

5.1 Selections of $r_1, r_2, \ldots, r_e$

The values of $r_1, r_2, \ldots, r_e$ can be determined as follows. In order to reduce the total number of symbols to be transmitted, in each transmission we should

[4] The source can notify the destination the number of transmitted symbols over plain text. Therefore, the event of symbols being dropped is similar to that of symbols being modified along the multi-hop paths.

add as few as possible redundancy that can correct one more error.[5] Therefore, noticing that r_1 symbols are added in order to correct the errors caused by one compromised path, we can determine r_1 as

$$\left(\frac{r_1}{m}+\frac{k}{m}\right) \leq \frac{r_1}{2} . \tag{2}$$

The left side of (2) is the total number of errors introduced by the compromised path. The right side of (2) is the error correction capability due to the transmission of the additional r_1 symbols.

Taking the smallest integer that satisfies the above inequality, we have

$$r_1 = \left\lceil \frac{2k}{m-2} \right\rceil . \tag{3}$$

In general, the value of r_ℓ, where $1 \leq \ell \leq e$, must satisfy the following inequality

$$\frac{\ell}{m}\left(k+\sum_{i=1}^{\ell} r_i\right) \leq \frac{1}{2}\sum_{i=1}^{\ell} r_i . \tag{4}$$

In order to reduce the total number of message transmissions, we choose the smallest r_ℓ that satisfies the above inequality. Therefore,

$$\begin{aligned} r_\ell &= \left\lceil \frac{2\ell k}{m-2\ell} - \sum_{i=1}^{\ell-1} r_i \right\rceil \\ &= \left\lceil \frac{2\ell k}{m-2\ell} \right\rceil - \sum_{i=1}^{\ell-1} r_i . \end{aligned} \tag{5}$$

(5) can be further rearranged to

$$r_\ell = \left\lceil \frac{2\ell k}{m-2\ell} \right\rceil - \left\lceil \frac{2(\ell-1)k}{m-2(\ell-1)} \right\rceil , \tag{6}$$

when $1 \leq \ell \leq e$.

Based on (5), when there are ℓ compromised paths between the source and the destination, the total additional symbols that should be transmitted is

$$\sum_{i=1}^{\ell} r_i = \left\lceil \frac{2\ell k}{m-2\ell} \right\rceil . \tag{7}$$

Since e is the maximum number of errors that can be corrected by the IRT scheme, the set of $r_1, r_2, \ldots, r_e$ should satisfy:

$$\sum_{i=1}^{e} r_i \leq n-k ,$$

[5] We neglect the overhead of sending such symbols in our analysis. Inclusion of overhead such as MAC layer headers and physical layer headers may affect the performance of our scheme.

which leads to (based on (7))

$$\left\lceil \frac{2ek}{m-2e} \right\rceil \leq n-k \,.$$

Therefore, e should satisfy

$$e < \frac{n-k}{n} \cdot \frac{m}{2} \,. \tag{8}$$

As a point of reference, when $n = 1024$, $k = 256$, $m = 15$, the maximum value of e is 5 due to (8). The array r_i, $0 \leq i \leq e$, is $\{256\ 39\ 54\ 77\ 122\ 220\}$.

5.2 Information Disclosure

We start our discussions on the security performance of information disclosure by presenting the attack model in the following: the adversary takes control of some compromised sensors and collects all information passing through them. An intelligent adversary may decide not to modify the information that is passing through in order to maximize the benefit of its eavesdropping effort.

We further assume that there are m_x paths that are compromised by the adversary among the m available paths.[6] We evaluate the *secret disclosure probability*, p_x, which is defined as the probability of disclosing enough symbols to the adversary so that it can obtain the key with relative ease.

In the SP scheme, the γ symbols are transmitted through one randomly chosen path among the m available paths. The secret disclosure probability can be calculated as

$$p_x^{(SP)} = \frac{m_x}{m} \,. \tag{9}$$

When the IRT scheme is used, the source transmits only $r(0) = k$ symbols and the destination gets all of these symbols successfully because no information is modified. Since such k symbols are transmitted through the m paths evenly, each path sends $\frac{k}{m}$ symbols. For a fair comparison between the IRT scheme and the SP scheme, we define the secret disclosure probability as the probability of at least αk symbols being disclosed to the adversary, where $0 < \alpha \leq 1$. Therefore, the secret disclosure probability can be calculated as

$$p_x^{(IRT)} = \frac{[\text{Number of cases where at least } \alpha k \text{ symbols are disclosed}]}{[\text{Total number of cases}]} \,. \tag{10}$$

The value of α depends largely on how the γ symbols of the secret link key information are encoded into the k symbols, i.e., it depends on the selection of function f in the scheme.

---

[6] We use a different variable than e in order to distinguish the two different kinds of compromises: information modification and information disclosure.

6 Performance Evaluation

Simulations were performed in Matlab to evaluate the efficiency of the proposed scheme. Unless specified otherwise, our simulations were set up with the following parameters: we randomly place $N = 400$ nodes on a square area of 1000 m by 1000 m. The radio transceiver range is 100 m. The MDS code is assumed to be $(n, k) = (1024, 256)$. We investigate the performance of the IRT scheme and the SP scheme for different γ (instead of varying (n, k)).

In Fig. 2, we show the number of paths with secure connections that are exactly h-hops from a source to a destination (assuming that they do not share a common key). The average number of paths is presented corresponding to various probabilities of any two neighboring nodes sharing a common key, p_{local}. We also present the number of paths for a similar network with half of the nodes, for comparison purposes. As can be observed from Fig. 2, the number of available paths increases with local connectivity, p_{local}. When nodal density increases, there are more paths as well. The number of h-hop paths also increases with h. Note that these paths may have common nodes besides the source and the destination.

When there are compromised nodes on the paths that are used to deliver the secret link key information and these compromised nodes modify the passing information, extra symbols need to be transmitted. Figure 3 shows the average number of symbols that need to be transmitted in order to allow the destination to re-generate the secret key. We use all of the available 2- and 3-hop paths in the IRT scheme. The SP scheme randomly chooses one out of these paths to send the secret link key. In Fig. 3, we fix (n, k) for the IRT scheme but vary γ. Therefore, the number of transmitted symbols in the SP scheme lowers as γ decreases. The relative transmission cost of the IRT scheme increases as γ

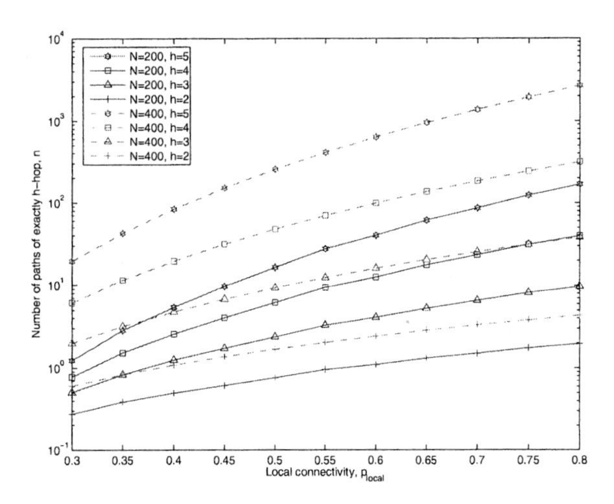

Fig. 2. Number of paths with exactly h-hops from source and destination

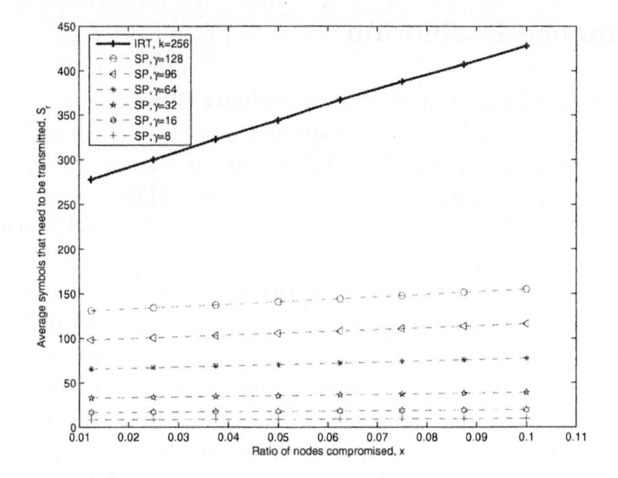

Fig. 3. Average symbols that should be transmitted

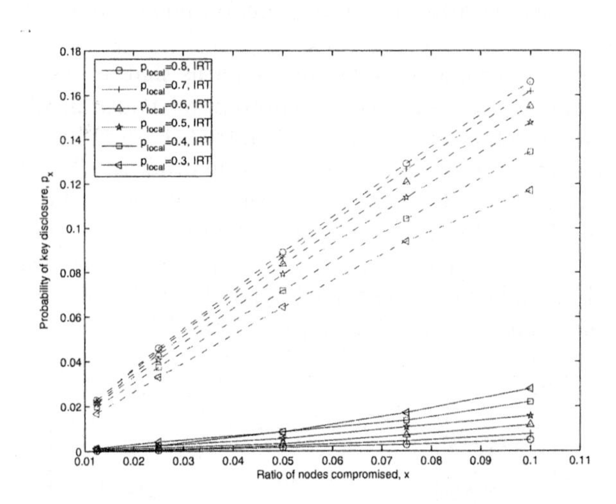

Fig. 4. Secret disclosure probability of the IRT and SP schemes. The red dashed lines represent the performance of the SP scheme.

decreases. Note that there is a slight increase of number of transmitted symbols in the SP scheme as the ratio of nodes compromised, x, increases. This is due to the higher probability of the used path being compromised.

In Fig. 4, we show the secret disclosure probability of the IRT scheme and the SP scheme. The value of γ is set to 64. The curves with solid lines represent p_x of the IRT scheme for different local connectivity, p_{local}. The red dashed lines show the p_x of the SP scheme. We can observe a much higher secret disclosure probability of the SP scheme. Furthermore, p_x increases with x, as expected.

The flexibility of secret disclosure probability of the IRT scheme is presented in Fig. 5. In this figure, we vary the value of α and showed the probability of

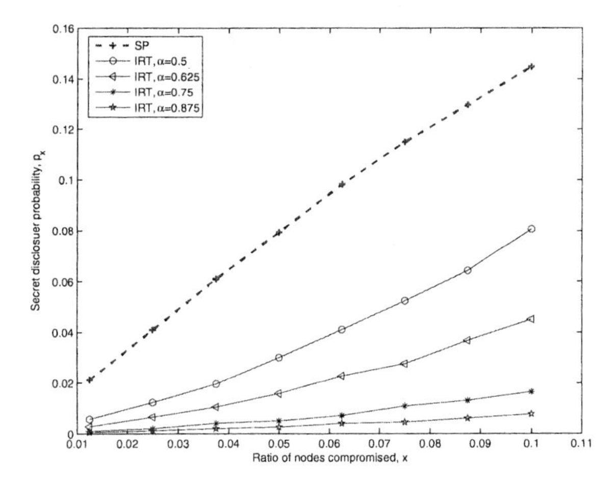

Fig. 5. The flexible secret disclosure probability of the IRT scheme

key disclosure for different node compromised ratio, x. It can be observed that the IRT scheme has a much lower p_x than the SP scheme when $\alpha < 1$. As α increases, the p_x value is smaller. Therefore, the IRT scheme provides a nice property of flexibility: a pre-defined threshold of probability of key disclosure can be guaranteed by varying (n, k).

7 Conclusions

We have proposed and investigated an Incremental Redundancy Transmission (IRT) scheme for the secret link key establishment process of key pre-distribution techniques. The IRT scheme uses the powerful MDS codes to encode the secret link key information and send through multiple multi-hop paths. The redundant symbols are transmitted only when they are needed to enable the destination to decode the secret information. One salient feature of the IRT scheme is the flexibility of trading transmission for lower information disclosure.

In the future work, we will consider sending different number of symbols along different paths based on the lengths of the paths in order to further reduce the overall transmission overhead and to improve the security performance. The effect of node disjoint paths will be investigated as well.

References

1. Eschenauer, L., Gligor, V.D.: A key-management scheme for distributed sensor networks. In: Proc. of the 9th ACM conference on Computer and communications security, Washington, DC, USA (2002) 41–47
2. Chan, H., Perrig, A., Song, D.: Random key predistribution schemes for sensor networks. In: Proc. of IEEE Symposium on Security and Privacy, Berkeley, California (2003) 197–213

3. Du, W., Deng, J., Han, Y.S., Varshney, P.K., Katz, J., Khalili, A.: A pairwise key pre-distribution scheme for wireless sensor networks. ACM Trans. on Information and System Security **8** (2005) 228–258
4. Liu, D., Ning, P.: Establishing pairwise keys in distributed sensor networks. In: Proc. of the 10th ACM Conference on Computer and Communications Security (CCS '03), Washington, DC, USA (2003) 52–61
5. Camtepe, S.A., Yener, B.: Combinatorial design of key distribution mechanisms for wireless sensor networks. In Samarati, P., Ryan, P., Gollmann, D., Molva, R., eds.: 9th European Symposium on Research Computer Security. Volume 3193 of Proceedings Series: Lecture Notes in Computer Science (LNCS)., Springer-Verlag (2004) 293–308
6. Lee, J., Stinson, D.R.: A combinatorial approach to key predistribution for distributed sensor networks. In: Proc. of IEEE Wireless Communications and Networking Conference (WCNC '05), New Orleans, LA, USA (2005)
7. Papadimitratos, P., Haas, Z.J.: Secure message transmission in mobile ad hoc networks. Elsevier Ad Hoc Networks **1** (2003) 193–209
8. Tsirigos, A., Haas, Z.J.: Multipath routing in the presence of frequent topological changes. IEEE Communication Magazine (2001) 132–138
9. Rabin, M.O.: Efficient dispersal of information for security, load balancing, and fault tolerance. Journal of the Association for Computing Machinery **36** (1989) 335–348
10. Pursley, M.B., Sandberg, S.D.: Incremental-redundancy transmission for meteor-burst communications. IEEE Trans. on Communications **39** (1991) 689–702
11. Wicker, S.B., Bartz, M.J.: Type-II hybrid-ARQ protocols using punctured MDS codes. IEEE Trans. on Communications **42** (1994) 1431–1440
12. Du, W., Deng, J., Han, Y.S., Chen, S., Varshney, P.K.: A key management scheme for wireless sensor networks using deployment knowledge. In: Proc. of the 23rd Conference of the IEEE Communications Society (Infocom '04), Hong Kong, China (2004) 586–597
13. Wicker, S.B., Bhargava, V.K.: Reed-Solomon Codes and Their Applications. Piscataway, NJ: IEEE Press (1994)
14. Lou, W., Fang, Y.: A multipath routing approach for secure data delivery. In: Proc. of IEEE Military Communications Conference (MILCOM '01), McLean, VA, USA (2001) 1467–1473

A Study on Efficient Key Management in Real Time Wireless Sensor Network

Sangchul Son[1], Miyoun Yoon[2], Kwangkyum Lee[1], and Yongtae Shin[1]

[1] Dept. of Computing, Graduate School, Soongsil University, Sangdo5-Dong, Dongjak-Gu, Seoul, Republic of Korea
{yelhorse, goodwin77, shin}@cherry.ssu.ac.kr
[2] Information Security Technology Division, Korea Information, Security Agency, Seoul, Korea
myyoon@kisa.or.kr

Abstract. Sensor network technique consists that collecting information, transfer data by sensor nodes and that control sink node. But this sensor node has limited energy power and computing capability, so it has big problem that apply traditional security technique for sensor network. Especially, Research of security technology in real time wireless sensor network is not consisting yet. So we propose about efficient key management mechanism to allow existent secret sharing mechanism and one-way hash algorithm in real time wireless sensor network environment.

1 Introduction

The ubiquitous sensor network is network that spread wide area or dense area for collecting information and transfer data of radio technique. Though sensor node receives data request from administration node that is sink node and send collected data, communication between sensor node and sensor node consists of radio and consists of radio sensor node and sink node interval at this process. Therefore, proceeding that each node certificate before receiving and sending, also this proceeding must consist and authentication key management is very important.

Real time sensor network gets completed information which of sensor node that scatter each other is conjoined. Dynamic adulterating and secession of sensor nodes happen by in real time wireless sensor network, key management energy of sensor node much depletive. Specially, when this situation or adulterating of node, secession that must collect information many long hours happens frequently, efficient key management is very important. We introduce about key management mechanism of sensor network that exist in existing in chapter 2 of this paper and analyze their advantage and disadvantage. Chapter 3 introduces mechanism that proposes in this paper and chapter 4 do comparative analyze with existing research [3] and analyze security. Finally, this paper presents about conclusion and explains hereafter research task in chapter 5.

X. Jia, J. Wu, and Y. He (Eds.): MSN 2005, LNCS 3794, pp. 745 – 752, 2005.

2 Related Works

[1] proposed mechanism that elect representation node of sensor node in sensor network. This mechanism is selecting representation node and divide sensor network as class by method to manage leaf node and use secret sharing mechanism and define certification method between sensor nodes in sensor network of schedule extent.

[2] is proposing secret sharing mechanism to use symmetric key algorithm. Secret sharing mechanism is if secret sharing mechanism does to keep dividing from single secret key to n's many key and schedule key is mixed, is algorithm that can reconstruct secret key of original. If it use this algorithm, user who has purpose that is enemy of evil invades and can prevent that when mixed the key in case $(n-1)$'s key flowed out, secret key is reconstructed.

[3] introduce light weight changed μ-TESLA techniques that improves TESLA that is using in existing network according to environment of sensor network. This mechanism must do that key creation and division consist to key chain per time and synchronization of always inaccurate time must consist. Key distribution is transmitted while when sender turns and key is created by one way hash algorithm.

Security mechanism that proposing [1] and [3] is efficient in security of sensor network but inefficient in real time sensor network environment. So we propose about efficient key management mechanism to use hierarchic certification method in real time sensor network in this paper.

3 Key Management Mechanism of Sensor Node

Each node is difficult to use various information and performance to use public key encryption algorithm in sensor network. Also, symmetric key encryption mode that exists in existing is inappropriate to store and achieves packet per necessity certification information at sensor node.[4] Therefore, we do light weight to be changed and present about stable symmetric key management mechanism taking advantage of existent secret sharing mechanism. Sensor node and reliable certification mechanism of sink node and key management mechanism to propose in this paper can be applied in efficient and some arrangement situation in time important matter that is real time.

3.1 Key Distribution and Key Management for Reliable Communication

As each sensor nodes form network through routing process of nodes if it is arranged in place that sensor nodes are specified to collect information and sink node and communication consist, sensor network is composed. After sensor network occurs so, do key value for first certification broadcasting to each sensor nodes at sink node for the first time. Elect area representation node of sensor network using algorithm that propose in [1] next. Representation node refers to this node as representation node as characteristic set node that can sense all sensors of each sensor node. Certification of representation node at sink node first time broadcasting consisting of done certification key representation node certification of leaf nodes take charge. After elect representation node, create secret key for certification when communicate with sink node using one way hash algorithm at sink node. After transmit key value k_i that is made write secret key value a_i and communication of representation nodes draw coefficient value a_i through secret sharing mechanism thereafter at the same time.

These two keys value are transmitted to sink node to representation node are known as master key. Figure 1 shows process that communicates certification key to master key and representation node to leaf node after first sensor network is composed.

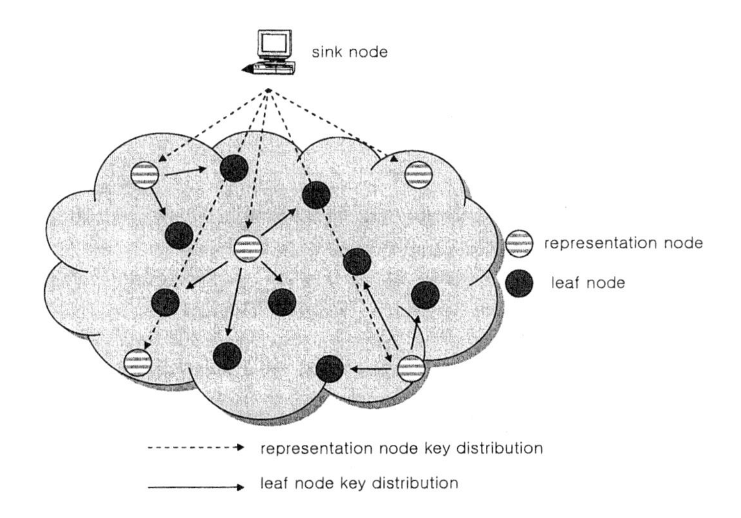

Fig. 1. When it forms sensor network, process of key distribution

Representation node that receive transmission stores coefficient value a_i to own buffer after create encryption key using one way hash algorithm by certification key that receive profit coefficient value a_i first time from sink node. Leaf node that deliver coefficient value a_i of next original to leaf node and charge a_i sends with encryption key that appear by incidental and value after do to do one way hash algorithm spending key value such as doing inceptive certification key value so that representation node may encrypt a_i that is storing when send information that collect by oneself while is storing the encryption key to representation node. Thus, representation node that receive information and encoded key value that is collected to leaf node can compare encryption key that receive to leaf node with encryption key in own buffer and do certification of leaf node. This time, representation node and limit sphere of leaf node decide by 1 hop unit of measure and leaf node belonged between different representation nodes is belonged to representation node that priority encryption key is received. And leaf node that certification key value is transmitted to several representation nodes takes charge role that is gateway between representation nodes.

Like this, Scatters node hierarchically and certification of each node consists by transmitting certification key. In case of mistaken value a_i or free value was passed while master key is transmitted to sink node to representation node first time, can reliable communicate between representation nodes because can not create key value that can quote sink node by secret sharing mechanism and authoritativeness can be offered because using one way hash algorithm between representation node and leaf node. Figure 2 schematizes sink node and representation node, certification key exchange between leaf nodes.

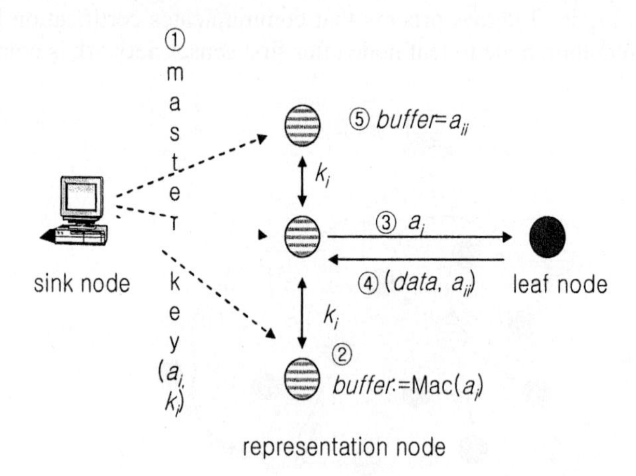

Fig. 2. Key distribution mechanism to consist at each node

---

input : Master key that is sent at sink node
output : Certification of representation node by master key

Procedure *leaf node authentication*
If receive *Master key*(k_i, a_i) from *sink node* **then**
Begin
buffer:= **Mac**(a_i)
Send a_i **to** *leaf node*
    If receive *data*, a_{ii} from *leaf node* **then**
    Begin
        If *buffer* = a_{ii}
        Begin
            Authenticate *leaf node*
            Send *data* **to** *next representation node*
        Else
        Listen *data*, a_{ii}
        End
    End
End
Else
Listen k_i, a_i

Fig. 3. Key management algorithm of representation node

Because it use secret sharing mechanism between sink node and representation node and unused a_i that is certification key of all representation nodes, and have used and draw key value a_i of n. Even if that's situation that reliable communication as that is available because it can not draw certification key value of representation node in case of a_i of n-1 piece collected if it become so and together what one node can not communicate happens, provide effect that breakdown of the node can be permitted. Also, one sensor node has low memory and low processing power, so sensor node requires more little processing because it uses one way hash algorithm once. Figure 3 is algorithm about key management mechanism that consists at representation node and table 1 explains about function and variable that is used in algorithm.

Table 1. Function, variable comment that is used in algorithm

Defined function and variable	Definition
Master key	Certification key value of representation node that is transmitted at sink node
k_i	One way hash encoded representation node key value
a_i	Representation node's certification key that use secret sharing mechanism
receive	Function that receive message or data
Mac	One way hash algorithm
a_{ii}	One way hash encoded key that is passed at leaf node
buffer	Sensor Node's temporary save space
Authenticate	Function that quote node
Listen	Function that wait data or variable
sink node	Generalization node that take charge data and certification of sensor node
next represe-ntation node	Node that representation subset of abutting sensor node
leaf node	Low level node belonging to representation node
Send	Function that transmit data or key
data	Information that sensor node collects

3.2 Key Refresh Cycle

Master key that is transmitted from sink node in real time sensor network environment should be updated periodically. This is processing ability special quality short life cycle and mobility of each sensor nodes, low processing ability of sensor network, there is all-important constituent to leave and enter of nodes that consist dynamically and communicate that is reliable from invasion of node that is not quoted. Specially, it is reason because position change and adulterating and secession of node consist frequently in sensor network environment that consist of real time, refresh cycle of master key is important. Element of key refresh cycle is change of representation node by first. Representation node in case this representation node does not achieve several function for some reason receiving key distribution that use secret sharing mechanism from sink node or in case adulterating and secession of network occurred sensor nodes again new representation node. Therefore, it should be always transmitted new master key value from sink node after this process is repeated. This time, elected representation node transmits existent certification key value to sink node after encipher one way hash algorithm and receive authentication from sink newly. Representation node that receive authentication so transmits new key value using mechanism that is proposed to leaf node.

Second, key value should be updated periodically regardless of change of sensor node by second. After first certification consists, if refresh of master key value does not consist, node that is enemy of evil which invade without permission can flow out information without notice routing process or information gathering process and can circulate wrong information. Specially, key refresh in real time environment is all-important to quote stream of collected information. Therefore, it does key refresh by cycle of time that divide time that whole data is collected in case of any change did not occur to sensor network itself as number of representation node. If it uses such key refresh cycle, it changes that master key change of whole sensor network gives group of representation node. Therefore, because master key value of whole network changes before find out certification key value of next time representation node even if node that is enemy of evil knows any mass of representation node certification key value invading within sensor network, it is no effectiveness of representation node certification key value that get in existing. It offers little more efficient authoritativeness in real time sensor network whole certification using this key refresh cycle. Next chapter 4 analyzes performance with existent sensor network through analysis of proposed mechanism.

4 Performance Analysis

It offers security that strong so much although if it make use of key management mechanism that is proposing in this paper, wastage increases than energy consumption of only routing process that consist in general sensor network. Specially, key management about data flowing is essential when it consider energy wastage in real time sensor network surrounding that adulterating and secession of sensor node are frequent. It analyzes efficiency of key management mechanism that proposes in paper that sees therefore about security required usually with analysis about extensity of representation node.

4.1 Extensity Analysis

Mechanism that is proposing in this paper is key management mechanism to divide and uses secret sharing mechanism and one way hash algorithm hierarchically. (Numerical formula 1, 2) is diet that express energy wastage of key management mechanism that propose in paper that see.

① When it enter and leave of representation node

$P \cdot (E_{receive} + N_{one\text{-}hop} \cdot (E_{send} + E_{receive}))$ (numerical formula 1)

② When it enter and leave of leaf node

$P \cdot (E_{send} + E_{receive})$ (numerical formula 2)

The other side, energy efficiency of key management mechanism that proposes in [3] is same with below.

$$P{\cdot}N_{total}{\cdot}E_{receive} \quad \text{(numerical formula 3)}$$

Variable P that is defined in (numerical formula 1, 2, 3) ratio of enter and leave and $E_{receive}$ of sensor nodes is unit consumption energy content of cosmic ray key value reception right time and E_{send} is unit consumption energy content of cosmic ray key value transmission of a message right time. $N_{one\text{-}hop}$ is number of representation node, and N_{total} displays number of whole node. As it appears in numerical formula, efficiency of key management mechanism that propose in [3] in situation that number of sensor node is not much good but it can know that efficiency of mechanism that propose in situation that number of node becomes much and adulterating and secession of node become frequent improved. Figure 4 expresses energy wastage that change P according to N_{total} in case.decided $N_{one\text{-}hop}$ by 5 by 0.1. When sensor node communicates, because consumption energy is much, processing energy did not consider in node and transmission of a message, reception unit consumption energy supposes by 210mW [5].

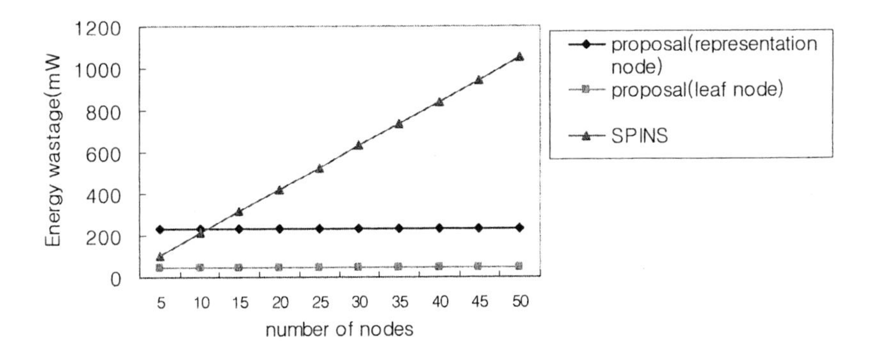

Fig. 4. Energy efficiency of mechanism that propose with SPINS

In case number of node is few in case of [3] with result of figure 4, energy wastage is few but adulterating and secession of node increase, consumption energy content of cosmic ray can see augmented thing as schedule ratio. The other side, after sensor network is composed first time in techniques to propose, because key value exchange process of nodes need, early energy wastage is big than mechanism of [3] that charge certification key value without this exchange process. However, it can see that energy wastage is fixed when adulterating and secession of node by breakdown of sensor node or replace of new node are frequent. This administrates key management at each node, and because whole key refresh is no necessity because have other key by area. Therefore, in case it did whole number of node by 50 that propose can know about quintuple than mechanism of [3] that energy efficiency is good.

4.2 Security Requirement Analysis

This paper explains about security of mechanism that is proposing in this paper. It must satisfy security such as confidentiality, integrity, certification, nonrepudiation and so on to secure sensor network from outside or interior attack.

Confidentiality can secure confidentiality of sensor node from first sink node before this certification key value is leaked because certification key value is transmitted into each sensor node.

Integrity can secure integrity between sink node and representation node even if a_i value that is limited that transmit coefficient value a_i for data certification between representation nodes of master key interior that is transmitted at sink node through secret sharing mechanism is outpoured.

Certification and nonrepudiation is other master key that is transmitted at sink node one way hash algorithm certification of each representation node through done value consist and because representation node and authentication of leaf node compare first certification key value each other with value that do one way hash algorithm, certification and nonrepudiation are guaranteed.

5 Conclusion and Further Works

This paper proposed efficient key management mechanism that can be applied to real time sensor network. After we divide sensor network as two classes greatly for certification of each node, certification between representation nodes consist and offer quotation using one way hash algorithm with leaf node dividing master key to representation node. It can offer efficient certification key management in real time environment using such mechanism. In hereafter, analysis of certification way of leaf nodes and atomization of refresh cycle, message overhead and so on will consist.

References

[1] Miyoun Yoon, Kwangkyum Lee, Sangcul Son, Yongtae Shin: “Robust Authentication Mechanism for Ad-hoc Sensor Network”, Korean institute of communication sciences, September 2004.

[2] Adi Shamir: "How to Share a Secret", Communications of the ACM, November 1979.

[3] A. Perrig, R. Szewczyk, V. Wen, D. Cullar, and J. D. Tygar: “SPINS: Security protocols for sensor networks”, Proceedings of the 7th Annual ACM/IEEE Internation Conference on Mobile Computing and Networking (MobiCom). Rome. Italy(2001) 189--199.

[4] Rosario Fennaro & Pankaj Rohatgi: “How to sign Digital Streams”, In CRYPTO'97, pages 180--197, 1997.

[5] Chalermek Intanagonwiwat, Ramesh Govindan, Deborah L. Estrin, John Heidemann, and Fabio Silva: "Directed Diffusion for Wireless Sensor Networking", IEEE/ACM Transactions on Networking, vol. 11, no. 1, pp. 2-16, February 2003.

Efficient Group Key Management for Dynamic Peer Networks

Wei Wang[1], Jianfeng Ma[1,2], and SangJae Moon[3]

[1] Key Laboratory of Computer Networks and Information Security(Ministry of Education), Xidian University, Xi'an 710071, China
{wwzwh, ejfma}@hotmail.com

[2] School of Computing and Automatization, Tianjin Polytechnic University, Tianjin 300160, China

[3] Mobile Network Security Technology Research Center, Kyungpook National University, Sankyuk-dong, Buk-ku, Daeyu 702-701, Korea
sjmoon@knu.ac.kr

Abstract. In dynamic peer networks, how to promote the performance of group key management without sacrificing the desired security is a critical and difficult problem. In this paper, a secure, efficient and distributed group key management scheme is presented and its security is proved. The scheme is based on hierarchical key tree and multi-party key agreement, and has the desired properties, suck as key independence and statelessness. The related analysis shows that the ternary key tree is most applicable to group key management, and the corresponding key management scheme is efficient in the computation cost, storage cost and feasibility.

1 Introduction

Many multicast applications require security services for information transmission. One basic approach to provide these security services is group key management. Current group key management schemes can be classified into two categories: the centralized approach [1] and the distributed approach [2]. The distributed approach is attractive for better scalability and resilience to intrusions, but the known approach is so complex that the serious performance degradation is caused. For example, in TGDH [3], which provides the best performance in peer-peer networks, the total costs of computation and storage are much worse than that of some schemes for non-peer networks, such as KEY GRAPH [4].

In this paper, we propose a secure, efficient key management scheme which has the desired properties of key independence [3] and statelessness [5], and exceeds TGDH in computation complexity, storage requirement and communication cost.

The rest of this paper is organized as follows. In Section 2, we briefly introduce the preliminary knowledge on contributory key agreement and multi-linear map. Section 3 presents our group key management protocols. In Section 4, the performance and security of the scheme is discussed. Then Section 5 draws conclusions and explores further work.

X. Jia, J. Wu, and Y. He (Eds.): MSN 2005, LNCS 3794, pp. 753 – 762, 2005.

2 Preliminaries

2.1 Contributory Key Agreement

Our scheme is based on the contributory key agreement, which means that the scheme must guarantee that:

(1) Each party who contributes one Con_i can calculate K;

(2) No information about K can be extracted from a group key management protocol without the knowledge of at least one of the contributions;

(3) All inputs Con_i are kept secret, i.e., if member N_i is honest, even a collusion of all other parties cannot extract any secret information about N_i from their combined view of the protocol.

2.2 Multi-linear Diffie-Hellman Assumption

Assume that the discrete logarithm problem in groups G_1 and G_2 is hard, where G_1 is an additive group and G_2 a multiplicative group.

***Definition* 1.** A map e: $G_1{}^d \rightarrow G_2$ is a *d multi-linear map* if it satisfies the following properties:

(1) G_1 and G_2 are groups of the same prime order p;

(2) For $\forall a_1,\ldots, a_d \in Z$ and $\forall x_1,\ldots, x_d \in G_1$, $e(a_1x_1,\ldots, a_dx_d) = e(x_1,\ldots, x_d)^{a_1 \ldots a_d}$;

(3) The map e is non-degenerate in the following sense: if $P \in G_1$ is a generator of G_1 then $e(P,\ldots,P)$ is a generator of G_2.

***Definition* 2.** The *Decisional Multi-linear Diffie-Hellman (DMDH) problem* is, given $(P, x_1P, x_2P,\ldots,x_{d+1}P)$ and $z \in G_2$, to decide whether $z=e(P,P,\ldots,P)^{x_1x_2\ldots x_{d+1}}$ or not.

***Definition* 3.** The *Decisional Multi-linear Diffie-Hellman Assumption* claims that for any polynomial time algorithm A and any $d > 1$, the advantage $\mathsf{AdvDHm}_{A,d}(t)$ of A in solving the Decisional Multi-linear Diffie-Hellman problem is negligible, where $\mathsf{AdvDHm}_{A,d}(t)$ is defined as the probability that A is able to distinguish $e(P,P,\ldots,P)^{x_1x_2\ldots x_{d+1}}$ from $z \in G_2$.

In Section 4.1, we illuminate the key tree is most applicable to group key management in dynamic peer groups when d=3, in which case the bilinear map is employed.

3 New Protocols

3.1 Notations and Lemmas

System setup: Let G_1 is an additive group and G_2 a multiplicative group, e: $G_1{}^d \rightarrow G_2$ is a *d multi-linear map* on G_1 and G_2. Choose a generator P of G_1 and a map H: $G_2 \rightarrow Z_p{}^*$. The system parameter is (G_1, G_2, P, H).

Key pairing generation: We employs a d-ary key tree whose nodes are denoted by $<l,v>$. Each node is associated with the key $K_{<l,v>}$ and the blinded key $BK_{<l,v>}=H(K_{<l,v>})P$. Assume that every member M_i(at node $<l,v>$) chooses a secret random number $r_i \pmod p$, and he knows every key along the path from $<l,v>$ to $<0,0>$, referred to as the key–path.

***Lemma* 1.** The key $K_{<l,v>}$ and blinded key $BK_{<l,v>}$ can be computed recursively as follows:

$$\begin{aligned} K_{<l,v>} &= e(BK_{<l+1,dv+1>}, BK_{<l+1,dv+2>}, \ldots, BK_{<l+1,dv+d-1>})^{H(K_{<l+1,dv>})} \\ &= e(BK_{<l+1,dv>}, BK_{<l+1,dv+2>}, \ldots, BK_{<l+1,dv+d-1>})^{H(K_{<l+1,dv+1>})} = \ldots \\ &= e(P, P, \ldots, P)^{H(K_{<l+1,dv>})H(K_{<l+1,dv+1>})\ldots H(K_{<l+1,dv+d-1>})} \end{aligned} \quad (1)$$

$$BK_{<l,v>}=H(K_{<l,v>})P$$

If a child node $<l+1,dv+i>$ is null, its blinded key $BK_{<l+1,dv+i>}$ should be replaced by a certain sibling node.

***Lemma* 2.** Any authorized group member M_i can compute the group key $K_{<0,0>}$ from its key and the blinded keys on the co-path [3], which is the set of siblings of each node on the M_i's key-path.

3.2 Protocols

In this section, protocols for join and leave events specified by a key tree are presented. Firstly, a framework of group applications is given in Fig 1.

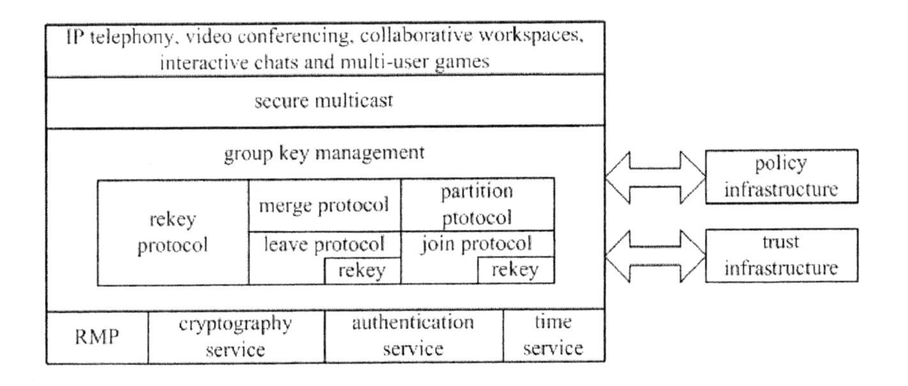

Fig. 1. Framework of group applications

Note that merge and partition protocols are based on join and leave protocols respectively, and the rekey operation is a special case of the leave protocol without any members leaving the group. Thus, we focus only on the join and leave protocols in this paper. In the following, we assume that (1) the underlying group communication system is resistant to fail-stop failures; (2) all communication channels are public and authentic; (3) any member can initiate the membership change protocols.

Join Protocol. Assume that there are N members $\{M_1,\ldots, M_N\}$ in the group. A new member M_{N+1} first initiates the protocol by sending a join request message that contains its blinded key to the group members.

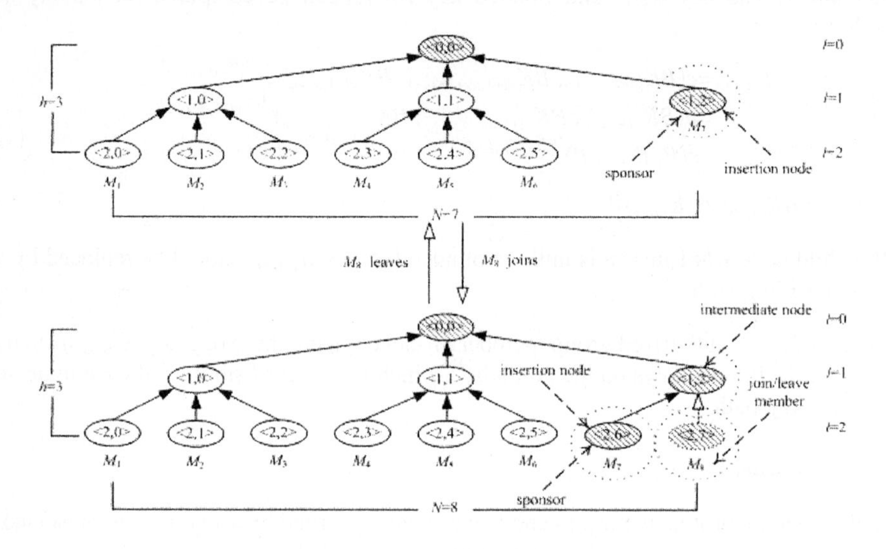

Fig. 2. Key trees before and after a join (leave)

After the request is accepted, if the key tree is full, M_{N+1} is inserted to the root node, else the insertion node and the sponsor should be determined. The insertion node is the topmost leftmost node where the join does not increase the height of the key tree and the sponsor is a random leaf node in the subtree rooted at the insertion node. When the insertion node has two or more child nodes, the sponsor inserts M_{N+1} under this insertion node. Otherwise, the sponsor creates a new intermediate node and a new member node, and promotes the new intermediate node to be the parent of the insertion node and M_{N+1}.

After updating the tree, the sponsor picks a new key, computes the new blinded keys on the key–path of M_{N+1} and multicasts the new tree to all group members.

An example of member M_8 joining to a group is given in Figure 2.

Leave Protocol. Assume that member M_l leaves the group. In this case, M_l's parent node is determined as the intermediate node and the sponsor is the random topmost leaf node of the subtree rooted at the intermediate node. If the degree of intermediate node is not two, the sponsor needs only to delete the leaf node corresponding to M_l. Otherwise, the leaving node is deleted and the sibling of M_l is promoted to replace M_l's parent node.

After updating the tree, the sponsor picks computes the new blinded keys on the key–path of itself and multicasts the new blinded keys to all group members which allows them recompute the new group key.

Assuming the setting of Figure 2, if member M_8 leaves the group, the sponsor M_7 deletes nodes <1,2> and <2,7>, and then renames <2,6> to <1,2>. After updating the tree, M_7 selects a new random secret key, computes the new blinded key $BK_{<1,2>}$, and multicasts it to the group.

4 Scheme Analysis

4.1 Performance Analysis

This sub-section devotes to the analysis of the computation, storage and communication costs for presented protocols. Consider a fully balanced key tree with degree d and height h and its corresponding secure group. Let $N = d^h$ be the number of group members.

Computation Costs. For each join/leave request, the member who requests the join/leave is called the *requesting member*, and the other members are *non-requesting members*. Apparently, for a join/leave request, a *requesting member* perform h+1 pairings and the sponsor h-1. The average computation overhead of a *non-requesting member* and the whole computation overhead of our scheme are given by the following expressions:

$$\frac{1}{N-1}\sum_{i=1}^{h} i(d-1)d^{h-i} = \frac{d}{d-1} - \frac{\log_d N}{N-1}$$
$$\left(\frac{d}{d-1} - \frac{\log_d N}{N-1}\right)(N-1) + (h-1) + \frac{1}{2}(h+1) = \frac{d(N-1)}{d-1} + \frac{\log_d N - 1}{2} \quad (2)$$

Storage Costs. The storage is measured by the number of keys and blinded keys that need to be stored. In our group key management scheme, each member M should store the keys on its key-path and the blinded keys on its co-path. Table 1 summarizes the computation and storage overheads for the presented scheme.

Table 1. Computation and storage costs

		requesting member	a non-requesting member	sponsor	whole scheme
computation cost	join	$O(\log_d N)$	$d/(d\text{-}1)\text{-}\log_d N/(N\text{-}1)$	$O(\log_d N)$	$O(dN/(d\text{-}1)+\log_d N/2)$
	leave	0	$d/(d\text{-}1)\text{-}\log_d N/(N\text{-}1)$	$O(\log_d N)$	
storage cost		$O(d\log_d N)$			$O(d\log_d N+N)$

The dependencies of the computation and storage costs on d are given in Fig 3, which show that the greater the degree d, the lower the computation complexity, but the storage requirement hits its lowest point at $d\approx 3$. Moreover, the bilinear pairing such as Weil pairing [6] corresponding to d=3 has been widely applied in cryptography literatures. Thus we draw the following conclusion.

Conclusion: The ternary tree is most applicable to group key management in dynamic peer networks.

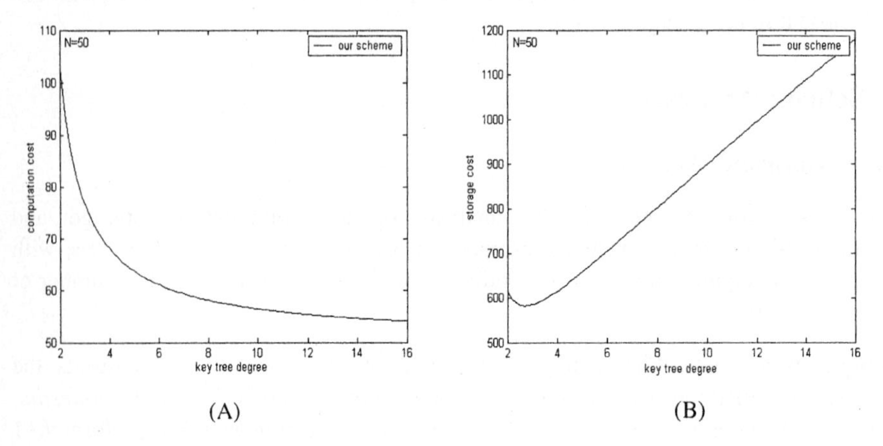

(A) (B)

Fig. 3. Computation and Storage Cost With Key Tree of Degree d. (A) Computation Cost and (B) Storage Cost.

Communication cost. Table 2 presents a numerical comparison among three kinds of the group key management schemes in terms of unicasts, multicasts, rounds and messages. In the leave protocol, three schemes have the same communication cost. But in the join protocol, the cost in our scheme provides the best performance with TGDH in second place, and KEY GRAPH in third place.

Table 2. Communication cost

		our scheme	TGDH	KEY GRAPH (group-oriented)
join	unicast	0	0	2
	multicast	2	2	1
	round	2	2	3
	message	2	3	3
leave	unicast	0	0	0
	multicast	1	1	1
	round	1	1	1
	message	1	1	1

Figure 4(A) and Figure 4(B) presents a comparison in the computation and storage overheads among our scheme, TGDH and KEY GRAPH with d=3, 4, 2, respectively. From the comparisons we can draw conclusions that our scheme approximates KEY GRAPH and is much better than TGDH in computation complexity, in addition it is a little better than TGDH but still much worse than KEY GRAPH in storage requirement. Although the pairings is computationally slower than modular exponentiation, fast implementation of it has been studied actively recently.

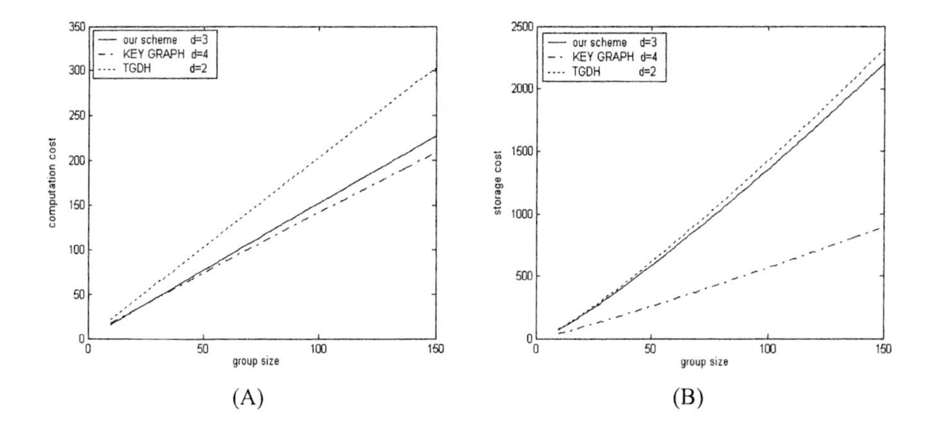

Fig. 4. Performance Comparison. (A) Computation Cost and (B) Storage Cost.

4.2 Security Analysis

In this section, we first prove the correctness and group key freshness of the presented scheme. Later, we prove that presented scheme provides the security requirements of group key secrecy, backward secrecy, forward secrecy and key independence.

Correctness. The correctness means that, in the presence of a passive adversary, all group members can compute the same group key.

***Theorem* 1.** All leaf nodes can get a same group key.

Proof. The proof is by induction on the height of key tree. We denote the key tree with height $h(=\lceil \log_d N \rceil)$ by T^h.

Basis: The case h=1 is trivial.

Induction Hypothesis: Assume that Theorem 1 holds for arbitrary trees of height $H<h$.

Induction Step: Consider a tree T^{H+1} and its corresponding group key K^{H+1}. From the induction hypothesis, in any sub-tree T_i^H, each leaf node can compute the same sub-group key K_i^H, i=0,…,d-1. According to Lemma 1, from the sub-group key K_i^H and all the other blinded sub-group keys of T_j^H where j=0,…,d-1, $j \neq i$, all leaf nodes of T_i^H can compute the group key: $K^{H+1} = e(H(K_0^H)P, H(K_1^H)P, \ldots, H(K_{d-2}^H)P)^{H(K_{d-1}^H)}$.

Therefore, by induction, the theorem holds for all possible cases. #

Group Key Freshness. A group key is fresh if it can be guaranteed to be new.

***Theorem* 2.** In our scheme, every group key is fresh.

Proof. In our scheme, the correctness is easily proved as follows: for every membership change, there is a member in the group generates its random key share, so the probability that new group key is same as any old group key is negligible, which guarantees the freshness of group key. #

TDMDH Problem. Next, we introduce the Tree based Decisional Multi-linear DH (TDMDH) problem and prove that it can be reduced to the DMDH problem.

For $(\Gamma, P, p) \leftarrow G(t, d)$, $N \in \mathbb{N}$, $\mathsf{R}=(R_1,R_2,\ldots,R_N)$ for $R_i \in G_2$, we define the following random variables:

—K_j^i: i-th level of j-th key (secret value, $K_{<i,j>}$), $K_j^h = R_k$ for some $k \in [1,N]$.

—BK_j^i: i-th level of j-th blinded key (public value, $BK_{<i,j>}$), *i.e.*, $H(K_j^i)P$.

Also we can define public and secret values as below:

$view(L, R, T)$: $=\{BK_j^L = H(K_j^L)P$ for some j defined according to $T\}$

$K(L, R, T) := e(BK_{dj+1}^{L+1}, \ldots, BK_{dj+d-1}^{L+1})^{H(K_{dj}^{L+1})}$

where L is some level in T, T is a subtree rooted at some node in level L and $R \subset \mathsf{R}$ corresponds to T.

Let the following two random variables are defined by generating $(\Gamma,P,p) \leftarrow G(t, d)$, choosing R randomly from G_2 and T randomly from all d-ary trees:

—$A_L := (view(L, R, T), y)$

—$B_L := (view(L, R, T), K(L, R, T))$

***Definition* 4.** Let $(\Gamma, P, p) \leftarrow G(t, d)$, and A_0 and B_0 be defined as above. *TDMDH algorithm* $\mathcal{A}$ is a probabilistic polynomial time 0/1 valued algorithm satisfying, for some fixed $k>0$ and sufficiently large m:

$$|\Pr[\mathcal{A}(A_0)=1]-\Pr[\mathcal{A}(B_0)=1]| > 1/m^k. \tag{3}$$

***Theorem* 3.** If the DMDH problem on the group G_1 and G_2 is hard, there is no probabilistic polynomial time algorithm that can distinguish A_0 from B_0 with non-negligible probability.

Proof. If $X_0=(R_{n_0},R_{n_0+1},\ldots, R_{n_0+n_1-1})$, and $X_{d-1}=(R_{n_{d-1}},R_{n_{d-1}+1},\ldots, R_{n_{d-1}+n_d-1})$, where each X_i is associated with some node $<L+1,dv+i>$ and a subtree T_i, and all of them possess the same parent node $<L,v>$. Firstly, with the knowledge of Lemma 1, A_L and B_L can be rewritten as:

$$\text{—}A_L := (view(L, R, T), y) \quad \text{for random } y \in G_2$$
$$=(view(L+1,X_0,T_0), view(L+1,X_1,T_1),\ldots, view(L+1,X_{d-1},T_{d-1}), H(K_{dv}^{L+1})p, H(K_{dv+1}^{L+1})p,\ldots,H(K_{dv+d-1}^{L+1})p, y)$$
$$\text{—}B_L := (view(L, R, T), K(L, R, T))$$
$$=(view(L+1,X_0,T_0), view(L+1,X_1,T_1),\ldots, view(L+1,X_{d-1}, T_{d-1}), H(K_{dv}^{L+1})p, H(K_{dv+1}^{L+1})p,\ldots,H(K_{dv+d-1}^{L+1})p,$$
$$e(P,P,\ldots,)^{H(K_{<l+1,dv>})H(K_{<l+1,dv+1>})\ldots H(K_{<l+1,dv+d-1>})})$$

Fact 1: There is no probabilistic polynomial time algorithm that can distinguish A_h from B_h, which is equivalent to solve the DMDH problem in G_1 and G_2.

The proof is by contradiction and induction.

Contradiction Hypothesis and *Induction Basis*: Assume that A_0 and B_0 can be distinguished in polynomial time.

Induction Hypothesis: Assume that there exists a polynomial algorithm that can distinguish A_L from B_L.

Induction Step: We will show that this algorithm can be used to distinguish A_{L+1} from B_{L+1} or can be used to solve the DMDH problem. Consider the following equations:

$$—A^0_L:=A_L:=(view(L+1,X_0,T_0),view(L+1,X_1,T_1),\ldots,view(L+1,X_{d-1},T_{d-1}), H(K^{L+1}_{dv})p, H(K^{L+1}_{dv+1})p, \ldots, H(K^{L+1}_{dv++d-1})p, y)$$

...

$$—A^d_L:= (view(L+1,X_0,T_0),view(L+1,X_1,T_1),\ldots,view(L+1,X_{d-1},T_{d-1}), r_0P, r_1P,\ldots, r_{d-1}P, y)$$

$$—B^d_L: = (view(L+1,X_0,T_0),view(L+1,X_1,T_1),\ldots,view(L+1,X_{d-1},T_{d-1}), r_0P, r_1P,\ldots, r_{d-1}P, e(P,P,\ldots,)^{H(K_{<l+1,dv>})H(K_{<l+1,dv+1>})\ldots H(K_{<l+1,dv+d-1>})})$$

...

$$—B^0_L:=B_H:=(view(L+1,X_0,T_0),view(L+1,X_1,T_1),\ldots,view(L+1,X_{d-1},T_{d-1}), H(K^{L+1}_{dv})p, H(K^{L+1}_{dv+1})p,\ldots,H(K^{L+1}_{dv++d-1})p, e(P,P,\ldots, P)^{H(K_{<l+1,dv>})H(K_{<l+1,dv+1>})\ldots H(K_{<l+1,dv+d-1>})})$$

Since A^0_L and B^d_L can be distinguished in polynomial time, the passive adversary can distinguish at least one of (A^i_L, A^j_L), (A^i_L, B^j_L),(B^i_L, B^j_L) for $i,j\in[0, d]$.

A^0_L and A^1_L (A^i_L and A^j_L, A^i_L and B^j_L, B^i_L and B^j_L, for $i,j\in[0, d]$): Suppose A^0_L and A^1_L can be distinguished in polynomial time. Suppose that the passive adversary want to decide whether $P'_{L+1}=(view(L+1, X_0, T_0), r')$ is an instance of TDMDH problem or r' is a random number. To solve this, the passive adversary generate trees $T_1, T_2,\ldots,T_{d-1}$ of level $L+1$ with distribution $X_1, X_2,\ldots,X_{d-1}$, respectively. Note that the passive adversary knows all secret and public information of $T_1, T_2,\ldots,T_{d-1}$. Then, by P'_{L+1} and (T_2, X_2), (T_3, X_3),..., (T_d, X_d) pairs, he can generate the distribution:

$$P'_L=(view(L+1,X_0,T_0),view(L+1,X_1,T_1),\ldots,view(L+1,X_{d-1},T_{d-1}),r'P, H(K^{L+1}_{dv+1})p,\ldots H(K^{L+1}_{dv+d-1})p, y)$$

Now the passive adversary input P'_L to the distinguisher $A_{A^0_L A^1_L}$. If P'_H is an instance of $A^0_L(A^1_L)$, then P'_{L+1} is an instance of A^0_{L+1} (A^1_{L+1}) , respectively.

From the above analysis, we get that the passive adversary can distinguish A_h from B. Consequently, there is no probabilistic polynomial time algorithm that can distinguish A_0 from B_0 with a non-negligible probability. #

Key Independence

***Theorem* 4.** In the presence of a passive adversary, our group key management scheme provides the security requirements of group key secrecy, backward secrecy, forward secrecy and key independence.

Proof. The relationship among the properties is given in [3]. Thus, we only need to show that the backward secrecy and forward secrecy is provided in the proposed scheme.

In the join event, the sponsor computes the new blinded keys and, consequently, previous root key is changed. Therefore, the view of the joining member M-with is exactly same as that of a passive adversary. Clearly, all the new keys will contain M's contribution. This reveals that the probability of M deriving any previous keys of the key tree is negligible. Hence, the backward secrecy can be provided in our scheme. Similarly, we can show that our protocols provide forward secrecy.

Thus, the presented scheme provides the security requirements of group key secrecy, backward secrecy, forward secrecy and key independence. #

4.3 Statelessness

Based on the interdependency of rekey messages, group key management algorithms can be classified into stateless and stateful schemes. In a stateless protocol, rekey messages are independent of each other. In our scheme, once a group member *M* was comes back on-line or misses previous group keys, the sponsor corresponding to *M* will send the current key tree message to it, from which *M* can compute the group key based on the key of itself and the blinded keys on its co-path. So, a member missing previous group keys needs not to contact other members to obtain keys that were transmitted in the past to gain the current group key.

5 Conclusions and Future Work

In this paper, we present an efficient scheme for distributed group key management in dynamic peer networks. The scheme supports dynamic membership group events, and satisfies the desired security requirements. Our analysis shows that the ternary key tree is most applicable to group key management in dynamic peer networks. As a result, we conclude that the new scheme is more efficient than the ones available in computation, storage and communication overheads.

The limitation of the proposed scheme is that it is not yet a satisfactory solution for the cases of merge and partition events. SDR [7] based approaches can deal with the problem of a center sending a message to a group of users. We are planning to extend the proposed scheme to all membership events by modifying the SDR model.

References

1. Harney, H. and C. Muckenhirn. Group Key Management Protocol (GKMP) Specification. RFC 2093, July 1997
2. M. Steiner, G. Tsudik, and M. Waidner. Key agreement in dynamic peer groups. IEEE Transactions on Parallel and Distributed Systems, August 2000
3. Y. Kim, A. Perrig, and G. Tsudik. Simple and fault-tolerant key agreement for dynamic collaborative groups. Technical Report 2, USC Technical Report 00-737, August 2000
4. C. Wong, M. Gouda, and S. Lam. Secure group communications using key graphs. IEEE/ACM TRANSACTIONS ON NETWORKING, VOL. 8, NO. 1, FEBRUARY 2000
5. D. Naor, M. Naor, and J. Lotspiech. Revocation and tracing schemes for stateless receivers. In Advances in cryptology - CRYPTO, Santa Barbara, CA, August 2001. Springer-Verlag Inc. LNCS 2139, 2001, pp. 41-62
6. D. Boneh and M. Franklin. Identity based encryption from the Weil pairing. SIAM J. of Computing, Vol. 32, No. 3, pp. 586-615, 2003
7. D. Naor, M. Naor, and J. Lotspiech. Revocation and tracing schemes for stateless receivers. Lecture Notes in Computer Science, 2139:41–62, 2001

Improvement of the Naive Group Key Distribution Approach for Mobile Ad Hoc Networks

Yujin Lim[1] and Sanghyun Ahn[2,*,**]

[1] Department of Information Media,
University of Suwon, Suwon, Korea
yujin@suwon.ac.kr
[2] School of Computer Science,
University of Seoul, Seoul, Korea
ahn@venus.uos.ac.kr

Abstract. Most of mobile ad hoc network (MANET) applications are based on the group communication and, because of the insecure characteristic of the wireless channel, multicast security is especially needed in MANET. Secure delivery of multicast data can be achieved with the use of a group key for data encryption. However, for the support of dynamic group membership, the group key has to be updated for each member join/leave and, consequently, a mechanism distributing an updated group key to members is required. The two major categories of the group key distribution mechanisms proposed for wired networks are the naive and the tree-based approaches. The naive approach is based on unicast, so it is not appropriate for large group communication environment. On the other hand, the tree-based approach is scalable in terms of the group size, but requires the reliable multicast mechanism for the group key distribution. In the sense that the reliable multicast mechanism requires a large amount of computing resources from mobile nodes, the tree-based approach is not that desirable for the small-sized MANET environment. However, recent studies on the secure multicast mechanism for MANET focus on the tree-based approach. Therefore, in this paper, we propose a new key distribution protocol, called the proxy-based key management protocol (PROMPT), which is based on the naive approach and reduces the message overhead of the naive by introducing the concept of the proxy node.

1 Introduction

The mobile ad hoc network (MANET) is a new paradigm for wireless communication among mobile devices (nodes). Within this network, mobile nodes can

* This research was supported by the MIC(Ministry of Information and Communication), Korea, under the Chung-Ang University HNRC-ITRC (Home Network Research Center) support program supervised by the IITA (Institute of Information Technology Assessment).

** This work was supported by grant No. R01-2004-10372-0 from the Basic Research Program of the Korea Science & Engineering Foundation.

X. Jia, J. Wu, and Y. He (Eds.): MSN 2005, LNCS 3794, pp. 763–772, 2005.

communicate without the help of fixed base stations and switching centers. Mobile nodes within their coverage can directly communicate via wireless channels, and those outside their coverage can also communicate with the help of other intermediate mobile nodes. MANET is originally researched for the communication in a military environment. Recently, with the advent of mobile devices like portable computers, MANET becomes a commercially possible solution.

Most of MANET applications like disaster relief and information exchange within a conference or a lecture room are based on the group communication. However, because of the insecure characteristic of the wireless channel, security becomes a big issue and makes the group communication within MANET more problematic. For security management within MANET, availability, confidentiality, integrity and authentication must be considered [1].

Secure delivery of multicast data can be achieved with the use of a group key for data encryption. However, for the support of dynamic group membership, the group key has to be updated for each member join/leave and, consequently, a mechanism distributing an updated group key to members is required. The two major categories of the group key distribution mechanisms proposed for wired networks are the naive and the tree-based approaches. The naive approach is based on unicast, so it is not appropriate for large group communication environment. On the other hand, the tree-based approach is scalable in terms of the group size, but it requires the reliable multicast mechanism for the group key distribution and this characteristic is not well suited for the MANET environment with energy-constraint. However, recent studies on the secure multicast mechanism for MANET focus on the tree-based approach (even most of them do not consider mobility). Since most of MANET applications are expected to be small-sized ones and MANET nodes are mobile and constrained with computing capability and power, a more simple secure multicast mechanism is required for the practicality of MANET.

Therefore, in this paper, we propose a simple secure multicast mechanism which adopts the naive approach and enhances it to reduce the group key update overhead of the naive by utilizing the broadcast capability of the wireless link and the neighbor information obtained from the MANET routing protocol such as AODV [2].

The rest of the paper is organized as follows; in Section 2, the schemes proposed for the multicast security in the wired network and MANET are described, and the motivation of our work is presented. In Section 3, we describe our newly proposed key distribution protocol, the proxy-based key management protocol (PROMPT). In Section 4, the performance of PROMPT is shown, and Section 5 concludes this paper.

2 Related Work

Recently, group communication applications like video conferencing and push-based information services over Internet have been attracted much attention. These group-based applications are closely related to multicast mechanisms and major issues in multicast include the provisioning of reliability and security. Multicast security implies only allowed senders/receivers can send/receive group messages. Secure delivery of multicast data can be achieved by encrypting data and, for this, the group key distribution among members is required.

In order to provide the multicast security, several aspects must be considered. First, group members must share a group key for data encryption and decryption. Second, the join/leave secrecy must be provided. The join secrecy implies that a newly joining member is not able to decrypt messages transmitted ahead of its join. For this, whenever a new member joins, the group controller must update the group key and encrypt the new group key with the old group key and multicast it to group members. The leave secrecy implies that a left member is not able to decrypt messages transmitted after its leave. Hence, the group key must be updated for each member leave. However, unlike the join secrecy, if the new group key is encrypted with the old group key, the left member can decrypt the new group key. Therefore, for the leave secrecy, a private key is assigned to each member and, whenever a member leaves, the new group key is encrypted with each private key and sent to the corresponding member (i.e., the new group key is not exposed to the left member).

For the distribution of an updated group key, two major approaches, the naive and the tree-based approaches, have been proposed [3]. In the naive approach, the group controller maintains one group key and N private keys for a group with N members. At the time of subscription, each member enters into the authentication procedure and receives the group key and its own private key from the group controller.In the naive approach, the leave secrecy is guaranteed by the group controller unicasting the new group key encrypted with each private key to each corresponding member. The advantages of the naive approach are simplicity, easiness of implementation, and no requirement on the reliable multicast mechanism. However, the larger the group size is, the more overhead it requires since the group controller transmits a new group key to each member via unicast.

The tree-based approach targets to reduce the key update overhead of the naive approach, by maintaining a logical key tree composed of the group key, subgroup keys and private keys at the group controller [4]. Subgroup keys help to reduce the number of messages generated for a key update at the group controller. A logical K-ary key tree structure is adopted for an efficient key management. A member of a group with N members has to maintain all of those keys on the path from the root to itself in the key tree. The group controller has to maintain all keys in the key tree, including the group key, subgroup keys and N private keys. Therefore, the number of keys maintained at the group controller is (KN - 1) / (K - 1), and that at a group member is $log_K N$.

As a summary, in the tree-based approach, the number of keys maintained at a member is increased and the number of messages transmitted by the group controller is reduced. Hence, the tree-based approach is more scalable in terms of the group size than the naive approach. Due to this advantage of the tree-based approach, recently proposed secure multicast mechanisms for MANET are focused on the tree-based approach and try to reduce the communication overhead of the tree-based approach [5] [6] [7] (some of the previously proposed MANET tree-based secure multicast mechanisms operate based on the GPS information which is not adequate for the MANET environment, and most of non-GPS based mechanisms do not consider mobility at all and only [7] provides moderate level of mobility). However, the key distribution of the tree-based approach is based on the multicast, a reliable multicast mechanism is required as a basic component since those members not receiving newly updated keys can not participate in the group communication any further.

In order to provide reliable multicast services, reliable multicast mechanisms have been extensively studied for the wired network. Recently, with the increase of the interest in MANET, reliable multicast mechanisms for MANET have been proposed by many researchers [8] [9] [10] [11] [12]. The reliable multicast mechanism requires buffering at sources and/or receivers for the recovery of lost packets, so even for the wired network reliable multicast mechanisms overburden nodes. Also, the reliable multicast mechanism can not be used for real-time applications due to the retransmission-based lost packet recovery. Especially in MANET which uses the wireless channel with high bit error rate (BER) and mobile nodes, providing a reliable multicast service is more difficult than that in the wired network. Moreover, with the energy and computing resource constraints of MANET nodes, using a complex reliable multicast mechanism is not preferable.

Therefore, instead of using the tree-based approach requiring the reliable multicast mechanism, using the naive approach is more practical for the small-sized MANET environment which does not have to be scalable. Therefore, we propose a new key distribution protocol, called the proxy-based key management protocol (PROMPT), which is based on the naive approach with utilizing the broadcast characteristic of the wireless channel in order to reduce the message overhead of the naive approach. In this paper, we focus only on the group key distribution mechanism itself and not on any other security related issues such as malicious proxy nodes.

3 Proxy-Based Key Management Protocol (PROMPT)

In this section, we propose a new key distribution protocol, the proxy-based key management protocol (PROMPT), which targets to reduce the message overhead of the naive approach.

In PROMPT, the join secrecy is easily provided by multicasting a new group key encrypted with the old group key for each member join. For the leave secrecy, the broadcast characteristic of the wireless channel is utilized to reduce the key update message overhead. Two basic operations, the first hop grouping and the last-hop grouping, are newly defined for PROMPT. The first-hop grouping is performed by the source which multicasts a new key information to its neighboring group members. This is feasible since each node maintains its own neighbor list. In this operation, the source encrypts an updated group key with private keys of its neighboring members and includes newly encrypted group keys in an update message and sends it via 1-hop flooding. The destination address of the update message has the group multicast address and the TTL field is set to 1 and the data field includes more than one [the IP address of a neighboring member, the new group key encrypted with the private key of the neighboring member] pairs.

The last-hop grouping is performed by a member with a number of neighboring members and uses the 1-hop flooding. For this operation, each member node has to know how many neighbor nodes are members. However, since the group membership information is managed only at the source, it is not possible for a node to know the membership of its neighbors. Therefore, to solve this problem, at the initial stage, the source unicasts the updated key information to each member not in the first-hop grouping as in the naive approach. The group address is included in the IP option field of this new key message to allow each member node to know the group membership of its neighbors. Once the source starts to send data after finishing the key update procedure, those member nodes with more than a pre-specified number, k, (which is a system parameter determined at the session set-up stage of a multicast session) of neighboring members send PROXY packets to the source to let the source know that it is a possible representative (i.e., proxy) of its neighboring members. Within a PROXY packet, the list of neighboring members is included. The proxy node selection problem is the set covering problem which is NP-hard [13] and, as a heuristic solution, a greedy approach is adopted in PROMPT. That is, the source receiving PROXY packets select those nodes with the largest number (which is greater than or equal to k) of neighboring members not covered yet as proxy nodes. Figure 1 shows the formal description on the first-hop grouping at the source and the last-hop grouping at a proxy node.

Figure 2 shows the format of the key update packet sent by the source. In the data field, the pair of the IP address and the group key encrypted with the private key of the proxy node and those of neighboring members of the proxy node are included. The destination field has the IP address of the proxy node. The IP option field has the following subfields:

- the multicast address to indicate the group information
- the proxy bit to let the receiving node know whether it is a proxy node or not
- the key bit to indicate whether the packet has the user data or the key information
- the number of [IP address, encrypted group key] pairs
- the length of an encrypted group key

Since the information used for the proxy node selection is collected during the previous group key update procedure, some of the neighboring nodes of a proxy node may not receive the key update message. In this case, the proxy node notifies the source of those non-receiving members in order to let the source unicast the update message to them.

```
At the source,
  select proxy nodes according to the proxy node selection algorithm

  // first-hop grouping
  destination address ← group address
  TTL ← 1
  IP options field ← a list of [IP address of a neighboring member,
                      the new group key encrypted with the private key of the neighboring member] pairs
  broadcast the key update packet

  for each proxy node {
      destination address ← IP address of the proxy node
      IP options field ← [IP address of the proxy node,
                          the new group key encrypted with the private key of the proxy node]
      IP options field ← a list of [IP address of a neighboring member of the proxy node,
                         the new group key encrypted with the private key of the neighboring member of the proxy node] pairs
      unicast the key update packet
  }

  for each non-proxy node involved neither in first-hop grouping nor in last-hop grouping {
      destination address ← IP address of the non-proxy node
      IP options field ← [IP address of the non-proxy node,
                          the new group key encrypted with the private key of the non-proxy node]
      unicast the key update packet
  }

At a proxy node,
  when a key update packet is received {   // last-hop grouping
      destination address ← group address
      IP options field ← a list of [IP address of a neighboring member,
                         the new group key encrypted with the private key of the neighboring member] pairs
      TTL ← 1
      broadcast the key update packet
  }
```

Fig. 1. The procedure of the first-hop grouping and the last-hop grouping

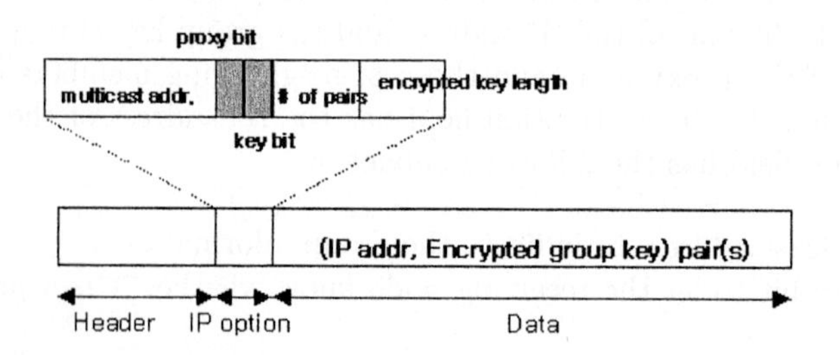

Fig. 2. The format of the key update packet

We can expect that PROMPT will show better performance for a dense group. For a dense group, the number of those nodes with more than a pre-specified number of neighboring members may be large and, if PROXY packets are sent almost simultaneously by members, the packet explosion problem may occur. To solve this problem, the node which has sent a PROXY packet to the source lets its neighbors know the fact in order to prevent them from sending their own PROXY packets to the source.

For a sparse group, the performance of PROMPT is expected to be similar to that of the naive approach. The reason for this is that in this case proxy nodes may not be selected and PROMPT works like the naive approach.

4 Performance Evaluation

To evaluate the performance of PROMPT, we have performed simulations using the GloMoSim simulator [14]. GloMoSim is a simulation package for wireless network systems written with the distributed simulation language from PARSEC. The simulation environment consists of 50 mobile nodes in the range of 1000 × 1000 m, and the transmission range of each node is set to 250 m in radius, and the channel capacity is set to 2 Mbps. The moving direction of each mobile node is chosen randomly. And we have applied the free space propagation model in which the power of the signal decreases $1/d^2$ for distance d, and assumed the IEEE 802.11 as the medium access control protocol. The AODV [2] is used as the underlying unicast routing protocol, and the source and receivers are randomly selected and the total simulation time is set to 1000 seconds.

Since those nodes with more than a pre-specified number (i.e., threshold k) of neighboring members send PROXY packets to the source, if the threshold is set to a small value, the PROXY packet implosion can happen. On the other hand, if the threshold is set to a large value, the number of generated PROXY packets decreases and PROMPT becomes the naive approach. Therefore, an appropriate threshold value needs to be determined.

Figure 3 shows the performance of PROMPT in terms of the number of transmitted packets (including PROXY and key update packets) with various threshold values (in this case, the group size is set to 15 and the pause time to 300 seconds). For smaller threshold values, more PROXY packets are generated since the possibility of being a proxy candidate node becomes higher ('PROMPT(control)' in figure 3 shows this result). For larger threshold val-

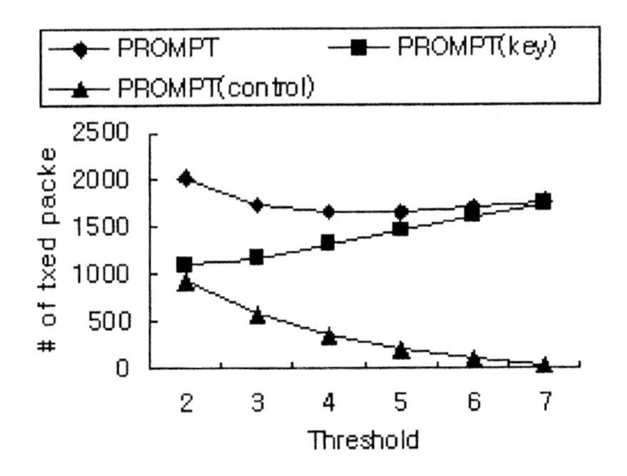

Fig. 3. The performance for various thresholds

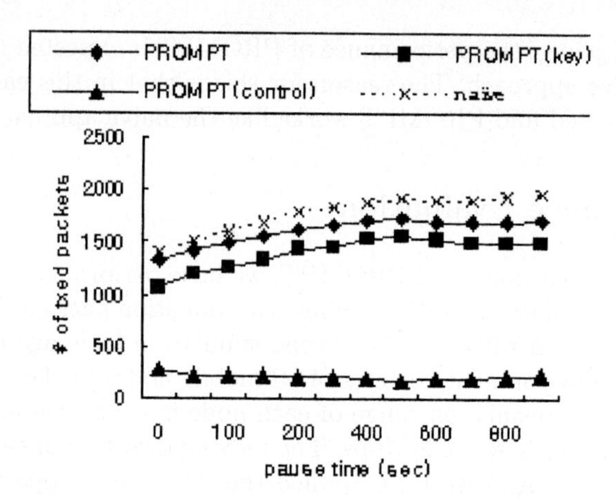

Fig. 4. The performance for various node mobility

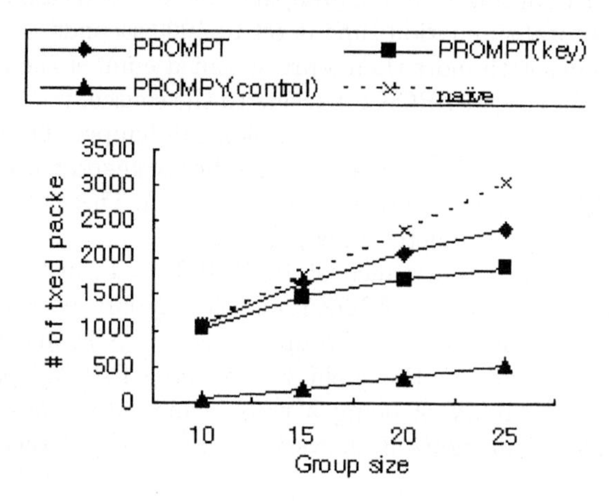

Fig. 5. The performance for various group sizes

ues, more key packets are generated by the source since the possibility of the last-hop grouping decreases ('PROMPT(key)' in figure 3 shows this result). In that figure, the plot labeled with 'PROMPT' is the result of summing up 'PROMPT(control)' and 'PROMPT(key)'. Overall, in this simulation environment the most appropriate threshold value is 5.

Figures 4 and 5 show the performance of PROMPT for various node mobility and group sizes with k = 5. As shown in figure 4, PROMPT outperforms the naive in terms of the number of transmitted packets and the control overhead due to PROXY packets is kept almost constant. As the node mobility decreases (i.e., as the pause time increases), the number of transmitted packets increases since only successful transmissions are considered (lower mobility gives a higher possibility of successful transmission).

Figure 5 shows that, in terms of the number of transmitted packets, PROMPT outperforms the naive for all group size cases and for larger group sizes PROMPT performs much better. The reason is that for larger group the possibility of having neighboring members increases. As a summary, PROMPT gives better performance in a dense group environment and even for the sparse group case it performs similar to that of the naive. Since most of the group communications in MANET happen in a dense environment, we can say that our PROMPT is appropriate for MANET.

5 Conclusion

MANET is a new paradigm for wireless communication among mobile devices. Most of MANET applications are based on the group communication. However, due to the insecure characteristic of the wireless channel, secure delivery of multicast data becomes a hot issue and, for this, the concept of the group key has been adopted. Since the group membership can change dynamically, the group key also has to be updated. Updated keys need to be distributed to group members, so the key distribution protocol is required for the secure multicast. The naive and the tree-based approaches are the most representative schemes for the group key management proposed for the wired network. However, since MANET is small-sized and dense in nature and hard to support the reliable multicast mechanism, the naive approach is more preferable than the tree-based approach in this kind of environments. Therefore, in this paper, we proposed a key distribution protocol, the proxy-based key management protocol (PROMPT), which is based on the naive and tries to reduce the number of key distribution-related packets. PROMPT introduces the concept of the proxy node which is a representative of its neighboring members and utilizes the broadcast characteristic of the wireless channel to reduce the key distribution overhead. From the performance evaluation, we have showed that PROMPT outperforms the naive in dense group and higher node mobility cases. Even in the worst case, PROMPT performs almost the same as the naive.

References

1. L. Zhou, and Z. J. Haas, "Securing Ad Hoc Networks", IEEE Network, pp24-30, Nov. 1999.
2. C. Perkins, E. Bolding-Royer, and S. Das, "Ad hoc On-Demand Distance Vector (AODV) Routing", IETF RFC 3561, July 2003.
3. M. J. Moyer, J. R. Rao, and P. Rohatgi, "A Survey of Security Issues in Multicast Communications", IEEE Network, pp12-23, Nov. 1999.
4. C. K. Wong, M. Gouda, and S. S. Lam, "Secure Group Communication Using Key Graphs", Proceedings of ACM SIGCOMM, 1998.
5. L. Lazos and R. Poovendran, "Energy-Aware Secure Multicast Communication in Ad-hoc Networks Using Geographic Location Information", IEEE International Conference on Acoustics Speech and Signal Processing (ICASSP '03), vol. 4, pp201-204, April 2003.

6. M. Moharrum, R. Mukkamala, and M. Eltoweissy, "CKDS: An Efficient Combinatorial Key Distribution Scheme for Wireless Ad-Hoc Networks", IEEE International Conference on Performance, Computing, and Communications (IPCCC '04), pp631-636, April 2004.
7. S. Zhu, S. Setia, S. Xu, and S. Jajodia, "GKMPAN: An Efficient Group Rekeying Scheme for Secure Multicast in Ad-Hoc Networks", International Conference on Mobile and Ubiquitous Systems: Networking and Services (MOBIQUITOUS '04), pp42-51, Aug. 2004.
8. R. Chandra, V. Ramasubramanian, and K. Birman, "Anonymouns Gossip: Improving Multicast Reliability in Mobile Ad-Hoc Networks", 21st International Conference on Distributed Computing Systems (ICDCS), pp275-283, April 2001.
9. S. Gupta and P. Srimani, "An Adaptive Protocol for Reliable Multicast in Mobile Multi-hop Radio Networks", IEEE WMCSA '99, pp111-122, Feb. 1999.
10. L. Klos and G. Richard III, "Reliable Group Communication in an Ad Hoc Network", IEEE International Conference on Local Computer Networks (LCN 2002), 2002.
11. A. Sobeih, H. Baraka, and A. Fahmy, "ReMHoc: A Reliable Multicast Protocol for Wireless Mobile Multihop Ad Hoc Networks", IEEE Consumer Communications and Networking Conference (CCNC), Jan. 2004.
12. K. Tang, K. Obraczka, S.-J. Lee, and M. Gerla, "Reliable Adaptive Lightweight Multicast Protocol", IEEE ICC 2003, May 2003.
13. T. H. Cormen, C. E. Leiserson, and R. L. Rivest, "Introduction to Algorithms", MIT Press, 1990.
14. UCLA Computer Science Department Parallel Computing Laboratory and Wireless Adaptive Mobility Laboratory, "GloMoSim: A Scalable Simulation Environment for Wireless and Wired Network Systems", `http://pcl.cs.ucla.edu/projects/domains/glomosim.html`

RAA: A Ring-Based Address Autoconfiguration Protocol in Mobile Ad Hoc Networks

Yuh-Shyan Chen and Shih-Min Lin

Department of Computer Science and Information Engineering,
National Chung Cheng University, Chiayi 621, Taiwan

Abstract. The problem for dynamic IP address assignment is manifest in mobile ad hoc networks (MANETs), especially in 4G all-IP-based heterogeneous networks. In this paper, we propose a ring-based address autoconfiguration protocol to configure node addresses. This work aims at the decentralized ring-based address autoconfiguration (DRAA) protocol, which has the capability to perform low latency and whose broadcast messages are reduced to lower control overhead. In addition, we introduce the centralized ring-based address autoconfiguration (CRAA) protocol to largely diminish control overhead and to serve as an even solution for IP address resource distribution. Both of DRAA and CRAA protocols are low-latency solutions because each node independently allocates partial IP addresses and does not need to perform the duplicate addresses detection (DAD) during the node-join operation. Communication overhead is significantly lessened in that RAA (DRAA and CRAA) protocols use the logical ring, thus utilizing fewer control messages solely by means of uni-cast messages to distribute address resources and to retrieve invalid addresses. Especially, the CRAA protocol reduces larger numbers of broadcast messages during network merging. The other important contribution is that our CRAA protocol also has an even capability so that address resources can be evenly distributed in each node in networks; this accounts for the reason our solution is suitable for large-scale networks. Finally, the performance analysis illustrates performance achievements of RAA protocols.

Keywords: Autoconfiguration, IP address assignment, MANET, RAA, wireless IP.

1 Introduction

Multiple functions of the fourth-generation (4G) communication system are envisioned to be extensively used in the near future. 4G networks are an all-IP-based heterogeneous network, exploiting IP-based technologies to achieve integration among multiple access network systems, such as 4G core networks, 3G core networks, wireless local area networks (WLANs) and MANETs. In IP-based MANETs, users communicate with others without infrastructures and service charges. A MANET is made up of identical mobile nodes, each node with a limited wireless transmission range to communicate with neighboring nodes. In

X. Jia, J. Wu, and Y. He (Eds.): MSN 2005, LNCS 3794, pp. 773–782, 2005.

order to link nodes through more than one hop, multi-hop routing protocols - such as DSDV, AODV, DSR, ZRP and OLSR - are designed. These multi-hop routing protocols require each node to have its own unique IP address to transmit packets hop by hop toward the destination. Hence to practice these routing protocols in a correct manner, a node must possess an IP address bearing no similarity to that of any other node.

In recent years, many solutions for dynamic IP address assignment in MANETs have been brought out. According to their dynamic addressing mechanisms, we organize these solutions into the following four categories: all agreement approaches [1][3][5], leader-based approaches [2][9], best-effort approaches [8] [10] and buddy system approaches [4][7]. The all agreement approach [5] featured a distributed, dynamic host configuration protocol for address assignment called MANETconf. The greatest communication overhead will be produced during network merging because nodes in networks must perform DAD. Among leader-based approaches, Y. Sun *et al.* [2] proposed a Dynamic Address Configuration Protocol (DACP). The biggest drawback Leader-based approaches bear is that the workload of the leader node is too heavy due to DAD for all joining nodes. Among best-effort approaches, the prophet address allocation protocol [10] makes use of an integer sequence consisting of random numbers through the stateful function $f(n)$ for conflict-free allocation. Prophet does not perform DAD to reduce communication overhead during network merging, but nodes with the smaller network identifier (NID) change their IP addresses no matter duplication occurs or not. Although Prophet brings the benefit of lower communication overhead, nodes break all on-going connections with the smaller NID. When two large networks merge, the impact of connection loss is significant. A.P. Tayal *et al.* [7] proposed an address assignment for the automatic configuration (named AAAC), to which the buddy system is applied whenever resources run out and new nodes seek to join a network. In AAAC, every node in merging networks broadcasts its IP address and address pool to whole networks for network merging, whose communication overhead is large.

To offer effective IP address assignment in a dynamic network environment, the solution provided by our addressing protocol presents three goals, aiming at efficient, rapid IP address distribution as well as applicable address resource maintenance: 1) low latency, 2) low communication overhead and 3) evenness. Namely, low latency produces the results that a requested node timely gets a unique address in the IP address assignment process, communication overhead is lessened to enhance network efficiency, and address resources are evenly distributed in each node.

The remaining sections of the paper are organized as follows. Section 2 proposes protocol comparisons and basic ideas of RAA protocols. In Section 3, we present the details of the decentralized RAA protocol (DRAA) with regard to address resource maintenance and node behavior handling. In Section 4, the centralized RAA protocol (CRAA) is introduced. Section 5 shows the performance analysis. Finally, Section 6 draws conclusions for the paper.

2 Basic Ideas

In this section, we proffer both conceptual discrepancies among various protocols and the basic idea of RAA protocols. The key to determining the effectiveness of a dynamic IP address assignment protocol mainly lies in the latency of node joining, and we illustrate the comparison among various protocols in terms of node joining in Fig. 1. We consider mobile wireless networks where all nodes use IP address to communicate with others. Such a network can be modeled as follows. A mobile wireless network is represented as a graph $G = (V, E)$ where V is the set of nodes and E is the edge set which gives available communications. In a given graph $G = (V, E)$, we denote by $n = |V|$ the number of nodes in the network. The identifier (ID) and the related list of node u are represented as N_u and RL_u respectively. The network identifier is represented as NID, which is 2-tuple: <IP address, Random number>, by whose uniqueness is distinguished in different networks. The ID of a node is 2-tuple: <NID, IP address>. The successor, the predecessor and the second predecessor of node u are represented as S_u, P_u and SP_u.

In Fig. 1(a) and (b), MANETconf and DACP are not conflicting free protocols, so they have to perform DAD during node joining, which increases the latency of node joining. On the contrary, the other protocols, such as Prophet, AAAC, DRAA and CRAA, are conflicting ones so that they can assign a unique address to new nodes without DAD. Furthermore, AAAC, DRAA and CRAA are categorized into buddy system approaches, whose main difference is the evenness of address blocks in all nodes during node joining. In AAAC and DRAA, a new node N_j broadcasts a one-hop address request (AREQ) to its neighbors and awaits the very first address reply (AREP). As Fig. 1(e) shown, the CRAA protocol awaits AREPs of all neighbors and picks up the biggest address block for use, which effectively improves the uneven distribution of address blocks shown in the former case.

In this paper, we draw on a novel technique developed in peer-to-peer (P2P) networks to provide a logical view for resource maintenance. It offers a logical network, allowing clients to share files in a peer-to-peer way. One of the distributed hash table (DHT) based approaches in P2P, known as Chord [6],

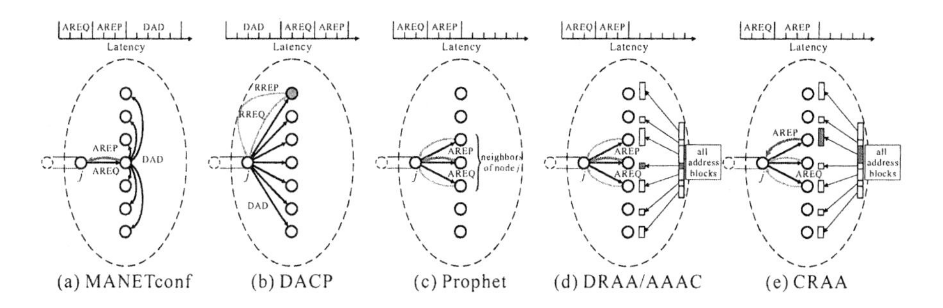

Fig. 1. The comparison of various protocols during node joining

inspires us to utilize its logical view to perform address resource management. In the prototype of Chord, in order to keep load balancing, each node uses the hash function to produce a node identifier. Unlike Chord, our solutions - RAA protocols - are not necessary to use any hash function to distribute address resources because IP address resources differ from file ones. If the hash function is applied to distribute address resources for the evenness in networks, the complexity of address management will be raised and not be suitable in MANETs. For this reason, the buddy system is combined in RAA protocols to manage resources. The binary buddy system is one common, famous type of buddy systems with a resource size of 2^m units and starting off this size. When the application issues a request, a 2^m-unit block is split into two with 2^{m-1} units respectively. However, resources in the binary buddy system are depleted rapidly. When a node owns 2^m addresses and allocates them to other nodes up to m times, address resources will be consumed. In the traditional buddy system approach (AAAC), address resources are retrieved solely during resource consumption. If a network changes with frequency, then resources usually run out. If so, the latency of a new node's requiring an available address block inevitably increases. In our solution, address resources are retrieved during both resource consumption and node joining.

In RAA protocols, each node records its logical neighbors' IDs on the related list (RL). Logical neighbors are the successor, the predecessor and the second predecessor. Notice that the RL is updated during node joining and network merging. If N_i exists in RAA protocols, the successor (S_i) is the node which allocates an address block to it and is also the first node in the clockwise course starting from N_i. The predecessor (P_i) and the second predecessor (SP_i) are the first node and the second node in the anticlockwise course starting from N_i respectively. The successor's, the predecessor's and the second predecessor's IDs can be represented as SID_i, PID_i and $SPID_i$. The RL of N_i is RL_i: $\{SID_i, PID_i, SPID_i\}$. Each node possesses its own anticlockwise-course free address block. During node leaving, only the successor in the clockwise course has to be informed; then it will retrieve the address block of the leaving node. Our solution achieves highly-efficient resource management and address allocation through the combination of logical ring and the buddy system.

3 Decentralized Ring-Based Address Autoconfiguration Protocol (DRAA)

Since nodes in MANETs have joining, leaving, partitioning and merging behaviors, this section sheds light on how the DRAA protocol handles these behaviors. Before we introduce it, the state diagram of RAA protocols should be shown first to help handle the node behaviors. There are totally five states in RAA protocols: INITIAL, STABLE, MERGE, HS and FINISH. When a new node intends to join a network, it enters the INITIAL state and waits for a free address block. A node enters the STABLE state when getting an address. If nodes are informed of network merging by some node, they enter the MERGE state for merging. If the holder leaves the network, other nodes enter the HS (holder selection) state to select a new holder. If leaving the network, a node enters the FINISH state. According to node behaviors, we realize how the DRAA protocol deal with joining and leaving. In following paragraphs, we will state the initiation of networks and specify node behaviors via the DRAA protocol.

3.1 Initial State

A node is in the INITIAL state before getting a usable address. At the start, the first node, N_i, enters a network and broadcasts a one-hop address request (AREQ) message. N_i awaits an address request timer (AREQ_Timer) to gather responses from other nodes. Once the timer expires and N_i does not get any response, N_i gets its identity as the first node (i.e. the holder) in the network. N_i randomly chooses an IP address, and sets it into its ID. The holder uses its ID and a random number as the network identifier (NID). The NID is periodically broadcast to the whole network by the holder, enabling nodes to get the NID of their locating network and to detect network partitioning and mergeing.

3.2 Node Joining

When aiming to enter a network, N_j needs to obtain an IP address and then checks whether P_j exists in the network in order to ensure the wholeness of address blocks. The joining procedure consists of two phases: 1) address requesting and 2) failed node checking. The address requesting phase is used to allocate an IP address to a new node whereas the failed node checking is used to check whether the predecessor of the new node is alive.

Address Requesting Phase. With N_j in the initial state and intending to join the network in DRAA, the address requesting phase will be triggered. The address requesting phase is applied to allocate an IP address and an address block to a new node, including an address request (AREQ) issued from the new node, an address reply (AREP) replied from neighbors of the new node and an address reply acknowledgement (AREP_ACK) sent by the new node to the neighboring node which is the first to transmit an AREP.

Failed Node Checking Phase. After the address requesting phase, N_j uses the IP address to communicate with other nodes in the MANET and enters the STABLE state from the INITIAL state. The Failed node checking phase is used to check whether the predecessor of the new node is alive. The new node first sends an alive checking (ACHK) message to its predecessor. If the predecessor alive, the predecessor replies with an ALIVE message to the new node. If the predecessor fails, the new node sends an address retrieve (ARET) message to its second predecessor to retrieve the address block of the predecessor. Then the second predecessor sends back an address retrieve acknowledgement (ARET_ACK) to the new node to inform it which ones are its new predecessor and second predecessor.

3.3 Node Leaving

When a node leaves the network genteelly, the leaving node sends a LEAVE message to its successor, and then the successor takes over the address block of the leaving node. Notice that during node leaving, the successor is unnecessary to be the one-hop neighbor of the leaving node, which has to procure the successor's routing path in the routing table. If nodes crash without a LEAVE message, the orphan block of the crashed node is retrieved in the failed node checking phase.

3.4 Network Partitioning

The only responsibility of the holder node in the DRAA protocol is to broadcast its NID. If there is a node not receiving the NID after one broadcasting period, the node may have been partitioned from the network. But wireless communication is not reliable; the NID may not be received due to packet loss. To reduce the impact, we define network partitioning as three broadcasting periods without the need to receive the NID. Once a node detects network partitioning, it performs the holder selection algorithm.

Holder Selection. In the holder selection algorithm, we borrow the random backoff in carrier sense multiple access with collision avoidance (CSMA/CA) protocol of IEEE 802.11 standard and slightly modify the idea. When network partitioning is detected by the node, the detecting node enters the holder selection (HS) state and chooses a random backoff timer (RB_Timer) which determines the amount of time the node must wait until it is allowed to broadcast its NID. The RB_Timer reduces the collision of NID messages, broadcast by nodes in the holder selection state.

3.5 Network Merging

Assume that there are two networks G (NID_G) and G' ($\text{NID}_{G'}$) intend to merge. Network G has n nodes and network G' has n' nodes. In the DRAA protocol, all nodes during network merging have to broadcast its ID for DAD. When all nodes get full information of networks, they choose a holder who broadcasts a bigger NID. If duplicate addressing occurs, duplicate nodes which have fewer TCP connections or are in the smaller network should rejoin the network. The benefits of choosing the node from a smaller network are quicker responses and lesser possible disconnections.

4 Centralized Ring-Based Address Autoconfiguration Protocol (CRAA)

With regard to low communication overhead and evenness, the centralized RAA protocol (named CRAA) is introduced. The differences between DRAA and CRAA protocols are in node joining and network merging. In the address requesting phase of node joining, the CRAA protocol selects the biggest free address block to use during the address requesting phase. Furthermore, the CRAA protocol in the failed node checking phase can recover more than one failed node because the holder in CRAA maintains the node list (NL) which contains all used IP addresses in the network. The NL is helpful when there are two or more continuous node failures in a ring, so that the lost address resources can be retrieved with the help of the holder. During networks merging, the NL reduces broadcast messages by exchanging the holder's NL only.

In the address requesting phase, the new node in the CRAA protocol waits for all neighboring nodes' AREP messages and selects the biggest free address block for use. The DRAA protocol can distribute valid addresses immediately, but the address block of each node will probably not be evenly for the long-term perspective. The advantage of the CRAA protocol is averaging the number of free address blocks managed by each node. Since the holder maintains the NL of the network in the CRAA protocol, each node sends an address registration request to the holder when a node takes over a free address block. After using a new node uses a new IP address, the new node sends a registration request (RREQ) to the holder that records the new node's address on the NL and the holder sends a registration reply (RREP) to the new node.

By contrast with the DRAA protocol during network merging, in the CRAA protocol, when the holder receives the merging request (MREQ), it broadcasts its own NL. Each node receives the holders' NLs, merges those NLs, modifies its own RL and chooses the bigger NID to be its network identifier. This method reduces large communication overhead during network merging. The NL becomes much significant in CRAA, so that we will introduce the replication of the NL. The NL incorporates in all used nodes in the network and is a helpful information source for both the lost address block retrieval and network merging. The NL replicas exist in the holder's successor, predecessor and second predecessor. When the holder gets an RREQ message, the holder copies the newest NL to these three nodes to ensure the freshness of the NL.

5 Performance Analysis

To make appropriate protocol simulation, all protocols are implemented with C++. We select AAAC [7] and Prophet [10] to compare with our protocols, and do not consider comparing with MANETconf and DACP. Both of AAAC, Prophet are conflicting free protocols, while MANETconf and DACP are not. The simulation parameter is described below. The simulation time is 1500 seconds. The speed of nodes is varied from 0 to 10 m/s, and the pause time is set to 0 for strict reasons. The mobile nodes move according to the random waypoint mobility model. The network size is set to 1000m $\times$ 1000m where the number of nodes is from 50 to 300. The size of address blocks is set to 2^m IP addresses and m are 10 and 24. The maximal size of the address block is 24-bit which makes a suitable selection for large networks. In particular, in our simulation, the transmission range is 250m. The underlying routing protocol applies the DSDV in all simulations. The network is initialized with a single node. Nodes join the network every 1 second and arriving nodes are placed randomly in the

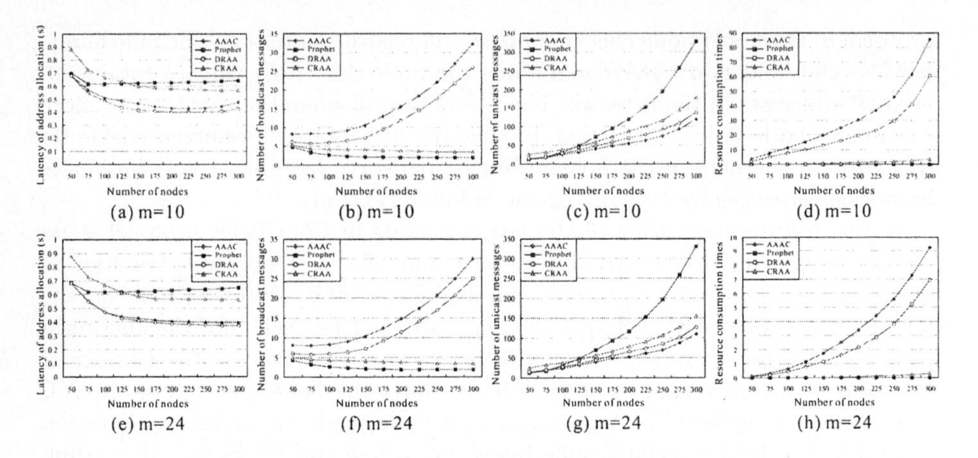

Fig. 2. Average latency of address allocation, the number of broadcast/unicast messages and resource consumption times with the varied number of nodes at m=10 and 24

rectangular region. The beacon interval for the holder to broadcast its NID is set to 10 seconds. The maximum retry times r of AREQ is set to 3 [10], and the AREQ_Timer is set to 2 seconds for quicker addressing. For strict reasons, when all nodes are stable, node leaving and rejoining, network partitioning and network merging are produced freely. The performance metrics of the simulation are given below.

- *Latency*: Latency of address allocation represents the average latency for a new node to obtain a unique IP address within the network. The shorter the latency, the better, since it means a new node can get a usable IP address more rapidly.
- *Communication overhead*: Communication overhead refers to the number of control messages transmitted during the simulation period, including unicast and broadcast messages. Normally, broadcast messages occupy more bandwidth than unicast messages do.
- *Evenness*: Evenness implies that address blocks should be evenly distributed in all nodes, which indicates that each node has the capability to assign address blocks to newly-joined nodes. The more evenly address blocks are allotted, the fewer the address resource consumption times in each node are, from which we are able to determine whether address blocks are evenly distributed.

Fig. 2(a) and (e) depict the average latency of address allocation versus the varied number of nodes, with two address block sizes: 2^{10} and 2^{24} respectively. In general, the DRAA protocol has the shortest latency because the DRAA protocol fast replies an address request and maintains orphan address blocks when a node joins the network. The CRAA protocol has longer latency than AAAC and DRAA protocols because it awaits AREPs of neighbors to select a bigger address block. Prophet has the longest latency when the number of nodes is more than 150 because during network merging, all nodes in the smaller NID have to rejoin the network and wait for the expiration of an AREQ_Timer. When the number of nodes is 50, all protocols have longer latency because the network size is 1000m $\times$ 1000m and the transmission range is 250m. When a new node joins the network, it has the higher probability that no node is within its transmission range, so that the new node needs to await the AREQ_Timer expiration. When the number of nodes is more than 150 at m=10 in Fig. 2(a), the latency of AAAC and DRAA increases with the node number because more resource consumption times in the whole network will expand the number of address requesting phase retries.

Fig. 2(b) and (f) show the number of broadcast messages versus the varied number of nodes, with two address block sizes: 2^{10} and 2^{24} respectively. Prophet has the fewest broadcast messages because it does not preform DAD during merging and all nodes with the smaller NID have to rejoin the network. The CRAA protocol has the second fewest broadcast messages because only the holder broadcasts NL for DAD during network merging. The DRAA protocol has fewer broadcast message than AAAC because the resource consumption times of DRAA are less than AAAC. Fig. 2(c) and (g) display the number of unicast messages versus the varied number of nodes, with two address block sizes: 2^{10} and 2^{24} respectively. AAAC has the fewest unicast messages because most of its control messages are handled with broadcast messages. In the CRAA protocol, more unicast messages are added to maintain the NL, so its unicast messages are more than those in DRAA as well as AAAC. Unicast messages in Prophet are the most because many nodes need to rejoin the network during network merging, which results in the large number of unicast messages.

Fig. 2(d) and (h) show the resource consumption times versus the varied number of nodes, with two address block sizes: 2^{10} and 2^{24} respectively. The DRAA protocol, during node joining, has the failed node checking phase to retrieve orphan blocks, so the resource consumption times decrease. The CRAA protocol has evenly-distributed resources because nodes in the CRAA protocol request address blocks from all neighbors and choose the biggest one to use. Although the way Prophet allots addresses is different from that of buddy system approaches and does not guarantee every allotted address is unique, $f(n)$, which distributes addresses, does not have the phenomenon of resource consumption. To be fair, the resource consumption times of Prophet are set to be 0. On the whole, when the address block size is big enough, the resource consumption times will be significantly reduced and it is the CRAA protocol whose resource consumption times are close to be 0.

6 Conclusions

This paper proposes two ring-based address autoconfiguration protocols in mobile ad hoc networks. Compared with existing address assignment protocols, the DRAA protocol successfully achieves low latency and low communication overhead, and the CRAA protocol further achieves low communication overhead and evenness of dynamic address assignment. RAA protocols use a logical ring to proceed address allocation and resource management. The ring provides unique address assignment without DAD. The DRAA protocol tolerates one node's invalidity and restores a failed node without help of the holder, and the CRAA protocol restores failed nodes with help of the holder. Based on the above advantages, RAA protocols shows high efficiency in address allocation as well as in resource management and suitability for the large scale mobile ad hoc network.

References

1. C. E. Perkins, J. T. Malinen, R.Wakikawa, E. M. Belding-Royer, and Y. Sun. Ad hoc Address Autoconfiguration. IETF Internet Draft, draft-ietf-manet-autoconf-01.txt, 2001. (Work in Progress).
2. Y. Sun and E. M. Belding-Royer. Dynamic Address Configuration in Mobile Ad hoc Networks. Technical report, Computer Science Department, UCSB, Mar. 2003.
3. Zero Configuration Networking. http://www.zeroconf.org/.
4. Mansoor Mohsin and Ravi Prakash. IP address assignment in a mobile ad hoc network. In *Proceedings of IEEE Military Communications Conference (MILCOM 2002)*, volume 2, pages 856–861, Anaheim, CA, United States, 7-10 Oct. 2002.
5. Sanket Nesargi and Ravi Prakash. MANETconf: Configuration of hosts in a mobile ad hoc network. In *Proceedings of the Twenty-first Annual Joint Conference of the IEEE Computer and Communications Societies (INFOCOM 2002)*, volume 2, pages 1059–1068, 23-27 Jun. 2002.
6. I. Stoica, R. Morris, D. Liben-Nowell, D. Karger, M. Frans Kaashoek, F. Dabek, and H. Balakrishnan. Chord: A scalable peer-to-peer lookup protocol for internet applications. *IEEE/ACM Transactions on Networking*, pages 149–160, 2002.
7. Abhishek Prakash Tayal and L. M. Patnaik. An address assignment for the automatic configuration of mobile ad hoc networks. *Personal Ubiquitous Computer*, volume 8, issue 1, pages 47–54, Feb. 2004.
8. Nitin H. Vaidya. Weak duplicate address detection in mobile ad hoc networks. In *Proceedings of the 3rd ACM international symposium on Mobile ad hoc networking and computing (MobiHoc 2002)*, pages 206–216, Lausanne, Switzerland, 9-11 Jun. 2002.
9. Kilian Weniger and Martina Zitterbart. IPv6 Autoconfiguration in Large Scale Mobile Ad-Hoc Networks. In *Proceedings of European Wireless 2002*, pages 142–148, Florence, Italy, Feb. 2002.
10. Hongbo Zhou, Lionel M. Ni, and Matt W. Mutka. Prophet address allocation for large scale MANETs. In *Proceedings of the Twenty-Second Annual Joint Conference of the IEEE Computer and Communications Societies (INFOCOM 2003)*, volume 2, pages 1304–1311, San Francisco, CA, United States, Mar. 30-Apr. 3 2003.

Dual Binding Update with Additional Care of Address in Network Mobility*

KwangChul Jeong, Tae-Jin Lee, and Hyunseung Choo**

School of Information and Communication Engineering,
Sungkyunkwan University, Suwon, 440-746 Korea
+82-31-290-7145
{drofcoms, tjlee, choo}@ece.skku.ac.kr

Abstract. In this paper, we propose an end-to-end route optimization scheme for nested mobile networks, which we refer to as *Dual Binding Update* (*DBU*). In general, the nested mobile networks easily suffer from a bi-directional pinball routing with hierarchically multiple mobile routers. To handle this matter, we provide a new binding update (BU) message to allow a Correspondent Node (CN) to keep an additional Care of Address (CoA). And we also allow intermediate Mobile Routers (MRs) maintain a routing table to forward packets inside the mobile network and replace a source address of the packet for reverse route optimization. We evaluate the DBU with existing schemes by analytical approaches. The results show that the DBU reduces the delay of route optimization significantly under various scenarios and also improves an average Round Trip Time (RTT) consistently for many nesting levels tested.

1 Introduction

As wireless networking technologies have drastically advanced and many electronic devices are given the capability of communications with their own IP addresses, users expect to be connected to the Internet from anywhere at anytime. The IETF has standardized protocols such as Mobile IPv4 (MIP) and Mobile IPv6 (MIPv6) [3] to support seamless connectivity to mobile hosts. Recently, more and more users require the seamless Internet services while they are on public transportation. And if one user moves with several wireless devices and communicates through the Internet, these moving devices with him (her) constitute a Wireless Personal Area Network. Unfortunately, because Mobile IP is designed for continuous accessibility to mobile hosts with mobility transparency on IPv4 or IPv6, it does not provide a solution in response to these new demands. The continuous mobility of a network will lead to movements of the nodes in the network. If all nodes are forced to run Mobile IP according to the movement of the network, the overhead causes futile consumption of network resources. Hence

* This work was supported in parts by IITA IT Research Center and University Fundamental Research Program of Ministry of Information and Communication, Korea.

** Corresponding author.

X. Jia, J. Wu, and Y. He (Eds.): MSN 2005, LNCS 3794, pp. 783–793, 2005.

an elected node called Mobile Router (MR) [1] should become a gateway instead of entire nodes in the network for efficient resource management.

The mobile network may have a complicated hierarchical architecture, and this situation is referred to as a *nested mobile network*. According to NEMO Basic Support Protocol (NBS) [8], it is required that all packets going through the nodes inside the nested mobile network should be tunneled to every Home Agent (HA) which they pass by. To avoid this, the packet should be delivered to the Top Level Mobile Router (TLMR) directly. However, because all Mobile Network Prefixes (MNP) of the MRs in the nested mobile network are different, the TLMR cannot route the packet to its destination. Hence we need to develop a mechanism for end-to-end routing optimization in the nested mobile network.

The rest of this paper is organized as follows. We introduce the bi-directional pinball routing problem of nested mobile networks and review the existing NEMO schemes for route optimization in Section 2. Section 3 describes the dual binding update method in the DBU for end-to-end route optimization of nested mobile networks. In Section 4, we evaluate the performance of the proposed scheme compared with the existing ones. Finally we conclude this paper in Section 5.

2 Related Works

2.1 NEMO Basic Support (NBS) [8]

In a NEMO network, a point of attachment can vary due to its movement. Since every NEMO network has its own home network, mobile networks configure addresses by using the prefix of its home. These addresses have a topological meaning while the NEMO network resides at home. When the NEMO network is away from home, a packet addressed to an Mobile Network Node (MNN) [1] still routes to the home network. The NBS is designed to preserve established communications between the MNNs and CNs while the NEMO network moves, and it creates bi-directional tunnels between the MNN_HA and MR_CoA to support network mobility .

The NBS, however, does not describe the route optimization [8], since the NBS is mainly designed with bi-directional tunnel. Hence if the mobile network has a nesting level (depth) of N, the packet is encapsulated N times. And when the MNN delivers the packet to the CN, all intermediate nodes replace the source addresses of the packet with their own CoAs to avoid ingress filtering and the destination addresses of the packet with their HAs for packet tunneling. Hence it causes the *reverse pinball routing problem*. The overhead of bi-directional pinball routing becomes more significant as the nesting level increases and a distance between the HAs of MRs becomes longer. Therefore the NBS lacks scalability and promptness with respect to a nested environment.

2.2 Recursive Binding Update Plus (RBU+) [2]

The RBU+ scheme is basically operated under the MIPv6 route optimization unlike the NBS. Thus in RBU+, any node receiving the packet via its HA performs

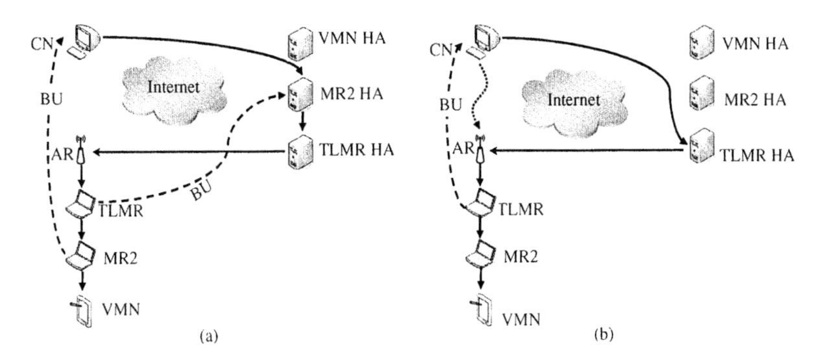

Fig. 1. Route optimization in RBU+

the binding update (BU). After the CN sends the first packet to the Visited Mobile Node (VMN) [1], both the TLMR and the MR*2* perform the BU as shown in Fig. 1(a). The RBU+ maintains the optimal route from the CN to the TLMR by updating its binding information recursively when it receives a BU message.

For example, the CN makes (*VMN_HoA:VMN_CoA*) as in Fig. 1(a) and (*MR2_prefix:MR2_CoA*) as in Fig. 1(a) out of (*VMN_HoA:MR2_CoA*) by the recursive binding update. This is because the VMN_CoA is configured with MR*2*_prefix. The TLMR also delivers the BU to the CN as in Fig. 1(b) and the CN also makes (*VMN_HoA:MR2_CoA*) and (*TLMR_prefix:TLMR_CoA*) out of (*VMN_HoA:MR2_CoA*) since the MR*2*_CoA is configured with TLMR_prefix. Therefore CN can deliver the packets to the TLMR in which the VMN resides.

However, The RBU+ should perform the recursive search for the recursive binding update whenever a BU message arrives, and the delay for route optimization becomes more serious as the nesting level increases.

2.3 Reverse Routing Header (RRH) [5]

The RRH scheme is basically based on a single tunnel between the first MR and its HA. The RRH records the addresses of intermediate MRs in the slots of routing header when the MNN sends the packet to the CN. Fig. 2(a) shows an example of the RRH operation. While a VMN sends the packet to a CN, an MR2 records the source address of the RRH with its CoA to avoid the ingress filtering. It also records the destination address of the RRH with its HA, and its HoA in the free slot of RRH. The TLMR performs the same tasks when it receives the packet from the MR*2*. Then the packet is delivered to the MR*2*_HA, which contains the multiple slots with the TLMR_CoA. Finally the MR*2* relays the packet to the CN according to the original routing header. Fig. 2(b) shows how to use the slot contents. When the CN sends the packets to the VMN, it is routed through MR*2*_HA. At this point, the MR*2*_HA records all intermediate nodes which the packets should traverse by using TLMR_CoA and multiple slots. In other words, the RRH performs source routing with multiple slots and it alleviates the pinball routing overhead by a single tunnel. However it requires

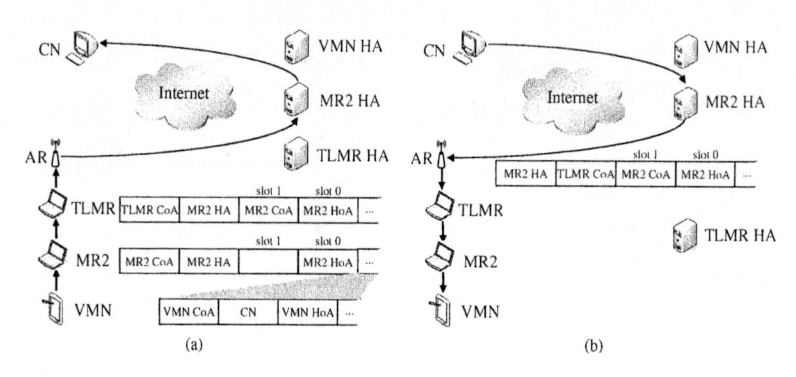

Fig. 2. Route optimization in RRH

more slots as the depth of nesting increases, and because the RRH scheme should suffer from inevitable single tunnel it still has potential overhead.

3 The Proposed Scheme

In this section, we propose a Dual Binding Update scheme to constitute the optimal routes by utilizing the BU message. And we focus on the *MIPv6-enabled VMN* which Home Address (HoA) does not consist of the MNP of parent-MR.

3.1 Dual Binding Update for End-to-End Route Optimization

The binding update that the MNN performs in MIPv6 can be divided into two types. The first one occurs when the MNN detects its movement, and the second one occurs when the packets are delivered via its HA. In both types, after the MNN sends the BU messages, each HA and CN records the (*MNN_HoA:MNN_CoA*) entry in the binding cache. However it cannot route packets through optimal route based on this binding entry. Hence the NBS incurs a pinball routing to support the nested mobile networks. In the DBU, the MNN sends binding entry of (*MNN_HoA:MNN_CoA*) when it performs BU to its HA, but it sends both (*VMN_HoA:VMN_CoA*) and *TLMR_CoA* when it performs BU

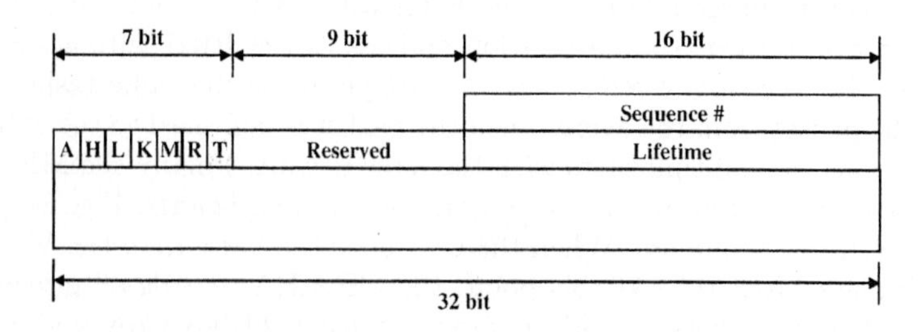

Fig. 3. The T bit defined in the BU message

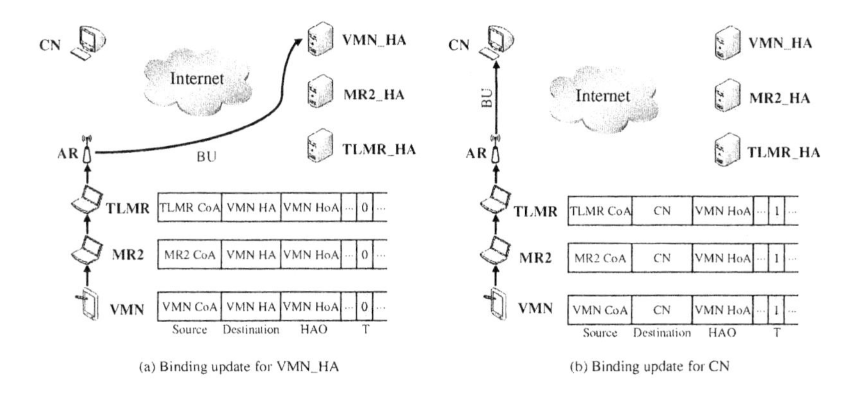

Fig. 4. Two types of BU message

to the CN. So the CN can keep the optimal route from the CN to the TLMR. Fig. 3 shows a newly defined T bit in the BU message. If T bit is set, a node receiving the BU message records the binding entry of (*VMN_HoA:VMN_CoA*) including the TLMR_CoA. In Fig. 4(a), the VMN sends the BU message to its HA in which T bit is unset, and then the VMN_HA records only (*VMN_HoA:VMN_CoA*) entry in the binding cache. Because when the mobile network with which the TLMR is associated moves, it generates a BU storm to the HAs of MR. Fig. 4(b) shows the situation that the VMN sends the BU to the CN. In this case T bit is set. The CN records the (*VMN_HoA:VMN_CoA:TLMR_CoA*) entry in the binding cache. Hence when the CN sends the packet to the VMN, it sends the packet to the TLMR directly not via intermediate MRs due to the additional TLMR_CoA entry. As shown in Fig. 5, every node which the BU message traverses replace the source address with its CoA. Unlike the NBS, it does not append an additional header. According to the NBS, when an MNN in a nested mobile network sends a packet to a CN, the packet is routed through each HA of the MRs including its HA. Because every MR including the VMN records the source address with its CoA and the destination address of the packet with its HoA to avoid the ingress filtering and to maintain the connection with the CN irrespective of the VMN's location. However this reverse pinball routing also causes serious overhead, and the overhead becomes more significant as the nesting level increases.

When the CN maintains the TLMR_CoA, it confirms the connection with the VMN since the source address of the packet is TLMR_CoA. And the VMN offers mobility transparency to transport layer of CN based on the Home Address Option (HAO). So the proposed DBU provides the solution for the reverse pinball routing problem. The following algorithm describes the operation while a certain node receives the BU message according to the mechanism mentioned above.

The route optimization schemes described above do not touch the route optimization from the TLMR to the VMN. While the TLMR receives a packet from the CN through the optimal route, it has no clue to route the packet to the VMN since it does not maintain any address information for the VMN. Hence we propose the route optimization scheme from TLMR to VMN by utilizing BU message. Every VMN entering a new mobile network sends BU message to its HA and CNs. All MRs above the VMN cache the VMN_HoA and the source address of the packet while BU messages go through them. So when the TLMR receives the packet with its destination specified by VMN_HoA, it is able to relay that packet to the proper intermediate child-MRs and finally to the right VMN.

In RBU+ scheme [2], we need an additional mechanism to send a packet from the TLMR to the VMN. Especially in the Route Request Broadcasting mechanism, all intermediate MRs should relay the request messages to the child nodes to find a destination node. Therefore the delay of packet delivery becomes larger as the nesting level increases.

3.2 More Mobility Concern

We can classify the mobility of the nested mobile network into three types. *1*) VMN moves; *2*) A partial network moves; *3*) An entire network including the TLMR moves. In types 2 and 3, the CN is disconnected temporarily from the VMN of the nested mobile network. Because the CN sends the packet to the TLMR_CoA according to the binding cache. In this paper, we consider this problem and propose a mechanism to handle it.

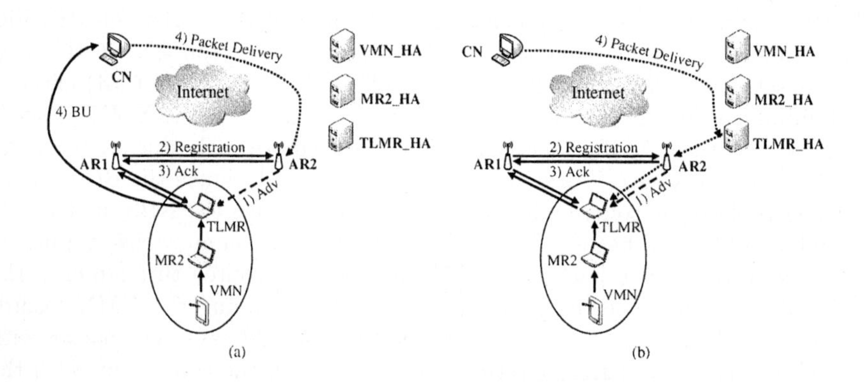

Fig. 5. In case of entire networks move

As you see in Fig. 5, the entire network including TLMR*1* moves to the network under AR*2*. The CN which communicates with the VMN cannot recognize the movement of the VMN, so when the CN sends the packet to the TLMR*1* and the VMN cannot receive it. We propose two methods to settle this problem.

The first one can be applied to the case that a partial network moves. That is the TLMR*1* sends the BU message to all CNs communicating with its child nodes. These messages are for exchanging the TLMR_CoA entry except for (*HoA:coA*). This method has an advantage of short delay, but we can expect high cost. The other one uses an extension of BU message. When the VMN delivers the BU message to the CN, it inserts an additional TLMR_HoA in the binding entry. Hence the CN maintains the TLMR_HoA together with the TLMR_CoA. In case that the CN cannot send the packet to the TLMR_CoA, it sends the packet to the TLMR_HoA temporarily. Then after the TLMR receives the packet through its HA, it can perform the BU to the CN or relay the packet for a while. This method prevents the TLMR from burst binding updates, but it suffers from relatively large delay and amount of packet losses.

4 Performance Evaluation

In this section, we evaluate our proposed scheme compared with the NBS, RBU+, and RRH analytically in terms of delay. At first, we consider the delay until a VMN receives a packet through an optimal route after it enters the mobile network with which a TLMR is associated. Secondly, we evaluate a Round Trip Time (RTT) including the NBS when a CN exchanges the amount of packets with the VMN for various nesting levels.

4.1 Analyses Environment

We assume an environment of performance evaluation as following:

Assumption 1) One MR has $\bar{t}$ nodes in average.

Assumption 2) Each node under the same parent_MR can be a MR with probability α and a VMN with probability 1-α.

Let T indicate the total number of nodes which a single TLMR constitutes.

$$T = \bar{t}(1-\alpha) + \bar{t}\cdot\alpha(\bar{t}(1-\alpha) + \bar{t}\cdot\alpha(\bar{t}(1-\alpha) + \cdots \quad = \frac{\bar{t}(\alpha-1)}{\bar{t}\alpha - 1}$$

When we Assume that $\bar{t} \geq 1$ and $T = \dfrac{\bar{t}(\alpha-1)}{\bar{t}\alpha-1} \geq 1$, *Hence*, $\bar{t}\alpha < 1$.

And since we also assume that the CN communicates with a voluntary VMN, we need to calculate an average number of VMNs for the nesting level. Equations below denote the average number of VMNs for each nesting level :

$$\text{Nesting level } i : \bar{t}^i\alpha^{i-1}\cdot(1-\alpha)$$

Therefore the probability of the voluntary VMN communicating with the CN resides in the nesting level of i is

$$P_{VMN}\ (i) = \frac{\bar{t}\alpha^{i-1}(1-\alpha)}{T} = (1-\bar{t}\alpha)(\bar{t}^{i-1}\cdot\alpha^{i-1}) = (1-\bar{t}\alpha)(\bar{t}\alpha)^{i-1}$$

When we consider the routing paths through which the CN communicates with the VMN, we can classify them into outside the TLMR and inside the TLMR. And if we assume the environment of evaluation as random networks, we can say the distances among any neighboring two entities are equal and define ω to represent the one hop delay of outside the TLMR. And we also assume that the distances among the nodes inside the TLMR are equal, and define φ which represents the one hop delay of inside the TLMR.

4.2 Comparison of Delay for Route Optimization

The users connecting the Internet in wireless networks expect to be provided various services with guaranteed delay. Therefore when we consider the network mobility which forms a nested mobile network frequently, it is especially important to deliver the packet via the optimal route as soon as it is possible. We evaluate the delay of route optimization when the CN communicates with the voluntary VMN for the given *Assumptions.* The delay of route optimization is calculated as the time until the VMN receives the packet through optimal route after the CN initiates the communication. And it consists of the delay from the the CN to TLMR, from the TLMR to VMN, and binding update delay.

The ω represents the one hop routing delay between the nodes outside the TLMR, $c(i)$ represents the number of hops which the packet traverses outside the TLMR until the optimal route constitutes. And φ represents the one hop routing delay between nodes inside the TLMR, $b(i)$ represents the number of hops which the packet traverses inside the TLMR until the optimal route constitutes.

4.2.1 The Delay of RBU+ Route Optimization

In the RBU+, any node performs the BU whenever a packet arrives via its HA since the it is operated based on the MIPv6 route optimization. Hence when the CN sends the packet to the VMN, several nodes perform the BU concurrently. However we focus on the delay, we only evaluate the BU delay for the VMN which is the largest among them. We can define the number of routing hops which the packet routes via outside the TLMR in terms of the nesting levels.

$$\begin{aligned} RBU_{CN-TLMR}(i) &= i + 2 + \left\lceil \frac{i+2}{2} \right\rceil + c(i-1) - (i+1) \\ &= \left\lceil \frac{i+2}{2} \right\rceil + c(i-1) + 1 \end{aligned}$$

After the TLMR receives the packet, all of intermediate MRs need relay it to the VMN. In this respect, we can also define the number of routing hops which the packet routes via inside the TLMR for the nesting levels.

$$RBU_{TLMR-MNN}(i) = i(i+2)$$

When we consider that the VMN sends the BU message, we can define the number of hops which it passes by for the nesting levels.

$$RBU_{MNN-TLMR}(i) = \frac{i(i+1)}{2}$$

$$RBU_{TLMR-CN}(i) = \frac{(i+2)(i+3)}{2} - 1$$

Finally, the average total delay for RBU+ route optimization is as follows.

$$RBU_{TOTAL} = \sum_{i=1}^{\infty} (1 - \bar{t}\alpha)(\bar{t}\alpha)^{i-1} \{\omega \cdot c(i) + b(i)\}$$

$$= \sum_{i=1}^{\infty}(1-\bar{t}\alpha)(\bar{t}\alpha)^{i-1}\Bigg[\omega\cdot\left\{\left\lceil\frac{i+2}{2}\right\rceil+c(i-1)+1\right\} + \varphi\cdot\left\{i(i+2)+\frac{i(i+1)}{2}\right\}+\omega\cdot\left\{\frac{(i+2)(i+3)}{2}-1\right\}\Bigg]$$

4.2.2 The Delay of RRH Route Optimization

The RRH is basically based on a single tunnel between the TLMR and the HA of the first MR, so it involves potential overheads continuously. We can simply define the number of routing hops outside the TLMR and inside the TLMR which the packet passes via for the nesting levels.

$$RRH_{CN-TLMR}(i) = (i+2)+2, \quad RRH_{TLMR-MNN}(i) = 2i$$

And we define the required number of routing hops when the VMN perform the BU for the nesting levels as follows.

$$RRH_{MNN-TLMR}(i) = i\ , \ \ RRH_{TLMR-CN}(i) = 2$$

Finally, the total average delay for RRH route optimization is

$$RRH_{TOTAL} = \sum_{i=1}^{\infty}(1-\bar{t}\alpha)(\bar{t}\alpha)^{i-1}\{\omega\cdot c(i)+b(i)\}$$

$$= \sum_{i=1}^{\infty}(1-\bar{t}\alpha)(\bar{t}\alpha)^{i-1}\cdot\Big[\omega\cdot\{(i+2)+2\}+\varphi\cdot(2i+i)+\omega\cdot 2\Big]$$

4.2.3 The Delay of DBU Route Optimization

The DBU shortens the delay of route optimization through the reverse route optimization compared with the RBU+. And since our proposed scheme does not have to pass by HA of the first MR after route optimization, it is more efficient than the RRH. We can define both the number of outside the TLMR hops and inside the TLMR which the packet passes through in terms of the nesting levels.

$$DBU_{CN-TLMR}(i) = i+3\ , \ \ DBU_{TLMR-MNN}(i) = 2i$$

And we define the required number of hops the packet passed by when the VMN perform the BU according to the nesting level.

$$DBU_{VMN-TLMR}(i) = i\ , \ DBU_{TLMR-CN}(i) = 1$$

As a result, the average total delay for proposed route optimization is

$$DBU_{TOTAL} = \sum_{i=1}^{\infty}(1-\bar{t}\alpha)(\bar{t}^{i-1}\cdot\alpha^{i-1})\{\omega\cdot c(i)+b(i)\}$$

$$= \sum_{i=1}^{\infty}(1-\bar{t}\alpha)(\bar{t}\alpha)^{i-1}\Big[\omega\cdot(i+3)+\{\varphi\cdot(2i+i)+\omega\}\Big]$$

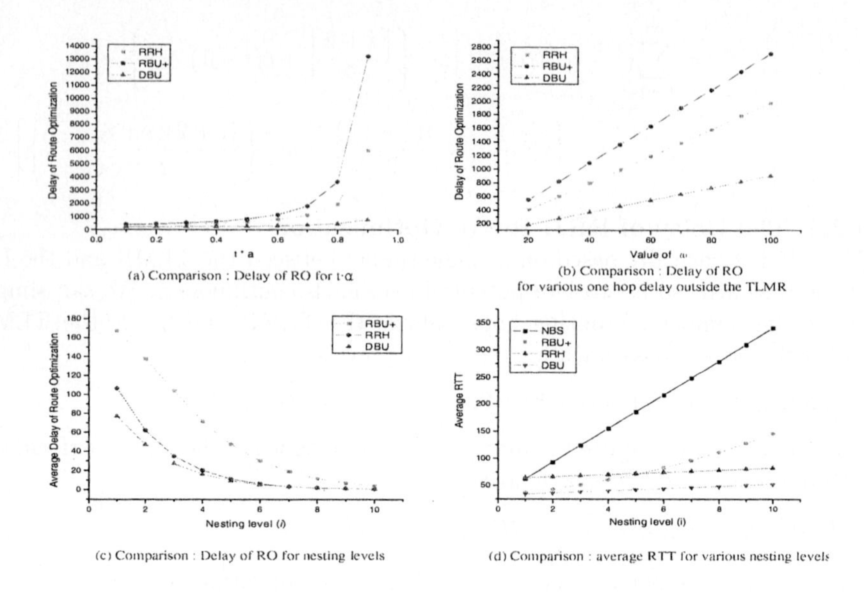

Fig. 6. Comparison : delay of route optimization for $t \cdot \alpha$

Fig. 6 shows the evaluation results under various parameters. In Fig. 6(a), when the $t \cdot \alpha$ is close to 1, the total delay of route optimization for the RRH and RBU+ is drastically augmented. And in general, a routing distance between the nodes at outside the TLMR is larger than that in inside the TLMR. We vary the value of ω and evaluate the delay of route optimization while we fix the value of φ to 1 and $t \cdot \alpha$ to 0.5 as in Fig. 6(b). The result shows that the delay gap becomes more significant as the value of ω increases.

Fig. 6(c) shows the average delay for route optimization when the MNN resides in the nesting level i. Since we fix the value of $t \cdot \alpha$ to 0.5, there is little chance that the nesting level is bigger than 10. The result indicates that the DBU is superior to the RRH and the RBU+ within reasonable scopes. We also evaluate average RTTs including the NBS when the CN exchanges the packet 30 times with the VMN under various nesting levels, and the value of ω is fixed to 30. The result shows that the average RTTs of all schemes increase linearly with different slopes. But as you see in Fig. 6(d), the average RTT of RBU+ exceeds the one of RRH starting from the nesting level 6. This is because the RBU+ takes more delay to constitute the optimal route as the nesting level increases.

5 Conclusion

This paper proposes a Dual Binding Update scheme for end-to-end route optimization in NEMO environment. The DBU defines a new T bit in a BU message, and CNs maintain an additional TLMR_CoA when the T bit is set. And another mechanism is needed to relay the packets to the correct MNN. So all nodes that the BU message traverses keep the MNN_HoA and CoA of a child-node for the end-to-end route optimization. The MNNs also can constitute reverse optimal routes by using the TLMR_CoA and HAO.

The RBU+ provides the end-to-end route optimization, but it does not offer reverse optimization and it takes more time as the nesting level increases. Although the RRH avoids the pinball routing, it requires additional multiple slots in the BU message for the source routing. Besides the RRH involves potential overhead due to the inevitable single tunnel. In our DBU, it is allowed that all nodes can communicate with MNNs with minimum delay. Our evaluation results the show advantages of the DBU compared to the other existing schemes.

References

1. C. Ng, P. Thubert, H. Ohnishi, E. Paik, "Taxonomy of Route Optimization models in the NEMO Context," IETF, draft-thubert-nemo-ro-taxonomy-04, February 2005, Work in progress.
2. C. Hosik, P. Eun Kyoung, and C. Yanghee, "RBU+: Recursive Binding Update for End-to-End Route Optimization in Nested Mobile Networks," HSNMC 2004, LNCS 3079, pp. 468-478, 2004.
3. D. Johnson, C. Perkins, and J. Arkko, "Mobility Support in IPv6," IETF, RFC 3775, June 2004.
4. M. Watari, T. Ernst, "Route Optimization with Nested Correspondent Nodes," IETF, draft-watari-nemo-nested-cn-01, February 2005, Work in progress.
5. P. Thubert, M. Molteni, "IPv6 Reverse Routing Header and its application to Mobile Networks," IETF, draft-thubert-nemo-reverse-routing-header-05, March 2005, Work in progress.
6. R. Wakikawa, S. Koshiba, K. Uehara, J. Murai, "ORC: Optimized Route Cache Management Protocol for Network Mobility," Telecommunications, ICT 2003, 10th International Conference, vol. 2, pp. 1194-1200, 23 Feb.-1 March 2003.
7. T. Clausen, E. Baccelli, R. Wakikawa, "NEMO Route Optimisation Problem Statement," IETF, draft-clausen-nemo-ro-problem-statement-00, October 2004.
8. V. Devarapalli, R. Wakikawa, A. Petrescu, P. Thubert, "Network Mobility (NEMO) Basic Support Protocol," IETF, January 2005, RFC 3963.

Optimistic Dynamic Address Allocation for Large Scale MANETs

Longjiang Li[1] and Xiaoming Xu[2]

[1] Department of Computer Science and Engineering, Shanghai Jiaotong University, Shanghai 200030, P.R. of China
[2] Department of Automatic Control, Shanghai Jiaotong University, Shanghai 200030, P.R. of China
{E_llj, xmxu}@sjtu.edu.cn

Abstract. In order to allow truly spontaneous and infrastructureless networking, the autoconfiguration algorithm of the mobile node addresses is important in the practical usage of most MANETs. The traditional methods such as DHCP can not be extended to MANETs because MANETs may operate in a stand-alone fashion and their topologies may change rapidly and unpredictably. The diversified schemes have been proposed to solve this problem. Some of them apply Duplicate Address Detection (DAD) algorithms to autoconfigure the address for each node in a MANET. However, the multi-hop broadcast used by DAD results in high communication overhead. Therefore, a new autoconfiguration algorithm is proposed in this article, which combines the enhanced binary split idea of Dynamic Address Allocation Protocol (DAAP) and the pseudo-random algorithm to construct the interface ID of IPv6 address. The allocation process is distributed and do not rely on the multi-hop broadcast, so our algorithm can be suitable for large scale MANETs through our simulation study.

1 Introduction

Mobile ad hoc networks (MANETs) are self-organizing wireless networks. Each mobile node has the capabilities to route data packets. Before the proper routing can be possible, every node needs to be configured a unique address. In MANETs, preconfiguration is not always possible and has some drawbacks. Furthermore, there is no central infrastructure managing all nodes in MANETs, so an autoconfiguration protocol is required to provide dynamic allocation of node's address [10].

The autoconfiguration protocols proposed presently can be classified in protocols utilizing stateless and stateful approaches. Most stateful approaches rely on the allocation table [1]. M. Günes and J. Reibel [6] have proposed an address autoconfiguration protocol utilizing a centralized allocation table. The protocols, MANETconf [5], Boleng's protocol [7], and the Prophet Allocation protocol [4], utilize a distributed common allocation table. The protocol, proposed by M. Mohsin and R. Prakash [9], utilizes the multiple disjoint allocation tables. The weakness is that most these approaches rely on reliable state synchronization in the presence of packet loss and network merging, which may consume a considerable amount of bandwidth [1]. In

X. Jia, J. Wu, and Y. He (Eds.): MSN 2005, LNCS 3794, pp. 794–803, 2005.

contrast, the stateless approaches usually need DAD algorithm to cope with address conflict. Perkins et al. [10] have proposed an autocofiguration protocol following a stateless approach. However, the broadcast used in DAD usually may result in high communication overhead and small scalability. The other two approaches, weak DAD (WDAD) [11] and passive DAD (PDAD) [12], need to change the routing protocol. And the protocol HCQA [13] combines elements of both stateful and stateless approaches, but incurs more complexities.

In order to overcome the aforementioned drawbacks, we propose a new address autoconfiguration algorithm, namely optimistic dynamic address allocation algorithm (ODAA), which combines the enhanced binary split idea of Dynamic Address Allocation Protocol (DAAP) [2][3] and the pseudo-random algorithm [15] to reduce the overhead of address configuration very much.

This paper is structured as follows. Next section introduces the basic idea of the proposed approach. Section 3 discusses its characteristics and gives a succinct comparison with two known algorithms, which depicts the superiority of the proposed algorithm over others. Section 4 presents some simulation results, which is equal to our analysis. Section 5 concludes the paper.

2 Basic Idea

We assume that each allocation process the MANET starts from a single node. Each such process is referred to as an address domain. The first node in a domain is called as domain initiator. In order to generate unique addresses in a domain, we define an allocation function based on the similar idea to Dynamic Address Allocation Protocol (DAAP) [2] [3]. Furthermore, a local decision policy is proposed to propagate address resource between neighboring nodes. Since each address holds particular ability, assigned by the allocation function, to generate new addresses, we refer to such address as meta address. When a node need leave the MANET, the node can transfer its address capsulated in an address message to one of any neighboring nodes for reuse.

2.1 Structure of Meta Address

We use a 3-tuble: (*address*, *power*, *DID*) to represent a meta address where *address* is its identification in the domain, *power* indicates its ability to generate new addresses and *DID* is the domain identification. Here the *address* is unique in a domain and the combination of *DID* and *address* identify a global address. Usually, DID must be generated with a pseudo-random algorithm consistent with [21] and is propagated to new nodes during the course of allocation. Because *DID* is a random number, if the number of bits for *DID* is large enough, two domains will have different *DID*s. The probability that two or more of *DID*s will collide can be approximated using the formula [15]:

$$P = 1 - \exp(-N^2 / 2^{(L+1)}),$$

where P is the probability of collision, N is the number of interconnected *DID*s, and L is the length of the *DID*. For example, in IPv6, the 64-bit interface ID can be coded using the combination of *DID* and *address*.

Suppose that the scope of *address* is [0, *M*-1] where *M* is the *power* of 2 ($=2^K$). If there is only one domain initiator in a MANET, every node in the MANET can obtain unique *address* belonging to one domain, i.e. *DID* plays a role of the network ID. Otherwise, multiple *DID*s may compete or coexist in the same MANET. When we only consider address allocation in a single domain, because all nodes have the same *DID*, for convenience, we may neglect *DID* and only use a 2-tuble: (*address*, *power*) to represent an address.

We denote the scope of *address* as S, i.e. S = [0, *M*-1]. The allocation function, say *f*, is defined as below.

1) *f*: S×T-> S×T, where T = [0, *K*] is a set of integer number scope;
2) For a input (*address*, *power*), *f* (*address*, *power*) = (*address*$|2^K$-*power*, *power*-1).
3) A meta address can call *f* to generate new address, if only its *power* is greater than zero. And a meta address reduce its *power* by 1 each time it call *f* successfully.

Note that "|" is a bit-wise OR operation, e.g. for a node with (*address*, *power*) = (010, 1) and *K*=3, *f* (010, 1) = (110, 0) (see Figure 1). Note that, here, *address* is coded in binary code.

A meta address will be prohibited from calling *f*, or we can write that *f* (*add ress*, <1) returns null value, if its *power* is smaller than 1. Ordinary address also can be regarded that its *power* equals 0.

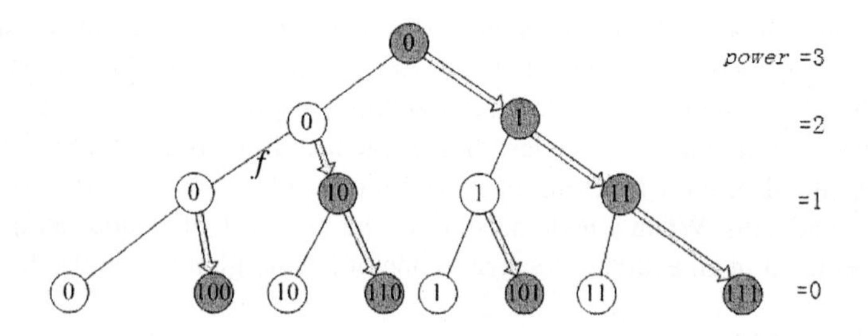

Fig. 1. A example to demonstrate the behavior of the allocation function with K=3

We designate that the meta address of domain initiator is (0, *K*, *DID*) from which all other meta addresses can be generated by directly or indirectly calling *f*. Obviously, the initial meta address can generate at most K new meta addresses by directly calling *f* and each generated address also can generate other meta addresses if only its *power* is not smaller than 1. It is easy to verify that the allocation function can generate all values in S from the initial address (0, *K*, *DID*) without duplication.

2.2 Process of Address Allocation

If a node has been configured address, we say that it is a configured node, otherwise an unconfigured node. Since an unconfigured node has no allocation ability, we may say that it has the lowest ability to simplify the process.

When a mobile node switches to the ad-hoc mode, it is still an unconfigured node. First of all, the node generates a link-local address by itself and broadcasts a query message to its neighbors within one hop scope. Its link-local address should be carried in the message, which may be used by the responder to build a unicast acknowledgement. When the node receives a reply, it may choose to send an address request to the responder. If the node receives an address message, it uses the meta address in the message as its address and becomes a configured node. Otherwise, if the node fails to become a configured node during a predefined period RETRIES_PERIOD, it resends the query messages until it becomes a configured node or the number of retires times exceeds a predefined maximal value MAX_RETRIES, say 1. If the number of retires times exceeds MAX_RETRIES, the node may choose to become a domain initiator with a special probability which should be small enough to reduce the number of domains in a MANET. A domain initiator should assign (0, *K*, *DID*) as its address and it becomes a configured node.

When a configured node receives a query message, it replies an acknowledgement message. If the node receives an address request, it executes a local decision to generate a meta address and sends the meta address capsulated in an address message to the requester.

2.3 Local Decision to Dispense Address

In order to support our algorithm, each node maintains three variables: *ADDR*, *LIST* and *lastHop*. Here *ADDR* is the meta address owned by the node, *LIST* is an array of pending meta addresses and *lastHop* records whom its addresses is assigned from. For a domain initiator, its *ADDR* is (0, *K*, *DID*), *LIST* is null and *lastHop* is null. For an unconfigured node, its *ADDR* is (-1,-1), its *LIST* is an empty *LIST* of length 0 and *lastHop* is null.

In order to evaluate total allocation ability that a node has, we sum up all *power* values of meta addresses just like below and write it as *heartpower*.

$$heartpower = \sum_{u=ADDR\, or\, u \in LIST} u.power$$

Note that *heartpower* of an unconfigured node equals -1.

Consider two neighboring nodes, say A and B. Assume that B has sent an address request to A and its *heartpower* is also carried in the request packet. Note that B should ensure that B.*heartpower* is smaller than A.*power* before B sends the request to A, which can reduce the useless requests. When A receives the request, if A.*power* is greater than B.*heartpower*, A may try to dispense a meta address to B in unicast, otherwise A does nothing. A always tries to dispense meta address from A.*LIST* firstly. Only when A.*LIST* has no element, A dispenses meta address from A.*ADDR*. Moreover, while dispensing meta address, A always can keep enough *power* for itself, unless A.*LIST* is empty and A.*ADDR*.*power* is smaller than 2. When B receives the meta address, B sets A as its *lastHop* firstly. If B is still unconfigured, B uses the meta address as its address which is stored in its *ADDR*. Otherwise, if B is a configured node, B inserts the meta address into its *LIST* as an item of pending address resource. After A dispenses a meta address successfully, if A.*LIST* has no element A also tries to send an address request to its *lastHop*. There is also a counter variable in each node. If A fails to receive reply from its *lastHop* before another request, the counter increases 1. When the counter exceeds special threshold, say 3, A sends in broadcast its neighbors a query message within one hop scope to update its *lastHop* and the counter is reset to zero. Note that this procedure is applied to all configured nodes.

2.4 Mechanism for Network Partition and Merger

Here we consider most common two kinds of scenarios about network partition and merger.

The first scenario is that a MANET partitions and then the partitions merge again. If there is only one domain initiator in a MANET, every node in the MANET can obtain unique *address* belonging to one domain, i.e. *DID* plays a role of the network ID. Since the allocation function always generates different addresses even if the MANET becomes partitioned, there is no conflict if the partitions become merged later. Otherwise, multiple *DID*s may compete or coexist in the same MANET. So long as multiple domains do not collide, there still is no conflict if the partitions become merged later. In case the domain collision is detected, an error message should be triggered.

The second scenario is that two separately configured MANETs merge. Since nodes in different MANETs belong to different domains, there is no conflict if only the domains do not collide. In case the domain collision is detected, an error message should be triggered.Any unconfigured node may choose to become a domain initiator with a special probability which should be small enough to reduce the number of domains in a MANET. However, as the number of domains decrease, the propagation of address allocation may cost more time because each node always obtains address from one of its neighbors. Therefore, our algorithm is not sensitive to network partition and merger and there is a trade-off between reliability and latency of address allocation.

3 Characteristics and Performance Comparison

In this section, we compare our algorithm with other two typical approaches.

The autoconfiguration protocol, proposed by Perkins et al. [10], follows a stateless approach. Each node selects an address by itself and employs DAD mechanism (query-based DAD) to determine whether the selected address is valid or not. No state is maintained. No address reclamation is needed. However, multi-hop broadcast adopted in conflict detection leads to high communication overhead, high latency, and small scalability. In addition, the protocol has not considered subnet merging.

MANETconf [5] is also an autoconfiguration protocol but following a stateful approach. The protocol maintains a global allocation state, so a new node can obtain a free address to join the subnet. However, the state management and synchronization to maintain global state incurs high complexity, high communication overhead, high latency, and low scalability.

In contrast, we argue that ODAA can achieve low communication overhead, low latency and high scalability. In ODAA, although the uniqueness of addresses is not rigorous, the probability of address collision may be extremely low. The feasibility follows the fact that when subnets merger occurs, address collision always is possible. Furthermore, the mechanism reporting the error messages can be employed to finally conquer the trouble incurred by address collision. Thus, the correctness of ODAA is promising. The configuration time is essentially proportional to the height of the virtual spanning tree. Since allocation process always starts from some domain initiator, the configuration time is proportional to the log of the number of nodes in the domain. In addition, multi-hop flooding is avoided and allocation processes in different domains can be parallel, so communication overhead is reduced very much. As a result, ODAA is suitable for large scale MANETs.

The comparison results are summarized in Table 1.

Table 1. Characteristics and performance comparison

	Perkins	MANETconf	ODAA
Network organization	Flat/hierachical	Flat/hierachical	Flat/hierachical
State maintanence	Stateless	Stateful	Stateful
Approach	query-based DAD	global allocation state	Allocation function
Address conflict	Yes	Yes	No
Support for subnet merging	No	Yes	Yes
Address reclamation	Unneeded	Needed	Needed
Complexity	Low	High	Low
Communication overhead	High	High	Low
Latency	High	High	low
Scalability	Low	Low	High

4 Simulation

We have implemented query-based DAD [10] together with ODAA to compare their performance. The simulation is based on Scalable Wireless Ad hoc Network Simulator (SWANS version 1.0.6) [19]. In order to investigate how well ODAA scales, we varied the number of nodes to investigate the performance of large scale MANETs. Statistics about communication overhead and latency in both experiments were collected, which shows that ODAA outperforms query-based DAD.

4.1 Simulation Parameters

The random waypoint mobility model [20] was adopted in the simulation. Node speeds are randomly distributed between zero and the maximum speed 10 m/s. The pause time is consistently 30 seconds.

As stated before, if the probability that an unconfigured node becomes a domain initiator is too small, high latency is incurred. Otherwise, if the probability is too large, too many domains emerge, which weakens the performance of the algorithm. We set that the typical value of the probability is 0.001 and K is 12. Thus, a domain can potentially accommodate at most 4096 nodes. During the simulation, mobile nodes join the network every 30s randomly. When all nodes are configured, the simulation program stops running and prints the result.

4.2 Simulation Results

In the first experiment, I vary the number of nodes, from 200 to 300, within a fixed field (7000 x 7000 meters). The results are equal to our analysis before. The results show that the number of domains is about 2 and the duplicate probability of domain DID is very low to 8.88×10^{-16}, that indicates there are almost no duplicate address generation during our first two simulations.

Because every successfully received packet or sent packet, either unicast packet or broadcast packet, must have consumed bandwidth (and *power* as well), we use the number of packets as the evaluation metric for communication overhead.

Figure 3 shows the average number of packets, including what a single node receives or sends, with different node numbers. The number of packets generated in query-based DAD is about 14.7 times of that of ODAA on average. And as the node density increases, the communication overhead of query-based DAD increases because the link number increases. However, the communication overhead of ODAA decreases because the neighboring nodes become more so that the number of retries to obtain addresses decreases.

In query-based DAD, nodes participating in the allocation try a maximum of three times for broadcast of duplicate address detection packet. While in ODAA, except domain initiators, every node tries infinitely to broadcast query packets until it receives a reply from its configured neighbor. In order to be easy to compare, the intervals for both are set to be the same and we can use their retry times as the evaluation metric for latency.

Figure 4 shows the average retry times within different sizes of fields. As the density decreases, the retry times in ODAA increases because the neighboring nodes decreases.

Figure 5 shows the relationship of retry times and the node number. As the node number increases, the node density increases but the retry times in ODAA decrease.

In fact, the interval for multi-hop broadcast in query-based DAD usually should be much longer than that for one-hop broadcast in ODAA, however, it seems difficult to compute them in advance because of dynamic topology. Thus, that the retry times in ODAA is more than that in query-based DAD does not means that latency in ODAA must exceed that in query-based DAD. From the results in Figure 4 and Figure 5, we can arrive at conclusion that that the latency in ODAA depends on the node density but it is not sensitive to the node number while the latency in query-based DAD increases for large scale MANETs.

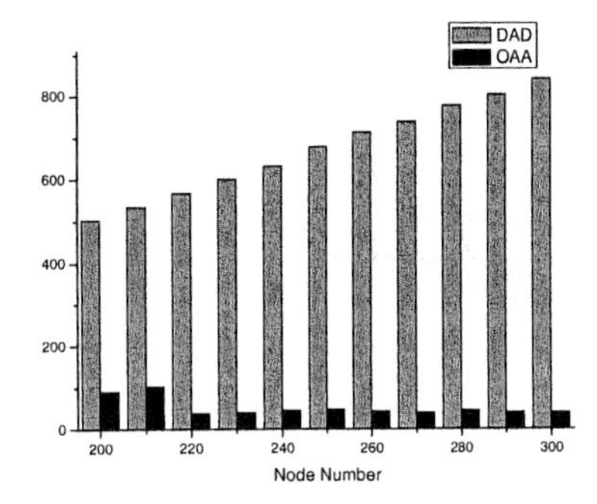

Fig. 3. Communication overhead for different node numbers

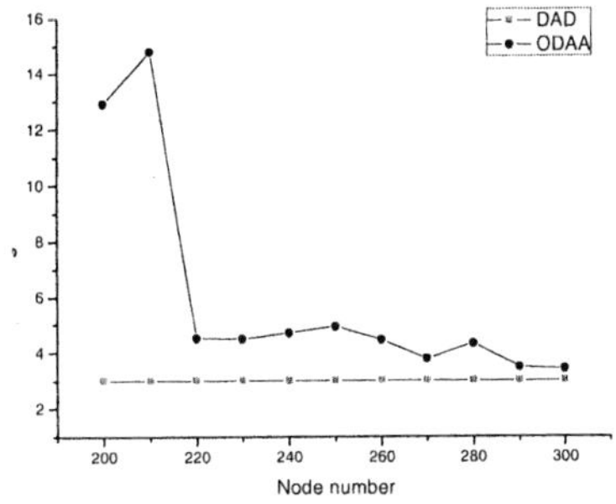

Fig. 4. Latency for different node numbers

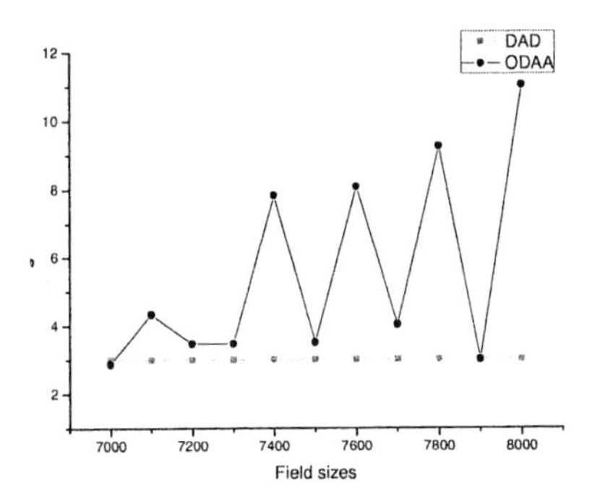

Fig. 5. Latency for 300 nodes

5 Conclusion and Future Work

We have presented an address autoconfiguration solution, which combines the enhanced binary split idea of Dynamic Address Allocation Protocol (DAAP) and the pseudo-random algorithm [15] to construct the interface ID of IPv6 address. Unlike query-based DAD algorithm, ODAA does not rely on the multi-hop flooding, so it can reduce the overhead of address configuration obviously. Each node runs locally a decision process to dispense address resources to other nodes, which may lead to the balanced distribution of address resources over the whole network. In the ideal case, an unconfigured node can obtain a new address from one of its neighboring nodes without retries after it switches to the ad-hoc mode. Furthermore, ODAA is not sensitive to both the network merger and the number of nodes and the allocation processes in different domains are parallel, so we believe that it may be suitable for large scale MANETs with low communication overhead, low latency, and high scalability.

A major issue that has been ignored in this paper is security. A faulty node perhaps can degrade the performance of address allocation, or even damage the consistency of address distribution. For instance, if a faulty node does not abide by the decision process or misbehaves in running the allocation function, a wrong address possibly generated by the node may impact many other nodes, which increases the probability of the address collision. This case may degrade the performance of address allocation though the mechanism reporting the error messages may correct the address collision at last. If an adversary node which knows the state of the system misbehaves deliberately, the robustness of ODAA may be suspicious in the worse case. Those problem illustrated above needs more research.

Acknowledgments

This paper is supported by the National Key Fundamental Research Program (2002cb312200), the National High Technology Research and Development Program of China (2002AA412010), and the National Natural Science Foundation of China (60174038).

References

1. K. Weniger and Z. Martina, "Mobile Ad Hoc Networks - Current Approaches and Future Directions", IEEE Network, Karlsruhe Univ., Germany, July 2004.
2. A. J. McAulley, and K. Manousakis: Self-Configuring Networks, MILCOM 2000, 21st Century Military Communications Conference Proceedings,Volume 1, Pages 315-319.
3. A. Misra, S. Das, A. McAulley, and S. K. Das: Autoconfiguration, Registration, and Mobility Management for Pervasive Computing, IEEE Personal Communications, Volume 8, Issue 4, August 2001, Pages 24-31.
4. Hongbo Zhou, Lionel M. Ni, Matt W. Mutka, "Prophet Address Allocation for Large Scale Manets," Proc. IEEE INFOCOM 2003, San Francisco, CA, Mar. 2003.
5. S. Nesargi and R. Prakash, "MANETconf: Configuration of Hosts in a Mobile Ad Hoc Network," Proc. IEEE INFOCOM 2002, New York, NY, June 2002.
6. M. Günes and J. Reibel, "An IP Address Configuration Algorithm for Zeroconf Mobile Multihop Ad Hoc Networks," Proc. Int'l. Wksp. Broadband Wireless Ad Hoc Networks and Services, Sophia Antipolis, France, Sept. 2002.
7. J. Boleng, "Efficient Network Layer Addressing for Mobile Ad Hoc Networks," Proc. Int'l. Conf. Wireless Networks, Las Vegas, NV, June 2002, pp. 271–77.
8. D. Simplot-Ryl, I. Stojmenovic, "Guest editorial - Ad hoc networking: data communications and topology control," IEEE Network, University of Lille, France, July 2004.
9. M. Mohsin and R. Prakash, "IP Address Assignment in a Mobile Ad Hoc Network," Proc. IEEE MILCOM 2002, Anaheim, CA, Oct. 2002.
10. C. Perkins et al., "IP Address Autoconfiguration for Ad Hoc Networks," IETF draft, 2001.
11. N. H. Vaidya, "Weak Duplicate Address Detection in Mobile Ad Hoc Networks," Proc. ACM MobiHoc 2002, Lausanne, Switzerland, June 2002, pp. 206–16.
12. K. Weniger, "Passive Duplicate Address Detection in Mobile Ad Hoc Networks," Proc. IEEE WCNC 2003, New Orleans, LA, Mar. 2003.

13. Y. Sun and E. M. Belding-Royer, "Dynamic Address Configuration in Mobile Ad Hoc Networks," UCSB tech. rep. 2003-11, Santa Barbara, CA, June 2003.
14. J. Jeong et al., "Ad Hoc IP Address Autoconfiguration", draft-jeong-adhoc-ip-*ADDR*-autoconf-04.txt, Feb. 2005.
15. R. Hinden, B. Haberman ,"Unique Local IPv6 Unicast Addresses", draft-ietf-ipv6-unique-local-*ADDR*-09.txt,Jan. 2005.
16. T. Narten et al., "Neighbor Discovery for IP Version 6 (IPv6)", draft-ietf-ipv6-2461bis-02.txt, Feb. 2005.
17. R.Droms et al., Dynamic Host Configuration Protocol for IPv6 (DHCPv6), Network Working Group RFC 3315 ,Jul. 2003.
18. Hongbo Zhou, Matt W. Mutka, Lionel M. Ni, "IP Address Handoff in the MANET", Proc. IEEE INFOCOM 2004, Mar. 2004.
19. R. Barr, JiST-Java in Simulation Time: User Guide and Tutorial. Mar. 2004.
20. J. Broch et al., A performance comparison of multi-hop wireless ad hoc routing protocols, in: Proceedings of the Fourth Annual ACM/IEEE Inter-national Conference on Mobile Computing and Networking, October 1998, pp. 85–97.
21. Eastlake, D. 3rd, S. Crocker, J. Schiller, "Randomness Recommendations for Security", RFC 1750, December 1994.

Boundary-Based Time Partitioning with Flattened R-Tree for Indexing Ubiquitous Objects*

Youn Chul Jung, Hee Yong Youn**, and Eun Seok Lee

School of Information and Communications Engineering,
Sungkyunkwan University, Suwon, Korea
{Jimmy4u, youn, eslee}@ece.skku.ac.kr

Abstract. The advances of wireless communication technologies, personal locator technology, and global positioning systems enable a wide range of location-aware services. To enable the services, a number of spatiotemporal access methods have been proposed for handling timestamp and time interval queries. However, the performance of the existing methods of a single index structure quickly degrades as time progresses. To overcome the problem, we propose to employ time-based partitioning on the R-tree called time boundary-based partitioning with flattened R-tree (BPR-Tree). The proposed scheme employs a new insertion policy to reduce the height of the tree and a time grouping method in order to minimize the search time of various queries. Extensive computer simulation reveals that the proposed scheme significantly outperforms the existing schemes.

Keywords: Indexing moving object, location based service, spatiotemporal database, time stamp and interval query, R-Tree, ubiquitous computing.

1 Introduction

The advances of wireless communication technologies, personal locator technology, and global positioning systems (GPS) enable a wide range of location-aware services. These are the key technologies allowing ubiquitous location-aware environments and services. Such services have the potential of improving the quality of life by adding location-awareness to virtually all objects of interests such as humans, cars, laptops, and buildings, etc. In addition, due to the progress of wireless networking with GPS and wide spread of personal communication devices such as mobile phone, concerns on contents services based on user location are increasing. The contents services are mainly based on location based services (LBS). In order to quickly search for the requested locations for a variety of services, the DBMS system in ubiquitous environment needs an access method capable of efficiently searching a large amount of objects.

---

* This research was supported in part by the Ubiquitous Autonomic Computing and Network Project, 21st Century Frontier R&D Program in Korea and the Brain Korea 21 Project in 2005.

** Corresponding author.

X. Jia, J. Wu, and Y. He (Eds.): MSN 2005, LNCS 3794, pp. 804 – 814, 2005.

The most recently developed spatiotemporal access methods such as HR-Tree [1], MV3R-Tree [3], 3D R-tree [4], TB-Tree, and STR-Tree [4] are for indexing the timestamp and time interval queries of moving objects. However, the spatiotemporal access methods can neither efficiently support a variety of queries nor reflect the change in the location of the objects. Accordingly, these spatiotemporal access methods are not suitable to the location aware systems in ubiquitous environment. As another access method, the TPR-Tree (Time-Based Partitioning R-tree) [12] was proposed which consists of multiple trees of two dimensions. As time passes, however, the height of the 2D R-tree of the TPR-tree gradually grows and dead space is generated. This causes the performance of the TPR-tree to degrade and thus it is not suitable for huge ubiquitous database system.

To solve the problems, we focus on reducing the cost of various queries, i.e. timestamp and time interval query in this paper. The proposed scheme also employs a new insertion policy to reduce the height of the tree in order to maximize the performance. Also, it employs a time grouping method in order to minimize the search time of various queries. The proposed scheme with the multiple-indexing and partitioning approach is called boundary-based time partitioning with flattened R-tree (BPR-Tree). It efficiently fits various queries (i.e. timestamp query and interval query) from the given spatial region of a fixed network and allows flattened tree. Computer simulation reveals that the proposed scheme outperforms other access methods for various combinations of queries, while the improvement is more substantial for time stamp query.

The rest of the paper is organized as follows. Section 2 reviews the existing access methods for timestamp and interval query. Section 3 describes the structure of the proposed BPR-Tree, and the insertion and search operation with the proposed structure are demonstrated. Section 4 presents the results of performance comparison. Finally, we conclude the paper in Section 5.

2 Related Works

In this section we introduce four typical access methods – HR-Tree, 3D R-Tree, MV3R-Tree, and TPR-Tree.

2.1 3D R-Tree

The 3D R-tree [4] simply considers time as another dimension of the R-tree. Whenever an object moves to another position or changes its shape, a new minimum bounding region (MBR) is created to represent the change of the object. The MBR containing the spatial extent and lifespan is inserted in the 3D R-tree.

Figure 1 shows an example of the 3D R-tree. In this figure, R_0, R_1, R_2, R_3, and R_4 represent the MBRs describing the movements of the objects, and R_1 contains R_3 and R_4. As time passes, thus, the time region enlarges accordingly. This causes the performance of the 3D R-tree to degrade because a single R-tree keeps the whole time region in the 3D R-tree. Also, long-lived records result in a huge dead space not covering any record but existing as a part of the MBR. Therefore, they degrade the performance of query with the 3D R-tree. However, the 3D R-tree takes a minimum space since it has no duplicate data.

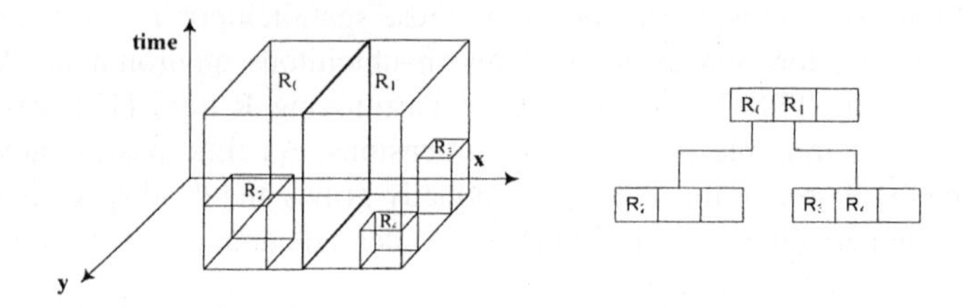

Fig. 1. The structure of the 3D R-Tree

2.2 MV3R-Tree

The MV3R-tree [3] combines the Multi-Version R-tree (MVR-tree) and a small auxiliary 3D R-tree built at the leaf nodes of the MVR-tree. The former is used to answer timestamp and short interval queries and the latter is to answer long interval queries. Figure 2 illustrates the structure of the MV3R-tree.

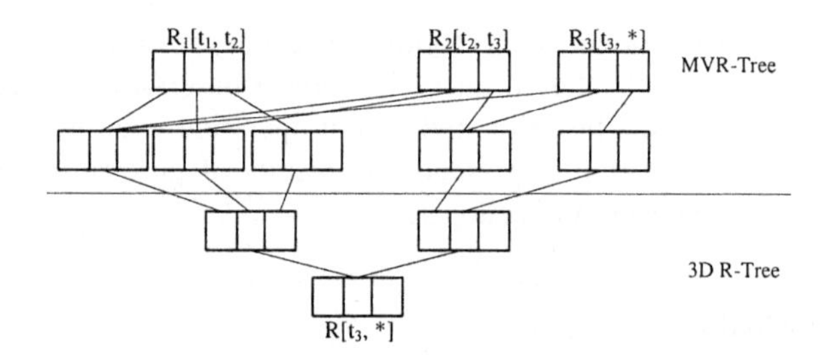

Fig. 2. The structure of the MV3R-Tree

As shown in the figure, the MVR-tree consists of multiple R-trees that have their own jurisdiction intervals. For example, $R_1[t_1,t_2]$ deals with the records whose time intervals belong to $[t_1,t_2]$, while $R_2[t_2,t_3]$ manipulates the records whose time intervals belong to $[t_2,t_3]$. Although the size of the auxiliary 3D R-tree is very small, it shares the leaf nodes of the MVR-tree. As a result, it improves the performance of interval queries and provides flexibility to the algorithm in processing other spatial queries such as join and k-nearest neighbor.

2.3 HR-Tree

The Historical R-tree (HR-tree) [1,2] creates an R-tree whenever the objects in the previous R-tree change their positions or shapes. Here common branches of consecutive R-trees are stored only once in order to save the space.

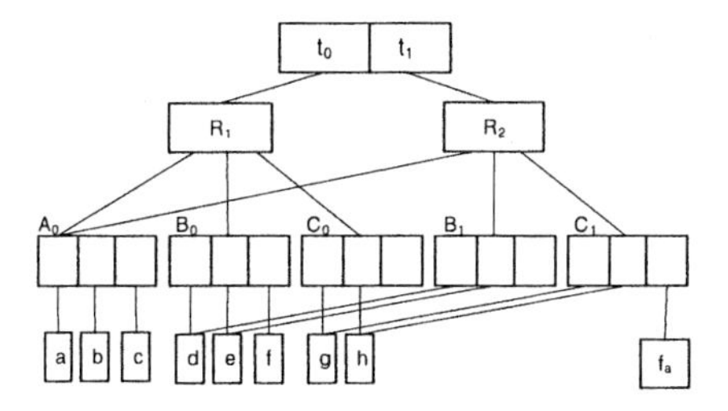

Fig. 3. The structure of the HR-Tree

The timestamp query is directed to the corresponding R-tree and search is performed only inside the tree. Thus, the timestamp query becomes an ordinary window query and it can be handled very efficiently. The interval query should search the corresponding R-trees of all the timestamps involved. Even when only one object has changed its position, the HR-tree may update the nodes contained in the path between the leaf node corresponding to the object and the root node. Therefore, the size of the HR-tree is several times larger than that of the 3D R-tree. Figure 3 shows an example of the HR-tree. Here $R_1[t_0,t_1]$ deals with the records whose time intervals are contained in time interval $[t_0,t_1]$, and $R_2[t_1, t_2]$ manipulates the records whose time intervals are contained in $[t_1,t_2]$. Suppose that object f changes its position to f_a from B_0 to C_1 at $[t_2,t_3]$. The three nodes B_0, B_1, and C_1 which are associated with f and f_a are updated and copied to node C_1.

2.4 TPR-Tree

TPR-Tree (Time-Based Partitioning R-tree) [12] consists of a single 2D R-Tree and multiple 1D R-Trees as shown in Figure 4. The 2D R-Tree is used to index the spatial data of the spatial region. The structure is similar to the original R-Tree [9] and each leaf node of the 2D R-Tree contains the pointers pointing the root of each 1D R-Tree

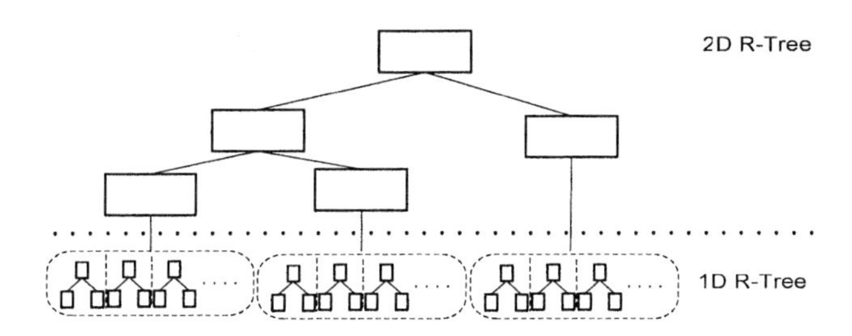

Fig. 4. The structure of the TPR-Tree

on the given spatial data of the region. Consequently, each leaf node of the 2D R-tree corresponds to a 1D R-tree. Each of the 1D R-Tree is responsible for a group of fixed time interval derived from various queries. Also, each 1D R-Tree is generated according to the time intervals and used to index various queries on the given spatial region.

It efficiently fits various queries (i.e. timestamp query and interval query). The TPR-Tree can be considered as a single 2D R-tree on top of multiple 1D R-trees.

As time passes, however, the height of the 2D R-tree grows and dead space is generated. Also, a 1D R-Tree is formed for a the time interval on the given spatial region. This causes the performance of the TPR-tree to degrade.

3 The Proposed Scheme

In ubiquitous environment the DBMS system needs an access method capable of efficiently searching, updating, and modifying the large amount of ubiquitous objects. However, most spatiotemporal access methods consist of a single index structure covering the entire time span of the network, and this degrades the performance as time progresses. Accordingly, they are not suitable to ubiquitous environment that has a large amount of continuously moving objects. To solve the problem, we propose the boundary-based time partitioning with flattened R-Tree (BPR-Tree) scheme. We first introduce the structure of the BPR-Tree. We then describe the insertion and search algorithm based on the BPR-Tree.

3.1 The Structure of the BPR-Tree

The structure of the BPR-tree is similar to that of the TPR-tree. It consists of a single 2D R-Tree and multiple 1D R-Trees. The 2D R-Tree is used to index the spatial data of the spatial region. Each 1D R-Tree is generated according to the time interval and they are grouped by the time intervals. It is used to index various queries on the given spatial region. The TPR-Tree [12] does not employ grouping policy based on time interval.

Figure 5 illustrates the structure of the 1D R-tree in the proposed BPR-Tree. Each leaf node entry of the 1D R-Tree is of the form <ID, MBR, GID, (t_{start}, t_{end})>. Here ID is the object identifier, and MBR covers the boundaries of the children nodes. GID is the group identifier of the time intervals combined together. (t_{start}, t_{end}) are the times when the object was inserted and deleted, respectively. If an entry has not yet been logically deleted, t_{end} is marked as "*". The time region boundary designates the life-span of a record of each 1D R-Tree, and L_i designates the interval of R_i.

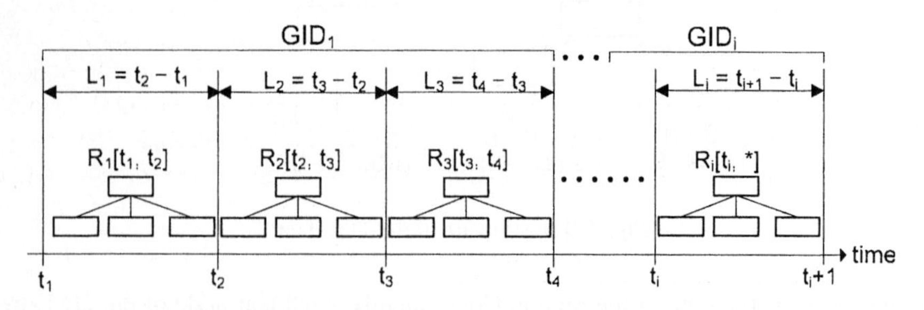

Fig. 5. The structure of the 1D R-Tree in the proposed BPR-Tree

We apply time partitioning of the records with time boundaries and grouping of several time intervals in the 1D R-Tree, which significantly enhance the performance of time queries. As a result, the BPR-Tree contains multiple 1D R-Trees in a given MBRs on a 2D R-Tree. The 1D R-Tree can dynamically support various queries and efficiently reflect the change of queries using the time based grouping. It outperforms other methods for time interval query as shown later. The proposed scheme is similar to the TPR-Tree [12]. However, since it applies grouping of time intervals, the proposed approach more efficiently fit ubiquitous environment and its applications than the TPR-tree scheme.

3.2 Insertion and Search with the BPR-Tree

In the 2D R-Tree, we use the R-Tree algorithm [10] for insertion, update, and search. Each leaf node of the 2D R-tree has a pointer that points the root node of the 1D R-Tree in the given network. Figure 6 shows the insertion and search procedure with the BPR-Tree.

The insertion and Search Procedure with the BPR-Tree

---

1. Execute the R-Tree search algorithm in the 2D R-Tree
2. Find the line segment leaving the moving object.
3. If the line segment is found, locate the leaf node that contains the given line segment and the corresponding 1D R-Tree
4. Execute insertion and search algorithm in the 1D R-Tree

Fig. 6. Insertion and search procedure with the BPR-Tree

To process the insertion and search operation, we use the search algorithm of the R-tree which is executed in four steps. An algorithm selects a leaf node in which a new index entry is placed. In case an insertion operation causes a node split, one of the three alternative split algorithms (Exhaustive, QuadraticSplit, and LinearSplit) is executed [10]. Here the algorithms are based on the least enlargement criterion, and they were designed to serve the spatial data inserted in the R-Tree. Accordingly, the proposed 2D R-Tree with the BPR-Tree formed using the algorithms usually renders small overlap between the nodes. If a line segment leaving the moving object is found, the leaf node containing the line segment and the corresponding 1D R-Tree are located. The insertion and search algorithm described in the next section are then executed.

3.3 Insertion and Search with the 1D R-Tree

When an insertion operation occurs at a specific spatial region with the BPR-Tree, the following steps are executed. In the 1st step, the R-Tree insertion algorithm is executed in the 2D R-Tree in order to insert the MBR covering the specific region of the leaf node of the 2D R-Tree. It then follows the pointer to the 1D R-Tree and finds the 1D R-trees with suitable GID_i and time interval $[T_{start}, T_{end}]$. The record is inserted into the right-most leaf node of the 1D R-tree. Accordingly, the 1D R-tree is kept as flat as possible. In the 2D R-Tree, we use the R-Tree algorithm [9] to index a spatial region. The next step is to execute the 1D R-tree algorithm.

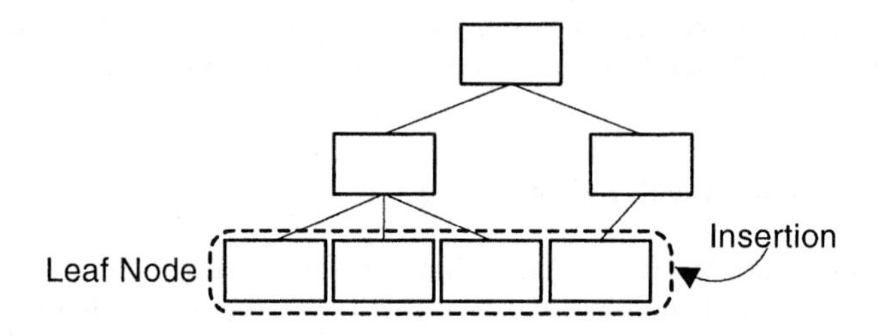

Fig. 7. The insertion policy with the 1D R-Tree

Figure 7 illustrates the insertion operation with the 1D R-tree. Here, every new object is inserted in the left-most leaf node of the 1D R-tree. When a leaf node of 1D R-tree is full, a new node is created at the right side of it and the new entry is inserted in that node. This approach of insertion makes the 1D R-tree have fully filled leaf nodes and very low overlapping. As a result, the structure of the 1D R-tree is flattened the height of the tree becomes low, which results in fast query operation.

```
Procedure insertion
 /*Find GIDi which contains the time interval */
 for GIDi(interval length){
    /*Find Ri which contains the record, and insert it
     into Ri*/
    for Ri(ti, ti+1){
        if (ti, ti+1) contains current time interval{
          insert record to Ri into right-most leaf node;
          break;
    }
    /* the logical deletion */
    if (current record=(ID, MBR, GIDi,[ti+1, *])){
        previous record=[ti, *] is changed into
        previous record=[ti, ti+1];
 }
    /*record intersect between time intervals*/
    for Ri(ti, ti+1){
       if(ti, ti+1) intersect the lifespan of the record{
           insert record to suitable time interval's Ri
          into right-most leaf node;
       }
    }
 }
```

Fig. 8. The insertion algorithm with the 1D R-Tree

Figure 8 shows the insertion algorithm for a record having GID_i as its group identifier with the 1D R-Tree. The algorithm of the 1D R-Tree employs the concept of time boundary and grouping of time intervals. Therefore, if the lifespan of a record does not cross the boundary, the record is inserted in the node of the R-Tree covering the lifespan of the record. However, if the record crosses the time intervals, the record is inserted in both the nodes in the 1D R-tree.

4 Performance Evaluation

This section describes the simulation environment and provides the results of computer simulation.

4.1 Simulation Environment

In our experiment we use a 2.4GHz Pentium IV machine with 1GB of main memory. Due to unavailability of movement data of actual objects, we use a synthetic dataset generated by the GSTD methods [8], which has been widely employed as a benchmarking environment for studying the access methods handling moving points and regions. It was designed to generate realistic data for the evaluation of database algorithms for indexing and storing dynamic location data.

The number of moving objects is 1,000 in the experiment. Spatial areas of queries are 1%, 3%, 5%, 7%, and 9% of the entire space. The simulation evaluates the cost of queries of combined timestamp and interval time in terms of the number of nodes accessed.

4.2 Simulation Results

Five workload data of different combination of timestamp and interval queries are used for simulation as shown in Table 1. Figure 9 shows the performance of the access methods compared for various queries of time interval and timestamp in terms of the number of nodes accessed. Here we compare the TPR-tree, MV3R-tree, 3DR-Tree, and the proposed BPR-Tree scheme.

Table 1. The workload data of timestamp and time interval queries

Workload	Timestamp	Time interval
1st Workload	100%	0%
2nd Workload	75%	25%
3rd Workload	50%	50%
4th Workload	25%	75%
5th Workload	0%	100%

As shown in Figure 9, the proposed BPR-Tree outperforms other methods for various queries of timestamp and interval time. Notice that performance of the 3D R-tree declines with timestamp query. This is because a single R-Tree keeps the entire time region in the 3DR-Tree while the time region gets large as time passes. The 3DR-Tree takes smallest space since it has no duplicate data. On the contrary, the

MV3R-Tree outperforms the 3DR-Tree. The TPR-Tree and BPR-tree display higher performance for timestamp queries than for time interval queries as other methods. Since they consist of multiple indexing trees and 1D R-Tree with the boundary-based partitioning method, it can dynamically fit various queries on the given interval time. Observe from the plots that the BPR-tree is more efficient than the TPR-tree for various workload data. This is because the 1D R-tree of the BPR-Tree uses time grouping technique and a new insertion policy. Also, it reduces dead space in the 1D R-Tree.

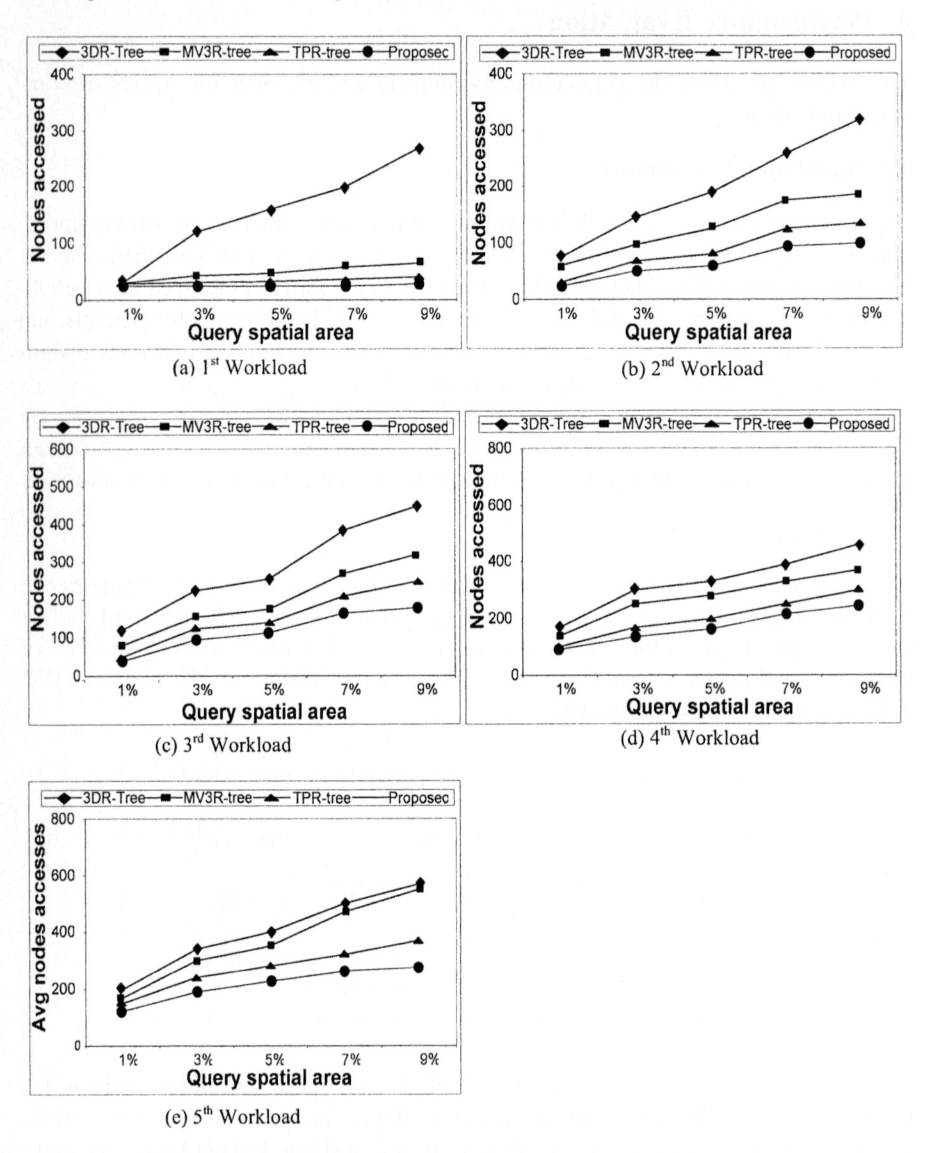

Fig. 9. The comparison of the costs of query of the four methods

5 Conclusion

Most existing spatiotemporal access methods consist of a single index structure covering the entire time span of the network. These index structures are not suitable to location awareness enabling the LBS system in ubiquitous environment. The performance may decrease as time passes since the size of indexed time gets very large. In this paper we have applied record partitioning with time boundary and grouping of time intervals to the 1D R-Tree, which splits the 1D R-tree according to the grouping of time intervals. We have also proposed a method consisting of multiple trees of two dimensions with boundary-based time partitioning with flattened 1D R-tree for indexing. It can efficiently handle various queries (i.e. timestamp query and interval query) from the given spatial region. The time interval length of the BPR-Tree well fits timestamp and interval of the query. Computer simulation reveals that the proposed BPR-Tree outperforms other methods for various combinations of queries of timestamp and interval time. The proposed approach is more suitable to ubiquitous environment and its applications than the earlier spatiotemporal methods. We will further investigate the performance of the proposed approach for other operational environment.

References

[1] Nascimento, M.A., Silva, J.R.O.: Towards historical R-Tree. In Proceesings of the 13th ACM Symposium on Applied Computing (ACM-SAC'98), 235-240, 1998.

[2] Nascimento, M.A., Silva, J.R.O, and Theodoridis, Y.: Evaluation for access Structures for discretely moving points. In Proceesings of the International Workshop on Spatio-Temporal Database Management (STDBM'99), Edinburgh, Scotland, 171-188, 1999.

[3] Tao, Y., and Papadias, D.: Mv3R-tree: a spatiotemporal access method for timestamp and intervalqueries. In Proceedings of the 27th International Conference on Very Large Databases, 431-440, 2001.

[4] Theodoridis, Y., Vazirgiannis, M., and Sellis, T.: Spatio-temporal Indexing for Large Multimedia Applications. In Proceedings of the 3rd IEEE Conference on Multimedia Computing and Systems, Hiroshima, Japan, 441-448, 1996.

[5] Pfoser D., Jensen C.S., and Theodoridis, Y.: Novel Approaches to the Indexing of Moving Object Trajectories. In Proceedings of the 26th International Conference on Very Large Databases, Cairo, Egypt, 395-406, 2000.

[6] Elias frentzos.: Indexing object moving on fixed networks. Advances in Spatial and Temporal Databases: 8th International Symposium, SSTD 2003 Santorini Island, Greece, 289-305, July 24-27, 2003 Proceedings.

[7] Brinkhoff, T.: Generation Network-Based Moving Objects. In Proceedings of the 12th Int'l Conference on Scientific and Statistical Database Management, SSDBM'00, Berlin, Germany, 253-255, 2000.

[8] Y.Theodoridis, J. R. O. silva, and M.A Nascimento: On the Generation of Spatiotemporal Datasets. SSD, 147-164, 1999.

[9] A.Guttman: R-Tree: A Dynamic Index Structure for Spatial Searching. ACM SIGMOD, 47-57, 1984.

[10] Pfoser, D., and Jensen, C.S.: Querying the Trajectories of On-Line Mobile Object. TIMECENTER Technical Report, TR-57, 2001
[11] Hadjieleftheriou, M., Kollios, G., Tsotras, V.J., Gunopulos, D.: Efficient Indexing of Spatiotemporal Objects. In Proceedings of 8th International Conference on Extending Database Technology (EDBT), Prague, Czech Republic, 251-268, 2002
[12] Jung, Y.C., Youn, H.Y., Kim, U.: Efficient Indexing of Moving Objects Using Time-based Partitioning with R-Tree: International Conference on Computational Science (ICCS), Atlanta, GA, USA, 2005

Authentication in Fast Handover of Mobile IPv6 Applying AAA by Using Hash Value

Hyungmo Kang and Youngsong Mun

School of Computing, Soongsil University
kangzzang@sunny.ssu.ac.kr, mun@computing.ssu.ac.kr

Abstract. The miniaturization of the computing machinery increases needs for using mobile computing of users. In this situation, Mobile IPv6 Working Group of Internet Engineering Task Force (IETF) suggests IPSec for Mobile IPv6 security. IPSec, however, has a long latency problem of dynamic security authentication due to the key management and a large amount of calculation. We therefore apply AAA in Mobile IPv6 to solve a dynamic security authentication problem of IPSec and simultaneously use Fast Handover technique for quick mobility. And we propose a method for reducing a long latency caused by authentication of mobile node (MN) during a handover. Old and new access routers authenticated by a hash value of an MN are able to reuse existing Security Authentication (SA) for a certain time and reduce latency for new SA. As a consequence, a mobile node can provide the real time services due to the reduction of authentication latency. Also, we can trust the buffered information through the tunnel between two access routers.

1 Introduction

The development of wireless communication and user's needs for new internet technique have improved a wireless mobile internet environment. While accomplishing the rapid growth of the mobile Internet technique, it is exposed to a lot of possibilities of attack. In this situation, Mobile IPv6 Working Group of Internet Engineering Task Force (IETF) suggests IPSec [2] for Mobile IPv6 [1] security. IPSec [2], the message authentication method that establishes Security Authentication (SA) [3], has strong authentication ability. But, in foreign network, bootstrapping causes IPSec to increase latency due to a lot of message switching. To solve bootstrapping problem, Mobile IPv6 Working Group suggests that Authentication, Authorization, Accounting (AAA) [4] as Mobile IPv6 security technique [5]. Although AAA infrastructure performs strong authentication functions, AAA has to do a lot of message switching to authenticate mobile node (MN) every movement. The latency happens due to lots of message switching and it is interfered with realization of real-time communication.

In this paper, we apply Dupont AAA [6] to using AAA infrastructure and Fast Handover [7] to minimizing the movement latency. Because the mobile node needs new SA that causes latency during handover, the aim of our proposal is to deduce the latency problem in handover situation and to solve it. The proposed method is not to use new SA after handover immediately, but it is to use the existing SA for a certain

X. Jia, J. Wu, and Y. He (Eds.): MSN 2005, LNCS 3794, pp. 815 – 824, 2005.

time until completion of new SA and then exchanges existing SA with new SA. For this, mobile node must have the characteristic hash value between old access router (PAR) and new access router (NAR). As a result, authentication latency is deleted by this proposed method and the reliance is also assured between PAR and NAR.

In section 2, in this paper, we analyze the scheme and procedure of a concurrent method [8] of Fast Handover [7] and Dupont AAA [6]. We introduce the proposed method and procedure in section 3. In section 4, we verify the performance improvement by cost comparison with other methods. In section 5, we reach a conclusion.

2 Related Works

2.1 A Concurrent Method [8] of Fast Handover [7] and Dupont AAA [6]

Fast Handover [7] supports the combination with AAA [4], [5] in Mobile IPv6 [1]. However, a general Fast Handover and AAA method causes long latency because it applies the AAA authentication after performing all handover process. To solve this latency problem, in reference [8], it suggests the simultaneous accomplishment of the Fast Handover and AAA authentication. Before the L2 handoff happens, a mobile node performs a series of the AAA authentication procedure in advance and subsequently it makes a reduction of the latency.

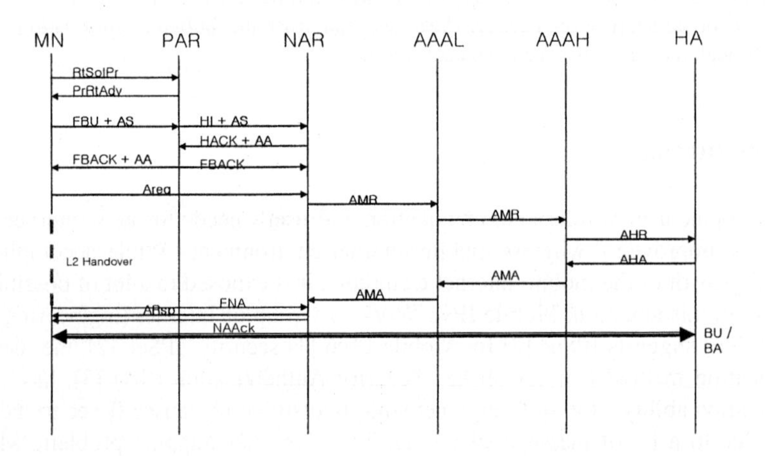

Fig. 1. When MN receives FBACK message before movement (Predictive mode in Fast Handover), the method [8] performs the rest of Dupont AAA procedures during L2 Handoff. Due to the simultaneous accomplishment of Hast Handover and Dupont AAA, the method [8] reduces latency.

Fig. 1 and 2 show the method in the reference [8] that performs simultaneously Fast Handover and Dupont AAA. After RtSolPr and PrRtAdv message exchange, MN sends FBU message including AS message to the PAR. According to this step, AAA authentication procedure starts in advance. AS message included into FBU and HI are sent to the NAR. In response to AS message, NAR sends HACK message including

AA message to PAR. At this moment, tunnel is created and PAR receives HACK message and then PAR sends FBACK message including AA message to the MN and NAR. If a MN receives FBACK message before movement, it sends AReq message to the local AAA server immediately and executes the AAA authentication procedure. During L2 Handoff time, a series of AAA authentication procedure are performed and the rest of AAA authentication procedure is performed after MN moves.

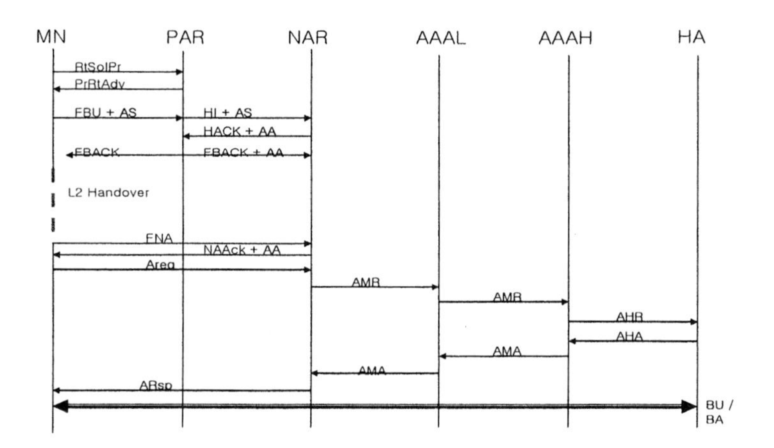

Fig. 2. When MN does not receive FBACK message before movement (Reactive mode in Fast Handover), the method [8] performs the rest of Dupont AAA procedures After movement. It makes latency.

2.2 Problem of Current System

If an MN performs the Fast Handover and Dupont AAA simultaneously, it is able to reduce the latency rather than sequentially performing two processes. But, if MN moves quickly and frequently, this method has a latency problem caused by authentication time and a cost problem of generating unnecessary messages. Also, the improvement of L2 Handoff and long distance between AAAL and AAAH cause the authentication latency. In these situations, it is difficult for MN to support the real time communication. To succeed in a real time communication and continuous connection, we have to analyze the reasons of problems and find the solution to minimize the latency.

3 Proposed Scheme

3.1 Proposed Method

Proposed method is to extend the reliable region (authenticated sections by AAA) from HA-PAR-MN to HA-PAR-NAR-MN while simultaneously performing Fast Handover and Dupont AAA. In other words, when exchanging the RtSolPr and PrRtAdv message in Fast Handover situation, MN creates characteristic value that is based on its own information between PAR and NAR. And, MN using the characteristic value authenticates the section between PAR and NAR. As a result, the reliable region is extended to the NAR by the characteristic value. To make a Hash value (H_value), MN uses NAI, HoA, HA, OCoA (old), NCoA (new), and aaa_key. Before movement, H_value generated by MN is sent to NAR through PAR and registered by NAR using MN's home address. NAR confirms MN's movement by FBACK message that contains MN's home address, and then set H_value flag of MN. After movement of MN, H_value recreated by MN is directly sent to the NAR and NAR checks the former H_value with latter H_value and H_value flag.

H_value = First (64, HMAC_SHA1(NAI,(HoA|OCoA|NCoA|HA|aaa_key)))

If authentication between the MN and NAR is completed, the scope of authentication extends to the NAR, and then MN is able to reuse existing SA (SA^+) for a while. Because the MN uses SA (SA^+) for a certain time until the new SA is establish, it is possible to make a continuous connection. (A certain time is average time about a vehicle moving quickly among routers.) If new SA fails or MN's life time of SA or SA^+ expires, MN stops its communication and requests re-authentication. If new SA succeeds while using the SA^+, new SA exchanges SA^+ and continues the communication. Authentication between PAR-NAR and MN is able to trust the information in the PAR-NAR tunnel. After H_value authentication procedure, NAR deletes MN's H_value registered by NAR to protect replay attack.

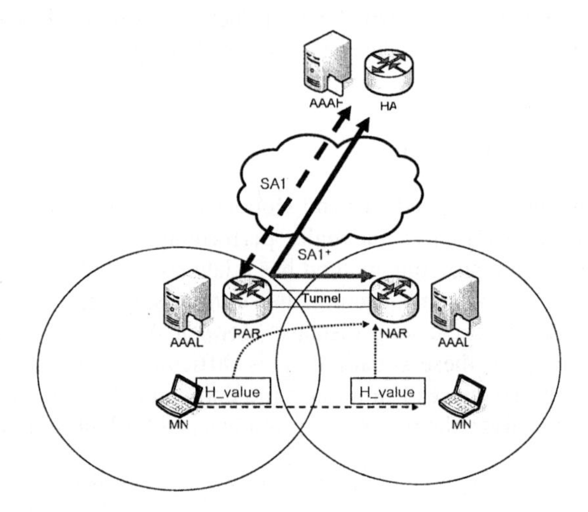

Fig. 3. Proposed scheme uses H_value. In PAR, MN first generates H_value and forwards it to NAR. After moving to NAR, MN regenerates H_value and compares with previous MN's H_value.

3.2 Procedure of Proposed Method

After exchanging the RtSolPr and PrRtAdv message, MN sends FBU message that contains AS message and H_value. PAR also sends that HI message contains AS

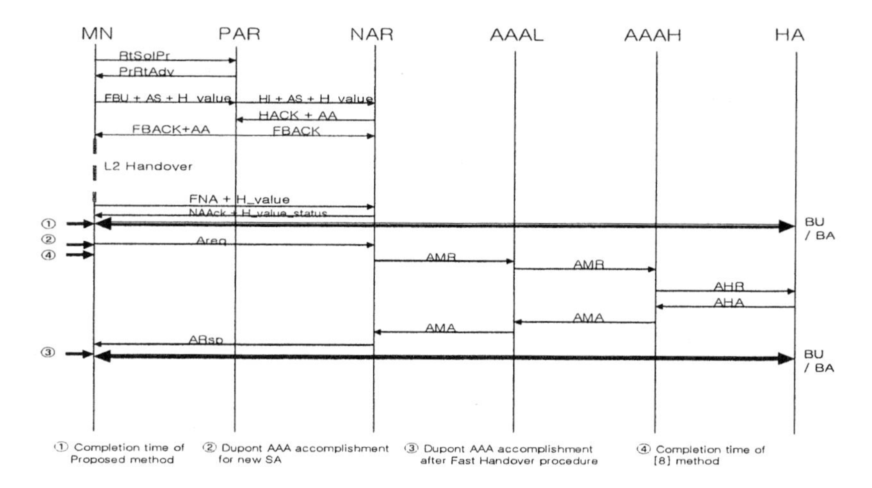

Fig. 4. When MN receives FBACK message before movement (Predictive mode in Fast Handover), it shows the proposed procedure. Arrows show a speed difference among the methods. The proposed method performs new SA after completing connection. FBU and FNA have to include H_value to authenticate MN.

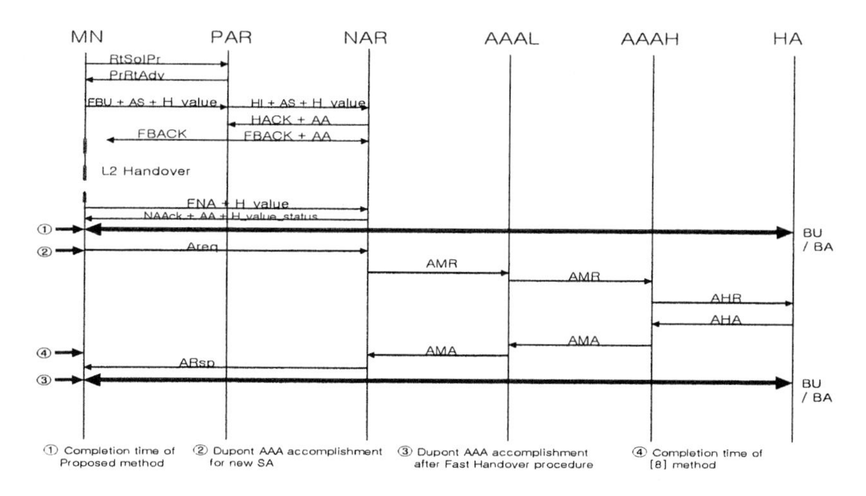

Fig. 5. When MN does not receive FBACK message before movement (Reactive mode in Fast Handover), it shows that the proposed procedure. Arrows show a speed difference among the methods. The proposed method has little latency.

message and H_value. NAR receives HI message, and then it registers H_value in a home address of MN. NAR confirms MN's movement by FBACK message that contains MN's home address, and then set H_value flag of MN. After L2 handoff, MN connects to the NAR and re-transmits H_value by FNA message. NAR receiving the FNA message compares re-generated H_value with registered H_value and check a flag. If the result of the comparison proves the H_value and flag to be right,

H_value authentication between PAR and NAR is completed. Since both HA-PAR section and PAR-NAR section are secure, existing SA (SA^+) is able to expand from HA-PAR to HA-PAR-NAR for a certain time. The proposed method extends time of using existing SA (SA^+) and it also tries to attempt new SA for MN's security. New SA is separately performed by Dupont AAA procedure during communication. If new SA is established, it exchanges a SA^+ and continues communication. In case of Fig. 5, comparison methods is difficult to continue the communication due to the latency caused by new SA establishment. But, proposed method is able to keep the continuity of communication.

4 Performance Evaluation

4.1 Cost Analysis of Proposed and Comparison Methods

This paper refers to the approach method in reference [9], [10] to evaluate a performance of a proposed system method. Fig. 6 shows the distance of entities.

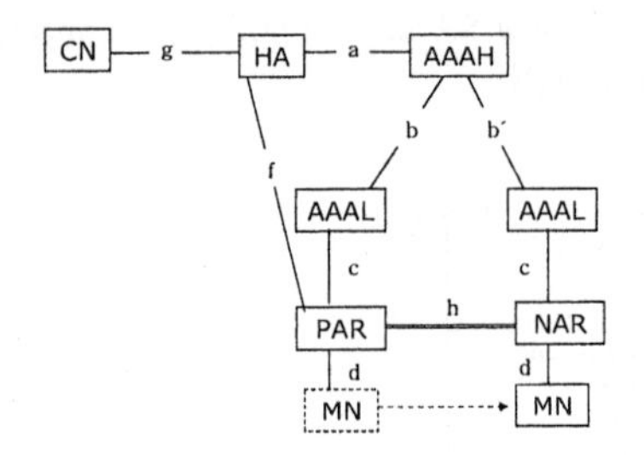

Fig. 6. System method for performance analysis

We define that a correspond node (CN) generates data packets destined for a mobile node (MN) at a mean rate λ, and a MN moves from one subnet to another at a mean rate μ. We also define the mean number of packets received by a MN from a CN per movement as Packet to Mobility Ratio (PMR $p = \lambda / \mu$), the each average packet length $l_c(control)$ (=200byte) , $l_d(data)$ (=1024 byte), and their ratio $l = l_d / l_c$. We assume the cost of processing control packets as r (all hosts).

In this paper, we compare a performance of proposed method with comparison methods and do not assume a fail in Fast Handover. The First comparison method is that performs Dupont AAA after Fast Handover procedure and the second is that refer to the method in reference [8].

The performance analysis of the first method shows below: When the MN moves a new domain in Mobile IPv6, its total cost is C_{total} (1).

$$C_{total} = C_{signal} + C_{packet} \tag{1}$$

Signaling cost consists of the cost of Fast Handover, AAA and registration with HA.

$$C_{signal} = M_{fast} + M_{AAA} + m = 2(a+b'+c) + 10d + 3h + m + 18r \quad (2)$$

M_{fast} and M_{AAA} represent the signaling cost of Fast Handover and AAA. m is a message cost for a registration with HA. Packet cost is C_{packet} (3) that consists of the packet loss cost (C_{loss}) and packet forwarding cost (C_{fwd}).

$$C_{packet} = C_{fwd} + C_{loss} = (\lambda \times t_{delay} \times C_{dt}) + C_{loss} \quad (3)$$

t_{delay} (4) is forwarding latency from receiving FBU message to finishing authentication.

$$t_{delay} = 2(t_a + t_{b'} + t_c) + 7t_d + 3t_h + 15t_r + t_{L2} + t_m \quad (4)$$

L2 handoff latency is (t_{L2}), the movement time of hops (t_x) and packet processing time on host (t_r). C_{dt} (5) is a single data packet cost delivered from CN to MN.

$$C_{dt} = l \times (g + f + h + d) + 3r \quad (5)$$

As s result, the total cost of AAA after Fast Handover procedure in Mobile IPv6 is

$$C_{total} = 2(a+b'+c) + 10d + 3h + m + 18r + (\lambda \times t_{delay} \times C_{dt}) + C_{loss} \quad (6)$$

The performance analysis of the second method (in reference [8]) shows below. It is divided by two situations.

- When MN receives FBACK message (Predictive),

$$C_{signal-p} = 2(a+b'+c) + 8d + 4h + m + 16r \quad (7)$$

$$t_{delay-p} = 2t_d + 3t_h + 4t_r + \mathrm{MAX}\{\ 2(t_a + t_{b'} + t_c) + t_d + 7t_r),\ (t_{L2} + 2t_d + 2t_r)\ \} + t_m \quad (8)$$

The latency of from sending AReq message to receiving ARsp message selects the maximum value between AAA authentication procedures:

$2(t_a + t_{b'} + t_c) + t_d + 7t_r$,and L2 Handoff : $t_{L2} + 2t_d + 2t_r$.

- When MN does not receive FBACK message (Reactive),

$$C_{signal-a} = 2(a+b'+c) + 8d + 3h + m + 16r \quad (9)$$

$$t_{delay-a} = 2(t_a + t_{b'} + t_c + 2t_d) + 3t_h + 13t_r + t_{L2} + t_m \quad (10)$$

The total cost of a concurrent method of Fast Handover and Dupont AAA,

$$C_{total-p} = 2(a+b'+c) + 8d + 4h + m + 16r + (\lambda \times t_{delay-p} \times C_{dt-p}) + C_{loss-p} \quad (11)$$

$$C_{total-a} = 2(a+b'+c) + 8d + 3h + m + 16r + (\lambda \times t_{delay-a} \times C_{dt-a}) + C_{loss-a} \quad (12)$$

A proposed method spends additional cost on processing the H_value at MN, PAR and NAR. We assume that the average of packet processing cost at these points is $2r$. A proposed method also divided by two situations.

- When MN receives FBACK message (predictive),

$$C_{signal-sug-p} = 2(a+b'+c)+8d+3h+m+20r \tag{13}$$

$$t_{delay-sug-p} = 3t_d + 2t_h + 7t_r + t_{L2} + t_m \tag{14}$$

The main focus of the proposed method is a reduction of packet cost and this is accomplished by a reduction of a packet forwarding latency ($t_{delay-sug-p}$).

- When MN does not receive FBACK message (reactive),

$$C_{signal-sug-a} = 2(a+b'+c)+8d+3h+m+20r \tag{15}$$

$$t_{delay-sug-a} = 2t_d + 3t_h + 7t_r + t_{L2} + t_m \tag{16}$$

The total cost of a proposed method,

$$C_{total-sug-p} = 2(a+b'+c)+8d+3h+m+20r+(\lambda \times t_{delay-sug-p} \times C_{dt-sug-p})+C_{loss-sug-p} \tag{17}$$

$$C_{total-sug-a} = 2(a+b'+c)+8d+3h+m+20r+(\lambda \times t_{delay-sug-a} \times C_{dt-sug-a})+C_{loss-sug-a} \tag{18}$$

4.2 Performance Evaluation of Proposed Method

Assume that the cost of message processing at any hop is same. We also assume the distance between two hosts on the same sub-network (a, c, d) is 1, between tunnel (h) is 2 and the other sub-network (b, b', g, f) is 5. We define the weighted value of loss cost when the tunnel fails in completion as two. To apply mobility, we refer to the method in reference [9] (uniform fluid model) and use the formula (19) of latency for wire section (msec).

$$t_{RT-wire}(h,k) = 3.63k + 3.21(h-1) \tag{19}$$

k is packet length (kbyte) and h is the number of hops. In Reference [10], it provides the latency (t_{L2}=84 msec and t_r=0.5 msec) for wireless section.

When a mobile node moves to the new domain, the rate, the total cost of proposed method about the total cost of the first method, shows below,

When MN receives FBACK message (Predictive),

$$\lim_{p\to\infty} \frac{C_{total-sug-p}}{C_{total}} = \lim_{p\to\infty} \frac{C_{signal-sug-p} + (\lambda \times t_{delay-sug-p} \times C_{dt-sug-p}) + C_{loss-sug-p}}{C_{signal} + (\lambda \times t_{delay} \times C_{dt}) + C_{loss}} \tag{20}$$

The rate converges at 0.734 (pedestrian) and at 0.733 (vehicle). The proposed method gets approximately 27% cost profit in comparison with first method.

When MN does not receive FBACK message (Reactive),

$$\lim_{p\to\infty} \frac{C_{total-sug-a}}{C_{total}} = \lim_{p\to\infty} \frac{C_{signal-sug-a} + (\lambda \times t_{delay-sug-a} \times C_{dt-sug-a}) + C_{loss-sug-a}}{C_{signal} + (\lambda \times t_{delay} \times C_{dt}) + C_{loss}} \tag{21}$$

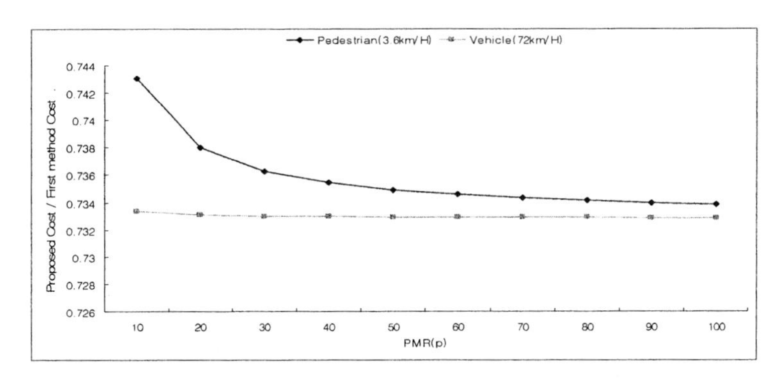

Fig. 7. The conclusion of exp. 20 (pedestrian: $\mu = 0.01$ (m/msec), vehicle: $\mu = 0.2$ (m/msec))

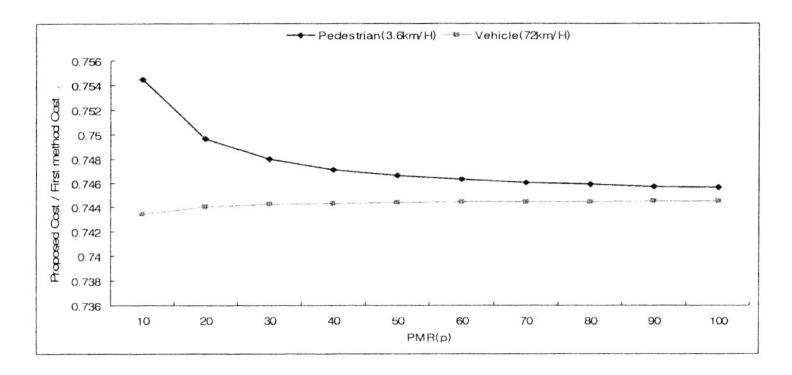

Fig. 8. The conclusion of exp. 21 (pedestrian: $\mu = 0.01$ (m/msec), vehicle: $\mu = 0.2$ (m/msec)).

The rate converges at 0.745 (pedestrian) and at 0.744 (vehicle). The proposed method gets approximately 25% cost profit in comparison with first method.

5 Conclusion

AAA authentication method causes long latency for its strong security. To make up for its defect of a long latency, several methods are suggested. In this paper, we propose the authentication by the Hash value to reduce the latency. The authentication of H_value is able to extend existing SA established with PAR to NAR. And the proposed method makes considerable reduction of latency for new SA and assurance of reliability between PAR and NAR. When MN receives a FBACK message, the proposed method get 27% cost profit in comparison with first method (existing method). When MN does not receive a FBACK message, proposed method get 25% cost profit in comparison with first method. Because the total packet cost is caused by latency of packet, the reduction of the cost implies the reduction of latency. In consequence, this proposal provides real-time communication.

Acknowledgment

This work was supported by the Soongsil University Research Fund.

References

1. Charles E. Perkins and David B. Johnson, "Mobility Support in IPv6", RFC 3775, June 2004.
2. J. Arkko, V. Devarapalli, F. Dupont, "Using IPSec to Protect Mobile IPv6 Signaling Between Mobile Nodes and Home Agents", RFC 3776, June 2004.
3. S. Kent, R. Atkinson, "Security Architecture for the Internet Protocol", RFC 2401, November 1998.
4. M. Faccin, E. Perkins, and etc. "Mobile IPv6 Authentication, Authorization, and Accounting Requirements", Internet Draft, draft-le-aaa-mipv6-requirements-03.txt February 2004
5. P. Flykt, C. Perkins, T. Eklund, "AAA for IPv6 Network Access", draft-perkins-ietf-aaav6-05.txt, May 2003
6. Francis Dupont "AAA for mobile IPv6", Internet Draft, draft-dupont-mipv6-aaa-01.txt, November 2001
7. R. Koodli et al, "Fast Handovers for Mobile IPv6", RFC 4068, July 2005.
8. Changnam Kim, Youngsong Mun and etc, "Performance Improvement in Mobile IPv6 Using AAA and Fast Handoff" ICCSA 2004
9. R. Jain, T.Raleigh and C, Graff, M. Bereschinsky, "Mobile Internet Access and Qos Guarantees using Mobile IP and RSVP with Local Register," in Proc. ICC'98 Conf,. pp. 1690-1695, Atlanta
10. Jon-Olov Vatn, "An experimental study of IEEE 802.11b handover performance and its effect on voice traffic", SE Telecommunication Systems Laboratory Department of Microelectronics and Information Technology (IMIT), July 2003.

The Tentative and Early Binding Update for Mobile IPv6 Fast Handover

Seonggeun Ryu and Youngsong Mun

School of Computing, Soongsil University, Seoul, Korea
sgryu@sunny.ssu.ac.kr, mun@computing.ssu.ac.kr

Abstract. In Mobile IPv6, a handover latency is an important issue. To reduce the handover latency, mipshop working group in IETF has studied the fast handover (FMIPv6) which creates and verifies a new care-of address (NCoA) in advance before a layer 2 handover resulting in reduced handover latency. Even in FMIPv6, the NCoA must be registered in a home agent (HA). This registration still creates a significant amount of delay. To reduce registration latency, we propose a tentative and early binding update (TEBU) scheme that the NCoA is registered in the HA in advance during the layer 2 handover based on FMIPv6. We use cost analysis for the performance evaluation. As a result, we found that the TEBU scheme guarantees lower handover latency than FMIPv6 as much as 21%.

1 Introduction

Since the requirements on the wireless Internet and mobility support are increased, the Internet engineering task force (IETF) has standardized Mobile IPv6 (MIPv6) [1]. MIPv6 supports maintaining an IP address of a mobile node (MN) while the MN moves from one subnet to another subnet. And hence, the MN can maintain sessions to correspondents. In MIPv6, when the MN moves to another subnet, it needs a certain process, a handover, which causes a long latency problem. To reduce the handover latency, mipshop working group in IETF has studied a fast handover (FMIPv6) [2] and a hierarchical MIPv6 (HMIPv6) [3], therefore the handover latency has been reduced. FMIPv6 creates and verifies a new care-of address (NCoA) in advance before a layer 2 handover and HMIPv6 administers mobility locally. However, there is a long delay that the NCoA is registered in a home agent (HA), and this delay remains a problem of the handover latency. To reduce registration latency, we suggest a tentative and early binding update (TEBU) scheme that the NCoA is registered in the HA in advance during the layer 2 handover.

In FMIPv6, the handover latency is shorter than the one of MIPv6, and packets from a correspondent node (CN) to the MN are not lost during the handover, because a tunnel is established between a previous access router (PAR) and a new access router (NAR) and the NAR will buffer packets destined to the MN. However, In FMIPv6, a registration latency of an NCoA still exists. To reduce the registration latency, the TEBU message is encapsulated in the FBU message when the MN sends

X. Jia, J. Wu, and Y. He (Eds.): MSN 2005, LNCS 3794, pp. 825–835, 2005.

a fast binding update (FBU) message. The NAR sends the TEBU message to the HA if it verifies the NCoA successfully. The HA receives the TEBU message and registers the NCoA with a binding cache entry (BCE). By using TEBU scheme the registration latency is reduced, since a registration of the NCoA is performed during a layer 2 handover.

The rest of this paper is organized as follows: in section 2, FMIPv6 scheme is presented as related works. The tentative and early binding update (TEBU) scheme is proposed in section 3, and in section 4, the TEBU scheme is evaluated using cost analysis. Finally, in section 5, we conclude discussion with future study.

2 Related Works

In MIPv6, when an MN moves from one subnet to another subnet, a certain process such as a handover is needed. The handover consists of layer 2 handover which is physical handover and layer 3 handover which is logical handover. In FMIPv6, two handovers are interleaved to reduce a handover latency.

FMIPv6 is a scheme which improves MIPv6 handover. In FMIPv6, there are two modes, such as predictive and reactive mode. In this paper, we only explain and use the predictive mode. In FMIPv6, several portions of the layer 3 handover are performed prior to the layer 2 handover. In other words, the MN performs the layer 3 handover while it is connected to a PAR, and in this case, the PAR must have known information about destined AR. The PAR establishes a tunnel between itself and a NAR, and then verifies MN's NCoA by exchanging a handover initiate (HI) message and a handover acknowledge (HAck) message. Packets that arrive at previous care-of address (PCoA) are sent to the NAR through an established tunnel during the handover. The tunnel is kept until MN's NCoA is registered to a HA.

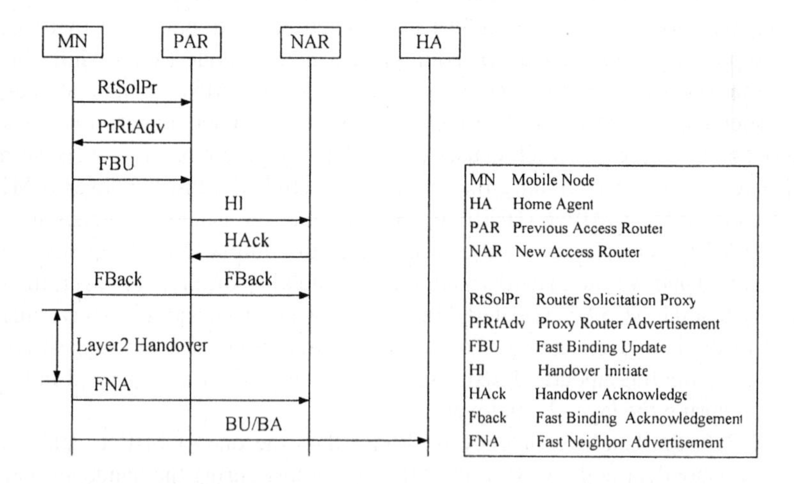

Fig. 1. MIPv6 fast handover Scheme (FMIPv6) – Predictive mode

Fig. 1 shows FMIPv6 scheme. The MN initiates FMIPv6 scheme when a layer 2 trigger take places. The MN sends an RtSolPr message to the PAR and receives a

PrRtAdv messages for information of the NAR. The MN creates the NCoA and sends a FBU message with the NCoA. When the PAR receives the FBU message, it sends a HI message and receives a HAck message in order to verify the NCoA in the NAR and to establish a tunnel between the PAR and the NAR. The PAR sends FBack messages to the MN and the NAR, respectively, after verification of the NCoA. The MN performs the layer 2 handover when it receives the FBack message. The MN sends a FNA message to NAR to inform its movement after the layer 2 handover. Then the MN interchanges a BU and a BA with the HA to register the NCoA. After registration, the handover is finished.

In FMIPv6, IEEE 802.11 [4] is practically selected as a layer 2 protocol. There is a study on MIPv6 fast handover for 802.11 networks [5], which defines an order that layer 2 and layer 3 handover processes are performed when the layer 2 protocol is the IEEE 802.11 wireless LAN and the layer 3 protocol is FMIPv6. Processes of IEEE 802.11 handover are a scan phase and an execution phase [4]. In [5], two phase of IEEE 802.11 handover can be performed separately, and the scan phase may be performed in advance in predictive mode of FMIPv6 to reduce latency of the layer 2 handover.

3 The Proposed Scheme (TEBU)

A handover process should be followed in order to preserve the connectivity when an MN moves from one subnet into another subnet. The handover process consists of layer 2 and layer 3 handovers. The layer 2 handover is performed by a scan phase and an execution phase consisting of an authentication phase and an association phase. The layer 3 handover is decomposed into creating, verifying and registering an NCoA, and detecting a movement. The layer 2 and layer 3 handovers are performed sequentially in MIPv6 while several messages related to layer 3 handover are exchanged prior to the layer 2 handover in FMIPv6. The registration of the NCoA still spends a long delay. The TEBU scheme is similar to FMIPv6. In the TEBU, the MN sends FBU encapsulating a TEBU message before the layer 2 handover. The registration process of the NCoA to the HA is performed during the layer 2 handover. Therefore, the registration latency of the NCoA is reduced.

We assume that the MN must be able to encapsulate the TEBU message to the FBU message. The TEBU lifetime is limited by 3 seconds which is referred to [6] to avoid a false binding. A PAR is assumed to derive the TEBU message from the FBU message, and to encapsulate the TEBU message to the HI message. The NAR is assumed to send the TEBU message to the HA after verification of the NCoA. If verification of the NCoA is failed, the NAR don't send the TEBU message to the HA. We assume that the HA must support dual CoA fields (CoA and tentative CoA fields) in a binding cache entry (BCE). Under this assumption, if the HA receives the TEBU message, then it stores the CoA with the tentative CoA field, also if the HA receives a general BU message, then it store the CoA with the CoA field and the tentative CoA field is cleared. If the tentative CoA field of the BCE is not empty, then the HA forwards packets to an IP address of tentative CoA field. If is not, then the HA forwards packets to an address of the CoA field. The TEBU scheme only can be performed when FMIPv6 scheme is the predictive mode. Otherwise, the MN performs a MIPv6 handover (or the reactive mode of FMIPv6). A return routability test (RR) and a registration to CNs are not considered in this paper, since FMIPv6 scheme and the TEBU scheme use a same scheme after the registration to the HA. We assume a limited lifetime of the TEBU message and dual CoA fields to protect a false binding update and a ping-pong condition.

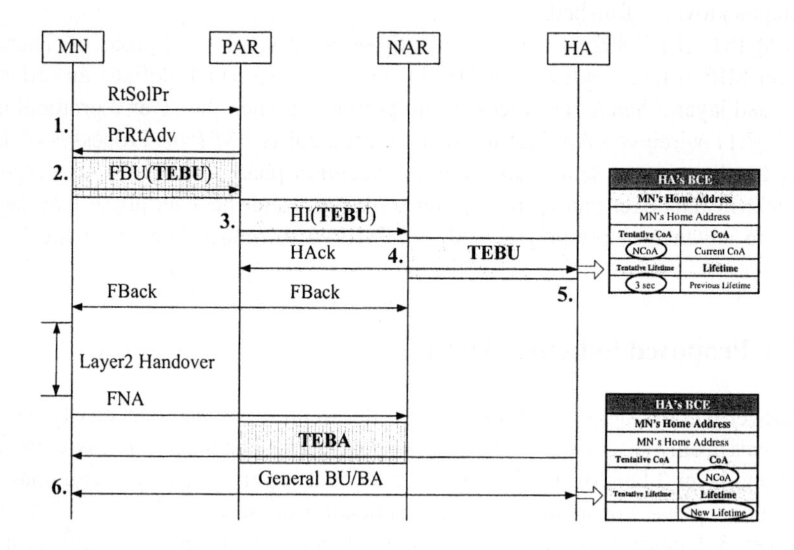

Fig. 2. Signals of the tentative and early binding update (TEBU) scheme

The TEBU scheme is as follow:

1. The MN initiates TEBU scheme from a layer 2 trigger. The MN exchanges RtSolPr and PrRtAdv messages with the PAR.
2. The MN sends the FBU message encapsulating the TEBU message to the PAR.
3. When the PAR receives the FBU message and it creates a HI message encapsulating the TEBU message
4. The NAR receives the HI message, verifies the NCoA and responds to the HI message with a HAck message. After the NAR verifies the NCoA, it sends the TEBU message to the HA to register the NCoA.
5. The HA receives the TEBU message and verifies authentication information. If the TEBU message is verified, a CoA is stored with a tentative CoA field in the BCE, and the HA sends a tentative and early BA (TEBA) message to the MN.
6. The MN initiate general registration of the NCoA with the HA to update a binding information after it receives TEBA message.

The tentative registration is completed when the MN receives the TEBA message. As a result, the handover latency is reduced.

4 Performance Evaluations

4.1 System Modeling

Fig. 3 shows a system model for evaluating performance of our TEBU scheme. In the system model, an MN moves from one subnet to another subnet, such as a handover. In this section, we analyze the TEBU scheme, MIPv6 and FMIPv6 in terms of handover latency, signaling cost and packet delivery cost. Finally, we will show that the performance of our scheme is more enhanced than those of MIPv6 and FMIPv6.

We assume that a CN generates data packets destined to the MN at a mean rate λ, and the MN moves from one subnet to another at a mean rate μ. We define packet to mobility ratio (PMR, p) as the mean number of packets received by the MN from the CN per movement. When the movement and packet generation processes are independent and stationary, the PMR is given by $p = \lambda / \mu$. We assume that the cost for transmitting a control packet is dependent on the distance between the sender and receiver. The cost for transmitting a data packet will be l times greater, where $l = l_d / l_c$, where l_d is the average length of a data packet, and l_c is the average length of a control packet. The average processing cost for control packets at any node is assumed r [7].

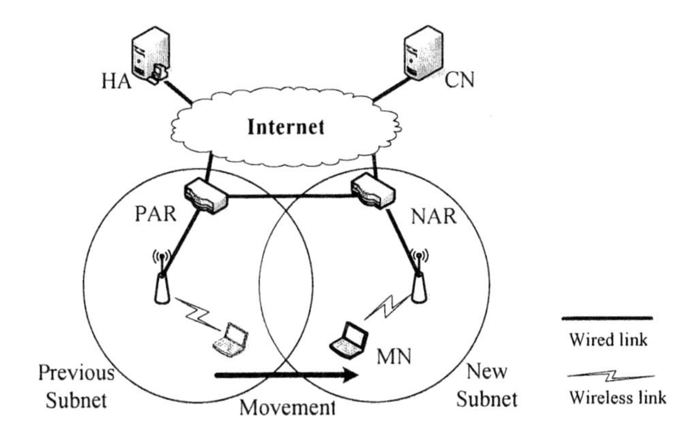

Fig. 3. System model

4.2 Handover Latency

The handover latency consists of three latencies such as a link switching latency, an IP connectivity latency and a location update latency. The link switching latency is due to a layer 2 handover. The IP connectivity latency is due to movement detection and new IP address configuration after the layer 2 handover. An MN can send packets from a new subnet after the IP connectivity latency and receive packets from a CN directly after the registration of the new IP address [8, 9].

Fig. 4 shows the timing diagram corresponding to MIPv6 handover, where T_{L2} is the link switching latency consisting of T_{l2scan} and T_{l2exec} which are respectively the scan phase and the execution phase of the layer 2 handover, $T_{IPbasic}$ is the IP connectivity latency consisting of T_{MD} and T_{AC} which are respectively the movement detection and the address configuration, respectively, and T_{BU} is the location update latency consisting of registration at the HA. In Fig. 4, the registration latency to the CN (T_{BUCN}) consisting of a RR test and a BU process is omitted after T_{BUHA}, since the latency is equal to FMIPv6 and the TEBU scheme. In MIPv6, the handover latency is $T_{L2} + T_{IPbasic} + T_{BUHA} + T_{BUCN}$.

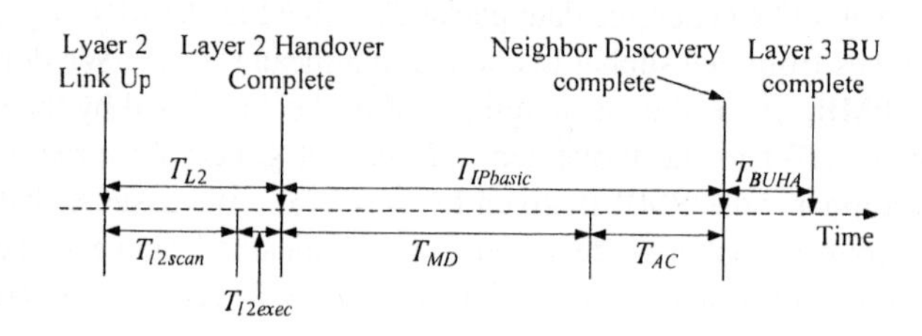

Fig. 4. Timing diagram in basic MIPv6

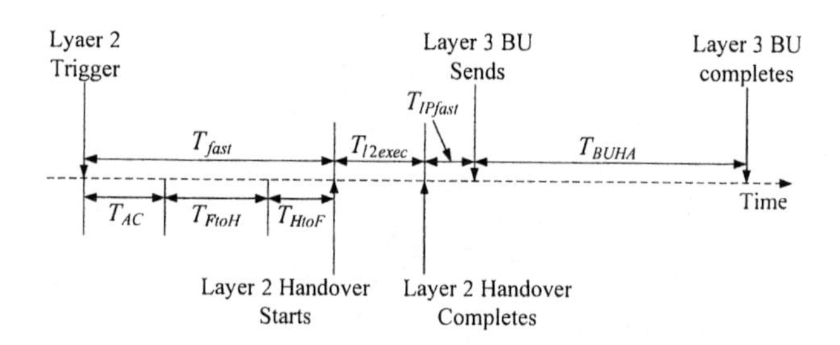

Fig. 5. Timing diagram in FMIPv6

In FMIPv6, a new IP address configuration (T_{AC}) and the movement detection (T_{MD}) are performed in advance before the layer 2 handover; where T_{MD} consists of T_{FtoH} and T_{HtoF} which are latencies between sending a FBU and receiving a HI and between sending a Hack and receiving a FBack, respectively. In predictive FMIPv6, the layer 2 handover is performed by an execution phase without a scan phase. Since the scan phase is performed periodically in advance before the handover, the link switching latency is T_{l2exec} [5]. T_{IPfast} is a latency during transmitting a FNA message. In FMIPv6, the handover latency is $T_{fast} + T_{l2exec} + T_{IPfast} + T_{BUHA} + T_{BUCN}$ and the timing diagram is shown by Fig. 5.

The Timing diagram in the TEBU scheme is similar to that in FMIPv6 except that T_{BUHA} is changed to T_{TEBU}. The TEBU message is sent to the HA after the NAR receives the HI message and verifies a NCoA. Therefore, the handover latency of the TEBU scheme is $(T_{fast} - T_{HtoF}) + \max(T_{TEBU}, T_{HtoF} + T_{l2exec} + T_{IPfast}) + T_{BUCN}$ and is shorter than that of FMIPv6. Fig. 6 shows a timing diagram in the TEBU scheme.

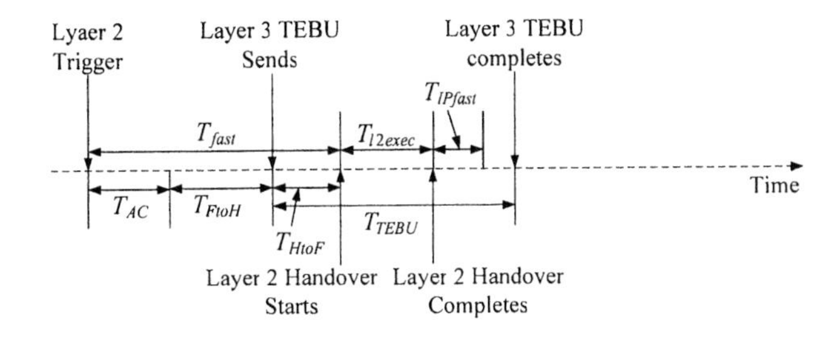

Fig. 6. Timing diagram in the TEBU scheme

4.3 Cost Analysis

In this paper, we compare the TEBU scheme with MIPv6 and FMIPv6, respectively. Overall handover cost (CO) consisting of signaling cost and delivery cost is given by

$$CO = CS + CD \ , \tag{1}$$

where CS is a sum of costs for signal messages, and CD is a sum of cost to deliver data packets during a handover. We don't consider layer 2 signals as the signaling cost during the handover. The signaling costs are given by

$$CS_{MIPv6} = CS_{IPbasic} + CS_{BUHA} + CS_{BUCN} \ ,$$

$$CS_{FMIPv6} = CS_{fast} + CS_{IPfast} + CS_{BUHA} + CS_{BUCN} \ ,$$

$$CS_{TEBU} = CS_{fast} + CS_{TEBUHA} + CS_{IPfast} + CS_{BUHA} + CS_{BUCN} \ ,$$

where CS_{MIPv6} , CS_{FMIPv6} and CS_{TEBU} are signaling costs during MIPv6, FMIPv6 and the TEBU scheme handovers, respectively. $CS_{IPbasic}$ is a signaling cost during a movement detection and the new IP address configuration, CS_{BUHA} is a signaling cost during registration to a HA, CS_{BUCN} is a signaling cost during a RR test and registration to a CN, CS_{fast} is a signaling cost during processes before the layer 2 handover in FMIPv6 and the TEBU scheme, CS_{IPfast} is a signaling cost during a FNA message, and CS_{TEBUHA} is a signaling cost during registration to the HA. In the TEBU scheme, registration to the HA is performed in twice. It is because that the TEBU message is a tentative BU message, and a general BU message must be sent after moving to a new subnet.

The delivery cost is estimated by both forwarding costs and lost cost. In this analysis, we consider the forwarding cost and the lost cost as the additional buffer space used by forwarding packets and the retransmission cost due to lost packets during the handover.

$$CD_{MIPv6} = \lambda \times CD_{newNet} \times T_{MIPv6Handover} \ ,$$

$$CD_{FMIPv6} = P_{suc.} \times \lambda \times \{CD_{preNet} \times (T_{AC} + T_{FtoH}) + CD_{newNetTunnel} \times T_{FMIPv6Tunnel}\} + \eta \times P_{fail} \times \lambda \times CD_{preNet} \times T_{MIPv6Handover} \ ,$$

$$CD_{TEBU} = P_{suc.} \times \lambda \times \{CD_{preNet} \times (T_{AC} + T_{FtoH}) + CD_{newNetTunnel} \times T_{TEBUTunnel}\} + \eta \times P_{fail} \times \lambda \times CD_{preNet} \times T_{MIPv6Handover} ,$$

where CD_{MIPv6} , CD_{FMIPv6} and CD_{TEBU} are delivery costs during the handover, CD_{newNet} , CD_{preNet} and $CD_{newNetTunnel}$ are the forwarding costs from the CN to the new subnet, from the CN to the previous subnet and from the CN to new subnet through a tunnel, respectively. The success ratio of FMIPv6 predictive mode is denoted by $P_{suc.}$ and η is a weight factor for retransmission due to the lost packets. In MIPv6, all packets are lost during the handover. If FMIPv6 is the predictive mode, all packets are tunneled to the NAR, and if not, then all packets are lost during the handover. The TEBU scheme is similar to FMIPv6. To calculate $P_{suc.}$ and P_{fail} , we refer to section 4.3 in [10] which shows a failure probability of the predictive mode of FMIPv6 with MN's velocities and radiuses of a cell, respectively [11].

4.4 Numerical Results and Analysis

In this paper, we use formulas derived from empirical communication delay model suitable for the scenario where a CN, a HA and ARs are connected to a wired enterprise network. Regression analysis of the collected data yields (2), where k is the length of the packet in KB, h is the number of hops, and $T_{wired-RT}$ is the round-trip time in milliseconds. A similar experiment over wireless link is shown in (3) [7].

$$T_{wired-RT}(h,k) = 3.63k + 3.21(h-1) . \quad (2)$$

$$T_{wireless-RT}(k) = 17.1k . \quad (3)$$

In Fig. 3, we assume that a message processing cost is equal to the cost for communication over a single hop ($r = 1$). A hop between an MN and an AP and a hop between the AP and a AR is 1, respectively. Hops between ARs are 2. Hops between the AR and a HA, hops between the AR and a CN and hops between the HA and the CN are 5, respectively. System parameters specified in Table 1 are based on formula (2), (3) [12, 13].

Table 1. Parameters for numerical analysis

Packet size (bytes)		Latency analysis (milliseconds)							
l_c	l_d	T_{fast}	T_{L2}	T_{l2exec}	$T_{IPbasic}$	T_{IPfast}	T_{BUHA}	T_{BUCN}	T_{proc}
200	1024	17.52	169.45	8.55	246.54	3.52	21.1	46.45	0.5

Dependences of signal costs and delivery costs related to the PMR

The signaling costs are independent of the PMR, since it is not affected by the PMR. The TEBU scheme costs a more signaling cost than FMIPv6 and MIPv6, because it performs BU process in twice. The delivery costs depends on the PMR, since it is dependent on λ. The TEBU scheme costs a less delivery cost than FMIPv6 and MIPv6, because its handover latency is shorter than FMIPv6 and MIPv6.

Cost ratio related to the PMR

Overall cost for a MIPv6 handover (CO_{MIPv6}), overall cost for FMIPv6 (CO_{FMIPv6}), and overall cost for the TEBU scheme (CO_{TEBU}) are calculated by (1), respectively.

On the other hand, we obtain that

$$\lim_{p \to \infty} \frac{CO_{TEBU}}{CO_{MIPv6}} = \lim_{p \to \infty} \frac{CS_{TEBU} + CD_{TEBU}}{CD_{MIPv6} + CD_{MIPv6}} \approx 0.21 \quad (4)$$

$$\lim_{p \to \infty} \frac{CO_{TEBU}}{CO_{FMIPv6}} \lim_{p \to \infty} \frac{CS_{TEBU} + CS_{TEBU}}{CD_{FMIPv6} + CD_{FMIPv6}} \approx 0.79 \quad (5)$$

where PMR > 450.

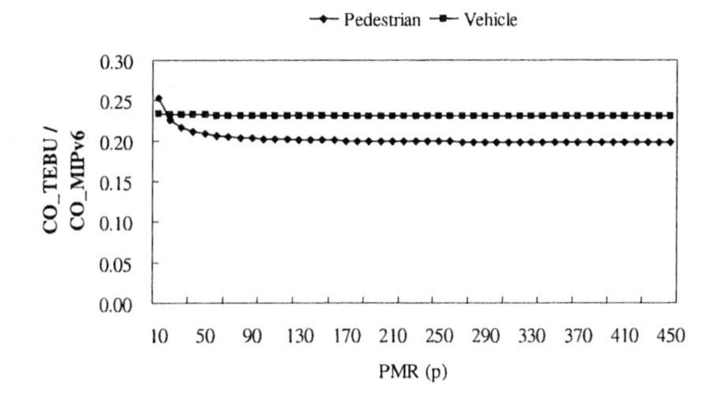

Fig. 7. Cost ratio of the TEBU scheme against MIPv6 handover

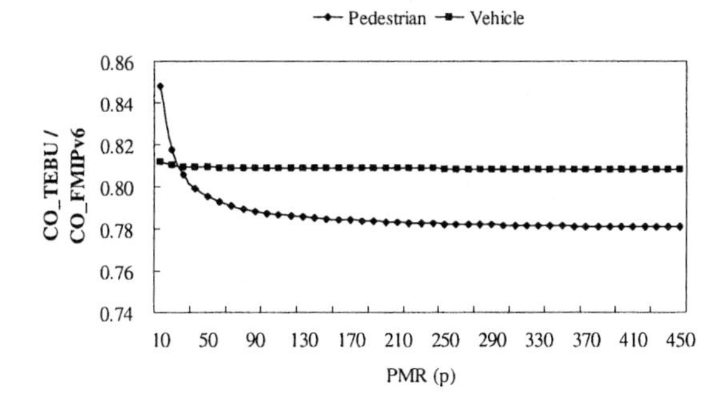

Fig. 8. Cost ratio of the TEBU scheme against FMIPv6

Fig. 7 and 8 show the variation of cost ratio against MIPv6 handover and FMIPv6, respectively when radius of a cell is 100 meters. In a case of the pedestrian, the variation of cost ratio is high at low PMR since the handover process is performed infrequently. In Fig. 7 and 8, the higher the PMR is, the higher the performance is, and values of those ratios are convergent to limits, respectively, when the PMR increases over 450. At (4), the mean cost ratio is 0.21, therefore, compared to MIPv6 handover, the TEBU scheme gains 79% improvement. The mean cost ratio is 0.79 by (5), where the TEBU scheme saves the cost by 21% against FMIPv6.

5 Conclusions and Future Study

A handover process occurs due to movements of an MN maintaining a home IP address in MIPv6. The handover process causes a long latency for the MN to receive or to send packets, such as a handover latency. In the MIPv6, the handover latency is a critical issue. FMIPv6 has studied to reduce the handover latency, but registration latency is still long. We propose a tentative and early binding update (TEBU) scheme to reduce the handover latency. Especially, the TEBU message is sent to HA in advance before a layer 2 handover.

We have showed an improved performance by means of comparison of costs, such as the cost ratio of the TEBU scheme against MIPv6 and FMIPv6. Compared to MIPv6 handover, the TEBU gains 79% improvement, while it gains 21% improvement compared to FMIPv6.

Currently, FMIPv6 is being standardized in IETF. The TEBU scheme can help improve the performance of FMIPv6. Therefore, when TEBU scheme is used with FMIPv6, a performance of the handover process will be improved.

Acknowledgments

This work was supported by the Soongsil University Research Fund.

References

1. Johnson, D., Perkins, C., Arkko, J.: Mobility Support in IPv6, RFC 3775 (2004)
2. Koodli, R.: Fast Handovers for Mobile IPv6, work in progress (2004)
3. Soliman, H., Malki, K. El: Hierarchical Mobile IPv6 mobility management (HMIPv6), work in progress (2004)
4. Ergen, M.: IEEE 802.11 Tutorial (2002)
5. McCann, P.: Mobile IPv6 Fast Handovers for 802.11 Networks, work in progress (2004)
6. Vogt, C., Bless, R., Doll, M.: Early Binding Updates for Mobile IPv6, work in progress (2004)
7. Jain, R., Raleigh, T., Graff, C., Bereschinsky, M.: Mobile Internet Access and QoS Guarantees using Mobile IP and RSVP with Location Registers, in Proc. ICC'98 Conf. (1998) 1690-1695
8. Koodli, R., Perkins, C.: Fast Handovers and Context Transfers in Mobile Networks, ACM Computer Communication Review, Vol. 31, No. 5 (2001)
9. Pack, S., Choi, Y.: Performance Analysis of Fast Handover in Mobile IPv6 Networks, work in progress, IFIP PWC 2003, Venice, Italy (2003)

10. Kim, J., Mun, Y.: A Study on Handoff Performance Improvement Scheme for Mobile IPv6 over IEEE 802.11 Wireless LAN, work in progress (2003)
11. McNair, J., Akyildiz, I.F., Bender, M.D.: An Inter-System Handoff Technique for the IMT-2000 System, IEEE INFOCOM, vol. 1 208-216 (2000)
12. Vatn, J.: An experimental study of IEEE 802.11b handover performance and its effect on voice traffic, SE Telecommunication Systems Laboratory Department of Microelectronics and Information Technology (IMIT) (2003)
13. Mishra, A., Shin, M., Arbaugh, W.: An Empirical Analysis of the IEEE 802.11 MAC Layer Handoff Process, ACM SIGCOMM Computer Communication Review (2003)

A Simulation Study to Investigate the Impact of Mobility on Stability of IP Multicast Tree*

Wu Qian[1], Jian-ping Wu[2], Ming-wei Xu[1], and Deng Hui[3]

[1] Department of Computer Science and Technology,
Tsinghua University, Beijing 100084, China
{wuqian, xmw}@csnet1.cs.tsinghua.edu.cn
http://netlab.cs.tsinghua.edu.cn

[2] Department of Computer Science and Technology,
Tsinghua University, Beijing 100084, China
jianping@cernet.edu.cn

[3] Hitachi (China) Investment, Ltd
Beijing 100004, China
hdeng@hitachi.cn

Abstract. Mobile users expect similar kinds of applications to static ones, including various IP multicast applications. In mobile environment, the multicast tree should face with not only the dynamic group membership problem but also the mobile node's position change issue. In this paper, we study the stability of IP multicast tree in mobile environment. We define a stability factor and investigate how the various elements of network and mobility impact on it. It is shown that the stability factor is mainly dominated by three elements, namely the ratio of the number of mobile nodes and network size, mobility model and the mobile multicast scheme. These results can give some useful references when we design a new mobile multicast scheme in the future.

1 Introduction

Mobile users expect similar kinds of applications to static ones, including attractive IP multicast applications. In the mean time, with the merit of efficient multi-destinations delivery, IP multicast can give the benefit of saving network bandwidth and releasing the burden of replications from the source. This kind of efficiency is especially valuable for mobile networks which usually use wireless infrastructure and face with the bandwidth scarce problem.

In mobile environment, multicast must deal with not only the dynamic group membership but also the dynamic member location. The current multicast protocols are developed implicitly for static members and do not consider the extra requirements to support mobile nodes. Every time a member changes its location, keeping track of it and reconstructing the multicast tree will involve extreme

* This work is supported by the Natural Science Foundation of China (No60373010), National 973 Project Fund of China (No 2003CB314801), and Cooperative research project on Mobile IPv6 Multicast between Hitachi (China) and Tsinghua University.

X. Jia, J. Wu, and Y. He (Eds.): MSN 2005, LNCS 3794, pp. 836–845, 2005.

overhead, while leaving this tree unchanged will result in inefficient sometimes incorrect delivery path. Mobile IP [5] provides two basic approaches to support mobile multicast, i.e., bi-directional tunneling (BT) and remote subscription (RS). Most of the other proposed solutions are based on them [2, 3].

The main problem of BT-based approaches is the poor multicast packet delivery efficiency [2, 3]. Sometimes the multicast would even degrade to unicast. On the contrary, RS-based approaches maintain most of the merits of multicast mechanism. The main issue of these approaches is the overhead to maintain the multicast delivery tree as joining and leaving behaviors occur much more frequently in mobile networks. The purpose of this paper is to investigate the impact of mobility on the stability of multicast tree. We implement our study both on plane and on hierarchical RS-based scheme. The study focuses on simulation-based evaluation because mobile multicast protocols are still an emerging area and deployment is relatively uneventful, and simulation can provide researchers with a number of significant benefits, including repeatable scenarios, isolation of parameters, and exploration of a variety of metrics.

In this paper, we investigate how the various elements of network and mobility impact on stability of multicast tree. These elements include the mobile multicast scheme, the network size, the move speed, the mobility model and the power range. From the abundant data obtained, it is shown that the stability factor defined in this paper as the average number of link changes per updating is mainly dominated by three elements, namely the ratio of the number of mobile nodes and network size, mobility model and the mobile multicast scheme. While the effect of speed and AR's power range remains slight. These results can give some useful references when we design new mobile multicast scheme in the future, and remind us to pay enough attention to the stability problem in our design procedure.

The rest of the paper is organized as follows. In section 2 we introduce former research on stability and our extension. In section 3 mobile multicast schemes are introduced. Section 4 presents our simulation environment and methodology. In section 5, a set of experiments are carried out and analytical results are presented. Finally, we conclude this paper and introduce future works.

2 Stability of Multicast Tree

One of the major points of interest in multicast tree is the stability problem. Besides the dynamics of topology changes, multicast also offers the possibility of joining and leaving a group at any time. This activity requires the multicast tree to be dynamically updated. If these changes occur too often, the tree may become unstable, resulting in undesirable routing overhead.

2.1 Former Research on Stability

Mieghem [1] implements a study on how the number of links on a multicast tree changes as the number of multicast users in a group change. In his paper, the stability of a multicast tree is defined as follows.

Definition 1. In a shortest path tree with m different group members uniformly distributed in the graph containing N nodes, the number of links in the tree that change after one multicast group member leaves the group has been chosen. If we denote this quantity by $\Delta_N(m)$, then, the average number of changes equals $E[\Delta_N(m)]$. The situation where $E[\Delta_N(m)] \leq 1$ may be regarded as a stable regime.

The definition is base on the assumption that either no or one group member can leave at a single instant of time. The author carried his research on a specific class of random graphs, called as RGU (random graphs with N nodes, independently chosen links with probability p and uniformly on [0,1] or exponentially distributed link metrics w) . The research shows for RGU when m is larger than $0.3161N \approx N/3$, the expression $E[\Delta_N(m)] \leq 1$ can be satisfied, and the multicast tree would be a stable one. In addition, it also quantifies the common belief that minimal spanning trees are more instable than shortest path trees.

2.2 Stability Problem in Mobile Networks and Our Motivation

With the character of mobility, the multicast tree will encounter much more joining and leaving events in mobile network than in static network. So it is significant to investigate the stability of multicast tree in mobile environment.

Because the coverage of wireless network can overlap, mobile node (MN) would change its position by three kinds of manner, *connect*, *disconnect* and *handoff* respectively.

In multicast application, these complex position change manner will result in complicated transform of multicast tree, so the assumption *Definition 1* can't comprehensively reflect the behavior of multicast tree in mobile environment. In this article we expand *Definition 1* and come to a more universal definition.

Definition 2. In a multicast tree with m different group members uniformly distributed in the graph containing N nodes, the number of links in the tree that change after one multicast group membership update (join, leave, or leave-join event) occurs has been chosen. We denote this quantity by $\Delta_N(m)$. The stability of the multicast tree may be measured by the average number of changes, $E[\Delta_N(m)]$. We call this value as Stability Factor α. The situation where $\alpha \leq 1$ may be regarded as a stable regime.

Clearly, the smaller α, the more stable the tree is.

Our study focuses on a simulation-based evaluation and qualitative analysis. There are three main reasons. First, as mentioned in [1], very few types of graphs can be computed to quantify the stability of multicast tree. Second, the topology of Internet is currently not sufficiently known to be categorized as a type or an instance of a class of graphs. Third, mobile multicast is still an emerging area and deployment is relatively uneventful, and simulation can provide researchers with a number of significant benefits, including repeatable scenarios, isolation of parameters, and exploration of a variety of metrics. Although it is impossible to simulate all kinds of graphs and there exist differences between the simulation and the real world, it can still give us the first order of estimates and useful references when designing a mobile multicast scheme.

3 Mobile Multicast Schemes

Bi-directional tunneling (BT) and remote subscription (RS) are the two basic multicast schemes introduced by Mobile IP to be used in mobile environment. Most of the other proposed solutions are based on them and attempt to solve some drawbacks of them [2,3]. In BT scheme, the MN implements multicast all through its home agent, including joining/leaving group and receiving multicast packets, so the routing inefficiency and bandwidth wasting become the main drawbacks. Because BT and its inheritors [2,3] weaken the primary characters of multicast, more attentions are paid on RS-based approaches and we focus our research just on RS-based approaches. We choose two typical ones among them, the original plane RS scheme and the hierarchical MobiCast [4] scheme.

3.1 Remote Subscription (RS)

In RS scheme, the MN would resubscribe to the multicast group whenever it changes the attachment to a new access network. The MN does the resubscription using its new care of address through the local multicast router just like the static one in the foreign network. Obviously, the multicast packets are delivered on the shortest paths. This scheme maintains the main merits of multicast mechanism.

RS faces with some new problems and the major one is the stability problem. In this scheme, both of the multicast delivery tree and the multicast group membership should be updated after the handoff, which would result in network overhead and computation overhead. If there are many mobile nodes moving quickly, this would cause many leaving and joining behaviors, and consequently bring the multicast tree serious stability problem.

3.2 MobiCast

MobiCast is a hierarchical RS-based scheme. This solution divides networks into domains, and each domain has a Domain Foreign Agent (DFA). MobiCast focus on the intra-domain multicast technique, while the inter-domain method is just directly chosen from RS or BT. DFA subscribes to the multicast group on behalf of the MNs and manages the multicast in its domain. For every multicast group, DFA provides a unique translated multicast address in the domain. Multicast packets are at first delivered to DFA, then DFA changes the group address to the translated multicast address. The Base Station (BS) in the domain subscribes to the translated multicast group, and forwards packets to the MN after retranslating them back. In order to achieve fast handoff within the domain, MN's affiliated BS will inform the physically adjacent BSs to subscribe to the corresponding translated multicast group and buffer multicast packets. These BSs form a virtual domain called Dynamic Virtual Macro-cells (DVM).

Comparing with RS, MobiCast hides the MN's movement from outside, and avoids the update of main multicast delivery tree. The other advantages are the reduced handoff latency and packets loss. But because a mass of unnecessary multicast packets are forwarded to adjacent BSs while there are maybe no group members, one of the main drawbacks of MobiCast would be bandwidth waste, which is critical for mobile environment. What's more, every time the mobile handoff occurs, it will cause several BSs to join the multicast group while others to leave. So the other main drawback is the significant multicast protocol cost and multicast tree update within the domain.

4 Simulation Environment and Methodology

In our simulation, multicast group members are all mobile nodes, and the update of multicast tree is absolutely caused by the position changes of mobile members. We record the total number of link changes of multicast tree and the total number of position change events. The ratio of them is the Stability Factor defined in *Definition 2.*

4.1 Investigating Elements

- **Mobile Multicast Scheme:** The multicast tree maintaining manner differs significantly in different mobile multicast schemes. In our simulation we investigate the RS scheme and MobiCast scheme. And for the MobiCast scheme, we study both the single-domain and multi-domain situations.
- **Number of MN & Network Size:** [1] shows that the number of MN, denoted by m, and the network size, denoted by N, are the most important elements impact the stability of multicast tree.
- **Mobility model:** The mobility model impacts when the MN would move and how it moves. The models used in the simulation are described below:
  - **Random Waypoint** (RPW for short) [7]: In this model an MN begins by staying in a randomly chosen location for a certain period of time. Once this time expires, it chooses a random destination in the simulation area and a speed that is uniformly distributed between [minSpeed, maxSpeed]. The MN then travels toward the new destination at the selected speed. Upon arrival, the MN pauses for a random period before starting the process again. This model can be used to mimic the wandering in an area action. In our simulation the pause time is uniformly chosen between 0 and 5s, and the speed is random between 0 and 40m/s.
  - **Gauss-Markov** [8]: This mobility model can be used to imitate the action of random movement while without sudden stops and sharp turns. MN's new speed and direction are correlated to its formal speed and direction and their mean values through the simulation. A tuning parameter, α, where $0 \leq \alpha \leq 1$, is used to vary the randomness. The smaller the , the greater the randomness is. We choose the value of to be 0.75 in our simulation, and the mean value of direction to be 90 degrees initially. The mean value of speed is fixes to 40m/s.
  - **Exhibition** [9]: MN in this model chooses a destination from among a fixed set of exhibition centers and then moves toward that center with

a fixed speed which uniformly chosen between [minSpeed, maxSpeed]. Once a node is within a certain distance of the center it pauses for a given time and then chooses a new center. This model can be used to mimic the action of people visiting a museum. Our simulation uses 10 centers placed uniformly. When a MN travels to a center, it stops when it is within 20 meters of the center and then pauses a time between 0 and 10s. The speed of a node is random between 0 and 40m/s.
- **Move Speed:** One of the main characters of mobility is the move speed. The faster the MN, the more probable the position change event will occur.
- **Power Range of AR:** The power range reflects the service area of an AR. When the power range is small, the MN is prone to change its serving AR more frequently.

4.2 Network Model and Methodology

The simulation is built on OMNET++ [6], a discrete event simulator.

The topology in our simulation is a mesh network in which each node acts as a multicast router of local network and also an AR for MN. Generally, the size of this mesh network is 10*10, the power range is set to a square of 100*100 meters for simplicity (only the *handoff* change manner would occur), and the distance between two nearby ARs is 100 meters. For MobiCast scheme, there are another 1 or 4 DFA routers in the topology. We will change the size of mesh network or the power range of AR when we investigate their impact on stability.

For simplicity, there is only one multicast group with one fixed source and the group members are all mobile nodes. We use a source based shortest path tree to deliver multicast packets. Because all the links have the same weight, the improved shortest path tree in our simulation is also a minimal spanning tree.

Originally, mobile nodes are randomly located at the mesh. The number of mobile nodes varies form 5 to 80. We run each simulation for 500 seconds.

5 Results

In this section, we illustrate how the mobility impacts on stability of IP multicast tree. The stability is measured by Stability Factor defined in *Definition 2.* The smaller , the more stable the tree is.

5.1 The Impact of Mobile Multicast Scheme

Fig 1 illustrates how the stability factor varies with the number of MN in different mobile multicast schemes. The schemes include RS, MobiCast with one domain (or DFA) and MobiCast with four domains (or DFAs). The size of mesh network is 10*10 and the speed of MN is random between 0 and 40m/s. Fig 1(a) is obtained under RWP model and 1(b) under Exhibition model.

All the curves in Fig 1 show that stability factor is decreasing with m (number of MN) which accords with our instinct. It also shows that the stability factor

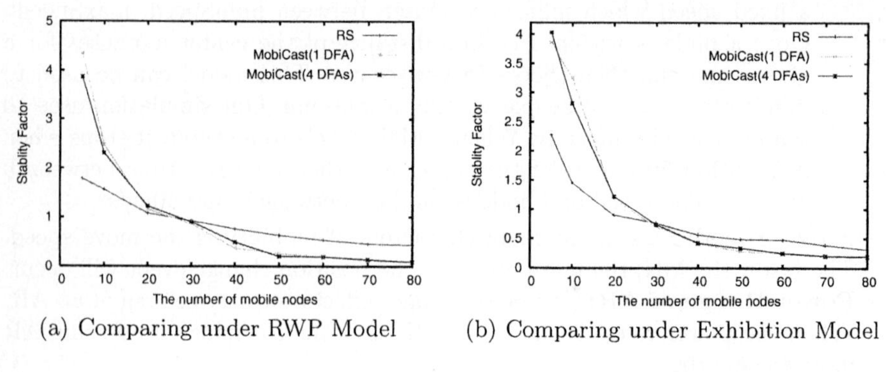

(a) Comparing under RWP Model (b) Comparing under Exhibition Model

Fig. 1. The Impact of Mobile Multicast schemes

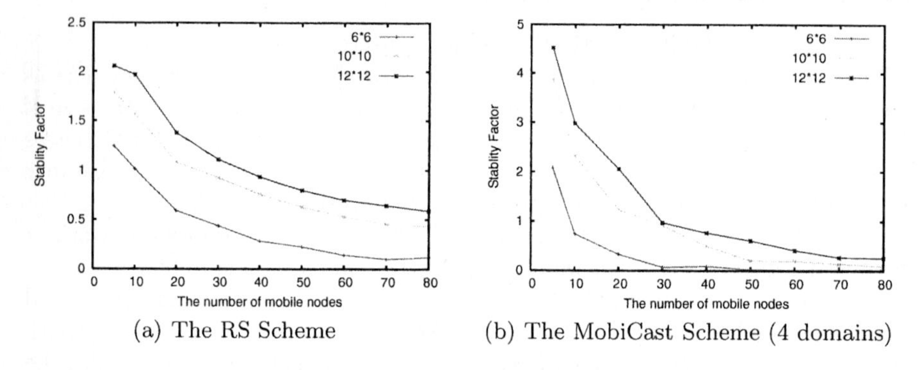

(a) The RS Scheme (b) The MobiCast Scheme (4 domains)

Fig. 2. The Impact of Network Size

of MobiCast scheme changes more intensively than that of RS, and MobiCast is more instable when m is small, but more stable when m is big enough. This phenomenon is due to the data redundancy mechanism used in MobiCast to achieve fast and nearly lossless intra-domain handoff. For MobiCast, when the number of MN is small, every time the mobile handoff occurs, it will cause several ARs to join the multicast group while others to leave. But when there are many mobile nodes in the network, most of the ARs will lie on the multicast tree, and a handoff event will not cause many varieties.

As an improved scheme, MobiCast achieves attractive performance in some aspects but aggravates the stability problem of multicast tree which embarrasses it from widely use.

5.2 The Impact of Network Size

In order to know how the network size impacts stability factor, we run simulation in three kinds of networks with different size. They are 6*6, 10*10 and 12*12. The other simulation environments are the same. Fig 2 illustrates the impact of network size in RS and MobiCast scheme. We can see that the bigger the network, the more instable the tree is. This is because MNs are more dispersive in big network, and the probability of moving to an AR already having MN in its range is small. Another result shown in Fig 2 is that although the absolute value of stability factor differs a lot, the curve of how it varies with the number of MN is much similar.

5.3 The Impact of Mobility Model

Fig 3 illustrates how the mobility model impacts stability factor. Fig 3(a) and 3(b) are the results gained in RS and MobiCast scheme respectively. We compare RWP, Gauss-Markov and Exhibition these three kinds of mobility models. As for the Gauss-Markov model, the tuning parameter is set to 0.75 which means the move track of MN is much influenced by its history movement.

As illustrated by Fig 3, mobility model does have some impact on stability. The stable order of multicast tree is Gauss-Markov>Exhibition>RWP, while the random character is just reverse, namely more random model would result in more instable. It can be explained that mobility model impacts how the MNs distributes in the network, and greater randomness will result in more dispersive.

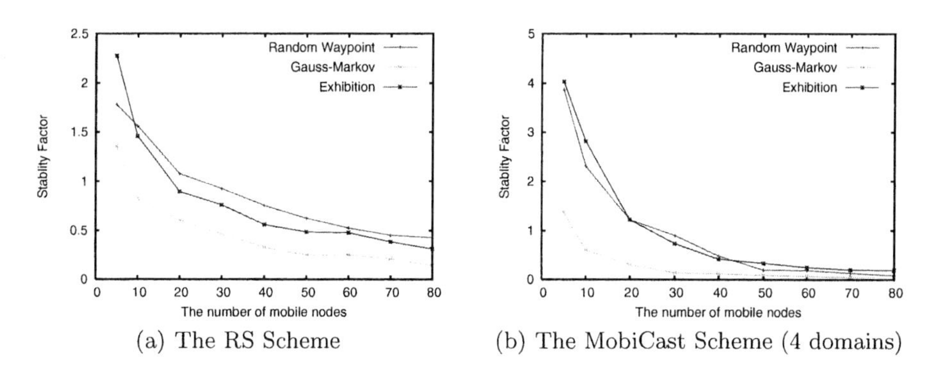

(a) The RS Scheme (b) The MobiCast Scheme (4 domains)

Fig. 3. The Impact of Mobility Model

5.4 The Impact of Move Speed

We investigate the impact of move speed on stability in RS scheme under different mobility models. Fig 4(a) shows when the number of MN is 30 how the stability factor varies with the average speed of MN. The variation manner is investigated in both RWP and Exhibition model. Fig 4(b) presents three curves about the stability factor's variation with the number of MN when the average move speed in Gauss-Markov model is 3, 10 and 40m/s.

We observe that the move speed does little impact on the stability factor. This result can be explained as follows. When the speed of MN is faster, the position change event will occur more often, which would cause the multicast

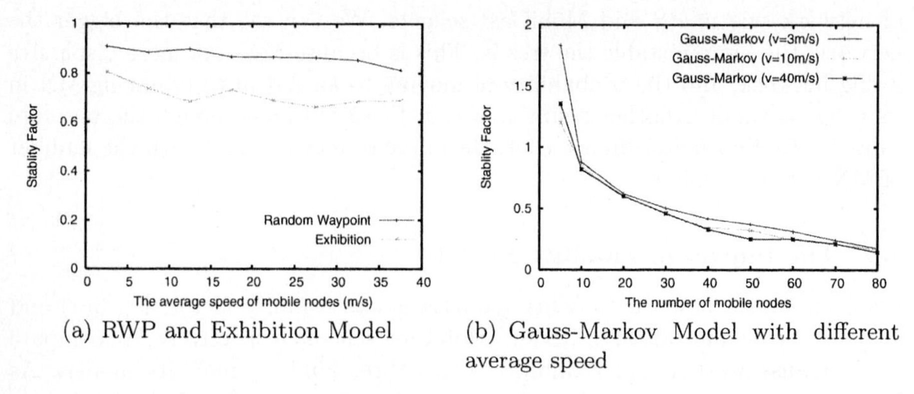

(a) RWP and Exhibition Model (b) Gauss-Markov Model with different average speed

Fig. 4. The Impact of Move Speed

tree to be updated more frequently. But the changes of link would increase too, and the average number of changes per updating, named as stability factor in our definition, would remain similar.

5.5 The Impact of the Power Range of AR

Fig 5 illustrates how the power range of AR impacts stability factor. We investigate RS scheme. The power range of AR is varied by 50 and 100 meters. Fig 5(a) and 5(b) are the results obtained under the RWP and Exhibition model respectively. From the figure we can see, the power range of AR also does little impact on the stability factor, yet when the power range is small the MN is prone to change its serving AR more frequently. The reason is the same as described in section 5.4.

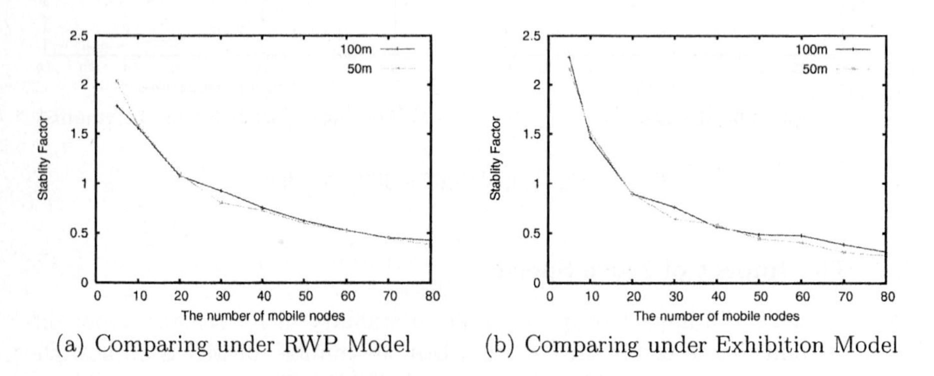

(a) Comparing under RWP Model (b) Comparing under Exhibition Model

Fig. 5. The Impact of the Power Range of AR

6 Conclusion and Future Work

In this paper, we study the stability of IP multicast tree in mobile environment. We implement our research both on the original RS scheme and the hierarchical

MobiCast scheme. The paper focuses on simulation-based evaluation and how the various elements of network and mobility impact on it. These elements include the mobile multicast scheme, the network size, the move speed, the mobility model and the power range. From the abundance data obtained, it is shown that the stability factor defined in this paper is mainly dominated by three elements, namely the ratio of the number of mobile nodes and network size, mobility model and the mobile multicast scheme. While the effect of speed and power range of AR remains slight. Although the simulation can't absolutely reflect the real world, the simulation results can still give us the first order of estimates and useful references when designing a mobile multicast scheme.

Sometimes the position change of a multicast source would cause the whole multicast tree to be updated. In the future, we will give more efforts to investigate the stability problem when multicast sender can be a mobile one.

References

1. P, V.Mieghem., M, Janic.: Stability of a Multicast Tree. IEEE INFOCOM 2002, Vol.2. Anchorage, Alaska (USA) (2002)1133-1142
2. I, Romdhani., M, Kellil., H-Y, Lach., A, Bouabdallah., H, Bettahar.: IP Mobile Multicast: Challenges and Solutions. IEEE Communications Surveys & Tutorials, vol.6, no.1. (2004) 18-41
3. Hrishikesh, Gossain., Carlos, De.M.Cordeiro., Dharma, P.Agrawal.: Multicast: Wired to Wireless. IEEE Communications Magazine, vol.40, no.6. (2002) 116-123
4. C, Lin.Tan., S, Pink.: MobiCast: a multicast scheme for wireless networks. ACM/Baltzer Mobile Networks and Applications, vol.5, no.4. (2000) 259-271
5. C, Perkins. (ed.): IP Mobility Support for IPv4. RFC 3344, August (2002)
6. Omnet++ Community Site. http://www.omnetpp.org
7. D, Johnson., D, Maltz.: Dynamic source routing in ad hoc wireless networks. In T. Imelinsky and H. Korth, editors, Mobile Computing. (1996) 153-181.
8. B, Liang., Z, Haas.: Predictive distance-based mobility management for PCS networks. Proceedings of the Joint Conference of the IEEE Computer and Communications Societies (INFOCOM). (1999) 1377-1384
9. P, Johansson. (ed.): Scenario Based Peformance Analysis of Routing Protcols for Mobile Ad Hoc Networks. IEEE MobiCom. (1999)

Fast Handover Method for mSCTP Using FMIPv6

Kwang-Ryoul Kim[1], Sung-Gi Min[1,*], and Youn-Hee Han[2]

[1] Dept. of Computer Science and Engineering,
Korea University, Seoul, Korea
{biofrog, sgmin}@korea.ac.kr
[2] Samsung AIT, Giheung-si, Kyounggi-do, Korea
yh21.han@samsung.com

Abstract. In this paper, we propose the fast handover method in mSCTP using FMIPv6. Using FMIPv6 as handover procedure in mSCTP, the performance of handover can be significantly improved. First, mSCTP can add new network address to the correspondent node quickly as FMIPv6 provides New Care of Address (CoA) without closing connection to current network. Second, mSCTP can determine when it has to change the primary IP address with the trigger from FMIPv6. This trigger indicates that the mobile node has completely joined the network of New Access Router (NAR) and confirms that the MN can receive data through the NAR. We present integrated handover procedures that maximizes the handover performance between mSCTP and FMIPv6. We implement the integration of mSCTP and FMIPv6 in our test bed and verify the operation of suggested handover procedures by analysis of the experimental result.

1 Introduction

There are increasing requirements for communication in wireless environment and mobility is also the most important aspect of future IP network. The functionality of mobile communication can be divided into three aspects that is the mobility management, the location management and the handover management.

The mobility management provides the mobile nodes (MN) to connect to the correspondent node (CN) while it is moving. When the MN enters the new networks, it finds the New Access Router (NAR) and obtains new address from the NAR. After this, the MN configures the new IP address for expected communication. The location management is used to find the MN and make connection from the CN to the MN. The location management keeps the location information of the MN, each time the MN enter the new wireless networks. The handover management provides continuous communication while the MN moves on communicating. Using the handover, the MN can change the wireless access networks seamlessly with minimizing service disruption and packet losses.

* Corresponding author.

X. Jia, J. Wu, and Y. He (Eds.): MSN 2005, LNCS 3794, pp. 846–855, 2005.

The Mobile IP (MIP) and Mobile IPv6 (MIPv6) is most representative one that provides mobility and location management with supporting handover during mobile communication. Basically, the management for mobile communication is processed in network layer, many currently used applications need not to be modified to support mobile communication. However, the MIP requires the specific routers like Home Agent (HA) and Foreign Agent (FA) that perform a mobility control and location management [1],[2].

The Session Initiation Protocol (SIP) can provide mobile communication in application layer [3]. The SIP is used for establishing or closing multimedia sessions and supports unicast and multicast connections. The SIP can provide handover by sending the updated INVITE message when it moved to new network and provide location management using the REGISTER message that register new location information to the SIP register server.

The Fast handover for Mobile IPv6 (FMIPv6) supports mobile communication in another aspect. The FMIPv6 is proposed to reduce handover latency of MIPv6 and it makes MIPv6 can be used for real time traffic such as voice and video [4]. Therefore, FMIPv6 is focused on handover management and does not provide any other functions for mobile communication.

The Mobile Stream Control Transmission Protocol (mSCTP) is recently proposed that provides mobility management and handover management in transport layer without requiring the specific routers such as HA or FA. However, the mSCTP does not provide location management and it is hard to provide low handover latency using mSCTP alone [5],[6].

The protocols such as MIP, SIP provide all mobile communication functionalities. However, some protocols are focused on specific aspects of mobile communication functionalities, so it must be used with other mobile protocols to support mobile communication. Currently, there are various requirements in mobile communication, especially in handover performance, using a single mobile communication protocols can't satisfy such requirements. The requirement for low handover latency is essential to support real time application in mobile environment. Therefore, it is expected that FMIPv6 is widely adopted to improve handover performance in mobile communication environment.

In this paper, we propose to use FMIPv6 with mSCTP to improve handover performance. We present the FMIPv6 can process handover with low latency and can be used as efficient under layer that support handover for mSCTP. We also describe the procedures of handover processing which is integrated both FMIPv6 and mSCTP procedure. We perform experiment to verify the operation of suggested handover procedures in our test bed. The handover performance of mSCTP using FMIPv6 is presented by comparing the handover latency.

2 Transport Layer Mobility Support in mSCTP

The SCTP has been approved by the IETF as a new reliable generic purpose transport layer protocol and it has many characteristic same as TCP that provide reliable session oriented transport service [7]. And SCTP has many additional features such as muti-streaming and muti-homing which support more variable needs of application.

In case of muti-homing feature, single SCTP endpoint can have multiple IP addresses for redundant purpose. In [8], multi-homing feature is extended with Dynamic Address Reconfiguration (DAR) which makes SCTP add or delete new IP address and change primary IP address while end to end session is active. The DAR extension makes SCTP can support mobility of moving terminal. The mSCTP is actually SCTP with DAR extension and support soft handover in the transport layer.

When the MN moves to the new network after it have been associated with the CN using the mSCTP, the MN obtains New Care of Address (NCoA) from the network, and mSCTP sends ASCONF-AddIP message to the CN to register NCoA. The mSCTP in MN change the primary IP address after it is considered that the MN joined the new network, and if a previous CoA goes inactive state, delete it from the CN by sending the ASCONF-DeleteIP message. By configuring address of the MN as it moves, session associated between the MN and the CN is not disturbed, and data can be transmitted continuously while the MN is moving.

The mobility management point of view which is focused on how the MN can connect to the CN for communication, mSCTP can handle it in an efficient way. If the MN needs to communicate with the CN, it can transfer data by initiating mSCTP session between them. The specific routers such as, HA and FA of the MIP are unnecessary and mSCTP provides mobility with an end to end principle. However, the mSCTP do not have a location manager by itself, if the CN needs to set up a session to the MN, it must find the information about MN using location manager. Once the session has been made on both cases, mSCTP control the handover and no other protocols need to support handover.

For handover perspective, mSCTP sends ASCONF-AddIP message after it obtain NCoA from the NAR, but mSCTP has no explicit information when it change the primary IP address which is much important to performance of handover. mSCTP resides in the transport layer, so it has not exact movement or location information of the MN. If mSCTP makes improper decision to change of primary IP address, we experience heavy processing overhead in handover and degraded handover performance. Therefore, to improve handover performance of mSCTP, much consideration about change the primary IP address is required. In current specification of mSCTP, the changing primary IP address is remained in main challenge issue [6].

If it is necessary to support real time application in mobile communication, the guarantee for low handover latency is required. And guarantee for low handover latency can't be satisfied in a single mobile protocol. In case of MIP, it is hard to satisfy the real time application for handover with MIP alone, the protocol such as FMIPv6 that provides much tight handover requirement is proposed.

mSCTP also can be used with other mobile protocols for location management or handover management. And it is required that mSCTP interoperates and uses the features of other protocols, if it can provide more handover performance or process a specific aspect of handover more efficiently.

3 Fast Handover in mSCTP Using FMIPv6

In this section, we discuss the handover procedures and the issues of handover in current mSCTP more specifically. After this, we describe the integrated procedure of handover using FMIPv6 with mSCTP for fast handover.

3.1 Current Handover Issues in mSCTP

When the MN moves from one location to others, the mSCTP adds new address sending the ASCONF-AddIP message after it acquires new IP address from NAR, but there are many possibilities when it sends ASCONF-DeleteIP message or more importantly when it changes the primary IP address using ASCONF-SetPrimaryAddress. Currently, the MN can change primary IP address as following cases [6].

1. Immediately after a MN received ASCONF-ACK for ASCONF-AddIP from the CN.
2. Using explicit handover information from layer2 or physical layer.
3. Using a handover trigger from upper layer.

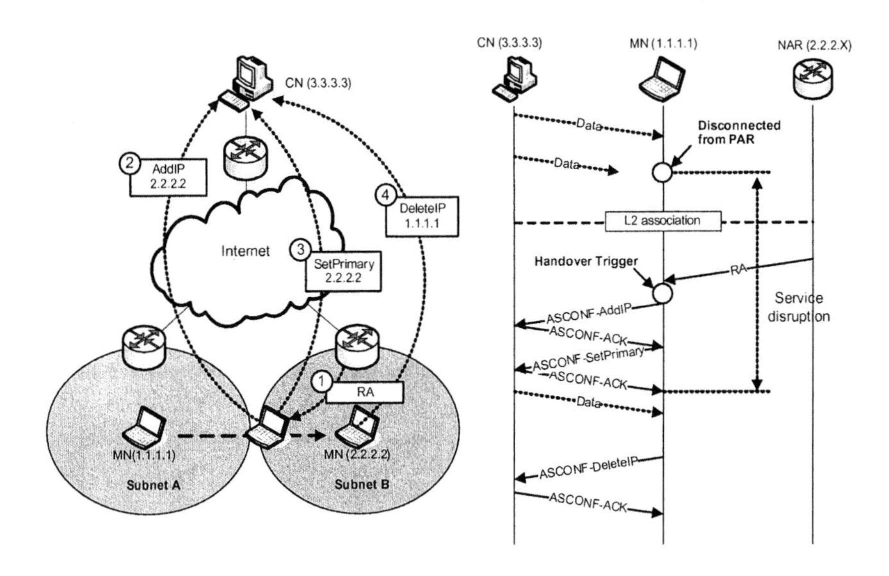

Fig. 1. The typical handover procures in mSCTP

In the first case, the mSCTP changes primary IP address after it registered NCoA to the CN. The performance of handover with this method depends on the state of handover at link layer. If the mSCTP changes the primary IP address after the link layer has successfully joined to the new network, then the mSCTP can receive data through the NAR and changing the primary path is valid. So, the handover in msCTP is also successful. However, mSCTP doesn't have correct movement information of the MN, it is possible that the MN obtains an NCoA actually it does not move to new location. In this case, receiving the ASCONF-ACK from CN does not confirm that the MN is in the new location and the changing primary IP address can makes heavy handover overloads.

The Fig. 1 describes the case1. The MN obtains NCoA from the NAR with Route Advertisement (RA) message, and registers the NCoA to the NAR with sending ASCONF-AddIP. If the MN receives the ASCONF-ACK for ADDIP from the CN, it sends the ASCONF-SetPrimaryAddress to change the primary IP address. In this case, the handover latency consists of the latency that the MN obtains the NCoA with receiving RA from the NAR and the latency that updates this address with sending ASCONF-AddIP and ASCONF-SetPrimaryAddress. The RA is distributed periodically from the NAR, and the latency for obtaining NCoA can be up to the distribution period of RA.

In the second case, it is shown that the handover performance of mSCTP which use the link layer information such as radio signal strength is better than the case without using it [9]. In this case, the mSCTP can process handover with much precise information about link layer, it provides improved handover performance. However, using the information comes from link layer or physical layer in the transport layer, the transport layer must have another independent component that receive information from under layer and store it. Then it evaluates under layer information and determines when it adds new address and changes the primary IP address. But receiving, storing and evaluating under layer handover information at the transport layer is not simple and it does not conform to layered architecture of protocol.

In the third case, the upper layer makes decision of the handover, and sends the trigger to the mSCTP. The upper layer has information about handover that is more than movement of the MN. Especially, if there are multiple wireless networks, the upper layer makes the handover decision from one network to another, considering with a bandwidth, cost, and other performance parameters. This kind of handover is called vertical handover and it can be triggered without moving of the MN. Thus, it is desirable to handover whenever the mSCTP receives handover request from the upper layer.

3.2 Using FMIPv6 for Fast Handover in mSCTP

The FMIPv6 is proposed to improve handover performance of MIPv6. When MN handovers using MIP, there are the IP connectivity latency and the binding update latency. The IP connectivity latency consist of movement detection latency and the NCoA configuration latency in the new subnet area. The binding update latency is the time that MN sends Binding Update to the HA and updates binding information on the HA.

With the FMIPv6, the MN can obtain NCoA without closing connection to current link when it discovers new access point. This process eliminates the IP connectivity latency. The RtSolPr message and PrRtAdv message are used for this purpose.

The FMIPv6 makes tunnel between the PAR and NAR using Fast Binding Update (FBU) message and the MN sends Fast Neighbor Advertisement (FNA) message to the NAR when it attached to the NAR. With this FBU and FNA process, the MN can reduce binding update latency significantly and minimize packet losses during handover.

If we consider the handover procedures of mSCTP, the IP connectivity latency still exists in the same way. The latency that updates the NCoA using ASCONF-AddIP and changes primary IP address is considered as binding update latency. So FMIPv6 can also be used with mSCTP to improve the handover performance.

As see in Fig. 2, when FMIPv6 is used for handover processing in mSCTP, the FMIPv6 processes handover information from link layer or physical layer. The FMIPv6 receives information from link layer and use this information to send RtSolPr message. If FMIPv6 acquires NCoA through RtSolPr and PrRtAdv messages exchange, this NCoA can be used directly in mSCTP to send the ASCONF-AddIP message. As a result, the IP connectivity latency in mSCTP can be reduced and the MN can start handover process without long IP connectivity latency.

After the FMIPv6 sent FNA message to the NAR which means that the MN has attached to the NAR, FMIPv6 immediately triggers mSCTP that it can

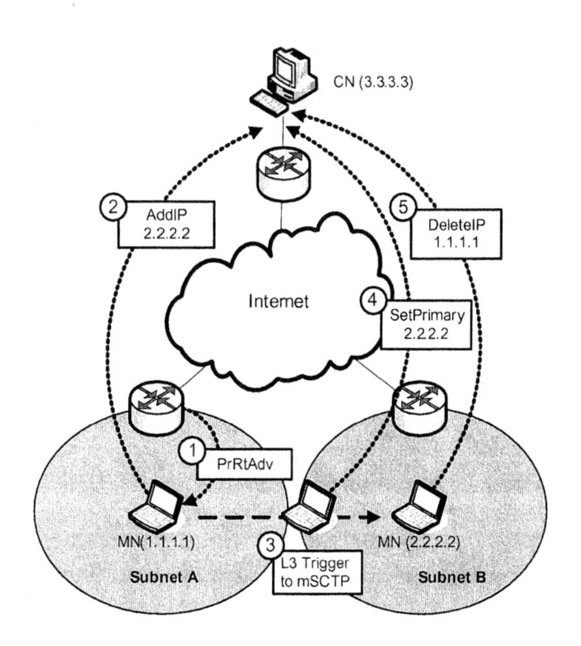

Fig. 2. Handover procedures in mSCTP with FMIPv6

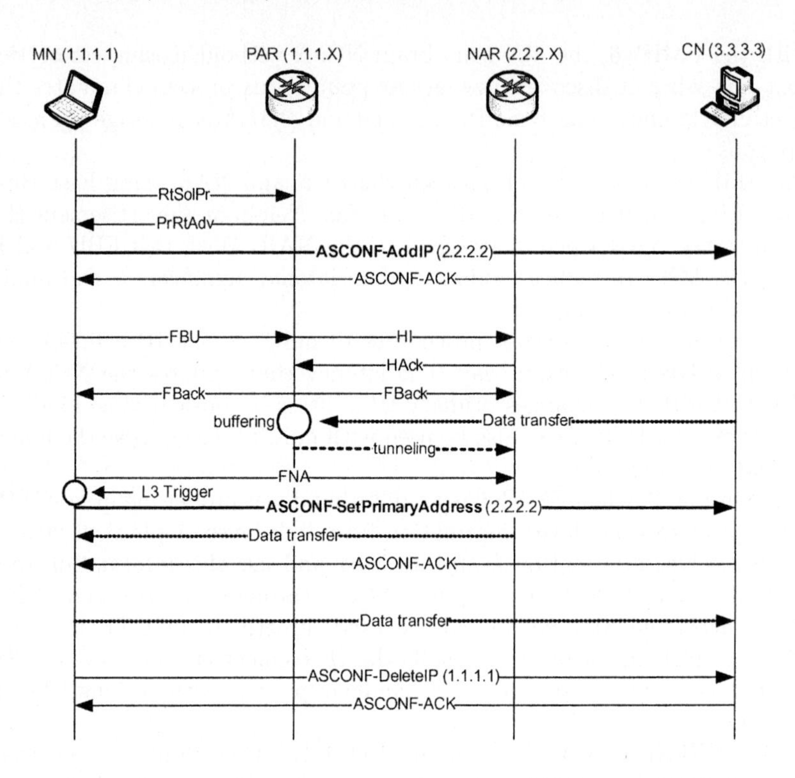

Fig. 3. Integrated handover procedures in mSCTP with FMIPv6

change primary IP address. Receiving trigger that the FMIPv6 has sent the FNA message to the NAR, confirms that the MN already has attached to the subnet of NAR. So, with this trigger, the MN can change the primary IP address always after it has attached to the subnet of NAR.

The integrated procedures of the handover using FMIPv6 with mSCTP which maximize the performance of the handover are shown in Fig. 3.

The moment that MN receive PrRtAdv message from PAR, MN can acquire network address of the NAR. And the MN can register network address of NAR to the address list of CN with ASCONF-AddIP message before moving to the location of NAR.

The FMIPv6 in MN generates trigger immediately after it sends FNA message to NAR. Using this trigger, mSCTP sends ASCONF-SetPrimaryAddress message to change the primary IP address.

In Fig. 3, using FMIPv6 with mSCTP, mSCTP can have more improved handover performance as following reasons.

1. mSCTP can obtain and register the NCoA to the CN without IP connectivity latency with the FMIPv6 RtSolPr, PrRtAdv address acquisition process.
2. mSCTP can change primary IP address with confirmation that the MN has attached the subnet of the NAR.

Actually, FMIPv6 processes handover sequences and sends trigger to mSCTP, mSCTP does not need to process information from link layer or physical layer to determine the time changing primary IP address. Through this, mSCTP can have simple and correct information about under layer handover state, and can make clear handover decision correspondent to under layer state.

4 Experiment and Result

In this section, we introduce experimental result of suggested handover procedure. Fig. 4 shows the network set up of experimental environment.

There are two client systems perform as MN and CN. The CN is fixed PC system and the MN has *Lucent Orinoco USB client silver* WLAN adapter. For both client systems, we ported Linux kernel 2.6.12.2 that includes *lksctp* [10] module for SCTP and DAR extension, and integrated it with FMIPv6 that have been ported our previous project. The PAR and NAR are typical routers based on Linux kernel 2.4.20 and support IPV6 routing. We also use two common *Linksys wireless-G* APs.

We performed experiment as following handover process shown in Fig. 1 and Fig. 2. Fig. 5 (a) shows the result of handover latency in mSCTP without FMIPv6. Each handover result shows three different latencies. The first bar (CONN-ADDIP) represents the latency from link up event to the event that MN sends ASCONF-AddIP message. The second bar (CONN-ACK) represents the latency from link up event to the event that MN receives ASCONF-ACK message for ASCONF-SetPrimaryAddress. And the third bar (DISC-ACK) rep-

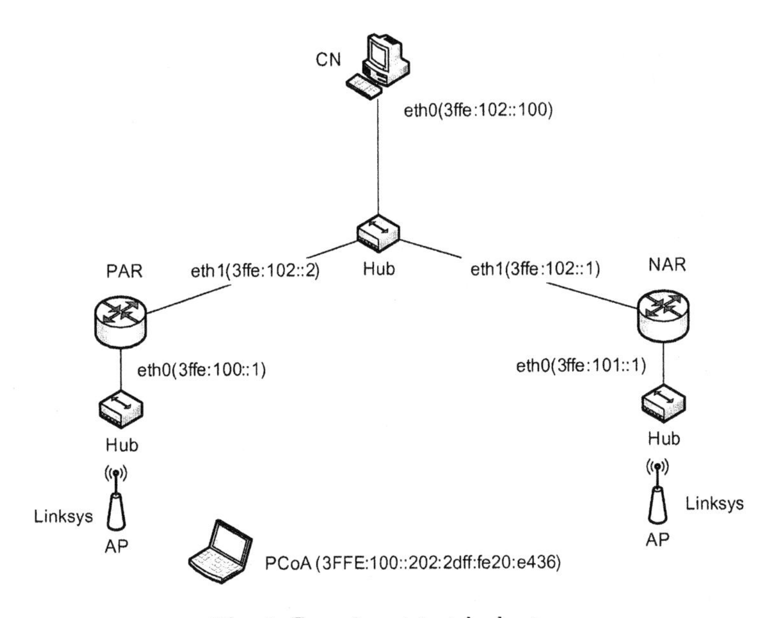

Fig. 4. Experiment test bed set up

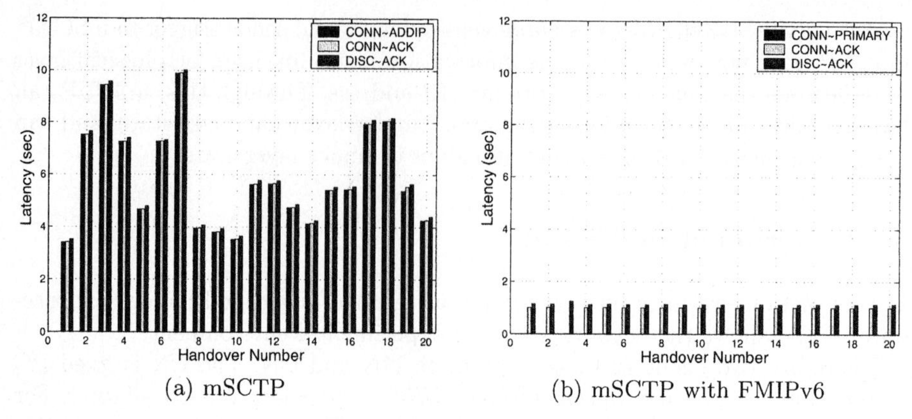

(a) mSCTP (b) mSCTP with FMIPv6

Fig. 5. Handover latencies

resents the total handover latency form link down event to the event that the MN receives ASCONF-ACK message for ASCONF-SetPrimaryAddress.

In handover result of mSCTP without FMIPv6, we can see that the total handover latency is mainly caused by CONN-ADDIP latency. In CONN-ADDIP latency, MN configures NCoA using RA that is periodically sent by NAR. We use RA sending period up to 10 seconds, therefore the CONN-ADDIP latency varies widely from few seconds to over than 10 seconds.

Fig. 5 (b) is the handover latency result from mSCTP using FMIPv6. The first bar (CONN-PRIMATY) represents the latency from link up to the event that MN sends ASCONF-SetPrimaryAddress. The second bar (CONN-ACK) represents the latency from link up to the event that MN receives ASCONF-ACK for ASCONF-SetPrimaryAddress message. And the third bar (DISC-ACK) represents the total handover latency from link down to the event that MN receives ASCONF-ACK for ASCONF-SetPrimaryAddress message.

We notice that CONN-PRIMARY latency is very low and takes only few *ms*. From this result, we can see that MN can acquire the NCoA very quickly by using FMIPv6 and it takes very small times to send ASCONF-SetPrimaryAddress message. However, the CONN-ACK latency takes more than 1 second. We found out that this latency was caused by the delay exist in IPv6 before sending first packet due to DAD, NUD, NS/NA processing, not by the round trip delay between ASCONF-SetPrimaryAddress to ASCONF-ACK. While implementing MN and CN for experiment, we did not optimize this processing for performance, so it exist both handover procedures in our experiment. Note that the handover performance of suggested procedure is greatly improved by pre-acquisition of NCoA and variation of handover latency is relatively small than that in mSCTP without FMIPv6.

5 Conclusion

We have introduced the fast handover method in mSCTP using FMIPv6. Using FMIPv6 as handover procedure, mSCTP can add NCoA to CN quickly without closing a connection to PAR and can change primary IP address after the MN has completely joined to the subnet of the NAR using the trigger from FMIPv6.

Therefore, handover latency can be reduced significantly and mSCTP can process handover with explicit trigger from FMIPv6. mSCTP does not have to maintain or evaluate the information from link layer and physical layer directly. And mSCTP resides in the transport layer and FMIPv6 is in the network layer, trigger handover between mSCTP and FMIPv6 are simple and efficient.

We have presented the integrated handover procedures of the FMIPv6 and the mSCTP to optimize handover performance and performed experiment on our test bed. The experimental result shows that the handover latency in mSCTP is mainly caused by NCoA acquisition process that depends on RA and FMIPv6 can improve the handover performance significantly by pre-acquisition of NCoA.

References

1. C. Perkins: IP Mobility Support for IPv4, RFC3344, Aug. 2002
2. D. Johnson, C. Perkins, J. Arkko: Mobility Support in IPv6, RFC3775, June 2004
3. Henning Schulzrinne, Elin Wedlund: Application-Layer Mobility Using SIP, Mobile Computing and Communications Review, Vol. 4, No. 3, July 2000
4. Rajeev Koodli: Fast Handover for Mobile IPv6, RFC 4068, Oct. 2004
5. Wei Xing, Holger Karl, Adam Wolisz: M-SCTP Design and Prototypical Implemetation of an End-to-End Mobility Concept, Proc. of 5th Intl. Workshop The Internet Challenge: Technology and Applications, Berlin, Germany, Oct. 2002
6. Seok. J. Koh, Qiaobing Xie: Mobile SCTP (mSCTP) for Internet Mobility, IETF Internet Draft, draft-sjkoh-msctp-00.txt, Mar. 2005, work in progress
7. R. Stewart, Q. Xie: Stream Control Transmission Protocol, RFC2960, Oct. 2000
8. R. Stewart, M. Ramalho, Q. Xie, M. Tuexen, P. Conrad: Stream Control Transmission Protocol (SCTP) Dynamic Address Reconfiguration, IETF Internet Draft, draft-ietf-tsvwg-addip-sctp-11.txt, Feb. 2005, work in progress
9. Moonjeong Chang, Meejeong Lee: A Transport Layer Mobility Support Mechanism, ICONS 2004 LNCS 3090
10. The Linux Kernel SCTP project, http://lksctp.sourceforge.net/

An Analytical Comparison of Factors Affecting the Performance of Ad Hoc Network*

Xin Li, Nuan Wen, Bo Hu, Yuehui Jin, and Shanzhi Chen

Broadband Network Research Center, State Key Laboratory of Networking and Switching, Beijing University of Posts and Telecommunications
lixin@bupt.edu.cn, ruby_218@163.com, hubo@bupt.edu.cn, yhjin@bupt.edu.cn, chenshanzhi@yahoo.com.cn

Abstract. In this paper an analytical model is proposed to investigate and quantify the effects and interactions of node mobility, network size and traffic load on the performance of ad hoc networks using AODV in terms of cost, average end-to-end delay and throughput. The analytical results reveal that contrary to the traditional concept, performance of ad hoc networks is much more sensitive to traffic load and network size than to node mobility. The capacity of ad hoc networks relies on the collective impact of all three factors but not any one alone. Furthermore, NS-2 based simulations are carried out to verify the theoretical model.

1 Introduction

Development of dynamic routing protocols is one of the central challenges in the design of ad hoc networks. Accurate understanding of the nature of network is the cornerstone of successful routing protocol design. But little work has been carried out to systematically analyze the impacts or interactions of factors such as node mobility, network size and traffic load. It has been a rooted belief that volatile topology could impose the most significant impact and as a result tremendous efforts have been focused on mechanisms that can catch up with the high degree of node mobility.

Researches[1][2][3] focused on comparison between different routing protocols have given some insights into the nature of ad hoc networks. It has been a common sense that there is no generally perfect ad hoc routing protocol. To achieve best performance existing routing protocols must be adapted to the given scenarios.

In this paper, an analytical model is proposed to facilitate qualifying and quantifying the impacts and correlation of node mobility, network size and traffic load Original version of AODV[4] is selected to represent the inherent nature of reactive routing protocols. But the framework of this model is not strictly AODV-oriented. It could also be utilized in analysis on other protocols.

* This project is supported by the state 863 high-tech program of China (the Project Num is 2005AA121630), National Natural Science Foundation of China (Grant No.90204003, Grant No. 60472067), National Basic Research Program of China (Grant No. 2003CB314806), and Beijing Municipal Key Disciplines (XK100130438) .

X. Jia, J. Wu, and Y. He (Eds.): MSN 2005, LNCS 3794, pp. 856 – 865, 2005.

NS-2[5] based simulations are also carried out to verify the theoretical analysis. Both analytical and simulation results show that contrary to the rooted belief that ad hoc networks suffer the most to node mobility, the collective impact of network size and traffic load are much more decisive to the network capacity. In the development of routing protocols more attention should be paid to improve the network scalability and alleviate congestion.

The rest of this paper is organized as follows. In section II, preliminary notions, assumptions and performance metrics are briefly presented. Section III gives an detailed introduction to the new model. Analytical and simulation results are presented in section IV and V respectively. Finally we conclude this paper in section VI.

2 Preliminaries

2.1 Basic Notions

Suppose node X and Y has been within the transmission range of each other during the time interval $(t_0, t]$ and at time t X needs to transmit a packet to Y, then the link (X, Y) can be in one of the following states:

Link Broken: Y has moved out of the transmission range of X.
Link Un-operational: Y is still within the transmission range of X, but the packet transmission from X to Y fails.
Def.1 Link Failure: link (X, Y) fails if it is broken or un-operational.
Def.2 Route Failure: a route $\{X_1,X_2,...X_k\}$ fails if a packet encounters link failure before it reaches X_k.

2.2 Assumptions

1) The ad hoc network is deployed within a square field. Network size is measured by border length b in terms of hops.
2) Route length in terms of hops between any two given nodes is uniformly distributed between 1 and the network diameter.
3) Traffic load is generated by a number of CBR sources with transmission rate λ and fixed packet size. UDP is selected as the transport layer protocol.
4) All nodes move in accordance with random waypoint model[6]. IEEE 802.11[7] is used as the MAC layer protocol. Due to the memoryless nature of random waypoint model, the hold time of links between adjacent nodes is exponentially distributed and has a mean μ.
5) To guarantee full connectivity in the network, the minimum number of neighbors of any node should be no less than π[8], which requires the number of nodes (N) should be no less than $(\pi+1)b^2/\pi$. Considering the density waves[6] of node distribution, N is defined to be:

$$N = 2ceil\left(\frac{(\pi+1)b^2}{\pi}\right) \tag{1}$$

6) Link-unoperation is mainly caused by noises, obstacles, network congestion, and multiple access interferences. In our model only high contention rates, congestion and hidden terminal problem that are proportional to the traffic load are taken into account. So the probability of link un-operational (p_u) is taken as the measure of traffic load.

7) The original version of AODV without expanding ring search is adopted in this model.

2.3 Performance Metrics

Cost—the average amount of data bytes and control bytes generated to successfully deliver a data byte. Each hop-wise transmission of routing and data bytes should be taken into account.

Throughput— ratio of the data packets delivered to the destination to those generated by the CBR sources.

Average end-to-end delay— this includes all possible delays caused by buffering during route discovery latency, retransmission delays caused by route failure and propagation time. For simplicity and clarity, average transfer time in one hop, D_L, is selected as the measure of delay.

3 Analytical Model

3.1 General Results

Lemma 1. The maximum route length within ad hoc network is:

$$L_{\max} = ceil\left(\sqrt{2}b\right) \tag{2}$$

The average route length within ad hoc network is:

$$E_L = ceil\left(\frac{1 + ceil\left(\sqrt{2}b\right)}{2}\right) \tag{3}$$

Function ceil(A) rounds the elements of A to the nearest integer greater than or equal to A.

Proof: according to assumptions 2) the result is straightforward.

Theorem 1. The probability that a packet is successfully transmitted over a link connecting two common nodes is:

$$p_L = (1 - p_u)e^{-\frac{1}{\lambda\mu}} \tag{4}$$

Proof: Due to the memoryless nature of exponential distribution, the hold time of a link between common nodes since the end of last packet transmission, say t, is independent of the link hold time before t and exponentially distributed with mean μ. Thus the probability that a link broken occurs before a transmission is given by:

$$p_B = P\left(t < \frac{1}{\lambda}\right) = 1 - e^{-\frac{1}{\lambda\mu}}$$

To successfully transmit a packet over a link connecting common nodes, two preconditions must be satisfied: the link is non-broken and operational. So the probability that a link is not failed is:

$$p_L = (1 - p_u)(1 - p_B) = (1 - p_u)e^{-\frac{1}{\lambda\mu}}$$

Lemma 2. The average number of links successfully passed by a packet along a route before an error occurs is:

$$q = \left[\frac{1}{L_{\max}} \sum_{H=1}^{L_{\max}} \sum_{k=1}^{H} kP_L^{k-1}(1 - P_L)\right] - 1 \tag{5}$$

Proof: For a route of H hops, let the number of links (including the failed link) passed by a packet to encounter a link failure be Q_H, thus

$$p(Q_H = k) = p_L^{k-1}(1 - p_L)$$

The expected value of Q_H is:

$$E[Q_H] = \sum_{k=1}^{H} kP(Q = k) = \sum_{k=1}^{H} kP_L^{k-1}(1 - P_L)$$

According to assumption 2), the probability that the route length between any pair of given nodes equals L is $1/L_{MAX}$. Hence for any packet in the network, the average number of links it passed to encounter an error (including the failed link), let it be Q, should be:

$$E[Q] = E\{E[Q_H]\} = \sum_{H=1}^{L_{MAX}} P(L)E[Q_H] = \frac{1}{L_{MAX}} \sum_{H=1}^{L_{MAX}} \sum_{k=1}^{H} kP_L^{k-1}(1 - P_L)$$

And the average number of links successfully passed by a packet along a route before an error occurs is:

$$q = Q - 1 = \left[\frac{1}{L_{\max}} \sum_{H=1}^{L_{\max}} \sum_{k=1}^{H} kP_L^{k-1}(1 - P_L)\right] - 1$$

Lemma 3. The average number of routing failures a given packet encountered before it is successfully delivered to the destination is:

$$z = \frac{1}{L_{MAX}} \sum_{H=1}^{L_{\max}} \frac{1}{p_L^H} - 1 \tag{6}$$

Proof: Let Z_H be an r.v. which describes the number of routing attempts needed to successfully deliver a packet to its final destination through a route with H hops. Z_H has a geometric distribution given by:

$$P(Z_H = k) = (1 - p_{SH})^{k-1} p_{SH}, \quad p_{SH} = p_L^H$$

p_{SH} is the probability that a packet is successfully routed to its final destination through H hops. Thus the expected value of Z_H is $1/p_{SH}$, and the expected value of Z should be:

$$E[Z] = E\{E[Z_H]\} = \sum_{H=1}^{L_{MAX}} P(H)\frac{1}{p_{SH}} = \frac{1}{L_{MAX}}\sum_{H=1}^{L_{MAX}}\frac{1}{p_L^H}$$

So the average number of routing failures before a packet is successfully delivered is:

$$z = E[Z] - 1 = \frac{1}{L_{MAX}}\sum_{H=1}^{L_{MAX}}\frac{1}{p_L^H} - 1$$

Theorem 2. Let the number of data bytes in a packet be η, the average traffic generated to successfully route a packet to its final destination using AODV is:

$$C = \frac{z(104q + 48E_L + 52N) + 64E_L}{\eta} \tag{7}$$

Proof: Let C_{LS}, C_{ERR}, C_{REQ} and C_{REP} be the one-hop cost at network layer of successfully delivering a data packet, route error message, route request message and route reply message respectively. Hence the traffic generated to successfully route a data packet is composed of the following costs:

The cost of successfully transmitting a data packet through the route: E_LC_{LS}.
The cost of failed data transmission: zqC_{LS}.
The cost of informing the source about route failure: zqC_{ERR}.
The cost of route discovery, including the cost incurred by the source to broadcast RREQ and the destination to send back RREP: $z(NC_{REQ}+E_LC_{REP})$.

We also assume that each source sends out packets without stop and the cost incurred by the first round of route discovery could be neglected. So the overall cost of successfully routing a packet is:

$$C' = z(qC_{ERR} + qC_{LS} + E_LC_{REP} + NC_{REQ}) + E_LC_{LS}$$

When using AODV, C_{ERR}, C_{REQ} and C_{REP} should be 40 bytes, 50 bytes and 48 bytes respectively. To calculate the byte efficiency, C' should be divided by the number of bytes in payload η. Thus the result is straightforward.

Theorem 3. When using AODV, throughput of the network is:

$$T = \frac{L_{MAX}}{\sum_{H=1}^{L_{max}}\frac{1}{p_L^H}} \tag{8}$$

Proof: By definition throughput is just the reciprocal of the average number of routing attempts needed to successfully deliver a packet to its final destination, i.e. 1/Z. Based on lemma 3, the result is straightforward.

Theorem 4. When using AODV, the average end-to-end delay is:

$$D = 2qzD_L + E_LD_L + 2zE_LD_L \tag{9}$$

Proof: When using AODV, before a packet is successfully delivered it may encounter several route failures. Using lemma 1, both the average time consumed to transmit a packet from source to the failed link and inform the source of the failure is qD_L. In route discovery process, although RREQ is flooded throughout the network, the average time for RREQ reaching destination is E_LD_L. The same is that for RREP being routed back. Thus the average cycle for one route discovery is $2E_LD_L$. Using lemma 3, the average delay incurred by handling route failures before a packet is delivered is $z(2qD_L+2E_LD_L)$. At last, the average time of successfully routing a packet—E_LD_L should be appended. So theorem 4 is proved.

4 Analytical Results

In this section the impacts imposed by node mobility, network size and traffic load on the performance of fast node are evaluated and compared using the model mentioned above. To facilitate quantifying the effects of different factors, CBR sources are assumed to send 64-byte packets with a transmission rate of 10 packets per second.

4.1 Analysis on Cost

Fig.1a gives cost as a function of average link hold time and network size. It is observed that cost keeps rather stable whenμvaries from 10 seconds to infinite. Steep rise occurs when$\mu$is less than 5 seconds. That is really an extremely high speed and not realistic, so the impact of node mobility is not so prominent as it has long been estimated. Both Fig.1a and Fig.1b demonstrate that network size and traffic load could impose much greater impact than node speed. Tiny variations in either factor could dramatically fluctuate the network performance. According to theorem 2, overhead incurred by flooding network with *RREQ* constitutes the greatest majority of cost. In the light of assumption 5 and theorem 2, route discovery overhead is proportional to z and b^2, while lemma 3 indicates that retransmission times per packet is mainly decided by b and p_u. So network size and traffic load could be much more influential than node mobility. Fig.1c gives more insights into the interaction of b and p_u. It shows that b or p_u alone could not dominate the network performance. In small networks even under heavy traffic load could cost still be kept at low level, while extremely low traffic load could remarkably improves the network scalability. But when both factors exceed some thresholds, network capacity could be highly vulnerable to any slight increment in either factor.

4.2 Analysis on Average End-to-End Delay and Throughput

Fig.2 depicts the relative impacts of node mobility, traffic load and network size on average end-to-end delay. They show that the average end-to-end delay is not sensible to node mobility too. Another interesting observation is that compared with cost, both the absolute value and variation pace in delay are kept at a moderate level. This is due to the reason that when handling a route failure, cost incurred by flooding *RREQ* across network has a magnitude of $2b^2$ bytes. But to delay, time interval incurred by

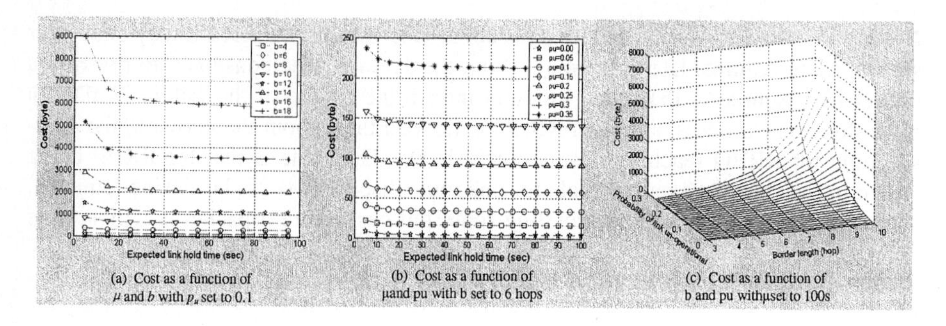

(a) Cost as a function of μ and b with p_u set to 0.1

(b) Cost as a function of μand pu with b set to 6 hops

(c) Cost as a function of b and pu withμset to 100s

Fig. 1. Analysis on cost

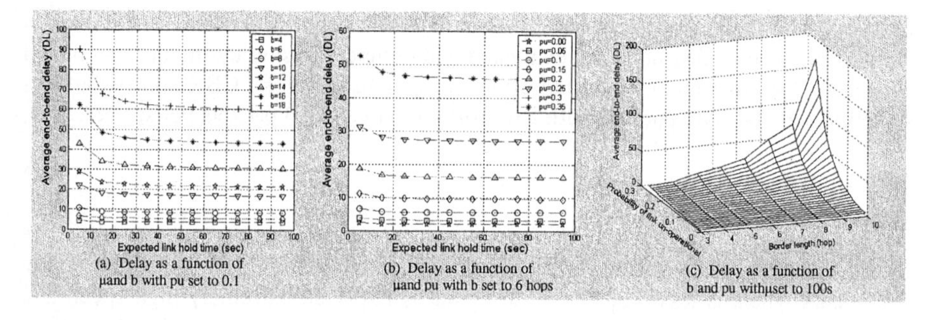

(a) Delay as a function of μand b with pu set to 0.1

(b) Delay as a function of μand pu with b set to 6 hops

(c) Delay as a function of b and pu withμset to 100s

Fig. 2. Analysis on the average end-to-end delay

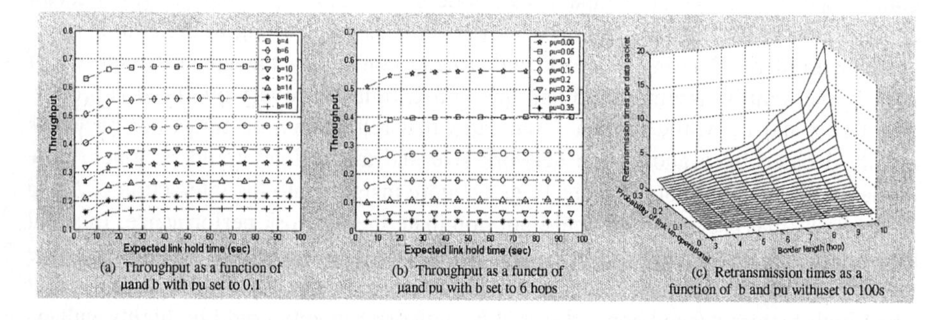

(a) Throughput as a function of μand b with pu set to 0.1

(b) Throughput as a functn of μand pu with b set to 6 hops

(c) Retransmission times as a function of b and pu withμset to 100s

Fig. 3. Analysis on throughput and retransmission times

one route discovery is just around $2E_L$, i.e. $b/\sqrt{2}$. This gives us an elicitation: by exploring the average retransmission times per packet, i.e. Z, more insights into the correlation between different factors could be achieved.

Just the same as we have observed in Fig.1 and Fig.2, Fig.3a and Fig.3b show that b and p_u could impose much greater impact on throughput than node mobility too. So we can conclud that node mobility is not a decisive factor to the network performance. Fig.3c illustrated that average retransmission times has almost the same characteristic as that of cost and delay. But the curves in this figure are steeper than their counterparts in Fig.1c and Fig.2c. The reason lies in that z is more directly influenced by network size and traffic load. Even within the scope where cost and delay keep rather stable, obvious variations in retransmission times could still be observed.

5 Simulation Results

5.1 The Simulation Model

NS-2 based simulations were also carried out to verify the theoretical model. Simulation configurations are the same as that mentioned in the theoretical analysis. Source-destination pairs are spread randomly over the network. For a network with N nodes, ceil(N/5), ceil(1.25N/5) and ceil(1.5N/5) sources are used to emulate three different traffic loads in the second set of simulations. Two more configurations: ceil(1.25N/5) and ceil(1.5N/5) are also adopted in the third set of experiments.

In our simulations node mobility is measured by the average speed with pause time set to 0 second. Five node speed configurations, 2m/s, 6m/s, 10m/s, 14m/s and 30m/s, are selected to represent a series of descending link hold time. But how to calculateμas a function of node speed is out of the scope of this paper.

Three network size configurations with network border length set to 4 hops, 5 hops and 6 hops respectively are adopted. The number of nodes in each field is decided according to assumption 5).

Simulations are run for 400 simulated seconds for each scenario. Each data point represents an average of at least five runs with identical traffic load, network size, average node speed, but different randomly generated mobility scenarios. It should be noticed that in all the simulations the average end-to-end delay is measured in seconds. But this will not affect the comparison between analytical and simulation results.

5.2 Simulation Results

The first set of experiments examines the effects of average node speed and network size. To keep traffic load at the same level, the number of sources is set to ceil (N/5) in each scenario. It could be observed in Fig.4 that even if the node speed has been raised to 14 times higher than the initial value, the network performance just suffer very little degradation of no more than 25%. But one hop increment in the network border length could lead to an increase of more than 80%.

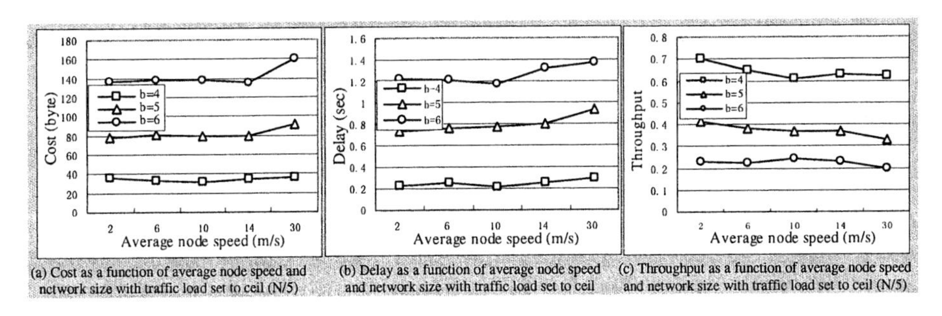

(a) Cost as a function of average node speed and network size with traffic load set to ceil (N/5)
(b) Delay as a function of average node speed and network size with traffic load set to ceil
(c) Throughput as a function of average node speed and network size with traffic load set to ceil (N/5)

Fig. 4. The relative impact of node speed and network size

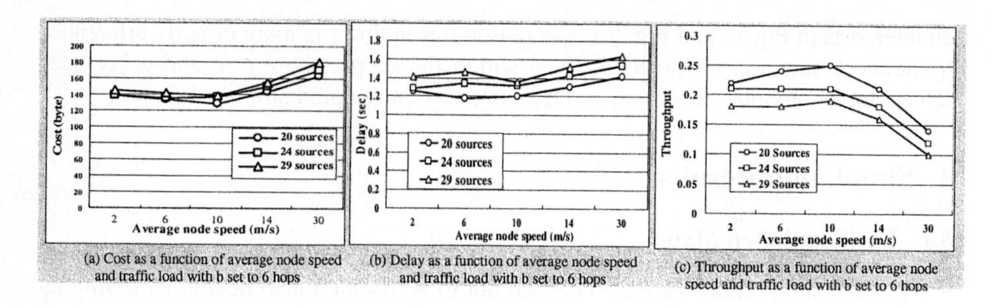

(a) Cost as a function of average node speed and traffic load with b set to 6 hops

(b) Delay as a function of average node speed and traffic load with b set to 6 hops

(c) Throughput as a function of average node speed and traffic load with b set to 6 hops

Fig. 5. The relative impact of node speed and traffic load

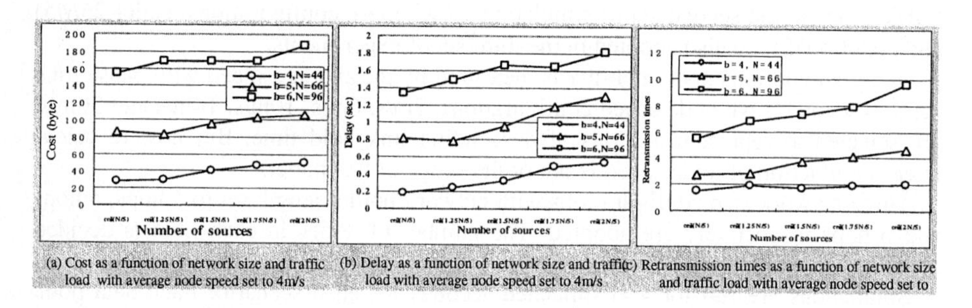

(a) Cost as a function of network size and traffic load with average node speed set to 4m/s

(b) Delay as a function of network size and traffic load with average node speed set to 4m/s

(c) Retransmission times as a function of network size and traffic load with average node speed set to

Fig. 6. The relative impact of network size and traffic load

In the second set of experiments correlation between traffic load and node mobility is investigated with the network border length uniformly set to 1500 meters in each scenario. Although we could not figure out which p_u should 20, 24 and 29 sources correspond to, Fig.5a, Fig.5b and Fig.5c still reveal that the network performance could suffer more to increased traffic load than boosted node speed. Fig.5 also shows that when node speed exceeds 10m/s, the network performance would be dramatically degraded by increased node mobility. The same trend could also be observed in Fig.4, although it is not so obvious when b is set to 4 and 5. In section Ⅳ, our analytical results have illustrated that the impact of node mobility could increase sharply as average link hold time decreases below certain threshold value and the pace is proportional to network size, which is coincident with the simulation results.

Fig.4 and Fig.5 also illustrate that the influence of traffic load are not so dramatic as network size. Fig.6 further confirms it. This also corresponds to the observations from Fig.1c and Fig.2c. Based on our analytical model the reason is rather straight forward: larger network means longer average route length, end-to-end delay and more routing overhead, especially the route discovery overhead, which directly contribute to the performance degradation. At the same time, just as shown in Fig.3c and Fig.6c, longer route length could lead to more route failures and retransmission times, thus further deteriorate the situation.

6 Conclusions

In this paper an analytical model is proposed to investigate and quantify the impacts and interactions of network size, node mobility and traffic load on the performance of ad hoc networks. The original version of AODV is selected to represent the inherent nature of reactive routing protocols. Ns-2 based simulations are also carried out to verify the validity of the theoretical model. Both the analytical and simulation results reveal that performance of ad hoc networks depends on the collective impact of different factors but not any one alone while network size and traffic load could impose much greater impact than node mobility. Network performance could start saturating even when the average link hold time is still at a very low level. These conclusions also reveal that routing protocol design in ad hoc networks is highly scenario-oriented. It is an art of balancing and compromising. Before setting about to solve problems it would be better to find out what the problem is.

References

1. Hsu. J, Bhatia. S, Takai. M, Bagrodia. R and Acriche. M.J, "Performance of mobile ad hoc networking routing protocols in realistic scenarios," MILCOM 2003. IEEE Volume 2, Page(s):1268 – 1273, 13-16 Oct. 2003
2. J. Broch, D. A. Maltz, D. B. Johnson, Y.-C. Hu, and J. Yetcheva, "A performance comparison of multi-hop wireless ad hoc network routing protocols", MOBICOM'98, 1998. Dallas, Texas.
3. D. Samir. R, R. Castañ, eda. J. Yan, "Simulation-based performance evaluation of routing protocols for mobile ad hoc networks," *Mobile Networks and Applications*, vol. 5 Issue 3, September 2000.
4. C. E. Perkins and E. M. Royer, "Ad Hoc On-demand Distance Vector Routing," Proc. 2nd IEEE Wksp. Mobile Comp. Sys. and Apps, pp. 90–100, Feb. 1999.
5. K. Fall and K. Varadhan, Eds, ns notes and ocumentation, 1999; available from http://www-mash.cs.berkeley.edu/ns/.
6. T. Camp, J. Boleng, and V. Davies. "A survey of mobility models for ad hoc network research", Wireless Comm. and Mobile Computing (WCMC), 2(5): 483--502, 2002
7. IEEE, "Wireless LAN Medium Access Control (MAC) and Physical Layer (PHY) Specifications," IEEE Std. 802.11-1997, 1997.
8. G. Ferrari and O.K. Tonguz, "Minimum number of neighbors for fully connected uniform ad hoc wireless networks", IEEE International Conference on Communications (ICC'04), Paris, France, 2004

Maximum Throughput and Minimum Delay in IEEE 802.15.4

Benoît Latré[1,*], Pieter De Mil[1], Ingrid Moerman[1], Niek Van Dierdonck[2], Bart Dhoedt[1], and Piet Demeester[1]

[1] Department of Information Technology (INTEC), Ghent University - IBBT - IMEC
Gaston Crommenlaan 8, bus 201, B-9050 Gent, Belgium
Tel. +32 9 331 49 00, Fax. +32 9 331 48 99
benoit.latre@intec.UGent.be
[2] Ubiwave NV, Warandestraat 3, B-9240 Zele, Belgium
Tel. +32 52 45 39 80, Fax. +32 52 45 39 89

Abstract. This paper investigates the maximum throughput and minimum delay of the new IEEE 802.15.4-standard. This standard was designed as a highly reliable and low-power protocol working at a low data rate and offers a beaconed and unbeaconed version. We will give the exact formulae for a transmission between one sender and one receiver for the unbeaconed version as this one has the least overhead. Further, the influence of the different address schemes, i.e. no addresses or the use of long and short addresses, is investigated. It is shown that the maximum throughput is not higher than 163 kbps when no addresses are used and that the maximum throughput drops when the other address schemes are used. Finally, we will measure the throughput experimentally in order to validate our theoretical analysis.

1 Introduction

The market of wireless devices has experienced a significant boost in the last few years and new applications are emerging rapidly. Several new protocols have been proposed such as IEEE 802.11g and IEEE 802.16. However, these protocols focus on achieving higher data rates in order to support high bit rate applications for as much users as possible. On the other hand, there is a growing need for low data rate solutions which provide high reliability for activities such as controlling and monitoring. Furthermore, these applications often use simple devices which are not capable of handling complex protocols. In order to cope with this problem, a new standard was defined in the end of 2003: IEEE 802.15.4 [1].

The goal of the IEEE 802.15.4 standard is to provide a low-cost, highly reliable and low-power protocol for wireless connectivity among inexpensive, fixed and portable devices such as sensor networks and home networks [2, 3]. This last type of networks is commonly referred to as Wireless Personal Area Networks (WPAN). The standard works in the 2.4 GHz range -the same range as 802.11b/g

* Corresponding author.

X. Jia, J. Wu, and Y. He (Eds.): MSN 2005, LNCS 3794, pp. 866–876, 2005.

and Bluetooth- and defines a physical layer and a MAC sub layer. The standard is used by the Zigbee Alliance [4] to build a reliable, cost-effective and low-power network.

In this paper, we will investigate the maximum throughput and minimum delay of 802.15.4. We will do this both analytically and experimentally. All the information needed for obtaining these results can be found in the standard [1]. This paper will offer the exact formulae for these calculations in order to give an overview and an easy way to calculate the maximum throughput without the need to completely understand the standard.

The paper is organized as follows. Section 2 will give a brief technical overview of 802.15.4. In section 3, the maximum throughput is calculated. The analysis of the results is given in section 4 and experimental validation is done in section 5. Finally, section 6 concludes the paper.

2 Technical Overview

The new IEEE 802.15.4 defines 16 channels in the 2.4 GHz band, 10 channels at 915 MHz and 1 channel at 868 MHz. The 2.4 GHz band is available worldwide and operates at a raw data rate of 250 kbps. The channel of 868 MHz is specified for operation in Europe with a raw data rate of 20 kbps and for North America the 915 MHz band is used at a raw data rate of 40 kbps. All of these channels use DSSS. The standard specifies further that each device shall be capable of transmitting at least 1 mW (0 dBm), but actual transmit power may be lower or higher. Typical devices are expected to cover a 10–20 m range.

The MAC sub layer supports different topologies: a *star topology* with a central network coordinator, a *peer to peer* topology (i.e. a tree topology) and a *combined topology* with interconnected stars (clustered stars). Both topologies use CSMA/CA to control access to the shared medium. All devices have 64-bit IEEE addresses, but short 16-bit addresses can be assigned.

In order to achieve low latencies, the IEEE 802.15.4 can operate in an optional superframe mode. In this mode, beacons are sent by a dedicated device, called a PAN-coordinator at predetermined intervals (PAN = Personal Area Network). These intervals can vary from 15 ms to 245 seconds. The time between these beacons is split in 16 slots of equal size and is divided in two groups: the contention access period (CAP) and the contention free period (CFP) in order to provide the data with quality of service (QoS). The time slots in the CFP are called guaranteed time slots (GTS) and are assigned by the PAN-coordinator. The channel access in the CAP is contention based (CSMA/CA). When a device wishes to transmit data, the device waits for a random number of back off periods. Subsequently, it checks if the medium is idle. If so, the data is transmitted, if not, the device backs off once again and so on.

As the MAC sub layer needs a finite amount of time to process data received from the PHY, the transmitted frames are followed by an Inter Frame Space (IFS) period. The length of the IFS depends on the size of the frame that has just been transmitted. Long frames will be followed by a Long IFS (LIFS) and

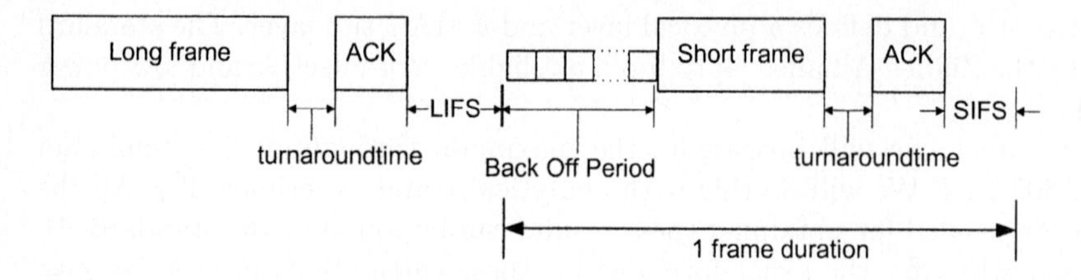

Fig. 1. Frame sequence in 802.15.4. We notice the back off period and that long frames are followed by a long inter frame space and short frames by a short inter frame space.

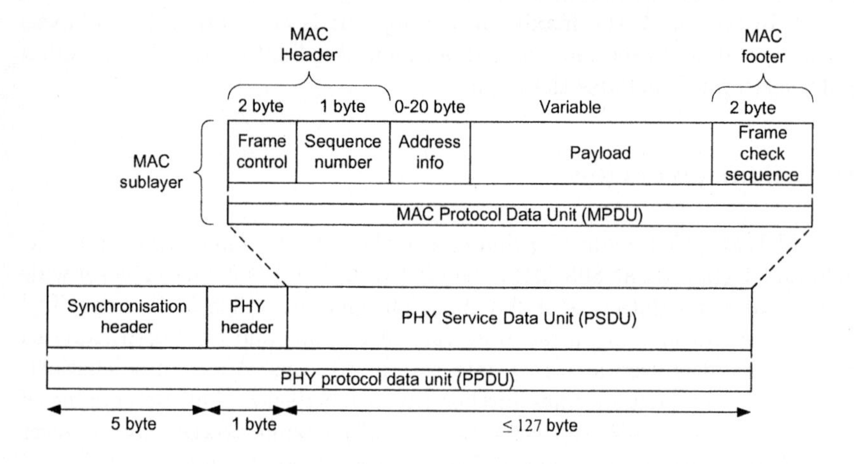

Fig. 2. Frame structure of IEEE 802.15.4

short frames by a Short IFS (SIFS). An example of a frame sequence, using acknowledgments (ACKs), is given in figure 1. If no ACKS are used, the IFS follows the frame immediately.

The packet structure of IEEE 802.15.4 is shown in figure 2. The size of the address info can vary between 0 and 20 bytes as both short and long addresses can be used and as a return acknowledgment frame does not contain any address information at all. Additionally, the address info field can contain the 16-bit PAN identifier, both from the sender and from the receiver. These identifiers can only be omitted when no addresses are sent. The payload of the MAC Protocol Data Unit is variable with the limitation that a complete MAC-frame (MPDU or PSDU) may not exceed 127 bytes.

3 Theoretical Calculations

3.1 Assumptions

The maximum throughput of IEEE 802.15.4 is determined as the number of data bits coming from the upper layer (i.e. the network layer) that can be transmitted.

Hence, we are only interested in the throughput of the MAC-layer according the OSI-protocol stack.

In this paper, only the unslotted version of the protocol (i.e. without the super frames) in the 2.4 GHz band is examined. Indeed, the 2.4 GHz band provides the most channels at the highest data rate and the unslotted version has the least overhead. Hence, CSMA with a back off scheme is used.

The maximum throughput is calculated between only 1 sender and only 1 receiver which are located close to each other. Therefore, we assume that there are no losses due to collisions, no packets are lost due to buffer overflow at either sender or receiver, the sending node has always sufficient packets to send and the BER is zero (i.e. we assume a perfect channel).

3.2 Calculations

The maximum throughput (TP) is calculated as follows. First the delay of a packet is determined. This overall delay accounts on the one hand for the delay of the data being sent and on the other hand for the delay caused by all the elements of the frame sequence, as is depicted in figure 1, i.e. back off scheme, sending of an acknowledgement, ... In other words, the overall delay is the time needed to transmit 1 packet. Subsequently, this overall delay is used to determine the throughput:

$$TP = \frac{8 \cdot x}{delay(x)} \qquad (1)$$

In this formula, x represents the number of bytes that has been received from the upper layer, i.e. the payload bytes from figure 1. The delay each packet experiences can be formulated as:

$$delay(x) = T_{BO} + T_{frame}(x) + T_{TA} + T_{ACK} + T_{IFS}(x) \qquad (2)$$

The following notations were used:

T_{BO} = Back off period
$T_{frame}(x)$ = Transmission time for a payload of x byte
T_{TA} = Turn around time (192 μs)
T_{ACK} = Transmission time for an ACK
T_{IFS} = IFS time

For the IFS, SIFS is used when the MPDU is smaller than or equal to 18 bytes. Otherwise, LIFS is used. (SIFS = 192 μs, LIFS = 640 μs). The different times are expressed as follows:

Back off period

$$T_{BO} = BO_{slots} \cdot T_{BO\,slot} \qquad (3)$$

BO_{slots} = Number of back off slots
$T_{BO\,slot}$ = Time for a back off slot (320 μs)

The number of back off slots is a random number uniformly in the interval (0, 2^{BE}-1) with BE the *back off exponent* which has a minimum of 3. As we only assume one sender and a BER of zero, the BE will not change. Hence, the number of back off slots can be represented as the mean of the interval: $\frac{2^3-1}{2}$ or 3.5.

Transmission time of a frame with a payload of x bytes

$$T_{frame}(x) = 8 \cdot \frac{L_{PHY} + L_{MAC\_HDR} + L_{address} + x + L_{MAC\_FTR}}{R_{data}} \tag{4}$$

L_{PHY} = Length of the PHY and synchronization header in bytes (6)
L_{MAC_HDR} = Length of the MAC header in bytes (3)
$L_{address}$ = Length of the MAC address info field
L_{MAC_FTR} = Length of the MAC footer in bytes (2)
R_{data} = Raw data rate (250 kbps)

$L_{address}$ incorporates the total length of the MAC address info field, thus including the PAN-identifier for both the sender as the destination if addresses are used. The length of one PAN-identifier is 2 bytes.

Transmission time for an acknowledgement

$$T_{ACK} = \frac{L_{PHY} + L_{MAC\_HDR} + L_{MAC\_FTR}}{R_{data}} \tag{5}$$

If no acknowledgements are used, T_{TA} and T_{ACK} are omitted in (2).
Summarizing, we can express the throughput using the following formula:

$$TP = \frac{8 \cdot x}{a \cdot x + b} \tag{6}$$

$$delay = a \cdot x + b \tag{7}$$

In this equations, a and b depends on the length of the data bytes (SIFS or LIFS) and the length of the address used (64 bit, 16 bit or no addresses). The parameter a expresses the delay needed for sending 1 data byte, parameter b is the time needed for the protocol overhead for sending 1 packet. The different values for a and b can be found in table 1.

Table 1. Overview of the parameters for equation 5

nr of address bits		a	b
0 bits	ACK	0.000032	0.002656
	no ACK	0.000032	0.002112
16 bits	ACK	0.000032	0.002912
	no ACK	0.000032	0.002368
64 bits	ACK	0.000032	0.003296
	no ACK	0.000032	0.002752

4 Analysis

In this section, we will analyze the throughput and bandwidth efficiency of IEEE 802.15.4 and we will discuss the lower delay limit. Several scenarios are considered: an address length of 64 bit address, of 16 bit address or without any address info and in all cases with or without the use of ACKs. The bandwidth efficiency is expressed as

$$\eta = \frac{TP}{R_{data}}. \tag{8}$$

The results can be found in figures 3 and 4 where figure 3 gives the useful bitrate and figure 4 the bandwidth efficiency. In the figures, the payload size represents the number of bits that are received from the upper layer. In section 2 it was mentioned that the maximum size of the MPDU is 127 bytes. Consequently, the number of data bytes that can be sent in one packet is limited. This can be seen in the figures: when the address length is set to 2 bytes (or 16 bits), the maximum payload size is 114 bytes. This can be calculated as follows: $MPDU = L_{MAC\_HDR} + L_{address} + L_{MAC\_FTR} + payload$, where $L_{address}$ equals to $2 \cdot 2$ bytes + $2 \cdot 2$ bytes for the PAN-identifiers and the short addresses respectively. Putting the correct values into the formula for MPDU, gives us 114 bytes as maximum payload length. When the long address structure is used (64 bits), 102 data bytes can be put into 1 packet. If no addresses are used, the PAN-identifiers can be omitted, which means that $L_{address}$ is zero. The maximum payload is now set to 122 bytes.

In general, we see that the number of useful bits or the bandwidth efficiency grows when the number of payload bits increases. The same remark was made when investigating the throughput of IEEE 802.11 [5] and is to be expected as all the packets have the same overhead irrespective of the length of the packet. Further, the small bump in the graph when the address length is 16 bits at 6 bytes, figure 3(b), is caused by the transition of the use of SIFS to LIFS: at that moment the MPDU will be larger than 18 bytes. In all cases, the bandwidth efficiency increases when no ACK is used, which is to be expected as less control traffic is being sent. In figure 3 and 4 we have only shown the graphs for short and long addresses. The graphs for the scenario without addresses are similar to the previous ones with the understanding that the maximum throughput is higher when no addresses are used. The graphs were omitted for reasons of clarity.

A summary can be found in table 2 where the maximum bit rate and bandwidth efficiency of the several scenarios are given.

We can see that we under optimal circumstances, i.e. using no addresses and with-out ACK, an efficiency of 64.9% can be reached. If acknowledgements are used, an efficiency of merely 59.5 % is obtained. Using the short address further lowers the maximum bit rate by about 4%. The worst result is an efficiency of only 49.8% which is reached when the long address is used with acknowledgements. The main reason for these low results is that the length of the MPDU is limited to 127 bytes. Indeed, the number of overhead bytes is relatively large compared to the number of useful bits (MPDU payload). This short packet

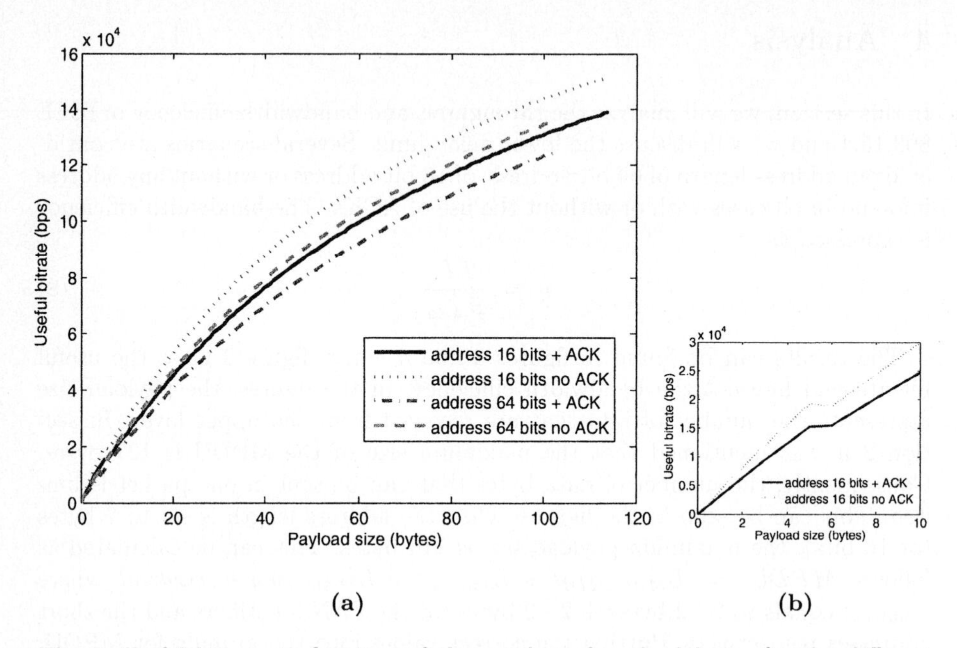

Fig. 3. Useful bitrate in function of the number of payload bytes for the different address schemes. The graph on the right (b) shows a snapshot of the left graph for an address size of 16 bits. The transition from SIFS to DIFS can be seen clearly.

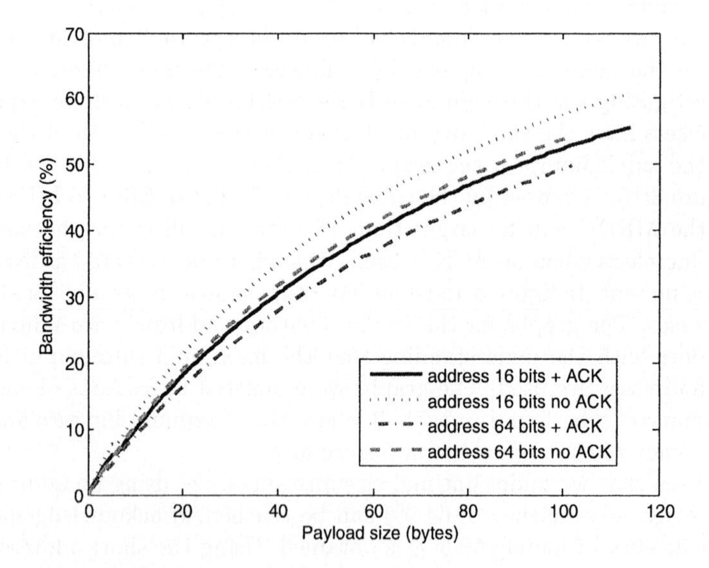

Fig. 4. Bandwidth efficiency of IEEE 802.15.4

length was chosen in order to limit the number of collisions (small packets are used) and to improve fair use of the medium. Further, the main application area

Table 2. Maximum bitrate and maximum efficiency of IEEE 802.15.4 for different address lengths

nr of address bits		maximum bitrate (bps)	maximum efficiency (%)
0 bits	ACK	147,780	59.5
	no ACK	162,234	64.9
16 bits	ACK	139,024	55.6
	no ACK	151,596	60.6
64 bits	ACK	124,390	49.8
	no ACK	135,638	54.8

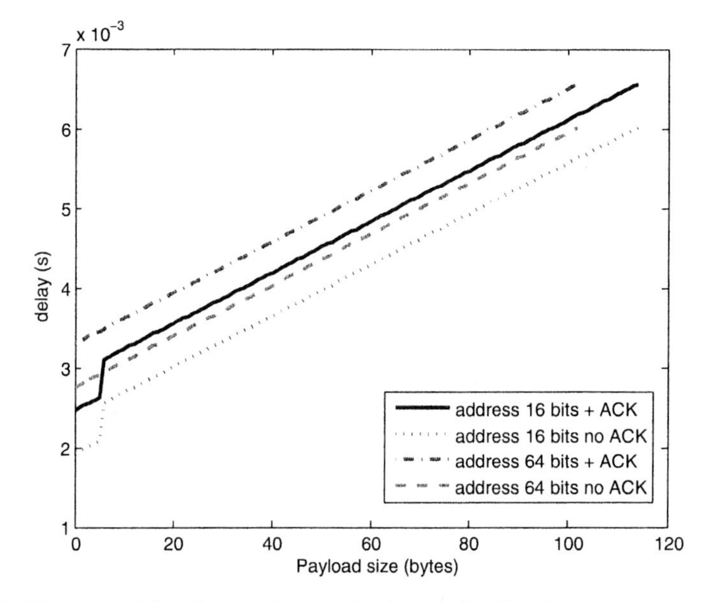

Fig. 5. Minimum delay for varying payload sizes for the short and long address

of this standard focuses on the transmission of small quantities of data, hence the small data packets.

Figure 5 gives the minimum delay each packet experiences. We immediately notice that the delay is a linear function of the number of payload bytes, as long as we assume a payload of more than 6 bytes for the short address scheme. The jump in the graph for the short address length is caused by the IFS-mechanism. In table 3, the minimum delay is given for the different scenarios. For the maximum payload, the minimum delay is the same for all the scenarios. Indeed, the MPDU is set to the maximum of 127 bytes. However, as can be seen in figure 5, the maximum number of payload bits differs when the short or long address is used.

5 Experimental Results

In order to validate our theoretically obtained maximum throughput, we will measure experimentally the throughput between 2 radios using the IEEE 802.15.4

Table 3. Minimum delay in ms for a payload of zero bits and a payload of a maximum number of bits

nr of address bits		delay (ms)	
		payload = 0 bits	payload = maximum
0 bits	ACK	2.21	6.56
	no ACK	1.66	6.02
16 bits	ACK	2.46	6.56
	no ACK	1.92	6.02
64 bits	ACK	3.30	6.56
	no ACK	2.75	6.02

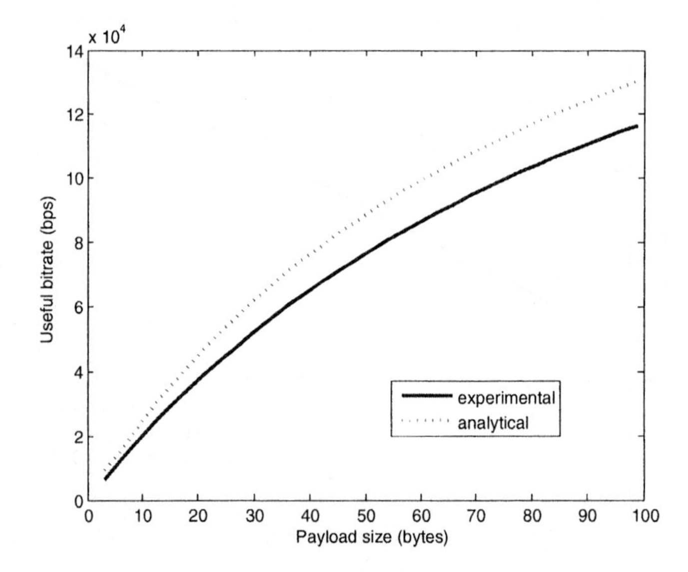

Fig. 6. Comparison between analytical and experimental results when short addressing and acknowledgments are used

specification. For our assays, we have used the 13192 DSK (Developer's Starter Kit) of Freescale Inc. This kit uses the MC13192 radio chip of Freescale Inc. [9]. This radio works at 2.4 GHz and software is included which implements the IEEE 802.15.4–standard. In order to minimize interference caused by other habitants of the 2.4 GHz-band, we have used channel 16 (highest channel) as this channel does not overlap with any of the channels of IEEE 802.11 [6]. Figure 6 gives a comparison between the theoretically and experimentally obtained results when a short address and an acknowledgment is used. We see that the experimental curve is lower than the one obtained analytically. The relative difference between the two curves is steady at about 11 %. However, we notice that the 2 graphs have the same curve. We have fitted the experimental curve with (7) and we obtained the following values for a and b respectively: 0.0000324 and 0.00359. The analytical values can be found in table 1 (16 bit address and ACK used):

0.000032 and 0.00291. We see that the main difference is in the part that is independent of the number of bytes sent. This is an indication that an extra delay or processing time needs to be added to each packet. The duration of this extra delay is about 680 μs (b is expressed in seconds: 0.00359-0.00291 = 0.00068 seconds). Another experiment was done where the long address was used without an ACK. Again a lower throughput than theoretically expected is achieved. Now we have a difference of about 9 %. The fitted values for a and b are 0.00003201 and 0.003271 respectively. As in the previous situation, the extra delay is independent of the number of bits sent and is about 520 μs. The time difference in the two situations is comparable. The extra delay is probably caused by processing the software on the devices.

6 Conclusion

The maximum throughput and minimum delay are determined under the condition that there is only 1 radio sending and 1 radio receiving. The next step in analyzing the performance of IEEE 802.15.4 would be introducing more transmitters and receivers which can hear each other. It is assumed that the maximum overall throughput, i.e. the throughput of all the radios achieved together, will fall as the different radios have to access the same medium. Indeed, this will result in collisions and longer back off periods. This also will cause lower throughput and larger delays. Another issue is the performance of the slotted version of IEEE 802.15.4 and the use of varying duty cycles. This study was done in [7]. As was mentioned in section 2 and 4, IEEE 802.15.4 works in the 2.4 GHz-band, the same band as WiFi (IEEE 802.11) and Bluetooth (IEEE 802.15.1). Consequently, these technologies will cause interference when used simultaneously. The interference between WiFi and 802.15.4 was investigated in [6] and [8]. It was concluded that WiFi interference is detrimental to a WPAN using 802.15.4. However, if the distance between the IEEE 802.15.4 and IEEE 802.11b radio exceeds 8 meter, the interference of IEEE 802.11b is almost negligible.

In this paper, we have presented the exact formulae for determining the maximum theoretical throughput of the unbeaconed version of IEEE 802.15.4. It was concluded this throughput varies according to the number of data bits in the packet and that a maximum throughput of 163 kbps can be achieved. Generally, the bandwidth efficiency is rather low due to the small packet size imposed in the standard.

Acknowledgements

This research is partly funded by the Belgian Science Policy through the IAP V/11 contract, by The Institute for the Promotion of Innovation through Science and Technology in Flanders (IWT-Vlaanderen) through the contracts No. 020152, No. 040286 and a PhD grant for B. Latré, by the Fund for Scientific Research - Flanders (F.W.O.-V., Belgium) and by the EC IST integrated project MAGNET (Contract no. 507102).

References

1. IEEE Std. 802.15.4: IEEE Standard for Wireless Medium Access Control and Physical Layer specifications for Low-Rate Wireless Personal Area Networks, 2003
2. Ed Callaway et al.,"Home Networking with IEEE 802.15.4: A developing Standard for Low-Rate Wireless Personal Area Networks", IEEE Communications Magazine, Vol 40 No. 8, pp 70-77, Aug. 2002
3. José A. Gutierrez, Marco Naeve, Ed Callaway, Monique Bourgeois, Vinay Mitter, Bob Heile, "IEEE 802. 15.4: A developing standard for low-power low-cost wireless personal area networks", IEEE Network, vol. 15, no. 5, September/October 2001 pp. 12-19
4. ZigBee Alliance, www.zigbee.org
5. Xiao Y. and Rosdahl J., "Throughput and delay limits of 802.11", IEEE Communications Letter, Vol. 6, No. 8, August 2002, pp. 355-357
6. Soo Young Shin, Sunghyun Choi, Hong Seong Park, Wook Hyun Kwon, "Packet Error Rate Analysis of IEEE 802.15.4 under IEEE 802.11b Interference", Wired/Wireless Internet Communications 2005, LCNS 3510, May 2005, pp.279-288
7. Jianliang Zheng and Myung J. Lee "Will IEEE 802.15.4 make Ubiquitous networking a reality?: A discussion on a potential low power, low bit rate standard", IEEE Communica-tions Magazine, Vol. 42, No. 6, Jun 2004, pp. 140 - 146
8. N. Golmie, D. Cypher, and O. Rebala, "Performance Evaluation of Low Rate WPANs for Sensors and Medical Applications", Proceedings of Military Communications Conference (MILCOM 2004), Oct. 31 - Nov. 3, 2004
9. Freescale Inc. http://www.freescale.com/ZigBee

On the Capacity of Hybrid Wireless Networks in Code Division Multiple Access Scheme

Qin-yun Dai[1], Xiu-lin Hu[1], Zhao Jun[2], and Yun-yu Zhang[1]

[1] Department of Electronic and Information Engineering,
Huazhong University of Science & Technology,Wuhan, Hubei 430074, China
daiqingyun929@163.com
[2] SHANGHAI Branch, China Netcom Corporation LTD.,
Pudong, Shanghai 201203, China
zhaojun4@china-netcom.com

Abstract. The hybrid wireless network is a kind of the novel network model, where a sparse network of base stations is placed within an ad hoc network. The problem on throughput capacity of hybrid wireless networks is considered to evaluate the performance of this network model. In this paper, we propose a general framework to analyze the capacity of hybrid wireless networks in code division multiple access scheme. Subsequently, we derive the mathematical analytical expressions of the capacity of hybrid wireless network systems under some assumptions. Finally, simulation results show that the hybrid wireless network could be a tradeoff between centrally controlled networks and ad hoc networks.

1 Introduction

Throughput capacity is an important parameter to evaluate the performance of wireless networks, which is denoted as the long-term achievable data transmission rate that a network can support. The network architecture is one of important factors influencing on the network capacity performance. The wireless network architecture can be roughly divided in two categories [1], as shown in Fig. 1. A widely used architecture is a network centrally controlled by base stations where every node communicates with others through the base stations, i.e., centrally controlled network. An alternative is the ad hoc architecture, where each node has the same capabilities. Two nodes wishing to communicate do so directly or use nodes lying in between them to route their packets. Recently, a hybrid wireless network model is proposed in [2], which is formed by placing a sparse network of base stations in an ad hoc network. Data may be forwarded in a multi-hop fashion or through base stations in a hybrid wireless network.

In the case of centrally controlled network, the capacity performance is analyzed on a cell basis by considering the uplink and downlink, which has been well studied over the last decades. Subsequently, the throughput capacity of an ad hoc network has been discussed widely. In fact, the physical layer of wireless networks is undergoing tremendous development because of the recent advances in signal processing and multiple antenna systems and has significantly impacts on the capacity performance.

X. Jia, J. Wu, and Y. He (Eds.): MSN 2005, LNCS 3794, pp. 877 – 885, 2005.

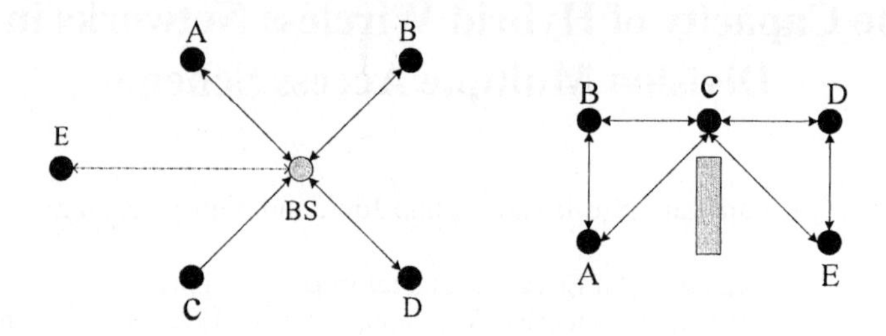

Fig. 1. Two wireless network architecture models are given. The left part shows the sketch map of a centrally controlled network and the right part is regards as an ad hoc network. A, B, C, D, E denote as different nodes in a network. BS is the abbreviation for a base station. The rectangle in the right part is considered as a barrier during the transmission process.

With a better coding scheme such as the complex network coding [3], including for example multiple-access and broadcast, network capacity could be improved under the same physical constraints. In spread spectrum, multi-packet reception (MPR) can be obtained by assigning multiple codes to a single receiver.

In this paper, we focus on the problem about the capacity of hybrid wireless networks. Liu [4] considered two different routing strategies and studied the scaling behavior of the throughput capacity of a hybrid wireless network under point-to-point model. It is well known that code division multiple access (CDMA) scheme is the possibility of receiving multiple packets at the same time. We propose a general framework to analyze the capacity of hybrid wireless networks in CDMA scheme by Markov chain approach and derive the mathematical analytical expressions of the hybrid wireless network capacity under some assumptions. Simulation experiments show that it is an attractive choice-of-technology for improving capacity performance of hybrid wireless networks.

2 The Hybrid Wireless Network Model

2.1 Network Component

A hybrid wireless network consists of two components. The first component is an ad hoc network including n_n nodes in a network region. The second component is a sparse network of n_B base stations, which are placed within the ad hoc network in a regular pattern.

The ad hoc network includes static and mobile nodes, where static nodes are distributed uniformly. The mobile nodes are randomly distributed at time t=0, and later, they move under the uniform mobility model so that the position of the mobile nodes at time t are independent of each other and the steady state distribution of the mobile nodes is uniform [5].

2.2 Routing Strategy

In a hybrid wireless network, there are two transmission modes: ad hoc mode and infrastructure mode. In the ad hoc mode, data are forwarded from the source to the destination in a multi-hop fashion without using base stations. In the infrastructure mode, data are forwarded through base stations. We adopt the routing strategy proposed by Liu [4]. If the destination is located in the same cell as the source node, data are forwarded in the ad hoc mode. Otherwise, data are forwarded in the infrastructure mode.

2.3 Transmission Mode Description in Code Division Multiple Access Scheme

In the infrastructure mode, multiple nodes transmit packets to each other through base stations. We assume a time division duplex (TDD) system with equal-sized uplink and downlink packets and each occupies one time slot. Nodes are half-duplex and are always in the receiving mode during the downlink period, and similarly, nodes are in the transmitting mode during the uplink period. A slotted aloha random access protocol is used by all nodes in the uplink: whenever a node has a new packet to transmit, it sends the packet in the earliest available uplink time slot. If the packet is not successfully received by the base station, the node will retransmit the packet with a fixed probability in each successive uplink slots until a successful transmission occurs.

Each node in the network transmits packets in the uplink using a unique spreading code which is assumed to be randomly generated. We assume that the receiver at the base station is a bank of matched filter, while the base station uses orthogonal codes for packets intended for different nodes in the downlink so that each receiving node always successfully receives its packets and the transmission success of a packet depends on the uplink reception alone.

In the ad hoc mode, nodes transmit to each other directly through a common channel by which all nodes are fully connected. The same slotted aloha random access protocol used in the infrastructure mode is also employed by all nodes. The transceiver at each node is also half-duplex. Every node uses a unique code to spread its transmitted packets. In order to receive packets from any potential nodes, we assume that each node has the knowledge of all possible spreading codes and the receiver at each node is also a bank of matched filters.

2.4 Assumptions

The classical analysis of slotted aloha by Kleinrock [6] is used in this paper: a node that needs to retransmit a packet is called as in the *backlogged* state; otherwise a node is in the *unbacklogged* state. To simplify analysis, we ignore noises and assume that errors in a packet are caused by multiple-access interference (MAI) alone. The following five assumptions about the hybrid wireless network are made [7].

Assumption 1: Nodes generate packets according to independent Poisson processes with equal arrival rate.

Assumption 2: There is an immediate feedback about the status of the transmission.

Assumption 3: There is no buffer at any node, i.e., each node can at most hold one packet at a time.

Assumption 4: Probability s_{ki} means the receiver detects successfully i out of k colliding packets in a time slot.

Assumption 5: Each node has equal probability to transmit to every other node.

3 Performance Analyses

For an n_n-node network, the Markov chain is characterized by $(n_n+1)\times(n_n+1)$ transition matrix $P=[p_{nk}]$ with p_{nk} being the probability that the network state goes from n to k in one transition. Next, we characterize the Markov chain for a hybrid wireless network by obtaining the transition matrix of the system.

3.1 Characterization of Hybrid Wireless Networks

In the infrastructure mode, the network state changes every two-time slots (packets are transmitted during the uplink time slot and received in the downlink time slot). p_{nk} represents the probability that the network state goes from n to k in two time slots. To obtain the state transition matrix $P_I=[p^I_{nk}]$, we first define the reception matrix S for the base station.

$$S=\begin{pmatrix} s_{10} & s_{11} & 0 & \cdots & 0 \\ s_{20} & s_{21} & s_{22} & \cdots & 0 \\ \vdots & & & & \vdots \\ s_{n_n0} & s_{n_n1} & s_{n_n2} & \cdots & s_{n_nn_n} \end{pmatrix}. \quad (1)$$

where s_{jk} is the probability that the base station successfully demodulates k out of j packets.

The computation of packet success probability follows [8]. Let k be the total number of packets in a slot, N be the spreading gain, the BER x is [9]: $x=Q\left(\sqrt{\frac{3N}{k-1}}\right)$, where $Q(y)=(1/\sqrt{2\pi})\int_y^\infty e^{-(t^2/2)}dt$. Under the assumption that errors occur independently in a packet, the packet success probability $p_I(k)=\sum_{i=0}^{e_b}\binom{L_p}{i}x^i(1-x)^{L_p-i}$, where e_b is the number of bit errors that can be corrected by coding and L_p is the length of a packet.

$$s_{kn}=\binom{k}{n}p_I(k)^n(1-p_I(k))^{k-n}. \quad (2)$$

Let $Q^I_a(k,n)$ be the probability that k unbacklogged nodes transmit packets in a given uplink slot and Q^I_r be the probability that k backlogged nodes transmit. $Q^I_a(k,n)=\binom{n_n-n}{k}(1-p^I_a)^{n_n-n-k}(p^I_a)^k$, $Q^I_r(k,n)=\binom{n}{k}(1-p_r)^{n-k}p_r^k$, where $p^I_a=1-e^{-(2\lambda/n_n)}$ is the probability that there is at least one packet arrives at an unbacklogged node during two slots for the Poisson arrival with rate λ and p_r is the retransmission probability for a backlogged node during the uplink slot. The transition probability with s_{00} defined to be one

$$p^{I}_{nk} = \begin{cases} \sum_{y=n-k}^{n} \sum_{x=0}^{n_n-n} s_{(x+y)[x+(n-k)]} Q^{I}_{r}(y,n) Q^{I}_{a}(x,n) & 0 \le k < n \\ \sum_{x=k-n}^{n_n-n} \sum_{y=0}^{n} s_{(x+y)[x-(k-n)]} Q^{I}_{a}(x,n) Q^{I}_{r}(y,n) & n \le k \le n_n \end{cases}. \quad (3)$$

The stationary distribution of the network state $\{q^I\}_{n=0}^{n_n}$ can be obtained by solving the following balance equation: $\vec{q}^I = \vec{q}^I P_I$, where $\vec{q}^I = [q^I_0, q^I_1, \cdots, q^I_{n_n}]$ and $\sum q^I_n = 1$.

In the ad hoc mode, the transition probability p_{nk} is defined for every time slot since the network state changes during one time slot. Define the network reception matrix R as

$$R = \begin{pmatrix} r_{10} & r_{11} & 0 & \cdots & 0 \\ r_{20} & r_{21} & r_{22} & \cdots & 0 \\ \vdots & & & & \vdots \\ r_{n_n 0} & r_{n_n 1} & r_{n_n 2} & \cdots & r_{n_n n_n} \end{pmatrix}. \quad (4)$$

where r_{jk} is the probability that k out of j packets in the time slot are received by their intended receivers in the network.

Theorem 1 [7]: Under assumption 1-5, given total $L \le n_n$ packets are transmitted in a time slot, the probability that there are $n \le L$ successfully received packets by their intended receivers in the network is given by

$$r_{Ln} = \sum_{l=n}^{L} \sum_{J=\min(J,l)}^{\min(l,n_n-L)} q_{Ll} \frac{\binom{n_n - L}{J}}{(n_n - L)^l} \times \sum_{\sum_{j=1}^{J} a_j = l} \frac{l!}{a_1! a_2! \cdots a_J!} \times \left(\sum_{\sum_{j=1}^{F} b_j = n} \prod_{i=1}^{J} d_{L,a_i,b_i} \right) \quad (5)$$

Where $a_j = 1,2 \cdots l,\ b_j = 0,1,\cdots a_j$ and $q_{Ll} = \binom{L}{l}\left(\frac{n_n - L}{n_n - 1}\right)^l \left(\frac{L-1}{n_n - 1}\right)^{L-l}$, $d_{L,a_i,b_i} = \sum_{k=b_i}^{L-(a_i-b_i)} \frac{\binom{a_i}{b_i}\binom{L-a_i}{k-b_i}}{\binom{L}{k}} s_{LK}$. Similar to the infrastructure mode, the transition probability

$$p^{a}_{nk} = \begin{cases} \sum_{y=n-k}^{n} \sum_{x=0}^{n_n-n} r_{(x+y)[x+(n-k)]} Q^{a}_{r}(y,n) Q^{a}_{a}(x,n) & 0 \le k < n \\ \sum_{x=k-n}^{n_n-n} \sum_{y=0}^{n} r_{(x+y)[x-(k-n)]} Q^{a}_{a}(x,n) Q^{a}_{r}(y,n) & n \le k \le n_n \end{cases}. \quad (6)$$

$Q^a_a(k,n) = \binom{n_n - n}{k}(1-p^a_a)^{n_n-n-k}(p^a_a)^k$, $Q^a_r(k,n) = \binom{n}{k}(1-p_r)^{n-k} p^k_r$ are probabilities that k packets are transmitted by unbacklogged and backlogged nodes in one time slot, respectively, and $p_a{}^a$ and p_r are packet transmission probabilities for unbacklogged and backlogged nodes in one time slot, respectively, r_{00} is also define to be one,

$p_a^a = 1 - e^{-(\lambda/n_n)}$. Similar to the infrastructure mode, the stationary distribution $\{q^a\}_{n=0}^{n_n}$ of the ad hoc mode can be obtained by solving the Markov-chain balance equation: $\vec{q}^a = \vec{q}^a P_a$, where $\vec{q}^a = [q_0^a, q_1^a, \cdots, q_{n_n}^a]$, and $\sum q_n^a = 1$.

3.2 Throughput of Hybrid Wireless Networks

In the infrastructure mode, given network state *n*, the number of packets successfully received by their intended receivers in two time slots is $\overline{N} = \sum_{k=1}^{n_n} p_k^I \sum_{l=0}^{k} ls_{kl}$, where $p_k^I = \sum_{x=0}^{k} Q_a^I(x,n) Q_r^I(k-x,n)$ is the probability that total *k* packets are transmitted in the uplink time slot. Because the throughput $\beta_I(n)$ and the average throughput $\overline{\beta}_I$ is defined per time slot, $\beta_I(n) = \overline{N}/2$, $\overline{\beta}_I = E(\beta_I(n)) = \sum_{n=0}^{n_n} \beta_I(n) q_n^I$, where q_n^I is the stationary distribution of the network state Markov chain.

In the ad hoc mode, $\beta_a(n)$ and the average throughput $\overline{\beta}_a$ are: $\beta_a(n) = \sum_{k=1}^{n_n} p_k^a \sum_{l=0}^{k} lr_{kl}$, $\overline{\beta}_a = E(\beta_a(n)) = \sum_{n=0}^{n_n} \beta_a(n) q_n^a$ where p_k^a is the probability that total *k* packets are transmitted in one time slot in the ad hoc mode.

Theorem 2: According to our model in section 2, the average throughput of the hybrid wireless network in CDMA scheme is $\overline{\beta}_H = (1 - 1/n_B)\overline{\beta}_I + 1/n_B \overline{\beta}_a$.

Proof: In our model, we assume the nodes of hybrid wireless network have the property that static nodes are distributed uniformly and the steady state distribution of the mobile nodes is uniform. According to the routing policy in section 2, the probability that the destination is located in the same cell as the source node is $1/n_B$, so that the probability that data are forwarded in the ad hoc mode is $1/n_B$, and similarly, the probability that data are forwarded in the infrastructure mode is $1-1/n_B$. Because the average throughput of infrastructure mode and ad hoc mode are $\overline{\beta}_I$ and $\overline{\beta}_a$ respectively in CDMA scheme, the result follows. ■

4 Experimental Results

In this section, the experimental results of throughput capacity of a hybrid wireless network in CDMA model and corresponding comparisons are given.

In three network models, i.e., hybrid wireless network, centrally controlled network, and ad hoc network, define n_B=3, n_n=50. Results on the throughput capacity of three network models are obtained when we simulate the traffic transmission within seconds under the same physical parameters such as the data arrival rate, the bit error rate and so on. Fig. 2 shows the hybrid wireless network model architecture would lead to being a tradeoff between a centrally controlled network and an ad hoc network, where the throughput capacity of a hybrid wireless network is slightly smaller than the centrally controlled network and far larger than the ad hoc network.

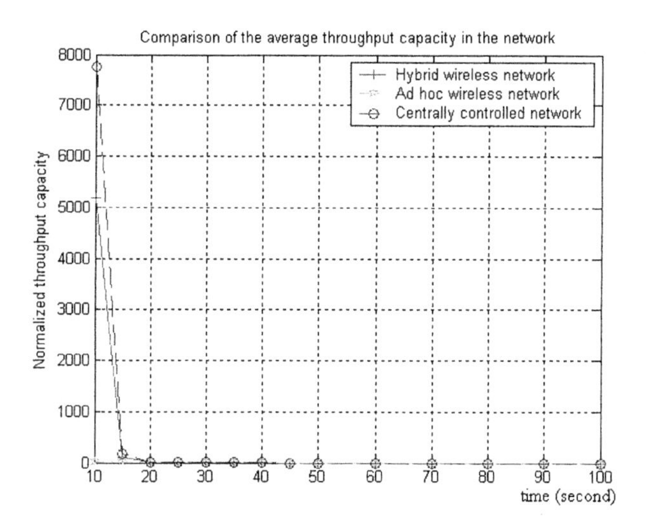

Fig. 2. Comparison of the average throughput capacity in three different network models under the same physical parameters

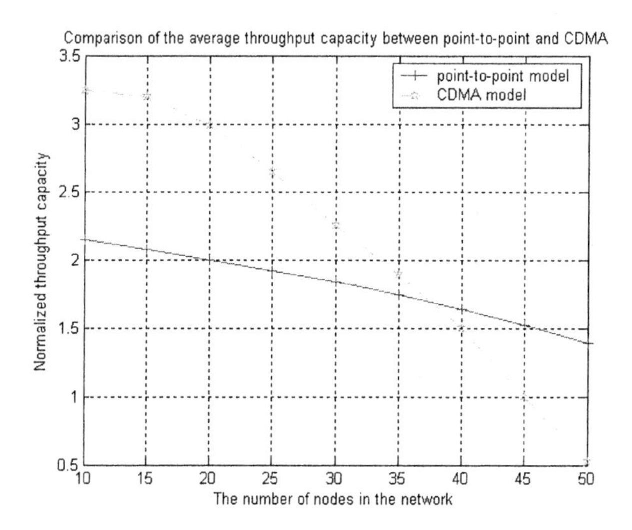

Fig. 3. Comparison of average throughput capacity between point-to-point and CDMA (n_B=1)

Fig. 3 gives the comparison of the throughput capacity of a hybrid wireless network between the point-to-point model and CDMA scheme. According to two different models, the variation trend of throughput could be analyzed as the number of nodes in the network changes from *10* to *50*. We observe that the throughput capacity in CDMA scheme is larger than those of the point-to-point model. Capacity performance in CDMA scheme is distinctly superior to those of the point-to-point model as the number of base stations increases. The throughput will decrease to the some value scale as the number of nodes increases. We consider the above results could be

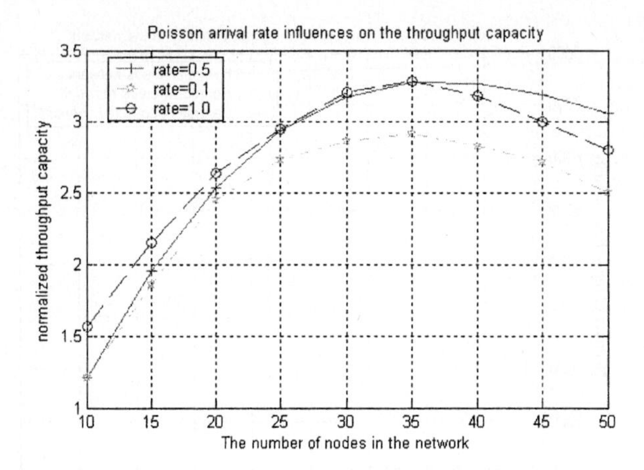

Fig. 4. Poisson arrival rate influences on the throughput capacity of a hybrid wireless network in CDMA scheme ($\lambda = 0.1, 0.5, 1.0$)

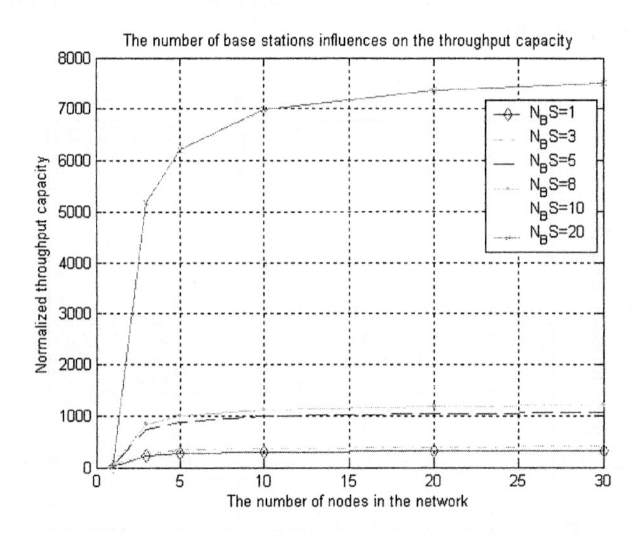

Fig. 5. The number of base stations influences on the throughput capacity of a hybrid wireless network in CDMA scheme (n_B=1,3,5,8,10,20)

improved on if the appropriate routing policy is selected, which could make throughput performance remain a high constant value and have no downtrend.

Subsequently, results of Poisson arrival rate λ influencing on the throughput performance of a hybrid wireless network in CDMA scheme are obtained. On the whole, the throughput performance would be improved as λ increases. From Fig. 4, we observe that the throughput capacity would keep the steady state at last after it is increased to some value as n_n is added, when n_B remains a constant. We guess that the hybrid wireless network architecture could exist some theory limit value of capacity perform ance on the given condition, which would be significantly influenced by the number of base stations n_B, though the quantitative conclusions are not presented in this paper.

Finally, different values of n_B influencing on the throughput capacity of a hybrid wireless network in CDMA scheme is also given. Experimental data from Fig. 5 are consistent with our intuitive imagination, which shows the throughput capacity will increase as n_B is added and will go to the steady state as the number of nodes in the network is added at last. Experiment results would forecast the conclusion that the capacity or the capability of the hybrid wireless network offering loads would be determined once n_B is determined, which could be improved farther through the advance signal process or others, but the capacity increment must be based on the theory limit value depending on n_B.

References

1. Toumpis, S., A.J.G.: Some Capacity Results for Ad Hoc Network. Allerton Conference on Communication, Control and Computing, Allerton II, Vol. 2 (2000) 775–784
2. Olivier Dousse, P.T., M. H.: Connectivity in Ad Hoc and Hybrid Network. IEEE INFOCOM'02, Vol. 2 (2002) 1079–1088
3. Gastpar, M., M.V.: On the Capacity of Wireless Networks: the Relay Case. IEEE INFOCOM'02, Vol. 3 (2002) 1577-1586
4. Liu, B., Liu, Z., Towsley, D.: On the Capacity of Hybrid Wireless Networks. IEEE INFOCOM'03, Vol. 2 (2003) 1543-1552
5. Nikhil Bansal, Z.L.: Capacity, Delay and Mobility in Wireless Ad Hoc Networks. IEEE INFOCOM'03 (2003) 1553-1563
6. Kleinrock, L., S.S.L.: Packet Switching in a Multi-access Broadcase Channel: Performance Evaluation. IEEE Transaction on Communication, Vol. 23 (1975) 410-423
7. Jeffrey, Q.Bao, L.T.: A Performance Comparison Between Ad Hoc and Centrally Controlled CDMA Wireless LANs. IEEE Transaction on Wireless Communication, Vol. 4 (2002) 829-841
8. Morrow Jr, R.K, J.S.L.: Bit-to-Bit Error Dependence in Slotted DS/SSMA Packet Systems with Random Signature Sequences. IEEE Transaction on Communication, Vol. 37 (1989) 1052-1061
9. Lehnert, J., M.P.: Error Probabilities for Binary Direct-sequence Spread-spretrum Communications with Random Signature Sequences. IEEE Transaction on Communication, Vol. 35 (1987) 87-98

Performance Evaluation of Existing Approaches for Hybrid Ad Hoc Networks Across Mobility Models

Francisco J. Ros, Pedro M. Ruiz, and Antonio Gomez-Skarmeta

DIIC, University of Murcia, Spain
{fjrm, pedrom, skarmeta}@dif.um.es

Abstract. There is being an on-going effort in the research community to efficiently interconnect Mobile Ad hoc Networks (MANET) to fixed ones like the Internet. Several approaches have been proposed within the MANET working group of the Internet Engineering Task Force (IETF), but there is still no clear evidence about which alternative is best suited for each mobility scenario, and how does mobility affect their performance. In this paper, we answer these questions through a simulation-based performance evaluation across mobility models. Our results show the performance trade-offs of existing proposals and the strong influence that the mobility pattern has on their behavior.

1 Introduction and Motivation

Mobile ad hoc networks consist of a number of mobile nodes which organize themselves in order to communicate one with each other wirelessly. These nodes have routing capabilities which allow them to create multihop paths connecting nodes which are not within radio range. These networks are extremely flexible, self-configurable, and they do not require the deployment of any infrastructure for their operation. However, the idea of facilitating the integration of MANETs and fixed IP networks has gained a lot of momentum within the research community. In such integrated scenarios, commonly known as hybrid ad hoc networks, mobile nodes are witnessed as an easily deployable extension to the existing infrastructure. Some ad hoc nodes are gateways which can be used by other nodes to seamlessly communicate with hosts in the fixed network.

Within the IETF, several solutions have been proposed to deal with the interconnection of MANETs to the Internet. One of the first proposals by Broch et al. [1] is based on an integration of Mobile IP and MANETs employing a source routing protocol. MIPMANET [2] followed a similar approach based on AODV, but it only works with Mobile IPv4 because it requires foreign agents (FA). In general, these approaches are tightly coupled with specific types of routing protocols, and therefore their applicability gets restricted.

The proposals which are receiving more attention within the IETF and the research community in general are those from Wakikawa et al. [3] and Jelger

X. Jia, J. Wu, and Y. He (Eds.): MSN 2005, LNCS 3794, pp. 886–896, 2005.

et al. [4], which define different gateway discovery functions and address allocation schemes. Another interesting proposal is that from Singh et al. [5], which proposes a hybrid gateway discovery procedure which is partially based on the previous schemes.

Many works in the literature have reported the strong impact that mobility has on the performance of MANETs. Thus, mobility will be a central aspect in our evaluations. In particular, we have employed three well-known mobility models (Random Waypoint, Gauss–Markov and Manhattan Grid) that we have used to deeply investigate the inter-relation between the Internet interconnection mechanism and the mobility of the network. An in-depth survey of the Random Waypoint and Gauss–Markov models (and others) can be found in [6], while the Manhattan Grid model is defined in [7].

The main novelty of this paper, is the investigation of the performance of the Internet connectivity solutions which are receiving more attention within the IETF. To the best of our knowledge, such kind of study has not been done before. In the authors' opinion, this paper sheds some light onto the performance implications of the main features of each approach, presenting simulation results which provide valuable information to interworking protocol designers. Moreover, these results can be used to properly tune parameters of a given solution depending on the mobility pattern of the network, what can also be useful for hybrid MANETs deployers.

The remainder of the paper is organized as follows: Sect. 2 provides a global sight of the most important current interworking mechanisms. The results of the simulations are shown in Sect. 3. Finally, Sect. 4 gives some conclusions and draws some future directions.

2 Analysis of Current Proposals

In this section we explore the most significant features of the main MANET interconnection mechanisms nowadays, namely those from Wakikawa et al., Jelger et al. and Singh et al. We refer to these solutions using the surname of their first author from now on. Table 1 summarizes the main features provided by each solution.

Table 1. Summary of features of well-known existing proposals

	Wakikawa	**Jelger**	**Singh**
Proactive/Reactive/Hybrid	P/R	P	H
Multiple Prefixes	Yes	Yes	No
Stateless/Stateful	Stateless	Stateless	n/a
DAD	Yes	No	n/a
Routing Header/Default Routing	RH	DR	RH/DR
Restricted Flooding	No	Yes	No
Load Balancing	No	No	Yes
Complete Specification	Yes	Yes	No

2.1 Address Allocation

Nodes requiring global connectivity need a globally routable IP address if we want to avoid other solutions like Network Address Translation (NAT). There are basically two alternatives to the issue of address allocation: they may be assigned by a centralized entity (stateful auto-configuration) or can be generated by the nodes themselves (stateless auto-configuration). The stateful approach is less suitable for ad hoc networks since partitions may occur, although it has also been considered in some works [8]. Both "Wakikawa" and "Jelger" specify a stateless auto-configuration mechanism which is based on network prefixes advertised by gateways. The nodes concatenate an interface identifier to one of those prefixes in order to generate the IP address. Currently, "Singh" does not deal with these issues.

2.2 Duplicate Address Detection

Once a node has an IP address, it may check whether the address is being used by other node. If that is the case, then the address should be deallocated and the node should try to get another one. This procedure is known as *Duplicate Address Detection* (DAD), and can be performed by asking the whole MANET if an address is already in use. When a node receives one of those messages requesting an IP address which it owns, then it replies to the originator in order to notify the duplication. This easy mechanism is suggested by "Wakikawa", but it does not work when network partitions and merges occur. Because of this and the little likelihood of address duplication when IPv6 interface identifiers are used, "Jelger" prefers avoiding the DAD procedure.

The main drawback of the DAD mechanism is the control overhead that it introduces in the MANET, specially if the procedure is repeated periodically to avoid address duplications when a partitioned MANET merges.

2.3 Gateway Discovery

The network prefix information is delivered within the messages used by the gateway discovery function. Maybe this is the hottest topic in hybrid MANETs research, since it has been the feature which has received more attention so far. Internet-gateways are responsible for disseminating control messages which advertise their presence in the MANET, and this can be accomplished in several different ways.

"Wakikawa" defines two mechanisms: a reactive and a proactive one. In the reactive version, when a node requires global connectivity it issues a request message which is flooded throughout the MANET. When this request is received by a gateway, then it sends a message which creates reverse routes to the gateway on its way back to the originator. The proactive approach of "Wakikawa" is based on the periodic flooding of gateway advertisement messages, allowing mobile nodes to create routes to the Internet in an unsolicited manner. Of course, this solution heavily increments the gateway discovery overhead because the gateway messages are sent to the whole MANET every now and then.

In order to limit that overhead of proactive gateway discovery, "Jelger" proposes a restricted flooding scheme which is based on the property of *prefix continuity*. A MANET node only forwards the gateway discovery messages which it uses to configure its own IP address. This property guarantees that every node shares the same prefix than its next hop to the gateway, so that the MANET gets divided in as many *subnets* as gateways are present. When "Jelger" is used with a proactive routing protocol, a node creates a default route when it receives a gateway discovery message and uses it to configure its own global address. But if the approach is integrated with a reactive routing protocol, then a node must perform a route discovery to avoid breaking the on-demand operation of the protocol.

Regarding "Singh" approach, it introduces a new scenario where gateways are mobile nodes which are one hop away from a wireless access router. Nodes employ a hybrid gateway discovery scheme, since they can request gateway information or receive it proactively. The first node which becomes a gateway is known as the "default gateway", and it is responsible for the periodic flooding of gateway messages. Remaining gateways are called "candidate gateways" and they only send gateway information when they receive a request message.

2.4 Routing Traffic to the Internet

The way traffic is directed to the Internet is also different across approaches. "Wakikawa" prefers using IPv6 routing headers to route data packets to the selected gateways. This introduces more overhead due to the additional header, but it is a flexible solution because nodes may dynamically vary the selected gateway without the need to change their IP address. This helps at maximizing the IP address lifetime. However, "Jelger" relies on *default routing*, i.e., nodes send Internet traffic using their default route and expect the remaining nodes to correctly forward the data packets to the suitable gateway. "Singh" uses both alternatives: default routing is employed when nodes want to route traffic through their "default gateway", but they can also use routing headers to send packets to a "candidate gateway".

2.5 Load Balancing

"Singh" depicts an interesting feature which does not appear in the rest of the proposals: a traffic balancing mechanism. Internet-gateways could advertise a metric of the load which passes across them within the gateway discovery messages. MANET nodes could use this information to take a more intelligent decision than what is taken when only the number of hops to the gateway is considered. Unfortunately, no detailed explanation of this procedure is provided in the current specification.

3 Performance Evaluation

To assess the performance of "Wakikawa" and "Jelger", we have implemented them within the version 2.27 of the *ns2* [1] network simulator. The gateway se-

[1] The Network Simulator, `http://www.isi.edu/nsnam/ns/`.

lection function uses in both cases the criterion of minimum distance to the gateway, in order to get a fair comparison between the two approaches. "Singh" has not been simulated because the current specification is not complete enough and therefore it has not captured the research community attention yet.

In addition, we have also implemented the OLSR protocol according to the latest IETF specification[2]. We have set up a scenario consisting of 25 mobile nodes using 802.11b at 2 Mb/s with a radio range of 250 m, 2 gateways and 2 nodes in the fixed network. These nodes are placed in a rectangular area of $1200x500m^2$. 10 active UDP sources have been simulated, sending out a constant bit rate of 20Kb/s using 512 bytes/packet. The gateways are located in the upper right and lower left corners, so that we can have long enough paths to convey useful information. In addition, we use the two different routing schemes which are being considered for standardization within the IETF: OLSR [9] as a proactive scheme, and AODV [10] as a reactive one. This will help us to determine not only the performance of the proposals, but the type of routing protocols for which they are most suitable under different mobility scenarios. The case of OLSR with a reactive gateway discovery has not been simulated because in OLSR all the routes to every node in the MANET (including the gateways) are already computed proactively. So, there is no need to reactively discover the gateway, because it is already available at every node. In both AODV and OLSR we activated the link layer feedback.

Movement patterns have been generated using the *BonnMotion* [3] tool, creating scenarios with the Random Waypoint, Gauss–Markov and Manhattan Grid mobility models. Random Waypoint is the most widely used mobility model in MANET research because of its simplicity. Nodes select a random speed and destination around the simulation area and move toward that destination. Then they stop for a given pause time and repeat the process. The Gauss–Markov model makes nodes movements to be based on previous ones, so that there are not strong changes of speed and direction. Finally, Manhattan Grid models the simulation area as a city section which is only crossed by vertical and horizontal streets. Nodes are only allowed to move through these streets.

All simulations have been run during 900 seconds, with speeds randomly chosen between 0 m/s and (5, 10, 15, 20) m/s. Random Waypoint and Manhattan Grid models have employed a mean pause time of 60 seconds, although the former has also been simulated with 0, 30, 60, 120, 300, 600 and 900 seconds of pause time in the case of 20 m/s as maximum speed. The Manhattan Grid scenarios have been divided into 8x3 blocks, what allows MAC layer visibility among nodes which are at opposite streets of a same block.

3.1 Packet Delivery Ratio

The Packet Delivery Ratio (PDR) is mainly influenced by the routing protocol under consideration, although Internet connectivity mechanisms also have an

[2] Code available at `http://ants.dif.um.es/masimum/`.

[3] Developed at the University of Bonn,
`http://web.informatik.uni-bonn.de/IV/Mitarbeiter/dewaal/BonnMotion/`.

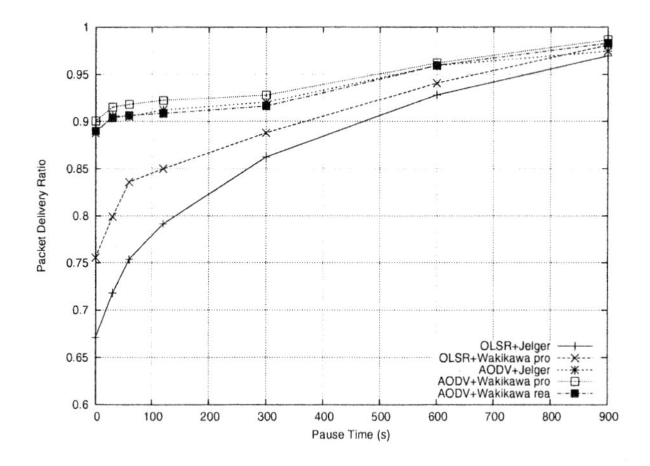

Fig. 1. PDR in Random Waypoint model using different pause times (maximum speed = 20 m/s)

impact. Similarly to previous simulations of OLSR in the literature, we can see in Fig. 1 that as the mobility increases in the Random Waypoint model, it offers a much lower performance compared to AODV. The reason is that OLSR has a higher convergence time compared to AODV as the link break rate increases. In addition, according to RFC 3626, when link layer feedback informs OLSR about a broken link to a neighbor, the link is marked as "lost" for 6 seconds. During this time packets using this link are dropped in OLSR. This behavior also affects the routes towards Internet gateways, which is the reason why the PDR is so low in OLSR simulations.

In the case of OLSR, "Jelger" performs surprisingly worse than the proactive version of "Wakikawa". Given that "Jelger" has a lower gateway discovery overhead we expected the results to be the other way around. The reason is that "Jelger" is strongly affected by the mobility of the network. After carefully analyzing the simulations we found out that the selection of next hops and gateways makes the topology created by "Jelger" very fragile to mobility. The problem is that the restrictions imposed by the prefix continuity in "Jelger" concentrates the traffic on a specific set of nodes. In AODV, this problem is not so dramatic because AODV, rather than marking a neighbor as lost, starts finding a new route immediately. So, we can conclude that although prefix continuity has very interesting advantages (as we will see), it has to be carefully designed to avoid data concentration and provide quick reactions to topological changes.

Regarding AODV, we can see how proactive "Wakikawa" offers a better PDR than the remaining solutions at high speeds. This is due to the proactive dissemination of information, what updates routes to the Internet as soon as they get broken. "Jelger" and reactive "Wakikawa" behave very much the same because the former is designed to create routes on-demand when it is integrated within a reactive routing protocol (although proactive flooding of gateway information is still performed).

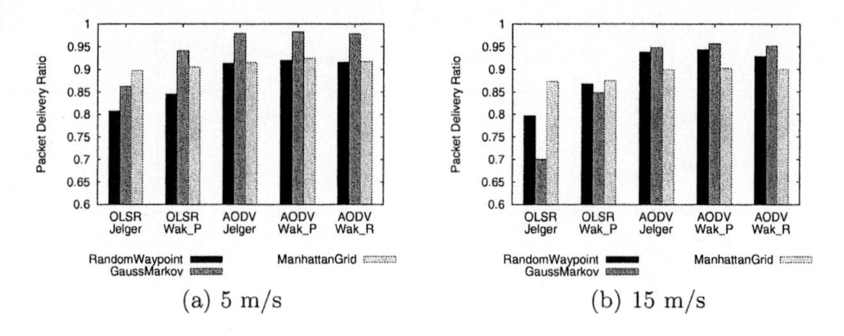

Fig. 2. PDR obtained from different mobility models for different maximum speeds

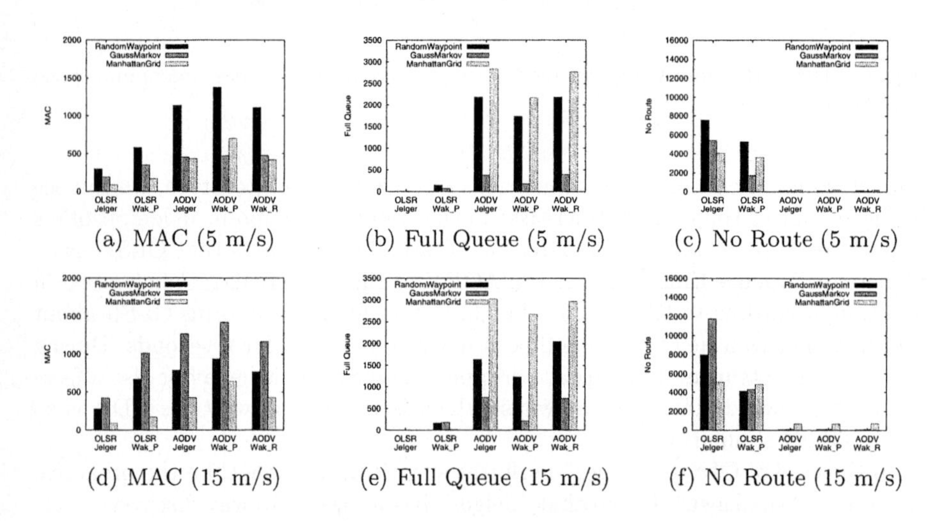

Fig. 3. Cause of packet drops for different mobility models

One of our goals is to analyze if the results are congruent across mobility models. Figure 2 shows a comparison between Random Waypoint, Gauss–Markov and Manhattan Grid mobility models with maximum speeds of 5 m/s and 15 m/s. Figures for other maximum speeds showed a similar trend (they are not included due to space constraints).

At first sight we can point out an interesting thing: mobility model can heavily influence the resulting PDR, but results seem to be consistent across mobility models. That is, "Jelger" continues offering a lower PDR than "Wakikawa" when they are integrated within an OLSR network, and AODV does not change its PDR very much regardless of the Internet interconnection mechanism and the mobility model used. But in fact, each mobility model influences in a different way every approach showing their strengths and drawbacks. We can better realize this if we make a more in-depth analysis of the causes of packet drops, as we will explain below.

The Gauss–Markov model presents the biggest link break rate of all the simulated mobility models when the maximum speed is high. However, it provokes very few link losses at low speeds. Because this mobility model does not perform strong changes in speed and direction, when a node picks a high speed then it is very likely that the node will continue travelling at high speeds, making links to break more often. Just the opposite occurs when the node initially chooses a low speed.

That sheds some light onto the results of Fig. 2, where it is worth pointing out that the PDR dramatically decreases in OLSR as the maximum available speed of the Gauss–Markov model increases. As we previously said, "Jelger" is less strong against frequent topology changes than "Wakikawa", and that is why this behavior of the Gauss–Markov model impacts more on its performance. Figure 3 clearly outlines this, because the number of drops due to the absence of a suitable route towards the Internet significantly grows at high speeds in Gauss–Markov model. Moreover, the number of packet drops due to the MAC layer not being able to deliver a packet to its destination (because of a link break) also increases. The mobility model has a lower influence in AODV than in OLSR, because the former is able to easily adapt to changing topologies.

On the other hand, Manhattan Grid model does not cause many link breaks because nodes have their mobility very restricted. Instead of that, nodes tend to form groups, increasing contention at link layer. This is why this model makes the PDR of OLSR and AODV very similar, enhancing results of the former. In addition, the performance of "Jelger" and "Wakikawa" also tend to equal since "Jelger" is very sensitive to those link breaks which this model lacks (see Fig. 3). Manhattan Grid mobility model fills up interface queues because of MAC layer contention, while it does not cause many drops due to link breaks (MAC and No Route drops). As a note, results obtained by this mobility model should depend on the number of blocks used (we have used a fixed configuration though).

In addition, we can ascertain from Fig. 3 that OLSR is not prone to packet drops due to filling up the interface queue, since it does not buffer data packets before sending them. Some of these types of drops appear in "Wakikawa" because of its non-controlled flooding, which creates more layer-2 contention than "Jelger". In the case of AODV, queues get full because data packets are buffered when a route is being discovered. But that is not so heavily evidenced in proactive "Wakikawa" because Internet routes are periodically refreshed.

3.2 Gateway Discovery Overhead

Finally, we evaluate the overhead of the gateway discovery function of each of the proposals. As we can see in Fig. 4, AODV simulations result in a higher gateway overhead as the mobility of the network increases in Random Waypoint model. This is due to the increase in the link break rate, which makes ad hoc nodes find a new route to the Internet as soon as their default route is broken. We can clearly see that proactive "Wakikawa" generates the biggest amount of Internet-gateway messages due to its periodic flooding through the whole

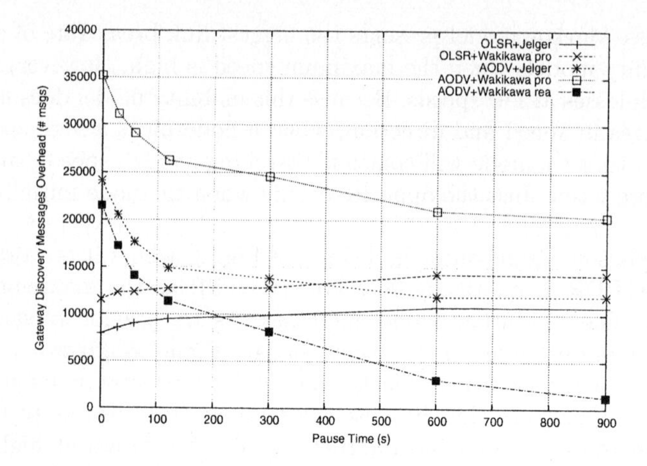

Fig. 4. Gateway discovery overhead in the Random Waypoint model using different pause times (maximum speed = 20 m/s)

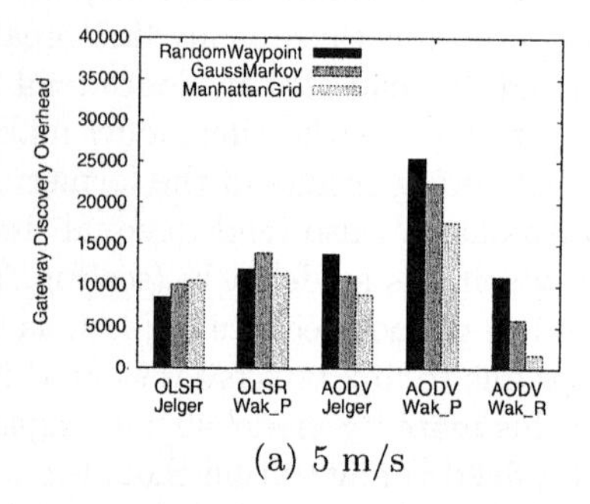

(a) 5 m/s

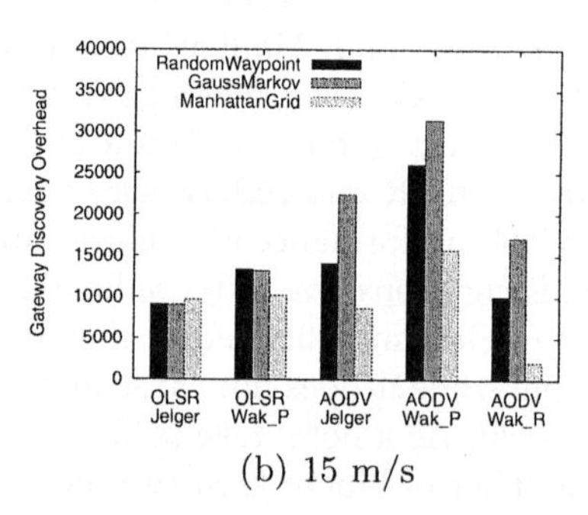

(b) 15 m/s

Fig. 5. Gateway discovery overhead obtained from different mobility models for different maximum speeds

network. Reactive "Wakikawa" shows the minimum gateway overhead thanks to its reactiveness. "Jelger" sits in between the other two, due to its limited periodical flooding.

As it was expected, the gateway discovery overhead for Internet connectivity mechanisms combined with OLSR remains almost unaffected by network mobility. This is due to the fact that Internet connectivity messages are periodically sent out by OLSR without reaction to link breaks. So, its gateway control overhead is not heavily affected by mobility. Figure 4 shows that "Jelger" always maintains a lower overhead than proactive "Wakikawa" due to the restriction of forwarding imposed by prefix continuity. The difference remains almost constant independently of the mobility of the network.

The number of messages due to the gateway discovery function in OLSR simulations does not vary very much regardless of the mobility model used (Fig. 5). The mobility model does not seem to significantly impact the overhead offered by all these approaches, except in the case of the Manhattan Grid model which tends to equal the results of "Jelger" and "Wakikawa" when they are integrated within OLSR . This is due to the higher contention caused by this mobility model, which reduces the number of control messages which can be sent in "Wakikawa".

The gateway discovery overhead of AODV gets very much affected by the influence of the mobility model, but as it happened with the PDR, it remains consistent across mobility models. The Manhattan Grid model offers the minimum amount of link breaks, and therefore there is a low overhead in all AODV solutions. The Gauss–Markov model causes little overhead at low speeds (few link breaks) but a lot of overhead at higher speeds (many link breaks). The Random Waypoint mobility model sits in between the others.

4 Conclusions, Discussion and Future Work

In this paper we have conducted a simulation-based study of the current approaches for interconnecting MANETs and fixed networks. This study has evaluated their performance, and it has shown how different mobility models influence in a different way the behavior of each solution.

Our results show that depending on the scenario we want to model, every solution has its strong and weak points. Hence, "Jelger" suits better for mobility patterns where few link breaks occur, like the Gauss–Markov (at low speeds) and Manhattan Grid mobility models. In those cases it offers a good PDR with a reduced gateway discovery overhead. However, we have seen that although prefix continuity offers an interesting mechanism of limited flooding, it has to be carefully designed in order to avoid routes which are fragile to changing topologies. On the other hand, reactive and proactive versions of "Wakikawa" are more suitable for high mobility scenarios. Random Waypoint and Gauss–Markov (at high speeds) mobility models generate a big number of link breaks, but "Wakikawa" solution is able to perform quite well under these circumstances. Nevertheless, it is also clear that proactive gateway discovery needs a constrained flooding mechanism to avoid the huge amount of overhead associated with the discovery of gateways.

In our opinion, this result opens up the need for new adaptive schemes being able to adapt to the mobility of the network. In addition to adaptive gateway discovery and auto-configuration, there are other areas in which we plan to focus our future work. These include among others improved DAD (Duplicate Address Detection) mechanisms, efficient support of DNS, discovery of application and network services, network authentication and integrated security mechanisms.

Acknowledgment

Part of this work has been funded by Spanish MCYT by means of the "Ramon y Cajal" workprogramme, the ICSI Call for Spanish Technologists and the SAM (MCYT, TIC2002-04531-C04-03) and I-SIS (2I04SU009) projects.

References

1. J. Broch, D.A. Maltz, D.B. Johnson, *"Supporting Hierarchy and Heterogeneous Interfaces in Multi-Hop Wireless Ad Hoc Networks."* Proceedings of the Workshop on Mobile Computing held in conjunction with the International Symposium on Parallel Architectures, Algorithms, and Networks, IEEE, Perth, Western Australia, June 1999.
2. U. Jonsson, F. Alriksson, T. Larsson, P. Johansson, G.M. Maquire, Jr. *"MIP-MANET:Mobile IP for Mobile Ad Hoc Networks."* IEEE/ACM Workshop on Mobile and Ad Hoc Networking and Computing, pp. 75–85. Boston, MA USA, August 1999.
3. R. Wakikawa, J. T. Malinen, C. E. Perkins, A. Nilsson, and A. Tuominen, *"Global Connectivity for IPv6 Mobile Ad Hoc Networks,"* Internet-Draft "draft-wakikawa-manet-globalv6-03.txt". Oct. 2003.
4. C. Jelger, T. Noel, and A. Frey, *"Gateway and Address Autoconfiguration for IPv6 Ad Hoc Networks,"* Internet-Draft "draft-jelger-manet-gateway-autoconf-v6-02.txt". Apr. 2004.
5. S. Singh, J.H. Kim, Y.G. Choi, K.L. Kang, and Y.S. Roh, *"Mobile Multi-gateway Support for IPv6 Mobile Ad Hoc Networks,"* Internet-Draft, "draft-sinhg-manet-mmg-00.txt", June 2004.
6. T. Camp, J. Boleng, and V. Davies, *"A Survey of Mobility Models for Ad Hoc Network Research,"* Wireless Communication & Mobile Computing (WCMC): Special issue on Mobile Ad Hoc Networking: Research, Trends and Applications, vol. 2, no. 5, pp. 483-502, 2002.
7. *"Selection Procedures for the Choice of Radio Transmission Technologies of the UMTS (TS30.03 v3.2.0)"*, TS 30.03 3GPP, April 1998.
8. H. W. Cha, J. S. Park, and H. J. Kim, *"Extended Support for Global Connectivity for IPv6 Mobile Ad Hoc Networks,"* Internet-Draft "draft-cha-manet-extended-support-globalv6-00.txt". Oct. 2003.
9. T. Clausen and P. Jacquet, Eds., *"Optimized Link State Routing Protocol (OLSR),"* IETF RFC 3626, October 2003.
10. C. Perkins, E. Belding-Royer, and S. Das, *"Ad hoc On Demand Distance Vector (AODV) Routing,"* IETF RFC 3561, July 2003.

UDC: A Self-adaptive Uneven Clustering Protocol for Dynamic Sensor Networks

Guang Jin and Silvia Nittel

Department of Spatial Information Science and Engineering,
University of Maine, Orono, ME 04469, USA
{jin, nittel}@spatial.maine.edu

Abstract. The constrained resources of sensor networks challenge researchers to design resource efficient protocols. Clustering protocols are efficient to support the aggregation queries in sensor databases. This paper presents a novel clustering protocol, named UDC (spatially Uneven Density Clustering), to prolong the lifetime of sensor networks. Unlike other clustering protocols, UDC forms distributed sensor nodes into spatially uneven clusters according to local network conditions. In short, clustered by UDC, the nodes nearby the central base are grouped into smaller clusters, while the distant nodes are clustered into larger groups to save resources. Our simulation results exemplify that UDC can extend the lifetime of sensor networks up to twice as long as the other clustering protocols do.

1 Introduction

Recent successes of nano-scale sensing devices, low-power wireless communicating devices and miniaturized manufacture of computing devices lead to the technology of sensor networks which have enabled us to explore the physical world in detail that cannot be obtained easily through other traditional ways. Consisting of a set of sensor nodes each of which have with different sensors attached, and connected to each other through wireless radio links, sensor networks provide a platform to monitor the physical world for different types of applications. Several resources, such as limited battery sources and limited communication bandwidth, however, constrain sensor networks. Those limitations require an efficient protocol to minimize the resource consumption of sensor networks and provide a reliable platform to other applications. For example, SMECN, SPIN-2,SAR, Directed Diffusion paradigm etc. are some of the proposed energy aware communication protocols designed for wireless sensor networks [1]. Sensor database management systems (sensor DBMS), such as TinyDB [2] and Cougar [3], also address the energy saving problems by minimizing the amount of data transmitted within networks. For example, in TAG [4] sensor nodes process raw readings into small amount of descriptive information in-network to save precious resources. Clustering protocols, such as LEACH [5] and HEED [6], are ideal to support the aggregation queries while prolonging the lifetime of sensor networks. By using clustering protocols, several sensor nodes are elected as cluster heads that collect raw readings from member nodes, process them and return processed results back to users. The rotation between

X. Jia, J. Wu, and Y. He (Eds.): MSN 2005, LNCS 3794, pp. 897–906, 2005.

two roles of head and non-head member reduces the average resource consumption of senor nodes and extends the lifetime of networks.

Based on a model for the energy consumption rate of sensor nodes, we propose a novel protocol, named UDC(spatially Uneven Density Clustering) to support aggregation queries, which adapts clusters to local network conditions. For example, those sensor nodes near to the central base can have direct communication links with the central base, while faraway nodes can establish larger clusters to save energy consumption. UDC forms clusters according to different conditions, such as sensor nodes' locations and cluster sizes. As a result, UDC can prolong the lifetime of networks while maintaining the high quality of networks compared with other approaches, which has been confirmed by our simulation results.

The remaining of the paper is organized as follows. Section 2 introduces related work on sensor databases and clustering protocols. A model of cluster characteristics is introduced in section 3, based on which we present the UDC protocol in section 4. We present our simulation results and compare UDC with other approaches in section 5. Finally, section 6 draws the conclusion and discusses the future work.

2 Background

In a typical scenario of sensor network applications, a powerful central base exists to receive queries from users, disseminate data into networks and receive readings. Although the central base can be characterized as a powerful machine, even with unlimited energy, the sensor nodes are constrained devices characterized by their constrained resources (e.g. limited energy sources, communication bandwidth and range)[7], which forces all applications over sensor networks to be resource conservative.

2.1 Sensor Database

Several "sensor databases" or "device databases", such as TinyDB and Cougar, have been recently established. So a sensor network cannot only monitor the physical world, but also respond to users' queries over the world. In most cases, users prefer a statistical or other descriptive information rather than raw readings from sensor networks, which inspires researches to study the in-network processing of aggregation queries. TAG[4] is a framework to support aggregation queries. In TAG, some sensor nodes take the responsibility to process raw readings from other nodes and aggregate them into a summarized information to avoid redundant resource overheads.

2.2 Protocol

Routing protocols in sensor networks is a widely researched area, but also are challenged by factors such as energy-awareness, lightweight computation, fault tolerance, scalability and topology of the network [8]. Main categories are Flat, Clustering (Hierarchical) and Adaptive protocols [9].

The first type includes routing mechanisms based on flat multi-hop communication. SAR [10] implements multi-cast approach creating a tree, initiated from source to destination nodes. Directed Diffusion [11] introduces a data-centric model, an alternative way to conventional address based routing methods. The second category of protocols, hierarchical or clustering approaches, is the basis for aggregation queries. LEACH [5] uses randomized election algorithm to cluster nodes in sensor networks. HEED [6] improves the efficiency by using additional cluster head election procedure among head candidates. In the third type of routing protocols, namely adaptive routing, all the information is disseminated among every node. The idea is to allow users to query any node in the network. SPIN-1 and SPIN-2[12] family of protocols utilizes data negotiation and adaptive algorithms. To reduce bandwidth overhead, nodes only exchange metadata instead of actual data for negotiation. As all the nodes exchange information, the current energy levels in the network are known.

3 Cluster Characteristic

Clustering methods have been studied over the decades and are useful in the many fields, such as data mining and artificial intelligence [13] to discover the relations within data. The computation costs of many well-established algorithms (e.g. k-means and g-mean), however, reduce their applicability in sensor networks. A model to analyze the energy consumption of sensor nodes can help us to design resource conservative clustering protocols.

3.1 A Energy Model of Cluster

Generally, clustering protocols divide time into time units, called *rounds*. In each round, a sensor node can either be a cluster head or be a non-head member of some cluster, based on the choice made by clustering protocols. A non-head member only sends its reading to the cluster head, and the cluster head aggregates raw readings from member nodes and returns the processed result back to the central base. Furthermore, since the size of message in most aggregation queries doesn't increase through processing procedures, the message will be assume constant (i.e. k bits) in our model.

Table 1. Parameters of energy consumption model

Parameter	Description
E_{tx}	Energy consumption rate for transmitting data.
ϵ_{fx}	Energy consumption rate for amplifying signal to nearby location.
ϵ_{amp}	Energy consumption rate for amplifying signal to faraway location.
d_0	Distance threshold for signal amplifiers.

The characteristics of low-energy radio [5, 6] enable an energy cost model, table 1, that helps us understand the behavior of clustered sensor nodes. To send k bits data and amplifies it to be detected by a remote sensor at distance of d, a node consumes a amount of $E_{SD}(k, d) = E_{tx} \cdot k + \epsilon_{fx} \cdot d^2$ energy, if the sink is located nearby. If the sink is far away (i.e. d is bigger than a threshold d_0) several sensor nodes have to act as routers to relay messages, and the distance has quartic effects on energy consumption as $E_{SD}'(k,d) = E_{tx} \cdot k + \epsilon_{amp} \cdot d^4$. A sensor node spends E_{RD} energy to receive k bits data from other nodes, as defined by $E_{RD}(k) = E_{tx} \cdot k$. If a cluster has m nodes, the distance between its cluster head and central base is d, and the message length is k bits, the energy consumption rate for its cluster head is,

$$ER_{CH}(d,m) = E_{RD}\left[(m-1)\cdot k\right] + E_{SD}^{'}(k,d). \tag{1}$$

In (1), the choice of E_{SD} or $E_{SD}^{'}$ is made based on cluster heads' locations. Although a cluster head consumes extra energy to receive messages from it members, the burden of a non-head node is lessened, since a non-head node only needs to return its reading to its cluster head located nearby. If the distance between a non-head node and its cluster head is d, we can define its energy consumption rate ER_{MEM} as

$$ER_{MEM}(d) = E_{SD}(k,d). \tag{2}$$

3.2 Benefit of Clustering

Above discussion excludes the applications of naïve methods (e.g. fixed cluster heads or all nodes with direct links to the central base) in the constrained environment of sensor networks. In a clustering protocol, on the other hand, all sensor nodes can rotate the roles of being a cluster head and a non-head member. For the whole life of a sensor node, the energy consumption rate can be averaged as,

$$ER_{SN}(d,m) = ER_{MEM}(\bar{D}) \cdot \frac{m-1}{m} + ER_{CH} \cdot \frac{1}{m}, \tag{3}$$

where m is the number of sensor nodes in the cluster, the distance between its cluster head and the central base is d, and $ER_{MEM}(\bar{D})$ indicates this node's average cost as being a non-head member while other nodes act as cluster heads. From (3) we can see that although a sensor node in a cluster with more nodes consumes more energy as being a cluster head, there is more chance that this node can act as a non-head member during its whole lifetime. The average energy consumption rate, ER_{SN}, should act as an important role in clustering protocols.

Similar models have been developed by [5, 6] to find an optimal probability of electing a cluster head, i.e. $1/m$ in (3). One of the deficiencies of previous approaches is that they assume the probability is predefined. Rather the optimal probability of being a cluster head depends on the sensor node's location and distances between neighbor nodes. For example, if a sensor node is located very near to the central base, it is optimal for the node to establish a direct communication link with the central base, since the cost of direct communication with the central base might be cheaper than using another cluster head as a mediator. The distant sensor nodes may create bigger clusters to lower ER_{SN}. However, while the cluster grows by adding more sensor nodes, the cost of being non-head members, ER_{MEM}, also increases because of the increasing distance between the cluster head and non-head members. The changing layout of mobile sensor networks brings more difficulties to clustering protocols based on fixed probability.

Besides the weakness that the global probability cannot catch local optimal choices, the fixed probability may result in an improper clustering pattern especially in a large network. If we set a global fixed probability, p_{CH}, for nodes to be cluster heads, the number of cluster heads actually follows the *binomial* distribution, $Prob(k) = C_n^k \cdot p_{CH} \cdot (1 - p_{CH})$, where k is the number of cluster heads after clustering procedures, and n is the number of total nodes. The variation of such distribution is given by, $n \cdot p_{CH} \cdot (1 - p_{CH})$. As we can see, the number of cluster heads after clustering procedure with fixed probability values varies more while the number of sensor nodes increases, which means a small chance to generate an optimal number of clusters in large networks.

4 UDC Protocol

A good clustering protocol should adapt node clusters to local network conditions. Expensive centralized algorithms can make a better clustering by analyzing all nodes' locations and requirements of queries with drawing on excessive communication overhead. Furthermore the changing layouts of mobile sensor networks bring more difficulties to any centralized approaches. Distributed approaches, like UDC, can be adaptive to local conditions at expense of exchanging collaborative messages between neighboring nodes. To lessen the burden of sensor nodes, a compact message to represent clusters rather than detailed information about individual sensor nodes is preferable to clustering protocols.

4.1 Cluster Bounding Circle

UDC uses a compact, fixed length message, named cluster bounding circle (CBC) to present clusters. A CBC is a circle centered at the cluster head, and the diameter is the longest distance between the cluster head and non-head members. A message of CBC i includes the location of its cluster head, CH_i, the diameter of CBC, d_i and the number of sensor nodes contained, m_i. Similar to (3), UDC estimates the energy consumption rate for the cluster head in CBC i as,

$$ER_{CH}(|CH_i|, m_i, d_i) = \frac{1}{m_i}[ER_{MEM}(d_i) \cdot (m_i - 1) + ER_{CH}(|CH_i|, m_i)], \quad (4)$$

where the $|CH_i|$ indicates the distance between the cluster head and the central base. Rather than the average cost $ER_{MEM}(\bar{D})$ in (3), UDC use a upper bound $ER_{MEM}(d_i)$ in (4) to estimate the cost when this cluster head acts as a non-head member, since in a dynamic environment, $ER_{MEM}(\bar{D})$ is hard to estimate.

After exchanging CBC messages, cluster heads may face different relations with other clusters. If CBC_1 contains CBC_2, head 1 uses $ER_{CH}(|CH_1|, m_1 + m_2, d_1)$ to estimate its new cost of combining cluster 2 into 1. Since adding cluster 2 into 1 do not increase the diameter of cluster 1, but decrease the chance that head 1 acts as a cluster head, which means the new cost is cheaper. In other situations where CBC_1 doesn't contain CBC_2 the diameter of cluster 1 will change by adding cluster 2. Sensor node 1 expects that adding cluster 2 increases the cost of being a non-head member due to the enlarged diameter of new cluster. Hence 1 uses the estimation as $ER_{CH}(|CH_1|, m_1 + m_2, d_2 + |CH_1 - CH_2|)$, where $|CH_1 - CH_2|$ indicates the distance between cluster head 1 and 2. The expected new cost might decrease if the changed chance of being a cluster head is more influential than the change of diameter.

4.2 Description of Algorithms

Reforming clusters may cause extra energy consumption, therefore UDC only allows cluster heads to reform clusters. In short words, if a cluster head chooses to join another cluster, all its member nodes will join the new cluster and change their cluster heads to the new head.

Figure 1 illustrates the clustering algorithm for the UDC protocol. In each round, all sensors choose themselves to be their own cluster heads at first. Then they enter a loop. In such a loop, all cluster head nodes prepare CBC messages based on local cluster information and exchange them in the beginning. Cluster head nodes then evaluate received CBC messages to find if it is worthy to enlarge the cluster by adding other clusters in, based on above discussions. If it is more expensive to enlarge itself, the cluster head node and all its members exit the loop and end the clustering procedure. Otherwise the cluster heads elect new cluster heads among them by a sub clustering procedure, *subclustering*() that can introduce the multi-hop idea to UDC. This procedure is not crucial to UDC, since the clusters can reform themselves based on CBC messages. In UDC, a cluster head can accept or reject join requests from other clusters by analyzing CBC messages. Further splitting operations may form better clusters, but UDC ignores them to avoid unnecessary energy consumption. After clusters are formed, no changes occur till next round.

While other protocols assign a predefined global probability to elect cluster heads, UDC adapts sensor nodes into different cluster densities based on local network conditions. Figure 2 shows an example of clustered layout based on the UDC protocol, where the star indicates the location of central base, the crosses indicate the locations of cluster heads, the dots indicate non-head nodes and the dotted lines indicate the com-

```
1.myHead=myID;
2.do{nextRound=true;
3.myCBC=prepare(myCurrentClusterInfo);
4.broadcast(myCBC);
5.othersCBCs[ ]=receive();
6.isWorthyToEnlarge=evaluate(othersCBCs[]);
7.if(!isWorthyToEnlarge) nextRound=false;
8.else{tentativeHead=subclustering();
9.if(myID==tentativeHead){
10.do{newMemCBC=receive();
11.if(isWorthyToAdd(newMemCBC)) confirmSender(i);
12.}until no one wants to join me;}
13.else{ sendTo(myCBC,tentativeHead);
14.waitForConfirm();
15.if(confirmed){informMembers(ChangeHeadTo,tentativeHead);
16.toBeMember();
17.nextRound=false;}}}
18.}while(nextRound)
```

Fig. 1. Algorithm description of the UDC protocol

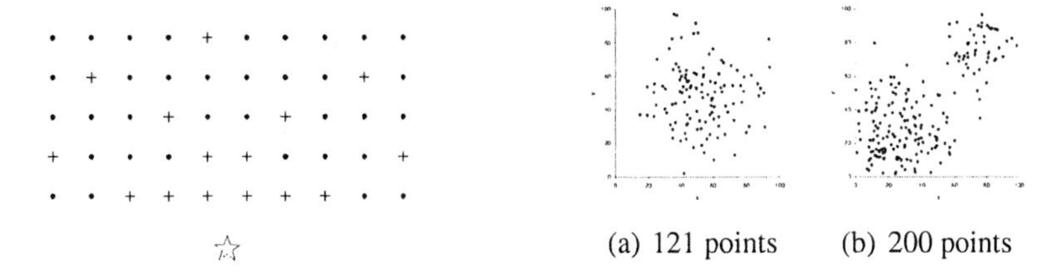
(a) 121 points (b) 200 points

Fig. 2. An example

Fig. 3. Plots of two skewed layouts

munication links within clusters. The sensor nodes nearby the central base tend to create small clusters, while the distant nodes need to create larger clusters to reduce the energy consumption. The self adaptive characteristic of UDC is more attractive to mobile sensor networks, since the layout of sensor nodes cannot be easily determined in advance in mobile sensor networks. Our simulation results validated our expectation that UDC outperforms current clustering protocols, as illustrated in section 5.

5 Simulation Result

We implemented UDC in Java, and simulated the sensor network behaviors by treating each sensor node as a thread. We tested the UDC protocol, LEACH and HEED in different network layouts. The first layout is a grid-like network with 121 nodes. In real applications, nevertheless, the mobile sensor networks cannot be in perfect grid forms all the time, but more realistically be skewed as showed by fig.3. In the first skewed layout, fig.3(a), we set a normal distribution along x and y axes both centered at 50 with variance 25 and generated 121 points. Similarly in the second skewed layout, two clustered locations with 150 and 50 points were generated centered at $(25, 25)$ and $(75, 75)$ respectively as shown in fig.3(b). We implemented a HEED-like algorithm as the $subclustering()$ in UDC. This $subclustering()$ uses a cluster head discovery procedure as HEED does, but just uses distance as the cost metric, while HEED also uses

Table 2. Parameter settings in simulation

Parameter	Value
Network region	from $(0, 0)$ to $(100, 100)$
Central base	at $(50, 125)$, $(50, 150)$ or $(50, 175)$
Threshold distance d_0	$75m$
E_{tx}	$5nj/bit$
ϵ_{fx}	$10pj/bit$
ϵ_{amp}	$0.0013pj/bit$
Data packet size (k)	2048 bits
Initial energy	$0.5J$
C_{prob}	5%
CH_{prob}	30%
$range$	$25m$

degree-based weights. In each round, all sensor nodes are clustered according to different clustering protocols, and then consume an amount of energy based on the simple model defined by fig.2. Most of our parameters are the same with [5, 6], among which several parameters are used only by LEACH and HEED. C_{prob} indicates the tentative cluster head probability within $range$ in HEED, and CH_{prob} is the global cluster head probability in LEACH. We set $CH_{prob} = 30\%$ according to the optimal choice defined in [5]. The reason that we ignore the data processing cost is that the communication cost is much more expensive than on-board computation. Furthermore, the data processing requirement depends on applications and query types, and can be generalized as E_{tx}. Our simulation program measures the remaining energy of each node, and records the node death rates based on different protocols. To be fair to all clustering protocols, the network scenario is set as single hop where all sensor nodes can establish direct links with the central base.

5.1 Lifespan of Sensor Networks

Figure 4 plots the rounds when the first and last sensor nodes die in the 121 grid sensor network for different locations of the central base. As we expected, UDC prolongs the time when the first node ceases working. UDC allows all sensor nodes to work up to twice longer than other clustering protocols do, as shown by fig.4(a). This metric is one of the most important requirements of sensor networks, since this time indicates how long a network can perform with *full* capacity. As the distance between the sensor network and the central base is enlarged, the lifetime of network decreases. Compared with other clustering methods, the decreasing trend of round when the first node dies in UDC is flatter than that in other protocols. But the decreasing trend of time when all nodes deplete their energy in UDC is steeper as illustrated by fig.4(b). Figure 5 shows the results in skewed layouts and compares them with the grid layout with the central base at $(50, 150)$. UDC still outperforms others in terms of the time when first one cease working as showed by fig.5(a). As we can see from fig.5, however, the predefined global probability does not work well for different network layouts. For example, LEACH works better than HEED in skewed network of 121 nodes, but is outperformed by HEED in the grid network with the same number of nodes and the same parameter settings.

On the other hand, the lifespan of a sensor network is a dynamic process as shown by fig.6 which records the number of alive nodes in each round in the grid-like network of 121 nodes with the central base at $(50, 175)$. It reveals that the sensor nodes in UDC

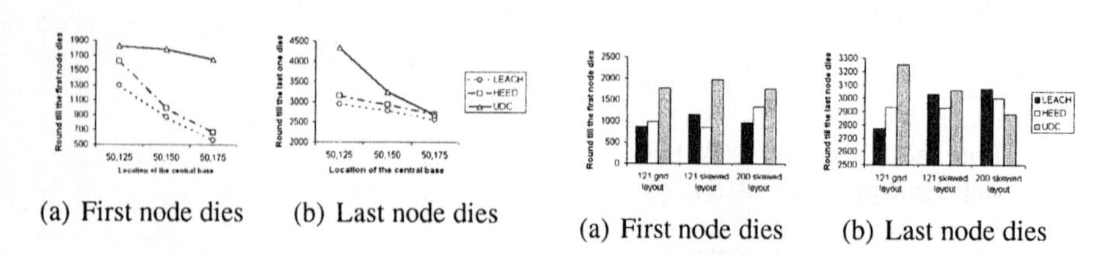

(a) First node dies (b) Last node dies

Fig. 4. Lifetime of 121 grid layout

(a) First node dies (b) Last node dies

Fig. 5. Lifetime of skewed sensor networks

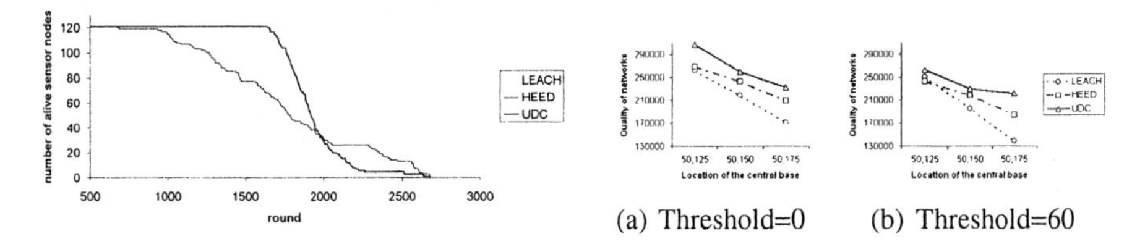

(a) Threshold=0 (b) Threshold=60

Fig. 6. The life span of 121 grid layout

Fig. 7. Quality of sensor networks

die "suddenly", while sensor nodes in other clustering protocols die more gradually. It is reasonable since sensor nodes in UDC spend a lot of energy to maintain all of the nodes alive, while in other clustering protocols alive nodes can alleviate their burden from the death of other nodes. In UDC, when the first one cease working the remaining nodes are also close to their ends. The "sudden death" property of UDC may cause a problem to a sensor network in a high sensor node density, since the redundant nodes may also be depleted in UDC protocol to maintain all sensor nodes alive as shown in fig.5(b).

5.2 Quality of Networks

It is hard to define the life time of a sensor network, since a network consists of more than one sensor node. The time when the first node or the last node stops functioning cannot describe the behavior of sensor networks perfectly. A quality description of sensor network may be more reasonable than the simple "lifetime" of networks. The quality of sensor networks, nevertheless, depends on applications. For example, some applications require all sensors to work, while half alive sensor nodes are enough for other applications. Hence a simple quality metric to describe the number of alive nodes in service can be defined as,

$$Q = \sum_{i=0}^{\infty} CountAliveNodes(i), \text{ if } CountAliveNodes(i) > threshold. \quad (5)$$

In (5), $CountAliveNodes(i)$ counts the number of alive sensor nodes in round i. This measurement Q tells us that "during its lifetime, how many alive sensor nodes can a sensor network totally provide to satisfy some services based on a *threshold* setting?". Compared with the lifetime of networks, the Q measure a quantity of service provided by sensor networks to satisfy applications. The quality of 121 grid network with different $threshold$ settings is measured for different central base locations, and shown by fig.7. UDC outperforms other protocols as shown by fig.7(a) for $threshold = 0$, where a single alive node is satisfactory. With a bigger and more realistic $threshold$, such an improvement would be more significant as shown by fig.7(b).

6 Conclusion and Future Work

This paper presents a novel protocol, UDC, designed for dynamic sensor networks to support aggregation queries. Compared with other clustering protocols, UDC benefits from adapting clusters to local network conditions, and significantly prolongs the lifetime of networks with a high quality. Although a sensor network in UDC protocol tends to undergo a "sudden death" after some time point, the "sudden death" property of UDC also throws a light for next version of UDC to reduce the "death rate" of nodes and enhance the quality of networks.

References

1. Akyildiz, I., Su, W., Sankarasubramaniam, Y., Cayirci, E.: A survey on sensor networks. IEEE Commun. Mag. **40** (2002) 102–114
2. Hellerstein, J.M., Hong, W., Madden, S., Stanek, K.: Beyond average: Toward sophisticated sensing with queries. In: IPSN. (2003) 63–79
3. Demers, A., Gehrke, J., Rajaraman, R., Trigoni, N., Yao, Y.: The cougar project: a work-in-progress report. SIGMOD Rec. **32** (2003) 53–59
4. Madden, S., Franklin, M.J., Hellerstein, J.M., Hong, W.: Tag: a tiny aggregation service for ad-hoc sensor networks. SIGOPS Oper. Syst. Rev. **36** (2002) 131–146
5. Heinzelman, W.R., Chandrakasan, A.P., Balakrishnan, H.: An application-specific protocol architecture for wireless microsensor networks. IEEE Transactions on Wireless Communications (2002) 660–670
6. Younis, O., Fahmy, S.: Distributed clustering in ad-hoc sensor networks:a hybrid, energy-efficient approach. IEEE INFOCOM 2004 (2004)
7. Culler, D., Estrin, D., Srivastava, M.: Overview of sensor networks. Computer (2004) 41–48
8. Tilak, S., Abu-Ghazaleh, N.B., Heinzelman, W.: A taxonomy of wireless micro-sensor network models. SIGMOBILE Mob. Comput. Commun. Rev. **6** (2002) 28–36
9. Karaki, A.J., Kamal, A.: 6. In: Handbook of Sensor Networks: Compact Wireless and Wired Sensing Systems. 1 edn. CRC Press (2005)
10. Conner, W.S., Chhabra, J., Yarvis, M., Krishnamurthy, L.: Experimental evaluation of synchronization and topology control for in-building sensor network applications. In: WSNA '03: Proceedings of the 2nd ACM international conference on Wireless sensor networks and applications, New York, NY, USA, ACM Press (2003) 38–49
11. Intanagonwiwat, C., Govindan, R., Estrin, D.: Directed diffusion for wireless sensor networks. IEEE/ACM Transactions on Networking **11** (2003)
12. Heinzelman, W.R., Kulik, J., Balakrishnan, H.: Adaptive protocols for information dissemination in wireless sensor networks. In: Mobile Computing and Networking. (1999) 174–185
13. Han, J., Kamber, M.: Data Mining: Concepts and Techniques. 1 edn. Morgan Kaufmann (2001)

A Backoff-Based Energy Efficient Clustering Algorithm for Wireless Sensor Networks

Yongtao Cao[1], Chen He[1], and Jun Wang[2]

[1] Department of Electronics Engineering,
Shanghai Jiao Tong University, China
ytcao@sjtu.edu.cn,
[2] Department of Communication Engineering,
Nanjing University of Posts and Telecommunications, China

Abstract. Wireless sensor networks have emerged recently as an effective way of monitoring remote or inhospitable physical environments. One of the major challenges in devising such networks is how to organize a large amount of sensor nodes without the coordination of any centralized access point. Clustering can not only conserve limited system resource, but also serve as an effective self-organization tool. In this paper, we present a distributed clustering algorithm based on adaptive backoff strategy. By adaptively adjusting the wakeup rate of the exponential distribution, a node with higher residual energy is more likely to be elected clusterhead. We also take advantage of the contention-based channel access method to ensure that clusterheads are well scattered. Simulation experiments illustrate that our algorithm is able to significantly prolong network life compared with the conventional approach.

1 Introduction

Wireless sensor networks have recently attracted intensive attention from both academic and industrial fields since their vast potential prospects in military and commercial applications, emergency relief *etc.* [1].Because of the absence of a centralized control, deployed sensors should be capable of self-organizing to construct the whole network topology.

Hierarchical (clustering) technique is an efficient tool to organize large-scale networks such as Internet or cellular networks. Clustering and data aggregation can also aid in reducing bandwidth and energy consumption, which is vital to resource-constrained sensor networks. Several clustering algorithms have been proposed for wireless sensor networks in recent years. Heinzelman *et al.*[2] have proposed a distributed algorithm LEACH to form one-hop clusters, in which sensors elect themselves as clusterheads with some probability and advertise their decisions to their one-hop neighbors. Assuming that sensors are uniformly distributed in the working region, the authors also present an analytical model to compute the optimal number of clusterheads. In [3], Bandyopadhyay *et al.* assume that sensors are distributed according to a homogeneous spatial Poisson

X. Jia, J. Wu, and Y. He (Eds.): MSN 2005, LNCS 3794, pp. 907–916, 2005.

process. By taking advantage of results in stochastic geometry, they obtain the optimal probability at which a sensor node choose itself as a clusterhead and the maximum number of hops allowed from a sensor to its clusters. Younis *et al.* [4] propose a clustering protocol HEED which periodically selects clusterheads according to a hybrid of residual energy and a second parameters. Among these protocols mentioned above, LEACH, due to its simplicity, effectiveness, low time complexity etc., has been well studied and become a referred baseline to evaluate the clustering performance in sensor networks.

In this paper, we first point out that LEACH may result in fast death of some sensor nodes because of its randomness. Then we propose a distributed load-balancing clustering algorithm based on an adaptive backoff strategy. Simulation experiments show that this algorithm is able to prolong the network life compared with LEACH.

2 Analysis on LEACH

In this section, we first define "the network life". The network life of wireless sensor networks can be defined as the time elapsed until the first node dies, the last node dies, or a fraction of nodes die. In many applications, each working node is critical for the whole system, so the network lifetime is defined as the shortest lifetime of any node in the network, i.e.

$$T_s = \min\{T_i, i \in V\} ,$$

where V is the set of nodes in the network, T_i is the lifetime of the node i and T_s represents the network lifetime. Since sensor networks often work in hazardous or hostile environments, it is difficult or impossible to recharge batteries for sensor nodes. How to prolong the network life is the first consideration in wireless sensor networks.

In clustered sensor networks, clusterheads are responsible for data fusion within each cluster and directly transmit the aggregated data to the remote base station (BS). With clustering and data compression, the network payload has been greatly reduced i.e. battery energy can be considerably saved. Among clustering protocols proposed recently, LEACH is a typical representative. LEACH is a dynamic distributed clustering protocol since it only depends on local information. Its clustering process can terminate in a constant number of iterations ($O(1)$ time complexity) regardless of network diameter or the number of nodes. Moreover, by rotating clusterheads, LEACH attempts to evenly distribute the energy loads among all nodes. However, because of some reasons, LEACH may result in faster death of some sensor nodes, i.e. shorten the network life. Our simulation experiments show LEACH is not load-balancing as expected. We think that two reasons are able to explain such result:

A. Some nodes become "forced clusterheads" and have to directly communicate with remote base station

Although [2] and [3] compute the optimal number of clusterheads for LEACH, the number of clusterheads produced by using LEACH doesn't always equal to the expected optimal value due to its randomness. For example, among N nodes, k clusterheads are expected. Then the self-electing probability is $p = {}^{k}/_{N}$. The event X is defined as " m clusterheads are elected", which is a discrete random variable following a Binomial distribution with parameters N and p, i.e.

$$\Pr(X = m) = \binom{N}{m} p^m (1-p)^{N-m}$$

When too few clusterheads are elected, it is very likely that there is no self-elected clusterhead in a certain node's proximity. So it has to become a "forced clusterhead" and communicates directly with BS which often locates far away from the node's working region. Even if LEACH results in the target number of the clusterheads, these clusterheads may scatter unevenly, i.e. most clusterheads clump in some regions while few clusterheads locate in other regions[5]. The unevenly distribution of clusterheads also leads to "forced clusterheads".Although LEACH attempts to rotate the CH's role among sensor nodes i.e. one node communicates directly with BS only once in a working cycle. But unfortunately, the randomness undermines the attempt. Simulation results shows that the probability that a node contact directly with BS only once in a 20-round cycle is little while the case a node communicates directly with BS more than eight times accounts for about 50%. It is well known that radio communication with low-lying antennae and near-ground channels has an exponential path loss, i.e. that the minimum output power required to transmit a signal over a distance d is proportional to $d^n, 2 \le n \le 4$. Since transmitting data directly to remote BS is much energy-consuming, the clustering algorithm should avoid too many nodes involving in such long-distance communications.

B. LEACH determines one node's "role" without taking into consideration its residual energy

LEACH determines clusterheads according to a predefined probability[1]. Although a sensor node, if it has been elected as a clusterhead once, has no chance to become a clusterhead again in a working cycle, it is still possible to consume much more energy than other nodes, for example, it has to serve as a "forced CH" in each round or manage too many nodes because of the uneven distribution of sensor nodes. However, in next working cycle these nodes with low residual energy have the same opportunity to be elected clusterheads, which will deplete their battery energy very soon.

3 A Clustering Algorithm with Adaptive Backoff Strategy

In this section, we present a distributed algorithm based on adaptive backoff strategy. The primary goal of our approach is to prolong network lifetime, i.e.

[1] Although LEACH provides an alternative way considering nodes' residual energy to determine the self-electing probability, it is unrealistic in large-scale sensor networks since each node needs to obtain other nodes' energy information throughout the whole network. So we ignore such method in this paper since it has no practical meaning.

the time elapsed until the first node dies. The scheme we proposed not only maintains the desirable features of LEACH but also helps to evenly distribute energy load among sensor nodes.

We model a sensor network by an unit disk graph $G = (V, E)$,the simplest model of ad hoc networks or sensor networks. Each node $v \in V$ represents a sensor node in this network. There is an edge$(u, v) \in E$ if and only if u and v locate in each other's transmission range. In this case we say that u and v are neighbors. The set of neighbors of a node v will be denoted by $N(v)$. Nodes can send and receive message from their neighbors. We also make the assumption that the radio channel is symmetric.

3.1 Clustering Algorithm

Our distributed clustering algorithm is shown in Fig.1, The parameters used in this algorithm are listed in Table 1. The operation of our algorithm is also divided into rounds, which is similar with LEACH. Each round begins with a cluster-forming phase. In this phase, nodes are initially in the waiting mode. Each node waits for a initiator timer according to an exponential random distribution, i.e. $f(t_i) = \lambda_i e^{-\lambda_i t_i}$, where $\lambda_i = \lambda_0 \frac{E^i_{residual}}{E_{\max}}.E^i_{residual}$ is the estimated current residual energy in the node i and $E^i_{\max}$ is a reference maximum energy. When the timer fires, the node first sends a bid to compete for the channel. If the node A win the channel, it elects itself as a clusterhead and broadcast an ADV_CH message to all its neighbors. The neighbors that receive the message, stop the timer and decide to join the cluster which A initializes. If one node simultaneously receives more than one ADV_CH message i.e. it falls within the range of more than one self-elected clusterheads, it uses the node ID or its distance to those clusterheads to break ties (The distance can be determined according to signal attenuation). When a node B decides to join a certain cluster, it broadcasts a JOIN message, JOIN(myID, myHEAD) and terminates the algorithm. On receiving a JOIN(u, t) message, , node v checks if it has previously sent a ADV_CH message. If this is the case, it checks if node u wants to join v's cluster $(v = t)$. If the node $v$has not sent a ADV_CH message, it record the u'sdecision. If all v's neighbors have decided their role, i.e. join a certain cluster before$v$'s timer fires, v has to become a forced CH and stop its timer. When T_{CF}, the maximal time of cluster forming elapses, each node which has not decided its role places a bid for the channel. The nodes which have successfully broadcast

Table 1. Parameters Used in the Algorithm

Γ	the set of ID's of my one-hop neighbors
$E_{residual}$	the estimated current residual energy in the node
$E_{\max}$	the fully charged battery energy
λ_0	the initial wakeup rate
T_{CF}	the maximal cluster-forming time

```
Clustering Algorithm(Γ, E_residual, E_max, λ_0, T_CF)
begin
 /*set the timer according to an exponential random distribution f(t_i) = λ_i e^(-λ_i t_i) */
  p_r =Random(0,1);  λ_i = λ_0 E'_residual / E_max ;
  t_i = -(1/λ_i) ln(p_r/λ_i);
  current_time=GetTime( );   /* get current   system time */
  set   timer=current_time + min(t_i, T_CF);
  while(current_time<timer)
  begin
    on receiving ADV_CH(ID)
    begin
      set myHead=ID;
      broadcast JOIN(myID,myHEAD);
          return;
    end
    on receiving JOIN(ID,clusteID)
    begin
      Γ=Γ - {ID};
      if (Γ==φ)
        break;
    end
    current_time=GetTime( );
  end
  Attempt to broadcast my ADV_CH(myID);
  Switch (Occurrence during the attempt do)
    Case the ADV_CH message is successfully broadcast:
        set myHEAD=myID;
        return;
    Case the ADV_CH(ID) message is received:
        set myHead=ID;
     broadcast JOIN(myID,myHEAD);
        return;
  end Switch
end.
```

Fig. 1. Algorithm Pesudo-code

the ADV_CH messages will become CHs while others will join a certain cluster. Then another round of cluster-formation will begin until all nodes are clustered.

3.2 Discussion

For our algorithm, we obtain the following properties:

Lemma 1. *Our algorithm is fully distributed.*

Proof. A node makes decision only depending on local information:it attempts to become a clusterhead when its timer fires or joins a cluster when it successfully receives an ADV_CH message.

Lemma 2. *The cluster-forming phase lasts for at most T_{CF}.*

Proof. When T_{CF} elapses, though a node's timer has not yet fire, it has to stop the timer and bid for a clusterhead. Thus T_{CF} is the upper bound of the execution time of our algorithm.

Lemma 3. *Our algorithm has $O(1)$ message complexity per node i.e. the total message comlexity is $O(1)$.*

Proof. During the execution of our algorithm, one node in the network at most sends one ADV_CH message or JOIN message.

Lemma 4. *There are no neighboring clusterheads i.e. the clusterheads are well scattered.*

Proof. Our approach that a sensor node contends to be a cluster head is based on a control channel broadcast access method, which is similar with [6], where only one node win in its neighborhood, and in consequence, the elected heads are well scattered.

Lemma 5. *when the wakeup rate $\lambda \geq \max\left\{e, -\frac{\ln(1-\varepsilon)}{T_{CF}}\right\}$, our clusterhead selection algorithm will perform dynamic load-balancing for wireless sensor networks.*

Proof. Derived from the exponential distribution $f(t) = \lambda e^{-\lambda t}$, When fixing $f(t) = \mu, \mu \in [0,1]$,we get $t(\lambda) = -\frac{1}{\lambda}\ln\frac{\mu}{\lambda}$. The first-order derivate of the function is $\frac{dt}{d\lambda} = \frac{1}{\lambda^2}(1 - \ln\frac{\lambda}{\mu})$, it is obvious that when $\lambda \geq \mu e$, $\frac{dt}{d\lambda} \leq 0$ i.e. the function is monotonously decreasing. So when we choose

$$\lambda \geq \max(\mu e) = e\} \quad (1)$$

the algorithm will ensure that the node with more residual energy have more opportunity to become a clusterhead since its timer is more likely to elapse before its neighbors with lower battery energy.

We also hope that the network elects enough number of clusterheads in one round before the time T_{CF} elapses, for instance 50% of nodes are expected to initialize cluster-formation within one minute. Thus the selection of λ should satisfy the following inequation: $p\{t > T_{CF}\} \leq 1 - \varepsilon$, where ε is the expected percentage. Therefore, in this case

$$\int_{T_{CF}}^{+\infty} f(t)dt \leq 1 - \varepsilon \Rightarrow \lambda \geq -\frac{\ln(1-\varepsilon)}{T_{CF}} \quad . \tag{2}$$

Based on (2), we can calculate that a λ of 0.012 ensures that 50% of the nodes initialize cluster-forming process within one minute ($\varepsilon = 0.5$, $T_{CF} = 60$sec) .

From (1) and (2), we can conclude:

$$\lambda_{\min} = \max\left\{e, -\frac{\ln(1-\varepsilon)}{T_{CF}}\right\} \tag{3}$$

4 Simulation Results

We conduct simulation experiments to compare performance of the proposed algorithm with LEACH. These two algorithms are implemented in Microsoft Visual C++. The entire simulation is conducted in a 100m∗100m region, which is between $(x = 0, y = 0)$ and $(x{=}100,\ y{=}100)$. 100 nodes with 2J initial energy are randomly spread in this region. Two nodes are said to have a wireless link between them if they are within communication range of each other. The performance is simulated with the communication range of the nodes set to 25 meters. Initially, each node is assigned a unique node ID and x, y coordinates within the region. The base station locates in the (50,175).

We assume the simple radio model proposed by Heinzelman et al.[2].The transmission range R is set to 25m.For LEACH, we set the optimal value $k = 5$. And for our algorithm, we set the initial wakeup rate $\lambda_0 = 56$ wakeup/sec. Moreover, the residual battery energy is discretized into 20 levels, so the minimum wakeup rate $\lambda_{\min}$equals 2.8, which satisfies the inequation (3) (we assume $\varepsilon = 0.5$, $T_{CF} = 60$sec). Simulation experiments proceed with rounds. In each round, one ordinary node, if it has enough residual energy to function properly, collects sensor data and sends a packet to its CH or BS. We call such packet **"the effective data packet"**.

We first measure how many times a node communicates directly with the BS during a 20-round cycle by using LEACH and our algorithm. Fig.2 shows that compared with LEACH, our algorithm greatly reduces the number of times one node contact directly with BS, since the contention-based head advertising ensures no neighboring clusterheads i.e. clusterheads are well-distributed.

We also compare network life with our algorithm to LEACH, where network life is the time until the first node dies. Fig.3 illustrates our algorithm greatly improves the network life over LEACH. This is because LEACH's randomness may lead to some "heavy-burdened" nodes whose battery energy is very likely to be depleted much faster. In contrast, the head selection in our algorithm is primarily based on nodes' residual energy. Those nodes with more residual energy have higher probability to become CHs, which provides good load-balance among sensor nodes. In addition, the contention-based head advertising ensures that clusterheads are well-distributed which further lessens some nodes' burden and prolongs the network life.

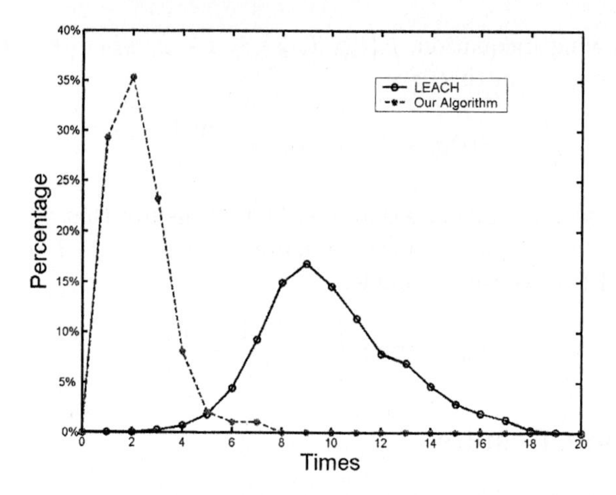

Fig. 2. Distribution of the number of times one node communicates directly with BS in a 20-round cycle using LEACH (100 nodes,$k_{opt} = 5$) vs. our algorithm(100 nodes,$\lambda_{min} = 2.8$)

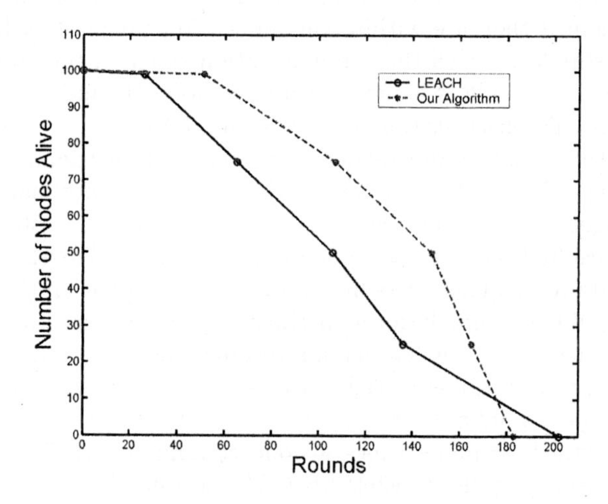

Fig. 3. Network Life using LEACH (100 nodes,$k_{opt} = 5$) vs. our algorithm(100 nodes,$\lambda_{min} = 2.8$)

Then we study the relationship of network life and effective sensor data. Observing the simulation results of Fig.4 shows that our algorithm will produce 50% effective sensor data more than LEACH over time since in the latter many nodes have to spend a large amount of energy communicating with BS. On the contrary, our algorithm effectively avoids this problem, hence the limited energy has been saved to send more effective sensor data.

Finally, we compare the energy-efficiency with LEACH to our algorithm. Fig.5 shows the total number of effective sensor data sent by network nodes for a

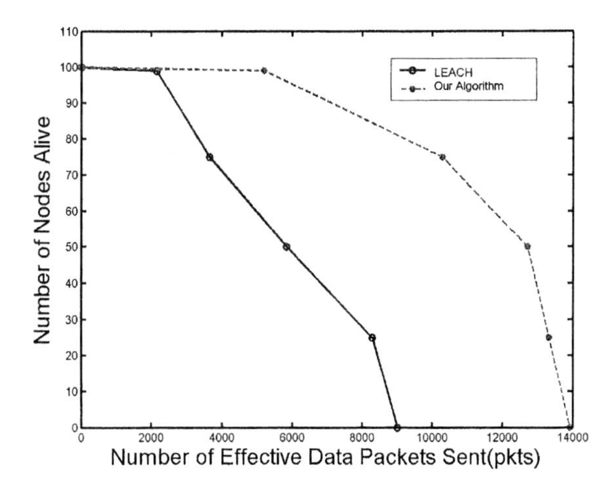

Fig. 4. Number of survival nodes per given amount of effective data packets sent using LEACH (100 nodes,$k_{opt} = 5$) vs. our algorithm(100 nodes,$\lambda_{min} = 2.8$)

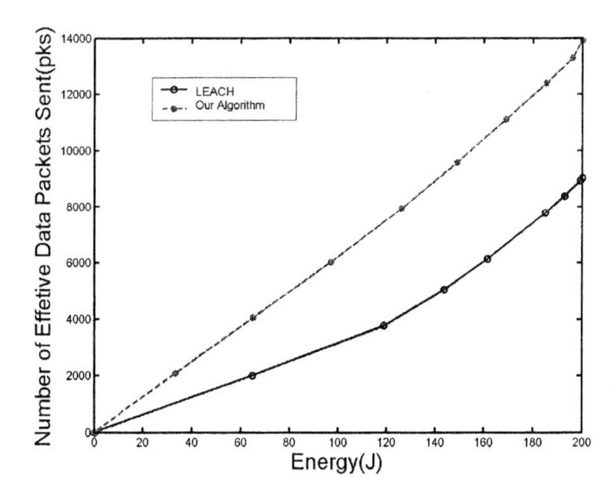

Fig. 5. Amount of effective data packets sent per given amount of energy using LEACH (100 nodes,$k_{opt} = 5$) vs. our algorithm(100 nodes,$\lambda_{min} = 2.8$)

given amount of energy. The result illustrates that our algorithm sends much more effective sensor data for a given amount of energy than LEACH i.e. our algorithm is more energy-efficient.

5 Conclusion

Clustering is one of the fundamental problems in wireless sensor networks. LEACH provides many advantageous features to meet the requirements of the severe resource constraints in sensor networks. However, due to its randomness, LEACH is not as load-balancing as expected. Both theoretical analysis and simulation results show that too much nodes have to communicate directly with the base station, which will consume a large amount of nodes' limited battery energy. To solve such problem, we propose a new distributed clustering algorithm which not only uses an adaptive backoff strategy to realize load balance among sensor node, but also introduces a contention-based message broadcast method to ensure there are no neighboring clusterheads. Simulation results also indicate that the proposed algorithm greatly reduces the number of "forced clusterheads" and prolongs the network life efficiently.

References

1. G. J. Pottie and W. J. Kaiser: Wireless Integrated Network Sensors. Commun. ACM, vol. 43, no. 5, May 2000, pp. 551-58.
2. W. B. Heinzelman, A. P. Chandrakasan, H. Balakrishnan: An Application-specific Protocol Architecture for Wireless Microsensor Networks. IEEE Tran. On Wireless Communications, Vol. 1, No. 4, pp.660-670, Oct. 2002.
3. Bandyopadhyay and E. J. Coyle: An Energy Efficient Hierarchical Clustering Algorithm for Wireless Sensor Networks. Proc. IEEE INFOCOM 2003.
4. O. Younis, and S. Fahmy: Distributed Clustering in Ad-hoc Sensor Networks: a Hybrid, Energy-efficient Approach. Proc.IEEE INFOCOM 2004.
5. L.Zhao, X. Hong and Q. Liang: Energy-Efficient Self-Organization for Wireless Sensor Networks: A Fully Distributed Approach. Poc. IEEE GLOBECOM 2004.
6. T.C. Hou and T.J. Tsai: An Access-Based Clustering Protocol for Multihop Wireless Ad Hoc Networks. IEEE JSAC, 2001,19 (7):120121210.

Energy-Saving Cluster Formation Algorithm in Wireless Sensor Networks

Hyang-tack Lee[1], Dae-hong Son[1], Byeong-hee Roh[1], S.W. Yoo[1], and Y.C. Oh[2]

[1] Graduate School of Information and Communication, Ajou University, San 5 Wonchon-dong, Youngtong-Gu, Suwon, 443-749, Korea
{hlee, ajouzam, bhroh, swyoo}@ajou.ac.kr
[2] Samsung Electronics Cooperation, Suwon, Korea
ycoh@samsung.com

Abstract. In this paper, we propose an efficient energy-saving cluster formation algorithm (ECFA) with sleep mode. ECFA can achieve energy efficient routing with the following two properties. First, ECFA reconfigures clusters with fair cluster formations, in which all nodes in a sensor network can consume their energies evenly. In order to achieve fair cluster regions, ECFA does not require any information on the location and energy of nodes. Second, by letting nodes very close to just elected cluster head be in sleep mode, ECFA can reduce unnecessary energy consumptions. Performances of ECFA are compared with LEACH and LEACH-C in the cases of the static and mobile nodes.

1 Introduction

Recently, sensor networks are becoming an important tool for monitoring a region of interest. To design sensor networks such that sensors utilize their energies in very effective ways is one of the most important issues to be solved for the efficient operation of the networks. Much of works for energy aware design for sensor networks have been carried out[1][2][3].

Among those, as a distributed cluster-based routing scheme, LEACH (Low-Energy Adaptive Clustering Hierarchy) has been proposed[4]. In LEACH, cluster heads that relay data from sensor nodes in a certain region called cluster to BS (base station) are periodically elected to prevent a specific sensor node from consuming its residual energy rapidly. However, since the cluster heads are elected in distributed and probabilistic way, there exist the possibilities of poor cluster formations, in which cluster heads are located very close to each other. LEACH-C (LEACH-Centralized)[5] has been proposed to solve the poor cluster formation problem in LEACH. In LEACH-C, each node sends the information on its location and residual energy to BS. By using the information, BS constructs clusters as optimal as it can, and broadcasts the cluster information to all nodes in the network. LEACH-C is more effective than LEACH from the cluster formation's viewpoints, but it consumes much energy compared with LEACH because all nodes have to communicate with BS at each round and it requires additional

X. Jia, J. Wu, and Y. He (Eds.): MSN 2005, LNCS 3794, pp. 917–926, 2005.

overhead for each sensor node to know its location information through additional communication technique such as GPS.

In this paper, we propose an efficient energy-saving cluster formation algorithm (ECFA) with sleep mode. ECFA can achieve energy efficient routing with the following two properties. First, ECFA can improve the energy efficiency by preventing the possibility of poor cluster formation as in LEACH. ECFA can produce clusters with fair cluster regions such that all the sensors in a sensor network can utilize their energies equally. In order to achieve fair cluster regions, ECFA does not require any information on the location and energy of each node. Second, by letting nodes very close to the cluster head just elected be in sleep mode, ECFA can reduce unnecessary active period which is one of the main causes of energy dissipation[6]. Performances of ECFA are compared with LEACH and LEACH-C in the cases of the static and mobile nodes.

The rest of the paper is organized as follows. In Section 2, some background on LEACH algorithm and its generic problem is explained. Then, our proposed scheme, ECFA, is illustrated in Section 3. In Section 4, some experimental results will be given. Finally, we make conclusions in Section 5.

2 Problem Definition

In LEACH, timeline is divided into rounds, and each round consists of Set-up Phase and Steady-state Phase[4]. Clusters are reconfigured in Set-up Phase. After then, in Steady-state Phase, actual data transmission can be done from the nodes to the cluster head, and then to the BS. Especially, Set-up Phase consists of three sub-phases such as Advertisement, Cluster Set-up and Schedule Creation Phases. In Advertisement Phase, each node decides whether it can be elected as a cluster head or not. Then, in Cluster Set-up Phase, all nodes except for cluster heads choose their cluster head, and then cluster reconfiguration is finished. Finally, TDMA schedule for data transmission in the network is arranged in Schedule Creation Phase.

For electing cluster heads in Advertisement Phase, n-th sensor node chooses a random number between 0 and 1, and compares the number with a threshold value $T(n)$ as following

$$T(n) = \begin{cases} \frac{P}{1-P\cdot(r \ mod \frac{1}{P})} & \text{if } n \in G \\ 0 & \text{otherwise} \end{cases} \quad (1)$$

where P is the desired percentage of the cluster heads, r is the current round, and G is the set of nodes that have not been cluster head in the last $1/P$ rounds. If the chosen random number by n-th sensor node is less than $T(n)$, the node is elected as a cluster head for the corresponding round. According to Eq.(1), LEACH ensures that all nodes become a cluster head exactly once during consecutive $1/P$ rounds.

There are two basic problems in LEACH with the above head election procedure.

First, since cluster heads are elected in a probabilistic way only using Eq.(1), there is no way to consider the formation of clusters. There exists some possibility of both good and poor cluster formations. In good cluster formations, since elected sensor nodes are evenly distributed in the region of the sensor network, all sensor nodes can consume their energy evenly in average. On the other hand, in some poor cluster formations that adjacent nodes can be elected as cluster heads, sensor nodes with longer distances to corresponding cluster head consume much more energies than those with shorter distances. In addition, collisions can be occurred frequently in the network due to short distance between cluster heads. Though LEACH-C[5] has been proposed to overcome the poor cluster formation problem, it requires an additional overhead for each node to know its location and to deliver its location and energy information to BS. The overhead results in consuming much energy compared with LEACH.

Second, in LEACH, all nodes have to keep awake in Advertisement Phase. This may cause a problem on unnecessary energy consumption. For example, some nodes that close to the first elected cluster head and will be a member of a cluster governed by the cluster head do not need to participate into following head election procedures and listen signals from other cluster heads.

3 Energy-Saving Cluster Formation Algorithm with Sleep Mode

In this paper, we propose an energy-saving cluster formation algorithm(ECFA) with sleep mode to solve the two problems described in the previous Section that LEACH has. Fig.1 shows the basic timeline of the proposed ECFA operation, in which timelines are divided into rounds as in LEACH. The Advertisement Phase of ECFA consists of K stages where K denotes the predefined number of cluster heads in the network. It is noted that there exists only one stage in Advertisement Phase in LEACH. This means that there are K chances for each node to become a cluster head at each round in ECFA, while only one in LEACH.

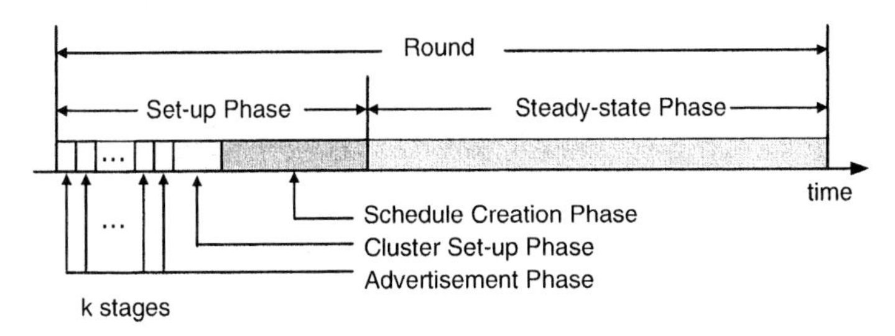

Fig. 1. Timeline of ECFA operation

In ECFA, nodes within a certain range from each cluster head elected at each stage do not take part in the cluster head process at next stages in the corresponding round. The detail procedure for the cluster head election will be explained later. If we let $C(r,s)$ be the number of nodes that can take part in the cluster head election process at a state s in a round r where s=0,1,$\cdots$,K-1 and r=0,1,2,$\cdots$. Ideally, the average number of nodes belong to the range covered by each cluster head can be $C(r,0)/K$ where $C(r,0)$ is the number of nodes that can be candidates of cluster heads at the beginning of the round r. Then, we have

$$C(r,s) = C(r,0) \cdot (1 - \frac{s}{K}) \tag{2}$$

where $C(r,0) = N - K \cdot (r \ mod \frac{N}{K})$ and N is the total number of nodes.

Let $P_n(r,s)$ and $E_{CH}(r,s)$ be the probability that n-th node becomes a cluster head and the expected number of nodes that can be elected as cluster heads, respectively. In order to get the ideal situation where one cluster head is elected at each stage, $P_n(r,s)$ and $E_{CH}(r,s)$ should satisfy the following condition. That is,

$$E_{CH}(r,s) = \sum_{n=1}^{N} P_n(r,s) = 1. \tag{3}$$

Let G be the set of nodes that have not been elected as cluster heads in the last N/K rounds, and $P_{n \in G}(r,s)$ be the probability that an arbitrary node n ($n \in G$) becomes a cluster head at time (r,s). The nodes that have not been a cluster head have identical $P_n(\cdot)$ at each stage. Then, we can rewrite Eq.(3) as

$$E_{CH}(r,s) = C(r,s) \cdot P_{n \in G}(r,s) = 1 \tag{4}$$

From Eq.(4), we have

$$P_{n \in G}(r,s) = \frac{1}{C(r,s)}. \tag{5}$$

It is noted that Eq.(5) is the probability that only one node can be elected as the cluster head at a certain stage. Substituting Eq.(2) into Eq.(5), we have the threshold for n-th node can become a cluster head at stage s in a round r as following

$$T(n,s) = \begin{cases} \frac{1}{\{N-K\cdot(r \ mod \frac{N}{K})\}\cdot(1-\frac{s}{K})} & \text{if } n \in G \\ 0 & \text{otherwise} \end{cases} \tag{6}$$

With the threshold of Eq.(6), the cluster head election procedure of the proposed ECFA for n-th sensor node at a stage s in a round r is shown in Fig.2. At the beginning of the stage, n-th node eligible for becoming a cluster head selects a value between 0 and 1 randomly, and then compares it with the threshold $T(n,s)$ as in Eq.(6). If the selected random number by the node is less than $T(n,s)$, the node is elected as a cluster head for the corresponding stage. After then, it broadcasts its advertisement message to neighbor nodes within a certain range. The neighbor nodes received the advertisement message estimate the distance d from itself to the cluster head by the received signal power. The operations of neighbor nodes are differentiated according to the distance d as following.

```
procedure ClusterHeadElection_ECFA(n,r,s) begin
 1:        if (s=0) then
 2:           initialize cluster head list
 3:           state = ACTIVE
 4:        end-if
 5:        if (state=ACTIVE) then
 6:           choose a random number v in [0,1]
 7:           if (v < T(n,s)) then
 8:               broadcast a cluster head advertisement message
 9:               state=SLEEP
10:           else
11:               goto S1
12:           end-if
13:        else if (state=LISTEN) then
14: S1:       wait and listen a cluster head advertisement message
15:           if (a cluster head advertisement message is received) then
16:               estimate the distance d from the cluster head
17:               add the cluster head to the cluster head list
18:               if (d ≤ R_o) then
19:                   state=SLEEP
20:               else if (R_o < d ≤ R) then
21:                   state=LISTEN
22:               end-if
23:           end-if
24:        end-if
end procedure
```

Fig. 2. Cluster head election procedure of ECFA

i) $0 \leq d \leq R_o$. The neighbor nodes within this area sleep until the beginning of Cluster Set-up Phase to reduce the unnecessary active periods. Since the nodes are in sleep, they do not take part in the next cluster head election processes and listen to the signal from other cluster heads.

ii) $R_o < d \leq R$. The neighbor nodes within this range do not take part in the cluster head election process at the next stages. Though these nodes do not participate into the head election process, they still listen to the signal from other cluster heads. After all cluster heads are selected, the nodes determine the cluster head with the strongest power as its cluster head during Cluster Set-up Phase.

iii) $d > R$. The nodes take part in the head election process at the next stage.

Determining R_o and R. Consider a sensor network with $M \times M$ size. Let us assume an ideal situation that sensor nodes are uniformly distributed on the sensor network and K elected cluster heads are evenly arranged. Then, we have the following approximation

$$M^2 = K\pi {R_o}^2 \tag{7}$$

where R_o is the radius for the advertisement region exclusively covered by each cluster head in the above ideal situation. However, when some of the cluster heads are located on the region near to edges of the network, it becomes unsuitable because the advertisement regions covered by those clusters are smaller than M^2/K, and it does not satisfy the ideal condition of Eq.(7).

In order to solve the problem, it needs to consider some larger value R than R_o of Eq.(7). For the upper limit of R, let assume an extreme case when K=1 and the cluster head is elected at a border of the sensor network. In this case, if $R = 2R_o$, the cluster head can cover the whole sensor network range. On the other hand, as K increases to infinite, R_o becomes small enough to satisfy the ideal situation of Eq.(7). Likewise, the radius R that we are trying to find out is highly related to the area of sensor networks as well as the number of clusters. Let R_a be the average radius of the sensor network. For sensor networks with $M \times M$ size, it can be approximated by $M^2 = \pi \cdot R_a^2$. if we define the radius $R \equiv (1 + \frac{R_o}{R_a})R_o$, then we have

$$R = (1 + \sqrt{1/K})R_o \quad (8)$$

It is noted that the factor in front of R_o of right side of Eq.(8) has the range between 1 and 2. This provides the consistency with the intuitions that we used for the derivation of R. That is, R=$2R_o$ when K=1, and $R \to R_o$ when $K \to \infty$.

4 Simulation Results

The simulation has been carried out using ns-2 network simulator [7][8]. For the simulation, a sensor network with size of 100m×100m and 100 arbitrarily distributed sensor nodes has been considered. We let the number of clusters be 5, the cluster formation be changed at every 20 seconds and the location of BS be (50,75). In addition, the same simulation environment and radio energy model as in [5] has been used. That is, we let the initial energy for each node be 2J and the spreading factor be 8. According to equations (3) and (4), as the values of R_o and R for ECFA, 25 and 37 have been used, respectively.

Effect of Sleep Range R_o on the System Time. To observe the effect of sleep mode, we carried out the simulations varying the sleep range(R_o) between 0 and R. Especially, in order to investigate how the energy of each node is consumed efficiently, in Fig.3, it is shown the system time that whole nodes are alive. That is, the y-axis of Fig.3 indicates the elapsed time until the fist dead node is happened. We can see from Fig.3 that as the sleep range R_o increases to around 25, the time also increases. However, as the sleep range increases over 25, it tends to decrease. This is because when the sleep range is larger than R_o, there are some nodes that use more energy for data transfer due to long distance between the cluster heads and them. This indicates that the value of 25 for the sleep range of R_o given from Eq.(7) is adequate. It is noted that the zero sleep range means that sleep mode is not used.

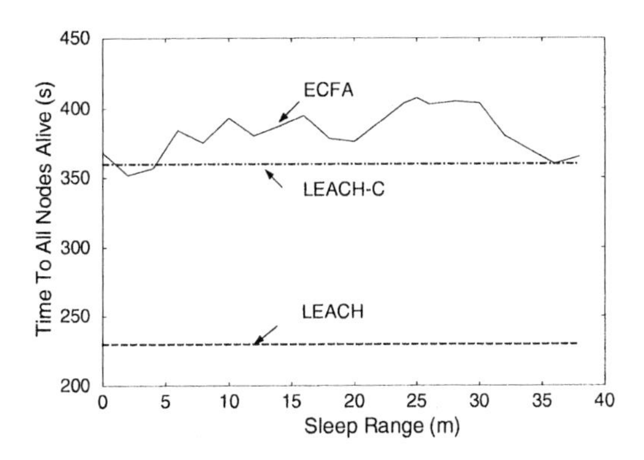

Fig. 3. Elapsed system time upto all the nodes are alive

Performances for Static Nodes' Case. Let define the system lifetime as the time until there is no active node in the network. In Fig.4(a), the system lifetime performances are compared. In Fig.4(a), ECFA(0) means ECFA without sleep mode, i.e. R_o=0. For ECFA, R_o=25 is used for the simulation. From Fig.4(a), we can see that LEACH-C shows the shortest system lifetime, because each node

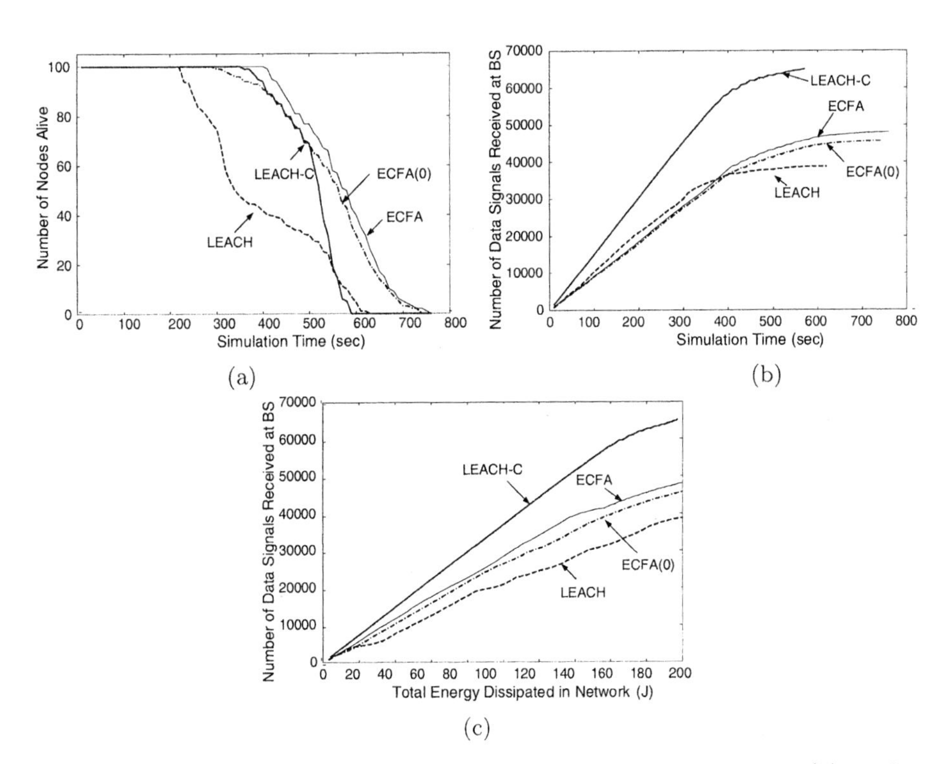

Fig. 4. Performances for static nodes: (a) number of nodes alive along time, (b) number of data signals received at BS over the time and (c) per given amount of energy

is required to maintain its location and energy information and to communicate the information with BS for forming clusters. On the other hand, ECFA and ECFA(0) show better system lifetime performances than LEACH. Especially, from the results that ECFA outperforms ECFA(0), we can see that the sleep mode can improve the system lifetime performances further.

Though the system lifetime of LEACH-C is shorter than other schemes, from the viewpoints of the amount of delivered data, LEACH-C shows the best performances as shown in Fig.4(b) and (c). In Fig.4(b), the total number of data signals received at BS during the system lifetime is compared. Though LEACH-C shows the shortest system lifetime, it can deliver much more data signals than other schemes. This is because LEACH-C can configure optimal cluster formations with the global knowledge of the network so that it can consume less energy for delivering data between cluster heads and their cluster member nodes once clusters are configured. This phenomenon can be illustrated by Fig.4(c), in which LEACH-C shows the largest delivered data signals per given energy. We can see also that ECFA and ECFA(0) show lower data deliver performances than LEACH-C, but much better than LEACH. And, ECFA show better than ECFA(0). Compared with LEACH, ECFA has more opportunities of delivering data in ECFA since ECFA keeps the nodes alive longer than LEACH by reducing the possibility of poor cluster formations. In addition, by using sleep mode, ECFA can achieve longer system lifetime and better data deliver performances. On the other hand, since ECFA does not require any knowledge of network, it can not get optimal cluster formations as in LEACH-C. However, ECFA can keep the system lifetime much longer than LEACH-C.

Performances for Mobile Nodes' Case. We also carried out simulations for the environment that sensor nodes are moving. For the simulation, we used Random Waypoint model presented in [9]. At every second, each node chooses a destination randomly and moves toward it with a velocity uniformly chosen from the range [0, V_{max}], where V_{max} is the maximum allowable velocity for every mobile nodes[10]. Same parameters used in [10] are applied to the simu-

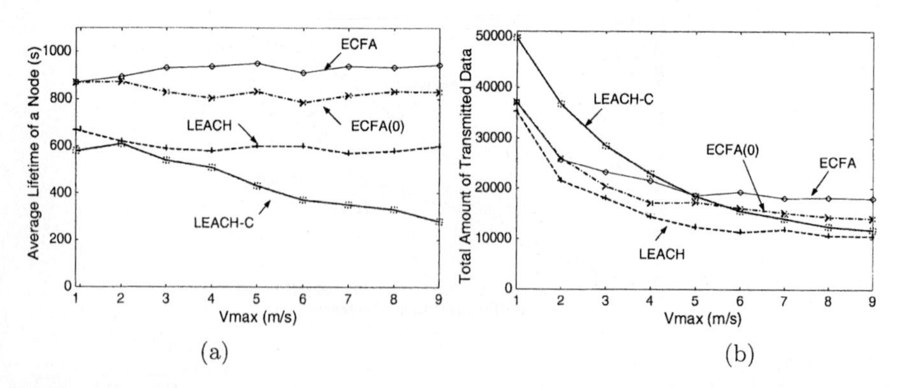

Fig. 5. Performances for mobile nodes: (a) average lifetime of a node (b) total amount of transmitted data

lation. The energy consumption due to mobility of each node is not considered in our simulation. By doing so, the intrinsic operational performances of those comparable schemes can be compared.

In Fig.5(a), system lifetime performances for mobile nodes' case are shown when V_{max} varies between 1 and 9 m/sec. Fig.5(a) shows that ECFA keeps the network longer than LEACH and LEACH-C. We can also see that ECFA with sleep mode outperforms ECFA(0) without sleep mode. Especially, the system lifetime of LEACH-C decreases as V_{max} increases, while other schemes are very few dependent on the variation of V_{max}. LEACH-C forms clusters as optimal as it can based on the global knowledge of the network. The cluster formation is very effective when the velocities of nodes are so slow that there are very slight changes in the whole network topology. However, it may not be kept when the nodes move so fast that there are large number of nodes very far from cluster heads and the global network topology is significantly changed from that the clusters were formed. Therefore, the lifetime of LEACH-C is getting shortened as the nodes move faster. In ECFA and LEACH, the nodes are not required to send the information at the beginning of the rounds. Thus, under the same condition, ECFA and LEACH dissipate less energy than LEACH-C. Moreover, ECFA makes the average lifetime of a node longer over all the velocities since ECFA is more energy-saving and forms the cluster more efficiently than LEACH.

Fig.5(b) shows the total number of data signals transmitted during the lifetimes of those comparable schemes. As the velocity increases, the amount of delivered data signals tends to decrease. At low velocities less than 5 m/s, LEACH-C shows the best data delivery performances. However, as the velocity increases larger than 5, ECFA shows better performances than LEACH-C because the lifetimes of LEACH-C are abruptly shortened for high velocities. Even though the lifetime of LEACH is longer than that of LEACH-C, LEACH cannot send more data because it cannot form more efficient cluster than LEACH-C.

5 Conclusion

In this paper, we proposed an efficient energy-saving cluster formation algorithm (ECFA) with sleep mode. ECFA can not only reconfigure clusters with fair cluster formations in which all nodes in a sensor network can consume their energies evenly, but also reduce unnecessary active period that is one of the main causes of energy dissipations. The simulation results showed that the system lifetime performance of ECFA outperforms those of other schemes such as LEACH and LEACH-C in both cases where the nodes are moved and not. For the data throughput, LEACH-C showed the best performance, and ECFA did better than LEACH in case the nodes are static. However, it was shown that ECFA outperforms the other schemes when the nodes are moving fast. Likewise, ECFA can achieve the improved lifetime performances better than LEACH-C as well as LEACH under both static and mobile nodes environments. Moreover, ECFA can be implemented with lower complexity than LEACH-C since ECFA does not require for nodes to maintain and send the information on their locations and energies to BS. We conclude that, ECFA can be well applied to the situation that longer system operation time is required and medium amount of sensing data is requested to be exchanged continuously under both environments where sensor nodes are moving or not.

Acknowledgement. This work was supported in part by MIC and IITA through IT Leading R&D Support Project, Korea.

References

1. Edger H. Callaway, *Wireless Sensor Networks: Architectures and Protocols*, Auerbach Publications, August 2003
2. Sameer Tilak, Nael B. Abu-Ghazaleh, and Wendi Heinzelman, "A Taxonomy of Wireless Micro-Sensor Network Models," ACM Mobile Computing and Communications Review, Vol. 6, No. 2, pp. 28-36, April 2002
3. I.F.Akyildiz, W.Su, Y.Sankarasubramaniam, E.Cayirci, "A survey on Sensor Networks," IEEE Communications Magazine, Vol. 40, No. 8, pp. 102-114, August 2002
4. W. Heinzelman, A. Chandrakasan, and H. Balakrishnan, "Energy-Efficient Communication Protocol for Wireless Microsensor Networks," Proceeding of Hawaii Conference on System Sciences, January 2000
5. W. Heinzelman, A. Chandrakasan, and H. Balakrishnan, "An Application-Specific Protocol Architecture for Wireless Microsensor Networks," IEEE Transactions on Wireless Communications, Vol. 1, No. 4, October 2002
6. W. Ye, J. Heidemann, and D. Estrin, "Medium Access Control With Coordinated Adaptive Sleeping for Wireless Sensor Networks," IEEE/ACM Transaction on Networking, Vol. 12, No. 3, June 2004
7. UCB/LBNL/VINT, "Network Simulator ns-2," `http://wwwmash.cs.berkeley.edu/ns.`
8. W.Heinzelman, A.Chandrakasan, and H.Balakrishnan, uAMPS ns Code Extensions, http://wwwmtl.mit.edu/ research/icsystems/uamps/leach
9. F.Bai, N.Sadagopan, A.Helmy, "The IMPORTANT Framework for Analyzing the impact of Mobility on Performance of Routing for Ad Hoc Networks," Ad-Hoc Networks, Vol. 1, No. 5, pp. 383 - 404, November 2003
10. L. Breslau, D. Estrin, K. Fall, S. Floyd, J. Heidemann, A. Helmy, P. Huang, S. McCanne, K. Varadhan, Y. Xu, H. Yu, "Advances in network simulation," IEEE Computer Magazine, Vol. 33, No. 5, pp. 59-67, May 2000

RECA: A Ring-Structured Energy-Efficient Cluster Architecture for Wireless Sensor Networks*

Guanfeng Li and Taieb Znati

Department of Computer Science,
University of Pittsburgh,
Pittsburgh, PA 15260, U.S.A
{ligf, znati}@cs.pitt.edu

Abstract. Clustering schemes have been proposed to prolong the lifetime of a Wireless Sensor Networks (WSNs). It is also desired that the energy consumption be evenly dispersed throughout the network. This paper presents RECA: a Ring-structured Energy-efficient Clustering Architecture. In RECA, nodes take turns to be the cluster-head and make local decisions on the fair-share length of their duty cycle according to their remaining energy and those of the rest of the nodes within the same cluster, consequently they deplete their energy supply at approximately the same time regardless of the initial amount of energy in their battery. RECA avoids the tight synchronization problem in LEACH and our primary results show that our scheme can achieve 50% - 150% longer network lifetime than LEACH depending on the initial energy level.

1 Introduction

Wireless Sensor Networks (WSNs) are autonomous systems which are often deployed in an uncontrolled manner such as dropping from an unmanned plane. The resulting network is characterized as an infrastructure-less, immobile wireless network. Applications employing these networks often take the form of one such network and one or a few data observers, called Base stations (BS's). Recent advances in VLSI technology have made individual sensor become smaller in size and more powerful in computation and radio communication, "anytime, anywhere" computing is becoming more and more realistic. However, the development of high-energy battery still lags far behind electronic progress. How to efficiently use the energy-constrained battery still remains a prominent research problem in WSNs.

Various techniques have been developed to address energy consumption problems in different aspects of WSNs. A comprehensive list of technologies to conserve energy in WSNs can be found in [1]. Among these technologies, clustering is

* This work is supported by NSF awards ANI-0325353, ANI-0073972, NSF-0524634 and NSF-0524429.

X. Jia, J. Wu, and Y. He (Eds.): MSN 2005, LNCS 3794, pp. 927–936, 2005.

particularly promising and has received much attention in the research community. In a clustered network, nodes are organized into non-overlapping groups. A special node in a cluster, called cluster-head, assumes extra responsibility, e.g. to fuse data collected by the nodes inside the same cluster and to route inter-cluster data packet. Because of the geographical proximity, the sensed data from nodes within one cluster usually exhibit high correlation, therefore a cluster-head can collect data from its member nodes, aggregate them to remove redundancy, and send one single packet to the BS. Number of transmission is reduced and hence the energy conservation.

Despite numerous advantages from clustering scheme, care must be taken to avoid its undesirable effects. If a node is fixed as the cluster-head during the whole network lifetime, its battery can be depleted very soon because of the extra work of a cluster-head has to perform. To prolong the network lifetime, it is therefore desired that the role of cluster-head rotate among nodes. To this end, there are two major schemes to switch the role of cluster-head among the network nodes. The first one is self-election and bully scheme. The operation of the system is divided into rounds, in each round, nodes randomly elect itself as the cluster-head and "bully" the neighboring nodes to be their cluster member. The second one is probe-and-switch. Each node sleeps for a random time and wakes up to probe the surrounding environment; if it perceives there are not enough cluster-heads around itself, it becomes a cluster-head. The inherent randomness in the cluster-head assignments in these schemes, however, results in uneven workload among all the network nodes thus leads to nodes dying prematurely. Clustering algorithms must strive to achieve balanced energy consumption in order to prolong the network lifetime.

Our proposed work, RECA, takes the node residual energy and its workload into consideration. The basic operation of RECA is divided into rounds. In each round, nodes within a cluster deterministically take turns in assuming the functionality of a the cluster-head for a time duration proportional to their residual energy. Energy dissipation is evenly distributed among all the nodes, thereby effectively prolonging the network lifetime.

2 Related Work

One pioneering work in this literature is LEACH [2]. In this work, operation of the network is divided into rounds. At the beginning of each round, each node generates a random number and compares to a threshold. If the generated random number is less than the threshold, this node becomes a cluster-head, otherwise, it joins the nearest self-elected cluster-head to form a cluster. The threshold is computed in such a way that nodes elect themselves to be cluster-heads with an increasing probability if they have not been a cluster-head in recent rounds. However, a recent study of the formation of clusters [3] in LEACH reveals that due to the stochastic way of the cluster-head election, the number of cluster-heads varies in a very large range. Specially, when the total number of nodes and the desired percentage of cluster-heads are small, there are no nodes electing themselves as the cluster-heads in some rounds. Furthermore, LEACH requires that nodes are at least loosely synchronized in order to start the cluster-head election at approximately the same time.

To improve LEACH performance, Handy, Haase, and Timmermann introduced energy factor into the cluster-head election process [4]. If a candidate for cluster-head has more residual energy in its battery, it has a better chance to be a cluster-head. Their addition to LEACH helps evenly distribute energy consumption to some extent, however, the election of a cluster-head remains largely in a random way.

Another improvement of LEACH was presented in PEGASIS [5]. In this work, nodes form a chain and send data only to the neighbor on the chain. Nodes take turns to be the cluster-head to fuse the data for the entire chain and send the result to the BS. Although PEGASIS can outperform LEACH in network lifetime, its performance on distributing energy consumption is worse as pointed out in [6].

GAF[7], ASCENT[8] and PEAS[9] are routing protocols employing the idea of a special node acting as a cluster-head in a geographical vicinity. Energy conservation is achieved through turning of the radio transmitter of some certain nodes. In GAF, nodes are supposed to be equipped with location-aware devices. The whole network field is divided into grids, sensors within the same grid are considered "equivalent" in routing a packet. As long as there is a node awake in a grid to route packets, other nodes can go to sleep to save energy. In ASCENT, nodes probe the surrounding environment to make local decision whether or not to switch state to be a cluster-head. Similar work can be found in PEAS. However, all of these schemes suffer from unbalanced workload among the sensors because of the lack of a schedule for active cluster-heads to be switch back to inactive state. In extreme cases, A cluster-head is forced to stay active until dead if no nodes are willing to take over the role of cluster-head.

Our proposed work, RECA differs from the above works in that cluster-head management is deterministic to ensure fairness among cluster nodes. Nodes assume cluster-head functionalities for a period of time proportional to their residual energy. Consequently, energy dissipation in the proposed scheme is evenly distributed among all the nodes, thereby effectively prolonging the network lifetime. Furthermore, cluster-head functionalities rotate deterministically among the cluster nodes. This deterministic aspect of the scheme eliminates the reclustering cost incurred by other clustering schemes. Finally, the requirement for tight synchronization is relaxed considerably since cluster formation happens only once at the initialization phase of the network.

3 RECA: A Ring-Structured Energy-Efficient Clustering Architecture

The basic operations of RECA are divided into two distinct phases: Energy-balanced Cluster Formation and Cluster-head Management. The first phase is executed only once. Its objective is to produce a set of energy-balanced clusters.

In the second phase, the role of cluster-head rotates among cluster nodes using the residual energy of a node to determine the length of the period during which it serves as a cluster-head.

3.1 Energy-Balanced Cluster Formation

In this phase, sensors are grouped into clusters with about the same total energy. We first obtain initial clusters using minimum transmission power, then we balance the energy among the initial clusters. To the contrary of other clustering schemes, this phase is executed only once throughout the whole lifetime of the network, re-clustering cost is therefore avoided.

Initial Cluster Formation. To minimize the intra-cluster communication cost later on in the system operation, one-hop clusters using minimum transmission power are formed. The algorithm used to obtain the initial clusters is outlined in Algo. 1.

First, the expected number of nodes in one cluster is estimated apriori as $\gamma = \frac{N \times \pi \times R^2}{A}$, where N is the total number of nodes in the network, A is the area that the network covers and R is the minimum transmission range. The cluster organizer election process is divided into time slots. In each slot, each node that has not elected itself as a cluster organizer or associated with a cluster generates a random number in [0, 1] and compares to a threshold $h = min(2^r \times$

Algorithm 1. Init_Clusters: Obtain the initial clusters

Define:
  γ: The expected number of nodes in one cluster.
  t_{slot}: Duration of one time slot.

Init_Clusters :executed at each node

for $r = 0$ to $\lceil log_2 \gamma \rceil$ **do**
  if not a cluster organizer nor a cluster member **then**
    generate a number α between 0 and 1
    if $\alpha < min(2^r \times \frac{1}{\gamma}, 1)$ **then**
      announce itself as a cluster organizer
      exit
    end if
  end if
  listen for a duration of t_{slot}
  if not a cluster organizer **then**
    if hears one or a few cluster organizer announcements **then**
      Choose the best SNR cluster to join
    end if
  end if
end for

$\frac{1}{\gamma}, 1)$, where r is the slot number. If the randomly generated number is less than h, this node becomes a cluster organizer and announces this information using minimum transmission power, if not, it listens to the cluster organizer announcements. If it hears any, it associates itself with the cluster from which the announcement has the best signal noise ratio (SNR). Nodes that associate with one cluster in previous slots also listen to the new announcements to adjust cluster membership. If one node finds a closer newly elected cluster organizer, it re-associates itself with the new one.

After $\lceil log_2\gamma \rceil$ time slots, each node in the network is either a cluster organizer or a cluster member. Each cluster organizer then compute the total energy in its cluster and exchange this information with other cluster organizers. After receiving this information from all the other clusters, a cluster organizer can compute the average energy E_{avg} for one cluster in the network.

Balancing Energy on Initial Clusters. Because the cluster organizer election is based on a random self-election mechanism, there is no guarantee on the total energy in one cluster. A group of energy balanced clusters can be achieved by applying the steps shown in Algo. 2 with the initial clusters.

A cluster is said to be acceptable if its total energy falls in the range $[E_{avg} - \delta, E_{avg} + \delta]$. δ is the maximum allowed variance of the total energy in a cluster. The algorithm starts from the cluster with the minimum total energy. If this cluster is not acceptable, the cluster organizer broadcast recruiting message to neighboring clusters. Nodes hearing this message reply with node ID along with its energy information if they belong to clusters that haven't had the chance to balancing their total energy yet. The cluster organizer then select the combination of nodes that makes the total energy fall in the acceptable range (or as close

---

Algorithm 2. Balance_Clusters: Balancing energy among clusters

---

Define:
- S: A set of clusters obtained from algorithm 1.
- c_i: a cluster, $c_i \in S$.
- E_{avg}: average total energy per cluster.
- δ: Maximum allowed variance of total energy for one cluster.

Balance_Clusters :executed at the cluster organizer for cluster c_i

if c_i has least total energy in S **then**
    recruit nodes from neighboring clusters in S s.t. total energy falls in $[E_{avg} - \delta, E_{avg} + \delta]$
    ask all the other cluster organizers to remove c_i from S
end if
if c_i has members deprived by other clusters **then**
    recompute the total energy
    update with all the other cluster organizers
end if

---

as possible) and broadcast the ID list of the selected nodes. The selected nodes then leave the current cluster and join the new cluster. To avoid oscillation, this cluster should not be touched again. Those clusters that have nodes deprived recompute the total energy and send it to the rest of the cluster organizers. Another cluster with least energy starts a new round until all the clusters have either had a chance to adapt its membership or they are acceptable in the first place. During this process, nodes drift from energy-rich clusters to energy-poor clusters.

After the cluster membership is stable, each node announces its cluster membership along with its initial energy level and records this information from the nodes within the same cluster. Nodes in the same cluster then organize themselves into a logic ring in the increasing order of their IDs.

3.2 Cluster-Head Management

The operation of a RECA system within one cluster is divided into rounds, each round lasts for a duration of R. R is a predefined system parameter. The value of R should be long enough to hide the system control overhead and short enough to avoid a node dying in its active duty cycle.

All nodes within a cluster take turns to be the cluster-head for a fair-share duration of a duty cycle in one round. Nodes that are not cluster-head use minimum transmission power to send data to the cluster-head. A node u computes its fair-share cluster-head duty cycle d_u as the following:

$$d_u = \frac{E_u}{\sum\limits_{\text{all w}} E_w} \times R.$$

where E_u are the energy level of node u. Node w is in the same cluster with u.

In addition to computing the duty cycle, each node computes and maintains two timers, t_0 and t_1. Timer t_0 is used to mark the beginning of the duty cycle as the cluster-head of a node. The value of t_0 for node u is:

$$t_0 = \sum_{\text{v's ID<u's ID}} d_v = \frac{\sum\limits_{\text{v's ID<u's ID}} E_v}{\sum\limits_{\text{all w}} E_w} \times R.$$

Timer t_1 is used to mark the end of one round, therefore its value is simply set to be R.

$$t_1 = R.$$

In each round, the node with the lowest ID starts to act as the cluster-head and announce the beginning of the new round. On hearing this message, each node starts the timers t_0 and t_1. When timer t_0 expires at node u, u signals the current cluster-head and takes over the duty as a new cluster-head for approximately its duty cycle d_u.

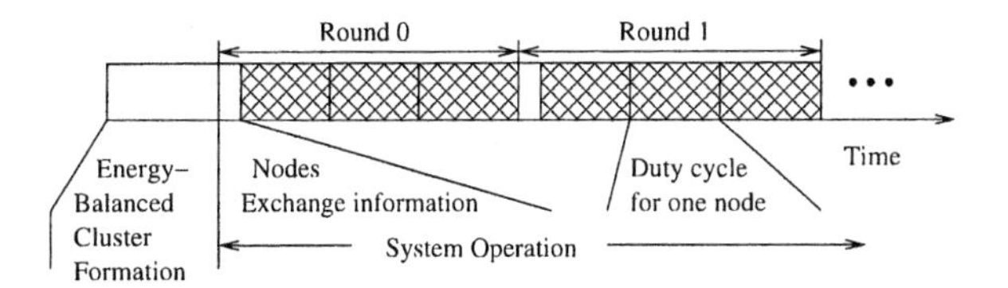

Fig. 1. RECA operation

When the timers t_1 expire at all nodes, each node reports its current energy level to and records this information from the rest of the nodes in the same cluster. This can be done in TDMA fashion to avoid collision. The node with the lowest ID in the cluster then announces and begins a new round.

The operation of RECA is summarized in Fig. 1.

Note that in RECA, each cluster acts independently of other clusters. Also a node will release its cluster-head duty only when signaled by the next cluster-head, this will prevent small clock skew within a period of R. Therefore, global tight synchronization in this phase is not necessary.

4 Performance Evaluation

Consider the case of a N-nodes cluster with initial energy value of e_i respectively.

Let $\Gamma = \sum_{i=1}^{N} e_i$. The variance of the energy level can be computed as:

$$\sigma^2 = \frac{\sum_{i=1}^{N}(e_i - \frac{\Gamma}{N})^2}{N}. \tag{1}$$

Assume same energy dissipation rate among the nodes in one round and the total energy consumed is E. The residual energy in node i's battery after this round is $e_i - \frac{e_i E}{\Gamma}$, therefore, the variance of energy level after this round is:

$$\sigma'^2 = \frac{\sum_{i=1}^{N}(e_i - \frac{e_i E}{\Gamma} - \frac{\Gamma - E}{N})^2}{N} = \frac{\sum_{i=1}^{N}((e_i - \frac{\Gamma}{N})(1 - \frac{E}{\Gamma}))^2}{N} = (1 - \frac{E}{\Gamma})^2 \sigma^2. \tag{2}$$

Compare (1) and (2), one can find that the variance of energy level is a strictly decreasing function. This proof shows that RECA can theoretically reduce the energy difference among nodes as the network runs.

4.1 Simulation and Results

Radio Model and Simulation Environments. To faithfully reproduce the results of LEACH and compare the performance, we adopt the same radio model and simulation environments used in [2]. Parameters used in this model are listed in table 1.

Table 1. Radio Model Parameters

Radio Parameter	Value
Transmitter / Receiver Electronics, E_{elec}	$50nJ/bit$
Transmitter Amplifier, ϵ_{amp}	$100pJ/bit/m^2$

Assuming path loss factor of 2, transmitting a k-bit message over a distance of d using this radio model causes a sensor to expend:

$$E_{Tx}(k,d) = E_{elec} \times k + \epsilon_{amp} \times k \times d^2.$$

To receive this message, a sensor expends:

$$E_{Rx}(k) = E_{elec} \times k.$$

In all the experiments, 100 sensor nodes are uniformly deployed on a [0,50]×[-25,25] field, the BS is at (150,0). The minimum radio range is set to be 15.

Simulation Results. First the balanced cluster formation algorithms are examined. Each node is given an initial energy uniformly distributed in [0.5J, 1J]. 11 clusters were obtained from Algo. 1. The distribution before and after energy balancing is shown in Fig. 2(a). The energy for each cluster before and after the balancing is shown in Fig. 2(b). As these figures illustrated, Algo. 2 can effectively balance the total energy per-cluster.

To compare the performance with LEACH, a similar scenario as in [2] is simulated. Initially each node is given 1J of energy in this experiment. The number of alive nodes in each round is shown in Fig. 3(a), the corresponding variance of residual energy level is shown in Fig. 3(b). These figures showed that RECA maintains smaller residual energy variance than that of LEACH all the

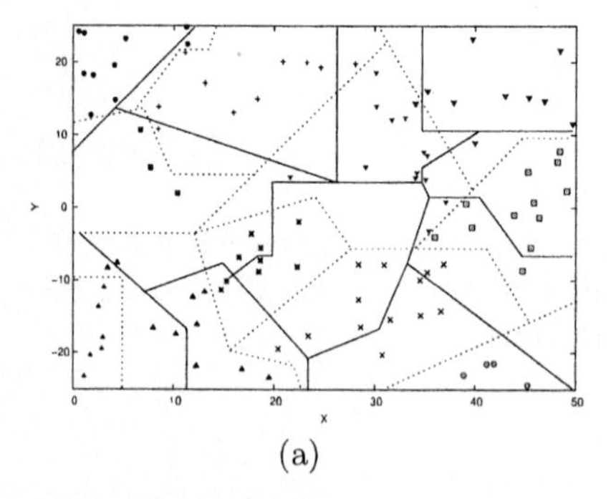

(a)

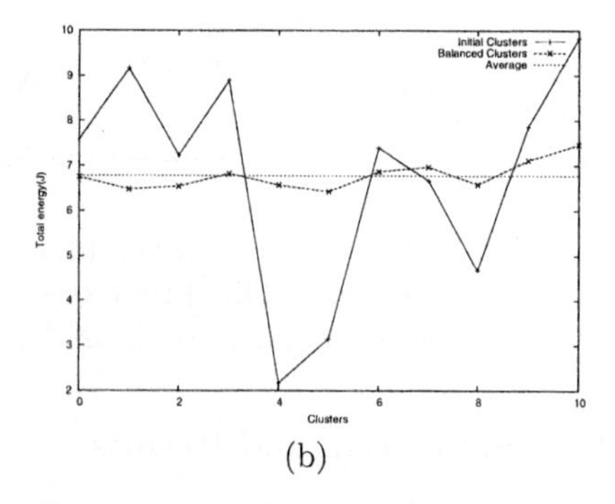

(b)

Fig. 2. (a) initial clusters obtained from Algo. 1 are separated by dotted lines clusters after energy balancing are grouped by solid lines, (b) Total energy for each cluster before and after energy balancing.

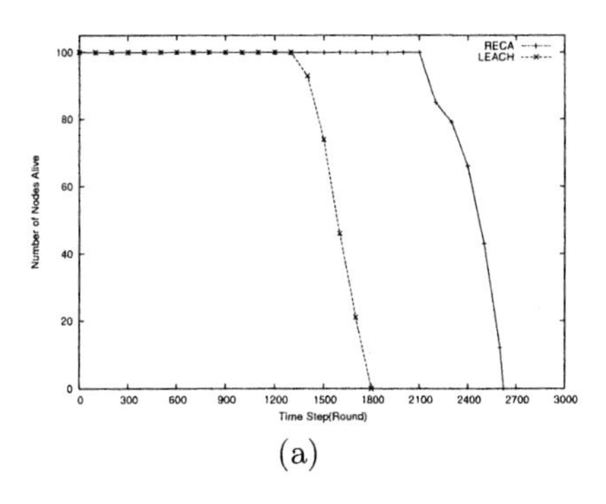

(a)

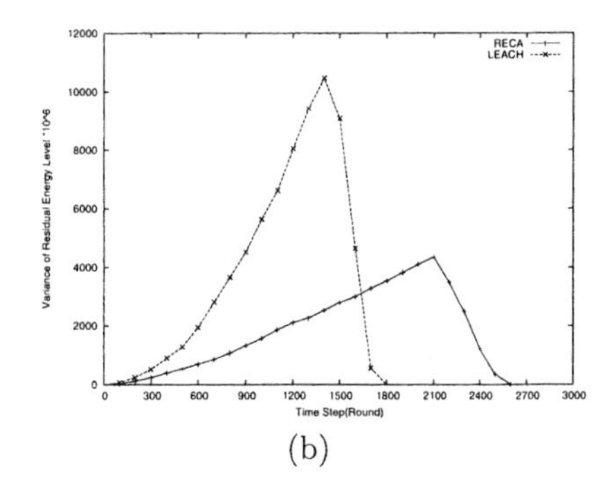

(b)

Fig. 3. (a) Number of nodes alive, (b) Variance of residual energy level with uniform initial energy level

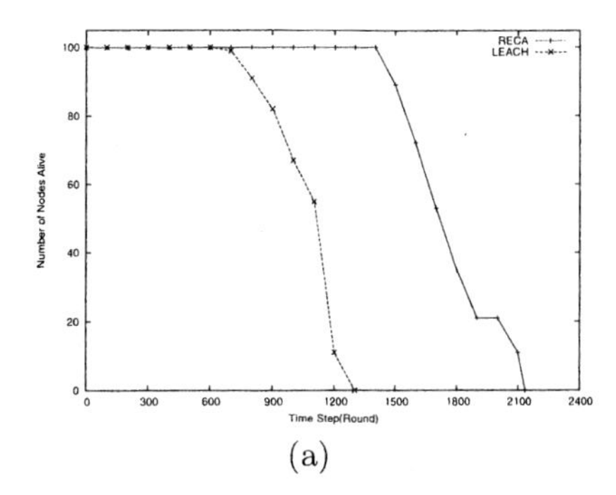

(a)

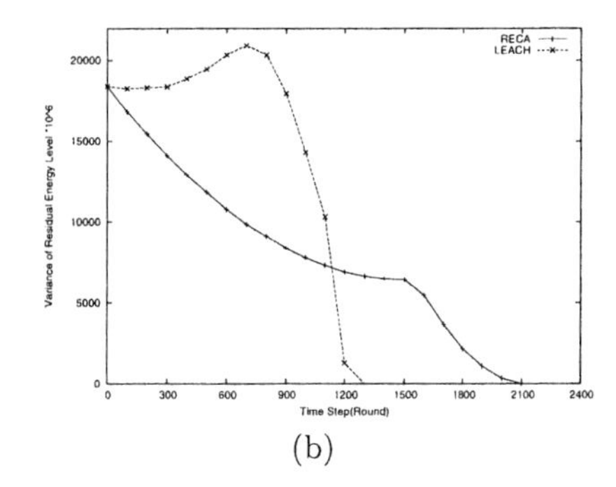

(b)

Fig. 4. (a)Number of nodes alive (b) Variance of residual energy level with variant initial energy level

time, consequently, RECA prolongs the network lifetime about 30% (time when first node dies).

Another experiment was set up to examine the impact of the initial energy level on both schemes. In this simulation, each node is given an initial energy uniformly distributed in [0.5J, 1J]. The results are shown in Fig. 4. RECA quickly reduced the difference of initial energy among all the nodes, thus a 150% longer network lifetime is achieved.

5 Conclusion

In this paper, a novel Ring-structured Energy-efficient Clustering Architecture for wireless sensor networks, referred to as RECA, is presented. RECA deterministically rotate cluster-head to ensure fairness among cluster nodes. Furthermore, nodes in RECA assume cluster-head functionalities for a period of time proportional to their residual energy with respect to other nodes in the same cluster. Consequently, energy dissipation in RECA is evenly distributed among all the nodes of a cluster. This effectively prolongs lifetime of the entire network. Finally, the requirement for tight synchronization is relaxed considerably since cluster formation happens only once at the initialization phase of the network and nodes in one cluster use hand-shake mechanism to circumvent the effect of small clock skew.

References

1. Min, R., Bhardwaj, M., Cho, S.H., Ickes, N., Shih, E., Sinha, A., Wang, A., Chandrakasan, A.: Energy-centric enabling technologies for wireless sensor networks. IEEE Wireless Communications Magazine (2002)
2. Heinzelman, W., Chandraksan, A., Balakrishnan, H.: Energy-efficient communication protocol for wireless microsensorr networks. Hawaiian International Conference on Systems Science (2000)
3. Vajanapoom, K.: A study of cluster formation in LEACH. Technical Report (2005)
4. Handy, M.J., Haase, M., Timmermann, D.: Low energy adaptive clustering hierarchy with deterministic cluster-head selection. IEEE International Conference on Mobile and Wireless Communications Networks, Stockholm. (2002)
5. Lindsey, S., Raghavendra, C.: Pegasis: Power-efficient gathering in sensor information systems. International Conference on Communications (2001)
6. Raicu, I.: Efficient even distribution of power consumption in wireless sensor networks. ISCA 18th International Conference on Computers and Their Applications, Honolulu, Hawaii, USA. (2003)
7. Xu, Y., Heidemann, J., Estrin, D.: Geography-informed energy conservation for ad hoc routing. Proceedings of 2001 ACM Mobicom, Rome, Italy (2001)
8. Cerpa, A., Estrin, D.: Ascent: Adaptive self-configuring sensor netwroks topologies. Proceedings of INFOCOM (2002)
9. Ye, F., Zhong, G., Lu, S., Zhang, L.: Peas: A robust energy conserving protocol for long-lived sensor network. Proceedings of the 10th IEEE International Conference on Network Protocols(ICNP'02) (2002)

A Distributed Efficient Clustering Approach for Ad Hoc and Sensor Networks

Jason H. Li[1], Miao Yu[2], and Renato Levy[1]

[1] Intelligent Automation, Inc., Rockville, MD 20855, USA*
{jli, rlevy}@i-a-i.com
[2] Department of Mechanical Engineering, University of Maryland, College Park, MD 20742, USA
mmyu@glue.umd.edu

Abstract. This paper presents a Distributed, Efficient Clustering Approach (DECA) for ad hoc and sensor networks. DECA can provide robustness against moderate node mobility and at the same time render energy-efficiency. The identified clusterheads cover the whole network and each node in the network can determine its cluster and only one cluster. The algorithm terminates in deterministic time without iterations, and each node transmits only one message during the algorithm. We prove analytically the correctness and complexity of the algorithm, and simulation results demonstrate that DECA is energy-efficient and resilient against node mobility.

1 Introduction

Communication between arbitrary endpoints in an *ad hoc network* typically requires routing over multiple-hop wireless paths due to the limited wireless transmission range. Without a fixed infrastructure, these paths consist of wireless links whose endpoints are likely to be moving independently of one another. Consequently, mobile end systems in an ad hoc network are expected to act cooperatively to route traffic and adapt the network to the dynamic state of its links and its mobility patterns. Unlike fixed infrastructure networks where link failures are comparatively rare events, the rate of link failure due to node mobility is the primary obstacle to routing in ad hoc networks [9].

A closely related area of ad hoc networks is *wireless sensor networks (WSNs)* [1], which comprise of a higher number of nodes (in the thousands and more) scattered over some region. Sensor nodes are typically less mobile, and more densely deployed than mobile ad hoc networks (MANETs). The sensor nodes gather data from the environment and can perform various kinds of activities—such as collaborative processing of the sensor data, and performing some synchronized actions based on the gathered sensor data. Sensor nodes are usually heavily resource-constrained (especially on power), irreplaceable, and become unusable

* This work was supported by the Air Force Research Laboratory, grant FA8750-05-C-0161.

X. Jia, J. Wu, and Y. He (Eds.): MSN 2005, LNCS 3794, pp. 937–949, 2005.

after failure or energy depletion. It is thus crucial to devise novel energy-efficient solutions for topology organization and routing that are scalable, efficient and energy conserving in order to increase the overall network longevity.

Given the potentially large number of mobile devices, scalability becomes a critical issue. To scale down networks with a large number of nodes, clustering protocols have been investigated for ad hoc and sensor networks in the literature [7][8][11]. While these strategies differ in the criteria used to organize the clusters, clustering decisions in each of these schemes are based on static views of the network topology; none of the proposed schemes, even equipped with some local maintenance schemes, is satisfactorily resistant to node mobility beyond rare and trivial node movement. One of the purposes of this work is to propose a clustering protocol that is resilient against mild to moderate mobility where each node can potentially move.

In the hybrid energy-efficient distributed clustering approach (HEED) proposed for ad hoc sensor networks [12], clusterhead selection is primarily based on the residual energy of each node. The clustering process entails a number of rounds of iterations; each iteration exploiting some probabilistic methods for nodes to elect to become a clusterhead. HEED is a fully distributed protocol and it ensures that each node can either elect to become a clusterhead or it joins a cluster within its range. While HEED is one of the most recognized energy-efficient clustering protocols, we argue that its clustering performance can be further enhanced. In this work, we will present a distributed, energy-efficient clustering approach (DECA) that outperforms HEED in terms of energy-efficiency and possesses the advantages of better clustering efficiency and resilience against node mobility.

Our contributions are as follows. DECA aims to prolong network lifetime by efficiently organizing nodes into clusters, with the clusters resistant to node mobility. The protocol terminates without rounds of iterations as required by HEED, which makes DECA a less complex and more efficient algorithm. Further, DECA's efforts of minimizing control overhead render even smaller overhead than HEED, which implies better energy-efficiency in sensor networks.

The remainder of this paper is organized as follows. Section 2 describes the network model and the clustering problem that we address in this work. Section 3 presents the DECA protocol with correctness and complexity analysis. Performance evaluation is presented in Section 4, followed by the descriptions on relevant work in Section 5. We conclude the paper in Section 6.

2 Problem Statement

An ad hoc wireless network is modeled as a set V of nodes that are interconnected by a set E of full-duplex directed communication links. Each node has a unique identifier and has at least one transmitter and one receiver. Two nodes are neighbors and have a link between them if they are in the transmission range of each other [5]. Neighboring nodes share the same wireless media, and each message is transmitted by a local broadcast. Nodes within the ad hoc network may move at any time without notice, but we assume that the node speed is moderate with respect to the packet transmission latency and wireless transmission range of the network hardware in use. It is our goal that the clustering protocol can still generate decent clusters under such mobility.

Let the clustering duration T_C be the time interval taken by the clustering protocol to cluster the network. Let the network operation interval T_O be the time needed to execute the intended tasks. In many applications, $T_O >> T_C$. In general, nodes that travel rapidly in the network may degrade the cluster quality because they alter the node distribution in their clusters and make the clusters unstable, possibly long before the end of T_O. However, research efforts on clustering should not be restricted only within the arena of static or quasi-stationary networks where node movements are rare and slow. Rather, for those applications where T_O is not much longer than T_C, we propose in this work an efficient protocol that generates clusters in *ad hoc networks* with mild to moderate node mobility. One such example is related to fast and efficient command and control in military applications, where nodes can frequently move.

In our model for *sensor networks*, though, the sensor nodes are assumed to be quasi-stationary and all nodes have similar capabilities. Nodes are location unaware and will be left unattended after deployment. Recharging is assumed not possible and therefore, energy-efficient sensor network protocols are required for energy conservation and prolonging network lifetime. For clustering, in particular, every node can act as both a source and a server (clusterhead). A node may fail if its energy resource is depleted, which motivates the need for rotating the clusterhead role in some fair manner among all neighboring nodes for load balancing and overall network longevity.

The problem of clustering is then defined as follows. For an ad hoc or sensor network with nodes set V, the goal is to identify a set of clusterheads that cover the whole network. Each and every node v in set V must be mapped into exactly one cluster, and each ordinary node in the cluster must be able to directly communicate to its clusterhead. The clustering protocol must be completely distributed meaning that each node independently makes its decisions based only on local information. Further, the clustering must terminate fast and execute efficiently in terms of processing complexity and message exchange. Finally, the clustering algorithm must be resistant to moderate mobility (in ad hoc networks) and at the same time renders energy-efficiency, especially for sensor networks.

3 DECA Clustering Algorithm

The DECA algorithm structure is somewhat similar to that presented by Lin and Gerla [8] in that each node broadcasts its decision as the clusterhead in the neighborhood based on some local information and score function. In [8] the score is computed based on node identifiers, and each node holds its message transmission until all its neighbors with better scores (lower ID) have done so. Each node stops its execution of the protocol if it knows that every node in its closed neighborhood (including itself) has transmitted. HEED [12] utilizes node residual energy as the first criterion and takes a cost function as the secondary criterion to compute the score, and each node probabilistically propagates tentative or final clusterhead announcement depending on its probability and connectivity. The execution of the protocol at each node will terminate when the probability of self-election, which gets doubled in every iteration, reaches 1.

It is assumed in [8] that the network topology does not change during the algorithm execution, and therefore it is valid for each node to wait until it overhears every higher-score neighbor transmitting. With some node mobility, however, this algorithm can halt since it is quite possible that an initial neighboring node leaves the transmission range for a node, say v, so that v cannot overhear its transmission. v then has to wait endlessly according to the stopping rule.

Similar assumption exists in HEED. Under node mobility, HEED will not halt though, since each node will terminate according to its probability-doubling procedure. However, we observe that the rounds of iterations are not necessary and can potentially harm the clustering performance due to the possibly excessive number of transmitted announcements.

We emphasize the important insights on distributed clustering: *those nodes with better scores should announce themselves earlier than those with worse scores.* In this work, we utilize a score function that captures node residual energy, connectivity and identifier. Each node does not need to hold its announcement until its better-scored neighbors have done so; each node does not need to overhear every neighbor in order to stop; and, each node only transmits one message, rather than going through rounds of iterations of probabilistic message announcement. Given the fact that it is communication that consumes far more energy in sensor nodes compared with sensing and computation, such saves on message transmissions lead to better energy efficiency.

3.1 DECA Operation

Each node periodically transmits a Hello message to identify itself, and based on such Hello messages, each node maintains a neighbor list. Define the score function at each node as score $= w_1 E + w_2 C + w_3 I$, where E stands for node residual energy, C stands for node connectivity, I stands for node identifier, and weights follow $\sum_{i=1}^{3} w_i = 1$. We put higher weight on node residual energy in our simulations. The computed score is then used to compute the delay for this node to announce itself as the clusterhead. The higher the score, the sooner the node will transmit. The computed delay is normalized between 0 and a certain upper bound D_{max}, which is a key parameter that needs to be carefully selected in practice, like the DIFS parameter in IEEE 802.11. In our simulation, we choose $D_{max} = 10$ms and the protocol works well. After the clustering starts, the procedure will terminate after time T_{stop}, which is another key parameter whose selection needs to take node computation capability and mobility into consideration. In the simulation, we choose $T_{stop} = 1$s.

The distributed clustering algorithm at each node is illustrated in the pseudo code fragments. Essentially, clustering is done periodically and at each clustering epoch, each node either immediately announces itself as a potential clusterhead or it holds for some delay time.

```
I. START-CLUSTERING-ALGORITHM()
1  myScore = w1E + w2C + w3I;
2  delay = (1000 – myScore)/100;
3  if (delay < 0)
4      then broadcastCluster (myId, myCid, myScore);
5      else
6            delayAnnouncement ();
7  Schedule clustering termination.
```

```
II. RECEIVING-CLUSTERING-MESSAGE(id, cid, score)
 1  if (id == cid)
 2      then if (myCid == UNKNOWN)
 3              then if (score > myScore)
 4                      then myCid = cid;
 5                           cancelDelayAnnouncement ();
 6                           broadcastCluster (myId, myCid,score);
 7           elseif (score > myScore)
 8              then if (myId == myCid)
 9                      then needConversion = true;
10                      else
11                           convertToNewCluster ();
```

```
III. FINALIZE-CLUSTERING-ALGORITHM()
1  if (needConversion)
2      then if (!amIHeadforAnyOtherNode ())
3              then convertToNewCluster ();
4  if (myCid == UNKNOWN)
5      then myCid = cid;
6            broadcastCluster (myId, myCid, score);
```

On receiving such clustering messages, a node needs to check whether the node ID and cluster ID embedded in the received message are the same; same node ID and cluster ID means that the message has been transmitted from a clusterhead. Further, if the receiving node does not belong to any cluster, and the received score is better than its own score, the node can simply join the advertised cluster and cancel its delayed announcement.

If the receiving node currently belongs to some other cluster, and the received score is better than its own score, two cases are considered. First, if the current node belongs to a cluster with itself as the head, receiving a better scored message means that this node may need to switch to the better cluster. However, cautions need to be taken here before switching since the current node, as a clusterhead, may already have other nodes affiliated with it. Therefore, inconsistencies can occur if it rushes to switch to another cluster. In our approach, we simply mark the necessity for switching (line 9 in Phase II) and defer it to finalizing phase, where it checks to make sure that no other nodes are affiliated with this node in the cluster as the head, before the switching can occur. But if the current node receiving a better-scored message is not itself a clusterhead, as an ordinary node, it can immediately convert to the new cluster, and this is the second case (line 11 in Phase II). It is critical to note that the switch process mandates that a node needs to leave a cluster first before joining a new cluster. In the finalizing phase, where each node is forced to enter after T_{stop}, each node checks to see if it needs to convert. Further, each node checks if it already belongs to a cluster and will initiate a new cluster with itself as the head if not so.

3.2 Correctness and Complexity

The protocol described above is completely distributed, and to prove the correctness of the algorithm, we need to show that 1) the algorithm terminates; 2) every node eventually determines its cluster; and 3) in a cluster, any two nodes are at most two-hops away.

Theorem 1. *Eventually DECA terminates.*

Proof. After the clustering starts, the procedure will stop receiving messages after time T_{stop}, and enter the finalizing phase, after which the algorithm will terminate. □

Note that in order for DECA to outperform related protocols presented in [8] and [12] under node mobility, it is critical to design the key parameters $D_{\max}$ and T_{stop} appropriately taking node computation and mobility patterns into considerations. With carefully designed parameters, node needs not to wait (possibly in vain as in [8]) to transmit or terminate, nor need it to go through rounds of probabilistic announcement. In HEED, every iteration takes time t_c, which should be long enough to receive messages from any neighbor within the transmission range. We can choose $D_{\max}$ to be roughly comparable to (probably slightly larger than) t_c and DECA can generally terminate faster than HEED.

Theorem 2. *At the end of Phase III, every node can determine its cluster and only one cluster.*

Proof. Suppose a node does not determine its cluster when entering Phase III. Then condition at line 4 holds and the node will create a new cluster and claims itself as the clusterhead. So every node can determine its cluster. Now we show that every node selects only one cluster. A node determines its cluster by one of the following three methods. First, it claims itself as the clusterhead; second, it joins a cluster with a better score when its cluster is undecided; and third, it converts from a cluster to another one. The first two methods do not make a node join more than one clusters, and the switch procedure checks for consistency and mandates that a non-responsible node (a node not serving as head for a cluster) can only leave the previous cluster first before joining the new cluster. As a result, no node can appear in two clusters. □

One may argue that Theorem 2 does not suffice for clustering purposes. For example, one can easily invent an algorithm such that every node creates a new cluster and claims itself as the clusterhead; obviously Theorem 2 holds. However, our algorithm does much better than such trivial clustering. Most of the clusters are formed executing line 4 to line 6 in Phase II, which means joining clusters with better-scored heads. This is due to the fact that the initial order of clusterhead announcements is strictly determined using the score function.

Theorem 3. *When clustering finishes, any two nodes in a cluster are at most two-hops away.*

Proof. The proof is based on the mechanisms by which a node joins a cluster. A node, say v, joins a cluster with head w only if v can receive an announcement from w with a better score. In other words, all ordinary nodes are within one-hop from the clusterhead and the theorem follows. □

To show that the algorithm is energy-efficient, we prove that the communication and time complexity is low.

Theorem 4. *In DECA, each node transmits only one message during the operation.*

Proof. In broadcastCluster method, a Boolean variable iAlreadySent (not shown in Pseudo code) ensures that each node cannot send more than once. Now we show that each node will eventually transmit. In Phase I execution when nodes start the clustering, each node either transmits immediately or schedules a delayed transmission, which will either get executed or cancelled at line 5 in Phase II. Note that the cancellation is immediately followed by a transmission so each node will eventually transmit. □

Theorem 5. *The time complexity of the algorithm is $O(|V|)$.*

Proof. From Phase II operations, each received message is processed by a fixed number of computation steps without any loop. By Theorem 4, each node only sends one message and therefore there are only $|V|$ messages in the system. Thus the time complexity is $O(|V|)$. □

4 Performance Evaluation

We evaluate the DECA protocol using an in-house simulation tool called agent-based ad-hoc network simulator (NetSim). In our simulations, random graphs are generated so that nodes are randomly dispersed in a 1000m × 1000m region and each node's transmission range is bound to 250m. We investigate the clustering performance under different node mobility patterns, and the node speed ranges from 0 to 50m/s. For each speed, each node takes the same maximum speed and a large number of random graphs get generated. Simulations are run and results are averaged over these random graphs.

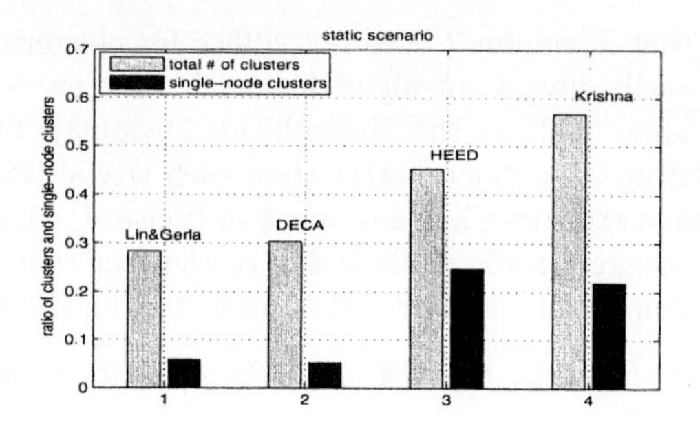

Fig. 1. Ratio of number of clusters. Static scenario.

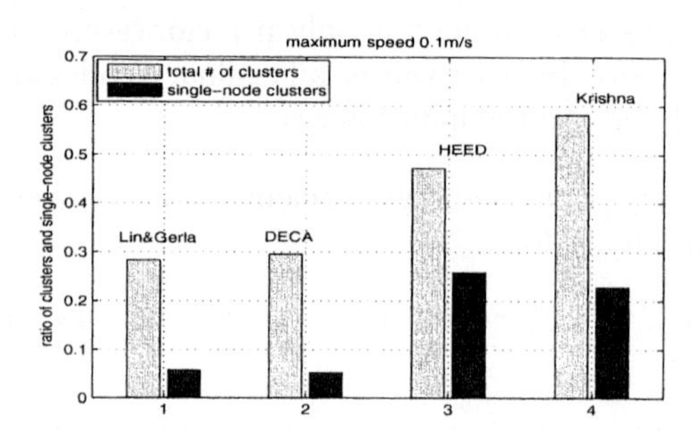

Fig. 2. Ratio of number of clusters. Maximum speed 0.1m/s.

In general, for any clustering protocol, it is undesirable to create single-node clusters. Single-node clusters arise when a node is forced to represent itself (because of not receiving any clusterhead messages). While many other protocols generate lots of single-node clusters as node mobility gets more aggressive, our algorithm shows much better resilience. We have considered the following metrics for performance comparisons: 1) the average overhead (in number of protocol messages); 2) the ratio of the number of clusters to the number of nodes in the network; 3) the ratio of the single-node clusters to the number of nodes in the network; and 4) the average residual energy of the selected clusterheads.

We first look at static scenarios where nodes do not move and the quasi-stationary scenarios where the maximum node speed is bounded at 0.1m/s. We choose [8] proposed by Lin & Gerla (LIN) as a representative for those general clustering protocols, and choose Krishna's algorithm (KRISHNA) [7] to represent dominating-set based clustering protocols. For energy-aware protocols, we choose HEED [12] to compare with DECA. From Fig. 1 (static scenario) and Fig. 2 (0.1m/s max. speed) it is easy to observe that KRISHNA has the worst clustering performance with the highest cluster-to-nodes ratio, while DECA and LIN possess the best performance. HEED performs in between.

Fig. 3, which combines Fig. 1 and Fig. 2, shows that all four protocols perform consistently under (very) mild node mobility. In fact, with maximum node speed set as 0.1m/s, both LIN and DECA perform exactly the same as their static scenarios, while HEED and KRISHNA degrade only to a noticeable extent.

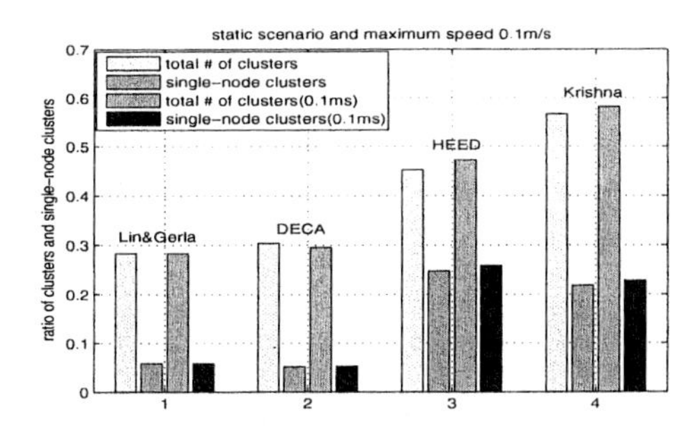

Fig. 3. Ratio of number of clusters. Put together.

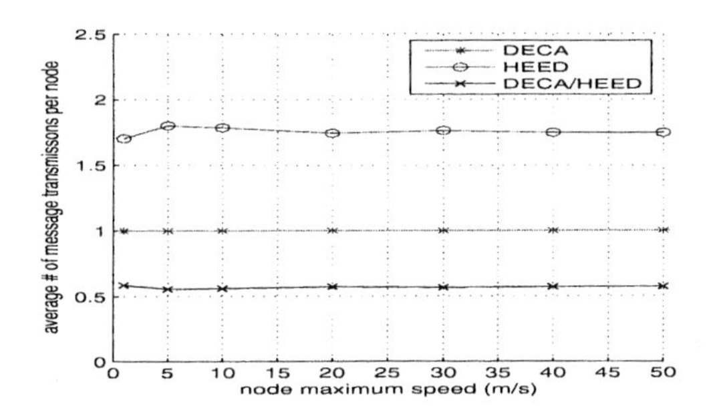

Fig. 4. Average number of transmissions per node and DECA/HEED ratio

During our simulations, as we increase the maximum node speed, both LIN and KRISHNA fail to generate clusters. This is expected. In LIN, a node will not transmit its message until all its better-scored neighbors have done so; the algorithm will not terminate if a node do not receive a message from each of its neighbors. Node mobility can make the holding node wait for ever. In KRISHNA, in order to compute clusters, each node needs accurate information of the entire network topology, facilitated by network-wide link state update which by itself is extremely vulnerable to node mobility. In contrast, we found that both HEED and DECA are quite resilient to node mobility in that they can generate decent clusters even when each node can potentially move independently of others. The following figures compare the performance of DECA and HEED under different node mobility.

Fig. 4 shows that for DECA, the number of protocol messages for clustering remains one per node, regardless of node speed, as proven in Theorem 4. For HEED, the number of protocol messages is roughly 1.8 for every node speed, and a node running DECA transmits about 56% number of messages as that in HEED (shown as DECA/HEED in Fig. 4). The fact that HEED incurs more message transmissions is due to the possibly many rounds of iterations (especially when node power is getting reduced), where each node in every iteration can potentially send a message to claim itself as the candidate clusterhead [12]. Reducing the number of transmissions is of great importance, especially in sensor networks, since it would render better energy efficiency and fewer packet colli-

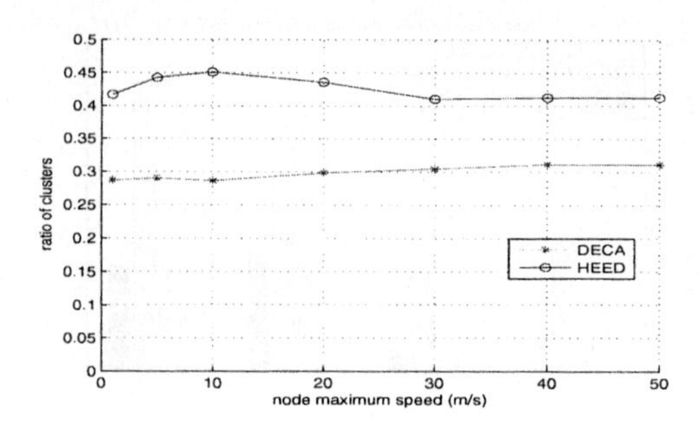

Fig. 5. Ratio of clusters to total number of nodes in network

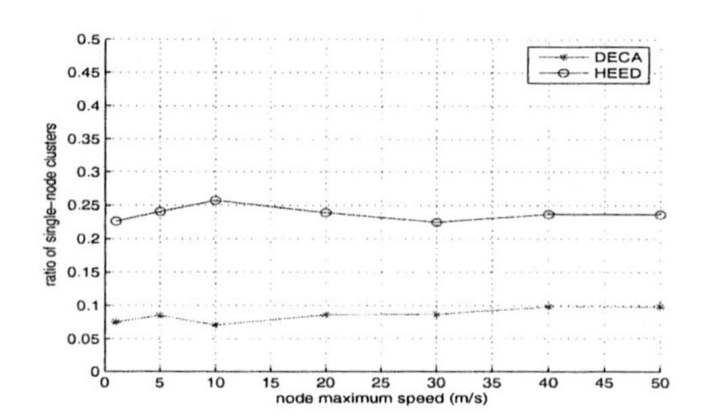

Fig. 6. Ratio of single-node clusters to total number of nodes in network

sions (e.g. CSMA/CA type MAC in IEEE 802.11). Fig. 5 and Fig. 6 illustrate the ratio of number of clusters and single node clusters to the total number of nodes in network. In both cases, DECA outperforms HEED.

Note that both DECA and HEED perform quite *consistently* under different maximum node speed and this is not coincident: a node in both DECA and HEED will stop trying to claiming itself as the potential clusterhead after some initial period (delayed announcement in DECA and rounds of iterations in HEED) and enters the finalizing phase. As a result, the local information gathered, which serves as the base for clustering, is essentially what can be gathered

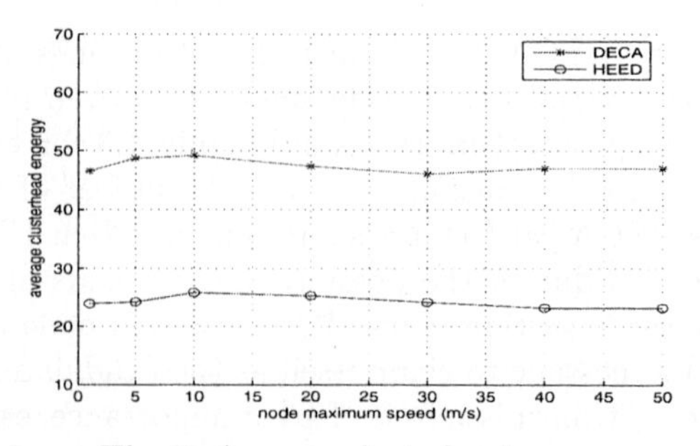

Fig. 7. Average clusterhead energy

within the somewhat invariant initial period which leads to consistent behaviors under different node mobility.

Further, we compare DECA and HEED with respect to the (normalized) average clusterhead energy in Fig. 7. Again both DECA and HEED perform quite consistently and DECA outperforms HEED with about **twice** the average clusterhead residual energy. This is in accordance with Fig. 5 where DECA consistently incurs fewer message transmissions than HEED. In sensor networks, sending fewer messages by each node in DECA while achieving the intended goal usually means energy-efficiency and longer node lifetime.

In addition, HEED may possess another undesirable feature in its protocol operation. Over time, each node's energy fades leading to a smaller probability of transmission in HEED for each node, which implies more rounds of iterations. As a result, more announcements could be sent and more energy could be consumed, which could lead to more messages sent and more energy consumed in the next round of clustering! In future work we will analyze HEED and execute more extensive simulations to see if such amplifying-effects really exist. DECA, on the contrary, does not posses this potential drawback even with energy fading, since each node only sends one message during the operation.

5 Related Work

Das and Sivakumar et al. [10] identified a subnetwork that forms a minimum connected dominating set (MCDS). Each node in the subnetwork is called a spine node and keeps a routing table that captures the topological structure of the whole network. The main drawback of this algorithm is that it still needs a non-constant number of rounds to determine a connected dominating set [11].

In [11] the authors proposed an efficient localized algorithms that can quickly build a backbone directly in ad hoc networks. This approach uses a localized algorithm called the *marking process* where hosts interact with others in restricted vicinity. This algorithm is simple, which greatly eases its implementation, with low communication and computation cost; but it tends to create small clusters.

Similar to [8], Basagni [3] proposed to use nodes' weights instead of lowest ID or node degrees in clusterhead decisions. Weight is defined by mobility related parameters, such as speed. Basagni [4] further generalized the scheme by allowing each clusterhead to have at most k neighboring clusterhead and described an algorithm for finding a maximal weighted independent set in wireless networks.

One of the first protocols that use clustering for network longevity is the Low-Energy Adaptive Clustering Hierarchy (LEACH) protocol [6]. In LEACH, a node elects to become a clusterhead randomly according to a target number of clusterheads in the network and its own residual energy, and energy load gets evenly distributed among the sensors in the network. In addition, when possible, data are compressed at the clusterhead to reduce the number of transmissions. A limitation of this scheme is that it requires all current clusterheads to be able to transmit directly to the sink.

6 Conclusion and Future Work

In this paper we present a distributed, efficient clustering algorithm that works with resilience to node mobility and at the same time renders energy efficiency. The algorithm terminates fast, has low time complexity and generates non-overlapping clusters with good clustering performance. Our approach is applicable to both mobile ad hoc networks and energy-constrained sensor networks. The clustering scheme provides a useful service that can be leveraged by different applications to achieve scalability.

It can be observed that in DECA the dispersed delay timers for clusterhead announcement assume the existence of a global synchronization system. While this might not be a problem for many (military) ad hoc network applications, for sensor networks synchronization can become trickier. It could be an interesting research to study time synchronization protocols combined with clustering protocol in sensor networks, with an effort to provide the maximum degree of functionality and flexibility with minimum energy consumption. Further, it could be interesting to observe how much improvement DECA can still maintain over HEED as transmission range varies.

References

1. I. F. Akyildiz, W. Su, Y. Sanakarasubramaniam, and E. Cayirci, "Wireless sensor networks: A survey," Computer Networks, vol. 38, no. 4, pp. 393-422, March 2002.
2. D. J. Baker, A. Ephremides, and J. A. Flynn, "The design and simulation of a mobile radio network with distributed control," IEEE Journal on Selected Areas in Communications, vol. SAC-2, no. 1, pp. 226-237, January 1984.
3. S. Basagni, "Distributed clustering for ad hoc networks," in Proceedings of the 1999 International Symposium on Parallel Architectures, Algorithms, and Networks (I-SPAN'99)
4. S. Basagni, D. Turgut, and S. K. Das, "Mobility-adaptive protocols for managing large ad hoc networks," in Proc of the IEEE International Conference on Communications, ICC 2001, June 11-14 2001, pp. 1539-1543.
5. B. N. Clark, C. J. Colburn, and D. S. Johnson, "Unit disk graphs," Discrete Mathematics, vol. 86, pp. 165-167, 1990.
6. W. R. Heinzelman, A. Chandrakasan, and H. Balakrishnan, "Energy efficient communication protocol for wireless microsensor networks," in Proceedings of the3rd Annual Hawaii International Conference on System Sciences, HICSS 2000,January 4-7 2000, pp. 3005-3014
7. P. Krishna, N.N. Vaidya, M. Chatterjee and D.K. Pradhan, A cluster-based approach for routing indynamic networks, ACM SIGCOMM Computer Communication Review 49 (1997) 49-64.
8. C. R. Lin and M. Gerla, "Adaptive clustering for mobile wireless networks," Journal on Selected Areas in Communications, vol. 15, no. 7, pp. 1265-1275, September 1997.
9. A. B. McDonald and T. Znati, "A mobility-based framework for adaptive clustering in wireless ad hoc networks," IEEE Journal on Selected Areas in Communications, vol. 17, no. 8, pp. 1466-1487, August 1999.

10. R. Sivakumar, B. Das, and B. V., "Spine-based routing in ad hoc networks," ACM/Baltzer Cluster Computing Journal, vol. 1, pp. 237-248, November 1998, special Issue on Mobile Computing.
11. J. Wu and H. Li, "On calculating connected dominating sets for efficient routing in ad hoc wireless networks," Telecommunication Systems, Special Issue on Mobile Computing and Wireless Networks, vol. 18,no. 1/3, pp. 13-36, September 2001.
12. O. Younis, S. Fahmy, "HEED: A Hybrid, Energy-Efficient,Distributed Clustering Approach for Ad Hoc Sensor Networks", IEEE TRANSACTIONS ON MOBILE COMPUTING, VOL. 3, NO. 4, OCTOBER-DECEMBER 2004

A Novel MAC Protocol for Improving Throughput and Fairness in WLANs

Xuejun Tian[1], Xiang Chen[2], and Yuguang Fang[2]

[1] Department of Information Systems,
Faculty of Information Science and Technology,
Aichi Prefectural University, Aichi, Japan
tan@ist.aichi-pu.ac.jp

[2] Department of Electrical and Computer Engineering,
University of Florida, Gainesville, FL, U.S.A
xchen@ecel.ufl.edu, fang@ece.ufl.edu

Abstract. Many schemes have been proposed to enhance throughput or fairness of the original IEEE 802.11 standard, however, they either fail to consider both throughput and fairness, or to do so with complicated algorithms. In this paper, we propose a new MAC scheme that dynamically optimizes each active node's backoff process. The key idea is to enable each node to adjust its Contention Window (CW) to approach the optimal one that will maximize the throughput. Meanwhile, when the network enters into steady state in saturated case, i.e., under heavy traffic load, all the nodes will maintain approximately identical CWs, which guarantees fair share of the channel among all nodes. Through simulation comparison with previous schemes, we show that our scheme can greatly improve the throughput no matter the network is in saturated or non-saturated case, while maintaining good fairness.

1 Introduction

Wireless local area networks (WLANs) have been increasingly popular and widely deployed in recent years. Currently, the IEEE 802.11 MAC standard includes two channel access methods: a mandatory contention based one called Distributed Coordination Function (DCF) and an optional centralized one called Point Coordination Function (PCF). Due to its inherent simplicity and flexibility, the DCF mode is preferred and has attracted most research attention. Meanwhile, PCF is not supported in most current wireless cards and may result in poor performance when working alone or together with DCF, as shown in [1][2]. In this paper, we focus on DCF.

Since all the nodes share a common wireless channel with limited bandwidth in the WLAN, it is highly desirable that an efficient and fair medium access control (MAC) scheme is employed. However, for the 802.11 DCF, there is room for improvement in terms of both efficiency([3][4][5]) and fairness. Cali et al. pointed out in [6] that depending on the network configuration, DCF may deliver a much lower throughput compared to the theoretical throughput limit. Meanwhile, as

X. Jia, J. Wu, and Y. He (Eds.): MSN 2005, LNCS 3794, pp. 950–965, 2005.

demonstrated in [7], the fairness as well as throughput of the IEEE 802.11 DCF could significantly deteriorate when the number of nodes increases.

Although extensive research has been conducted to improve throughput ([8] [9][10][6][11][12][13][14]) or fairness ([6] [15]), except in [11], these two performance indexes are rarely considered together. In this paper, we aim to enhance both throughput and fairness for DCF at the same time by proposing a novel MAC scheme called DOB. Compared to the original 802.11 DCF and previous enhancement approaches, this scheme has the following distinguishing features:

- Unlike [6] that relies on the accurate on-line estimation of the number of active nodes, we use a simply and accurate measure called average idle interval, which is easily obtained and reflects network traffic load, to Dynamically Optimizing the Backoff algorithm.
- It is known that in the 802.11 DCF, each node exponentially increases it contention window (CW) in case of collision and reset it after successful transmission. Although this is designed to avoid collisions, the fact that the CWs change drastically lead to neither fast collision resolution nor high throughput [11]. In contrast, DOB enables each node to keep a quasi-stable CW that oscillates around an *optimal* value that leads to a throughput close to maximum. More specifically, the current CW will be decreased if it is greater than the optimal CW and be increased otherwise.
- Since each node in the network maintains its CW around the optimal value, all nodes will have equal opportunities to seize the channel. As a result, the fairness is improved compared to the original DCF.

The remainder of this paper is organized as follows. In Section 2, we describe the IEEE 802.11 MAC protocol and then discuss the related work. We elaborate on our key idea and the theoretical analysis for improvement in Section 3. Then, we present in detail our proposed DOB scheme in Section 4. Section 5 gives performance evaluation and the discussions on the simulation results. Finally, concluding remarks are given in Section 6.

2 Preliminaries

In this section, we discuss the related work. Especially, we focus on Cali's work ([6]) and FCR (Fast Collision Resolution, [11]), as these two schemes resolve MAC collisions through dynamically adjusting the contention window.

2.1 Related Work

Considerable research efforts on IEEE 802.11 DCF have been expended on either theoretical analysis or throughput improvement ([7][9][10][6][11][16]). In [7], Bianchi used a Markov chain to model the binary exponential backoff procedure. By assuming the collision probability of each node's transmission is constant and independent of the number of retransmission, he derived the saturated throughput for the IEEE 802.11 DCF. In [8], Bharghvan analyzed and improved the performance of the IEEE 802.11 MAC. Although the contention information appended to the transmitted packets can help in collision resolution, its transmission increases the traffic load and the delay results in insensitivity to the traffic changes. Kim and Hou developed a model-based frame scheduling algorithm to improve the protocol capacity of the 802.11 [16]. In this scheme, each node sets its backoff timer in the same way as in the IEEE 802.11; however, when the backoff timer reaches zero, it waits for an additional amount of time before accessing the medium. Though this scheme improves the efficiency of medium access, the calculation of the additional time is complicated since the number of active nodes must be accurately estimated.

Cali et al. [6] studied the 802.11 protocol capacity by using a p-persistent backoff strategy to approximate the original backoff in the protocol. In addition, they showed that given a certain number of active nodes and average frame length, there exists an average contention window that maximizes throughput. Basing on this analysis, they proposed a dynamic backoff tuning algorithm to approach the maximum throughput. It is important to note that the performance of the tuning algorithm depends largely on the accurate estimation of the number of active nodes. However, in practice, there is no simple and effective run-time estimation algorithm due to the distributed nature of the IEEE 802.11 DCF. Meanwhile, a complicated algorithm ([6]) would impose a significant computation burden on each node and be insensitive to the changes in traffic load.

Fairness is another important issue in MAC protocol design for WLANs. In a shared channel wireless network, throughput and fairness essentially conflict with each other as shown in [17]. The analysis in [7] demonstrated that the fairness as well as the throughput of IEEE 802.11 DCF could significantly deteriorate when the number of nodes increases. Several research works addressed this issue[6] [15]. In [6], the number of active nodes needs to be estimated as mentioned and in [15], only initial contention window is adjusted and thus the contention window is not optimized.

As will be shown later, our proposed DOB preserves the advantages and overcomes the deficiencies of the work [6] and FCR. While relying on dynamic tuning of CW, it needs not estimate the number of active nodes, as is the case in [6]. Compared to the original IEEE 802.11 or FCR, since each node, without initializing its CW with the minimum value, keeps its CW close to the same optimal value, DOB can maintain fairness and keep the network operating with less fluctuation. Consequently, the network always works in a quasi-stable state. In other words, the nodes with a smaller CW than the optimal CW will increase CW and the nodes with a greater CW will decrease CW.

3 Design Motivation and Analysis

3.1 Motivation

In the IEEE 802.11 MAC, an appropriate CW is the key to providing throughput and fairness. A small CW results in high collision probability, whereas a large CW results in wasted idle time slots. In [6], Cali et al. showed that given the number of active nodes, there exists an optimal CW that leads to the theoretical throughput limit and when the number of active nodes changes, so does this optimal CW. Since in practice, the number of active nodes always changes, to let each node attain and keep using the corresponding optimal CW requires the estimation of the number of active node. However, this is not an easy task in the network environment where a contention-based MAC protocol is used. To get around this difficulty, we are thus motivated to find other effective measures that also lead us to the optimal CW and hence the maximal throughput. Therefore, we focuses on the average idle interval in the channel between two consecutive busy periods (due to transmissions or collisions) that each node locally observes. It has two merits. One is that without complex computation, each node can obtain the average idle interval online by observation, which is quite simple since the DCF is in fact built on the basis of physical and virtual carrier sensing mechanisms.

In the following, we derive the relationship between average idle interval and throughput through analysis. For the purpose of simplicity, we assume the frame length is constant. Later on, we will show that the performance is not sensitive to a variable frame length.

3.2 Analytical Study

In [6], IEEE 802.11 DCF is analyzed based on an assumption that, in each time slot, each node contends the medium with the same probability p subjected to $p = 1/(E[B]+1)$, where $E[B]$ is the average backoff timer and equals $(E[CW]-1)/2$ for DCF. In the strict sense, this assumption is not true because every node maybe use different CW. Since our DOB will enable each node to settle on a quasi-stable CW, we assume that all the nodes use the same and fixed CW for simplicity reason. Consequently, we have

$$p = 2/(CW+1) \tag{1}$$

as all the expectation signs E can be removed.

We assume every node is an active one, i.e., always having packets to transmit and for every packet transmission, the initial backoff timer is uniformly selected from $[0, CW-1]$. For each virtual backoff time slot, it may be idle, busy due to a successful transmission, busy due to collision. Accordingly, we denote by t_{slt}, T_s, and T_{col} the time durations of the three types of virtual slots, respectively, and denote by p_{idl}, p_s, and p_{col} the associated probabilities, respectively. Thus, we can express the above probabilities as follows.

$$\begin{aligned} p_{idl} &= (1-p)^n \\ p_s &= np(1-p)^{(n-1)} \\ p_{col} &= 1 - p_{idl} - p_s \end{aligned} \tag{2}$$

where n is the number of active nodes. Thus, the throughput is expressed as

$$\rho = \frac{Tp_s}{t_{slt}p_{idl} + T_{col}p_{cll} + T_s p_s}$$

$$= \frac{T}{t_{slt}p_{idl}/p_s + T_{col}p_{col}/p_s + T_s} \quad (3)$$

where T is the transmission time of one packet, which can be obtained by subtracting overhead from T_s. In the above formula, the term p_{idl}/p_s can be thought of as the average number of idle slots for every successful transmission and the term p_{col}/p_s the average number of collisions for every successful transmission. If we denote by L_{idl} the average idle interval, it can be expressed as

$$L_{idl} = p_{idl}/(1 - p_{idl}) \quad (4)$$

Considering Equation (1) and (2), this equation can be further written as

$$L_{idl} = \frac{1}{(1 + 2/(CW-1))^n - 1}$$
$$= \frac{1}{n\frac{2}{CW-1} + ... + \binom{n}{i}(\frac{2}{CW-1})^{(n-i)} + ... + (\frac{2}{CW-1})^n} \quad (5)$$

In Equation (5), we can see that when CW is large enough, $L_{idl} = (CW - 1)/(2n)$. As a matter of fact, this is the case when the network traffic load is heavy. In this case, to effectively avoid collisions, the optimal CW is large enough for the approximation $L_{idl} = (CW - 1)/(2n)$ in our DOB, which is also verified through simulations.

With Equation (2), (3), and (5), we can express the throughput as a function of L_{idl}, as shown in Fig. 1. Several important observations are made. First, we find that every curve follows the same pattern; namely, as the average idle

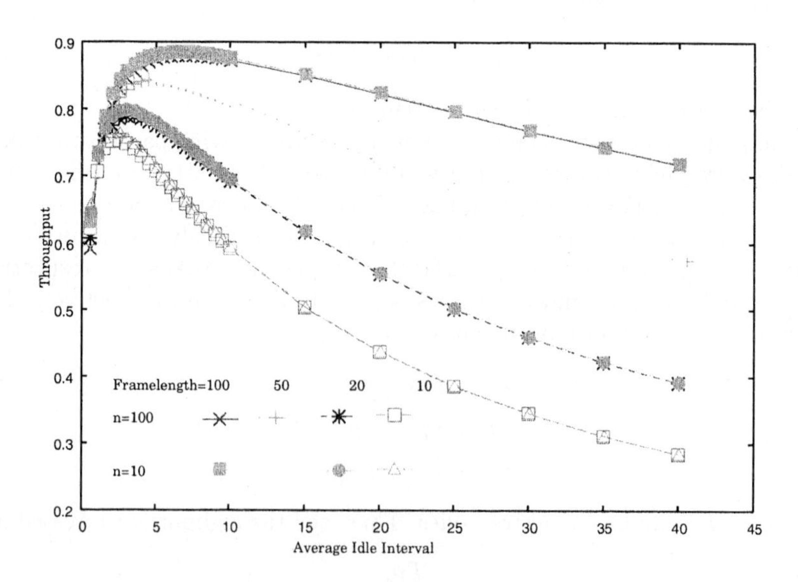

Fig. 1. Throughput vs. average idle interval

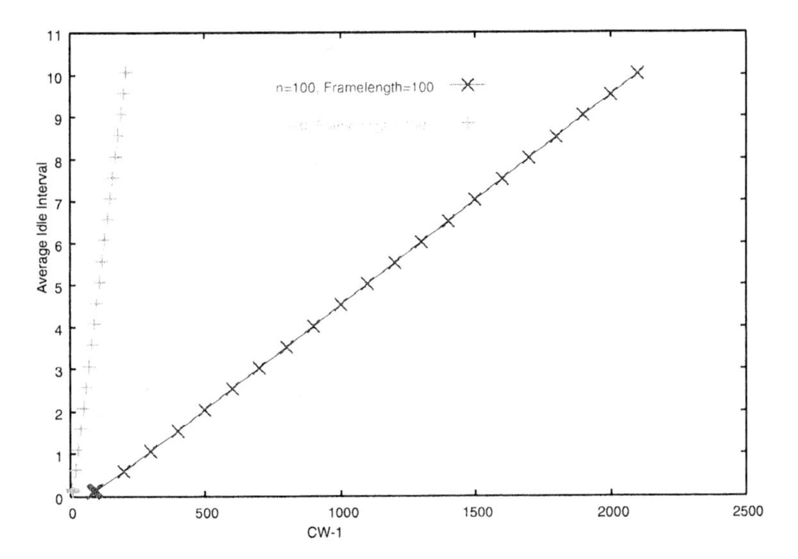

Fig. 2. Average idle interval vs. Contention Window

interval L_{idl} increases, the throughput first rises quickly, and then decreases relatively slowly after reaching its peak. Second, although the optimal value of L_{idl} that maximizes throughput is different in cases of different frame lengths, it varies in a very small range, which hereafter is called the optimal range of L_{idl}. Finally, this optimal value is almost independent of the number of active nodes. Therefore, L_{idl} is a suitable measure that indicates the network throughput.

$$L_{idl} = (CW - 1)/(2n) - \alpha \tag{6}$$

Fig. 2 shows the relationship between L_{idl} and the contention window CW, as revealed in Equation (5). It can be observed that L_{idl} is almost a linear function of CW when CW is larger than a certain value. Specifically, in the optimal range of L_{idl} for aggregate throughput, say L_{idl}=[3,8], we can estimate L_{idl} using an approximation linear formula:

Where, α is a constant. Since we are interested in tuning the network to work with maximal throughput, given the nice linear relationship, we can achieve this goal by adjusting the size of CW. In other words, each node can observe the average idle interval locally and adjust its backoff window accordingly such that the network throughput is maximized.

Clearly, the above results hold when all nodes have the same CW. In reality, different nodes may have different CWs that fluctuates around the optimal CW. Next, we give a theorem that given the average idle interval, i.e., given the idle probability p_{idl}, the achieved throughput will not be lower than the minimum throughput which is obtained under the condition that all nodes have the same CW.

Theorem: Given the probability that a slot is idle is P and the number of active nodes n, the throughput is minimal in the case of $p_1 = p_2 = ... = p_i = ... = p_n$. Due to the fact that $CW = \frac{2}{p_i} - 1$, it follows that each node has the same CW.

This Theorem can be proved by mathematical induction, here we omit it because of page limit. The *Theorem* reveals two important points. First, if we keep the average idle interval as the one corresponding to the peak throughput shown in Fig. 1, the throughput will not be less than the peak throughput, as the peak throughput is derived under the condition that all nodes have the same CW. Second, it shows that there is a tradeoff between throughput and fairness. If all nodes keep the same CW, which means the channel is fairly shared among all nodes, the throughput is sacrificed.

By detecting the average idle interval, each node can adjust its current CW around the optimal CW at runtime. Assume the observed current average idle interval is l_{idl}, the optimal CW corresponding to the optimal L_{io} is CW_o, given Equation (6), we can estimate CW_o as

$$CW_o = (CW_c - 1)\frac{L_{io} + \alpha}{l_{idl} + \alpha} + 1 \tag{7}$$

where CW_c is the current CW. Clearly, we obtain the optimal CW while avoiding the difficult task of estimating the number of active nodes. Then, we can adjust the current CW basing on Equation (7) so as to approach the optimal CW and hence tune the network to deliver high throughput. In the following, we give the tuning algorithm in detail.

4 DOB Scheme

In this section, we describe the dynamically optimizing backoff (DOB) scheme in which each node adapt its backoff process according to the observed average idle interval l_{idl}, which reflects the network traffic load. The goal is that each node always uses a contention window close to the optimal CW.

Unlike the IEEE 802.11, DOB divides a node's backoff process into three stages, namely stage 0, 1, and 2. When a node starts backoff, depending on the initially chosen backoff timer (BT), it may enter stage 1 or 2. After a collision, a node may enter into stage 0. At each stage, the node decrements its backoff timer when an idle time slot is detected. The node contends for the channel only when its BT reaches 0 and it is at stage 2. At the end of stage 0 and stage 1, the backoff nodes refresh their CWs basing on the average idle interval l_{idl} observed in previous stages according to the formula revised from (7) so that a high aggregate throughput can be achieved. We introduce two parameters K_h and K_l as thresholds for observed channel idle interval l_{idl}. Specifically, K_h is the threshold corresponding to high traffic load and K_l the threshold corresponding to low traffic load. When observed l_{idl} is lower than K_h , the node increases its CW and when l_{idl} is larger than K_l the node decreases its CW. K_h and K_l are defined as follows:

$$K_h = K_h^{'} - (CW_i - 1)/CW_{ct} \tag{8}$$

$$K_l = K_l^{'} - (CW_i - 1)/CW_{ct} \tag{9}$$

Where, the range defined by $[K_h^{'}, K_l^{'}]$ is correspondent to the optimal range around to L_{io} shown in formula (7). The term $(CW_i - 1)/CW_{ct}$, where CW_i

is the current CW of node i and CW_{ct} is the same constant for each node, is introduced to enhance the fairness in short term. We can understand this term as follows. In Equation (7), new CW depends on the old one for every node. Ideally each node may have the same initial CW when the network enters into steady state in saturated case, in reality, when a new active node initializes its CW as CW_{min} different from CWs used in other nodes and begins to transmit or the traffic load changes, CWs of different nodes changes and maybe different for a short term, which degrades fairness. Independent from average idle interval, the term $[-(CW_i - 1)/CW_{ct}]$ allows nodes with larger CW are likely to decrease its CW and the nodes with smaller CW likely to increase its CW. We also define a modified version of L_{io} denoted by L_c:

$$L_c = L_{io} + 0.5 - (CW_i - 1)/CW_{ct} \tag{10}$$

Here, we approximate α as 0.5. Obviously, we have $K_h^{'} < L_{io} < K_l^{'}$. Then the Equation (7) becomes

$$CW_o = (CW_c - 1)\frac{L_c}{l_{idl} + 0.5} + 1 \tag{11}$$

If we assume that finally $CW_o = CW_c$ for stable state and CWs converge to the same value, we can express this value as

$$CW = \frac{L_{io} + 0.5}{\frac{1}{2n} + \frac{1}{CW_{ct}}} + 1 \tag{12}$$

which is verified through average CW obtained from the following simulations. From the above equation, we can find that CW slips from the optimal value shown in Equation (6) because of introduction of CW_{ct} which is a tradeoff between fairness and aggregate throughput. For instance, $CW_{ct} = 250$ when the number of nodes is less than or equal to 100. When the number of active nodes in the network is more than 100, as far as throughput is concerned, we can set a larger CW_{ct} to keep the CW in Equation (12) closer to the optimal value in order to increase throughput ; however, this could lead to a little degradation in fairness as each node adapts its CW in a slower pace.

To avoid measuring the average channel idle interval in a short term, a node observes the idle slots to calculate the average idle interval at least more than Observation Window (OW) which is a certain number of idle slots. DOB adopts different backoff processes for new transmissions and retransmissions after a collision. In the case new transmission, a node, say node A, follows the following algorithm (Algorithm I):

1. Node A uses its current CW if it has one; otherwise it selects CW_{min} as current CW and set its backoff timer (BT) to uniform[0,CW-1]. Node A enters into backoff stage 2 if BT<OW or stage 1 otherwise.

2. Node A in stage 2 decreases BT by 1 whenever it detects an idle time slot. If BT=0, it begins transmission. If node A is in stage 1, it counts the number of consecutive idle slots while decreasing BT by 1 for each idle slot. When BT

reaches zero, it calculates l_{idl}. Then, it compares l_{idl} with K_l and K_h. If l_{idl} is in the range between K_h and K_l, i.e., $K_h < l_{idl} < K_l$, node A begins transmission immediately without changing CW; otherwise it acts as follows.

i) if $l_{idl} > K_l$, node A starts transmission immediately and decreases its CW as $newCW = (CW - 1)\frac{L_c}{l_{idl}+0.5} + 1$, as derived from Equation (7).

ii) if $l_{idl} < K_h$, node A increases its CW as $newCW = (CW - 1)\frac{L_c}{l_{idl}+0.5} + 1$ and resets its BT as BT=uniform[0, newCW-CW]. Then node A enters backoff stage 2.

In the case of collision, node A follows the following algorithm (Algorithm II):

1. Node A sets its BT as BT=uniform[0,2CW+1] without changing its CW.

2. If BT<OW, node A enters into backoff stage 2 and begins decreasing BT by 1 for every idle slot until BT=0; then it starts transmission. If BT≥OW, node A enters into backoff stage 0, while decreasing BT by 1 for each idle slot, it starts the calculation process of l_{idl} as stated previously. After OW idle slots, node A calculates l_{idl}. If l_{idl} is in the range between K_h and K_l, i.e., $K_h < l_{idl} < K_l$, node A enters into backoff stage 2 without any other adjustment. Otherwise, node A acts as follows.

i) if $l_{idl} > K_l$, node A still uses the current CW and resets BT as BT= uniform[0,CW-1].

ii) if $l_{idl} < K_h$, node A increases its CW as $newCW = (CW - 1)\frac{L_c}{l_{idl}+0.5} + 1$ and resets its BT as BT=uniform[0, newCW-1].

Then node A enters into stage 1 and follows step 2 in Algorithm I.

5 Performance Evaluation

In this section, we focus on evaluating the performance of our DOB through simulations, which are carried out on OPNET [18]. For comparison purpose, we also present the simulation results for the IEEE 802.11 DCF and FCR. In all the simulations, we consider the basic MAC scheme. In other words, the RTS/CTS mechanism is not used. The DCF-related and DOB-related parameters are set as the TABLE.

We assume each node is Poison process source with the same arrival rate which increases till to saturation in simulations. As shown below, DOB exhibits a better performance than the IEEE 802.11 and FCR in terms of throughput and fairness.

Table 1. Network configuration and Backoff parameters

Parameter	Value	Parameter	Value
SIFS	10 μsec	$K_h^{'}$	5.8
DIFS	50 μsec	$K_l^{'}$	6.0
Slot Length	20 μsec	L_{io}	5.9
aPreamblLength	144 bits	OW	15
aPLCPHeaderLength	48 bits	CW_{ct}	250
Bit rate	1 Mbps		

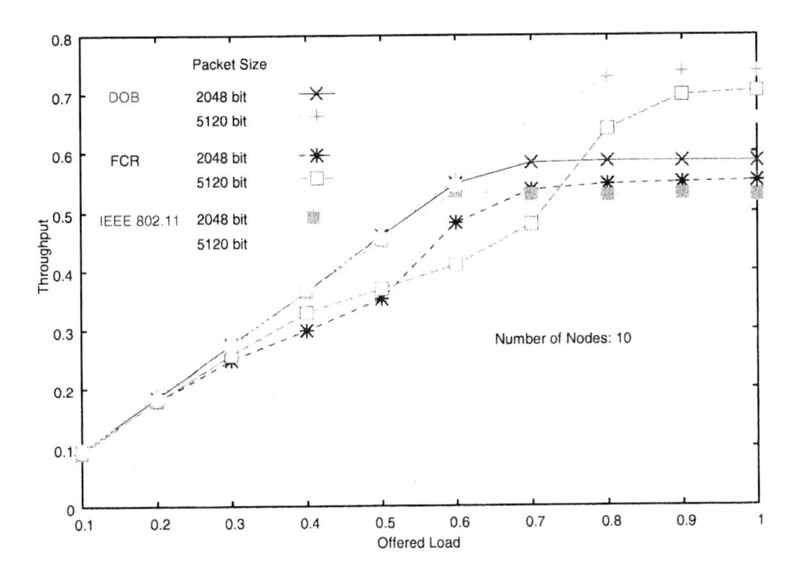

Fig. 3. Throughput vs. offered traffic with 10 nodes

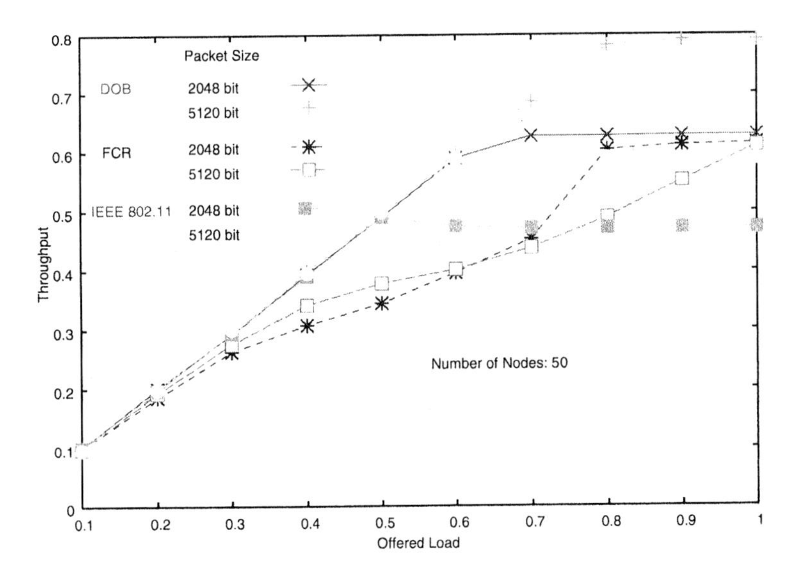

Fig. 4. Throughput vs. offered traffic with 50 nodes

5.1 Throughput

Firstly, we present the throughput obtained for the three schemes, i.e., DOB, FCR and the IEEE 802.11, under different offered load. Fig. 3, 4 show the throughput results when the number of nodes is 10 and 50. In each figure, we also consider two different frame lengths: 2048 bits and 5120 bits. Note that the frame length is the size of payload data and does not include MAC overhead, which is a reason that the simulation results are lower than the theoretical values.

It can be observed that when the traffic load is low, say lower than 0.6, the throughputs of DOB and the IEEE 802.11 are almost the same and equal to the offered load, whereas the throughput of FCR is lower. It can be explained as follows. For both DOB and the 802.11, since the offered load is low, the MAC collisions are slight and all the offered traffic can get through. For FCR, a node sets its CW as minimum after successful transmission, while the other nodes enlarge their CWs. In this way, once a node obtains medium, probably it can transmit continuously. But FCR is not efficient in the case of nonsaturation since a node has not enough packets for continuous transmission. This observation is more pronounced as the number of nodes increases. When the traffic load becomes heavy and the network enters into saturation, we see that the throughput of the 802.11 first increases with the traffic load, then slightly decreases after reaching the peak, and finally stabilizes at a certain value. This phenomenon is due to the fact that the maximum throughput of the 802.11 is larger than its saturated throughput. For both DOB and FCR, their throughputs first increase with the traffic load and then becomes stable. In the stable state, it can be seen that the 802.11 yields the lowest throughput among the three schemes. FCR is much better than the 802.11 because it can resolve the collisions in saturated case faster and more efficiently; and when a node seizes the channel, it will continuously transmit with a very high probability, resulting in high channel utilization. Even so, our DOB outperforms FCR though the difference in the throughput dwindles as the number of nodes become large.

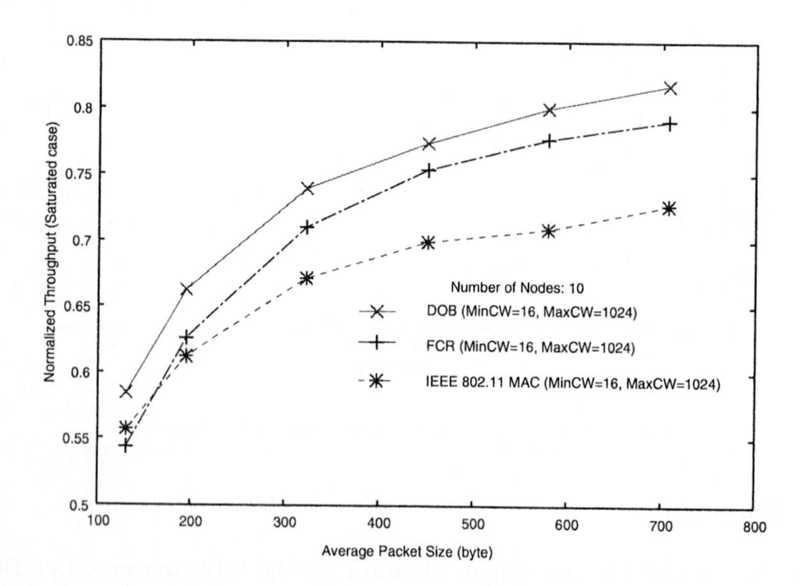

Fig. 5. Simulation results of throughput with 10 nodes

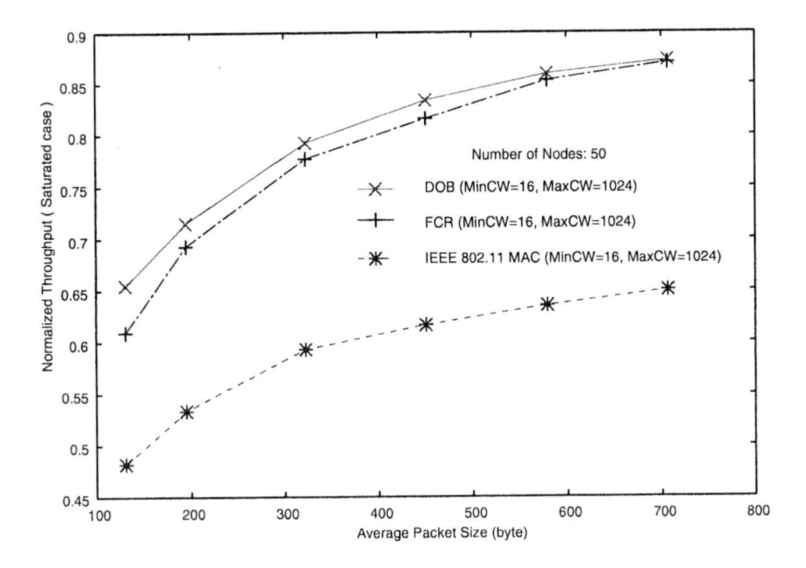

Fig. 6. Simulation results of throughput with 50 nodes

We also compare the throughputs of these three scheme as a function of the average packet size in saturated case, where each node always has packets in its buffer waiting for transmission, as shown in Fig. 5 and Fig. 6, which corresponds to the cases where the number of nodes is 10 and 50, respectively. In all cases, we find that the throughput increases along with the average packet size. This is because in the saturated case, given the number of nodes, the probabilities p_{idl}, p_s, and p_{col} are constant. Accordingly, according to Equation (3), the throughput gets larger if T gets larger while the overhead is the same. We also find that in all cases, our DOB achieves the highest throughput.

To sum up, the throughput performance of DOB is the best either in non-saturated case or saturated case. This is attributed to the fact that our DOB uses the average idle interval to adapt CW to approach the optimal CW, which can efficiently resolve collisions and lead to high throughput. Compared to FCR, DOB overcomes FCR's inefficiency in non-saturated case while being slightly more efficient in dealing with collisions in saturated case.

5.2 Fairness

To evaluate the fairness of DOB, we adopt the following Fairness Index (FI) [19] that is commonly accepted:

$$FI = \frac{(\sum_i T_i/\phi_i)^2}{n\sum_i (T_i/\phi_i)^2} \tag{13}$$

where T_i is throughput of flow i, ϕ_i is the weight of flow i (here we assume all nodes have the same weight in simulation). According to Equation (13), $FI \leq 1$,

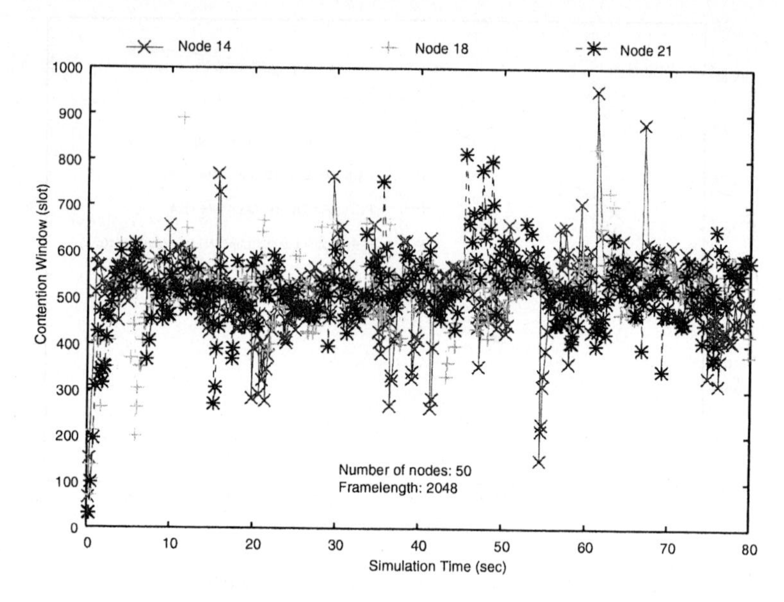

Fig. 7. Changes of CW in simulation with 50 nodes

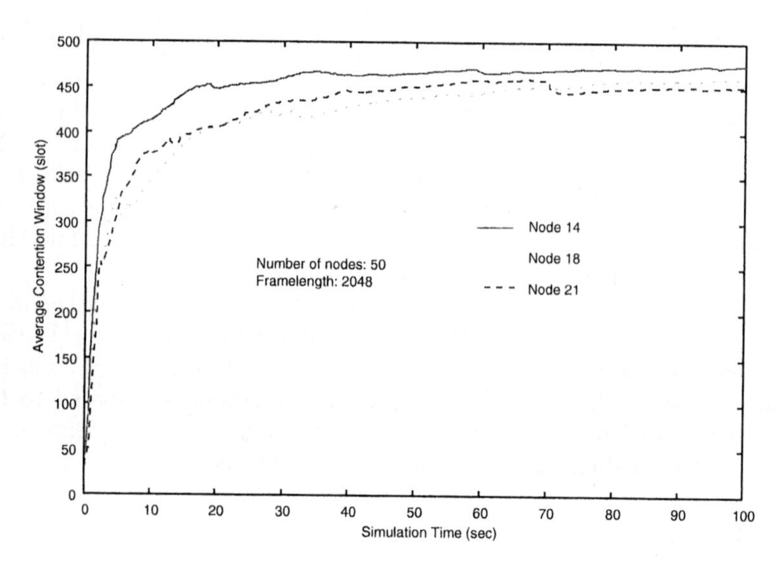

Fig. 8. Average CW in simulation with 50 nodes

where the equation holds only when all T_i/ϕ_i are equal. Normally, a higher FI means a better fairness.

Before comparing the fairness indexes between DOB and the IEEE 802.11 (Note we do not include FCR since it depends on an additional scheduling algorithm to achieve good fairness), we show how each node's CW changes in the course of simulation. Fig. 7 show the instantaneous change of the CW for three

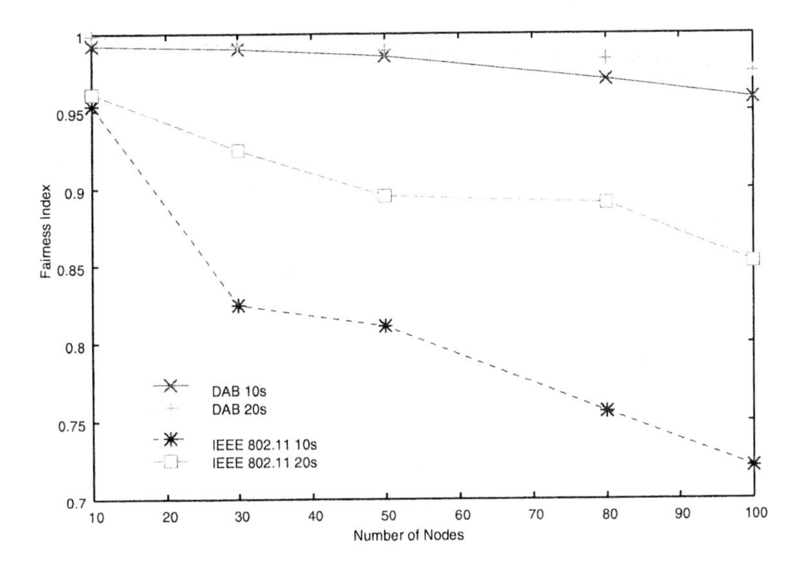

Fig. 9. Fairness index

nodes that are randomly selected from a network of 50 nodes. We see that while the CWs fluctuates from time to time, they have close average values, which are illustrated in Fig. 8. In Fig. 8, the average CW is about 450 close to the value 464.286 obtained by Equation (12).

Since DOB ensures that the all the nodes use about the same CW that is around the optimal value, it is expected that it can achieve a better fairness. (Because of high probability of consecutive transmission after gaining right of medium for FCR, FCR itself can inversely affect the fairness without additional fair scheduling mechanism.) Here, we compare fairness index of DOB with IEEE 802.11 as shown in Fig. 9. Fig. 9 shows that the fairness of DOB within 10s and 20s periods are significantly improved over that of the IEEE 802.11. It can also be seen that as the number of nodes rises, the fairness drops quickly for the 802.11, whereas for DOB, the fairness only slightly decreases. This is because the CWs of DOB used in every nodes are close as shown in Fig. 7 and 8.

6 Conclusion

In this paper, we first show that under the assumption that all the nodes use the same and fixed contention window, an index called average channel idle interval can indicate network throughput and has a simple relationship with the optimal contention window CW that leads to the maximal throughput. Meanwhile, if all the nodes use the same CW, they will fairly share the common channel.

Through both analysis and simulation, our scheme has the following advantages. First, as shown in the analysis, the average idle interval of channel is a suitable index for each node to grasp the network traffic situation and is insensitive to the change in packet length or the number of active nodes. Each node only needs to adjust its backoff process based on the observed average channel idle interval, avoiding the difficult task of estimating the number of active nodes in a changing network environment as required in [6]. Second, compared with the original IEEE 802.11, DOB achieves much higher throughput in saturated case; compared with FCR, DOB overcomes its inefficiency in non-saturated case. This is attributed to the fact that in DOB, each node always adjust its backoff to approach the optimal CW and thus leads to high throughput. Finally, DOB achieves good fairness since when the network becomes stable, all the nodes maintains almost identical average CW.

This research was partially supported by the Ministry of Education, Culture, Sports, Science and Technology, Grant-in-Aid for Exploratory Research, 17650020, 2005.

References

1. M. A. Visser, and M. E. Zarki: Voice and data transmission over an 802.11 wireless network. IEEE PIMRC (1995)
2. J.Y. Yeh and C. Chen: Support of multimedia services with the IEEE 802-11 MAC protocol. Proc. ICCE2 (2002)
3. L. Bononi and M. Conti and E. Gregori: Runtime Optimization of IEEE 802.11 Wireless LANs Performance. IEEE Trans. Parallel Distrib. Syst. **15(1)** (2004) 66-80
4. R. Bruno and M. Conti and E. Gregori: Distributed Contention Control in Heterogeneous 802.11b WLANs. In WONS, St Moritz, Switzerland, January 2005.
5. Q. Pang and S. C. Liew and J. Y. B. Lee and V. C. M. Leung. Performance evaluation of an adaptive backoff scheme for WLAN. Wireless Communication and Mobile Computing. **5(8)** (2004) 867-879
6. F. Cali, M. Conti and E. Gregori: Dynamic tuning of the IEEE 802.11 protocol to achieve a theoretical throughput limit. IEEE/ACM Transactions on Networking. **8(6)** (2000) 785-799
7. G. Bianchi: Performance analysis of the IEEE 802.11 distributed coordination function. IEEE JSAC. **18 (3)** (2000)
8. V. Bharghvan: Performance evaluation of algorithms for wireless medium access. IEEE International Computer Performance and Dependability Symposium IPDS'98. (1998) 142-149
9. Y. C. Tay and K. C. Chua: A capacity analysis for the IEEE 802.11 MAC protocol. ACM/Baltzer Wireless Networks **7 (2)** (2001)
10. J. H. Kim and J. K. Lee: Performance of carrier sense multiple access with collision avoidance protocols in wireless LANs. Wireless Personal Communications **11 (2)**, (1999) 161-183
11. Y. Kwon, Y. Fang and H. Latchman: A novel MAC protocal with fast collosion resolution for wireless Lans. IEEE INFOCM. (2003)
12. J. Weinmiller, H. Woesner. J. P. Ebert and A. Wolisz: Analyzing and tuning the distributed coordination function in the IEEE 802.11 DFWMAC draft standard. Proc. MASCOT (1996)
13. H. Wu, Y. Peng, K. Long, S. Cheng and J. Ma: Performance of reliable transport protocol over IEEE 802.11 wireless LAN: analysis and enhancement. IEEE INFOCOM, **2** (2002) 599-607

14. H. S. Chhaya and S. Gupta: Performance modeling of asynchronous data transfer methods of IEEE 802.11 MAC protocol. Wireless Networks. **3** (1997) 217-234
15. P. Yong, H. Wu, S. Cheng and K. Long: A new Self-Adapt DCF Algorithm. IEEE GLOBECOM **2** (2002)
16. H. Kim and J. Hou: Improving protocol capacity with model-based frame scheduling in IEEE 802.11-operated WLANs. Proc. of ACM MobiCom. (2003)
17. T. Nandagopal, T-E Kim, X. Gao and V. Bharghavan: Achieving MAC layer fairness in wireless packet networks. Proc. ACM MOBICOM (2000) 87-98
18. OPNET Modeler. http://www.opnet.com.
19. R. Jain, A. Durresi, and G. Babic: Throughput Fairness Index: An Explanation. ATM Forum/99-0045 (1999)

Optimal Control of Packet Service Access State for Cdma2000-1x Systems

Cai-xia Liu[1], Yu-bo Tan[2], and Dong-nian Cheng

[1] PLA Information Engineering University, Henan, P.R. China, 450002
[2] PDL, Computer School, NUDT, Changsha, Hunan, P.R. China, 410073
lcx@mail.ndsc.com.cn

Abstract. Packet data service accessing state control is an effective wireless resource control scheme for 3G systems. This paper proposed an optimal control mechanism for packet data service accessing state by taking an integrated performance function of the mean packets waiting time W, the mean saving in the signaling overhead S and the mean channel utilization U as target function. By introducing a specific *IBP* model to better capture the characteristics of WWW traffic over cdma2000-1x systems, a system performance model was established. By analyzing the relationships between service model parameters and the state transition control timer, this paper presents a reference for project development to set some system parameters.

1 Introduction

The ratio of data service is gradually increasing in the new generation mobile communication systems. In order to effectively utilize the wireless resource and satisfy the need of new services, IS-2000 systems presented the new packet data accessing control mechanism according to packet data services' burst characteristic, which uses four states to control the accessing process and the states transition process is shown in Fig.1.

Hereinto, new-arrival packets trigger the transitions from the dormancy, suspended or control hold state to the active state, and the upper-layer signaling together with the expiration timer control the transitions of any other two states. The expiration timer includes active timer T_{active}, control hold timer T_{hold} and suspended timer $T_{suspended}$.

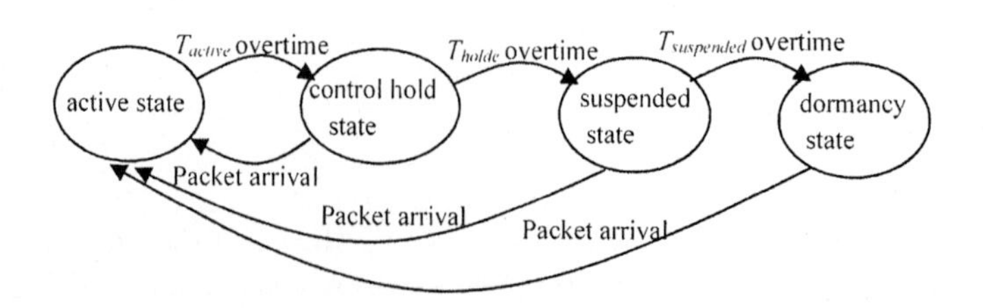

Fig. 1. Packet service states transition process

X. Jia, J. Wu, and Y. He (Eds.): MSN 2005, LNCS 3794, pp. 966–975, 2005.

If a data user is in active state, the base station (BS) will start the T_{active} after finishing transmitting the user's packets. If new packets have arrived before T_{active} expired, the BS stopped the T_{active}. Otherwise, if T_{active} expired, the system will release the traffic channels and the user will transfer into control hold state, at the same time,T_{hold} is started. If T_{hold} expired, the user will be in suspended state and $T_{suspended}$ is started. Similarly, if $T_{suspended}$ expired, the user will transfer into dormancy state.

Thus, the value of the three timers determines the BS to mark the users' accessing states. For example, if the value of T_{active} is too large, the user will occupy the traffic channels in a longer time, which may reduce the channel utilization. Whereas, a larger timing value can avoid rebuilding traffic channels repeatedly before sending packets, which not only may reduce the packets waiting time, also may save the signaling overheads for rebuilding traffic channels. Contrarily, a too small timing value may decrease the traffic channel occupying time and advance the channels' utilization, but may increase the packets waiting time and signaling overheads. So, it's very important to choose an appropriate timing value based on synthetically considering the critical system performance factors, such as the channel utilization, packets waiting time and the signaling overhead etc.

At present, there are no rules to determine the timing value of T_{active}, T_{hold} and $T_{suspended}$, also no methods to choose the timing value in the technology specification for cdma2000-1x systems. Reference [3] analyzed the threshold of the MAC states transition timers based on the self-similarity packet traffic model. While, according to the mathematics description of "self-similarity", only the traffic flow with "lots of" ON/OFF sources in the network systems may approach to self-similarity characteristic, and the total service flux in a BS evidently doesn't possess the "lots of" characteristic.

This paper introduced a new method to determine the timing value based on a specific *IBP*(*Interrupted Bernoulli Process*) model by taking an integrated performance function of the mean packets waiting time W, the mean saving in the signaling overhead S and the mean channel utilization U as target function. *IBP* model may better capture the characteristics of packet data traffic over cdma2000-1x systems.

2 Packet Service Model

2.1 Packet Service Characteristics

Generally, with different traffic sources, packet data service may be separated into many categories and 80 percent of the packet data service will be WWW service, so this paper mainly introduces WWW forward data service characteristics.

WWW forward service is generally described using three-layer model: session layer, packet call layer and packet layer. A user's one time WWW service process is named by "a session" and one time page-request is named by "a packet call". During the transmission, each page-request's objects are packed into many

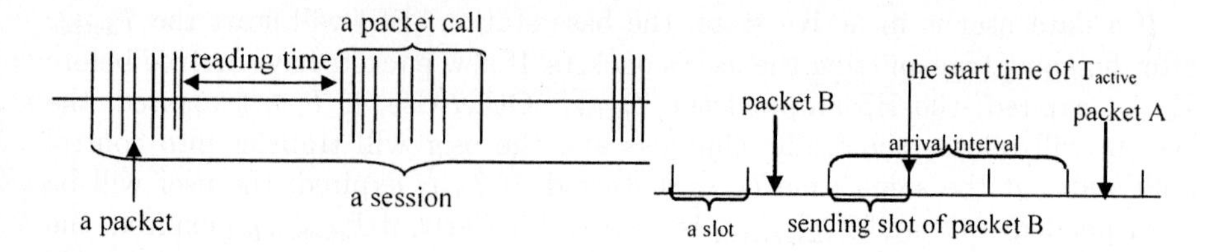

Fig. 2. Three-layer model of WWW service **Fig. 3.** Sketch map of packet arrival

transmission layer packets. The interval between two packet calls is the user's reading time as showed in Fig.2.

Usually, the state during a page-request being sended and the page being downloaded to a user is named by active state. The state when a user is reading the page downloaded is named by dormancy state. The other packet services, such as FTP, WAP etc., also have the similar active and dormancy characteristics.

2.2 Dacket Service Model

In virtue of reference [4] and [5] which analyzed the traffic characteristics in cdma2000-1x systems, we induct the *IBP* Process to describe the arrival characteristics of packet service. *IBP* is the discrete form of *IPP* (*Interrupted Poisson Process*)[7], and in which the time is dispersed into slots with equal length and any slot is either in ON period or in OFF period. In any slot of ON period, the packets arrive with probability a and no packets arrive in OFF period. Denoting p is the transferring probability that a slot is in ON period and its next slot is in OFF period and$q$ is the transferring probability that a slot is in OFF period and the next slot is in ON period. If D_{on} and D_{off} respectively denote the number of slots in ON period and in OFF period, then $D_{on}$and D_{off} may be expressed by formula (1).

$$D_{on} = \frac{1}{p}, \; D_{off} = \frac{1}{q} \tag{1}$$

We use the following *IBP* process to describe the WWW traffic flow in cdma2000-1x systems:

(1)A packet call's duration conforms to geometry distribution with mean being D_{on} and unit being the length of one slot;

(2)The average packets number in a packet call is N_d;

(3)The length of the reading period between two adjacent packet calls, denoted by t_{pc}, conforms to geometry distribution with mean being D_{pc}, and supposing no packets arrive in reading periods;

(4)The arrival interval of two adjacent packets in one packet call, denoted byt_p, approximatively conforms to geometry distribution with mean being D_d When the number of slots during one packet call is large enough, where the arrival interval denotes the number of slots in which there are no packets arrive during two adjacent packets(showed as Fig.3).

Hereinto, the length of a slot may be 20ms, 40ms or 80ms, which depends on the size of the packet.

From the *IBP* process mentioned above, we can deduce the following:

(1)The probability of the reading period (OFF period) including $k$slots, denoted by $f_{tpc}(k)$,$f_{tpc}(k) = p_1(1-p_1)^{k-1}$.

Hereinto, $p_1$is the transferring probability that a slot is in OFF period and the next slot is in ON period, from formula (1), $p_1 = 1/D_{pc}$.

(2)The probability of the packet arrival interval being k slots, denoted by $f_{tp}(k)$, is $f_{tp}(k) = p_2(1-p_2)^k$.

Hereinto, p_2 is the packet arrival probability in each slot of ON period and $p_2 = 1/(D_d+1)$.

3 Performance Model

This section will build the integrated performance target function which depends on the mean packets waiting time (W), the mean saving in the signaling overhead (S) and the mean channel utilization (U).

It's obvious that U,W and $S$are all relative with T_{active}, where$0 \leq U \leq 1$ and T_{active}denotes the number of the slots. For simple analysis, we define S being the probability of not requiring to rebuild traffic channels when one packet arrives, so$0 \leq S \leq 1$. Usually, when a packet arrives at a queue in *BS*, if the traffic channel for it already exists and no other packets are waiting in the queue, the BS will transmit the packet at the beginning of the next slot[2].Since a packet's arrival instant is random in one slot, it's reasonable to assume the arrival instant is uniformly distributed in the interval of a slot. So, the average minimum waiting time for any packet in a BS is a half slot length. If defining W is the average waiting time with unit the length of a slot, then we get $0 < \frac{1}{2W} \leq 1$. So, Formula (2) restricts the value of F in [0,1],where ω_1, ω_2 and ω_3 are three weights and they satisfy $\omega_1 + \omega_2 + \omega_3 = 1$.

$$F = \varpi_1 U + \varpi_2 S + \varpi_3 \frac{1}{2W} \tag{2}$$

Formula (2) will be used as the system target function and by tuning the value of ω_1,ω_2 and ω_3, the target function will be suitable to different systems that emphasize particularly on different performance parameters.

Our objective is to maximize U, maximize$S$ and minimize W (that is to maximize 1/(2W)). While optimizing the three parameters individually would result in three different (perhaps conflicting) operating points. A better optimization function can be derived by combining all these three parameters, so our objective is to:

$$Maximize(\omega_1 U + \omega_2 S + \omega_3 \frac{1}{2W})$$

This ensures all the three parameters collectively optimize the overall system performance. In the following, we derive expressions for the three parameters,U,S and W. Then the T_{active} that make the overall system performance optimize is also derived.

3.1 Expression for S

Supposing A and B are two packets belonging to the same session and B arrived at BS just before A. If t_{A-B} is the arrival interval between A and B and $t_B$is the time of system needing to send B, then, when $t_{A-B} - t_B \leq T_{active}$, the traffic channel for the session is active when A arrives. We assume the packets in BS needn't be split when being sent, which means the system will finish sending a packet in a slot, that is t_B=1. The relationships among the various time parameters are showed in Fig.3.

Packet A may be the first-arrival packet in a ON period, which means the ON period is activated due to the arrival of A, then packet B would be the last-arrived packet before the previous ON period ended, which means the next slot would enter into a OFF period after B arrived. A also may be the packet arrived in any slot after a ON period was activated, and packet B would be the packet arrived before the ON period ended.

If A is the first-arrival packet of a ON period, then t_{A-B} will equal to the length of a OFF period, t_{pc}, and if $(t_{pc} - 1) \leq T_{active}$, the traffic channel for the session is active when B arrives; If A is the packet arrived in any slot after a ON period was activated, then t_{A-B} =t_p, and if $(t_p - 1) \leq T_{active}$, the traffic channel is also active for B. Moreover, a packet arrives to the BS with a probability of $1/N_d$ being the first packet of a ON period and with the probability of $(N_d-1)/N_d$ being other one. So, S, the probability of not requiring rebuilding traffic channels when a packet arrives, may be expressed as formula (3).

$$S = \frac{N_d - 1}{N_d}(1 - (1 - p_2)^{T_{active}+2}) + \frac{1}{N_d}(1 - (1 - p_1)^{T_{active}+1}) \quad (3)$$

3.2 Expression for U

Assuming that:

All sessions have the same bandwidth requirement which is marked with C (bps);

All packets arriving to a BS have the same size which is marked with $\frac{C \cdot T_f}{8}$(bytes);

In the duration of a session, the mean number of packet calls is N_{pc}.

Then, a session's average duration time, denoted by L, is:$L = N_{pc} \cdot (D_{on} + D_{pc})$.

Defining U is a traffic channel's utility during the traffic channel being occupied by a session. From the foregoing analysis, the average number of slots occupied by the system for sending packets during a session, denoted by T_u, is : $T_u = N_d \cdot N_{pc}$.

The data quantity sent in various slots will be equal if there are packets to be sent in these slots, and the quantity is $C \cdot T_f$bits. While there may be no packets sent in some slots during the active period of a session, and the session may still occupy the traffic channel and the channels' utility will be affected. It's very obvious the value of T_{active} determines the channels' utility. (1) T_{active}=0 means the system will release the traffic channel occupied by a session when it just finished sending a packet of the session, so if not considering the channel release time, the additional slots occupied will equal to zero, that is T_e =0; (2) $T_{active} > 0$ means the session will additionally occupy the traffic channel in $(D_{d1} - 1)$ slots if $1 < t_p \leq T_{active} + 1$,and will additionally occupy the traffic channel in T_{active} slots if $t_p > T_{active}+1$,where D_{d1} is the average packets arrival interval when $1 < t_p \leq T_{active} + 1$.

So, the average number of the slots being additionally occupied in an ON period for a session, denoted by T_{on}, is:

$$T_{on} = (N_d - 1) \cdot \{(1-p_2)^2[1-(1-p_2)^{T_{active}}] \cdot (D_{d1} - 1) + (1 - p_2)^{T_{active}+2} \cdot T_{active})\}, \quad (T_{active} \geq 1)$$

where D_{d1} is expressed as the following:

$$D_{d1} = (1-p_2)^2 + \frac{(1-p_2)^2(1-(1-p_2)^{T_{active}})}{p_2} - (T_{active} + 1)(1 - p_2)^{T_{active}+2}, (T_{active} \geq 1)$$

In the same way, the average number of the slots being additionally occupied in an OFF period for a session, denoted by T_{off}, is:

$$T_{off} = (1-p_1)[1-(1 - p_1)^{T_{active}}](D_{pc1}-1) + (1-p_1)^{T_{active}+1} \cdot T_{active} \;, \quad (T_{active} \geq 1)$$

Where $D_{pc1}$is the average length of an OFF period when $1 < t_{pc} \leq (T_{active}+1)$ and it is expressed as:

$$D_{pc1} = (1-p_1) + \frac{(1-p_1)(1-(1-p_1)^{T_{active}})}{p_1} - (T_{active} + 1)(1 - p_1)^{T_{active}+1}, \quad (T_{active} \geq 1)$$

It's easy to get the average number of the slots being additionally occupied in the duration of a session when $T_{active} \geq 1$, denoted by T_e.

$$T_e = N_{pc} \cdot (T_{on} + T_{off})$$

So, the traffic channel's utility, U, may be expressed as formula (4).

$$U = \frac{T_u}{T_u + T_e} = \begin{cases} \frac{N_d \cdot N_{pc}}{N_d \cdot N_{pc} + N_{pc} \cdot (T_{on}+T_{off})} \\ 1 \end{cases} = \begin{cases} \frac{N_d}{N_d+T_{on}+T_{off}}, & (T_{active} \geq 1) \\ 1\;, & (T_{active} = 0) \end{cases} \quad (4)$$

3.3 Expression for *W*

As the foregoing depiction of this paper: (1) A packet has to wait for average 1/2 slot in BS before it is sent with the precondition of the buffer being vacant and the traffic channel being active when the packet arrives;(2) If the traffic channels have been released, the packet has to request to rebuild them if there are free resource in the system.

From session 3.2, the occupying ratio of a traffic channel during a session being activeis:

$$p3 = \frac{T_u + T_e}{L} = \begin{cases} \frac{N_d+T_{on}+T_{off}}{D_{on}+D_{pc}}, & (T_{active} \geq 1) \\ \frac{N_d}{D_{on}+D_{pc}}, & (T_{active} = 0) \end{cases}$$

On the assumption that the wireless resources are all occupied by M sessions when a packet belonging to a dormancy session arrives, then the probability that there are traffic channels being released in the follow-up any slot is $p_4=1-p_3^M$. From the definition of geometry distribution, the probability that there are no free resources until the $(k+1)$th (k=0,1,2,...) slots after a packet has arrived is $p_4\ (1-\ p_4)^k$, which means the packet's mean waiting time for free resources is $(1-p_4)/p_4$ slots. Supposing T_s slots are needed to rebuild a traffic channel, then the mean waiting time of a packet belonging to a dormancy session is: $0.5+(1-p_4)/p_4+T_s$

Thus, the mean waiting time for any packet arriving at BS is:

$$W = 0.5S + (0.5 + \frac{1-p_4}{p_4} + Ts)(1-S) \tag{5}$$

3.4 Determination of the Optimal T_{active}

From formula (3), (4) and (5), the integrated performance function of U, S and 1/ $(2W)$ can be obtained. It's obvious that the right side of formula (2) is correlative with T_{active}, and the value of T_{active} that makes Fmaximize can be deduced by differential coefficient, which is showed in formula (6).

$$\frac{d(\omega_1 U + \omega_2 S + \omega_3 \frac{1}{2W})}{dT_{active}} = 0 \tag{6}$$

4 Performance Analysis

For convenience, we assume the other parameters, with the exception of T_{active}, in the expressions of F,S,U and Wtake values according to Table 1.

Figure 4-6 show the relationships between S, U, $1/(2W)$and T_{active} respectively. Intuitively analyzing, the relationships between the three parameters and T_{active} are directly affected by N_d,D_d and D_{pc}, where N_d,D_d and D_{pc} are determined by the statistical characteristic of packet service in cdma2000-1x systems. Figure 4-6 conclude D_d has greater effect on the value of S, U and $1/(2W)$ when T_{active} takes value in $(0 \sim 50)$ and when $T_{active} > 20$, the value of U and $1/(2W)$ are great affected by N_d.

Table 1. Parameters Evaluation

parameters	value
N_d(unit : slot)	20,30,40
D_d(unit : slot)	2, 4, 6
D_{pc} (unit : slot)	250,350,500
D_{on}(unit : slot)	500
T_s (unit: slot)	5
M(the maximum session number)	5
ω_1, ω_2 ω_3	1/3,1/3,1/3

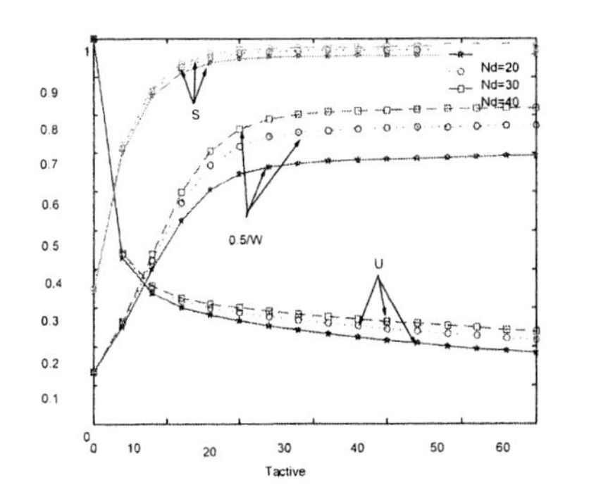

Fig. 4. N_d varies, S,U, $0.5/W$ as functions of T_{active}

Fig. 5. D_d varies, S,U, $0.5/W$ as functions of T_{active}

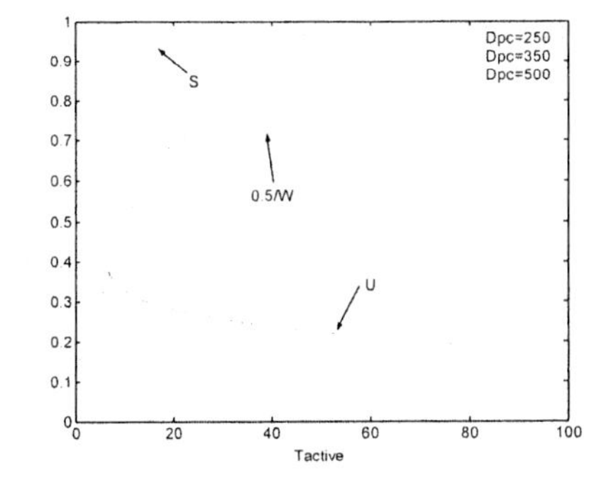

Fig. 6. D_{pc} varies, S,U,$0.5/W$ as functions of T_{active}

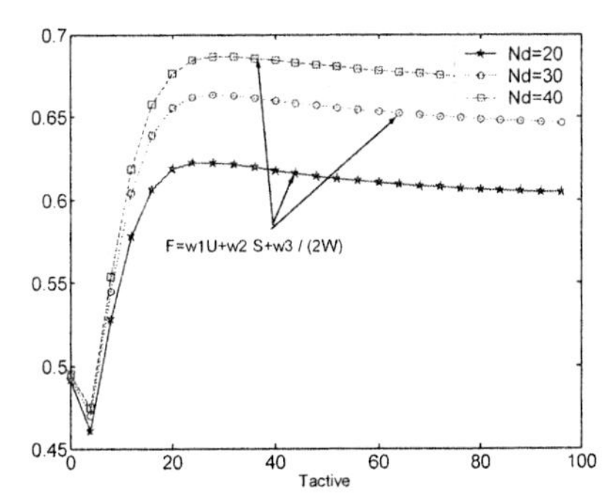

Fig. 7. N_d varies, F as a function of T_{active}

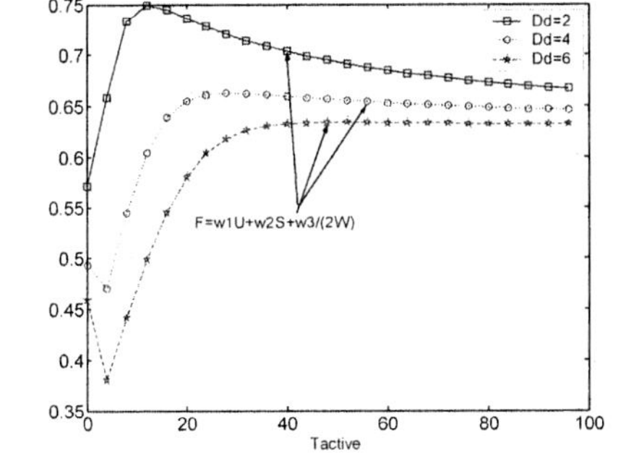

Fig. 8. D_d varies, F as a function of T_{active}

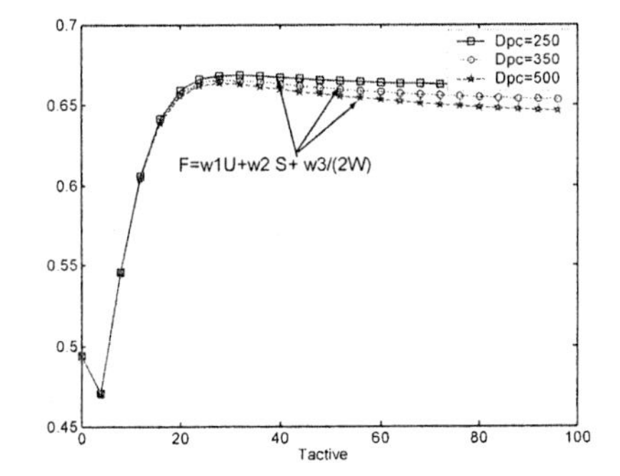

Fig. 9. D_{pc} varies, F as a function of T_{active}

Figure7-9 depict the relationships between system performance target functions and T_{active}, respectively when N_d, D_d and D_{pc} vary. Not losing universality, we temporarily assume ω_1, ω_2,and$\omega_3$all take value 1/3.

Figure7-9 show the same phenomena as the following:

(1)When T_{active} is smaller, F increases as T_{active} increasing and arrives to a maximum value;

(2)After F arrived to maximum, F decreases as T_{active} unceasingly increasing.

From the foregoing analysis, the value of T_{active} that makes F maximize is the target value that optimize the system performance.

Figure 7-9 also show that the system performance and the optimal T_{active} are obviously affected by N_d and D_d, which is mainly represented as:

When T_{active} takes values in the area of less than 20, the system performance improves as N_d increasing, and the optimal T_{active} varies little;

When T_{active} takes values in the area of less than 100,the system performance decreases as D_d increasing, and the optimal T_{active} increases;

In the all value area, the system performance varies little as D_{pc} varies, especially, when $T_{active} < 20$, the system performace is almost unacted on D_{pc}, as well as the optimal T_{active} does.

Thus, it can be seen: when the packet service statistical parameters N_d and D_d change in different cdma2000-1x systems, the optimal T_{active} should be newly chosen, while the other parameters change, the optimal T_{active} needs not be adjusted.

Otherwise, intuitively analyzing formula (2), we can see that the optimal T_{active} will take different values when the weights ω_1,ω_2 or ω_3 change.

5 Conclusions

Packet data service accessing state control is an effective wireless resource control scheme for 3G systems according to the burst characteristic of packet data service. This paper proposed an optimizing control mechanism for packet data service accessing state by taking the integrated performance function of the mean packets waiting time W, the mean saving in the signaling overhead S and the mean channel utilization U as target function. By introducing a specific *IBP* model to better capture the characteristics of WWW traffic over cdma2000-1x systems, we established a system performance model based on the target function, introduced a calculation method for T_{active} and analyzed the relationships between service model parameters and the T_{active}. According to performance analysis, we educed several conclusions:(1) in different value areas, N_d and D_d have greater effect on the system performance and the optimal T_{active} ;(2) When N_d or D_d changes, especially Dd changes, the system need to newly select the optimal T_{active} value.

The service model adopted in this paper is accord with cdma2000-1x system WWW service characteristics, and using the integrated performance of U,S and W as the system performance target function has certain a prevalent significance.

The analysis method in this paper is also the same with T_{hold} and $T_{suspended}$.

References

1. B.Mah," An Empirical Model of HTTP Network Traffic," *Proceeding of INFOCOM'97*, 1997.
2. 3GPP2 A.S0001-A,"3GPP2 Access Network Interfaces Interoperability Specification," November 30, 2000.
3. Mainak Chatterjee and Sajal K. Das," Optimal MAC State Switching for cdma2000 Networks," Proceedings of IEEE INFOCOM 2002, Vol. 1, pp. 400-406.
4. ITU: US TG 8/1, Radio Communication Study Group," The radio cdma2000 RTT Candidate submission," Technical Report TR 45-5, Jun. 1998.
5. C. Comaniciu, NB Mandayam, D. Famolari, P. Agrawal,"Wireless Access to the World Wide Web in an Integrated CDMA System,"IEEE Transactions on Wireless Communications, May 2003.
6. M.E. Crovela and A. Bestavros," Self-Similarity in World Wide Web Traffic: Evidence and Possible Causes," IEEE/ACM Transaction on Networking, vol 5, No 6, December 1997, pp. 835-846.
7. R.O.Onvural,"Asynchronous Transfer Mode Network: performance Issues," Artech House, 1994.
8. Y.S.Rao and A.Kripalani,"cdma2000 mobile radio access for IMT2000," IEEE International Conference on Personal Wireless Communication, 1999, pp 6-15.
9. C. Comaniciu, NB Mandayam, D. Famolari, P. Agrawal," Wireless Access to the World Wide Web in an Integrated CDMA System," IEEE Transactions on Wireless Communications, May 2003.
10. Peng Peng," A Critical Review of Packet Data Services in Wireless Personal Communication System," Virginia Tech ECPE 6504: Wireless Networks and Mobile Computing , Spring 2000. http:// fiddle.visc.vt.edu/courses/ecpe6504-wireless/ projects.

A Cross-Layer Optimization for Ad Hoc Networks*

Yuan Zhang, Wenwu Wu, and Xinghai Yang

School of Information Science and Engineering, Jinan University, Jinan 250022, China
{yzhang, wuww, ise_yangxh}@ujn.edu.cn

Abstract. The lack of an established infrastructure and the hostile nature of the wireless channel make the design of ad hoc networks a challenging task. The cross-layer design methodology, which has been strongly advocated in recent years, essentially aims to overcome the sub-optimality introduced by designing each layer in isolation. This paper explores one aspect of such optimizations, namely using multiple antennas at each node for receiving and transmitting data using directional beams. By jointly analyzing both the MAC layer parameters and choosing a suitable routing protocol afterwards, improvements over earlier published schemes are obtained. We also perform in-depth simulations to characterize the better performance of various scenarios.

1 Introduction

The layering principle, which was originally proposed for the wired line networks, simplifies design and implementation and provides the possibility of alternative layer implementations. After its huge success in the Internet, it was naturally extended for the wireless networks too. However, the characteristics of wireless networks, caused by their low link capacity and high bit error rates, differ from wired networks in several ways. In particular, some of the concerns unique to a wireless ad hoc mobile environment include energy efficiency, link stability under mobility, routing in the absence of global location knowledge and scalability of network. As is mentioned in [1], due to the many unique challenges posed by such networks, the traditional approach to optimizing performance by separately optimizing different layers of the OSI-model may not be optimal. In order to obtain the best results, it might be necessary to perform optimization using the information available across many layers. Such techniques are known as cross-layer design.

Cross-layer design is able to improve the network performance [2]. One example of the coupling, which has been addressed in [3,4,5], is between routing in the network layer and the access control in the medium access control sub-layer. By using theoretical upper bounds on the performance that are given in terms of capacity regions, [6] evaluates the effects of various design choices, with an emphasis on the MAC sub-layer. In [7] the cross-layer design addresses the joint problem of power control and scheduling for multi-hop wireless networks with QoS. It takes SINR and minimum rate as constraints to minimize the total transmit over the links.

---

* This work is supported by Science & Technology Foundation of Jinan University under Grant No. Y0520 and Y0519.

X. Jia, J. Wu, and Y. He (Eds.): MSN 2005, LNCS 3794, pp. 976 – 985, 2005.

This paper in particular explores one aspect of such optimizations, which is the use of multiple antennas at each node for receiving and transmitting data using directional beams. With the current move towards higher frequency bands, the size of such multiple-element antennas is no longer a constraint, making these systems an interesting and viable alternative to the current single-element omnidirectional antennas. In addition, to obtain the best performance, selecting the correct parameters at the MAC layer and a suitable routing protocol are also essential.

The rest of this paper is organized as follows. Section 2 discusses the physical layer and some of the characteristics and problems of having multiple antenna systems. Similarly, section 3 looks at the MAC layer while section 4 focuses on routing. Section 5 provides simulation results and characterizes the performance of various schemes. Finally, section 6 presents the conclusions and a brief statement of the future work.

2 Physical Layer (PHY)

At the physical layer, two distinct sets of decisions need to be made – choice of antenna system and choice of the air interface.

The choice of air interface for an ad hoc wireless system is overviewed in [8], and in numerous papers in the literature. However, this will not be discussed here since the selection of such an air interface is based on a number of factors other than those considered in this paper. It will be assumed that the factors taken into consideration in this paper are orthogonal to those involved in selecting the air interface. In this paper, a more detailed discussion of the antenna systems will be presented.

Directional antennas can be classified into two main categories – switched beam antennas and steered beam antennas. In [9], the author presents some of the first results in using beamforming to achieve improved throughput and reduced end-to-end delay in a static ad hoc setting. The author argues that the performance observed is similar for static and dynamic environments since the same routing and MAC protocols are used. However, this is not true.

At the physical layer, in a static environment, once the direction of the transmission has been determined, a node can cache this information. For any subsequent transmissions to/from this node, the beam can be directed in this direction (except in the event of node failure). In a dynamic environment, however, any of the scenarios below can occur –

1. *The position of the receiver changes between different transmissions*
This is a likely scenario even in a low-mobility, low-data rate network. One of the methods to overcome stale directions to receivers in a node's cache is to use HELLO packets as described in [9, 10]. However, this can quickly impose an unnecessary overhead on the entire system. In addition, in a high data-rate or otherwise densely populated network, this can lead to congestion of the network and consequently a reduction in throughput. An alternative (which is used in the simulations conducted for this paper) is to attempt to communicate with the receiver using the direction stored in the cache. After a predetermined number of failed attempts (4 in the simulations), the antenna reverts to either an omnidirectional transmission, or it can start sweeping each sector sequentially. Thus, if a node wants to communicate with a neighbor and does not

find its direction in the cache, it initiates a RREQ by sweeping through each sector (for both beam switching and beam steering antennas).

2. *The position of the receiver changes between different packets for the same transmission*
This is a possibility in a moderate-to-high mobility environment with moderate-to-large data transfers. The approach described in (1.) above can be used to combat this problem. In addition, the ACKs transmitted by the receiver can be used to determine the direction of arrival and thus select the appropriate antenna for transmission.

3. *The position of the receiver changes during a single packet transmission*
Given the transfer rate (11 Mbps) and the typical packet size in 802.11 networks, this is not a very likely scenario for most ad hoc networks and so will not be considered.

3 Medium Access Control Layer (MAC)

Implementing a directional antenna system is intimately connected to the MAC layer, which should not only be able to support such directional transmission, but also leverage the additional information to improve performance. A number of papers in the literature address this problem [10,11,12,13]. However, most of the protocols discussed in these papers are adapted from the IEEE 802.11 wireless standard. In addition, it is noted that in general, most of these papers do not differ much in their proposed schemes. Consequently, leveraging the work done in these papers, the aspects of the MAC implemented in our simulations will be discussed and its merits and demerits evaluated.

For the purposes of supporting directional transmission, two main modifications need to be made to the MAC protocol.

1. Every node needs to maintain a list of its neighbors as well as the directions they are present at, with respect to the node. This list could be maintained proactively using HELLO packets or performing a 360° sweep to locate the node when there is a need to transmit to it.
2. A Directional NAV table needs to be included in the MAC layer [13]. This maintains a list of those directions to which transmissions need to be differed due to existing communications in that area. In [13], this angle was set to be equal to the beamwidth of each directional antenna based on some assumptions made by the authors.

Some of the salient features of the MAC protocol implemented in this paper include –

• *Directional RTS/CTS*
The source node transmits its RTS using a directional beam in the supposed direction of the receiver. The receiver transmits the CTS back using a directional beam. The source repeats the RTS up to 4 times (a design parameter), before transmitting in omni-directional mode.

• *Omnidirectional Idle/Receive*
All nodes listen to the channel in an omni-directional mode when not transmitting or receiving. This ensures that they can receive RTS/CTS from any direction and also maintain an up-to-date DNAV table.

• *Directional Data Exchange*
Once the RTS/CTS handshaking procedure has been established, the data is exchanged using directional beams.

In [13], a number of problems have been identified stemming from the use of a directional MAC protocol. While many of these problems, such as deafness and tradeoffs including those between spatial reuse and collisions are inherent with directional transmissions, one problem identified there can be resolved using steerable antennas. By choosing a beam pattern such that it has a gain just enough to reach the destination, but not much farther, the interference caused to nodes in the vicinity of the destination can be significantly reduced. This also has the added advantage of savings in terms of reduced energy required for transmission, as will be illustrated in the simulations.

4 Routing

As has been mentioned earlier, most of the effort in relation to directional antennas has been done at the MAC layer. Much less focus has been done developing routing protocols that leverage the directional information at the routing level. Below, some of the work done in this area is presented.

In [10], the authors evaluate the performance of DSR over their Directional MAC (called DiMAC). They choose to use the greater range of directional antennas, instead of their energy conservation property. In keeping with this, they propose a few optimizations to the traditional DSR algorithm. If DSR is implemented directly over a Directional MAC, then it is likely that the destination node will reply to the first route request packet which reaches it. The authors optimize DSR by making the destination wait for a short period to hear from multiple routes, before choosing the path with the lowest hop count. It can be argued, however, that this is not necessarily an optimal solution to the problem. It is because the lowest hop count might imply having to transmit over longer distances. This in turn has the dual disadvantage of consuming more energy and also causing more interference to other nodes in the neighborhood of these transmissions. Alternatively, it could be argued that using a lower hop count implies fewer nodes are involved in communications, leaving more nodes free to process other routes. Hence, choosing which route to reply to is a tradeoff which depends on factors such as energy constraints and traffic density in the network.

Another optimization proposed in [10] is to have every intermediate node forward Route Request packets to only a fraction of the nodes in its neighborhood. This can reduce congestion in the system and also conserve energy without adversely affecting the throughput or delay. However, in a highly dynamic situation, this strategy may not be effective and could lead to performance degradation.

In [11], the authors describe a network aware MAC and routing protocol to achieve load balancing using directional antennas. The concept of zone-disjoint transmissions is used to reduce overlap of communications (both in terms of nodes as well as antenna beams) and hence improve transmission. However, for this protocol to be implemented effectively, a number of tables containing location and routing-related information need to be proactively exchanged. The authors have shown that their combined MAC and routing protocol has good performance characteristics in a static scenario as compared to the DSR protocol. However, they noted that the performance dropped under mobility conditions.

Based on some of the results discussed above, it would seem that a suboptimal, but easy, approach would be to use an appropriate routing protocol with directional antennas and a directional MAC to achieve similar performance to that of routing protocols which have been specially modified for such situations. A number of papers [14,15] have been written comparing various routing protocols and analyzing their characteristics. Some of the conclusions derived by the authors are included below.

- For most scenarios, one of AODV or DSR typically performs best or at least on the same order as any protocol specifically tailor-made for such a scenario.
- In low-mobility, low-moderate traffic conditions, DSR tends to have the best performance due to its route caching and promiscuous listening feature.
- In high-mobility or heavily congested networks, AODV works well since it has a very low control overhead.
- In large networks, DSR does not scale well. Consequently, protocols such as LANMAR or ZRP which are based on hierarchical routing are preferable.

Based on these conclusions, in the simulations performed, AODV and/or DSR will be used as the routing protocols without any modifications.

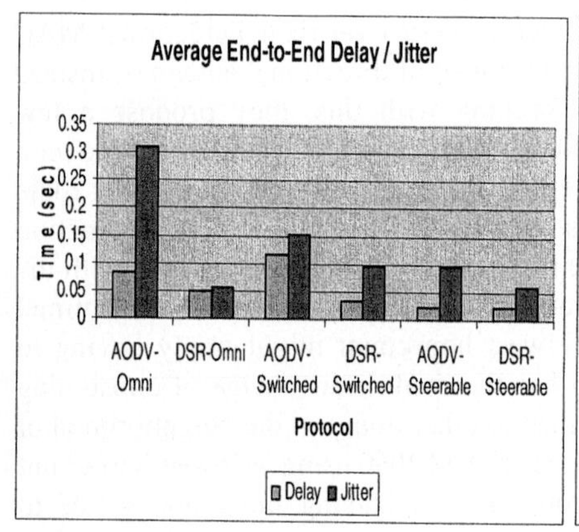

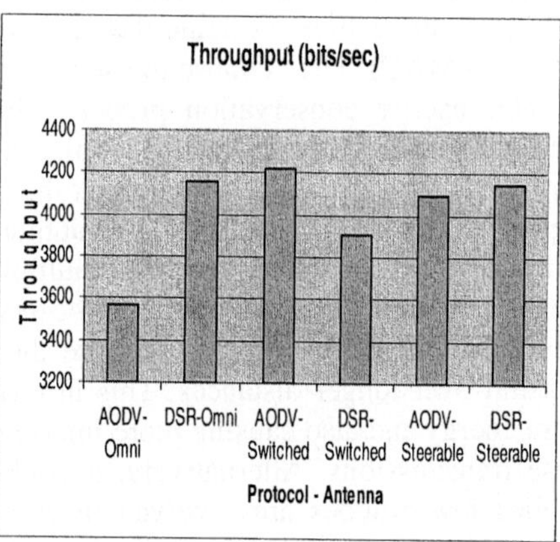

Fig. 1. Delay/Jitter and Throughput for Scenario 1

5 Performance Evaluation

In this section, the performance of the designed discussed in the past sections is presented. Based on these results, the advantages and disadvantages of this system will be presented. The simulations were performed using the QualNet simulator, version 3.7 [16].

5.1 Simulation Model

In these simulations, it is assumed that the gain of the directional antenna (as compared to an omnidirectional antenna) is 15dB. The 802.11b radio was used in these simulations, which has an omni-directional range of 250m. The 2-ray propagation model has been used. Node mobility is simulated using the random way-point model. Constant shadowing with an average of 4dB is assumed. Fading has not been taken into account for any of the simulations. The 802.11 DCF was used as the MAC layer protocol.

5.2 Performance Criteria

The performance criteria used to evaluate the various schemes include –

1) Average end-to-end delay: This includes all possible delays caused by buffering during route discovery latency, queuing at the interface queue, retransmission delays at the MAC, and propagation and transmission delays. This parameter helps in determining whether packets will be delivered within given time constraints.
2) Throughput: This is number of bits received by the server per second (excluding control data).
3) Signals Transmitted: This is the total number of signals transmitted per node. It can be used as an approximate indicator of energy consumed.
4) Packets Dropped due to Retransmission Limit: This is the number of packets that were dropped due to lack of acknowledgement from the receiver. This can be used as a measure of how well the next-hop nodes are chosen to avoid interference with other communications. The retransmission limit was set to 4 for these simulations.

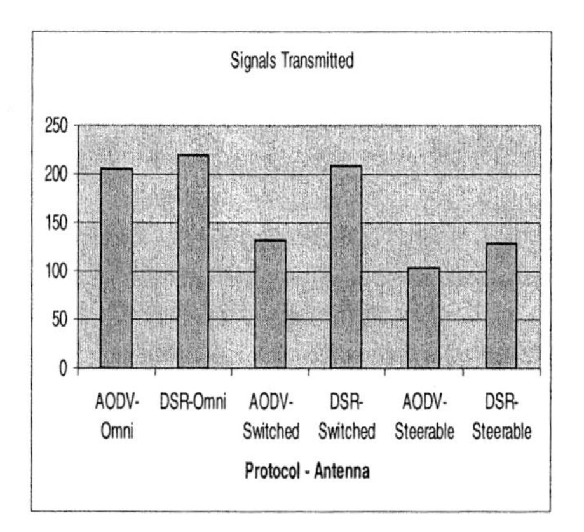

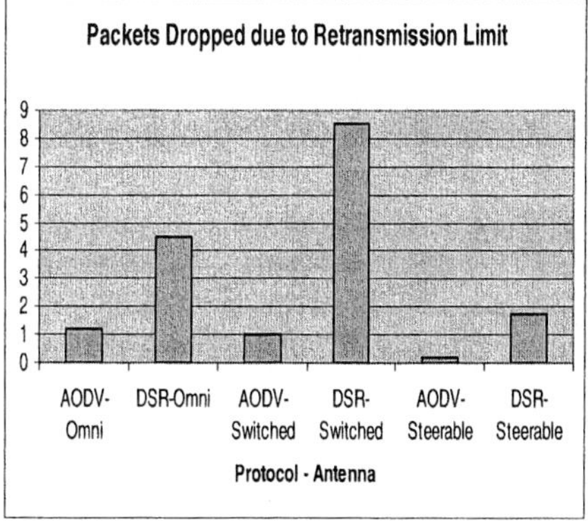

Fig. 2. Signals Transmitted and Packets dropped due to Retransmission Limit

5.3 Simulation Results

Scenario 1

This scenario was used to model a low-density, moderate mobility and moderate-traffic condition environment. 36 nodes were randomly placed in a 1500x1500m area. Up to

40% of the nodes were involved in transmissions at any given time. A random waypoint model was used as the mobility model with the speeds of nodes varying between 0-10 m/sec with up to 30sec pause time. The DNAV angle was chosen to be 45°. The results are illustrated in fig. 1.

A number of deductions can be drawn from figure 4 –

- DSR tends to outperform AODV for any type of antenna system, and for both delay/jitter and throughput. This can be explained by the nature of this scenario, as described earlier, which enables DSR to perform better since it has the ability to cache routes.
- Steerable antennas provide the best performance in terms of both delay and throughput. This can be explained by the fact that steerable antennas have many patterns to choose from. Hence, it is possible to choose an antenna pattern which either allows transmissions to more distant nodes, or reduces interference with neighboring transmissions.
- The throughput of directional antennas is at least on the order of the omni-directional antennas, though it generally performs better.
- Beam switching using AODV seems to have the best throughput in this scenario (though it has the second-highest delay/jitter characteristics). This can be explained by the fact that some of the gains of DSR over steerable beams were negated due to the low control overhead imposed by AODV.

Fig. 2 illustrates the signals transmitted per node as well as the number of packets dropped due to retransmission limit. As can be seen, beam steering requires the lowest number of total signals to be transmitted. This again can be explained by the fact that of the 3 antenna techniques, beam steering typically provides the best pattern to a particular destination. As a result, fewer packets are dropped due to mobility or interference. In addition, the pattern of signals transmitted is supported by the number of packets dropped due to retransmission limit. DSR, which uses both larger control packets, as well as route caching, has a greater number of packets dropped. This is because, route caching can cause the wrong node to be chosen as the next hop (problem of stale routes in DSR, as discussed in [9, 10]). In short, DSR is not as energy-efficient as AODV, even though it has better delay-throughput characteristics.

Scenario 2

This scenario was used to model a moderate-to-heavy density, moderate-to-high mobility and a moderate-to-heavy traffic condition environment. 100 nodes were randomly placed in a 1500x1500m area. Between 10 - 50% of the nodes were involved in transmissions at any given time. A random waypoint model was used as the mobility model with the speeds of nodes varying between 0-10 m/sec with up to 10sec pause time. The DNAV angle was chosen to be 45°.

The trends observed in fig. 3 are similar to those observed for scenario 1, with one main exception. In most cases, the performance of AODV is very similar to that of DSR. This is because, as mobility increases, the low overhead of AODV overcomes the route caching advantage of DSR.

It is also interesting to note that both directional antenna systems have almost similar performance, which is almost twice as good as the omnidirectional antenna case. It was expected that the ability of a beam-steering antenna to "follow" the destination would enable it to perform better than beam switching. However, this is not the case. This could be hypothesized by the fact that the time chosen for expiry of direction-of-neighbor information at the MAC layer was optimized for this particular scenario. Hence, the beam-switched antenna did not transmit too many signals in the wrong direction results in similar throughputs. This hypothesis is further borne out by fig. 4 for signals transmitted and packets dropped due to retransmission limit.

As can be seen from fig. 4, the number of signals transmitted and packets dropped due to the retransmission limit are almost the same for the beam steering and beam switching cases, supporting the argument made earlier.

Also, these graphs clearly illustrate the advantages that AODV has over DSR in a high mobility situation. For all cases, AODV requires fewer signals to be transmitted and has only a fraction the number of packets dropped as DSR, reasons for which have been given earlier. Also, both the directional antenna require half or less signals to be transmitted as compared to the omnidirectional case, thus illustrating the energy savings that can be obtained from using these schemes.

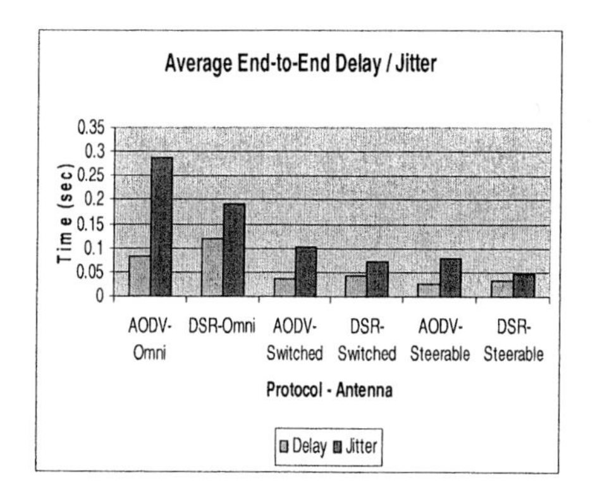

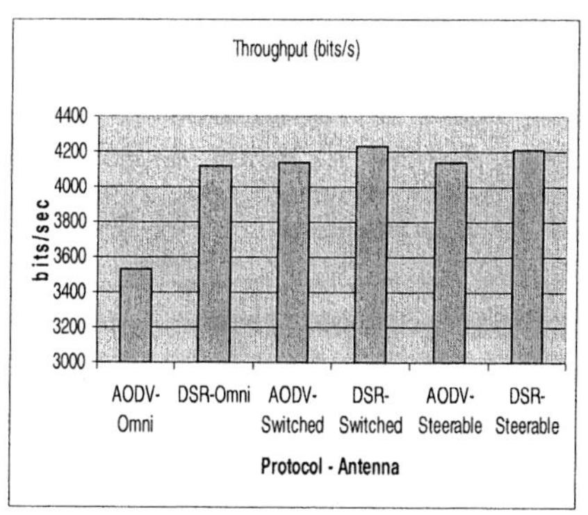

Fig. 3. Delay/Jitter and Throughput for Scenario 2

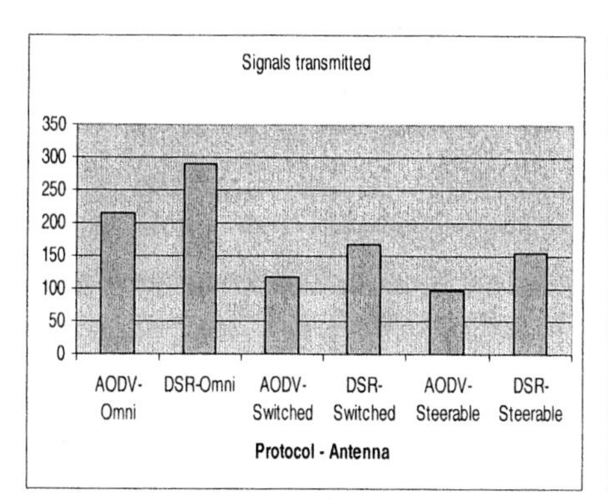

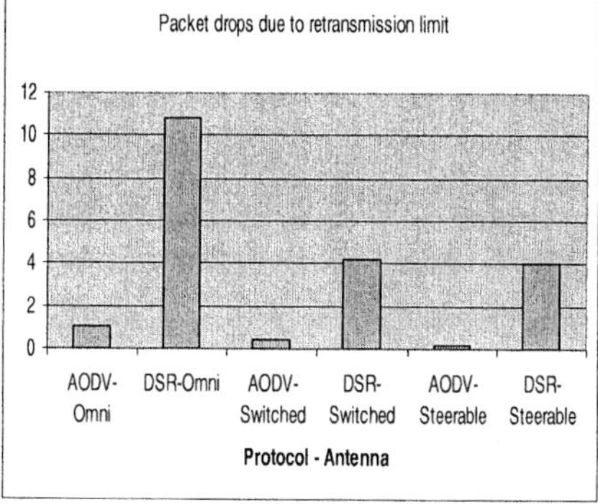

Fig. 4. Signals Transmitted and Packets Dropped for Scenario 2

6 Conclusions

Based on the simulations conducted, a number of conclusions can be drawn,

- Considerable gains can be obtained from the deployment of directional antennas.
- Directional antennas can be used either for the purpose of bridging network partitions or reducing energy consumption
- In [10], it was concluded that in static environments, directional antennas do not perform well in dense networks. However, the results obtained from this scenario do not bear out this conclusion. As was shown above, considerable gains can be obtained by deploying directional antennas even in dense networks and in mobility conditions.
- The choice of routing protocol, based on the network's characteristics, is important for extracting maximum gains.
- Bean switching typically performs a little worse than beam steering, though the overall performance is on the same order as beam steering. However, since it is easier to implement, beam switching could be considered as a "cheaper alternative" to implementing a full adaptive beam steering solution.
- The DNAV angle to be used for directional 802.11-based MACs should be set close to the beam-width of the antenna pattern for optimal performance.
- Conventional routing protocols for ad hoc networks using directional MACs have performance gains similar to those reported for routing protocols optimized for directional transmissions.

While a number of useful results have been obtained, many research challenges still exist in this area. Foremost would be the development of a routing protocol that is able to harness the power of directional antennas without significantly increasing the overhead (and thus negating any gains obtained). In addition, the MAC protocol can be optimized to better support long-range transmissions to bridge network partitions. It would also be interesting to see if it would be possible for a single node to support multiple communications using different elements of a sectored antenna. The tradeoffs in terms of energy consumption and interference versus reduced latency will be useful to observe. Finally, efforts should be made to translate information on network congestion, as determined by packets dropped or other metrics, into useable information at the application layer. Consequently, in order to meet constraints such as latency or throughput, other factors such as coding and compression could be altered accordingly.

References

[1] Goldsmith, A. J., Wicker, S.B.: Design Challenges for Energy-Constrained Ad Hoc Wireless Networks. IEEE Wireless Commun. Mag., 9 (2002) 8-27

[2] Shakkottai, S., Rappaport, T. S., Karlsson, P. C.: Cross-layer Design for Wireless Networks. IEEE Commun. Mag., 41 (2003) 74-80

[3] Ayyagari, D., Michail, A., Ephremides A.: Unified Approach to Scheduling, Access Control and Routing for Ad-hoc Wireless Networks. IEEE VTC, 1 (2000) 380-384

[4] Barrett, C., Marathe, A., Marathe, M. V., Drozda, M.: Characterizing the Interaction between Routing and MAC Protocols in Ad-hoc Networks. ACM MobiHoc (2002) 92-103
[5] Santhanam, A. V., Cruz, R. L.: Optimal Routing, Link Scheduling and Power Control in Multi-hop Wireless Networks. IEEE INFOCOM, 1 (2003) 702-711
[6] Toumpis, S., Goldsmith, A. J.: Performance, Optimization and Cross-layer Design of Media Access Protocols for Wireless Ad hoc Networks. IEEE ICC, 3 (2003) 2234-2240
[7] Kozat, U. C., Koutsopoulos I., Tassiulas, L.: A Framework for Cross-layer Design of Energy-efficient Communication with QoS Provisioning in Multi-hop Wireless Networks. IEEE INFOCOM, 23 (2004) 1446-1456
[8] Leiner, B., Nielson, D., Tobagi, F.: Issues in Packet Radio Network Design. Proceedings of the IEEE, 75 (1987) 6-20
[9] Ramanathan, R.: On the Performance of Ad Hoc Networks with Beamforming Antennas. ACM MobiHoc (2001) 95-105
[10] Choudhury R.R., Vaidya, N. H.: Impact of Directional Antennas on Ad Hoc Routing. ICPWC (2003) 590-600
[11] Roy, S., Saha, D., Bandyopadhyay, S., Ueda T., Tanaka, S.: A Network-Aware MAC and Routing Protocol for Effective Load Balancing in Ad Hoc Wireless Networks with Directional Antenna. MobiHoc (2003) 88~97
[12] Nasipuri, A., Ye, S., You J., Hiromoto, R.E.: A MAC Protocol for Mobile Ad Hoc Networks Using Directional Antennas. IEEE WCNC, 3 (2000) 23-28
[13] Choudhury, R. R., Yang, X., Ramanathan R., Vaidya, N.: Using Directional Antennas for Medium Access Control in Ad Hoc Networks. MobiCom (2002) 59-70
[14] Royer, E.M., Toh, C.K.: A Review of Current Routing Protocols for Ad Hoc Mobile Wireless Networks. IEEE Personal Communication, 6 (1999) 46-55
[15] Perkins, C.E., Royer, E.M., Das, S.R., Marina, M.K.: Performance Comparison of Two On-demand Routing Protocols for Ad Hoc Networks. IEEE Personal Communication, 8 (2001) 16-28
[16] QualNet Simulator, Version 3.7, Scalable Network Technologies, www.scalable-networks.com

A Novel Media Access Control Algorithm Within Single Cluster in Hierarchical Ad Hoc Networks

Dongni Li[1] and Yasha Wang[2]

[1] School of Computer Science Technology, Beijing Institute of Technology,
Beijing 100081, China
ldn@bit.edu.cn
[2] Institute of Software Engineering, School of Electronics Engineering and Computer Science,
Peking University, Beijing 100871, China
wangys@sei.pku.edu.cn

Abstract. Media Access Control (MAC) is one the most critical issues for wireless ad hoc networks. In a hierarchical ad hoc network, all the nodes in the same cluster share the wireless channel. The current MAC algorithms can hardly adapt to both light and heavy traffic loads, thus cannot perform well with remarkably and frequently changing traffic load in ad hoc networks in which nodes keep moving in and out of clusters. A multi-token MAC (MTM) algorithm is proposed in this paper. It is used within single cluster in hierarchical ad hoc networks. It can automatically compromise between CSMA/CA in IEEE 802.11 and the token scheduling in cluster-head-gateway switching routing (CGSR) algorithm. Simulation results show that in the ad hoc network with active nodes moving in and out of clusters, MTM gives better throughput ratio and average packet delay.

1 Introduction

In hierarchical ad hoc networks, nodes are aggregated in clusters. All nodes in a cluster can communicate with a cluster-head and possibly with each other [17]. Media Access Control (MAC) is one of the most important issues in ad hoc networks. Across clusters, frequency or code division can be adopted to access the wireless channel. Within a cluster, special MAC algorithms should be used to allocate the shared channel among competing nodes [6], [17]. Usually, there are two types of MAC algorithms in wired or wireless local area networks, referred to as competition algorithms [8], [9], [10], [11] and queue algorithms [2], [4], [5] in this paper. In a competition algorithm, nodes attempting to transmit randomly compete for the shared channel. If a node wins the competition, it has the right to transmit data at once; otherwise it will participate in the next competition after waiting for a random amount of time. A typical example of the competition algorithms is carrier-sense multiple access with collision avoidance (CSMA/CA) in IEEE 802.11. In a queue algorithm, there is a polling among all the nodes according to some order, and the node transmits only when in its turn. A typical example of the queue algorithms is the one proposed for the cluster-head-gateway switching routing (CGSR) algorithm [17], hereafter referred to as TSC (token scheduling in CGSR), which is similar to the Token Ring in IEEE 802.5.

X. Jia, J. Wu, and Y. He (Eds.): MSN 2005, LNCS 3794, pp. 986–995, 2005.

Generally, when the network load is light, the competition algorithms perform better than the queue type with respect to the aggregate throughput over all flows in the network [3], [1], because under this situation the channel is mostly idle. In a competition algorithm, the nodes can transmit without any delay [14], [7], however, the nodes in a queue algorithm have to wait for its turn even if the channel is available. On the other hand, when the network load is heavy, the queue algorithms give better aggregate throughput. This is because heavy load results in many collisions in competition algorithms. However, the queue algorithms enable the nodes to transmit in turn under the control of some scheduling schemes, thus avoiding collisions and increasing the channel utilization [12], [13].

Generally, when we design a local area network, the number of nodes and network traffic are estimated first, and then the result is used to decide which type of MAC algorithms should be deployed. However, unfortunately, this method is not suitable for an ad hoc network, mainly due to the following two reasons:

1. The distribution of nodes in an ad hoc network is often uneven, and this results in different node densities in different clusters (i.e. the number of nodes contained in a cluster differs from each other.). A cluster with more nodes generally has heavier load on it. If we select the same MAC algorithm for all clusters in advance, no matter competition or queue, the performance of the whole network cannot be optimized. On the other hand, if we select different MAC algorithms for different clusters according to their loads, the problem cannot be solved either. This is because in ad hoc networks, nodes may move to other clusters, two clusters may incorporate into one cluster, or one cluster may split into two clusters, all of above situations may lead to changes in cluster node densities. The MAC algorithm chosen in advance cannot always meet the requirements of such changes.
2. If we implement two algorithms on every node, i.e. a competition algorithm and a queue algorithm, and make each node switch the algorithms according to the current cluster load, there will be some new problems. First, the two-MAC-algorithm configuration enhances the algorithm stack complexity of nodes; second, when the node switches from one algorithm to another, this sudden change in MAC layer may results in negative impact on the application's QoS (quality of service); third, when the load of a cluster is between light and heavy, the decision of choosing algorithms is hard to make.

In order to solve the above-mentioned problems in ad hoc networks, a novel MAC algorithm is proposed in this paper. It compromise between CSMA/CA and TSC according to the varying network loads. It can be deployed by all the clusters in a hierarchical ad hoc network. Under light traffic load, it is more like CSMA/CA; as the traffic load increases, it tends towards TSC.

It should be noticed that MTM is used within single cluster, while across clusters CDMA or FDMA can be deployed, and this is beyond the scope of this paper.

The rest of the paper is organized as follows. Section II presents our proposed multi-token MAC algorithm (hereafter referred as MTM). Section III discusses some key elements of MTM. Section IV describes the simulation model and discusses the simulation results. Section V concludes the paper in the end.

2 Proposed Multi-token MAC Algorithm

MTM is proposed to solve the aforementioned problems within a single cluster where all the nodes share the same channel. We model a hierarchical ad hoc network as described in [17]. All nodes in a cluster can communicate with the cluster-head and they are all 1-hop from the cluster-head. Only the nodes holding tokens have the chances to transmit and the tokens are controlled by the cluster-head. Initially, every node in the cluster is allocated a token and the nodes (include cluster-head and cluster-members) begin to detect the network load in the cluster. There are three states with respect to the network load: *heavy*, *normal* and *light*. According to the detection results of the cluster-head, if the network load is *heavy*, the cluster-head will remove some tokens until the network load state turns to be *normal* or only one token is active in the cluster; otherwise, if the network load is *light*, the cluster-head will insert more tokens in the cluster until the network load turns to be *normal*, or every node holds a token. Suppose the extreme situations: if every node has a token, MTM is actually CSMA/CA; while if there is only one node in this cluster, MTM is actually TSC. MTM can be divided into two parts: determination of network load state and management of tokens.

2.1 Determination of Network Load State

MTM allows multiple tokens exist at the same time, and CSMA/CA is deployed among these competing tokens. As specified in CSMA/CA, in case of collision, a node backs off for a random amount of time. If collisions happen more and more frequently, the back-off time gets longer as a consequence. This means that the network load is getting heavier and the transmission delay is getting longer. So we calculate for each node the weighted average of the number of collisions it encountered recently, and take it as the measure of the network load state.

Every node in the cluster maintains the following set of variables:

1. Collision number (denoted a_i), which records the number of collisions it encountered while attempting to transmit a frame. For instance, a_0 denotes the number of collisions it encountered during transmitting the current frame; a_1 denotes the number of collisions during transmitting the last frame; a_i denotes the number of collisions during transmitting the last i-th frame.
2. Collision number window (denoted w), which denotes the number of a_is contributing to the weighted average of the number of collisions.
3. Weight vector (denoted V). $V = (\theta_0, \theta_1, \cdots, \theta_{w-1})$, where $\sum_{i=0}^{w-1} \theta_i = 1$, $\theta_0 \geq \theta_1 \geq \cdots \geq \theta_{w-1} \geq 0$.
4. Weighted average of the number of collisions (denoted C), denotes while attempting to transmit a frame, how many collisions the node encountered averagely.

$$C = \sum_{i=0}^{w-1} (a_0 \theta_0) \tag{1}$$

5. Threshold of light load (denoted T_l) and threshold of heavy load (denoted T_h). It is obvious that $T_h \geq T_l > 0$. They both denote the number of collisions during transmitting a frame.

While transmitting frames, the node counts the number of collisions. Whenever the node has performed a successful transmission, or the transmission has been canceled (this is because the number of retransmissions goes beyond a limit), it calculates C according to (1). If $C > T_h$, then the network load state is considered heavy; if $C < T_l$, then the network load is considered light; if $T_h \geqslant C \geqslant T_l$, then the network load is considered normal.

2.2 Token Management

MTM defines two types of tokens: time-limiting token and non-time-limiting token. The holder of a time-limiting token can hold the token for a finite amount of time, denoted τ. When τ expires, the node must return the token to the cluster-head

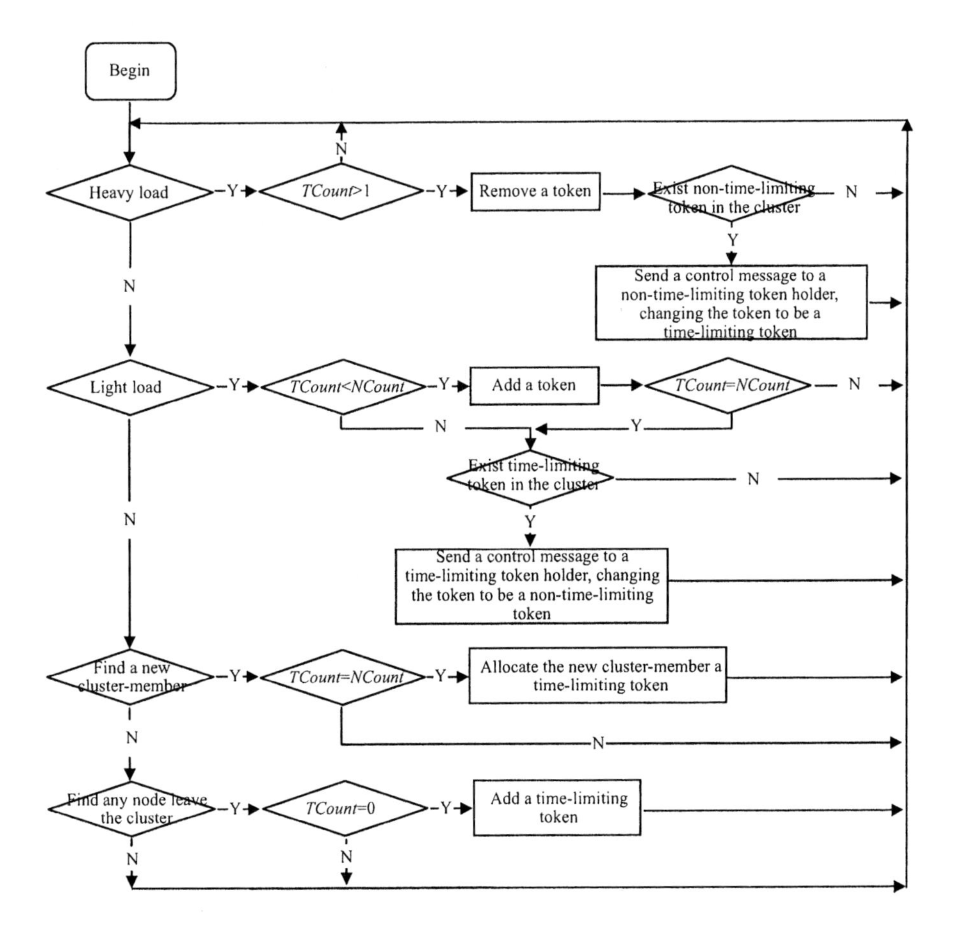

Fig. 1. Management of tokens performed by a cluster-head

regardless of whether it still wants to transmit. The finite amount of time τ is called token time limit. The holder of a non-time-limiting token can always hold the token (no matter whether it wants to transmit) until it receives a token taking-back packet (hereafter referred as TTP) from the cluster-head.

The process of the multi-token management is shown in Fig.1, where *TCount* denotes the total number of tokens (held either by the cluster-head or by the cluster-members) in the cluster and *NCount* denotes the total number of nodes. Initially, the cluster-head allocates each cluster-member a non-time-limiting token, and for each node in the cluster (including the cluster-head), C and the set of a_i are set to zero. So at the beginning, every node has a token and they compete for the channel by means of CSMA/CA. It is necessary to point out that, on allocated a token, a cluster-member cannot add or remove any tokens. Only the cluster-head can do this. It means that the cluster-head can change the total number of tokens in the cluster (by the operations of removing or adding tokens), while the cluster-members cannot.

As shown in Fig.1, the cluster-head removes a token whenever it found the network load is heavy and the current number of tokens is more than one. The process of removing a token is indicated in Fig. 2. As shown in Fig. 2, the cluster-head checks whether there exist any tokens that have not been allocated yet. If there exist any such tokens, the cluster-head will remove one of them. If such tokens do not exist, the cluster-head will wait until a time-limiting token returns, and then remove it. If time-limiting tokens do not exist, the cluster-head will send a TTP to a non-time-limiting token holder, asking it to return the token, and then remove it. Such sequence performed in the process can reduce the overhead of control messages and the algorithm complexity.

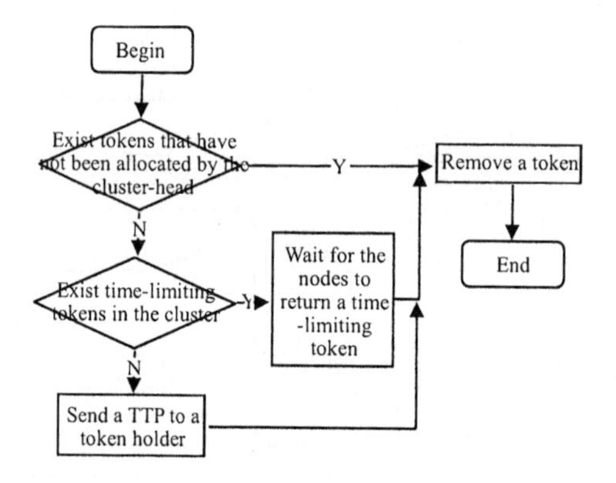

Fig. 2. The process of the cluster-head removing a token

3 Discussions

3.1 Why Choosing *C* to Measure the Network Load State

Taking C as the measure of network load state benefits from the following three aspects:

1. Simplicity. The nodes only need to record the number of collisions it encountered while transmitting the last w frames, i.e. (a_0, a_1,...,a_w), and then calculate C through simple computation, see formula (1).
2. Low overhead. The nodes do not need to transmit any special massages to detect the network load, saving bandwidth overhead. Moreover, the nodes do not need to sense the channel all along, so that they can go to the doze state when necessary, saving battery energy consumption.
3. Proper process. While attempting to transmit, the node updates C, and determines the network load state based on the updated C. This process benefits from two aspects: (a) A node begins the process of determining the network load state only when it is about to transmit, and it transmits only when it is holding a token, thus on one hand it can handle the token duly, on the other hand, for those nodes that are not holding tokens, they do not need to determine the network load state, saving their computation overhead and battery energy; (b) The more frequently a node transmits, the more frequently it updates C, thus it can reflect the current network load state better than a node transmitting less frequently. As specified in CGSR, for a cluster, all the packets going across it must be forwarded via the cluster-head, meanwhile the cluster-head need to periodically advertise some control messages, such as routing advertisements, cluster-member advertisements, etc. A cluster-head is usually the node transmitting the most frequently and as the manager of tokens, it has the highest probability to get tokens, so that it can indicate the current network load state better than other nodes in the cluster. This characteristic very matches the requirement in MTM that as the controller of tokens, the cluster-head must learn the real-time network load information as much as possible.

3.2 Why the Cluster-Members Also Need to Determine the Network Load State

One of the most important characteristics of ad hoc networks is the frequent dynamic reorganization. In a hierarchical ad hoc network, the situations such as the cluster-head becoming invalid, the cluster-head departing from the current cluster, clusters reorganizing, etc., occur from time to time. These may incur the re-election of cluster-head. Letting all the cluster-members perform determining the network load state is to prepare for the quick cluster-head switch.

3.3 Why Defining Two Types of Token

In order to explain this point, we assume the following two situations:

1. If all the tokens are non-time-limiting tokens
When the number of tokens is less than the number of nodes, in order to give the nodes not holding tokens the chances to transmit, the current token holders must release tokens according to some rules. If all the tokens are set to non-time-limiting, the cluster-head must send TTPs, asking the holders to return tokens. In case of heavy load, the number of tokens is relatively less, thus the cluster-head must transmit TTPs frequently and the tokens transfer frequently, increasing transmission overhead. In contrast, time-limiting tokens require the holders to automatically return them after a finite amount of time, thus simplifying the cluster-head's work.

2. If all the tokens are time-limiting tokens
Assume the situation that the network load is light enough to enable every node to hold a token. If all the tokens are time limiting, after some time the cluster-members must return it to the cluster-head. In this case the only thing the cluster-head need to do is to send them back to the cluster-members again. This kind of interactions is meaningless. On the contrary, it will increase transmission delay and overhead. In contrast, if in this situation the tokens are non-time-limiting, the nodes will compete for the channel by means of CSMA/CA, thus reducing the transmission overhead and shortening the transmission delay.

Time-limiting tokens and non-time-limiting tokens fit heavy load and light load respectively, so that it is necessary to define the two types of tokens.

4 Performance Evaluation To

We have simulated the proposed MTM and the CSMA/CA, as well as the TSC. Two metrics are used to evaluate the MAC algorithms: throughput ratio over all flows and average packet delay over all flows in the simulated cluster.

4.1 Simulation Model

We use ns-2 for our simulations. The simulator runs on a SUN workstation, and the operating system is SunOs5.7. Each simulation is performed for a duration of 20 seconds. The simulated cluster's maximal size is 50 nodes for a hierarchical ad hoc network. The nodes uses omni-directional antenna and the transmission mode is two-ray ground reflection. The channel bit rate is 11 Mbps and the transmission range is 400 m. The nodes in the simulated cluster are randomly placed in a 260 X 260 m^2 area, and continuously move at the rate of up to 20 *m/s*. To monitor the global throughput at the application level□we set application-level data flow from the sources to the destinations. The routing protocol on network level is set to dynamic source routing (DSR) protocol. A source node generates and transmits const bit rate (CBR) traffic. The packet size of each flow is fixed at 512 bytes unless otherwise specified. The parameters in MTM are set as follows:

- Collision number window: $w = 6$
- Weight vector: $V = (0.4, 0.2, 0.1, 0.1, 0.1, 0.1)$
- Threshold of light load: $T_l = 0.1$
- Threshold of heavy load: $T_h = 1$

All the simulation results are averages over 20 runs.

4.2 Simulation Results

The graphs in this paper show three curves labeled as CSMA/CA, TSC and MTM. The curve labeled as CSMA/CA corresponds to IEEE 802.11 CSMA/CA. the curve labeled as TSC indicates the token scheduling in CGSR. The curve labeled as MTM is for the scheme proposed in this paper.

Varying the Number of Flows. In order to simulate the changes in node density caused by the nodes' movement, we put 30 nodes in our simulated cluster, and make them silent initially. After that, we let 20 source nodes enter this cluster one by one gradually. Each source node transmits a packet every 10ms, and all destinations are in this cluster. Fig. 3 shows the throughput ratio varying the network load. As indicated in Fig. 3, when the network load is low, for instance initially each node holds a token, the throughput ratio of MTM is quite similar to that of CSMA/CA. As the network load increases, the throughput ratio of CSMA/CA degrades severely. This is mainly due to the overhead of competition. The throughput ratio of MTM also degrades, but the degradation is not as significant as CSMA/CA. This is because as the number of tokens reduces, the competition becomes less severe, and more bandwidth can be used for data transmissions. As the network load go on increasing, the number of tokens reduces until finally to 1, and MTM performs more and more similarly to TSC. This is because as the tokens are removed by the cluster-head, MTM tends towards TSC.

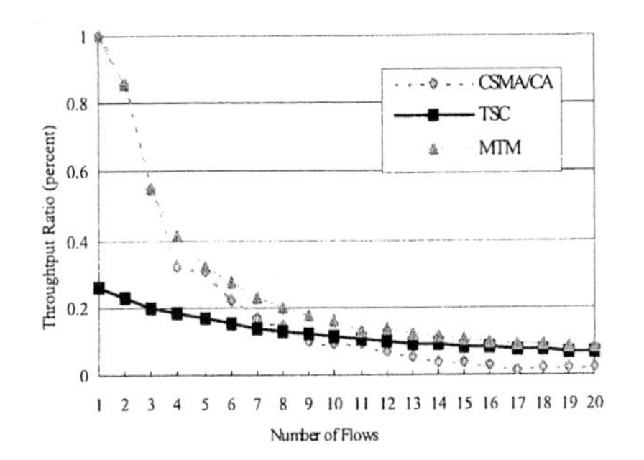

Fig. 3. Throughput ratio with fixed packet size

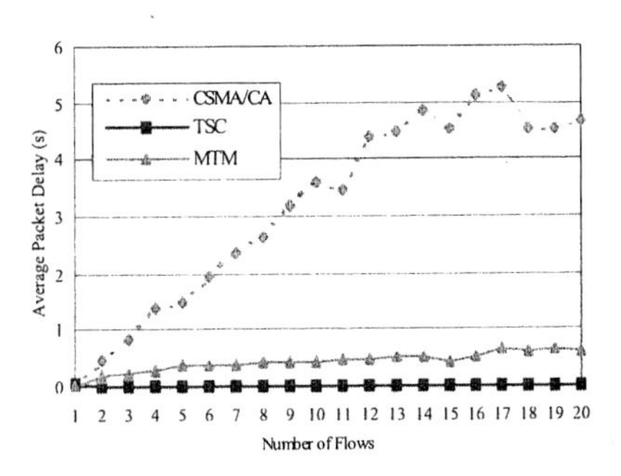

Fig. 4. Average delay with fixed packet size

Fig. 4 shows the average packet delay varying number of flows. As shown in Fig. 4, the packet delay of CSMA/CA increases significantly as the number of flows increases, while the packet delay of MTM changes moderately all along. Especially when the number of flows is more than two, the packet delay of MTM is much shorter than that of CSMA/CA. It is observed from Fig. 4 that although MTM and TSC perform similarly, the packet delay of MTM is always slightly longer than that of TSC. This is because on receipt of a token, in TSC the node can transmit at once, while in MTM, the node has to sense the channel for a while.

Varying the Traffic of Each Flow. If the number of flows is fixed, varying the traffic of each flow also results in the network load changing, thus we change the traffic of each flow by increasing the packet size, and see how the throughput ratio and the average packet delay tend in this case. Fig. 5 and 6 show the throughput ratio and the average packet delay varying packet sizes, respectively. The simulated packet sizes are 128, 256, 512 and 1024 bytes. Each source node transmits 20 packets per seconds. We simulated three schemes with 10, 15 and 20 flows respectively. As the packet size increases in Fig.5, the throughput ratio of all schemes degrades. However, the throughput ratio of MTM is always much higher than, or similar to, that of CSMA/CA or TSC. Especially under lower network load, MTM performs similarly to CSMA/CA, and much better than TSC (see Fig.5 (a)); under higher network load, it performs similarly to TSC, and much better than CSMA/CA (see Fig. 5 (c)).

As the packet size increases in Fig. 6, the average packet delays of CSMA/CA and TSC increase, however as the network load becomes higher, the increasing rate of MTM becomes much slower than that of CSMA/CA. This is mainly because in MTM the number of collisions is restricted with the number of tokens, which is dynamically changed with the network load. The average packet delay of TSC always keeps very short regardless of the network load. This is mainly because the existence of the single token avoids collisions effectively.

This kind of situation (the number of flows is relatively fixed, but the traffic of each flow changes, leading to the total load of the channel changing) also happens to the wireless networks without dynamic topology changes, thus in addition to ad hoc networks, MTM also performs well in other wireless networks.

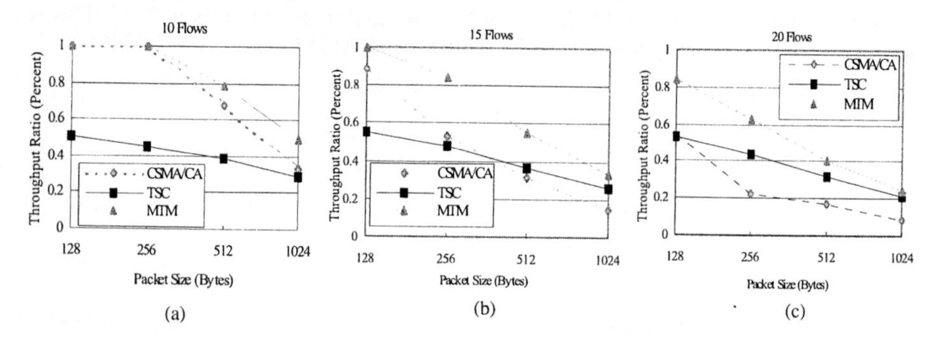

Fig. 5. Throughput ratio with different packet sizes

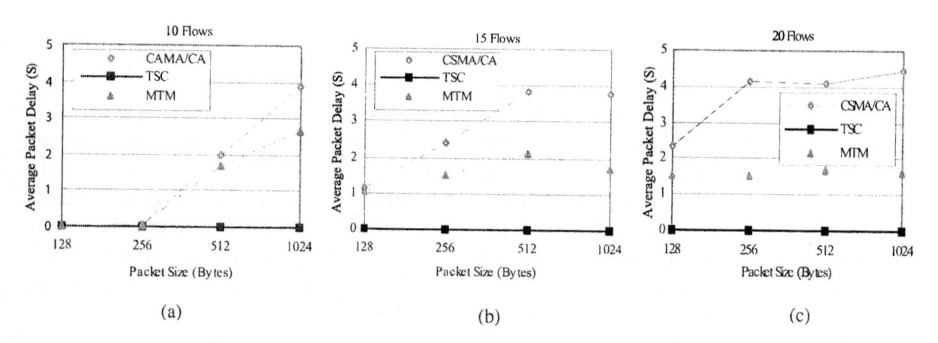

Fig. 6. Average packet delay with different packet sizes

5 Conclusion

We have presented the multi-token MAC algorithm. It can automatically adjust itself between the competition algorithm (taking IEEE 802.11 CSMA/CA as an example in this paper) and the queue algorithm (taking TSC as an example) according to the varying network load. In case of light load, MTM tends toward CSMA/CA. As the network load increases, MTM tends toward TSC. Simulation results confirm that in a hierarchical ad hoc network with significantly varying traffic loads, MTM gives better performance on the whole, as compared to CSMA/CA and TSC.

References

1. J. Broch et al.: A Performance Comparison of Multi-Hop Wireless Ad Hoc Network Routing Protocols,Proc. Mobicom '98.
2. N. Malpani, N. Vaidya, and J. Welch: Distributed token circulation in mobile ad hoc networks. In Proc. 9th International Conference on Network Protocols (2001).
3. Mustafa Ergen, Duke Lee, Raja Sengupta, Pravin Variaya: Wireless Token Ring Protocol (WTRP) Performance Comparison with IEEE 802.11 In ISCC, Antalya (2003)
4. D. Lee, R. Attias, A. Puri, R. Sengupta, S. Tripakis, P.Varaiya: A Wireless Token Ring Protocol For Ad-Hoc Networks, 2002 IEEE Aerospace Conference Proceedings, BigSky, Montana, USA.(2002)
5. M. Ergen, D. Lee, R. Attias, A. Puri, R. Sengupta, S. Tripakis, P. Varaiya: Wireless Token Ring Protocol, SCI Orlando(2002)
6. D. Maniezzo, G. Pau, M. Gerla, G. Mazzini, K. Yao: T-MAH: A Token Passing MAC protocol for Ad Hoc Networks, MedHocNet2002, Chia, Sardegna, Italy(2002)
7. G. Bianchi: Performance Analysis of the IEEE 802.11 Distributed Coordination Function, IEEE Journal on Selected Areas in Communications, Vol.18, No. 3 (2000).
8. Y.Wang: Achieving Fairness in IEEE 802.11 DFWMAC with Variable Packet Size, Global Telecommunications Conference (2001).
9. T. Pagtzis, P. Kirstein, S. Hailes: Operational and Fairness Issues with Connection-less Traffic over IEEE802.11b, ICC (2001).
10. R. Akester and S. Hailes: LA-DCF: A New Multiple Access Protocol for Ad-Hoc Wireless Networks, London Communications Symposium 14 th – 15 th (2000).
11. R. Bruno, M. Conti, E. Gregori: WLAN technologies for Mobile ad hoc Networks, In Proceedings of the 34th Hawaii International Conference on System Sciences, Hawaii(2001)
12. F.Calì, M. Conti, E. Gregori: IEEE 802.11 protocol: design and performance evaluation of an adaptive backoff mechanism. IEEE Journal Selected Areas in Comm. (2000).
13. L. Bononi, M. Conti, E. Gregori: Design and Performance Evaluation of an Asymptotically Optimal Backoff Algorithm for IEEE 802.11 Wireless LANs, Proc. HICSS-33, Maui, Hawaii (2000).
14. Bianchi G, Fratta L, Oliveri M.: Performance evaluation and enhancement of the CSMA/CA MAC protocol for 802.11 wireless LANs[A], Proc. PIMRC[C] (1996):392-396.
15. IEEE Standard for Information technology--Telecommunications and information exchange between systems--Local and metropolitan area networks--Specific requirements--Part 5: Token Ring Access Method and Physical Layer Specification, ISO/IEC 8802-5:1998 (IEEE Std 802.5), 1998.
16. Chiang C C, Gerla M.: Routing and Multicast in Multihop, Mobile Wireless Networks, Proc. ICUPC'97[C] (1997):197-211.

IEE-MAC: An Improved Energy Efficient MAC Protocol for IEEE 802.11-Based Wireless Ad Hoc Networks

Bo Gao[1], Yuhang Yang[1], and Huiye Ma[2]

[1] Dept. of Electronic Engineering, Shanghai Jiao Tong University, Shanghai, China
gaobo@sjtu.edu.cn

[2] Dept. of Computer Science and Engineering, The Chinese University of Hong Kong
hyma@cse.cuhk.edu.hk

Abstract. Power consumption becomes a primary issue since wireless devices are often powered by batteries. This paper presents IEE-MAC, an improved energy efficient MAC protocol designed for IEEE 802.11-Based wireless Ad Hoc networks. In IEE-MAC, every node constructs a scheduling table in the distributed way by exchanging control information in ATIM window. Every node will only wake up to transmit or receive data frames and then enter doze state following the scheduling table to conserve energy consumption. Using this mechanism, IEE-MAC can eliminate the collision, overhearing, and idle listening. Moreover, IEE-MAC can adjust the size of ATIM window adaptively according to the actual traffic load in order to reduce unnecessary energy consumption without degrading the throughput of the network. Simulation results show that our protocol attains the better energy efficiency as well as throughput than other protocols in the literature.

1 Introduction

In recent years, wireless networks have received more and more attention because they can provide more flexible and convenient connection than wired networks. Wireless devices are often powered by batteries. Since batteries can provide a finite amount of energy, it is important that each device reduces energy consumption to increase its lifetime. Past research has investigated energy conserving mechanisms at various layers of the protocol stack, including work on routing [1, 2, 3], medium access control (MAC)[4, 5, 6], and transport protocols [7, 8]. Along with many approaches addressing energy conserving issue at various layers of the network protocols stack, this paper designs an improved energy-efficient medium access control (MAC) protocol for wireless networks.

The IEEE 802.11 MAC [9] is one of the most widely used medium access protocols in wireless LANs. In addition to specify Distributed Coordination Function (DCF) for contention-based medium access methods, it also provides a Power Saving Mode (PSM) for ad hoc mode called Independent Basic Service

X. Jia, J. Wu, and Y. He (Eds.): MSN 2005, LNCS 3794, pp. 996–1005, 2005.

Set (IBSS). In IBSS, a set of mobile nodes can communicate each other under no underlying infrastructure such as base station. In the IEEE 802.11 Power Saving Mode (PSM), a node can be in one of two different power modes i.e., active mode when a node can transmit or receive frames at any time and power save mode (PS) when a node is mainly in low-power state and transits to full powered state subject to the rules described next. The low-power state usually consumes at least an order of magnitude less power than in the active state.

In the Power Saving Mode (PSM), time is divided into beacon intervals and all nodes are synchronized with each other. At the start of each beacon interval, each node contends to send a beacon for time synchronization with each other and wakes up for an ATIM window (Ad-hoc Traffic Indication Message window) interval. If a node has buffered frames to PS nodes, it will send an ATIM frame to the PS nodes within the ATIM window period. On receiving the ATIM frame, the PS node responses an ATIM_ACK and both the sender and receiver will keep awake for the whole beacon interval. The other nodes which do not send nor receive any ATIM frame during an ATIM window, will go to the doze state for the rest of the beacon interval. After the end of ATIM window, all nodes which successfully exchange the ATIM and ATIM_ACK frames in ATIM window will follow the normal DCF access procedure to transmit their data frames.

Figure 1 illustrates an operating example of the IEEE 802.11 Power Saving Mode (PSM) in IBSS. Station A, B and C wake up almost at the same time, TBTT (Target Beacon Transmission Time) which is the beginning of each beacon interval. If station A wants to transmit frames to station B, an ATIM frame is sent from A to B during the ATIM window. Station B replies an ATIM_ACK to A. After the ATIM window ends, both station A and B remain in the active state and exchange their data frames using the DCF procedure. Because station C does not send or receive any ATIM frame, it stays in the doze state after the ATIM window. All dozing stations wake up again at the beginning of the next ATIM window.

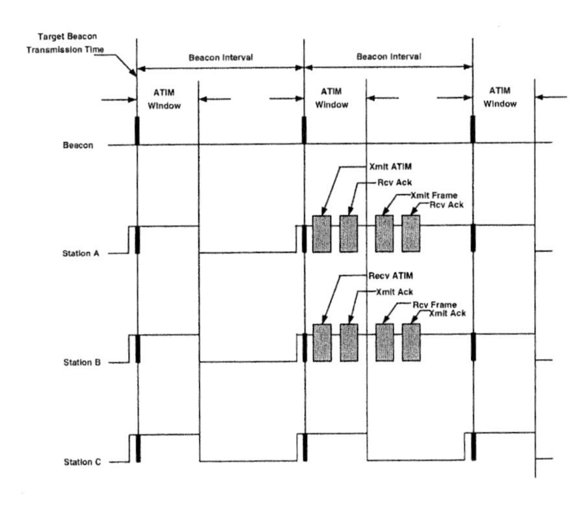

Fig. 1. Basic Operation of the IEEE 802.11 PSM in IBSS

However, when running IEEE 802.11 PSM MAC protocol on an Ad Hoc network, much energy is wasted due to the following source of overhead. The first one is collision. If two nodes transmit at the same time and interfere with each other's transmission, packets are corrupted. Hence the energy used during transmission and reception is wasted. The second source is overhearing, meaning that a node picks up packets that are destined to other nodes. The third source is idle listening. In PSM specified in IEEE 802.11, when a node transmits or receives an ATIM frame during an ATIM window, it must stay awake during the entire beacon interval. At low loads, this approach results in a much higher energy consumption than necessary. Moreover, in PSM specified in IEEE 802.11, all nodes use the fixed ATIM window size. Since the ATIM window size critically affects throughput and energy consumption, a fixed ATIM window does not perform well in all situations, as shown in [10].

In this paper, we propose an improved energy efficient MAC protocol (IEE-MAC) explicitly designed for IEEE 802.11-Based Ad Hoc networks. IEE-MAC not only eliminates the collision, overhearing, and idle listening, but also adjusts the ATIM window size adaptively. While reducing energy consumption is the primary goal in our design, IEE-MAC also achieves good throughput. This mechanism is achieved by building a scheduling table at each node and keeping a separate timer by each node.

The rest of the paper is organized as follows. In Section 2, we briefly review the related work. Section 3 presents our proposed protocol. Section 4 evaluates performance of the proposed protocol by simulations. At last, we make some conclusions in Section 5.

2 Related Work

Woesner *et al.* [10] presented simulation results for the power saving mechanisms of two wireless LAN standards, IEEE 802.11 and HIPERLAN. It showed the different sizes of beacon intervals and ATIM windows in IEEE 802.11 had a significant impact on throughput and energy consumption.

Krashinsky *et al.* [8] showed that the IEEE 802.11 PSM increased round trip times and proposed a PSM that dynamically adapted to network activity. Jung *et al.* [11] proposed a method that dynamically adjusted the size of ATIM window and allowed nodes to stay awake for only a fraction of the beacon interval to save more energy. In [12], Choi *et al.* presented another variation of IEEE 802.11 PSM.

In [13], Wu *et al.* proposed an energy efficiency MAC protocol for IEEE 802.11 networks by scheduling transmission after the ATIM window and adjusting the ATIM window dynamically to adapt to the traffic states. But they did not consider the energy consumption caused by overhearing among neighboring nodes.

PAMAS [1] made an improvement on energy saving by trying to avoid the overhearing among neighboring nodes. To achieve this goal, it required two independent radio channels, which in most cases indicated two independent radio systems on each node. Similar to PAMAS, S-MAC [14] allowed nodes to sleep during neighbors' transmissions; nodes went to sleep after hearing an RTS or a CTS destined for neighbors. However, it is dedicated to sensor network with low load and long latency.

3 Proposed IEE-MAC (Improved Energy Efficient MAC) Protocol

As noted previously, the power saving mode of IEEE 802.11 has four disadvantages that are collision, overhearing, idle listening, and fixed ATIM window size. Our proposed IEE-MAC protocol reduces energy waste from the above four aspects. We construct a scheduling table at each node by exchanging control messages in the ATIM window. Each entry of the scheduling table includes the starting time and the duration time. The transmission period corresponding to each entry is defined to begin at the starting time and last for the duration time. After building the tables of all nodes, the transmission period of each entry after the ATIM window is guaranteed not to be overlapped by others. Thus collision among different nodes is eliminated. Each node wakes up at the beginning of its transmission period, transmits or receives data in the period, and enters doze state after the period. By this mechanism, we can avoid collision, overhearing, and idle listening. As to fixed ATIM window size, our method is to end the ATIM window in advance if the status of the channel continues to be idle in a pre-specified period of time. Utilizing IEE-MAC protocol, we solve the collision, overhearing, idle listening, and fixed ATIM window size in IEEE 802.11 PSM. Figure 2 is an example to illustrate our main idea. This mechanism is explained in the following three phases.

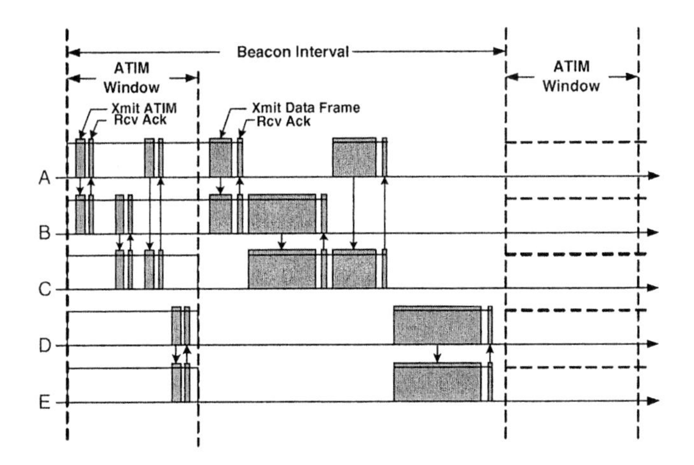

Fig. 2. An example of IEE-MAC protocol

A

Direction	Action	Starting time	Duration time
A->B	Sending	0	10+Ack_Duration
A->C	Sending	40+2Ack_Duration	20+Ack_Duration

B

Direction	Action	Starting time	Duration time
A->B	Receiving	0	10+Ack_Duration
B->C	Sending	10+Ack_Duration	30+Ack_Duration

C

Direction	Action	Starting time	Duration time
B->C	Receiving	10+Ack_Duration	30+Ack_Duration
A->C	Receiving	40+2Ack_Duration	20+Ack_Duration

D

Direction	Action	Starting time	Duration time
D->E	Sending	60+3Ack_Duration	40+Ack_Duration

E

Direction	Action	Starting time	Duration time
D->E	Receiving	60+3Ack_Duration	40+Ack_Duration

Fig. 3. Scheduling tables of Nodes

3.1 Distributed Arrangement Transmission Period of Nodes

In this phase, we build a scheduling table at each node in order to arrange the transmission period of all nodes. We assume that all nodes are fully connected and time synchronization so that all PS nodes can wake up at almost the same TBTT. At the TBTT, each node wakes up for an ATIM window interval. In the ATIM window, each node can send two kinds of frames. One kind of frames is the ATIM frame which is defined as ATIM(sender, receiver, the duration time). The other kind of frames is the ATIM_ACK frame which is defined as ATIM_ACK(the starting time, the duration time). Each node has a separate timer to record the beginning of available transmission time.

If a node with buffered frames to a PS node, it will send an ATIM frame containing duration field which indicates how long the remaining transmission will be to the PS node within the ATIM window period. On receiving the ATIM frame, the indicated PS node responses an ATIM_ACK with starting time and duration to the sender of the ATIM frame and completes the reservation of the data frames transmissions.

Our main idea is illustrated in detail by Fig.2. In the ATIM window, five PS nodes send the ATIM frames. Only four ATIM frames are announced successfully sequentially, i.e., ATIM(A, B, 10), ATIM(B, C, 30), ATIM(A, C, 20), and ATIM(D, E, 40). On receiving ATIM(A, B, 10), node B constructs the ATIM_ACK(0, 10+ACK_duration) where ACK_duration is equal to 2SIFS+ACK and replies it to A. Meanwhile, all nodes know the channel is busy during (0, 10+ACK_duration) and update their timers to 10+ACK_duration. On receiving ATIM(B, C, 30), C constructs the ATIM_ACK(10+ACK_duration, 30+ACK_duration) and sends it back to B. All nodes update their timers to 40+2ACK_duration. Then, on receiving ATIM(A, C, 20), C constructs the ATIM_ACK(40+ 2ACK_duration, 20+ACK_duration) and responses it to A. All nodes update their timers to 60+3ACK_duration. At last, on receiving ATIM(D, E, 40), E constructs the ATIM_ACK(60+3ACK_duration, 40+ACK_duration) and send it to D. All nodes update their timers to 100+4ACK_duration. Now, all nodes can decide their transmission periods and build their own scheduling tables as shown in Fig.3.

3.2 Transmission Data Frames for PS Nodes

At the end of ATIM window, each PS node which successfully transmitted or received an ATIM frame during an ATIM window wakes up to exchange its data frames and then enters doze state again according to its individual scheduling table. But if the duration between the current ending time and the next starting time is small, specifically, in our simulation, node will not enter doze state if the remaining duration is less than two SIFS durations. We continue the scenario as the above example to explain this operation. After ATIM window and a SIFS time, A and B wake up and exchange their frames announced in ATIM window. When completing the exchange of its traffic, A goes to doze state. Then C wakes up according to its schedule and B continues to transmit data frames to C. As noted above, after receiving data frames from A, B doesn't enter doze state because the duration between its current ending time and its next starting time in B's schedule table is too small. After completing the transmission to C, B goes to doze state. At the same time, A wakes up again and exchanges data frames with C follows the same rule as B and C. Then, A and C enter doze state until the end of the beacon interval since there is no entry in their schedule table. At last, D and E wake up and exchange their data frames transmission. When completing the exchange of their traffic, D and E go to doze state.

3.3 Adjust ATIM Window Size

In PSM of IEEE 802.11, there is a fixed value of ATIM window. Thus, nodes have to stay awake for the whole ATIM Window unnecessarily because they are not allowed to enter doze state until this fixed awake duration ends. In the worst case when no nodes have data packets to send, all nodes have to awake and just waste their energies. In our proposed scheme, each node measures how long the channel was idle continuously during the ATIM window. If nodes sense the channel is idle more than $DIFS + \frac{1}{2}CW_{max}$, we assume that no nodes with buffered frames to send. As a result, all nodes can end the ATIM window and enter the data transmission or doze state according to their individual schedule. Utilizing above method, we can dynamically adjust the ATIM window according to the actual traffic load of the network, conserve more power of PS nodes, and improve the network throughput.

4 Performance Evaluation

To evaluate the performance of the proposed protocol IEE-MAC, we implement IEE-MAC, Wu *et al.* in [13] proposed energy efficient MAC protocol, hereafter referred as Wu-MAC, the dynamic power saving mechanism (DPSM) proposed in [11], the PSM scheme in IEEE 802.11, and IEEE 802.11 without power saving mechanism in ns-2 with the CMU wireless extensions [15]. We name the last two schemes as PSM and NO-PSM respectively. In our simulation, we evaluate the above protocols in terms of the following two metrics.

- Aggregate throughput in the network.
- Total data delivered per unit of energy consumption (Kbits delivered per joule): This metric measures the amount of data delivered per joule of energy.

4.1 Simulation Model

For conveniently comparison, we use the similar simulation model with [11] for our simulations. The duration of each simulation is 20 seconds in a wireless LAN. Each flow transmits CBR (Constant Bit Rate) traffic, and the rate of traffic is varied in different simulations. The channel bit rate is 2Mbps and the packet size is fixed at 512 bytes.

To measure energy consumption, we use 1.65w, 1.4w, 1.15w, and 0.045w as values of power consumed by the wireless network interface in transmit, receive, idle modes and the doze state, respectively.

Each node starts with enough energy so that it will not run out of its energy during the simulations. All the simulation results are averages over 30 runs. The length of beacon interval is 100 ms; the ATIM window size in 802.11 is 5 ms [9]. Our protocol uses flexible ATIM window size.

Based on the above experimental setting, we carry out two groups of experiments. In the first group, we vary the total network load in order to observe the effect of the network load on throughput and energy consumption. Simulated network loads are 10%, 20%, 40%, and 60%, measured as a fraction of the channel bit rate of 2 Mbps. In the second group, we change the number of nodes to explore the effect of the number of nodes on throughput and energy consumption. The number of nodes is chosen to be 10, 20, 30, 40 and 50. In all cases, half of the nodes transmit packets to the other half.

4.2 Simulation Results

Figure 4 plots the aggregate throughput against the network load with a fixed number of network nodes (30 nodes). NO-PSM has the best throughput since the nodes keep awake at all time and has not extra overhead of ATIM scheme. The throughput of IEE-MAC and Wu-MAC are very close to that of NO-PSM, because both of methods decrease the possibility of frame collision. However, the throughput of PSM degrades when the load of the network is high since highly loaded network needs more time for data transmission but PSM use the extra channel capacity for the ATIM window. In addition, the performance gap between NO-PSM and DPSM becomes big with the load of the network increasing due to the impact of collision increasing too.

Figure 5 shows the dependence of the total data delivered per joule upon the network load. The same as Fig.4, the number of nodes is fixed as 30. We can see clearly that our protocol has the best energy efficiency among all these protocols. In particular, the energy efficiency of NO-PSM is the worst in all protocols since all nodes are always awake and there is no energy savings from doze state. However, the energy efficiency of NO-PSM increases with the network load grows due to more energy used to transmitting frames rather than

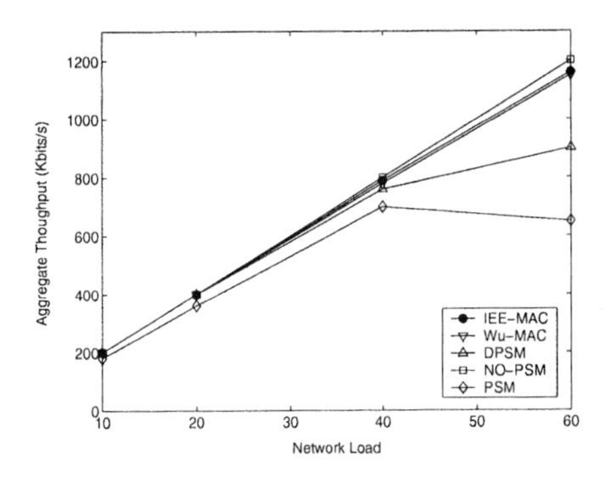

Fig. 4. Aggregate throughput vs. network load

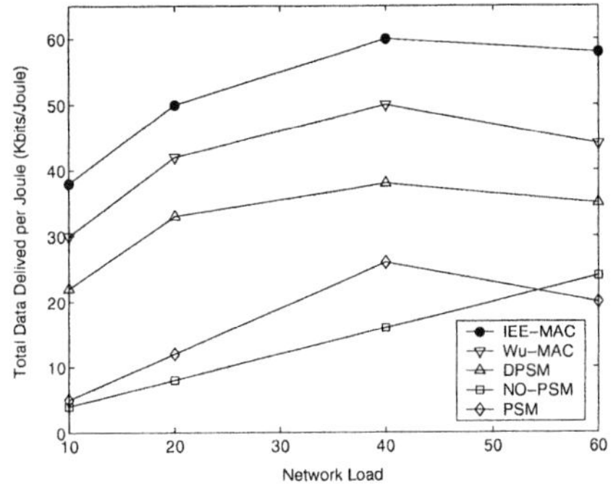

Fig. 5. Aggregate data delivered per joule vs. network load

wasted at idle state. The energy efficiency of PSM is better than that of NO-PSM in the case of low and medium network load. When the network load is low, the performance improvement of PSM compared with NO-PSM is slight since the node cannot enter doze state even if it has at least one packet to transmit or receive. As the load of the network increases, the energy gain from PSM become larger because more time is used to transmit or receive data frames during the nodes are awake. Finally, when the network load is high, the energy efficiency of PSM degrades even lower than that of NO-PSM. The reason is that highly loaded network needs more time for data transmission, but PSM use extra channel capacity for ATIM window. Energy gain from Wu-MAC and DPSM also become smaller when the network load gets higher for the same reason. However, our proposed IEE-MAC always has the best performance in all of the cases because our mechanism avoids overhearing and collision and thus decreases the energy consumption. In addition, our mechanism adjusts the ATIM Window adaptively so that we can avoid PS nodes keeping awake unnecessarily.

Figure 6 illustrates the effect of the number of the nodes in the network on the throughput with the fixed network load (network load = 40%) when using our proposed IEE-MAC, Wu-MAC, DPSM, PSM, and NO-PSM schemes, respectively. NO-PSM has the best performance since the nodes keep awake at all time. The throughput of PSM and DPSM decrease rapidly as the number of nodes in the network increases especially in the case of the large network size, since more contending nodes result in more packets collisions and the ATIM window uses extra channel capacity. On the contrary, as shown in Fig.6, the throughput of our IEE-MAC and Wu-MAC only degrade slightly with the number of nodes increases. The reason lying behind is that these two schemes employ the scheduling transmission mechanism to decrease significantly the possibility of frames collision.

The aggregate data delivered per joule of these five protocols versus the number of the nodes in the network is depicted in Fig.7. Our proposed IEE-MAC always performs better than other four protocols. Particularly, the data deliv-

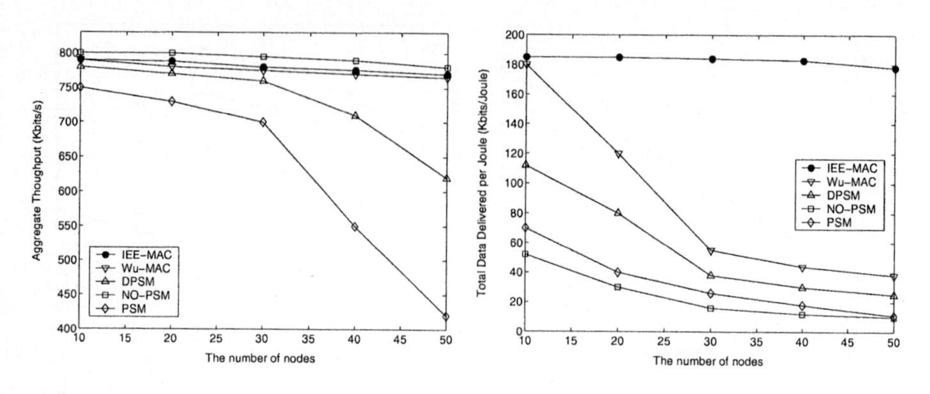

Fig. 6. Aggregate throughput vs. the number of nodes

Fig. 7. Aggregate data delivered per joule vs. the number of the nodes

ered per joule of PSM and NO-PSM significantly decrease with the increase of the number of the nodes because there are more collisions among the nodes and nodes need to take more time to finish data transmission. The energy efficiency of Wu-MAC also experiences similarly degradation with the number of the nodes increase. However, the reason is different from the case of PSM and NO-PSM. In Wu-MAC, there is more energy wasting on overhearing with the number of nodes increase. On the contrary, the energy efficiency of IEE-MAC remains stable in most of the cases and only slightly degrades in the very large network size. This occurs because our mechanism eliminates overhearing and collision and conserves the energy consumption efficiently when the network size is large.

5 Conclusions

We have presented an improved energy efficient MAC (IEE-MAC) protocol for IEEE 802.11-Based Ad Hoc networks. The proposed protocol has four main contributions. First, to avoid unnecessary frame collisions and backoff waiting time in data frame transmissions, the protocol can schedule all nodes to transmit their frames to PS nodes sequentially. Secondly, the protocol can avoid overhearing of nodes in order to save their energy when other nodes are transmitting. Thirdly, IEE-MAC also allows node to enter doze state whenever it has finished transmitting packets other than stay awake until the end of the beacon interval. Lastly, to conserve more power of PS nodes and improve the channel utilization, the protocol utilizes an adaptive strategy to dynamically adjust the ATIM window size according to the actual traffic load. Results of simulation show that IEE-MAC obtains best performance in terms of throughput and energy efficiency compared with other energy efficient MAC protocols mentioned previously.

References

1. Singh, S., Woo, M., Raghavendra, C.S.: Power-aware routing in mobile ad hoc networks. In: MobiCom '98: Proceedings of the 4th annual ACM/IEEE international conference on Mobile computing and networking, ACM Press (1998) 181–190
2. Xu, Y., Heidemann, J., Estrin, D.: Geography-informed energy conservation for ad hoc routing. In: Proceedings of the ACM/IEEE International Conference on Mobile Computing and Networking, Rome, Italy, USC/Information Sciences Institute, ACM (2001) 70–84
3. Chang, J.H., Tassiulas, L.: Energy conserving routing in wireless ad-hoc networks. In: INFOCOM (1). (2000) 22–31
4. Monks, J., Bharghavan, V., mei W. Hwu, W.: A power controlled multiple access protocol for wireless packet networks. In: INFOCOM. (2001) 219–228
5. Feeney, L.M., Nilsson, M.: Investigating the energy consumption of a wireless network interface in an ad hoc networking environment. In: IEEE INFOCOM. (2001)
6. Tseng, Y.C., Hsu, C.S., Hsieh, T.Y.: Power-saving protocols for ieee 802.11-based multi-hop ad hoc networks. Comput. Networks **43** (2003) 317–337
7. Zorzi, M., Rao, R.: Is tcp energy efficient? In: Proceedings of Sixth IEEE International Workshop on Mobile Multimedia Communications. (1999)
8. Krashinsky, R., Balakrishnan, H.: Minimizing energy for wireless web access with bounded slowdown. In: MobiCom 2002, Atlanta, GA (2002)
9. LAN MAN Standards Committee of the IEEE Computer Society: IEEE Std 802.11-1999, Wireless LAN Medium Access Control (MAC) and Physical Layer (PHY) Specification. IEEE (1999)
10. Woesner, H., Ebert, J.P., Schlager, M., Wolisz, A.: Power saving mechanisms in emerging standards for wireless lans: The mac level perspecitve. IEEE Personal Communications **5** (1998) 40–48
11. Jung, E.S., Vaidya, N.H.: An energy efficient mac protocol for wireless lans. In: INFOCOM. (2002)
12. Choi, J.M., Ko, Y.B., Kim, J.H.: Enhanced power saving scheme for ieee 802.11 dcf based wireless networks. In: PWC. (2003) 835–840
13. Wu, S.L., Tseng, P.C.: An energy efficient mac protocol for ieee 802.11 wlans. In: CNSR 2004. (2004) 137–145
14. Ye, W., Heidemann, J., Estrin, D.: Medium access control with coordinated adaptive sleeping for wireless sensor networks. IEEE/ACM Trans. Netw. **12** (2004) 493–506
15. The CMU Monarch Project: (The cmu monarch project's wireless and mobility extension to ns)

POST: A Peer-to-Peer Overlay Structure for Service and Application Deployment in MANETs*

Anandha Gopalan and Taieb Znati

Department of Computer Science,
University of Pittsburgh, Pittsburgh, PA 15260, U.S.A
{axgopala, znati}@cs.pitt.edu

Abstract. Ad-hoc networks are an emerging technology with immense potential. Providing support for large-scale service and application deployment in these networks, however is crucial to make them a viable alternative. The lack of infrastructure, coupled with the time-varying characteristics of ad-hoc networks, brings about new challenges to the design and deployment of applications. This paper addresses these challenges and presents a unified, overlay-based service architecture to support large-scale service and application deployment in ad-hoc networks. We discuss the main functionalities of the architecture and describe the algorithms for object registration and discovery. Finally, the proposed architecture was evaluated using simulations and the results show that the architecture performs well under different network conditions.

1 Introduction

Advances in wireless technology and portable computing along with demands for greater user mobility have provided a major impetus towards development of an emerging class of self-organizing, rapidly deployable network architectures referred to as ad-hoc networks. Ad-hoc networks, which have proven useful in military applications, are expected to play an important role in commercial settings where mobile access to a wired network is either ineffective or impossible. Despite their advantages, large-scale deployment of services and applications over these networks has been lagging. This is due to the lack of an efficient and scalable architecture to support the basic functionalities necessary to enable node interaction. Several challenges must be addressed in order to develop an effective service architecture to support the deployment of applications in a scalable manner. These challenges are related to the development of several capabilities necessary to support the operations of nodes, which include: Object registration and discovery, Mobile node location, and Traffic routing and forwarding.

Node mobility, coupled with the limitation of computational and communication resources, brings about a new set of challenges that need to be addressed

* This work has been supported by NSF award ANI-0073972.

X. Jia, J. Wu, and Y. He (Eds.): MSN 2005, LNCS 3794, pp. 1006–1015, 2005.

in order to enable an efficient and scalable architecture for service and application deployment in MANETs. In addition to object information, peers must also register their location and mobility information to facilitate peer[1] interaction. This information, however, changes dynamically due to peer mobility. Efficient mechanisms must, therefore be in place to update this information as peers move.

The main contribution of this paper is in providing a novel Peer-to-peer Overlay STructure (*POST*) that allows for service and application deployment in MANETs. This paper takes a unique peer-to-peer based approach to provide a scalable, robust and efficient framework along with the protocols and algorithms that allow for large-scale service and application deployment in MANETs. POST is efficient, by not requiring nodes to maintain routing information. Bootstrapping in POST does not require the knowledge of other POST nodes. A node only needs to know the *hash* function used and the location of the *zone virtual manager service* to register and query for objects.

The main components of POST are: *zones*, *virtual homes* and *mobility profile.* A zone is a physical area in the network that acts as an information database for objects in the network. The zones are organized as a virtual DHT-based structure that enables object location through distributed indexing. The DHT uses a virtual structure that is tightly coupled to the physical structure of the network to locate peers where object information is stored. A *virtual home* of a node is the physical area where the node is most likely located. In the case when a node is mobile, it leaves behind its mobility information with select *proxy* nodes within its virtual home. This information constitutes the *mobility profile* of the node and consists of its expected direction and speed of travel.

The rest of the paper is organized as follows: Section 2 details the related work in this area while Section 3 details the network characteristics used in POST, Section 4 details the different components of the system architecture and the algorithms used, Section 5 details the simulations and the ensuing results and Section 6 concludes the paper and identifies areas of future work.

2 Related Work

Service discovery for ad-hoc networks is still a very new area of research. There have been some protocols for service location and discovery that have been developed for LANs, namely: Service Location Protocol [2] and Simple Service Discovery Protocol [16]. SLP relies on agents to search for and locate services in the network; an *user agent* is used on behalf of users to search for services, while a *service agent* advertises services on behalf of a server and finally a *directory agent* collects the advertisements sent out by the *server agent.* SSDP uses a specific protocol and port number to search for and locate services in the network. HTTP UDP is used on the reserved local multicast address *239.255.255.250* along with the *SSDPport* while searching for services. Both SLP and SSDP cannot be directly used for MANETs due to their reliance on an existing network structure.

[1] We will use the term node and peer interchangeably in this paper.

CAN [12] provides a distributed, Internet-scale hash table. The network is divided into zones according to a virtual co-ordinate system, where each node is responsible for a virtual zone. Given (key,value) pairs, CAN maps the key to a point P in the co-ordinate system using a uniform hash function. The corresponding (key,value) pair is stored at the node that owns the zone containing P. We differ from CAN by having more nodes in a zone to hold object information. Also, a zone is not split when a new node arrives and the overhead is avoided. Mobility is also incorporated by using the *mobility profile management base*.

The Landmark routing hierarchy [11] provides a set of algorithms for routing in large, dynamic networks. Nodes in this hierarchy have a permanent node ID and a *landmark* address that is used for routing. The *landmark* address consists of a list of the IDs of nodes along the path from this node to a well known *landmark* node. Location service is provided in the landmark hierarchy by mapping node IDs to addresses. A node X chooses its address server by hashing its node id. The node whose value matches or is closest to the hash value is chosen as X's address server.

GLS [6] provides distributed location information service in mobile ad-hoc networks. GLS combined with geographic forwarding can be used to achieve routing in the network. A node X "recruits" a node that is "closest" to its own ID in the ID space to act as its location server. POST differs from GLS and the Landmark scheme by using a group of nodes (that are selected from within a zone) to act as the object location server. The information is stored in a manner such that only k out of N fragments are necessary to *re-construct* it. This increases the robustness of POST, since the information is still available, even after the failure of some nodes.

[5, 4, 14] use the concept of home regions. Each node is mapped to an area (using a hash function) in the network that is designated as its home region. The home region holds the location information about the mobile nodes which map to this location. A node updates its location information by sending updates to its home region. In our scheme, a node does not keep updating its *virtual home*, rather it leaves a trail behind that can be used by other nodes to locate it.

Ekta [3] integrates distributed hash tables into MANETs and provides an architecture for constructing distributed applications and services. POST differs by not using Pastry [13] as the DHT. The DHT is constructed in a manner that allows it to take advantage of the location information provided. POST also takes node mobility into consideration. [8, 15] provide basis for resource discovery in MANETs. Our main contribution when compared to these works is the use of a DHT-based system for providing resource discovery in MANETs.

3 POST: Network Characteristics

3.1 Zones and Virtual Homes (VHs)

Consider an ad-hoc network covering a specific geographical area, denoted by Λ. We perceive this area to be divided into zones as shown in fig. 1. A zone for example, can be an administrative domain where geographical proximity facilitates

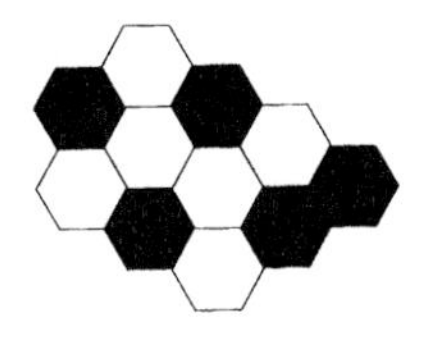

Fig. 1. Partitioning the network into zones

communication between nodes of the zone. A zone Z_i, is uniquely characterized by an identifier and contains one or more neighborhoods. A neighborhood is defined as a physical area within a zone in the network.

Each zone also has a *zone virtual manager service* ($ZVMS$) that returns the physical co-ordinates of a zone given the *zone id*. The $ZVMS$ is collaboratively provided by a group of nodes within a zone. The *zones* form a virtual DHT-based distributed database that holds object information. The virtual DHT maps the objects into zones, where object information is stored. The novelty of this approach is that the virtual DHT-structure is tightly coupled to the physical structure of the network, where each physical zone is responsible for holding information about those objects that map to it. The database is thus, a fixed size hash table with each entry corresponding to a *zone* in the network. To ensure that there are no *hot spots* in the network due to the hashing of object keys to zones, the hash function is chosen to be a $uniform$ hash function.

The *virtual home* of a node is the neighborhood within a zone where the node is most likely to be located and is defined by its physical co-ordinates. The VH of a node can change over time depending on the mobility of the node. For example, a user may be at home at time t_1 and at time t_2, he/she may be at their office, in which case their *new* virtual home is characterized by the co-ordinates of the office. A user can choose to provide this information while registering, thus facilitating location of the user. Hence, a virtual home of a node refers to the actual physical location of the node and also reflects user behavior.

A node is characterized by a unique $identifier$ and its *virtual home*. Each node knows the physical location of the neighborhood where the $ZVMS$ is provided. A node, A, upon departure from its VH leaves behind some information so that other nodes can use this information to locate A. This involves building the $MPMB$, that comprises of a set of *proxy* nodes that are responsible for holding A's mobility information. This information contains the expected direction and speed of travel of A and is called the *mobility profile.*

Each object is characterized by its object id and a *unique* key. Any node wishing to register (or) query for an object (O) in the network, must first calculate the *hash value* (using a system-wide hash function), based on the key associated with O. This *hash value* provides a *zone id* that maps to a zone that is responsible for holding information about O.

Zone Virtual Manager Service: Within a zone, the *zone virtual manager service* is mapped to a neighborhood and is collaboratively provided by the set of nodes whose VHs map to this neighborhood. Given a *zone id*, the $ZVMS$ returns the physical co-ordinates of the corresponding zone.

Table 1. Key range

Zone	Interval	Lower Limit	Upper Limit
.	.	.	.
z_i	I_i	$(i-1)\cdot\frac{\tau}{\alpha_z}$	$(i\cdot\frac{\tau}{\alpha_z})-1$
.	.	.	.
z_{α_z}	I_{α_z}	$(\alpha_z-1)\cdot\frac{\tau}{\alpha_z}$	$(\alpha_z\cdot\frac{\tau}{\alpha_z})-1$

Zone Key Management: A zone is responsible for holding information about objects whose keys hash to that zone. There is a need for an algorithm to assign keys to zones. Let the entire key space be of size τ. τ is divided into α_z intervals; α_z = total number of zones in the network. Each interval contains α keys; $\alpha = \frac{\tau}{\alpha_z}$. Let the zones be: $z_1, z_2, ..., z_{\alpha_z}$. Let $I_1, I_2, .., I_{\alpha_z}$ be the intervals, where each interval contains a set of keys. In POST, each zone z_i is responsible for the keys corresponding to interval I_i. This is expressed in Table 1.

4 POST: System Architecture and Services

4.1 Object Registration and Discovery

Object Registration: A peer A, managing a collection of objects registers its objects by hashing the *object key* to obtain a *hash value*, which maps to a *zone id* of a zone (Z) that A must register with. Node A queries its $ZVMS$ to get the physical co-ordinates of Z. A sends a message with its *virtual home*, object key and other attributes related to O to Z. Nodes in Z collaboratively register this. Algorithm 1 details the process by which node A, containing an object O with key k_i, registers this information, (H = hash function).

Object Discovery: Object discovery is performed in a method, similar to object registration. A peer A, wishing to locate the object(s) in the network, first calculates the *hash value* by hashing the object key. This *hash value* maps to a *zone id* of a zone (Z). Node A queries its $ZVMS$ to get the physical co-ordinates of Z. Using directional routing, peer A sends a request to Z for information about the location of the object. Nodes in Z reply with a list of peers that own this resource. Algorithm 2 details the process by which node A, discovers information about an object O with key k_i.

Algorithm 1. Object Registration

(1) Calculate $Z_{id} = H(k_i)$
(2) Query $ZVMS$ to get co-ordinates of zone, Z with id Z_{id}
(3) Use directional routing to send a message to Z to register
(4) Nodes in Z register the object information

Algorithm 2. Object Discovery

(1) Calculate $Z_{id} = H(k_i)$
(2) Query $ZVMS$ to get co-ordinates of zone, Z with id Z_{id}
(3) Use directional routing to send a query to Z with k_i
(4) Nodes in Z reply with a list of peers that *own* the object

4.2 Mobile Node Location and Peer Interaction

Mobile Node Location: Node mobility is incorporated by using a node's *mobility profile.* Consider the situation when a node A leaves its current *virtual home.* Node A has some knowledge about its intended destination and its direction and speed of travel and leaves behind this information called the *mobility profile* with select *proxy* nodes that act as the *MPMB.* Using this mobility profile, other nodes can locate node A. To recruit *proxy* nodes, node A sends out a broadcast message *once* at its highest power (to ensure maximum broadcast radius) and waits for replies from the other nodes. The *mobility profile* is encoded in a manner such that k out of N (number of replies) fragments are enough to re-construct it. This ensures that the mobility profile is available even after a few proxy nodes have left the *VH.* [10] Defines an encoding scheme that can be used for this purpose. Algorithm 3 details the steps involved in building the *MPMB.*

Algorithm 3. Building the MPMB

(1) Broadcast-Msg to recruit
(2) Encode *mobility profile*
(3) Send encoded parts to the nodes that replied

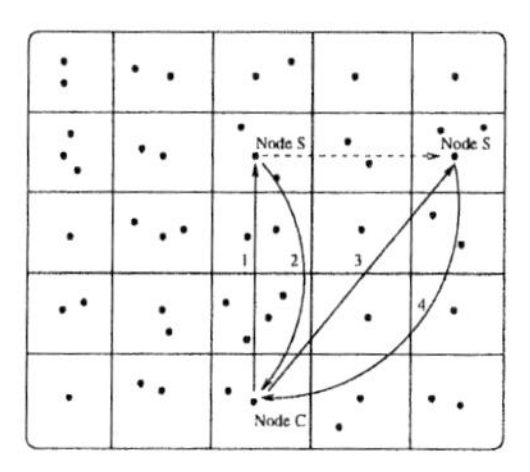

Fig. 2. Mobile Peer Location

Peer Interaction: Once a node discovers the resources available in the network, it tries to interact with the peer containing the required resource. Algorithm 4 and figure 2 detail the procedure by which a node C obtains a resource from a node S that is mobile.

Algorithm 4. Handling peer mobility

(1) C receives VH(S) from the zone holding object information
(2) Using directional routing, C sends messages towards VH(S) (msg 1 in the figure)
(3) **case** S is in VH(S)
(4) Connection is established between C and S
(5) **case** S is currently not in VH(S)
(6) The nodes in VH(S) reply with the *mobility profile* of S, a metric: $[t_0, V(t_0), D(t_0), P_V(t), P_D(t)]$ (msg 2)
(7) C uses the *mobility profile* of S to determine the current position of S and sends messages in this direction (msg 3)
(8) S upon receiving the messages by C acknowledges it (msg 4) and initiates a conversation with C

The components of the *mobility profile* in algorithm 4 are: t_0: starting time, $V(t_0)$: expected starting speed, $D(t_0)$: expected initial direction, $P_v(t)$: Predictor for speed after t time units since departure, $P_d(t)$: Predictor for direction after t time units since departure.

4.3 Traffic Forwarding Algorithm

This section gives a brief description of the forwarding algorithm used in POST (for more details, refer to [1]). Consider the scenario when a source S attempts to route traffic to a destination D. To limit flooding in the network, traffic is sent in a truncated cone-shaped manner towards D as shown in figure 3. Nodes in zone 1 have the highest priority to forward the traffic. If no nodes are currently available in zone 1, the transmission area is expanded to include zone 2, after a timeout. This prioritizes the neighboring nodes in such a way that nodes that are more in line with the direction of the destination have higher priority to forward the message, thereby reducing the delay traffic suffers on its way towards the destination. Nodes that receive the message sent by S calculate their priorities based on which they decide whether to forward the message. Furthermore, upon hearing a transmission within the zone, the remaining eligible nodes drop the message. As the message progresses toward its destination, the highest priority node calculates a new cone and re-iterates the process.

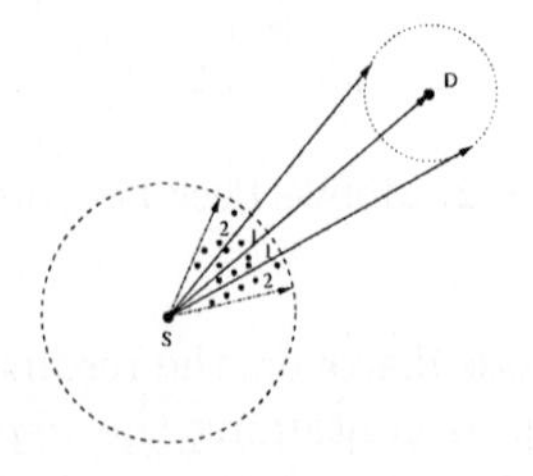

Fig. 3. Directional Routing

5 Simulation and Results

The protocol was implemented in the Glomosim network simulator [9] and was tested under different network scenarios. The set of tests were conducted as part of the sensitivity analysis of POST. In the experiments, the hit rate and the response times for object registration and discovery were measured for a network of static, low mobility and high mobility nodes, while varying the network density. The channel characteristics used were: *TWO-RAY* (receiving antenna sees two signals, a direct path signal and a signal reflected off the ground) and *FREE-SPACE* (signal propagation is in the absence of any reflections or multipath). The mobility models used were: *NO-MOBILITY* (static) and the *Random Trip* mobility model [7]. The speed of the nodes was varied to simulate a network of slow moving nodes (5 m/s) and a network of fast moving nodes (25 m/s). The number of nodes was varied from 100 to 500 and these nodes were placed in a network grid of size 2800x2800m. The network grid was further divided into zones of size 400x400m. Traffic statistics were collected at the sender to evaluate the *response time.* To evaluate the "hit rate", we measure the % of packets that reach the destination. Each data point represented was the value averaged over 10 independent experimental runs. The first set of experiments (figures 4(a), 4(b)) were performed by varying the channel characteristic. The second set of experiments (figures 5(a), 5(b)) were performed to measure the response time for object registration and discovery.

From figure 4(a), we conclude that in all the cases the hit rate increases and reaches 100% as the network density increases. The hit rate is always lower for the high mobility case as when compared to the static and low mobility cases. This can be attributed to the fact that node mobility has an impact on the number of nodes in a zone that are available to hold information about the objects and the mobility profiles. We can further observe that due to this the hit rate for a sparse network is lower.

From figure 4(b), we conclude that the hit rate is higher. This is because the transmission range for a node using the channel *FREE-SPACE* is much higher when compared to *TWO-RAY.* This increase in transmission radius leads to a

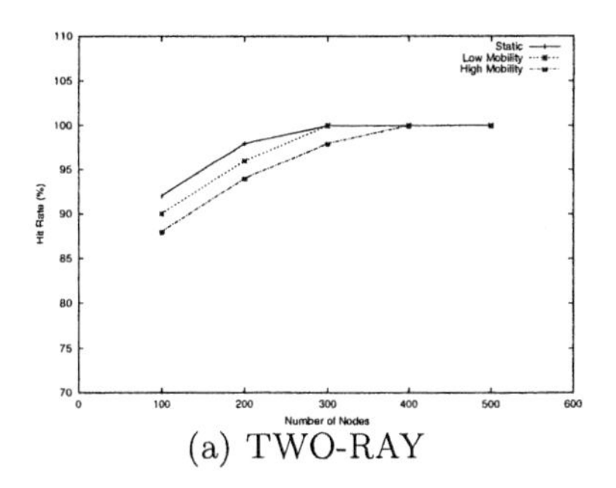

(a) TWO-RAY

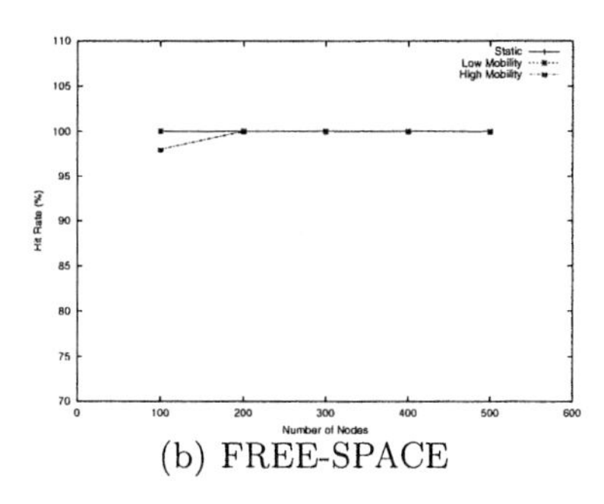

(b) FREE-SPACE

Fig. 4. Hit Rate using different Channel Characteristics

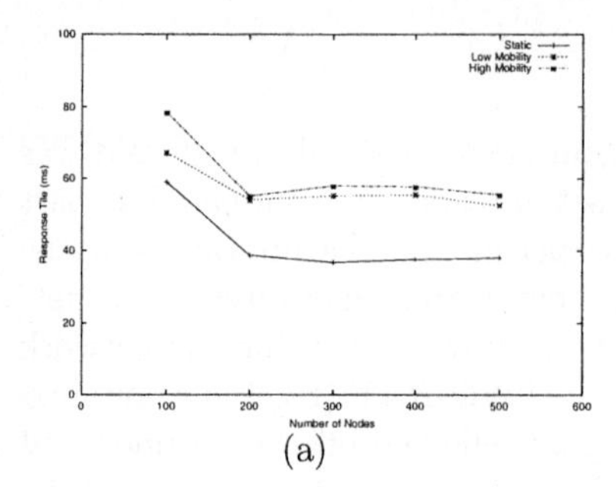

(a)

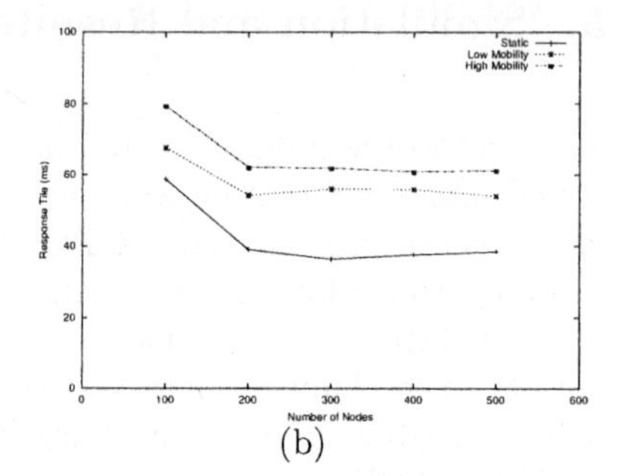

(b)

Fig. 5. Object Registration and Discovery

higher number of nodes that can be reached and hence the availability of more nodes that hold the information about the objects and the mobility profiles.

From figure 5(a), we conclude that for a slightly dense to a very dense network, the response time remains almost a constant. For a sparsely populated network, the response time is significantly higher. This is due to the fact that the paucity of nodes in the network leads to a lower probability of finding a node in the direction of the destination to forward the traffic. We can also observe that the response time for a high mobility network is higher (though not by much). This is because node mobility causes frequent changes to the zone membership.

From figure 5(b), we conclude that the response time is very similar to the response times observed for object registration. This can be attributed to the fact that the protocol followed for object discovery is very similar that of object registration.

6 Conclusion and Future Work

The major contribution of this paper is in providing a novel Peer-to-peer Overlay STructure (*POST*) for service and application deployment in MANETs. The proposed framework is scalable, robust and efficient. POST is efficient since nodes do not maintain and update routing tables. Bootstrapping does not require the knowledge of other POST nodes. A node needs to know the *hash* function used in the network and the $ZVMS$, that maps zone ids to the zones to register and query for objects in the network. Object registration and discovery are achieved by hashing the key associated with an object to obtain the *zone id* of the zone associated with it. POST is scalable, since the hash function ensures that the database containing the object information is spread across various *zone*s in the network. A zone key management protocol was developed to map object keys to the corresponding zones in the network. The mobility of the nodes in the network is incorporated into POST by using the *mobility profile* of a node.

There is a lot of potential for future work in this area. Security needs to be incorporated into this service-architecture at various levels, be it object registration and discovery or handling the *mobility profile*s using the *MPMB*.

References

1. A. Gopalan, T. Znati and P. K. Chrysanthis. Structuring Pervasive Services in Infrastructureless Networks. In *Proc. IEEE International Conference on Pervasive Services (ICPS '05)*, 2005.
2. E. Guttmann and C. Perkins and J. Veizades and M. Day. Service Location Protocol, Version 2. *Internet Draft, RFC 2608*, 1999.
3. H. Pucha, S. M. Das and Y. Charlie Hu. Ekta: An Efficient DHT Substrate for Distributed Applications in Mobile Ad Hoc Networks. In *Proc. of the 6th IEEE Workshop on Mobile Computing Systems and Applications*, December 2004.
4. J. P. Hubaux, J. Y. Le Boudec, and M. Vetterli Th. Gross. Towards self-organizing mobile ad-hoc networks: the terminodes project. *IEEE Comm Mag*, 39(1):118 -124, January 2001.
5. I. Stojmenovic. Home agent based location update and destination search schemes in ad hoc wireless networks. Technical Report TR-99-10, Computer Science, SITE, University of Ottawa, Sep. 1999.
6. J. Li and J. Jannotti and D. De Couto and D. Karger and R. Morris. A Scalable Location Service for Geographic Ad-Hoc Routing. In *Proc. Mobicom*, pages 120–130, August 2000.
7. Jean-Yves Le Boudec and Milan Vojnovic. Perfect Simulation and Stationarity of a Class of Mobility Models. In *Proceedings of IEEE INFOCOM*, 2005.
8. Jivodar B. Tchakarov and Nitin H. Vaidya. Efficient Content Location in Mobile Ad Hoc Networks. In *Proc. IEEE International Conference on Mobile Data Management (MDM 2004)*, January 2004.
9. L. Bajaj, M. Takai, R. Ahuja, R. Bagrodia and M. Gerla. Glomosim: A scalable network simulation environment. Technical Report 990027, UCLA, 13, 1999.
10. Gretchen Lynn. ROMR: Robust Multicast Routing in Mobile Ad-Hoc Networks. *PhD. Thesis, University of Pittsburgh*, December 2003.
11. Paul F. Tsuchiya. The Landmark Hierarchy: A New Hierarchy for Routing in very large Networks. In *Proc. ACM SIGCOMM*, pages 35–42, August 1988.
12. Sylvia Ratnasamy, Paul Francis, Mark Handley, Richard Karp, and Scott Shenker. A scalable content-addressable network. In *Proceedings of ACM SIGCOMM*, 2001.
13. Antony Rowstron and Peter Druschel. Pastry: Scalable, decentralized object location, and routing for large-scale peer-to-peer systems. *Lecture Notes in Computer Science*, 2218:329–350, 2001.
14. Seung-Chul M. Woo and Suresh Singh. Scalable Routing Protocol for Ad Hoc Networks. *Journal of Wireless Networks*, 7(5), Sep. 2001.
15. U. Kozat and L. Tassiulas. Network Layer Support for Service Discovery in Mobile Ad Hoc Networks. In *Proceedings of IEEE INFOCOM*, 2003.
16. Yaron Y. Goland, Ting Cai, Paul Leach, Ye Gu. Simple Service Discovery Protocol/1.0. *Internet Draft*, Oct. 1999.

An Efficient and Practical Greedy Algorithm for Server-Peer Selection in Wireless Peer-to-Peer File Sharing Networks*

Andrew Ka Ho Leung and Yu-Kwong Kwok**

Department of Electrical and Electronic Engineering,
The University of Hong Kong, Pokfulam, Hong Kong
{khleung, ykwok}@eee.hku.hk

Abstract. Toward a new era of "Ubiquitous Networking" where people are interconnected in anywhere and at anytime via the wired and wireless Internet, we have witnessed an increasing level of impromptu interactions among human beings in recent years. One important aspect of these interactions is the ***Peer-to-Peer (P2P) Networking*** that is becoming a dominant traffic source in the wired Internet. In these Internet overlay networks, users are allowed to exchange information through instant messaging and file sharing. Unfortunately, most of the previous work proposed in the literature on P2P networking is designed for the traditional wired Internet, without much regard to important issues pertinent to wireless communications. In this paper, we attempt to provide some insight into P2P networking with respect to a wireless environment. We focus on P2P file sharing, already a hot application in the wired Internet, and will be equally important in the wireless counterpart. We propose a greedy server-peer selection algorithm to decide from which peer should a client download files so that the level of fairness of the whole network is increased and expected service life of the whole file sharing network is extended. We also propose a new performance metric called ***Energy-Based Data Availability, EBDA,*** which is an important performance metric for improving the effectiveness of a wireless P2P file sharing network.

Keywords: wireless networking, P2P systems, file sharing, energy efficiency, fairness, greedy algorithm.

1 Introduction

It is reported in a recent survey [32] that Peer-to-Peer (P2P) applications generate one-fifth of the total Internet traffic, and it is believed that it will continue to grow. Furthermore, an ISP (Internet Services Provider) solution company [29] reported that the top-four hottest P2P file sharing applications are BitTorrent

* This research was supported by a grant from the Research Grants Council of the HKSAR under project number HKU 7157/04E.

** Corresponding author.

X. Jia, J. Wu, and Y. He (Eds.): MSN 2005, LNCS 3794, pp. 1016–1025, 2005.

[1] (occupies 53 % of all P2P traffic), eDonkey2000 [5] (occupies 24 %), FastTrack [6] (occupies 19 %) and Gnutella [10] (occupies 4 %). Indeed, given the recent rapid development of high speed wireless communication technologies, including 3G, post-3G and WLAN, it is widely envisioned that file sharing over wireless P2P network will naturally be the next trend. An efficient wireless P2P network is reckoned as a key component of our next generation ubiquitous and pervasive mobile computing platform. However, such a wireless P2P network is likely to be *energy constrained* in nature, since mobile wireless devices are mostly battery operated and the battery has limited life. Running P2P applications on top of it requires network developers to incorporate energy conservation idea in their design.

In this paper, We propose a server-peer selection rule to increase the performance of a wireless P2P file sharing network by using a greedy algorithm to select a suitable peer to download the requested files, when there are multiple peers who have the file we request for. Different from previous works, mobility factor and transmit power constraint in wireless network are explicitly incorporated into our design. The concept of "data availability" has been used in [22] and [20] to study the provision of information resource in P2P networks. We introduce a new performance metric, namely "energy-based data availability" for *wireless* P2P network. Our view is that, in an energy constrained network, "data availability" should not only refer to the amount of file resources possessed by somebody or being shared, but also refer to the energy levels of entities who hold the file resources. It makes no sense to have a wireless P2P network where some users hold a large amount of popular file objects for which other users always request (e.g., popular music or latest movie trailer) but these users who possess resources are of low battery level. All the valuable resources would be lost when these users' energy exhausted. Our greedy algorithm can significantly increase the energy efficiency in terms of fairness. We avoid any single peer drop too much in energy level and has its energy level deviate too far from the others. In Section 2, we provide some background information on P2P networking. In Section 3, we describe our approach to increasing the energy efficiency by using a greedy server-peer selection algorithm. In Sections 4 and 5 we describe our simulation platform and the simulation results, respectively. We summarize our conclusions in Section 6.

2 Background and Related Work

Apart from data availability, architecture of file sharing system is another aspect of P2P networking which receives most attention, including the famous "first generation" centralized system Napster [23], which uses a centralized database to maintain a directory of file resources in the Internet, Gnutella [10] and KaZaA [14] which use decentralized searching. More advanced searching algorithm can be found in so-called "second generation" P2P system architectures, including Chord [17], Pastry [28], Tapestry [33] and CAN [27] which use distributed hash table (DHT) for searching and ask each peer to hold a subset of the whole routing information set. BitTorrent(BT) [1] is another latest common decentralized P2P protocol deployed in our Internet. However, all the above designs are not tailor-made for a wireless medium. Therefore, in order to achieve the goal of having ubiquitous wireless P2P file sharing, new design criteria must be taken into account. Our work focus on one of the key design issue, which is the energy constraint. Different from above work, our goal is *not* to find a searching algorithm nor routing protocol, but to propose a "server-peer selection algorithm" to select the server-client pairs in file sharing, assume that file searching result is already available. Our work is believed to be effective in extending P2P file sharing from traditional wired Internet to a wireless environment in the future.

3 Our Proposed Approach

3.1 Definition

Consider a P2P file sharing application running on top of a mobile wireless ad hoc network with N users. We denote each user by n_i:

$$U = \{n_i | i = 1, 2, 3, \ldots, N\} \tag{1}$$

We use E_i to denote the remaining energy level (battery level) of a mobile user n_i. We then use a binary vector V_i to denote the file objects that user n_i possesses:

$$V_i = \{\delta_{ik} | k = 1, 2, 3, \ldots, M\},\ i = 1, 2, 3, \ldots, N \tag{2}$$

where

$$\delta_{ik} = \begin{cases} 1 & \text{If user } i \text{ has file object } f_k \\ 0 & \text{If user } i \text{ does not have file object } f_k \end{cases} \tag{3}$$

We assume that there are totally M different file objects in the network:

$$F = \{f_j | j = 1, 2, 3, \ldots, M\} \tag{4}$$

In this P2P file sharing platform, we quantify the popularity of each file object and represent it using a weight, with respect to a particular user. Different files are of different popularity in the network, some file objects (e.g., latest pop music) are of higher ranking in the mind of some users. Thus, we represent the weight (rank) of an object f_k in user n_i's mind by a weight w_{ik}. Different file objects could have different weight values for different people, thus we need a N by M weight matrix W to denote weights of all file objects for all users:

$$W = \begin{bmatrix} w_{11} \ldots w_{1M} \\ \ldots\ldots \\ w_{N1} \ldots w_{NM} \end{bmatrix} \tag{5}$$

The weight measure, which is related to the taste of a particular user such as, "favorite music" or "favorite singer" are something quite subjective. We can make a more objective ranking of files by restricting each user to give their weight values using some scale, say, "How many stars (five stars is maximum) do you give for this song?" ranking like this are commonly found in online forums and fans sites. It should be noted that, although the calculation of W require heavy computation and its value can be dynamically changing when users' taste and community's trends are changing, a single user need *not* to calculate it. As shown in later sections, each user only has to do local calculation on client-peer's and server-peer's weight values on file objects when running our greedy algorithm. Now we define the energy-based data availability, EBDA, of a user as:

$$D_i = E_i \sum_{k=1}^{M} w_{ik} \delta_{ik} \tag{6}$$

and the EBDA of the whole network is:

$$D = \sum_{i=1}^{N} (E_i \sum_{k=1}^{M} w_{ik} \delta_{ik}) \tag{7}$$

where

$$\delta_{ik} = \begin{cases} 1 & \text{If user } i \text{ has file object } f_k \\ 0 & \text{If user } i \text{ does not has file object } f_k \end{cases} \tag{8}$$

We interpret this performance metric in the following way. Firstly, "availability" of file objects in a P2P network depends on how many copies of each file are there in the network, this is represented by δ terms in our metric. Secondly, different files are of different popularity in the network, some file objects (e.g., latest pop music) are of higher ranking in the mind of some users, our metric take this into account and represent this by a weight w.

In the definition of EBDA, the sum of all file object weights of each user is scaled (multiplied) by the amount of energy of each user, energy level of user n_i by E_i. This multiplicative factor corresponds to the fact that all file objects an user hold would be rendered as nothing when the energy of this user exhausts. Consider a P2P network, given the same number of copies of the same set of file objects distributed over two group of users, each have the N users. We see that a higher value of $\sum_{i=1}^{N} D_i$ for a group has the implication that (1) the file object resources possessed by the whole group is more "durable", more future sharing is possible before the energy of the holders exhaust and (2) on average, a holder of a file can keep their favorite file longer.

3.2 Greedy Algorithm for Server-Peer Selection

In the last section we see that EBDA is more meaningful performance metric for a wireless P2P network compare to traditional metrics used in wired networks. But how can we control this performance metric in a P2P network? Firstly, we note that it is undesirable to control which user could request for a file, this affects the freedom of users and possibly discourages peers' participation. Secondly, we should not prohibit a peer to give response to a request posted by another peer, this restricts the free flow of information and data. At least we allow all searching results and response to keep going. Thus, we let the "client-peer" who ask for the file to decide from which "server-peer" to download the file, we control *who delivers* the file object. Given that there are a number of peers reply to a file object request and possess the file object being requested for, we select the one which leads to a more positive or less negative change of EBDA to transfer this file to the requesting peer.

In particular, when a peer n_c issues a request for a file object f_j, let n_s be the peer which possess f_j and transmit this file to n_c, we determine the change of total EBDA after the sharing action as follows:
Let $D_c^{'}$ and $D_s^{'}$ denote the EBDA of n_c and n_s *before* sharing f_j respectively,

$$\begin{cases} D_c^{'} = E_c \sum_{k=1}^{M} w_{ck}\delta_{ck} = E_c O_c \\ \\ D_s^{'} = E_s \sum_{k=1}^{M} w_{sk}\delta_{sk} = E_s O_s \end{cases} \tag{9}$$

where $O_c = \sum_{k=1}^{M} w_{ck}\delta_{ck}$ and $O_s = \sum_{k=1}^{M} w_{sk}\delta_{sk}$.

n_s transmits a file f_j to n_c, both transmit and receive action consumes power. Let P be the transmit power of n_s, αP be the power consumption of the receiver on n_c, let r be the transmission bit rate (which is assumed to be the same for all users), $S(f_j)$ be the file size of f_j, then the energy reduced on n_s and n_c are $e = PS(f_j)/r$ and $\alpha e = \alpha PS(f_j)/r$ respectively. Let $D_c^{'}$ and $D_s^{'}$ denote the EBDA of n_c and n_s *after* sharing f_j,

$$\begin{cases} D_c = (E_c - e\alpha)(O_c + w_{cj}) \\ \\ D_s = (E_s - e)O_s \end{cases} \tag{10}$$

Thus, change of total EBDA after this sharing is:

$$\begin{aligned} \Delta D &= D_c + D_s - D_c^{'} - D_s^{'} \\ &= (E_c - e\alpha)(O_c + w_{cj}) + (E_s - e)O_s \\ &\quad - E_c O_c - E_s O_s \\ &= E_c w_{cj} - e(O_s + \alpha O_c + \alpha w_{cj}) \end{aligned} \tag{11}$$

Traditionally, a node is selected as the server-peer if it owns the requested file object and it is closest to the requesting node (smallest hop count). Our idea is to add the EBDA metric in server-peer selection process:

We construct a server-set S which consists of all nodes which owns the requested file object and we find out the node which is closest to the requesting node, in terms of number of hop count and, if the file is sent from this node to the requesting node, the change of EBDA would be least negative or most positive. Then we select from this server-set a node n_s This node n_s would be the selected server-peer.

Simply put, our idea is to add the EBDA metric into the traditional server-peer process.

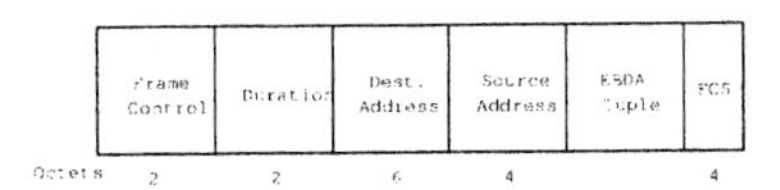

Fig. 1. Format of Request Packet

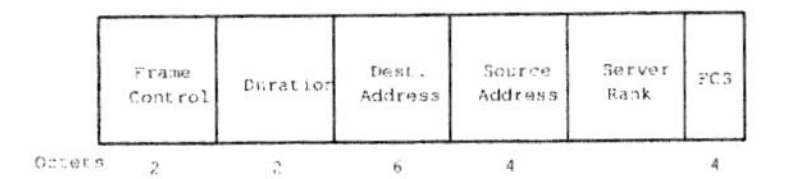

Fig. 2. Format of Response Packet

3.3 Implementation Issues

We use a common wireless ad hoc network standard, IEEE 802.11b Wireless LAN [13], as the platform to investigate the implementation aspect of our algorithm. Firstly, the client-peer n_c who request for the file f_j should transmit a request packet containing a *EBDA tuple* $<$ file ID of $f_j, E_c, O_c, w_{cj}, S(f_j, w_{cj}) >$ so that each peer n_d who receive this request packet *and* possess f_j can evaluate the value of $\Delta D(n_d, n_c)$ if it acts as the server-peer. In order to calculate e, n_d needs the file size of f_j (which is given in the request packet), transmit power P and bit rate r. In IEEE 802.11b, there is no power control such as those in CDMA IS-95 [31], each peer (e.g., a user holding a PDA with WLAN adapter) uses a known fixed power. Common transmit power ranges from 13 dBm to 30 dBm, depends on the brand of the WLAN adapter [15, 3]. The formats of the request packet and the response packet are shown in Figure 1 and Figure 2. They are modified from IEEE 802.11 MAC ACK packet [13]. After calculation of $\Delta D(n_d, n_c)$, each user n_d replies n_c by transmitting a response packet, containing a server rank, which value is given by the ΔD value calculated in Section 3.2. The client n_c would decide who to download the file from and directly request that server-peer. The download process afterward is the same as ordinary file download in conventional WLAN network.

For the seek of simplicity, we assume a one-hop search scenario, where the file request is not re-transmitted by the receiver within transmission range of n_c to other multiple–hop peers from n_c, this rules out any influence of scalability problem or route-failure problem. Contention algorithms like CSMA/CA and random back-off in IEEE 802.11b is a prerequisite for running our algorithm so that the response packets from $n_d \in N_j$ would not collided upon receiving file object request from n_c.

This technique of modifying the MAC layer packet has already been used by Zhu *et al.* to enhance WLAN to a be relay-enabled with a *rDCF* MAC layer [34].

Trustworthiness of nodes is also an important issue. "What if a user cheating?" or "What if a user giving wrong values of remaining energy levels or weight values, etc?" To address this issue we rely on other research effort made on "Reputation Systems" which are designed for evaluating the trust-worthiness and behaviour of nodes in ad hoc network [8]. Modification on these system could be used to increase the efficiency of the greedy server-peer selection protocol.

4 Simulation Platform

4.1 Mobility Model and Energy Consumption Model

In our simulations, 100 mobile users are assumed to be scattered randomly in a 400 m × 400 m area initially, they are allowed to move according to a random-walk-like model. The original idea of random walk was firstly investigated by K. Pearson, in a letter to English journal *Nature* in 1905 [25, 11] and has been used to study P2P network in [9]. Briefly speaking, it investigates the position of an moving object after n equal length movements, each with an independent, random direction. In our simulations we modified the random walk model in such a way that each user moves in a particular direction with random period of time until it change to another random direction. The velocities of all users are not changed simultaneously, at a particular time instant only a random number of users decide to change their direction of motion. Also, the velocities of users are assumed to be distributed in a Gaussian manner with mean equals $0.83ms^{-1}$ [21].

Now we define the energy model used in this paper. Firstly, the wireless devices are assumed to have three possible modes of operation: Transmit, Receive and Idle. The energy consumption ratio of the three modes is set as $1 : 0.6 : 0.4$, as indicated by the experimental measurements done by Feeney and Nilsson [7]. So the energy consumption on a node is set as:

$$P_{\mathrm{Tx}}T_{\mathrm{Tx}} + P_{\mathrm{Rx}}T_{\mathrm{Rx}} + P_{\mathrm{IDLE}}T_{\mathrm{IDLE}}$$

where the first three P terms represent power consumption in Transmit, Receive and Idle modes, respectively (exclude the exchange of control packets).

4.2 Wireless Channel Characteristics

The mobile devices in the simulations have similar wireless transmission parameters as those commonly found in IEEE 802.11b WLAN adapters available in the market which operate at 2.4 GHz ISM band, where the transmit power generally ranges from 13 dBm to 30 dBm, depending on the brand of the WLAN adapter. We set the transmit power as 20 dBm in our simulations. For simplicity we assume that the transmission rate of all users is fixed at 1 Mbps. At this bit rate the receiver sensitivity is usually around −90 dBm. We set it as −91 dBm, the same as Orinoco 802.11b Gold PCu card [24]. For radio propagation, we adopt the *Okumura-Hata Model* which is commonly used in the literature [26] to estimate the path loss.

4.3 File Searching and Fetching Model

The sizes of file objects are less than 5 MB. File objects can be MP3 songs, ring tones for mobile phones, short movie trailers in relatively low resolution, etc. Each peer is assumed to be a PDA running WLAN. We assume a file searching engine which could search the file and give the path as long as there exist a path between the requesting-node and the sever-peer. There are totally 60 different file objects in the network.

About behavior of users in simulations: a weight matrix for different objects and an aggressiveness matrix for peers. "Who request a file" and "What file object is being request" are decided according to these two matrices. It is *not* a random generation but use Zipf distribution for popularity of objects and Roulette's wheel selection for generating the requests.

5 Performance Results

The performance gain of using the greedy algorithm is of two-fold. First, we observe that the standard deviation of energy levels of all nodes become smaller (see Figure 3). We interpret this as follows. Consider the algorithm aforementioned, among all the potential server-peer, we are selecting the one with most positive or least negative change of EBDA. This means we are, in a way, selecting the sever-peer with smallest number of objects on hands to share (consider Equation (11)). This server-peer is expected to be less frequently asked to share files (because it has less files to donate). This could implicitly allocate the sever-role among different nodes more evenly, as long as the request is satisfied. The more even distribution results in a less deviated set of energy levels of different nodes. We regarded this as an energy-sense fairness.

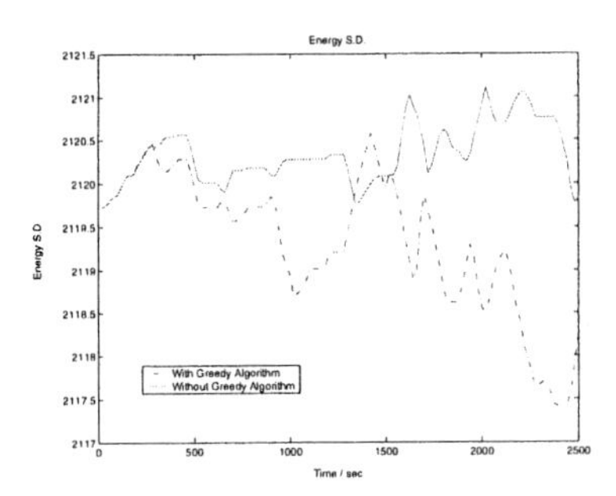

Fig. 3. Standard deviation of energy levels of nodes

Secondly, the more even distribution of server task could also avoid the existence of busy-server and thus slightly increase the file request successful ratio.

The only drawback of using the greedy algorithm is that the delay would be longer (on average 15 seconds more) since the greedy algorithm represent an extra criteria besides hop count in selection of server-peer.

6 Conclusions

We have proposed a greedy sever-peer selection algorithm which can increase the energy efficiency of the network. Our work shed some insight into a yet-to-be explored area, *wireless* P2P file sharing network which is believed to be an important interaction platform in next generation wireless communication. A new performance metric, namely energy-based data availability, has been presented. With simulations we demonstrate that the performance of the network is improved in terms of fairness and file request successful ratio.

References

1. BitTorrent, `http://bitconjurer.org/BitTorrent/`, 2004
2. C. Bram, "Incentives build robustness in BitTorrent," available at: `http://bitconjurer.org/BitTorrent/bittorrentecon.pdf`, 2004
3. Chris De Herrera, "802.11b Wireless LAN PC Cards," `http://www.cewindows.net/peripherals/ pccardwirelesslan.htm`, 2004.
4. E. Cohen, A. Fiat and H. Kaplan, "Associative search in peer to peer networks: harnessing latent semantics," *Proc. INFOCOM 2003*, vol. 2, pp. 1261–1271, Mar.-Apr. 2003.
5. eDonkey2000, `http://www.edonkey2000.com/`, 2004.
6. FastTrack, `http://www.slyck.com/ft.php`, 2004.
7. L. M. Feeney and M. Nilsson, "Investigating the energy consumption of a wireless network interface in an ad hoc networking environment," *Proc. IEEE INFOCOM 2001*, vol. 3, pp. 1548–1557, Apr. 2001.
8. S. Ganeriwal and M. B. Srivastava, "Reputation-based framework for high integrity sensor networks," *Proc. ACM SASN 2004*, pp. 66–77, Oct. 2004.
9. C. Gkantsidis, M. Mihail and A. Saberi, "Random Walks in Peer-to-Peer Networks," *Proc. IEEE INFOCOM 2004*, vol. 1, pp. 120–130, Mar. 2004.
10. Gnutella, `http://www.gnutella.com/`, 2004.
11. B. D. Huges, *Random walks and random environments*, Oxford Science Publications, 1995.
12. ICQ, `http://www.icq.com/`, 2004
13. IEEE, "Wireless LAN Medium Access Control (MAC) and Physical Layer (PHY) specifications," *IEEE Std 802.11b - 1999*, 1999.
14. KaZaA, `http://www.kazaa.com/`, 2004
15. S. Kishore, J. C. Chen, K. M. Sivalingam and P. Agrawal, "A comparison of MAC protocols for wireless local networks based on battery power consumption," *Proc. IEEE INFOCOM 1998*, vol. 1, pp. 150–157, Mar.-Apr. 1998.
16. J. Kulik, W. Rabiner and H. Balakrishnan ,"Adaptive protocols for information dissemination in wireless sensor networks ," *Proc. 5th ACM/IEEE MobiCom*, pp. 174–185, Aug. 1999.
17. I. Stoica, R. Morris, D. Liben-Nowell, D. R. Karger, M. F. Kaashoek, F. Dabek and H. Balakrishnan, "Chord: a scalable peer-to-peer lookup protocol for Internet applications," *IEEE/ACM Trans. Networking*, vol. 11, issue. 1, pp. 17–32, Feb. 2003.
18. Z. Li and M. H. Ammar, "A file-centric model for peer-to-peer file sharing systems," *Proc. 11th Int'l Conf. Network Protocols*, pp. 28–37, Nov. 2003.
19. Z. Li, E. W. Zegura and M. H. Ammar "The effect of peer selection and buffering strategies on the performance of peer-to-peer file sharing systems," *Proc. IEEE MASCOTS 2002*, pp. 63–70, Oct. 2002.
20. X. Liu, G. Yang and D. Wang, "Stationary and adaptive replication approach to data availability in structured peer-to-peer overlay networks," *Proc. ICON2003*, pp.265–270, Sept.–Oct. 2003.

21. J. G. Markoulidakis, G. L. Lyberopoulos, D. F. Tsirkas and E. D. Sykas, "Mobility modeling in third-generation mobile telecommunications systems," *IEEE Personal Communications*, vol. 4, no. 4, pp. 41–56, Aug. 1997.
22. M. D. Mustafa, B. Nathrah, M. H. Suzuri and M. T. Abu Osman, "Improving data availability using hybrid replication technique in peer-to-peer environments," AINA 2004, vol. 1, pp.593–598, Mar. 2004.
23. "Napster", http://www.napster.com/, 2004
24. Orinoco, "Orinoco 802.11b Gold PC card data sheet", http://www.proxim.com/products/ wifi/11b/, 2004.
25. K. Pearson, "The problem of the random walk," *Nature*, vol. 72, p. 294, July 1905.
26. T. S. Rappaport, *Wireless communication, principle and practice*, Prentice Hall, 1996.
27. S. Ratnasamy, P. Francis, M. Handley, R. Karp, and S. Shenker, "A scalable content-addressable network," *Proc. ACM SIGCOMM 2001*, Aug. 2001.
28. A. Rowstron and P. Druschel, "Pastry: Scalable, distributed object location and routing for large-scale peer-to-peer systems," *IFIP/ACM Int'l Conf. Distributed Systems Platforms (Middleware)*, pp. 329–350, Nov. 2001.
29. Slyck.com, http://www.slyck.com/news.php?story=574, 2004.
30. Sony CLIE handheld, http://sonyelectronics.sonystyle.com /micros/clie/, 2004.
31. TIA/EIA Interim Standard 95, July 1993.
32. The Washington Times Online, http://www.washtimes.com/technology/20040303-094741-3574r.htm, 2004.
33. B. Y. Zhao, L. Huang J. Stribling, S. C. Rhea, A. D. Joseph and J. D. Kubiatowicz, "Tapestry: a resilient global-scale overlay for service deployment," *IEEE. Journal Selected Areas in Communications*, vol. 22, issue. 1, pp. 41–53, Jan. 2004.
34. H. Zhu and G. Cao, "rDCF: A relay-enabled medium access control protocol for wireless ad hoc networks," *Proc. IEEE INFOCOM 2005*, Mar. 2005.

Can P2P Benefit from MANET? Performance Evaluation from Users' Perspective

Lu Yan

Turku Centre for Computer Science (TUCS),
FIN-20520 Turku, Finland
lu.yan@ieee.org

Abstract. With the advance in mobile wireless communication technology and the increasing number of mobile users, peer-to-peer computing, in both academic research and industrial development, has recently begun to extend its scope to address problems relevant to mobile devices and wireless networks. This paper is a performance study of peer-to-peer systems over mobile ad hoc networks. We show that cross-layer approach performs better than separating the overlay from the access networks with the comparison of different settings for the peer-to-peer overlay and underlying mobile ad hoc network.

1 Introduction

Peer-to-Peer (P2P) computing is a networking and distributed computing paradigm which allows the sharing of computing resources and services by direct, symmetric interaction between computers. With the advance in mobile wireless communication technology and the increasing number of mobile users, peer-to-peer computing, in both academic research and industrial development, has recently begun to extend its scope to address problems relevant to mobile devices and wireless networks.

Mobile Ad hoc Networks (MANET) and P2P systems share a lot of key characteristics: self-organization and decentralization, and both need to solve the same fundamental problem: connectivity. Although it seems natural and attractive to deploy P2P systems over MANET due to this common nature, the special characteristics of mobile environments and the diversity in wireless networks bring new challenges for research in P2P computing.

Currently, most P2P systems work on wired Internet, which depends on application layer connections among peers, forming an application layer overlay network. In MANET, overlay is also formed dynamically via connections among peers, but without requiring any wired infrastructure. So, the major differences between P2P and MANET that we concerned in this paper are (a) P2P is generally referred to the application layer, but MANET is generally referred to the network layer, which is a lower layer concerning network access issues. Thus, the immediate result of this layer partition reflects the difference of the packet transmission methods between P2P and MANET: the P2P overlay is a unicast network with virtual broadcast consisting of numerous single unicast packets; while the MANET overlay always performs physical broadcasting. (b) Peers in P2P overlay is usually referred to static node

X. Jia, J. Wu, and Y. He (Eds.): MSN 2005, LNCS 3794, pp. 1026 – 1035, 2005.

though no priori knowledge of arriving and departing is assumed, but peers in MANET is usually referred to mobile node since connections are usually constrained by physical factors like limited battery energy, bandwidth, computing power, etc.

The above similarities and differences between P2P and MANET lead to an interesting but challenging research on P2P systems over MANET. Although both P2P and MANET have recently becoming popular research areas due to the widely deployment of P2P applications over Internet and rapid progress of wireless communication, few research has been done for the convergence of the two overlay network technologies. In fact, the scenario of P2P systems over MANET seems feasible and promising, and possible applications for this scenario include car-to-car communication in a field-range MANET, an e-campus system for mobile e-learning applications in a campus-range MANET on top of IEEE 802.11, and a small applet running on mobile phones or PDAs enabling mobile subscribers exchange music, ring tones and video clips via Bluetooth, etc.

2 Background and State-of-the-Art

Since both P2P and MANET are becoming popular only in recent years, the research on P2P systems over MANET is still in its early stage. The first documented system is Proem [1], which is a P2P platform for developing mobile P2P applications, but it seems to be a rough one and only IEEE 802.11b in ad hoc mode is supported. 7DS [2] is another primitive attempt to enable P2P resource sharing and information dissemination in mobile environments, but it is rather a P2P architecture proposal than a practical application. In a recent paper [3], Passive Distributed Indexing was proposed for such kind of systems to improve the search efficiency of P2P systems over MANET, and in ORION [4], a Broadcast over Broadcast routing protocol was proposed. The above works were focused on either P2P architecture or routing schema design, but how efficient is the approach and what is the performance experienced by users are still in need of further investigation.

Previous work on performance study of P2P over MANET was mostly based on simulative approach and no concrete analytical mode was introduced. Performance issues of this kind of systems were first discussed in [5], but it simply shows the experiment results and no further analysis was presented. There is a survey of such kind of systems in [6] but no further conclusions were derived, and a sophisticated experiment and discussion on P2P communication in MANET can be found in [7]. Recently, B. Bakos etc. with Nokia Research analyzed a Gnutella-style protocol query engine on mobile networks with different topologies in [8], and T. Hossfeld etc. with Siemens Labs conducted a simulative performance evaluation of mobile P2P file-sharing in [9]. However, all above works fall into practical experience report category and no performance models are proposed.

3 Performance Evaluation of P2P over MANET

There have been many routing protocols in P2P networks and MANET respectively. For instance, one can find a very substantial P2P routing scheme survey from HP

Labs in [10], and US Navy Research publish ongoing MANET routing schemes in [11]; but all above schemes fall into two basic categories: broadcast-like and DHT-like. More specifically, most early P2P search algorithms, such as in Gnutella [12], Freenet [13] and Kazaa [14], are broadcast-like and some recent P2P searching, like in eMule [15] and BitTorrent [16], employs more or less some feathers of DHT. On the MANET side, most on-demand routing protocols, such as DSR [17] and AODV [18], are basically broadcast-like. Therefore, we here introduce different approaches to integrate these protocols in different ways according to categories.

3.1 Broadcast over Broadcast

The most straight forward approach is to employ a broadcast-like P2P routing protocol at the application layer over a broadcast-like MANET routing protocol at the network layer. Intuitively, in the above settings, every routing message broadcasting to the *virtual* neighbors at the application layer will result to a full broadcasting to the corresponding *physical* neighbors at the network layer.

The scheme is illustrated in Figure 1 with a searching example: peer *A* in the P2P overlay is trying to search for a particular piece of information, which is actually available in peer *B*. Due to broadcast mechanism, the search request is transmitted to its neighbors, and recursively to all the members in the network, until a match is found or timeout. Here we use the blue lines represent the routing path at this application layer. Then we map this searching process into the MANET overlay, where node *A0* is the corresponding mobile node to the peer *A* in the P2P overlay, and *B0* is related to *B* in the same way. Since the MANET overlay also employs a broadcast-like routing protocol, the request from node *A0* is flooded (broadcast) to directly connected neighbors, which themselves flood their neighbors etc., until the request is answered or a maximum number of flooding steps occur. The route establishing lines in that network layer is highlighted in red, where we can find that there are few overlapping routes between these two layers though they all employ broadcast-like protocols.

We have studied a typical broadcast-like P2P protocol, Guntella [19], in the previous work [20]. This is a pure P2P protocol, as shown in Figure 2, in which no advertisement of shared resources (e.g. directory or index server) occurs. Instead, each request from a peer is broadcasted to directly connected peers, which themselves

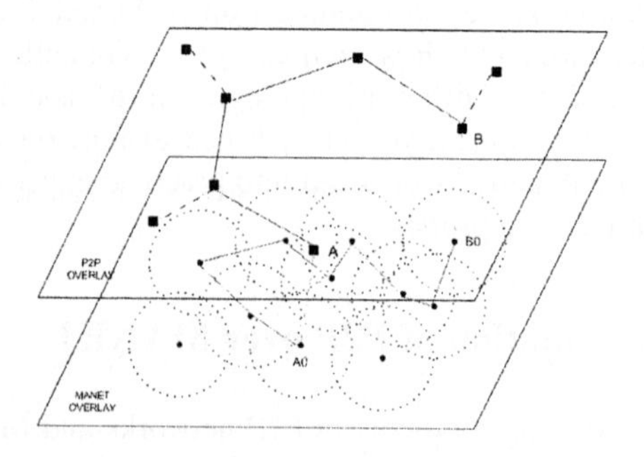

Fig. 1. Broadcast over Broadcast

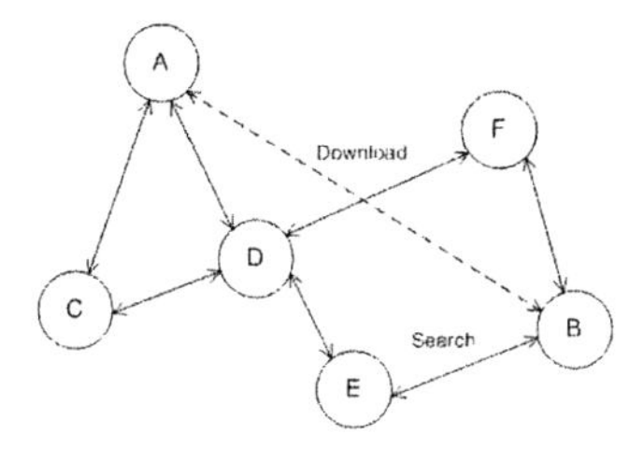

Fig. 2. Broadcast-like P2P Protocol

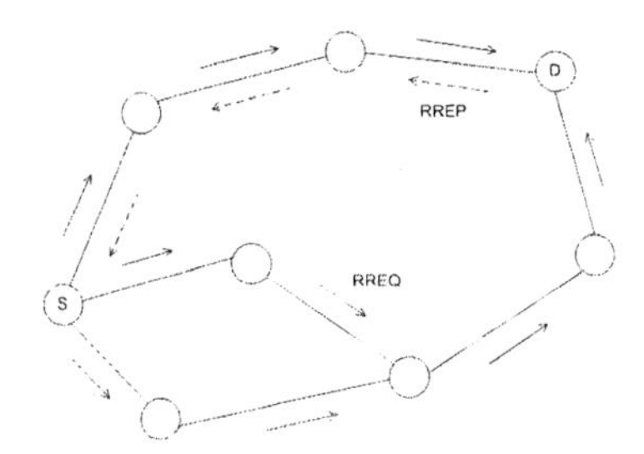

Fig. 3. Broadcast-like MANET Protocol

broadcast this request to their directly connected peers etc., until the request is answered or a maximum number of broadcast steps occur. It is easy to see that this protocol requires a lot of network bandwidth, and it does not prove to be very scalable. The complexity of this routing algorithm is $O(n)$. [21, 22]

Generally, most on-demand MANET protocols, like DSR [23] and AODV [24], are broadcast-like in nature [25]. Previously, one typical broadcast-like MANET protocol, AODV, was studied in [26]. As shown in Figure 3, in that protocol, each node maintains a routing table only for active destinations: when a node needs a route to a destinations, a path discovery procedure is started, based on a RREQ (route request) packet; the packet will not collect a complete path (with all IDs of involved nodes) but only a hop count; when the packet reaches a node that has the destination in its routing table, or the destination itself, a RREP (route reply) packet is sent back to the source (through the path that has been set-up by the RREQ packet), which will insert the destination in its routing table and will associate the neighbour from which the RREP was received as preferred neighbour to that destination. Simply speaking, when a source node wants to send a packet to a destination, if it does not know a valid route, it initiates a route discovery process by flooding RREQ packet through the network. AODV is a pure on-demand protocol, as only nodes along a path maintain routing information and exchange routing tables. The complexity of that routing algorithm is $O(n)$. [27]

This approach is probably the easiest one to implement, but the drawback is also obvious: the routing path of the requesting message is not the shortest path between source and destination (e.g. the red line in Figure 1), because the *virtual* neighbors in the P2P overlay are not necessarily also the *physical* neighbors in the MANET overlay, and actually these nodes might be physically far away from each other.

Therefore, the resulting routing algorithm complexity of this broadcast over broadcast scheme is unfortunately $O(n^2)$ though each layer's routing algorithm complexity is $O(n)$ respectively.

It is not practical to deploy such kind of scheme for its serious scalability problem due to the double broadcast; and taking the energy consumption portion into consideration, which is somehow critical to mobile devices, the double broadcast will also cost a lot of energy consumption, and make it infeasible in cellular wireless date networks.

3.2 DHT over Broadcast

The scalability problem of broadcast-like protocols has long been observed and many revisions and improvement schemas are proposed in [28, 29, 30]. To overcome the scaling problems in broadcast-like protocols where data placement and overlay network construction are essentially random, a number of proposals are focused on structured overlay designs. Distributed Hash Table (DHT) [31] and its varieties [32, 33, 34] advocated by Microsoft Research seem to be promising routing algorithms for overlay networks. Therefore it is interesting to see the second approach: to employ a DHT-like P2P routing protocol at the application layer over a broadcast-like MANET routing protocol at the network layer.

The scheme is illustrated in Figure 4 with the same searching example. Comparing to the previous approach, the difference lies in the P2P overlay: in a DHT-like protocol, files are associated to keys (e.g. produced by hashing the file name); each node in the system handles a portion of the hash space and is responsible for storing a certain range of keys. After a lookup for a certain key, the system returns the identity (e.g. the IP address) of the node storing the object with that key. The DHT functionality allows nodes to put and get files based on their key, and each node handles a portion of the hash space and is responsible for a certain key range. Therefore, routing is location-deterministic distributed lookup (e.g. the blue line in Figure 4).

DHT was first proposed by Plaxton etc. in [35], without intention to address P2P routing problems. DHT soon proved to be a useful substrate for large distributed systems and a number of projects are proposed to build Internet-scale facilities

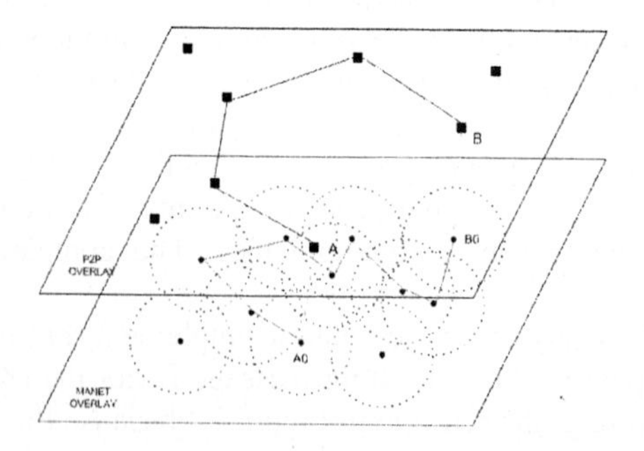

Fig. 4. DHT over Broadcast

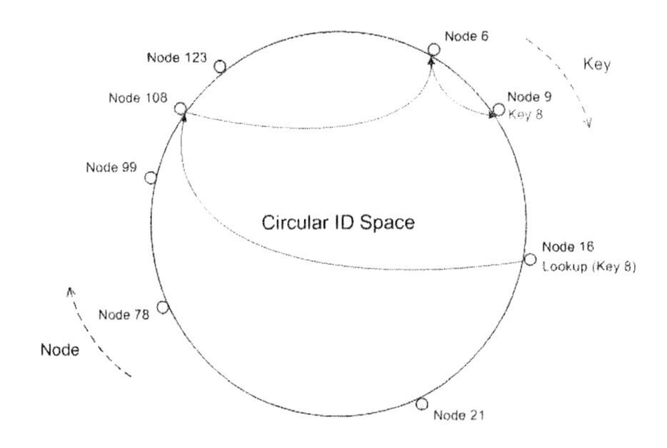

Fig. 5. DHT-like P2P Protocol

layered above DHTs, among them are Chord [31], CAN [32], Pastry [33], Tapestry [34] etc. As illustrated in Figure 5, all of them take a key as input and route a message to the node responsible for that key. Nodes have identifiers, taken from the same space as the keys. Each node maintains a routing table consisting of a small subset of nodes in the system. When a node receives a query for a key for which it is not responsible, the node routes the query to the *hashed* neighbor node towards resolving the query. In such a design, for a system with n nodes, each node has $O(log\ n)$ neighbors, and the complexity of the DHT-like routing algorithm is $O(\log n)$. [36]

Additional work is required to implement this approach, partly because DHT requires a periodical maintenance (i.e. it is just like an Internet-scale hash table, or a large distributed database); since each node maintains a routing table (i.e. hashed keys) to its neighbors according to DHT algorithm, following a node join or leave, there is always a nearest key reassignment between nodes.

This DHT over Broadcast approach is obviously better than the previous one, but it still does not solve the shortest path problem as in the Broadcast over Broadcast scheme. Though the P2P overlay algorithm complexity is optimized to $O(log\ n)$, the mapped message routing in the MANET overlay is still in the broadcast fashion with complexity $O(n)$; the resulting algorithm complexity of this approach is as high as $O(n\ log\ n)$.

This approach still requires a lot of network bandwidth, and hence does not prove to be very scalable, but it is efficient in limited communities, such as a company network.

3.3 Cross-Layer Routing

A further step of the Broadcast over Broadcast approach would be a Cross-Layer Broadcast. Due to similarity of Broadcast-like P2P and MANET protocols, the second broadcast could be skipped if the peers in the P2P overlay would be mapped directly into the MANET overlay, and the result of this approach would be the merge of application layer and network layer (i.e. the *virtual* neighbors in P2P overlay overlaps the *physical* neighbors in MANET overlay).

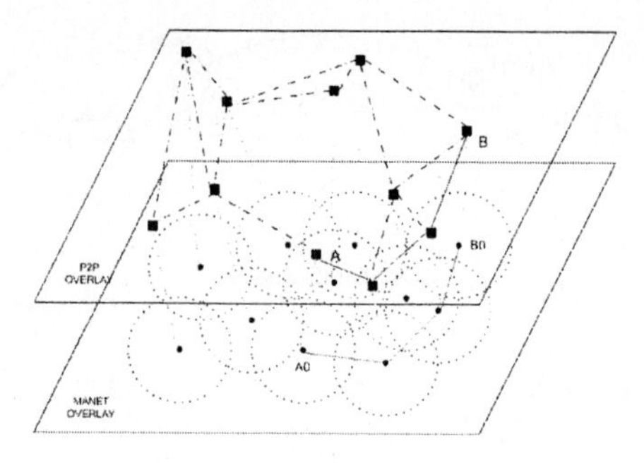

Fig. 6. Cross-Layer Broadcast

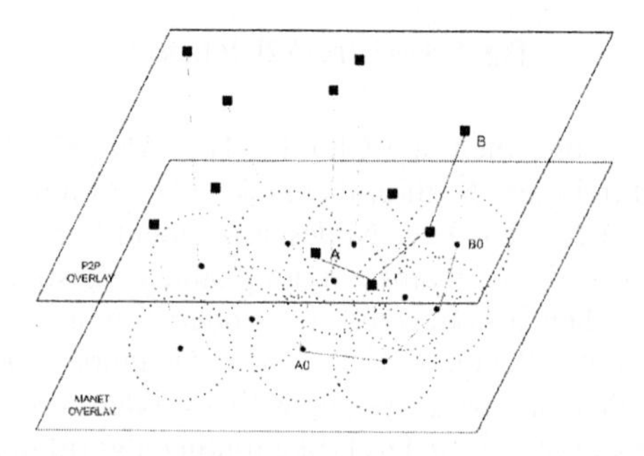

Fig. 7. Cross-Layer DHT

The scheme is illustrated in Figure 6, where the advantage of this cross-layer approach is obvious: the routing path of the requesting message is the shortest path between source and destination (e.g. the blue and red lines in Figure 6), because the *virtual* neighbors in the P2P overlay are *de facto physical* neighbors in the MANET overlay due to the merge of two layers. Thanks to the nature of broadcast, the algorithm complexity of this approach is $O(n)$, making it suitable for deployment in relatively large scale networks, but still not feasible for Internet scale networks.

It is also possible to design a Cross-Layer DHT in Figure 7 with the similar inspiration, and the algorithm complexity would be optimized to $O(log\ n)$ with the merit of DHT, which is advocated to be efficient even in Internet scale networks. The difficulty in that approach is implementation: there is no off-the-shelf DHT-like MANET protocol as far as know, though recently, some research projects, like Ekta [37], towards a DHT substrate in MANET are proposed.

Table 1. How efficient does a user try to find a specific piece of data?

	Efficiency	Scalability	Implementation
Broadcast over Broadcast	$O(n^2)$	N.A.	Easy
DHT over Broadcast	$O(n\ log\ n)$	Bad	Medium
Cross-Layer Broadcast	$O(n)$	Medium	Difficult
Cross-Layer DHT	$O(log\ n)$	Good	N.A.

As an answer to the question in our paper title, we show that the cross-layer approach performs better than separating the overlay from the access networks, with the comparison of different settings for the peer-to-peer overlay and underlying mobile ad hoc network in above four approaches in Table 1.

4 Concluding Remarks

In this paper, we studied the peer-to-peer systems over mobile ad hoc networks with a comparison of different settings for the peer-to-peer overlay and underlying mobile ad hoc network. We show that cross-layer approach performs better than separating the overlay from the access networks. We hope our results would potentially provide useful guidelines for mobile operators, value-added service providers and application developers to design and dimension mobile peer-to-peer systems, and as a foundation for our long term goal [38].

References

1. G. Kortuem, J. Schneider, D. Preuitt, T. G. C. Thompson, S. Fickas, Z. Segall. When Peer-to-Peer comes Face-to-Face: Collaborative Peer-to-Peer Computing in Mobile Ad hoc Networks. In *Proc. 1st International Conference on Peer-to-Peer Computing (P2P 2001)*, Linkoping, Sweden, August 2001.
2. M. Papadopouli and H. Schulzrinne. A Performance Analysis of 7DS a Peer-to-Peer Data Dissemination and Prefetching Tool for Mobile Users. In *Advances in wired and wireless communications, IEEE Sarnoff Symposium Digest*, 2001, Ewing, NJ.
3. C. Lindemann and O. Waldhorst. A Distributed Search Service for Peer-to-Peer File Sharing in Mobile Applications. In *Proc. 2nd IEEE Conf. on Peer-to-Peer Computing (P2P 2002)*, 2002.
4. A. Klemm, Ch. Lindemann, and O. Waldhorst. A Special-Purpose Peer-to-Peer File Sharing System for Mobile Ad Hoc Networks. In *Proc. IEEE Vehicular Technology Conf.*, Orlando, FL, October 2003.
5. S. K. Goel, M. Singh, D. Xu. Efficient Peer-to-Peer Data Dissemination in Mobile Ad-Hoc Networks. In *Proc. International Conference on Parallel Processing (ICPPW '02)*, IEEE Computer Society, 2002.
6. G. Ding, B. Bhargava. Peer-to-peer File-sharing over Mobile Ad hoc Networks. In *Proc. 2nd IEEE Conf. on Pervasive Computing and Communications Workshops*. Orlando, Florida, 2004.
7. H.Y. Hsieh and R. Sivakumar. On Using Peer-to-Peer Communication in Cellular Wireless Data Networks. In *IEEE Transaction on Mobile Computing*, vol. 3, no. 1, January-March 2004.
8. B. Bakos, G. Csucs, L. Farkas, J. K. Nurminen. Peer-to-peer protocol evaluation in topologies resembling wireless networks. An Experiment with Gnutella Query Engine. In *Proc. International Conference on Networks*, Sydney, Oct., 2003.
9. T. Hossfeld, K. Tutschku, F. U. Andersen, H. Meer, J. Oberender. Simulative Performance Evaluation of a Mobile Peer-to-Peer File-Sharing System. *Research Report 345*, University of Wurzburg, Nov. 2004.
10. D. S. Milojicic, V. Kalogeraki, R. Lukose, K. Nagaraja, J. Pruyne, B. Richard, S. Rollins, Z. Xu. Peer-to-Peer Computing. *Technical Report HPL-2002-57*, HP Labs.

11. *MANET Implementation Survey*. Available at http://protean.itd.nrl.navy.mil/manet/survey/survey.html
12. Gnutella: http://www.gnutella.com/
13. Freenet: http://freenet.sourceforge.net/
14. Kazaa: http://www.kazaa.com/
15. eMule: http://www.emule-project.net/
16. BitTorrent: http://bittorrent.com/
17. *DSR IETF draft v1.0*. Available at http://www.ietf.org/internet-drafts/draft-ietf-manet-dsr-10.txt
18. *AODV IETF draft v1.3*. Available at http://www.ietf.org/internet-drafts/draft-ietf-manet-aodv-13.txt
19. Clip2. *The gnutella protocol specification v0.4 (document revision 1.2)*. Available at http://www9.limewire.com/developer /gnutella protocol 0.4.pdf, Jun 2001.
20. L. Yan and K. Sere. Stepwise Development of Peer-to-Peer Systems. In *Proc. 6th International Workshop in Formal Methods (IWFM'03)*. Dublin, Ireland, July 2003.
21. M. Ripeanu, I. Foster and A. Iamnitch. Mapping the Gnutella Network: Properties of Large-Scale Peer-to-Peer Systems and Implications for System Design. In *IEEE Internet Computing*, vol. 6(1) 2002.
22. Y. Chawathe, S. Ratnasamy, L. Breslau, S. Shenker. Making Gnutella-like P2P Systems Scalable. In *Proceedings of ACM SIGCOMM*, 2003.
23. D. B. Johnson, D. A. Maltz. Dynamic Source Routing in Ad-Hoc Wireless Networks. In *Mobile Computing*, Kluwer, 1996.
24. C. E. Perkins and E. M. Royer. The Ad hoc On-Demand Distance Vector Protocol. In *Ad hoc Networking*. Addison-Wesley, 2000.
25. F. Kojima, H. Harada and M. Fujise. A Study on Effective Packet Routing Scheme for Mobile Communication Network. In *Proc. 4th Intl. Symposium on Wireless Personal Multimedia Communications*, Denmark, Sept. 2001.
26. L. Yan and J. Ni. Building a Formal Framework for Mobile Ad Hoc Computing. In *Proc. International Conf. on Computational Science (ICCS 2004)*. Krakow, Poland, June 2004. LNCS 3036, Springer-Verlag.
27. E. M. Royer and C. K. Toh. A Review of Current Routing Protocols for Ad-Hoc Mobile Wireless Networks. In *IEEE Personal Communications*, April 1999.
28. Q. Lv, S. Ratnasamy and S. Shenker. Can Heterogeneity Make Gnutella Scalable? In *Proc. 1st International Workshop on Peer-to-Peer Systems (IPTPS '02)*, Cambridge, MA, March 2002.
29. B. Yang and H. Garcia-Molina. Improving Search in Peer-to-Peer Networks. In *Proc. Intl. Conf. on Distributed Systems (ICDCS)*, 2002.
30. Y. Chawathe, S. Ratnasamy, L. Breslau, and S. Shenker. Making Gnutella-like P2P Systems Scalable. In *Proc. ACM SIGCOMM 2003*, Karlsruhe, Germany, August 2003.
31. I. Stoica, R. Morris, D. Karger, F. Kaashoek and H. Balakrishnan. Chord: A Scalable Peer-To-Peer Lookup Service for Internet Applications. In *Proc. ACM SIGCOMM*, 2001.
32. S. Ratnasamy, P. Francis, M. Handley, R. Karp and S. Schenker. A scalable content-addressable network. In *Proc. Conf. on applications, technologies, architectures, and protocols for computer communications*, ACM, 2001.
33. A. Rowstron and P. Druschel. Pastry: Scalable, distributed object location and routing for large-scale peer-to-peer systems. In *Proc. IFIP/ACM International Conference on Distributed Systems Platforms (Middleware)*, Heidelberg, Germany, pages 329-350, November, 2001.

34. B. Y. Zhao, L. Huang, J. Stribling, S. C. Rhea, A. D. Joseph, and J. Kubiatowicz. Tapestry: A Resilient Global-scale Overlay for Service Deployment. In *IEEE Journal on Selected Areas in Communications*, January 2004, Vol. 22, No. 1.
35. C. Plaxton, R. Rajaraman, A. Richa. Accessing nearby copies of replicated objects in a distributed environment. In *Proc. ACM SPAA*, Rhode Island, June 1997.
36. S. Ratnasamy, S. Shenker, I. Stoica. Routing Algorithms for DHTs: Some Open Questions. In *Proc. 1st International Workshop on Peer-to-Peer Systems*, March 2002.
37. H. Pucha, S. M. Das and Y. C. Hu. Ekta: An Efficient DHT Substrate for Distributed Applications in Mobile Ad Hoc Networks. In *Proc. 6th IEEE Workshop on Mobile Computing Systems and Applications*, December 2004, UK.
38. L. Yan, K. Sere, X. Zhou, and J. Pang. Towards an Integrated Architecture for Peer-to-Peer and Ad Hoc Overlay Network Applications. In *Proc. 10th IEEE International Workshop on Future Trends of Distributed Computing Systems (FTDCS 2004)*, May 2004.

Research on Dynamic Modeling and Grid-Based Virtual Reality

Luliang Tang[1,2,*] and Qingquan Li[1,2]

[1] State Key Laboratory of Information Engineering in Surveying, Mapping and Remote Sensing (LIESMRS), Wuhan University, Wuhan, China
tll@whu.edu.cn

[2] Research Center of Spatial Information and Network Communication (SINC), Wuhan University, Wuhan, China
qqli@whu.edu.cn

Abstract. It is publicly considered that the next generational Internet technology is grid computing, which supports the sharing and coordinated use of diverse resources in dynamic virtual organizations from geographically and organizationally distributed components. Grid computing characters strong computing ability and broad width information exchange[1]. Globus presented Open Grid Services Architecture (OGSA), which centered on grid services[3]. According to the characteristic of Grid-based Virtual Reality (GVR) and the development of current grid computing, this paper put forward the Orient-Grid Distributed Network Model for GVR, whose dynamic Virtual group is corresponding with the Virtual Organization in OGSA service. The GDNM is of more advantage to the distributed database consistency management, and is more convenient to the virtual group users acquiring the GVR data information, and the dynamic virtual groups in GDNM are easier and more directly to utilize the grid source and communication each other. The architecture of GVR designed in this paper is based on OGSA and web services, which is based on the OGSA. This architecture is more convenient to utilizing grid service and realizing the GVR. This paper put forward the method of virtual environment Object-oriented Dynamic Modeling (OODM) based on Problem Solving (PS), which is applied with dynamic digital terrain and dynamic object modeling. This paper presents the implementation of GVR and the interfaces of Grid Service.

1 Introduction

It plays an increasingly important role in modern society that the useful information is processed, visualized and integrated from various sources, while the information sources may be widely distributed, and the data processing requirements can be highly variable, and the type of resources are required, and the processing demands are put upon the researchers [5].

Grid technology is a major cornerstone of today's computational science and engineering, and provides a powerful medium to achieve the integration of large amounts

[*] Corresponding author.

X. Jia, J. Wu, and Y. He (Eds.): MSN 2005, LNCS 3794, pp. 1036–1042, 2005.

of experimental data and computational resources, from simple parameters and highly distributed networks, into complex interactive operation that allow the researchers to efficiently run their experiments, by optimizing overhead and performance.

The next generational internet based on grid computing makes it possible to build the Distributed Virtual Reality (DVR), which can realizes the mass data dynamic modeling, distributed virtual geographic environment, and collaborative GIS.

2 Grid Computing and Middleware

2.1 Grid Computing

The Grid has been defined as "flexible, secure, coordinated resource sharing among dynamic collections of individuals, institutions and resources (A virtual organization)" and "A computational grid is a hardware and software infrastructure that provides dependable, consistent, pervasive, and inexpensive access to high-end computational capabilities."[1].

The first grid test bed was put up with 1000M bandwidth in 1990s, more and more great grid projects were started such as I-WAY, Globus, Legion and the Golobal Grid Forum (CGF) in America, CERN DataGrid, UNICORE and MOL in Europe, Nimrod/G and EcoGrid in Austria, Ninf and Bricks in Japan [1]. In China, Li Sanli started his "ACI" grid research, and Chinese Academy of Sciences started state "863" project named "Vega Grid" in 1999. China National Grid (CNGrid) was started with many grid nodes in 2002, the China Grid Forum (CGF) came into existence in 2003[6].

2.2 Middleware and Globus Toolkit

The Globus Toolkit is a software toolkit that allows develop to program grid-based applications, which is one of the biggest corporations in research on grid computing [2,8], and is being the practical standard for grid system researches and applications.

Resource manager is most important function in Globus Toolkit as a grid middleware, which is divided into two services, the resource broker and the job manager. Figure 1 displays the workflow of Globus Toolkit Grid middleware.

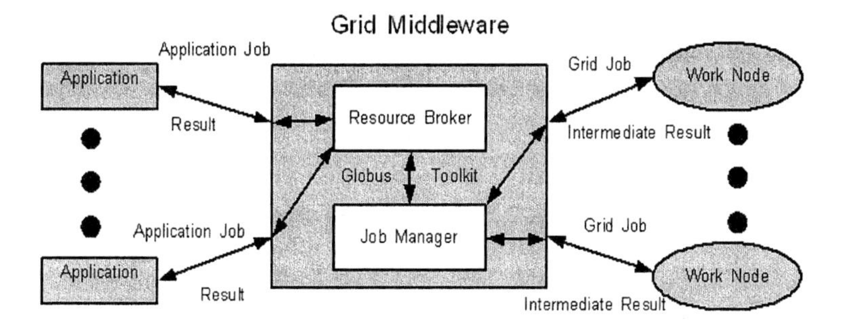

Fig. 1. Basic principle of Globus Toolkit Grid Middleware

3 Distributed Virtual Reality and Grid-Based Virtual Reality

3.1 Distributed Virtual Reality

With the development of the internet, the demand of widely distributed resource and data processing requirements, DVR (Distributed Virtual Reality, DVR) becomes one of hot research topics. The standard of DVR has been changed from the Distributed Information System (DIS) interactive simulation of SIMUNET to High Level Architecture (HLA), which includes three parties such as OMT (Object Model Temple), Rules and Interface Specification. WorldToolKit (WTK) is a develop tool for virtual reality. Vega is another developing tool kit for virtual reality. Canada Alberta University developed MR, and English Division corporation developed dVS, and Sweden distributed information system lab develops DIVE (Distributed Interactive Virtual Environment, DIVE), and America Naval Postgraduate School develops NPSNET-IV [4].

The network data model of DVR has three types, such as concentrated network model, distributed network model and duplicate network model.

3.2 Grid-Based Virtual Reality

The development of GVR depends on the development of network technology and distributed computing as same as the development of GIS depends on the development of computer science and technology in a certain extent, and.

The network based on grid computing makes it possible to build GVR, which can realize the mass data dynamic modeling, and build the distributed virtual geographic environment, and fulfill the collaborative GIS and interactive cooperation, and make it come into the fact that the tele-immersion and visual systems are built.

GVR would change the old "humans/computers" interaction mode into "humans/ computers/ humans cooperation" interaction mode.

GVR can be applied to distributed dynamic modeling, military simulation, collaborative virtual environment (CVE), tele-immersion, such as distributed military environment dynamic modeling, cooperative training, fight scene distribution, resource layout, information service, Log service, virtual remote library, this paper analyses and put forward the application of GVR.

4 The Architecture of GVR

4.1 Grid-Oriented Distributed Network Model of GVR

In this paper, a network model of GVR is designed, named Grid-Oriented Distributed Network Model (GDNM), and Figure 2 displays the Grid-Oriented Distributed Network Model of GVR. GDNM is composed of virtual groups and GVR server, and virtual groups can access each other directly by the grid, and the GVR server is answer for the database consistency of the distributed virtual group server. Virtual group is composed of group client and group server, and group clients acquire the scene database from the self-group server no accessing the database from other group servers or GVR server. Virtual group is dynamic group and corresponding with the Virtual Organization in grid OGSA service, group clients can access the group server scene database from GVR server.

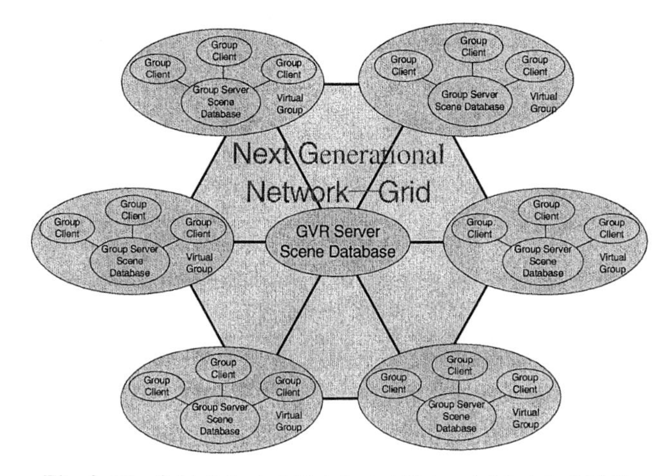

Fig. 2. The Grid-Oriented Distributed Network Model of GVR

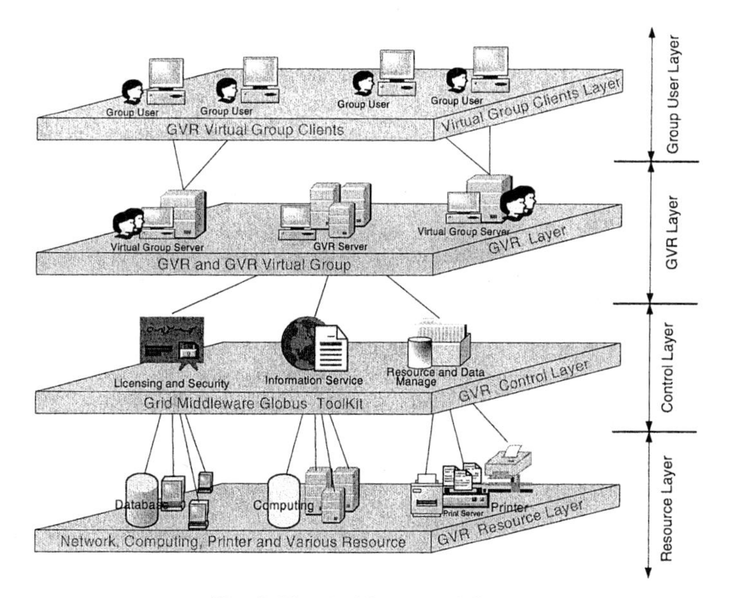

Fig. 3. The Architecture of GVR

4.2 The Architecture of GVR

The architecture of GVR is composed of four tiers such as GVR Resource Layer, GVR Control Layer, GVR Layer and Virtual Group Clients Layer. GVR Resource layer includes the network grid-based, high performance computer server, scene database and various resources. GVR Control Layer answers for the security management, information service, data management and resource management. GVR Layer is composed with GVR server and GVR Virtual Group server, and GVR server answer for the GVR data management and the databases consistency management that are distributed in virtual group servers, and GVR Virtual Groups acquire the GVR scene data from the GVR server and answer for the GVR group client users fulfilling the distributed simulation (Figure 3).

The architecture of GVR is based on the Open Grid Services Architecture (OGSA), which is centered with Web Services. This architecture of GVR is OGSA-oriented, which can utilize grid service efficiency, and decrease the conflict with the grid environment.

4.3 Object-Oriented Dynamic Modeling Based on Problem Solving

Modeling is an important in Virtual Reality (VR), and there are much software which can fulfill the static object modeling, such as 3DMax, Maya, MultiGen Creator and so on. GVR is based on grid computing, and makes it possible that the dynamic object modeling is practiced in distributed environment.

Terrain models can be classified into two types: Digital Elevation Model (DEM) and Triangulated Irregular Network (TIN) [9]. Generally speaking, a DEM is easier to be constructed than a TIN, but for a multi-resolution model, DEM structure may have the problem of terrain tearing between tiles with different resolutions. Although a TIN works well in the multi-resolution case, its complex algorithms of the transformation between different resolution models impede its use in real-time applications.

The architecture of GVR is OGSA-oriented which is a service-oriented architecture, the researchers pay attention to the problem more than the resource type and the resource coming, because the grid computing supplies a Problem Solving Environment. A Problem Solving Environment (PSE) provides the user a complete and integrated environment for problem composition, solution, and analysis [10]. The environment should provide an intuitive interface to the available Grid resource, which abstracts the complexities of accessing grid resource by providing a complete suite of high-level tools designed to tackle a particular problem [11].

Dynamic digital terrain in GVR is applied for the method of Object-oriented Dynamic Modeling (OODM) based on the Problem Solving (PS). Digital terrain model is presented by triangulated irregular network (TIN), and the TIN creation becomes the problem. GVR put forward this problem to the grid middleware, and the Grid Service find and get the computing resource in the Problem Solving Environment.

5 Implementation of GVR

It is important for GVR that the task management middleware and resource management middleware, which decide the success or failure of the whole system, now we will discuss these two parts' technical implementation in details. Task management middleware takes charge of all the tasks including task management and dispatch, disintegration, distribution, result merge, report generation, etc.

The mission of task management middleware in this system is mainly completed by task object. After receiving users' application job, task management middleware dispatch task object to job manager, every task object has the object-oriented characteristic. Users cannot operate with resources directly except by task object. Different task objects have different missions, including searching, displaying, suspending, resuming, stopping the operation, which were decided by task management middleware when they were generated.

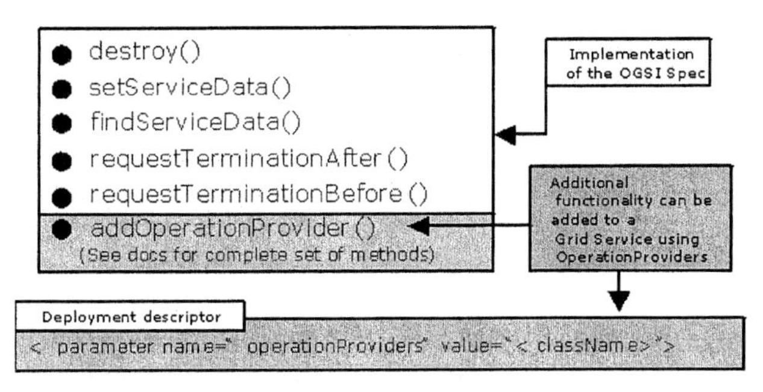

Fig. 4. Implementation of the GridService port Type

Table 1. The Interface of GridService

Interface	Operation
GridService	FindServiceData()
	SetTerminationTime()
	Destroy()
NotificationSource	SubscribeToNotificationTopic()
	UnSubscibeToNotificationTopic()
NotificationSink	DeliverNotification()
Registry	RegisterService()
	UnRegisterService()
Factory	CreateService()
PrimaryKey	FindByPrimaryKey()
	DestroyPrimaryKey()
HandleMap	FindByHandle()

Resource management middleware deals with the problems how to describe resources in resource registry center and how to publish, find and bind service resources. Figure 4 displays the implementation API interface of grid service, users may fulfill the grid application in "C", Java, VS.NET. The OGSI supplies the GridServiec interface using API operations such as FindServiceData(), SetTerminationTime() and Destroy().

OGSI supplies user other interfaces such as NotificationSource, NotificationSink, Registry, Factory, PrimaryKey and HandleMap (Table 1).

6 Conclusion

According to the characteristic of Grid-based Virtual Reality (GVR) and the development of current grid computing, this paper put forward the Orient-Grid Distributed Network

Model for GVR, whose dynamic Virtual group is corresponding with the Virtual Organization in OGSA service. The GDNM is of more advantage to the distributed database consistency management, and is more convenient to the virtual group users acquiring the GVR data information, and the dynamic virtual groups in GDNM are easier and more directly to utilize the grid source and communication each other.

The architecture of GVR designed in this paper is based on OGSA and web services, which is keep to "the five-tiers sandglass structure" of the OGSA. This architecture is more convenient to utilizing grid service and decreasing the conflict with the grid environment. This paper presents the implementation of GVR and the interfaces of Grid Service.

The next network based on grid computing bring us much more new idea and new change to our social, and many grid network researchers describe the blueprint of grid network infrastructure and good foreground of application in GIS field. While GVR is a new issue that faces many researchers in computer, virtual reality and GIS fields, and it is only in experimental period at present. GVR is a sort of project that involves in different regions, different users, different platforms, different subjects and different fields' researchers, and will bring us more challenge.

Acknowledgement

This work has been supported by the Open Research Fund of LIESMARS, No. (03)0404.

References

1. I. Foster and C. Kesselman. The Grid: Blueprint for a New Computing Infrastructure. Morgan Kaufmann, Los Altos, CA, 1998. www.mkp.com/book_catalog/1-55860-475-8.asp
2. What is the Globus Toolkit? http://www.globus.org/
3. I. Foster, C. Kesselman, J. Nick, S. Tuecke, "The Physiology of the Grid: An Open Grid Services Architecture for Distributed Systems Integration," June 22, 2002.
4. Zhang Maojun. Virtual reality system. Beijing, Science Publisher.2001
5. William E. Johnston. Computational and data Grids in large-scale science and engineering. Future Generation Computer Systems 18 (2002) 1085–1100.
6. Dou Zhihui, Chen Yu, Liu Peng. Grid computing. Beijing, Tsinghua University Publisher. 2002
7. Gong Jianhua, Lin Hui. Study on distributed virtual Geo-environments. Journal of Image and Graphics. Vol. 6 (A), No. 9. 2001
8. I. Foster, C. Kesselman. The Globus Project: A Status Report. Proc. IPPS/SPDP '98 Heterogeneous Computing Workshop, pp. 4-18, 1998.
9. Abdelguerfi, M., Wynne, C., Cooper, E. Roy, L. and Shaw, K., Representation of 3-D elevation in terrain database using hierarchical triangulated irregular networks: a comparative analysis, INT. J. Geographical Information Science, 1998, Vol. 12, No. 8, pp. 853-873.
10. Walker, D.W., Li, M., Rana, O.F., Shields, M.S., Huang, Y,: The software architecture of a distributed problem-solving environment. Concurrency: Practice and Experience 12(15) (2000) 1455-1480. http://www.cs.cf.ac.uk/User/David.W. Walker/papers/psearch01.ps
11. Hakki Eres, Graeme Pound, Zhouan Jiao, et al. Implementation of a Grid-Enabled Problem Solving Environment in Matlab. P.M.A. Sloot et al. (Eds.): ICCS 2003, LNCS 2660, pp. 420–429, 2003.

Design of Wireless Sensors for Automobiles

Olga L. Diaz–Gutierrez and Richard Hall

Dept. Computer Science & Engineering,
La Trobe University, Melbourne 3086, Victoria, Australia
oldiazgutierrez@students.latrobe.edu.au,
richard.hall@latrobe.edu.au

Abstract. Automobile manufacturers require sense data to analyse and improve the driving experience. Currently, sensors are physically wired to both data collectors and the car battery, thus the number of wires scale linearly with sensors. We design alternative power and communications subsystems to minimise these wires' impact on the test environment.

1 Introduction

Like many other industries, the multinational automobile industry uses radio frequency (RF) applications [1], specifically to assist driving and vehicle diagnosis [2, 3]. Such applications are supported in modern vehicles via the standard provision of hardware infrastructure consisting of the two–wire serial controller area network (CAN) bus platform [4]. All vehicle applications are required by international standards to have undergone exhaustive testing under statistically sound empirical experiments that involve the collection of sense data [5]. The development of wireless sensor networks in automobiles is thus a natural progression of typical sensing and RF devices for this industry.

Wireless sensors (WS) are desirable in an automobile test environment for three reasons. First, WS are smaller since cable ports are unnecessary. Second, they eliminate the time required to painstakingly connect all sensors to power and data cables in vehicles. Consequently, WS can be more quickly deployed and easily moved around, improving both the spatial resolution of sense data and fault tolerance of the sensor network [6, 7]. Finally, the number of cables that can be safely crammed into an automobile cabin (along with sensors) during test driving is limited (see Figure 1) .

Fig. 1. Example test vehicle cabin view (courtesy Siemens VDO Automotive Pty. Ltd.)

This paper is organised as follows. In Section 2 we present our WS hardware design. Our proposed WS collaboration method for the automobile-testing-

X. Jia, J. Wu, and Y. He (Eds.): MSN 2005, LNCS 3794, pp. 1043–1050, 2005.

environment sensor network will be discussed in Section 3. This network lifetime is bounded by the average power consumption of individual WS, discussed in Section 4. Finally we describe future work in Section 5.

2 Wireless Sensor Architecture: Hardware Design

We adopted a typical WS architecture [6, 8] consisting of a power unit, sensing module, radio communications module, and a microcontroller which performs analog–to–digital conversion (A/D) as well as software operations including aggregating data into packets at the transmitting end, and filtering and interpolates data at the receiving WS (see Figure 2).

WS *power units* typically have operating voltages ranging from 3.0–3.6V. We decided to give the tester the option to power the WS either from small batteries as per normal usage [6] or from the automobile battery, if convenient. Thus we incorporated a TLE4274 regulator to convert this battery voltage to within the WS operating voltage (see Figure 3).

In order to minimise power drain, the *sensing module* (see Figure 4) is switched on just before a measurement is performed then off immediately afterwards (rapid switching has no negative impact on the components [9]). As usual, this measurement, represented as an analog signal (produced by a world phenomenon sensing device) is stabilised for input to the A/D [10, 11]. The WS is designed to be phenomenon-independent: automobile manufacturers can study any phenomen of interest with these WS so long as they have a measuring device that outputs an appropriate analog signal.

Being unsure about the RF operating environment within automobiles of different makes, we wanted the *radio communications module* to meet two design

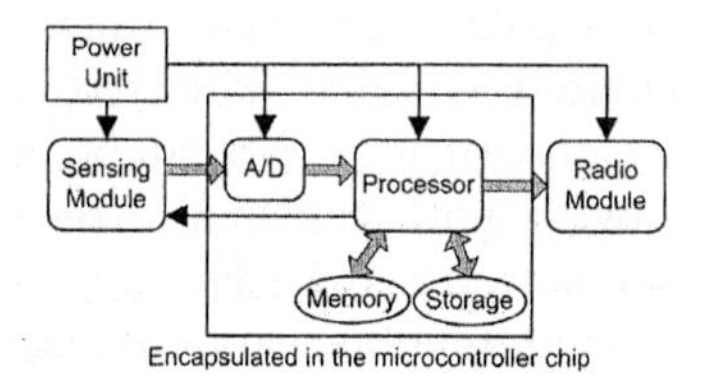

Fig. 2. Architecture of a Wireless Sensor

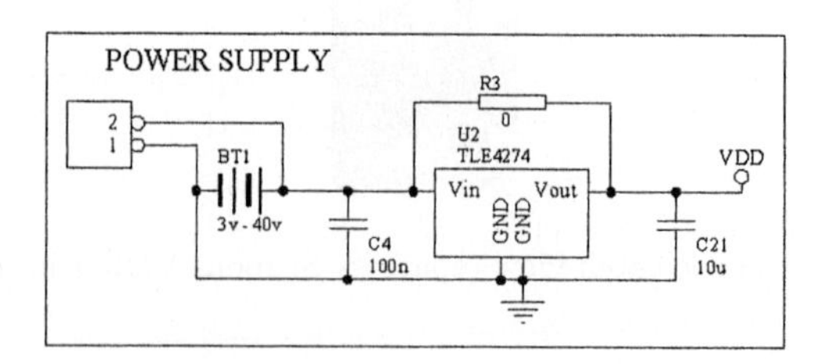

Fig. 3. Power Module Schematics

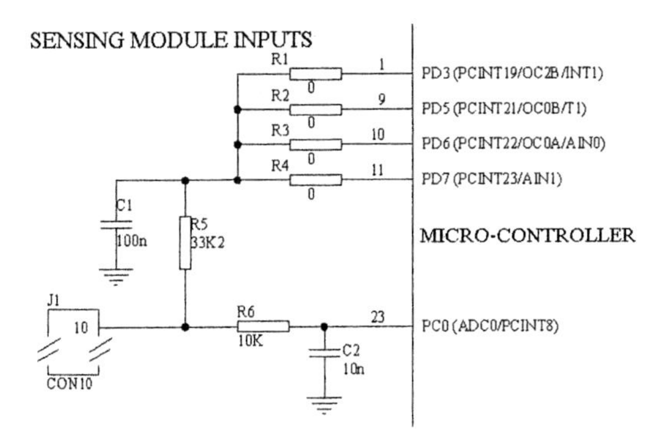

Fig. 4. Sensing Module Design

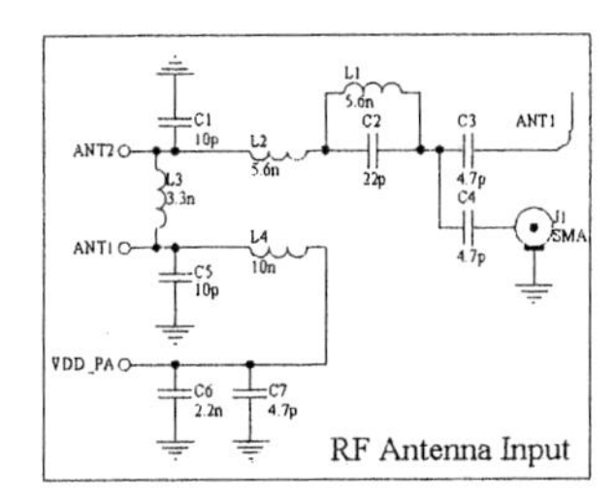

Fig. 5. Wireless Sensor RF Antenna Schematics

criteria. First, we wanted the option of being able to communicate across multiple bandwidths, so we selected two compact and low–power RF chips - the nRF905 transceiver for the 433/868/915MHz band [12] and the nRF2401 transceiver for the 2.4GHz band [13]. Second, we wanted two antenna options that could be selected where necessary (see Figure 5): both PCB (compact and lower power requirements but hard to tune [14, 15, 16]) and external antennas (vice versa [17]). We hope that the ability to switch between different RF frequencies and antennas will assist communications particularly between the engine bay and the cabin across the thick metal plate firewall standard in all vehicles.

The two requirements of our *microcontroller* were that it had minimal power drain (to prolong WS life) and that it had a built–in Serial Peripheral Interface (SPI) functionality (to assist data logging onto the CAN bus [18]). The chip we selected was the Atmel ATmega88: it is low–powered, and has fast stand–by–mode to active–mode transition time (so the WS draws power for the shortest amount of time during measurements and transmitting). It has a built in brown-out detector circuit, to alert the sensor network user as to when WS batteries need replacing. It also has a useful built–in *A/D* which we will use to convert sensor input to a 10-bit digital value (zero represents 0 volts and 2^{10} - 1 represents the power source voltage/maximum possible voltage [18]). The meaning of this value is derived by mapping the voltage into the measurement scale for the real-world phenomena being sensed.

In this section we discussed the design of an individual wireless sensor. In the automobile test environment however, it is necessary to consider how many of these sensors will simultaneously sense and communicate data.

3 Wireless Sensor Network

Our WS will use the communications protocol Time Division Multiple Access (TDMA), as there is an insufficient frequency band to support FDMA and the chosen transceivers do not support CDMA [19]. The use of TDMA means that only the actively data transmitting and receiving WS are in (the relatively high–power) communications on-mode thus all-but-two of the WS are in communications stand-by-mode at any instant. We assume that a single WS (master) operates as a constant data receiver (star topology network) and is powered by the vehicle battery (so it is unnecessary to consider power usage for the master WS). In addition to acting as a central receiver, it synchronises when all slave WS can begin data transmission using a unique req-trans message to each slave.

Our two transceiver options both produce RF packets with a bit length of 240 [12, 13], and in the first instance we decided to use fifteen 16-bit packets (3 bits to id the 5 A/D channels, 10 bits sense data/packet, 3 bits left over). A window of 10ms has been allowed for simple packet processing by the receiver and a minimum TDMA slot of 5ms has been allowed.

We now calculate the maximum number of WS that a TDMA network with these values can support.

$NoDatabits_{packet}$ = No. of data bits in a packet = 240
$Measurement_{size}$ = Size of measurement (number of bits) = 16
$Samples_{interval}$ = Sample time interval, user defined (in seconds)
$Sensors_{number}$ = Number of sensors, user defined (in seconds)
$TDMA_{threshold}$ = Minimum time needed for each transmitter when using TDMA protocol (ms) = 5ms
$MCU_{threshold}$ = Minimum time needed for the master (MCU) to process each packet received (ms) = 10ms

Using the variables above, the following data can be calculated:

$Samples_{packet}$ = No. Samples / Packet $= \frac{NoDatabits_{packet}}{Measurement_{bitsize}} = \frac{240}{16} = 15$
$TXInterval_{avg}$ = Average Transmission Interval
$TDMA_{MCU}$ = Average window time for each sensor, for both TDMA transmission and processing at master (ms)

Every sensor will transmit a packet when 15 samples are taken, which are measured at a rate of $Samples_{interval}$. The average transmission interval is thus:

$$TXInterval_{avg} = Samples_{packet} \times Samples_{interval}$$

If sensors transmit a packet every $TXInterval_{avg}$, then each sensor is allocated a window in that interval to implement TDMA. To calculate the size of the time slot, the interval is divided by the number of nodes in the network. The 1,000 is to convert seconds to milli–seconds:

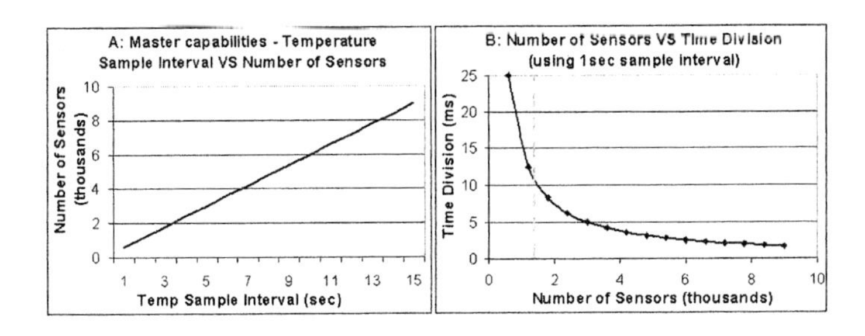

Fig. 6. System Bottleneck: Master VS Protocol Capabilities

$$TDMA_{MCU} = \frac{TXInterval_{avg}}{Sensors_{number}} \times 1000 \tag{1}$$

where $TDMA_{MCU} > TDMA_{threshold}$ and $TDMA_{MCU} > MCU_{threshold}$

The maximum number of WS our network could support depends on the interval at which samples are taken ($Samples_{interval}$). Once a suitable interval has been chosen (based on input rate of change), we use equation (1) to determine the number of nodes ($Sensors_{number}$) and TDMA slot.

For instance, if a sample interval of 1 second is required, then the maximum number of nodes the master can process is 1,400 (with a TDMA_{MCU} of 10.714ms) since the master's 10ms threshold has been reached (Figure 6B).

In this section we discussed the way that our wireless sensors would collaborate and showed that relatively high spatial resolution (>1000 WS nodes with respect to the size of an automobile) could be achieved using the proposed design. However, if sensors that run off their own batteries draw power too quickly, the task of regularly replacing over a thousand batteries might be prohibitive in the testing environment, thus practically limiting spatial resolution. Therefore, we consider sensor network lifetime with respect to power consumption.

4 Sensor Network Lifetime

The three main sources of current consumption in our WS are the sensing module, the radio module and the processor module (see Figure 2). These modules are considered with respect to both WS operating modes: on–mode and off–mode. The current averages over time for each module and mode were analysed (calculations not shown) then all results are summed to generate an overall WS average power consumption.

Average Current per Module (μA)	On–Mode	Off–Mode
Sensing Module	47.475	0
Radio Module	5.369	2.232
Processor Module	400.737	15.247
Total Average Current = $AvI_{ON/OFF}$	453.581	17.479

In addition, since we intend to use a standard 500mAh battery, which is incapable of supplying the current required by a WS as its charge decreases, an estimated 20% of the theoretical power is deducted [20]. Thus, to calculate how average WS lifetime ($Hours$), the following formula is used:

$$Bat_{size} = \text{Electric current provided by the battery for an hour} = 500\text{mAh} \quad (2)$$

$$AvI_{ON} = \text{Average WS current during on–mode} = 453.581\mu\text{A} \quad (3)$$

$$AvI_{OFF} = \text{Average WS current during off–mode} = 17.479\mu \quad (4)$$

$$T_{ON} = \text{Percentage of time that WS are in on–mode (\%)} \quad (5)$$

$$Hours = \left[\frac{Bat_{size}}{\left(AvI_{ON} \times \frac{T_{ON}}{100}\right) + \left(AvI_{OFF} \times \frac{100-T_{ON}}{100}\right)} \right] \times 0.8 \quad (6)$$

Assuming WS are active constantly, battery life is 1.2 months. More realistically, if the sensor network was made active in an automobile testing environment for 10% of the time, the network could operate for a little more than 9 months.

5 Conclusion

In this paper we presented the hardware design of a wireless sensor network that could be used to assist automobile manufacturers to collect large amounts of real-world data in an automobile testing environment. This design is relatively flexible; our WS will be able to use multiple bandwidths, antennas, power sources, and sensor types. There are a number of questions remaining however about how well this design will perform in the field.

The automobile testing environment is much less than ideal to implement a wireless star network topology. There may be multiple RF devices transmitting simultaneously, causing intermodulation distortion. The same signal will be received multiple times as the signals wave propagate (reflect, diffract etc.) around objects in the vehicle. The antennas may also be placed close to conducting surfaces, which may cause noisy signals. It will take some experimenting to characterise the operational range for a WS in an automobile environment. We intend to test this design in the first instance using electronic temperature sensors (thermistors), placing 60 WS in a Siemens VDO test vehicle. These tests will allow us to assess whether WS deliver reductions in automobile test setup time with respect to wiring, and whether WS lead to improvements with respect to data spatial resolution.

It is essential that TDMA time slots are allocated accurately and WS transmit only during their respective slot to avoid interference. However, since external oscillators are expensive in terms of power consumption, a less accurate, internal microcontroller oscillator will be used. Thus, due to higher clock error rate, re-synchronisation must be performed periodically in order to achieve the expected communication performance. Several re-synchronisation algorithms will be trialled to determine the most suitable.

Acknowledgement. We would like to thank Siemens VDO Automotive Pty Ltd for sponsoring this project, particularly Shaun Murray and Martin Gonda, whose assistance with design development has been exhaustive. We would also like to thank Darrell Elton and Paul Main from the Department of Electronic Engineering LTU, for providing enormous practical assistance in electronics design.

References

1. CHANG, K. In: RF and Microwave Wireless Systems. John Wiley & Sons, Inc (2000)
2. KOPETZ, H.: (Automotive electronics – present state and future prospects) http://www.nzdl.org/cgi-bin/cstrlibrary?e=d-0cstr--00-0-0-014-Document---0-11--1-en-50---20-about-RF+sensors+automotive--001-001-0isoZz-8859Zz-1-0&cl=search& d=HASH015d59d7e6c34232e3473dcd.1&hl=0&gc=0>=1.
3. ERIKSSON, L., S.BRODEN: High performance automotive radar. In: Microwave Journal. Volume 39. (1996) 24 – 38
4. VRBA, M.S.P.B.R., ZEZULKA, F.: Chapter 10: Introduction to industrial sensor networking. In: Handbook of Sensor Networks: Compact Wireless and Wired Sensing Systems, CRC Press (2005) 10:5
5. FLINK, J. In: The Automobile Age. The MIT Press (1988) 1 – 26, 358 – 376
6. HASSANEIN, Q.W.H., XU, K.: Chapter 9: A practical perspective on wireless sensor networks. In: Handbook of Sensor Networks: Compact Wireless and Wired Sensing Systems, CRC Press (2005)
7. ROMER, K.: Tracking real–worls phenomena with smart dust. In: Wireless Sensor Networks – First European Workshop, Berlin, Germany, John Wiley & Sons, Inc (2004)
8. PAPAVASSILIOU, S., ZHU, J.: Chapter 15: Architecture and modeling of dynamic wireless sensor networks. In: Handbook of Sensor Networks: Compact Wireless and Wired Sensing Systems, CRC Press (2005) 15:3
9. Corporation, B.B.: MicroPower, Single–Supply OPERATIONAL AMPLIFIERS MicroAmplifier Series. (1999) http://www.fulcrum.ru/Read/CDROMs/TI-2001.June/docs/sbos088.pdf.
10. BOYLESTAD, R.: 21.5: R–C Low–Pass Filter. In: Introductory Circuit Analysis. 9 edn. Prentice Hall (2000) 916 – 923
11. SEDRA, A., SMITH, K.: 3.8: Limiting and Clamping Circuits. In: Microelectronic Circuits. 4 edn. Oxford University Press, Inc (1998) 195
12. ASA, N.S.: Single Chip 433/868/915 MHz Transceiver nRF905 Product Specification. 1.2 edn. (2005) http://www.nordicsemi.no/files/Product/data_sheet/nRF905rev1_2.pdf.
13. ASA, N.S.: Single Chip 2.4 GHz Transceiver nRF2401A Product Specification. (2004) http://www.nordicsemi.no/files/Product/data_sheet/nRF2401A_rev1_0.pdf.
14. ASA, N.S.: Quarterwave printed monopole antenna for 2.45ghz. Technical report (2003) http://www.nordicsemi.no/files/Product/white_paper/PCB-quarterwave-2_4GHz-monopole-jan05.pdf.
15. ASA, N.S.: nAN900–04: nRF905 RF and antenna layout. Technical report (2004) http://www.nordicsemi.no/files/Product/applications/nAN900-04_nRF905_RF_and_antenna_layout_rev2_0.pdf.

16. ASA, N.S.: nAN24–01: nRF2401 RF layout. Technical report (2004) http://www.nordicsemi.no/files/Product/applications/nAN24-01 rev2_0.pdf.
17. ASA, N.S.: nAN24–05: nRF24E1 wireless hands–free demo. Technical report (2003) http://www.nordicsemi.no/files/Product/applications/nAN24-05 _rev1_2.pdf.
18. Corporation, A.: ATmega48/88/168 Product Specification. E edn. (2005) http://www.atmel.com/dyn/resources/prod_documents/doc2545.pdf.
19. RAZAVI, B.: 4: Multiple Access Techniques and Wireless Standards. Prentice Hall Communications Engineering and Emerging Technologies. In: RF Microelectronics. PRENTICE HALL PTR (1998) 103 – 110
20. WRIGHT, S.R.D.S.L.F.P., RABAEY, J.: Power sources for wireless sensor networks. In: Wireless Sensor Networks – First European Workshop, Berlin, Germany, John Wiley & Sons, Inc (2004)

Mobile Tracking Using Fuzzy Multi-criteria Decision Making

Soo Chang Kim[1], Jong Chan Lee[2], Yeon Seung Shin[1], and Kyoung-Rok Cho[3]

[1] Converged Access Network Research Team, ETRI, Korea
sckim@etri.re.kr
[2] Dept. of Computer Information Science, Kunsan National Univ., Korea
[3] Dept. of Information & Comm. Eng., Chungbuk National Univ., Korea

Abstract. In the microcell- or picocell-based system the frequent movements of the mobile bring about excessive traffics into the networks. A mobile location estimation mechanism can facilitate both efficient resource allocation and better QoS provisioning through handoff optimization. In this study, we propose a novel mobile tracking method based on Multi-Criteria Decision Making (MCDM), in which uncertain parameters such as PSS (Pilot Signal Strength), the distance between the mobile and the base station, the moving direction, and the previous location are used in the decision process using the aggregation function in fuzzy set theory. Through numerical results, we show that our proposed mobile tracking method provides a better performance than the conventional method using the received signal strength.

1 Introduction

There will be a strong need for the mobile terminal tracking in the next generation mobile communication systems. The location of a Mobile Terminal must be found out, e.g., in wireless emergency calls already in the near future. It is of great importance to the efficiency of next generation mobile communication systems to know the exact position of the moving mobile user in order to reduce the number of paging messages and cell handover messages. Handover efficiency will be an important aspect in next generation mobile communication systems because it affects directly to the switching road and QoS, particularly in combined micro cell of Pico cell networks.

Many methods and systems have been proposed based on radio signal strength measurement of a mobile object's transmitter by a set of base stations. Time of arrival (TOA) of a signal from a mobile to neighboring base stations are used in [1], but this scheme has two problems. First, an accurate synchronization is essential between all sending endpoints and all receiving ones in the system. An error of 1 μs in synchronization results to 300 m error in location. Secondly this scheme is not suitable for the microcellular environment because it also assumes LOS environment. Time difference of arrival (TDOA) of signals from two base stations is considered in [2]. TOA scheme and TDOA scheme have been studied for IS-95B where PN code of CDMA system can be used for the location estimation. Enhanced Observed Time Difference (E-OTD) is a TDOA positioning method based on OTD feature already existing in GSM. The mobile measures arrival time of signals from three or more cell sites in a

X. Jia, J. Wu, and Y. He (Eds.): MSN 2005, LNCS 3794, pp. 1051 – 1058, 2005.

network. In this method the position of mobile is determined by trilateration [3]. E-OTD, which relies upon the visibility of at least three cell sites to calculate it, is not a good solution for rural areas where cell-site separation is large. However, it promises to work well in areas of high cell-site density and indoors.

In this study, to enhance estimation accuracy, we propose a scheme based on MCDM which considers multiple parameters: the signal strength, the distance between the base station and mobile, the moving direction, and the previous location. This process is based on three step location estimations which can determine the mobile position by gradually reducing the area of the mobile position [4]. Using MCDM, the estimator first estimates the locating sector in the sector estimation step, then estimates the locating zone in the zone estimation step, and then finally estimates the locating block in the block estimate step.

2 Estimation Procedure

Figure 1 shows how our scheme divides a cell into many blocks based on the signal strength and then estimates the optimal block stepwise where the mobile is located using MCDM.

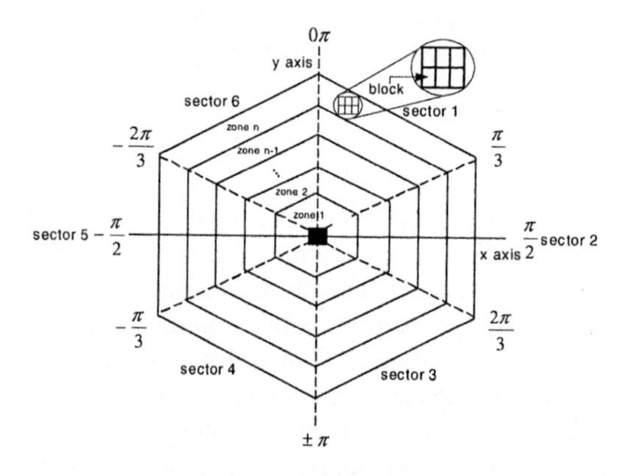

Fig. 1. Sector, Zone and Block

The location of a mobile within a cell can be defined by dividing each cell into sectors, zones and blocks and relating these to the signal level received by it at that point. It is done automatically in three phases of sector definition, zone definition and block definition. Then the location definition block is constructed with these results. They are performed at the system initialization before executing the location estimation. The sector definition phase divides a cell into sectors, and assigns a sector number to blocks belonging to each sector. The zone definition phase divides each sector into zones, and assigns a zone number to blocks belonging to each zone. The block definition phase assigns a block number to each block. In order to indicate the location of each block within a cell, 2-dimensional vector (d, a) is assigned to each block. After the completion of this phase each block has a set of block information.

The set of block information is called the *block object*. The block object contains the following information: the sector number, the zone number, the block number, the vector data (d, a), the maximum and the minimum values of the average PSS for the LOS block, the compensated value for the NLOS block, and a bit for indicating "node" or "edge".

```
class blockObject
{
  private:
        unsigned int sector_num;
        unsigned int zone_num;
        unsigned int block_num;
        double vector(double d, double a);
        unsigned int los_sig(int min, int max);
        unsigned int nlos_sig;
        unsigned int node;
        unsigned int edge;
  public:
              .
              .
              .
}
```

3 Mobile Tracking Based on MCDM

In our study, the received signal strength, the distance between the mobile and the base station, the previous location, and the moving direction are considered as decision parameters. The received signal strength has been used in many schemes, but it has very irregular profiles due to the effects of radio environments. The distance is considered because it can explain the block allocation plan; however, it may also be inaccurate due to the effect of multi-path fading, etc. It is not sufficient by itself. We consider the previous location. It is normally expected that the estimated location should be near the previous one. Therefore, if the estimated location is too far from the previous one, the estimation may be regarded as inaccurate. We also consider the moving direction. Usually the mobile is most likely to move forward, less likely to move rightward or leftward, and least likely to move backward more than one block. The low-speed mobile (a pedestrian) has a smaller moving radius and a more complex moving pattern, while the high-speed mobile (a motor vehicle) has a larger radius and a simpler pattern.

3.1 Membership Function

The membership function with a trapezoidal shape is used for determining the membership degree of the mobile because it provides a more versatile degree between the upper and the lower limits than the membership function with a step-like shape. Let us define the membership functions for the pilot signal strengths (PSSs) from neighboring base stations. The membership function of PSS_i, $\mu_R(PSS_i)$, is given by Eq. (1). PSS_i is the signal strength received from the base station i, s_1 is the lower limit, and s_2 is the upper limit.

$$\mu_R(PSS_i) = \begin{bmatrix} 0, & PSS_i < s_1 \\ 1 - \dfrac{PSS_i - s_1}{|s_2 - s_1|}, & s_1 \le PSS_i \le s_2 \\ 1, & PSS_i > s_2 \end{bmatrix} \quad (1)$$

Now we define the membership function of the distance. The membership function of the distance $\mu_R(D_i)$ is given by Eq. (2), where D_i is the distance between the base station i and the mobile [4].

$$\mu_R(D_i) = \begin{bmatrix} 1, & D_i < d_1 \\ 1 - \dfrac{|D_i - d_2|}{|d_1 - d_2|}, & d_1 \le D_i \le d_2 \\ 0, & D_i > d_2 \end{bmatrix} \quad (2)$$

The membership function of the previous location of the mobile $\mu_R(L_i)$ is given by Eq. (3), where L_i is its current location, $E_1,\cdots,E_4$ is the previous location, and g_i is the physical difference between them [4].

$$\mu_R(L_i) = \begin{bmatrix} 0, & L_i < E_1 \\ 1 - \dfrac{L_i - E_1}{|g_i|}, & E_1 \le L_i \le E_2 \\ 1, & E_2 \le L_i \le E_3 \\ 1 - \dfrac{L_i - E_3}{|g_i|}, & E_3 \le L_i \le E_4 \\ 0, & L_i > E_4 \end{bmatrix} \quad (3)$$

The membership function of the moving direction $\mu_R(C_i)$ is given by Eq. (4). C_i is the moving direction of the mobile, $PSS_1,\cdots,PSS_4$ is the pilot signal strength, and o_i the physical difference between the previous location and the current one.

$$\mu_R(C_i) = \begin{bmatrix} 0, & C_i < PSS_1 \\ 1 - \dfrac{C_i - PSS_1}{|o_i|}, & PSS_1 \le C_i \le PSS_2 \\ 1, & PSS_2 \le C_i \le PSS_3 \\ 1 - \dfrac{C_i - PSS_3}{|o_i|}, & PSS_3 \le C_i \le PSS_4 \\ 0, & C_i > PSS_4 \end{bmatrix} \quad (4)$$

3.2 Location Estimation

Most of the MCDM approaches face the decision problem in two consecutive steps: aggregating all the judgments with respect to all the criteria and per decision alternative and ranking the alternatives according to the aggregated criterion. Also our approach uses this two-steps decomposition [5, 6].

Let J_i (i ∈ {1, 2,..., n} be a finite number of alternatives to be evaluated against a set of criteria K_j (j=1, 2,..., m). Subjective assessments are to be given to determine (a) the degree to which each alternative satisfies each criterion, represented as a fuzzy matrix referred to as the decision matrix, and (b) how important each criterion is for the problem evaluated, represented as a fuzzy vector referred to as the weighting vector. Each decision problem involves n alternatives and m linguistic attributes corresponding to m criteria. Thus, decision data can be organized in a $m \times n$ matrix. The decision matrix for alternatives is given by Eq. (5).

$$\mu = \begin{bmatrix} \mu_R(PSS_{11}) & \mu_R(D_{12}) & \mu_R(L_{13}) & \mu_R(C_{14}) \\ \mu_R(PSS_{21}) & \mu_R(D_{22}) & \mu_R(L_{23}) & \mu_R(C_{24}) \\ \mu_R(PSS_{31}) & \mu_R(D_{32}) & \mu_R(L_{33}) & \mu_R(C_{34}) \\ \cdots & \cdots & \cdots & \cdots \\ \mu_R(PSS_{n1}) & \mu_R(D_{n2}) & \mu_R(L_{n3}) & \mu_R(C_{n4}) \end{bmatrix} \tag{5}$$

The weighting vector for evaluation criteria can be given by using linguistic terminology with fuzzy set theory [5, 6]. It is a finite set of ordered symbols to represent the weights of the criteria using the following linear ordering: very high $\geq$ high $\geq$ medium $\geq$ low $\geq$ very low. Weighting vector W is represented as Eq. (6).

$$W = (w_i^{PSS}, w_i^{D}, w_i^{L}, w_i^{C}) \tag{6}$$

3.2.1 Sector Estimation Based on Multi –criteria Parameters

The decision parameters considered in the Sector Estimation step are the signal strength, the distance and the previous location. The mobile is estimated to be located at the sector neighboring to the base station whose total membership degree is the largest. The sector estimation is performed as follows.

Procedure 1. Membership degrees are obtained using the membership function for the signal strength, the distance and the previous location.

Procedure 2. Membership degrees obtained in Procedure 1 for the base station neighboring to the present station are totalized using the fuzzy connective operator as shown in Eq. (7).

$$\mu_i = \mu_R(PSS_i) \cdot \mu_R(D_i) \cdot \mu_R(L_i) \tag{7}$$

We obtain Eq. (8) by imposing the weight on μ_i. The reason for weighting is that the parameters used may differ in their importance.

$$\omega\mu_i = \mu_R(PSS_i) \cdot W_{PSS} + \mu_R(D_i) \cdot W_D + \mu_R(L_i) \cdot W_L \tag{8}$$

where W_{PSS} is the weight for the received signal strength, W_D for the distance, and W_L for the location. Also $W_{PSS} + W_D + W_L = 1$, and W_{PSS} =0.5, W_D =0.3 and W_L=0.2 respectively.

Procedure 3. Blocks with the sector number estimated are selected from all the blocks within the cell for the next step of the estimation. Selection is done by examining sector number in the block object information.

3.2.2 Track Estimation Based on Multi -criteria Parameters

The decision parameters considered in the Track Estimation step are the signal strength, the distance and the moving direction. From the blocks selected in the sector estimation step, this step estimates the track of blocks at one of which the mobile locates using the following algorithm.

Procedure 1. Membership degrees are obtained using the membership function for the signal strength, the distance and the moving direction.

Procedure 2. Membership degrees obtained in Procedure 1 is totalized using the fuzzy connective operator as shown in Eq. (9).

$$\mu_i = \mu_R(PSS_i) \cdot \mu_R(D_i) \cdot \mu_R(C_i) \tag{9}$$

We obtain Eq. (10) by imposing the weight on μ_i .

$$\omega\mu_i = \mu_R(PSS_i) \cdot W_{PSS} + \mu_R(D_i) \cdot W_D + \mu_R(C_i) \cdot W_C \tag{10}$$

where W_{PSS} is assumed to be 0.6, W_D 0.2 and W_C 0.2 respectively.

Procedure 3. Blocks which belong to the track estimated above are selected for the next step. It is done by examining the track number of the blocks selected in the sector estimation.

3.2.3 Block Estimation Based on Multi –criteria Parameters

From the blocks selected in the zone estimation step, this step uses the following algorithm to estimate the block in which the mobile may be located.

Procedure 1. Membership degrees are obtained using the membership function for the signal strength, the distance and the moving direction.

Procedure 2. Membership degrees obtained in Procedure 1 are totalized using the fuzzy connective operator as shown in Eq. (11).

$$\mu_i = \mu_R(PSS_i) \cdot \mu_R(D_i) \cdot \mu_R(C_i) \tag{11}$$

We obtain Eq. (12) by imposing the weight on μ_i .

$$\omega\mu_i = \mu_R(PSS_i) \cdot W_{PSS} + \mu_R(D_i) \cdot W_D + \mu_R(C_i) \cdot W_C \tag{12}$$

where W_{PSS} is assumed to be 0.6, W_D 0.1 and W_C 0.3 respectively.

Procedure 3. The selection is done by examining the block number of the blocks selected in the track estimation.

4 Performance Analysis

In our paper we assume that low speed mobiles, pedestrians, occupy 60% of the total population in the cell and high-speed mobiles, vehicles, 40%. One half of the pedestrians are assumed to be still and another half moving. Also the private owned cars occupy 60% of the total vehicle, the taxi 10% and the public transportation 30%. Vehicles move forward, leftward/rightward and U- Turn. The moving velocity is assumed to have a uniform distribution. The walking speed of pedestrians is 0~5 Km/hr, the speed of private cars and taxis 30~100 Km/hr, and buses 10~70 Km/hr.

The speed is assumed to be constant during walking or driving. In order to reflect more realistic information into our simulation, it is assumed that the signal strength is sampled every 0.5 sec, 0.2 sec, 0.1 sec, 0.1 sec and 0.05 sec for the speed of ≤10km/h, ≤ 20km/h, ≤50km/h, ≤70km/h and ≤ 100km/h, respectively. If CT is too small, we cannot obtain enough samples to calculate the average signal strength.

Figure 2 shows estimation results of three schemes for the situation where the high speed mobile moves along a straight or curved sector boundary area. In this figure the horizontal and vertical axes represent the relative location of the area observed and the path generated in this simulation. Results are shown for AP (Area Partitioning), VA (Virtual Area) [4] and MCDM from left to right in this figure respectively. As can be seen AP sometimes selects faulty locations far away from the generated path. That is because inaccurate results in sector estimation stage are escalated into track and block estimation. VA has better accuracies for curved path. In our understanding it may be attributed to the fact that the average value of pilot signal strengths sampled by high speed mobile passing through two sectors falls into the range of PSS values of the sector boundary area. The performance of MCDM is less affected during a left turn or right turn. A left or right turn causes an abrupt signal distortions, but their effects on estimation can be compensated for by using information on previous location and distance to base station.

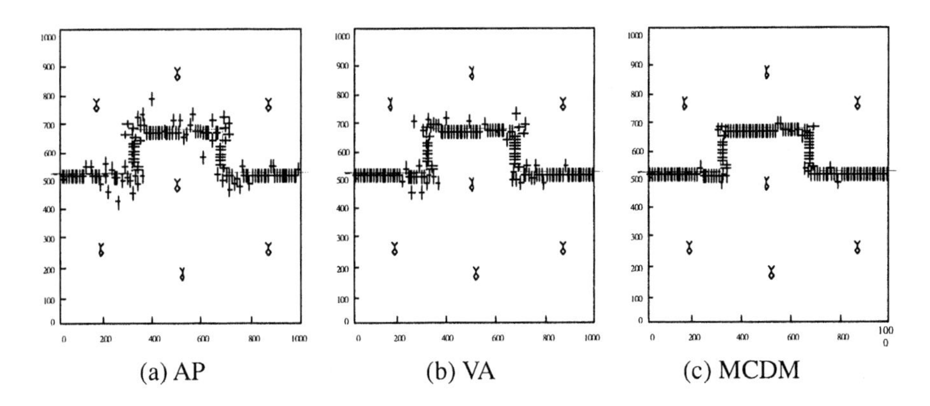

Fig. 2. The estimation results on the move

We compare our scheme, MCDM with VA [4], E-OTD and TDOA in Figure 3. In this figure the mobile maintains its y position at 1000m and traverse x axis from $x =$ 0m to $x =$ 2000m. The y axis means how the drms varies with x position of mobile. Drms (Distance Root Mean Square) stands for the root-mean-square value of the distances from the true location point of the position fixes in a collection of measurements. In order to get the estimated values for comparison we take an average of 20 values for each mobile position. We assume NLOS environment and the signal level of mobile may change abruptly due to shadowing. It shows that the performance of MCDM is least affected by abrupt change of signal level. MCDM has the most accurate result. This may well be attributed to the fact that it imposes less weight on

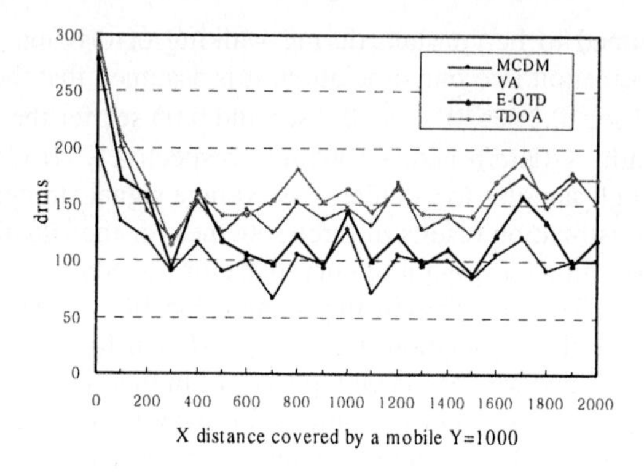

Fig. 3. Comparison of the estimation accuracy

the received signal strength in NLOS area and, instead, greater weights on other parameters such as the distance between mobile and base station, previous location, and moving direction are considered as decision parameters.

5 Conclusions

In this study, we proposed a MCDM-based mobile tracking method for estimating more accurately the mobile location by considering multiple parameters, such as the signal strength, the distance between the base station and mobile, the moving direction, and previous location. We have demonstrated that our scheme increases the estimation accuracy when the mobile moves along a boundary area. The effect of weight factor variations on the estimation performance of our scheme and the determination of the optimal weight should be the subject of a future study. Also further researches are required on their implementation and applications to the handoff and channel allocation strategies.

References

1. T. Nypan and O. Hallingstad, "Cellular Positioning by Database Comparison and Hidden Markov Models.," PWC2002:, pp. 277-284, Oct. 2002
2. T. S. Rappaport, J. H. Reed and B. D. Woerner, "Position Location Using Wireless Communications on Highways of the Future," IEEE Communications Magazine, pp. 33-41, Oct.
3. Y. A. Spirito, "On the Accuracy of Cellular Mobile Station Location Estimation," IEEE Trans. Veh. Technol., vol. 50, no. 3, pp. 674-685, 2001.
4. J. C. Lee and Y. S. Mun, "Mobile Location Estimation Scheme," SK telecommunications Review, Vol. 9, No. 6, pp. 968-983, Dec. 1999.
5. C. Naso and B. Turchiano, "A Fuzzy Multi-Criteria Algorithm for Dynamic Routing in FMS," IEEE ICSMC'1998, Vol. 1, pp. 457-462, Oct. 1998.
6. C. H. Yeh and H. Deng, "An Algorithm for Fuzzy Multi-Criteria Decisionmaking," IEEE ICIPS'1997, pp. 1564-1568, 1997.

Pitfall in Using Average Travel Speed in Traffic Signalized Intersection Networks

Bongsoo Son[1], Jae Hwan Maeng[1], Young Jun Han[1], and Bong Gyou Lee[2]

[1] Dept. of Urban Planning and Eng., Yonsei Univ., Seoul, Korea
{sbs, mjray, hizune}@yonsei.ac.kr
[2] Graduate School of Information, Yonsei Univ., Seoul, Korea
bglee@yonsei.ac.kr

Abstract. For the effective use and management of urban traffic control systems, it is necessary to collect and process traffic data and produce traffic information. Average travel speed is typically used for classifying traffic conditions of traffic signalized intersection networks. The purpose of this paper is to solve the pitfall caused by the usage of average travel speed estimated by conventional technique for the signalized intersection networks. To do this, this paper has suggested the basis of criteria for selecting the speed data to be used for final estimation of travel speed. The key point is to check the relevancy of travel time of the vehicles traveled during the same evaluation time period.

1 Introduction

A number of delays such as stopped delay, approach delay, travel-time delay and time-in-queue delay may occurred at a traffic signalized intersection during the same time period. These delays are mainly dependent upon the coordination of green times of traffic signals relatively closely spaced in the signalized intersection network [1]. Thus, it is important for the traffic engineers to coordinate their green times in order to improve the vehicles' movement through the set of signals. It is often called as "signal progression" in traffic engineering.

Figure 1 shows the time-space diagram for the signal progression. In the figure the yellow interval are not shown due to the scale. If a vehicle were to travel at the speed limit (or free-flow travel speed), it would arrive at each of the signals just as they turn green; this is indicated by the heavy dashed line. There is a window of green in the figure and this window is called the "bandwidth" that is a measure of how large a platoon of vehicles can be passed without stopping. It is noteworthy that the efficiency of a bandwidth is defined as the ratio of the bandwidth to the cycle length, expressed as a percentage, and an efficiency of 40% to 55% is considered good. In fact, it is almost impossible to coordinate all green times for all approaches at traffic signal intersection network such as shown in Figure 1. Due to this fact, some vehicles inherently experience the delays under the same traffic condition.

More conventionally, average travel speed is typically used for classifying traffic conditions of traffic signalized intersection networks. Theoretic methods for travel

X. Jia, J. Wu, and Y. He (Eds.): MSN 2005, LNCS 3794, pp. 1059 – 1064, 2005.

speed estimation are common in the traffic literature, but treatments of how to account delay caused by traffic signal are less common. Conventional method is to simply calculate the average value of travel speed of all vehicles traveled during the same time period [2]. However, if we estimate the average travel speed of all vehicles traveled during the same time period, it may lead to misunderstanding for traffic condition as well as traffic information.

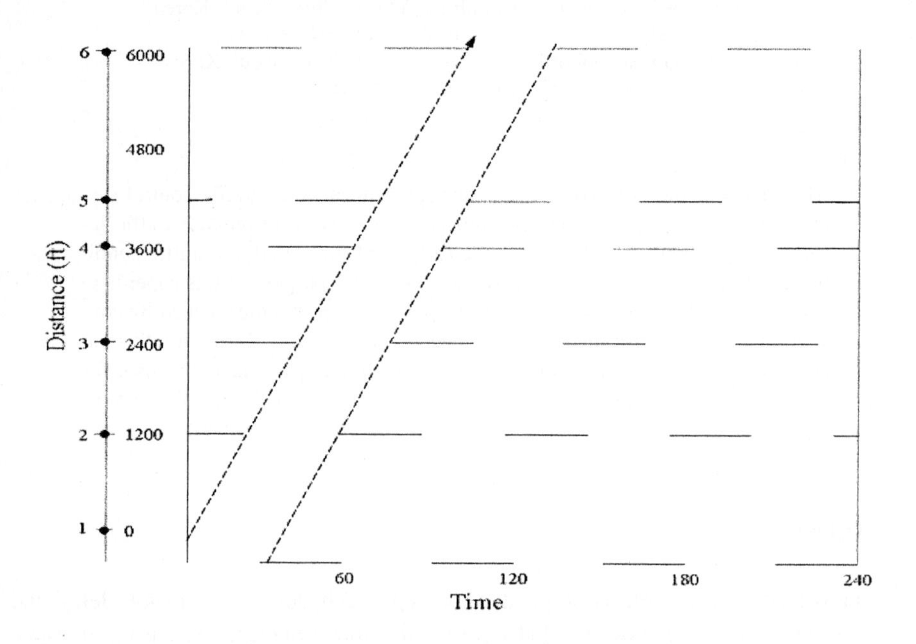

Fig. 1. Vehicle trajectory and "bandwidth" in a signal progression

This paper is an attempt to solve the pitfall caused by the usage of average travel speed in urban traffic signalized intersection network. To do so, this paper employs the vehicles' trajectories in the time-space diagram. Son *et al.* [3] have first employed the trajectories for predicting the bus arrival time in the signalized intersection networks. Most recently, Kim *et al.* [4] have proposed a heuristic algorithm for estimating travel speed in the signalized intersection networks, which was developed based on the trajectories. However, the algorithm has limitation to avoid the pitfall caused by the usage of average travel speed during non-congested time period.

2 Vehicles' Trajectories

Figure 2 illustrates the trajectories that five vehicles take as time passes in traffic signalized intersection network. In the figure, trajectory type I is associated with the vehicles arrived at traffic signal *i* during the red time period. Trajectory II represents the vehicles arrived at the signal during the period between ending of red time and beginning of green time and experienced delay for passing the traffic signal.

Trajectory type III is related to the vehicles passed the traffic signal without any delay. The three types of trajectories indicate that the travel speeds on traffic signalized intersection network are widely different and greatly dependent upon whether or not vehicles await at traffic signals. In other words, the travel speed in traffic signalized intersection network significantly vary depending upon the state of signals (i.e., green time or red time of traffic signal) as well as the coordination of green times of traffic signals [4].

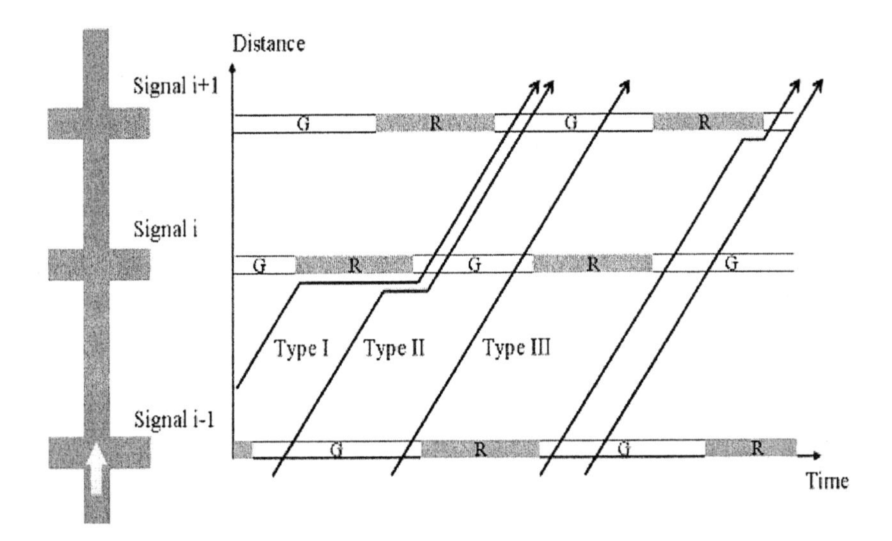

Fig. 2. Time-space diagram for vehicles' trajectories in traffic signalized network

As above-mentioned, the conventional method estimates the average travel speed of all vehicles passing a roadway section in traffic signalized intersection network over some specified time period (e.g., every 5 to 15 minutes), where the speed is the inverse of the time taken by the vehicles to traverse the roadway distance. However, as can be seen from Figure 2, the waiting times occurred at traffic signalized intersections would cause major differences in estimating of each vehicle's travel speed in traffic signalized intersection network. The conventional method definitely has limitation in reflecting the variation of vehicles' trajectories for the estimation of the travel speed.

3 Pitfalls of Average Travel Speed

Some field speed data were collected by floating car method for both congested and non-congested time periods. A total of ten passenger cars were assigned to depart from the upstream of "Intersection 1" to "Intersection 5" every 1 minute time interval. The study site consisted of five traffic signal intersections is 2.5 km long a major arterial that links between the western region and the old downtown area in Seoul. The study site is consisted of 6 to 8-lane sections. The speed limit of the study site is 60km/h.

Table 1 shows a set of sample speed data which were calculated based on each vehicle's travel time measured from the field. In the table, the speed data were grouped for 5-minute evaluation time period.

Table1. Travel speed data

Speed values of ten floating vehicles	
Non-congested period	Congested period
{45, 43}	{16, 13, 12, 14, 12}
{22, 25, 29, 35}	{13, 10, 11}
{29, 35, 43, 46}	{8, 11}

With respect to non-congested traffic condition, the speed values measured under non-congested traffic condition were fluctuated from 22km/h to 46km/h. With the exception of the first 5-minute evaluation time period, the fluctuation was severe during the other periods. These results are not surprising, since the speed variation would be occurred by both the state of traffic signals and the coordination of green times of traffic signals. (In order to better understand some of the discussion, refer to Figure 2.)

For the first time period, the average value of speed data during the time period is reasonable for representing the reality, but the average values of the other two time periods are not good, since the vehicles' speeds vary even under the same traffic condition. It is logical to not wonder why travel speed of each vehicle does result in sufficient discrepancy from the average speed estimated during the same evaluation time period. However, if we estimate the average travel speed of all vehicles traveled during the same evaluation time period by using the same manner of conventional method, it may lead to misunderstanding for traffic condition as well as traffic information.

More specifically, for an example, there is enough unused green time during a given green phase under non-congested traffic condition. If some vehicles arrived just after the ending of green time during the same evaluation time period, they should wait until the signal turns green again. Thus, the average travel speed would result in lower value due to the red time of signals included during the same time period. Consequently, the average travel speed may not be appropriate for representing the non-congested traffic conditions. For this case, somewhat higher speed value among all the speeds measured during the same time period is more appropriate for representing the real traffic conditions rather than the average value.

Most of the vehicles traveled on congested signalized intersection network experienced all of stopped delay, approach delay, travel-time delay and time-in-queue delay at the signalized intersections. This case will not be misinterpreted if we use the average travel speed estimated by the conventional method, since most of the speeds collected during the same time period would be maintained at the similar speed level. Therefore, the data associated with the congested traffic condition are not severely fluctuated so that the average speed of all vehicles traveled during the same evaluation time period seems to be reasonable for representing the reality.

4 Method for Selecting Reasonable Speed Values

The most significance shortcoming of conventional technique is that the delays caused by traffic signals are overlooked in estimating the average travel speed. For the congested traffic condition, we can simply figure out reasonable speed as the average value of all vehicles' speeds, since each vehicle's speed does not much differ from the average travel speed of all the vehicles. However, the problem occurs when the traffic condition is not congested and traffic signals do not well coordinated all green times for the vehicles. The main task to do in this paper is how to figure out reasonable speed value for representing the real traffic condition as well as traffic information disseminated to the travelers. To do this, this paper has carefully reviewed two cases of traffic situations for the illustrative pitfall being used.

Case 1

Traffic signals well coordinated all green times for the vehicles so that they moved efficiently through all the traffic signals without any unnecessary stopping and delay. In this case, the difference of travel times under non-congested traffic condition would be mainly caused by stopped delay occurred at the signalized intersection rather than approach delay, travel-time delay and time-in-queue delay.

All of the speed values collected in the case during the same time period would be less than the speed limit and the speed values might be severely fluctuated, since at most 40% to 55% of all green times could be coordinated. Consequently, all travel times of the vehicles traveled would be larger than free flow travel time (that is equal to travel distance divided by speed limit), but would be smaller than the sum of the free flow travel time and total red times of all the traffic signals passed. Theoretically, the reasonable travel time of the case seems to be close to the sum of the free flow travel time and the half of total red times of all the traffic signals passed.

Therefore, in order to define a reasonable speed value, we have to first check whether the speed values collected during the same time period meet the above-mentioned travel time range; and then we have to select relatively higher speed values among them, since lower speed values would be caused by the signal progression, but not caused by traffic condition.

Case 2

Traffic signals have not well coordinated all green times for the vehicles so that they could not move efficiently through all the traffic signals even though traffic condition was not congested. In this case, the vehicles could not maintain higher speed so that their speed values collected during the same time period would be not severely fluctuated. However, this case makes things complex since some of the speed values would be very low even though traffic condition is not congested. Such case will be misinterpreted if we use the average travel speed estimated by the conventional method.

In order to solve this problem, we have to check whether all travel times of the vehicles traveled during the same time period are reasonable. For this case, the reasonable travel time should be somewhat less than the sum of free flow travel time and total red times of all the traffic signals passed, since the efficiency of a bandwidth would be much less than 40%. Therefore, it seems to be reasonable to select the highest value among the speed values which are relevant to the above-mentioned travel time range. Otherwise, we may misunderstand this non-congested traffic condition as congested by using the lower speed values.

5 Conclusions

Careful investigation of the field speed data, it was confirmed that the travel speed values measured under non-congested traffic condition were severely fluctuated and the speed variation would be occurred by both the state of signals and the coordination of green times of signals. Thus, it is not reasonable to estimate the travel speed for representing the reality simply by using the average speed of all vehicles traveled during the same evaluation time period. This paper presents two cases for the illustrative pitfall being used and suggests a method for selecting reasonable speed value among all speed data collected during the same time period. The key point is to check the relevancy of travel time of the vehicles traveled during the same time period. To do this, this paper has suggested the basis of criteria for selecting the speed data to be used for final estimation of travel speed. The criteria seem to be theoretically sound and promising.

References

1. W. R. McShane and R .P. Roess, Traffic Engineering, Prentice-Hall, Inc. 1990
2. B. Son and S. Lee, "A Development of Evaluation Method for Road Performance," The 4th Conference of Eastern Asia Society for Transportation Studies, 24-27 October, Hanoi, Vietnam, 2001
3. B. Son, H. Kim, C. Shin, and S. Lee, "Bus Arrival Time Prediction Method for ITS Application," KES 2004, LNAI Vol. 3215, pp.88-94, 2004
4. H. Kim, B. Son, S. Lee and S. Oh, "Heuristic Algorithm for Estimating Travel Speed in Traffic Signalized Networks," LNCS Vol. 3415, 2005

Static Registration Grouping Scheme to Reduce HLR Traffic Cost in Mobile Networks

Dong Chun Lee

Dept. of Computer Science Howon Univ., South Korea,
727, Wolha-Ri, Impi, KunSan, ChonBuk, Korea
ldch@sunny.howon.ac.kr

Abstract. This paper proposes the static registration grouping scheme that solves the Home Location Register (HLR) bottleneck due to the terminal's frequent registration area (RA) crossings and that distributes the registration traffic to each of the local signaling transfer point (LSTP) area. The RAs in LSTP area are grouped statically. It is to remove the signaling overhead and to mitigate the Regional STP (RSTP) bottleneck. The proposed scheme solves the HLR bottleneck due to the terminal's frequent RA crossings to each of the LSTP area.

1 Introduction

Universal Mobile Telecommunication System (UMTS) and cdma2000 are two major standards for third generation (3G) mobile telecommunication or IMT-2000 [13, 21]. Many operators commit to deploy UMTS and/or cdma2000-based 3G networks. Evolving from the existing 2G networks, construction of effective 3G networks is critical for provisioning future mobile services. The standard commonly used in North America is the Electronics Industry Association/Telecommunications Industry Association (EIA / TIA) Interim Standard 95 (IS-95), and in Europe the GSM [2]. IS-95 and GSM have a structural drawback: as the number of user increase, HLR becomes the bottleneck.

A number of works have been reported to reduce the bottleneck of the HLR problems. In [8], a Location Forwarding Strategy is proposed to reduce the signaling costs for location registration. A Local Anchoring Scheme is introduced in [5]. Under these schemes, signaling traffic due to location registration is reduced by elimination the need to report location changes to the HLR. Location update and paging subject to delay constraints is considered in [9]. When an incoming call arrives, the residing area of the terminal is partitioned into a number of sub-areas, and then these sub-areas are polled sequentially. With increasing the delay time needed to connect a call, the cost of location update is reduced. Hierarchical database system architecture is introduced in [1]. A queuing model of three-level hierarchical database system is illustrated in [6], [10]. These schemes can reduce both signaling traffics due to location registration and call delivery using the properties of call locality and local mobility.

Above schemes are proposed to reduce the costs for location registration or call tracking. The general caching scheme is effective when the call requests to the callee

X. Jia, J. Wu, and Y. He (Eds.): MSN 2005, LNCS 3794, pp. 1065 – 1072, 2005.

from one RA are very much. This implies that its effectiveness is god enough only when the degree of the call locality is extremely high. However, it is not real in wireless environments. Also, there exists a consistency problem between the cached information and the entry in the callee's VLR. It also limits the cached information and the entry in the calee's VLR. It also limits the mobility patterns. In aspect of the registration, it is same as the IS-95 scheme. As for the LA scheme, they are cost effective in reducing the HLR access traffic. However, there is a trade-off between the registration cost and call tracking cost.

2 Proposed Scheme

I define post VLR, PVLR which keeps the callee's current location information as long as the callee moves within its LSTP area. If a terminal crosses the LSTP area, the VLR which serves the new RA is set to a new PVLR. If the terminal moves within the LSTP area, it is registered at its own PVLR not HLR. If the terminal moves out from the area, it is registered at the HLR. In case that a terminal is switched on, the VLR which serves the terminal's current RA is PVLR and the VLRs which serve the intermediate RAs in terminal's moving route within the LSTP area report the terminal's location information to the PVLR. We note that we don't have to consider where the callee is currently. It is because the PVLR keeps the callee's current location as long as the callee moves within its LSTP area. Therefore, without the terminal movements into a new LSTP area, the registration at HLR does not occur.

I statically group the VLRs in LSTP area in order to localize the HLR traffic. It is also possible to group the RAs dynamically regardless of the LSTP area. Suppose that the P_{VLR} and the VLR which serves the callee's RA belong to the same dynamic group but are connected to the physically different LSTPs. In this case, I should tolerate the additional signaling traffic even though the caller and callee belong to the same dynamic group. A lot of signaling messages for registering user locations and tracking calls is transmitted via RSTP instead of LSTP. If the cost of transmitting the signaling messages via RSTP is large enough compared to that via LSTP, dynamic grouping method may be degrade the performance although it solves the Ping-Pong effect. Furthermore, it is critical in case that the RSTP is bottlenecked. A location registration in HLR occurs only when a terminal moves out of its LSTP area. The RA crossings within LSTP area generate the registrations in P_{VLR}. The followings describe the registration steps.

(1) The terminal which has moved to a new RA requests a registration to the VLR which serves the new RA.

(2) The new VLR inquires the id. of the current P_{VLR} of the old VLR and the old VLR replies to the new VLR with ACK message where the id. is piggybacked. If the terminal sends to the new VLR a message containing the id. of its P_{VLR}, the query messages can be omitted. The new VLR only sends a registration message to P_{VLR}. If the id. of its P_{VLR} does not exist in the entry, the new VLR regards the terminal as having moved out of its LSTP area. The new VLR becomes the new P_{VLR} of the terminal. Therefore, the new VLR should determine whether the old RA belongs to its LSTP area or not.

If belongs: The new VLR sends the location in formation of the terminal to P_{VLR}. And then a registration cancellation message is sent to the old VLR by new VLR or P_{VLR}.

If not: After the new VLR transmits the location information of the terminal to HLR, HLR transmits a registration cancellation message to the old P_{VLR}. And then the old P_{VLR} sends a cancellation message to the old VLR, subsequently. In this case, the new VLR plays a role of the new P_{VLR}.

Fig. 1 shows the message flow due to the location registration according to the status of P_{VLR} change.

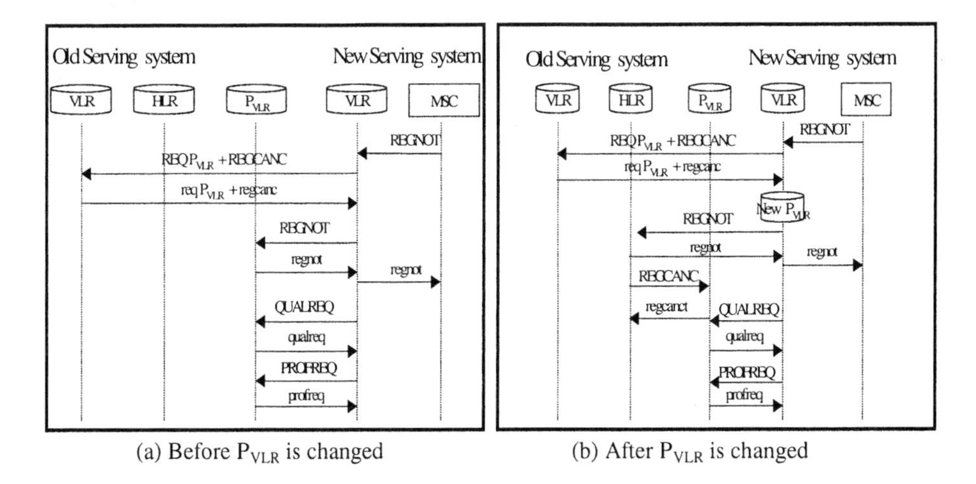

(a) Before P_{VLR} is changed (b) After P_{VLR} is changed

Fig. 1. Location registration in proposed scheme

If the caching scheme is applied to call tracking, the most important thing is how to maintain the ICR low to some extent that the call tracking using the cache is cost effective compared to otherwise. If the proposed scheme is used in the caching scheme, the consistency can be maintained as long as the terminal crosses the RAs within the LSTP area. As for the hierarchical level in which the cache is located, the MSC level is not desirous considering the general call patterns. It is effective only for the users of which working and resident areas are regionally very limited, e.g., one or two RAs.

3 Performance Analysis

For numerical analysis, terminal-moving probability should be computed. I adopt hexagon model as geometrical RA model. That model is considered to be common for modeling the RA. Generally, it is assumed that a VLR serves a RA.

3.1 RA Model

As shown in Fig. 2, RAs in LSTP area can be grouped. There are 1, 7, and 19 RAs in circle 0, circle 1, and circle 2 areas, respectively. The terminals in RAs inside circle n area still exist in circle n area after their first RA crossing. That is, the terminals inside circle area cannot cross their LSTP area when they cross the RA one time. While the terminals in RAs which meet the line of the circle in figure can move out from their LSTP area. I can simply calculate the number of moving out terminals. Intuitively, I can consider that the terminals in arrow marked areas move out in Fig. 2. Using the number of outside edges in arrow marked polygons. I can compute the number of terminals which move out from the LSTP area as follows.

$$\left(\frac{\text{Total No. of outside edges in arrow marked polygons}}{\text{No. of edges of hexagon} \times \text{No. of RAs in LSTP area}}\right) \times \text{No. of terminals in LSTP area} \qquad (1)$$

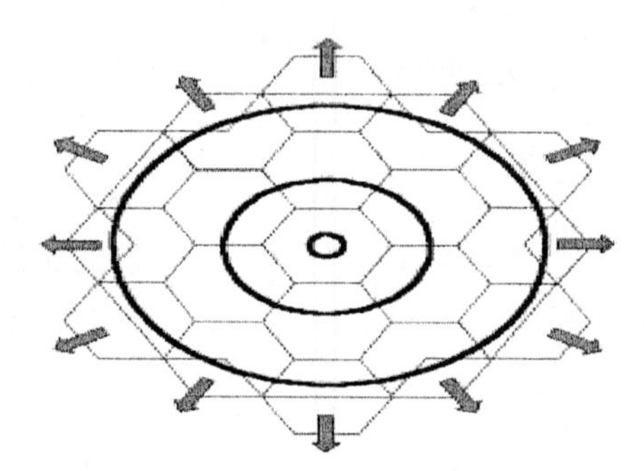

Fig. 2. The RA Hexagons model

For example, there are 2 or 3 outside edges in arrow marked area in hexagon model. So 2/6 or 3/6 of terminals in corresponding RA which meets the line of the circle move out the LSTP area. In case of circle 2 in Fig.2 , the number of RAs is 19 and the number of terminals which move out from LSTP area is given by the terminal in LSTP area × (5/19).

The number of VLRs in LSTP area represented as circle n can be generalized as follows.

$$\text{No. of VLRs in LSTP area} = 1+3n(n+1) \quad (\text{where } n = 1, 2, \ldots) \qquad (2)$$

The rate of terminals which move out from LSTP area can be generalized as follows.

$$R_{\text{move\_out,No.of VLRs in LSTP area}} = \frac{2n+1}{1+3n(n+1)} \quad (\text{where } n = 1, 2, \ldots) \qquad (3)$$

In the RA model, the rate of terminal movements within the LSTP area, $R_{\text{move\_out,No.of VLRs in LSTP area}}$ is ($R_{\text{move\_out,No.of VLRs in LSTP area}}$). Two RAs are said to be locally related when they belong to the same LSTP area and remotely related when the one of them belongs to the different LSTP area. The terminal's RA crossings should be classified according to the local and remote relations in the following schemes.

1) Proposed Scheme

- Relation of PVLR serving RA and callee's last visiting RA
- Relation of RA where callee's last visiting RA and callee's current RA

2) LA Scheme

- Relation of LA serving RA and callee's last visiting RA
- Relation of callee's last visiting RA and callee's current RA

I define the probabilities that a terminal moves within the LSTP area and crosses the LSTP area as P (local) and P (remote), respectively. The above relations are determined according to the terminal's moving patterns. P (local) and P (remote) can be written as follows.

$$P\ (\text{local}) = R_{\text{move\_in,No.of VLRs in LSTP area}},\quad P\ (\text{remote}) = R_{\text{move\_out,No.of VLRs in LSTP area}} \tag{4}$$

P (local) and P (remote) are varied according to the number of VLRs in LSTP area. Table 1 shows the probabilities in case of the one time RA crossing. P (local) and P (remote) when the number of VLRs in LSTP area is 7, 9, 19, and 25 are shown in Table 1.

Suppose that the number of RAs in LSTP area is 7. If a terminal moves into a new LSTP area in its nth movement, the terminal is located in one of outside RAs - 6 RAs - in the new LSTP area. If the terminal moves into a new RA in its $(n + 1)^{th}$ movement, $R_{\text{move\_in,No.of VLRs in LSTP area}}$ and $R_{\text{move\_out,No.of VLRs in LSTP area}}$ are both 3/6. If the terminal's $(n + 1)^{th}$ movement occurs within the LSTP area, the two probabilities according to its $(n + 2)^{th}$ movement are 4/7 and 3/7, respectively. Otherwise, the terminal moves into a new LSTP area again. Therefore, the two probabilities are both 3/6.

Table 1. P (local) & P (remote) according to the number of VLRs in LSTP area

Probability	Number of VLRs in group (LSTP area)			
	7	9	19	25
P(local)	4/7	6/9	14/19	20/25
P(remote)	3/7	3/9	5/15	5/25

Once a terminal moves into a new LSTP area, the pairs of two probabilities due to the next movement of the terminal according to the number of RAs in LSTP area are (3/6,3/6), (5/8,3/8), (7/12,5/12), and (11/16, 5/16), respectively. Therefore, I should classify the terminal moving patterns and compute the probabilities to evaluate the traffic costs correctly. To evaluate the performance, I define the Signaling Costs (SCs) as follows.

SC1: Cost of transmitting a message from one VLR to another VLR through HLR
SC2: Cost of transmitting a message from one VLR to another VLR through RSTP
SC3: Cost of transmitting a message from one VLR to another VLR through LSTP

I evaluate the performance according to the relative values of SC1, SC2, and SC3 which are needed for location registration. Intuitively, we can assume $SC3 \leq SC2 < SC1$ or $SC3 \leq SC2 << SC1$. Even though it is difficult to determine the exact values, the difference of relative values can be computed.

To evaluate the registration traffic, I define Registration Cost Set (RCS). RCS is composed of the SC1, SC2, and SC3. SC2 may not belong to the registration cost according to the applied schemes. In IS-95 scheme, only SC1 is used. I assume the various RCSs. The following 4 cases are considered to evaluate the performance.

Case 1: $SC3 < SC2 < SC1$
Case 2: $SC3, SC2 << SC1$
Case 3: $SC3 << SC2, SC1$
Case 4: $SC3, SC2 < SC1$

In following figures, I assume that the larger RCS number is, SC2 and SC3 are larger and SC3/SC2 is larger in above 4 cases. It is to investigate how much SC2 and SC3 affect the registration cost. Because SC1 is fixed regardless of the RCS number, the traffic dependency on SC1 is reduced with larger RCS number. Using the RCS, I can compare the registration costs in several cases with that in IS-95 scheme. Because SC1, SC2, and SC3 are dependent on system environment, several cases of RCS should be considered to evaluate the traffic fairly.

If a terminal moves within the LSTP area - in case of the local relation -, the P_{VLR} is updated and the entry in old VLR is cancelled. If a terminal moves into a new LSTP area, the HLR is updated. Subsequently, the entries in old P_{VLR} and old VLR are cancelled. The registration cost is computed as follows.

$$C_{\text{proposed scheme , Loc.Reg}} = R_{\text{move\_in, No. of VLRs in LSTP area}} \cdot (2 \cdot 2SC3) + R_{\text{move\_out, No. of VLRs in LSTP area}} \cdot 2(SC1 + SC3) \tag{5}$$

As you see in Eq. (5, the registration cost in proposed scheme is independent of SC2. In above 4 cases, I assume that SC3 in case 4 - SC3, SC2 < SC1 - is larger than those in the other cases. The registration traffic dependency on SC1 in proposed scheme is smaller than that in IS-95 scheme. Therefore, with a common SC1, case 1, 2, and 3 shows the better results compared to case 4.

3.2 Location Registration Cost Comparison

1) IS- 95 Scheme

Whenever a terminal moves into a new RA, it is registered at the HLR. The registration cost in IS-95 scheme is computed as follows.

$$C_{\text{IS-95,Loc.Reg.}} = 2 \times SC1 \tag{6}$$

2) LA Scheme

The local and remote relations between the RA into which a terminal moves and LA serving RA should be considered. When a terminal moves into a new RA, the LA is updated or a new LA is selected, and subsequently the entry in old VLR is cancelled.

Suppose that a call is requested to a terminal after terminal's n time movements. I define the number of terminal's RA crossings before call request as *M_no.*. The location registration cost in LA scheme can be written as follows.

$$C_{LA,Loc.Reg.} = (1 - (1/M\_no.))\{2 \cdot 2SC3 \cdot P(local) + 2 \cdot 2SC2 \cdot P(remote)\} + (1/M\_no.)\{2(SC1+SC3) \cdot P(local) + 2(SC1 + SC2) \cdot P(remote)\} \quad (7)$$

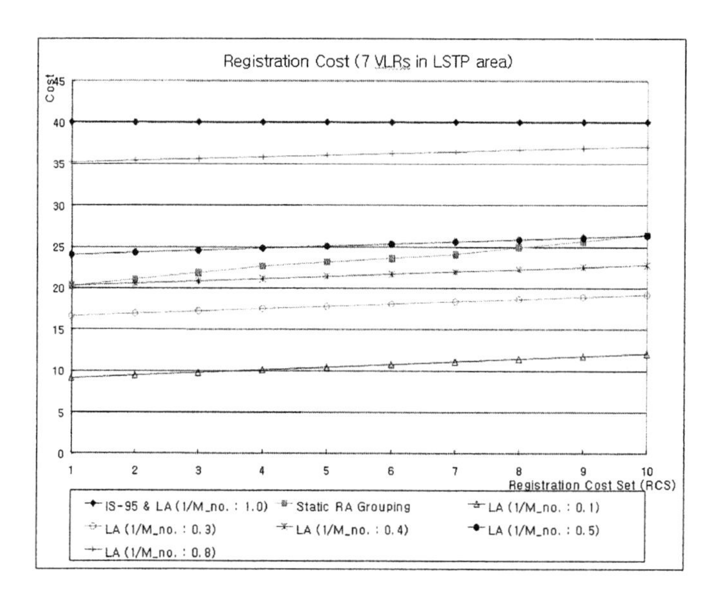

Fig. 3. Location registration cost

In Fig. 3, the proposed static RA grouping scheme shows the better results compare to the LA scheme in case of 1/M_no. ≥ 0.5, where *M_no.* is the number of terminal's RA crossings before call request. The smaller 1 over 1/M_no. is in LA scheme, the more VLRs in LSTP area can be taken for the similar performance. This implies that proposed scheme with large number of VLRs show the similar performance to LA scheme in high mobility environments.

4 Conclusions

In this letter, we presented the static registration grouping scheme that is mainly focused on reducing the HLR bottleneck traffic and distributing the HLR traffic to each of the LSTP area. The proposed scheme decreases the additional signaling traffic compare to the dynamic one, and is relatively insensitive to the user's call-to-mobility ratio, compared to other schemes. As a result of cost evaluation, the more VLRs in LSTP area are, the performance is improved greater than previous schemes.

Acknowledgements

This work was supported by the Howon University Fund, 2005.

References

1. A. D. Malyan, Leslie J. Ng, Victor C.M. Leung, and Robert W. Donaldson, "Network Architecture and Signaling for Wireless Personal Communications", IEEE JSAC., Vol. 11, No. 6, August 1993, pp. 830-840.
2. EIA/TIA, "Cellular Radio-telecommunications Intersystem Operations: Automatic Roaming", Technical Report IS-95 (Revision D), EIA/TIA, July 1999.
3. G.P. Pollini, "Signaling Traffic Volume Generated by Mobile and Personal Communications", IEEE Comm. Mag., Vol.33 No. 6, June 1995, pp. 60-65.
4. G.P. Pollini and D.J.Goodman, "Signaling System Performance Evaluation for Personal Communications", IEEE Trans. on Veh. Tech., May 1994.
5. J.S.M. Ho and I.F.Akyildiz, "Local Anchor scheme for Reducing Location Tracking Cost in PCNs", Proceeding of ACM MOBICOM'95, 1995.
6. J.Z. Wang, "A Fully Distributed Location Registration Strategy for Universal Personal Communication Systems", IEEE JSAC., Vol. 11, No. 6, August 1993, pp. 850-860.
7. R. Jain and Y.B Lin, "A Caching Strategy to Reduce Network Impacts of PCS, IEEE JSAC.", Vol. 12, No. 8, Oct. 1994, pp. 1434-1444.
8. R. Jain and Y.B.Lin, "An Auxiliary User Location Strategy Employing Forwarding Pointers to Reduce Network Impacts of PCS", Proceeding of IEEE ICC'95, 1995.
9. S. Mohan and R. Jain, "Two User Location Strategies for Personal Communications Services", IEEE Personal Comm. Mag., Vol. 1, No.1, First Quarter, 1994, pp. 42-50.
10. SeungJoon Park, DongChun Lee, and JooSeok Song, "Locality Based Location Tracking Using Virtually Hierarchical Link in Personal Communications Services", IEICE Trans. Commun., Vol. E81-B, No. 9, Sep. 1998, pp. 1779- 981.

Towards Security Analysis to Binding Update Protocol in Mobile IPv6 with Formal Method*

Jian-xin Li, Jin-peng Huai, Qin Li, and Xian-xian Li

School of Computer Science, Beihang University, Beijing, China
{lijx, huaijp, liqin, lixx}@act.buaa.edu.cn

Abstract. Mobile IPv6 (MIPv6) is one key protocol for IPv6 enabled computers and handsets providing always-on capabilities and seamless mobility between wireless and wired networks. Binding Update protocol, which has resolved the triangle routing problem in MIPv6, was fraught with vulnerabilities due to mobility. In this paper, we presented three typical lightweight unilateral authentication protocols for securing Binding Update, and compared their security features. We proposed one approach based on BAN logic, a famous formal method has been successfully used in several projects, with extended rules and definitions that are capable of specifying security properties and goals of these protocols. Finally, and the effectiveness of this approach was demonstrated. We hope these works can contribute to the ongoing design and deployment of MIPv6 and pave the way for future research in the security evaluation of MIPv6.

1 Introduction

As the ubiquitous and pervasive computing paradigms gain popularity, data and service interchanges among mobile equipments throughout the Internet became not only possible but essential. The rapid development of the mobile technologies is having a dramatic impact on how individuals and institutions communicate and do business with one another. Mobile IP [1] enables Mobile Nodes to remain reachable and keep connections of transport and high-layer communications alive while moving into other network links from home link. That is, mobile equipments are always accessible via their home network address regardless of whether they are at home link or not. Many key issues in MIPv6 have been extensively studied, such as Routing Optimization (RO), Binding Update [1][2]. Unfortunately, the MIPv6 standardization has ever halted for a long time because of vulnerabilities in Binding Update. Hence, it is desirable to explore solutions enabling the receivers to authenticate the origin of Binding Update message.

MIPv6 is a wide global environment, it is impractical to build a global PKI and other centralized security infrastructures for any Mobile Node. Previously, many researchers attempted to use IPSec [3] as security mechanism for Binding Update in Mobile IPv6. Scott Bradner et al proposed a novel approach in [4] to authenticate the

* This work is supported by the National Natural Science Foundation of China under Grant 91412011.

X. Jia, J. Wu, and Y. He (Eds.): MSN 2005, LNCS 3794, pp. 1073 – 1080, 2005.

origin of Binding Update message with temporarily generated public-private key pair, which is known as PBK (Purpose Built Key) protocol. In 2001, IETF Mobile Work Group ever planned to adopt it as standard, but due to the man-in-the-middle attack occurred in it, the standardization halted for long time. Furthermore, [5] introduces another protocol named CAM based on public-private key pair. Return-Routability protocol presented in [2] mainly relies on the premise that the routing path is semi-reliable among non-mobile nodes, which can defend against bombing and refection attacks etc. Beside the approaches of designing special security protocols, [6] also presents an advanced scheme known as trust management, to secure the Binding Update request. To the best of our knowledge, there no previous investigation on the synthetically comparison and particularly analysis for these protocols in the sense of using formal method, yet the analysis manually is infeasible and the results also is imprecise, thus it is necessary to apply formal method for the analysis of security properties to these mobile protocols.

We proposed a new formal method based on BAN logic to analyze the security of Binding Update protocols in MIPv6. The major contributions of this paper as follows:

1. We provide insight into the security research trends in the area of MIPv6, and give the analysis of security issues in MIPv6.
2. We provide information about the current state of the art in authentication protocols for Binding Update in MIPv6, as well as point out some problems that need careful attention from the research community. Then compare their aspects with respect to their security mechanisms and the rational behind them.
3. We give a comprehensive and reasonable method, which has extended definitions and rules based on BAN logic, to support the formal analysis for Binding Update protocols.

The rest of this paper is organized as followings. In Section 2, we will review basic concepts in MIPv6. Then introduce three typical protocols designed for Binding Update authentication, and give their differences (Section 3). Next, we will show how to depict these protocols running over mobile network with BAN logic [7] and illustrate how to analyze their security properties (Section 4), and finally conclude in Section 5.

2 Basic Concepts

To state the approach of security analysis to Binding Update protocols more precisely, we briefly review some basic notations used to depict MIPv6 scenario, more details discussed in [1][2]:

MN (Mobile Node): A node can change its point of attachment from one link to another, while still being reachable via its home address.
HA (Home Agent): A router on a Mobile Node's home link with which the Mobile Node has registered its current care-of address. *CN* (Correspond Node) : A peer node, may be either mobile or stationary, with which a Mobile Node is communicating.
BU (Binding Update): Message that *MN* used to notify its current binding to *HA* or *CN*.
HoA(*MN*) (Home of Address): The link address of *MN* attached at *HA*.
CoA(*MN*) (Care-of Address): A unicast routable address associated with a Mobile Node while visiting a foreign link.

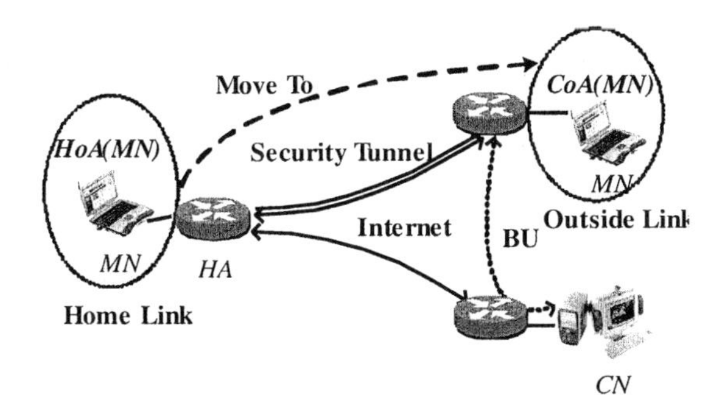

Fig. 1. The basic model of Binding Update communication in MIPv6, with a Secure Tunnel established between *MN* and *HA*

During the design of MIPv6 protocol, some principles obeyed are as follows:

1. The mobile host *MN*, as long as it connects to the Internet somewhere in the world, should be reachable via *HA* routing address [8] [9].
2. All transport-layer and higher-layer connections and Security Associations (SAs) between the *MN* and its *CN* should survive the address change [9]. For example, a mobile host receiving video, when switching to new network link and be able to continue to receive the data stream without re-connection.
3. The stated goal in the IETF working group was that the MIPv6 protocol should not be lower than current non-mobile IPv4 Internet in respect of security [9].

In the MIPv6 network environment (shown in Fig.1.), every Mobile Node is assigned to a unique *HA*, and *MN*'s home link address is *HoA*(*MN*). While *MN* visiting a foreign link, the current address becomes *CoA*(*MN*), and the corresponding data packets will be transmitted from *MN* to *CN* via *HA*, where the "triangle routing" problem occurred, which results in many disadvantages such as low efficiency. To optimize the routing path and offer directly communication between *MN* and *CN*, *MN* must notify *CN* of its current binding address *CoA*(*MN*) by Binding Update protocol between *MN* and *CN*. But any communicating node can acts as one *MN* to forge Binding Update messages so that the normal communication is corrupted. The routing from *HA* to *CN* is semi-reliable, no node outside this routing can wiretap the communicating information. In Addition, secure tunnel would be bridged between *HA* and *MN* with IPSec etc [2][9]. Therefore, we only consider the unilateral authentication of Binding Update from *CN* to *MN*, according to principle 3, the design and applicability of these protocols should be lightweight and flexible.

In summary, Binding Update authentication protocols in MIPv6 maintain three features: ① The running process is rather concise and lightweight; ②The appropriate security level is required, refer to principle(3);③The security goal is perspicuous, only need to authenticate the origin of *BU*.

3 Three Binding Update Authentication Protocols

3.1 PBK Protocol

PBK protocol intends to establish authentication of Binding Update message relying on a pair of temporary public-private keys, including two procedures:

(1) **Generation of *PBID* (Purpose Built ID, PBID)**:
Firstly, *MN* generates temporary public-private key pair, then computes relevant hash code $PBID = Hash(K_{MN})$ from public key K_{MN} , sends *PBID* to *CN* via *HA* routing path:

1. $MN \rightarrow HA \rightarrow CN : PBID$

After *CN* receives *PBID*, *CN* caches a tuple <*HoA*(*MN*),*PBID*> into its local table.

(2) **Procedure of Binding Update Request**：
Once *MN* moves to foreign network link, it should provide its public key K_{MN} and the message *BU* signed by its own private key K_{MN}^{-1} to *CN*:

2. $MN \rightarrow CN : K_{MN}, BU, (BU)_{K_{MN}^{-1}}$

Upon receiving Binding Update request message from *MN*, the *CN* will verify the signed *BU* by comparing the K_{MN} with *PBID* in the caching table to authenticate the origin identity of *BU*.

3.2 CAM Protocol

In CAM protocol, *MN* also generates its own public-private key pair respectively before moving away from home link, and *CN* transmits K_{MN} to *CN*. A novel aspect of this protocol is that *MN*'s Home Agent address associated with the hash code of its temporary public key. The Binding Update request in CAM protocol is as follows:

$$M \rightarrow C\text{: } CoA(MN), HoA(CN), HoA(MN), i, T_m, \{Hash(CoA(MN), HoA(CN), HoA(MN), T_m)\} K_{MN}^{-1}$$
$$HostPart(HoA(MN)) = Hash(K_{MN}, i)$$

In it, the modifier *i* used to resolve name clashes, and *HoA*(*MN*) plays a two-fold purposes here:① As the proof of *MN*'s identity, with which *CN* can verify *MN*; ② As the suffix of network address, providing the routing path from *HA* to *MN*.

3.3 RR (Return Routability) Protocol

IETF Mobile Group has recommended RR protocol as Binding Update authentication standard.

(1) **RR Test Procedure:**
As shown below, *MN* will send the term *init* of test message to *CN* via both *HA* routing path and directly respectively, and request *CN* to generate a key token:

$$MN \to HA \to CN : init_1$$
$$MN \to CN : init_2$$

Here, *Init* is used to distinguish different sessions, other terms include the *HoA* address (or *CoA* address) and session *Cookie* etc.

Upon receiving request from *MN*, *CN* will generate two key tokens K_{HoA} and K_{CoA} respectively with hash of its key K_{cn}, fresh nonce and *HoA* (or *CoA*), then responds them to *MN* as follows:

$$CN \to HA \to MN : init_1, K_{HoA}$$
$$CN \to MN : init_2, K_{CoA}$$

As such, *MN* can receive the acknowledgements from *CN* via both routing path, yet generates the key $K_{BU} = Hash(K_{HoA}, K_{CoA})$ by which *CN* can authorize *MN*'s binding.

(2) **Binding Update Request Procedure:**
MN can send Binding Update request message *BU* signed by K_{BU} to *CN*:

$$MN \to CN : HoA(MN), CoA(MN), BU, MAC\{(CoA(MN), HoA(MN), BU), K_{BU}\}$$

3.4 Comparisons of Three Protocols

The above protocols mentioned such as PBK protocol, CAM protocol and RR protocol are all designed elaborately for the purpose of unilateral authentication of Binding Update in MIPv6. According to the introduction and analysis above, we show the main differences (see **Table 1**) among them:

Table 1. Comparisons of PBK CAM and RR protocol. Interaction column shows their different characteristics. Recently, IETF Mobile Team has recommended the RR protocol as standard for Binding Update.

Protocol	Interaction	Security Mechanisms	Applicability	Security
PBK	initialization: 1 BU request: 1	Hash and Signature	Easy	Low
CAM	initialization: 1 BU request:1	Hash, Timestamp and Signature	Easy, but trusted time server needed	A Little high
RR	initialization: 4 BU request:1	MAC, Symmetric Key	Some simple, standard proposed	Medium

4 Formal Analysis to These Protocols

Prior work on the analysis of Binding Update protocols lacks an adequate notion of security, Fundamental questions such as "what needs to protected in theses protocols?" and "what are the security requirements?" are not adequately answered. We will show one approach based on BAN logic with extended rules and appropriate definitions to specify running environment and security goal of MIPv6 protocol rather than merely manual work.

4.1 Description of Binding Update Protocols

The logic, named BAN [10] after the authors, is a formal logic used to reason about beliefs, encryption, and protocols. BAN logic is simple and has been successfully used in analyzing several projects. Due to paper space limitation, we will not introduce the BAN logic here; the interested reader can refer to [10].

Basic Notations:
The *principal* set: Entities referred to as principals, e.g., *P*, *Q* in capital. In the basic communication model in MIPv6, the principals form a set $ID=\{MN, CN, HA\}$;The *key* terms: The temporary public key K_{TP} and private key K_{TP}^{-1} , by which *P* convince others.In the basic MIPv6 model, other messages terms include *BU*, *HoA*(*P*), *CoA*(*P*) and *Nonce* that generated randomly by principals and timestamp *T* chosen in protocols.

Definition 1 (Binding Update Security Goal): In MIPv6 Binding Update protocol, *CN* must authorize the binding message. However, it is unnecessary to authenticate the identity of *MN* (*MN* can keep anonymous), merely verify the origin of *BU* really sent by *MN*, which is a unilateral authentication. We denote the goal of Binding Update protocol as:

$$CN \mid\equiv MN \mid\equiv (BU)$$

In Dolev-Yao model [11] where defines the internet is entirely open and every participants' public key is known to others, whereas in practical mobile internet environment, the public-private key pair is generated temporary, the scope of known is limited.

Definition 2 (Public Key Condition): Assuming that node *MN* secures the communication using temporary public-private key pair over mobile internet environment. If another node *P* can obtain the public key K_{MN} owned by *MN*, that is, *P* can see the key K_{MN}, we postulate:

$$P \lhd K_{MN} \quad (\textit{Public Key Condition}).$$

To facilitate the description and evaluation of security of MIPv6, *Public Key Condition* restricts the ability of attackers, as to PBK, CAM etc, where using temporary key, we can analyze their security under such condition.

In addition, the *Message Meaning Rule* for public key in BAN logic should makes relevant rectification as follows:

$$\frac{CN \mid\equiv K_{MN} \mapsto MN,\ \ CN \lhd \{X\}_{K^{-1}},\ \ CN \lhd K_{MN}}{CN \mid\equiv MN \mid\sim X} \quad (\textit{Restricted Message Meaning Rule}).$$

In Dolev-Yao model [11], where attackers can arbitrarily intercept, modify and forge communicating messages, but in some special situations, such as VPN and private network tunnel, the attackers' ability to destroy the communication among participants is also limited.

Definition 3 (Secure Tunnel Condition): Assume that secure tunnel, with IPSec technologies etc., can be bridged between participant *A* and *B* as to the transmitting $A \to B : m$ or $B \to A : m$, that means nobody can acquired message term *m* except for *A* and *B*, we denote:

$$A \overset{m}{\rightleftarrows} B \quad (\textit{Secure Tunnel Condition}).$$

Because of the secure tunnel established between *MN* and *CN* in MIPv6, Defination3 can properly security depict the exact security properties and server appropriate analysis method for special internet environment.

4.2 Analysis to These Protocols

Now we show how to analyze the security of these protocols with two propositions with BAN logic under the conditions assumption for the mobile network environment.

Proposition 1: PBK protocol cannot satisfy Binding Update security goal.

Proof: To start, from PBK protocol we give following assumptions:

$$(1) MN \mid\sim BU$$
$$(2) CN \triangleleft (BU, \{BU\}_{K_{MN}^{-1}})$$
$$(3) CN \triangleleft K_{MN} (\textit{Public Key Condition})$$

From the above assumptions and signature rule, we can infer the belief relation: $CN \mid\equiv MN \mid\sim BU$, but due to be absence of the assumption: $CN \mid\equiv \# BU$, the security goal $CN \mid\equiv MN \mid\equiv BU$ for Binding Update cannot be inferred. □

Since *CN* cannot verify freshness of *BU* so that PBK protocol easily suffers from man-in-the-middle attack. Assuming that *MN* can establish sessions with both *CN* and *P*(hostile attacker) respectively, *P* will easily forge *MN*'s Binding Update request to *CN* so *CN* mistakes that *MN*'s *CoA* address has changed, which will spoil the latter communication:

$$MN \rightarrow P : BU, (BU)_{K_{MN}^{-1}}$$
$$P(MN) \rightarrow CN : BU, (BU)_{K_{MN}^{-1}}$$

According to above principle, *P* also can simultaneously forge Binding Update requests from both *MN* and *CN*, so that it's messages will redirect to *P* :

$$P(MN) \rightarrow CN : BU, (BU)_{K_{MN}^{-1}}$$
$$P(CN) \rightarrow MN : BU, (BU)_{K_{CN}^{-1}}$$

Scott Bradner et al. [4] proposed to prevent such threat of man-in-the-middle attack by adding challenge-response process. Besides, replay attack can also be defended against with timestamp or fresh.

Proposition 2: CAM protocol cannot satisfy Binding Update security goal without Timestamp T_m.

Proof: First, we write the assumptions according to CAM protocol:

$$(1) CN \mid\equiv K_{MN} \mapsto MN$$
$$(2) CN \triangleleft K_{MN} (\textit{Public Key Condition})$$

From CAM protocol, we also can obtain following result:

$$(3) CN \triangleleft \{Hash(CoA(MN), HoA(CN), HoA(MN))\}_{K_{MN}^{-1}}$$

From (1), (2), (3) and *Restricted Message Meaning Rule*, the final result is obtained:

$$(4) CN \mid\equiv MN \mid\sim CoA(MN), HoA(CN), HoA(MN)$$

Due to the absence of timestamp T_m implying freshness of message, we cannot infer the following belief:

$$MN \models \#(CoA(MN), HoA(CN), HoA(MN))$$

Certainly, the security goal of $CN \models MN \models CoA(MN)$ cannot be satisfied. □

5 Conclusions

We summarize three typical authentication protocols for Binding Update in previous publications, together with their comparison. A dedicated approach based on a formal method, BAN logic, and some extended definitions and rules were given. From corresponding propositions' proof, we show why our formal definitions are effective to the security analysis of Binding Update protocols in MIPv6.

This paper aimed to identify and solve certain security analysis problems in MIPv6 protocol, but much remains to be done. Our ongoing work includes the design and implementation of new protocols, and the ultimate goal is build feasible formal methods for the security analysis to the whole MIPv6 protocol.

References

1. Perkins C. IP mobility support. IETF RFC,RFC 2002,http://www.ietf.org/ rfc/, 1996.
2. Johnson D, Perkins C, Arkko J. IETF RFC 3775,Mobility Support in IPv6. 2004
3. Arkko J, Devarapalli V, Dupont F. IETF RFC 3776, Using IPsec to Protect Mobile IPv6 Signaling between Mobile Nodes and Home Agents. 2004.
4. Scott Bradner, Allison Mankin, Jeffrey I. Schiller .A Framework for Purpose-Built Keys (PBK), draft-bradner-pbk-frame-06.txt,IETFhttp://www.ietf.org/internet-drafts/draft-bradner-pbk-frame-06.txt, 2003.
5. Greg O'Shea and Michael Roe. Child-proof authentication for MIPv6 (CAM). Computer Communications Review, April 2001.
6. Holly Xiao. Trust Management for Mobile IPv6 Binding Update. Proceedings of the International Conference on Security and Management, Las Vegas, SAM '03, 23-26, June , 2003.
7. Michael Burrows, Martín Abadi, and Roger Needham. A logic of authentication. Proceedings of the Royal Society of London A, 426, 233~271, 1989.
8. Tuomas Aura. Mobile IPv6 Security. In Proc. Security Protocols, 10th International Workshop, LNCS, Cambridge, UK, April 2002.
9. Pekka Nikander, Jari Arkko,Tuomas Aura and Gabriel Montenegro. Mobile IP version 6 (MIPv6) Route Optimization Security Design. In Proc. IEEE Vehicular Technology Conference Fall 2003, Orlando, FL USA. IEEE Press, October 2003.
10. J. Arkko and P. Nikander. Limitations of IPsec Policy Mechanisms. Security Protocols Workshop, Cambridge, UK, April ,2003.
11. Dolev D, Yao A. On the security of public key protocols. IEEE Transactions on Information Theory, 29(2): 198-208, 1983

Enhancing of the Prefetching Prediction for Context-Aware Mobile Information Services

In Seon Choi[1], Hang Gon Lee[2], and Gi Hwan Cho[3]

[1] Dept. of Computer Statistics & Information, Chonbuk Univ., Korea
ischoi@dcs.chonbuk.ac.kr
[2] Korea Institute of Science and Technology Information, Korea
hglee@kisti.re.kr
[3] Division of Electronics and Information Engineering, Chonbuk Univ., Korea
ghcho@dcs.chonbuk.ac.kr

Abstract. This paper deals with a prefetching method to enhance its prediction for context-aware services in mobile information system. This method aims to reduce the latency time to get the refreshed information appropriated to the current location of mobile users. To achieve this, our approach is to effectively limit the prefetched information into the most next location context. It makes use of the prefetching zone that reflects the user's mobility speed and direction, and the mobile reference count that stands for the user's visiting frequency to a given area. Then it considers the residence time, in order to further predict the prefetching candidates.

1 Introduction

The cellular technology is basically based on the voice communication, but the new network environment is required gradually to ease data communication by increasing the bandwidth. As the information communication industry is developed rapidly, it is expected to be increased the number of wireless Internet service. The application services that are general in the Internet should be offered to a similar extent in mobile IP networks. However technologies applied in existent network environment have a lot of restrictions to apply directly to wireless network environment [1, 2]. Namely, a mobile information service requires quick context aware conversion in movement, so the mobile user must get new information when moved to new location. The low bandwidth, high latency, traffic and frequent connection due to the characteristics of mobile environment are remained the obstacles to users.

Therefore, it is required to find a solution by utilizing the existing bandwidth instead of increasing the bandwidth that would cause additional expense. To resolve this, the most prominent method is prefetching. The basic idea is to prefetch a set of information in advance and use them again to accommodate wireless network property. But the prefetched data expected to be referred in the near future has the flaw that needs lots of memory and spends much computing time. So, to improve the performance of mobile information service, it is important to enhance the prediction level of prefetching zone, in order to give maximum effectiveness.

X. Jia, J. Wu, and Y. He (Eds.): MSN 2005, LNCS 3794, pp. 1081 – 1087, 2005.

This paper presents a prefetching method to enhance its prediction for context-aware services in mobile information system. It aims to reduce the latency time to get the refreshed information appropriated to the current location of mobile users. To achieve this, our approach is to effectively limit the prefetched information into the most next location context. It makes first use of the prefetching zone that reflects the user's mobility speed and direction, and the mobile reference count that stands for the user's visiting frequency to a given area. Then it considers the residence time, in order to further predict the prefetching candidates.

2 Related Work

In mobile information systems, servers can no longer be conscious about the data available on a mobile user or even about the mobile users connected to the network. It is thus the clients' responsibility to initiate the validity procedure. Prefetching method is a well established technique to improve performance in tradition distributed systems. Based on fixed nodes, and several papers exist about this topic. Some works have also considered the utility of this technique in the framework of mobile computing, in general from the view point of improving the access to remote file systems; the use of mobility prediction has been also considered for this purpose.

Jiang and Kleinrock [3] proposed an adaptive network prefetching scheme. This scheme predicts the files' future access probabilities based on the access history and the network condition. The scheme allows the prefetching of a file only if the access probability of the file is greater than a function of the system bandwidth, delay and retrieval time. Dar et al. [4] proposed to invalidate the set of data that is semantically furthest away from the current user context. This includes the current location, but also moving behaviors like speed, direction of the user. Ye et al. [5] makes use of predefined routes to detect the regions of interest for which data is required. In such a way, they have location information for the whole ongoing trip and do not have to compute the target areas while on the move. Cho [1] suggested a prefetching approach by considering the speed and moving direction of the mobile user. The speed provides about the velocity with which a user changes locations. Moreover, the size of the user's area is largely dependent on the speed. Whenever the user crosses the borders of the current zone, new prefetching zones is computed. Depending on the speed in the moment that the user leaves the scope of a zone, the new one considers more or less adjacent network cells.

Choi [2] proposed a frequency based prefetching method to analyze mobility pattern of user accumulated during fixed period. Thus, frequency is based on the speed. If predict with data that is accumulated during given period, there is problem to itself. To resolve this problem, this paper presents a progressively refined scheme, named in Temporal Locality in Spatial (TLS).

3 TLS Scheme

The proposed TLS scheme makes use of two localities. The first locality is the spatial locality which considers visiting frequency for prefetching. The second locality is the temporal locality which counts how many times should be residence in an area. Fig. 1 shows an illustrative example of the proposed scheme.

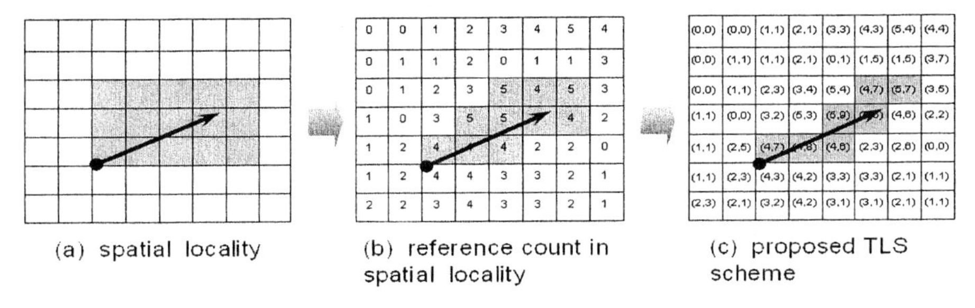

(a) spatial locality
(b) reference count in spatial locality
(c) proposed TLS scheme

Fig. 1. An example of the proposed scheme

3.1 Spatial Locality Model

A frequency-based function [1, 2, 6] is utilized to apply the reference count in spatial locality. The reference count stands for the visiting frequency of each area. It reflects the moving pattern with user's interesting. The following notations are used in our description:

i : The number of cell
SP : The velocity based prefetching
rc : Reference counts of cell
f : Frequency
N : The number of total cell established by velocity
rcw : Critical weight of the reference count [0, 1]

To make an effective restriction of prefetching area, a prefetching threshold can be defined based on the frequency-based function. It would be applied to reduce prefetching area with a space based criteria. Space-based prefetching threshold value, PT_s, is calculated as following.

$$PT_s = \left(\frac{1}{N} \sum_{i \in SP} rc(i) \cdot f(i) \right) \cdot rcw$$

Fig. 1 (a) shows an example which is initially defined with the velocity (speed and direction) based prefetching area, borrowed from reference [1]. Fig. 1 (b) shows a result by applying PT_s. The dark parts depict the newly confined areas among the velocity based prefetching areas.

3.2 Temporal Locality in Spatial Locality

In addition to the frequency-based function, a time-based function can be utilized to minimize the number of prefetching area. That is, if a user resides for a long time in a given area, this means that the user has a lot of interest in the area. We suppose for mean residence time T_i in free cell. Accordingly, the time function can be written as

$$T_i = \frac{\sum_{j=1}^{rc} sum_j}{rc}$$

where n is the reference count of arbitrary cell and sum_j is residence time sum-whenever a mobile user visits. The following notations are used in our description:

RP: The reference based prefetching;
T : Average residence time of any cell;
f : Frequency;
N : The number of total cell established by reference;
rw : Critical weight of the residence time [0, 1];

To make a prefetching area decision, we again determine a prefetching threshold for temporal locality. Temporal-based prefetching critical value PT_t is as following.

$$PT_t = \left(\frac{1}{N} \sum_{i \in RP} T(i) \cdot f(i) \right) \cdot rw$$

Fig. 1 (c) shows an example of applying PT_t. The dark parts depict the newly confined areas among the reference based prefetching areas.

3.3 TLS Process Algorithm

Due to power (bandwidth) limitations of a wireless device (network), a prefetching scheme must be used prudently. If the number of prefetched information is full or the utilization rate of the prefetched information is low, wireless resource will be wasted. To avoid to be wasted of valuable resources, it is important that the prefetched information should have high frequency of use. Therefore, the TLS scheme decides these information areas based on the time-based function. Fig. 2 shows the details of the TLS process algorithm.

Notations:

- V_x, V_y : user's speed values.
- S_{area} : the estimated velocity-based area.
- R_{area} : the estimated reference-based area.
- T_{area} : the estimated time-based area.
- PT_s, PT_t : spatial based prefetching threshold, and temporal based prefetching threshold.
- rcw, rw : critical weight of the reference count, and critical weight of the residence time.
- T_i : average residence time of cell.
- rc : reference count.

While(x and y coordinate aren't the end)
obtain value from accumulated information;
// Spacial Locality

$$S_{area} = \left| \sqrt{V_{x^2} + V_{y^2}} \right| * \left| (|V_x| + |V_y|) / 2 \right|;$$

$$PT_s = \left(\frac{1}{N} \sum_{i \in S_{area}} rc(i) \cdot f(i) \right) \cdot rcw \quad ;$$

$$R_{area} \leftarrow Select\ area\ with\ more\ than\ PT_s\ from\ S_{area}\ ;$$

// Temporal Locality

$$T_i = \left(\sum_{j=1}^{n} sum_j \right) / rc \ ;$$

$$PT_t = \left(\frac{1}{N} \sum_{i \in R_{area}} T(i) \cdot f(i) \right) \cdot rw \quad ;$$

$$T_{area} \leftarrow Select\ area\ with\ more\ than\ PT_t\ from\ R_{area}\ ;$$

End while

Fig. 2. TLS process algorithm

4 Performance Evaluation

4.1 Simulation Model

In order to analyze the performance of proposed scheme, the typical moving scenario is utilized for the velocity-based mobility model. A user moves in a two-dimensional portion's area with the constant speed and direction during any given unit time period. In this simulation, a mobile user is assumed to move around 25 by 25 portions. The user repeats process that move to position preserve of following destination 20 times according to given coordinate value beforehand.

The simulator has been implemented using event-based simulator CSIM [7]. A set of system parameters was given from the other works [6]: the bandwidth of the wired line is assumed to very from 800 kbps to 1.2 Mbps, the bandwidth of wireless medium is a tenth of that of wired line, the time of move detection is assumed to take 100ms. The mobility reference count has been obtained by accumulating for 30 days in simulation time. The proposed TLS scheme has been compared with SL (Spatial Locality) method from [1], and RC (Reference Count) in SL from [6].

4.2 The Numerical Results

With given the user mobility scenario, the number of prefetched portion information has been evaluated to show the communication and storage overhead of three strategies. Fig. 3 shows the number of prefetched portion information on user's move sequence. As a result, proposed scheme TLS was shown its performance improvement about 0.394 in compared with RL.

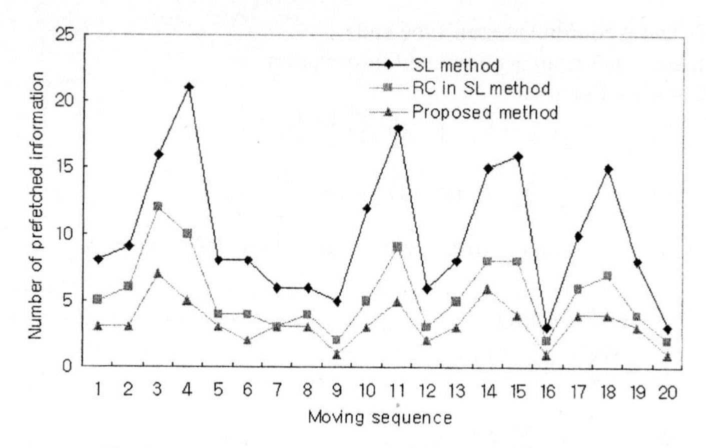

Fig. 3. The number of prefetched portion information

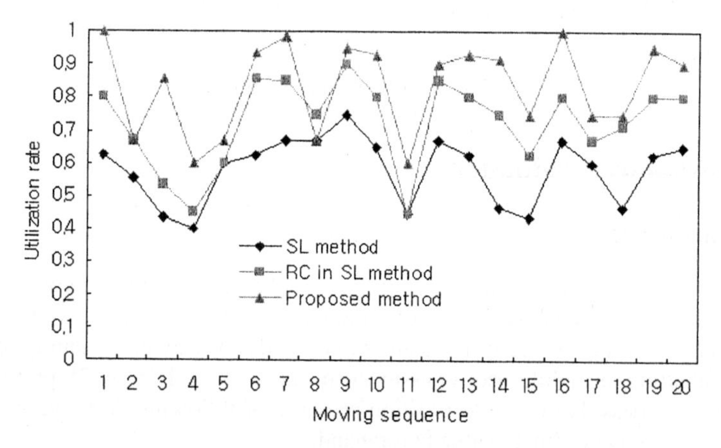

Fig. 4. Utilization rate of the prefetched portion information

Now, it is meaningful to figure out the utilization rate. The utilization rate can be achieved with the number of prefetched information actually participated for the user's location-aware service out of all the number of prefetched information. It just reflects the predictability degree for the given prefetching strategy. With given the user mobility model, the utilization rate with RC in SL method is much better than that with SL method; the former shows that over 0.723 of prefetched portion information is utilized for real service, while the later figures around 0.581. With TLS, the rate is getting to 0.836.

5 Conclusions

In this paper we proposed TLS scheme so that reduces to latency time in prefetching information. The proposed scheme considers the temporal locality in addition to the spatial locality. Simulation results verified that the proposed scheme can reduce the number of prefetched information, and improve the utilization rate of the prefetched information in compared with the other schemes. That is to improve the effectiveness of the limited memory of mobile computing by maximizing the utilization rate of prefetching information.

References

1. G. Cho, "Using Predictive Prefetching to Improve Location Awareness of Mobile Information Service," Lecture Notes in Computer Science, vol. 2331, pp. 1128-1136, 2002
2. I. Choi, "Applying Mobility Pattern to Location-aware Mobile Information Services," Master Thesis, 2003
3. Z. Jiang and L. Kleinrock, "An Adaptive Network Prefetch Scheme," IEEE Journal on Selected Areas in Communications, vol 16, no 3, pp. 1–11, April 1998
4. S. Dar, M. J. Franklin, B. T. Jónsson, D. Srivastava and M. Tan, "Semantic Data Caching and Replacement," Proc. of VLDB, pp. 330-341, 2002
5. T. Ye, H.-A. Jacobsen and R. Katz, "Mobile Awareness in a Wide Area Wireless Network of Info-stations," Proc. of MobiCom'98, pp. 109-120, 1998
6. S. M. Park, D. Y. Kim and G. H. Cho, "Improving Prediction Level of Prefetching for Location-Aware Mobile Information Service," Future Generation Computer Systems, vol 20, pp. 197-203, 2004
7. CSIM18 Simulation Engine, Mesquite Software Inc., 1997

A Mobile Multimedia Database System for Infants Education Environment

Keun Wang Lee[1], Hyeon Seob Cho[2], Jong Hee Lee[3], and Wha Yeon Cho[4]

[1] Dept. of Multimedia Science, Chungwoon Univ., Chungnam, Korea
Kwlee@chungwoon.ac.kr
[2] Dept. of Electronics Eng., Chungwoon Univ., Chungnam, Korea
[3] Principal Research Engineer, Retail Tech Co., LTD., Seoul, Korea
[4] Dept. of Early Childhood Edu., Hyejeon College, Chungnam, Korea

Abstract. To effectively deal with video data, a semantic-based retrieval scheme that allows for processing diverse user queries and saving them on the database is required. This paper proposes a semantic-based multimedia database system that enables users to search the meaning of video data in a diverse manner. It uses both semantic and dependency weights to perform video retrieval for the environment education of infants. In the proposed system, the user searches multimedia data for environment education through entering keywords. The mobile agent then computes both semantic and dependency weights, ensuring the accuracy of data retrieved using calculated weights as annotative information of key frames. As a result of implementing and testing the prototype of the proposed system, a higher precision of approximately 96.5% was obtained.

1 Introduction

In recent years, in order to meet user needs, rapid progress has been made in e-learning technology, multimedia content service technology, and accurate retrieval technology. As a result, user demand for large-capacity video data in mobile environments is growing. To meet the diverse needs of different users, a vast amount of video data is required to be effectively managed [1]. The effective and efficient management of video data requires a technology that enables systematic classification and integration of large-capacity video data. In addition, a system allowing for effective retrieval and storage of video data should be in place to provide users with information on video data according to diverse user environments, for example, on mobile terminals [2].

However, compared with desktop computers, mobile terminals have many inherent disadvantages such as low CPU processing rate/bandwidth/battery capacity, and small screens [3, 4]. In particular, low CPU processing rates and bandwidths are key inhibitors to servers that aim to provide seamless multimedia data. To ensure the effective retrieval and playout of video data on mobile terminals, CPU performance of terminals must be improved, and network technologies and systematic video data indexing technologies should be developed. Currently, studies that address such restraints and investigate methods to effectively index video data are being actively conducted [5, 6]. However, such video indexing methods are based on simply classifying video genres or types, rather than on reflecting user needs.

X. Jia, J. Wu, and Y. He (Eds.): MSN 2005, LNCS 3794, pp. 1088 – 1094, 2005.

To attach various types of information to video data is difficult since video data contain no textualized information relative to typical text data. As such, there is a need for semantic-based retrieval using additional information such as frames, key frames, and annotations in video. It is very important to make information on video data more systematic and concrete so that the user can more accurately perform content-based retrieval of such video data [7]. In this paper, a wireless multimedia database system is presented that supports mobile-learning for infants education environment using mobile terminals in wireless environments.

2 Related Work

Annotation-based retrieval, which is being widely used in performing content-based retrieval of multimedia data, is a retrieval scheme that performs a comparative retrieval of user-searched annotations by entering user annotations for different key frames and then saving the information [5-6].

The AVIS (Advanced Video Information System), developed by the University of Maryland in the United States, defines metadata as objects, events, and behaviors that are displayed in a video, and proposes an effective retrieval scheme by linking the metadata with video segments [7].

VideoSTAR (Video Storage and Retrieval), developed by the Norwegian University of Science and Technology, is a database system based on the relational database model in which various attributes such as characters, locations, and events make up a metadata, a structured video data. Those attributes are classified back into basic, primary, and secondary contexts that allow for the easy reusing and sharing of metadata, and enable the user to smoothly configure queries through the use of fixed attributes [8].

This scheme offers the advantage of correctly expressing and retrieving video contents because the user can process video contents with annotations while watching a video [9]. However, the user must attach annotations to each individual video one-by-one through the use of characters [10]. This not only involves a lot of time and effort, but also causes a vast amount of unnecessary annotations. In addition, it cannot achieve retrieval accuracy since many different annotators attach their own meanings to videos.

This paper aims to learn user questions and answers, and proposes a mobile multimedia database system that uses an automatic indexing agent in a mobile environment where the metadata of multimedia data can be updated by semantic and dependency weights in an automatic and continuous manner. This research also presents retrieval results using data for environment education of infants as test contents.

3 Multimedia Database System Using Auto-annotation Method

3.1 System Architecture

The indexing agent extracts keywords from the queries submitted from the user for video retrieval, and matches them with basic annotation information stored in the metadata. It then detects the key frames that have the same keywords as annotation information, and sends them to the user.

Figure 1 presents the architecture of annotation-based retrieval performed by the indexing agent in the Agent Middleware.

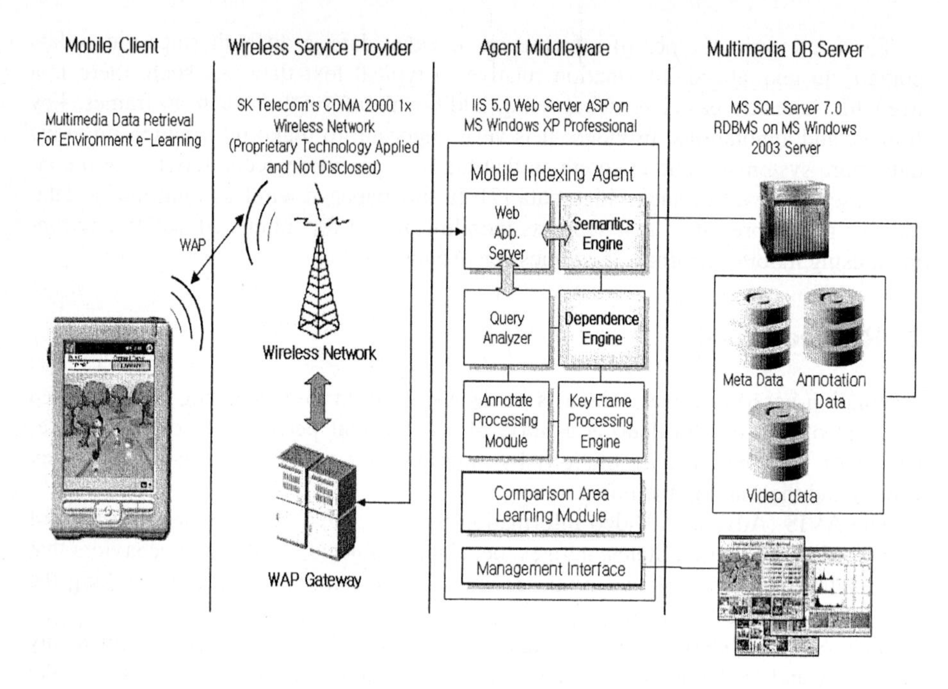

Fig. 1. System Architecture

Once entered, queries submitted from the user are analyzed and keywords are extracted. The keywords extracted are matched with the annotation information of metadata stored in the annotation database. As a result of matching, the key frames having exactly matched keywords are detected in the database and sent to the user. Additionally, the keywords that do not exactly match the annotation information among queries received from the user are defined as potential keywords.

If the mobile agent receives queries on retrieval of video contents from the user, it extracts keywords among query words and matches them to the basic annotation information of the metadata. It then detects key frames that have the same keywords as the annotation information and sends them to the user.

3.2 Semantic Weight Value

Once a user query consisting of one or more words is inputted, a corresponding keyword is extracted. The key frames containing user keywords are searched by the extracted keyword inputted by the user. The user keywords are then classified into "real keywords" and "potential keywords". The accurately matched keywords among keywords in the annotation information are classified as the "same keywords", while the accurately unmatched keywords are classified as "difference keywords".

The agent extracts key frames containing identical keywords and displays key frame lists to the user. If the user selects a specific key frame among the key frame lists, semantic weight values are calculated for each individual keyword that belongs to the specific key frame.

Where annotation keywords in the key frame are the same keywords, a new semantics weight is calculated as in Equ. (1):

$$W_{Keyword\_new} = W_{Keyword\_old} + \frac{1}{N_{Kframe\_SK}} \tag{1}$$

Where $W_{Keyword\_new}$ is the new semantics weight for annotation keywords and $W_{Keyword\_old}$ is the previous semantics weight for annotation keywords. N_{Kframe_SK} is the number of key frames with the same keywords.

In the meantime, where annotation keywords in the key frame are difference keywords, a new semantics weight is calculated as in Equ. (2).

$$W_{Keyword\_new} = W_{Keyword\_old} - \frac{1}{N_{Kframe\_SK}} \tag{2}$$

3.3 Dependence Weight Value

A dependence weight value indicates the dependence between annotations. If the user inputs a keyword, the inputted keyword and the existing dependence weight value are calculated in the order of dependence to assist the user in inputting the next keyword. The dependence weight value is added and/or subtracted by Equ. (3) and (4).

3.3.1 Addition of Dependence Weight Values

As shown in Formula 1, the dependence is increased due to the effect of learning if there are more than two keywords inputted by the user, and if there is an existing dependence between two annotations. This is useful when a keyword inputted by another user is inputted improperly, or when a wrong keyword is selected.

$$W_{Dep\_ab} = W_{Dep\_ab} + \frac{N_{Kframe\_a \cap b}}{N_{Kframe\_a} + N_{Kframe\_b}} \tag{3}$$

- W_{Dep_ab} : Dependence weight values of the annotations a and b;
- W_{Kframe_a} : Number of key frames that have the annotation a in the whole annotation DB;
- W_{Kframe_b} : Number of key frames that have the annotation b in the whole annotation DB;
- $W_{Kframe\_a \cap b}$: Number of key frames that have annotations **a** and **b** in the whole annotation.

3.3.2 Addition of Dependence Weight Values

In the case where the keyword inputted by the user contains annotation a without any annotation b, and there are dependence weight values ( ) of annotations a and b, these dependence weight values should be decreased as shown in Equ. (4).

$$W_{Dep\_ab} = W_{Dep\_ab} - \frac{1}{N_{Kframe\_a} + N_{Kframe\_b}} \tag{4}$$

($Kframe\_b$ is a DB annotation, not a keyword inputted by the user.)

- W_{Dep_ab} : Dependence weight values of the annotations **a** and **b**;
- W_{Kframe_a} : Number of key frames that have the annotation a in the whole annotation DB;
- W_{Kframe_b} : Number of key frames that have the annotation **b** in the whole annotation DB;
- $W_{Kframe\_a \cap b}$: Number of key frames that have annotations **a** and **b** in the whole annotation.

If an inputted keyword combination isn't included in the DB, the addition of a dependence weight value allows for decreasing the dependence weight value in such a way as to search dependence weight values corresponding to each keyword. In this case, subtracted dependence weight values should be lower than the added ones.

4 Implementation and Experimental Results

4.1 Implementation

Figure 2 shows the interface that allows the user to perform semantic-based retrieval on a mobile terminal. The user can search key frames for his/her desired scenes by inputting several search words.

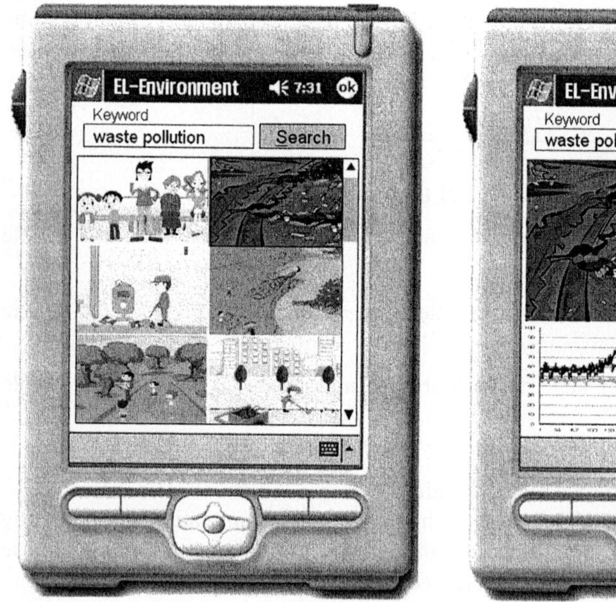

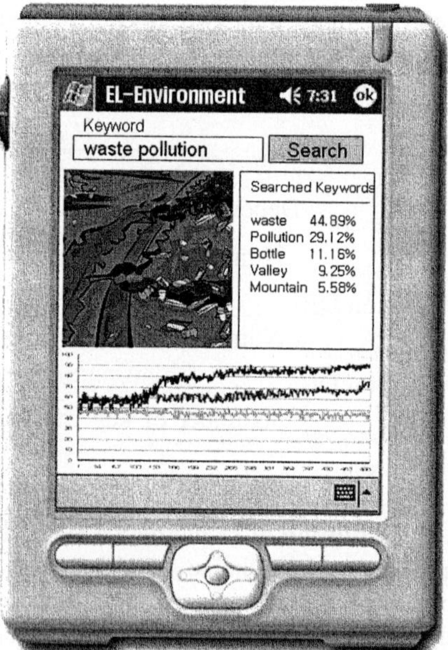

Fig. 2. Mobile Terminal Interface for Annotation-based Retrieval

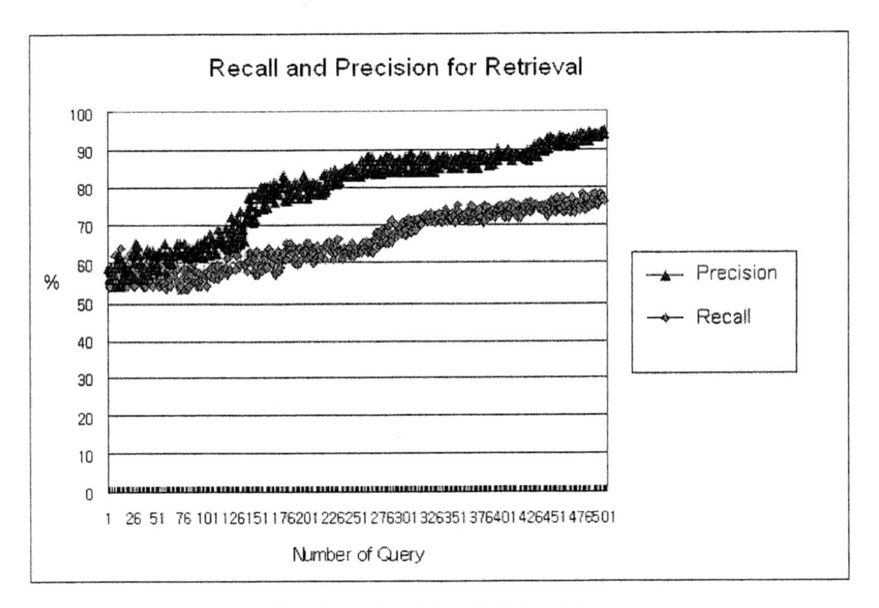

Fig. 3. Retrieval Recall & Precision

4.2 Experimental Results

MPEG formatted movie video files were used as domains for video data in order to evaluate the proposed system. For the purpose of this study, we used around 20 video clips for environment education that detected a total of 10,341 key frames. By default, a single annotation was attached to 1,524 key frames, except in the case where a key frame had duplicate portions due to successive cuts or when individual objects of a key frame were indistinguishable.

5 Conclusions

In this paper, a mobile multimedia database system that allows the user to search various meanings through the use of semantics and dependence weight value of environment mobile-learning data is proposed. In the performance of annotation-based retrieval, the proposed system reduces keyword errors by the user through the use of semantics and dependence weight value, thus allowing the user to make queries using more accurate and concrete keywords.

It is believed that future research should focus on a method for extracting voice from video data and performing similarity retrieval, thus allowing the search for video scenes using both keywords and the user's voice through speech recognition technologies.

Acknowledgements

This subject was supported by Ministry of Environment as "The Eco-technopia 21 Project".

References

1. Sibel Adali et al., "The Advanced Video Information System: data structure and query processing," Multimedia System, pp.172-186, 1996.
2. B. Y. Ricardo and R. N. Berthier, Modern Information Retrieval, ACM press, 1999.
3. T. Kamba, S. A. et al., "Using small screen space more efficiently," In Proceedings of CHI'96, ACM Press, 1996.
4. O. Buyukkokten, H. et al., "Power Browser: Efficient Web Browsing for PDAs," In Proceedings of CHI'2000, ACM Press, 2000.
5. Jiang Li et al., "Scalable portrait video for mobile video communication," IEEE Trans on CSVT, Vol. 13, No. 5 pp.376-384, 2003.
6. C. Dubuc et al., "The design and simulated performance of a mobile video telephony application for satellite third-generation wireless systems," IEEE Trans on Multimedia, Vol. 3, No. 4 pp.424-431, 2001.
7. S. Adali, "The Advanced Video Information System: Data Structure and Query Processing," ACM Multimedia System, Vol. 4, No. 4, pp. 172-186, 1996.
8. R. Hjelsvold, "VideoSTAR-A Database for Video Information Sharing," Ph.D. Thesis, Norwegian Institute of Technology, 1995.
9. Chong-Wah Ngo, Ting-Chuen Pong, Hong-Jiang ZhangOn, "Clustering and retrieval of video shots through temporal slices analysis," IEEE Trans on Multimedia, Vol.04, No.04, pp. 0446-0458, 2002.
10. Jonghee Lee, Keunwang Lee, "A Semantics-based Video Retrieval System using Indexing Agent", Conference on KMMS '03, Vol. 6, No. 2, pp.281-284, 2003.

Towards a Symbolic Bisimulation for the Spi Calculus

Yinhua Lü[1], Xiaorong Chen[1], Luming Fang[2], and Hangjun Wang[2]

[1] Department of Computer Science,
Guizhou University, Guiyang 550025, China
[2] School of Information Science and Technology,
Zhejiang Forestry University, Lin'an 311300, China
yhlv@sohu.com, xrchen@tmail.gzu.edu.cn

Abstract. Observational equivalence is a powerful means for formulating the security properties of cryptographic protocols. However suffering from the infinite quantifications over contexts, its proof becomes notoriously troublesome. This paper addresses the problem with a symbolic technique. We propose a symbolic bisimulation for spi calculus based on an environmental sensitive label transition system semantics, which restrict the infinite inputs of a process to only finite transitions. We also prove that the symbolic bisimulation is sound to the traditional concrete bisimulation, and furthermore is a promising means to automatically verifying the security protocols.

1 Introduction

In the recent decade, formal analysis of cryptographic protocols has been greatly developed under the use of process calculus, such as Spi calculus ([AG99], [LGCF05]), CSP [Hoa85] and CCS [Mil89], which provide good foundations for modelling protocols and reasoning interactions between them.

Among these models, Abadi and Gordon's spi calculus receives high research interests. It is extended from pi calculus [MPW92] with cryptographic primitives and some relations. Previously, protocols' security properties are generally formulated in the *partial order* relation $\ll$ ([Sch98]). For example, to state that protocol P holds the secrecy property, one can construct a typical observer O paralleling with P in an environment ε and formalize it as $(\varepsilon, P \mid O) \models \perp \ll \overline{err}\langle M\rangle$ indicating that the message M should never be leaked out in all executing traces of $P \mid O$. One problem of this formations is that the observer process O can not include all possible invaders.

However, in spi calculus by virtue of the *observational equivalence* $\simeq$ [AG99], the secrecy of a security protocol is successfully defined as the *syntactical indistinguishability*, that is $P(d) \simeq P(d')$, for any data d, d' which means the behavior of P can not be distinguished under all contexts. This formalization is attractive because the quantification over all contexts directly captures an arbitrary attacker rather than explicitly building it.

X. Jia, J. Wu, and Y. He (Eds.): MSN 2005, LNCS 3794, pp. 1095–1102, 2005.

Although intuitive, the infinite quantifications make proofs of such equivalence quite difficult. *Environmental bisimulations* are consequently developed to approach it (e.g. Abadi and Gordon's *framed bisimulation* [AG98], Borgström et al.'s *hedged bisimulation* [BN03] and Boreale et.al.'s *environment sensitive labelled bisimulation* [BDP02], comparisons of these bisimulation relations can be found in [BN03]). Basically, two processes simulate each other if it can build a correspondent operation between their interactions with each environment and these correspondences should always keep two environments equivalent. Nevertheless, that an input action may potentially receive arbitrary messages from its environment makes automatical proof of a bisimulation relation impractical.

Central to our approach is the development of *symbolic operational semantics* where symbolic actions such as input $\bar{c}\langle t\rangle$, output $c(x)$ and their residuals are associated with *variable messages.* More generally, symbolic actions are related to boolean formulaes indicating under which conditions they can be performed. An environment sensitive label transition semantics is given to the configuration, which is the role of our bisimulation game. Important to our semantics is that an input configuration can transit to another only if the input process is ready to receive the terms derivable from its symbolic environment. Our *symbolic bisimulation* is built on this environment sensitive lts.

Related work. Symbolic technique can be found in *model checking* to deal with the state explosion problem and formed a technical branch–*symbolic model checking*[CGP99]. Hennessy and Lin, to our knowledge, firstly employed it in the value-passing process algebra to realize the proof of standard bisimulation equivalence mechanically [HL95].

To cope with the bisimulation checking problem in spi calculus as stated above, Borgström et al. as we do, constructed a symbolic bisimulation in spi calculus [BBN04]. Comparing with our work, it has a complicated semantics, e.g. much consideration on transferable channels and scope restrictions, which seem to us can be simplified.

In their definition of symbolic environment, each component in the environment is parameterized with a time stamp in order to guarantee that substituted messages for an input variable at time t must be synthesized from the environment before that time, and during each step the environment's time value must be carefully computed. Yet, we use an environment *Knowledge Deductive System* to ensure that the symbolic substitution for an input is derivable from the environment.

Moreover, one of their conditions on equating two environments only requires if two messages both belong to the clear text set then they can be regarded as indistinguishable to the environments while in ours their value equivalence is also required.

Summary. This paper is structured as follows: Section 2 presents the syntax of the Spi calculus and operational semantics used in this paper. Section 3 takes insight into a paired process's environments in which they interact. Section 4 mainly describes our bisimulation relations under the concrete and symbolic contexts. Finally, section 5 offers some conclusion remarks.

2 Spi Calculus

Spi calculus is defined in [AG99] as a process algebraic language developed from pi calculus [MPW92]. Equipped with the primitives for encryption and decryption, hashing, it is capable of modelling the security protocols. In addition, we have primitive constructs for pairing, pair splitting and match(including mismatch) to condition the execution of a process.

We adopt the same assumption on the perfect encryption and hashing as stated in ([AG99], [BDP02]). Moreover, we require channels of input and output are just constant names under the assumption of global communications in a protocol, and encryption keys are atomic.

Syntax. The syntax of spi calculus is given in table 1. Names are constants used for plain-texts, keys and nonces. Variables are unknown names to be concretized. Messages are names, together with compound names built from pairing, encryptions and hashes. Terms are composed of variables, messages, pair spitting and decryptions. In boolean formulas, the predicate $[t : \mathcal{N}]$ judges for whether t belongs to the name set $\mathcal{N}$ while the formulae $[ev \vdash t]$ tests for whether t can be deduced from the environment ev, and it will be further explained in section 3.

Process P includes input, output, parallel, choice, restriction, pair splitting and decryption. In particular, we replace the traditional unbound replication process $!P$ with multiplication process P^n which represents finite n Ps in parallel.

To treat asymmetric encryption, we assume a relation $\mathcal{K} \subseteq \mathcal{N} \times \mathcal{N}$, s. t. $(k, k') \in \mathcal{K}$ iff messages encrypted with k can be decrypted with k'. We define $I(k) = \{k' | (k, k') \in \mathcal{K}\}$. A substitution σ is a partial function s. t. $\sigma(\cdot) : \mathcal{V} \to \mathcal{T}$. Substitutions are applied to processes, terms, formulae and actions in the straightforward way.

Semantics. First, we present two evaluation functions to evaluate the terms and boolean formulaes. The evaluation function for terms, $\rho(\cdot) : \mathcal{T} \to \mathcal{M}$ is defined by induction on term $\mathcal{T}$, while the evaluation function for boolean formulaes, $\rho(\cdot) : \mathcal{B} \to \{tt, ff\}$ is defined by induction on formulaes $\mathcal{B}$.

Formal semantics formalizing inference rules of process P in this paper is structure operational semantics. Our concrete semantics is similar to the one

Table 1. Syntax of the Spi Calculus

$n ::= a, c, \cdots, k, n \cdots$	Names $\mathcal{N}$
$x ::= x, y, z \cdots$	Variables $\mathcal{V}$
$M ::= n \mid (M_1, M_2) \mid E(M, n) \mid H(M)$	Messages $\mathcal{M}$
$t ::= x \mid M \mid (t_1, t_2) \mid E(t_1, t_2) \mid H(t) \mid \pi_i(t_1, t_2) \mid D(t_1, t_2)$	Terms $\mathcal{T}$
$b ::= tt \mid b_1 \wedge b_2 \mid \neg b \mid [t_1 = t_2] \mid [t : \mathcal{N}] \mid [ev \vdash t]$	Booleans $\mathcal{B}$
$P ::= \mathbf{0} \mid c(x).P \mid \bar{c}\langle t \rangle.P \mid P \vert Q \mid b : P, Q \mid (v\, a)P \mid$ let $x = D(t, t')$ in $P \mid$ case x of $\pi_i(t_1, t_2)$ in P	Processes $\mathcal{P}$

used in [BDP02], apart from that we do not have the internal τ-action when two processes are in communication. The concrete semantics characterizes input and output terms with (concrete) messages. In the concrete semantics, actions $\mu \in \mathcal{A}_c$, where $\mathcal{A}_c = \{c(x),\ \bar{c}\langle M\rangle\}$.

Different from the concrete semantics, when executing an input action of a process P in the symbolic semantics, messages for concretizing input variables are delayed until possibly wanted, and the variables are substituted with symbolic terms. Moreover, boolean conditions guarded for a transition are collected in transition constrains ([HL95], [BBN04]). The symbolic transition takes the form of $P \overset{\lambda}{\underset{b}{\longmapsto}} P'$, where $\lambda \in \mathcal{A}_s = \{c(x),\ \bar{c}\langle t\rangle\}$, $b \in \mathcal{B}$. Symbolic operational semantics for spi calculus are listed in table 2.

The symbolic transition mimics the concrete transition with true constrains.

Lemma 1. *$P \overset{\mu}{\longrightarrow} P'$ iff $\exists\ \rho,\ \sigma,\ b$ s.t. $P \overset{\lambda}{\underset{b}{\longmapsto}} Q$ where $P' = Q \cdot \sigma$, $\mu = \lambda \cdot \sigma$, and $\rho(b \cdot \sigma) = tt$.*

Table 2. Symbolic Operational Semantics

$$(\text{SIN})\ \frac{--}{c(x).P \overset{c(y)}{\underset{tt}{\longmapsto}} P[y/x]}\quad y = \text{new}\{\mathcal{T}\} \qquad (\text{SOUT})\ \frac{--}{\bar{c}\langle t\rangle.P \overset{\bar{c}\langle t\rangle}{\underset{tt}{\longmapsto}} P}$$

$$(\text{SPAR})\ \frac{P \overset{\lambda}{\underset{b}{\longmapsto}} P'}{P \mid Q \overset{\lambda}{\underset{b}{\longmapsto}} P' \mid Q} \qquad (\text{SCHS})\ \frac{P \overset{\lambda}{\underset{b}{\longmapsto}} P'}{b' : P, Q \overset{\lambda}{\underset{b \wedge b'}{\longmapsto}} P'}$$

$$(\text{SRES})\ \frac{P \overset{\lambda}{\underset{b}{\longmapsto}} P'}{(\nu\, a)P \overset{\lambda}{\underset{b}{\longmapsto}} (\nu\, a)P'}\quad a \notin \{c\} \cup n(b) \qquad (\text{SSPL})\ \frac{P[t/x] \overset{\lambda}{\underset{b}{\longmapsto}} P'}{case\ x\ of\ \pi_1(t_1, t_2)\ in\ P \overset{\lambda}{\underset{b}{\longmapsto}} P'}$$

$$(\text{SDEC})\ \frac{P[t/x] \overset{\lambda}{\underset{b}{\longmapsto}} P'}{let\ x = D(E(t, t'), t'')\ in\ P \overset{\lambda}{\underset{b \wedge b' \wedge b''}{\longmapsto}} P'}\quad b' = (I(t'') = t'),\ b'' = \{t', t''\} : \mathcal{N}$$

3 Process Environment

In spi calculus, a process is supposed to act in a malicious environment able to observe and forge any message from the process. Two types of environments are discussed in this paper according to the different semantics. The *concrete environment cev* of a process P is a *message set* composed from P's outputs and synthesis rules, while the *symbolic environment ev* (just refer as the *environment ev* for convenience) of P is a *term set*. This section provides insights into process's symbolic environment analysis and hope readers have not much difficulties in understanding those about the concrete environment.

As we can see, when a process exports a message, its environment will duplicate and store it into its own message set. On the other hand, when the process needs to import a message, its environment will provide with all possible messages it can make. However, as mentioned in the introduction, this will give rise to infinite branches, which is hard to cope with.

However this problem can be well solved by taking into account substituting the input virtually rather than concretely. That is to substitute the input with a term derivable from its environment. We start with an example for this development.

$$P \equiv \bar{c}\langle E(M,k)\rangle \mid c(x).\ \text{let } u = \text{Dec}(x, I(k)) \text{ in } \bar{c}\langle u\rangle, \quad ev = \emptyset$$

In the symbolic executions, P may receive a term $t = E(y,k)$ from its ev. Because the k is still unknown to ev, t can only be synthesized under the constrains of substitution $\sigma[M/y]$ if ev knows $E(M,k)$.

There is a *Knowledge Deductive System* (KDS)[FA01] for deciding the judgement of $\sigma : ev \vdash t$, which means under the substitution of σ the term t may be derived from ev.

A *configuration* $C = (ev,\ P)$ is a tuple of a process P with its symbolic environment ev. Transitions on a configuration $C = (ev, P)$ corresponds to the movements of interactions between P and ev. Now, we are ready to define the lts semantics for a configuration in table 3.

Table 3. LTS Semantics for a Configuration

$$(\text{FIN})\frac{P \overset{c(x)}{\underset{b}{\longmapsto}} P'[t/x]}{(ev, P) \overset{c(x)}{\underset{b\wedge[ev\vdash t]}{\longmapsto}} (ev\cdot\sigma, P'[t/x]\cdot\sigma)}\ \sigma : ev \vdash t \qquad (\text{FOUT})\frac{P \overset{\bar{c}\langle t\rangle}{\underset{b}{\longmapsto}} P'}{(ev, P) \overset{\bar{c}\langle t\rangle}{\underset{b}{\longmapsto}} (ev \cup \{t\}, P')}$$

3.1 Coherent Environments

In this part, we reason on a paired-environment from two processes and offer conditions to equate these two environments. Similar discussion can be found in ([AG98], [BN03]).

Suppose process P's and Q's environments are ev_1, ev_2. A paired-environment ε is the set of $ev_1 \times ev_2$. The synthesis function $\mathcal{S}(\cdot)$ and analysis function $\mathcal{A}(\cdot)$ are closure operations over the power set of ε, which simulate the composition and decomposition actions conducted by an attacker.

The kernel of the paired-environment ε is $\mathcal{K}(\varepsilon)$ excluding redundant paired-terms. We assume that ε is always refined to $\mathcal{K}(\varepsilon)$. Next, we equate two environments of ε in the term of *coherent environment.* This leads to the important idea of *indistinguishability* of two environments in the eye of a bystander process(always an invader).

Definition 1 (Coherent Environments). *A paired-environment ε is semi-coherent if whenever$(t_1, t_2) \in \varepsilon$*

a. if $t_1 \in \mathcal{N}$ then $t_2 \in \mathcal{N}$ s.t. $t_1 = t_2$;
b. if $t_1 \in \mathcal{V}$ then $t_2 \in \mathcal{V}$;

c. *if* $(t_1', t_2') \in$ *!varepsilon s.t.* $t_1 = t_1'$ *then* $t_2 = t_2'$;
d. *if* $t_1 = E(u_1, u_1')$ *and* $I(u_1') \in \pi_1(\mathcal{S}(\varepsilon))$ *then* $t_2 = E(u_2, u_2')$ *s.t.* $(I(u_1'), I(u_2')) \in \mathcal{S}(\varepsilon)$ *and* $(u_1, u_2) \in \mathcal{S}(\varepsilon)$;
e. *if* $t_1 = (u_1, u_2)$ *then* $t_2 = (u_1', u_2')$ *and for* $(u_1, u_1'), (u_2, u_2') \in \varepsilon$ *satisfying conditions (a)–(e).*

ε is *coherent* iff both ε and ε^T are semi-coherent. A relation $\mathcal{R}$ is a subset of $\varepsilon \times \mathcal{P} \times \mathcal{P}$ and is coherent if $\varepsilon \vdash (P, Q)$ implies that ε is coherent.

Now, we relate it to the concrete environment under the help of substitutions.

Definition 2 (Well-Formed Substitutions). *A substitution pair* (σ_1, σ_2) *is called well-formed, written as* $\varepsilon \vdash (t_1, t_2)$ *iff*

a. *if* $(v_1, v_2) \in \varepsilon$, *then* $\varepsilon \vdash \sigma_1(v_1) \leftrightarrow \sigma_2(v_2)$;
b. *if* $(t_1, t_2) \in \varepsilon$, *then* $\varepsilon \vdash \rho(t_1 \cdot \sigma_1) \leftrightarrow \rho(t_2 \cdot \sigma_2), \rho(t_i \cdot \sigma_i) \neq \perp$;
c. *For two processes* P, Q, *if* $b_i \in \mathcal{B}(P)$, $b_i' \in \mathcal{B}(Q)$, *then* $\rho(b_i \cdot \sigma_1)$, $\rho(b_i' \cdot \sigma_2)$ *is defined.*

When applying a well-formed substitution pair (σ_1, σ_2) to a coherent symbolic environment pair ε, then we will get the corresponding coherent concrete environment pair $\varepsilon^c_{\sigma_1, \sigma_2}$.

Lemma 2. *A coherent paired-environment* ε *can be converted into a concrete coherent paired-environment* $\varepsilon^c_{\sigma_1, \sigma_2}$ *under two well-formed substitutions,* σ_1, σ_2, *and an evaluation* ρ.

4 Reefed Bisimulations

Bisimulation equivalence is of great importance in relating two processes in spi calculus. To verify a bisimulation between P and Q, it is to build corresponding steps between them under an equivalent paired environment. In the following, we give definitions for both concrete and symbolic bisimulations–nicknamed *reefed bisimulations*.

Concrete reefed bisimulation is given in the traditional style of an environmental bisimulation ([BN03], [BDP02]).

Definition 3. *A concrete reefed relation* $\mathcal{R}^c$ *is coherent relation if* $(ev_1, ev_2) \vdash P\mathcal{R}^cQ$ *implies whenever*

a. $(cev_1, P) \xrightarrow{c(x)} (cev_1', P')$, *there is a configuration* (cev_2', Q') *with* $(cev_2, Q) \xrightarrow{c(y)} (cev_2', Q')$ *and for all messages* M *and* N, $(cev_1, cev_2) \vdash (M \leftrightarrow N)$, *such that* $(cev_1', cev_2') \vdash P'[M/x]\mathcal{R}^cQ'[N/y]$;
b. $(cev_1, P) \xrightarrow{\bar{c}\langle M_1 \rangle} (cev_1', P')$, *where* $cev_1' = cev_1 \cup \{M_1\}$, *there exits a configuration* (cev_2', Q') *with* $(cev_2, Q) \xrightarrow{\bar{c}\langle M_2 \rangle} (cev_2', Q')$, $cev_2 = cev_2 \cup \{M_2\}$ *and* $(cev_1, cev_2) \vdash P'\mathcal{R}cQ'$.

A concrete reefed bisimulation is a reefed relation $\mathcal{R}^c$ such that both $\mathcal{R}^c$ and $\mathcal{R}^{cT}$ are concrete reefed simulations. *Concrete reefed bisimilarity*($\sim_r^c$) is the union of all concrete reefed bisimulations.

In the definition above, as for inputs of two processes, it must take into account of all paired indistinguishable messages constructed by each *cev*. This is the main problem for automated bisimulation checks, since the set of potential inputs is infinite.

Based on previous preparations and inspired by the symbolic bisimulation in [HL95], we now define a symbolic bisimulation for spi calculus, with the property that every symbolic input action only lead to finite new process pair. Different from the concrete bisimulation, symbolic transition in our definition is guarded by boolean formulaes (or constrains). These constrains are collected and satisfied during the simulation transitions. Intuitively, for two processes to be bisimilar under a given coherent environment, every guarded transition of one of the processes must be simulated by a guarded transition of the other process such that the new environment is coherent and so on.

Definition 4. *A symbolic reefed relation $\mathcal{R}^b$ is a coherent relation if $(ev_1, ev_2) \vdash P\mathcal{R}^b Q$ implies whenever*

i. $(ev_1, P) \xrightarrow[b_1 \wedge [ev_1 \vdash t]]{c(x)} (ev_1', P'[t/x] \cdot \sigma_1)$, *where* $\sigma_1 : ev_1 \vdash t$ *and* $ev_1' = ev_1 \cdot \sigma_1$, *there exists a substitution* σ_2 *and boolean set* B, *satisfying (1).* $\sigma_2 : ev_2 \vdash t$ *and* $ev_2' = ev_2 \cdot \sigma_2$; *(2).* $b \wedge b_1 \Rightarrow \vee B, \forall\, b' \in B, b' \Rightarrow b_2$, *such that* $(ev_2, Q) \xrightarrow[b_2 \wedge [ev_2 \vdash t]]{c(y)} (ev_2', Q'[t/y] \cdot \sigma_2)$ *and* $(ev_1', ev_2') \vdash P'[t/x] \cdot \sigma_1 \, \mathcal{R}^{b'} \, Q'[t/y] \cdot \sigma_2$

ii. $(ev_1, P) \xrightarrow[b_1]{\bar{c}\langle t_1 \rangle} (ev_1', P')$, *where* $ev_1' = ev_1 \cup t_1$, *there exists a boolean set* B, *where* $b \wedge b_1 \Rightarrow \vee B, \forall\, b' \in B, b' \Rightarrow b_2$, *such that* $(ev_2, Q) \xrightarrow[b_2]{\bar{c}\langle t_2 \rangle} (ev_2', Q')$, *where* $ev_2' = ev_2 \cup t_2$, *and* $(ev_1', ev_2') \vdash P' \mathcal{R}^{b'} Q'$.

A symbolic reefed bisimulation is a reefed relation $\mathcal{R}^b$ such that both $\mathcal{R}^b$ and $\mathcal{R}^{bT}$ are reefed simulations.*Symbolic reefed bisimilarity*($\sim_r^b$) is the union of all reefed bisimulations.

We have that symbolic reefed bisimilarity is a sound approximation to concrete one.

Theorem 1 (Soundness). *If* $\varepsilon \vdash P \sim_r^b Q$, $\varepsilon = (ev_1, ev_2)$ *then there exists well-formed substitutions* σ_1, σ_2 *and an evaluation function* ρ, *such that* $\varepsilon^c_{\sigma_1,\sigma_2} \vdash P \cdot \sigma_1 \sim_r^c Q \cdot \sigma_2$.

An Example. Consider two processes:

$$P \equiv \bar{c}\langle E(M,k)\rangle \mid c(x).let\ u = Dec(x, I(k))\ in\ \bar{c}\langle M\rangle$$
$$Q \equiv \bar{c}\langle E(M,k)\rangle \mid c(y).let\ u = Dec(y, I(k))\ in\ [u = M]\bar{c}\langle u\rangle, \bar{c}\langle M\rangle$$

where k is a public key and the initial environments for both processes are $ev_1 = ev_2 = \{k, E(M_{old}, k), x\}$. Let $b = tt$, to prove that $P \sim_r^b Q$.

Proposition 1. $(ev_1, ev_2) \vdash P \sim_r^{tt} Q$

5 Conclusions

We have developed a symbolic environment sensitive lts semantics for the spi calculus with full encryption system under a clear and rich formalization. We also propose a symbolic notion of a bisimulation, the symbolic reefed bisimilarity for spi calculus, and prove it a sound approximation to concrete bisimilarity. Based on these work, it is promising for the automatic checking of observational equivalence in security properties formulation.

Future work will focus on an implementation of this bisimilarity in mechanizing protocol analysis, also will try to reason about the completeness of symbolic reefed bisimilarity.

References

[AG98] M. Abadi and A. D. Gordon. A bisimulation method for cryptographic protocols. *Nordic Journal of Computing*, 5(4): 267-303, 1998.

[AG99] M. Abadi and A. D. Gordon. A calculus for cryptographic protocols: The spi calculus. *Information and Computation*, 148(1):1-70, 1999.

[BDP02] M. Boreale, R. De Nicola, and Rosario Pugliese. Proof techniques for cryptographic processes. *SIAM Journal on Computing*, 31(3): 947-986, 2002.

[BN03] J. Borgström, U. Nestmann. On bisimulations for the spi-calculus. Technical Report, EPFL, Switzerland, 2003.

[BBN04] J. Borgström, S. Briais and U. Nestmann. Symbolic bisimulations in the spi calculus. *Proc. of CONCUR 2004, volume 3170 of LNCS*: 161-170, 2004.

[CGP99] Edmund M. Clarke, Jr., Orna Grumberg and Doror A. Peled. *Model Checking.* MIT Press, Page:61-87, 1999.

[FA01] M. Fiore and M. Abadi. Computing symbolic models for verifying cryptographic protocols. *Proc. of 14th IEEE CSFW*: 160-173, 2001.

[Hoa85] C.Hoare. *Commnunicating Sequential Processes.* Prentice-Hall International, 1985.

[HL95] M. Hennessy and H. Lin. Symbolic bisimulations. *Theoretical Computer Science*, 138(1995): 353-389, 1995.

[LGCF05] Y.H. Lü, Y.G. Gu, X.R. Chen, and Y.Fu. Analyzing security protocols by a bisimulation method based on environmental knowledge. *Proc. of ICCCAS 2005*:79-83, 2005.

[MPW92] R. Milner, J. Parrow, and D. Walker. A calculus of mobile processes, I and II. *Information and Computation*, 100(1):1-77, 1992.

[Mil89] R. Milner. *Communication and Concurrency.* Prentice-Hall International, 1989.

[Sch98] S. Schneider. Verifying authentication protocols in CSP. *IEEE Tran. Softw. Eng. 24*, 9, 741-758, 1998

Mobile Agent-Based Framework for Healthcare Knowledge Management System

Sang-Young Lee and Yun-Hyeon Lee

Dept. of Health Administration, Namseoul University, Korea
21 Maeju-ri, Seongwan-eup, Cheonan
{sylee, skylee}@snu.ac.kr

Abstract. Induction of knowledge management system about healthcare enterprise is required. To address this issue of the lack of true knowledge management in healthcare enterprises, we propose a framework for common healthcare knowledge management. This framework is made up of two areas of applications and services, i.e. the mobile agent-based knowledge management application area and the strategic visualization, planning and coalition information service area.

1 Introduction

Importance of effective information and knowledge management in enterprise has been emphasized [1]. Now, knowledge is realized with the biggest competition weapon, and not a part of property possessed by the organization simply.

Therefore, necessity of Knowledge Management System(KMS) that can change the data and information stored transfer to the form of knowledge, and store these into software, database and manage efficiently have been embossed[2, 3]. Especially, various attempts to process the inefficient and lack of knowledge management problem in healthcare sectors applied in this paper have been demonstrated [4, 5]. However, study on framework for knowledge management in healthcare field that consisted of fundamental basis has not been exploited yet [6]. Mainly, researches on the healthcare knowledge management tool have been done [7, 8].

This paper was presented to develop healthcare knowledge management framework through a general architecture scheme. This healthcare knowledge management framework processes a special knowledge management process and has a target of common application. And is consisted of application of mobile agent based and service to achieve the knowledge-based general process [9] efficiently.

2 Healthcare Knowledge Management System Based on Mobile Agent

Proposed healthcare knowledge management framework is consisted of the application and the service area.

X. Jia, J. Wu, and Y. He (Eds.): MSN 2005, LNCS 3794, pp. 1103 – 1109, 2005.

2.1 Application Area of Mobile Agent

In the application area to process important knowledge management process on knowledge acquisition, identification, interchange, organization, and reusability. The following components are involved.

- Knowledge Acquisition Tool: Basic knowledge gets from healthcare expert.
- Identification and Interchange Tool for Knowledge: To share knowledge provided by healthcare expert through mutual interchange with Mobile agent and to provide health care knowledge for individual use.
- Organization and Reusability Tool for Knowledge: Systematization and reformatting for other purpose and circumstance can be achieved since constructed the repository of knowledge.

2.2 Service Area of Dynamic Information

This area is carried out for to facilitate strategic plan and judgment in healthcare department. This area is consisted of following components.

- Dynamic Healthcare Knowledge Visualize Tool: To achieve effective visualization and browsing of healthcare knowledge acquired from the repository.
- Dynamic Healthcare Planning Tool: To provide for individual use and provide customizing possible.
- Healthcare Coalition Information Tool: To utilize plan, schedule and resources to form the optimal team to execute work in healthcare part.

Components about this application and service are consisted of 4 structures as shown in Figure 1.

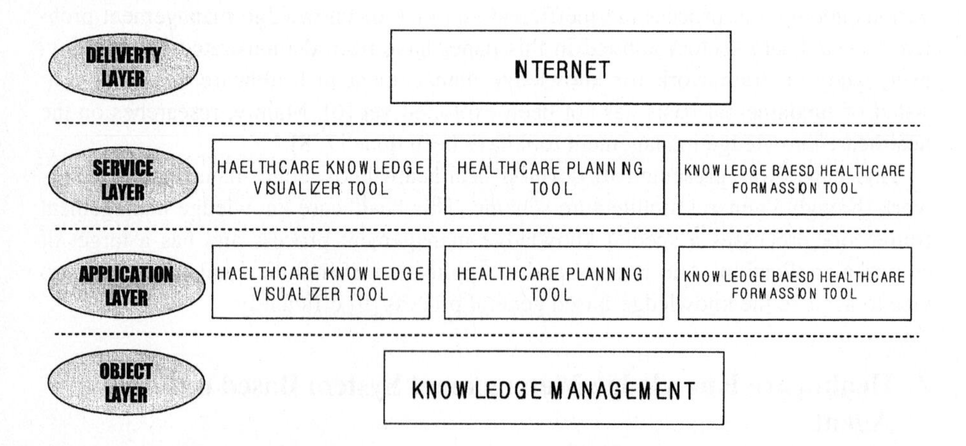

Fig. 1. Layer Architecture of the Healthcare Framework

3 Mobile Agent-Based Application Area

This is used to create knowledge for efficient decision making of healthcare part and manages, and does effective judgment through link with service area and to lay out strategic plan.

3.1 Knowledge Acquisition Tool

This is consisted of mail server, application server and mail (knowledge) repository as shown in Figure 2.

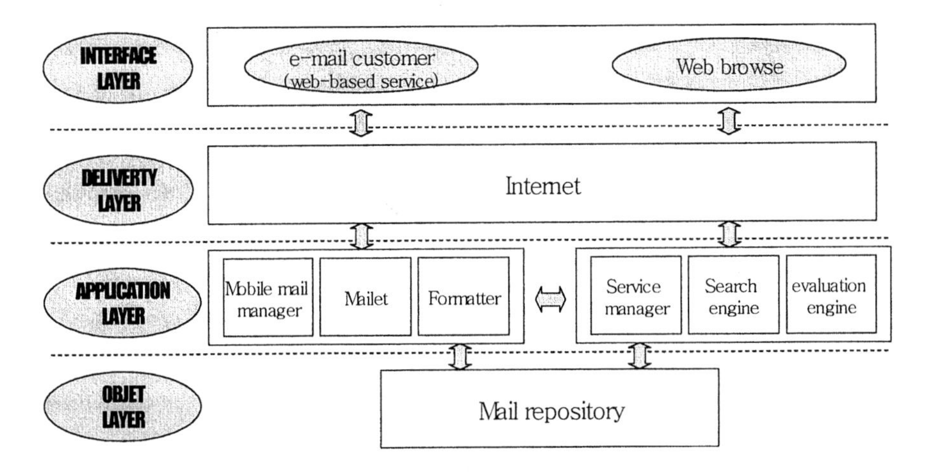

Fig. 2. Four-Layered Architecture of the Knowledge Acquisition Tool

1) Mail Server

The mail server supports the basic function in sending and receiving to and from various e-mail customers. The main serve components of mail server are listed as followings.

- Intellectual e-mail Administrator: Handles e-mail transaction.
- Mailet: Mailet make e-mail forwarding or forming an auditor group specialized for health issues controlled by the intellectual e-mail administrator.
- E-mail Formatter: e-mail proceeds to be re-formatted before it saved to repository or forwarded to receiver. To capture suitable messages i.e. sender, receiver, date and time, the intellectual parsing function is adopted.

2) Application Server

- Service Manager: Receive services required by user.
- Evaluation Engine: Evaluation of the e-mail qualities (usage, accurate, pertinence).

3) Mail Repository: Main save mechanism

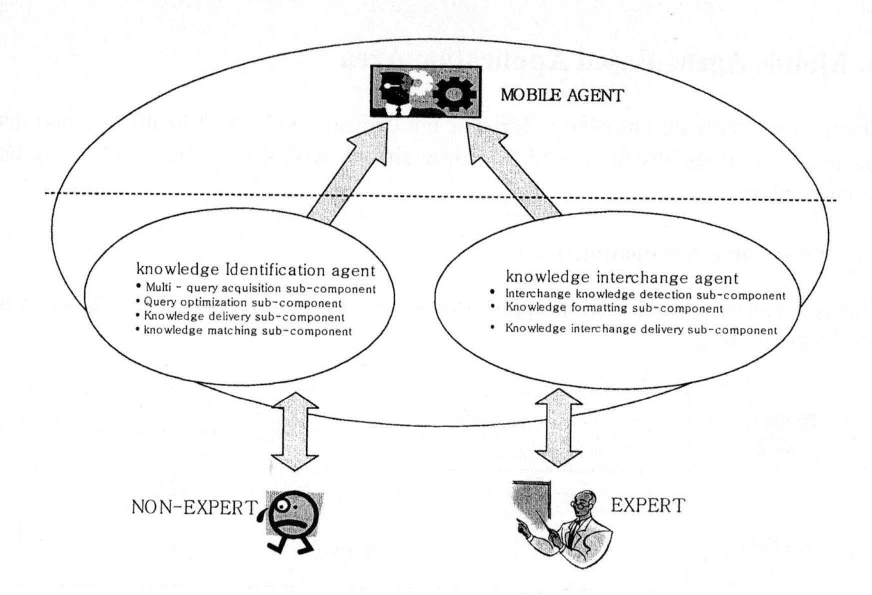

Fig. 3. Knowledge Identification and Interchange Tool

3.2 Identification and Interchange Tool for Knowledge

The knowledge identification and knowledge interchange tool is consisted of the knowledge management framework of mobile agent base, the agent based knowledge confirmation agent and the agent based knowledge interchange agent. Figure 3 shows these structures.

1) Knowledge Management Framework of Mobile Agent Base
Knowledge management framework of mobile agent base offers the basic and general architecture of mobile agent sub-component.

2) Knowledge Identification Agent
Knowledge identification agent involves the search protocol that those who nonprofessional healthcare knowledge can locate search possibility easily. The search protocol applied the search classification for the algorithm to bring the most exact result of presentation. The main composition is as following.

- Multi-Query Acquisition Sub-Component: This acquires query of users dynamically for user to take spare query intellectually for actualization.
- Query Optimization Sub-Component: Query optimization sub-component involves the process that formatting receiving query again as concise and easy forms.
- Knowledge Delivery Sub-Component: To take responsibility on conductions of agents to deliver query to other agent.
- Knowledge Matching Sub-Component: To confirm and amend knowledge required for user.

3) Knowledge Interchange Agent

- Interchange Knowledge Detection Sub-Component: Use language analysis method to search document and e-mail including data connected with healthcare experience.
- Knowledge Formatting Sub-Component: Interchanged knowledge is reformed by mean of format for confirmation and searching effectively.
- Knowledge Interchange Delivery Sub-Component: Roles of share and transmit the knowledge are achieved through agent.

3.3 Organization and Reusability Tool for Knowledge

The knowledge organization and reusability tool is consisted of 3 components. These are the knowledge management framework of Mobile agent base, knowledge composition of agent base and the knowledge reusability agent base.

1) Knowledge Organization Agent
Categorization of mobile agent to approach easy to various medical treatment repository.

- Repository Organization Sub-Component: Knowledge in repository is organized automatically by a special protocol or an algorithm composes.
- Garbage Collection Sub-Component: Garbage collection sub-component maintains an up to date knowledge in the repository, and remove the old and least usable from the knowledge items.

2) Knowledge Reusability Agent
To solve current problem provide to reuse healthcare knowledge effectively. This paper focused on knowledge application, knowledge personalization and knowledge combination.

- Knowledge Application Sub-Component: Apply the used healthcare knowledge for new problem or query based on application algorithm.
- Knowledge Personalization Sub-Component: Personalize the applied know-how or solutions considering user's preference. Age, economical background, health, past experience and family history provide influences here.
- Knowledge Combination Sub-Component: Make for user to master easily and effectively through all knowledge that have been applied and personalized.

4 Service Area About Dynamic Information

The visualization, planning and coalition of information service area support knowledge management application. For the creation of healthcare value, provide for to assist for knowledge management application area of mobile agent base.

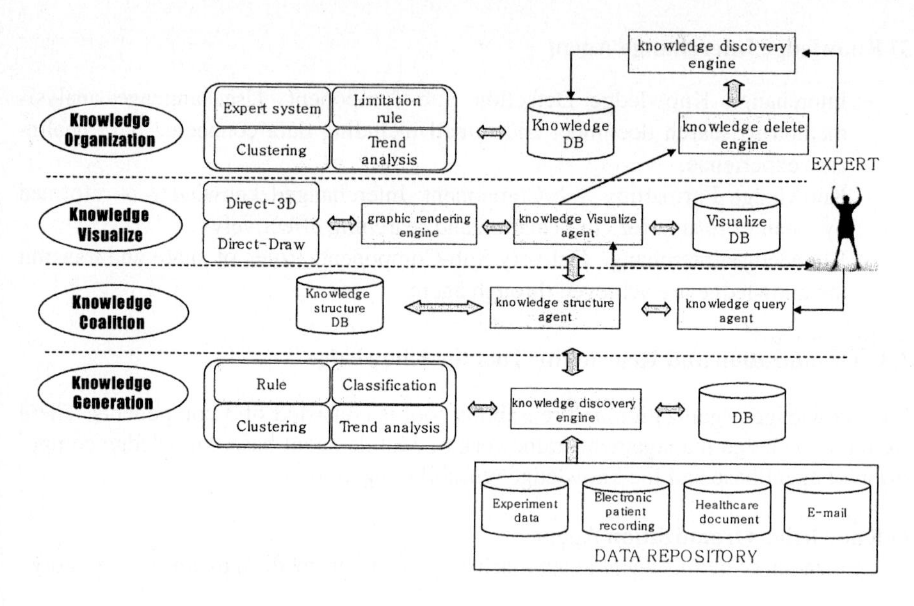

Fig. 4. Dynamic Healthcare Knowledge Visualize Tool

4.1 Dynamic Healthcare Knowledge Visualize Tool

Dynamic visualize tool of healthcare knowledge include knowledge discovery engine, knowledge query agent, knowledge structure engine, knowledge structure DB, graphic rendering engine, knowledge Visualize agent, knowledge delete engine and new knowledge discovery agent. Figure 4 shows these components.

These components play an important role in visualization and use of life cycle on knowledge.

4.2 Dynamic Healthcare Planning Tool

Three sub-components are participated in the establishment of healthcare planning for healthcare part and individual person.

- General Planning Repository: This is a repository which store general plan for healthcare part and health plan for individual person.
- Resource and Schedule Repository: This is a repository which stores related information resource and save each individual health condition.
- Mobile Dynamic Planning Agent: This is a center engine that takes the responsibility for fundamental plan. In here, the progress is achieves in cooperation with the repositories of general plan, resources and schedule.

5 Conclusions

Knowledge management system is a nuclear rising with core in all enterprises or departments presently. However, it can not but take doubt in practical effectiveness by the existing solutions leaned on documents management. This has shortcoming in it can not present the necessary application to makes knowledge management process and hardly manage common knowledge and experience properly.

Therefore, in this paper, the common healthcare knowledge management framework that composed knowledge management tool of Mobile agent base was presented. These frameworks presented in the existed general architecture. Also, healthcare knowledge management framework of this paper achieves core areas of knowledge management, and presented the supporting framework of knowledge management processes.

References

1. Thomas A. Stewart, Intellectual Capital: The New Wealth of Organization, New York: Currency Doubleday (1997)
2. Kathy A. Stewart, Richard Baskerville, Veda C.Storey, James A. Senn, Arjan Raven, CherieLong, Confronting the Assumptions Underlying the Management of Knowledge: An Agenda for Understanding and Investigating Knowledge Management, The Database for Advances in Information systems, Vol. 31(4).(2000)123-132
3. Atreyi Kankanhalli, Fransiska Tanudidjaja, Juliana Sutanto, Bernard C.Y. Tan, The Role of IT in Successful Knowledge Management Initiatives, Communications of the ACM, Vol. 46(9).(2003)69-73
4. Hansen, Morten T. Nohria, Nitin Tierney, Thomas, What's Your Strategy for Managing Knowledge, Harvard Business Review, Vol. 77(2). (1999)106-116
5. Despres, Charles Chauvel, Dabiel, Knowledge management, Journal of Knowledge Management, Vol. 3(2). (1999)110-120
6. Duff L. A., Casey A., Using informatics to Help Implement Clinical Guidelines, Health Informatics Journal, No. 5. (1999)90-97
7. Heathfield H., Louw G., New Challenges for Clinical Informatics: Knowledge Management Tools, Health Informatics Journal, Vol. 5(4). (1999)67-73
8. Hanka R., O'Brien Claire, Heathfield H., Buchan I. E., WAX ActiveLibrary: A Tool to Manage Information Overload, Topics in Health Informatics Management, Vol. 20(2). (1999)69-82
9. O'Dell, C., Grayson, C. J., If We Only Knew What We Know: Identification and Transfer of Internal Best Practices, American Productivity and Quality Center White Paper (1997)

Assurance Method of High Availability in Information Security Infrastructure System

SiChoon Noh [1], JeomGu Kim [2], and Dong Chun Lee [3]

[1] Dept. of General Education, Namseoul Univ., Korea
nsc321@nsu.ac.kr
[2] Dept. of Computer Science, Namseoul Univ., Korea
[3] Dept. of Computer Science, Howon Univ., Korea
ldch@sunny.howon.ac.kr

Abstract. It is very important for the information protection system to maintain high availability at each moment as a variety of intrusions occur continuously. The high availability of information protection system shall be primarily studied in relation to the infrastructure. The high availability on the infrastructure is assured by letting the fail over mechanism operate upon the entire structure through the structural design and the implementation of functions. The proposed method reduces the system overload rating due to trouble packets and improves the status of connection by SNMP Polling Trap and the ICMP transport factor by ping packet.

1 Introduction

High Availability is a broad concept for the computer system function designed to allow the user to continuously use specific application or device in spite of H/W or S/W failure. With the explosive growth of the Internet and it's increasingly important role on our lives, the traffic on the Internet is increasing dramatically, which has been rowing at over 100% annual rate. The load on popular Internet sites is growing rapidly [3]. Thus the performance bottleneck problem of the security systems, and with the increasing access requests the servers will be easily overloaded for a short time. Nowadays, more and more companies are moving their businesses on the Internet, any interrupt/stop of services on the security system means business lose, and high availability of these servers becomes increase-singly important [4, 5].

However, various information security systems of today have focused on improving the efficiency while giving importance to detecting and cutting off harmful traffic by the information security system itself. As the result, the threat to availability is found primarily from the infrastructure in the actual fields operating information security systems [6]. This paper proposes the method to assure the safe operation of overall network and information system by implementing enhanced high availability infrastructure concerning the network infrastructure.

X. Jia, J. Wu, and Y. He (Eds.): MSN 2005, LNCS 3794, pp. 1110–1116, 2005.

2 High Availability Infrastructure Architecture

The basic structure of the infra is divided into the core, transmission and access layers according to the architecture design principle. Apply the high availability structure and function to this hierarchy structure based on the methodology suggested in the infrastructure configuration method. For external point of interface, locate the exterior router in the front end to constitute WAN edge layer.

WAN edge layer is connected to LAN edge and up to the access layer, LAN connection section through the core layer, LAN Backbone section. For the infra-structure, the improved topology with high availability structure in all layers was configured. The design principle of high availability infrastructure is the auto dual structure in all layers in terms of structure and the introduction of full mesh in terms of function.

Availability is assured in three aspects of service, client and session status availabilities. Service availability is the capacity that allows clients to continuously access services. Session availability is the capacity that maintains the session statuses while the failure occurs.

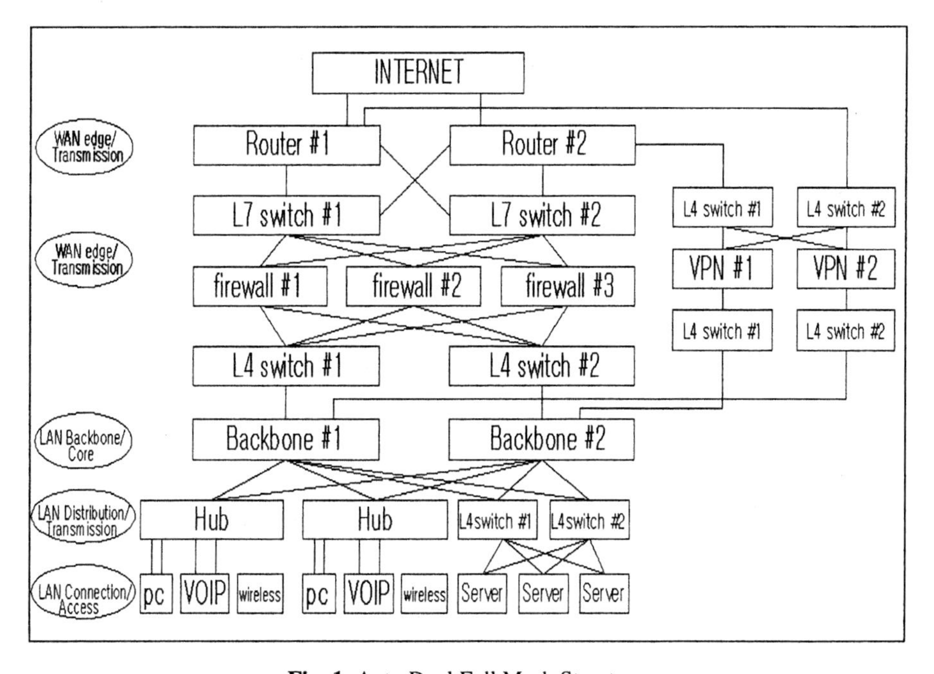

Fig. 1. Auto Dual Full Mesh Structure

3 Failover Algorithm

STEP1. Securing the Automated Recoverability of Core and Transmission Layers: Recoverability is the function that promptly recovers the system to the original state from failure. Maximize the recoverability of the network interworking in the equipment and link, protocol and application levels for the core, transmission and access layers. The recoverability of the equipment level is developed with the technology derived from duplicated processors and IOS software such as SSO (Stateful Switchover). The recoverability in the link level can be provided using the technology such as recoverability packet ring (RPR), multi-link point to point protocol (PPP), or the spanning tree enhancement boards like EtherChannel(r), IEEE802.1w or 802.1s. The recoverability in the protocol level can be realized by adapting gateway load balancing protocol (GLBP) or the protocol integration and optimization for routing and access protocol. The recoverability in the application level can be realized by applying the stateful NAT (Network Address Translation) or server load balancing.

STEP 2. Process Control: The control module based on the high availability daemon process control the duplication for the duplication of the system is placed in both systems to determine the system status through continuous monitoring of the system. It determines the active and standby statuses between two systems and synchronizes the data that need to be synchronized between two systems. This is applied to initial start up and restarts and assures the least packet loss in system switch over that occurs in delivery the traffic.

Continuous monitoring of the system activates the periodical Heart Bit through Ethernet based TCP/IP connection between high availability, and this monitors not only the process status of the other system but also the Ethernet connection line for high availability.

The most important element in the status change between systems is the prevention of data loss due to quick change over. The precise load sharing for all data on high availability connection lines between systems can greatly reduce the loss of data and the probability of congestion.

STEP 3. Hot Sparing and Heartbeat Connection: Core systems apply the hot sparing function through perfect redundancy. For core switch and router, design switching fabric and supervisor engine with the hot sparing with zero loss rates. The interface port module and multi service blade shall also have internal spares that can turn to service within few seconds. The core system shall also have sufficient processing capacity and scalability to be able to accommodate additional IP-based services such as videophone meeting and public radio access within the network.

Prepare the heartbeat connection to exchange the server status data and session data between primary server and backup server. If the heartbeat connection is not available, the back up system will not transfer its role to the primary system even after the primary system is recovered.

STEP 4. Fail Over: Both primary and secondary functions of routing, L7 switching, firewall packet filtering, L4 switching and backbone networking operate in active mode to handle the networking traffic and VPN traffic passing through each equipment. When one function fails, the other function becomes the master to take over 100% of the traffic. When fail over occurs in full mesh mode, adjust the traffic volume per section in order to protect equipment performance.

Retain a couple of exclusive high availability interfaces to achieve availability and assure the synchronization between two equipments. If the interface connection is lost due to some causes use other interfaces to fail over the synchronization information.

Send heartbeat messages by the configurable interval (minimum 200ms) to check whether the fail over has started. In this case, the fail over mechanism is activated through the events of heartbeat loss, interface link loss, and the loss of configured IP address or monitored IP address set.

4 Configuration Method

Structure of Backbone and Distribution Layer - Auto Dual Structure:
The first requirement in designing the high availability network is to duplicate the structure and function of the backbone and distribution layer. Considering the intranet system, if the core layer is LAN backbone section, the transmission layer is the LAN distribution layer and the access layer is the LAN connection layer.

Auto dual hierarchy structure means moving away from traditional semi dual structure.

The traditional semi dual structure partially applies the dual structure centering on the firewall without applying the availability structure for the core and distribution layers.

Partially dual structure has limit in assuring the system availability. The reason is that the threat to the network availability has increased due to the diversification of network interfaces, the massive expansion of traffic volume, and the diversification of data types all of which hinder the partially dual structure to achieve high availability.

Traffic Exit Point Distributed Layout: Traffic Exit Point Distribution is designed to improve the speed and stability of the network using a number of internet connection lines.

Traffic distribution joins a variety of high speed internet circuits such as the private lines, ADSL and cable modems from a number of ISPs and use as a single private line. It also manages the internet traffic efficiently and optimizes traffic handling efficiency by relieving traffic congestion to unnecessary lines.

The IP band of internal network can either use the IP band of the private lines or set PAT (Port Address Translation) configuration by setting private IP band. It also allows the internet network to share IPs using PAT especially when there are small numbers of IPs allocated from internet lines or ADSL/VDSL lines. Existing round-robin, hashing, least connection, maximum response time, maximum bandwidth methods are used as load distribution method.

Use of Multi-layer Switching Structure for Packet Handling: Supplement the limit of the security system by retaining the high availability through the load balancing with the multi layer switch equal to or above layer 7.

Handle packets and act as the virus wall and perform multi-layer switching. Multi-layer switch will act as the control for protocols such as TCP and UDP in upper layers and will have additional functions such as traffic control. And the addition of the application layer, the 7th layer, will enable direct packet filtering and QoS.
It uses exclusive ASIC and will have additional load balancing functions of security devices such as the firewall, VPN and IDS other than the basic functions like SLB(Server Load Balancing) or LLB(Line Load Balancing).

Information Security System Clustering Structure: Bind 2 or more security systems into one using clustering method to have the systems virtually operated as one system. In this case, develop a security network for the mesh structure so that the one side can cover up the capacity of the other when the other side fails in order to maintain the maximum availability. The goal is to make the system change over and all security processes for system recovery activate within few seconds (maintaining 99.999% of availability, dropping less than 10 pings) in case of unexpected downtime. This type of technology is obtained by clustering a number of independent security systems through the network and interconnecting with security software with high availability.

5 Performance Analysis

The system used for measurement is the system of corporation A which has 30 or more buildings, desktops, data centers, and research centers. The system is composed of over 800 switches, over 200 routers, over 250 console servers, over 800 access points and contents switching devices.

General structure has semi dual structure and performs partial traffic load distribution with its infrastructure. High availability structure has adapted the hierarchy of the network structure and the traffic load distribution structure suggested in this study and applied Layer7 multi-layer switching and clustering, and fail over mechanism.

The record of availability is the matrix of the spreadsheet including all types of device lists, and it is collected through the analysis software that provides the information about the number of devices, total device failure time, total failure fighting time, device availability activation time, and the down time.

The most important data are MTBT (Mean Time between Failures) and MTTR (Mean Time to Recovery) and these are collected and analyzed from all devices and paths in the network.

(1) System Overload Rating Derived by Trouble Ticket

Switch is the section where the overload trouble occurs primarily by the packets passing the traffic routing process through the external router. The following table shows the overload trouble traffic amount occurring at the application switch before and after improving the structure. In the proposed structure, the possibility of application switch bottleneck due to the congestion of balancing amount appears to be reduced by 28%. Also, the occurrence of such will be limited to the main section and will not affect other sections.

Table 1. Switching Overload Rating

Unit: mill. Packets/day

Division	General Structure	Proposed Structure	
Function	Main	Primary	Backup
Switching Type	Layer4	Layer7	Layer7
Throughput	36	46	9.2
Share Rate	100%	72%	28%

Table 2. Packet Filtering Connection and Equipment Status

Unit: mill. Packets/day

Division	General Structure	Proposed Structure				
Firewall	Main	A	B	C	D	Total
Number of Firewall	6	5	1	1	1	8
Share Rate	100%	72%	12%	12%	5%	100%
Packets Handled	36	46	7.6	7.6	3.2	

Table 3. ICMP Transport Factor

Division	General Structure			Proposed Structure		
	Transmission	Reception	Loss	Transmission	Reception	Loss
Daum	1,768	1,296	26.7%	1,296	1,214	6.4%
Yahoo!	1,721	1,519	11.8%	1,253	1,195	4.7%
Naver	1,974	1,900	3.75%	1,476	1,461	1.02%
Altavista	1,699	1,598	5.94%	1,572	1,529	2.74%
Average	1,791	1,578	12.0%	1,399	1,350	3.68%

(2) **Connection through SNMP Polling Trap and Equipment Status**
This is the measurement related to how the traffic load-balanced by the application switch is dispersed in each system. The firewall which becomes the bottleneck factor for the gateway section before structural improvement is the main firewall, and the load derating of the main firewall is the point of measurement, and the traffic whose load is reduced through the main firewall shows varied results in the firewall for each function.

100% of the total traffic load handled by the main firewall in the existing structure reduced down to 72% by 28% in the high availability structure. Therefore the possibility of bottleneck on the firewall is reduced by 28%.

(3) **ICMP Transport Factor by Ping Packet**
If the traffic exceeds CPU's capacity limit while the amount of transmission packets is gradually increasing, the network delay will increase and the packets will be lost. The enhancement of capacity after the structural conversion reduced system overload and also the information loss. The following table shows the result of ping test using specific site in the transmission section which passes through the firewall in the internal network at the lower end of the firewall.

6 Conclusions

In spite of the massive scale and the complexity of the system used for measurement, the network with 100% UPS and emergency power generator performed 99.999% of availability. This performance proved that the system can operate without suspension with high-availability infrastructure and precise fail over algorithm even when the failure occurred. In handling the fail over, the back up unit includes the essential network configuration, session status and security connection beforehand in order to almost immediately handle the existing traffic continuously within the fail over time. Using the built-in fail over protocol and dynamic routing, it's possible to lay the security system in the connected network environment or load sharing environment.

High availability and technology enhances the function and flexibility of the network through which the information is accessed and delivered and also can prevent financial loss and productivity loss through the continuous service of the security system. The high availability of the security system can be realized with the topology and configuration properly defined to fully utilize the recovery function of the security system in the thoroughly planned optimized method.

References

1. Willam Stallings, "Network and Internetwork Security". Prentice Hall, 1995.
2. D. Dias, W. Kish, R. Mukherjee and R. Tewari, "A Scalable and Highly Available Server", COMPCON 1996, pp. 85-92, 1996.
3. Highly Available and Scalable Cluster-based Web Servers, Xuehong gan, Trevor Schroeder, Steve Goddard and Byrav Ramamurthy, in submission to The Eighth IEEE international Conference on Computer Communications and Networks , 1999
4. G. Hunt, G. Goldszmid, R. King, and R. Mukherjee: Network Dispatcher: A Connection Router for Scalable Internet Service, Computer Networks and ISDN Systems, Vol.30, pp.347 357, 1998.
5. D. Dias, W. Kish, R. Mukherjee and R. Tewari : A scaleable and highly available Web server. IEEE international Conference on Data Engineering. Now Orleans, February 1996.
6. G. Trent and M. Sake, WebStone: the First Generation in HTTP Benchmarking, MTS Silicon Graphics, February 1995.
7. "IA-LVS: Design of the Improving Availability for Linus Virtual Server: 2nd International Conference on Software Engineering, Artificial Intelligence, Networking & Parallel/Distributed Computing August 20-22, 2001 Nagoya Institute of Technology, Japan
8. Sookheon Lee, Geunyoung Chun, Myongsoon Park, Load balancing mechanism for dispatchers in Web server Cluster, International Conference on ITCC 2002, Orleans, Las Vegas, U.S.A, 2002
9. Rainer Link, "Server-based Virus-protection on Unix/Linux", University of Applied Science Frut wangen, 2003.

Fuzzy-Based Prefetching Scheme for Effective Information Support in Mobile Networks

Jin Ah Yoo[1] and Dong Chun Lee[2]

[1] Dept. of Information Statistics, Jeonju Univ., South Korea
gina@jj.ac.kr
[2] Dept. of Computer Science, Howon Univ., South Korea
ldch@sunny.howon.ac.kr

Abstract. This paper proposes a fuzzy-based Prefetching method which obtains information to be referred in the near future with using the properties of mobile movement. We consider a velocity-based fuzzy model to find a pattern of movement for location-awareness service which provides the effective information about location change of mobile user or mobile computing on the context of Mobile Information Services. This procedure provides basic solutions which are optimal for a present situation with restricting the number of considered subject by defining prefetching area based on the movement length and direction.

1 Introduction

We need to minimize the delayed time for collecting new information that meets the needs of the mobile user when providing mobile information service for location-awareness. There are, however, many problems in applying wire communication technologies to wireless communication due to its own characteristics. The information should be updated continually according to a user's position responding to requests for contents issued by a user. The low bandwidth, high delay, traffic, and frequent disconnection are inevitable barrier in mobile environment. Increasing the bandwidth might be the solution if the cost is not crucial. But to save expense, the optimal solution should be to utilize the existing bandwidth efficiently.

When we use a prefetching method to overcome these problems, which collects, in advance, information that a user may request in the future on the user's mobile device, the better quality of wireless communication and fitness of information can be provided. So, we can confront the properties of wireless communication by storing information and reusing them. But data prefetching has two main defects, lots of memory and processing time. So, we need to set a suitable prefetching area so that we use the prefetching data effectively.

This paper proposes a fuzzy-based prefetching method which adopts the pattern of mobile movement for location-awareness service.

X. Jia, J. Wu, and Y. He (Eds.): MSN 2005, LNCS 3794, pp. 1117 – 1124, 2005.

2 Related Work

While a mobile user moves from a location to the other location connected with fixed network, the delay time on mobile service contains the time to aware location-change and to collect information. We should reduce the time for collecting data so as to take enough time for effective information service. Mobile users want to get data which fit in his/her mobile circumstance without any delay. so we need enhanced methods applying to the mobile states as like location, velocity and direction so that we can get the most suitable information independent of time and place in a mobile environment[18, 25].

Some prefetching methods considering those conditions are suggested, the velocity-based prefetching methods are most popular.

The velocity-prefetching [1] is a technique to reduce the load of local area server due to pre-stored data, which uses square or rectangle shaped area distinction instead of the ring shaped so as to confine an extent of downloading from information might be needed in the future. Namely the idea of movement velocity prefetching is to confine the distance and width of prefetching area with its velocity and direction.

But the delay is inevitable because it is difficult to decide whether we should prefetch general information or specific data. Since the prefeching method depends only on the past behavior, it has a limits of its own and lack of flexibility. To solve this problem, we propose a fuzzy-based prefetching method in the next section.

3 Fuzzy-Based Prefetching Method

A mobile user is assumed to have a prefetching method to get suitable information matching the environment. The location of a mobile user refers the predefined geographical regions using the exclusive device location of mobile user or terminal. For simplification of expansion, a mobile user prefetches data at the same pattern of movement in a given unit time. The given (present) location, the relative distance and the direction would be the basic elements to derive the location of a mobile model.

We prefetches the information partition associated with Prefeching Zone (PZ), the set of location partition, on the area within distance d from the present location. The prefetching scheme selects PZ and prefetch data matching with the zone, which includes all location partitions within distance d from the present point. The zone can be set with the partitions which located in radius d at the centered by present location. If the mobile user is moving at location partitions in PZ, when prefetching is done, we don't have to reload information of the partition from the remote server. When, however, the user moves out the PZ, the new location partition becomes the starting point for selecting a new PZ. As soon as information partition associated with the location partition being loaded, the effective information service is made for the user, and the service structure is used to load information partition with background work in the rest partition of new zone at the same time.

The mobile motion of user applies to the calculation of velocity-based model [1], the PZ is derived from the fuzzy-based model. If we select the location partition as a ring shaped, it would result the prefetching of unnecessary information partitions, so we propose the scheme to set PZ as rectangular shaped structure. According to the pattern of movement of mobile user, we set up PZ as wider when faster and narrower when slower. Given the present location, the future location of user is based on the event and a new location is estimated while the direction and the velocity of movement are changed.

Velocity-based prefetching [1] is based on time, so we have to respond the change of velocity or direction in the case of occurrence of new mobile environment or accident.

Let the velocity-based prefetching zone $V_{X,Y}$, and define k as the number of cell that a mobile user moves and t as the movement time. The equation for D_t is as follows

$$D_t = \left| \frac{k}{t} \times V_{X,Y} \right| \tag{1}$$

When D_t is 0, it implies no movement, and "D_t is 1" means keeping up the speed, otherwise, the mobile user moves slower or faster than before, the pattern changes. We can draw the membership function as shown in Fig. 1 using the fuzzy triangular function.

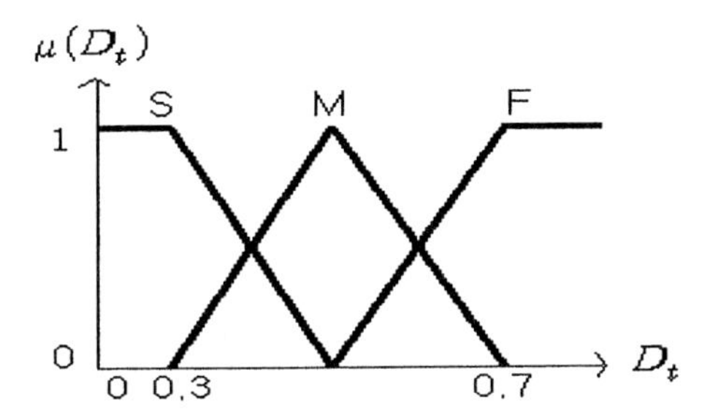

Fig. 1. Fuzzy membership function

We define what slow S, maintenance M, and fast F is for each fuzzy set.

$$\mu(D_t) = \begin{cases} \mu_p(D_t) > \mu_q(D_t), -1 < \mu(D_t) < 0 \\ \mu_p(D_t) = \mu_q(D_t), \mu(D_t) = 1 \\ \mu_p(D_t) < \mu_q(D_t), 0 < \mu(D_t) < 1 \end{cases}$$

where $\mu_p(D_t)$ is the past location of the user and $\mu_q(D_t)$ is the present location. "$\mu_q(D_t)$=0" means stillness, the zone is getting narrower, so we have to derive the PZ again. "$\mu_p(D_t) > \mu_q(D_t)$" implies getting slower and on the contrary "$\mu_p(D_t) < \mu_q(D_t)$" does getting faster to get wider the PZ. "$\mu_p(D_t) = \mu_q(D_t)$" has no change in the velocity, neither does the PZ.

For example, a mobile user reduces the speed on the point (4,5) ($\mu_p(D_t) > \mu_q(D_t)$). Since the velocity-based prefetching zone comes from the average of V_x and V_y, the PZ would be the Fig. 2 (a) with V=<3,5>, d=6, and w=4[1]. But applying the change of the velocity, the fuzzy based PZ is as shown in Fig. 2 (b).

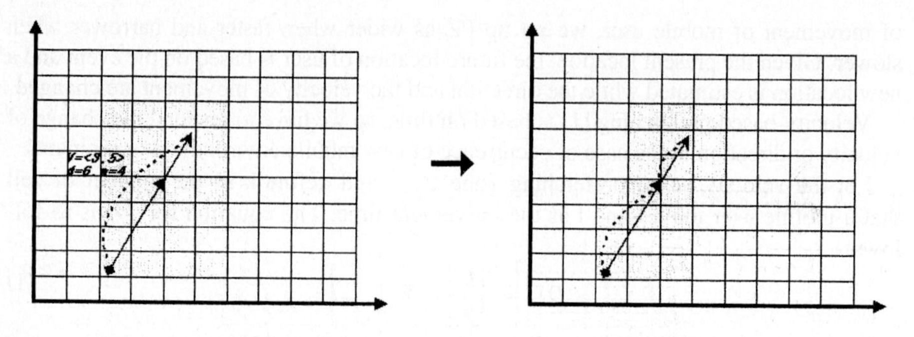

Fig. 2. Velocity-based PZ **Fig. 3.** Fuzzy-based PZ

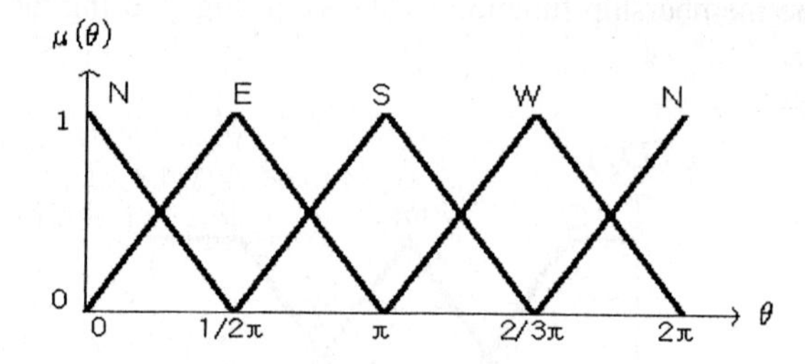

Fig. 4. Fuzzy membership function for movement direction

Also we use the angles so as to derive the direction of the PZ as following

$$\theta_t = \left| \theta_{t-1} + \Theta_t \right| \tag{2}$$

where θ_t is the newly estimated movement direction, θ_{t-1} is the prior estimated movement direction, and Θ is the current observed movement.

From Eq. (2), the estimated movement direction θ_t is calculated that turns to clockwise from the north 0°. We use membership functions to select a candidate prefetching zone for estimating movement direction. The estimated movement direction is mapped to a basic fuzzy set by the membership function that has larger membership value as shown in Fig. 4.

For the example of Fig. 5, if we assume that the user was occurred at cell A, a movement direction is C, and an estimated movement direction is π/2. For the estimated direction, the membership value of the membership function is $\mu_1(\pi/2)=1.0$.

A movement direction is B, and an estimated movement direction is π/6. For the estimated direction, the membership value of the membership function is $\mu_1(\pi/6)=0.5$.

The PZ based on fuzzy prefetching using the distance and the direction is as follows,

$$PZ = \arg\min(P(D, \theta))$$

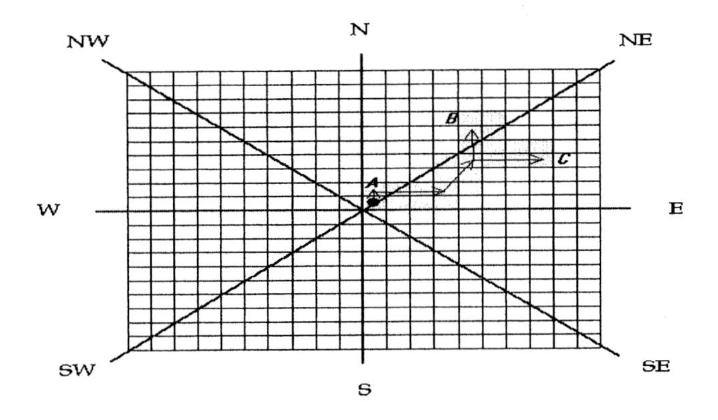

Fig. 5. Example of zone

This structure considering the velocity and direction may result to include the surrounding area located in the future as well as the location on the movement path. We can design a similar fuzzy simulation process for computing. When input variable are the followings,

θ_t = {N, NE, E, SE, S, SW, W, NW}, the fuzzy sets of movement direction

D_t = {slow, maintenance, fast} the fuzzy sets for movement velocity

the process would be,

Step 1. Set PZ = 0.
Step 2. Randomly generate $V_{X1}; V_{X2}; ¢ ¢ ¢ ; V_{Xn}$ and $V_{Y1}; V_{Y2}; ¢ ¢ ¢ ; V_{Yn}$, and set the rectangular zone from the velocity V_{XY}.
Step 3. Check the velocity and the direction.
If D_p 6= D_q then we calculate $'(D_t)$
If $\mu_{t;1}$ 6= $£_t$ then we calculate $'(\mu_t)$
Else Set PZ and go to step2
Step 4. $D_1 \_ D_2 \_ ¢ ¢ ¢ \_ D_T, \mu_1 \_ \mu_2 \_ ¢ ¢ ¢ \_ \mu_T$ and $'(D; \mu) = '(D_1) \wedge '(D_2) \wedge ¢ ¢ ¢ \wedge '(D_T) \wedge '(\mu_1) \wedge '(\mu_2) \wedge ¢ ¢ ¢ \wedge '(\mu_T)$.
Step 5. Set PZ and go to step2.

When a mobile user demands information, the acquiring process is formed as follows. The request information links to a base station from the mobile device, the base station connects to the server, the server picks the corresponding data stored in Database, and they links to the server and the mobile terminal of the user eventually.

Prefetching is performed with the application of VPT (Virtual Prefetching Table) and APT(Actual Prefetching Table) on the server before the request data transmit to the user. We add up our proposed prefetching on APT so that we can collect, in advance, information referred in the future.

4 Performance Evaluation

The comparison study is conducted in the view point of the usage rate of memories that respond the user's request. The simulation program is implemented in C combining a powerful simulation engine, CSIM18 [3]. To make a fuzzy-based model for the motion of a mobile user, we assumed the followings. Even though a mobile user goes around on three dimensional spaces in the real life, for simplification, a mobile user is assumed to move on a two-dimensional area at fixed velocity and direction for fixed period under the movement model.

In the simulation the user starts at the point (5,6) and move to next point in a sequence. We repeat the process 20 times according to the given point value so that we obtain the virtual prefetching zone based on velocity.

We gave the restrictions that the movement path must contain specific points in the order, from D0 (5, 6), to D1(22, 7), D2 (12, 17), D3(22, 18), D4 (5, 23), and D5 (5, 6) so as to control the missing rate of the users. The number of cells lying on the path is 106 and the virtual zone is set under the assumption of a fixed velocity. As shown Fig. 6, the portion of prefetched data of fuzzy-based method compared with the velocity-based scheme. The proposed scheme prefetched 109 portion as a total and 5.45 as the average, it is the improvement of 32% throughout prefetching data having high probability of user's request. The velocity-based prefetching shows the usage rate of 57.6% on the average, the proposed scheme shows that of 89.7%, which made improvements of 16%.

We find the proposed scheme is very efficient in the number of prefetched data and the rate of usage of information, and need to improve the failure rate of information search. So, in this simulation study, we considered the fuzzy-based and found out the improvement of 15% in aspect of the missing rate of information retrieval compared to the velocity-based prefetching.

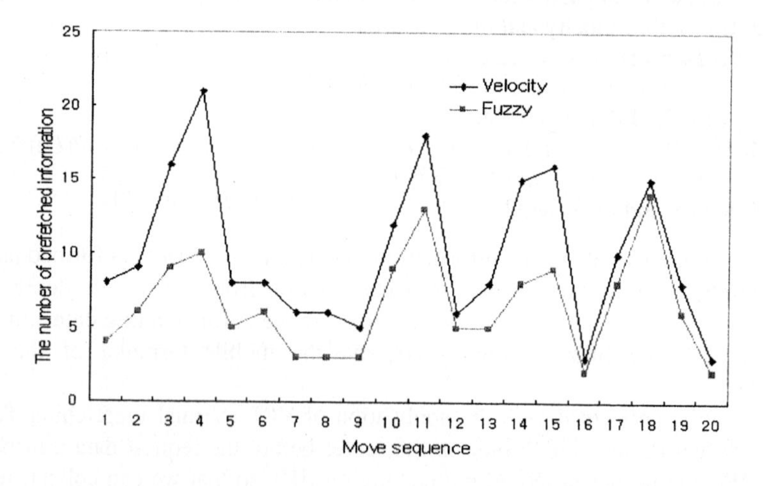

Fig. 6. The number of prefetched portion information

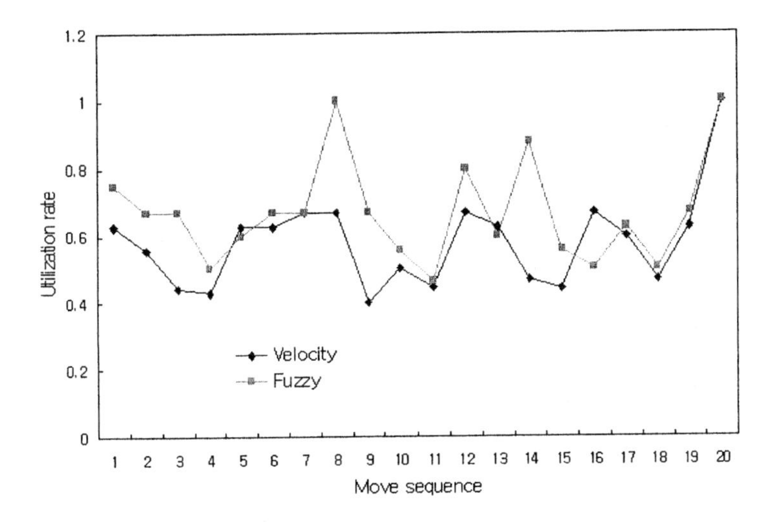

Fig. 7. Utilization rate of the prefetched portion information

5 Conclusions

We proposed the fuzzy-based prefetching method considering the movement pattern of mobile user. The proposed method made an improvement comparing with the velocity-based prefetching and the frequency-based one, which raise the usage rate of information with the effectiveness of pre-stored data.

A mobile user requests information about the location to go soon on the mobile environments. The existing prefetching methods are confined to prefetch uniformly, but flexible methods are inevitable to satisfy the needs of mobile users under the various kind of situations. The prefetching zone should be set up flexibly according as velocity, stopping or moving.

Acknowledgements

This work was supported by grand from ITRC Support Program in 2005.

References

1. Cho, G. Using Predictive Prefetching to Improve Location Awareness of Mobile Information Service, Lecture Notes in Computer Science Vol. 2331/2002 pp. 1128-1136, 2002.
2. CSIMI18 Simulation Engine, Mesquite Software Inc.,1997.
3. G. Liu. Exploitation of location-dependent caching and prefetching techniques for supporting mobile computing and communications. Proc. WIRELESS-94, 1994.
4. Gitzenis, S.& N. Bambos. Power-Controlled Data Prefetching/Caching in Wireless Packet Networks.Proc. IEEE INFOCOM 2002, 2002.

5. G.Y. Liu, G.Q. Maguire. A predictive mobility management scheme for supporting wireless mobile computing. Proc.ICUP-95, 1995.
6. Imai, N., H. Morikawa & T. Aoyama. Prefetching Architecture for Hot-Spotted Network. Proc. ICC2001, 2001.
7. R. Hugo Patterson , Garth A. Gibson , M. Satyanarayanan, A status report on research in transparent informed prefetching, ACM SIGOPS Operating Systems Review, v.27 n.2, p.21-34, April. 1993.
8. S.M. Park, D.Y. Kim, G.H. Cho, Improving prediction level of prefetching for location-aware mobile information service, Future Generation Computer Systems 20, 2004, pp. 197-203
9. Q.Ren and M.H.Dunham. "Using semantic caching to manage location dependent data in mobile computing". In Proc. ICMCN 2000, 2000.
10. V.N. Padmanabhan, J. C. Mogul, Using Predictive prefetching to improve World WideWeb Latency. ACM SIGCOM Computer Communication Review, Vol. 26(3), 1998.
11. Z. Jiang and L. Kleinrock. An Adaptive Network Prefetch Scheme. IEEE Journal on Selected Areas in Communications, Vol. 16(3):1-11, 1998.

Duplex Method for Mobile Communication Systems

Gi Sung Lee

Dept. of Computer Science Howon Univ., Korea
ygslee@sunny.howon.ac.kr

Abstract. High reliability and real-time response are required in the Main Control Processor of Radio Network Controller for mobile communication systems. In spite of its robustness some fault rates is inevitable, so the processors are duplicated for non-interrupted service and services are switched to the standby processor in case of faults. The hot-standby sharing scheme has advantages of no data loss and non-proliferation of error data in comparison with the warm-standby sharing scheme but it has difficulties with implementation due to synchronization problem. This paper proposes an expanded warm-standby sharing scheme based on a novel massage processing and error detection mechanism for improving performances.

1 Introduction

The importance of communication system stability has been increasingly recognized as various types of high performance systems have been developed and industry gets highly dependent on these systems. Particularly, In RNC of mobile communication systems, MCP (Main Control Processors) is responsible for call control function, so that the level of fault tolerance in MCP is considered as the key factor measuring the stability of whole communication systems. The MCP must be implemented with high tolerance, but it is inevitable for them to have small fault. So, in order to reduce fault rate to acceptable level and maintain the system stability, two or more processors are redundant, and as an active processor suffers from faults, the other standby processor must take over its position and provide seamless services.

The present systems supporting fault-tolerant capability adopt basically warm-standby scheme or hot-standby scheme [1-3]. The implementation of the warm-standby scheme is easier but it has drawbacks that some data losses occur depending on the types and the scopes of faults. It also propagates to the entire system false data that have not been detected by currently active unit. Most of commercial systems used currently such as AT&T ESS-5 and ETRI TDX-10 adopt the warm standby sparing scheme. The systems that adopt hot-standby sparing scheme have advantage of no data loss. But they are not easily able to obtain theoretical performance when implementing the real-time switching system because of difficulties in synchronization between fault-tolerant modules and restoration of recovered system module [4-6].

This paper aims to obtain simultaneously both the readiness and lower cost of the implementation and the performance enhancements by avoiding loss of data and the proliferation of false data effectively.

X. Jia, J. Wu, and Y. He (Eds.): MSN 2005, LNCS 3794, pp. 1125 – 1132, 2005.

2 Expanded Warm-Standby Sharing

Fault is proliferated to the entire system and places serious impacts to call processing and other processing functions if the Standby side just turned over to active state performs the recovery procedure based on faulty data generated after the fault has occurred during call processing in the Active side. We can say that data loss and the proliferation of fault data occur due to following two situations. We attempt to solve these problems in both hardware and software aspects. An error detection scheme is proposed in hardware aspect and a novel message processing mechanism proposed in software aspect.

2.1 Task Based Duplex Architecture

First of all the task based duplex architecture of our proposed system is described. In our task based duplex architecture, the task is activated per each message, changed data in memory are transmitted to the Standby side and messages are logged for recovery when switching is done. This architecture does not require status information for registers and processor. The synchronization between the processors is performed per message basis, simple and easy to implement. Also the dual-down problem can be avoided as no information about mal-function is transferred to Standby side. Figure 1 shows the duplex architecture embedding our proposed warm-standby sharing scheme. Each task loaded into Active side and Standby side generates TCB (Task Control Block) and registers its storage area in initialization. This step is performed in Active side and Standby side in the same procedure. After this step both Active side and Standby side receive Call message simultaneously. The IPC (Inter-process Communication) stage in Active side transfer this message to the corresponding task for processing. If anything has been changed as a result of processing the message, backup data (which is content changed and information on that message) is transmitted to the duplex module. Then the duplex module maintains this backup data in TCB and transmits it to the duplex module of the Standby side when the task calls checkpoint.

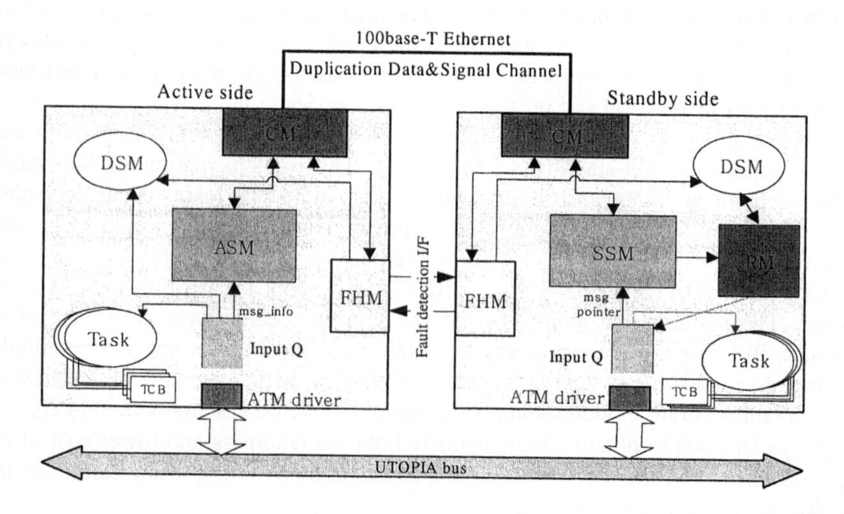

Fig. 1. Duplex Architecture

The DSM (Duplex Support Module) has information on the overall state of the duplex block, controls other module and communicates with upper level processor by using this information. The synchronization module for synchronizing the state of Active side and Standby side is composed of ASM (Active Synchronization Module) and SSM (Standby Synchronization Module) The FHM (Fault Handling Module) detects and handles any faults The RM (Recovery Module) is performed in Standby side after the fault is detected in Active side, and is responsible for following up through the point where the former Active side is executed.

2.2 Hardware-Based Control Mechanism

Figure 2 shows the diagram for connecting our duplex boards that can cope with data loss and the proliferation of faulty data. This mechanism dramatically reduces the possibility of not detecting faults by two boards checking not only themselves but also each other. In other words they periodically check their own status and monitor any fault of each other in the hardware level by transmitting fault signals concerned with duplex in TTL through the backplane. In this way they find out whether they are Active or Standby respectively each other and then they are switched by software.

The duplex switching determines active state by analyzing both the internal status information detected from themselves and the status information of the decoder logic reported from each other. For this purpose we use the memory mapped registers, MFSR (Mate Fault State Register) and MSSR (My Fault Status Register). The status of each own board is written in MFSR and the status of each other written in MSSR by hardware. Figure 3 shows the fault detection in hardware level and the switching in software level. If any fault occurs at the mate board performing Active side, the fault detection logic of the mate board generates Mate Fault Detection Interrupt and writes the reason of fault in MFSR. If Mate Fault Detection Interrupt occurs, the ISR (Interrupt Service Routine) is called for reading MFSR, figuring out the reason of the fault and executing the appropriate handling process. In such a situation as system initialization process where the interrupt is disabled, it cannot occur even if there are faults in the mate board. In this case the control module finds out faults of the mate board by accessing MFSR with no interrupt event.

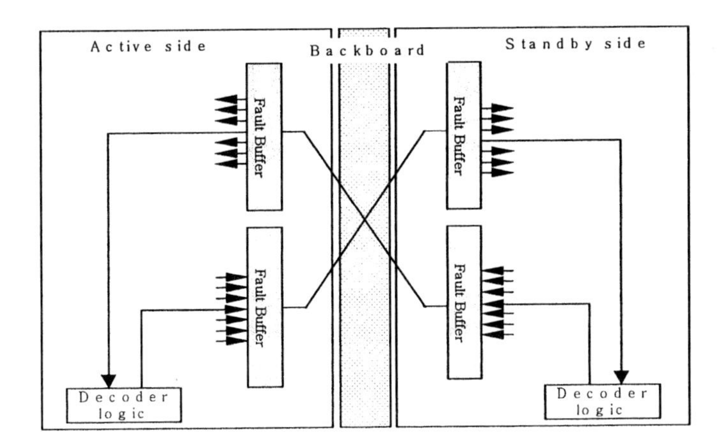

Fig. 2. The diagram for connecting duplex boards

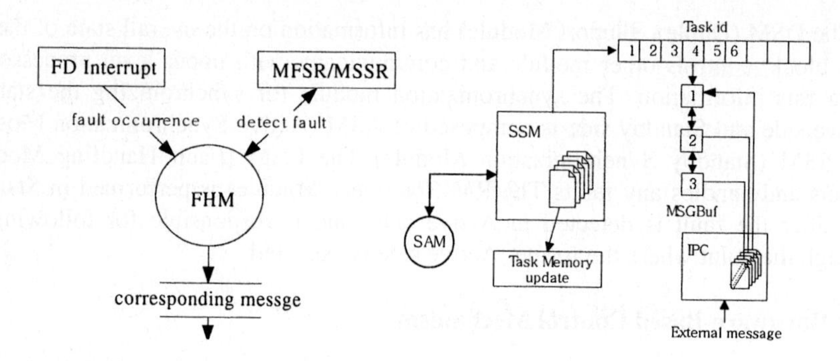

Fig. 3. Fault Detection Mechanism **Fig. 4.** Message Processing Mechanism

Each board has an ACTIVE register for use of ACTIVE indicator and it is set in hardware level immediately when any fault occurs in each board. The ACTIVE/STANDBY status bit is set, the corresponding handling routine is executed and the Active side is turned into sleep state. In this way we can prevent any faulty data from being transferred in the beginning.

2.3 Software-Based Control Mechanism

The proliferation of faulty data is avoided effectively by using the TCB of each task as the buffered area. Storing the backup data that is generated as the result of processing the message is delayed until the final data arrives, and it is stored in the corresponding task memory when the final data arrive at the Standby side. For this purpose the backup data is managed by each task with the start address and the size of the area to be backed up in TCB. Information about the recent external input message is stored for collecting garbage in the Standby side. For transmitting a series of backup data registered in TCB to the duplex path, the ASM divides them into 1024 byte sized packets, attaches the sequence number for reassembling and transfers them onto the COM. The SSM manages those packets received from the Active side by using TCB of the duplex module in the same way as in the ASM for reassembling them later on. That is, packets from the Active side are held in TCB until they are reassembled in the SSM. When the last packet of backup data arrives, the synchronization is done by storing this backup data in the same memory where the start address is stored. If the switching occurs before the last packet is received, the proliferation of faulty data is avoided by deleting from TCB incomplete data relating to this message. This mechanism for deleting and reassembling is shown in Fig. 4.

In the step of data backup, all the messages concerning the backup process should be deleted from MSGbuf when the Standby side receives backup message described in the header of the last packet. For this to be done, TASK ID relating to data backed up is sought and then the corresponding MSGbuf is located. From all the messages connected to this MSGbuf, the message involved with the backup is sought and the messages succeeding to this message are freed. If the switching is done without receiving backup message, data is recovered by re-processing the corresponding message.

To avoid data loss during the data backup, only after all the data generated as the result of processing the message have stored in the memory of the corresponding task in the Standby side, the matching message in the Standby side can be deleted. Any lost data is recovered by reprocessing this message if the switching occurs without backup done. While switching, the messages in MSGbuf are deleted after they are transferred to the appropriate task. The Standby side is initiated as the new Active side only when no message is in MSGbuf.

3 Markov Model for the Duplicated Processors

The probability of the fault occurrence in a board, λ, corresponds to the hazard function $Z(t)=\lambda$ with the exponential distribution. D_1 denotes Mate Fault Detection Probability, D_0 My Fault Detection Probability, Δt the time increment factor, and β the probability of the fault recovery. The probability of the fault recovery means the probability that the system can be recovered to the normal state without loss after the fault has been detected. The Figure 5 shows Markov model of our proposed system.

The probability that the system is in the state S at time $t+\Delta t$ is defined by the probability that the system will transit from the arbitrary state to state S and the probability that the system will remain at the current state.

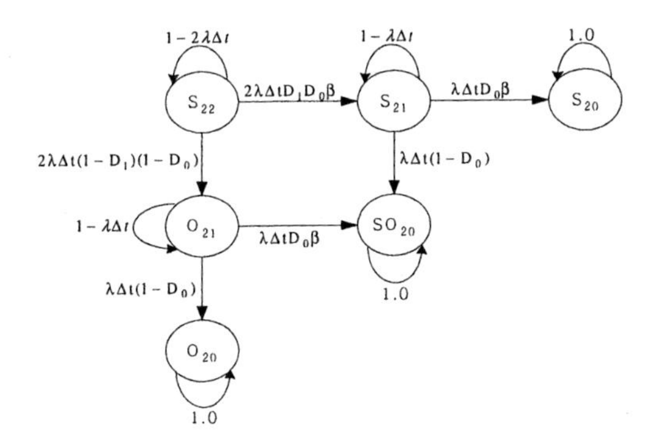

Fig. 5. Markov model of the expanded Warm-Standby Scheme

At state $P_{S_{22}}(t+\Delta t)$ where the two boards are normal in hardware level, the call processing and the backup are under way. From the Markov level the following holds.

$$P_{S_{22}}(t+\Delta t) = (1-2\lambda\Delta t)P_{S_{22}}(t) \tag{1}$$

The state $P_{S_{21}}(t+\Delta t)$ indicates that the fault is detected at the Standby side after the Standby side has been successfully switched to the Active due to the fault of the Active side.

$$P_{S_{21}}(t+\Delta t) = 2\lambda\Delta t D_1 D_0 \beta P_{S_{22}}(t) + (1-\lambda\Delta t)P_{S_{21}}(t) \tag{2}$$

The state $P_{S_{20}}(t+\Delta t)$ indicates that the faults have occurred and been detected in both the Active side and the Standby side.

$$P_{S_{20}}(t+\Delta t) = \lambda \Delta t D_0 \beta P_{s_{21}}(t) + P_{s_{20}}(t) \quad (3)$$

The state $P_{O_{21}}(t+\Delta t)$ indicates that the faults has occurred in the Active side or the Standby side but that it has not been detected.

$$P_{O_{21}}(t+\Delta t) = 2\lambda \Delta t(1-D_1)(1-D_0)P_{s_{22}}(t) + (1-\lambda \Delta t)P_{O_{21}}(t) \quad (4)$$

The state $P_{O_{20}}(t+\Delta t)$ indicates that the faults has occurred and not been detected in both the Active side and the Standby side, and that the Standby side cannot be used.

$$P_{O_{20}}(t+\Delta t) = \lambda \Delta t(1-D_0)P_{O_{21}}(t) + P_{O_{20}}(t) \quad (5)$$

The state $P_{SO_{20}}(t+\Delta t)$ indicates that the faults has occurred in both side and only one of them been detected.

$$P_{SO_{20}}(t+\Delta t) = \lambda \Delta t(1-D_0)P_{s_{21}}(t) + \lambda \Delta t D_0 \beta P_{O_{21}}(t) + P_{SO_{20}}(t) \quad (6)$$

The reliability means the degree that the system is able to operate safely and it can be represented by the probability which is given in (7).

The safety means the degree that the board keeps stable by handling the fault effectively when the fault has occurred.

$$R_W(t) = P_{S_{22}}(t) + P_{S_{21}}(t) + P_{O_{21}}(t) \quad (7)$$

$$S_W(t) = P_{S_{22}}(t) + P_{S_{21}}(t) + P_{S_{20}}(t) + P_{O_{21}}(t) + P_{O_{20}}(t) \quad (8)$$

4 Performance Analysis

In this section we analyze how the ability of detecting the hardware faults and coping with them by such a standby scheme affects the system reliability and the safety. In our system the fault detection and recovery relies on the sum of both the probability of the fault detection in each mate board and the probability of the fault detection in each own board. The result of performance analysis by the numerical analysis is described in this section.

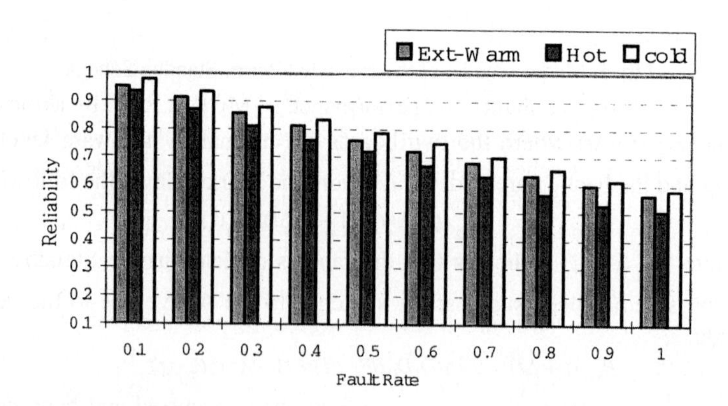

Fig. 6. Comparison of reliability for proposed scheme

In Figure 6 the reliability of our expanded warm-standby scheme is compared with those of the existing schemes. We can find that the reliability of our expanded warm-standby scheme is better than that of the hot-standby scheme. The cold-standby system has the best reliability, where the sub module is waiting or performing the diagnostics while the main module is performing the normal operation.

Figure 7 compares the safety of our scheme with that of the existing scheme. Our scheme has the best safety among three schemes. The cold-standby scheme has the worst because its fault detection and system reconfiguration rely only on the self diagnostics. In the hot-standby scheme where each processor detects errors by executing the self-diagnostic routine for finding out the faulty processor, its safety is affected severely if the fault occurs in one module and it is not detected by self-diagnostic. On the other hand our scheme is best because both the real time fault detection of the child processor and the mate processor by hardware level and the periodical fault detection by software level is performed simultaneously.

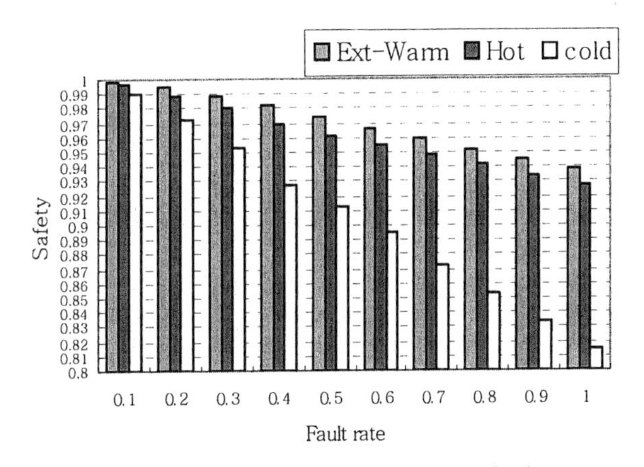

Fig. 7. Comparison of safety for proposed scheme

5 Conclusions

In this paper we tried to enhance the performance of MCP in RNC by avoiding effectively data loss and the proliferation of faulty data which is the major problem of the existing warm-standby scheme. We proposed the task based duplex architecture in which the task is activated per each message, changed data in memory are transmitted to the Standby side and messages are logged for recovery when switching is done. This architecture does not require status information for registers and processor. The synchronization between the processors is performed per message basis, simple and easy to implement. In hardware aspect the probability of detecting the fault is increased by checking the state of each own board and each mate board in real time. And the transfer of the faulty data is avoided in the beginning by setting the status bit in hardware level and turning the Active side into sleep state. In software aspect the message processing mechanism is proposed for avoiding data loss and the proliferation of faulty data. The performance analysis shows that our expanded warm-standby scheme has better reliability and safety. Later on the software fault detection mechanism has to be studied in order to cope with the failure of the hardware fault detection mechanism.

Acknowledgements

This work was supported by the Howon University Fund, 2005.

References

1. D. P. Siewiorek, "Architecture of fault - tolerant computers: an historical perspective," Proc. IEEE, Vol. 79, No. 12, 1710 ~ 1734, Dec. 1991.
2. B. W. Johnson, Design and Analysis of Fault Tolerant Digital Systems, Addis on - Wesley Pub. Co., 1989.
3. J. C, Laprie, "Definition and Analysis of Hardware and Software Fault-Tolerant Architectures," Computer, pp.39 ~ 51, July 1990.
4. A.M. Tyrrell, "CSP Method for Identifying Atomic Actions in the Design of Fault Tolerant Concurrent Systems," IEEE Trans. on SE, Vol. 21, No. 7, pp. 59-68, 1995.
5. L. Young, "Software Fault Tolerance at the Operating System Level," Proc. of COMPSAC '95, pp.393, 1995.
6. M. Russinovich, and Z. Segall, "Fault Tolerance for Off-The-Shelf Applications and Hardware," Proc. of FTCS-25, pp. 67-71, June 1995.

Author Index

Lecture Notes in Computer Science

For information about Vols. 1–3719

please contact your bookseller or Springer

Vol. 3835: G. Sutcliffe, A. Voronkov (Eds.), Logic for Programming, Artificial Intelligence, and Reasoning. XIV, 744 pages. 2005. (Subseries LNAI).

Vol. 3833: K.-J. Li, C. Vangenot (Eds.), Web and Wireless Geographical Information Systems. XI, 309 pages. 2005.

Vol. 3824: L.T. Yang, M. Amamiya, Z. Liu, M. Guo, F.J. Rammig (Eds.), Embedded and Ubiquitous Computing. XXIII, 1204 pages. 2005.

Vol. 3823: T. Enokido, L. Yan, B. Xiao, D. Kim, Y. Dai, L.T. Yang (Eds.), Embedded and Ubiquitous Computing. XXXII, 1317 pages. 2005.

Vol. 3821: R. Ramanujam, S. Sen (Eds.), FSTTCS 2005: Foundations of Software Technology and Theoretical Computer Science. XIV, 566 pages. 2005.

Vol. 3818: S. Grumbach, L. Sui, V. Vianu (Eds.), Advances in Computer Science – ASIAN 2005. XIII, 294 pages. 2005.

Vol. 3814: M. Maybury, O. Stock, W. Wahlster (Eds.), Intelligent Technologies for Interactive Entertainment. XV, 342 pages. 2005. (Subseries LNAI).

Vol. 3810: Y.G. Desmedt (Ed.), Cryptology and Network Security. XI, 349 pages. 2005.

Vol. 3809: S. Zhang, R. Jarvis (Eds.), AI 2005: Advances in Artificial Intelligence. XXVII, 1344 pages. 2005. (Subseries LNAI).

Vol. 3808: C. Bento, A. Cardoso, G. Dias (Eds.), Progress in Artificial Intelligence. XVIII, 704 pages. 2005. (Subseries LNAI).

Vol. 3807: M. Dean, Y. Guo, W. Jun, R. Kaschek, S. Krishnaswamy, Z. Pan, Q.Z. Sheng (Eds.), Web Information Systems Engineering – WISE 2005 Workshops. XV, 275 pages. 2005.

Vol. 3806: A.H. H. Ngu, M. Kitsuregawa, E.J. Neuhold, J.-Y. Chung, Q.Z. Sheng (Eds.), Web Information Systems Engineering – WISE 2005. XXI, 771 pages. 2005.

Vol. 3805: G. Subsol (Ed.), Virtual Storytelling. XII, 289 pages. 2005.

Vol. 3804: G. Bebis, R. Boyle, D. Koracin, B. Parvin (Eds.), Advances in Visual Computing. XX, 755 pages. 2005.

Vol. 3803: S. Jajodia, C. Mazumdar (Eds.), Information Systems Security. XI, 342 pages. 2005.

Vol. 3799: M. A. Rodríguez, I.F. Cruz, S. Levashkin, M.J. Egenhofer (Eds.), GeoSpatial Semantics. X, 259 pages. 2005.

Vol. 3798: A. Dearle, S. Eisenbach (Eds.), Component Deployment. X, 197 pages. 2005.

Vol. 3797: S. Maitra, C. E. V. Madhavan, R. Venkatesan (Eds.), Progress in Cryptology - INDOCRYPT 2005. XIV, 417 pages. 2005.

Vol. 3796: N.P. Smart (Ed.), Cryptography and Coding. XI, 461 pages. 2005.

Vol. 3795: H. Zhuge, G.C. Fox (Eds.), Grid and Cooperative Computing - GCC 2005. XXI, 1203 pages. 2005.

Vol. 3794: X. Jia, J. Wu, Y. He (Eds.), Mobile Ad-hoc and Sensor Networks. XX, 1136 pages. 2005.

Vol. 3793: T. Conte, N. Navarro, W.-m.W. Hwu, M. Valero, T. Ungerer (Eds.), High Performance Embedded Architectures and Compilers. XIII, 317 pages. 2005.

Vol. 3792: I. Richardson, P. Abrahamsson, R. Messnarz (Eds.), Software Process Improvement. VIII, 215 pages. 2005.

Vol. 3791: A. Adi, S. Stoutenburg, S. Tabet (Eds.), Rules and Rule Markup Languages for the Semantic Web. X, 225 pages. 2005.

Vol. 3790: G. Alonso (Ed.), Middleware 2005. XIII, 443 pages. 2005.

Vol. 3789: A. Gelbukh, Á. de Albornoz, H. Terashima-Marín (Eds.), MICAI 2005: Advances in Artificial Intelligence. XXVI, 1198 pages. 2005. (Subseries LNAI).

Vol. 3788: B. Roy (Ed.), Advances in Cryptology - ASIACRYPT 2005. XIV, 703 pages. 2005.

Vol. 3785: K.-K. Lau, R. Banach (Eds.), Formal Methods and Software Engineering. XIV, 496 pages. 2005.

Vol. 3784: J. Tao, T. Tan, R.W. Picard (Eds.), Affective Computing and Intelligent Interaction. XIX, 1008 pages. 2005.

Vol. 3781: S.Z. Li, Z. Sun, T. Tan, S. Pankanti, G. Chollet, D. Zhang (Eds.), Advances in Biometric Person Authentication. XI, 250 pages. 2005.

Vol. 3780: K. Yi (Ed.), Programming Languages and Systems. XI, 435 pages. 2005.

Vol. 3779: H. Jin, D. Reed, W. Jiang (Eds.), Network and Parallel Computing. XV, 513 pages. 2005.

Vol. 3778: C. Atkinson, C. Bunse, H.-G. Gross, C. Peper (Eds.), Component-Based Software Development for Embedded Systems. VIII, 345 pages. 2005.

Vol. 3777: O.B. Lupanov, O.M. Kasim-Zade, A.V. Chaskin, K. Steinhöfel (Eds.), Stochastic Algorithms: Foundations and Applications. VIII, 239 pages. 2005.

Vol. 3775: J. Schönwälder, J. Serrat (Eds.), Ambient Networks. XIII, 281 pages. 2005.

Vol. 3773: A. Sanfeliu, M.L. Cortés (Eds.), Progress in Pattern Recognition, Image Analysis and Applications. XX, 1094 pages. 2005.

Vol. 3772: M. Consens, G. Navarro (Eds.), String Processing and Information Retrieval. XIV, 406 pages. 2005.

Vol. 3771: J.M.T. Romijn, G.P. Smith, J. van de Pol (Eds.), Integrated Formal Methods. XI, 407 pages. 2005.

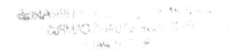

Vol. 3770: J. Akoka, S.W. Liddle, I.-Y. Song, M. Bertolotto, I. Comyn-Wattiau, W.-J. van den Heuvel, M. Kolp, J. Trujillo, C. Kop, H.C. Mayr (Eds.), Perspectives in Conceptual Modeling. XXII, 476 pages. 2005.

Vol. 3768: Y.-S. Ho, H.J. Kim (Eds.), Advances in Multimedia Information Processing - PCM 2005, Part II. XXVIII, 1088 pages. 2005.

Vol. 3767: Y.-S. Ho, H.J. Kim (Eds.), Advances in Multimedia Information Processing - PCM 2005, Part I. XXVIII, 1022 pages. 2005.

Vol. 3766: N. Sebe, M.S. Lew, T.S. Huang (Eds.), Computer Vision in Human-Computer Interaction. X, 231 pages. 2005.

Vol. 3765: Y. Liu, T. Jiang, C. Zhang (Eds.), Computer Vision for Biomedical Image Applications. X, 563 pages. 2005.

Vol. 3764: S. Tixeuil, T. Herman (Eds.), Self-Stabilizing Systems. VIII, 229 pages. 2005.

Vol. 3762: R. Meersman, Z. Tari, P. Herrero (Eds.), On the Move to Meaningful Internet Systems 2005: OTM 2005 Workshops. XXXI, 1228 pages. 2005.

Vol. 3761: R. Meersman, Z. Tari (Eds.), On the Move to Meaningful Internet Systems 2005: CoopIS, DOA, and ODBASE, Part II. XXVII, 653 pages. 2005.

Vol. 3760: R. Meersman, Z. Tari (Eds.), On the Move to Meaningful Internet Systems 2005: CoopIS, DOA, and ODBASE, Part I. XXVII, 921 pages. 2005.

Vol. 3759: G. Chen, Y. Pan, M. Guo, J. Lu (Eds.), Parallel and Distributed Processing and Applications - ISPA 2005 Workshops. XIII, 669 pages. 2005.

Vol. 3758: Y. Pan, D.-x. Chen, M. Guo, J. Cao, J.J. Dongarra (Eds.), Parallel and Distributed Processing and Applications. XXIII, 1162 pages. 2005.

Vol. 3757: A. Rangarajan, B. Vemuri, A.L. Yuille (Eds.), Energy Minimization Methods in Computer Vision and Pattern Recognition. XII, 666 pages. 2005.

Vol. 3756: J. Cao, W. Nejdl, M. Xu (Eds.), Advanced Parallel Processing Technologies. XIV, 526 pages. 2005.

Vol. 3754: J. Dalmau Royo, G. Hasegawa (Eds.), Management of Multimedia Networks and Services. XII, 384 pages. 2005.

Vol. 3753: O.F. Olsen, L.M.J. Florack, A. Kuijper (Eds.), Deep Structure, Singularities, and Computer Vision. X, 259 pages. 2005.

Vol. 3752: N. Paragios, O. Faugeras, T. Chan, C. Schnörr (Eds.), Variational, Geometric, and Level Set Methods in Computer Vision. XI, 369 pages. 2005.

Vol. 3751: T. Magedanz, E.R. M. Madeira, P. Dini (Eds.), Operations and Management in IP-Based Networks. X, 213 pages. 2005.

Vol. 3750: J.S. Duncan, G. Gerig (Eds.), Medical Image Computing and Computer-Assisted Intervention – MICCAI 2005, Part II. XL, 1018 pages. 2005.

Vol. 3749: J.S. Duncan, G. Gerig (Eds.), Medical Image Computing and Computer-Assisted Intervention – MICCAI 2005, Part I. XXXIX, 942 pages. 2005.

Vol. 3748: A. Hartman, D. Kreische (Eds.), Model Driven Architecture – Foundations and Applications. IX, 349 pages. 2005.

Vol. 3747: C.A. Maziero, J.G. Silva, A.M.S. Andrade, F.M.d. Assis Silva (Eds.), Dependable Computing. XV, 267 pages. 2005.

Vol. 3746: P. Bozanis, E.N. Houstis (Eds.), Advances in Informatics. XIX, 879 pages. 2005.

Vol. 3745: J.L. Oliveira, V. Maojo, F. Martín-Sánchez, A.S. Pereira (Eds.), Biological and Medical Data Analysis. XII, 422 pages. 2005. (Subseries LNBI).

Vol. 3744: T. Magedanz, A. Karmouch, S. Pierre, I. Venieris (Eds.), Mobility Aware Technologies and Applications. XIV, 418 pages. 2005.

Vol. 3742: J. Akiyama, M. Kano, X. Tan (Eds.), Discrete and Computational Geometry. VIII, 213 pages. 2005.

Vol. 3740: T. Srikanthan, J. Xue, C.-H. Chang (Eds.), Advances in Computer Systems Architecture. XVII, 833 pages. 2005.

Vol. 3739: W. Fan, Z.-h. Wu, J. Yang (Eds.), Advances in Web-Age Information Management. XXIV, 930 pages. 2005.

Vol. 3738: V.R. Syrotiuk, E. Chávez (Eds.), Ad-Hoc, Mobile, and Wireless Networks. XI, 360 pages. 2005.

Vol. 3735: A. Hoffmann, H. Motoda, T. Scheffer (Eds.), Discovery Science. XVI, 400 pages. 2005. (Subseries LNAI).

Vol. 3734: S. Jain, H.U. Simon, E. Tomita (Eds.), Algorithmic Learning Theory. XII, 490 pages. 2005. (Subseries LNAI).

Vol. 3733: P. Yolum, T. Güngör, F. Gürgen, C. Özturan (Eds.), Computer and Information Sciences - ISCIS 2005. XXI, 973 pages. 2005.

Vol. 3731: F. Wang (Ed.), Formal Techniques for Networked and Distributed Systems - FORTE 2005. XII, 558 pages. 2005.

Vol. 3729: Y. Gil, E. Motta, V. R. Benjamins, M.A. Musen (Eds.), The Semantic Web – ISWC 2005. XXIII, 1073 pages. 2005.

Vol. 3728: V. Paliouras, J. Vounckx, D. Verkest (Eds.), Integrated Circuit and System Design. XV, 753 pages. 2005.

Vol. 3727: M. Barni, J. Herrera Joancomartí, S. Katzenbeisser, F. Pérez-González (Eds.), Information Hiding. XII, 414 pages. 2005.

Vol. 3726: L.T. Yang, O.F. Rana, B. Di Martino, J.J. Dongarra (Eds.), High Performance Computing and Communications. XXVI, 1116 pages. 2005.

Vol. 3725: D. Borrione, W. Paul (Eds.), Correct Hardware Design and Verification Methods. XII, 412 pages. 2005.

Vol. 3724: P. Fraigniaud (Ed.), Distributed Computing. XIV, 520 pages. 2005.

Vol. 3723: W. Zhao, S. Gong, X. Tang (Eds.), Analysis and Modelling of Faces and Gestures. XI, 4234 pages. 2005.

Vol. 3722: D. Van Hung, M. Wirsing (Eds.), Theoretical Aspects of Computing – ICTAC 2005. XIV, 614 pages. 2005.

Vol. 3721: A.M. Jorge, L. Torgo, P.B. Brazdil, R. Camacho, J. Gama (Eds.), Knowledge Discovery in Databases: PKDD 2005. XXIII, 719 pages. 2005. (Subseries LNAI).

Vol. 3720: J. Gama, R. Camacho, P.B. Brazdil, A.M. Jorge, L. Torgo (Eds.), Machine Learning: ECML 2005. XXIII, 769 pages. 2005. (Subseries LNAI).